P9-CQG-806

# COMMUNICATION SYSTEMS

4TH EDITION

# COMMUNICATION SYSTEMS

## 4TH EDITION

**Simon Haykin**
*McMaster University*

JOHN WILEY & SONS, INC.
New York ■ Chichester ■ Weinheim ■ Brisbane ■ Singapore ■ Toronto

| | |
| --- | --- |
| *Editor* | Bill Zobrist |
| *Marketing Manager* | Katherine Hepburn |
| *Associate Production Director* | Lucille Buonocore |
| *Senior Production Editor* | Monique Calello |
| *Cover Designer* | Madelyn Lesure |
| *Illustration Coordinator* | Gene Aiello |
| *Illustration Studio* | Wellington Studios |

*Cover Photo* NASA/Photo Researchers, Inc.

This book was set in 10/12 Times Roman by UG / GGS Information Services, Inc. and printed and bound by Hamilton Printing Company. The cover was printed by Phoenix Color Corporation.

This book is printed on acid-free paper. 

The paper in this book was manufactured by a mill whose forest management programs include sustained yield harvesting of its timberlands. Sustained yield harvesting principles ensure that the numbers of trees cut each year does not exceed the amount of new growth.

Copyright © 2001, John Wiley & Sons, Inc. All rights Reserved.

No part of this publication may be reproduced, stored in a retrieval system or transmitted in any form or by any means, electronic, mechanical, photocopying, recording, scanning or otherwise, except as permitted under Sections 107 or 1089 of the 1976 United States Copyright Act, without either the prior written permission of the Publisher, or authorization through payment of the appropriate per-copy fee to the Copyright Clearance Center, 222 Rosewood Drive, Danvers, MA 01923, (508) 750-8400, fax (508) 750-4470. Requests to the Publisher for permission should be addressed to the Permissions Department, John Wiley & Sons, Inc., 605 Third Avenue, New York, NY 10158-0012, (212) 850-6011, fax (212) 850-6008, E-Mail: PERMREQ@WILEY.COM.

To order books or for customer service call 1-800-CALL-WILEY (225-5945).

***Library of Congress Cataloging-in-Publication Data***

Haykin, Simon
Communication systems / Simon Haykin.—4th ed.
p. cm.
ISBN 0-471-17869-1 (cloth : alk. paper)
1. Telecommunication. 2. Signal theory (Telecommunication) I. Title.

TK5101 .H37 2000
621.382—dc21 99-042977

Printed in the United States of America

10 9 8 7 6 5 4 3 2 1

*In loving memory of Vera*

# PREFACE

Electrical engineering education has undergone some radical changes during the past couple of decades and continues to do so. A modern undergraduate program in electrical engineering includes the following two introductory courses:

- *Signals and Systems*, which provides a balanced and integrated treatment of continuous-time and discrete-time forms of signals and systems. The Fourier transform (in its different forms), Laplace transform, and $z$-transform are treated in detail. Typically, the course also includes an elementary treatment of communication systems.
- *Probability and Random Processes*, which develops an intuitive grasp of discrete and continuous random variables and then introduces the notion of a random process and its characteristics.

Typically, these two introductory courses lead to a senior-level course on communication systems.

The fourth edition of this book has been written with this background and primary objective in mind. Simply put, the book provides a modern treatment of communication systems at a level suitable for a one- or two-semester senior undergraduate course. The emphasis is on the statistical underpinnings of communication theory with applications.

The material is presented in a logical manner, and it is illustrated with examples, with the overall aim being that of helping the student develop an intuitive grasp of the theory under discussion. Except for the Background and Preview chapter, each chapter ends with numerous problems designed not only to help the students test their understanding of the material covered in the chapter but also to challenge them to extend this material. Every chapter includes notes and references that provide suggestions for further reading. Sections or subsections that can be bypassed without loss of continuity are identified with a footnote.

A distinctive feature of the book is the inclusion of eight computer experiments using MATLAB. This set of experiments provides the basis of a "Software Laboratory", with each experiment being designed to extend the material covered in the pertinent chapter. Most important, the experiments exploit the unique capabilities of MATLAB in an instructive manner. The MATLAB codes for all these experiments are available on the Wiley Web site: **http://www.wiley.com/college/haykin/.**

The Background and Preview chapter presents introductory and motivational material, paving the way for detailed treatment of the many facets of communication systems in the subsequent 10 chapters. The material in these chapters is organized as follows:

- Chapter 1 develops a detailed treatment of *random*, or *stochastic*, *processes*, with particular emphasis on their partial characterization (i.e., second-order statistics). In effect, the discussion is restricted to wide-sense stationary processes. The correlation properties and power spectra of random processes are described in detail. Gaussian processes and narrowband noise feature prominently in the study of communication systems, hence their treatment in the latter part of the chapter. This treatment naturally leads to the consideration of the Rayleigh and Rician distributions that arise in a communications environment.

- Chapter 2 presents an integrated treatment of *continuous-wave (CW) modulation* (i.e., *analog communications*) and their different types, as outlined here:
  (i) *Amplitude modulation*, which itself can assume one of the following forms (depending on how the spectral characteristics of the modulated wave are specified):
    - Full amplitude modulation
    - Double sideband-suppressed carrier modulation
    - Quadrature amplitude modulation
    - Single sideband modulation
    - Vestigial sideband modulation

  (ii) *Angle modulation*, which itself can assume one of two interrelated forms:
    - Phase modulation
    - Frequency modulation

  The time-domain and spectral characteristics of these modulated waves, methods for their generation and detection, and the effects of channel noise on their performances are discussed.

- Chapter 3 covers *pulse modulation* and discusses the processes of sampling, quantization, and coding that are fundamental to the digital transmission of analog signals. This chapter may be viewed as the transition from analog to digital communications. Specifically, the following types of pulse modulation are discussed:
  (i) *Analog pulse modulation*, where only time is represented in discrete form; it embodies the following special forms:
    - Pulse amplitude modulation
    - Pulse width (duration) modulation
    - Pule position modulation

  The characteristics of pulse amplitude modulation are discussed in detail, as it is basic to all forms of pulse modulation, be they of the analog or digital type.

  (ii) *Digital pulse modulation*, in which both time and signal amplitude are represented in discrete form; it embodies the following special forms:
    - Pulse-code modulation
    - Delta modulation
    - Differential pulse-code modulation

  In delta modulation, the sampling rate is increased far in excess of that used in pulse-code modulation so as to simplify implementation of the system. In contrast, in differential pulse-code modulation, the sampling rate is reduced through the use of a predictor that exploits the correlation properties of the information-bearing signal.

  (iii) MPEG/audio coding standard, which includes a psychoacoustic model as a key element in the design of the encoder.

- Chapter 4 covers *baseband pulse transmission*, which deals with the transmission of pulse-amplitude modulated signals in their baseband form. Two important issues are discussed: the effects of channel noise and limited channel bandwidth on the performance of a digital communication system. Assuming that the channel noise is additive and white, this effect is minimized by using a *matched filter*, which is basic to the design of communication receivers. As for limited channel bandwidth, it manifests itself in the form of a phenomenon known as *intersymbol interference*. To combat the degrading effects of this signal-dependent interference, we may use either a *pulse-shaping filter* or *correlative encoder/decoder*; both of these approaches are discussed. The chapter includes a discussion of digital subscriber lines for direct communication between a subscriber and an Internet service provider. This is followed by a derivation of the optimum linear receiver for combatting the combined effects of channel noise and intersymbol interference, which, in turn, leads to an introductory treatment of *adaptive equalization*.

- Chapter 5 discusses *signal-space analysis* for an additive white Gaussian noise channel. In particular, the foundations for the geometric representation of signals with finite energy are established. The correlation receiver is derived, and its equivalence with the matched filter receiver is demonstrated. The chapter finishes with a discussion of the probability of error and its approximate calculation.
- Chapter 6 discusses *passband data transmission*, where a sinusoidal carrier wave is employed to facilitate the transmission of the digitally modulated wave over a band-pass channel. This chapter builds on the geometric interpretation of signals presented in Chapter 5. In particular, the effect of channel noise on the performance of digital communication systems is evaluated, using the following modulation techniques:
  - (i) *Phase-shift keying*, which is the digital counterpart to phase modulation with the phase of the carrier wave taking on one of a prescribed set of discrete values.
  - (ii) Hybrid amplitude/phase modulation schemes including *quadrature-amplitude modulation* (QAM), and *carrierless amplitude/phase modulation* (CAP).
  - (iii) *Frequency-shift keying*, which is the digital counterpart of frequency modulation with the frequency of the carrier wave taking on one of a prescribed set of discrete values.
  - (iv) Generic multichannel modulation, followed by *discrete multitone,* the use of which has been standardized in asymmetric digital subscriber lines.

  In a digital communication system, timing is everything, which means that the receiver must be *synchronized* to the transmitter. In this context, we speak of the receiver being coherent or noncoherent. In a *coherent receiver,* provisions are made for the recovery of both the carrier phase and symbol timing. In a *noncoherent receiver* the carrier phase is ignored and provision is only made for symbol timing. Such a strategy is dictated by the fact that the carrier phase may be random, making phase recovery a costly proposition. Synchronization techniques are discussed in the latter part of the chapter, with particular emphasis on discrete-time signal processing.
- Chapter 7 introduces *spread-spectrum modulation.* Unlike traditional forms of modulation discussed in earlier chapters, channel bandwidth is purposely sacrificed in spread-spectrum modulation for the sake of security or protection against interfering signals. The direct-sequence and frequency-hop forms of spread-spectrum modulation are discussed.
- Chapter 8 deals with *multiuser radio communications*, where a multitude of users have access to a common radio channel. This type of communication channel is well represented in satellite and wireless communication systems, both of which are discussed. The chapter includes a presentation of link budget analysis, emphasizing the related antenna and propagation concepts, and noise calculations.
- Chapter 9 develops the *fundamental limits in information theory*, which are embodied in Shannon's theorems for data compaction, data compression, and data trans-

mission. These theorems provide upper bounds on the performance of information sources and communication channels. Two concepts, basic to formulation of the theorems, are (1) the *entropy* of a source (whose definition is analogous to that of entropy in thermodynamics), and (2) channel capacity.

- Chapter 10 deals with *error-control coding*, which encompasses techniques for the encoding and decoding of digital data streams for their reliable transmission over noisy channels. Four types of error-control coding are discussed:
  - (i) *Linear block codes*, which are completely described by sets of linearly independent code words, each of which consists of message bits and parity-check bits. The parity-check bits are included for the purpose of error control.
  - (ii) *Cyclic codes*, which form a subclass of linear block codes.
  - (iii) *Convolutional codes*, which involve operating on the message sequence continuously in a serial manner.
  - (iv) *Turbo codes*, which provide a novel method of constructing good codes that approach Shannon's channel capacity in a physically realizable manner.

  Methods for the generation of these codes and their decoding are discussed.

The book also includes supplementary material in the form of six appendices as follows:

- Appendix 1 reviews probability theory.
- Appendix 2, on the representation of signals and systems, reviews the Fourier transform and its properties, the various definitions of bandwidth, the Hilbert transform, and the low-pass equivalents of narrowband signals and systems.
- Appendix 3 presents an introductory treatment of the Bessel function and its modified form. Bessel functions arise in the study of frequency modulation, noncoherent detection of signals in noise, and symbol timing synchronization.
- Appendix 4 introduces the confluent hypergeometric function, the need for which arises in the envelope detection of amplitude-modulated signals in noise.
- Appendix 5 provides an introduction to cryptography, which is basic to secure communications.
- Appendix 6 includes 12 useful tables of various kinds.

As mentioned previously, the primary purpose of this book is to provide a modern treatment of communication systems suitable for use in a one- or two-semester undergraduate course at the senior level. The make-up of the material for the course is naturally determined by the background of the students and the interests of the teachers involved. The material covered in the book is both broad and deep enough to satisfy a variety of backgrounds and interests, thereby providing considerable flexibility in the choice of course material. As an aid to the teacher of the course, a detailed solutions manual for all the problems in the book is available from the publisher.

## *Acknowledgments*

I wish to express my deep gratitude to Dr. Gregory J. Pottie (University of California, Los Angeles), Dr. Santosh Venkatesh (University of Pennsylvania), Dr. Stephen G. Wilson (University of Virginia), Dr. Gordon Stüber (Georgia Institute of Technology), Dr. Venugopal Veeraralli (Cornell University), and Dr. Granville E. Ott (University of Texas at Austin) for critical reviews of an earlier version of the manuscript and for making numerous suggestions that have helped me shape the book into its present form. The treatment of the effect of noise on envelope detection presented in Chapter 2 is based on course notes made available to me by Dr. Santosh Venkatesh, for which I am grateful. I am grateful to Dr. Gordon Stüber for giving permission to reproduce Figure 6.32.

I am indebted to Dr. Michael Moher (Communications Research Centre, Ottawa) for reading five chapters of an earlier version of the manuscript and for making many constructive comments on turbo codes. I am equally indebted to Dr. Brendan Frey (University of Waterloo, Ontario) for his invaluable help in refining the material on turbo codes, comments on low-density parity-check codes, for providing the software to plot Fig. 9.18, and giving me the permission to reproduce Figures 10.27 and 10.33. I am grateful to Dr. David Conn (McMaster University, Ontario) for his critical reading of the Background and Preview Chapter and for making suggestions on how to improve the presentation of the material therein.

I also wish to thank Dr. Jean-Jacque Werner (Lucent Technologies, Holmdel), Dr. James Mazo (Lucent Technologies, Murray Hill), Dr. Andrew Viterbi (Qualcom, San Diego), Dr. Radford Neal (University of Toronto, Ontario), Dr. Yitzhak (Irwin) Kalet (Technion, Israel), Dr. Walter Chen (Motorola), Dr. John Cioffi (Stanford University), Dr. Jon Mark (University of Waterloo, Ontario), and Dr. Robert Dony (University of Guelph, Ontario); I thank them all for their helpful comments on selected sections in the book. Corrections and suggestions for improvements to the book made by Dr. Donald Wunsch II (University of Missouri) are also appreciated.

I am grateful to my graduate student Mathini Sellathurai (McMaster University) for performing the computer experiments in the book, and Hugh Pasika (McMaster University) for many useful comments on the Background and Preview Chapter and for doing the computations on some graphical plots in the book. Proofreading of the page proofs by Mathini Sellathurai and Alpesh Patel is much appreciated.

I am particularly grateful to my editor at Wiley, Bill Zobrist, for his strong support and help throughout the writing of the book. I am indebted to Monique Calello, Senior Production Editor at Wiley, for her tireless effort in overseeing the production of the book in its various stages. I thank Katherine Hepburn for advertising and marketing the book. I thank Karen Tongish for her careful copyediting of the manuscript, Katrina Avery for her careful proofreading of the page proofs, and Kristen Maus for composing the index of the book.

Last but by no means least, as always, I am grateful to my Technical Coordinator, Lola Brooks, for her tireless effort in typing the manuscript of the book. I also wish to record my gratitude to Brigitte Maier, Assistant Librarian, and Regina Bendig, Reference Librarian, at McMaster University, for helping me on numerous occasions in tracing references for the bibliography.

*Simon Haykin*
*Ancaster, Ontario*
*January, 2000*

# Contents

## Background and Preview 1



## Chapter 1 *Random Processes* 31

# BACKGROUND AND PREVIEW

The background and preview material presented herein sets the stage for a statistical treatment of communication systems in subsequent chapters. In particular, we describe the following:

- *The communication process.*
- *Primary communication resources, namely, transmitted power and channel bandwidth.*
- *Sources of information.*
- *The two primary types of switching: circuit switching and packet switching.*
- *Communication channels for the transportation of information-bearing signals from the transmitter to the receiver.*
- *The modulation process, which is basic to communication systems.*
- *Analog and digital types of communication systems.*
- *Shannon's information capacity theorem.*
- *A digital communications problem.*

The chapter concludes with some historical notes, as a source of motivation for the reader.

## The Communication Process[1]

Today, *communication* enters our daily lives in so many different ways that it is very easy to overlook the multitude of its facets. The telephones at our hands, the radios and televisions in our living rooms, the computer terminals with access to the Internet in our offices and homes, and our newspapers are all capable of providing rapid communications from every corner of the globe. Communication provides the senses for ships on the high seas, aircraft in flight, and rockets and satellites in space. Communication through a wireless telephone keeps a car driver in touch with the office or home miles away. Communication keeps a weather forecaster informed of conditions measured by a multitude of sensors. Indeed, the list of applications involving the use of communication in one way or another is almost endless.

In the most fundamental sense, communication involves implicitly the transmission of *information* from one point to another through a succession of processes, as described here:

1. The generation of a *message signal*: voice, music, picture, or computer data.
2. The description of that message signal with a certain measure of precision, by a set of *symbols*: electrical, aural, or visual.
3. The *encoding* of these symbols in a form that is suitable for transmission over a physical medium of interest.
4. The *transmission* of the encoded symbols to the desired destination.
5. The *decoding* and *reproduction* of the original symbols.
6. The *re-creation* of the original message signal, with a definable degradation in quality; the degradation is caused by imperfections in the system.

There are, of course, many other forms of communication that do not directly involve the human mind in real time. For example, in *computer communications* involving communication between two or more computers, human decisions may enter only in setting up the programs or commands for the computer, or in monitoring the results.

Irrespective of the form of communication process being considered, there are three basic elements to every communication system, namely, *transmitter*, *channel*, and *receiver*, as depicted in Figure 1. The transmitter is located at one point in space, the receiver is located at some other point separate from the transmitter, and the channel is the physical medium that connects them. The purpose of the transmitter is to convert the *message signal* produced by the *source of information* into a form suitable for transmission over the channel. However, as the transmitted signal propagates along the channel, it is distorted due to channel imperfections. Moreover, noise and interfering signals (originating from other sources) are added to the channel output, with the result that the *received signal* is a corrupted version of the *transmitted signal*. The receiver has the task of operating on the received signal so as to reconstruct a recognizable form of the original message signal for a user.

There are two basic modes of communication:

1. *Broadcasting*, which involves the use of a single powerful transmitter and numerous receivers that are relatively inexpensive to build. Here information-bearing signals flow only in one direction.
2. *Point-to-point communication*, in which the communication process takes place over a link between a single transmitter and a receiver. In this case, there is usually a bidirectional flow of information-bearing signals, which requires the use of a transmitter and receiver at each end of the link.

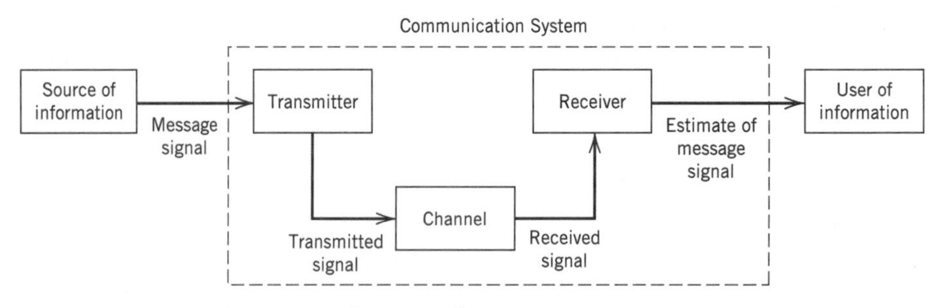


**FIGURE 1** Elements of a communication system.

The broadcasting mode of communication is exemplified by radio and television, and the ubiquitous telephone provides the means for one form of point-to-point communication. Another example of point-to-point communication is the link between an Earth station and a robot navigating the surface of a distant planet.

All these different communication systems as well as others not mentioned here share a common feature: The underlying communication process in each and every one of them is *statistical* in nature. Indeed, it is for this important reason that much of this book is devoted to the statistical underpinnings of communication systems. In so doing, we develop an exposition of the fundamental issues involved in the study of different communication methodologies and thereby provide a natural forum for their comparative evaluations.

## Primary Communication Resources

In a communication system, two primary resources are employed: *transmitted power* and *channel bandwidth*. The transmitted power is the average power of the transmitted signal. The channel bandwidth is defined as the band of frequencies allocated for the transmission of the message signal. A general system design objective is to use these two resources as efficiently as possible. In most communication channels, one resource may be considered more important than the other. We may therefore classify communication channels as *power limited* or *band limited*. For example, the telephone circuit is a typical band-limited channel, whereas a space communication link or satellite channel is typically power limited.

When the spectrum of a message signal extends down to zero or low frequencies, we define the bandwidth of the signal as that upper frequency above which the spectral content of the signal is negligible and therefore unnecessary for transmitting information. For example, the average voice spectrum extends well beyond 10 kHz, though most of the average power is concentrated in the range of 100 to 600 Hz, and a band from 300 to 3100 Hz gives good articulation. Accordingly, we find that telephone circuits that respond well to this latter range of frequencies give quite satisfactory commercial telephone service.

Another important point that we have to keep in mind is the unavoidable presence of noise in a communication system. *Noise* refers to unwanted waves that tend to disturb the transmission and processing of message signals in a communication system. The sources of noise may be internal or external to the system.

A quantitative way to account for the effect of noise is to introduce *signal-to-noise ratio* (SNR) as a system parameter. For example, we may define the SNR at the receiver input as *the ratio of the average signal power to the average noise power*, both being measured at the same point. The customary practice is to express the SNR in *decibels* (dBs), defined as 10 times the logarithm (to base 10) of the power ratio. For example, signal-to-noise ratios of 10, 100, and 1,000 correspond to 10, 20, and 30 dBs, respectively.

## Sources of Information

The telecommunications environment is dominated by four important sources of information: *speech*, *music*, *pictures*, and *computer data*. A source of information may be characterized in terms of the signal that carries the information. A *signal* is defined as a single-valued function of *time* that plays the role of the independent variable; at every instant of time, the function has a unique value. The signal can be *one-dimensional*, as in the case of speech, music, or computer data; *two-dimensional*, as in the case of pictures;

*three-dimensional*, as in the case of video data; and *four-dimensional*, as in the case of volume data over time. In the sequel, we elaborate on different sources of information.

(i) *Speech* is the primary method of human communication. Specifically, the speech communication process involves the transfer of information from a speaker to a listener, which takes place in three successive stages:

- *Production.* An intended message in the speaker's mind is represented by a *speech signal* that consists of sounds (i.e., pressure waves) generated inside the vocal tract and whose arrangement is governed by the rules of language.
- *Propagation.* The sound waves propagate through the air at a speed of 300 m/s, reaching the listener's ears.
- *Perception.* The incoming sounds are deciphered by the listener into a received message, thereby completing the chain of events that culminate in the transfer of information from the speaker to the listener.

The speech-production process may be viewed as a form of filtering, in which a *sound source excites a vocal tract filter*. The vocal tract consists of a tube of nonuniform cross-sectional area, beginning at the *glottis* (i.e., the opening between the vocal cords) and ending at the *lip*. As the sound propagates along the vocal tract, the spectrum (i.e., frequency content) is shaped by the frequency selectivity of the vocal tract; this effect is somewhat similar to the resonance phenomenon observed in organ pipes. The important point to note here is that the power spectrum (i.e., the distribution of long-term average power versus frequency) of speech approaches zero for zero frequency and reaches a peak in the neighborhood of a few hundred hertz. To put matters into proper perspective, however, we have to keep in mind that the hearing mechanism is very sensitive to frequency. Moreover, the type of communication system being considered has an important bearing on the band of frequencies considered to be "essential" for the communication process. For example, as mentioned previously, a bandwidth of 300 to 3100 Hz is considered adequate for commercial telephonic communication.

(ii) The second source of information, *music*, originates from instruments such as the piano, violin, and flute. The note made by a musical instrument may last for a short time interval as in the pressing of a key on a piano, or it may be sustained for a long time interval as in the example of a flute player holding a prolonged note. Typically, music has two structures: a *melodic* structure consisting of a time sequence of sounds, and a *harmonic* structure consisting of a set of simultaneous sounds. Like a speech signal, a musical signal is bipolar. However, a musical signal differs from a speech signal in that its spectrum occupies a much wider band of frequencies that may extend up to about 15 kHz. Accordingly, musical signals demand a much wider channel bandwidth than speech signals for their transmission.

(iii) The third source of information, *pictures*, relies on the human visual system for its perception. The picture can be *dynamic*, as in television, or *static*, as in facsimile. Taking the case of television first, the pictures in motion are converted into electrical signals to facilitate their transport from the transmitter to the receiver. To do so, each complete picture is *sequentially scanned*. The scanning process is carried out in a TV *camera*. In a black-and-white TV, the camera contains optics designed to focus an image on a *photocathode* consisting of a large number of photosensitive elements. The charge pattern so generated on the photosensitive surface is scanned by an *electron beam*, thereby producing an output current that varies *temporally* with the way in which the brightness of the original picture varies *spatially* from one point to another. The resulting output current is called a *video signal*. The type of scanning used in television is a form of spatial sampling called *raster scanning*, which converts a two-dimensional image intensity into a one-dimensional waveform; it is somewhat analogous to the manner in which we read a printed paper in that the scanning is performed from left to right on a line-by-line basis. In North American analog television, a picture is divided into 525 lines, which constitute a *frame*. Each frame is decomposed into two *interlaced fields*, each of which consists of 262.5 lines. For convenience of presentation, we will refer to the two fields as I and II. The scanning procedure is illustrated in Figure 2. The lines of field I are depicted as solid lines, and those of field II are depicted as dashed lines. The *start* and *end* of each field are also included in the figure. Field I is scanned first. The scanning spot of the TV camera moves with constant velocity across each line of the field from left to right, and the image intensity at the center of the spot is measured; the scanning spot itself is partly responsible for local spatial averaging of the image. When the end of a particular line is reached, the scanning spot quickly flies back (in a horizontal direction) to the start of the next line down in the field. This flyback is called the *horizontal retrace*. The scanning process described here is continued until the whole field has been accounted for. When this condition is reached, the scanning spot moves quickly (in a vertical direction) from the end of field I to the start of field II. This second flyback is called the *vertical retrace*. Field II is treated in the same fashion as field I. The time taken for each field to be scanned is 1/60 s. Correspondingly, the time taken for a frame or a complete picture to be scanned is 1/30 s. With 525 lines in a frame, the *line-scanning frequency* equals 15.75 kHz. Thus, by flashing 30 still pictures per second on the display tube of the TV receiver, the human eye perceives them to be moving pictures. This effect is due to a phenomenon known as the *persistence of vision*. During the horizontal- and vertical-retrace intervals, the picture tube is made inoperative by means of *blanking pulses* that are generated at the transmitter. Moreover, synchronization between the various scanning operations at both transmitter and receiver is accomplished by means of special pulses that are transmitted during the blanking periods; thus, the synchronizing pulses do not show on the reproduced picture. The reproduction quality of a TV picture is limited by two basic factors:

1. The number of lines available in a raster scan, which limits resolution of the picture in the vertical direction.
2. The channel bandwidth available for transmitting the video signal, which limits resolution of the picture in the horizontal direction.

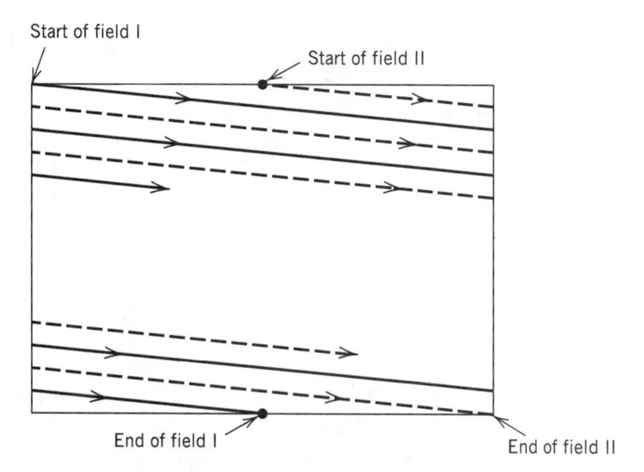


**FIGURE 2** Interlaced raster scan.

For each direction, *resolution* is expressed in terms of the maximum number of lines alternating between black and white that can be resolved in the TV image along the pertinent direction by a human observer. In the NTSC (National Television System Committee) system, which is the North American standard, the parameter values used result in a *video bandwidth* of 4.2 MHz, which extends down to zero frequency. This bandwidth is orders of magnitude larger than that of a speech signal. Note also that whereas a speech signal is bipolar, a video (television) signal is inherently positive (i.e., unipolar).

In color TV, the perception of color is based on the three types of color receptors (cones) in the human eye: red, green, and blue, whose wavelengths are 570 nm, 535 nm, and 445 nm, respectively. These three colors are referred to as *primary colors* because any other color found in nature can be approximated by an additive mixture of them. This physical reality is indeed the basis for the transmission of color in commercial TV broadcasting. The three primary colors are represented by the video signals $m_R(t)$, $m_G(t)$, and $m_B(t)$, respectively. To conserve bandwidth and produce a picture that can be viewed on a conventional black-and-white (monochrome) television receiver, the transmission of these three primary colors is accomplished by observing that they can be uniquely represented by any three signals that are independent linear combinations of $m_R(t)$, $m_G(t)$, and $m_B(t)$. The three signals are as follows:

- A *luminance signal*, $m_L(t)$, which produces a black-and-white version of the color picture when it is received on a conventional monochrome television receiver.
- A pair of signals, $m_I(t)$ and $m_Q(t)$, called the *chrominance signals*, which indicate the way the color of the picture departs from shades of gray.

The luminance signal $m_L(t)$ is assigned the entire 4.2 MHz bandwidth. Owing to certain properties of human vision, tests show that if the nominal bandwidths of the chrominance signals $m_I(t)$ and $m_Q(t)$ are 1.6 MHz and 0.6 MHz, respectively, satisfactory color reproduction is possible.

Turning next to a *facsimile (fax) machine*, the purpose of this machine is to transmit still pictures over a communication channel (most notably, a telephone channel). Such a machine provides a highly popular facility for the transmission of handwritten or printed text from one point to another; transmitting text by facsimile is treated simply like transmitting a picture. The basic principle employed for signal generation in a facsimile machine is to scan an original document (picture) and use an image sensor to convert the light to an electrical signal.

(iv) Finally, *personal computers* (PCs) have become an integral part of our daily lives. We use them for electronic mail, exchange of software, and sharing of resources. The text transmitted by a PC is usually encoded using the *American Standard Code for Information Interchange* (ASCII), which is the first code developed specifically for computer communications. Each character in ASCII is represented by seven *data bits* constituting a unique binary pattern made up of 0s and 1s; bit is acronym for binary digit. Thus a total of $2^7 = 128$ different characters can be represented in ASCII. The characters are various lowercase and uppercase letters, numbers, special control symbols, and punctuation symbols commonly used such as @, $, and %. Some of the special "control" symbols, such as BS (backspace) and CR (carriage return), are used to control the printing of characters on a page. Other symbols, such as ENQ (enquiry) and ETB (end of transmission block), are used for communication purposes. (A complete listing of ASCII characters is given in Table A6.1.) The seven data bits are ordered starting with the most significant bit $b_7$ down to the least significant bit $b_1$,

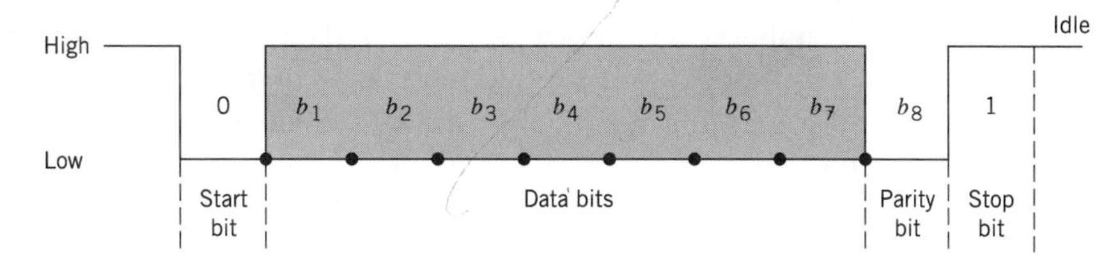


**FIGURE 3** The bit format for sending asynchronous serial data used in the RS-232 standard.

as illustrated in Figure 3. At the end of the data bits, an extra bit $b_8$ is appended for the purpose of *error detection*. This error-detection bit is called a *parity bit*. A sequence of eight bits is referred to as a *byte*, or an *octet*. The parity bit is set in such a way that the total number of 1s in each byte is odd for odd parity and even for even parity. Suppose, for example, the communicators agree to use even parity; then the parity bit will be a 0 when the number of 1s in the data bit is even and a 1 when it is odd. Hence, if a single bit in a byte is received in error and thereby violates the even parity rule, it can be detected and then corrected through retransmission. Personal computers are often connected via their RS (recommended standard)-232 ports. When ASCII data (in fact, all character data) are transmitted through these ports, a start bit, set to 0, and one or more stop bits, set to 1, as shown in Figure 3, are added to provide character framing. When the transmission is idle, a long series of 1s is sent so as to keep the circuit connection alive. In Figure 3, symbols 0 and 1 are designated as "low" and "high," respectively. They are also sometimes referred to as "space" and "mark," respectively; the latter terminology comes from the days of telegraphy. The text prepared on a PC is usually stored and then transmitted over a communication channel (e.g., a telephone channel) with a single character being sent at a time. This form of data transmission is called *asynchronous transmission*, as opposed to *synchronous transmission* in which a whole sequence of encoded characters is sent over the channel in one long transmission. Encoded characters produced by a mixture of asynchronous and synchronous terminals are combined by means of *data multiplexers*. The multiplexed stream of data so formed is then applied to a device called a *modem* (modulator-demodulator) for the purpose of transmission over the channel.

In summary, computer-generated data and television signals are both *wideband signals*, in that their power content occupies a wide range of frequencies. Another important characteristic of data communication between personal computers is *burstiness*, which means that information is usually transmitted from one terminal to another in bursts with silent periods between bursts. Indeed, data traffic involving computers in one form or another tends to be of a bursty nature. This is to be contrasted with traffic in a digital transmission network due to voice or interactive video, which, relatively speaking, is *continuous*.

Another way in which we use the computer is to *download* compressed forms of text, audio, and video data from a service provider at a remote location. *Data compression* provides a practical means for the efficient storage and transmission of these kinds of data. A data compression system consists of an *encoder* and a *decoder*, where the compression of an incoming data stream and its reconstruction are performed, respectively. Basically, there are two forms of data compression:

1. *Lossless compression* operates by removing the *redundant* information contained in the data of interest. The compression is said to be lossless because it is com-

pletely reversible in that the original data can be reconstructed exactly. Lossless compression is also referred to as *data compaction.*

2. *Lossy compression* involves the loss of information in a controlled manner; the compression may therefore not be completely reversible. Lossy compression is, however, capable of achieving a compression ratio higher than that attainable with lossless methods.

For digital text, lossless compression is required. In this context, we mention the *Lempel-Ziv algorithm*, which is intrinsically adaptive and capable of encoding groups of source symbols that occur frequently. It achieves a compression of approximately 55 percent on ordinary English text, which, loosely speaking, corresponds to the compression that would be achieved by encoding pairs of letters. The Lempel-Ziv algorithm is a form of *entropic coding*, or *source coding*, which is discussed in Chapter 9.

In many other applications, lossy compression is usually the preferred approach as its use can substantially reduce the data size without significantly altering the perceptual quality of an image or audio signal. For such applications, this form of data compression is acceptable, and in high-throughput data-transmission applications such as the Internet, it is a necessity. But in some other applications such as a clinical setting, the quality of a medical image (e.g., digital x-ray radiograph) must not be degraded on reconstruction.

For digital audio and video applications involving storage or transmission to be viable in today's marketplace, we need *standard* compression algorithms that enable the interoperability of equipment produced by different manufacturers. In this context, we mention three prominent standard compression algorithms that cater to different needs:

- The *JPEG image coding standard*[2] is designed to compress full-color or grayscale images of natural, real-world scenes by exploiting known limitations of the human visual system; JPEG stands for Joint Photographic Experts Group. At the input to the encoder, *picture elements*, or *pixels*, are grouped into $8 \times 8$ blocks, which are applied to a relative of the Fourier transform known as the *discrete cosine transform* (DCT)[3]. The DCT decomposes each block of pixels into a set of 64 coefficients that closely satisfy two related objectives:

  1. The coefficients should be as *uncorrelated* as possible.
  2. The energy of the input signal should be packed into the *smallest number of coefficients possible.*

  The next operation in the encoder is that of *quantization*, where each of the 64 DCT coefficients is *rounded off*. In JPEG, quantization is performed in conjunction with a *quantization table* supplied by the user as an input to the encoder. Each element of the table is an integer from 1 to 255 that specifies the step size of the DCT coefficients, which, in turn, permits the representation of each quantized DCT coefficient by an 8-bit code word. Basically, the purpose of quantization is to discard information that is not perceptually discernible. Quantization is a *many-to-one mapping* and therefore the principal source of lossiness in the encoder. The final operation in the encoder is that of *Huffman coding*, which is a form of entropic (source) coding also discussed in Chapter 9. Huffman coding achieves additional data compression in a lossless manner by encoding the quantized DCT coefficients in accordance with their statistical characteristics. At the decoder, data reconstruction is performed through a sequence of operations that are the inverse of those in the encoder, namely, Huffman decoding, dequantization in accordance with the quantization table, and finally the inverse DCT.

- The *MPEG-1/video coding standard*[4] is designed primarily to compress video signals at 30 frames per second (fps) into bit streams running at the rate of 1.5 megabits per second (Mb/s); MPEG stands for Motion Photographic Experts Group. The MPEG-1 video coding standard achieves this design goal by exploiting four basic forms of redundancy inherently present in video data:

  1. Interframe (temporal) redundancy.
  2. Interpixel redundancy within a frame.
  3. Psychovisual redundancy.
  4. Entropic coding redundancy.

  It is the exploitation of interframe redundancy that distinguishes MPEG-1 from JPEG. In principle, neighboring frames in typical video sequences are highly *correlated*. The meaning of this high correlation is that, in an average sense, a video signal does not change rapidly from one frame to the next, and as a result, the difference between adjacent frames has a variance (i.e., average power) that is much smaller than the variance of the video signal itself. Accordingly, the interframe redundancy can be significantly reduced to produce a more efficiently compressed video signal. This reduction is achieved through the use of *prediction* to estimate each frame from its neighbors; the resulting prediction error is transmitted for motion estimation and compensation. The prediction is nonlinear by virtue of the nature of the problem. As with JPEG, the interpixel redundancy is reduced through the combined use of the DCT, quantization, and lossless entropic coding. The net result is that full-motion video becomes a 1.5 Mb/s stream of computer data that can be stored on compact discs or integrated with texts and graphics. Most important, the full-motion video and associated audio can be delivered over existing computer and telecommunication networks, which, in turn, makes it possible to fulfill the need for *video-on-demand* on the Internet.

- The *MPEG-1/audio coding standard*[5] is based on *perceptual coding*, which is a waveform-preserving process; that is, the amplitude-time waveform of the decoded audio signal closely approximates that of the original audio signal. In basic terms, the encoding process encompasses four distinct operations:

  1. *Time-frequency mapping*, whereby the input audio signal is decomposed into multiple subbands.
  2. *Psychoacoustic modeling*, which simultaneously operates on the input audio signal to compute certain thresholds using known rules from the psychoacoustic behavior of the human auditory system.
  3. *Quantization and coding*, which, in conjunction with the psychoacoustic model, works on the output of the time-frequency mapper so as to maintain the noise resulting from quantization process at an inaudible level.
  4. *Frame-packing*, which is used to format the quantized audio samples into a decodable bit stream.

  The psychoacoustic model builds on a perceptual phenomenon known as *auditory masking*. Specifically, the human ear does *not* perceive quantization noise in a given frequency band if the average noise power lies below the *masking threshold* (i.e., the threshold of just noticeable distortion). For a given frequency band of interest, the masking threshold varies with frequency across that band. The *min-*

*imum masking threshold* is the one that is employed in the psychoacoustic model on a band-by-band basis. For example, the net result of using the MPEG-1 standard on the two audio channels of a stereo program is that each digitized audio signal, coming in at the rate of 768 kilobits per second (kb/s), is compressed to a rate as low as 16 kb/s. (The incoming data rate of 768 kb/s corresponds to a sampling rate of 48 kHz, with each sample being represented by a 16-bit code word.) Thus the MPEG-1/audio coding standard is suitable for the storage of audio signals in inexpensive media or their transmission over channels with limited bandwidth, while at the same time maintaining perceptual quality.

## Communication Networks[6]

A *communication network* (or simply *network*), illustrated in Figure 4, consists of an interconnection of a number of *routers* made up of intelligent processors (e.g., microprocessors). The primary purpose of these processors is to route data through the network, hence the name. Each router has one or more *hosts* attached to it; hosts are devices that communicate with one another. The network is designed to serve as a shared resource for moving data exchanged between hosts in an efficient manner and to provide a framework to support new applications and services.

The telephone network is an example of a communication network in which *circuit switching* is used to provide a dedicated communication path, or *circuit*, between two hosts. The circuit consists of a connected sequence of links from source to destination. For example, the links may consist of time slots for which a common channel is available for access by a multitude of users. The circuit, once in place, remains uninterrupted for the duration of transmission. Circuit switching is usually controlled by a centralized hierarchical control mechanism with knowledge of the network's organization. To establish a circuit-switched connection, an available path through the network is seized and then dedicated to the exclusive use of the two hosts wishing to communicate. In particular, a call-request signal must propagate all the way to the destination and be acknowledged before transmission can begin. Then, the network is effectively transparent to the users. This means that during the connection time, the bandwidth and resources allocated to the circuit are essentially "owned" by the two hosts until the circuit is disconnected. The circuit

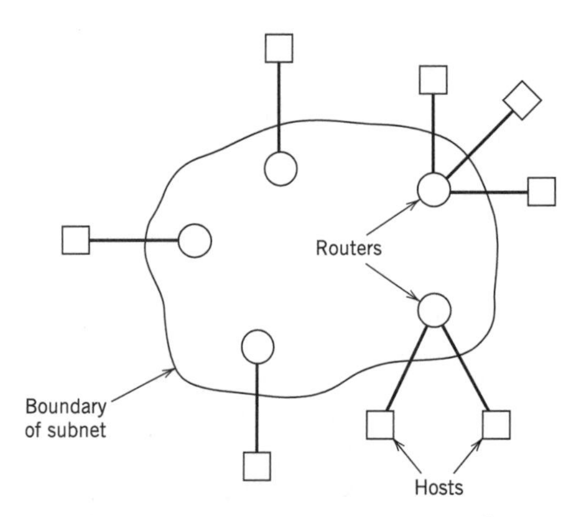


**FIGURE 4** Communication network.

thus represents an efficient use of resources only to the extent that the allocated bandwidth is properly used. Although the telephone network is used to transmit data, voice constitutes the bulk of the network's traffic. Indeed, circuit switching is well suited to the transmission of voice signals, since voice gives rise to a stream traffic and voice conversations tend to be of long duration (about 2 minutes on the average) compared to the time required for setting up the circuit (about 0.1 to 0.5 seconds). Moreover, in most voice conversations, there is information flow for a relatively large percentage of the connection time, which makes circuit switching all the more suitable for voice conversations.

In circuit switching, a communication link is shared between the different sessions using that link on a *fixed* allocation basis. In *packet switching*, on the other hand, the sharing is done on a *demand* basis, so it has an advantage over circuit switching in that when a link has traffic to send, the link may be more fully utilized.

The network principle of packet switching is "store and forward." Specifically, in a *packet-switched network*, any message larger than a specified size is subdivided prior to transmission into segments not exceeding the specified size. The segments are commonly referred to as *packets*. The original message is reassembled at the destination on a packet-by-packet basis. The network may be viewed as a distributed pool of *network resources* (i.e., channel bandwidth, buffers, and switching processors) whose capacity is *shared dynamically* by a community of competing hosts wishing to communicate. In contrast, in a circuit-switched network, resources are dedicated to a pair of hosts for the entire period they are in session. Accordingly, packet switching is far better suited to a computer-communication environment in which bursts of data are exchanged between hosts on an occasional basis. The use of packet switching, however, requires that careful *control* be exercised on user demands; otherwise, the network may be seriously abused.

The design of a *data network* (i.e., a network in which the hosts are all made up of computers and terminals) may proceed in an orderly way by looking at the network in terms of a *layered architecture*, regarded as a hierarchy of nested layers. *Layer* refers to a process or device inside a computer system, designed to perform a specific function. Naturally, the designers of a layer will be intimately familiar with its internal details and operation. At the system level, however, a user views the layer merely as a "black box" that is described in terms of inputs, outputs, and the functional relation between outputs and inputs. In a layered architecture, each layer regards the next lower layer as one or more black boxes with some given functional specification to be used by the given higher layer. Thus, the highly complex communication problem in data networks is resolved as a manageable set of well-defined interlocking functions. It is this line of reasoning that has led to the development of the *open systems interconnection* (OSI)[7] *reference model* by a subcommittee of the International Organization for Standardization. The term *open* refers to the ability of any two systems conforming to the reference model and its associated standards to interconnect.

In the OSI reference model, the communications and related-connection functions are organized as a series of layers, or *levels*, with well-defined *interfaces*, and with each layer built on its predecessor. In particular, each layer performs a related subset of primitive functions, and it relies on the next lower layer to perform additional primitive functions. Moreover, each layer offers certain services to the next higher layer and shields the latter from the implementation details of those services. Between each pair of layers, there is an *interface*. It is the interface that defines the services offered by the lower layer to the upper layer.

The OSI model is composed of seven layers, as illustrated in Figure 5; this figure also includes a description of the functions of the individual layers of the model. Layer $k$ on system $A$, say, communicates with layer $k$ on some other system $B$ in accordance with a

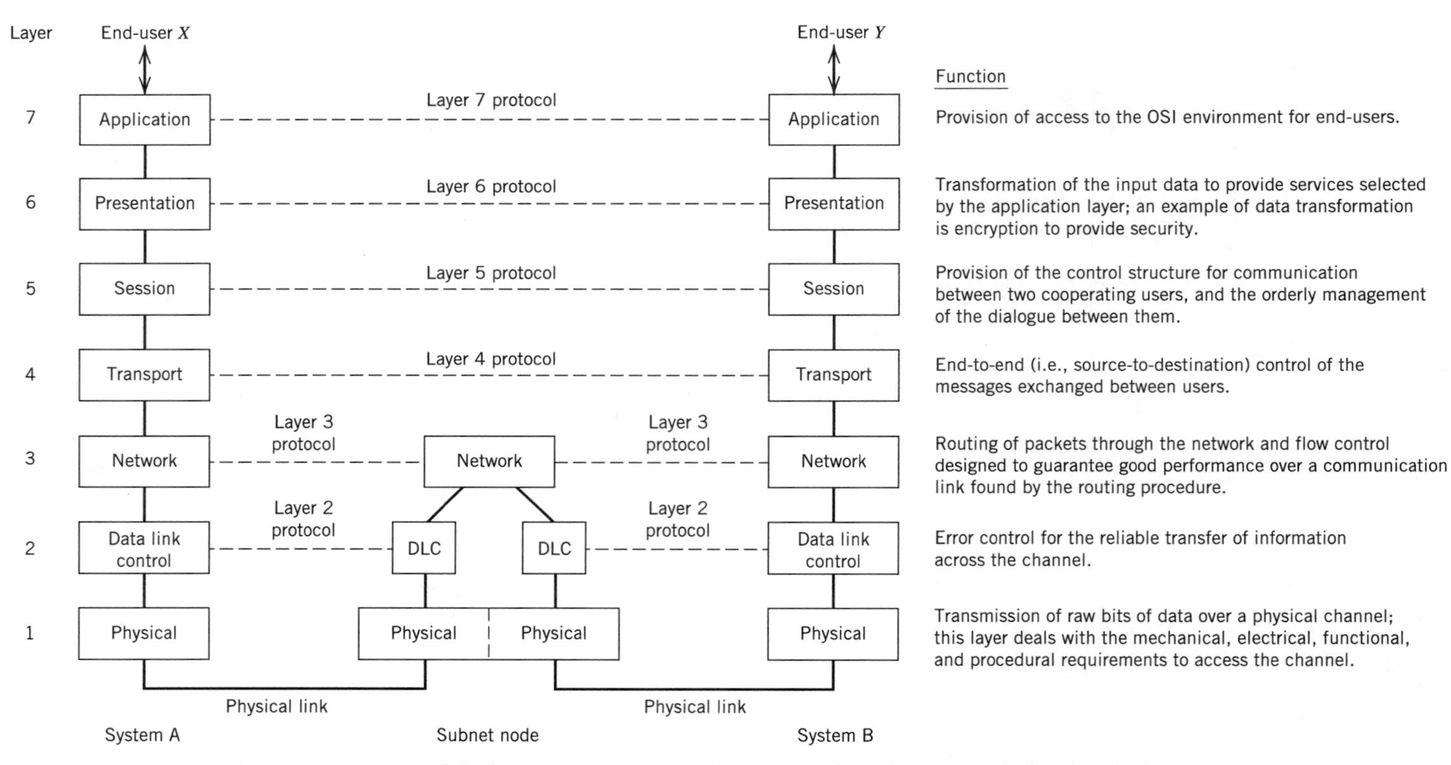


**FIGURE 5** OSI model; the acronym DLC in the middle of the figure stands for *data link control*.

set of rules and conventions, collectively constituting the layer *k protocol*, where $k = 1, 2, \ldots 7$. (The term *protocol* has been borrowed from common usage, describing conventional social behavior between human beings.) The entities that comprise the corresponding layers on different systems are referred to as *peer processes*. In other words, communication is achieved by having the peer processes in two different systems communicate via a protocol, with the protocol itself being defined by a set of rules of procedure. Physical communication between peer processes exists only at layer 1. On the other hand, layers 2 through 7 are in *virtual communication* with their distant peers. However, each of these six layers can exchange data and control information with its neighboring layers (below and above) through layer-to-layer interfaces. In Figure 5, physical communication is shown by solid lines and virtual communication by dashed lines.

Our primary interest in this book is in the physical layer of the OSI model.

## INTERNET

Any discussion of computer networks naturally leads to the *Internet*. In the Internet paradigm, the underlying network technology is decoupled from the applications at hand by adopting an abstract definition of network service. In more specific terms, we may say the following:

- The applications are carried out independently of the technology employed to construct the network.
- By the same token, the network technology is capable of evolving without affecting the applications.

The Internet architecture, depicted in Figure 6, has three functional blocks: hosts, subnets, and routers. The hosts constitute nodes of the network, where data originate or where they are delivered. The routers constitute intermediate nodes that are used to cross subnet boundaries. Within a subnet, all the hosts belonging to that subnet exchange data directly; see, for example, subnets 1 and 3 in Figure 6.

Like other computer networks, the Internet has a layered set of protocols. In particular, the exchange of data between the hosts and routers is accomplished by means of the *Internet protocol* (IP), as illustrated in Figure 7. The IP is a universal protocol that resides in the network layer (i.e., layer 3 of the OSI reference model). It is simple, defining an addressing plan with a built-in capability to transport data in the form of packets from node to node. In crossing a subnetwork boundary, the routers make the decisions as to how the packets addressed for a specified destination should be routed. This is done on the basis of routing tables that are developed through the use of custom protocols for

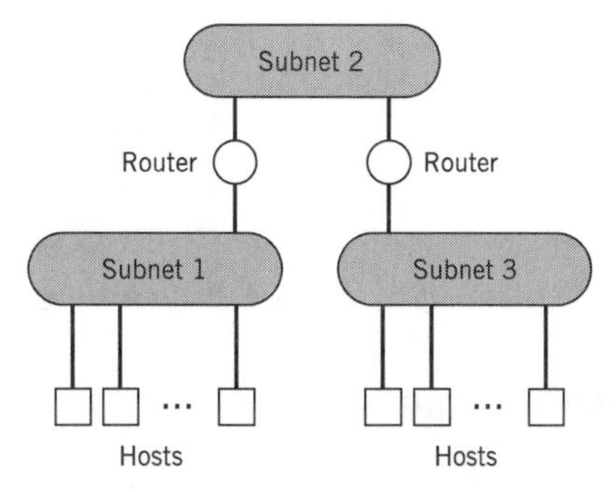


**FIGURE 6** An interconnected network of subnets.

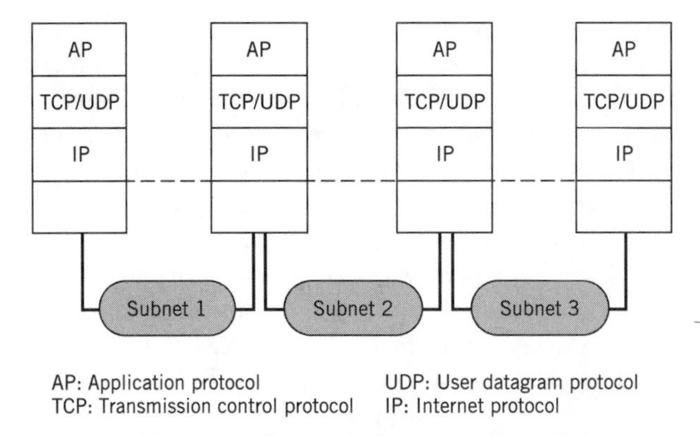


**FIGURE 7** Illustrating the network architecture of the Internet.

exchanging pertinent information with other routers. The net result of using the layered set of protocols is the delivery of *best effort service*. That is, the Internet offers to deliver each packet of data, but there are no guarantees on the transit time experienced in delivery or even whether the packets will be delivered to the intended recipient.

## ■ BROADBAND NETWORKS

With the ever-increasing demand for new services (e.g., video on demand, multimedia communications) and the availability of key enabling technologies (e.g., optical fibers, digital switches), the telephone network is evolving into an all-purpose broadband network known as the *broadband integrated services digital network* (B-ISDN). The underlying technology that makes B-ISDN possible is a user-network interface protocol called the *asynchronous transfer mode (ATM)*. ATM is a high-bandwidth, low-delay, packet-like technique used for switching and multiplexing; it is independent of the physical means of transport. The low-delay feature of the technique is needed to support real-time services such as voice. The high-bandwidth feature is required to handle video on demand. Simply put, ATM is both a technology that is hidden from the users and a connection-oriented service that is visible to the users.

As the name implies, ATM is not synchronous (i.e., tied to a master clock). It allows for the transport of digital information in the form of small, fixed-size packets called *cells*. The key feature of ATM to note here is that the connection-oriented service preserves call sequencing, which means that no reassembly of cells is needed prior to presenting the traffic stream to the destination host. The deployment of a cell-switching technology in B-ISDN is a gigantic break with the traditional use of circuit switching in the telephone network.

The primary purpose of ATM is to allocate *network resources* (i.e., bandwidth, buffers, and processing horsepower) efficiently so as to guarantee the expected *quality of service* (QoS) for each connection. QoS is measured in terms of three parameters:

- *Cell loss ratio*, defined as the ratio of the number of cells lost in transport across the network to the total number of cells pumped into the network.
- *Cell delay*, defined as the time taken for a cell of a particular connection to transit across the network.
- *Cell delay variation*, defined as the dispersion or jitter about the mean cell delay.

Quality of service offered in B-ISDN is to be contrasted with *best effort service* offered by the Internet.

**TABLE 1** *Hierarchy of SONET data rates*

| *Level*[a] | *Data Rate (Mb/s)* |
|---|---|
| OC-1 | 51.84 |
| OC-3 | 155.52 |
| OC-9 | 466.56 |
| OC-12 | 622.08 |
| OC-18 | 933.12 |
| OC-24 | 1,244.16 |
| OC-36 | 1,866.24 |
| OC-48 | 2,488.32 |

[a]OC stands for optical carrier level.

After their generation, the ATM cells are structured for transport across the network. The cells in B-ISDN are placed on an optical transmission system called the *synchronous optical network* (SONET);[8] optical fibers are discussed in the next section. SONET uses time-division multiplexing, whereby the entire bandwidth of an optical fiber is devoted to different incoming data streams on a time-shared basis, hence the need for a synchronous operation. SONET is controlled by a master clock with an accuracy of about 1 part in $10^9$. Thus bits of data are sent on a SONET line at extremely precise intervals, controlled by the master clock. Nevertheless, SONET permits the irregular time arrivals of ATM cells.

The basic SONET frame is a block of 810 bytes put out every 125 $\mu$s for an overall data rate of 51.84 Mb/s. Having 8000 frames every second exactly matches the sampling rate of 8 kHz, which is the standard sampling rate for the digital transmission of voice signals across the telephone network. The basic data rates of 51.84 Mb/s are synchronously byte-interleaved to generate a hierarchy of data rates, as summarized in Table 1.

## Communication Channels

The transmission of information across a communication network is accomplished in the physical layer by means of a *communication channel.* Depending on the mode of transmission used, we may distinguish two basic groups of communication channels: channels based on *guided propagation* and those based on *free propagation.* The first group includes telephone channels, coaxial cables, and optical fibers. The second group includes wireless broadcast channels, mobile radio channels, and satellite channels. These six channels are described in what follows.

(i) As mentioned earlier, a typical telephone network uses circuit switching to establish an end-to-end communication link on a temporary basis. The primary purpose of the network is to ensure that the telephone transmission between a speaker at one end of the link and a listener at the other end is an acceptable substitute for face-to-face conversation. In this form of communication, the message source is the sound produced by the speaker's voice, and the ultimate destination is the listener's ear. The telephone channel, however, supports only the transmission of electrical signals. Accordingly, appropriate transducers are used at the transmitting and receiving ends of the system. Specifically, a *microphone* is placed near the speaker's mouth to convert sound waves into an electrical signal, and the electrical signal is converted back into acoustic form by means of a *moving-coil receiver* placed near the listener's ear. Present-day designs of these two transducers have been perfected so as to respond well to frequencies ranging from 20 to 8000 Hz; moreover, a pair of them can be compactly packaged inside a single telephone set that is easy to speak into or listen from. The telephone channel is a *bandwidth-limited channel.* The restriction on bandwidth arises from the requirement of sharing the channel among a multitude of users at any one time. A practical solution to the telephonic communication problem must therefore minimize the channel bandwidth requirement, subject to a satisfactory transmission of human voice. To meet this requirement, the transducers and channel specifications must conform to standards based on subjective tests that are performed on the intelligibility, or articulation, of telephone signals by representative male and female speakers. A speech signal (male or female) is essentially limited to a band from 300 to 3100 Hz in the sense that frequencies outside this band do not contribute much to articulation efficiency. This frequency band may therefore be viewed as a rough guideline for the passband of a telephone channel that provides a satisfactory service, as illustrated in Figure 8 for a typical toll connection. Figure 8*a* shows the insertion loss of the channel plotted versus frequency; *insertion loss* (in dB) is defined as $10 \log_{10}(P_0/P_L)$, where $P_L$ is the power delivered to a load from a source via the channel and $P_0$ is the power delivered to the same load when it is connected directly to the source. Figure 8*b* shows the corresponding plot of the envelope (group) delay (in milliseconds) versus frequency; *envelope delay* is defined as the negative of the derivative of the phase response with respect to the angular frequency $\omega = 2\pi f$. The plots of Figure 8 clearly illustrate the *dispersive* nature of the telephone channel.

The telephone channel is built using twisted pairs for signal transmission. A *twisted pair* consists of two solid copper conductors, each of which is encased in a polyvinylchloride (PVC) sheath. Typically, each pair has a twist rate of 2 to 12 twists per foot, and a characteristic impedance of 90 to 110 ohms. Twisted pairs are usually made up into cables, with each cable consisting of many pairs in close proximity to

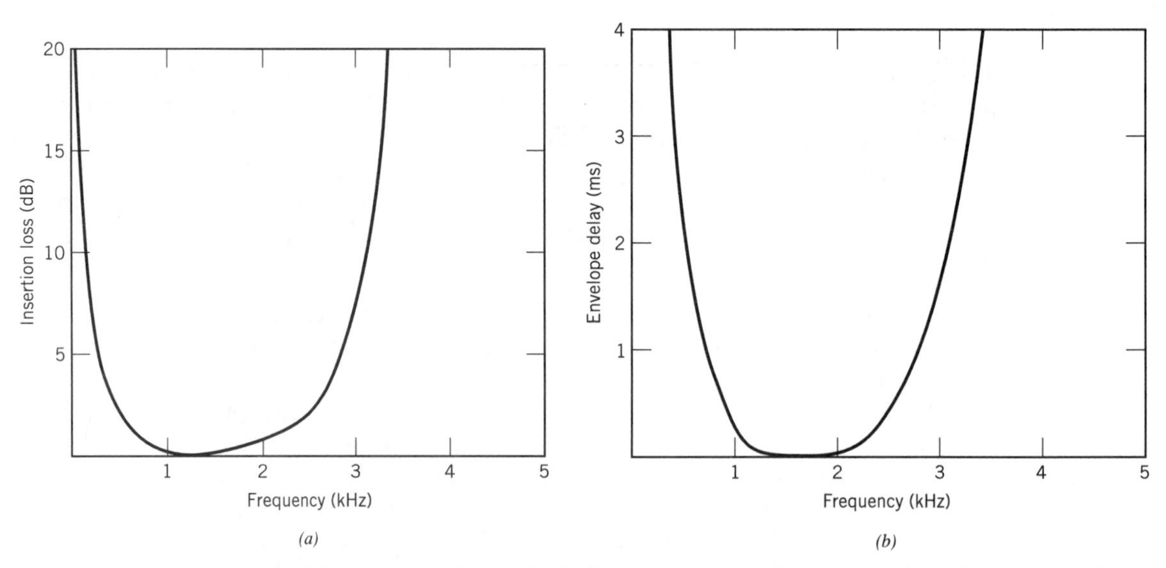


**FIGURE 8** Characteristics of typical telephone connection: (*a*) Insertion loss. (*b*) Envelope delay. (Adapted from Bellamy, 1991.)

each other. Twisted pairs are naturally susceptible to *electromagnetic interference* (EMI), the effects of which are mitigated through twisting the wires.

(ii) A *coaxial cable* consists of an inner conductor and an outer conductor, separated by a dielectric insulating material. The inner conductor is made of a copper wire encased inside the dielectric material. As for the outer conductor, it is made of copper, tinned copper, or copper-coated steel. Typically, a coaxial cable has a characteristic impedance of 50 or 75 ohms. Compared to a twisted-pair cable, a coaxial cable offers a greater degree of immunity to EMI. Moreover, because of their much higher bandwidth, coaxial cables can support the transmission of digital data at much higher bit rates than twisted pairs. Rates up to 20 Mb/s are feasible using coaxial cables, with 10 Mb/s being the standard.

Whereas the use of a twisted pair has been confined mainly to point-to-point service, a coaxial cable can operate as a multiple-access medium by using high-impedance taps. A common application of coaxial cables is as the transmission medium for local area networks in an office environment.

Another common application of coaxial cables is in *cable-television systems*, also known as *community-antenna television (CATV) systems*. In this application coaxial cables are used to distribute television, audio, and data signals from the head end to the subscribers. The *head end* is the central originating unit of the CATV system, where all signals are carried and processed.

(iii) An *optical fiber* is a dielectric wave guide that transports light signals from one place to another just as a twisted-wire pair or a coaxial cable transports electrical signals. It consists of a central *core* within which the propagating electromagnetic field is confined and which is surrounded by a *cladding* layer, which is itself surrounded by a thin protective *jacket*.[9] The core and cladding are both made of pure silica glass, whereas the jacket is made of plastic. Optical fibers have unique characteristics that make them highly attractive as a transmission medium. In particular, they offer the following unique characteristics:

- *Enormous potential bandwidth*, resulting from the use of optical carrier frequencies around $2 \times 10^{14}$ Hz; with such a high carrier frequency and a bandwidth roughly equal to 10 percent of the carrier frequency, the theoretical bandwidth of a lightwave system is around $2 \times 10^{13}$ Hz, which is very large indeed.
- *Low transmission losses*, as low as 0.1 dB/km.
- *Immunity to electromagnetic interference*, which is an inherent characteristic of an optical fiber viewed as a dielectric waveguide.
- *Small size and weight*, characterized by a diameter no greater than that of a human hair.
- *Ruggedness and flexibility*, exemplified by very high tensile strengths and the possibility of being bent or twisted without damage.

Last, but by no means least, optical fibers offer the potential for low-cost line communications since they are fabricated from sand, which, unlike the copper used in metallic conductors, is not a scarce resource. The unique properties of optical fibers have fuelled phenomenal advances in lightwave systems technology, which have, in turn, revolutionized long-distance communications and continue to do so.

(iv) *Wireless broadcast channels* support the transmission of radio and television signals. The information-bearing signal, representing speech, music, or pictures, is modulated onto a carrier frequency that identifies the transmitting station; modulation is described in the next section. The transmission originates from an *antenna* that acts as the transition or matching unit between the source of the modulated signal and electromagnetic waves in free space. The objective in designing the antenna is to excite the waves in the required direction or directions, as efficiently as possible. Typically, the transmitting antenna is mounted on a tower to provide an unobstructed view of the surrounding area, as far afield as possible. By virtue of the phenomenon of *diffraction*, which is a fundamental property of wave motion, radio waves are bent around the earth's surface. Propagation beyond the line of sight is thereby made possible, albeit with somewhat greater loss than is incurred in free space.

At the receiving end, an antenna is used to pick up the radiated waves, establishing a communication link to the transmitter. Most radio receivers are of the *superheterodyne* type. This technique consists of down-converting the received signal to some convenient frequency band, called the *intermediate frequency (IF) band*, and then recovering the original information-bearing signal by means of an appropriate detector.

(v) A *mobile radio channel* extends the capability of the public telecommunications network by introducing *mobility* into the network by virtue of its ability to broadcast. The term *mobile radio* is usually meant to encompass terrestrial situations where a radio transmitter or receiver is capable of being moved, regardless of whether it actually moves or not. The major propagation effects encountered in the use of a mobile radio in built-up areas are due to the fact that the antenna of the mobile unit may lie well below the surrounding buildings. Simply put, there is no "line-of-sight" path for communication; rather, radio propagation takes place mainly by way of scattering from the surfaces of the surrounding buildings and by diffraction over and/or around them. The end result is that energy reaches the receiving antenna via more than one path. In a mobile radio environment, we thus speak of a *multipath phenomenon* in that the various incoming radio waves reach their destination from different directions and with different time delays. Indeed, there may be a multitude of propagation paths with different electrical lengths, and their contributions to the received signal could combine in a variety of ways. Consequently, the received signal strength varies with location in a very complicated fashion, and so a mobile radio channel may be viewed as a *linear time-varying channel* that is statistical in nature.

(vi) Finally, a *satellite channel* adds another invaluable dimension to the public telecommunications network by providing *broad-area coverage* in both a continental and an intercontinental sense. Moreover, access to remote areas not covered by conventional cable or fiber communications is also a distinct feature of satellites. In almost all satellite communication systems, the satellites are placed in *geostationary orbit*. For the orbit to be geostationary, it has to satisfy two requirements. First, the orbit is *geosynchronous*, which requires the satellite to be at an altitude of 22,300 miles; a geosynchronous satellite orbits the Earth in 24 hours (i.e., the satellite is synchronous with the Earth's rotation). Second, the satellite is placed in orbit directly above the equator on an eastward heading (i.e., it has zero inclination). Viewed from Earth, a satellite in geostationary orbit appears to be stationary in the sky. Consequently, an Earth station does *not* have to track the satellite; rather, it merely has to point its antenna along a fixed direction, pointing toward the satellite. By so doing, the system design is simplified considerably. Communications satellites in geostationary orbit offer the following unique system capabilities:

- Broad-area coverage.
- Reliable transmission links.
- Wide transmission bandwidths.

In terms of services, satellites can provide fixed point-to-point links extending over long distances and into remote areas, communication to mobile platforms (e.g., aircraft, ships), or broadcast capabilities. Indeed, communications satellites play a key role in the notion of the whole world being viewed as a "global village." In a typical satellite communication system, a message signal is transmitted from an Earth station via an *uplink* to a satellite, amplified in a *transponder* (i.e., electronic circuitry) on board the satellite, and then retransmitted from the satellite via a *downlink* to another Earth station. With the satellite positioned in geostationary orbit, it is always visible to all the Earth stations located inside the satellite antenna's coverage zones on the Earth's surface. In effect, the satellite acts as a powerful *repeater* in the sky. The most popular frequency band for satellite communications is 6 GHz for the uplink and 4 GHz for the downlink. The use of this frequency band offers the following attributes:

- Relatively inexpensive microwave equipment.
- Low attenuation due to rainfall; rainfall is a primary atmospheric cause of signal loss.
- Insignificant sky background noise; the sky background noise (due to random noise emissions from galactic, solar, and terrestrial sources) reaches its lowest level between 1 and 10 GHz.

In the 6/4-GHz band, a typical satellite is assigned a 500 MHz bandwidth that is divided among 12 transponders on board the satellite. Each transponder, using approximately 36 MHz of the satellite bandwidth, corresponds to a specific radio channel. A single transponder can carry at least one color television signal, 1200 voice circuits, or digital data at a rate of 50 Mb/s.

To summarize, a communication channel is central to the operation of a communication system. Its properties determine both the information-carrying capacity of the system and the quality of service offered by the system. We may classify communication channels in different ways:

- A channel may be *linear* or *nonlinear*; a wireless radio channel is linear, whereas a satellite channel is usually (but not always) nonlinear.
- A channel may be *time invariant* or *time varying*; an optical fiber is time invariant, whereas a mobile radio channel is typically time varying.
- A channel may be *bandwidth limited* or *power limited* (i.e., limited in the available transmitted power); a telephone channel is bandwidth limited, whereas an optical fiber link and a satellite channel are both power limited.

Now that we have some understanding of sources of information and communication channels, we may return to the block diagram of a communication system shown in Figure 1.

## Modulation Process

The purpose of a communication system is to deliver a message signal from an information source in recognizable form to a user destination, with the source and the user being physically separated from each other. To do this, the transmitter modifies the message signal into a form suitable for transmission over the channel. This modification is achieved by means of a process known as *modulation*, which involves varying some parameter of a *carrier wave* in accordance with the message signal. The receiver re-creates the original message signal from a degraded version of the transmitted signal after propagation through the channel. This re-creation is accomplished by using a process known as *demodulation*, which is the reverse of the modulation process used in the transmitter. However, owing to the unavoidable presence of noise and distortion in the received signal, we find that the receiver cannot re-create the original message signal exactly. The resulting degradation in overall system performance is influenced by the type of modulation scheme used. Specifically, we find that some modulation schemes are less sensitive to the effects of noise and distortion than others.

We may classify the modulation process into *continuous-wave modulation* and *pulse modulation*. In continuous-wave (CW) modulation, a sinusoidal wave is used as the carrier. When the amplitude of the carrier is varied in accordance with the message signal, we have *amplitude modulation* (AM), and when the angle of the carrier is varied, we have *angle modulation*. The latter form of CW modulation may be further subdivided into *frequency modulation* (FM) and *phase modulation* (PM), in which the instantaneous frequency and phase of the carrier, respectively, are varied in accordance with the message signal.

In pulse modulation, on the other hand, the carrier consists of a periodic sequence of rectangular pulses. Pulse modulation can itself be of an analog or digital type. In *analog pulse modulation*, the amplitude, duration, or position of a pulse is varied in accordance with sample values of the message signal. In such a case, we speak of *pulse-amplitude modulation* (PAM), *pulse-duration modulation* (PDM), and *pulse-position modulation* (PPM).

The standard digital form of pulse modulation is known as *pulse-code modulation* (PCM) that has no CW counterpart. PCM starts out essentially as PAM, but with an important modification: the amplitude of each modulated pulse (i.e., *sample* of the original message signal) is *quantized* or *rounded off* to the nearest value in a prescribed set of *discrete* amplitude levels and then *coded* into a corresponding sequence of binary symbols. The binary symbols 0 and 1 are themselves represented by pulse signals that are suitably shaped for transmission over the channel. In any event, as a result of the quantization process, some information is always lost and the original message signal cannot therefore be reconstructed exactly. However, provided that the number of quantizing (discrete amplitude) levels is large enough, the distortion produced by the quantization process is not discernible to the human ear in the case of a speech signal or the human eye in the case of a two-dimensional image. Among all the different modulation schemes, pulse-code modulation has emerged as the preferred method of modulation for the transmission of analog message signals for the following reasons:

- *Robustness* in noisy environments by *regenerating* the transmitted signal at regular intervals
- *Flexible* operation
- *Integration* of diverse sources of information into a common format
- *Security* of information in its transmission from source to destination

In introducing the idea of modulation, we stressed its importance as a process that ensures the transmission of a message signal over a prescribed channel. There is another important benefit, namely, *multiplexing*, that results from the use of modulation. Multiplexing is the process of combining several message signals for their simultaneous transmission over the same channel. Three commonly used methods of multiplexing are as follows:

- *Frequency-division multiplexing* (FDM), in which CW modulation is used to translate each message signal to reside in a specific frequency slot inside the passband of the channel by assigning it a distinct carrier frequency; at the receiver, a bank of filters is used to separate the different modulated signals and prepare them individually for demodulation.

- *Time-division multiplexing* (TDM), in which pulse modulation is used to position samples of the different message signals in nonoverlapping time slots.
- *Code-division multiplexing* (CDM), in which each message signal is identified by a distinctive code.

In FDM the message signals overlap with each other at the channel input; hence the system may suffer from *crosstalk* (i.e., interaction between message signals) if the channel is nonlinear. In TDM the message signals use the full passband of the channel, but on a time-shared basis. In CDM the message signals are permitted to overlap in both time and frequency across the channel.

Mention should also be made of *wavelength-division multiplexing* (*WDM*), which is special to optical fibers. In WDM, wavelength is used as a new degree of freedom by concurrently operating distinct portions of the wavelength spectrum (i.e., distinct *colors*) that are accessible within the optical fiber. However, recognizing the reciprocal relationship that exists between the wavelength and frequency of an electromagnetic wave, we may say that WDM is a form of FDM.

## Analog and Digital Types of Communication

Typically, in the design of a communication system the information source, communication channel, and information sink (end user) are all specified. The challenge is to design the transmitter and the receiver with the following guidelines in mind:

- Encode/modulate the message signal generated by the source of information, transmit it over the channel, and produce an "estimate" of it at the receiver output that satisfies the requirements of the end user.
- Do all of this at an affordable cost.

We have the option of using a digital or analog communication system.

Consider first the case of a *digital communication* system represented by the block diagram of Figure 9, the rationale for which is rooted in information theory. The functional blocks of the transmitter and the receiver, starting from the far end of the channel, are paired as follows:

- *Source encoder-decoder.*
- *Channel encoder-decoder.*
- *Modulator-demodulator.*

The source encoder removes redundant information from the message signal and is responsible for the efficient use of the channel. The resulting sequence of symbols is called the *source code word*. The data stream is processed next by the channel encoder, which produces a new sequence of symbols called the *channel code word*. The channel code word is longer than the source code word by virtue of the *controlled* redundancy built into its construction. Finally, the modulator represents each symbol of the channel code word by a corresponding analog symbol, appropriately selected from a finite set of possible analog symbols. The sequence of analog symbols produced by the modulator is called a *waveform*, which is suitable for transmission over the channel. At the receiver, the channel output (received signal) is processed in reverse order to that in the transmitter, thereby recon-

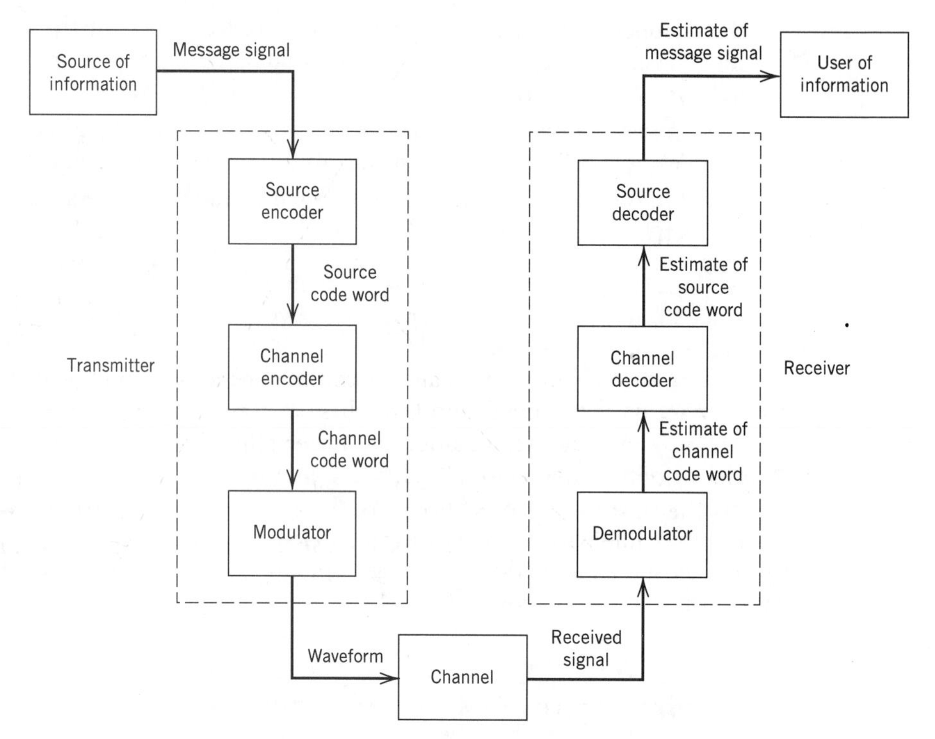


**FIGURE 9** Block diagram of digital communication system.

structing a recognizable version of the original message signal. The reconstructed message signal is finally delivered to the user of information at the destination. From this description it is apparent that the design of a digital communication system is rather complex in conceptual terms but easy to build. Moreover, the system is *robust*, offering greater tolerance of physical effects (e.g., temperature variations, aging, mechanical vibrations) than its analog counterpart.

In contrast, the design of an *analog communication system* is simple in conceptual terms but difficult to build because of stringent requirements on linearity and system adjustment. For example, voice communication requires nonlinear distortion products at least 40 dB below the wanted message signal. In signal-processing terms, the transmitter consists of a modulator and the receiver consists of a demodulator, the details of which are determined by the type of CW modulation used.

The conceptual simplicity of analog communications is due to the fact that analog modulation techniques, exemplified by their wide use in radio and television, make relatively superficial changes to the message signal in order to prepare it for transmission over the channel. More specifically, there is no significant effort made by the system designer to tailor the waveform of the transmitted signal to suit the channel at any deeper level. On the other hand, digital communication theory endeavors to find a finite set of waveforms that are closely matched to the characteristics of the channel and which are therefore more tolerant of channel impairments. In so doing, reliable communication is established over the channel. In the selection of good waveforms for digital communication over a noisy channel, the design is influenced solely by the channel characteristics. However, once the appropriate set of waveforms for transmission over the channel has been selected, the source information can be encoded into the channel waveforms, and the efficient transmission of information from the source to the user is thereby ensured. In summary, the use of digital communications provides the capability for information transmission that is both *efficient* and *reliable*.

From this discussion, it is apparent that the use of digital communications requires a considerable amount of electronic circuitry, but nowadays electronics are inexpensive, due to the ever-increasing availability of very-large-scale integrated (VLSI) circuits in the form of silicon chips. Thus although cost considerations used to be a factor in selecting analog communications over digital communications in the past, that is no longer the case.

Despite the trend toward the ever-increasing use of digital communications, a strong case can be made for the study of analog communications for two important reasons:

1. As long as we hear and see analog communications around us via radio and television, we need to understand how these communications systems work. Moreover, the study of analog modulation motivates other digital modulation schemes.
2. Analog devices and circuits have a natural affinity for operating at very high speeds and they consume very little power compared to their digital counterparts. Accordingly, the implementation of very high-speed or very low-power communication systems dictates the use of an analog approach.

## Shannon's Information Capacity Theorem

The goal of a communication system designer is to configure a system that transports a message signal from a source of interest across a noisy channel to a user at the other end of the channel with the following objective:

> The message signal is delivered to the user both efficiently and reliably, subject to certain design constraints: allowable transmit power, available channel bandwidth, and affordable cost of building the system.

In the case of a digital communication system, reliability is commonly expressed in terms of *bit error rate* (BER) or *probability of bit error* measured at the receiver output. Clearly, the smaller the BER, the more reliable the communication system is. A question that comes to mind in this context is whether it is possible to design a communication system that operates with zero BER even through the channel is noisy. In an *ideal* setting, the answer to this question is an emphatic *yes*. The answer is embodied in one of Shannon's celebrated theorems,[10] which is called the information capacity theorem.

Let $B$ denote the channel bandwidth, and let SNR denote the received signal-to-noise ratio. The *information capacity theorem* states that ideally these two parameters are related as

$$C = B \log_2(1 + \text{SNR}) \text{ b/s} \tag{1}$$

where $C$ is the information capacity of the channel. The *information capacity* is defined as the maximum rate at which information can be transmitted across the channel without error; it is measured in *bits per second* (b/s). For a prescribed channel bandwidth $B$ and received SNR, the information capacity theorem tells us that a message signal can be transmitted through the system without error even when the channel is noisy, provided that the *actual signaling rate* $R$ in bits per second, at which data are transmitted through the channel, is less than the information capacity $C$.

Unfortunately, Shannon's information capacity theorem does not tell us how to design the system. Nevertheless, from a design point of view, the theorem is very valuable for the following reasons:

1. The information capacity theorem provides a *bound* on what rate of data transmission is theoretically attainable for prescribed values of channel bandwidth $B$ and received SNR. On this basis, we may use the ratio

$$\eta = \frac{R}{C}$$

as a measure of the *efficiency* of the digital communication system under study. The closer $\eta$ is to unity, the more efficient the system is.

2. Equation (1) provides a basis for the trade-off between channel bandwidth $B$ and received SNR. In particular, for a prescribed signaling rate $R$, we may reduce the required SNR by increasing the channel bandwidth $B$, hence the motivation for using a wideband modulated scheme (e.g., pulse-code modulation) for improved noise performance.
3. Equation (1) provides an idealized framework for comparing the noise performance of one modulation scheme against another.

## A Digital Communication Problem

When we speak of a digital communication system having a low bit error rate, say, the implication is that only a small fraction in a long stream of binary symbols is decoded in error by the receiver. The issue of the receiver determining whether a binary symbol sent over a noisy channel is decoded in error or not is of fundamental importance to the design of digital communication systems. It is therefore appropriate briefly to discuss this basic issue so as to motivate the study of communication systems.

Suppose we have a random binary signal, $m(t)$, consisting of symbols 1 and 0 that are equally likely. Symbol 1 is represented by a constant level $+1$, and symbol 0 is represented by a constant level $-1$, each of which lasts for a duration $T$. Such a signal may represent the output of a digital computer or the digitized version of a speech signal. To facilitate the transmission of this signal over a communication channel, we employ a simple modulation scheme known as *phase-shift keying*. Specifically, the information bearing signal $m(t)$ is multiplied by a sinusoidal carrier wave $A_c \cos(2\pi f_c t)$, where $A_c$ is the carrier amplitude, $f_c$ is the carrier frequency, and $t$ is time. Figure 10*a* shows a block diagram of the transmitter, the output of which is defined by

$$s(t) = \begin{cases} A_c \cos(2\pi f_c t) & \text{for symbol 1} \\ -A_c \cos(2\pi f_c t) & \text{for symbol 0} \end{cases} \tag{2}$$

where $0 \le t \le T$. The carrier frequency $f_c$ is a multiple of $1/T$.

The channel is assumed to be distortionless but noisy, as depicted in Figure 10*b*. The received signal $x(t)$ is thus defined by

$$x(t) = s(t) + w(t) \tag{3}$$

where $w(t)$ is the additive channel noise.

The receiver consists of a correlator followed by a decision-making device, as depicted in Figure 10*c*. The *correlator* multiplies the received signal $x(t)$ by a locally generated

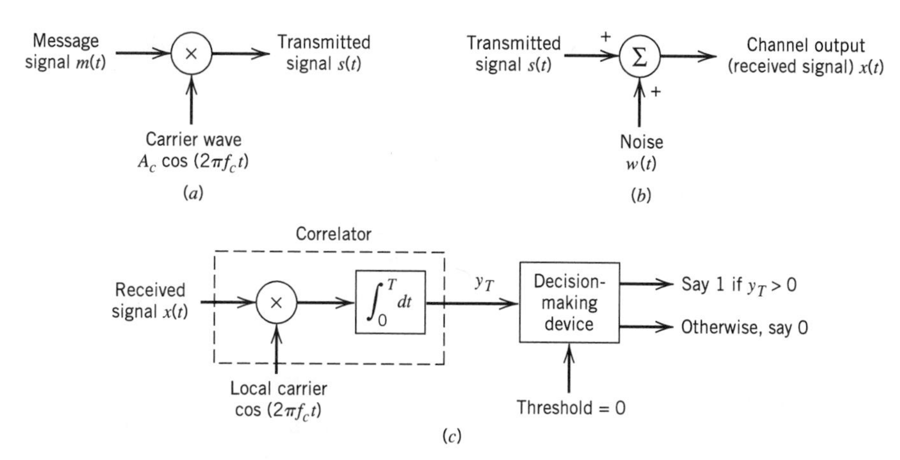


**FIGURE 10** Elements of a digital communication system. (*a*) Block diagram of transmitter. (*b*) Block diagram of channel. (*c*) Block diagram of receiver.

carrier $\cos(2\pi f_c t)$ and then integrates the product over the symbol interval $0 \leq t \leq T$, producing the output

$$y_T = \int_0^T x(t)\cos(2\pi f_c t)\, dt \tag{4}$$

Substituting Equations (2) and (3) into (4) and invoking the assumption that the carrier frequency $f_c$ is a multiple of $1/T$, we obtain (after the simplification of terms)

$$y_T = \begin{cases} +\dfrac{A_c}{2} + w_T & \text{for symbol 1} \\ -\dfrac{A_c}{2} + w_T & \text{for symbol 0} \end{cases} \tag{5}$$

where $w_T$ is the contribution of the correlator output due to the channel noise $w(t)$. To reconstruct the original binary signal $m(t)$, the correlator output $y_T$ is compared against a threshold of zero volts by the *decision-making device*, the operation of which is based on the following rule:

> If the correlator output $y_T$ is greater than zero, the receiver outputs symbol 1; otherwise, it outputs symbol 0.

With this background, we may now discuss/raise some basic issues. First, from Fourier analysis we find that the *time-bandwidth product* of a pulse signal is constant. This means that the bandwidth of a rectangular pulse of duration $T$ is inversely proportional to $T$. The transmitted signal in Figure 10*a* consists of the product of this rectangular signal and the sinusoidal carrier $A_c \cos(2\pi f_c t)$. The multiplication of a signal by a sinusoid has the effect of shifting the Fourier transform of the signal to the right by $f_c$ and to the left by an equal amount, except for the scaling factor of 1/2. It follows therefore that the bandwidth of the transmitted signal $m(t)$, and therefore the required channel bandwidth, is inversely proportional to the reciprocal of the symbol duration $T$. For the problem at hand, the reciprocal of $T$ is also the signaling rate of the system in b/s.

There are, however, some other issues that require theoretical considerations:

1. What is the justification for the receiver structure of Figure 10*c*?
2. The noise contribution $w_T$ is the value of a random variable $W$ produced by sampling a certain realization $w(t)$ of the channel noise at time $t = T$ in accordance with Equations (3) and (4). How do we relate the statistics of the random variable $W$ to the statistical characteristics of the channel noise?
3. The receiver of Figure 10*c* makes occasional errors due to the random nature of the correlator output. That is, the receiver decides in favor of symbol 0 given that symbol 1 was actually transmitted, and vice versa. What is the probability of decision errors?

Moreover, there are some important practical issues that need attention:

1. Channel bandwidth is a highly valuable resource. How do we choose a modulation scheme that conserves bandwidth in a cost-effective manner?
2. The binary signal $m(t)$ may include redundant symbols introduced into it through the use of channel encoding so as to provide protection against channel noise. How do we design the channel encoder in the transmitter and the channel decoder in the receiver so as to come very close to Shannon's information capacity theorem in a physically realizable manner?
3. The locally generated carrier in the receiver of Figure 10*c* is physically separate from the carrier source used for modulation in the transmitter. How do we *synchronize* the receiver to the transmitter with respect to both the carrier phase and symbol timing so as to justify the use of Equation (4) as the basis of decision-making in the reconstruction of the original binary signal?

The theoretical and practical issues raised here in the context of the simple digital communication system of Figure 10 are addressed in the following chapters of the book.

## *Historical Notes*[11]

A preview of communications would be incomplete without a history of the subject. In this final section of this introductory chapter we present some historical notes on communications; each paragraph focuses on some important and related events. It is hoped that this material will provide a sense of inspiration and motivation for the reader.

In 1837, the telegraph was perfected by Samuel Morse, a painter. With the words "What hath God wrought," transmitted by Morse's electric telegraph between Washington, D.C., and Baltimore, Maryland, in 1844, a completely revolutionary means of real-time, long-distance communications was triggered. The telegraph is the forerunner of digital communications in that the *Morse code* is a *variable-length* ternary code using an alphabet of four symbols: a dot, a dash, a letter space, and a word space; short sequences represent frequent letters, whereas long sequences represent infrequent letters. This type of signaling is ideal for manual keying. Subsequently, Emile Baudot developed a *fixed-length* binary code for telegraphy in 1875. In Baudot's telegraphic code, well-suited for use with teletypewriters, each code word consists of five equal-length code elements, and each element is assigned one of two possible states: a *mark* or a *space* (i.e., symbol 1 or 0 in today's terminology).

In 1864, James Clerk Maxwell formulated the electromagnetic theory of light and predicted the existence of radio waves; the underlying set of equations bears his name. The existence of radio waves was established experimentally by Heinrich Hertz in 1887. In 1894, Oliver Lodge demonstrated wireless communication over a relatively short distance (150 yards). Then, on December 12, 1901, Guglielmo Marconi received a radio signal at Signal Hill in Newfoundland; the radio signal had originated in Cornwall, England, 1700 miles away across the Atlantic. The way was thereby opened toward a tremendous broadening of the scope of communications. In 1906, Reginald Fessenden, a self-educated academic, made history by conducting the first radio broadcast.

In 1875, the *telephone* was invented by Alexander Graham Bell, a teacher of the deaf. The telephone made real-time transmission of speech by electrical encoding and replication of sound a practical reality. The first version of the telephone was crude and weak, enabling people to talk over short distances only. When telephone service was only a few years old, interest developed in automating it. Notably, in 1897, A. B. Strowger, an undertaker from Kansas City, Missouri, devised the automatic *step-by-step switch* that bears his name; of all the electromechanical switches devised over the years, the Strowger switch was the most popular and widely used.

In 1904, John Ambrose Fleming invented the *vacuum-tube diode*, which paved the way for the invention of the *vacuum-tube triode* by Lee de Forest in 1906. The discovery of the triode was instrumental in the development of transcontinental telephony in 1913 and signaled the dawn of wireless voice communications. Indeed, until the invention and perfection of the transistor, the triode was the supreme device for the design of electronic amplifiers.

In 1918, Edwin H. Armstrong invented the *superheterodyne radio* receiver; to this day, almost all radio receivers are of this type. In 1933, Armstrong demonstrated another revolutionary concept, namely, a modulation scheme that he called *frequency modulation* (FM); Armstrong's paper making the case for FM radio was published in 1936.

The first all-electronic *television* system was demonstrated by Philo T. Farnsworth in 1928, and then by Vladimir K. Zworykin in 1929. By 1939, the British Broadcasting Corporation (BBC) was broadcasting television on a commercial basis.

In 1928, Harry Nyquist published a classic paper on the theory of signal transmission in telegraphy. In particular, Nyquist developed criteria for the correct reception of telegraph signals transmitted over dispersive channels in the absence of noise. Much of Nyquist's early work was applied later to the transmission of digital data over dispersive channels.

In 1937, Alec Reeves invented *pulse-code modulation* (PCM) for the digital encoding of speech signals. The technique was developed during World War II to enable the encryption of speech signals; indeed, a full-scale, 24-channel system was used in the field by the United States military at the end of the war. However, PCM had to await the discovery of the transistor and the subsequent development of large-scale integration of circuits for its commercial exploitation.

In 1943, D. O. North devised the *matched filter* for the optimum detection of a known signal in additive white noise. A similar result was obtained in 1946 independently by J. H. Van Vleck and D. Middleton, who coined the term *matched filter*.

In 1947, the geometric representation of signals was developed by V. A. Kotel'nikov in a doctoral dissertation presented before the Academic Council of the Molotov Energy Institute in Moscow. This method was subsequently brought to full fruition by John M. Wozencraft and Irwin M. Jacobs in a landmark textbook published in 1965.

In 1948, the theoretical foundations of digital communications were laid by Claude Shannon in a paper entitled "A Mathematical Theory of Communication." Shannon's paper was received with immediate and enthusiastic acclaim. It was perhaps this response that emboldened Shannon to amend the title of his paper to "The Mathematical Theory of Communication" when it was reprinted a year later in a book co-authored with Warren Weaver. It is noteworthy that prior to the publication of Shannon's 1948 classic paper, it was believed that increasing the rate of information transmission over a channel would increase the probability of error; the communication theory community was taken by surprise when Shannon proved that this was not true, provided that the transmission rate was below the channel capacity. Shannon's 1948 paper was followed by some significant advances in coding theory, which include the following:

- Development of the first nontrivial *error-correcting codes* by M. J. E. Golay in 1949 and Richard W. Hamming in 1950.
- Development of *turbo codes* by C. Berrou, A. Glavieux, and P. Thitimajshima in 1993; turbo codes provide near-optimum error-correcting coding and decoding performance in the Shannon sense.

The *transistor* was invented in 1948 by Walter H. Brattain, John Bardeen, and William Shockley at Bell Laboratories. The first silicon integrated circuit (IC) was produced by Robert Noyce in 1958. These landmark innovations in solid-state devices and integrated circuits led to the development of *very-large-scale integrated* (VLSI) circuits and single-chip *microprocessors*, and with them the nature of signal processing and the telecommunications industry changed forever.

The invention of the transistor in 1948 spurred the application of electronics to switching and digital communications. The motivation was to improve reliability, increase capacity, and reduce cost. The first call through a stored-program system was placed in March 1958 at Bell Laboratories, and the first commercial telephone service with digital switching began in Morris, Illinois, in June 1960. The first *T-1 carrier system* transmission was installed in 1962 by Bell Laboratories.

During the period 1943 to 1946, the first electronic digital computer, called the ENIAC, was built at the Moore School of Electrical Engineering of the University of Pennsylvania under the technical direction of J. Presper Eckert, Jr., and John W. Mauchly. However, John von Neumann's contributions were among the earliest and most fundamental to the theory, design, and application of digital computers, which go back to the first draft of a report written in 1945. Computers and terminals started communicating with each other over long distances in the early 1950s. The links used were initially voice-grade telephone channels operating at low speeds (300 to 1200 b/s). Various factors have contributed to a dramatic increase in data transmission rates; notable among them are the idea of *adaptive equalization*, pioneered by Robert Lucky in 1965, and efficient modulation techniques, pioneered by G. Ungerboeck in 1982. Another idea widely employed in computer communications is that of *automatic repeat-request* (ARQ). The ARQ method was originally devised by H. C. A. van Duuren during World War II and published in 1946. It was used to improve radio-telephony for telex transmission over long distances.

From 1950 to 1970, various studies were made on *computer networks*. However, the most significant of them in terms of impact on computer communications was the Advanced Research Project Agency Network (ARPANET), first put into service in 1971. The development of ARPANET was sponsored by the Advanced Research Projects Agency of the U.S. Department of Defense. The pioneering work in *packet switching* was done on ARPANET. In 1985, ARPANET was renamed the *Internet*. The turning point in the evolution of the Internet occurred in 1990 when Tim Berners-Lee proposed a hypermedia software interface to the Internet, which he named the *World Wide Web*.[12] Thereupon, in the space of only about two years, the Web went from nonexistence to worldwide popularity, culminating in its commercialization in 1994. How do we explain the explosive growth of the Internet? We may answer this question by offering these reasons:[13]

- Before the Web exploded into existence, the ingredients for its creation were already in place. In particular, thanks to VLSI, personal computers (PCs) had already become ubiquitous in homes throughout the world, and they were increasingly equipped with modems for interconnectivity to the outside world.
- For about two decades, the Internet had grown steadily (albeit within a confined community of users), reaching a critical threshold of user-value based electronic mail and file transfer.
- Standards for document description and transfer, hypertext markup language (HTML), and hypertext transfer protocol (HTTP) had been adopted.

Thus, everything needed for creating the Web was already in place except for two critical ingredients: a simple user interface and a brilliant service concept.

In 1955, John R. Pierce proposed the use of satellites for communications. This proposal was preceded, however, by an earlier paper by Arthur C. Clark that was published in 1945, also proposing the idea of using an *Earth-orbiting* satellite as a relay point for communication between two Earth stations. In 1957, the Soviet Union launched Sputnik I, which transmitted telemetry signals for 21 days. This was followed shortly by the launching of Explorer I by the United States in 1958, which transmitted telemetry signals for about five months. A major experimental step in communications satellite technology was taken with the launching of Telstar I from Cape Canaveral on July 10, 1962. The Telstar satellite was built by Bell Laboratories, which had acquired considerable knowledge from pioneering work by Pierce. The satellite was capable of relaying TV programs across the Atlantic; this was made possible only through the user of maser receivers and large antennas.

The use of optical means (e.g., smoke and fire signals) for the transmission of information dates back to prehistoric times. However, no major breakthrough in optical communications was made until 1966, when K. C. Kao and G. A. Hockham of Standard Telephone Laboratories, U.K., proposed the use of a clad glass fiber as a dielectric waveguide. The *laser* (an acronym for light amplification by stimulated emission of radiation) had been invented and developed in 1959 and 1960. Kao and Hockham pointed out that (1) the attenuation in an optical fiber was due to impurities in the glass, and (2) the intrinsic loss, determined by Rayleigh scattering, is very low. Indeed, they predicted that a loss of 20 dB/km should be attainable. This remarkable prediction, made at a time when the power loss in a glass fiber was about 1000 dB/km, was to be demonstrated later. Nowadays, transmission losses as low as 0.1 dB/km are achievable.

The spectacular advances in microelectronics, digital computers, and lightwave systems that we have witnessed to date, and that will continue into the future, are all responsible for dramatic changes in the telecommunications environment; many of these changes are already in place, and more changes will evolve as time goes on.

## Notes and References

1. For essays on an early account of communications and other related disciplines (e.g., electronics, computers, radar, radio astronomy, satellites), see Overhage (1962); in particular, see the chapter on "Communications" by L. V. Berkner, pp. 35–50.

2. The JPEG image coding standard is discussed in the papers by Wallace (1991); see also the article by T. A. Ramstad in the handbook edited by Madisetti and Williams (1998).

3. The discrete cosine transform (DCT) and its inverse for a block of 8 × 8 source image samples are respectively defined by

$$F(u, v) = \frac{1}{4} C(u)C(v)\left[\sum_{x=0}^{7}\sum_{y=0}^{7} f(x, y) \cos\left(\frac{(2x + 1)u\pi}{16}\right) \cos\left(\frac{(2y + 1)v\pi}{16}\right)\right]$$

$$f(x, y) = \frac{1}{4}\left[\sum_{u=0}^{7}\sum_{v=0}^{7} C(u)C(v)F(u, v) \cos\left(\frac{(2x + 1)u\pi}{16}\right) \cos\left(\frac{(2y + 1)v\pi}{16}\right)\right]$$

where

$$C(u), C(v) = \begin{cases} \dfrac{1}{\sqrt{2}} & \text{for } u = 0 \text{ and } v = 0 \\ 1 & \text{otherwise} \end{cases}$$

For a full treatment of the DCT, see Rao and Yip (1990).

4. The MPEG-1 video coding standard is discussed in the paper by Gall (1991); see also the article by A. M. Tekalp in the handbook edited by Madisetti and Williams (1998), which discusses the follow-up versions of the MPEG video coding standard.

5. The MPEG-1 audio coding standard is discussed in the papers by Brandenburg and Stoll (1994) and Pan (1993); see also the article by P. Noll in the handbook edited by Madisetti and Williams (1998), which also discusses the follow-up versions of the MPEG audio coding standard. In particular, the widespread use of the more current standard, MPEG-3 audio, is resulting in a level of piracy that may dwarf the earlier problems of "bootleg" cassette tapes.

6. For a detailed discussion of communication networks, see Tanenbaum (1996).

7. The OSI reference model was developed by a subcommittee of the International Organization for Standardization (ISO) in 1977. For a discussion of the principles involved in arriving at the seven layers of the OSI model and a description of the layers themselves, see Tanenbaum (1996).

8. SONET was originally proposed by Telcordia Technologies Inc. (then known as Bellcore) and standardized by the American National Standards Institute (ANSI). Later, CCITT approved a SONET standard and issued a set of parallel recommendations called *synchronous digital hierarchy* (SDH). The differences between SONET and SDH are of a minor nature.

9. For a thorough and precise analysis of the propagation of light waves in an optical fiber, we need to treat it as a dielectric waveguide and use Maxwell's equations to carry out the analysis; such an analysis is highly mathematical in nature. For a readable account of the analysis, see Chapter 3 of Green, Jr. (1993).

10. For a semitechnical overview of Shannon's theorems on information theory presented in a highly readable fashion, see the book entitled *Silicon Dreams* by Lucky (1989).

11. For a readable account of the history of communications, see Lebow (1995).

12. For a historical account of the development of the Internet, see Leiner et al. (1997).

13. For an insightful essay on new telecommunications services and how society reacts to their development, see Lucky (1997). This paper points to *Metcalf's law*, according to which it seems as if any new telecommunications service must take a long time for it to build to universal acceptance. Lucky cites the World Wide Web as a startling counterexample to Metcalf's law and gives the reasons why.

# Chapter 1

# Random Processes

This chapter presents an introductory treatment of stationary random processes with emphasis on second-order statistics. In particular, it discusses the following issues:

- *The notion of a random process.*
- *The requirement that has to be satisfied for a random process to be stationary.*
- *The partial description of a random process in terms of its mean, correlation, and covariance functions.*
- *The conditions that have to be satisfied for a stationary random process to be ergodic, a property that enables us to substitute time averages for ensemble averages.*
- *What happens to a stationary random process when it is transmitted through a linear time-invariant filter?*
- *The frequency-domain description of a random process in terms of power spectral density.*
- *The characteristics of an important type of random process known as a Gaussian process.*
- *Sources of noise and their narrowband form.*
- *Rayleigh and Rician distributions, which represent two special probability distributions that arise in the study of communication systems.*

## 1.1 *Introduction*

The idea of a *mathematical model* used to describe a physical phenomenon is well established in the physical sciences and engineering. In this context, we may distinguish two classes of mathematical models: deterministic and stochastic. A model is said to be *deterministic* if there is no uncertainty about its time-dependent behavior at any instant of time. However, in many real-world problems the use of a deterministic model is inappropriate because the physical phenomenon of interest involves too many unknown factors. Nevertheless, it may be possible to consider a model described in probabilistic terms in that we speak of the *probability* of a future value lying between two specified limits. In such a case, the model is said to be *stochastic* or *random*. A brief review of probability theory is presented in Appendix 1.

Consider, for example, a radio communication system. The received signal in such a system usually consists of an *information-bearing signal* component, a random *interference* component, and *channel noise*. The information-bearing signal component may represent, for example, a voice signal that, typically, consists of randomly spaced bursts of energy of random duration. The interference component may represent spurious electromagnetic waves produced by other communication systems operating in the vicinity of the radio receiver. A major source of channel noise is *thermal noise*, which is caused by the random motion of the electrons in conductors and devices at the front end of the receiver. We thus find that the received signal is random in nature. Although it is not possible to predict the exact value of the signal in advance, it is possible to describe the signal in terms of statistical parameters such as average power and power spectral density, as discussed in this chapter.

## 1.2 Mathematical Definition of a Random Process

In light of these introductory remarks, it is apparent that random processes have two properties. First, they are functions of time. Second, they are random in the sense that before conducting an experiment, it is not possible to exactly define the waveforms that will be observed in the future.

In describing a random experiment it is convenient to think in terms of a sample space. Specifically, each outcome of the experiment is associated with a *sample point*. The totality of sample points corresponding to the aggregate of all possible outcomes of the experiment is called the *sample space*. Each sample point of the sample space is a function of time. The sample space or ensemble composed of functions of time is called a *random* or *stochastic process*.[1] As an integral part of this notion, we assume the existence of a probability distribution defined over an appropriate class of sets in the sample space, so that we may speak with confidence of the probability of various events.

Consider, then, a random experiment specified by the outcomes $s$ from some *sample space* $\mathsf{S}$, by the events defined on the sample space $\mathsf{S}$, and by the probabilities of these

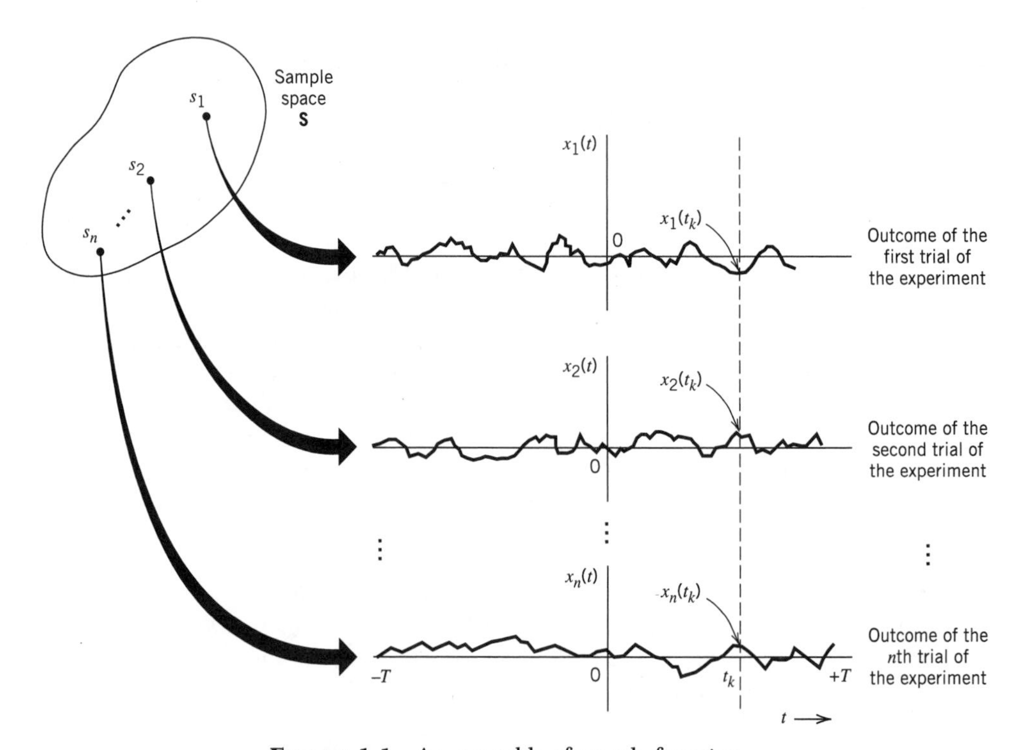


**FIGURE 1.1** An ensemble of sample functions.

events. Suppose that we assign to each sample point $s$ a function of time in accordance with the rule:

$$X(t, s), \qquad -T \leq t \leq T \tag{1.1}$$

where $2T$ is the *total observation interval*. For a fixed sample point $s_j$, the graph of the function $X(t, s_j)$ versus time $t$ is called a *realization* or *sample function* of the random process. To simplify the notation, we denote this sample function as

$$x_j(t) = X(t, s_j) \tag{1.2}$$

Figure 1.1 illustrates a set of sample functions $\{x_j(t) \mid j = 1, 2, \ldots, n\}$. From this figure, we note that for a fixed time $t_k$ inside the observation interval, the set of numbers

$$\{x_1(t_k), x_2(t_k), \ldots, x_n(t_k)\} = \{X(t_k, s_1), X(t_k, s_2), \ldots, X(t_k, s_n)\}$$

constitutes a *random variable*. Thus we have an indexed ensemble (family) of random variables $\{X(t, s)\}$, which is called a *random process*. To simplify the notation, the customary practice is to suppress the $s$ and simply use $X(t)$ to denote a random process. We may now formally define a random process $X(t)$ as *an ensemble of time functions together with a probability rule that assigns a probability to any meaningful event associated with an observation of one of the sample functions of the random process.* Moreover, we may distinguish between a random variable and a random process as follows:

- For a random variable, the outcome of a random experiment is mapped into a number.
- For a random process, the outcome of a random experiment is mapped into a waveform that is a function of time.

## 1.3 *Stationary Processes*

In dealing with random processes encountered in the real world, we often find that the statistical characterization of a process is independent of the time at which observation of the process is initiated. That is, if such a process is divided into a number of time intervals, the various sections of the process exhibit essentially the same statistical properties. Such a process is said to be *stationary*. Otherwise, it is said to be *nonstationary*. Generally speaking, a stationary process arises from a stable physical phenomenon that has evolved into a steady-state mode of behavior, whereas a nonstationary process arises from an unstable phenomenon.

To be more precise, consider a random process $X(t)$ that is initiated at $t = -\infty$. Let $X(t_1), X(t_2), \ldots, X(t_k)$ denote the random variables obtained by observing the random process $X(t)$ at times $t_1, t_2, \ldots, t_k$, respectively. The joint distribution function of this set of random variables is $F_{X(t_1), \ldots, X(t_k)}(x_1, \ldots, x_k)$. Suppose next we shift all the observation times by a fixed amount $\tau$, thereby obtaining a new set of random variables $X(t_1 + \tau)$, $X(t_2 + \tau), \ldots, X(t_k + \tau)$. The joint distribution function of this latter set of random variables is $F_{X(t_1+\tau), \ldots, X(t_k+\tau)}(x_1, \ldots, x_k)$. The random process $X(t)$ is said to be *stationary in the strict sense* or *strictly stationary* if the following condition holds:

$$F_{X(t_1+\tau), \ldots, X(t_k+\tau)}(x_1, \ldots, x_k) = F_{X(t_1), \ldots, X(t_k)}(x_1, \ldots, x_k) \tag{1.3}$$

for all time shifts $\tau$, all $k$, and all possible choices of observation times $t_1, \ldots, t_k$. In other words, *a random process $X(t)$, initiated at time $t = -\infty$, is strictly stationary if the joint distribution of any set of random variables obtained by observing the random process $X(t)$ is invariant with respect to the location of the origin $t = 0$.* Note that the finite-dimensional distributions in Equation (1.3) depend on the relative time separation between random variables but not on their absolute time. That is, the random process has the same probabilistic behavior through all time.

Similarly, we may say that two random processes $X(t)$ and $Y(t)$ are *jointly strictly stationary* if the joint finite-dimensional distributions of the two sets of random variables $X(t_1), \ldots, X(t_k)$ and $Y(t'_1), \ldots, Y(t'_j)$ are invariant with respect to the origin $t = 0$ for all $k$ and $j$ and all choices of observation times $t_1, \ldots, t_k$ and $t'_1, \ldots, t'_j$.

Returning to Equation (1.3), we may distinguish two situations of special interest:

1. For $k = 1$, we have

$$F_{X(t)}(x) = F_{X(t+\tau)}(x) = F_X(x) \qquad \text{for all } t \text{ and } \tau \tag{1.4}$$

   That is, *the first-order distribution function of a stationary random process is independent of time.*

2. For $k = 2$ and $\tau = -t_1$, we have

$$F_{X(t_1),X(t_2)}(x_1, x_2) = F_{X(0),X(t_2-t_1)}(x_1, x_2) \qquad \text{for all } t_1 \text{ and } t_2 \tag{1.5}$$

   That is, *the second-order distribution function of a stationary random process depends only on the time difference between the observation times* and not on the particular times at which the random process is observed.

These two properties have profound implications for the statistical parameterization of a stationary random process; this issue is discussed in Section 1.4.

### ▶ EXAMPLE 1.1

Consider Figure 1.2, depicting three spatial windows located at times $t_1$, $t_2$, $t_3$. We wish to evaluate the probability of obtaining a sample function $x(t)$ of a random process $X(t)$ that passes through this set of windows, that is, the probability of the joint event

$$A = \{a_i < X(t_i) \leq b_i\}, \qquad i = 1, 2, 3$$

In terms of the joint distribution function, this probability equals

$$P(A) = F_{X(t_1),X(t_2),X(t_3)}(b_1, b_2, b_3) - F_{X(t_1),X(t_2),X(t_3)}(a_1, a_2, a_3)$$

Suppose now the random process $X(t)$ is known to be strictly stationary. An implication of strict stationarity is that the probability of the set of sample functions of this process passing through the windows of Figure 1.3*a* is equal to the probability of the set of sample functions passing through the corresponding time-shifted windows of Figure 1.3*b*. Note, however, that it is not necessary that these two sets consist of the same sample functions. ◀

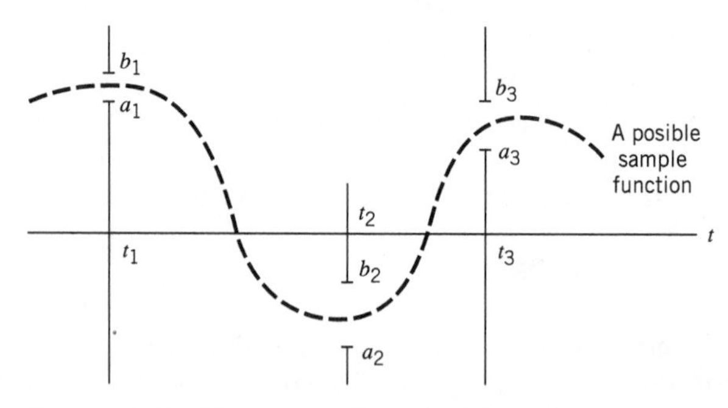


**FIGURE 1.2** Illustrating the probability of a joint event.

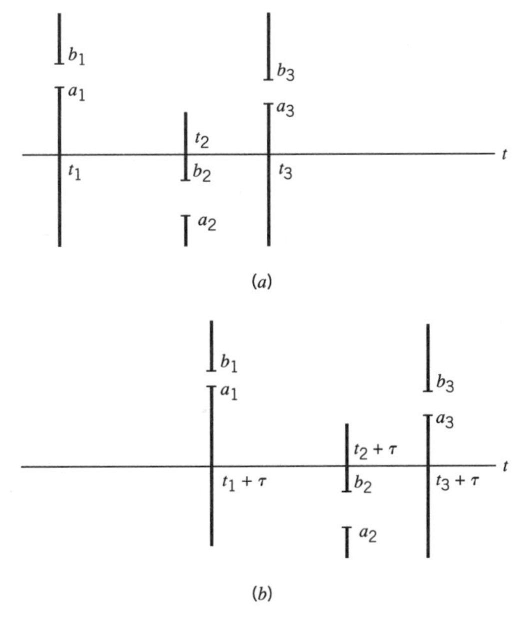


**FIGURE 1.3** Illustrating the concept of stationarity in Example 1.1.

## 1.4 *Mean, Correlation, and Covariance Functions*

Consider a strictly stationary random process $X(t)$. We define the *mean* of the process $X(t)$ as the expectation of the random variable obtained by observing the process at some time $t$, as shown by

$$\begin{aligned} \mu_X(t) &= E[X(t)] \\ &= \int_{-\infty}^{\infty} x f_{X(t)}(x)\, dx \end{aligned} \tag{1.6}$$

where $f_{X(t)}(x)$ is the first-order probability density function of the process. From Equation (1.4) we deduce that for a strictly stationary random process, $f_{X(t)}(x)$ is independent of time $t$. Consequently, *the mean of a strictly stationary process is a constant*, as shown by

$$\mu_X(t) = \mu_X \qquad \text{for all } t \tag{1.7}$$

We define the *autocorrelation function* of the process $X(t)$ as the expectation of the product of two random variables, $X(t_1)$ and $X(t_2)$, obtained by observing the process $X(t)$ at times $t_1$ and $t_2$, respectively. Specifically, we write

$$\begin{aligned} R_X(t_1, t_2) &= E[X(t_1)X(t_2)] \\ &= \int_{-\infty}^{\infty} \int_{-\infty}^{\infty} x_1 x_2 f_{X(t_1),X(t_2)}(x_1, x_2)\, dx_1 dx_2 \end{aligned} \tag{1.8}$$

where $f_{X(t_1),X(t_2)}(x_1, x_2)$ is the second-order probability density function of the process. From Equation (1.5), we deduce that for a strictly stationary random process, $f_{X(t_1),X(t_2)}(x_1, x_2)$ depends only on the difference between the observation times $t_1$ and $t_2$.

This, in turn, implies that *the autocorrelation function of a strictly stationary process depends only on the time difference* $t_2 - t_1$, as shown by

$$R_X(t_1, t_2) = R_X(t_2 - t_1) \qquad \text{for all } t_1 \text{ and } t_2 \tag{1.9}$$

Similarly, the *autocovariance function* of a strictly stationary process $X(t)$ is written as

$$\begin{aligned} C_X(t_1, t_2) &= E[(X(t_1) - \mu_X)(X(t_2) - \mu_X)] \\ &= R_X(t_2 - t_1) - \mu_X^2 \end{aligned} \tag{1.10}$$

Equation (1.10) shows that, like the autocorrelation function, the autocovariance function of a strictly stationary process $X(t)$ depends only on the time difference $t_2 - t_1$. This equation also shows that if we know the mean and autocorrelation function of the process, we can uniquely determine the autocovariance function. The mean and autocorrelation function are therefore sufficient to describe the first two moments of the process.

However, two important points should be carefully noted:

1. The mean and autocorrelation function only provide a *partial description* of the distribution of a random process $X(t)$.
2. The conditions of Equations (1.7) and (1.9), involving the mean and autocorrelation function, respectively, are *not* sufficient to guarantee that the random process $X(t)$ is strictly stationary.

Nevertheless, practical considerations often dictate that we simply limit ourselves to a partial description of the process given by the mean and autocorrelation function. The class of random processes that satisfy Equations (1.7) and (1.9) has been given various names, such as *second-order stationary*, *wide-sense stationary*, or *weakly stationary* processes. Henceforth, we shall simply refer to them as *stationary processes*.[2]

A stationary process is not necessarily strictly stationary because Equations (1.7) and (1.9) obviously do not imply the invariance of the joint ($k$-dimensional) distribution of Equation (1.3) with respect to the time shift $\tau$ for all $k$. On the other hand, a strictly stationary process does not necessarily satisfy Equations (1.7) and (1.9) as the first- and second-order moments may not exist. Clearly, however, the class of strictly stationary processes with finite second-order moments forms a subclass of the class of all stationary processes.

## ■ PROPERTIES OF THE AUTOCORRELATION FUNCTION

For convenience of notation, we redefine the autocorrelation function of a stationary process $X(t)$ as

$$R_X(\tau) = E[X(t + \tau)X(t)] \qquad \text{for all } t \tag{1.11}$$

This autocorrelation function has several important properties:

1. The *mean-square value* of the process may be obtained from $R_X(\tau)$ simply by putting $\tau = 0$ in Equation (1.11), as shown by

$$R_X(0) = E[X^2(t)] \tag{1.12}$$

2. The autocorrelation function $R_X(\tau)$ is an even function of $\tau$, that is,

$$R_X(\tau) = R_X(-\tau) \tag{1.13}$$

This property follows directly from the defining equation (1.11). Accordingly, we may also define the autocorrelation function $R_X(\tau)$ as

$$R_X(\tau) = E[X(t)X(t - \tau)]$$

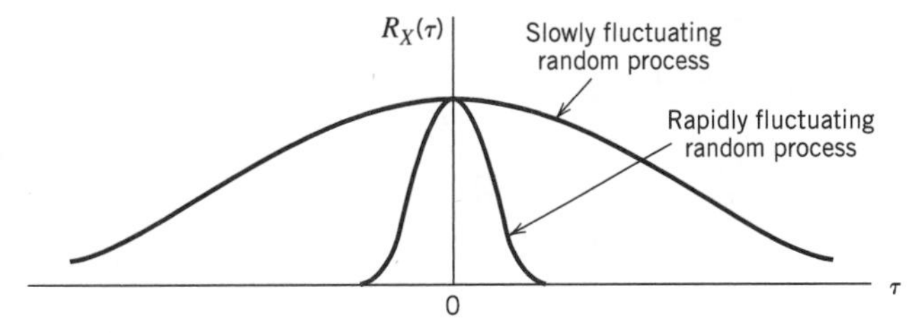


**FIGURE 1.4** Illustrating the autocorrelation functions of slowly and rapidly fluctuating random processes.

3. The autocorrelation function $R_X(\tau)$ has its maximum magnitude at $\tau = 0$, that is,

$$|R_X(\tau)| \leq R_X(0) \tag{1.14}$$

To prove this property, consider the nonnegative quantity

$$E[(X(t + \tau) \pm X(t))^2] \geq 0$$

Expanding terms and taking their individual expectations, we readily find that

$$E[X^2(t + \tau)] \pm 2E[X(t + \tau)X(t)] + E[X^2(t)] \geq 0$$

which, in light of Equations (1.11) and (1.12), reduces to

$$2R_X(0) \pm 2R_X(\tau) \geq 0$$

Equivalently, we may write

$$-R_X(0) \leq R_X(\tau) \leq R_X(0)$$

from which Equation (1.14) follows directly.

The physical significance of the autocorrelation function $R_X(\tau)$ is that it provides a means of describing the interdependence of two random variables obtained by observing a random process $X(t)$ at times $\tau$ seconds apart. It is therefore apparent that the more rapidly the random process $X(t)$ changes with time, the more rapidly will the autocorrelation function $R_X(\tau)$ decrease from its maximum $R_X(0)$ as $\tau$ increases, as illustrated in Figure 1.4. This decrease may be characterized by a *decorrelation time* $\tau_0$, such that for $\tau > \tau_0$, the magnitude of the autocorrelation function $R_X(\tau)$ remains below some prescribed value. We may thus define the decorrelation time $\tau_0$ of a stationary process $X(t)$ of zero mean as the time taken for the magnitude of the autocorrelation function $R_X(\tau)$ to decrease to 1 percent, say, of its maximum value $R_X(0)$.

### ▶ EXAMPLE 1.2 Sinusoidal Wave with Random Phase

Consider a sinusoidal signal with random phase, defined by

$$X(t) = A \cos(2\pi f_c t + \Theta) \tag{1.15}$$

where $A$ and $f_c$ are constants and $\Theta$ is a random variable that is *uniformly distributed* over the interval $[-\pi, \pi]$, that is,

$$f_\Theta(\theta) = \begin{cases} \dfrac{1}{2\pi}, & -\pi \leq \theta \leq \pi \\ 0, & \text{elsewhere} \end{cases} \tag{1.16}$$

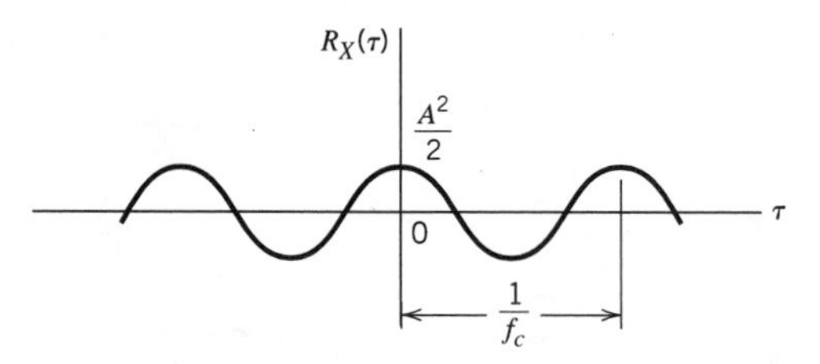


**FIGURE 1.5** Autocorrelation function of a sine wave with random phase.

This means that the random variable $\Theta$ is equally likely to have any value $\theta$ in the interval $[-\pi, \pi]$. Each value of $\Theta$ corresponds to a sample in the sample space of the random process $X(t)$.

The process $X(t)$ defined by Equations (1.15) and (1.16) may represent a locally generated carrier in the receiver of a communication system, which is used in demodulation of the received signal. In particular, the random variable $\Theta$ denotes the phase difference between this locally generated carrier and the sinusoidal carrier wave used to modulate the message signal in the transmitter.

The autocorrelation function of $X(t)$ is

$$\begin{aligned} R_X(\tau) &= E[X(t+\tau)X(t)] \\ &= E[A^2 \cos(2\pi f_c t + 2\pi f_c \tau + \Theta) \cos(2\pi f_c t + \Theta)] \\ &= \frac{A^2}{2} E[\cos(4\pi f_c t + 2\pi f_c \tau + 2\Theta)] + \frac{A^2}{2} E[\cos(2\pi f_c \tau)] \\ &= \frac{A^2}{2} \int_{-\pi}^{\pi} \frac{1}{2\pi} \cos(4\pi f_c t + 2\pi f_c \tau + 2\theta)\, d\theta + \frac{A^2}{2} \cos(2\pi f_c \tau) \end{aligned}$$

The first term integrates to zero, and so we get

$$R_X(\tau) = \frac{A^2}{2} \cos(2\pi f_c \tau) \tag{1.17}$$

which is plotted in Figure 1.5. We see therefore that the autocorrelation function of a sinusoidal wave with random phase is another sinusoid at the same frequency in the "$\tau$ domain" rather than the original time domain. ◀

## ▶ EXAMPLE 1.3 Random Binary Wave

Figure 1.6 shows the sample function $x(t)$ of a process $X(t)$ consisting of a random sequence of *binary symbols* 1 and 0. The following assumptions are made:

1. The symbols 1 and 0 are represented by pulses of amplitude $+A$ and $-A$ volts, respectively, and duration $T$ seconds.
2. The pulses are not synchronized, so the starting time $t_d$ of the first complete pulse for positive time is equally likely to lie anywhere between zero and $T$ seconds. That is, $t_d$ is the sample value of a uniformly distributed random variable $T_d$, with its probability density function defined by

$$f_{T_d}(t_d) = \begin{cases} \frac{1}{T}, & 0 \le t_d \le T \\ 0, & \text{elsewhere} \end{cases}$$

3. During any time interval $(n-1)T < t - t_d < nT$, where $n$ is an integer, the presence of a 1 or a 0 is determined by tossing a fair coin; specifically, if the outcome is heads,

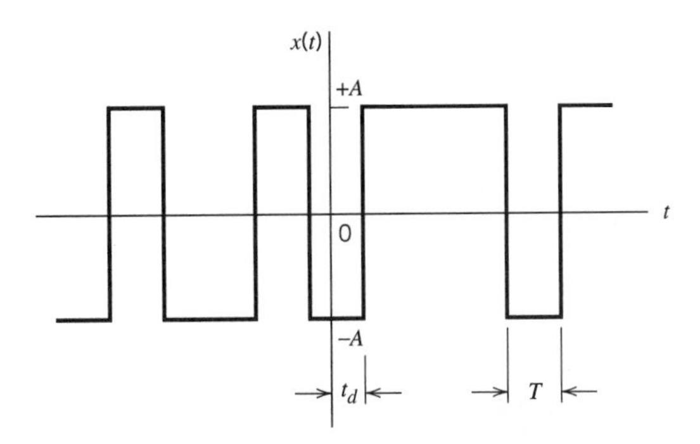


**FIGURE 1.6** Sample function of random binary wave.

we have a 1 and if the outcome is tails, we have a 0. These two symbols are thus equally likely, and the presence of a 1 or 0 in any one interval is independent of all other intervals.

Since the amplitude levels $-A$ and $+A$ occur with equal probability, it follows immediately that $E[X(t)] = 0$ for all $t$, and the mean of the process is therefore zero.

To find the autocorrelation function $R_X(t_k, t_i)$, we have to evaluate $E[X(t_k)X(t_i)]$, where $X(t_k)$ and $X(t_i)$ are random variables obtained by observing the random process $X(t)$ at times $t_k$ and $t_i$, respectively.

Consider first the case when $|t_k - t_i| > T$. Under this condition the random variables $X(t_k)$ and $X(t_i)$ occur in different pulse intervals and are therefore independent. We thus have

$$E[X(t_k)X(t_i)] = E[X(t_k)]E[X(t_i)] = 0, \qquad |t_k - t_i| > T$$

Consider next the case when $|t_k - t_i| < T$, with $t_k = 0$ and $t_i < t_k$. In such a situation we observe from Figure 1.6 that the random variables $X(t_k)$ and $X(t_i)$ occur in the same pulse interval if and only if the delay $t_d$ satisfies the condition $t_d < T - |t_k - t_i|$. We thus obtain the *conditional expectation*:

$$E[X(t_k)X(t_i)|t_d] = \begin{cases} A^2, & t_d < T - |t_k - t_i| \\ 0, & \text{elsewhere} \end{cases}$$

Averaging this result over all possible values of $t_d$, we get

$$\begin{aligned} E[X(t_k)X(t_i)] &= \int_0^{T-|t_k-t_i|} A^2 f_{T_d}(t_d)\, dt_d \\ &= \int_0^{T-|t_k-t_i|} \frac{A^2}{T}\, dt_d \\ &= A^2\left(1 - \frac{|t_k - t_i|}{T}\right), \qquad |t_k - t_i| < T \end{aligned}$$

By similar reasoning for any other value of $t_k$, we conclude that the autocorrelation function of a random binary wave, represented by the sample function shown in Figure 1.6, is only a function of the time difference $\tau = t_k - t_i$, as shown by

$$R_X(\tau) = \begin{cases} A^2\left(1 - \dfrac{|\tau|}{T}\right), & |\tau| < T \\ 0, & |\tau| \geq T \end{cases} \tag{1.18}$$

This result is plotted in Figure 1.7. ◀

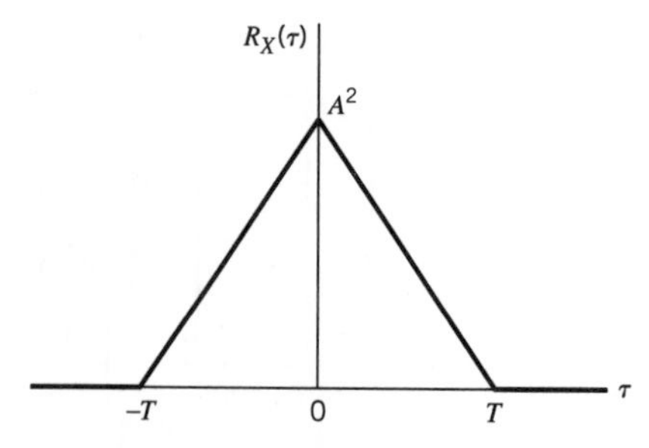


**FIGURE 1.7** Autocorrelation function of random binary wave.

## ■ CROSS-CORRELATION FUNCTIONS

Consider next the more general case of two random processes $X(t)$ and $Y(t)$ with autocorrelation functions $R_X(t, u)$ and $R_Y(t, u)$, respectively. The two *cross-correlation functions* of $X(t)$ and $Y(t)$ are defined by

$$R_{XY}(t, u) = E[X(t)Y(u)] \tag{1.19}$$

and

$$R_{YX}(t, u) = E[Y(t)X(u)] \tag{1.20}$$

where $t$ and $u$ denote two values of time at which the processes are observed. In this case, the correlation properties of the two random processes $X(t)$ and $Y(t)$ may be displayed conveniently in matrix form as follows:

$$\mathbf{R}(t, u) = \begin{bmatrix} R_X(t, u) & R_{XY}(t, u) \\ R_{YX}(t, u) & R_Y(t, u) \end{bmatrix}$$

which is called the *correlation matrix* of the random processes $X(t)$ and $Y(t)$. If the random processes $X(t)$ and $Y(t)$ are each stationary and, in addition, they are jointly stationary, then the correlation matrix can be written as

$$\mathbf{R}(\tau) = \begin{bmatrix} R_X(\tau) & R_{XY}(\tau) \\ R_{YX}(\tau) & R_Y(\tau) \end{bmatrix} \tag{1.21}$$

where $\tau = t - u$.

The cross-correlation function is not generally an even function of $\tau$ as was true for the autocorrelation function, nor does it have a maximum at the origin. However, it does obey a certain symmetry relationship as follows (see Problem 1.9):

$$R_{XY}(\tau) = R_{YX}(-\tau) \tag{1.22}$$

### ▶ EXAMPLE 1.4 Quadrature-Modulated Processes

Consider a pair of *quadrature-modulated processes* $X_1(t)$ and $X_2(t)$ that are related to a stationary process $X(t)$ as follows:

$$X_1(t) = X(t)\cos(2\pi f_c t + \Theta)$$
$$X_2(t) = X(t)\sin(2\pi f_c t + \Theta)$$

where $f_c$ is a carrier frequency, and the random variable $\Theta$ is uniformly distributed over the interval $[0, 2\pi]$. Moreover, $\Theta$ is independent of $X(t)$. One cross-correlation function of $X_1(t)$ and $X_2(t)$ is given by

$$\begin{aligned}
R_{12}(\tau) &= E[X_1(t)X_2(t-\tau)] \\
&= E[X(t)X(t-\tau)\cos(2\pi f_c t + \Theta)\sin(2\pi f_c t - 2\pi f_c \tau + \Theta)] \\
&= E[X(t)X(t-\tau)]E[\cos(2\pi f_c t + \Theta)\sin(2\pi f_c t - 2\pi f_c \tau + \Theta)] \\
&= \tfrac{1}{2}R_X(\tau)E[\sin(4\pi f_c t - 2\pi f_c \tau + 2\Theta) - \sin(2\pi f_c \tau)] \\
&= -\tfrac{1}{2}R_X(\tau)\sin(2\pi f_c \tau)
\end{aligned} \tag{1.23}$$

where, in the last line, we have made use of the uniform distribution of the random variable $\Theta$ representing phase. Note that at $\tau = 0$, the factor $\sin(2\pi f_c \tau)$ is zero and therefore

$$\begin{aligned}
R_{12}(0) &= E[X_1(t)X_2(t)] \\
&= 0
\end{aligned}$$

This shows that the random variables obtained by simultaneously observing the quadrature-modulated processes $X_1(t)$ and $X_2(t)$ at some fixed value of time $t$ are orthogonal to each other. ◀

## 1.5 *Ergodic Processes*

The *expectations* or *ensemble averages* of a random process $X(t)$ are averages "across the process." For example, the mean of a random process $X(t)$ at some fixed time $t_k$ is the expectation of the random variable $X(t_k)$ that describes *all possible values* of the sample functions of the process observed at time $t = t_k$. Naturally, we may also define *long-term sample averages*, or *time averages* that are averages "along the process." We are therefore interested in relating ensemble averages to time averages, for time averages represent a practical means available to us for the *estimation* of ensemble averages of a random process. The key question, of course, is: When can we substitute time averages for ensemble averages? To explore this issue, consider the sample function $x(t)$ of a stationary process $X(t)$, with the observation interval defined as $-T \le t \le T$. The DC *value* of $x(t)$ is defined by the time average

$$\mu_x(T) = \frac{1}{2T}\int_{-T}^{T} x(t)\, dt \tag{1.24}$$

Clearly, the time average $\mu_x(T)$ is a random variable, as its value depends on the observation interval and which particular sample function of the random process $X(t)$ is picked for use in Equation (1.24). Since the process $X(t)$ is assumed to be stationary, the mean of the time average $\mu_x(T)$ is given by (after interchanging the operation of expectation and integration):

$$\begin{aligned}
E[\mu_x(T)] &= \frac{1}{2T}\int_{-T}^{T} E[x(t)]\, dt \\
&= \frac{1}{2T}\int_{-T}^{T} \mu_X\, dt \\
&= \mu_X
\end{aligned} \tag{1.25}$$

where $\mu_X$ is the mean of the process $X(t)$. Accordingly, the time average $\mu_x(T)$ represents an *unbiased estimate* of the ensemble-averaged mean $\mu_X$. We say that the process $X(t)$ is *ergodic in the mean* if two conditions are satisfied:

- The time average $\mu_x(T)$ approaches the ensemble average $\mu_X$ in the limit as the observation interval $T$ approaches infinity; that is,

$$\lim_{T\to\infty} \mu_x(T) = \mu_X$$

- The variance of $\mu_x(T)$, treated as a random variable, approaches zero in the limit as the observation interval $T$ approaches infinity; that is,

$$\lim_{T\to\infty} \mathrm{var}[\mu_x(T)] = 0$$

The other time average of particular interest is the autocorrelation function $R_x(\tau, T)$ defined in terms of the sample function $x(t)$ observed over the interval $-T \le t \le T$. Following Equation (1.24), we may formally define the *time-averaged autocorrelation function* of a sample function $x(t)$ as follows:

$$R_x(\tau, T) = \frac{1}{2T}\int_{-T}^{T} x(t + \tau)x(t)\, dt \tag{1.26}$$

This second time-average should also be viewed as a random variable with a mean and variance of its own. In a manner similar to ergodicity of the mean, we say that the process $x(t)$ is *ergodic in the autocorrelation function* if the following two limiting conditions are satisfied:

$$\lim_{T\to\infty} R_x(\tau, T) = R_X(\tau)$$
$$\lim_{T\to\infty} \mathrm{var}[R_x(\tau, T)] = 0$$

We could, of course, go on in a similar way to define ergodicity in the most general sense by considering higher-order statistics of the process $X(t)$. In practice, however, ergodicity in the mean and ergodicity in the autocorrelation function, as described here, are often (but not always) considered to be adequate. Note also that the use of Equations (1.24) and (1.26) to compute the time averages $\mu_x(T)$ and $R_x(t, T)$ requires that the process $X(t)$ be stationary. In other words, for a random process to be ergodic, it has to be stationary; however, the converse is not necessarily true.

## 1.6 *Transmission of a Random Process Through a Linear Time-Invariant Filter*

Suppose that a random process $X(t)$ is applied as input to a *linear time-invariant filter of impulse response* $h(t)$, producing a new random process $Y(t)$ at the filter output, as in Figure 1.8. In general, it is difficult to describe the probability distribution of the output random process $Y(t)$, even when the probability distribution of the input random process $X(t)$ is completely specified for $-\infty < t < \infty$.

In this section, we determine the time-domain form of the input–output relations of the filter for defining the mean and autocorrelation functions of the output random process $Y(t)$ in terms of those of the input $X(t)$, assuming that $X(t)$ is a stationary process.

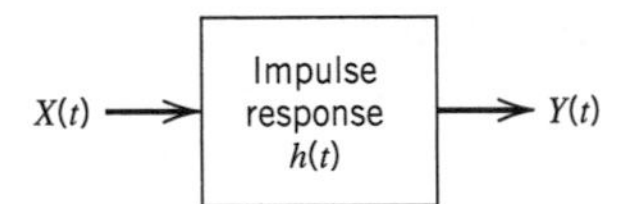


**FIGURE 1.8** Transmission of a random process through a linear time-invariant filter.

The transmission of a process through a linear time-invariant filter is governed by the *convolution integral*; for a review of this operation, see Appendix 2. For the problem at hand, we may thus express the output random process $Y(t)$ in terms of the input random process $X(t)$ as

$$Y(t) = \int_{-\infty}^{\infty} h(\tau_1)X(t - \tau_1)\, d\tau_1$$

where $\tau_1$ is the integration variable. Hence, the mean of $Y(t)$ is

$$\begin{aligned} \mu_Y(t) &= E[Y(t)] \\ &= E\left[\int_{-\infty}^{\infty} h(\tau_1)X(t - \tau_1)\, d\tau_1\right] \end{aligned} \tag{1.27}$$

Provided that the expectation $E[X(t)]$ is finite for all $t$ and the system is stable, we may interchange the order of expectation and integration in Equation (1.27) and so write

$$\begin{aligned} \mu_Y(t) &= \int_{-\infty}^{\infty} h(\tau_1)E[X(t - \tau_1)]\, d\tau_1 \\ &= \int_{-\infty}^{\infty} h(\tau_1)\mu_X(t - \tau_1)\, d\tau_1 \end{aligned} \tag{1.28}$$

When the input random process $X(t)$ is stationary, the mean $\mu_X(t)$ is a constant $\mu_X$, so that we may simplify Equation (1.28) as follows:

$$\begin{aligned} \mu_Y &= \mu_X \int_{-\infty}^{\infty} h(\tau_1)\, d\tau_1 \\ &= \mu_X H(0) \end{aligned} \tag{1.29}$$

where $H(0)$ is the zero-frequency (DC) response of the system. Equation (1.29) states that the mean of the random process $Y(t)$ produced at the output of a linear time-invariant system in response to $X(t)$ acting as the input process is equal to the mean of $X(t)$ multiplied by the DC response of the system, which is intuitively satisfying.

Consider next the autocorrelation function of the output random process $Y(t)$. By definition, we have

$$R_Y(t, u) = E[Y(t)Y(u)]$$

where $t$ and $u$ denote two values of the time at which the output process is observed. We may therefore use the convolution integral to write

$$R_Y(t, u) = E\left[\int_{-\infty}^{\infty} h(\tau_1)X(t - \tau_1)\, d\tau_1 \int_{-\infty}^{\infty} h(\tau_2)X(u - \tau_2)\, d\tau_2\right] \tag{1.30}$$

Here again, provided that the mean-square value $E[X^2(t)]$ is finite for all $t$ and the system is stable, we may interchange the order of the expectation and the integrations with respect to $\tau_1$ and $\tau_2$ in Equation (1.30), obtaining

$$\begin{aligned} R_Y(t, u) &= \int_{-\infty}^{\infty} d\tau_1 h(\tau_1) \int_{-\infty}^{\infty} d\tau_2 h(\tau_2) E[X(t - \tau_1)X(u - \tau_2)] \\ &= \int_{-\infty}^{\infty} d\tau_1 h(\tau_1) \int_{-\infty}^{\infty} d\tau_2 h(\tau_2) R_X(t - \tau_1, u - \tau_2) \end{aligned} \tag{1.31}$$

When the input $X(t)$ is a stationary process, the autocorrelation function of $X(t)$ is only a function of the difference between the observation times $t - \tau_1$ and $u - \tau_2$. Thus, putting $\tau = t - u$ in Equation (1.31), we may write

$$R_Y(\tau) = \int_{-\infty}^{\infty} \int_{-\infty}^{\infty} h(\tau_1)h(\tau_2)R_X(\tau - \tau_1 + \tau_2)\, d\tau_1\, d\tau_2 \tag{1.32}$$

On combining this result with that involving the mean $\mu_Y$, we see that if the input to a stable linear time-invariant filter is a stationary process, then the output of the filter is also a stationary process.

Since $R_Y(0) = E[Y^2(t)]$, it follows that the mean-square value of the output random process $Y(t)$ is obtained by putting $\tau = 0$ in Equation (1.32). We thus get the result

$$E[Y^2(t)] = \int_{-\infty}^{\infty} \int_{-\infty}^{\infty} h(\tau_1)h(\tau_2)R_X(\tau_2 - \tau_1)\, d\tau_1\, d\tau_2 \tag{1.33}$$

which is a constant.

## 1.7 *Power Spectral Density*

Thus far we have considered the characterization of stationary processes in linear systems in the time domain. We turn next to the characterization of random processes in linear systems by using frequency-domain ideas. In particular, we wish to derive the frequency-domain equivalent to the result of Equation (1.33) defining the mean-square value of the filter output.

By definition, the impulse response of a linear time-invariant filter is equal to the inverse Fourier transform of the frequency response of the system; a review of the Fourier transform is presented in Appendix 2. Using $H(f)$ to denote the *frequency response* of the system, we may thus write

$$h(\tau_1) = \int_{-\infty}^{\infty} H(f) \exp(j2\pi f\tau_1)\, df \tag{1.34}$$

Substituting this expression for $h(\tau_1)$ into Equation (1.33), we get

$$\begin{aligned} E[Y^2(t)] &= \int_{-\infty}^{\infty} \int_{-\infty}^{\infty} \left[\int_{-\infty}^{\infty} H(f) \exp(j2\pi f\tau_1)\, df\right] h(\tau_2)R_X(\tau_2 - \tau_1)\, d\tau_1\, d\tau_2 \\ &= \int_{-\infty}^{\infty} df\, H(f) \int_{-\infty}^{\infty} d\tau_2 h(\tau_2) \int_{-\infty}^{\infty} R_X(\tau_2 - \tau_1) \exp(j2\pi f\tau_1)\, d\tau_1 \end{aligned} \tag{1.35}$$

In the last integral on the right-hand side of Equation (1.35), define a new variable

$$\tau = \tau_2 - \tau_1$$

Then we may rewrite Equation (1.35) in the form

$$E[Y^2(t)] = \int_{-\infty}^{\infty} df H(f) \int_{-\infty}^{\infty} d\tau_2 h(\tau_2) \exp(j2\pi f \tau_2) \int_{-\infty}^{\infty} R_X(\tau) \exp(-j2\pi f \tau)\, d\tau \quad (1.36)$$

However, the middle integral on the right-hand side in Equation (1.36) is simply $H^*(f)$, the complex conjugate of the frequency response of the filter, and so we may simplify this equation as

$$E[Y^2(t)] = \int_{-\infty}^{\infty} df |H(f)|^2 \int_{-\infty}^{\infty} R_X(\tau) \exp(-j2\pi f \tau)\, d\tau \quad (1.37)$$

where $|H(f)|$ is the *magnitude response* of the filter. We may further simplify Equation (1.37) by recognizing that the last integral is simply the Fourier transform of the autocorrelation function $R_X(\tau)$ of the input random process $X(t)$. This prompts us to introduce the definition of a new parameter

$$S_X(f) = \int_{-\infty}^{\infty} R_X(\tau) \exp(-j2\pi f \tau)\, d\tau \quad (1.38)$$

The function $S_X(f)$ is called the *power spectral density*, or *power spectrum*, of the stationary process $X(t)$. Thus substituting Equation (1.38) into (1.37), we obtain the desired relation:

$$E[Y^2(t)] = \int_{-\infty}^{\infty} |H(f)|^2 S_X(f)\, df \quad (1.39)$$

Equation (1.39) states that *the mean-square value of the output of a stable linear time-invariant filter in response to a stationary process is equal to the integral over all frequencies of the power spectral density of the input process multiplied by the squared magnitude response of the filter*. This is the desired frequency-domain equivalent to the time-domain relation of Equation (1.33).

To investigate the physical significance of the power spectral density, suppose that the random process $X(t)$ is passed through an ideal narrowband filter with a magnitude response centered about the frequency $f_c$, as shown in Figure 1.9; that is,

$$|H(f)| = \begin{cases} 1, & |f \pm f_c| < \frac{1}{2}\Delta f \\ 0, & |f \pm f_c| > \frac{1}{2}\Delta f \end{cases} \quad (1.40)$$

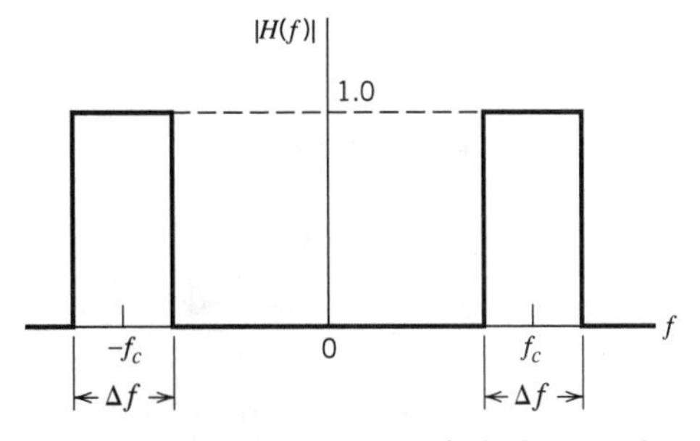


**FIGURE 1.9** Magnitude response of ideal narrowband filter.

where $\Delta f$ is the bandwidth of the filter. Then from Equation (1.39) we find that if the filter bandwidth $\Delta f$ is sufficiently small compared to the midband frequency $f_c$ and $S_X(f)$ is a continuous function, the mean-square value of the filter output is approximately

$$E[Y^2(t)] \simeq (2\Delta f)S_X(f_c) \tag{1.41}$$

The filter, however, passes only those frequency components of the input random process $X(t)$ that lie inside a narrow frequency band of width $\Delta f$ centered about the frequency $\pm f_c$. Thus $S_X(f_c)$ represents the frequency density of the average power in the random process $X(t)$, evaluated at the frequency $f = f_c$. The dimensions of the power spectral density are therefore in watts per Hertz (W/Hz).

## ■ PROPERTIES OF THE POWER SPECTRAL DENSITY

The power spectral density $S_X(f)$ and the autocorrelation function $R_X(\tau)$ of a stationary process $X(t)$ form a Fourier-transform pair with $\tau$ and $f$ as the variables of interest, as shown by the pair of relations

$$S_X(f) = \int_{-\infty}^{\infty} R_X(\tau) \exp(-j2\pi f\tau)\, d\tau \tag{1.42}$$

$$R_X(\tau) = \int_{-\infty}^{\infty} S_X(f) \exp(j2\pi f\tau)\, df \tag{1.43}$$

Equations (1.42) and (1.43) are basic relations in the theory of spectral analysis of random processes, and together they constitute what are usually called the *Einstein–Wiener–Khintchine relations.*[3]

The Einstein–Wiener–Khintchine relations show that if either the autocorrelation function or power spectral density of a random process is known, the other can be found exactly. But these functions display different aspects of the correlation information about the process. It is commonly accepted that for practical purposes, the power spectral density is the more useful "parameter."

We now wish to use this pair of relations to derive some general properties of the power spectral density of a stationary process.

**Property 1**

*The zero-frequency value of the power spectral density of a stationary process equals the total area under the graph of the autocorrelation function; that is,*

$$S_X(0) = \int_{-\infty}^{\infty} R_X(\tau)\, d\tau \tag{1.44}$$

This property follows directly from Equation (1.42) by putting $f = 0$.

**Property 2**

*The mean-square value of a stationary process equals the total area under the graph of the power spectral density; that is,*

$$E[X^2(t)] = \int_{-\infty}^{\infty} S_X(f)\, df \tag{1.45}$$

This property follows directly from Equation (1.43) by putting $\tau = 0$ and noting that $R_X(0) = E[X^2(t)]$.

**Property 3**

*The power spectral density of a stationary process is always nonnegative; that is,*

$$S_X(f) \geq 0 \qquad \text{for all } f \tag{1.46}$$

This property is an immediate consequence of the fact that, in Equation (1.41), the mean-square value $E[Y^2(t)]$ must always be nonnegative.

**Property 4**

*The power spectral density of a real-valued random process is an even function of frequency; that is,*

$$S_X(-f) = S_X(f) \tag{1.47}$$

This property is readily obtained by substituting $-f$ for $f$ in Equation (1.42):

$$S_X(-f) = \int_{-\infty}^{\infty} R_X(\tau) \exp(j2\pi f\tau)\, d\tau$$

Next, substituting $-\tau$ for $\tau$, and recognizing that $R_X(-\tau) = R_X(\tau)$, we get

$$S_X(-f) = \int_{-\infty}^{\infty} R_X(\tau) \exp(-j2\pi f\tau)\, d\tau = S_X(f)$$

which is the desired result.

**Property 5**

*The power spectral density, appropriately normalized, has the properties usually associated with a probability density function.*

The normalization we have in mind here is with respect to the total area under the graph of the power spectral density (i.e., the mean-square value of the process). Consider then the function

$$p_X(f) = \frac{S_X(f)}{\int_{-\infty}^{\infty} S_X(f)\, df} \tag{1.48}$$

In light of Properties 2 and 3, we note that $p_X(f) \geq 0$ for all $f$. Moreover, the total area under the function $p_X(f)$ is unity. Hence, the normalized form of the power spectral density, as defined in Equation (1.48), behaves similar to a probability density function.

### ▶ Example 1.5 Sinusoidal Wave with Random Phase (continued)

Consider the random process $X(t) = A \cos(2\pi f_c t + \Theta)$, where $\Theta$ is a uniformly distributed random variable over the interval $[-\pi, \pi]$. The autocorrelation function of this random process is given by Equation (1.17), which is reproduced here for convenience:

$$R_X(\tau) = \frac{A^2}{2} \cos(2\pi f_c \tau)$$

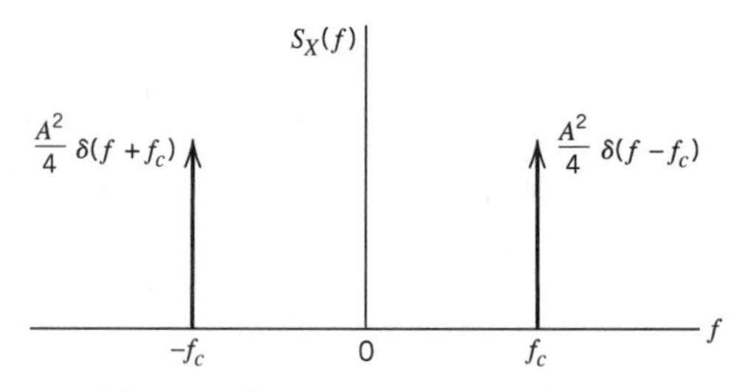


**FIGURE 1.10** Power spectral density of sine wave with random phase; $\delta(f)$ denotes the delta function at $f = 0$.

Let $\delta(f)$ denote the *delta function* at $f = 0$; for the definition of the delta function and its properties, see Appendix 2. Taking the Fourier transform of both sides of the relation defining $R_X(\tau)$, we find that the power spectral density of the sinusoidal process $X(t)$ is

$$S_X(f) = \frac{A^2}{4}\,[\delta(f - f_c) + \delta(f + f_c)] \tag{1.49}$$

which consists of a pair of delta functions weighted by the factor $A^2/4$ and located at $\pm f_c$, as illustrated in Figure 1.10. We note that the total area under a delta function is one. Hence, the total area under $S_X(f)$ is equal to $A^2/2$, as expected. ◀

### ▶ EXAMPLE 1.6 Random Binary Wave (continued)

Consider again a random binary wave consisting of a sequence of 1s and 0s represented by the values $+A$ and $-A$, respectively. In Example 1.3 we showed that the autocorrelation function of this random process has a triangular waveform, as shown by

$$R_X(\tau) = \begin{cases} A^2\left(1 - \dfrac{|\tau|}{T}\right), & |\tau| < T \\ 0, & |\tau| \geq T \end{cases}$$

The power spectral density of the process is therefore

$$S_X(f) = \int_{-T}^{T} A^2\left(1 - \frac{|\tau|}{T}\right) \exp(-j2\pi f\tau)\, d\tau$$

Using the Fourier transform of a triangular function (see Table A6.3), we obtain

$$S_X(f) = A^2T\,\mathrm{sinc}^2(fT) \tag{1.50}$$

which is plotted in Figure 1.11. Here again we see that the power spectral density is nonnegative for all $f$ and that it is an even function of $f$. Noting that $R_X(0) = A^2$ and using Property 2, we find that the total area under $S_X(f)$, or the average power of the random binary wave described here, is $A^2$, which is intuitively satisfying. ◀

The result of Equation (1.50) may be generalized as follows. We note that the *energy spectral density* (i.e., the squared magnitude of the Fourier transform) of a rectangular pulse $g(t)$ of amplitude $A$ and duration $T$ is given by

$$\mathcal{E}_g(f) = A^2T^2\,\mathrm{sinc}^2(fT) \tag{1.51}$$

We may therefore rewrite Equation (1.50) in terms of $\mathcal{E}_g(f)$ simply as

$$S_X(f) = \frac{\mathcal{E}_g(f)}{T} \tag{1.52}$$

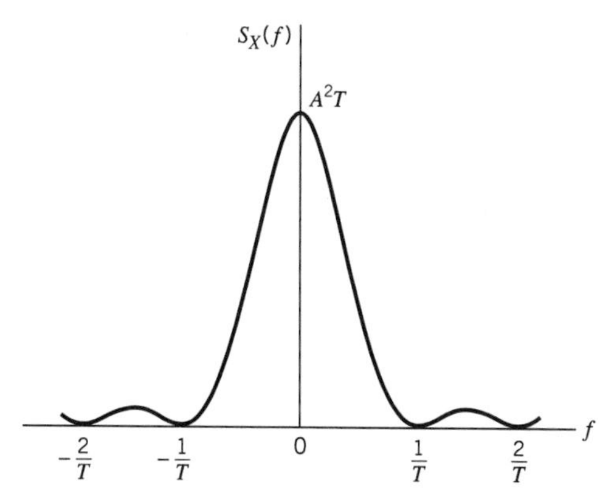


**FIGURE 1.11** Power spectral density of random binary wave.

Equation (1.52) states that for a random binary wave in which binary symbols 1 and 0 are represented by pulses $g(t)$ and $-g(t)$, respectively, the power spectral density $S_X(f)$ is equal to the energy spectral density $\mathcal{E}_g(f)$ of the *symbol shaping pulse* $g(t)$, divided by the *symbol duration* $T$.

### ▶ EXAMPLE 1.7 Mixing of a Random Process with a Sinusoidal Process

A situation that often arises in practice is that of *mixing* (i.e., multiplication) of a stationary process $X(t)$ with a sinusoidal wave $\cos(2\pi f_c t + \Theta)$, where the phase $\Theta$ is a random variable that is uniformly distributed over the interval $[0, 2\pi]$. The addition of the random phase $\Theta$ in this manner merely recognizes the fact that the time origin is arbitrarily chosen when $X(t)$ and $\cos(2\pi f_c t + \Theta)$ come from physically independent sources, as is usually the case. We are interested in determining the power spectral density of the random process $Y(t)$, defined by

$$Y(t) = X(t) \cos(2\pi f_c t + \Theta) \tag{1.53}$$

Using the definition of autocorrelation function of a stationary process and noting that the random variable $\Theta$ is independent of $X(t)$, we find that the autocorrelation function of $Y(t)$ is given by

$$\begin{aligned} R_Y(\tau) &= E[Y(t + \tau)Y(t)] \\ &= E[X(t + \tau) \cos(2\pi f_c t + 2\pi f_c \tau + \Theta)X(t) \cos(2\pi f_c t + \Theta)] \\ &= E[X(t + \tau)X(t)]E[\cos(2\pi f_c t + 2\pi f_c \tau + \Theta) \cos(2\pi f_c t + \Theta)] \\ &= \tfrac{1}{2}R_X(\tau)E[\cos(2\pi f_c \tau) + \cos(4\pi f_c t + 2\pi f_c \tau + 2\Theta)] \\ &= \tfrac{1}{2}R_X(\tau) \cos(2\pi f_c \tau) \end{aligned} \tag{1.54}$$

Because the power spectral density is the Fourier transform of the autocorrelation function, we find that the power spectral densities of the random processes $X(t)$ and $Y(t)$ are related as follows:

$$S_Y(f) = \tfrac{1}{4}[S_X(f - f_c) + S_X(f + f_c)] \tag{1.55}$$

According to Equation (1.55), the power spectral density of the random process $Y(t)$ defined in Equation (1.53) is obtained as follows: We shift the given power spectral density $S_X(f)$ of random process $X(t)$ to the right by $f_c$, shift it to the left by $f_c$, add the two shifted power spectra, and divide the result by 4. ◀

## ■ RELATION AMONG THE POWER SPECTRAL DENSITIES OF THE INPUT AND OUTPUT RANDOM PROCESSES

Let $S_X(f)$ denote the power spectral density of the output random process $Y(t)$ obtained by passing the random process $X(t)$ through a linear filter of frequency response $H(f)$. Then, recognizing by definition that the power spectral density of a random process is equal to the Fourier transform of its autocorrelation function and using Equation (1.32), we obtain

$$\begin{aligned} S_Y(f) &= \int_{-\infty}^{\infty} R_Y(\tau) \exp(-j2\pi f\tau)\, d\tau \\ &= \int_{-\infty}^{\infty}\int_{-\infty}^{\infty}\int_{-\infty}^{\infty} h(\tau_1)h(\tau_2)R_X(\tau - \tau_1 + \tau_2) \exp(-j2\pi f\tau)\, d\tau_1\, d\tau_2\, d\tau \end{aligned} \tag{1.56}$$

Let $\tau - \tau_1 + \tau_2 = \tau_0$, or, equivalently, $\tau = \tau_0 + \tau_1 - \tau_2$. Then by making this substitution in Equation (1.56), we find that $S_X(f)$ may be expressed as the product of three terms: the frequency response $H(f)$ of the filter, the complex conjugate of $H(f)$, and the power spectral density $S_X(f)$ of the input random process $X(t)$. We may thus simplify Equation (1.56) as

$$S_Y(f) = H(f)H^*(f)S_X(f) \tag{1.57}$$

Finally, since $|H(f)|^2 = H(f)H^*(f)$, we find that the relationship among the power spectral densities of the input and output random processes is expressed in the frequency domain by writing

$$S_Y(f) = |H(f)|^2 S_X(f) \tag{1.58}$$

Equation (1.58) states that *the power spectral density of the output process* $Y(t)$ *equals the power spectral density of the input process* $X(t)$ *multiplied by the squared magnitude response of the filter*. By using this relation, we can therefore determine the effect of passing a random process through a stable, linear, time-invariant, filter. In computational terms, Equation (1.58) is usually easier to handle than its time-domain counterpart of Equation (1.32), involving the autocorrelation function.

## ■ RELATION AMONG THE POWER SPECTRAL DENSITY AND THE MAGNITUDE SPECTRUM OF A SAMPLE FUNCTION

We now wish to relate the power spectral density $S_X(f)$ directly to the spectral properties of a sample function $x(t)$ of a stationary process $X(t)$ that is ergodic. For the sample function $x(t)$ to be Fourier transformable, however, it must be absolutely integrable; that is

$$\int_{-\infty}^{\infty} |x(t)|\, dt < \infty \tag{1.59}$$

This condition can never be satisfied by any stationary sample function $x(t)$ of infinite duration. In order to use the Fourier transform technique, we consider a truncated segment of $x(t)$, defined over the observation interval $-T \le t \le T$, say. Thus, using $X(f, T)$ to denote the Fourier transform of the truncated sample function so defined, we may write

$$X(f, T) = \int_{-T}^{T} x(t) \exp(-j2\pi ft)\, dt \tag{1.60}$$

Assuming that the process $x(t)$ is also ergodic, we may evaluate the autocorrelation function $R_X(\tau)$ of $X(t)$ using the time-average formula (see Section 1.5)

$$R_X(\tau) = \lim_{T\to\infty} \frac{1}{2T}\int_{-T}^{T} x(t+\tau)x(t)\,dt \tag{1.61}$$

It is customary to view the sample function $x(t)$ as a *power signal* (i.e., a signal with finite average power). Hence, we may formulate the following Fourier-transform pair:

$$\frac{1}{2T}\int_{-T}^{T} x(t+\tau)x(t)\,dt \rightleftharpoons \frac{1}{2T}|X(f, T)|^2 \tag{1.62}$$

The parameter on the left-hand side is a time-averaged autocorrelation function. The parameter on the right-hand side is called the *periodogram*, whose dimensions are the same as those of the power spectral density. This terminology is a misnomer, however, since the periodogram is a function of frequency, not period. Nevertheless, it has wide usage. The quantity was first used by statisticians to look for periodicities such as seasonal trends in data.

Using the formula for the inverse Fourier transform in the Fourier-transform pair of Equation (1.62), we may express the time-averaged autocorrelation function of the sample function $x(t)$ in terms of the periodogram as

$$\frac{1}{2T}\int_{-T}^{T} x(t+\tau)x(t)\,dt = \int_{-\infty}^{\infty} \frac{1}{2T}|X(f, T)|^2 \exp(j2\pi fT)\,df \tag{1.63}$$

Hence, substituting Equation (1.63) into (1.61), we get

$$R_X(\tau) = \lim_{T\to\infty} \int_{-\infty}^{\infty} \frac{1}{2T}|X(f, T)|^2 \exp(j2\pi f\tau)\,df \tag{1.64}$$

For a fixed value of the frequency $f$, the periodogram is a random variable in that its value varies in a random manner from one sample function of the random process to another. Thus, for a given sample function $x(t)$, the periodogram does not converge in any statistical sense to a limiting value as $T$ tends to infinity. As such, it would be incorrect to interchange the order of the integration and limiting operations in Equation (1.64). Suppose, however, that we take the expectation of both sides of Equation (1.64) over the ensemble of all sample functions of the random process and recognize that for an ergodic process the autocorrelation function $R_X(\tau)$ is unchanged by such an operation. Then, since each sample function of an ergodic process eventually takes on nearly all the modes of behavior of each other sample function, we may thus write

$$R_X(\tau) = \lim_{T\to\infty} \int_{-\infty}^{\infty} \frac{1}{2T} E[|X(f, T)|^2] \exp(j2\pi f\tau)\,df \tag{1.65}$$

Now we may interchange the order of the integration and limiting operations and so obtain

$$R_X(\tau) = \int_{-\infty}^{\infty} \left\{\lim_{T\to\infty} \frac{1}{2T} E[|X(f, T)|^2]\right\} \exp(j2\pi f\tau)\,df \tag{1.66}$$

Hence, comparing Equations (1.66) and (1.43), we obtain the desired relation between the power spectral density $S_X(f)$ of an ergodic process and the squared magnitude spectrum $|X(f, T)|^2$ of a truncated sample function of the process:

$$\begin{aligned} S_X(f) &= \lim_{T\to\infty} \frac{1}{2T} E[|X(f, T)|^2] \\ &= \lim_{T\to\infty} \frac{1}{2T} E\left[\left|\int_{-T}^{T} x(t)\exp(-j2\pi ft)\,dt\right|^2\right] \end{aligned} \tag{1.67}$$

It is important to note that in Equation (1.67) it is not possible to let $T \to \infty$ before taking the expectation. Equation (1.67) provides the mathematical basis for *estimating*[4] the power spectral density of an ergodic random process, given a sample function $x(t)$ of the process observed over the interval $[-T, T]$.

## ■ CROSS-SPECTRAL DENSITIES

Just as the power spectral density provides a measure of the frequency distribution of a single random process, cross-spectral densities provide a measure of the frequency interrelationship between two random processes. In particular, let $X(t)$ and $Y(t)$ be two jointly stationary processes with their cross-correlation functions denoted by $R_{XY}(\tau)$ and $R_{YX}(\tau)$. We then define the *cross-spectral densities* $S_{XY}(f)$ and $S_{YX}(f)$ of this pair of random processes to be the Fourier transforms of their respective cross-correlation functions, as shown by

$$S_{XY}(f) = \int_{-\infty}^{\infty} R_{XY}(\tau) \exp(-j2\pi f\tau)\, d\tau \tag{1.68}$$

and

$$S_{YX}(f) = \int_{-\infty}^{\infty} R_{YX}(\tau) \exp(-j2\pi f\tau)\, d\tau \tag{1.69}$$

The cross-correlation functions and cross-spectral densities thus form Fourier-transform pairs. Accordingly, using the formula for inverse Fourier transformation we may also write

$$R_{XY}(\tau) = \int_{-\infty}^{\infty} S_{XY}(f) \exp(j2\pi f\tau)\, df \tag{1.70}$$

and

$$R_{YX}(\tau) = \int_{-\infty}^{\infty} S_{YX}(f) \exp(j2\pi f\tau)\, df \tag{1.71}$$

The cross-spectral densities $S_{XY}(f)$ and $S_{YX}(f)$ are not necessarily real functions of the frequency $f$. However, substituting the relationship

$$R_{XY}(\tau) = R_{YX}(-\tau)$$

into Equation (1.68) and then using Equation (1.69) we find that $S_{XY}(f)$ and $S_{YX}(f)$ are related by

$$S_{XY}(f) = S_{YX}(-f) = S_{YX}^*(f) \tag{1.72}$$

### ▶ EXAMPLE 1.8

Suppose that the random processes $X(t)$ and $Y(t)$ have zero mean, and they are individually stationary. Consider the sum random process

$$Z(t) = X(t) + Y(t)$$

The problem is to determine the power spectral density of $Z(t)$.

The autocorrelation function of $Z(t)$ is given by

$$\begin{aligned} R_Z(t, u) &= E[Z(t)Z(u)] \\ &= E[(X(t) + Y(t))(X(u) + Y(u))] \\ &= E[X(t)X(u)] + E[X(t)Y(u)] + E[Y(t)X(u)] + E[Y(t)Y(u)] \\ &= R_X(t, u) + R_{XY}(t, u) + R_{YX}(t, u) + R_Y(t, u) \end{aligned}$$

Defining $\tau = t - u$, we may therefore write

$$R_Z(\tau) = R_X(\tau) + R_{XY}(\tau) + R_{YX}(\tau) + R_Y(\tau) \tag{1.73}$$

when the random processes $X(t)$ and $Y(t)$ are also jointly stationary. Accordingly, taking the Fourier transform of both sides of Equation (1.73), we get

$$S_Z(f) = S_X(f) + S_{XY}(f) + S_{YX}(f) + S_Y(f) \tag{1.74}$$

We thus see that the cross-spectral densities $S_{XY}(f)$ and $S_{YX}(f)$ represent the spectral components that must be added to the individual power spectral densities of a pair of correlated random processes in order to obtain the power spectral density of their sum.

When the stationary processes $X(t)$ and $Y(t)$ are uncorrelated, the cross-spectral densities $S_{XY}(f)$ and $S_{YX}(f)$ are zero, and so Equation (1.74) reduces as follows:

$$S_Z(f) = S_X(f) + S_Y(f) \tag{1.75}$$

We may generalize this latter result by stating that when there is a multiplicity of zero-mean stationary processes that are uncorrelated with each other, the power spectral density of their sum is equal to the sum of their individual power spectral densities. ◀

### ▶ EXAMPLE 1.9

Consider next the problem of passing two jointly stationary processes through a pair of separate, stable, linear, time-invariant filters, as shown in Figure 1.12. In particular, suppose that the random process $X(t)$ is the input to the filter of impulse response $h_1(t)$ and that the random process $Y(t)$ is the input to the filter of impulse response $h_2(t)$. Let $V(t)$ and $Z(t)$ denote the random processes at the respective filter outputs. The cross-correlation function of $V(t)$ and $Z(t)$ is therefore

$$\begin{aligned} R_{VZ}(t, u) &= E[V(t)Z(u)] \\ &= E\left[\int_{-\infty}^{\infty} h_1(\tau_1)X(t - \tau_1)\, d\tau_1 \int_{-\infty}^{\infty} h_2(\tau_2)Y(u - \tau_2)\, d\tau_2\right] \\ &= \int_{-\infty}^{\infty}\int_{-\infty}^{\infty} h_1(\tau_1)h_2(\tau_2)E[X(t - \tau_1)Y(u - \tau_2)]\, d\tau_1\, d\tau_2 \\ &= \int_{-\infty}^{\infty}\int_{-\infty}^{\infty} h_1(\tau_1)h_2(\tau_2)R_{XY}(t - \tau_1, u - \tau_2)\, d\tau_1\, d\tau_2 \end{aligned} \tag{1.76}$$

where $R_{XY}(t, u)$ is the cross-correlation function of $X(t)$ and $Y(t)$. Because the input random processes are jointly stationary (by hypothesis), we may set $\tau = t - u$ and so rewrite Equation (1.76) as follows:

$$R_{VZ}(\tau) = \int_{-\infty}^{\infty}\int_{-\infty}^{\infty} h_1(\tau_1)h_2(\tau_2)R_{XY}(\tau - \tau_1 + \tau_2)\, d\tau_1\, d\tau_2 \tag{1.77}$$

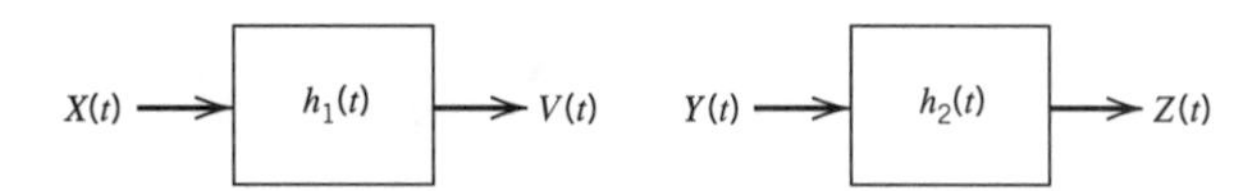


**FIGURE 1.12** A pair of separate linear time-invariant filters.

Taking the Fourier transform of both sides of Equation (1.77) and using a procedure similar to that which led to the development of Equation (1.39), we finally get

$$S_{VZ}(f) = H_1(f)H_2^*(f)S_{XY}(f) \tag{1.78}$$

where $H_1(f)$ and $H_2(f)$ are the frequency responses of the respective filters in Figure 1.12, and $H_2^*(f)$ is the complex conjugate of $H_2(f)$. This is the desired relationship between the cross-spectral density of the output processes and that of the input processes. ◀

## 1.8 *Gaussian Process*

The material we have presented on random processes up to this point in the discussion has been of a fairly general nature. In this section, we consider an important family of random processes known as Gaussian processes.[5]

Let us suppose that we observe a random process $X(t)$ for an interval that starts at time $t = 0$ and lasts until $t = T$. Suppose also that we weight the random process $X(t)$ by some function $g(t)$ and then integrate the product $g(t)X(t)$ over this observation interval, thereby obtaining a random variable $Y$ defined by

$$Y = \int_0^T g(t)X(t)\,dt \tag{1.79}$$

We refer to $Y$ as a *linear functional* of $X(t)$. The distinction between a function and a functional should be carefully noted. For example, the sum $Y = \sum_{i=1}^{N} a_i X_i$, where the $a_i$ are constants and the $X_i$ are random variables, is a linear *function* of the $X_i$; for each observed set of values for the random variables $X_i$, we have a corresponding value for the random variable $Y$. On the other hand, in Equation (1.79) the value of the random variable $Y$ depends on the course of the *argument function* $g(t)X(t)$ over the entire observation interval from 0 to $T$. Thus a functional is a quantity that depends on the entire course of one or more functions rather than on a number of discrete variables. In other words, the domain of a functional is a set or space of admissible functions rather than a region of a coordinate space.

If in Equation (1.79) the weighting function $g(t)$ is such that the mean-square value of the random variable $Y$ is finite, and if the random variable $Y$ is a *Gaussian-distributed* random variable for every $g(t)$ in this class of functions, then the process $X(t)$ is said to be a *Gaussian process*. In other words, the process $X(t)$ is a Gaussian process if every linear functional of $X(t)$ is a Gaussian random variable.

We say that the random variable $Y$ has a Gaussian distribution if its probability density function has the form

$$f_Y(y) = \frac{1}{\sqrt{2\pi}\sigma_Y} \exp\left[-\frac{(y-\mu_Y)^2}{2\sigma_Y^2}\right] \tag{1.80}$$

where $\mu_Y$ is the mean and $\sigma_Y^2$ is the variance of the random variable $Y$. A plot of this probability density function is given in Figure 1.13 for the special case when the Gaussian random variable $Y$ is *normalized* to have a mean $\mu_Y$ of zero and a variance $\sigma_Y^2$ of one, as shown by

$$f_Y(y) = \frac{1}{\sqrt{2\pi}} \exp\left(-\frac{y^2}{2}\right)$$

Such a normalized Gaussian distribution is commonly written as $\mathcal{N}(0, 1)$.

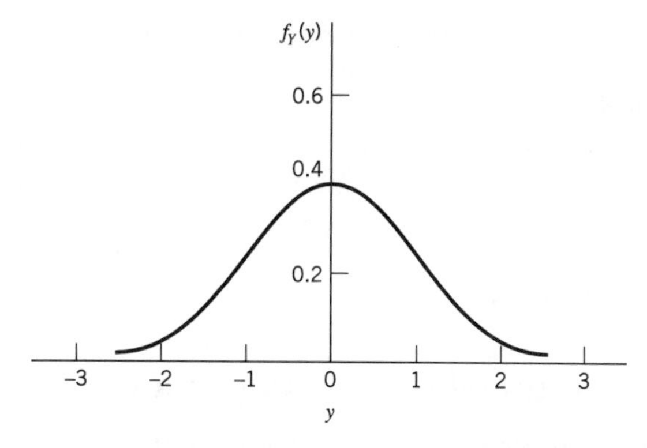


**FIGURE 1.13** Normalized Gaussian distribution.

A Gaussian process has two main virtues. First, the Gaussian process has many properties that make analytic results possible; we will discuss these properties later in the section. Second, the random processes produced by physical phenomena are often such that a Gaussian model is appropriate. Furthermore, the use of a Gaussian model to describe the physical phenomena is usually confirmed by experiments. Thus the frequent occurrence of physical phenomena for which a Gaussian model is appropriate, together with the ease with which a Gaussian process is handled mathematically, make the Gaussian process very important in the study of communication systems.

## ■ CENTRAL LIMIT THEOREM

The *central limit theorem* provides the mathematical justification for using a Gaussian process as a model for a large number of different physical phenomena in which the observed random variable, at a particular instant of time, is the result of a large number of individual random events. To formulate this important theorem, let $X_i$, $i = 1, 2, \ldots,$ $N$, be a set of random variables that satisfies the following requirements:

1. The $X_i$ are statistically independent.
2. The $X_i$ have the same probability distribution with mean $\mu_X$ and variance $\sigma_X^2$.

The $X_i$ so described are said to constitute a set of *independently and identically distributed* (i.i.d.) random variables. Let these random variables be *normalized* as follows:

$$Y_i = \frac{1}{\sigma_X}(X_i - \mu_X), \qquad i = 1, 2, \ldots, N$$

so that we have

$$E[Y_i] = 0$$

and

$$\text{var}[Y_i] = 1$$

Define the random variable

$$V_N = \frac{1}{\sqrt{N}} \sum_{i=1}^{N} Y_i$$

*The central limit theorem states that the probability distribution of $V_N$ approaches a normalized Gaussian distribution $\mathcal{N}(0, 1)$ in the limit as the number of random variables N approaches infinity.*

It is important to realize, however, that the central limit theorem gives only the "limiting" form of the probability distribution of the normalized random variable $V_N$ as $N$ approaches infinity. When $N$ is finite, it is sometimes found that the Gaussian limit gives a relatively poor approximation for the actual probability distribution of $V_N$ even though $N$ may be quite large.

## ■ PROPERTIES OF A GAUSSIAN PROCESS

A Gaussian process has some useful properties that are described in the sequel.

### Property 1

*If a Gaussian process $X(t)$ is applied to a stable linear filter, then the random process $Y(t)$ developed at the output of the filter is also Gaussian.*

This property is readily derived by using the definition of a Gaussian process based on Equation (1.79). Consider the situation depicted in Figure 1.8, where we have a linear time-invariant filter of impulse response $h(t)$, with the random process $X(t)$ as input and the random process $Y(t)$ as output. We assume that $X(t)$ is a Gaussian process. The random processes $Y(t)$ and $X(t)$ are related by the convolution integral

$$Y(t) = \int_0^T h(t - \tau)X(\tau)\, d\tau, \qquad 0 \le t < \infty \tag{1.81}$$

We assume that the impulse response $h(t)$ is such that the mean-square value of the output random process $Y(t)$ is finite for all $t$ in the range $0 \le t < \infty$ for which $Y(t)$ is defined. To demonstrate that the output process $Y(t)$ is Gaussian, we must show that any linear functional of it is a Gaussian random variable. That is, if we define the random variable

$$Z = \int_0^\infty g_Y(t) \int_0^T h(t - \tau)X(\tau)\, d\tau\, dt \tag{1.82}$$

then $Z$ must be a Gaussian random variable for every function $g_Y(t)$, such that the mean-square value of $Z$ is finite. Interchanging the order of integration in Equation (1.82), we get

$$Z = \int_0^T g(\tau)X(\tau)\, d\tau \tag{1.83}$$

where

$$g(\tau) = \int_0^\infty g_Y(t)h(t - \tau)\, d\tau \tag{1.84}$$

Since $X(t)$ is a Gaussian process by hypothesis, it follows from Equation (1.83) that $Z$ must be a Gaussian random variable. We have thus shown that if the input $X(t)$ to a linear filter is a Gaussian process, then the output $Y(t)$ is also a Gaussian process. Note, however, that although our proof was carried out assuming a time-invariant linear filter, this property is true for any arbitrary stable linear system.

**Property 2**

*Consider the set of random variables or samples $X(t_1), X(t_2), \ldots, X(t_n)$, obtained by observing a random process $X(t)$ at times $t_1, t_2, \ldots, t_n$. If the process $X(t)$ is Gaussian, then this set of random variables is jointly Gaussian for any $n$, with their $n$-fold joint probability density function being completely determined by specifying the set of means*

$$\mu_{X(t_i)} = E[X(t_i)], \qquad i = 1, 2, \ldots, n$$

*and the set of covariance functions*

$$C_X(t_k, t_i) = E[(X(t_k) - \mu_{X(t_k)})(X(t_i) - \mu_{X(t_i)})], \qquad k, i = 1, 2, \ldots, n$$

Let the $n$-by-1 vector $\mathbf{X}$ denote the set of random variables $X(t_1), \ldots, X(t_n)$ derived from the Gaussian process $X(t)$ by sampling it at times $t_1, \ldots, t_n$. Let $\mathbf{x}$ denote a value of $\mathbf{X}$. According to Property 2, the random vector $\mathbf{X}$ has a *multivariate Gaussian distribution* defined in matrix form as

$$f_{X(t_1),\ldots,X(t_2)}(x_1, \ldots, x_2) = \frac{1}{(2\pi)^{n/2}\Delta^{1/2}} \exp\left(-\frac{1}{2}(\mathbf{x} - \boldsymbol{\mu})^T \boldsymbol{\Sigma}^{-1}(\mathbf{x} - \boldsymbol{\mu})\right) \qquad (1.85)$$

where the superscript $T$ denotes transposition and

$$\begin{aligned} \boldsymbol{\mu} &= \text{mean vector} \\ &= [\mu_1, \mu_2, \ldots, \mu_n]^T \\ \boldsymbol{\Sigma} &= \text{covariance matrix} \\ &= \{C_X(t_k, t_i)\}_{k,i=1}^{n} \\ \boldsymbol{\Sigma}^{-1} &= \text{inverse of covariance matrix} \\ \Delta &= \text{determinant of covariance matrix } \boldsymbol{\Sigma} \end{aligned}$$

Property 2 is frequently used as the definition of a Gaussian process. However, this definition is more difficult to use than that based on Equation (1.79) for evaluating the effects of filtering on a Gaussian process.

We may extend Property 2 to two (or more) random processes as follows. Consider the composite set of random variables $X(t_1), X(t_2), \ldots, X(t_n), Y(u_1), Y(u_2), \ldots, Y(u_m)$ obtained by observing a random process $X(t)$ at times $\{t_i, i = 1, 2, \ldots, n\}$, and a second random process $Y(t)$ at times $\{u_k, k = 1, 2, \ldots, m\}$. We say that the processes $X(t)$ and $Y(t)$ are *jointly Gaussian* if this composite set of random variables is jointly Gaussian for any $n$ and $m$. Note that in addition to the mean and correlation functions of the random processes $X(t)$ and $Y(t)$ individually, we must also know the cross-covariance function

$$E[(X(t_i) - \mu_{X(t_i)})(Y(u_k) - \mu_{Y(u_k)})] = R_{XY}(t_i, u_k) - \mu_{X(t_i)}\mu_{Y(u_k)}$$

for any pair of observation instants $(t_i, u_k)$. This additional knowledge is embodied in the cross-correlation function, $R_{XY}(t_i, u_k)$, of the two processes $X(t)$ and $Y(t)$.

**Property 3**

*If a Gaussian process is stationary, then the process is also strictly stationary.*

This follows directly from Property 2.

**Property 4**

*If the random variables $X(t_1), X(t_2), \ldots, X(t_n)$, obtained by sampling a Gaussian process $X(t)$ at times $t_1, t_2, \ldots, t_n$, are uncorrelated, that is,*

$$E[(X(t_k) - \mu_{X(t_k)})(X(t_i) - \mu_{X(t_i)})] = 0, \qquad i \neq k$$

*then these random variables are statistically independent.*

The uncorrelatedness of $X(t_1), \ldots, X(t_n)$ means that the covariance matrix $\mathbf{\Sigma}$ is a diagonal matrix as shown by

$$\mathbf{\Sigma} = \begin{bmatrix} \sigma_1^2 & & & \mathrm{O} \\ & \sigma_2^2 & & \\ & & \ddots & \\ \mathrm{O} & & & \sigma_n^2 \end{bmatrix}$$

where

$$\sigma_i^2 = E[(X(t_i) - E[X(t_i)])^2], \qquad i = 1, 2, \ldots, n$$

Under this condition, the multivariate Gaussian distribution of Equation (1.85) simplifies to

$$f_{\mathbf{X}}(\mathbf{x}) = \prod_{i=1}^{n} f_{X_i}(x_i)$$

where $X_i = X(t_i)$ and

$$f_{X_i}(x_i) = \frac{1}{\sqrt{2\pi}\sigma_i} \exp\left(-\frac{(x_i - \mu_{X_i})^2}{2\sigma_i^2}\right)$$

In words, if the Gaussian random variables $X(t_1), \ldots, X(t_n)$ are uncorrelated, then they are statistically independent, which, in turn, means that the joint probability density function of this set of random variables can be expressed as the product of the probability density functions of the individual random variables in the set.

## 1.9 *Noise*

The term *noise* is used customarily to designate unwanted signals that tend to disturb the transmission and processing of signals in communication systems and over which we have incomplete control. In practice, we find that there are many potential sources of noise in a communication system. The sources of noise may be external to the system (e.g., atmospheric noise, galactic noise, man-made noise), or internal to the system. The second category includes an important type of noise that arises from *spontaneous fluctuations* of current or voltage in electrical circuits.[6] This type of noise represents a basic limitation on the transmission or detection of signals in communication systems involving the use of electronic devices. The two most common examples of spontaneous fluctuations in electrical circuits are *shot noise* and *thermal noise*, which are described in the sequel.

### ■ SHOT NOISE

Shot noise arises in electronic devices such as diodes and transistors because of the discrete nature of current flow in these devices. For example, in a *photodetector* circuit a current pulse is generated every time an electron is emitted by the cathode due to incident light from a source of constant intensity. The electrons are naturally emitted at random times denoted by $\tau_k$, where $-\infty < k < \infty$. It is assumed that the random emissions of electrons have been going on for a long time. Thus, the total current flowing through the photodetector may be modeled as an infinite sum of current pulses, as shown by

$$X(t) = \sum_{k=-\infty}^{\infty} h(t - \tau_k) \tag{1.86}$$

where $h(t - \tau_k)$ is the current pulse generated at time $\tau_k$. The process $X(t)$ defined by Equation (1.86) is a stationary process called *shot noise.*

The number of electrons, $N(t)$, emitted in the time interval $[0, t]$ constitutes a discrete stochastic process, the value of which increases by one each time an electron is emitted. Figure 1.14 shows a sample function of such a process. Let the mean value of the number of electrons, $\nu$, emitted between times $t$ and $t + t_0$ be defined by

$$E[\nu] = \lambda t_0 \tag{1.87}$$

The parameter $\lambda$ is a constant called the *rate* of the process. The total number of electrons emitted in the interval $[t, t + t_0]$, that is,

$$\nu = N(t + t_0) - N(t)$$

follows a *Poisson distribution* with a mean value equal to $\lambda t_0$. In particular, the probability that $k$ electrons are emitted in the interval $[t, t + t_0]$ is defined by

$$P(\nu = k) = \frac{(\lambda t_0)^k}{k!} e^{-\lambda t_0} \qquad k = 0, 1, \ldots \tag{1.88}$$

Unfortunately, a detailed statistical characterization of the shot-noise process $X(t)$ defined in Equation (1.86) is a difficult mathematical task. Here we simply quote the results pertaining to the first two moments of the process:

- The mean of $X(t)$ is

$$\mu_X = \lambda \int_{-\infty}^{\infty} h(t)\, dt \tag{1.89}$$

where $\lambda$ is the rate of the process and $h(t)$ is the waveform of a current pulse.

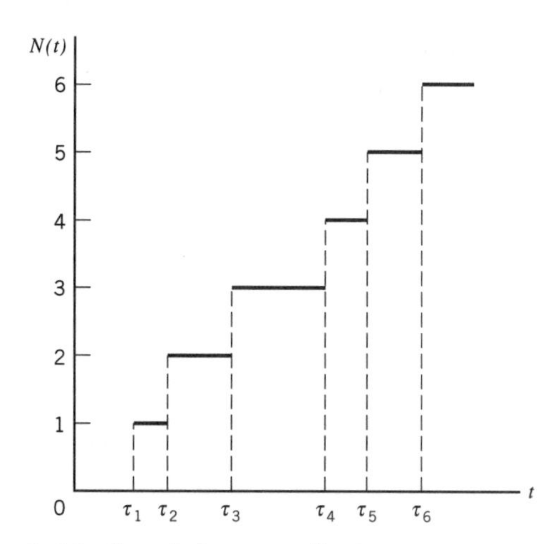


**FIGURE 1.14** Sample function of a Poisson counting process.

- The autocovariance function of $X(t)$ is

$$C_X(\tau) = \lambda \int_{-\infty}^{\infty} h(t)h(t + \tau)\, dt \tag{1.90}$$

This second result is known as *Campbell's theorem.*

For the special case of a waveform $h(t)$ consisting of a rectangular pulse of amplitude $A$ and duration $T$, the mean of the shot-noise process $X(t)$ is $\lambda AT$, and its autocovariance function is

$$C_X(\tau) = \begin{cases} \lambda A^2(T - |\tau|), & |\tau| < T \\ 0, & |\tau| \geq T \end{cases}$$

which has a triangular form similar to that shown in Figure 1.7.

## ■ THERMAL NOISE

*Thermal noise* is the name given to the electrical noise arising from the random motion of electrons in a conductor. The mean-square value of the thermal noise voltage $V_{TN}$ appearing across the terminals of a resistor, measured in a bandwidth of $\Delta f$ Hertz, is, for all practical purposes, given by

$$E[V_{TN}^2] = 4kTR\ \Delta f \text{ volts}^2 \tag{1.91}$$

where $k$ is *Boltzmann's constant* equal to $1.38 \times 10^{-23}$ joules per degree Kelvin, $T$ is the *absolute temperature* in degrees Kelvin, and $R$ is the resistance in ohms. We may thus model a noisy resistor by the *Thévenin equivalent circuit* consisting of a noise voltage generator of mean-square value $E[V_{TN}^2]$ in series with a noiseless resistor, as in Figure 1.15*a*. Alternatively, we may use the *Norton equivalent circuit* consisting of a noise current generator in parallel with a noiseless conductance, as in Figure 1.15*b*. The mean-square value of the noise current generator is

$$\begin{aligned} E[I_{TN}^2] &= \frac{1}{R^2} E[V_{TN}^2] \\ &= 4kTG\ \Delta f \text{ amps}^2 \end{aligned} \tag{1.92}$$

where $G = 1/R$ is the conductance. It is also of interest to note that because the number of electrons in a resistor is very large and their random motions inside the resistor are statistically independent of each other, the central limit theorem indicates that thermal noise is Gaussian distributed with zero mean.

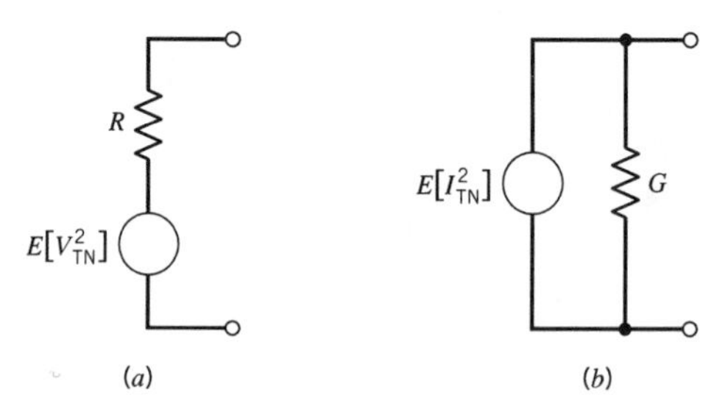


**FIGURE 1.15** Models of a noisy resistor. (*a*) Thévenin equivalent circuit. (*b*) Norton equivalent circuit.

Noise calculations involve the transfer of power, and so we find that the use of the *maximum-power transfer theorem* is applicable to such calculations. This theorem states that the maximum possible power is transferred from a source of internal resistance $R$ to a load of resistance $R_l$ when $R_l = R$. Under this *matched condition*, the power produced by the source is divided equally between the internal resistance of the source and the load resistance, and the power delivered to the load is referred to as the *available power*. Applying the maximum-power transfer theorem to the Thévenin equivalent circuit of Figure 1.15*a* or the Norton equivalent circuit of Figure 1.15*b*, we find that a noisy resistor produces an *available noise power* equal to $kT\,\Delta f$ watts.

## ■ White Noise

The noise analysis of communication systems is customarily based on an idealized form of noise called *white noise*, the power spectral density of which is independent of the operating frequency. The adjective *white* is used in the sense that white light contains equal amounts of all frequencies within the visible band of electromagnetic radiation. We express the power spectral density of white noise, with a sample function denoted by $w(t)$, as

$$S_W(f) = \frac{N_0}{2} \tag{1.93}$$

which is illustrated in Figure 1.16*a*. The dimensions of $N_0$ are in watts per Hertz. The parameter $N_0$ is usually referenced to the input stage of the receiver of a communication system. It may be expressed as

$$N_0 = kT_e \tag{1.94}$$

where $k$ is Boltzmann's constant and $T_e$ is the *equivalent noise temperature* of the receiver.[7] The equivalent noise temperature of a system is defined as *the temperature at which a noisy resistor has to be maintained such that, by connecting the resistor to the input of a noiseless version of the system, it produces the same available noise power at the output of the system as that produced by all the sources of noise in the actual system.* The important feature of the equivalent noise temperature is that it depends only on the parameters of the system.

Since the autocorrelation function is the inverse Fourier transform of the power spectral density, it follows that for white noise

$$R_W(\tau) = \frac{N_0}{2}\,\delta(\tau) \tag{1.95}$$

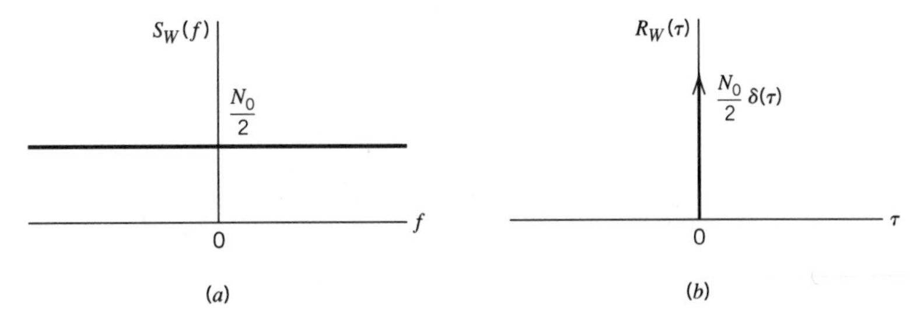


**Figure 1.16** Characteristics of white noise. (*a*) Power spectral density. (*b*) Autocorrelation function.

That is, the autocorrelation function of white noise consists of a delta function weighted by the factor $N_0/2$ and occurring at $\tau = 0$, as in Figure 1.16*b*. We note that $R_W(\tau)$ is zero for $\tau \neq 0$. Accordingly, any two different samples of white noise, no matter how closely together in time they are taken, are uncorrelated. If the white noise $w(t)$ is also Gaussian, then the two samples are statistically independent. In a sense, white Gaussian noise represents the ultimate in "randomness."

Strictly speaking, white noise has infinite average power and, as such, it is not physically realizable. Nevertheless, white noise has simple mathematical properties exemplified by Equations (1.93) and (1.95), which make it useful in statistical system analysis.

The utility of a white noise process is parallel to that of an impulse function or delta function in the analysis of linear systems. Just as we may observe the effect of an impulse only after it has been passed through a system with a finite bandwidth, so it is with white noise whose effect is observed only after passing through a similar system. We may state, therefore, that as long as the bandwidth of a noise process at the input of a system is appreciably larger than that of the system itself, then we may model the noise process as white noise.

### ▶ EXAMPLE 1.10 Ideal Low-Pass Filtered White Noise

Suppose that a white Gaussian noise $w(t)$ of zero mean and power spectral density $N_0/2$ is applied to an ideal low-pass filter of bandwidth $B$ and passband magnitude response of one. The power spectral density of the noise $n(t)$ appearing at the filter output is therefore (see Figure 1.17)

$$S_N(f) = \begin{cases} \dfrac{N_0}{2}, & -B < f < B \\ 0, & |f| > B \end{cases} \tag{1.96}$$

The autocorrelation function of $n(t)$ is the inverse Fourier transform of the power spectral density shown in Figure 1.17*a*:

$$\begin{aligned} R_N(\tau) &= \int_{-B}^{B} \frac{N_0}{2} \exp(j2\pi f\tau)\, df \\ &= N_0 B \operatorname{sinc}(2B\tau) \end{aligned} \tag{1.97}$$

This autocorrelation function is plotted in Figure 1.17*b*. We see that $R_N(\tau)$ has its maximum value of $N_0B$ at the origin, and it passes through zero at $\tau = \pm k/2B$, where $k = 1, 2, 3, \cdots$.

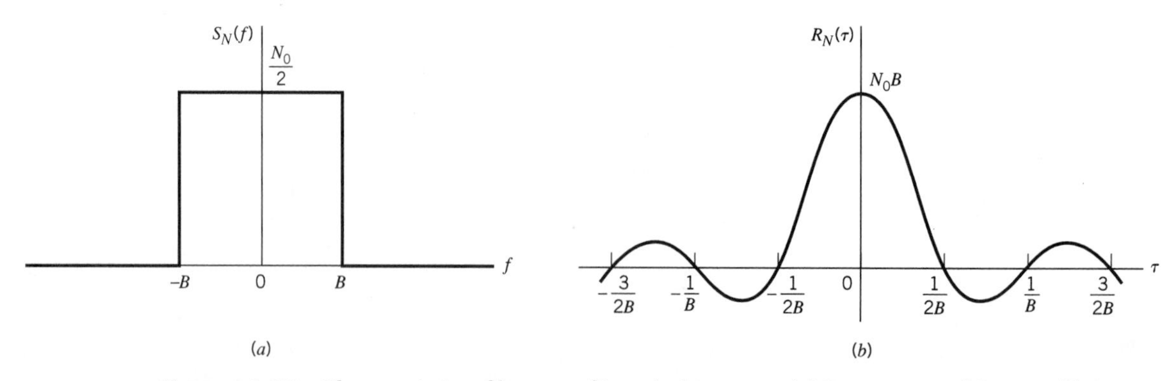


**FIGURE 1.17** Characteristics of low-pass filtered white noise. (*a*) Power spectral density. (*b*) Autocorrelation function.

Since the input noise $w(t)$ is Gaussian (by hypothesis), it follows that the band-limited noise $n(t)$ at the filter output is also Gaussian. Suppose now that $n(t)$ is sampled at the rate of $2B$ times per second. From Figure 1.17*b*, we see that the resulting noise samples are uncorrelated and, being Gaussian, they are statistically independent. Accordingly, the joint probability density function of a set of noise samples obtained in this way is equal to the product of the individual probability density functions. Note that each such noise sample has a mean of zero and variance of $N_0B$. ◀

### ▶ Example 1.11 Correlation of White Noise with a Sinusoidal Wave

Consider the sample function

$$w'(t) = \sqrt{\frac{2}{T}} \int_0^T w(t) \cos(2\pi f_c t)\, dt \tag{1.98}$$

which is the output of a correlator with white Gaussian noise $w(t)$ and sinusoidal wave $\sqrt{2/T} \cos(2\pi f_c t)$ as inputs; the scaling factor $\sqrt{2/T}$ is included here to make the sinusoidal wave input have unit energy over the interval $0 \le t \le T$. (This problem was encountered in the Background and Preview chapter but was not elaborated on at that time.) With the noise $w(t)$ having zero mean, it immediately follows that the correlator output $w'(t)$ has zero mean, too. The variance of the correlator output is defined by

$$\begin{aligned}
\sigma^2 &= E\left[\frac{2}{T}\int_0^T \int_0^T w(t_1) \cos(2\pi f_c t_1) w(t_2) \cos(2\pi f_c t_2)\, dt_1\, dt_2\right] \\
&= \frac{2}{T}\int_0^T \int_0^T E[w(t_1)w(t_2)] \cos(2\pi f_c t_1) \cos(2\pi f_c t_2)\, dt_1\, dt_2 \\
&= \frac{2}{T}\int_0^T \int_0^T R_W(t_1, t_2) \cos(2\pi f_c t_1) \cos(2\pi f_c t_2)\, dt_1\, dt_2
\end{aligned}$$

where $R_W(t_1, t_2)$ is the autocorrelation function of the white noise $w(t)$. But from Equation (1.95):

$$R_W(t_1, t_2) = \frac{N_0}{2}\, \delta(t_1 - t_2)$$

where $N_0/2$ is the power spectral density of the white noise $w(t)$. Accordingly, we may simplify the expression for the variance $\sigma^2$ as

$$\sigma^2 = \frac{N_0}{2} \cdot \frac{2}{T} \int_0^T \int_0^T \delta(t_1 - t_2) \cos(2\pi f_c t_1) \cos(2\pi f_c t_2)\, dt_1\, dt_2$$

We now invoke the *sifting property* of the delta function, namely,

$$\int_{-\infty}^{\infty} g(t)\, \delta(t)\, dt = g(0)$$

where $g(t)$ is a continuous function of time, assuming the value $g(0)$ at time $t = 0$. Hence, we may further simplify $\sigma^2$ as

$$\begin{aligned}
\sigma^2 &= \frac{N_0}{2} \cdot \frac{2}{T} \int_0^T \cos^2(2\pi f_c t)\, dt \\
&= \frac{N_0}{2}
\end{aligned} \tag{1.99}$$

where it is assumed that the frequency $f_c$ of the sinusoidal wave input is an integer multiple of the reciprocal of $T$. ◀

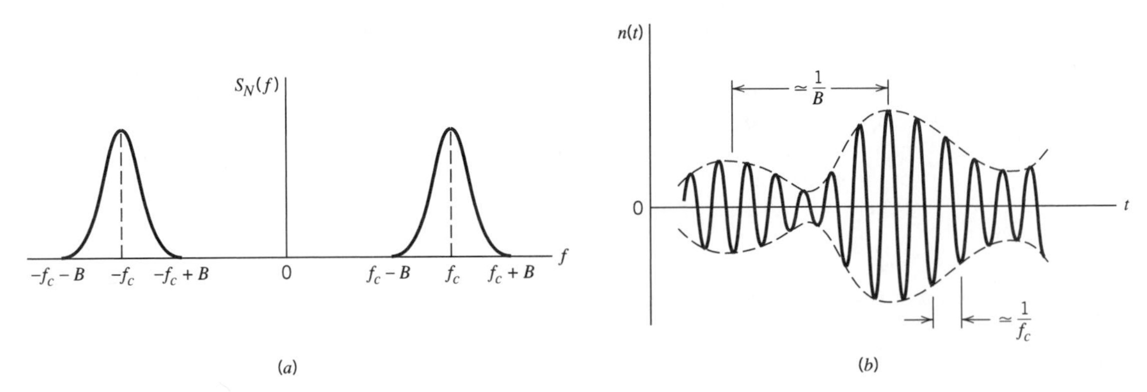


**FIGURE 1.18** (*a*) Power spectral density of narrowband noise. (*b*) Sample function of narrowband noise.

## 1.10 *Narrowband Noise*

The receiver of a communication system usually includes some provision for *preprocessing* the received signal. The preprocessing may take the form of a narrowband filter whose bandwidth is just large enough to pass the modulated component of the received signal essentially undistorted but not so large as to admit excessive noise through the receiver. The noise process appearing at the output of such a filter is called *narrowband noise*. With the spectral components of narrowband noise concentrated about some midband frequency $\pm f_c$ as in Figure 1.18*a*, we find that a sample function $n(t)$ of such a process appears somewhat similar to a sine wave of frequency $f_c$, which undulates slowly in both amplitude and phase, as illustrated in Figure 1.18*b*.

To analyze the effects of narrowband noise on the performance of a communication system, we need a mathematical representation of it. Depending on the application of interest, there are two specific representations of narrowband noise:

1. The narrowband noise is defined in terms of a pair of components called the *in-phase* and *quadrature* components.
2. The narrowband noise is defined in terms of two other components called the *envelope* and *phase*.

These two representations are described in what follows. For now it suffices to say that given the in-phase and quadrature components, we may determine the envelope and phase components, and vice versa. Moreover, in their own individual ways, the two representations are not only basic to the noise analysis of communication systems but also to the characterization of narrowband noise itself.

## 1.11 *Representation of Narrowband Noise in Terms of In-Phase and Quadrature Components*

Consider a narrowband noise $n(t)$ of bandwidth $2B$ centered on frequency $f_c$, as illustrated in Figure 1.18. In light of the theory of band-pass signals and systems presented in Appendix 2, we may represent $n(t)$ in the canonical (standard) form:

$$n(t) = n_I(t)\cos(2\pi f_c t) - n_Q(t)\sin(2\pi f_c t) \tag{1.100}$$

where $n_I(t)$ is called the *in-phase component* of $n(t)$, and $n_Q(t)$ is called the *quadrature component* of $n(t)$. Both $n_I(t)$ and $n_Q(t)$ are low-pass signals. Except for the midband frequency $f_c$, these two components are fully representative of the narrowband noise $n(t)$.

Given the narrowband noise $n(t)$, we may extract its in-phase and quadrature components using the scheme shown in Figure 1.19*a*. It is assumed that the two low-pass filters used in this scheme are ideal, each having a bandwidth equal to $B$ (i.e., one-half the bandwidth of the narrowband noise $n(t)$). The scheme of Figure 1.19*a* follows from the representation of Equation (1.100). We may, of course, use this equation directly to generate the narrowband noise $n(t)$, given its in-phase and quadrature components, as shown in Figure 1.19*b*. The schemes of Figures 1.19*a* and 1.19*b* may thus be viewed as narrowband noise *analyzer* and *synthesizer*, respectively.

The in-phase and quadrature components of a narrowband noise have important properties that are summarized here:

1. The in-phase component $n_I(t)$ and quadrature component $n_Q(t)$ of narrowband noise $n(t)$ have zero mean.
2. If the narrowband noise $n(t)$ is Gaussian, then its in-phase component $n_I(t)$ and quadrature component $n_Q(t)$ are jointly Gaussian.
3. If the narrowband noise $n(t)$ is stationary, then its in-phase component $n_I(t)$ and quadrature component $n_Q(t)$ are jointly stationary.
4. Both the in-phase component $n_I(t)$ and quadrature component $n_Q(t)$ have the same power spectral density, which is related to the power spectral density $S_N(f)$ of the narrowband noise $n(t)$ as

$$S_{N_I}(f) = S_{N_Q}(f) = \begin{cases} S_N(f - f_c) + S_N(f + f_c), & -B \le f \le B \\ 0, & \text{otherwise} \end{cases} \tag{1.101}$$

where it is assumed that $S_N(f)$ occupies the frequency interval $f_c - B \le |f| \le f_c + B$, and $f_c > B$.

5. The in-phase component $n_I(t)$ and quadrature component $n_Q(t)$ have the same variance as the narrowband noise $n(t)$.
6. The cross-spectral density of the in-phase and quadrature components of narrowband noise $n(t)$ is purely imaginary, as shown by

$$\begin{aligned} S_{N_I N_Q}(f) &= -S_{N_Q N_I}(f) \\ &= \begin{cases} j[S_N(f + f_c) - S_N(f - f_c)], & -B \le f \le B \\ 0, & \text{otherwise} \end{cases} \end{aligned} \tag{1.102}$$

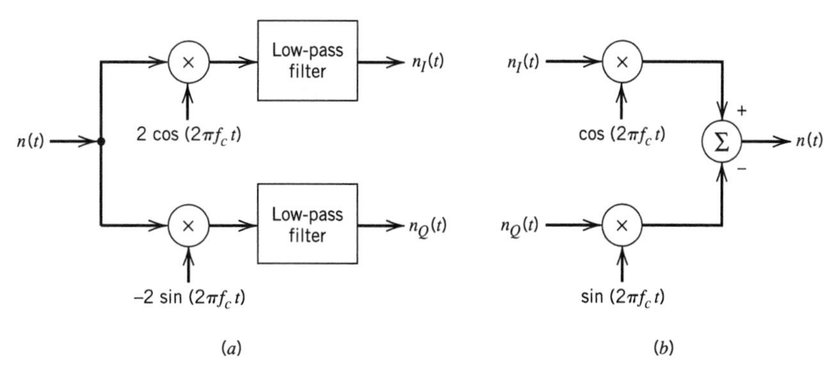


**FIGURE 1.19** (*a*) Extraction of in-phase and quadrature components of a narrowband process. (*b*) Generation of a narrowband process from its in-phase and quadrature components.

7. If the narrowband noise $n(t)$ is Gaussian and its power spectral density $S_N(t)$ is symmetric about the mid-band frequency $f_c$, then the in-phase component $n_I(t)$ and quadrature component $n_Q(t)$ are statistically independent.

For further discussions of these properties, the reader is referred to Problems 1.28 and 1.29.

### ▶ EXAMPLE 1.12 Ideal Band-Pass Filtered White Noise

Consider a white Gaussian noise of zero mean and power spectral density $N_0/2$, which is passed through an ideal band-pass filter of passband magnitude response equal to one, mid-band frequency $f_c$, and bandwidth $2B$. The power spectral density characteristic of the filtered noise $n(t)$ will therefore be as shown in Figure 1.20*a*. The problem is to determine the autocorrelation functions of $n(t)$ and its in-phase and quadrature components.

The autocorrelation function of $n(t)$ is the inverse Fourier transform of the power spectral density characteristic shown in Figure 1.20*a*:

$$\begin{aligned} R_N(\tau) &= \int_{-f_c-B}^{-f_c+B} \frac{N_0}{2} \exp(j2\pi f\tau)\, df + \int_{f_c-B}^{f_c+B} \frac{N_0}{2} \exp(j2\pi f\tau)\, df \\ &= N_0 B \operatorname{sinc}(2B\tau)[\exp(-j2\pi f_c\tau) + \exp(j2\pi f_c\tau)] \\ &= 2N_0 B \operatorname{sinc}(2B\tau) \cos(2\pi f_c\tau) \end{aligned} \tag{1.103}$$

which is plotted in Figure 1.20*b*.

The spectral density characteristic of Figure 1.20*a* is symmetric about $\pm f_c$. Therefore, we find that the corresponding spectral density characteristic of the in-phase noise component

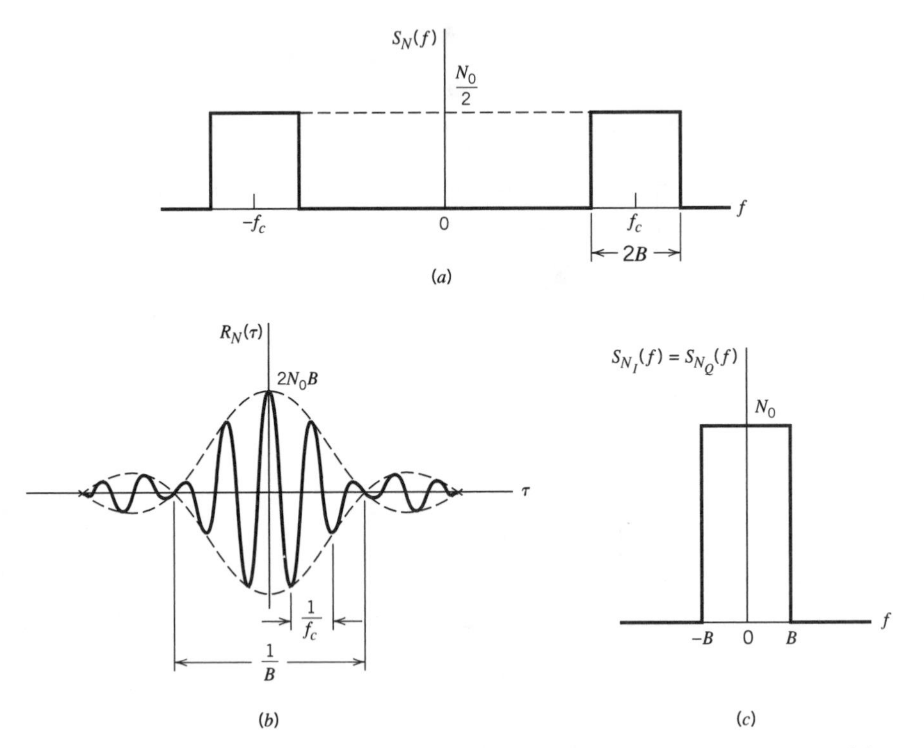


**FIGURE 1.20** Characteristics of ideal band-pass filtered white noise. (*a*) Power spectral density. (*b*) Autocorrelation function. (*c*) Power spectral density of in-phase and quadrature components.

$n_I(t)$ or the quadrature noise component $n_Q(t)$ is as shown in Figure 1.21*c*. The autocorrelation function of $n_I(t)$ or $n_Q(t)$ is therefore (see Example 1.10):

$$R_{N_I}(\tau) = R_{N_Q}(\tau) = 2N_0B \,\text{sinc}(2B\tau) \tag{1.104}$$

◀

## 1.12 *Representation of Narrowband Noise in Terms of Envelope and Phase Components*

In Section 1.11 we considered the representation of a narrowband noise $n(t)$ in terms of its in-phase and quadrature components. We may also represent the noise $n(t)$ in terms of its envelope and phase components as follows:

$$n(t) = r(t) \cos[2\pi f_c t + \psi(t)] \tag{1.105}$$

where

$$r(t) = [n_I^2(t) + n_Q^2(t)]^{1/2} \tag{1.106}$$

and

$$\psi(t) = \tan^{-1}\left[\frac{n_Q(t)}{n_I(t)}\right] \tag{1.107}$$

The function $r(t)$ is called the *envelope* of $n(t)$, and the function $\psi(t)$ is called the *phase* of $n(t)$.

The envelope $r(t)$ and phase $\psi(t)$ are both sample functions of low-pass random processes. As illustrated in Figure 1.18*b*, the time interval between two successive peaks of the envelope $r(t)$ is approximately $1/B$, where $2B$ is the bandwidth of the narrowband noise $n(t)$.

The probability distributions of $r(t)$ and $\psi(t)$ may be obtained from those of $n_I(t)$ and $n_Q(t)$ as follows. Let $N_I$ and $N_Q$ denote the random variables obtained by observing (at some fixed time) the random processes represented by the sample functions $n_I(t)$ and $n_Q(t)$, respectively. We note that $N_I$ and $N_Q$ are independent Gaussian random variables of zero mean and variance $\sigma^2$, and so we may express their joint probability density function by

$$f_{N_I,N_Q}(n_I, n_Q) = \frac{1}{2\pi\sigma^2} \exp\left(-\frac{n_I^2 + n_Q^2}{2\sigma^2}\right) \tag{1.108}$$

Accordingly, the probability of the joint event that $N_I$ lies between $n_I$ and $n_I + dn_I$ and that $N_Q$ lies between $n_Q$ and $n_Q + dn_Q$ (i.e., the pair of random variables $N_I$ and $N_Q$ lies jointly inside the shaded area of Figure 1.21*a*) is given by

$$f_{N_I,N_Q}(n_I, n_Q)\, dn_I\, dn_Q = \frac{1}{2\pi\sigma^2} \exp\left(-\frac{n_I^2 + n_Q^2}{2\sigma^2}\right) dn_I\, dn_Q \tag{1.109}$$

Define the transformation (see Figure 1.21*a*)

$$n_I = r \cos \psi \tag{1.110}$$

$$n_Q = r \sin \psi \tag{1.111}$$

In a limiting sense, we may equate the two incremental areas shown shaded in Figures 1.21*a* and 1.21*b* and thus write

$$dn_I\, dn_Q = r\, dr\, d\psi \tag{1.112}$$

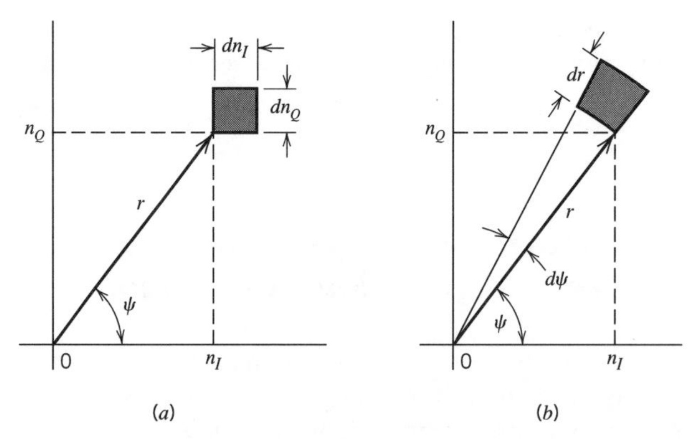


**FIGURE 1.21** Illustrating the coordinate system for representation of narrowband noise: (*a*) in terms of in-phase and quadrature components, and (*b*) in terms of envelope and phase.

Now, let $R$ and $\Psi$ denote the random variables obtained by observing (at some time $t$) the random processes represented by the envelope $r(t)$ and phase $\psi(t)$, respectively. Then, substituting Equations (1.110)–(1.112) into (1.109), we find that the probability of the random variables $R$ and $\Psi$ lying jointly inside the shaded area of Figure 1.21*b* is equal to

$$\frac{r}{2\pi\sigma^2}\exp\left(-\frac{r^2}{2\sigma^2}\right) dr\, d\psi$$

That is, the joint probability density function of $R$ and $\Psi$ is

$$f_{R,\Psi}(r, \psi) = \frac{r}{2\pi\sigma^2}\exp\left(-\frac{r^2}{2\sigma^2}\right) \tag{1.113}$$

This probability density function is independent of the angle $\psi$, which means that the random variables $R$ and $\Psi$ are statistically independent. We may thus express $f_{R,\Psi}(r, \psi)$ as the product of $f_R(r)$ and $f_\Psi(\psi)$. In particular, the random variable $\Psi$ representing phase is *uniformly distributed* inside the range 0 to $2\pi$, as shown by

$$f_\Psi(\psi) = \begin{cases} \dfrac{1}{2\pi}, & 0 \le \psi \le 2\pi \\ 0, & \text{elsewhere} \end{cases} \tag{1.114}$$

This leaves the probability density function of the random variable $R$ as

$$f_R(r) = \begin{cases} \dfrac{r}{\sigma^2}\exp\left(-\dfrac{r^2}{2\sigma^2}\right), & r \ge 0 \\ 0, & \text{elsewhere} \end{cases} \tag{1.115}$$

where $\sigma^2$ is the variance of the original narrowband noise $n(t)$. A random variable having the probability density function of Equation (1.115) is said to be *Rayleigh distributed*.[8]

For convenience of graphical presentation, let

$$v = \frac{r}{\sigma} \tag{1.116}$$

$$f_V(v) = \sigma f_R(r) \tag{1.117}$$

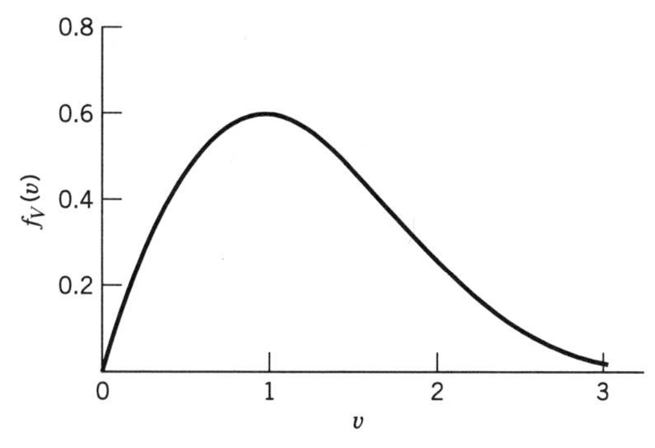


**FIGURE 1.22** Normalized Rayleigh distribution.

Then we may rewrite the Rayleigh distribution of Equation (1.115) in the *normalized* form

$$f_V(v) = \begin{cases} v \exp\left(-\dfrac{v^2}{2}\right), & v \geq 0 \\ 0, & \text{elsewhere} \end{cases} \tag{1.118}$$

Equation (1.118) is plotted in Figure 1.22. The peak value of the distribution $f_V(v)$ occurs at $v = 1$ and is equal to 0.607. Note also that, unlike the Gaussian distribution, the Rayleigh distribution is zero for negative values of $v$. This is because the envelope $r(t)$ can assume only nonnegative values.

## 1.13 *Sine Wave Plus Narrowband Noise*

Suppose next that we add the sinusoidal wave $A\cos(2\pi f_c t)$ to the narrowband noise $n(t)$, where $A$ and $f_c$ are both constants. We assume that the frequency of the sinusoidal wave is the same as the nominal carrier frequency of the noise. A sample function of the sinusoidal wave plus noise is then expressed by

$$x(t) = A\cos(2\pi f_c t) + n(t) \tag{1.119}$$

Representing the narrowband noise $n(t)$ in terms of its in-phase and quadrature components, we may write

$$x(t) = n_I'(t)\cos(2\pi f_c t) - n_Q(t)\sin(2\pi f_c t) \tag{1.120}$$

where

$$n_I'(t) = A + n_I(t) \tag{1.121}$$

We assume that $n(t)$ is Gaussian with zero mean and variance $\sigma^2$. Accordingly, we may state the following:

1. Both $n_I'(t)$ and $n_Q(t)$ are Gaussian and statistically independent.
2. The mean of $n_I'(t)$ is $A$ and that of $n_Q(t)$ is zero.
3. The variance of both $n_I'(t)$ and $n_Q(t)$ is $\sigma^2$.

We may therefore express the joint probability density function of the random variables $N_I'$ and $N_Q$, corresponding to $n_I'(t)$ and $n_Q(t)$, as follows:

$$f_{N_I',N_Q}(n_I', n_Q) = \frac{1}{2\pi\sigma^2}\exp\left[-\frac{(n_I' - A)^2 + n_Q^2}{2\sigma^2}\right] \tag{1.122}$$

Let $r(t)$ denote the envelope of $x(t)$ and $\psi(t)$ denote its phase. From Equation (1.120), we thus find that

$$r(t) = \{[n_I'(t)]^2 + n_Q^2(t)\}^{1/2} \tag{1.123}$$

and

$$\psi(t) = \tan^{-1}\left[\frac{n_Q(t)}{n_I'(t)}\right] \tag{1.124}$$

Following a procedure similar to that described in Section 1.12 for the derivation of the Rayleigh distribution, we find that the joint probability density function of the random variables $R$ and $\Psi$, corresponding to $r(t)$ and $\psi(t)$ for some fixed time $t$, is given by

$$f_{R,\Psi}(r, \psi) = \frac{r}{2\pi\sigma^2}\exp\left(-\frac{r^2 + A^2 - 2Ar\cos\psi}{2\sigma^2}\right) \tag{1.125}$$

We see that in this case, however, we cannot express the joint probability density function $f_{R,\Psi}(r, \psi)$ as a product $f_R(r)f_\Psi(\psi)$. This is because we now have a term involving the values of both random variables multiplied together as $r\cos\psi$. Hence, $R$ and $\Psi$ are dependent random variables for nonzero values of the amplitude $A$ of the sinusoidal wave component.

We are interested, in particular, in the probability density function of $R$. To determine this probability density function, we integrate Equation (1.125) over all possible values of $\psi$ obtaining the marginal density

$$\begin{aligned} f_R(r) &= \int_0^{2\pi} f_{R,\Psi}(r, \psi)\, d\psi \\ &= \frac{r}{2\pi\sigma^2}\exp\left(-\frac{r^2 + A^2}{2\sigma^2}\right)\int_0^{2\pi}\exp\left(\frac{Ar}{\sigma^2}\cos\psi\right) d\psi \end{aligned} \tag{1.126}$$

The integral in the right-hand side of Equation (1.126) can be identified in terms of the defining integral for the *modified Bessel function of the first kind of zero order* (see Appendix 3); that is,

$$I_0(x) = \frac{1}{2\pi}\int_0^{2\pi}\exp(x\cos\psi)\, d\psi \tag{1.127}$$

Thus, letting $x = Ar/\sigma^2$, we may rewrite Equation (1.126) in the compact form:

$$f_R(r) = \frac{r}{\sigma^2}\exp\left(-\frac{r^2 + A^2}{2\sigma^2}\right)I_0\left(\frac{Ar}{\sigma^2}\right) \tag{1.128}$$

This relation is called the *Rician distribution*.[9]

As with the Rayleigh distribution, the graphical presentation of the Rician distribution is simplified by putting

$$v = \frac{r}{\sigma} \tag{1.129}$$

$$a = \frac{A}{\sigma} \tag{1.130}$$

$$f_V(v) = \sigma f_R(r) \tag{1.131}$$

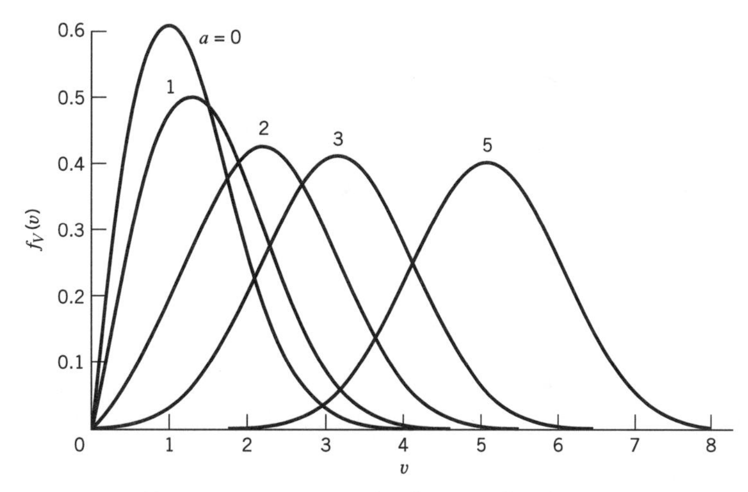


**FIGURE 1.23** Normalized Rician distribution.

Then we may express the Rician distribution of Equation (1.128) in the *normalized* form

$$f_V(v) = v \exp\left(-\frac{v^2 + a^2}{2}\right) I_0(av) \tag{1.132}$$

which is plotted in Figure 1.23 for the values 0, 1, 2, 3, 5, of the parameter $a$. Based on these curves, we may make the following observations:

1. When $a$ is zero, the Rician distribution reduces to the Rayleigh distribution.
2. The envelope distribution is approximately Gaussian in the vicinity of $v = a$ when $a$ is large, that is, when the sine-wave amplitude $A$ is large compared with $\sigma$, the square root of the average power of the noise $n(t)$.

## 1.14 *Computer Experiments: Flat-Fading Channel*

In this section we use computer simulations to study a *multipath channel* characterized by *Rayleigh fading*, examples of which arise in wireless communications and long-range radio transmission via the ionosphere. Fading occurs because of interference between different versions of the transmitted signal, which reach the receiver at correspondingly different times. The net result is that the received signal can vary widely in both amplitude and phase. Under certain conditions, the statistical time-varying nature of the received signal's envelope is closely described by a Rayleigh distribution as demonstrated herein.

Figure 1.24 presents a model of a multipath channel. It consists of a large collection of *scatterers* randomly positioned in space, whereby a single incident beam is converted into a correspondingly large number of scattered beams at the receiving antenna. The transmitted signal is set equal to $A \cos(2\pi f_c t)$. It is assumed that all the scattered beams travel at the same mean velocity. However, they differ from each other in amplitude and phase by virtue of differences in path loss and path delay. Thus the $k$th scattered beam is given by $A_k \cos(2\pi f_c t + \Theta_k)$, where the amplitude $A_k$ and phase $\Theta_k$ are random variables that vary slowly with time. Moreover, the $\Theta_k$ are all independent of one another and

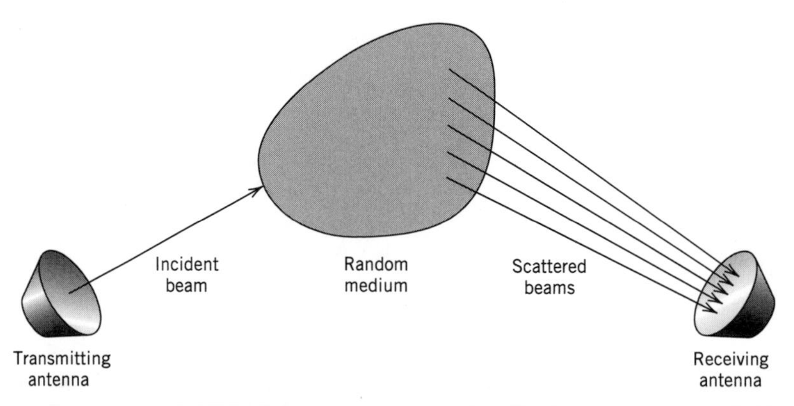


**FIGURE 1.24** Model of a multipath channel.

uniformly distributed inside the interval $[0, 2\pi]$. The type of fading exhibited by the multipath channel described herein is referred to as "flat fading" because the spectral characteristics of the transmitted signal are completely preserved at the channel output. However, the strength of the channel output changes with time due to random fluctuations in the gain of the channel caused by the multipath phenomenon.

Summing the contributions of all the scatterers, assumed to be $N$ in number, we may express the random process representing the received signal as

$$X(t) = \sum_{k=1}^{N} A_k \cos(2\pi f_c t + \Theta_k) \tag{1.33}$$

which may be rewriten in the equivalent form

$$X(t) = X_I \cos(2\pi f_c t) - X_Q \sin(2\pi f_c t) \tag{1.134}$$

where the $X_I$ and $X_Q$ are respectively defined by

$$X_I = \sum_{k=1}^{N} A_k \cos \Theta_k \tag{1.135}$$

and

$$X_Q = \sum_{k=1}^{N} A_k \sin \Theta_k \tag{1.136}$$

For convenience of presentation and without loss of generality, we may assume that $A_k$ lies in the closed interval $[-1, 1]$ for all $k$.

**Experiment 1. Gaussian Distributions**

From the central limit theorem we note that as the number of scatterers, $N$, approaches infinity, both $X_I$ and $X_Q$ should approach Gaussian random variables. To test the validity of this statement, the probability distributions of the in-phase component $X_I$ and quadrature component $X_Q$ are computed for $N$ = 10, 100, 1000, and 10,000. To test the validity of the central limit theorem, we need a measure of the goodness-of-fit that tests the equivalence of the measured probability distribution of the sampled data for varying

$N$ to the theoretical Gaussian distribution. One way of performing such a test is to use central moments of a distribution (up to order 4) to define the following two parameters:

$$\beta_1 = \frac{\mu_3^2}{\mu_2^3} \tag{1.137}$$

and

$$\beta_2 = \frac{\mu_4}{\mu_2^2} \tag{1.138}$$

where $\mu_2$, $\mu_3$, and $\mu_4$ are the second, third, and fourth central moments, respectively. The parameters $\beta_1$ and $\beta_2$ together provide a measure of the *skewness* of the distribution under test. The closer the values $\beta_1$ and $\beta_2$ for the measured distribution are to the corresponding ones for the theoretical distribution, the better is the goodness-of-fit. For a Gaussian random variable $X$ of mean $\mu_X$ and variance $\sigma_X^2$ we have

$$\mu_2 = \sigma_X^2$$
$$\mu_3 = 0$$
$$\mu_4 = 3\sigma_X^4$$

which yield

$$\beta_1 = 0$$

and

$$\beta_2 = 3$$

Table 1.1 presents the values of $\beta_1$ and $\beta_2$ computed for both the in-phase component $X_I$ and quadrature component $X_Q$ for varying $N$. Comparing these values with the corresponding ones for a Gaussian distribution, we clearly see that as the number of scatterers, $N$, increases the distributions of both $X_I$ and $X_Q$ do approach a zero-mean Gaussian distribution in accordance with the central limit theorem.

**TABLE 1.1** ***β Values for in-phase and quadrature components***

*(a) Measured Distribution*

| | | *Number of Scatterers,* N | | | |
|---|---|---|---|---|---|
| | | 10 | 100 | 1000 | 10,000 |
| In-phase component $X_I$ | $\beta_1$ | 0.2443 | 0.0255 | 0.0065 | 0.0003 |
| | $\beta_2$ | 2.1567 | 2.8759 | 2.8587 | 3.0075 |
| Quadrature component $X_Q$ | $\beta_1$ | 0.0874 | 0.0017 | 0.0004 | 0.0000 |
| | $\beta_2$ | 1.9621 | 2.7109 | 3.1663 | 3.0135 |

*(b) Theoretical Distribution: Gaussian*

$\beta_1 = 0$
$\beta_2 = 3$

### Experiment 2. Rayleigh Distribution

In Equation (1.134) the random process $X(t)$ is expressed in terms of its in-phase and quadrature components. Equivalently, we may express $X(t)$ in terms of its envelope and phase as

$$X(t) = R\cos(2\pi f_c t + \Psi) \tag{1.139}$$

where

$$R = \sqrt{X_I^2 + X_Q^2} \tag{1.140}$$

and

$$\Psi = \tan^{-1}\left(\frac{X_Q}{X_I}\right) \tag{1.141}$$

Note that in the experiments considered here the in-phase component $X_I$, quadrature component $X_Q$, envelope $R$, and phase $\Psi$ are all independent of time.

If $X_I$ and $X_Q$ approach Gaussian random variables for increasing $N$, then from the theory presented in Section 1.12 we note that the envelope $R$ will approach a Rayleigh distribution, and the phase $\Psi$ will approach a uniform distribution. In Figure 1.25 we present the actual probability density function of $r$ for data generated for the case of $N = 10{,}000$, with 100 histograms and 100 ensemble averages being computed. This figure also includes the theoretical curve. There is close agreement between these two curves, substantiating the assertion that the envelope $R$ of the received signal approaches a Rayleigh distribution.

Figure 1.26 illustrates the effect of Rayleigh fading on the waveform of the received signal $x(t)$, a sample function of $X(t)$, for the case of a sinusoidal transmitted signal with unit amplitude (i.e., $A = 1$) and frequency $f_c = 1$ MHz. Specifically, the transmitted signal and the corresponding received signal are shown in parts *a* and *b* of Figure 1.26, respectively. Comparing these two waveforms, we see that transmission through the multipath

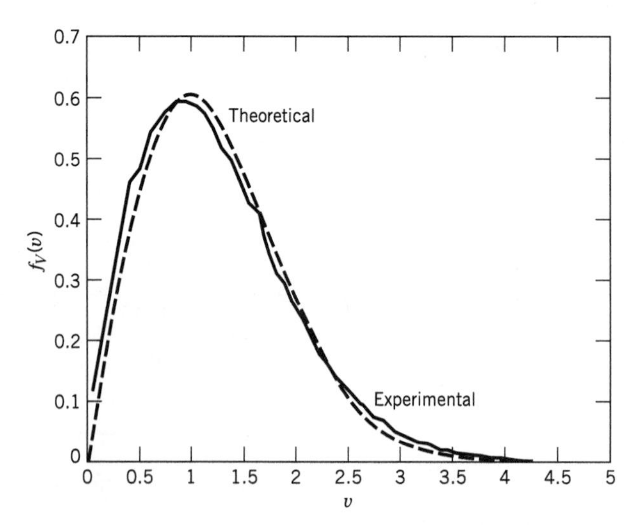


**FIGURE 1.25** Probability density function of the envelope of random process $X(t)$: comparing theory and experiment.

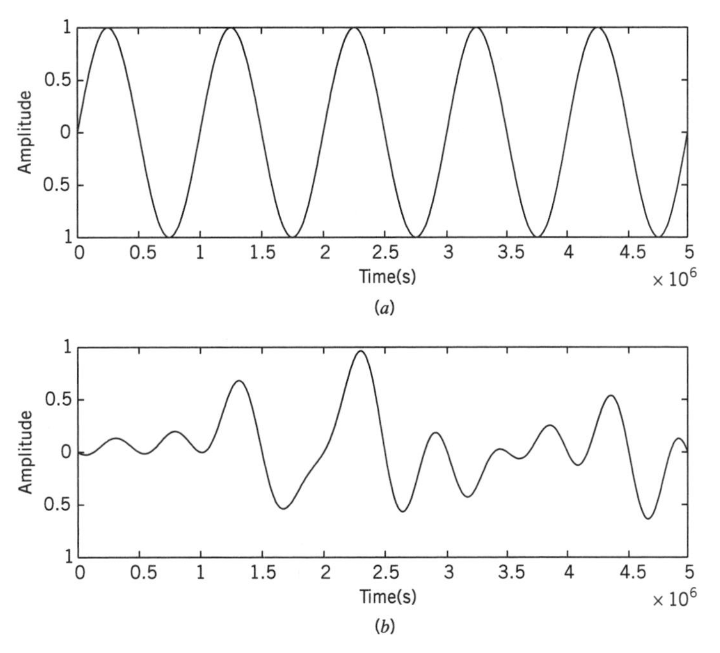


**FIGURE 1.26** Effect of Rayleigh fading on a sinusoidal wave. (*a*) Input sinusoidal wave. (*b*) Waveform of the resulting signal.

channel of Figure 1.24 has resulted in a received signal whose amplitude and phase components vary randomly with time, as expected.

## 1.15 *Summary and Discussion*

Much of the material presented in this chapter has dealt with the characterization of a particular class of random processes known to be stationary and ergodic. The implication of (wide-sense) stationarity is that we may develop a partial description of a random process in terms of two ensemble-averaged parameters: (1) a mean that is independent of time, and (2) an autocorrelation function that depends only on the difference between the times at which two observations of the process are made.[10] Ergodicity enables us to use time averages as "estimates" of these parameters. The time averages are computed using a sample function (i.e., single realization) of the random process.

Another important parameter of a random process is the power spectral density. The autocorrelation function and the power spectral density constitute a Fourier-transform pair. The formulas that define the power spectral density in terms of the autocorrelation function and vice versa are known as the Einstein-Wiener-Khintchine relations.

In Table 1.2 we present a graphical summary of the autocorrelation functions and power spectral densities of important random processes. All the processes described in this table are assumed to have zero mean and unit variance. This table should give the reader a feeling for (1) the interplay between the autocorrelation function and power spectral density of a random process, and (2) the role of linear filtering in shaping the autocorrelation function or, equivalently, the power spectral density of a white noise process.

The latter part of the chapter dealt with a noise process that is Gaussian and narrowband, which is the kind of filtered noise encountered at the front end of an idealized form of communication receiver. Gaussianity means that the random variable obtained by

**TABLE 1.2** ***Graphical summary of autocorrelation functions and power spectral densities of random processes of zero mean and unit variance***

| *Type of Process, $X(t)$* | *Autocorrelation Function, $R_X(\tau)$* | *Power Spectral Density, $S_X(f)$* |
|---|---|---|
| Sinusoidal process of unit frequency and random phase | 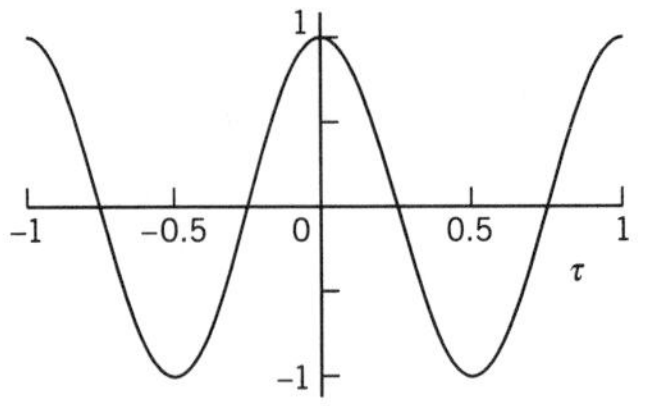 | 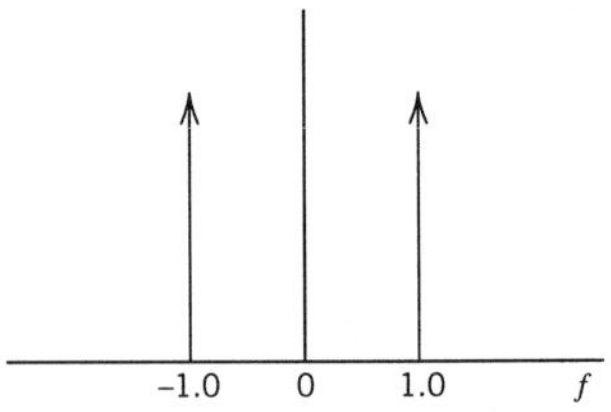 |
| Random binary wave of unit symbol-duration | 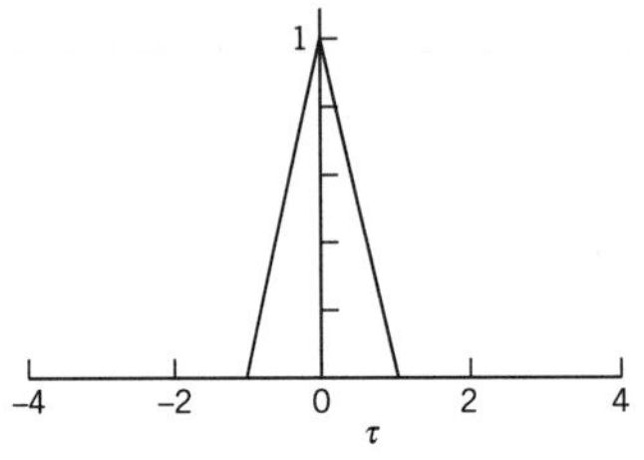 | 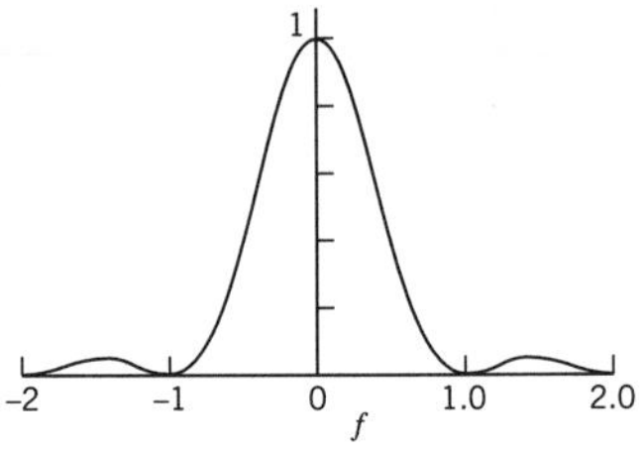 |
| RC low-pass filtered white noise | 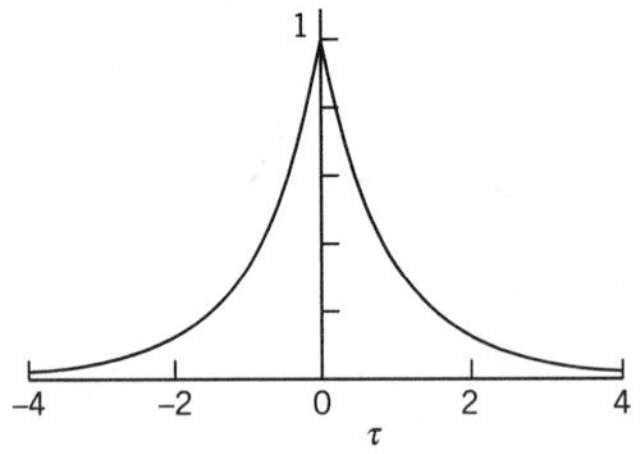 | 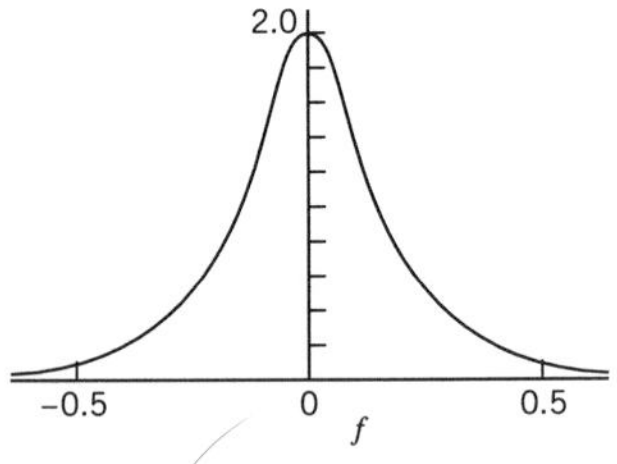 |
| Ideal low-pass filtered white noise | 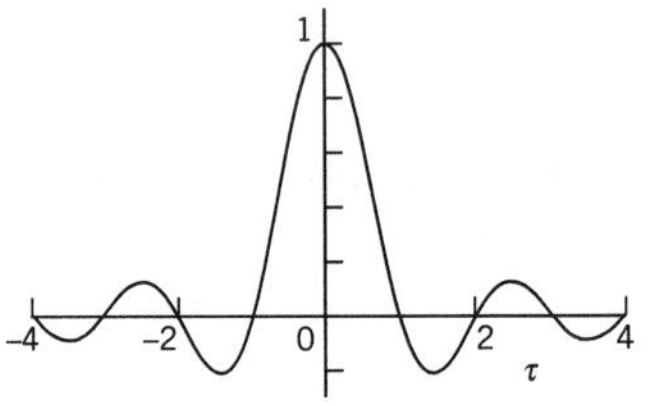 | 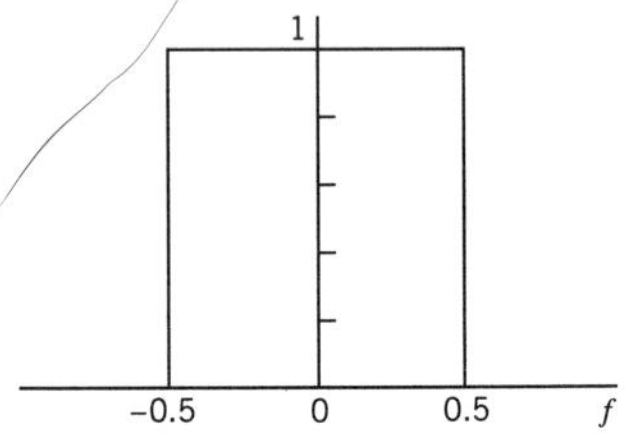 |
| Ideal band-pass filtered white noise | 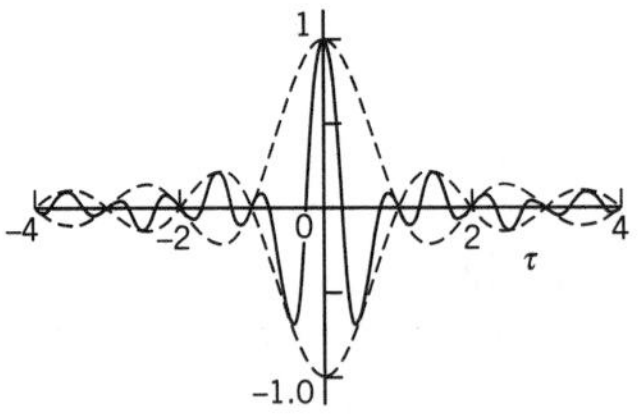 | 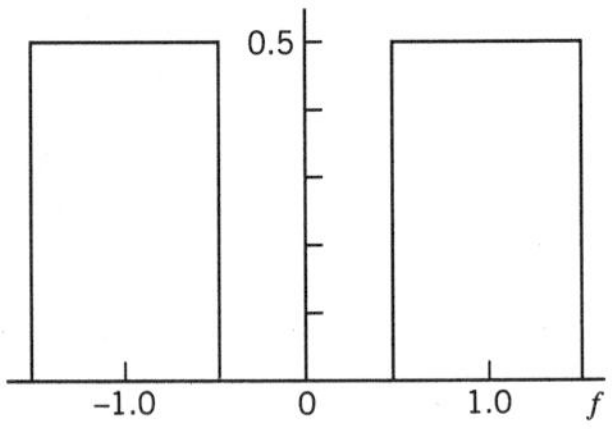 |
| RLC-filtered white noise | 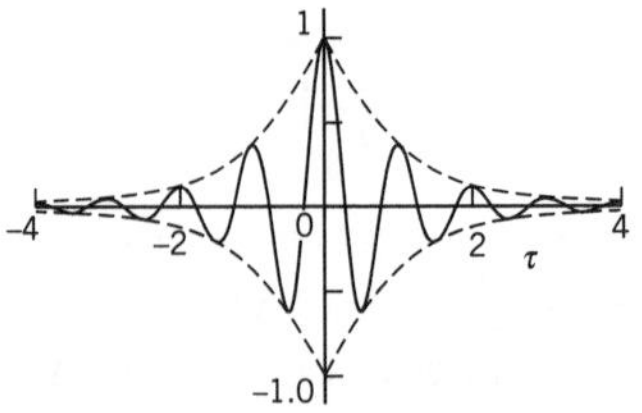 | 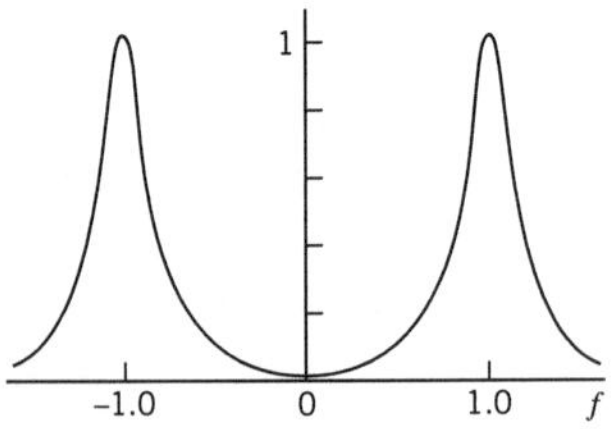 |

observing the output of the filter at some fixed time has a Gaussian distribution. The narrowband nature of the noise means that it may be represented in terms of an in-phase and a quadrature component. These two components are both low-pass, Gaussian processes, each with zero mean and a variance equal to that of the original narrowband noise. Alternatively, a Gaussian narrowband noise may be represented in terms of a Rayleigh-distributed envelope and a uniformly distributed phase. Each of these representations has its own specific area of application, as shown in subsequent chapters of the book.

## Notes and References

1. For a rigorous treatment of random processes, see the classic books of Doob (1953), Loève (1963), and Cramér and Leadbetter (1967).

2. There is another important class of random processes commonly encountered in practice, the mean and autocorrelation function of which exhibit *periodicity*, as in

$$\mu_X(t_1 + T) = \mu_X(t_1)$$
$$R_X(t_1 + T, t_2 + T) = R_X(t_1, t_2)$$

   for all $t_1$ and $t_2$. A random process $X(t)$ satisfying this pair of conditions is said to be *cyclostationary* (in the wide sense). Modeling the process $X(t)$ as cyclostationary adds a new dimension, namely, period $T$ to the partial description of the process. Examples of cyclostationary processes include a television signal obtained by raster-scanning a random video field, and a modulated process obtained by varying the amplitude, phase, or frequency of a sinusoidal carrier. For detailed discussion of cyclostationary processes, see Franks (1969), pp. 204–214, and the paper by Gardner and Franks (1975).

3. Traditionally, Equations (1.42) and (1.43) have been referred to in the literature as the Wiener-Khintchine relations in recognition of pioneering work done by Norbert Wiener and A. I. Khintchine; for their original papers, see Wiener (1930) and Khintchine (1934). A discovery of a forgotten paper by Albert Einstein on time-series analysis (delivered at the Swiss Physical Society's February 1914 meeting in Basel) reveals that Einstein had discussed the autocorrelation function and its relationship to the spectral content of a time series many years before Wiener and Khintchine. An English translation of Einstein's paper is reproduced in the *IEEE ASSP Magazine*, vol. 4, October 1987. This particular issue also contains articles by W. A. Gardner and A. M. Yaglom, which elaborate on Einstein's original work.

4. For further details of power spectrum estimation, see Blackman and Tukey (1958), Box and Jenkins (1976), Marple (1987), and Kay (1988).

5. The Gaussian distribution and associated Gaussian process are named after the great mathematician C. F. Gauss. At age 18, Gauss invented *the method of least squares* for finding the best value of a sequence of measurements of some quantity. Gauss later used the method of least squares in fitting orbits of planets to data measurements, a procedure that was published in 1809 in his book entitled *Theory of Motion of the Heavenly Bodies*. In connection with the error of observation, he developed the *Gaussian distribution*. This distribution is also known as the *normal distribution*. Partly for historical reasons, mathematicians commonly use the term normal, while engineers and physicists commonly use the term Gaussian.

6. For a detailed treatment of electrical noise, see Van der Ziel (1970) and the collection of papers edited by Gupta (1977).

   An introductory treatment of shot noise is presented in Helstrom (1990). For a more detailed treatment, see the paper by Yue, Luganani, and Rice (1978).

Thermal noise was first studied experimentally by J. B. Johnson in 1928, and for this reason it is sometimes referred to as the *Johnson noise*. Johnson's experiments were confirmed theoretically by Nyquist (1928).

7. The noisiness of a receiver may also be measured in terms of the so-called *noise figure*. The relationship between the noise figure and the equivalent noise temperature is developed in Chapter 8.
8. The Rayleigh distribution is named after the English physicist J. W. Strutt, Lord Rayleigh.
9. The Rician distribution is named in honor of Stephen O. Rice for the original contribution reported in a pair of papers published in 1944 and 1945, which are reproduced in Wax (1954).
10. The statistical characterization of communication systems presented in this book is confined to the first two moments, mean and autocorrelation function (equivalently, autocovariance function) of the pertinent random process. However, when a random process is transmitted through a nonlinear system, valuable information is contained in higher-order moments of the resulting output process. The parameters used to characterize higher-order moments in the time domain are called *cumulants*, and their multidimensional Fourier transforms are called *polyspectra*. For a discussion of higher-order cumulants and polyspectra and their estimation, see the paper by Nikias and Raghuveer (1987).

## PROBLEMS

### Stationarity and Ergodicity

**1.1** Consider a random process $X(t)$ defined by

$$X(t) = \sin(2\pi f_c t)$$

in which the frequency $f_c$ is a random variable uniformly distributed over the interval $[0, W]$. Show that $X(t)$ is nonstationary. *Hint:* Examine specific sample functions of the random process $X(t)$ for the frequency $f = W/4$, $W/2$, and $W$, say.

**1.2** Consider the sinusoidal process

$$X(t) = A\cos(2\pi f_c t)$$

where the frequency $f_c$ is constant and the amplitude $A$ is uniformly distributed:

$$f_A(a) = \begin{cases} 1, & 0 \le a \le 1 \\ 0, & \text{otherwise} \end{cases}$$

Determine whether or not this process is strictly stationary.

**1.3** A random process $X(t)$ is defined by

$$X(t) = A\cos(2\pi f_c t)$$

where $A$ is a Gaussian-distributed random variable of zero mean and variance $\sigma_A^2$. This random process is applied to an ideal integrator, producing the output

$$Y(t) = \int_0^t X(\tau)\, d\tau$$

(a) Determine the probability density function of the output $Y(t)$ at a particular time $t_k$.

(b) Determine whether or not $Y(t)$ is stationary.

(c) Determine whether or not $Y(t)$ is ergodic.

**1.4** Let $X$ and $Y$ be statistically independent Gaussian-distributed random variables, each with zero mean and unit variance. Define the Gaussian process

$$Z(t) = X\cos(2\pi t) + Y\sin(2\pi t)$$

(a) Determine the joint probability density function of the random variables $Z(t_1)$ and $Z(t_2)$ obtained by observing $Z(t)$ at times $t_1$ and $t_2$, respectively.

(b) Is the process $Z(t)$ stationary? Why?

## Correlation and Spectral Density Functions

**1.5** Prove the following two properties of the autocorrelation function $R_X(\tau)$ of a random process $X(t)$:

(a) If $X(t)$ contains a DC component equal to $A$, then $R_X(\tau)$ will contain a constant component equal to $A^2$.

(b) If $X(t)$ contains a sinusoidal component, then $R_X(\tau)$ will also contain a sinusoidal component of the same frequency.

**1.6** The square wave $x(t)$ of Figure P1.6 of constant amplitude $A$, period $T_0$, and delay $t_d$, represents the sample function of a random process $X(t)$. The delay is random, described by the probability density function

$$f_{T_d}(t_d) = \begin{cases} \dfrac{1}{T_0}, & -\dfrac{1}{2}T_0 \le t_d \le \dfrac{1}{2}T_0 \\ 0, & \text{otherwise} \end{cases}$$

(a) Determine the probability density function of the random variable $X(t_k)$ obtained by observing the random process $X(t)$ at time $t_k$.

(b) Determine the mean and autocorrelation function of $X(t)$ using ensemble-averaging.

(c) Determine the mean and autocorrelation function of $X(t)$ using time-averaging.

(d) Establish whether or not $X(t)$ is stationary. In what sense is it ergodic?

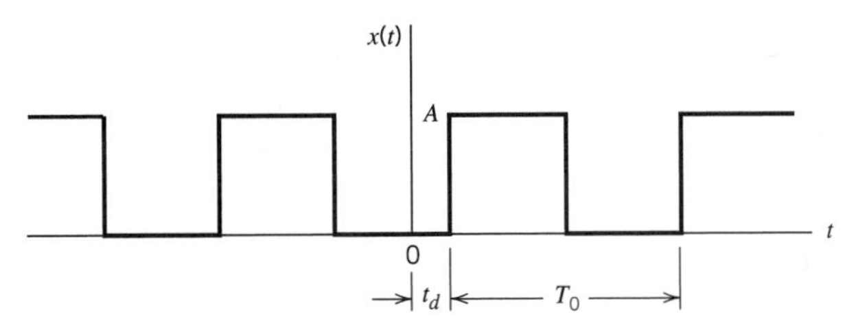


**FIGURE P1.6**

**1.7** A binary wave consists of a random sequence of symbols 1 and 0, similar to that described in Example 1.3, with one basic difference: symbol 1 is now represented by a pulse of amplitude $A$ volts and symbol 0 is represented by zero volts. All other parameters are the same as before. Show that for this new random binary wave $X(t)$:

(a) The autocorrelation function is

$$R_X(\tau) = \begin{cases} \dfrac{A^2}{4} + \dfrac{A^2}{4}\left(1 - \dfrac{|\tau|}{T}\right), & |\tau| < T \\ \dfrac{A^2}{4}, & |\tau| \ge T \end{cases}$$

(b) The power spectral density is

$$S_X(f) = \frac{A^2}{4}\,\delta(f) + \frac{A^2 T}{4}\,\text{sinc}^2(fT)$$

What is the percentage power contained in the DC component of the binary wave?

**1.8** A random process $Y(t)$ consists of a DC component of $\sqrt{3/2}$ volts, a periodic component $g(t)$, and a random component $X(t)$. The autocorrelation function of $Y(t)$ is shown in Figure P1.8.

(a) What is the average power of the periodic component $g(t)$?

(b) What is the average power of the random component $X(t)$?

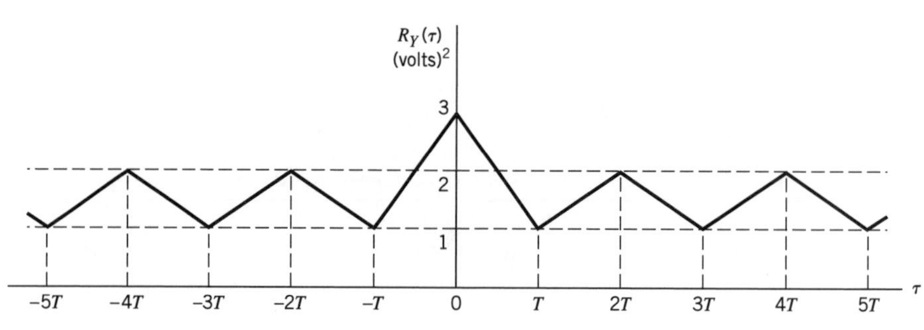


**FIGURE P1.8**

**1.9** Consider a pair of stationary processes $X(t)$ and $Y(t)$. Show that the cross-correlations $R_{XY}(\tau)$ and $R_{YX}(\tau)$ of these processes have the following properties:

(a) $R_{XY}(\tau) = R_{YX}(-\tau)$

(b) $|R_{XY}(\tau)| \leq \frac{1}{2}[R_X(0) + R_Y(0)]$
where $R_X(\tau)$ and $R_Y(\tau)$ are the autocorrelation functions of $X(t)$ and $Y(t)$, respectively.

**1.10** Consider two linear filters connected in cascade as in Figure P1.10. Let $X(t)$ be a stationary process with autocorrelation function $R_X(\tau)$. The random process appearing at the first filter output is $V(t)$ and that at the second filter output is $Y(t)$.

(a) Find the autocorrelation function of $Y(t)$.

(b) Find the cross-correlation function $R_{VY}(\tau)$ of $V(t)$ and $Y(t)$.

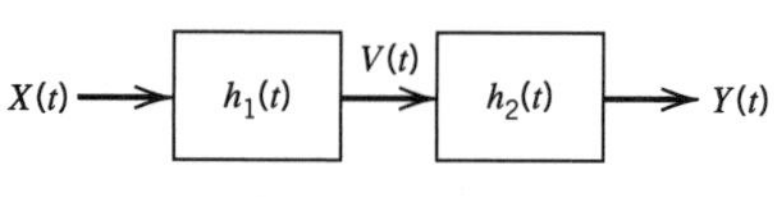


**FIGURE P1.10**

**1.11** A stationary process $X(t)$ is applied to a linear time-invariant filter of impulse response $h(t)$, producing an output $Y(t)$.

(a) Show that the cross-correlation function $R_{YX}(\tau)$ of the output $Y(t)$ and the input $X(t)$ is equal to the impulse response $h(\tau)$ convolved with the autocorrelation function $R_X(\tau)$ of the input, as shown by

$$R_{YX}(\tau) = \int_{-\infty}^{\infty} h(u) R_X(\tau - u)\, du$$

Show that the second cross-correlation function $R_{XY}(\tau)$ equals

$$R_{XY}(\tau) = \int_{-\infty}^{\infty} h(-u) R_X(\tau - u)\, du$$

(b) Find the cross-spectral densities $S_{YX}(f)$ and $S_{XY}(f)$.

(c) Assuming that $X(t)$ is a white noise process with zero mean and power spectral density $N_0/2$, show that

$$R_{YX}(\tau) = \frac{N_0}{2} h(\tau)$$

Comment on the practical significance of this result.

**1.12** The power spectral density of a random process $X(t)$ is shown in Figure P1.12. It consists of a delta function at $f = 0$ and a triangular component.

(a) Determine and sketch the autocorrelation function $R_X(\tau)$ of $X(t)$.

(b) What is the DC power contained in $X(t)$?

(c) What is the AC power contained in $X(t)$?

(d) What sampling rates will give uncorrelated samples of $X(t)$? Are the samples statistically independent?

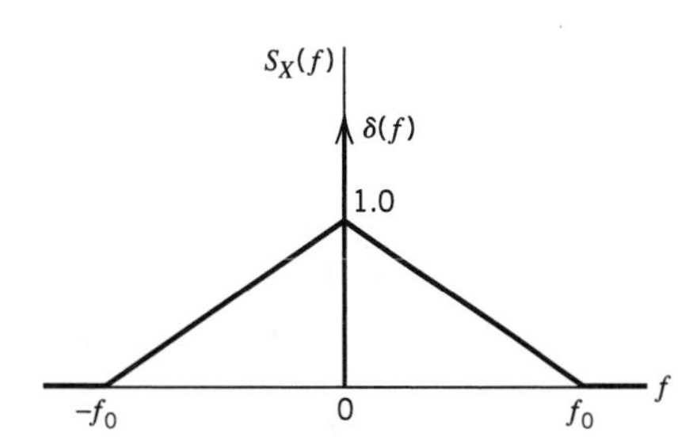


**FIGURE P1.12**

**1.13** A pair of noise processes $n_1(t)$ and $n_2(t)$ are related by

$$n_2(t) = n_1(t) \cos(2\pi f_c t + \theta) - n_1(t) \sin(2\pi f_c t + \theta)$$

where $f_c$ is a constant, and $\theta$ is the value of a random variable $\Theta$ whose probability density function is defined by

$$f_\Theta(\theta) = \begin{cases} \dfrac{1}{2\pi}, & 0 \leq \theta \leq 2\pi \\ 0, & \text{otherwise} \end{cases}$$

The noise process $n_1(t)$ is stationary and its power spectral density is as shown in Figure P1.13. Find and plot the corresponding power spectral density of $n_2(t)$.

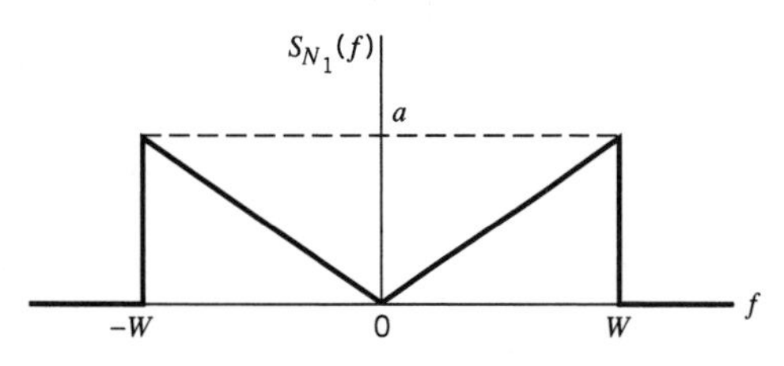


**FIGURE P1.13**

**1.14** A *random telegraph signal* $X(t)$, characterized by the autocorrelation function

$$R_X(\tau) = \exp(-2v|\tau|)$$

where $v$ is a constant, is applied to the low-pass RC filter of Figure P1.14. Determine the power spectral density and autocorrelation function of the random process at the filter output.

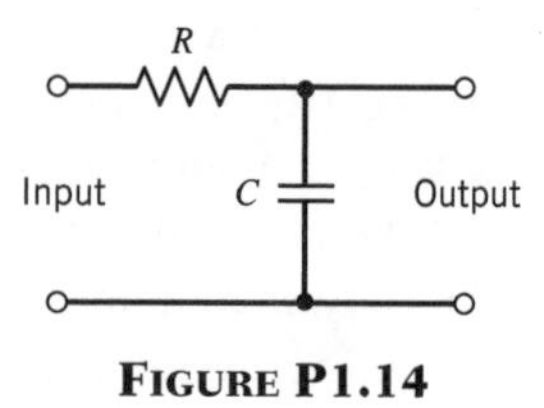


FIGURE P1.14

**1.15** A *running integrator* is defined by

$$y(t) = \int_{t-T}^{t} x(\tau)\, d\tau$$

where $x(t)$ is the input, $y(t)$ is the output, and $T$ is the integration period. Both $x(t)$ and $y(t)$ are sample functions of stationary processes $X(t)$ and $Y(t)$, respectively. Show that the power spectral density of the integrator output is related to that of the integrator input as

$$S_Y(f) = T^2 \operatorname{sinc}^2(fT) S_X(f)$$

**1.16** A zero-mean stationary process $X(t)$ is applied to a linear filter whose impulse response is defined by a truncated exponential:

$$h(t) = \begin{cases} ae^{-at}, & 0 \le t \le T \\ 0, & \text{otherwise} \end{cases}$$

Show that the power spectral density of the filter output $Y(t)$ is defined by

$$S_Y(f) = \frac{a^2}{a^2 + 4\pi^2 f^2}(1 - 2\exp(-aT)\cos(2\pi fT) + \exp(-2aT))S_X(f)$$

where $S_X(f)$ is the power spectral density of the filter input.

**1.17** The output of an oscillator is described by

$$X(t) = A\cos(2\pi ft - \Theta)$$

where $A$ is a constant, and $f$ and $\Theta$ are independent random variables. The probability density function of $\Theta$ is defined by

$$f_\Theta(\theta) = \begin{cases} \dfrac{1}{2\pi}, & 0 \le \theta \le 2\pi \\ 0, & \text{otherwise} \end{cases}$$

Find the power spectral density of $X(t)$ in terms of the probability density function of the frequency $f$. What happens to this power spectral density when the frequency $f$ assumes a constant value?

## Gaussian Processes

**1.18** A stationary, Gaussian process $X(t)$ has zero mean and power spectral density $S_X(f)$. Determine the probability density function of a random variable obtained by observing the process $X(t)$ at some time $t_k$.

**1.19** A Gaussian process $X(t)$ of zero mean and variance $\sigma_X^2$ is passed through a full-wave rectifier, which is described by the input–output relation of Figure P1.19. Show that the probability density function of the random variable $Y(t_k)$, obtained by observing the random process $Y(t)$ at the rectifier output at time $t_k$, is as follows:

$$f_{Y(t_k)}(y) = \begin{cases} \sqrt{\dfrac{2}{\pi}} \dfrac{1}{\sigma_X} \exp\left(-\dfrac{y^2}{2\sigma_X^2}\right), & y \geq 0 \\ 0, & y < 0 \end{cases}$$

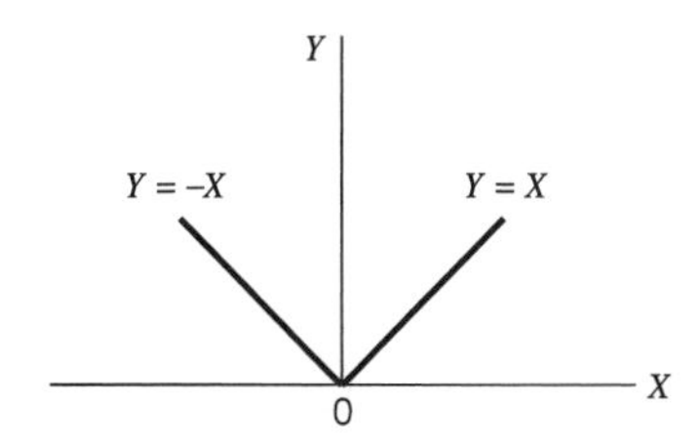


**Figure P1.19**

**1.20** Let $X(t)$ be a zero-mean, stationary, Gaussian process with autocorrelation function $R_X(\tau)$. This process is applied to a square-law device, which is defined by the input–output relation

$$Y(t) = X^2(t)$$

where $Y(t)$ is the output.

(a) Show that the mean of $Y(t)$ is $R_X(0)$.

(b) Show that the autocovariance function of $Y(t)$ is $2R_X^2(\tau)$.

**1.21** A stationary, Gaussian process $X(t)$ with mean $\mu_X$ and variance $\sigma_X^2$ is passed through two linear filters with impulse responses $h_1(t)$ and $h_2(t)$, yielding processes $Y(t)$ and $Z(t)$, as shown in Figure P1.21.

(a) Determine the joint probability density function of the random variables $Y(t_1)$ and $Z(t_2)$.

(b) What conditions are necessary and sufficient to ensure that $Y(t_1)$ and $Z(t_2)$ are statistically independent?

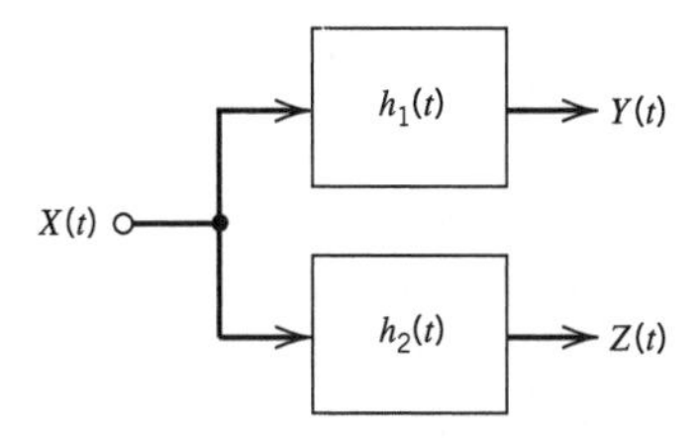


**Figure P1.21**

1.22 A stationary, Gaussian process $X(t)$ with zero mean and power spectral density $S_X(f)$ is applied to a linear filter whose impulse response $h(t)$ is shown in Figure P1.22. A sample $Y$ is taken of the random process at the filter output at time $T$.

(a) Determine the mean and variance of $Y$.

(b) What is the probability density function of $Y$?

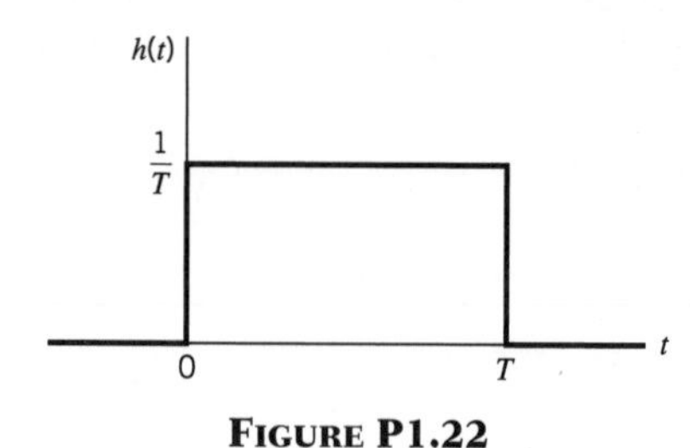


FIGURE P1.22

## Noise

1.23 Consider a white Gaussian noise process of zero mean and power spectral density $N_0/2$ that is applied to the input of the high-pass RL filter shown in Figure P1.23.

(a) Find the autocorrelation function and power spectral density of the random process at the output of the filter.

(b) What are the mean and variance of this output?

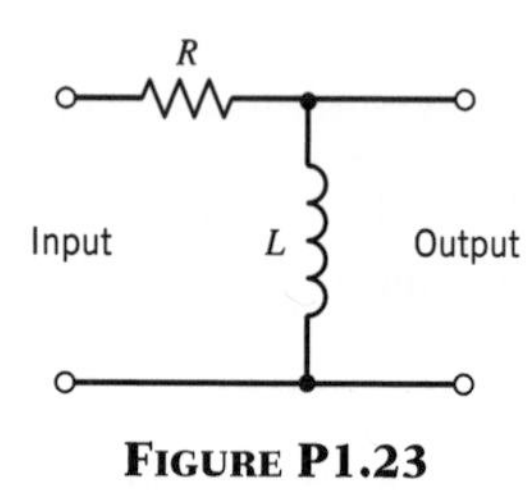


FIGURE P1.23

1.24 A white noise $w(t)$ of power spectral density $N_0/2$ is applied to a *Butterworth* low-pass filter of order $n$, whose magnitude response is defined by

$$|H(f)| = \frac{1}{[1 + (f/f_0)^{2n}]^{1/2}}$$

(a) Determine the noise equivalent bandwidth for this low-pass filter. (See Appendix 2 for the definition of noise equivalent bandwidth.)

(b) What is the limiting value of the noise equivalent bandwidth as $n$ approaches infinity?

1.25 The shot-noise process $X(t)$ defined by Equation (1.86) is stationary. Why?

1.26 White Gaussian noise of zero mean and power spectral density $N_0/2$ is applied to the filtering scheme shown in Figure P1.26*a*. The frequency responses of these two filters are shown in Figure P1.26*b*. The noise at the low-pass filter output is denoted by $n(t)$.

(a) Find the power spectral density and the autocorrelation function of $n(t)$.

(b) Find the mean and variance of $n(t)$.

(c) What is the rate at which $n(t)$ can be sampled so that the resulting samples are essentially uncorrelated?

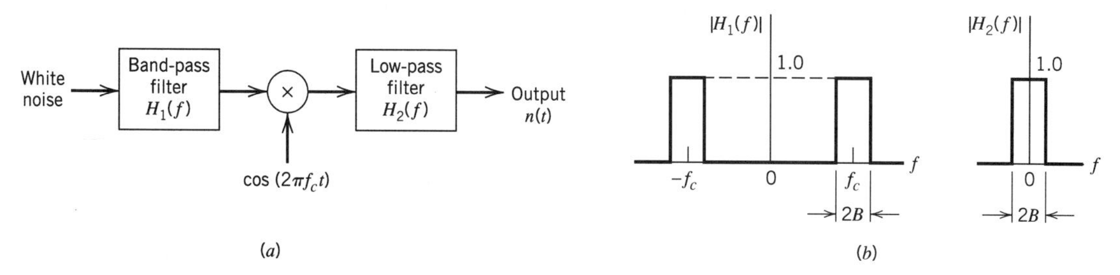


**FIGURE P1.26**

**1.27** Let $X(t)$ be a stationary process with zero mean, autocorrelation function $R_X(\tau)$, and power spectral density $S_X(f)$. We are required to find a linear filter with impulse response $h(t)$, such that the filter output has the same statistical characteristics as $X(t)$ when the input is white noise of power spectral density $N_0/2$.

(a) Determine the condition which the impulse response $h(t)$ must satisfy to achieve this requirement.

(b) What is the corresponding condition on the frequency response $H(f)$ of the filter?

## Narrowband Noise

**1.28** In the noise analyzer of Figure 1.19*a*, the low-pass filters are ideal with a bandwidth equal to one-half that of the narrowband noise $n(t)$ applied to the input. Using this scheme, derive the following results:

(a) Equation (1.101), defining the power spectral densities of the in-phase noise component $n_I(t)$ and quadrature noise component $n_Q(t)$ in terms of the power spectral density of $n(t)$.

(b) Equation (1.102), defining the cross-spectral densities of $n_I(t)$ and $n_Q(t)$.

**1.29** Assume that the narrowband noise $n(t)$ is Gaussian and its power spectral density $S_N(f)$ is symmetric about the midband frequency $f_c$. Show that the in-phase and quadrature components of $n(t)$ are statistically independent.

**1.30** The power spectral density of a narrowband noise $n(t)$ is as shown in Figure P1.30. The carrier frequency is 5 Hz.

(a) Find the power spectral densities of the in-phase and quadrature components of $n(t)$.

(b) Find their cross-spectral densities.

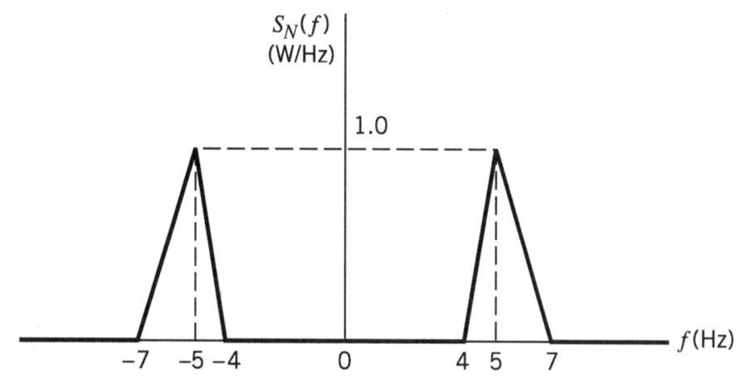


**FIGURE P1.30**

**1.31** Consider a Gaussian noise $n(t)$ with zero mean and the power spectral density $S_N(f)$ shown in Figure P1.31.

(a) Find the probability density function of the envelope of $n(t)$.

(b) What are the mean and variance of this envelope?

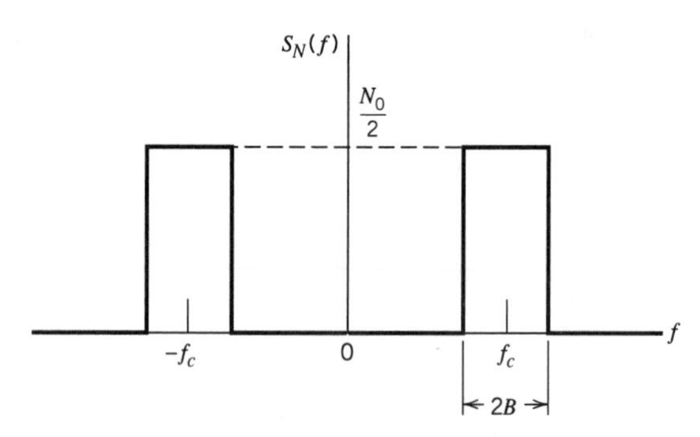


**FIGURE P1.31**

## Computer Experiments

**1.32** In this computer experiment we study the statistical characterization of a random process $X(t)$ defined by

$$X(t) = A\cos(2\pi f_c t + \Theta) + W(t)$$

where the phase $\Theta$ of the sinusoidal component is a uniformly distributed random variable over the interval $[-\pi, \pi]$, and $W(t)$ is a white Gaussian noise component of zero mean and power spectral density $N_0/2$. The two components of $X(t)$ are statistically independent; hence the autocorrelation function of $X(t)$ is

$$R_X(\tau) = \frac{A^2}{2}\cos(2\pi f_c \tau) + \frac{N_0}{2}\delta(\tau)$$

This equation shows that for $|\tau| > 0$ the autocorrelation function $R_X(\tau)$ has the same sinusoidal waveform as the signal component of $X(t)$.

The purpose of this computer experiment is to perform the computation of $R_X(\tau)$ using two different methods:

(a) *Ensemble averaging.* Generate $M = 50$ randomly picked realizations of the process $X(t)$. Hence compute the product $x(t + \tau)x(t)$ for some fixed time $t$, where $x(t)$ is a realization of $X(t)$. Repeat the computation of $x(t + \tau)x(t)$ for the $M$ realizations of $X(t)$, and thereby compute the average of these computations over $M$. Repeat this sequence of computations for different values of $\tau$.

(b) *Time averaging.* Compute the time-averaged autocorrelation function

$$R_x(\tau, T) = \frac{1}{2T}\int_{-T}^{T} x(t + \tau)x(t)\,dt$$

where $x(t)$ is a particular realization of $X(t)$, and $2T$ is the total observation interval. For this computation, use the Fourier-transform pair:

$$R_x(\tau, T) \rightleftharpoons \frac{1}{2T}|X_T(f)|^2$$

where $|X_T(f)|^2/2T$ is the periodogram of the process $X(t)$. Specifically, compute the Fourier transform $X_T(f)$ of the time-windowed function

$$x_T(t) = \begin{cases} x(t), & -T \leq t \leq T \\ 0, & \text{otherwise} \end{cases}$$

Hence compute the inverse Fourier transform of $|X_T(f)|^2/2T$.

Compare the results of your computation of $R_X(\tau)$ using these two approaches.

**1.33** In this computer experiment we continue the study of the multipath channel described in Section 1.14. Specifically, consider the situation where the received signal includes a line-of-sight component, as shown by

$$X(t) = \sum_{k=1}^{N} A_k \cos(2\pi f_c t + \Theta_k) + a \cos(2\pi f_c t)$$

where $a \cos(2\pi f_c t)$ is the directly received component. Following the material presented in Section 1.14, compute the envelope of $X(t)$ for $N = 10{,}000$, and $a = 0, 1, 2, 3, 5$. Compare your results with the Rician distribution studied in Section 1.13.

CHAPTER 2

# CONTINUOUS-WAVE MODULATION

In this chapter we study continuous-wave modulation, which is basic to the operation of analog communication systems. The chapter is divided into two related parts. In the first part we study the time-domain and frequency-domain descriptions of two basic families of continuous-wave modulation:

- *Amplitude modulation, in which the amplitude of a sinusoidal carrier is varied in accordance with an incoming message signal.*
- *Angle modulation, in which the instantaneous frequency or phase of the sinusoidal carrier is varied in accordance with the message signal.*

The second part of the chapter focuses on the effects of channel noise on the performance of the receivers pertaining to these modulation schemes.

Advantages and disadvantages of the different methods of continuous-wave modulation are highlighted in light of the material presented herein.

## 2.1 *Introduction*

The purpose of a communication system is to transmit *information-bearing signals* through a communication channel separating the transmitter from the receiver. Information-bearing signals are also referred to as *baseband signals*. The term *baseband* is used to designate the band of frequencies representing the original signal as delivered by a source of information. The proper use of the communication channel requires a shift of the range of baseband frequencies into other frequency ranges suitable for transmission, and a corresponding shift back to the original frequency range after reception. For example, a radio system must operate with frequencies of 30 kHz and upward, whereas the baseband signal usually contains frequencies in the audio frequency range, and so some form of frequency-band shifting must be used for the system to operate satisfactorily. A shift of the range of frequencies in a signal is accomplished by using *modulation*, which is defined as *the process by which some characteristic of a carrier is varied in accordance with a modulating wave (signal)*. A common form of the carrier is a *sinusoidal wave*, in which case we speak of a *continuous-wave modulation*[1] process. The baseband signal is referred to as the *modulating wave*, and the result of the modulation process is referred to as the *modulated wave*. Modulation is performed at the transmitting end of the communication system. At the receiving end of the system, we usually require the original baseband signal to be restored. This is accomplished by using a process known as *demodulation*, which is the reverse of the modulation process.

In basic signal-processing terms, we thus find that the transmitter of an analog communication system consists of a modulator and the receiver consists of a demodulator, as

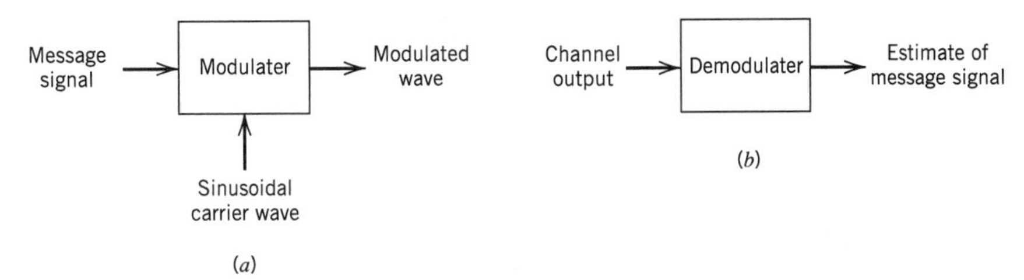


**FIGURE 2.1** Components of a continuous-wave modulation system: (*a*) transmitter, and (*b*) receiver.

depicted in Figure 2.1. In addition to the signal received from the transmitter, the receiver input includes channel noise. The degradation in receiver performance due to channel noise is determined by the type of modulation used.

In this chapter we study two families of continuous-wave (CW) modulation systems, namely, *amplitude modulation* and *angle modulation*. In amplitude modulation, the amplitude of the sinusoidal carrier wave is varied in accordance with the baseband signal. In angle modulation, the angle of the sinusoidal carrier wave is varied in accordance with the baseband signal. Figure 2.2 displays the waveforms of amplitude-modulated and angle-modulated signals for the case of sinusoidal modulation. Parts (*a*) and (*b*) of the figure show the sinusoidal carrier and modulating waves, respectively. Parts (*c*) and (*d*) show the

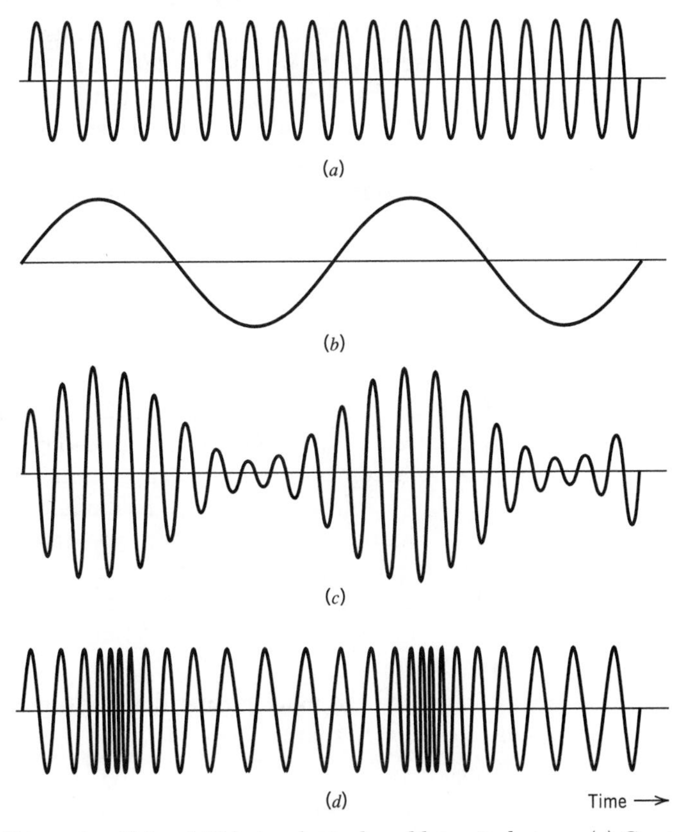


**FIGURE 2.2** Illustrating AM and FM signals produced by a single tone. (*a*) Carrier wave. (*b*) Sinusoidal modulating signal. (c) Amplitude-modulated signal. (*d*) Frequency-modulated signal.

corresponding amplitude-modulated and frequency-modulated waves, respectively; frequency modulation is a form of angle modulation. This figure clearly illustrates the basic differences between amplitude modulation and angle modulation, which are discussed in what follows.

## 2.2 Amplitude Modulation

Consider a *sinusoidal carrier wave* $c(t)$ defined by

$$c(t) = A_c \cos(2\pi f_c t) \tag{2.1}$$

where $A_c$ is the *carrier amplitude* and $f_c$ is the *carrier frequency*. To simplify the exposition without affecting the results obtained and conclusions reached, we have assumed that the phase of the carrier wave is zero in Equation (2.1). Let $m(t)$ denote the baseband signal that carries the specification of the message. The source of carrier wave $c(t)$ is physically independent of the source responsible for generating $m(t)$. *Amplitude modulation (AM) is defined as a process in which the amplitude of the carrier wave c(t) is varied about a mean value, linearly with the baseband signal m(t).* An amplitude-modulated (AM) wave may thus be described, in its most general form, as a function of time as follows:

$$s(t) = A_c[1 + k_a m(t)] \cos(2\pi f_c t) \tag{2.2}$$

where $k_a$ is a constant called the *amplitude sensitivity* of the modulator responsible for the generation of the modulated signal $s(t)$. Typically, the carrier amplitude $A_c$ and the message signal $m(t)$ are measured in volts, in which case $k_a$ is measured in volt$^{-1}$.

Figure 2.3*a* shows a baseband signal $m(t)$, and Figures 2.3*b* and 2.3*c* show the corresponding AM wave $s(t)$ for two values of amplitude sensitivity $k_a$ and a carrier amplitude $A_c = 1$ volt. We observe that the *envelope* of $s(t)$ has essentially the same shape as the baseband signal $m(t)$ provided that two requirements are satisfied:

1. The amplitude of $k_a m(t)$ is always less than unity, that is,

$$|k_a m(t)| < 1 \quad \text{for all } t \tag{2.3}$$

This condition is illustrated in Figure 2.3*b*; it ensures that the function $1 + k_a m(t)$ is always positive, and since an envelope is a positive function, we may express the envelope of the AM wave $s(t)$ of Equation (2.2) as $A_c[1 + k_a m(t)]$. When the amplitude sensitivity $k_a$ of the modulator is large enough to make $|k_a m(t)| > 1$ for any $t$, the carrier wave becomes *overmodulated*, resulting in carrier phase reversals whenever the factor $1 + k_a m(t)$ crosses zero. The modulated wave then exhibits *envelope distortion*, as in Figure 2.3*c*. It is therefore apparent that by avoiding overmodulation, a one-to-one relationship is maintained between the envelope of the AM wave and the modulating wave for all values of time—a useful feature, as we shall see later on. The absolute maximum value of $k_a m(t)$ multiplied by 100 is referred to as the *percentage modulation*.

2. The carrier frequency $f_c$ is much greater than the highest frequency component $W$ of the message signal $m(t)$, that is

$$f_c \gg W \tag{2.4}$$

We call $W$ the *message bandwidth*. If the condition of Equation (2.4) is not satisfied, an envelope cannot be visualized (and therefore detected) satisfactorily.

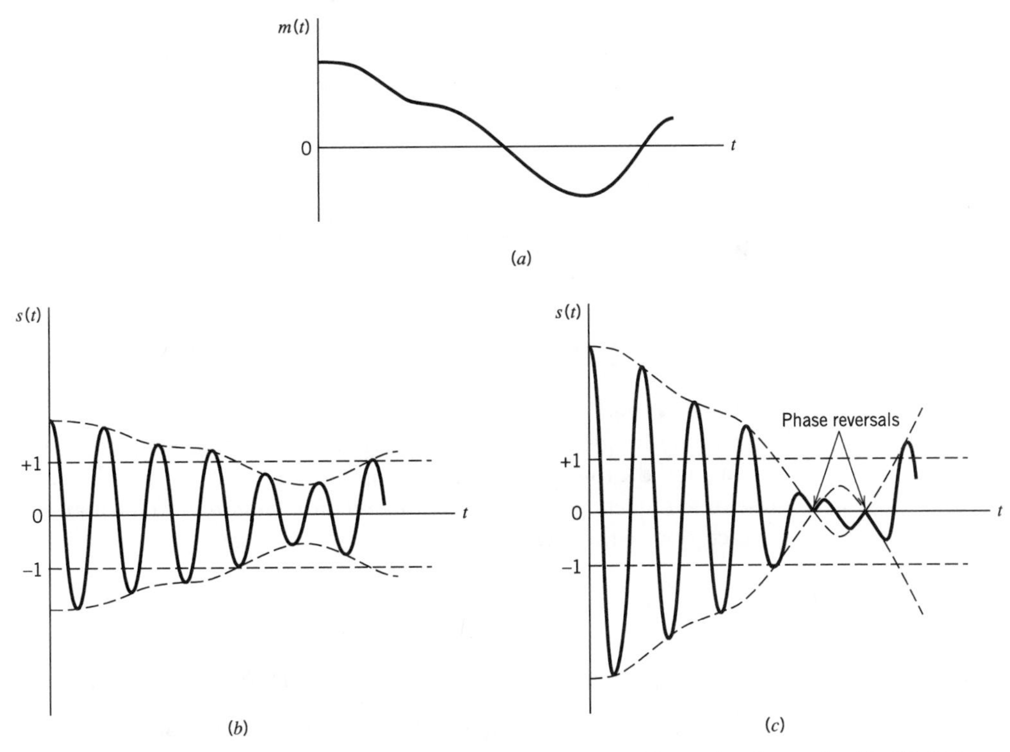


**FIGURE 2.3** Illustrating the amplitude modulation process. (*a*) Baseband signal $m(t)$. (*b*) AM wave for $|k_a m(t)| < 1$ for all $t$. (c) AM wave for $|k_a m(t)| > 1$ for some $t$.

From Equation (2.2), we find that the Fourier transform of the AM wave $s(t)$ is given by

$$S(f) = \frac{A_c}{2}[\delta(f - f_c) + \delta(f + f_c)] + \frac{k_a A_c}{2}[M(f - f_c) + M(f + f_c)] \tag{2.5}$$

Suppose that the baseband signal $m(t)$ is band-limited to the interval $-W \leq f \leq W$, as in Figure 2.4*a*. The shape of the spectrum shown in this figure is intended for the purpose of illustration only. We find from Equation (2.5) that the spectrum $S(f)$ of the AM wave is as shown in Figure 2.4*b* for the case when $f_c > W$. This spectrum consists of two delta functions weighted by the factor $A_c/2$ and occurring at $\pm f_c$, and two versions of the baseband spectrum translated in frequency by $\pm f_c$ and scaled in amplitude by $k_a A_c/2$. From the spectrum of Figure 2.4*b*, we note the following:

1. As a result of the modulation process, the spectrum of the message signal $m(t)$ for negative frequencies extending from $-W$ to 0 becomes completely visible for positive (i.e., measurable) frequencies, provided that the carrier frequency satisfies the condition $f_c > W$; herein lies the importance of the idea of "negative" frequencies.
2. For positive frequencies, the portion of the spectrum of an AM wave lying above the carrier frequency $f_c$ is referred to as the *upper sideband*, whereas the symmetric portion below $f_c$ is referred to as the *lower sideband*. For negative frequencies, the upper sideband is represented by the portion of the spectrum below $-f_c$ and the lower sideband by the portion above $-f_c$. The condition $f_c > W$ ensures that the sidebands do not overlap.

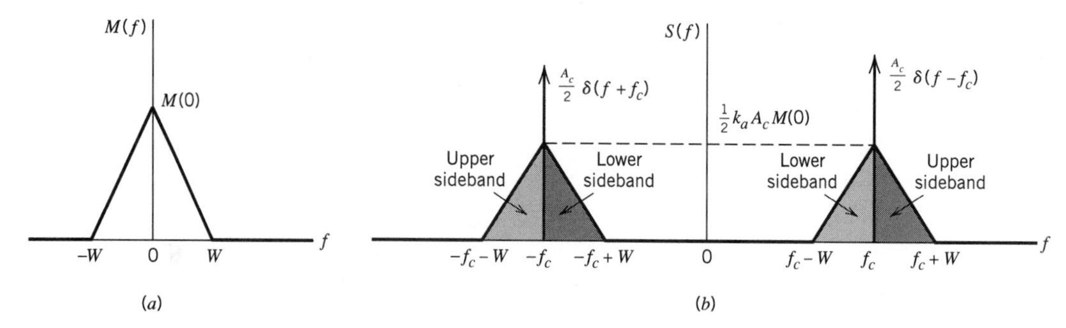


**FIGURE 2.4** (*a*) Spectrum of baseband signal. (*b*) Spectrum of AM wave.

3. For positive frequencies, the highest frequency component of the AM wave equals $f_c + W$, and the lowest frequency component equals $f_c - W$. The difference between these two frequencies defines the *transmission bandwidth* $B_T$ for an AM wave, which is exactly twice the message bandwidth $W$, that is,

$$B_T = 2W \tag{2.6}$$

## ■ VIRTUES AND LIMITATIONS OF AMPLITUDE MODULATION

Amplitude modulation is the oldest method of performing modulation. Its greatest virtue is the simplicity of implementation:

- In the transmitter, amplitude modulation is accomplished using a nonlinear device. For example, in the *switching modulator* discussed in Problem 2.3, the combined sum of the message signal and carrier wave is applied to a diode, with the carrier amplitude being large enough to swing across the characteristic curve of the diode. Fourier analysis of the voltage developed across a resistive load reveals the generation of an AM component, which may be extracted by means of a band-pass filter.
- In the receiver, amplitude demodulation is also accomplished using a nonlinear device. For example, we may use a simple and yet highly effective circuit known as the *envelope detector*, which is discussed in Problem 2.5. The circuit consists of a diode connected in series with the parallel combination of a capacitor and load resistor. Some version of this circuit is found in most commercial AM radio receivers. Provided that the carrier frequency is high enough and the percentage modulation is less than 100 percent, the demodulator output developed across the load resistor is nearly the same as the envelope of the incoming AM wave, hence the name "envelope detector."

Recall, however, that transmitted power and channel bandwidth are our two primary communication resources, and they should be used efficiently. In this context, we find that the standard form of amplitude modulation defined in Equation (2.2) suffers from two major limitations:

1. *Amplitude modulation is wasteful of power.* The carrier wave $c(t)$ is completely independent of the information-bearing signal $m(t)$. The transmission of the carrier wave therefore represents a waste of power, which means that in amplitude modulation only a fraction of the total transmitted power is actually affected by $m(t)$.

2. *Amplitude modulation is wasteful of bandwidth.* The upper and lower sidebands of an AM wave are uniquely related to each other by virtue of their symmetry about the carrier frequency; hence, given the magnitude and phase spectra of either sideband, we can uniquely determine the other. This means that insofar as the transmission of information is concerned, only one sideband is necessary, and the communication channel therefore needs to provide only the same bandwidth as the baseband signal. In light of this observation, amplitude modulation is wasteful of bandwidth as it requires a transmission bandwidth equal to twice the message bandwidth.

To overcome these limitations, we must make certain modifications: suppress the carrier and modify the sidebands of the AM wave. These modifications naturally result in increased system complexity. In effect, we trade system complexity for improved use of communication resources. The basis of this trade-off is linear modulation, which is discussed in the next section. In a strict sense, full amplitude modulation does not qualify as linear modulation because of the presence of the carrier wave.

## 2.3 *Linear Modulation Schemes*

In its most general form, linear modulation is defined by

$$s(t) = s_I(t)\cos(2\pi f_c t) - s_Q(t)\sin(2\pi f_c t) \tag{2.7}$$

where $s_I(t)$ is the *in-phase component* of the modulated wave $s(t)$, and $s_Q(t)$ is its *quadrature component*. Equation (2.7) is recognized as the canonical representation of a narrowband signal, which is discussed in detail in Appendix 2. In linear modulation, both $s_I(t)$ and $s_Q(t)$ are low-pass signals that are linearly related to the message signal $m(t)$.

Indeed, depending on how these two components of $s(t)$ are defined, we may identify three types of linear modulation involving a single message signal:

1. *Double sideband-suppressed carrier (DSB-SC) modulation*, where only the upper and lower sidebands are transmitted.
2. *Single sideband (SSB) modulation*, where only one sideband (the lower sideband or the upper sideband) is transmitted.
3. *Vestigial sideband (VSB) modulation*, where only a vestige (i.e., trace) of one of the sidebands and a correspondingly modified version of the other sideband are transmitted.

Table 2.1 presents a summary of the definitions of these three special forms of linear modulation. There are two important points to note from Table 2.1:

1. The in-phase component $s_I(t)$ is solely dependent on the message signal $m(t)$.
2. The quadrature component $s_Q(t)$ is a *filtered* version of $m(t)$. The spectral modification of the modulated wave $s(t)$ is solely due to $s_Q(t)$.

To be more specific, the role of the quadrature component (if present) is merely to interfere with the in-phase component, so as to reduce or eliminate power in one of the sidebands of the modulated signal $s(t)$, depending on how the quadrature component is defined.

**TABLE 2.1** *Different forms of linear modulation*

| *Type of Modulation* | *In-Phase Component* $s_I(t)$ | *Quadrature Component* $s_Q(t)$ | *Comments* |
|---|---|---|---|
| DSB-SC | $m(t)$ | 0 | $m(t)$ = message signal |
| SSB:[a] | | | |
| (a) Upper sideband transmitted | $\frac{1}{2}m(t)$ | $\frac{1}{2}\hat{m}(t)$ | $\hat{m}(t)$ = Hilbert transform of $m(t)$ |
| (b) Lower sideband transmitted | $\frac{1}{2}m(t)$ | $-\frac{1}{2}\hat{m}(t)$ | |
| VSB: | | | |
| (a) Vestige of lower sideband transmitted | $\frac{1}{2}m(t)$ | $\frac{1}{2}m'(t)$ | $m'(t)$ = output of the filter of frequency response $H_Q(f)$ due to $m(t)$. For the definition of $H_Q(f)$, see Eq. (2.16) |
| (b) Vestige of upper sideband transmitted | $\frac{1}{2}m(t)$ | $-\frac{1}{2}m'(t)$ | |

[a]For the mathematical description of single sideband modulation, see Problem 2.16.

## ■ DOUBLE SIDEBAND-SUPPRESSED CARRIER (DSB-SC) MODULATION

This form of linear modulation is generated by using a *product modulator* that simply multiplies the message signal $m(t)$ by the carrier wave $A_c \cos(2\pi f_c t)$, as illustrated in Figure 2.5*a*. Specifically, we write

$$s(t) = A_c m(t) \cos(2\pi f_c t) \tag{2.8}$$

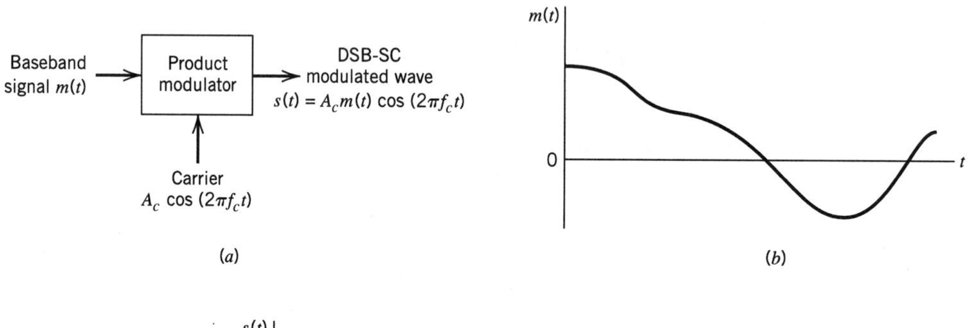


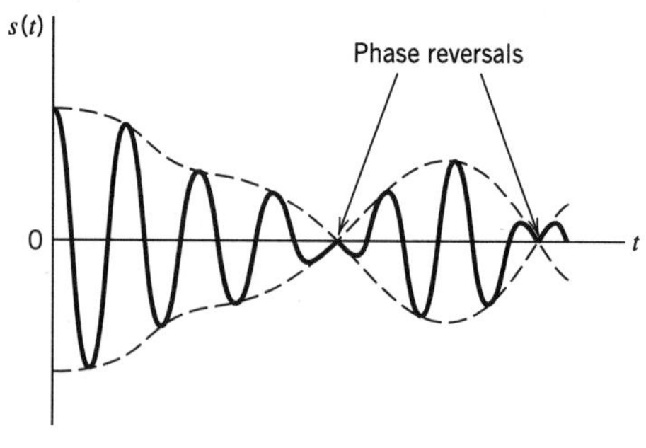


**FIGURE 2.5** (*a*) Block diagram of product modulator. (*b*) Baseband signal. (*c*) DSB-SC modulated wave.

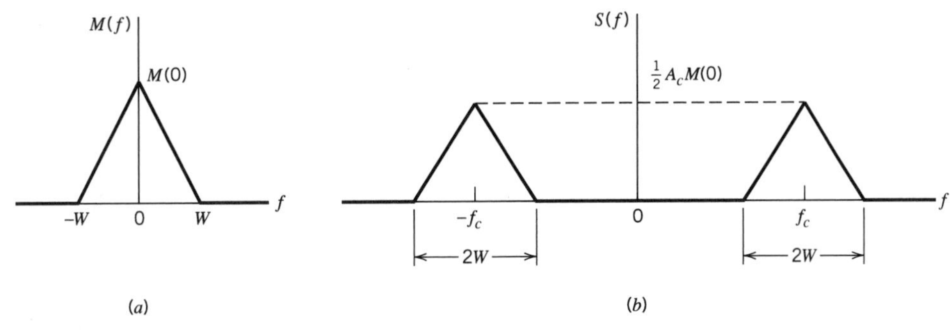


**FIGURE 2.6** (*a*) Spectrum of baseband signal. (*b*) Spectrum of DSB-SC modulated wave.

Figure 2.5*c* shows the modulated signal $s(t)$ for the arbitrary message waveform of Figure 2.5*b*. The modulated signal $s(t)$ undergoes a *phase reversal* whenever the message signal $m(t)$ crosses zero. Consequently, the envelope of a DSB-SC modulated signal is different from the message signal; this is unlike the case of an AM wave that has a percentage modulation less than 100 percent.

From Equation (2.8), the Fourier transform of $s(t)$ is obtained as

$$S(f) = \frac{1}{2} A_c[M(f - f_c) + M(f + f_c)] \tag{2.9}$$

For the case when the baseband signal $m(t)$ is limited to the interval $-W \leq f \leq W$, as in Figure 2.6*a*, we thus find that the spectrum $S(f)$ of the DSB-SC wave $s(t)$ is as illustrated in Figure 2.6*b*. Except for a change in scale factor, the modulation process simply *translates* the spectrum of the baseband signal by $\pm f_c$. Of course, the transmission bandwidth required by DSB-SC modulation is the same as that for amplitude modulation, namely, $2W$.

## ■ COHERENT DETECTION

The baseband signal $m(t)$ can be uniquely recovered from a DSB-SC wave $s(t)$ by first multiplying $s(t)$ with a locally generated sinusoidal wave and then low-pass filtering the product, as in Figure 2.7. It is assumed that the local oscillator signal is exactly coherent or synchronized, in both frequency and phase, with the carrier wave $c(t)$ used in the product modulator to generate $s(t)$. This method of demodulation is known as *coherent detection* or *synchronous demodulation.*

It is instructive to derive coherent detection as a special case of the more general demodulation process using a local oscillator signal of the same frequency but arbitrary phase difference $\phi$, measured with respect to the carrier wave $c(t)$. Thus, denoting the local

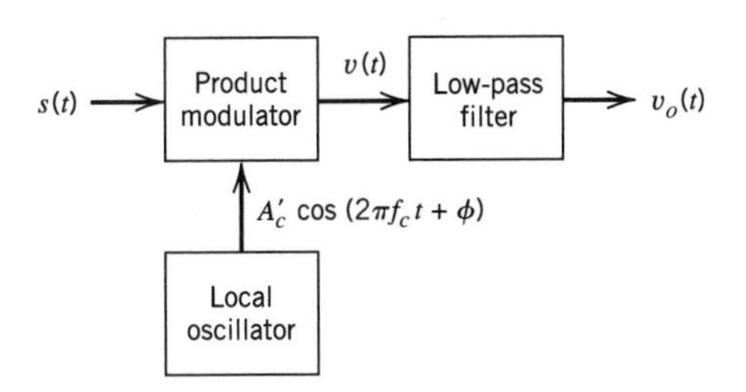


**FIGURE 2.7** Coherent detector for demodulating DSB-SC modulated wave.

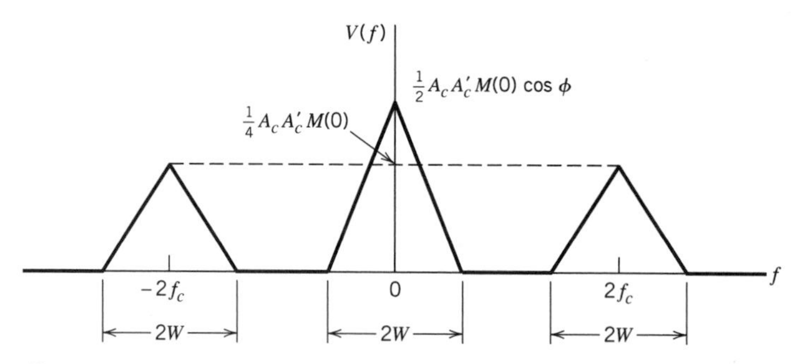


**FIGURE 2.8** Illustrating the spectrum of a product modulator output with a DSB-SC modulated wave as input.

oscillator signal by $A_c' \cos(2\pi f_c t + \phi)$, and using Equation (2.8) for the DSB-SC wave $s(t)$, we find that the product modulator output in Figure 2.7 is

$$\begin{aligned} v(t) &= A_c' \cos(2\pi f_c t + \phi) s(t) \\ &= A_c A_c' \cos(2\pi f_c t) \cos(2\pi f_c t + \phi) m(t) \\ &= \frac{1}{2} A_c A_c' \cos(4\pi f_c t + \phi) m(t) + \frac{1}{2} A_c A_c' \cos\phi\, m(t) \end{aligned} \tag{2.10}$$

The first term in Equation (2.10) represents a DSB-SC modulated signal with a carrier frequency $2f_c$, whereas the second term is proportional to the baseband signal $m(t)$. This is further illustrated by the spectrum $V(f)$ shown in Figure. 2.8, where it is assumed that the baseband signal $m(t)$ is limited to the interval $-W \le f \le W$. It is therefore apparent that the first term in Equation (2.10) is removed by the low-pass filter in Figure 2.7, provided that the cut-off frequency of this filter is greater than $W$ but less than $2f_c - W$. This requirement is satisfied by choosing $f_c > W$. At the filter output we then obtain a signal given by

$$v_o(t) = \frac{1}{2} A_c A_c' \cos\phi\, m(t) \tag{2.11}$$

The demodulated signal $v_o(t)$ is therefore proportional to $m(t)$ when the phase error $\phi$ is a constant. The amplitude of this demodulated signal is maximum when $\phi = 0$, and it is minimum (zero) when $\phi = \pm\pi/2$. The zero demodulated signal, which occurs for $\phi = \pm\pi/2$, represents the *quadrature null effect* of the coherent detector. Thus the phase error $\phi$ in the local oscillator causes the detector output to be attenuated by a factor equal to $\cos\phi$. As long as the phase error $\phi$ is constant, the detector provides an undistorted version of the original baseband signal $m(t)$. In practice, however, we usually find that the phase error $\phi$ varies randomly with time, due to random variations in the communication channel. The result is that at the detector output, the multiplying factor $\cos\phi$ also varies randomly with time, which is obviously undesirable. Therefore, provision must be made in the system to maintain the local oscillator in the receiver in perfect synchronism, in both frequency and phase, with the carrier wave used to generate the DSB-SC modulated signal in the transmitter. The resulting system complexity is the price that must be paid for suppressing the carrier wave to save transmitter power.

## ■ COSTAS RECEIVER

One method of obtaining a practical synchronous receiver system, suitable for demodulating DSB-SC waves, is to use the *Costas receiver*[2] shown in Figure 2.9. This receiver

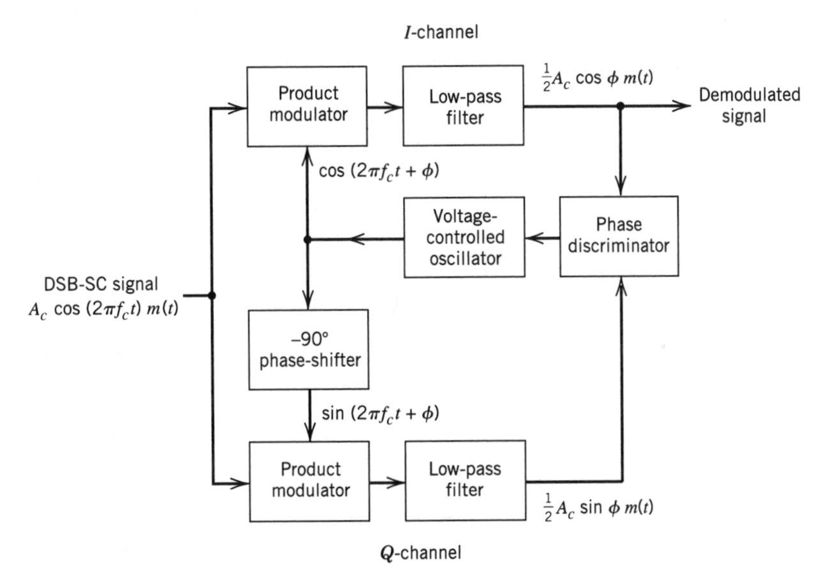


**FIGURE 2.9** Costas receiver.

consists of two coherent detectors supplied with the same input signal, namely, the incoming DSB-SC wave $A_c \cos(2\pi f_c t)m(t)$, but with individual local oscillator signals that are in phase quadrature with respect to each other. The frequency of the local oscillator is adjusted to be the same as the carrier frequency $f_c$, which is assumed known *a priori*. The detector in the upper path is referred to as the *in-phase coherent detector* or *I-channel*, and that in the lower path is referred to as the *quadrature-phase coherent detector* or *Q-channel*. These two detectors are coupled together to form a *negative feedback* system designed in such a way as to maintain the local oscillator synchronous with the carrier wave.

To understand the operation of this receiver, suppose that the local oscillator signal is of the same phase as the carrier wave $A_c \cos(2\pi f_c t)$ used to generate the incoming DSB-SC wave. Under these conditions, we find that the $I$-channel output contains the desired demodulated signal $m(t)$, whereas the $Q$-channel output is zero due to the quadrature null effect of the $Q$-channel. Suppose next that the local oscillator phase drifts from its proper value by a small angle $\phi$ radians. The $I$-channel output will remain essentially unchanged, but there will now be some signal appearing at the $Q$-channel output, which is proportional to $\sin \phi \simeq \phi$ for small $\phi$. This $Q$-channel output will have the same polarity as the $I$-channel output for one direction of local oscillator phase drift and opposite polarity for the opposite direction of local oscillator phase drift. Thus, by combining the $I$- and $Q$-channel outputs in a *phase discriminator* (which consists of a multiplier followed by a low-pass filter), as shown in Figure 2.9, a DC control signal is obtained that automatically corrects for local phase errors in the *voltage-controlled oscillator*.

It is apparent that phase control in the Costas receiver ceases with modulation and that phase-lock has to be reestablished with the reappearance of modulation. This is not a serious problem when receiving voice transmission, because the lock-up process normally occurs so rapidly that no distortion is perceptible.

## ■ QUADRATURE-CARRIER MULTIPLEXING

The quadrature null effect of the coherent detector may also be put to good use in the construction of the so-called *quadrature-carrier multiplexing* or *quadrature-amplitude*

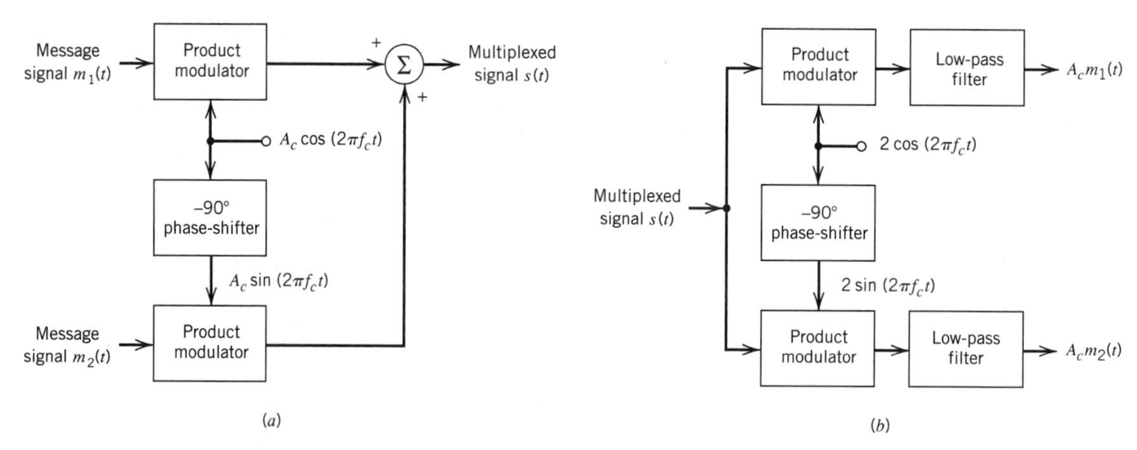


**FIGURE 2.10** Quadrature-carrier multiplexing system. (*a*) Transmitter. (*b*) Receiver.

*modulation* (QAM). This scheme enables two DSB-SC modulated waves (resulting from the application of two physically *independent* message signals) to occupy the same channel bandwidth, and yet it allows for the separation of the two message signals at the receiver output. It is therefore a *bandwidth-conservation scheme.*

A block diagram of the quadrature-carrier multiplexing system is shown in Figure 2.10. The transmitter part of the system, shown in Figure 2.10*a*, involves the use of two separate product modulators that are supplied with two carrier waves of the same frequency but differing in phase by −90 degrees. The transmitted signal $s(t)$ consists of the sum of these two product modulator outputs, as shown by

$$s(t) = A_c m_1(t) \cos(2\pi f_c t) + A_c m_2(t) \sin(2\pi f_c t) \tag{2.12}$$

where $m_1(t)$ and $m_2(t)$ denote the two different message signals applied to the product modulators. Thus $s(t)$ occupies a channel bandwidth of $2W$ centered at the carrier frequency $f_c$, where $W$ is the message bandwidth of $m_1(t)$ or $m_2(t)$. According to Equation (2.12), we may view $A_c m_1(t)$ as the in-phase component of the multiplexed band-pass signal $s(t)$ and $-A_c m_2(t)$ as its quadrature component.

The receiver part of the system is shown in Figure 2.10*b*. The multiplexed signal $s(t)$ is applied simultaneously to two separate coherent detectors that are supplied with two local carriers of the same frequency but differing in phase by −90 degrees. The output of the top detector is $A_c m_1(t)$, whereas the output of the bottom detector is $A_c m_2(t)$. For the system to operate satisfactorily, it is important to maintain the correct phase and frequency relationships between the local oscillators used in the transmitter and receiver parts of the system.

To maintain this synchronization, we may send a *pilot signal* outside the passband of the modulated signal. In this method, the pilot signal typically consists of a low-power sinusoidal tone whose frequency and phase are related to the carrier wave $c(t)$; at the receiver, the pilot signal is extracted by means of a suitably tuned circuit and then translated to the correct frequency for use in the coherent detector.

## ■ SINGLE-SIDEBAND MODULATION

In single-sideband modulation, only the upper or lower sideband is transmitted. We may generate such a modulated wave by using the *frequency-discrimination method* that consists of two stages:

- The first stage is a product modulator, which generates a DSB-SC modulated wave.
- The second stage is a band-pass filter, which is designed to pass one of the sidebands of this modulated wave and suppress the other.

From a practical viewpoint the most severe requirement of SSB generation using the frequency discrimination method arises from the unwanted sideband. The nearest frequency component of the unwanted sideband is separated from the desired sideband by twice the lowest frequency component of the message (modulating) signal. The implication here is that for the generation of an SSB modulated signal to be possible, the message spectrum must have an energy gap centered at the origin, as illustrated in Figure 2.11*a*. This requirement is naturally satisfied by voice signals, whose energy gap is about 600 Hz wide (i.e., it extends from $-300$ to $+300$ Hz). Thus, assuming that the upper sideband is retained, the spectrum of the SSB modulated signal is as shown in Figure 2.11*b*.

In designing the band-pass filter used in the frequency-discriminator for generating a SSB-modulated wave, we must meet the three basic requirements:

- The desired sideband lies inside the passband of the filter.
- The unwanted sideband lies inside the stopband of the filter.
- The filter's transition band, which separates the passband from the stopband, is twice the lowest frequency component of the message signal.

This kind of frequency discrimination usually requires the use of highly selective filters, which can only be realized in practice by means of crystal resonators.

To demodulate a SSB modulated signal $s(t)$, we may use a coherent detector, which multiplies $s(t)$ by a locally generated carrier and then low-pass filters the product. This method of demodulation assumes perfect synchronism between the oscillator in the coherent detector and the oscillator used to supply the carrier wave in the transmitter. This requirement is usually met in one of two ways:

- A low-power *pilot carrier* is transmitted in addition to the selected sideband.
- A highly stable oscillator, tuned to the same frequency as the carrier frequency, is used in the receiver.

In the latter method, it is inevitable that there would be some phase error $\phi$ in the local oscillator output with respect to the carrier wave used to generate the incoming SSB modulated wave. The effect of this phase error is to introduce a *phase distortion* in the demodulated signal, where each frequency component of the original message signal undergoes a constant phase shift $\phi$. This phase distortion is tolerable in voice communications,

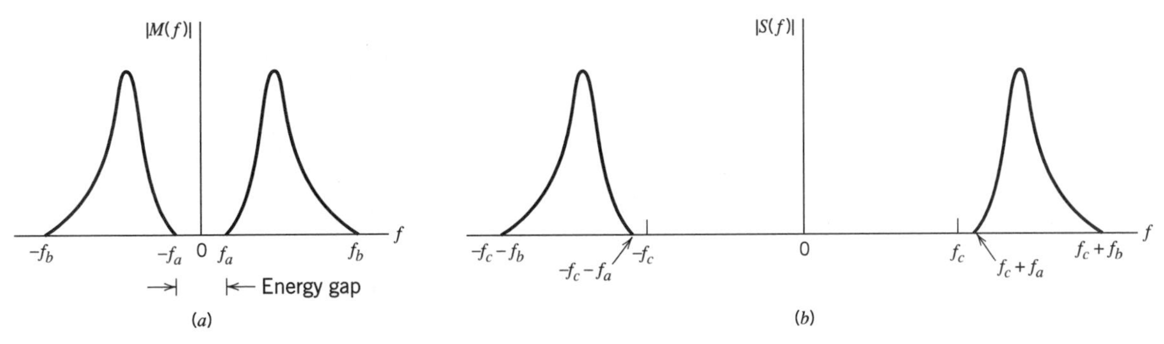


**FIGURE 2.11** (*a*) Spectrum of a message signal $m(t)$ with an energy gap of width $2f_a$ centered on the origin. (*b*) Spectrum of corresponding SSB signal containing the upper sideband.

because the human ear is relatively insensitive to phase distortion. In particular, the presence of phase distortion gives rise to a *Donald Duck voice effect*. In the transmission of music and video signals, on the other hand, the presence of this form of waveform distortion is utterly unacceptable.

## ■ VESTIGIAL SIDEBAND MODULATION

In *vestigial sideband (VSB) modulation*, one of the sidebands is partially suppressed and a vestige of the other sideband is transmitted to compensate for that suppression. A popular method for generating a VSB-modulated wave is to use the *frequency discrimination method*. First, we generate a DSB-SC modulated wave and then pass it through a band-pass filter, as shown in Figure 2.12; it is the special design of the band-pass filter that distinguishes VSB modulation from SSB modulation. Assuming that a vestige of the lower sideband is transmitted, the frequency response $H(f)$ of the band-pass filter takes the form shown in Figure 2.13. To simplify matters, only the response for positive frequencies is shown here. This frequency response is normalized, so that at the carrier frequency $f_c$ we have $|H(f_c)| = 1/2$. The important feature to note from Figure 2.13 is that the cutoff portion of the frequency response around the carrier frequency $f_c$ exhibits *odd symmetry*. That is, inside the transition interval $f_c - f_v \leq |f| \leq f_c + f_v$ the following two conditions are satisfied:

1. The sum of the values of the magnitude response $|H(f)|$ at any two frequencies equally displaced above and below $f_c$ is unity.
2. The phase response $\arg(H(f))$ is linear. That is, $H(f)$ satisfies the condition

$$H(f - f_c) + H(f + f_c) = 1 \qquad \text{for } -W \leq f \leq W \tag{2.13}$$

Note also that outside the frequency band of interest (i.e., $|f| > f_c + W$), the frequency response $H(f)$ may have an arbitrary specification. Accordingly, the transmission bandwidth of VSB modulation is

$$B_T = W + f_v \tag{2.14}$$

where $W$ is the message bandwidth, and $f_v$ is the width of the vestigial sideband.

According to Table 2.1, the VSB modulated wave is described in the time domain as

$$s(t) = \frac{1}{2} A_c m(t) \cos(2\pi f_c t) \pm \frac{1}{2} A_c m'(t) \sin(2\pi f_c t) \tag{2.15}$$

where the plus sign corresponds to the transmission of a vestige of the upper sideband, and the minus sign corresponds to the transmission of a vestige of the lower sideband. The signal $m'(t)$ in the quadrature component of $s(t)$ is obtained by passing the message signal

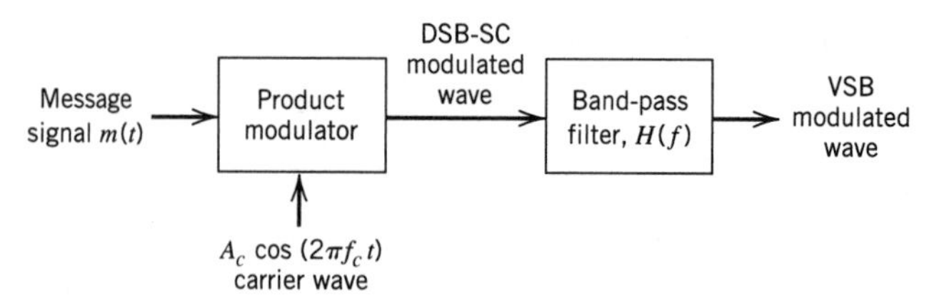


FIGURE 2.12 Filtering scheme for the generation of VSB modulated wave.

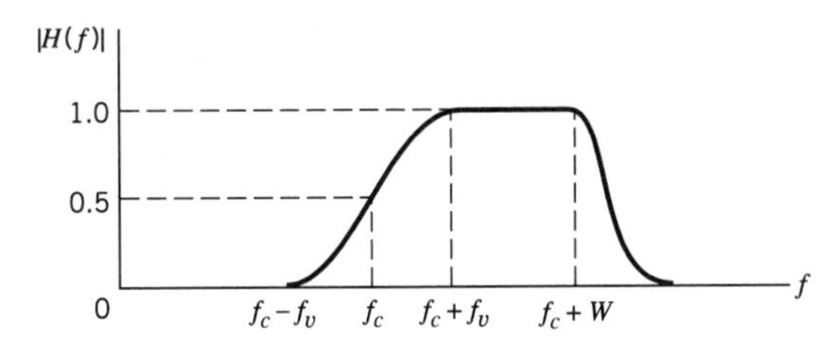


**FIGURE 2.13** Magnitude response of VSB filter; only the positive-frequency portion is shown.

$m(t)$ through a filter whose frequency response $H_Q(f)$ satisfies the following requirement (see Problem 2.20):

$$H_Q(f) = j[H(f - f_c) - H(f + f_c)] \qquad \text{for } -W \le f \le W \tag{2.16}$$

Figure 2.14 displays a plot of the frequency response $H_Q(f)$, scaled by $1/j$. The role of the quadrature component determined by $H_Q(f)$ is to interfere with the in-phase component in Equation (2.15) so as to partially reduce power in one of the sidebands of the modulated wave $s(t)$ and retain simply a vestige of the other sideband, as desired.

It is of interest to note that SSB modulation may be viewed as a special case of VSB modulation. Specifically, when the vestigial sideband is reduced to zero (i.e., we set $f_v = 0$), the modulated wave $s(t)$ of Equation (2.15) takes the limiting form of a single-sideband modulated wave.

## TELEVISION SIGNALS

A discussion of vestigial sideband modulation would be incomplete without a mention of its role in commercial television (TV) broadcasting. The exact details of the modulation format used to transmit the video signal characterizing a TV system are influenced by two factors:

1. The video signal exhibits a large bandwidth and significant low-frequency content, which suggest the use of vestigial sideband modulation.
2. The circuitry used for demodulation in the receiver should be simple and therefore inexpensive; this suggests the use of envelope detection, which requires the addition of a carrier to the VSB-modulated wave.

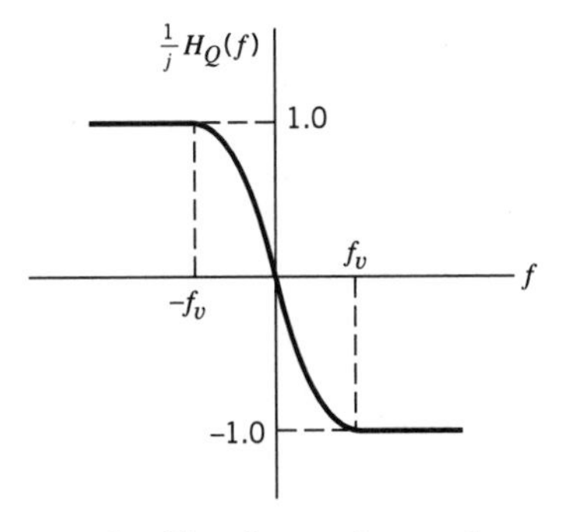


**FIGURE 2.14** Frequency response of a filter for producing the quadrature component of the VSB modulated wave.

With regard to point 1, however, it should be stressed that although there is indeed a basic desire to conserve bandwidth, in commercial TV broadcasting the transmitted signal is not quite VSB modulated. The reason is that at the transmitter the power levels are high, with the result that it would be expensive to rigidly control the filtering of sidebands. Instead, a *VSB filter* is inserted in each receiver, where the power levels are low. The overall performance is the same as conventional vestigial-sideband modulation, except for some wasted power and bandwidth. These remarks are illustrated in Figure 2.15. In particular, Figure 2.15*a* shows the idealized spectrum of a transmitted TV signal. The upper sideband, 25 percent of the lower sideband, and the picture carrier are transmitted. The frequency response of the VSB filter used to do the required spectrum shaping in the receiver is shown in Figure 2.15*b*.

The channel bandwidth used for TV broadcasting in North America is 6 MHz, as indicated in Figure 2.15*b*. This channel bandwidth not only accommodates the bandwidth requirement of the VSB modulated video signal but also provides for the accompanying sound signal that modulates a carrier of its own. The values presented on the frequency axis in Figures 2.15*a* and 2.15*b* pertain to a specific TV channel. According to this figure, the picture carrier frequency is at 55.25 MHz, and the sound carrier frequency is at 59.75 MHz. Note, however, that the information content of the TV signal lies in a *baseband spectrum* extending from 1.25 MHz below the picture carrier to 4.5 MHz above it.

With regard to point 2, the use of envelope detection (applied to a VSB modulated

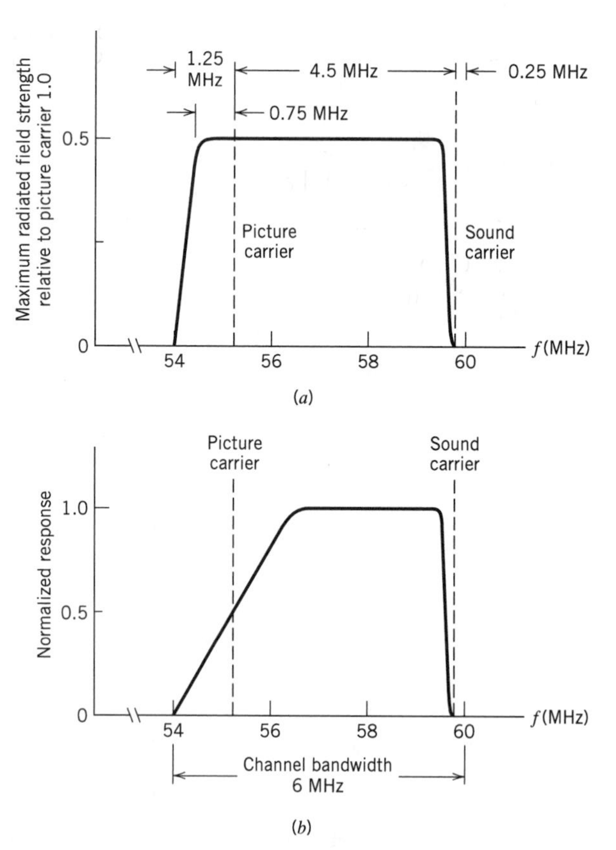


**FIGURE 2.15** (*a*) Idealized magnitude spectrum of a transmitted TV signal. (*b*) Magnitude response of VSB shaping filter in the receiver.

wave plus carrier) produces *waveform distortion* in the video signal recovered at the detector output. The distortion is produced by the quadrature component of the VSB modulated wave; this issue is discussed next.

The use of the time-domain description given in Equation (2.15) enables the determination of the waveform distortion caused by the envelope detector. Specifically, adding the carrier component $A_c \cos(2\pi f_c t)$ to the VSB-modulated wave of Equation (2.15), the latter being scaled by a factor $k_a$, modifies the modulated signal applied to the envelope detector input as

$$s(t) = A_c\left[1 + \frac{1}{2}k_a m(t)\right]\cos(2\pi f_c t) \pm \frac{1}{2}k_a A_c m'(t)\sin(2\pi f_c t) \tag{2.17}$$

where the constant $k_a$ determines the percentage modulation. The envelope detector output, denoted by $a(t)$, is therefore

$$\begin{aligned} a(t) &= A_c\left\{\left[1 + \frac{1}{2}k_a m(t)\right]^2 + \left[\frac{1}{2}k_a m'(t)\right]^2\right\}^{1/2} \\ &= A_c\left[1 + \frac{1}{2}k_a m(t)\right]\left\{1 + \left[\frac{\frac{1}{2}k_a m'(t)}{1 + \frac{1}{2}k_a m(t)}\right]^2\right\}^{1/2} \end{aligned} \tag{2.18}$$

Equation (2.18) indicates that the distortion is contributed by $m'(t)$, which is responsible for the quadrature component of the incoming VSB-modulated signal. This distortion can be reduced by using two methods:

- Reducing the percentage modulation to reduce the amplitude sensitivity $k_a$.
- Increasing the width of the vestigial sideband to reduce $m'(t)$.

Both methods are in fact used in practice. In commercial TV broadcasting, the width of the vestigial sideband (which is about 0.75 MHz, or one-sixth of a full sideband) is determined to keep the distortion due to $m'(t)$ within tolerable limits when the percentage modulation is nearly 100.

## 2.4 Frequency Translation

The basic operation involved in single-sideband modulation is in fact a form of *frequency translation*, which is why single-sideband modulation is sometimes referred to as *frequency changing*, *mixing*, or *heterodyning*. This operation is clearly illustrated in the spectrum of the signal shown in Figure 2.11*b* compared to that of the original message signal in Figure 2.11*a*. Specifically, we see that a message spectrum occupying the band from $f_a$ to $f_b$ for positive frequencies in Figure 2.11*a* is shifted upward by an amount equal to the carrier frequency $f_c$ in Figure 2.11*b*, and the message spectrum for negative frequencies is translated downward in a symmetric fashion.

The idea of frequency translation described herein may be generalized as follows. Suppose that we have a modulated wave $s_1(t)$ whose spectrum is centered on a carrier frequency $f_1$, and the requirement is to translate it upward in frequency such that its carrier frequency is changed from $f_1$ to a new value $f_2$. This requirement may be accomplished using the *mixer* shown in Figure 2.16. Specifically, the *mixer* is a device that consists of a product modulator followed by a band-pass filter.

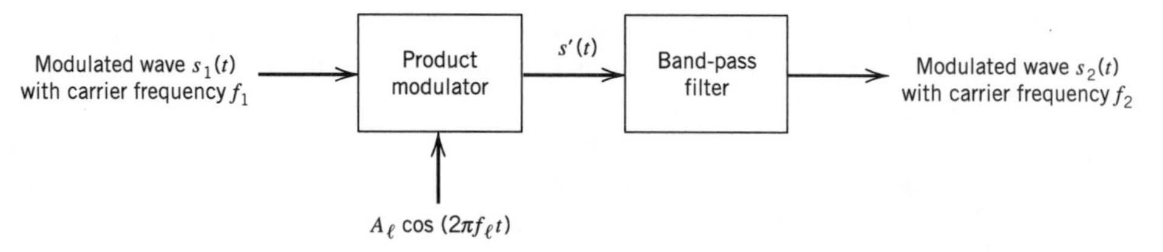


**FIGURE 2.16** Block diagram of mixer.

To explain the action of the mixer, consider the situation depicted in Figure 2.17, where, for the purpose of illustration, it is assumed that the mixer input $s_1(t)$ is an AM signal with carrier frequency $f_1$ and bandwidth $2W$. Part (*a*) of Figure 2.17 displays the AM spectrum $S_1(f)$ assuming that $f_1 > W$. Part (*b*) of the figure displays the spectrum $S'(f)$ of the resulting signal $s'(t)$ at the product modulator output.

The signal $s'(t)$ may be viewed as the sum of two modulated components: one component represented by the shaded spectrum in Figure 2.17*b*, and the other component represented by the unshaded spectrum in this figure. Depending on whether the incoming carrier frequency $f_1$ is translated upward or downward, we may identify two different situations, as described here:

*Up conversion.* In this case the translated carrier frequency $f_2$ is greater than the incoming carrier frequency $f_1$, and the required local oscillator frequency $f_l$ is therefore defined by

$$f_2 = f_1 + f_l$$

or

$$f_l = f_2 - f_1$$

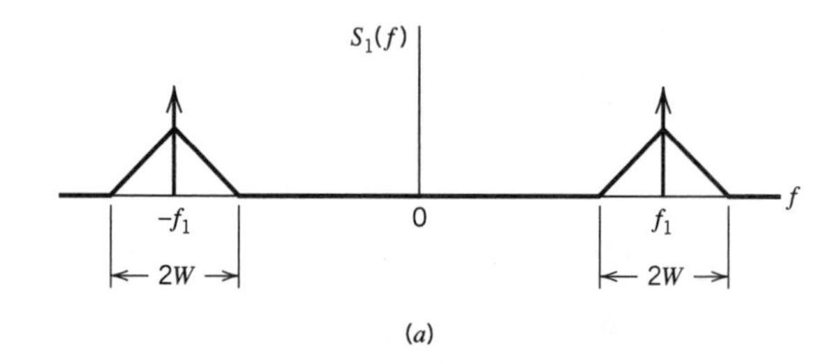


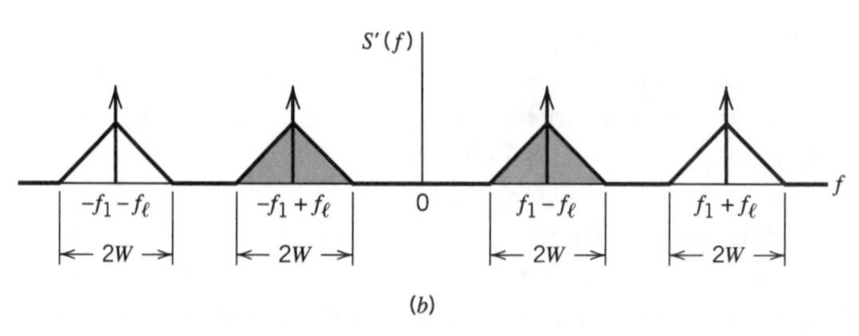


**FIGURE 2.17** (*a*) Spectrum of modulated signal $s_1(t)$ at the mixer input. (*b*) Spectrum of the corresponding signal $s'(t)$ at the output of the product modulator in the mixer.

The unshaded part of the spectrum in Figure 2.17*b* defines the wanted modulated signal $s_2(t)$, and the shaded part of this spectrum defines the *image signal* associated with $s_2(t)$. For obvious reasons, the mixer in this case is referred to as a *frequency-up converter*.

*Down conversion.* In this second case the translated carrier frequency $f_2$ is smaller than the incoming carrier frequency $f_1$, and the required oscillator frequency $f_l$ is therefore defined by

$$f_2 = f_1 - f_l$$

or

$$f_l = f_1 - f_2$$

The picture we have this time is the reverse of that pertaining to up conversion. In particular, the shaded part of the spectrum in Figure 2.17*b* defines the wanted modulated signal $s_2(t)$, and the unshaded part of this spectrum defines the associated image signal. The mixer is now referred to as a *frequency-down converter*. Note that in this case the translated carrier frequency $f_2$ has to be larger than $W$ (i.e., one half of the bandwidth of the modulated signal) to avoid sideband overlap.

The purpose of the band-pass filter in the mixer of Figure 2.16 is to pass the wanted modulated signal $s_2(t)$ and eliminate the associated image signal. This objective is achieved by aligning the midband frequency of the filter with the translated carrier frequency $f_2$ and assigning it a bandwidth equal to that of the incoming modulated signal $s_1(t)$.

It is important to note that mixing is a *linear* operation. Accordingly, the relation of the sidebands of the incoming modulated wave to the carrier is completely preserved at the mixer output.

## 2.5 *Frequency-Division Multiplexing*

Another important signal processing operation is *multiplexing*, whereby a number of independent signals can be combined into a composite signal suitable for transmission over a common channel. Voice frequencies transmitted over telephone systems, for example, range from 300 to 3100 Hz. To transmit a number of these signals over the same channel, the signals must be kept apart so that they do not interfere with each other, and thus they can be separated at the receiving end. This is accomplished by separating the signals either in frequency or in time. The technique of separating the signals in frequency is referred to as *frequency-division multiplexing* (FDM), whereas the technique of separating the signals in time is called *time-division multiplexing* (TDM). In this section, we discuss FDM systems, and TDM systems are discussed in Chapter 3.

A block diagram of an FDM system is shown in Figure 2.18. The incoming message signals are assumed to be of the low-pass type, but their spectra do not necessarily have nonzero values all the way down to zero frequency. Following each signal input, we have shown a low-pass filter, which is designed to remove high-frequency components that do not contribute significantly to signal representation but are capable of disturbing other message signals that share the common channel. These low-pass filters may be omitted only if the input signals are sufficiently band limited initially. The filtered signals are applied

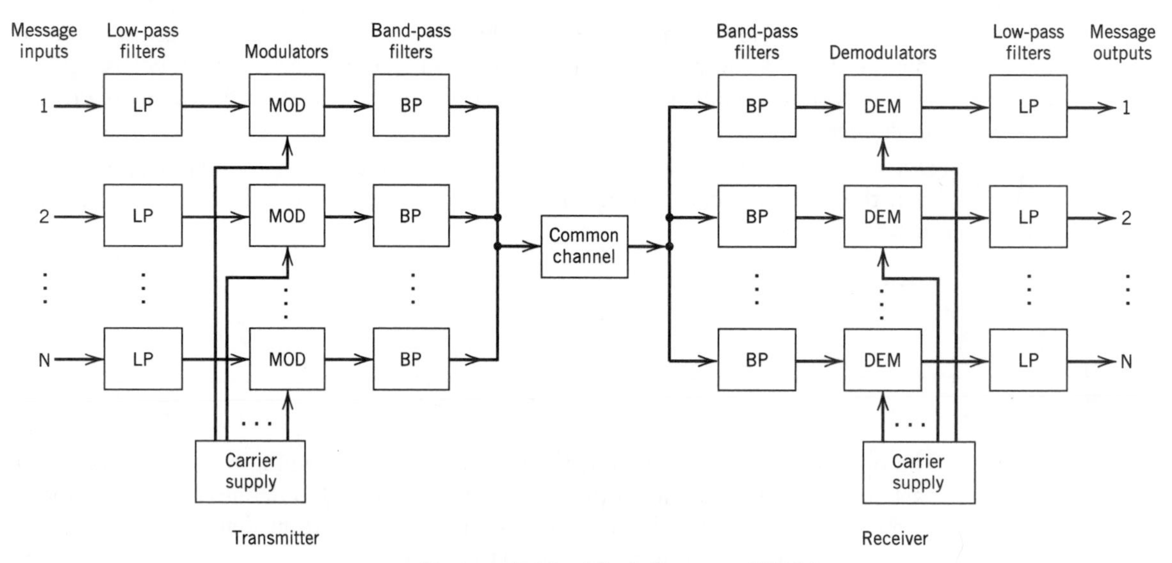


**FIGURE 2.18** Block diagram of FDM system.

to modulators that shift the frequency ranges of the signals so as to occupy mutually exclusive frequency intervals. The necessary carrier frequencies needed to perform these frequency translations are obtained from a carrier supply. For the modulation, we may use any one of the methods described in previous sections of this chapter. However, the most widely used method of modulation in frequency-division multiplexing is single sideband modulation, which, in the case of voice signals, requires a bandwidth that is approximately equal to that of the original voice signal. In practice, each voice input is usually assigned a bandwidth of 4 kHz. The band-pass filters following the modulators are used to restrict the band of each modulated wave to its prescribed range. The resulting band-pass filter outputs are next combined in parallel to form the input to the common channel. At the receiving terminal, a bank of band-pass filters, with their inputs connected in parallel, is used to separate the message signals on a frequency-occupancy basis. Finally, the original message signals are recovered by individual demodulators. Note that the FDM system shown in Figure 2.18 operates in only one direction. To provide for two-way transmission, as in telephony, for example, we have to completely duplicate the multiplexing facilities, with the components connected in reverse order and with the signal waves proceeding from right to left.

### ▶ EXAMPLE 2.1

The practical implementation of an FDM system usually involves many steps of modulation and demodulation, as illustrated in Figure 2.19. The first multiplexing step combines 12 voice inputs into a *basic group*, which is formed by having the $n$th input modulate a carrier at frequency $f_c = 60 + 4n$ kHz, where $n = 1, 2, \ldots, 12$. The lower sidebands are then selected by band-pass filtering and combined to form a group of 12 lower sidebands (one for each voice input). Thus the basic group occupies the frequency band 60 to 108 kHz. The next step in the FDM hierarchy involves the combination of five basic groups into a *supergroup*. This is accomplished by using the $n$th group to modulate a carrier of frequency $f_c = 372 + 48n$ kHz, where $n = 1, 2, \ldots, 5$. Here again the lower sidebands are selected by filtering and then

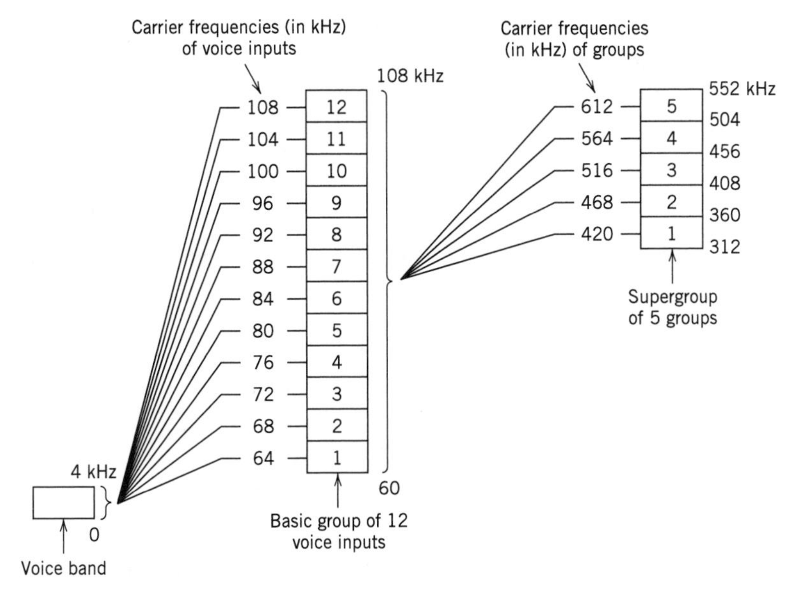


**FIGURE 2.19** Illustrating the modulation steps in an FDM system.

combined to form a supergroup occupying the band 312 to 552 kHz. Thus a supergroup is designed to accommodate 60 independent voice inputs. The reason for forming the supergroup in this manner is that economical filters of the required characteristics are available only over a limited frequency range. In a similar manner, supergroups are combined into *mastergroups*, and mastergroups are combined into *very large groups.* ◀

## 2.6 *Angle Modulation*

In the previous sections of this chapter, we investigated the effect of slowly varying the amplitude of a sinusoidal carrier wave in accordance with the baseband (information-carrying) signal. There is another way of modulating a sinusoidal carrier wave, namely, *angle modulation* in which the angle of the carrier wave is varied according to the baseband signal. In this method of modulation, the amplitude of the carrier wave is maintained constant. An important feature of angle modulation is that it can provide better discrimination against noise and interference than amplitude modulation. As will be shown later in Section 2.7, however, this improvement in performance is achieved at the expense of increased transmission bandwidth; that is, angle modulation provides us with a practical means of exchanging channel bandwidth for improved noise performance. Such a trade-off is *not* possible with amplitude modulation, regardless of its form.

### ■ BASIC DEFINITIONS

Let $\theta_i(t)$ denote the *angle* of a modulated sinusoidal carrier, assumed to be a function of the message signal. We express the resulting *angle-modulated wave* as

$$s(t) = A_c \cos[\theta_i(t)] \tag{2.19}$$

where $A_c$ is the carrier amplitude. A complete oscillation occurs whenever $\theta_i(t)$ changes by $2\pi$ radians. If $\theta_i(t)$ increases monotonically with time, the average frequency in Hertz, over an interval from $t$ to $t + \Delta t$, is given by

$$f_{\Delta t}(t) = \frac{\theta_i(t + \Delta t) - \theta_i(t)}{2\pi \, \Delta t} \tag{2.20}$$

We may thus define the *instantaneous frequency* of the angle-modulated signal $s(t)$ as follows:

$$\begin{aligned} f_i(t) &= \lim_{\Delta t \to 0} f_{\Delta t}(t) \\ &= \lim_{\Delta t \to 0} \left[ \frac{\theta_i(t + \Delta t) - \theta_i(t)}{2\pi \, \Delta t} \right] \\ &= \frac{1}{2\pi} \frac{d\theta_i(t)}{dt} \end{aligned} \tag{2.21}$$

Thus, according to Equation (2.19), we may interpret the angle-modulated signal $s(t)$ as a rotating phasor of length $A_c$ and angle $\theta_i(t)$. The angular velocity of such a phasor is $d\theta_i(t)/dt$ measured in radians per second, in accordance with Equation (2.21). In the simple case of an unmodulated carrier, the angle $\theta_i(t)$ is

$$\theta_i(t) = 2\pi f_c t + \phi_c$$

and the corresponding phasor rotates with a constant angular velocity equal to $2\pi f_c$. The constant $\phi_c$ is the value of $\theta_i(t)$ at $t = 0$.

There are an infinite number of ways in which the angle $\theta_i(t)$ may be varied in some manner with the message (baseband) signal. However, we shall consider only two commonly used methods, phase modulation and frequency modulation, defined as follows:

1. *Phase modulation* (PM) *is that form of angle modulation in which the angle $\theta_i(t)$ is varied linearly with the message signal $m(t)$*, as shown by

$$\theta_i(t) = 2\pi f_c t + k_p m(t) \tag{2.22}$$

The term $2\pi f_c t$ represents the angle of the *unmodulated* carrier; and the constant $k_p$ represents the *phase sensitivity* of the modulator, expressed in radians per volt on the assumption that $m(t)$ is a voltage waveform. For convenience, we have assumed in Equation (2.22) that the angle of the unmodulated carrier is zero at $t = 0$. The phase-modulated signal $s(t)$ is thus described in the time domain by

$$s(t) = A_c \cos[2\pi f_c t + k_p m(t)] \tag{2.23}$$

2. *Frequency modulation* (FM) *is that form of angle modulation in which the instantaneous frequency $f_i(t)$ is varied linearly with the message signal $m(t)$*, as shown by

$$f_i(t) = f_c + k_f m(t) \tag{2.24}$$

The term $f_c$ represents the frequency of the unmodulated carrier, and the constant $k_f$ represents the *frequency sensitivity* of the modulator, expressed in Hertz per volt

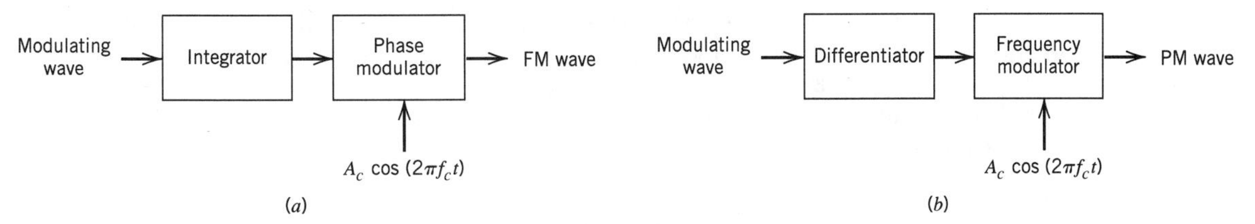


**FIGURE 2.20** Illustrating the relationship between frequency modulation and phase modulation. (*a*) Scheme for generating an FM wave by using a phase modulator. (*b*) Scheme for generating a PM wave by using a frequency modulator.

on the assumption that $m(t)$ is a voltage waveform. Integrating Equation (2.24) with respect to time and multiplying the result by $2\pi$, we get

$$\theta_i(t) = 2\pi f_c t + 2\pi k_f \int_0^t m(\tau)\, d\tau \tag{2.25}$$

where, for convenience, we have assumed that the angle of the unmodulated carrier wave is zero at $t = 0$. The frequency-modulated signal is therefore described in the time domain by

$$s(t) = A_c \cos\left[2\pi f_c t + 2\pi k_f \int_0^t m(\tau)\, d\tau\right] \tag{2.26}$$

A consequence of allowing the angle $\theta_i(t)$ to become dependent on the message signal $m(t)$ as in Equation (2.22) or on its integral as in Equation (2.25) is that the *zero crossings* of a PM signal or FM signal no longer have a perfect regularity in their spacing; zero crossings refer to the instants of time at which a waveform changes from a negative to a positive value or vice versa. This is one important feature that distinguishes both PM and FM signals from an AM signal. Another important difference is that the envelope of a PM or FM signal is constant (equal to the carrier amplitude), whereas the envelope of an AM signal is dependent on the message signal.

Comparing Equation (2.23) with (2.26) reveals that an FM signal may be regarded as a PM signal in which the modulating wave is $\int_0^t m(\tau)\, d\tau$ in place of $m(t)$. This means that an FM signal can be generated by first integrating $m(t)$ and then using the result as the input to a phase modulator, as in Figure 2.20*a*. Conversely, a PM signal can be generated by first differentiating $m(t)$ and then using the result as the input to a frequency modulator, as in Figure 2.20*b*. We may thus deduce all the properties of PM signals from those of FM signals and vice versa. Henceforth, we concentrate our attention on FM signals.

## 2.7 *Frequency Modulation*

The FM signal $s(t)$ defined by Equation (2.26) is a nonlinear function of the modulating signal $m(t)$, which makes frequency modulation a *nonlinear modulation process*. Consequently, unlike amplitude modulation, the spectrum of an FM signal is not related in a simple manner to that of the modulating signal; rather, its analysis is much more difficult than that of an AM signal.

How then can we tackle the spectral analysis of an FM signal? We propose to provide an empirical answer to this important question by proceeding in the following manner:

- We consider the simplest case possible, namely, that of a single-tone modulation that produces a narrowband FM signal.
- We next consider the more general case also involving a single-tone modulation, but this time the FM signal is wideband.

We could, of course, go on and consider the more elaborate case of a multitone FM signal. However, we propose not to do so, because our immediate objective is to establish an empirical relationship between the transmission bandwidth of an FM signal and the message bandwidth. As we shall subsequently see, the two-stage spectral analysis described here provides us with enough insight to propose a solution to the problem.

Consider then a sinusoidal modulating signal defined by

$$m(t) = A_m \cos(2\pi f_m t) \tag{2.27}$$

The instantaneous frequency of the resulting FM signal equals

$$\begin{aligned} f_i(t) &= f_c + k_f A_m \cos(2\pi f_m t) \\ &= f_c + \Delta f \cos(2\pi f_m t) \end{aligned} \tag{2.28}$$

where

$$\Delta f = k_f A_m \tag{2.29}$$

The quantity $\Delta f$ is called the *frequency deviation*, representing the maximum departure of the instantaneous frequency of the FM signal from the carrier frequency $f_c$. A fundamental characteristic of an FM signal is that the frequency deviation $\Delta f$ is proportional to the amplitude of the modulating signal and is independent of the modulation frequency.

Using Equation (2.28), the angle $\theta_i(t)$ of the FM signal is obtained as

$$\begin{aligned} \theta_i(t) &= 2\pi \int_0^t f_i(\tau)\, d\tau \\ &= 2\pi f_c t + \frac{\Delta f}{f_m} \sin(2\pi f_m t) \end{aligned} \tag{2.30}$$

The ratio of the frequency deviation $\Delta f$ to the modulation frequency $f_m$ is commonly called the *modulation index* of the FM signal. We denote it by $\beta$, and so write

$$\beta = \frac{\Delta f}{f_m} \tag{2.31}$$

and

$$\theta_i(t) = 2\pi f_c t + \beta \sin(2\pi f_m t) \tag{2.32}$$

From Equation (2.32) we see that, in a physical sense, the parameter $\beta$ represents the phase deviation of the FM signal, that is, the maximum departure of the angle $\theta_i(t)$ from the angle $2\pi f_c t$ of the unmodulated carrier; hence, $\beta$ is measured in radians.

The FM signal itself is given by

$$s(t) = A_c \cos[2\pi f_c t + \beta \sin(2\pi f_m t)] \tag{2.33}$$

Depending on the value of the modulation index $\beta$, we may distinguish two cases of frequency modulation:

- *Narrowband* FM, for which $\beta$ is small compared to one radian.
- *Wideband* FM, for which $\beta$ is large compared to one radian.

These two cases are considered next, in that order.

## NARROWBAND FREQUENCY MODULATION

Consider Equation (2.33), which defines an FM signal resulting from the use of a sinusoidal modulating signal. Expanding this relation, we get

$$s(t) = A_c \cos(2\pi f_c t) \cos[\beta \sin(2\pi f_m t)] - A_c \sin(2\pi f_c t) \sin[\beta \sin(2\pi f_m t)] \quad (2.34)$$

Assuming that the modulation index $\beta$ is small compared to one radian, we may use the following approximations:

$$\cos[\beta \sin(2\pi f_m t)] \simeq 1$$

and

$$\sin[\beta \sin(2\pi f_m t)] \simeq \beta \sin(2\pi f_m t)$$

Hence, Equation (2.34) simplifies to

$$s(t) \simeq A_c \cos(2\pi f_c t) - \beta A_c \sin(2\pi f_c t) \sin(2\pi f_m t) \quad (2.35)$$

Equation (2.35) defines the approximate form of a narrowband FM signal produced by a sinusoidal modulating signal $A_m \cos(2\pi f_m t)$. From this representation we deduce the modulator shown in block diagram form in Figure 2.21. This modulator involves splitting the carrier wave $A_c \cos(2\pi f_c t)$ into two paths. One path is direct; the other path contains a $-90$ degree phase-shifting network and a product modulator, the combination of which generates a DSB-SC modulated signal. The difference between these two signals produces a narrowband FM signal, but with some distortion.

Ideally, an FM signal has a constant envelope and, for the case of a sinusoidal modulating signal of frequency $f_m$, the angle $\theta_i(t)$ is also sinusoidal with the same frequency.

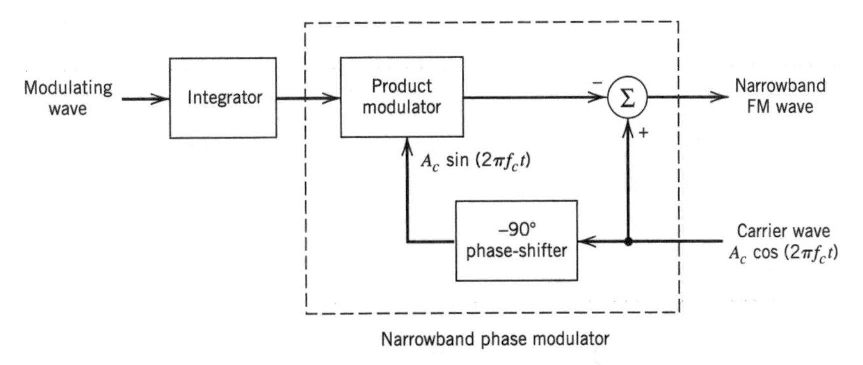


**FIGURE 2.21** Block diagram of a method for generating a narrowband FM signal.

But the modulated signal produced by the narrowband modulator of Figure 2.21 differs from this ideal condition in two fundamental respects:

1. The envelope contains a *residual* amplitude modulation and, therefore, varies with time.
2. For a sinusoidal modulating wave, the angle $\theta_i(t)$ contains *harmonic distortion* in the form of third- and higher-order harmonics of the modulation frequency $f_m$.

However, by restricting the modulation index to $\beta \leq 0.3$ radians, the effects of residual AM and harmonic PM are limited to negligible levels.

Returning to Equation (2.35), we may expand it as follows:

$$s(t) \simeq A_c \cos(2\pi f_c t) + \frac{1}{2} \beta A_c \{\cos[2\pi(f_c + f_m)t] - \cos[2\pi(f_c - f_m)t]\} \tag{2.36}$$

This expression is somewhat similar to the corresponding one defining an AM signal, which is as follows:

$$s_{\mathrm{AM}}(t) = A_c \cos(2\pi f_c t) + \frac{1}{2} \mu A_c \{\cos[2\pi(f_c + f_m)t] + \cos[2\pi(f_c - f_m)t]\} \tag{2.37}$$

where $\mu$ is the modulation factor of the AM signal. Comparing Equations (2.36) and (2.37), we see that in the case of sinusoidal modulation, the basic difference between an AM signal and a narrowband FM signal is that the algebraic sign of the lower side frequency in the narrowband FM is reversed. Thus, a narrowband FM signal requires essentially the same transmission bandwidth (i.e., $2f_m$) as the AM signal.

We may represent the narrowband FM signal with a phasor diagram as shown in Figure 2.22*a*, where we have used the carrier phasor as reference. We see that the resultant

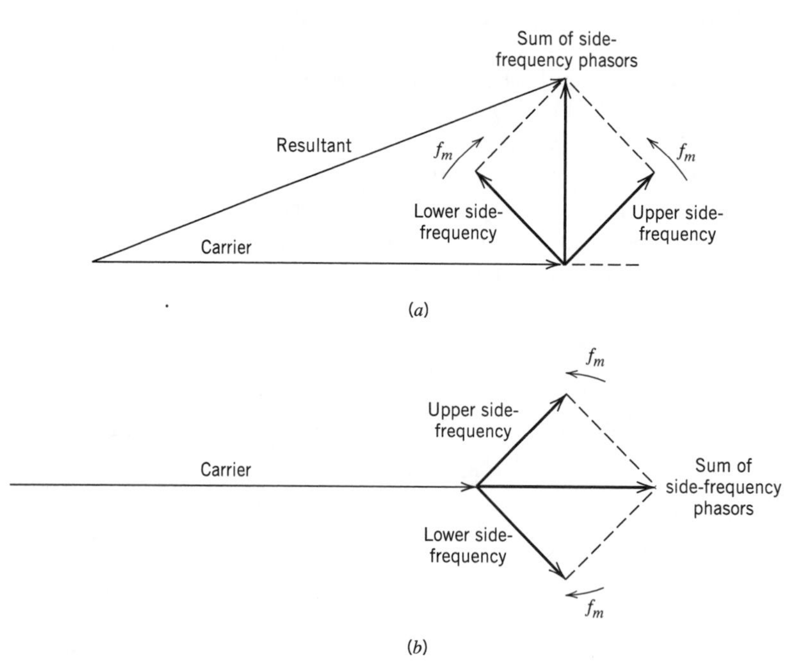


**FIGURE 2.22** A phasor comparison of narrowband FM and AM waves for sinusoidal modulation. (*a*) Narrowband FM wave. (*b*) AM wave.

of the two side-frequency phasors is always at right angles to the carrier phasor. The effect of this is to produce a resultant phasor representing the narrowband FM signal that is approximately of the same amplitude as the carrier phasor, but out of phase with respect to it. This phasor diagram should be contrasted with that of Figure 2.22*b*, representing an AM signal. In this latter case we see that the resultant phasor representing the AM signal has an amplitude that is different from that of the carrier phasor but always in phase with it.

## ■ WIDEBAND FREQUENCY MODULATION

We next wish to determine the spectrum of the single-tone FM signal of Equation (2.33) for an arbitrary value of the modulation index $\beta$. In general, an FM signal produced by a sinusoidal modulating signal, as in Equation (2.33), is in itself nonperiodic unless the carrier frequency $f_c$ is an integral multiple of the modulation frequency $f_m$. However, we may simplify matters by using the complex representation of band-pass signals described in Appendix 2. Specifically, we assume that the carrier frequency $f_c$ is large enough (compared to the bandwidth of the FM signal) to justify rewriting this equation in the form

$$\begin{aligned} s(t) &= \text{Re}[A_c \exp(j2\pi f_c t + j\beta \sin(2\pi f_m t))] \\ &= \text{Re}[\tilde{s}(t) \exp(j2\pi f_c t)] \end{aligned} \tag{2.38}$$

where $\tilde{s}(t)$ is the *complex envelope* of the FM signal $s(t)$, defined by

$$\tilde{s}(t) = A_c \exp[j\beta \sin(2\pi f_m t)] \tag{2.39}$$

Thus, unlike the original FM signal $s(t)$, the complex envelope $\tilde{s}(t)$ is a periodic function of time with a fundamental frequency equal to the modulation frequency $f_m$. We may therefore expand $\tilde{s}(t)$ in the form of a complex Fourier series as follows:

$$\tilde{s}(t) = \sum_{n=-\infty}^{\infty} c_n \exp(j2\pi n f_m t) \tag{2.40}$$

where the complex Fourier coefficient $c_n$ is defined by

$$\begin{aligned} c_n &= f_m \int_{-1/2f_m}^{1/2f_m} \tilde{s}(t) \exp(-j2\pi n f_m t)\, dt \\ &= f_m A_c \int_{-1/2f_m}^{1/2f_m} \exp[j\beta \sin(2\pi f_m t) - j2\pi n f_m t]\, dt \end{aligned} \tag{2.41}$$

Define a new variable:

$$x = 2\pi f_m t \tag{2.42}$$

Hence, we may rewrite Equation (2.41) in the new form

$$c_n = \frac{A_c}{2\pi} \int_{-\pi}^{\pi} \exp[j(\beta \sin x - nx)]\, dx \tag{2.43}$$

The integral on the right-hand side of Equation (2.43), except for a scaling factor, is recognized as the *nth order Bessel function of the first kind*[3] and argument $\beta$. This function is commonly denoted by the symbol $J_n(\beta)$, as shown by

$$J_n(\beta) = \frac{1}{2\pi} \int_{-\pi}^{\pi} \exp[j(\beta \sin x - nx)]\, dx \tag{2.44}$$

Accordingly, we may reduce Equation (2.43) to

$$c_n = A_c J_n(\beta) \tag{2.45}$$

Substituting Equation (2.45) in (2.40), we get, in terms of the Bessel function $J_n(\beta)$, the following expansion for the complex envelope of the FM signal:

$$\tilde{s}(t) = A_c \sum_{n=-\infty}^{\infty} J_n(\beta) \exp(j2\pi n f_m t) \tag{2.46}$$

Next, substituting Equation (2.46) in (2.38), we get

$$s(t) = A_c \cdot \text{Re}\left[\sum_{n=-\infty}^{\infty} J_n(\beta) \exp[j2\pi(f_c + nf_m)t]\right] \tag{2.47}$$

Interchanging the order of summation and evaluation of the real part in the right-hand side of Equation (2.47), we finally get

$$s(t) = A_c \sum_{n=-\infty}^{\infty} J_n(\beta) \cos[2\pi(f_c + nf_m)t] \tag{2.48}$$

This is the desired form for the Fourier series representation of the single-tone FM signal $s(t)$ for an arbitrary value of $\beta$. The discrete spectrum of $s(t)$ is obtained by taking the Fourier transforms of both sides of Equation (2.48); we thus have

$$S(f) = \frac{A_c}{2} \sum_{n=-\infty}^{\infty} J_n(\beta)[\delta(f - f_c - nf_m) + \delta(f + f_c + nf_m)] \tag{2.49}$$

In Figure 2.23 we have plotted the Bessel function $J_n(\beta)$ versus the modulation index $\beta$ for different positive integer values of $n$. We can develop further insight into the behavior

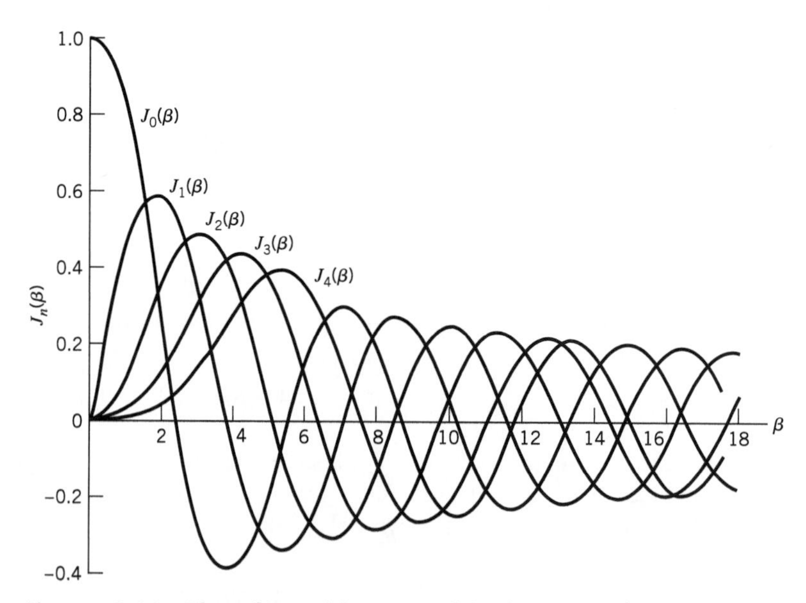


**FIGURE 2.23** Plots of Bessel functions of the first kind for varying order.

of the Bessel function $J_n(\beta)$ by making use of the following properties (see Appendix 3 for more details):

1.
$$J_n(\beta) = (-1)^n J_{-n}(\beta) \qquad \text{for all } n \text{, both positive and negative} \tag{2.50}$$

2. For small values of the modulation index $\beta$, we have

$$\left.\begin{aligned} J_0(\beta) &\simeq 1 \\ J_1(\beta) &\simeq \frac{\beta}{2} \\ J_n(\beta) &\simeq 0, \qquad n > 2 \end{aligned}\right\} \tag{2.51}$$

3.
$$\sum_{n=-\infty}^{\infty} J_n^2(\beta) = 1 \tag{2.52}$$

Thus, using Equations (2.49)–(2.52) and the curves of Figure 2.23, we may make the following observations:

1. The spectrum of an FM signal contains a carrier component and an infinite set of side frequencies located symmetrically on either side of the carrier at frequency separations of $f_m$, $2f_m$, $3f_m$, $\cdots$. In this respect, the result is unlike that which prevails in an AM system, since in an AM system a sinusoidal modulating signal gives rise to only one pair of side frequencies.
2. For the special case of $\beta$ small compared with unity, only the Bessel coefficients $J_0(\beta)$ and $J_1(\beta)$ have significant values, so that the FM signal is effectively composed of a carrier and a single pair of side frequencies at $f_c \pm f_m$. This situation corresponds to the special case of narrowband FM that was considered earlier.
3. The amplitude of the carrier component varies with $\beta$ according to $J_0(\beta)$. That is, unlike an AM signal, the amplitude of the carrier component of an FM signal is dependent on the modulation index $\beta$. The physical explanation for this property is that the envelope of an FM signal is constant, so that the average power of such a signal developed across a 1-ohm resistor is also constant, as shown by

$$P = \frac{1}{2} A_c^2 \tag{2.53}$$

When the carrier is modulated to generate the FM signal, the power in the side frequencies may appear only at the expense of the power originally in the carrier, thereby making the amplitude of the carrier component dependent on $\beta$. Note that the average power of an FM signal may also be determined from Equation (2.48), obtaining

$$P = \frac{1}{2} A_c^2 \sum_{n=-\infty}^{\infty} J_n^2(\beta) \tag{2.54}$$

Substituting Equation (2.52) into (2.54), the expression for the average power $P$ reduces to Equation (2.53), and so it should.

### ▶ EXAMPLE 2.2

In this example, we wish to investigate the ways in which variations in the amplitude and frequency of a sinusoidal modulating signal affect the spectrum of the FM signal. Consider

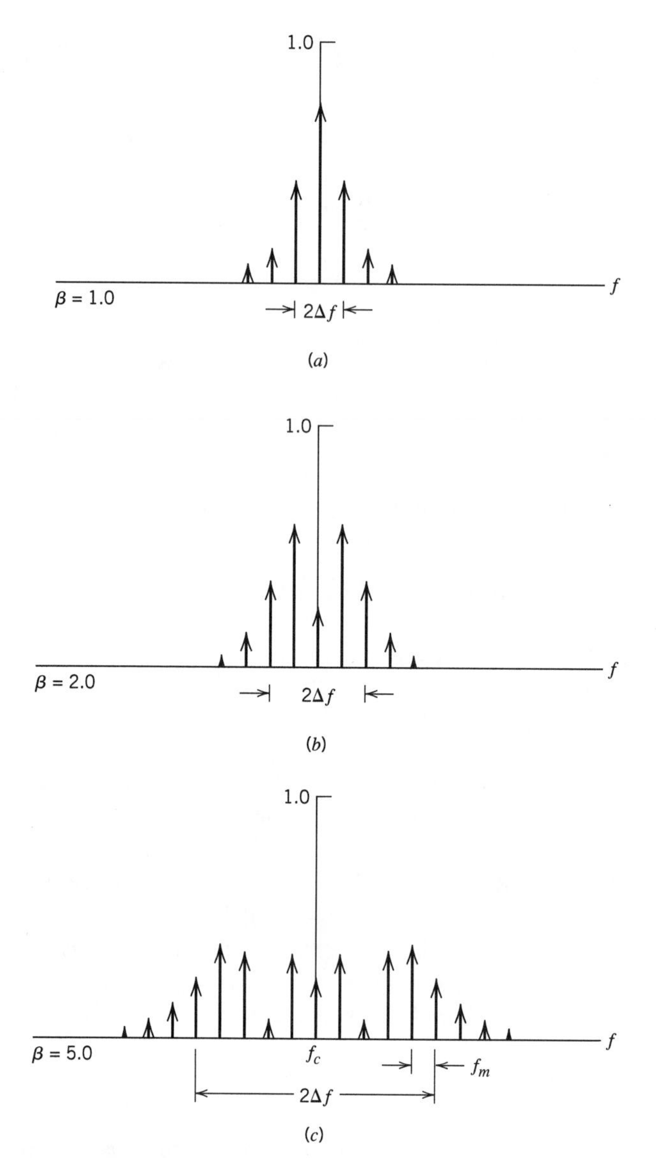


**FIGURE 2.24** Discrete amplitude spectra of an FM signal, normalized with respect to the carrier amplitude, for the case of sinusoidal modulation of fixed frequency and varying amplitude. Only the spectra for positive frequencies are shown.

first the case when the frequency of the modulating signal is fixed, but its amplitude is varied, producing a corresponding variation in the frequency deviation $\Delta f$. Thus, keeping the modulation frequency $f_m$ fixed, we find that the amplitude spectrum of the resulting FM signal is as shown plotted in Figure 2.24 for $\beta = 1$, 2, and 5. In this diagram we have normalized the spectrum with respect to the unmodulated carrier amplitude.

Consider next the case when the amplitude of the modulating signal is fixed; that is, the frequency deviation $\Delta f$ is maintained constant, and the modulation frequency $f_m$ is varied. In this case we find that the amplitude spectrum of the resulting FM signal is as shown plotted in Figure 2.25 for $\beta = 1$, 2, and 5. We see that when $\Delta f$ is fixed and $\beta$ is increased, we have an increasing number of spectral lines crowding into the fixed frequency interval $f_c - \Delta f < |f| < f_c + \Delta f$. That is, when $\beta$ approaches infinity, the bandwidth of the FM wave

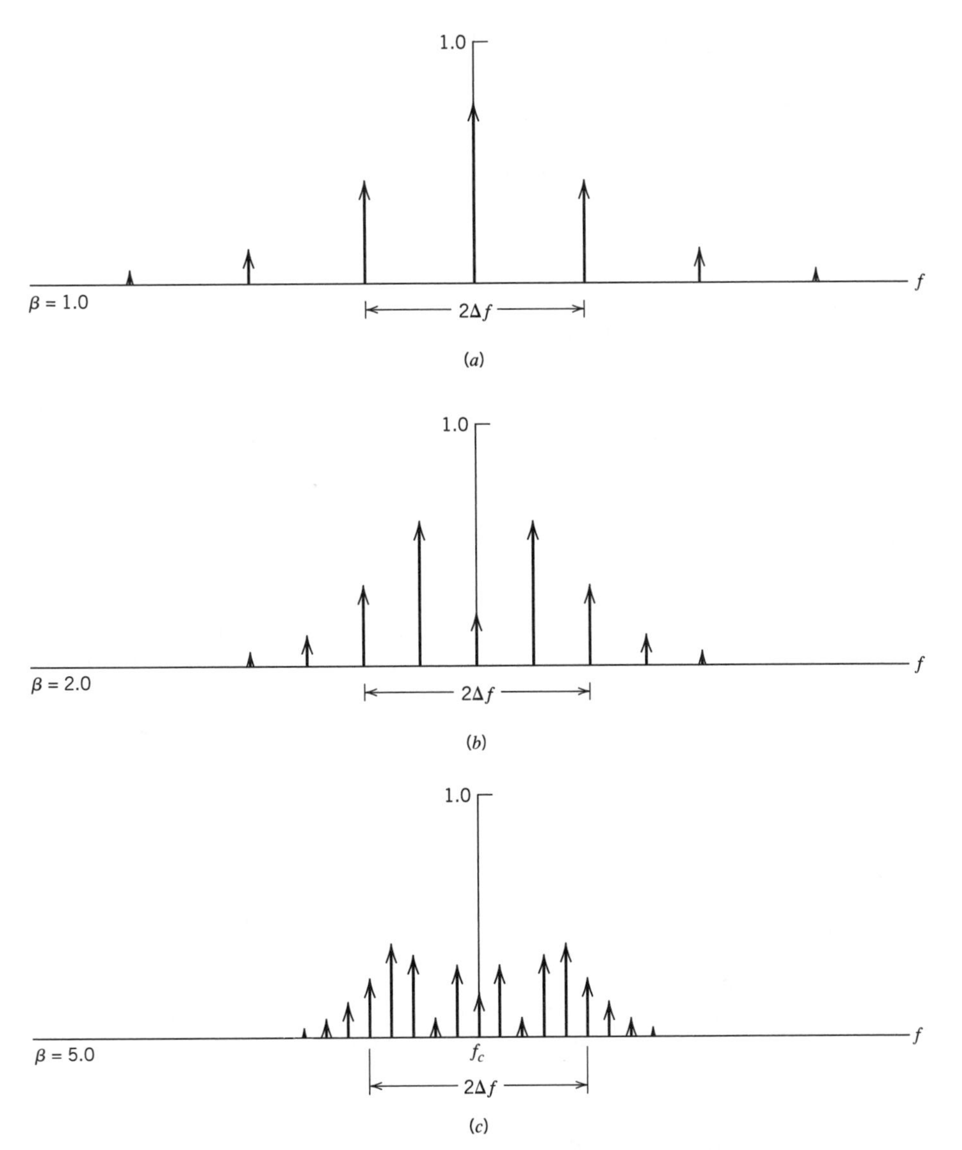


**FIGURE 2.25** Discrete amplitude spectra of an FM signal, normalized with respect to the carrier amplitude, for the case of sinusoidal modulation of varying frequency and fixed amplitude. Only the spectra for positive frequencies are shown.

approaches the limiting value of $2\Delta f$, which is an important point to keep in mind for later discussion. ◀

## ■ TRANSMISSION BANDWIDTH OF FM SIGNALS

In theory, an FM signal contains an infinite number of side frequencies so that the bandwidth required to transmit such a signal is similarly infinite in extent. In practice, however, we find that the FM signal is effectively limited to a finite number of significant side frequencies compatible with a specified amount of distortion. We may therefore specify an effective bandwidth required for the transmission of an FM signal. Consider first the case of an FM signal generated by a single-tone modulating wave of frequency $f_m$. In such an FM signal, the side frequencies that are separated from the carrier frequency $f_c$ by an amount greater than the frequency deviation $\Delta f$ decrease rapidly toward zero, so that the bandwidth always exceeds the total frequency excursion, but nevertheless is limited. Specifically, for large values of the modulation index $\beta$, the bandwidth approaches, and is only slightly greater than, the total frequency excursion $2\Delta f$ in accordance with the situation shown in Figure 2.25. On the other hand, for small values of the modulation index $\beta$, the spectrum of the FM signal is effectively limited to the carrier frequency $f_c$ and one pair of side frequencies at $f_c \pm f_m$, so that the bandwidth approaches $2f_m$. We may thus define an approximate rule for the transmission bandwidth of an FM signal generated by a single-tone modulating signal of frequency $f_m$ as follows:

$$B_T \simeq 2\Delta f + 2f_m = 2\Delta f\left(1 + \frac{1}{\beta}\right) \tag{2.55}$$

This empirical relation is known as *Carson's rule*.[4]

For an alternative assessment of the bandwidth requirement of an FM signal, we may use a definition based on retaining the maximum number of significant side frequencies whose amplitudes are all greater than some selected value. A convenient choice for this value is 1 percent of the unmodulated carrier amplitude. We may thus define *the transmission bandwidth of an FM wave as the separation between the two frequencies beyond which none of the side frequencies is greater than 1 percent of the carrier amplitude obtained when the modulation is removed.* That is, we define the transmission bandwidth as $2n_{\max}f_m$, where $f_m$ is the modulation frequency and $n_{\max}$ is the largest value of the integer $n$ that satisfies the requirement $|J_n(\beta)| > 0.01$. The value of $n_{\max}$ varies with the modulation index $\beta$ and can be determined readily from tabulated values of the Bessel function $J_n(\beta)$. Table 2.2 shows the total number of significant side frequencies (including both the upper and lower side frequencies) for different values of $\beta$, calculated on the 1 percent basis explained herein. The transmission bandwidth $B_T$ calculated using this procedure can be presented in the form of a *universal curve* by normalizing it with respect to the frequency deviation $\Delta f$ and then plotting it versus $\beta$. This curve is shown in Figure 2.26, which is drawn as a best fit through the set of points obtained by using Table 2.2. In Figure 2.26 we note that as the modulation index $\beta$ is increased, the bandwidth occupied

**TABLE 2.2** ***Number of significant side frequencies of a wideband FM signal for varying modulation index***

| *Modulation Index* $\beta$ | *Number of Significant Side Frequencies* $2n_{max}$ |
|---|---|
| 0.1 | 2 |
| 0.3 | 4 |
| 0.5 | 4 |
| 1.0 | 6 |
| 2.0 | 8 |
| 5.0 | 16 |
| 10.0 | 28 |
| 20.0 | 50 |
| 30.0 | 70 |

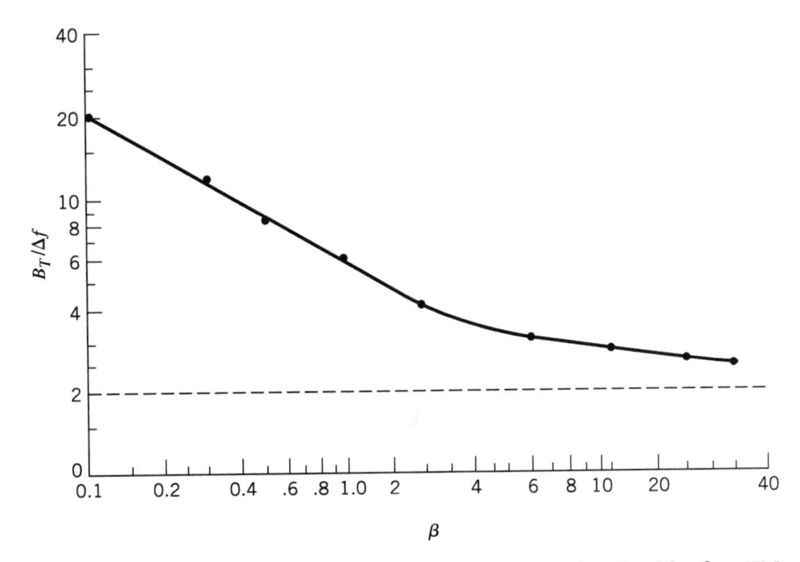


**FIGURE 2.26** Universal curve for evaluating the 1 percent bandwidth of an FM wave.

by the significant side frequencies drops toward that over which the carrier frequency actually deviates. This means that small values of the modulation index $\beta$ are relatively more extravagant in transmission bandwidth than are the larger values of $\beta$.

Consider next the more general case of an arbitrary modulating signal $m(t)$ with its highest frequency component denoted by $W$. The bandwidth required to transmit an FM signal generated by this modulating signal is estimated by using a worst-case tone-modulation analysis. Specifically, we first determine the so-called *deviation ratio* $D$, defined as the ratio of the frequency deviation $\Delta f$, which corresponds to the maximum possible amplitude of the modulation signal $m(t)$, to the highest modulation frequency $W$; these conditions represent the extreme cases possible. *The deviation ratio $D$ plays the same role for nonsinusoidal modulation that the modulation index $\beta$ plays for the case of sinusoidal modulation.* Then, replacing $\beta$ by $D$ and replacing $f_m$ with $W$, we may use Carson's rule given by Equation (2.55) or the universal curve of Figure 2.26 to obtain a value for the transmission bandwidth of the FM signal. From a practical viewpoint, Carson's rule somewhat underestimates the bandwidth requirement of an FM system, whereas using the universal curve of Figure 2.26 yields a somewhat conservative result. Thus, the choice of a transmission bandwidth that lies between the bounds provided by these two rules of thumb is acceptable for most practical purposes.

### ▶ EXAMPLE 2.3

In North America, the maximum value of frequency deviation $\Delta f$ is fixed at 75 kHz for commercial FM broadcasting by radio. If we take the modulation frequency $W = 15$ kHz, which is typically the "maximum" audio frequency of interest in FM transmission, we find that the corresponding value of the deviation ratio is

$$D = \frac{75}{15} = 5$$

Using Carson's rule of Equation (2.55), replacing $\beta$ by $D$, and replacing $f_m$ by $W$, the approximate value of the transmission bandwidth of the FM signal is obtained as

$$B_T = 2(75 + 15) = 180 \text{ kHz}$$

On the other hand, use of the curve of Figure 2.26 gives the transmission bandwidth of the FM signal to be

$$B_T = 3.2\Delta f = 3.2 \times 75 = 240 \text{ kHz}$$

In practice, a bandwidth of 200 kHz is allocated to each FM transmitter. On this basis, Carson's rule underestimates the transmission bandwidth by 10 percent, whereas the universal curve of Figure 2.26 overestimates it by 20 percent. ◀

## ■ GENERATION OF FM SIGNALS

There are essentially two basic methods of generating frequency-modulated signals, namely, *direct* FM and *indirect* FM. In the direct method the carrier frequency is directly varied in accordance with the input baseband signal, which is readily accomplished using a *voltage-controlled oscillator*. In the indirect method, the modulating signal is first used to produce a narrowband FM signal, and frequency multiplication is next used to increase the frequency deviation to the desired level. The indirect method is the preferred choice for frequency modulation when the stability of carrier frequency is of major concern as in commercial radio broadcasting, as described next.

### *Indirect FM*[5]

A simplified block diagram of an indirect FM system is shown in Figure 2.27. The message (baseband) signal $m(t)$ is first integrated and then used to phase-modulate a crystal-controlled oscillator; the use of crystal control provides *frequency stability*. To minimize the distortion inherent in the phase modulator, the maximum phase deviation or modulation index $\beta$ is kept small, thereby resulting in a narrowband FM signal; for the implementation of the narrow-band phase modulator, we may use the arrangement described in Figure 2.21. The narrowband FM signal is next multiplied in frequency by means of a frequency multiplier so as to produce the desired wideband FM signal.

A *frequency multiplier* consists of a nonlinear device followed by a band-pass filter, as shown in Figure 2.28. The implication of the nonlinear device being memoryless is that it has no energy-storage elements. The input-output relation of such a device may be expressed in the general form

$$v(t) = a_1 s(t) + a_2 s^2(t) + \cdots + a_n s^n(t) \tag{2.56}$$

where $a_1, a_2, \ldots, a_n$ are coefficients determined by the operating point of the device, and $n$ is the *highest order of nonlinearity*. In other words, the memoryless nonlinear device is an $n$th power-law device. The input $s(t)$ is an FM signal defined by

$$s(t) = A_c \cos\left[2\pi f_c t + 2\pi k_f \int_0^t m(\tau)\, d\tau\right]$$

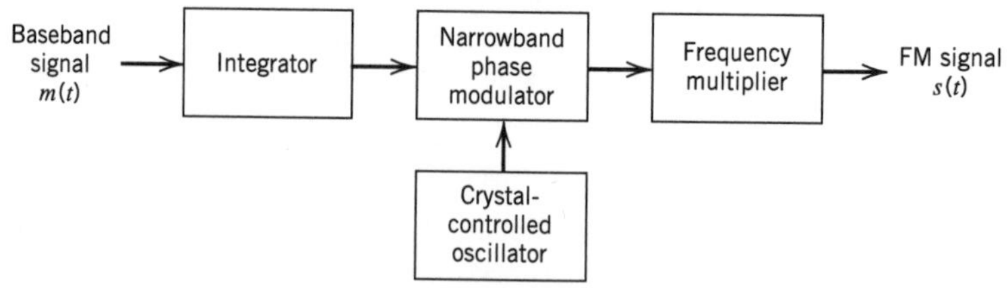

**FIGURE 2.27** Block diagram of the indirect method of generating a wideband FM signal.

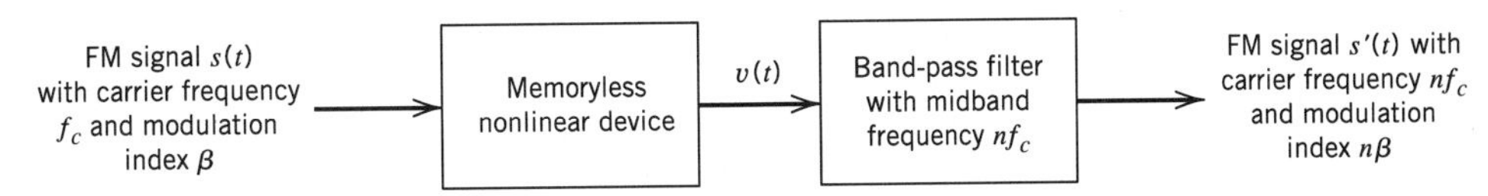


**FIGURE 2.28** Block diagram of frequency multiplier.

whose instantaneous frequency is

$$f_i(t) = f_c + k_f m(t) \tag{2.57}$$

The mid-band frequency of the band-pass filter in Figure 2.28 is set equal to $nf_c$, where $f_c$ is the carrier frequency of the incoming FM signal $s(t)$. Moreover, the band-pass filter is designed to have a bandwidth equal to $n$ times the transmission bandwidth of $s(t)$. In Section 2.8 dealing with nonlinear effects in FM systems, we describe the spectral contributions of such nonlinear terms as the second- and third-order terms in the input–output relation of Equation (2.56). For now it suffices to say that after band-pass filtering of the nonlinear device's output $v(t)$, we have a new FM signal defined by

$$s'(t) = A_c' \cos\left[2\pi n f_c t + 2\pi n k_f \int_0^t m(\tau)\, d\tau\right] \tag{2.58}$$

whose instantaneous frequency is

$$f_i'(t) = nf_c + nk_f m(t) \tag{2.59}$$

Thus, comparing Equation (2.59) with (2.57), we see that the nonlinear processing circuit of Figure 2.28 acts as a frequency multiplier. The frequency multiplication ratio is determined by the highest power $n$ in the input–output relation of Equation (2.56), characterizing the memoryless nonlinear device.

## DEMODULATION OF FM SIGNALS

*Frequency demodulation* is the process that enables us to recover the original modulating signal from a frequency-modulated signal. The objective is to produce a transfer characteristic that is the inverse of that of the frequency modulator, which can be realized directly or indirectly. Here we describe a direct method of frequency demodulation involving the use of a popular device known as a frequency discriminator, whose instantaneous output amplitude is directly proportional to the instantaneous frequency of the input FM signal. In Section 2.14, we describe an indirect method of frequency demodulation that uses another popular device known as a phase-locked loop.

Basically, the *frequency discriminator* consists of a *slope circuit* followed by an *envelope detector*. An ideal slope circuit is characterized by a frequency response that is purely imaginary, varying linearly with frequency inside a prescribed frequency interval. Consider the frequency response depicted in Figure 2.29*a*, which is defined by

$$H_1(f) = \begin{cases} j2\pi a\left(f - f_c + \dfrac{B_T}{2}\right), & f_c - \dfrac{B_T}{2} \le f \le f_c + \dfrac{B_T}{2} \\ j2\pi a\left(f + f_c - \dfrac{B_T}{2}\right), & -f_c - \dfrac{B_T}{2} \le f \le -f_c + \dfrac{B_T}{2} \\ 0, & \text{elsewhere} \end{cases} \tag{2.60}$$

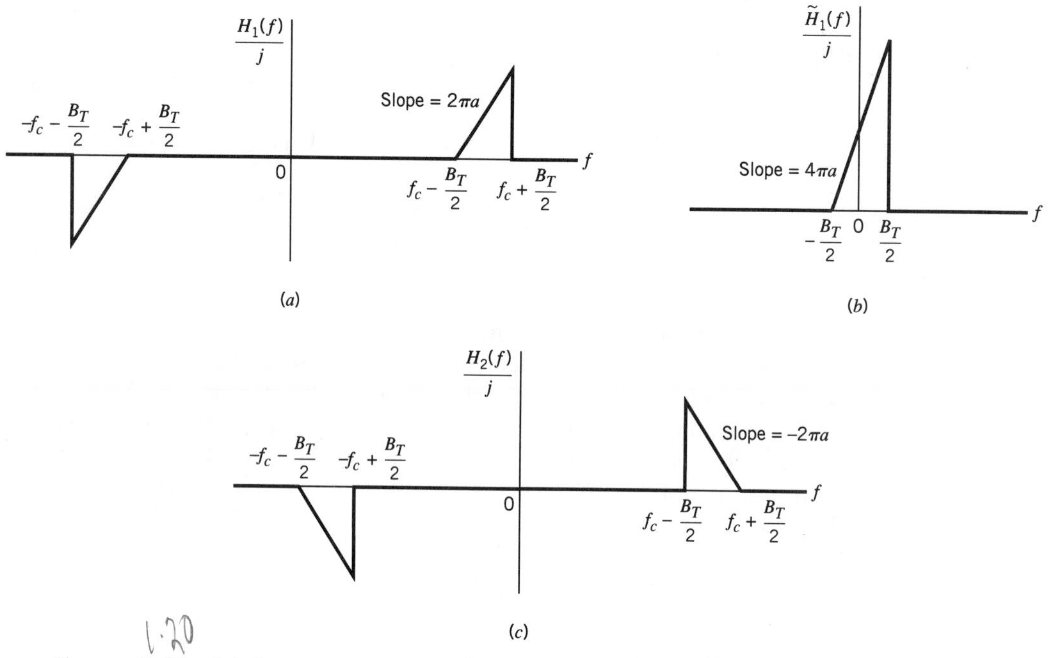


1.20

**FIGURE 2.29** (*a*) Frequency response of ideal slope circuit. (*b*) Frequency response of the slope circuit's complex low-pass equivalent. (*c*) Frequency response of the ideal slope circuit complementary to that of part (*a*).

where $a$ is a constant. We wish to evaluate the response of this slope circuit, denoted by $s_1(t)$, which is produced by an FM signal $s(t)$ of carrier frequency $f_c$ and transmission bandwidth $B_T$. It is assumed that the spectrum of $s(t)$ is essentially zero outside the frequency interval $f_c - B_T/2 \le |f| \le f_c + B_T/2$.

We may simplify the analysis of the frequency discriminator by invoking the isomorphism between a real-valued band-pass filter and a corresponding complex-valued low-pass filter. This isomorphism is discussed in Appendix 2. According to the material presented in that appendix, we may replace the band-pass filter with frequency response $H_1(f)$ with an equivalent low-pass filter with frequency response $\tilde{H}_1(f)$ by doing two things:

1. We shift $\tilde{H}_1(f)$ to the right by $f_c$, where $f_c$ is the midband frequency of the band-pass filter; this operation aligns the translated frequency response of the equivalent low-pass filter with that of the band-pass filter.
2. We set $\tilde{H}_1(f - f_c)$ equal to $2H_1(f)$ for $f > 0$.

Thus for the problem at hand we get

$$\tilde{H}_1(f - f_c) = 2H_1(f), \qquad f > 0 \tag{2.61}$$

Hence, using Equations (2.60) and (2.61), we get

$$\tilde{H}_1(f) = \begin{cases} j4\pi a\left(f + \dfrac{B_T}{2}\right), & -\dfrac{B_T}{2} \le f \le \dfrac{B_T}{2} \\ 0, & \text{elsewhere} \end{cases} \tag{2.62}$$

which is plotted in Figure 2.29*b*.

The incoming FM signal $s(t)$ is defined by Equation (2.26), which is reproduced here for convenience:

$$s(t) = A_c \cos\left[2\pi f_c t + 2\pi k_f \int_0^t m(\tau)\, d\tau\right]$$

Given that the carrier frequency $f_c$ is high compared to the transmission bandwidth of the FM signal $s(t)$, the complex envelope of $s(t)$ is

$$\tilde{s}(t) = A_c \exp\left[j2\pi k_f \int_0^t m(\tau)\, d\tau\right] \tag{2.63}$$

Let $\tilde{s}_1(t)$ denote the complex envelope of the response of the slope circuit defined by Figure 2.29*a* due to $\tilde{s}(t)$. Then, following the material presented in Appendix 2, we may express the Fourier transform of $\tilde{s}_1(t)$ as follows:

$$\begin{aligned} \tilde{S}_1(f) &= \frac{1}{2}\tilde{H}_1(f)\tilde{S}(f) \\ &= \begin{cases} j2\pi a\left(f + \dfrac{B_T}{2}\right)\tilde{S}(f), & -\dfrac{B_T}{2} \le f \le \dfrac{B_T}{2} \\ 0, & \text{elsewhere} \end{cases} \end{aligned} \tag{2.64}$$

where $\tilde{S}(f)$ is the Fourier transform of $\tilde{s}(t)$. From Fourier analysis we know that multiplication of the Fourier transform of a signal by $j2\pi f$ is equivalent to differentiating the signal in the time domain; see item 8 of Table A6.2. Hence, from Equation (2.64) we deduce

$$\tilde{s}_1(t) = a\left[\frac{d\tilde{s}(t)}{dt} + j\pi B_T \tilde{s}(t)\right] \tag{2.65}$$

Substituting Equation (2.63) into (2.65), we get

$$\tilde{s}_1(t) = j\pi B_T a A_c\left[1 + \frac{2k_f}{B_T}m(t)\right]\exp\left[j2\pi k_f \int_0^t m(\tau)\, d\tau\right] \tag{2.66}$$

The desired response of the slope circuit is therefore

$$\begin{aligned} s_1(t) &= \text{Re}[\tilde{s}_1(t)\exp(j2\pi f_c t)] \\ &= \pi B_T a A_c\left[1 + \frac{2k_f}{B_T}m(t)\right]\cos\left[2\pi f_c t + 2\pi k_f \int_0^t m(\tau)\, d\tau + \frac{\pi}{2}\right] \end{aligned} \tag{2.67}$$

The signal $s_1(t)$ is a hybrid-modulated signal in which both amplitude and frequency of the carrier wave vary with the message signal $m(t)$. However, provided that we choose

$$\left|\frac{2k_f}{B_T}m(t)\right| < 1 \qquad \text{for all } t$$

then we may use an envelope detector to recover the amplitude variations and thus, except for a bias term, obtain the original message signal. The resulting envelope-detector output is therefore

$$|\tilde{s}_1(t)| = \pi B_T a A_c\left[1 + \frac{2k_f}{B_T}m(t)\right] \tag{2.68}$$

The bias term $\pi B_T a A_c$ in the right-hand side of Equation (2.68) is proportional to the slope $a$ of the transfer function of the slope circuit. This suggests that the bias may be

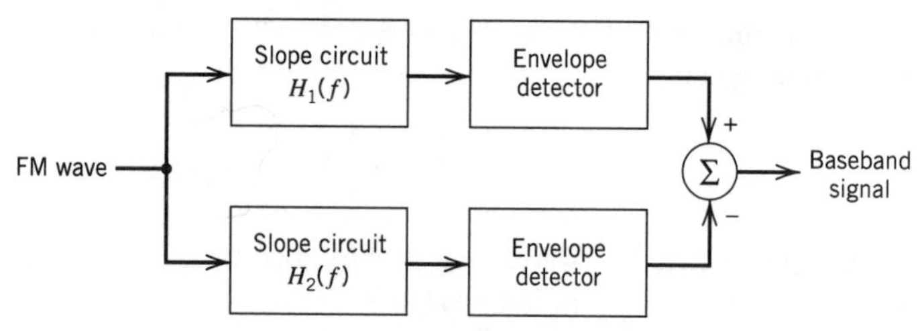


FIGURE 2.30 Block diagram of frequency discriminator.

removed by subtracting from the envelope-detector output $|\tilde{s}_1(t)|$ the output of a second envelope detector preceded by the *complementary slope circuit* with a frequency response $H_2(f)$ as described in Figure 2.29*c*. That is, the two slope circuits are related by

$$\tilde{H}_2(f) = \tilde{H}_1(-f) \tag{2.69}$$

Let $s_2(t)$ denote the response of the complementary slope circuit produced by the incoming FM signal $s(t)$. Then, following a procedure similar to that just described, we may write

$$|\tilde{s}_2(t)| = \pi B_T a A_c \left[1 - \frac{2k_f}{B_T} m(t)\right] \tag{2.70}$$

where $\tilde{s}_2(t)$ is the complex envelope of the signal $s_2(t)$. The difference between the two envelopes in Equations (2.68) and (2.70) is

$$\begin{aligned} s_o(t) &= |\tilde{s}_1(t)| - |\tilde{s}_2(t)| \\ &= 4\pi k_f a A_c m(t) \end{aligned} \tag{2.71}$$

which is a scaled version of the original message signal $m(t)$ and free from bias.

We may thus model the *ideal frequency discriminator* as a pair of slope circuits with their complex transfer functions related by Equation (2.69), followed by envelope detectors and finally a summer, as in Figure 2.30. This scheme is called a *balanced frequency discriminator*.

## ■ FM STEREO MULTIPLEXING[6]

*Stereo multiplexing* is a form of frequency-division multiplexing (FDM) designed to transmit two separate signals via the same carrier. It is widely used in FM radio broadcasting to send two different elements of a program (e.g., two different sections of an orchestra, a vocalist and an accompanist) so as to give a spatial dimension to its perception by a listener at the receiving end.

The specification of standards for FM stereo transmission is influenced by two factors:

1. The transmission has to operate within the allocated FM broadcast channels.
2. It has to be compatible with monophonic radio receivers.

The first requirement sets the permissible frequency parameters, including frequency deviation. The second requirement constrains the way in which the transmitted signal is configured.

Figure 2.31*a* shows the block diagram of the multiplexing system used in an FM stereo transmitter. Let $m_l(t)$ and $m_r(t)$ denote the signals picked up by left-hand and

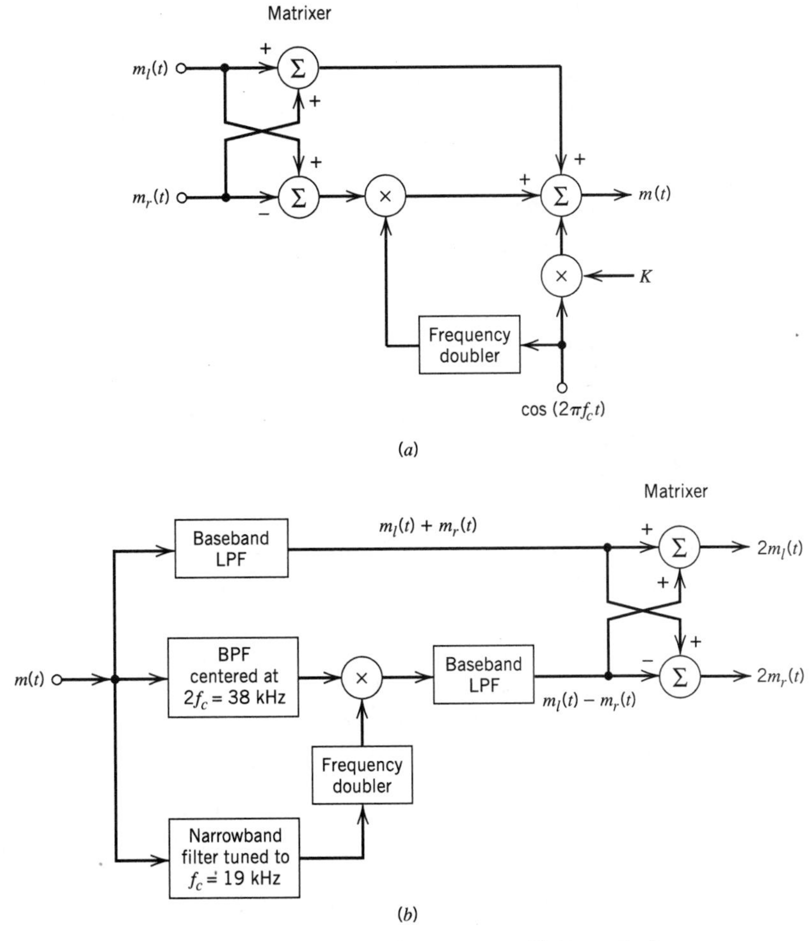


**FIGURE 2.31** (*a*) Multiplexer in transmitter of FM stereo. (*b*) Demultiplexer in receiver of FM stereo.

right-hand microphones at the transmitting end of the system. They are applied to a simple *matrixer* that generates the *sum signal*, $m_l(t) + m_r(t)$, and the *difference signal*, $m_l(t) - m_r(t)$. The sum signal is left unprocessed in its baseband form; it is available for monophonic reception. The difference signal and a 38-kHz subcarrier (derived from a 19-kHz crystal oscillator by frequency doubling) are applied to a product modulator, thereby producing a DSB-SC modulated wave. In addition to the sum signal and this DSB-SC modulated wave, the multiplexed signal $m(t)$ also includes a 19-kHz pilot to provide a reference for the coherent detection of the difference signal at the stereo receiver. Thus the multiplexed signal is described by

$$m(t) = [m_l(t) + m_r(t)] + [m_l - m_r(t)] \cos(4\pi f_c t) + K \cos(2\pi f_c t) \qquad (2.72)$$

where $f_c = 19$ kHz, and $K$ is the amplitude of the pilot tone. The multiplexed signal $m(t)$ then frequency-modulates the main carrier to produce the transmitted signal. The pilot is allotted between 8 and 10 percent of the peak frequency deviation; the amplitude $K$ in Equation (2.72) is chosen to satisfy this requirement.

At a stereo receiver, the multiplexed signal $m(t)$ is recovered by frequency demodulating the incoming FM wave. Then $m(t)$ is applied to the *demultiplexing system* shown in Figure 2.31*b*. The individual components of the multiplexed signal $m(t)$ are separated by the use of three appropriate filters. The recovered pilot (using a narrowband filter tuned to 19 kHz) is frequency doubled to produce the desired 38-kHz subcarrier. The availability of this subcarrier enables the coherent detection of the DSB-SC modulated wave, thereby recovering the difference signal, $m_l(t) - m_r(t)$. The baseband low-pass filter in the top path of Figure 2.31*b* is designed to pass the sum signal, $m_l(t) + m_r(t)$. Finally, the simple matrixer reconstructs the left-hand signal $m_l(t)$ and right-hand signal $m_r(t)$, except for scaling factors, and applies them to their respective speakers.

## 2.8 Nonlinear Effects in FM Systems

In the preceding two sections, we studied frequency modulation theory and methods for its generation and demodulation. We complete the discussion of frequency modulation by considering nonlinear effects in FM systems.

*Nonlinearities*, in one form or another, are present in all electrical networks. There are two basic forms of nonlinearity to consider:

1. The nonlinearity is said to be *strong* when it is introduced intentionally and in a controlled manner for some specific application. Examples of strong nonlinearity include square-law modulators, hard-limiters, and frequency multipliers.
2. The nonlinearity is said to be *weak* when a linear performance is desired, but non-linearities of a parasitic nature arise due to imperfections. The effect of such weak nonlinearities is to limit the useful signal levels in a system and thereby become an important design consideration.

In this section we examine the effects of weak nonlinearities on frequency modulation.

Consider a communications channel, the transfer characteristic of which is defined by the nonlinear input–output relation

$$v_o(t) = a_1 v_i(t) + a_2 v_i^2(t) + a_3 v_i^3(t) \tag{2.73}$$

where $v_i(t)$ and $v_o(t)$ are the input and output signals, respectively, and $a_1$, $a_2$, and $a_3$ are constants; Equation (2.73) is a truncated version of Equation (2.56) used in the context of frequency multiplication. The channel described in Equation (2.73) is said to be *memoryless* in that the output signal $v_o(t)$ is an instantaneous function of the input signal $v_i(t)$ (i.e., there is no energy storage involved in the description). We wish to determine the effect of transmitting a frequency-modulated wave through such a channel. The FM signal is defined by

$$v_i(t) = A_c \cos[2\pi f_c t + \phi(t)]$$

where

$$\phi(t) = 2\pi k_f \int_0^t m(\tau)\, d\tau$$

For this input signal, the use of Equation (2.73) yields

$$\begin{aligned} v_o(t) = {} & a_1 A_c \cos[2\pi f_c t + \phi(t)] + a_2 A_c^2 \cos^2[2\pi f_c t + \phi(t)] \\ & + a_3 A_c^3 \cos^3[2\pi f_c t + \phi(t)] \end{aligned} \tag{2.74}$$

Expanding the squared and cubed cosine terms in Equation (2.74) and then collecting common terms, we get

$$\begin{aligned} v_o(t) &= \frac{1}{2} a_2 A_c^2 + \left( a_1 A_c + \frac{3}{4} a_3 A_c^3 \right) \cos[2\pi f_c t + \phi(t)] \\ &+ \frac{1}{2} a_2 A_c^2 \cos[4\pi f_c t + 2\phi(t)] \\ &+ \frac{1}{4} a_3 A_c^3 \cos[6\pi f_c t + 3\phi(t)] \end{aligned} \tag{2.75}$$

Thus the channel output consists of a DC component and three frequency-modulated signals with carrier frequencies $f_c$, $2f_c$, and $3f_c$; the latter components are contributed by the linear, second-order, and third-order terms of Equation (2.73), respectively.

To extract the desired FM signal from the channel output $v_o(t)$, that is, the particular component with carrier frequency, $f_c$, it is necessary to separate the FM signal with this carrier frequency from the one with the closest carrier frequency, $2f_c$. Let $\Delta f$ denote the frequency deviation of the incoming FM signal $v_i(t)$, and $W$ denote the highest frequency component of the message signal $m(t)$. Then, applying Carson's rule and noting that the frequency deviation about the second harmonic of the carrier frequency is doubled, we find that the necessary condition for separating the desired FM signal with the carrier frequency $f_c$ from that with the carrier frequency $2f_c$ is

$$2f_c - (2\Delta f + W) > f_c + \Delta f + W$$

or

$$f_c > 3\Delta f + 2W \tag{2.76}$$

Thus, by using a band-pass filter of midband frequency $f_c$ and bandwidth $2\Delta f + 2W$, the channel output is reduced to

$$v_o'(t) = \left( a_1 A_c + \frac{3}{4} a_3 A_c^3 \right) \cos[2\pi f_c t + \phi(t)] \tag{2.77}$$

We see therefore that the only effect of passing an FM signal through a channel with amplitude nonlinearities, followed by appropriate filtering, is simply to modify its amplitude. That is, unlike amplitude modulation, frequency modulation is not affected by distortion produced by transmission through a channel with amplitude nonlinearities. It is for this reason that we find frequency modulation used in microwave radio systems: It permits the use of highly nonlinear amplifiers and power transmitters, which are particularly important to producing a maximum power output at radio frequencies.

An FM system is extremely sensitive to *phase nonlinearities*, however, as we would intuitively expect. A common type of phase nonlinearity that is encountered in microwave radio systems is known as *AM-to-PM conversion*. This is the result of the phase characteristic of repeaters or amplifiers used in the system being dependent on the instantaneous amplitude of the input signal. In practice, AM-to-PM conversion is characterized by a constant $K$, which is measured in degrees per dB and may be interpreted as the peak phase change at the output for a 1-dB change in envelope at the input. When an FM wave is transmitted through a microwave radio link, it picks up spurious amplitude variations due to noise and interference during the course of transmission, and when such an FM wave is passed through a repeater with AM-to-PM conversion, the output will contain unwanted phase modulation and resultant distortion. It is therefore important to keep the AM-to-PM conversion at a low level. For example, for a good microwave repeater, the AM-to-PM conversion constant $K$ is less than 2 degrees per dB.

## 2.9 *Superheterodyne Receiver*[7]

In a *broadcasting* system, irrespective of whether it is based on amplitude modulation or frequency modulation, the receiver not only has the task of demodulating the incoming modulated signal, but it is also required to perform some other system functions:

- *Carrier-frequency tuning*, the purpose of which is to select the desired signal (i.e., desired radio or TV station).
- *Filtering*, which is required to separate the desired signal from other modulated signals that may be picked up along the way.
- *Amplification*, which is intended to compensate for the loss of signal power incurred in the course of transmission.

The *superheterodyne receiver*, or *superhet* as it is often referred to, is a special type of receiver that fulfills all three functions, particularly the first two, in an elegant and practical fashion. Specifically, it overcomes the difficulty of having to build a tunable highly selective and variable filter. Indeed, practically all radio and TV receivers now being made are of the superheterodyne type.

Basically, the receiver consists of a radio-frequency (RF) section, a mixer and local oscillator, an intermediate-frequency (IF) section, demodulator, and power amplifier. Typical frequency parameters of commercial AM and FM radio receivers are listed in Table 2.3. Figure 2.32 shows the block diagram of a superheterodyne receiver for amplitude modulation using an envelope detector for demodulation.

The incoming amplitude-modulated wave is picked up by the receiving antenna and amplified in the RF section that is tuned to the carrier frequency of the incoming wave. The combination of mixer and local oscillator (of adjustable frequency) provides a *heterodyning* function, whereby the incoming signal is converted to a predetermined fixed *intermediate frequency*, usually lower than the incoming carrier frequency. This frequency translation is achieved without disturbing the relation of the sidebands to the carrier; see Section 2.4. The result of the heterodyning is to produce an intermediate-frequency carrier defined by

$$f_{\mathrm{IF}} = f_{\mathrm{LO}} - f_{\mathrm{RF}} \tag{2.78}$$

where $f_{\mathrm{LO}}$ is the frequency of the local oscillator and $f_{\mathrm{RF}}$ is the carrier frequency of the incoming RF signal. We refer to $f_{\mathrm{IF}}$ as the intermediate frequency (IF), because the signal

**TABLE 2.3** *Typical frequency parameters of AM and FM radio receivers*

| | *AM Radio* | *FM Radio* |
|---|---|---|
| RF carrier range | 0.535–1.605 MHz | 88–108 MHz |
| Midband frequency of IF section | 0.455 MHz | 10.7 MHz |
| IF bandwidth | 10 kHz | 200 kHz |

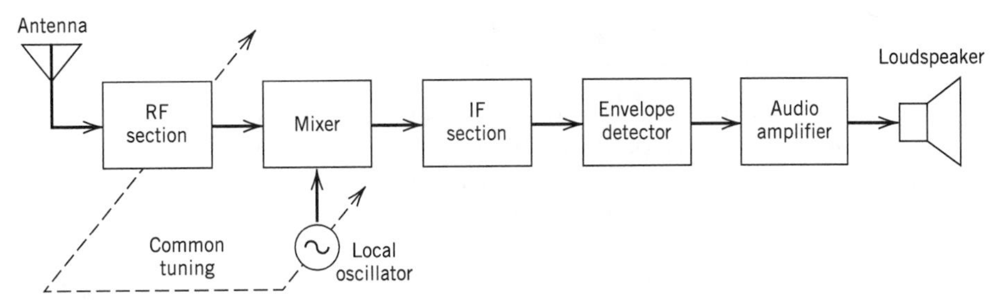


**FIGURE 2.32** Basic elements of an AM radio receiver of the superheterodyne type.

is neither at the original input frequency nor at the final baseband frequency. The mixer–local oscillator combination is sometimes referred to as the *first detector*, in which case the demodulator is called the *second detector*.

The IF section consists of one or more stages of tuned amplification, with a bandwidth corresponding to that required for the particular type of modulation that the receiver is intended to handle. The IF section provides most of the amplification and selectivity in the receiver. The output of the IF section is applied to a demodulator, the purpose of which is to recover the baseband signal. If coherent detection is used, then a coherent signal source must be provided in the receiver. The final operation in the receiver is the power amplification of the recovered message signal.

In a superheterodyne receiver the mixer will develop an intermediate frequency output when the input signal frequency is greater or less than the local oscillator frequency by an amount equal to the intermediate frequency. That is, there are two input frequencies, namely, $|f_{LO} \pm f_{IF}|$, which will result in $f_{IF}$ at the mixer output. This introduces the possibility of simultaneous reception of two signals differing in frequency by twice the intermediate frequency. For example, a receiver tuned to 0.65 MHz and having an IF of 0.455 MHz is subject to an *image interference* at 1.56 MHz; indeed, any receiver with this value of IF, when tuned to any station, is subject to image interference at a frequency of 0.910 MHz higher than the desired station. Since the function of the mixer is to produce the difference between two applied frequencies, it is incapable of distinguishing between the desired signal and its image in that it produces an IF output from either one of them. The only practical cure for image interference is to employ highly selective stages in the RF section (i.e., between the antenna and the mixer) in order to favor the desired signal and discriminate against the undesired or *image signal.* The effectiveness of suppressing unwanted image signals increases as the number of selective stages in the RF section increases, and as the ratio of intermediate to signal frequency increases.

The basic difference between AM and FM superheterodyne receivers lies in the use of an FM demodulator such as limiter-frequency discriminator. In an FM system, the message information is transmitted by variations of the instantaneous frequency of a sinusoidal carrier wave, and its amplitude is maintained constant. Therefore, any variations of the carrier amplitude at the receiver input must result from noise or interference. An *amplitude limiter*, following the IF section, is used to remove amplitude variations by clipping the modulated wave at the IF section output almost to the zero axis. The resulting rectangular wave is rounded off by a band-pass filter that suppresses harmonics of the carrier frequency. Thus the filter output is again sinusoidal, with an amplitude that is practically independent of the carrier amplitude at the receiver input (see Problem 2.42).

## 2.10 *Noise in CW Modulation Systems*

Up to this point in our discussion we have focused attention on the characterization of continuous-wave (CW) modulation techniques, entirely from a deterministic perspective. In the remainder of the chapter, we study the effects of *channel noise* on the reception of CW modulated signals and thereby develop a deeper understanding of the behavior of analog communications.

To undertake such a study we follow the customary practice by formulating two models:

1. *Channel model*, which assumes a communication channel that is distortionless but perturbed by *additive white Gaussian noise* (AWGN).
2. *Receiver model*, which assumes a receiver consisting of an ideal band-pass filter followed by an ideal demodulator appropriate for the application at hand; the band-pass filter is used to minimize the effect of channel noise.

These simplifying assumptions are made in order to obtain a basic understanding of the way in which noise affects the performance of the receiver. Moreover, they provide a framework for the comparison of different CW modulation-demodulation schemes.

Figure 2.33 shows the *noisy receiver model* that combines the above two assumptions. In this figure, $s(t)$ denotes the incoming modulated signal and $w(t)$ denotes the channel noise. The *received signal* is therefore made up of the sum of $s(t)$ and $w(t)$; this is the signal that the receiver has to work on. The *band-pass filter* in the model of Figure 2.33 represents the combined filtering action of the tuned amplifiers used in the actual receiver for the purpose of signal amplification prior to demodulation. The bandwidth of this band-pass filter is just wide enough to pass the modulated signal $s(t)$ without distortion. As for the *demodulator* in the model of Figure 2.33, its details naturally depend on the type of modulation used.

### ■ SIGNAL-TO-NOISE RATIOS: BASIC DEFINITIONS

Let the power spectral density of the noise $w(t)$ be denoted by $N_0/2$, defined for both positive and negative frequencies; that is, $N_0$ *is the average noise power per unit bandwidth measured at the front end of the receiver*. We also assume that the band-pass filter in the receiver model of Figure 2.33 is ideal, having a bandwidth equal to the transmission bandwidth $B_T$ of the modulated signal $s(t)$ and a midband frequency equal to the carrier frequency $f_c$. The latter assumption is justified for double sideband–suppressed carrier (DSB-SC) modulation, full amplitude modulation (AM), and frequency modulation (FM); the cases of single sideband (SSB) modulation and vestigial sideband (VSB) modulation require special considerations. Taking the midband frequency of the band-pass filter to be the same as the carrier frequency $f_c$, we may model the power spectral density $S_N(f)$ of the noise $n(t)$, resulting from the passage of the white noise $w(t)$ through the filter, as shown

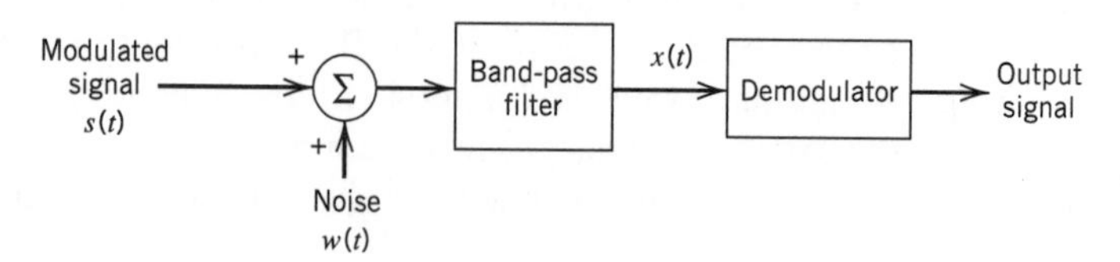


**FIGURE 2.33** Receiver model.

in Figure 2.34. Typically, the carrier frequency $f_c$ is large compared to the transmission bandwidth $B_T$. We may therefore treat the *filtered noise* $n(t)$ as a narrowband noise represented in the canonical form

$$n(t) = n_I(t)\cos(2\pi f_c t) - n_Q(t)\sin(2\pi f_c t) \tag{2.79}$$

where $n_I(t)$ is the *in-phase noise component* and $n_Q(t)$ is the *quadrature noise component*, both measured with respect to the carrier wave $A_c \cos(2\pi f_c t)$. The filtered signal $x(t)$ available for demodulation is defined by

$$x(t) = s(t) + n(t) \tag{2.80}$$

The details of $s(t)$ depend on the type of modulation used. In any event, the average noise power at the demodulator input is equal to the total area under the curve of the power spectral density $S_N(f)$. From Figure 2.34 we readily see that this average noise power is equal to $N_0 B_T$. Given the format of $s(t)$, we may also determine the average signal power at the demodulator input. With the demodulated signal $s(t)$ and the filtered noise ($n(t)$ appearing additively at the demodulator input in accordance with Equation (2.80), we may go on to define an *input signal-to-noise ratio*, $(\text{SNR})_I$, *as the ratio of the average power of the modulated signal* $s(t)$ *to the average power of the filtered noise* $n(t)$.

A more useful measure of noise performance, however, is the *output signal-to-noise ratio*, $(\text{SNR})_O$, *defined as the ratio of the average power of the demodulated message signal to the average power of the noise, both measured at the receiver output.* The output signal-to-noise ratio provides an intuitive measure for describing the fidelity with which the demodulation process in the receiver recovers the message signal from the modulated signal in the presence of additive noise. For such a criterion to be well defined, the recovered message signal and the corruptive noise component must appear *additively* at the demodulator output. This condition is perfectly valid in the case of a receiver using coherent detection. On the other hand, when the receiver uses envelope detection as in full AM or frequency discrimination as in FM, we have to assume that the average power of the filtered noise $n(t)$ is relatively low to justify the use of output signal-to-noise ratio as a measure of receiver performance.

The output signal-to-noise ratio depends, among other factors, on the type of modulation used in the transmitter and the type of demodulation used in the receiver. Thus it is informative to compare the output signal-to-noise ratios for different modulation-demodulation systems. However, for this comparison to be of meaningful value, it must be made on an equal basis as described here:

- The modulated signal $s(t)$ transmitted by each system has the same average power.
- The channel noise $w(t)$ has the same average power measured in the message bandwidth $W$.

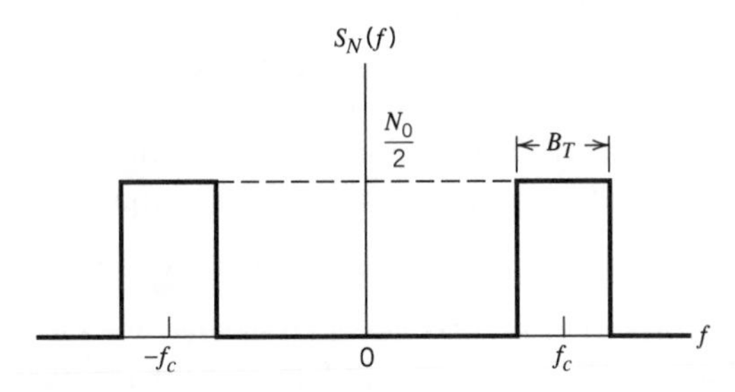


**FIGURE 2.34** Idealized characteristic of band-pass filtered noise.

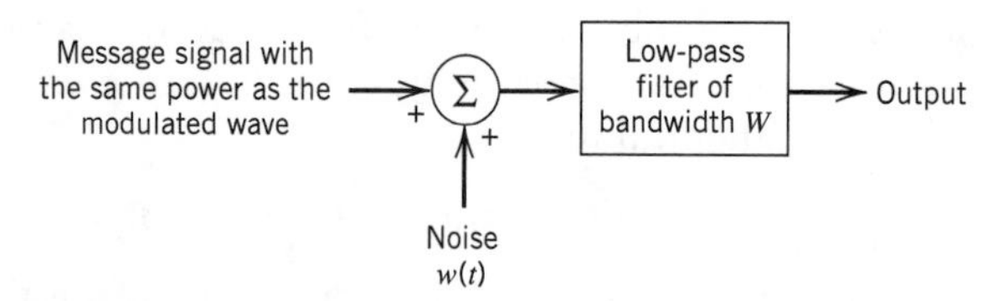


**FIGURE 2.35** The baseband transmission model, assuming a message signal of bandwidth $W$, used for calculating the channel signal-to-noise ratio.

Accordingly, as a frame of reference we define the *channel signal-to-noise ratio*, $(\text{SNR})_C$, as *the ratio of the average power of the modulated signal to the average power of channel noise in the message bandwidth, both measured at the receiver input*. This definition is illustrated in Figure 2.35.

For the purpose of comparing different continuous-wave (CW) modulation systems, we *normalize* the receiver performance by dividing the output signal-to-noise ratio by the channel signal-to-noise ratio. We thus define a *figure of merit* for the receiver as follows:

$$\text{Figure of merit} = \frac{(\text{SNR})_O}{(\text{SNR})_C} \tag{2.81}$$

Clearly, the higher the value of the figure of merit, the better will the noise performance of the receiver be. The figure of merit may equal one, be less than one, or be greater than one, depending on the type of modulation used, which will become apparent from the discussion that follows.

## 2.11 *Noise in Linear Receivers Using Coherent Detection*

From Sections 2.2 and 2.3 we recall that the demodulation of an amplitude-modulated wave depends on whether the carrier is suppressed or not. When the carrier is suppressed we usually require the use of coherent detection, in which case the receiver is *linear*. On the other hand, when the amplitude modulation includes transmission of the carrier, demodulation is accomplished simply by using an envelope detector, in which case the receiver is *nonlinear*. In this section we study the effect of noise on the performance of a linear receiver. The more difficult case of a nonlinear receiver is deferred to Section 2.12.

Consider the case of DSB-SC modulation Figure 2.36 shows the model of a DSB-SC receiver using a coherent detector. The use of coherent detection requires multiplication of the filtered signal $x(t)$ by a locally generated sinusoidal wave $\cos(2\pi f_c t)$ and then low-pass filtering the product. To simplify the analysis, we assume that the amplitude of the locally generated sinusoidal wave is unity. For this demodulation scheme to operate satisfactorily, however, it is necessary that the local oscillator be synchronized both in phase and in frequency with the oscillator generating the carrier wave in the transmitter. We assume that this synchronization has been achieved.

The DSB-SC component of the filtered signal $x(t)$ is expressed as

$$s(t) = CA_c \cos(2\pi f_c t)m(t) \tag{2.82}$$

where $A_c \cos(2\pi f_c t)$ is the sinusoidal carrier wave and $m(t)$ is the message signal. In the expression for $s(t)$ in Equation (2.82) we have included a *system-dependent scaling factor* $C$, the purpose of which is to ensure that the signal component $s(t)$ is measured in the same units as the additive noise component $n(t)$. We assume that $m(t)$ is the sample function of

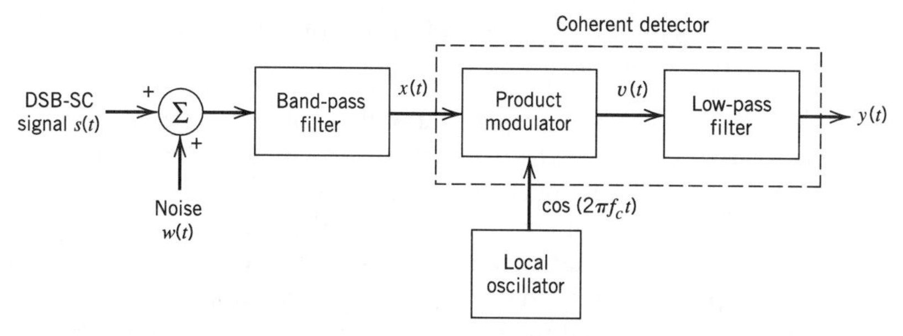


**FIGURE 2.36** Model of DSB-SC receiver using coherent detection.

a stationary process of zero mean, whose power spectral density $S_M(f)$ is limited to a maximum frequency $W$; that is, $W$ is the *message bandwidth*. The average power $P$ of the message signal is the total area under the curve of power spectral density, as shown by

$$P = \int_{-W}^{W} S_M(f)\, df \tag{2.83}$$

The carrier wave is statistically independent of the message signal. To emphasize this independence, the carrier should include a random phase that is uniformly distributed over $2\pi$ radians. In the defining equation for $s(t)$ this random phase angle has been omitted for convenience of presentation. Using the result of Example 1.7 of Chapter 1 on a modulated random process, we may express the average power of the DSB-SC modulated signal component $s(t)$ as $C^2A_c^2P/2$. With a noise spectral density of $N_0/2$, the average noise power in the message bandwidth $W$ is equal to $WN_0$. The channel signal-to-noise ratio of the DSB-SC modulation system is therefore

$$(\text{SNR})_{\text{C,DSB}} = \frac{C^2A_c^2P}{2WN_0} \tag{2.84}$$

where the constant $C^2$ in the numerator ensures that this ratio is dimensionless.

Next, we wish to determine the output signal-to-noise ratio of the system. Using the narrowband representation of the filtered noise $n(t)$, the total signal at the coherent detector input may be expressed as

$$\begin{aligned} x(t) &= s(t) + n(t) \\ &= CA_c \cos(2\pi f_c t)m(t) + n_I(t)\cos(2\pi f_c t) - n_Q(t)\sin(2\pi f_c t) \end{aligned} \tag{2.85}$$

where $n_I(t)$ and $n_Q(t)$ are the in-phase and quadrature components of $n(t)$ with respect to the carrier. The output of the product–modulator component of the coherent detector is therefore

$$\begin{aligned} v(t) &= x(t)\cos(2\pi f_c t) \\ &= \tfrac{1}{2}CA_c m(t) + \tfrac{1}{2}n_I(t) \\ &\quad + \tfrac{1}{2}[CA_c m(t) + n_I(t)]\cos(4\pi f_c t) - \tfrac{1}{2}n_Q(t)\sin(4\pi f_c t) \end{aligned}$$

The low-pass filter in the coherent detector in Figure 2.36 removes the high-frequency components of $v(t)$, yielding the receiver output

$$y(t) = \tfrac{1}{2}CA_c m(t) + \tfrac{1}{2}n_I(t) \tag{2.86}$$

Equation (2.86) indicates the following:

1. The message signal $m(t)$ and in-phase noise component $n_I(t)$ of the filtered noise $n(t)$ appear additively at the receiver output.

2. The quadrature component $n_Q(t)$ of the noise $n(t)$ is completely rejected by the coherent detector.

These two results are independent of the input signal-to-noise ratio. Thus, coherent detection distinguishes itself from other demodulation techniques in an important property: The output message component is unmutilated and the noise component always appears additively with the message, irrespective of the input signal-to-noise ratio.

The message signal component at the receiver output is $CA_c m(t)/2$. Therefore, the average power of this component may be expressed as $C^2 A_c^2 P/4$, where $P$ is the average power of the original message signal $m(t)$ and $C$ is the system-dependent scaling factor referred to earlier.

In the case of DSB-SC modulation, the band-pass filter in Figure 2.36 has a bandwidth $B_T$ equal to $2W$ in order to accommodate the upper and lower sidebands of the modulated signal $s(t)$. It follows therefore that the average power of the filtered noise $n(t)$ is $2WN_0$. From the discussion of narrowband noise presented in Section 1.11, we know that the average power of the (low-pass) in-phase noise component $n_I(t)$ is the same as that of the (band-pass) filtered noise $n(t)$. Since from Equation (2.86) the noise component at the receiver output is $n_I(t)/2$, it follows that the average power of the noise at the receiver output is

$$(\tfrac{1}{2})^2 2WN_0 = \tfrac{1}{2}WN_0$$

The output signal-to-noise for a DSB-SC receiver using coherent detection is therefore

$$\begin{aligned}(\text{SNR})_{\text{O, DSB-SC}} &= \frac{C^2 A_c^2 P/4}{WN_0/2} \\ &= \frac{C^2 A_c^2 P}{2WN_0}\end{aligned} \tag{2.87}$$

Using Equations (2.84) and (2.87), we obtain the figure of merit

$$\left.\frac{(\text{SNR})_\text{O}}{(\text{SNR})_\text{C}}\right|_{\text{DSB-SC}} = 1 \tag{2.88}$$

Note that the factor $C^2$ is common to both the output and channel signal-to-noise ratios, and therefore cancels out in evaluating the figure of merit.

Following through the noise analysis of a coherent detector for SSB, we find that, despite the fundamental differences between it and the coherent detector for DSB-SC modulation, the figure of merit is exactly the same for both of them; see Problem 2.49.

The important conclusions to be drawn from the discussions presented in this section and Problem 2.49 are two-fold:

1. For the same average transmitted or modulated signal power and the same average noise power in the message bandwidth, a coherent SSB receiver will have exactly the same output signal-to-noise ratio as a coherent DSB-SC receiver.
2. In both cases, the noise performance of the receiver is exactly the same as that obtained by simply transmitting the message signal in the presence of the same channel noise. The only effect of the modulation process is to translate the message signal to a different frequency band to facilitate its transmission over a band-pass channel.

Simply put, neither DSB-SC modulation nor SSB modulation offers the means for a trade-off between improved noise performance and increased channel bandwidth. This is a serious problem when high quality of reception is a requirement.

## 2.12 Noise in AM Receivers Using Envelope Detection

The next noise analysis we perform is for an amplitude modulation (AM) system using an envelope detector in the receiver, as shown in the model of Figure 2.37. In a full AM signal, both sidebands and the carrier wave are transmitted, as shown by

$$s(t) = A_c[1 + k_a m(t)] \cos(2\pi f_c t) \tag{2.89}$$

where $A_c \cos(2\pi f_c t)$ is the carrier wave, $m(t)$ is the message signal, and $k_a$ is a constant that determines the percentage modulation. In the expression for the amplitude-modulated signal component $s(t)$ given in Equation (2.89), we see no need for the use of a scaling factor, because it is reasonable to assume that the carrier amplitude $A_c$ has the same units as the additive noise component.

The average power of the carrier component in the AM signal $s(t)$ is $A_c^2/2$. The average power of the information-bearing component $A_c k_a m(t) \cos(2\pi f_c t)$ is $A_c^2 k_a^2 P/2$, where $P$ is the average power of the message signal $m(t)$. The average power of the full AM signal $s(t)$ is therefore equal to $A_c^2(1 + k_a^2 P)/2$. As for the DSB-SC system, the average power of noise in the message bandwidth is $WN_0$. The channel signal-to-noise ratio for AM is therefore

$$(\text{SNR})_{C,\text{AM}} = \frac{A_c^2(1 + k_a^2 P)}{2WN_0} \tag{2.90}$$

To evaluate the output signal-to-noise ratio, we first represent the filtered noise $n(t)$ in terms of its in-phase and quadrature components. We may therefore define the filtered signal $x(t)$ applied to the envelope detector in the receiver model of Figure 2.37 as follows:

$$\begin{aligned} x(t) &= s(t) + n(t) \\ &= [A_c + A_c k_a m(t) + n_I(t)] \cos(2\pi f_c t) - n_Q(t) \sin(2\pi f_c t) \end{aligned} \tag{2.91}$$

It is informative to represent the components that comprise the signal $x(t)$ by means of phasors, as in Figure 2.38*a*. From this phasor diagram, the receiver output is readily obtained as

$$\begin{aligned} y(t) &= \text{envelope of } x(t) \\ &= \{[A_c + A_c k_a m(t) + n_I(t)]^2 + n_Q^2(t)\}^{1/2} \end{aligned} \tag{2.92}$$

The signal $y(t)$ defines the output of an ideal envelope detector. The phase of $x(t)$ is of no interest to us, because an ideal envelope detector is totally insensitive to variations in the phase of $x(t)$.

The expression defining $y(t)$ is somewhat complex and needs to be simplified in some manner to permit the derivation of insightful results. Specifically, we would like to approximate the output $y(t)$ as the sum of a message term plus a term due to noise. In general, this is quite difficult to achieve. However, when the average carrier power is large com-

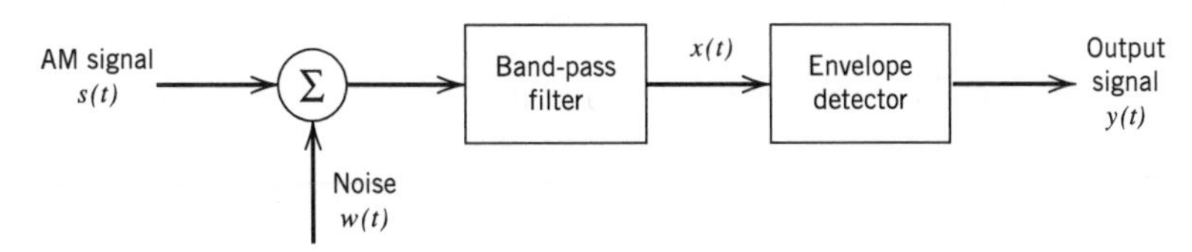


**FIGURE 2.37** Model of AM receiver.

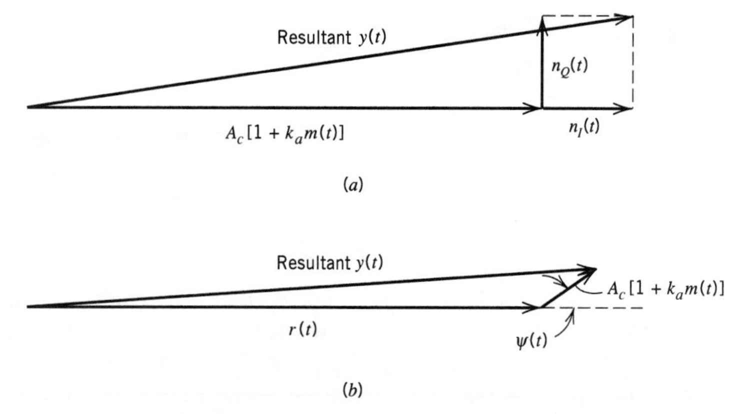


**FIGURE 2.38** (*a*) Phasor diagram for AM wave plus narrowband noise for the case of high carrier-to-noise ratio. (*b*) Phasor diagram for AM wave plus narrowband noise for the case of low carrier-to-noise ratio.

pared with the average noise power, so that the receiver is operating satisfactorily, then the signal term $A_c[1 + k_a m(t)]$ will be large compared with the noise terms $n_I(t)$ and $n_Q(t)$, at least most of the time. Then we may approximate the output $y(t)$ as (see Problem 2.51):

$$y(t) \simeq A_c + A_c k_a m(t) + n_I(t) \tag{2.93}$$

The presence of the DC or constant term $A_c$ in the envelope detector output $y(t)$ of Equation (2.93) is due to demodulation of the transmitted carrier wave. We may ignore this term, however, because it bears no relation whatsoever to the message signal $m(t)$. In any case, it may be removed simply by means of a blocking capacitor. Thus if we neglect the DC term $A_c$ in Equation (2.93), we find that the remainder has, except for scaling factors, a form similar to the output of a DSB-SC receiver using coherent detection. Accordingly, the output signal-to-noise ratio of an AM receiver using an envelope detector is approximately

$$(\text{SNR})_{\text{O,AM}} \simeq \frac{A_c^2 k_a^2 P}{2WN_0} \tag{2.94}$$

Equation (2.94) is, however, valid only if the following two conditions are satisfied:

1. The average noise power is small compared to the average carrier power at the envelope detector input.
2. The amplitude sensitivity $k_a$ is adjusted for a percentage modulation less than or equal to 100 percent.

Using Equations (2.90) and (2.94), we obtain the following figure of merit for amplitude modulation:

$$\left.\frac{(\text{SNR})_{\text{O}}}{(\text{SNR})_{\text{C}}}\right|_{\text{AM}} \simeq \frac{k_a^2 P}{1 + k_a^2 P} \tag{2.95}$$

Thus, whereas the figure of merit of a DSB-SC receiver or that of an SSB receiver using coherent detection is always unity, the corresponding figure of merit of an AM receiver using envelope detection is always less than unity. In other words, *the noise performance of a full AM receiver is always inferior to that of a DSB-SC receiver.* This is due to the wastage of transmitter power, which results from transmitting the carrier as a component of the AM wave.

### ▶ EXAMPLE 2.4 Single-Tone Modulation

Consider the special case of a sinusoidal wave of frequency $f_m$ and amplitude $A_m$ as the modulating wave, as shown by

$$m(t) = A_m \cos(2\pi f_m t)$$

The corresponding AM wave is

$$s(t) = A_c[1 + \mu \cos(2\pi f_m t)] \cos(2\pi f_c t)$$

where $\mu = k_a A_m$ is the modulation factor. The average power of the modulating wave $m(t)$ is (assuming a load resistor of 1 ohm)

$$P = \tfrac{1}{2}A_m^2$$

Therefore, using Equation (2.95), we get

$$\left.\frac{(\text{SNR})_O}{(\text{SNR})_C}\right|_{\text{AM}} = \frac{\frac{1}{2}k_a^2 A_m^2}{1 + \frac{1}{2}k_a^2 A_m^2} = \frac{\mu^2}{2 + \mu^2} \tag{2.96}$$

When $\mu = 1$, which corresponds to 100 percent modulation, we get a figure of merit equal to 1/3. This means that, other factors being equal, an AM system (using envelope detection) must transmit three times as much average power as a suppressed-carrier system (using coherent detection) to achieve the same quality of noise performance. ◀

## ■ THRESHOLD EFFECT

When the carrier-to-noise ratio is small compared with unity, the noise term dominates and the performance of the envelope detector changes completely from that just described. In this case it is more convenient to represent the narrowband noise $n(t)$ in terms of its envelope $r(t)$ and phase $\psi(t)$, as shown by

$$n(t) = r(t) \cos[2\pi f_c t + \psi(t)] \tag{2.97}$$

The corresponding phasor diagram for the detector input $x(t) = s(t) + n(t)$ is shown in Figure 2.38*b*, where we have used the noise envelope as reference, because it is now the dominant term. To the noise phasor $r(t)$ we have added a phasor representing the signal term $A_c[1 + k_a m(t)]$, with the angle between them being equal to the phase $\psi(t)$ of the noise $n(t)$. In Figure 2.38*b* it is assumed that the carrier-to-noise ratio is so low that the carrier amplitude $A_c$ is small compared with the noise envelope $r(t)$, at least most of the time. Then we may neglect the quadrature component of the signal with respect to the noise, and thus find from Figure 2.38*b* that the envelope detector output is

$$y(t) \simeq r(t) + A_c \cos[\psi(t)] + A_c k_a m(t) \cos[\psi(t)] \tag{2.98}$$

This relation reveals that when the carrier-to-noise ratio is low, the detector output has no component strictly proportional to the message signal $m(t)$. The last term of the expression defining $y(t)$ contains the message signal $m(t)$ multiplied by noise in the form of

$\cos[\psi(t)]$. From Section 1.11 we recall that the phase $\psi(t)$ of the narrowband noise $n(t)$ is uniformly distributed over $2\pi$ radians. It follows therefore that we have a complete loss of information in that the detector output does not contain the message signal $m(t)$ at all. The loss of a message in an envelope detector that operates at a low carrier-to-noise ratio is referred to as the *threshold effect*. By *threshold* we mean *a value of the carrier-to-noise ratio below which the noise performance of a detector deteriorates much more rapidly than proportionately to the carrier-to-noise ratio*. It is important to recognize that every nonlinear detector (e.g., envelope detector) exhibits a threshold effect. On the other hand, such an effect does *not* arise in a coherent detector.

A rigorous mathematical analysis of the threshold effect for the general case of an AM wave is beyond the scope of this book. In the next subsection we simplify matters by considering the case of an unmodulated carrier. Despite this simplification, we can still develop a great deal of insight into the threshold effect experienced in an envelope detector.

### *General Formula for $(SNR)_O$ in Envelope Detection**

Consider an envelope detector whose input signal is defined by

$$x(t) = A_c \cos(2\pi f_c t) + n(t) \tag{2.99}$$

where $A_c \cos(2\pi f_c t)$ is the unmodulated carrier and $n(t)$ is the sample function of band-limited, zero-mean, white Gaussian noise $N(t)$. The power spectral density of $N(t)$ is

$$S_N(t) = \begin{cases} \dfrac{N_0}{2} & \text{for } |f - f_c| \leq W \\ 0 & \text{otherwise} \end{cases} \tag{2.100}$$

Representing the narrowband noise $n(t)$ in terms of its in-phase component $n_I(t)$ and quadrature component $n_Q(t)$, we may express the noisy signal at the detector input as

$$x(t) = (A_c + n_I(t)) \cos(2\pi f_c t) - n_Q(t) \sin(2\pi f_c t) \tag{2.101}$$

The noise components $n_I(t)$ and $n_Q(t)$ are zero-mean, jointly Gaussian, mutually independent low-pass random processes with identical power spectral densities (see Equation 1.101):

$$S_{N_I}(f) = S_{N_Q}(f) = \begin{cases} S_N(f - f_c) + S_N(f + f_c) & \text{for } |f| \leq W \\ 0 & \text{otherwise} \end{cases} \tag{2.102}$$

For the problem at hand, the input signal consists of an unmodulated carrier with average power equal to $A_c^2/2$. The average noise power at the detector input is

$$\sigma_N^2 = 2WN_0 \tag{2.103}$$

The *carrier-to-noiser ratio* is therefore defined by

$$\begin{aligned} \rho &= \frac{A_c^2/2}{\sigma_N^2} \\ &= \frac{A_c^2}{4WN_0} \end{aligned} \tag{2.104}$$

We may think of $\rho$ as an input signal-to-noise ratio for the problem described herein.

---

*A reader who is not interested in the mathematical details of how noise affects the envelope detection of an AM signal may skip the material up to Eq. (2.124) and read the two limiting cases of the formula in that equation.

However, determination of the output signal-to-noise ratio is a more difficult undertaking because the envelope detector output

$$y(t) = \sqrt{(A_c + n_I(t))^2 + n_Q^2(t)} \tag{2.105}$$

is a nonlinear combination of signal and noise terms. With no clear-cut separation between signal and noise at the detector output $y(t)$, how then do we isolate the contribution of the signal $s(t)$ to $y(t)$ from the contribution due to the noise $n(t)$? To resolve this issue, we adopt a *heuristic* approach based on signal averaging. Specifically, we introduce the following two definitions:

1. The *mean output signal*, $s_o$, is the difference between the expectation of $y(t)$ in the combined presence of signal and noise and the expectation of $y(t)$ in the presence of noise alone, as shown by

$$s_o = E[y(t)] - E[y_o(t)] \tag{2.106}$$

where $y(t)$ is itself defined by Equation (2.105) and $y_o(t)$ is defined by

$$y_o(t) = \sqrt{n_I^2(t) + n_Q^2(t)} \tag{2.107}$$

2. The *mean output noise power* is the difference between the mean-square value of the detector output $y(t)$ and the square of the mean value of $y(t)$, as shown by

$$\text{var}[y(t)] = E[y^2(t)] - (E[y(t)])^2 \tag{2.108}$$

On this basis, we define the *output signal-to-noise ratio* as

$$(\text{SNR})_O = \frac{s_o^2}{\text{var}[y(t)]} \tag{2.109}$$

From Section 1.12, we recall that the envelope detector output due to noise alone is Rayleigh distributed; that is

$$f_{Y_o}(y) = \begin{cases} \dfrac{y}{\sigma_N^2}\exp\left(-\dfrac{y^2}{2\sigma_N^2}\right), & y \geq 0 \\ 0 & \text{otherwise} \end{cases} \tag{2.110}$$

The expectation of $y_o(t)$ is therefore

$$\begin{aligned} E[y_o(t)] &= \int_{-\infty}^{\infty} y f_{Y_o}(y)\, dy \\ &= \int_0^{\infty} \frac{y^2}{\sigma_N^2}\exp\left(-\frac{y^2}{2\sigma_N^2}\right) dy \end{aligned} \tag{2.111}$$

From the definition of the *gamma function* for real positive values of the argument $x$, we have

$$\Gamma(x) = \int_0^{\infty} z^{x-1}\exp(-z)\, dz \tag{2.112}$$

We may therefore rewrite Equation (2.111) as

$$\begin{aligned} E[y_o(t)] &= \sqrt{2}\sigma_N \Gamma\left(\frac{3}{2}\right) \\ &= \sqrt{\frac{\pi}{2}}\,\sigma_N \end{aligned} \tag{2.113}$$

where we have used the value $\Gamma(3/2) = \sqrt{\pi}/2$. To calculate the mean signal $s_o$ at the detector output, we also need the expectation of $y(t)$. Due to the combined presence of signal and noise, we recall from Section 1.13 that $y(t)$ is Rician distributed, as shown by

$$f_Y(y) = \begin{cases} \dfrac{y}{\sigma_N^2} \exp\left(-\dfrac{y^2 + A_c^2}{2\sigma_N^2}\right) I_0\left(\dfrac{A_c y}{\sigma_N^2}\right) & \text{for } y \geq 0 \\ 0 & \text{otherwise} \end{cases} \tag{2.114}$$

where $I_0(\cdot)$ is the modified Bessel function of the first kind of zero order (see Appendix 3). Hence,

$$E[y(t)] = \int_0^\infty \frac{y^2}{\sigma_N^2} \exp\left(-\frac{y^2 + A_c^2}{2\sigma_N^2}\right) I_0\left(\frac{A_c y}{\sigma_N^2}\right) dy \tag{2.115}$$

Putting $A_c y/\sigma_N^2 = u$ and recognizing that $\rho = A_c^2/2\sigma_N^2$, we may recast this expectation in the form

$$E[y(t)] = \frac{\sigma_N}{(2\rho)^{3/2}} \exp(-\rho) \int_0^\infty u^2 \exp\left(-\frac{u^2}{4\rho}\right) I_0(u)\, du \tag{2.116}$$

The integral in Equation (2.116) can be written in a concise form by using *confluent hypergeometric functions*; see Appendix 4. In particular, using the integral representation

$$\int_0^\infty u^{m-1} \exp(-b^2u^2) I_0(u)\, du = \frac{\Gamma(m/2)}{2b^m}\left({}_1F_1\left(\frac{m}{2};1;\frac{1}{4b^2}\right)\right) \tag{2.117}$$

with $m = 3$, $\Gamma(m/2) = \sqrt{\pi}/2$ and $b^2 = 1/4\rho$, we may express the expectation of $y(t)$ in terms of the confluent hypergeometric function ${}_1F_1(3/2;1;\rho)$ as

$$E[y(t)] = \sqrt{\frac{\pi}{2}}\, \sigma_N \exp(-\rho)\left({}_1F_1\left(\frac{3}{2};1;\rho\right)\right) \tag{2.118}$$

We may further simplify matters by using the following identity:

$$\exp(-u)({}_1F_1(\alpha;\beta;u)) = {}_1F_1(\beta - \alpha;\beta;-u) \tag{2.119}$$

and so finally express the expectation of $y(t)$ in the concise form

$$E[y(t)] = \sqrt{\frac{\pi}{2}}\, \sigma_N\left({}_1F_1\left(-\frac{1}{2};1;-\rho\right)\right) \tag{2.120}$$

Thus using Equations (2.113) and (2.120) in Equation (2.106) yields the mean output signal as

$$s_o = \sqrt{\frac{\pi}{2}}\, \sigma_N\left({}_1F_1\left(-\frac{1}{2};1;-\rho\right) - 1\right) \tag{2.121}$$

whose dependence on the standard deviation $\sigma_N$ of the noise $n(t)$ is testimony to the intermingling of signal and noise at the detector output.

Following a similar procedure, we may express the mean-square value of the detector output $y(t)$ as

$$\begin{aligned} E[y^2(t)] &= \int_0^\infty \frac{y^3}{\sigma_N^2} \exp\left(-\frac{y^2 + A_c^2}{2\sigma_N^2}\right) I_0\left(\frac{A_c y}{\sigma_N^2}\right) dy \\ &= 2\sigma_N^2({}_1F_1(-1;1;-\rho)) \end{aligned} \tag{2.122}$$

Hence using Equations (2.120) and (2.122) in Equation (2.108) yields the mean output noise power as

$$\operatorname{var}[y(t)] = 2\sigma_N^2\left({}_1F_1(-1;1;-\rho) - \frac{\pi}{4}\left({}_1F_1\left(-\frac{1}{2};1;-\rho\right)\right)^2\right) \tag{2.123}$$

Finally, using Equations (2.121) and (2.123) in Equation (2.109) yields the output signal-to-noise ratio for the envelope detection problem at hand as

$$(\mathrm{SNR})_O = \frac{\left({}_1F_1\left(-\frac{1}{2};1;-\rho\right) - 1\right)^2}{(4/\pi)\left({}_1F_1(-1;1;-\rho)\right) - \left({}_1F_1\left(-\frac{1}{2};1;-\rho\right)\right)^2} \tag{2.124}$$

Equation (2.124) is the general formula for the output signal-to-noise of an envelope detector whose input consists of an unmodulated carrier and band-limited, white Gaussian noise. Two limiting cases of this general formula are of particular interest:

1. *Large carrier-to-noise ratio.* For large $\rho$, we may use the following asymptotic formula (see Appendix 4)

$$ {}_1F_1\left(-\frac{1}{2};1;-\rho\right) \simeq 2\sqrt{\frac{\rho}{\pi}} \qquad \text{for } \rho \to \infty \tag{2.125}$$

Moreover, the following identity

$$ {}_1F_1(-1;1;-\rho) = 1 + \rho \tag{2.126}$$

holds exactly for all $\rho$. Accordingly, the use of Equations (2.125) and (2.126) in Equation (2.124) yields the following approximate formula for the output signal-to-noise ratio:

$$(\mathrm{SNR})_O \simeq \rho \qquad \text{for } \rho \to \infty \tag{2.127}$$

where we have ignored contributions due to $\rho^{1/2}$ and $\rho^0$ in the numerator of Equation (2.124) as being subdominant compared to $\rho$ for large $\rho$. Equation (2.127) shows that for large carrier-to-noise $\rho$ the envelope detector behaves like a coherent detector, in that the output signal-to-noise ratio is proportional to the input signal-to-noise ratio.

2. *Small carrier-to-noise ratio.* For small $\rho$, we have (see Appendix 4)

$$ {}_1F_1(a;c;-\rho) \simeq 1 - \frac{a}{c}\rho \qquad \text{for } \rho \to 0 \tag{2.128}$$

Hence, using this asymptotic formula, we may approximate the output signal-to-noise ratio for small $\rho$ as

$$\begin{aligned}(\mathrm{SNR})_O &\simeq \frac{\pi\rho^2}{16 - 4\pi} \\ &\simeq 0.91\rho^2 \qquad \text{for } \rho \to 0\end{aligned} \tag{2.129}$$

where, in the denominator, we have ignored contributions due to $\rho$ and $\rho^2$ as being subdominant compared to $\rho^0$ for small $\rho$. Equation (2.129) shows that for a small carrier-to-noise ratio, the output signal-to-noise ratio of the envelope detector is proportional to the squared input signal-to-noise ratio.

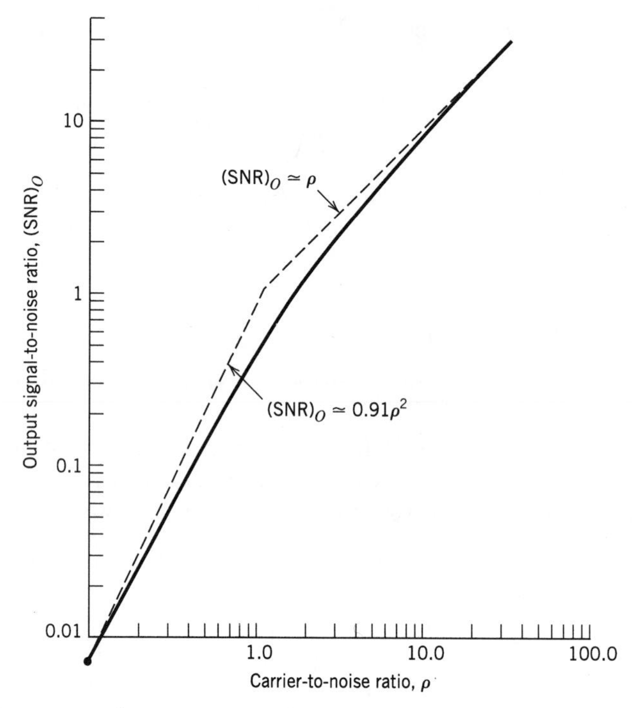


**FIGURE 2.39** Output signal-to-noise ratio of an envelope detector for varying carrier-to-noise ratio.

The conclusions drawn from the two limiting cases considered herein are that an envelope detector favors strong signals and penalizes weak signals. The phenomenon of weak signals being penalized by the detector is referred to as *weak signal suppression*, which is a manifestation of the threshold effect.

Using the formula of Equation (2.124), in Figure 2.39 we have plotted the output signal-to-noise ratio $(\text{SNR})_O$ of the envelope detector versus the carrier-to-noise ratio $\rho$ using tabulated values of confluent hypergeometric functions. This figure also includes the two asymptotes for large $\rho$ and small $\rho$. From Figure 2.39 we see that the output signal-to-noise ratio deviates from a linear behavior around a carrier-to-noise ratio of 10 dB (i.e., $\rho \simeq 10$).

## 2.13 *Noise in FM Receivers*

Finally, we turn our attention to the noise analysis of a frequency modulation (FM) system, for which we use the receiver model shown in Figure 2.40. As before, the noise $w(t)$ is modeled as white Gaussian noise of zero mean and power spectral density $N_0/2$. The

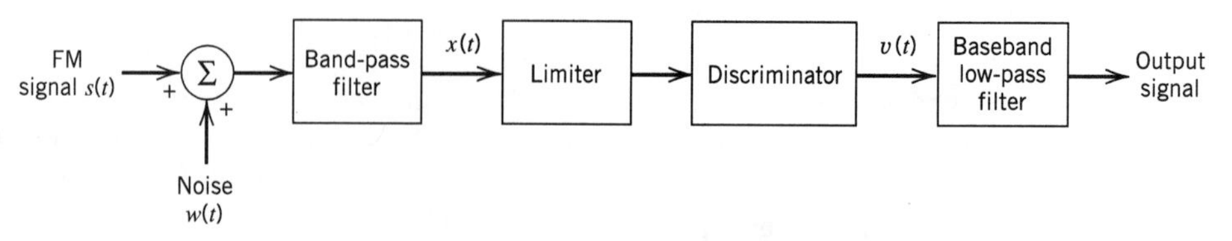


**FIGURE 2.40** Model of an FM receiver.

received FM signal $s(t)$ has a carrier frequency $f_c$ and transmission bandwidth $B_T$, such that only a negligible amount of power lies outside the frequency band $f_c \pm B_T/2$ for positive frequencies, and similarly for negative frequencies.

As in the AM case, the band-pass filter has a midband frequency $f_c$ and bandwidth $B_T$ and therefore passes the FM signal essentially without distortion. Ordinarily, $B_T$ is small compared with the midband frequency $f_c$, so that we may use the narrowband representation for $n(t)$, the filtered version of channel noise $w(t)$, in terms of its in-phase and quadrature components.

In an FM system, the message signal is transmitted by variations of the instantaneous frequency of a sinusoidal carrier wave, and its amplitude is maintained constant. Therefore, any variations of the carrier amplitude at the receiver input must result from noise or interference. The amplitude *limiter*, following the band-pass filter in the receiver model of Figure 2.40, is used to remove amplitude variations by clipping the modulated wave at the filter output almost to the zero axis. The resulting rectangular wave is rounded off by another band-pass filter that is an integral part of the limiter, thereby suppressing harmonics of the carrier frequency. Thus, the filter output is again sinusoidal, with an amplitude that is practically independent of the carrier amplitude at the receiver input.

The discriminator in the model of Figure 2.40 consists of two components:

1. A *slope network* or *differentiator* with a purely imaginary frequency response that varies linearly with frequency. It produces a hybrid-modulated wave in which both amplitude and frequency vary in accordance with the message signal.
2. An envelope detector that recovers the amplitude variation and thus reproduces the message signal.

The slope network and envelope detector are usually implemented as integral parts of a single physical unit.

The *postdetection filter*, labeled "baseband low-pass filter" in Figure 2.40, has a bandwidth that is just large enough to accommodate the highest frequency component of the message signal. This filter removes the out-of-band components of the noise at the discriminator output and thereby keeps the effect of the output noise to a minimum.

The filtered noise $n(t)$ at the band-pass filter output in Figure 2.40 is defined in terms of its in-phase and quadrature components by

$$n(t) = n_I(t)\cos(2\pi f_c t) - n_Q(t)\sin(2\pi f_c t)$$

Equivalently, we may express $n(t)$ in terms of its envelope and phase as

$$n(t) = r(t)\cos[(2\pi f_c t) + \psi(t)] \tag{2.130}$$

where the envelope is

$$r(t) = [n_I^2(t) + n_Q^2(t)]^{1/2} \tag{2.131}$$

and the phase is

$$\psi(t) = \tan^{-1}\left[\frac{n_Q(t)}{n_I(t)}\right] \tag{2.132}$$

The envelope $r(t)$ is Rayleigh distributed, and the phase $\psi(t)$ is uniformly distributed over $2\pi$ radians (see Section 1.12).

The incoming FM signal $s(t)$ is defined by

$$s(t) = A_c \cos\left[2\pi f_c t + 2\pi k_f \int_0^t m(\tau)\, d\tau\right] \tag{2.133}$$

where $A_c$ is the carrier amplitude, $f_c$ is the carrier frequency, $k_f$ is the frequency sensitivity, and $m(t)$ is the message signal. Note that, as with the standard AM, in FM there is no need to introduce a scaling factor in the definition of the modulated signal $s(t)$, since it is reasonable to assume that its amplitude $A_c$ has the same units as the additive noise component $n(t)$. To proceed, we define

$$\phi(t) = 2\pi k_f \int_0^t m(\tau)\, d\tau \tag{2.134}$$

We may thus express $s(t)$ in the simple form

$$s(t) = A_c \cos[2\pi f_c t + \phi(t)] \tag{2.135}$$

The noisy signal at the band-pass filter output is therefore

$$\begin{aligned} x(t) &= s(t) + n(t) \\ &= A_c \cos[2\pi f_c t + \phi(t)] + r(t)\cos[2\pi f_c t + \psi(t)] \end{aligned} \tag{2.136}$$

It is informative to represent $x(t)$ by means of a phasor diagram, as in Figure 2.41. In this diagram we have used the signal term as reference. The phase $\theta(t)$ of the resultant phasor representing $x(t)$ is obtained directly from Figure 2.41 as

$$\theta(t) = \phi(t) + \tan^{-1}\left\{\frac{r(t)\sin[\psi(t) - \phi(t)]}{A_c + r(t)\cos[\psi(t) - \phi(t)]}\right\} \tag{2.137}$$

The envelope of $x(t)$ is of no interest to us, because any envelope variations at the band-pass filter output are removed by the limiter.

Our motivation is to determine the error in the instantaneous frequency of the carrier wave caused by the presence of the filtered noise $n(t)$. With the discriminator assumed ideal, its output is proportional to $\theta'(t)/2\pi$ where $\theta'(t)$ is the derivative of $\theta(t)$ with respect to time. In view of the complexity of the expression defining $\theta(t)$, however, we need to make certain simplifying approximations, so that our analysis may yield useful results.

We assume that the carrier-to-noise ratio measured at the discriminator input is large compared with unity. Let $R$ denote the random variable obtained by observing (at some fixed time) the envelope process with sample function $r(t)$ [due to the noise $n(t)$]. Then, at least most of the time, the random variable $R$ is small compared with the carrier amplitude $A_c$, and so the expression for the phase $\theta(t)$ simplifies considerably as follows:

$$\theta(t) \simeq \phi(t) + \frac{r(t)}{A_c}\sin[\psi(t) - \phi(t)] \tag{2.138}$$

or, using the expression for $\phi(t)$ given in Equation (2.134),

$$\theta(t) \simeq 2\pi k_f \int_0^t m(\tau)\, d\tau + \frac{r(t)}{A_c}\sin[\psi(t) - \phi(t)] \tag{2.139}$$

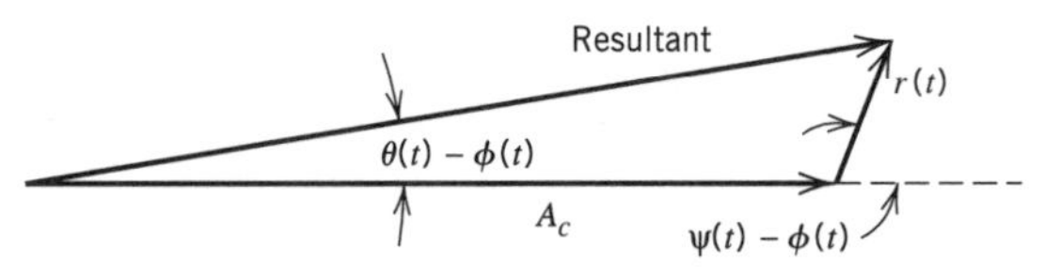


**FIGURE 2.41** Phasor diagram for FM wave plus narrowband noise for the case of high carrier-to-noise ratio.

The discriminator output is therefore

$$\begin{aligned} v(t) &= \frac{1}{2\pi}\frac{d\theta(t)}{dt} \\ &= k_f m(t) + n_d(t) \end{aligned} \tag{2.140}$$

where the noise term $n_d(t)$ is defined by

$$n_d(t) = \frac{1}{2\pi A_c}\frac{d}{dt}\{r(t)\sin[\psi(t) - \phi(t)]\} \tag{2.141}$$

We thus see that provided the carrier-to-noise ratio is high, the discriminator output $v(t)$ consists of the original message signal $m(t)$ multiplied by the constant factor $k_f$, plus an additive noise component $n_d(t)$. Accordingly, we may use the output signal-to-noise ratio as previously defined to assess the quality of performance of the FM receiver. Before doing this, however, it is instructive to see if we can simplify the expression defining the noise $n_d(t)$.

From the phasor diagram of Figure 2.41, we note that the effect of variations in the phase $\psi(t)$ of the narrowband noise appear referred to the signal term $\phi(t)$. We know that the phase $\psi(t)$ is uniformly distributed over $2\pi$ radians. It would therefore be tempting to assume that the phase difference $\psi(t) - \phi(t)$ is also uniformly distributed over $2\pi$ radians. If such an assumption were true, then the noise $n_d(t)$ at the discriminator output would be independent of the modulating signal and would depend only on the characteristics of the carrier and narrowband noise. Theoretical considerations show that this assumption is justified provided that the carrier-to-noise ratio is high.[9] Then we may simplify Equation (2.141) as:

$$n_d(t) \simeq \frac{1}{2\pi A_c}\frac{d}{dt}\{r(t)\sin[\psi(t)]\} \tag{2.142}$$

However, from the defining equations for $r(t)$ and $\psi(t)$, we note that the quadrature component $n_Q(t)$ of the filtered noise $n(t)$ is

$$n_Q(t) = r(t)\sin[\psi(t)] \tag{2.143}$$

Therefore, we may rewrite Equation (2.142) as

$$n_d(t) \simeq \frac{1}{2\pi A_c}\frac{dn_Q(t)}{dt} \tag{2.144}$$

This means that *the additive noise $n_d(t)$ appearing at the discriminator output is determined effectively by the carrier amplitude $A_c$ and the quadrature component $n_Q(t)$ of the narrowband noise $n(t)$.*

The output signal-to-noise ratio is defined as the ratio of the average output signal power to the average output noise power. From Equation (2.140), we see that the message component in the discriminator output, and therefore the low-pass filter output, is $k_f m(t)$. Hence, the average output signal power is equal to $k_f^2 P$, where $P$ is the average power of the message signal $m(t)$.

To determine the average output noise power, we note that the noise $n_d(t)$ at the discriminator output is proportional to the time derivative of the quadrature noise component $n_Q(t)$. Since the differentiation of a function with respect to time corresponds to multiplication of its Fourier transform by $j2\pi f$, it follows that we may obtain the noise process $n_d(t)$ by passing $n_Q(t)$ through a linear filter with a frequency response equal to

$$\frac{j2\pi f}{2\pi A_c} = \frac{jf}{A_c}$$

This means that the power spectral density $S_{N_d}(f)$ of the noise $n_d(t)$ is related to the power spectral density $S_{N_Q}(f)$ of the quadrature noise component $n_Q(t)$ as follows:

$$S_{N_d}(f) = \frac{f^2}{A_c^2} S_{N_Q}(f) \tag{2.145}$$

With the band-pass filter in the receiver model of Figure 2.40 having an ideal frequency response characterized by bandwidth $B_T$ and midband frequency $f_c$, it follows that the narrowband noise $n(t)$ will have a power spectral density characteristic that is similarly shaped. This means that the quadrature component $n_Q(t)$ of the narrowband noise $n(t)$ will have the ideal low-pass characteristic shown in Figure 2.42*a*. The corresponding power spectral density of the noise $n_d(t)$ is shown in Figure 2.42*b*; that is,

$$S_{N_d}(f) = \begin{cases} \dfrac{N_0 f^2}{A_c^2}, & |f| \le \dfrac{B_T}{2} \\ 0, & \text{otherwise} \end{cases} \tag{2.146}$$

In the receiver model of Figure 2.40, the discriminator output is followed by a low-pass filter with a bandwidth equal to the message bandwidth $W$. For wideband FM, we usually find that $W$ is smaller than $B_T/2$, where $B_T$ is the transmission bandwidth of the FM signal. This means that the out-of-band components of noise $n_d(t)$ will be rejected. Therefore, the power spectral density $S_{N_o}(f)$ of the noise $n_o(t)$ appearing at the receiver output is defined by

$$S_{N_o}(f) = \begin{cases} \dfrac{N_0 f^2}{A_c^2}, & |f| \le W \\ 0, & \text{otherwise} \end{cases} \tag{2.147}$$

as shown in Figure 2.42*c*. The average output noise power is determined by integrating the power spectral density $S_{N_o}(f)$ from $-W$ to $W$. We thus get the following result:

$$\begin{aligned} \text{Average power of output noise} &= \frac{N_0}{A_c^2} \int_{-W}^{W} f^2 \, df \\ &= \frac{2N_0 W^3}{3A_c^2} \end{aligned} \tag{2.148}$$

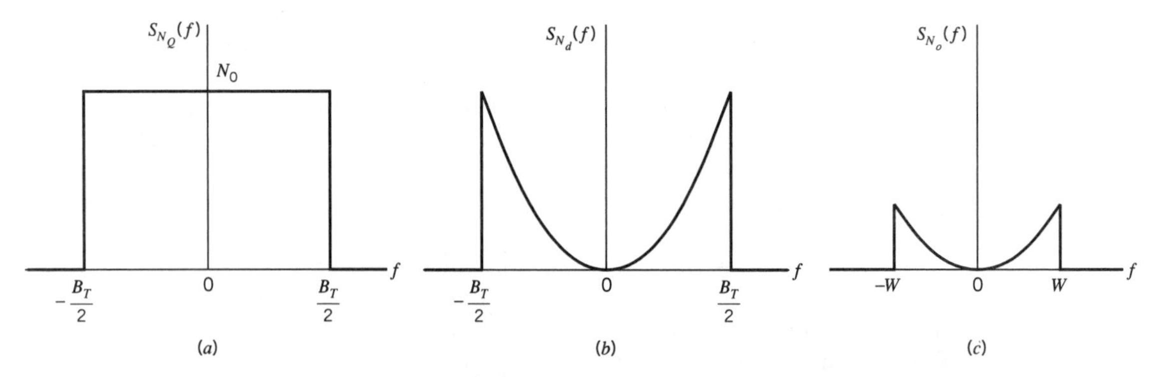


**FIGURE 2.42** Noise analysis of FM receiver. (*a*) Power spectral density of quadrature component $n_Q(t)$ of narrowband noise $n(t)$. (*b*) Power spectral density of noise $n_d(t)$ at the discriminator output. (*c*) Power spectral density of noise $n_o(t)$ at the receiver output.

Note that the average output noise power is inversely proportional to the average carrier power $A_c^2/2$. Accordingly, in an FM system, increasing the carrier power has a *noise-quieting effect.*

Earlier we determined the average output signal power as $k_f^2 P$. Therefore, provided the carrier-to-noise ratio is high, we may divide this average output signal power by the average output noise power of Equation (2.148) to obtain the output signal-to-noise ratio

$$(\mathrm{SNR})_{\mathrm{O,FM}} = \frac{3A_c^2 k_f^2 P}{2N_0 W^3} \tag{2.149}$$

The average power in the modulated signal $s(t)$ is $A_c^2/2$, and the average noise power in the message bandwidth is $WN_0$. Thus the channel signal-to-noise ratio is

$$(\mathrm{SNR})_{\mathrm{C,FM}} = \frac{A_c^2}{2WN_0} \tag{2.150}$$

Dividing the output signal-to-noise ratio by the channel signal-to-noise ratio, we get the following figure of merit for frequency modulation:

$$\left.\frac{(\mathrm{SNR})_{\mathrm{O}}}{(\mathrm{SNR})_{\mathrm{C}}}\right|_{\mathrm{FM}} = \frac{3k_f^2 P}{W^2} \tag{2.151}$$

From Section 2.7 we recall that the frequency deviation $\Delta f$ is proportional to the frequency sensitivity $k_f$ of the modulator. Also, by definition, the deviation ratio $D$ is equal to the frequency deviation $\Delta f$ divided by the message bandwidth $W$. In other words, the deviation ratio $D$ is proportional to the ratio $k_f P^{1/2}/W$. It follows therefore from Equation (2.151) that the figure of merit of a wideband FM system is a quadratic function of the deviation ratio. Now, in wideband FM, the transmission bandwidth $B_T$ is approximately proportional to the deviation ratio $D$. Accordingly, we may state that *when the carrier-to-noise ratio is high, an increase in the transmission bandwidth $B_T$ provides a corresponding quadratic increase in the output signal-to-noise ratio or figure of merit of the FM system.* The important point to note from this statement is that, unlike amplitude modulation, the use of frequency modulation does provide a practical mechanism for the exchange of increased transmission bandwidth for improved noise performance.

### ▶ EXAMPLE 2.5 Single-Tone Modulation

Consider the case of a sinusoidal wave of frequency $f_m$ as the modulating signal, and assume a peak frequency deviation $\Delta f$. The modulated FM signal is thus defined by

$$s(t) = A_c \cos\left[2\pi f_c t + \frac{\Delta f}{f_m}\sin(2\pi f_m t)\right]$$

Therefore, we may write

$$2\pi k_f \int_0^t m(\tau)\, d\tau = \frac{\Delta f}{f_m}\sin(2\pi f_m t)$$

Differentiating both sides with respect to time and solving for $m(t)$, we get

$$m(t) = \frac{\Delta f}{k_f}\cos(2\pi f_m t)$$

Hence, the average power of the message signal $m(t)$, developed across a 1-ohm load, is

$$P = \frac{(\Delta f)^2}{2k_f^2}$$

Substituting this result into the formula for the output signal-to-noise ratio given in Equation (2.149), we get

$$(\text{SNR})_{O,\text{FM}} = \frac{3A_c^2(\Delta f)^2}{4N_0W^3} = \frac{3A_c^2\beta^2}{4N_0W}$$

where $\beta = \Delta f/W$ is the modulation index. Using Equation (2.151) to evaluate the corresponding figure of merit, we get

$$\left.\frac{(\text{SNR})_O}{(\text{SNR})_C}\right|_{\text{FM}} = \frac{3}{2}\left(\frac{\Delta f}{W}\right)^2 = \frac{3}{2}\beta^2 \tag{2.152}$$

It is important to note that the modulation index $\beta = \Delta f/W$ is determined by the bandwidth $W$ of the postdetection low-pass filter and is not related to the sinusoidal message frequency $f_m$, except insofar as this filter is usually chosen so as to pass the spectrum of the desired message; this is merely a matter of consistent design. For a specified system bandwidth $W$, the sinusoidal message frequency $f_m$ may lie anywhere between 0 and $W$ and would yield the same output signal-to-noise ratio.

It is of particular interest to compare the noise performance of AM and FM systems. An insightful way of making this comparison is to consider the figures of merit of the two systems based on a sinusoidal modulating signal. For an AM system operating with a sinusoidal modulating signal and 100 percent modulation, we have (from Example 2.4):

$$\left.\frac{(\text{SNR})_O}{(\text{SNR})_C}\right|_{\text{AM}} = \frac{1}{3}$$

Comparing this figure of merit with the corresponding result described in Equation (2.152) for an FM system, we see that the use of frequency modulation offers the possibility of improved noise performance over amplitude modulation when

$$\tfrac{3}{2}\beta^2 > \tfrac{1}{3}$$

that is,

$$\beta > \frac{\sqrt{2}}{3} = 0.471$$

We may therefore consider $\beta = 0.5$ as defining roughly *the transition between narrowband FM and wideband FM*. This statement, based on noise considerations, further confirms a similar observation that was made in Section 2.7 when considering the bandwidth of FM waves. ◀

## ■ CAPTURE EFFECT

The inherent ability of an FM system to minimize the effects of unwanted signals (e.g., noise, as just discussed) also applies to *interference* produced by another frequency-modulated signal whose frequency content is close to the carrier frequency of the desired FM wave. However, interference suppression in an FM receiver works well only when the interference is weaker than the desired FM input. When the interference is the stronger one of the two, the receiver locks onto the stronger signal and thereby suppresses the desired FM input. When they are of nearly equal strength, the receiver fluctuates back and forth between them. This phenomenon is known as the *capture effect*, which describes another distinctive characteristic of frequency modulation.

## FM THRESHOLD EFFECT

The formula of Equation (2.149), defining the output signal-to-noise ratio of an FM receiver, is valid only if the carrier-to-noise ratio, measured at the discriminator input, is high compared with unity. It is found experimentally that as the input noise power is increased so that the carrier-to-noise ratio is decreased, the FM receiver *breaks*. At first, individual clicks are heard in the receiver output, and as the carrier-to-noise ratio decreases still further, the clicks rapidly merge into a *crackling* or *sputtering sound*. Near the breaking point, Equation (2.149) begins to fail by predicting values of output signal-to-noise ratio larger than the actual ones. This phenomenon is known as the *threshold effect*.[10] The threshold is defined as the minimum carrier-to-noise ratio yielding an FM improvement that is not significantly deteriorated from the value predicted by the usual signal-to-noise formula assuming a small noise power.

For a qualitative discussion of the FM threshold effect, consider first the case when there is a no signal present, so that the carrier wave is unmodulated. Then the composite signal at the frequency discriminator input is

$$x(t) = [A_c + n_I(t)] \cos(2\pi f_c t) - n_Q(t) \sin(2\pi f_c t) \tag{2.153}$$

where $n_I(t)$ and $n_Q(t)$ are the in-phase and quadrature components of the narrowband noise $n(t)$ with respect to the carrier wave. The phasor diagram of Figure 2.43 displays the phase relations between the various components of $x(t)$ in Equation (2.153). As the amplitudes and phases of $n_I(t)$ and $n_Q(t)$ change with time in a random manner, the point $P_1$ [the tip of the phasor representing $x(t)$] wanders around the point $P_2$ (the tip of the phasor representing the carrier). When the carrier-to-noise ratio is large, $n_I(t)$ and $n_Q(t)$ are usually much smaller than the carrier amplitude $A_c$, and so the wandering point $P_1$ in Figure 2.43 spends most of its time near point $P_2$. Thus the angle $\theta(t)$ is approximately $n_Q(t)/A_c$ to within a multiple of $2\pi$. When the carrier-to-noise ratio is low, on the other hand, the wandering point $P_1$ occasionally sweeps around the origin and $\theta(t)$ increases or decreases by $2\pi$ radians. Figure 2.44 illustrates how in a rough way the excursions in $\theta(t)$, depicted in Figure 2.44*a*, produce impulselike components in $\theta'(t) = d\theta/dt$. The discriminator output $v(t)$ is equal to $\theta'(t)/2\pi$. These impulselike components have different heights depending on how close the wandering point $P_1$ comes to the origin O, but all have areas nearly equal to $\pm 2\pi$ radians, as illustrated in Figure 2.44*b*. When the signal shown in Figure 2.44*b* is passed through the postdetection low-pass filter, corresponding but wider impulselike components are excited in the receiver output and are heard as clicks. The clicks are produced only when $\theta(t)$ changes by $\pm 2\pi$ radians.

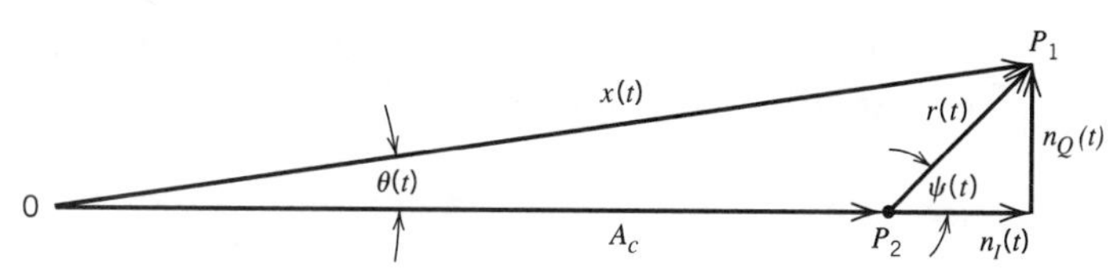


**FIGURE 2.43** Phasor diagram interpretation of Equation (2.153).

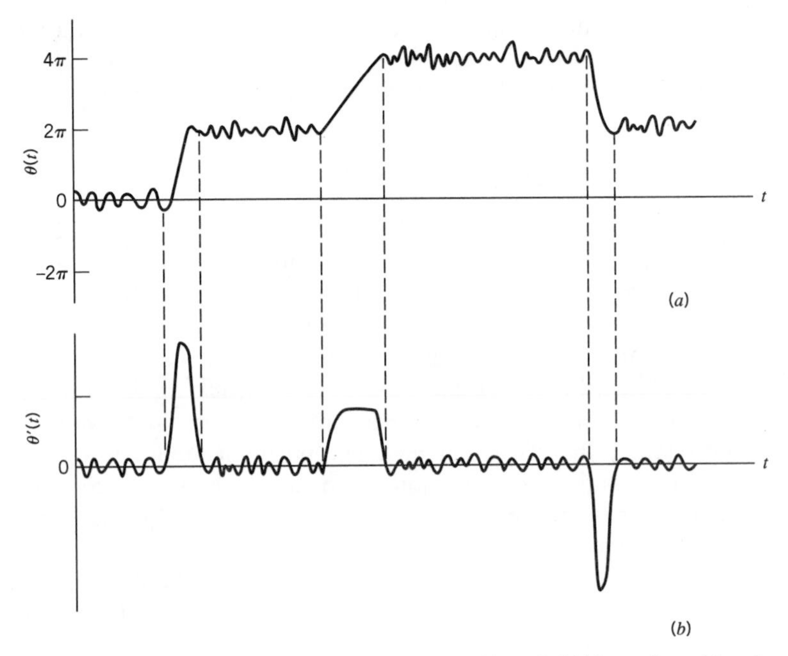


**FIGURE 2.44** Illustrating impulselike components in $\theta'(t) = d\theta(t)/dt$ produced by changes of $2\pi$ in $\theta(t)$; (*a*) and (*b*) are graphs of $\theta(t)$ and $\theta'(t)$, respectively.

From the phasor diagram of Figure 2.43, we may deduce the conditions required for clicks to occur. A positive-going click occurs when the envelope $r(t)$ and phase $\psi(t)$ of the narrowband noise $n(t)$ satisfy the following conditions:

$$r(t) > A_c$$
$$\psi(t) < \pi \le \psi(t) + d\psi(t)$$
$$\frac{d\psi(t)}{dt} > 0$$

These conditions ensure that the phase $\theta(t)$ of the resultant phasor $x(t)$ changes by $2\pi$ radians in the time increment $dt$, during which the phase of the narrowband noise increases by incremental amount $d\psi(t)$. Similarly, the conditions for a negative-going click to occur are as follows:

$$r(t) > A_c$$
$$\psi(t) > -\pi > \psi(t) + d\psi(t)$$
$$\frac{d\psi(t)}{dt} < 0$$

These conditions ensure that $\theta(t)$ changes by $-2\pi$ radians during the time increment $dt$.

The *carrier-to-noise ratio* is defined by

$$\rho = \frac{A_c^2}{2B_T N_0} \tag{2.154}$$

As $\rho$ is decreased, the average number of clicks per unit time increases. When this number becomes appreciably large, threshold is said to occur.

The output signal-to-noise ratio is calculated as follows:

1. The output signal is taken as the receiver output measured in the absence of noise. The average output signal power is calculated assuming a sinusoidal modulation that produces a frequency deviation $\Delta f$ equal to $B_T/2$, so that the carrier swings back and forth across the entire input frequency band.
2. The average output noise power is calculated when there is no signal present; that is, the carrier is unmodulated, with no restriction imposed on the value of the carrier-to-noise ratio $\rho$.

On this heuristic basis, theory[11] yields Curve I of Figure 2.45 presenting a plot of the output signal-to-noise ratio versus the carrier-to-noise ratio when the ratio $B_T/2W$ is equal to 5. This curve shows that the output signal-to-noise ratio deviates appreciably from a linear function of the carrier-to-noise ratio $\rho$ when $\rho$ is less than about 10 dB. Curve II of Figure 2.45 shows the effect of modulation on the output signal-to-noise ratio when the modulating signal (assumed sinusoidal) and the noise are present at the same time. The average output signal power pertaining to curve II may be taken to be effectively the same as for curve I. The average output noise power, however, is strongly dependent on the presence of the modulating signal, which accounts for the noticeable deviation of curve II from curve I. In particular, we find that as $\rho$ decreases from infinity, the output signal-to-

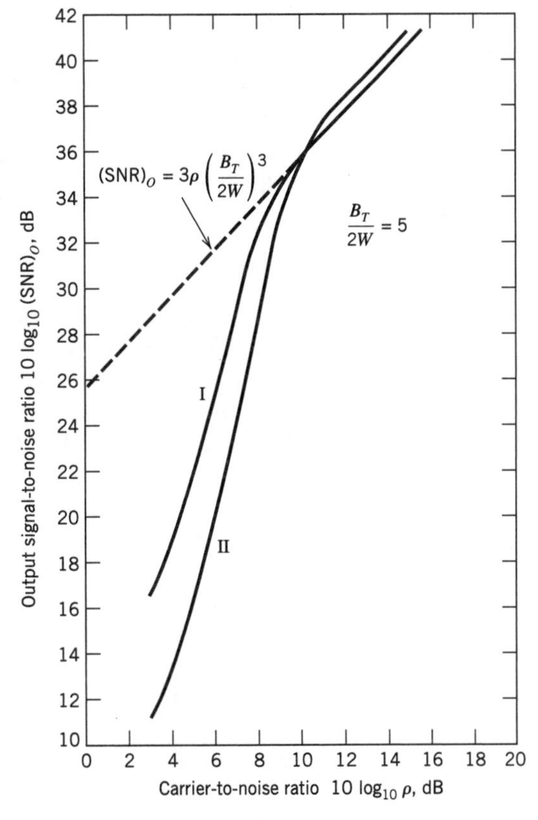


**FIGURE 2.45** Dependence of output signal-to-noise ratio on input carrier-to-noise ratio for FM reciever. In curve I, the average output noise power is calculated assuming an unmodulated carrier. In curve II, the average output noise power is calculated assuming a sinusoidally modulated carrier. Both curves I and II are calculated from theory.

noise deviates appreciably from a linear function of $\rho$ when $\rho$ is about 11 dB. Also when the signal is present, the resulting modulation of the carrier tends to increase the average number of clicks per second. Experimentally, it is found that occasional clicks are heard in the receiver output at a carrier-to-noise ratio of about 13 dB, which appears to be only slightly higher than what theory indicates. Also it is of interest to note that the increase in the average number of clicks per second tends to cause the output signal-to-noise ratio to fall off somewhat more sharply just below the threshold level in the presence of modulation.

From the foregoing discussion we may conclude that threshold effects in FM receivers may be avoided in most practical cases of interest if the carrier-to-noise ratio $\rho$ is equal to or greater than 20 or, equivalently, 13 dB. Thus using Equation (2.154) we find that the loss of message at the discriminator output is negligible if

$$\frac{A_c^2}{2B_T N_0} \geq 20$$

or, equivalently, if the average transmitted power $A_c^2/2$ satisfies the condition

$$\frac{A_c^2}{2} \geq 20 B_T N_0 \tag{2.155}$$

To use this formula, we may proceed as follows:

1. For a specified modulation index $\beta$ and message bandwidth $W$, we determine the transmission bandwidth of the FM wave, $B_T$, using the universal curve of Figure 2.26 or Carson's rule.
2. For a specified average noise power per unit bandwidth, $N_0$, we use Equation (2.155) to determine the minimum value of the average transmitted power $A_c^2/2$ that is necessary to operate above threshold.

## ■ FM THRESHOLD REDUCTION

In communication systems using frequency modulation, there is particular interest in reducing the noise threshold in an FM receiver so as to satisfactorily operate the receiver with the minimum signal power possible. Threshold reduction in FM receivers may be achieved by using an FM demodulator with negative feedback[12] (commonly referred to as an *FMFB demodulator*), or by using a *phase-locked loop demodulator*. Such devices are referred to as *extended-threshold demodulators*, the idea of which is illustrated in Figure 2.46. The threshold extension shown in this figure is measured with respect to the standard frequency discriminator (i.e., one without feedback).

The block diagram of an FMFB demodulator[13] is shown in Figure 2.47. We see that the local oscillator of the conventional FM receiver has been replaced by a voltage-controlled oscillator (VCO) whose instantaneous output frequency is controlled by the demodulated signal. In order to understand the operation of this receiver, suppose for the moment that the VCO is removed from the circuit and the feedback loop is left open. Assume that a wideband FM signal is applied to the receiver input, and a second FM signal, from the same source but whose modulation index is a fraction smaller, is applied to the VCO terminal of the mixer. The output of the mixer would consist of the difference frequency component, because the sum frequency component is removed by the band-pass filter. The frequency deviation of the mixer output would be small, although the frequency deviation of both input FM waves is large, since the difference between their instantaneous deviations is small. Hence, the modulation indices would subtract and the resulting FM wave at the mixer output would have a smaller modulation index. The FM wave with

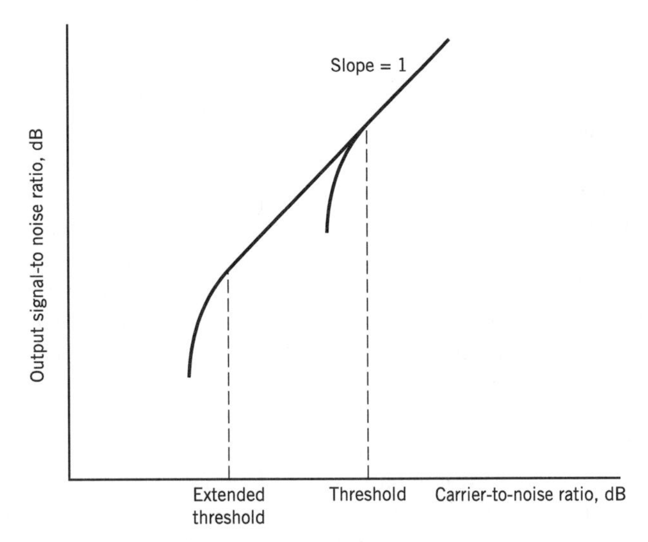


**FIGURE 2.46** FM threshold extension.

reduced modulation index may be passed through a band-pass filter, whose bandwidth need only be a fraction of that required for either wideband FM, and then frequency demodulated. It is now apparent that the second wideband FM signal applied to the mixer may be obtained by feeding the output of the frequency discriminator back to the VCO.

It will now be argued that the signal-to-noise ratio of an FMFB receiver is the same as that of a conventional FM receiver with the same input signal and noise power if the carrier-to-noise ratio is sufficiently large. Assume for the moment that there is no feedback around the demodulator. In the combined presence of an unmodulated carrier $A_c \cos(2\pi f_c t)$ and narrowband noise

$$n(t) = n_I(t) \cos(2\pi f_c t) - n_Q(t) \sin(2\pi f_c t)$$

the phase of the composite signal $x(t)$ at the limiter-discriminator input is approximately equal to $n_Q(t)/A_c$, assuming that the carrier-to-noise ratio is high. The envelope of $x(t)$ is of no interest to us, because the limiter removes all variations in the envelope. Thus the composite signal at the frequency discriminator input consists of a small index phase-modulated wave with the modulation derived from the component $n_Q(t)$ of noise that is in phase quadrature with the carrier. When feedback is applied, the VCO generates a frequency-modulated signal that reduces the phase-modulation index of the wave in the band-pass filter output, that is, the quadrature component $n_Q(t)$ of noise. Thus we see that as long as the carrier-to-noise ratio is sufficiently large, the FMFB receiver does not respond to the in-phase noise component $n_I(t)$, but that it would demodulate the quadrature noise component $n_Q(t)$ in exactly the same fashion as it would demodulate signal modulation.

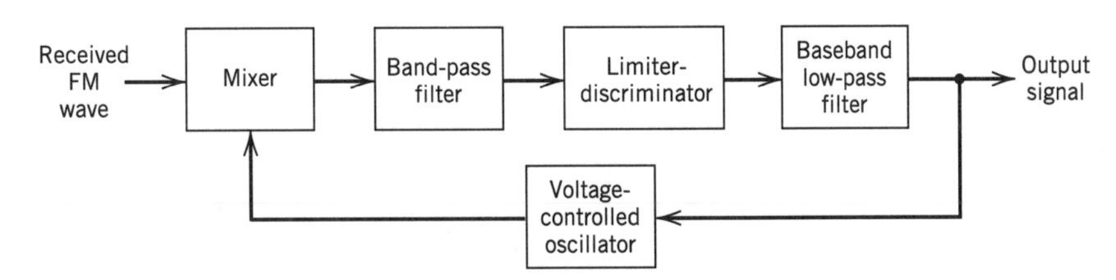


**FIGURE 2.47** FM demodulator with negative feedback.

Signal and quadrature noise are reduced in the same proportion by the applied feedback, with the result that the baseband signal-to-noise ratio is independent of feedback. For large carrier-to-noise ratios, the baseband signal-to-noise ratio of an FMFB receiver is then the same as that of a conventional FM receiver.

The reason that an FMFB receiver is able to extend the threshold is that, unlike a conventional FM receiver, it uses a very important piece of *a priori* information, namely, that even though the carrier frequency of the incoming FM wave will usually have large frequency deviations, its rate of change will be at the baseband rate. An FMFB demodulator is essentially a *tracking filter* that can track only the slowly varying frequency of a wideband FM signal, and consequently it responds only to a narrowband of noise centered about the instantaneous carrier frequency. The bandwidth of noise to which the FMFB receiver responds is precisely the band of noise that the VCO tracks. The end result is that an FMFB receiver is capable of realizing a threshold extension on the order of 5–7 dB, which represents a significant improvement in the design of minimum power FM systems.

Like the FMFB demodulator, the phase-locked loop (discussed later in Section 2.14) is also a tracking filter and, as such, the noise bandwidth to which it responds is precisely the band of noise tracked by the VCO. Indeed, the phase-locked loop demodulator offers a threshold extension capability with a relatively simple circuit. Unfortunately, the amount of threshold extension is not predictable by any existing theory, and it depends on signal parameters. Roughly speaking, improvement by a few (on the order of 2 to 3) decibels is achieved in typical applications, which is not as good as an FMFB demodulator.

## ■ PRE-EMPHASIS AND DE-EMPHASIS IN FM

Equation (2.147) shows that the power spectral density of the noise at the output of an FM receiver has a square-law dependence on the operating frequency; this is illustrated in Figure 2.48*a*. In Figure 2.48*b*, we have included the power spectral density of a typical message source; audio and video signals typically have spectra of this form. In particular, we see that the power spectral density of the message usually falls off appreciably at higher frequencies. On the other hand, the power spectral density of the output noise increases rapidly with frequency. Thus around $f = \pm W$, the relative spectral density of the message is quite low, whereas that of the output noise is quite high in comparison. Clearly, the message is not using the frequency band allotted to it in an efficient manner. It may appear that one way of improving the noise performance of the system is to slightly reduce the bandwidth of the postdetection low-pass filter so as to reject a large amount of noise power while losing only a small amount of message power. Such an approach, however, is usually not satisfactory because the distortion of the message caused by the reduced filter bandwidth, even though slight, may not be tolerable. For example, in the case of music, we find that although the high-frequency notes contribute only a very small fraction of the total power, nonetheless, they contribute a great deal from an esthetic viewpoint.

A more satisfactory approach to the efficient use of the allowed frequency band is based on the use of *pre-emphasis* in the transmitter and *de-emphasis* in the receiver, as

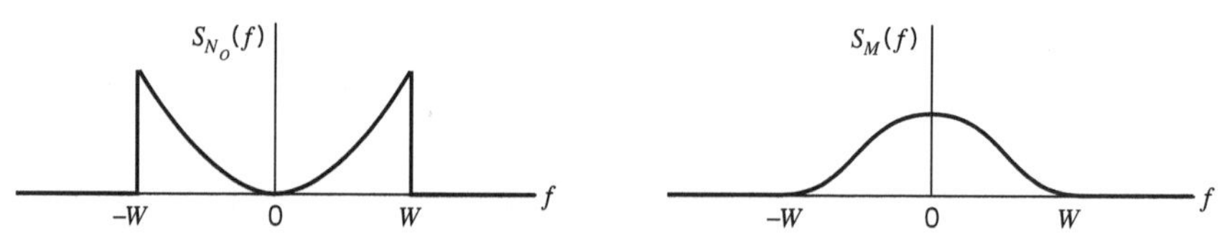


**FIGURE 2.48** (*a*) Power spectral density of noise at FM receiver output. (*b*) Power spectral density of a typical message signal.

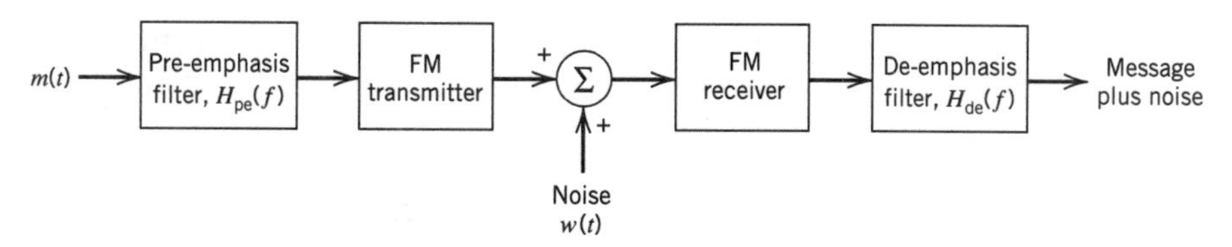


**FIGURE 2.49** Use of pre-emphasis and de-emphasis in an FM system.

illustrated in Figure 2.49. In this method, we artificially emphasize the high-frequency components of the message signal prior to modulation in the transmitter, and therefore before the noise is introduced in the receiver. In effect, the low-frequency and high-frequency portions of the power spectral density of the message are equalized in such a way that the message fully occupies the frequency band allotted to it. Then, at the discriminator output in the receiver, we perform the inverse operation by de-emphasizing the high-frequency components, so as to restore the original signal-power distribution of the message. In such a process, the high-frequency components of the noise at the discriminator output are also reduced, thereby effectively increasing the output signal-to-noise ratio of the system. Such a pre-emphasis and de-emphasis process is widely used in commercial FM radio transmission and reception.

In order to produce an undistorted version of the original message at the receiver output, the pre-emphasis filter in the transmitter and the de-emphasis filter in the receiver must ideally have frequency responses that are the inverse of each other. That is, if $H_{pe}(f)$ designates the frequency response of the pre-emphasis filter, then the frequency response $H_{de}(f)$ of the de-emphasis filter must ideally be (ignoring transmission delay)

$$H_{de}(f) = \frac{1}{H_{pe}(f)}, \qquad -W \le f \le W \tag{2.156}$$

This choice of frequency responses makes the average message power at the receiver output independent of the pre-emphasis and de-emphasis procedure.

From our previous noise analysis in FM systems, assuming a high carrier-to-noise ratio, the power spectral density of the noise $n_d(t)$ at the discriminator output is given by Equation (2.146). The modified power spectral density of the noise at the de-emphasis filter output is therefore

$$|H_{de}(f)|^2 S_{N_d}(f) = \begin{cases} \dfrac{N_0 f^2}{A_c^2} |H_{de}(f)|^2, & |f| \le \dfrac{B_T}{2} \\ 0, & \text{otherwise} \end{cases} \tag{2.157}$$

Recognizing, as before, that the postdetection low-pass filter has a bandwidth $W$ that is, in general, less than $B_T/2$, we find that the average power of the modified noise at the receiver output is as follows:

$$\begin{pmatrix} \text{Average output noise} \\ \text{power with de-emphasis} \end{pmatrix} = \frac{N_0}{A_c^2} \int_{-W}^{W} f^2 |H_{de}(f)|^2 \, df \tag{2.158}$$

Because the average message power at the receiver output is ideally unaffected by the combined pre-emphasis and de-emphasis procedure, it follows that the improvement in output signal-to-noise ratio produced by the use of pre-emphasis in the transmitter and de-emphasis in the receiver is defined by

$$I = \frac{\text{average output noise power without pre-emphasis and de-emphasis}}{\text{average output noise power with pre-emphasis and de-emphasis}} \tag{2.159}$$

Earlier we showed that the average output noise power without pre-emphasis and de-emphasis is equal to $(2N_0W^3/3A_c^2)$; see Equation (2.148). Therefore, after cancellation of common terms, we may express the improvement factor $I$ as

$$I = \frac{2W^3}{3\int_{-W}^{W} f^2 |H_{\text{de}}(f)|^2 \, df} \tag{2.160}$$

It must be emphasized that this improvement factor assumes the use of a high carrier-to-noise ratio at the discriminator input in the receiver.

## ▶ EXAMPLE 2.6

A simple pre-emphasis filter that emphasizes high frequencies and is commonly used in practice is defined by the frequency response

$$H_{\text{pe}}(f) = 1 + \frac{jf}{f_0}$$

which is closely realized by the RC-amplifier network shown in Figure 2.50*a*, provided that $R \ll r$ and $2\pi f Cr \ll 1$ inside the frequency band of interest. The amplifier in Figure 2.50*a* is intended to make up for the attenuation introduced by the RC network at low frequencies. The frequency parameter $f_0$ is $1/(2\pi Cr)$.

The corresponding de-emphasis filter in the receiver is defined by the frequency response

$$H_{\text{de}}(f) = \frac{1}{1 + jf/f_0}$$

which can be realized using the simple RC network of Figure 2.50*b*.

The improvement in output signal-to-noise ratio of the FM receiver, resulting from the combined use of the pre-emphasis and de-emphasis filters of Figure 2.50, is therefore

$$\begin{aligned} I &= \frac{2W^3}{3\int_{-W}^{W} \frac{f^2\, df}{1 + (f/f_0)^2}} \\ &= \frac{(W/f_0)^3}{3[(W/f_0) - \tan^{-1}(W/f_0)]} \end{aligned} \tag{2.161}$$

In commercial FM broadcasting, we typically have $f_0 = 2.1$ kHz, and we may reasonably assume $W = 15$ kHz. This set of values yields $I = 22$, which corresponds to an improvement of 13 dB in the output signal-to-noise ratio of the receiver. The output signal-to-noise

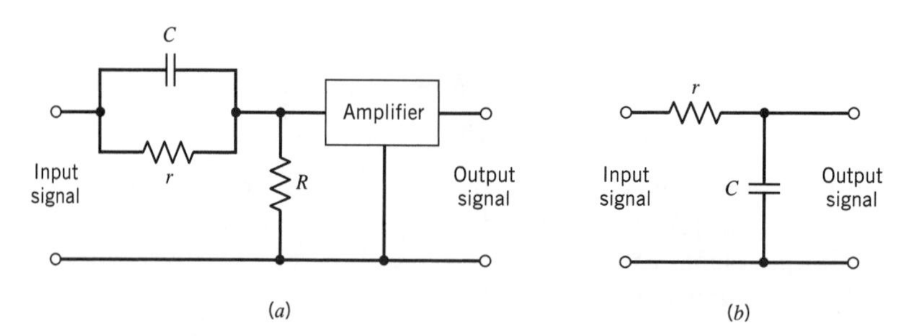


**FIGURE 2.50** (*a*) Pre-emphasis filter. (*b*) De-emphasis filter.

ratio of an FM receiver without pre-emphasis and de-emphasis is typically 40–50 dB. We see, therefore, that by using the simple pre-emphasis and de-emphasis filters shown in Figure 2.50, we can realize a significant improvement in the noise performance of the receiver. ◀

The use of the simple *linear* pre-emphasis and de-emphasis filters just described is an example of how the performance of an FM system may be improved by using the differences between characteristics of signals and noise in the system. These simple filters also find application in audio tape-recording. Specifically, *nonlinear* pre-emphasis and de-emphasis techniques have been applied successfully to tape recording. These techniques[14] (known as *Dolby-A*, *Dolby-B*, and *DBX* systems) use a combination of filtering and dynamic range compression to reduce the effects of noise, particularly when the signal level is low.

## 2.14 *Computer Experiments: Phase-Locked Loop*

The experimental study presented in this section focuses on the use of a phase-locked loop for the demodulation of a frequency modulated signal. Before proceeding with the experiments, however, we first present a brief exposition of phase-locked loop theory.

Basically, the *phase-locked loop* consists of three major components: a *multiplier*, a *loop filter*, and a *voltage-controlled oscillator* (VCO) connected together in the form of a feedback system, as shown in Figure 2.51. The VCO is a sinusoidal generator whose frequency is determined by a voltage applied to it from an external source. In effect, any frequency modulator may serve as a VCO. We assume that initially we have adjusted the VCO so that when the control voltage is zero, two conditions are satisfied:

1. The frequency of the VCO is precisely set at the unmodulated carrier frequency $f_c$.
2. The VCO output has a 90 degree phase-shift with respect to the unmodulated carrier wave.

Suppose then that the input signal applied to the phase-locked loop is an FM signal defined by

$$s(t) = A_c \sin[2\pi f_c t + \phi_1(t)]$$

where $A_c$ is the carrier amplitude. With a modulating signal $m(t)$, the angle $\phi_1(t)$ is related to $m(t)$ by the integral

$$\phi_1(t) = 2\pi k_f \int_0^t m(\tau)\, d\tau$$

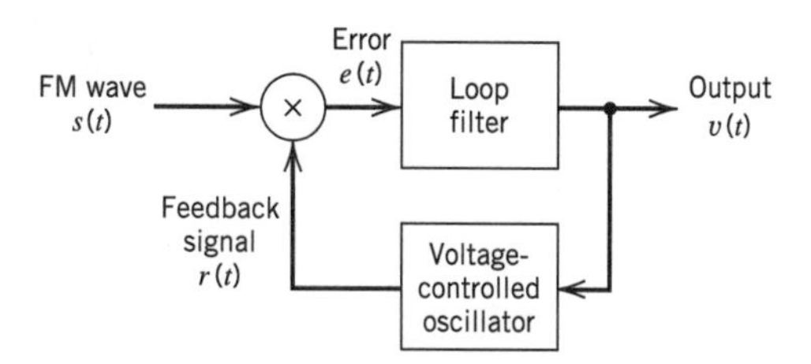


**FIGURE 2.51** Phase-locked loop.

where $k_f$ is the frequency sensitivity of the frequency modulator. Let the VCO output in the phase-locked loop be defined by

$$r(t) = A_v \cos[2\pi f_c t + \phi_2(t)]$$

where $A_v$ is the amplitude. With a control voltage $v(t)$ applied to the VCO input, the angle $\phi_2(t)$ is related to $v(t)$ by the integral

$$\phi_2(t) = 2\pi k_v \int_0^t v(\tau)\, d\tau \tag{2.162}$$

where $k_v$ is the frequency sensitivity of the VCO, measured in Hertz per volt. The object of the phase-locked loop is to generate a VCO output $r(t)$ that has the same phase angle (except for the fixed difference of 90 degrees) as the input FM signal $s(t)$. The time-varying phase angle $\phi_1(t)$ characterizing $s(t)$ may be due to modulation by a message signal $m(t)$, in which case we wish to recover $\phi_1(t)$ and thereby produce an estimate of $m(t)$. In other applications of the phase-locked loop, the time-varying phase angle $\phi_1(t)$ of the incoming signal $s(t)$ may be an unwanted phase shift caused by fluctuations in the communication channel; in this latter case, we wish to *track* $\phi_1(t)$ so as to produce a signal with the same phase angle for the purpose of coherent detection (synchronous demodulation).

## ■ MODEL OF THE PHASE-LOCKED LOOP[15]

To develop an understanding of the phase-locked loop, it is desirable to have a *model* of the loop. We start by developing a nonlinear model, which is subsequently linearized to simplify the analysis. According to Figure 2.51, the incoming FM signal $s(t)$ and the VCO output $r(t)$ are applied to the multiplier, producing two components:

1. A high-frequency component, represented by the *double-frequency* term

$$k_m A_c A_v \sin[4\pi f_c t + \phi_1(t) + \phi_2(t)]$$

2. A low-frequency component represented by the *difference-frequency* term

$$k_m A_c A_v \sin[\phi_1(t) - \phi_2(t)]$$

where $k_m$ is the *multiplier gain*, measured in volt$^{-1}$.

The loop filter in the phase-locked loop is a low-pass filter, and its response to the high-frequency component will be negligible. The VCO also contributes to the attenuation of this component. Therefore, discarding the high-frequency component (i.e., the double-frequency term), the input to the loop filter is reduced to

$$e(t) = k_m A_c A_v \sin[\phi_e(t)] \tag{2.163}$$

where $\phi_e(t)$ is the *phase error* defined by

$$\begin{aligned} \phi_e(t) &= \phi_1(t) - \phi_2(t) \\ &= \phi_1(t) - 2\pi k_v \int_0^t v(\tau)\, d\tau \end{aligned} \tag{2.164}$$

The loop filter operates on the error $e(t)$ to produce an output $v(t)$ defined by the convolution integral:

$$v(t) = \int_{-\infty}^{\infty} e(\tau) h(t - \tau)\, d\tau \tag{2.165}$$

where $h(t)$ is the impulse response of the loop filter. Using Equations (2.164) and (2.165) to relate $\phi_e(t)$ and $\phi_1(t)$, we obtain the following nonlinear integro-differential equation as the descriptor of the dynamic behavior of the phase-locked loop:

$$\frac{d\phi_e(t)}{dt} = \frac{d\phi_1(t)}{dt} - 2\pi K_0 \int_{-\infty}^{\infty} \sin[\phi_e(\tau)]h(t - \tau)\, d\tau \tag{2.166}$$

where $K_0$ is as *loop-gain parameter* defined by

$$K_0 = k_m k_v A_c A_v \tag{2.167}$$

The amplitudes $A_c$ and $A_v$ are both measured in volts, the multiplier gain $k_m$ in volt$^{-1}$ and the frequency sensitivity $k_v$ in Hertz per volt. Hence, it follows from Equation (2.167) that $K_0$ has the dimensions of frequency. Equation (2.166) suggests the model shown in Figure 2.52 for a phase-locked loop. In this model we have also included the relationship between $v(t)$ and $e(t)$ as represented by Equations (2.163) and (2.165). We see that the model of Figure 2.52 resembles the actual block diagram of Figure 2.51. The multiplier at the input of the phase-locked loop is replaced by a subtracter and a sinusoidal nonlinearity, and the VCO by an integrator.

The sinusoidal nonlinearity in the model of Figure 2.52 complicates the task of analyzing the behavior of the phase-locked loop. It would be helpful to *linearize* this model to simplify the analysis and yet give a good approximate description of the loop's behavior in certain modes of operation. When the phase error $\phi_e(t)$ is zero, the phase-locked loop is said to be in *phase-lock*. When $\phi_e(t)$ is at all times small compared with one radian, we may use the approximation

$$\sin[\phi_e(t)] \simeq \phi_e(t)$$

which is accurate to within 4 percent for $\phi_e(t)$ less than 0.5 radians. In this case, the loop is said to be *near phase-lock*, and the sinusoidal nonlinearity of Figure 2.52 may be disregarded. Under this condition, $v(t)$ is approximately equal to $m(t)$, except for the scaling factor $k_f/k_v$.

The complexity of the phase-locked loop is determined by the frequency response $H(f)$ of the loop filter. The simplest form of a phase-locked loop is obtained when $H(f) = 1$; that is, there is no loop filter, and the resulting phase-locked loop is referred to as a *first-order phase-locked loop*. A major limitation of a first-order phase-locked loop is that the loop gain parameter $K_0$ controls both the loop bandwidth as well as the hold-in frequency range of the loop; the *hold-in frequency range* refers to the range of frequencies

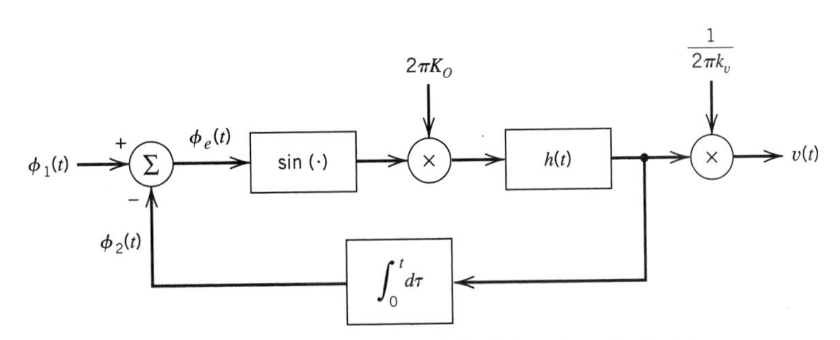


**FIGURE 2.52** Nonlinear model of the phase-locked loop.

for which the loop remains phase-locked to the input signal. We may overcome this limitation by using a loop filter with the frequency response

$$H(f) = 1 + \frac{a}{jf} \tag{2.168}$$

where $a$ is a constant. Then with this loop filter in place and the phase-locked loop operating in its linear mode, we find from Equation (2.166) that the phase-locked loop behaves as a *second-order feedback system*, as shown by the standard frequency response

$$\frac{\Phi_e(f)}{\Phi_1(f)} = \frac{(jf/f_n)^2}{1 + 2\,\zeta(jf/f_n) + (jf/f_n)^2} \tag{2.169}$$

where $\Phi_e(f)$ and $\Phi_1(f)$ are the Fourier transforms of $\phi_e(t)$ and $\phi_1(t)$, respectively. The system is parameterized by the *natural frequency*, $f_n$, and *damping factor*, $\zeta$, which are respectively defined by

$$f_n = \sqrt{aK_0} \tag{2.170}$$

and

$$\zeta = \sqrt{\frac{K_0}{4a}} \tag{2.171}$$

The second-order phase-locked loop so described is the subject of the computer experiments presented next.

### Experiment 1: Acquisition Mode

When a phase-locked loop is used for coherent detection (synchronous demodulation), the loop must first lock onto the input signal and then follow the variations of its phase angle with time. The process of bringing a loop into phase-lock is called *acquisition*, and the ensuing process of following angular variations in the input signal is called *tracking*. In the acquisition mode and quite possibly the tracking mode, the phase error $\phi_e(t)$ between the input signal $s(t)$ and the VCO output $r(t)$ will certainly be large, thereby mandating the use of the nonlinear model of Figure 2.52. However, a nonlinear analysis of the acquisition process based on this latter model is beyond the scope of this book. In this experiment, we use computer simulations to study the acquisition process and thereby develop insight into some of its features.

Consider a second-order phase-locked loop using the loop filter of Equation (2.168) and having the following parameters:

$$\text{Natural frequency } f_n = \frac{1}{2\pi}\text{ Hz}$$

$$\text{Damping factor } \zeta = 0.3,\ 0.707,\ 1.0$$

To accommodate variation in $\zeta$, the filter parameter $a$ is varied in accordance with the formula

$$a = \frac{f_n}{2\zeta}$$

which follows from Equations (2.170) and (2.171). Figure 2.53 presents the variation in the phase error $\phi_e(t)$ with time for each of the three specified values of damping factor $\zeta$,

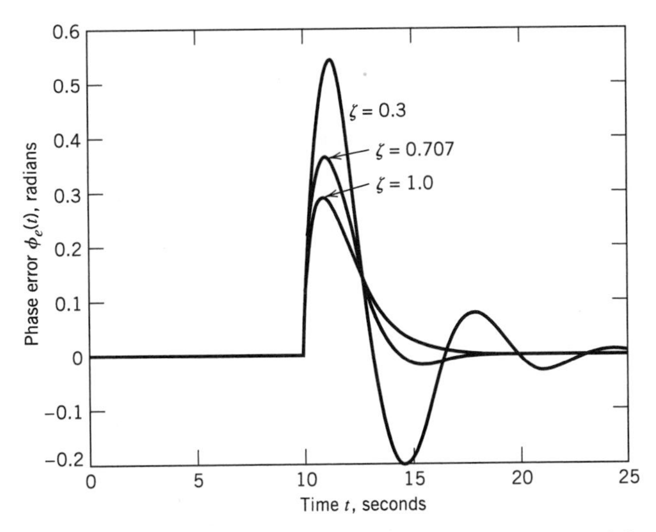


**FIGURE 2.53** Variation of the phase error for three different values of damping factor.

assuming a frequency step of 0.125 Hz. These results show that the damping factor $\zeta = 0.707$ gives the best compromise between a fast response time and an underdamped oscillatory behavior.

## Experiment 2: Phase-Plane Portrait

A *phase-plane portrait* is a family of trajectories, with each trajectory representing a single solution of Equation (2.166). For the second experiment we plot the phase-plane portrait of a second-order phase-locked loop for the case of sinusoidal modulation. The system parameters of the loop are as follows:

$$\text{Loop-gain parameter } K_0 = \frac{50}{2\pi} \text{ Hz}$$

$$\text{Loop-natural frequency } f_n = \frac{50}{2\sqrt{2}} \text{ Hz}$$

$$\text{Sinusoidal modulation frequency } f_m = \frac{50}{2\pi\sqrt{2\pi}} \text{ Hz}$$

Figure 2.54 presents the phase-plane portrait of the phase-locked loop adjusted for critical damping, where the trajectories (frequency error versus phase error) are plotted for different starting points. From this portrait we make the following observations:

1. For a sinusoidal nonlinearity, the phase-plane portrait is itself periodic with period $2\pi$ in the phase error $\phi_e$, but it is aperiodic in $d\phi_e/dt$.
2. For an initial frequency error

$$\frac{1}{K}\frac{d\phi_e}{dt}$$

with an absolute value less than or equal to 1, the phase-locked loop is assured of attaining a *stable (equilibrium) point* at (0, 0) or (0, $2\pi$); the multiplicity of equilibrium points is a manifestation of periodicity of the phase-plane portrait.

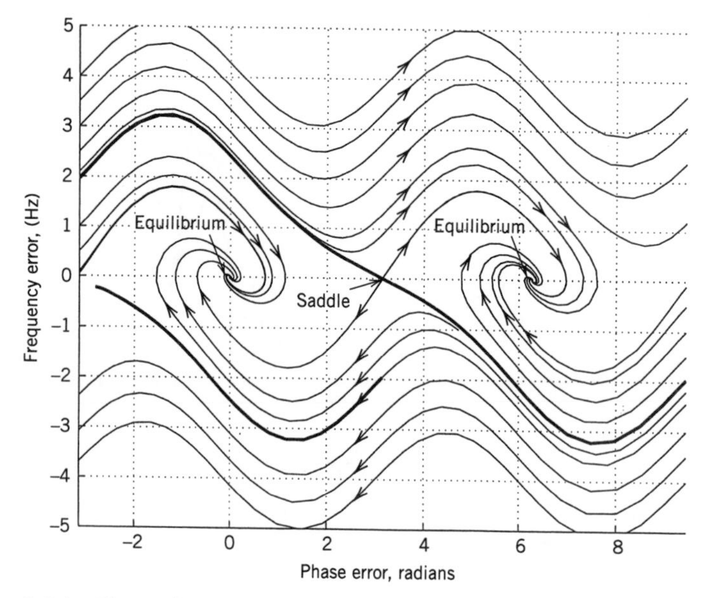


**FIGURE 2.54** Phase-plane portrait for critical damping and sinusoidal modulation.

3. For an initial frequency error

$$\frac{1}{K}\frac{d\phi_e}{dt}$$

with an absolute value equal to 2, we have a *saddle point* at $(0, \pi)$ where the slightest perturbation applied to the phase-locked loop causes it to shift to the equilibrium point $(0, 0)$ or $(0, 2\pi)$.

## 2.15 *Summary and Discussion*

In this chapter we studied the principles of continuous-wave (CW) modulation. This analog form of modulation uses a sinusoidal carrier whose amplitude or angle is varied in accordance with a message signal. We may thus distinguish two families of CW modulation: amplitude modulation and angle modulation.

### ■ AMPLITUDE MODULATION

Amplitude modulation may itself be classified into four types, depending on the spectral content of the modulated signal. The four types of amplitude modulation and their practical merits are as follows:

1. *Full amplitude modulation (AM), in which the upper and lower sidebands are transmitted in full, accompanied by the carrier wave.*
Accordingly, demodulation of an AM signal is accomplished rather simply in the receiver by using an envelope detector, for example. It is for this reason we find that full AM is commonly used in commercial AM *radio broadcasting*, which involves a single powerful transmitter and numerous receivers that are relatively inexpensive to build.

2. *Double sideband-suppressed carrier (DSB-SC) modulation, in which only the upper and lower sidebands are transmitted.*
The suppression of the carrier wave means that DSB-SC modulation requires much less power than full AM to transmit the same message signal; this advantage of DSB-SC modulation over full AM is, however, attained at the expense of increased receiver complexity. DSB-SC modulation is therefore well suited for *point-to-point communication* involving one transmitter and one receiver; in this form of communication, transmitted power is at a premium and the use of a complex receiver is therefore justifiable.

3. *Single sideband (SSB) modulation, in which only the upper sideband or lower sideband is transmitted.*
SSB modulation is the optimum form of CW modulation in the sense that it requires the minimum transmitted power and the minimum channel bandwidth for conveying a message signal from one point to another. However, its use is limited to message signals with an energy gap centered on zero frequency.

4. *Vestigial sideband modulation, in which almost all of one sideband and a vestige of the other sideband are transmitted in a prescribed complementary fashion.*
VSB modulation requires a channel bandwidth that is between that required for SSB and DSB-SC systems, and the saving in bandwidth can be significant if modulating signals with large bandwidths are being handled, as in the case of television signals and high-speed data.

DSB-SC, SSB, and VSB are examples of linear modulation, whereas, strictly speaking, full AM is nonlinear. However, the deviation of full AM from linearity is of a mild sort. Accordingly, all four forms of amplitude modulation lend themselves readily to spectral analysis using the Fourier transform.

## Angle Modulation

Angle modulation may be classified into frequency modulation (FM) and phase modulation (PM). *In FM, the instantaneous frequency of a sinusoidal carrier is varied in proportion to the message signal. In PM, on the other hand, it is the phase of the carrier that is varied in proportion to the message signal.* The instantaneous frequency is defined as the derivative of the phase with respect to time, except for the scaling factor $1/(2\pi)$. Accordingly, FM and PM are closely related to each other; if we know the properties of the one, we can determine those of the other. For this reason, and because FM is commonly used in broadcasting, much of the material on angle modulation in the chapter was devoted to FM.

Unlike amplitude modulation, FM is a nonlinear modulation process. Accordingly, spectral analysis of FM is more difficult than for AM. Nevertheless, by studying single-tone FM, we were able to develop a great deal of insight into the spectral properties of FM. In particular, we derived an empirical rule known as Carson's rule for an approximate evaluation of the transmission bandwidth $B_T$ of FM. According to this rule, $B_T$ is controlled by a single parameter: the modulation index $\beta$ for sinusoidal FM, or the deviation ratio $D$ for nonsinusoidal FM.

## Noise Analysis

We conclude the chapter on CW modulation systems by presenting a comparison of their noise performances. For this comparison, we assume that the modulation is produced by

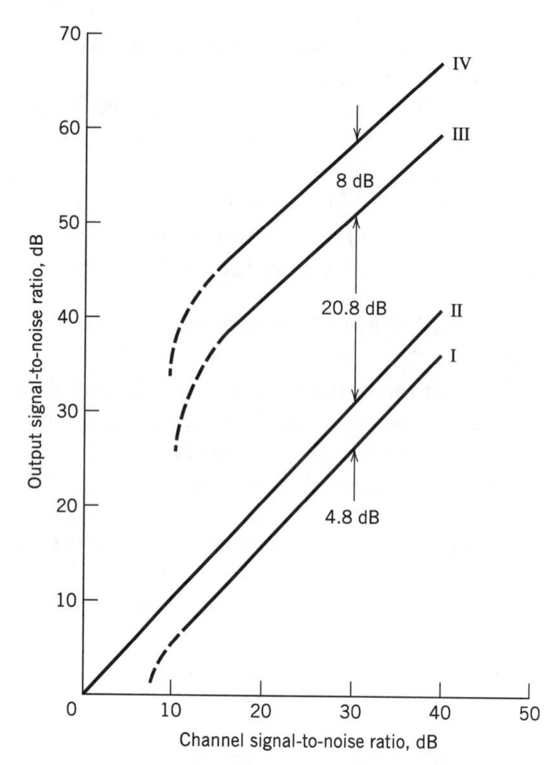


**FIGURE 2.55** Comparison of the noise performance of various CW modulation systems. Curve I: Full AM, $\mu = 1$. Curve II: DSB-SC, SSB. Curve III: FM, $\beta = 2$. Curve IV: FM, $\beta = 5$. (Curves III and IV include 13-dB pre-emphasis, de-emphasis improvement.)

a sinusoidal wave. For the comparison to be meaningful, we also assume that the modulation systems operate with exactly the same channel signal-to-noise ratio. We may thus plot the output signal-to-noise ratio versus the channel signal-to-noise ratio as in Figure 2.55 for the following modulation schemes:

- Full AM with 100 percent modulation
- Coherent DSB-SC, SSB
- FM with $\beta = 2$ and $\beta = 5$

Figure 2.55 also includes the AM and FM threshold effects. In making the comparison, it is informative to keep in mind the transmission bandwidth requirement of the modulation system in question. In this regard, we use a *normalized transmission bandwidth* defined by

$$B_n = \frac{B_T}{W}$$

where $B_T$ is the transmission bandwidth of the modulated signal, and $W$ is the message bandwidth. Table 2.4 presents the values of $B_n$ for the different CW modulation schemes.

From Figure 2.55 and Table 2.4 we make the following observations:

- Among the family of AM systems, SSB modulation is optimum with regard to noise performance as well as bandwidth conservation.
- The use of FM improves noise performance but at the expense of an excessive transmission bandwidth. This assumes that the FM system operates above threshold for the noise improvement to be realized.

**TABLE 2.4** *Values of $B_n$ for various CW modulation schemes*

| | | | *FM* | |
|---|---|---|---|---|
| | *AM, DSB-SC* | *SSB* | $\beta = 2$ | $\beta = 5$ |
| $B_n$ | 2 | 1 | 8 | 16 |

On an important point to conclude the discussion on CW modulation, only frequency modulation offers the capability to trade off transmission bandwidth for improved noise performance. The trade-off follows a square law, which is the best that we can do with CW modulation (i.e., analog communications). In Chapter 3 we describe pulse-code modulation, which is basic to the transmission of analog information-bearing signals by a digital communication system, and which can indeed do much better.

## NOTES AND REFERENCES

1. It appears that the terms *continuous wave* and *heterodyning* were first used by Reginald Fessenden in the early 1900s.
2. The Costas receiver is named in honor of its inventor; see the paper by Costas (1956).
3. Bessel functions play an important role in the study of both analog and digital communication systems. They can be of the so-called first kind or second kind. Appendix 3 discusses mathematical details and properties of both kinds of Bessel functions. A table of Bessel functions of the first kind is presented in Table A6.5.
4. Carson's rule for the bandwidth of FM signals is named in honor of its originator; Carson and Fry (1937) wrote one of the early classic papers on frequency modulation theory.
5. The indirect method of generating a wideband FM wave was first proposed by Armstrong (1936). Armstrong was also the first to recognize the noise-robustness properties of frequency modulation.
6. Stereo multiplexing usually involves the use of frequency modulation for radio transmission. However, it may also be transmitted using amplitude modulation as discussed in Problem 2.14; for more details, see the paper by Mennie (1978).
7. For detailed description of the superheterodyne receiver, see the *Radio Engineering Handbook* edited by Henney (1958, pp. 19-34–19-41).
8. The qualitative study of threshold in envelope detection presented here follows Downing (1964, p. 71).
9. For a justification of the critical assumption on which the simplification presented in Equation (2.142) rests, see Rice (1963).
10. For a detailed discussion of the threshold effect in FM receivers, see the paper by Rice (1963) and the book by Schwartz, Bennett, and Stein (1966, pp. 129–163).
11. Figure 2.45 is adapted from Rice (1963). The validity of the theoretical curve II in this figure has been confirmed experimentally; see Schwartz, Bennett, and Stein (1966, p. 153). For some earlier experimental work on the threshold phenomenon in FM, see the paper by Crosby (1937).
12. The idea of using feedback around an FM demodulator was originally proposed by Chaffee (1939).

13. The treatment of the FMFB demodulator presented in Section 2.13 is based on the paper by Enloe (1962); see also Roberts (1977, pp. 166–181).
14. For a detailed discussion of Dolby systems mentioned in the latter part of Section 2.13, see Stremler (1990, pp. 732–734).
15. For a full treatment of the nonlinear analysis of a phase-locked loop, see Gardner (1979), Lindsey (1972), and Viterbi (1966).

## PROBLEMS

### Amplitude Modulation

2.1 Suppose that nonlinear devices are available for which the output current $i_o$ and input voltage $v_i$ are related by

$$i_o = a_1 v_i + a_3 v_i^3$$

where $a_1$ and $a_3$ are constants. Explain how these devices may be used to provide: (a) a product modulator and (b) an amplitude modulator.

2.2 Figure P2.2 shows the circuit diagram of a *square-law modulator*. The signal applied to the nonlinear device is relatively weak, such that it can be represented by a square law:

$$v_2(t) = a_1 v_1(t) + a_2 v_1^2(t)$$

where $a_1$ and $a_2$ are constants, $v_1(t)$ is the input voltage, and $v_2(t)$ is the output voltage. The input voltage is defined by

$$v_1(t) = A_c \cos(2\pi f_c t) + m(t)$$

where $m(t)$ is a message signal and $A_c \cos(2\pi f_c t)$ is the carrier wave.

(a) Evaluate the output voltage $v_2(t)$.

(b) Specify the frequency response that the tuned circuit in Figure P2.2 must satisfy in order to generate an AM signal with $f_c$ as the carrier frequency.

(c) What is the amplitude sensitivity of this AM signal?

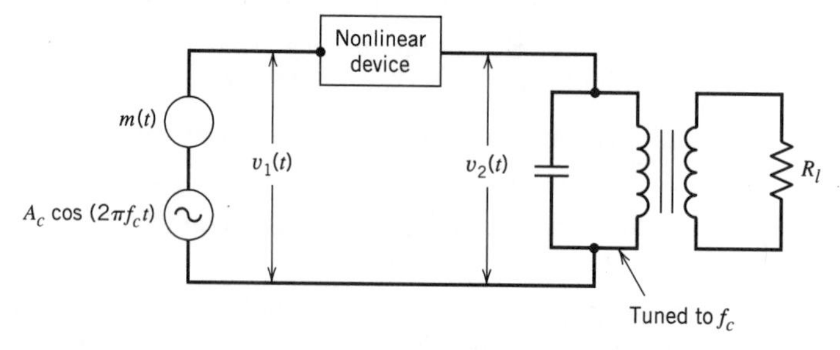


**FIGURE P2.2**

2.3 Figure P2.3a shows the circuit diagram of a *switching modulator*. Assume that the carrier wave $c(t)$ applied to the diode is large in amplitude, so that the diode acts like an ideal switch: it presents zero impedance when forward biased (i.e., $c(t) > 0$). We may thus approximate the transfer characteristic of the diode-load resistor combination by a piece-wise-linear characteristic defined as (see Figure P2.3*b*)

$$v_2(t) = \begin{cases} v_1(t), & c(t) > 0 \\ 0, & c(t) < 0 \end{cases}$$

That is, the load voltage $v_2(t)$ varies periodically between the values $v_1(t)$ and zero at a rate equal to the carrier frequency $f_c$. Hence, we may write

$$v_2(t) \simeq [A_c \cos(2\pi f_c t) + m(t)]g_{T_0}(t)$$

where $g_{T_0}(t)$ is a periodic pulse train defined by

$$g_{T_0}(t) = \frac{1}{2} + \frac{2}{\pi}\sum_{n=1}^{\infty} \frac{(-1)^{n-1}}{2n-1} \cos[2\pi f_c t(2n-1)]$$

(a) Find the AM wave component contained in the output voltage $v_2(t)$.

(b) Specify the unwanted components in $v_2(t)$ that need to be removed by a band-pass filter of suitable design.

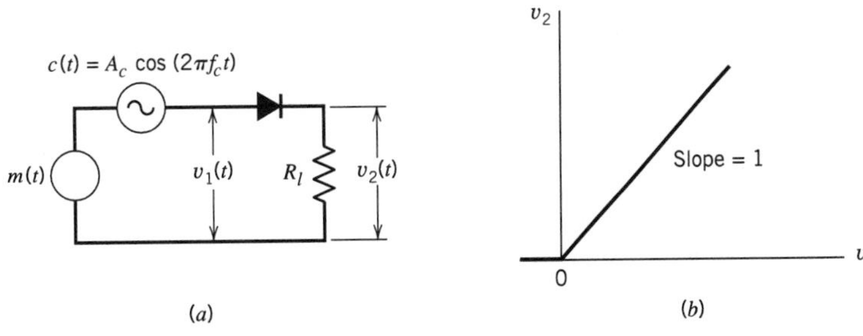


**FIGURE P2.3**

2.4 Consider the AM signal

$$s(t) = A_c[1 + \mu \cos(2\pi f_m t)] \cos(2\pi f_c t)$$

produced by a sinusoidal modulating signal of frequency $f_m$. Assume that the modulation factor is $\mu = 2$, and the carrier frequency $f_c$ is much greater than $f_m$. The AM signal $s(t)$ is applied to an ideal envelope detector, producing the output $v(t)$.

(a) Determine the Fourier series representation of $v(t)$.

(b) What is the ratio of second-harmonic amplitude to fundamental amplitude in $v(t)$?

2.5 Figure P2.5 shows the circuit diagram of an *envelope detector*. It consists simply of a diode and resistor-capacitor (RC) filter. On a positive half-cycle of the input signal, the diode is forward-biased and the capacitor C charges up rapidly to the peak value of the input signal. When the input signal falls below this value, the diode becomes reverse-biased and the capacitor C discharges slowly through the load resistor $R_l$. The discharging process continues until the next positive half-cycle. Thereafter, the charging-discharging routine is continued.

(a) Specify the condition that must be satisfied by the capacitor C for it to charge rapidly and thereby follow the input voltage up to the positive peak when the diode is conducting.

(b) Specify the condition which the load resistor $R_l$ must satisfy so that the capacitor C discharges slowly between positive peaks of the carrier wave, but not so long that the capacitor voltage will not discharge at the maximum rate of change of the modulating wave.

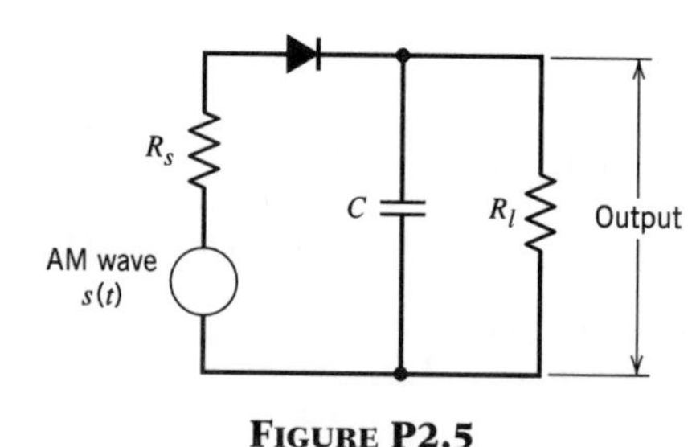


FIGURE P2.5

2.6 Consider a *square-law detector*, using a nonlinear device whose transfer characteristic is defined by

$$v_2(t) = a_1 v_1(t) + a_2 v_1^2(t)$$

where $a_1$ and $a_2$ are constants, $v_1(t)$ is the input, and $v_2(t)$ is the output. The input consists of the AM wave

$$v_1(t) = A_c[1 + k_a m(t)] \cos(2\pi f_c t)$$

(a) Evaluate the output $v_2(t)$.
(b) Find the conditions for which the message signal $m(t)$ may be recovered from $v_2(t)$.

2.7 The AM signal

$$s(t) = A_c[1 + k_a m(t)] \cos(2\pi f_c t)$$

is applied to the system shown in Figure P2.7. Assuming that $|k_a m(t)| < 1$ for all $t$ and the message signal $m(t)$ is limited to the interval $-W \le f \le W$ and that the carrier frequency $f_c > 2W$ show that $m(t)$ can be obtained from the square-rooter output $v_3(t)$.

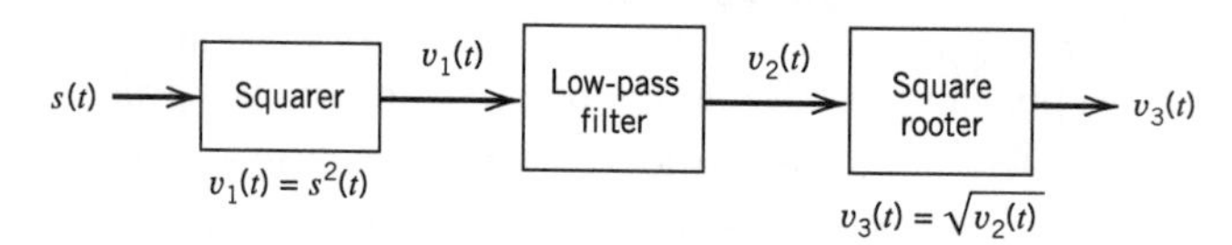


FIGURE P2.7

2.8 Consider a message signal $m(t)$ with the spectrum shown in Figure P2.8. The message bandwidth $W = 1$ kHz. This signal is applied to a product modulator, together with a carrier wave $A_c \cos(2\pi f_c t)$, producing the DSB-SC modulated signal $s(t)$. The modulated signal is next applied to a coherent detector. Assuming perfect synchronism between the carrier waves in the modulator and detector, determine the spectrum of the detector output when: (a) the carrier frequency $f_c = 1.25$ kHz and (b) the carrier frequency $f_c = 0.75$ kHz. What is the lowest carrier frequency for which each component of the modulated signal $s(t)$ is uniquely determined by $m(t)$?

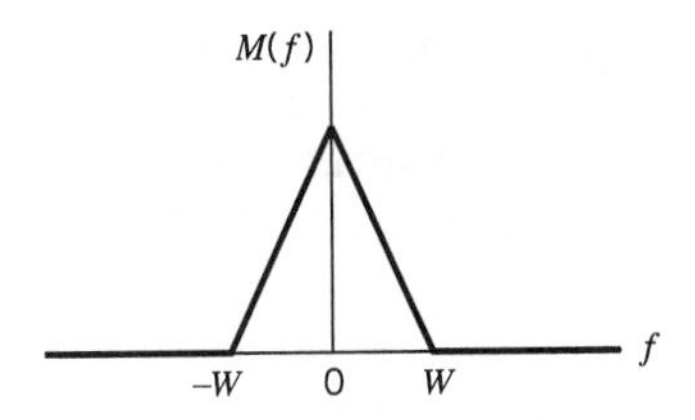


FIGURE P2.8

**2.9** Figure P2.9 shows the circuit diagram of a *balanced modulator*. The input applied to the top AM modulator is $m(t)$, whereas that applied to the lower AM modulator is $-m(t)$; these two modulators have the same amplitude sensitivity. Show that the output $s(t)$ of the balanced modulator consists of a DSB-SC modulated signal.

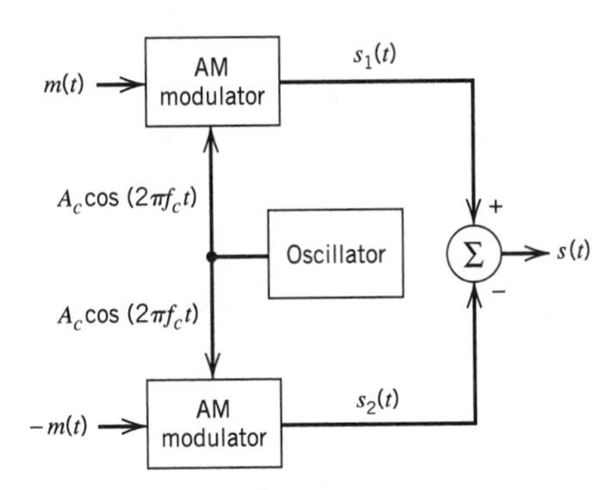


FIGURE P2.9

**2.10** A DSB-SC modulated signal is demodulated by applying it to a coherent detector.

(a) Evaluate the effect of a frequency error $\Delta f$ in the local carrier frequency of the detector, measured with respect to the carrier frequency of the incoming DSB-SC signal.

(b) For the case of a sinusoidal modulating wave, show that because of this frequency error, the demodulated signal exhibits *beats* at the error frequency. Illustrate your answer with a sketch of this demodulated signal.

**2.11** Consider the DSB-SC signal

$$s(t) = A_c \cos(2\pi f_c t) m(t)$$

where $A_c \cos(2\pi f_c t)$ is the carrier wave and $m(t)$ is the message signal. This modulated signal is applied to a square-law device characterized by

$$y(t) = s^2(t)$$

The output $y(t)$ is next applied to a narrowband filter with a passband magnitude response of one, midband frequency $2f_c$, and bandwidth $\Delta f$. Assume that $\Delta f$ is small enough to treat the spectrum of $y(t)$ as essentially constant inside the passband of the filter.

(a) Determine the spectrum of the square-law device output $y(t)$.

(b) Show that the filter output $v(t)$ is approximately sinusoidal, given by

$$v(t) \simeq \frac{A_c^2}{2} E \, \Delta f \cos(4\pi f_c t)$$

where $E$ is the energy of the message signal $m(t)$.

**2.12** Consider the quadrature-carrier multiplex system of Figure 2.10. The multiplexed signal $s(t)$ produced at the transmitter output in Figure 2.10*a* is applied to a communication channel of frequency response $H(f)$. The output of this channel is, in turn, applied to the receiver input in Figure 2.10*b*. Prove that the condition

$$H(f_c + f) = H^*(f_c - f), \qquad 0 \leq f \leq W$$

is necessary for recovery of the message signals $m_1(t)$ and $m_2(t)$ at the receiver outputs; $f_c$ is the carrier frequency, and $W$ is the message bandwidth. *Hint:* Evaluate the spectra of the two receiver outputs.

**2.13** Suppose that in the receiver of the quadrature-carrier multiplex system of Figure 2.10*b* the local carrier available for demodulation has a phase error $\phi$ with respect to the carrier source used in the transmitter. Assuming a distortionless communication channel between transmitter and receiver, show that this phase error will cause *cross-talk* to arise between the two demodulated signals at the receiver outputs. By cross-talk we mean that a portion of one message signal appears at the receiver output belonging to the other message signal, and vice versa.

**2.14** A particular version of *AM stereo* uses quadrature multiplexing. Specifically, the carrier $A_c \cos(2\pi f_c t)$ is used to modulate the sum signal

$$m_1(t) = V_0 + m_\ell(t) + m_r(t)$$

where $V_0$ is a DC offset included for the purpose of transmitting the carrier component, $m_\ell(t)$ is the left-hand audio signal, and $m_r(t)$ is the right-hand audio signal. The quadrature carrier $A_c \sin(2\pi f_c t)$ is used to modulate the difference signal

$$m_2(t) = m_\ell(t) - m_r(t)$$

(a) Show that an envelope detector may be used to recover the sum $m_r(t) + m_\ell(t)$ from the quadrature-multiplexed signal. How would you minimize the signal distortion produced by the envelope detector?

(b) Show that a coherent detector can recover the difference $m_\ell(t) - m_r(t)$.

(c) How are the desired $m_\ell(t)$ and $m_r(t)$ finally obtained?

**2.15** Using the message signal

$$m(t) = \frac{1}{1 + t^2}$$

determine and sketch the modulated waves for the following methods of modulation:

(a) Amplitude modulation with 50 percent modulation.

(b) Double sideband-suppressed carrier modulation.

(c) Single sideband modulation with only the upper sideband transmitted.

(d) Single sideband modulation with only the lower sideband transmitted.

**2.16** The *Hilbert transform* of a Fourier transformable signal $m(t)$, denoted by $\hat{m}(t)$, is defined by (see Appendix 2)

$$\hat{m}(t) = \frac{1}{\pi} \int_{-\infty}^{\infty} \frac{m(\tau)}{t - \tau} \, d\tau$$

In the frequency domain, we have

$$\hat{M}(f) = -j \, \text{sgn}(f) M(f)$$

where $m(t) \rightleftharpoons M(f)$, $\hat{m}(t) \rightleftharpoons \hat{M}(f)$, and $\text{sgn}(f)$ is the signum function.

Using the definition of the Hilbert transform, show that a single-sideband modulated signal resulting from the message signal $m(t)$ and carrier $\cos(2\pi f_c t)$ of unit amplitude is given by (see Table 2.1)

$$s(t) = \frac{1}{2} m(t) \cos(2\pi f_c t) \pm \frac{1}{2} \hat{m}(t) \sin(2\pi f_c t)$$

where the minus sign corresponds to the transmission of the upper sideband and the plus sign corresponds to the transmission of the lower sideband.

**2.17** The local oscillator used for the demodulation of an SSB signal $s(t)$ has a frequency error $\Delta f$ measured with respect to the carrier frequency $f_c$ used to generate $s(t)$. Otherwise, there is perfect synchronism between this oscillator in the receiver and the oscillator supplying the carrier wave in the transmitter. Evaluate the demodulated signal for the following two situations:

(a) The SSB signal $s(t)$ consists of the upper sideband only.

(b) The SSB signal $s(t)$ consists of the lower sideband only.

**2.18** Figure P2.18 shows the block diagram of *Weaver's method* for generating SSB modulated waves. The message (modulating) signal $m(t)$ is limited to the band $f_a \le |f| \le f_b$. The auxiliary carrier applied to the first pair of product modulators has a frequency $f_0$, which lies at the center of this band, as shown by

$$f_0 = \frac{f_a + f_b}{2}$$

The low-pass filters in the in-phase and quadrature channels are identical, each with a cutoff frequency equal to $(f_b - f_a)/2$. The carrier applied to the second pair of product modulators has a frequency $f_c$ that is greater than $(f_b - f_a)/2$. Sketch the spectra at the various points in the modulator of Figure P2.18, and hence show that:

(a) For the lower sideband, the contributions of the in-phase and quadrature channels are of opposite polarity, and by adding them at the modulator output, the lower sideband is suppressed.

(b) For the upper sideband, the contributions of the in-phase and quadrature channels are of the same polarity, and by adding them, the upper sideband is transmitted.

(c) How would you modify the modulator of Figure P2.18 so that only the lower sideband is transmitted?

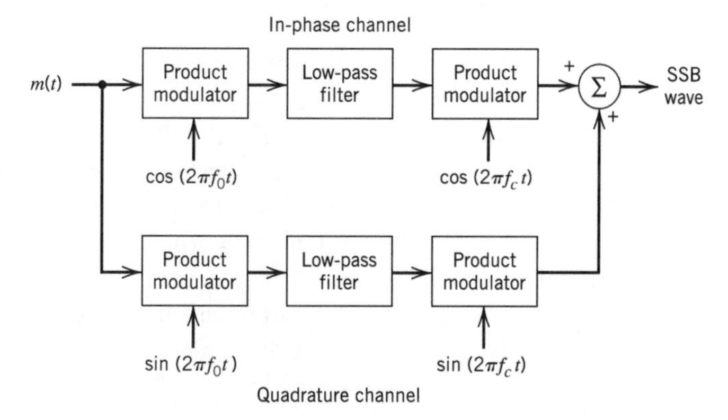


**FIGURE P2.18**

**2.19** The spectrum of a voice signal $m(t)$ is zero outside the interval $f_a \leq |f| \leq f_b$. To ensure communication privacy, this signal is applied to a *scrambler* that consists of the following cascade of components: product modulator, high-pass filter, second product modulator, and low-pass filter. The carrier wave applied to the first product modulator has a frequency equal to $f_c$, whereas that applied to the second product modulator has a frequency equal to $f_b + f_c$; both of them have unit amplitude. The high-pass and low-pass filters have the same cutoff frequency at $f_c$. Assume that $f_c > f_b$.

(a) Derive an expression for the scrambler output $s(t)$, and sketch its spectrum.

(b) Show that the original voice signal $m(t)$ may be recovered from $s(t)$ by using an *unscrambler* that is identical to the unit just described.

**2.20** In this problem we derive Equation (2.16) that defines the frequency response $H_Q(f)$ of the filter that operates on the message signal $m(t)$ to produce $m'(t)$ for VSB modulation. The signal $m'(t)$, except for a scaling factor, constitutes the quadrature component of $s(t)$. To do the derivation, we apply $s(t)$ to the coherent detector of Figure P2.20 so as to recover a scaled version of the original message signal $m(t)$.

(a) Starting with the block diagram of Figure 2.12 for the generation of a VSB modulated wave, determine the Fourier transform $V(f)$ of the product modulator output $v(t)$ in Figure P2.20 in terms of the Fourier transform of the message signal $m(t)$ and the frequency response $H(f)$ of the band-pass filter in Figure 2.12.

(b) Hence, by evaluating the Fourier transform of the low-pass filter output in Figure P2.20, determine the conditions that must be satisfied by $H_Q(f)$ in terms of $H(f)$ to assure perfect recovery of the original message signal $m(t)$, except for a scaling factor.

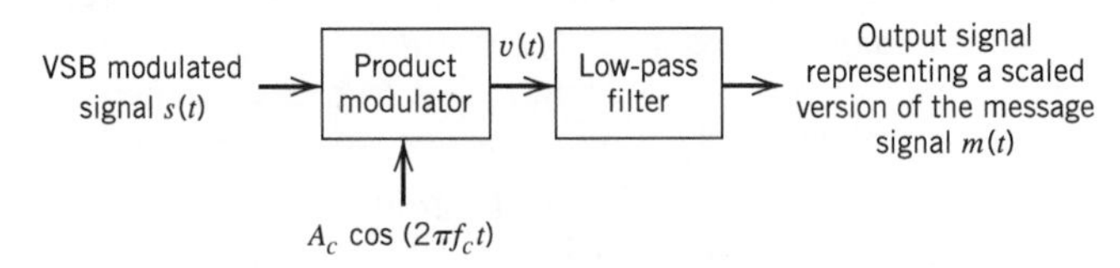


**FIGURE P2.20**

**2.21** The single-tone modulating signal $m(t) = A_m \cos(2\pi f_m t)$ is used to generate the VSB signal

$$s(t) = \frac{1}{2} aA_m A_c \cos[2\pi(f_c + f_m)t] + \frac{1}{2} A_m A_c(1 - a) \cos[2\pi(f_c - f_m)t]$$

where $a$ is a constant, less than unity, representing the attenuation of the upper side frequency.

(a) Find the quadrature component of the VSB signal $s(t)$.

(b) The VSB signal, plus the carrier $A_c \cos(2\pi f_c t)$, is passed through an envelope detector. Determine the distortion produced by the quadrature component.

(c) What is the value of constant $a$ for which this distortion reaches its worst possible condition?

**2.22** In this problem we study the idea of mixing in a superheterodyne receiver. To be specific, consider the block diagram of the *mixer* shown in Figure P2.22 that consists of a product modulator with a local oscillator of *variable frequency* $f_l$, followed by a band-pass filter. The input signal is an AM wave of bandwidth 10 kHz and carrier frequency that may lie anywhere in the range of 0.535 to 1.605 MHz; these parameters are typical of AM radio broadcasting. It is required to translate this signal to a frequency band centered at a fixed *intermediate frequency* (IF) of 0.455 MHz. Find the range of tuning that must be provided in the local oscillator to achieve this requirement.

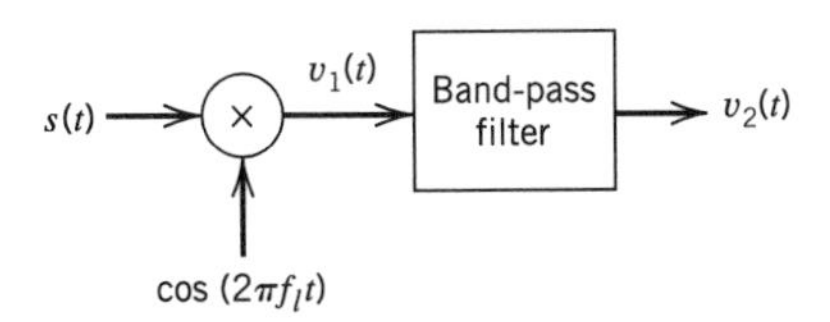


**FIGURE P2.22**

2.23 Figure P2.23 shows the block diagram of a *heterodyne spectrum analyzer*. It consists of a variable-frequency oscillator, multiplier, band-pass filter, and root mean square (RMS) meter. The oscillator has an amplitude $A$ and operates over the range $f_0$ to $f_0 + W$, where $f_0$ is the midband frequency of the filter and $W$ is the signal bandwidth. Assume that $f_0 = 2W$, the filter bandwidth $\Delta f$ is small compared with $f_0$ and that the passband magnitude response of the filter is one. Determine the value of the RMS meter output for a low-pass input signal $g(t)$.

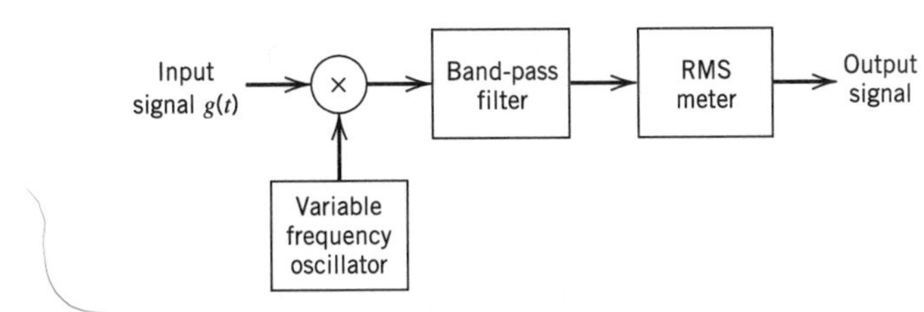


**FIGURE P2.23**

## Angle Modulation

2.24 Sketch the PM and FM waves produced by the sawtooth wave shown in Figure P2.24.

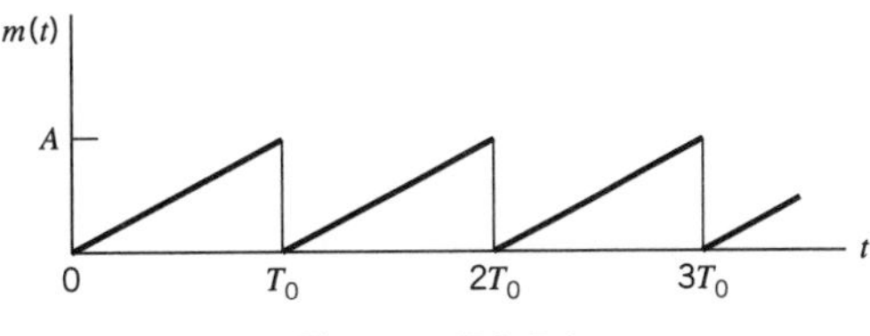


**FIGURE P2.24**

2.25 In a *frequency-modulated radar*, the instantaneous frequency of the transmitted carrier is varied as in Figure P2.25, which is obtained by using a triangular modulating signal. The instantaneous frequency of the received echo signal is shown dashed in Figure P2.25, where $\tau$ is the round-trip delay time. The transmitted and received echo signals are applied to a mixer, and the difference frequency component is retained. Assuming that $f_0\tau \ll 1$, determine the number of beat cycles at the mixer output, averaged over one second, in terms of the peak deviation $\Delta f$ of the carrier frequency, the delay $\tau$, and the repetition frequency $f_0$ of the transmitted signal.

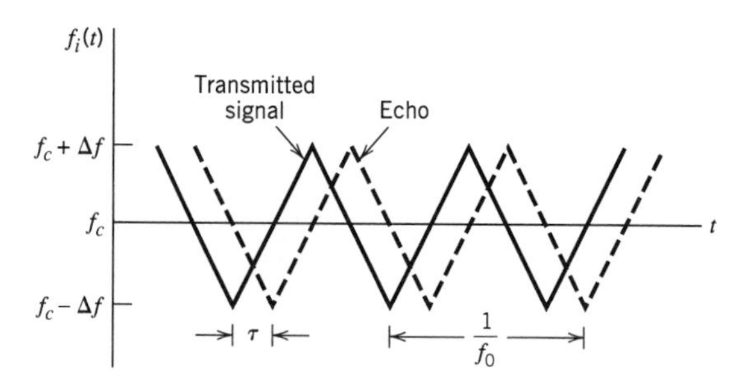


**FIGURE P2.25**

**2.26** The instantaneous frequency of a sine wave is equal to $f_c - \Delta f$ for $|t| \leq T/2$, and $f_c$ for $|t| > T/2$. Determine the spectrum of this frequency-modulated wave. *Hint:* Divide up the time interval of interest into three regions: $-\infty < t < -T/2$, $-T/2 \leq t \leq T/2$, and $T/2 < t < \infty$.

**2.27** Single-sideband modulation may be viewed as a hybrid form of amplitude modulation and frequency modulation. Evaluate the envelope and instantaneous frequency of an SSB wave for the following two cases:

(a) When only the upper sideband is transmitted.

(b) When only the lower sideband is transmitted.

**2.28** Consider a narrowband FM signal approximately defined by

$$s(t) \simeq A_c \cos(2\pi f_c t) - \beta A_c \sin(2\pi f_c t) \sin(2\pi f_m t)$$

(a) Determine the envelope of this modulated signal. What is the ratio of the maximum to the minimum value of this envelope? Plot this ratio versus $\beta$, assuming that $\beta$ is restricted to the interval $0 \leq \beta \leq 0.3$.

(b) Determine the average power of the narrowband FM signal, expressed as a percentage of the average power of the unmodulated carrier wave. Plot this result versus $\beta$, assuming that $\beta$ is restricted to the interval $0 \leq \beta \leq 0.3$.

(c) By expanding the angle $\theta_i(t)$ of the narrow-band FM signal $s(t)$ in the form of a power series, and restricting the modulation index $\beta$ to a maximum value of 0.3 radians, show that

$$\theta_i(t) \simeq 2\pi f_c t + \beta \sin(2\pi f_m t) - \frac{\beta^3}{3} \sin^3(2\pi f_m t)$$

What is the power ratio of third harmonic to fundamental component for $\beta = 0.3$?

**2.29** The sinusoidal modulating wave

$$m(t) = A_m \cos(2\pi f_m t)$$

is applied to a phase modulator with phase sensitivity $k_p$. The unmodulated carrier wave has frequency $f_c$ and amplitude $A_c$.

(a) Determine the spectrum of the resulting phase-modulated signal, assuming that the maximum phase deviation $\beta_p = k_p A_m$ does not exceed 0.3 radians.

(b) Construct a phasor diagram for this modulated signal, and compare it with that of the corresponding narrowband FM signal.

**2.30** Suppose that the phase-modulated signal of Problem 2.29 has an arbitrary value for the maximum phase deviation $\beta_p$. This modulated signal is applied to an ideal band-pass filter with midband frequency $f_c$ and a passband extending from $f_c - 1.5 f_m$ to $f_c + 1.5 f_m$.

Determine the envelope, phase, and instantaneous frequency of the modulated signal at the filter output as functions of time.

**2.31** A carrier wave is frequency-modulated using a sinusoidal signal of frequency $f_m$ and amplitude $A_m$.

(a) Determine the values of the modulation index $\beta$ for which the carrier component of the FM signal is reduced to zero. For this calculation you may use the values of $J_0(\beta)$ given in Table A6.5.

(b) In a certain experiment conducted with $f_m = 1$ kHz and increasing $A_m$ (starting from 0 volts), it is found that the carrier component of the FM signal is reduced to zero for the first time when $A_m = 2$ volts. What is the frequency sensitivity of the modulator? What is the value of $A_m$ for which the carrier components is reduced to zero for the second time?

**2.32** An FM signal with modulation index $\beta = 1$ is transmitted through an ideal band-pass filter with midband frequency $f_c$ and bandwidth $5f_m$, where $f_c$ is the carrier frequency and $f_m$ is the frequency of the sinusoidal modulating wave. Determine the magnitude spectrum of the filter output.

**2.33** A carrier wave of frequency 100 MHz is frequency-modulated by a sinusoidal wave of amplitude 20 volts and frequency 100 kHz. The frequency sensitivity of the modulator is 25 kHz per volt.

(a) Determine the approximate bandwidth of the FM signal, using Carson's rule.

(b) Determine the bandwidth by transmitting only those side frequencies whose amplitudes exceed 1 percent of the unmodulated carrier amplitude. Use the universal curve of Figure 2.26 for this calculation.

(c) Repeat your calculations, assuming that the amplitude of the modulating signal is doubled.

(d) Repeat your calculations, assuming that the modulation frequency is doubled.

**2.34** Consider a wideband PM signal produced by a sinusoidal modulating wave $A_m \cos(2\pi f_m t)$, using a modulator with a phase sensitivity equal to $k_p$ radians per volt.

(a) Show that if the maximum phase deviation of the PM signal is large compared with one radian, the bandwidth of the PM signal varies linearly with the modulation frequency $f_m$.

(b) Compare this characteristic of a wideband PM signal with that of a wideband FM signal.

**2.35** Figure P2.35 shows the block diagram of a real-time *spectrum analyzer* working on the principle of frequency modulation. The given signal $g(t)$ and a frequency-modulated signal $s(t)$ are applied to a multiplier and the output $g(t)s(t)$ is fed into a filter of impulse response $h(t)$. The $s(t)$ and $h(t)$ are *linear FM signals* whose instantaneous frequencies vary linearly with time at opposite rates, as shown by

$$s(t) = \cos(2\pi f_c t - \pi k t^2)$$
$$h(t) = \cos(2\pi f_c t + \pi k t^2)$$

where $k$ is a constant. Show that the envelope of the filter output is proportional to the magnitude spectrum of the input signal $g(t)$ with $kt$ playing the role of frequency $f$. *Hint:* Use the complex notations described in Appendix 2 for the analysis of band-pass signals and band-pass filters.

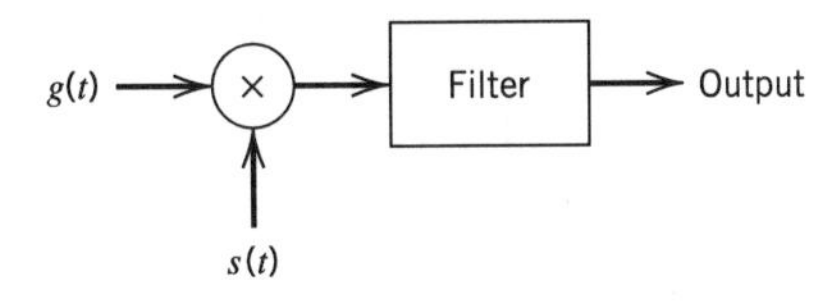


**FIGURE P2.35**

2.36 An FM signal with a frequency deviation of 10 kHz at a modulation frequency of 5 kHz is applied to two frequency multipliers connected in cascade. The first multiplier doubles the frequency and the second multiplier triples the frequency. Determine the frequency deviation and the modulation index of the FM signal obtained at the second multiplier output. What is the frequency separation of the adjacent side frequencies of this FM signal?

2.37 Figure P2.37 shows the block diagram of a wideband frequency modulator using the indirect method. This modulator is used to transmit audio signals containing frequencies in the range of 100 Hz to 15 kHz. The narrowband phase modulator is supplied with a carrier of frequency $f_1 = 0.1$ MHz by a crystal-controlled oscillator. A second crystal-controlled oscillator supplies a sinusoidal wave of frequency 9.5 MHz to the mixer. The system specifications are as follows:
Carrier frequency at the transmitter output, $f_c = 100$ MHz
Minimum frequency deviation, $\Delta f = 75$ kHz
Maximum modulation index in the phase modulator = 0.3 radians

(a) Calculate the frequency multiplication ratios $n_1$ and $n_2$ (preceding and following the mixer), which will satisfy these specifications.

(b) Specify the values of the carrier frequency and frequency deviation at the various points in the modulator of Figure P2.37.

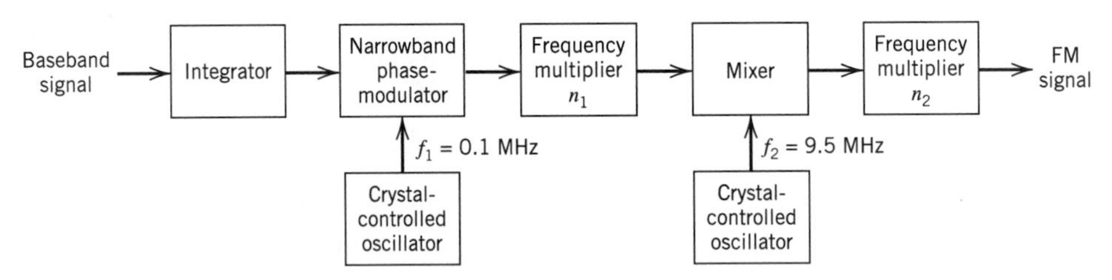


FIGURE P2.37

2.38 Figure P2.38 shows the frequency-determining network of a voltage-controlled oscillator. Frequency modulation is produced by applying the modulating signal $A_m \sin(2\pi f_m t)$ plus a bias $V_b$ to a pair of varactor diodes connected across the parallel combination of a 200-$\mu$H inductor and 100-pF capacitor. The capacitor of each varactor diode is related to the voltage $V$ (in volts) applied across its electrodes by

$$C = 100V^{-1/2} \text{ pF}$$

The unmodulated frequency of oscillation is 1 MHz. The VCO output is applied to a frequency multiplier to produce an FM signal with a carrier frequency of 64 MHz and a modulation index of 5. Determine (a) the magnitude of the bias voltage $V_b$ and (b) the amplitude $A_m$ of the modulating wave, given that $f_m = 10$ kHz.

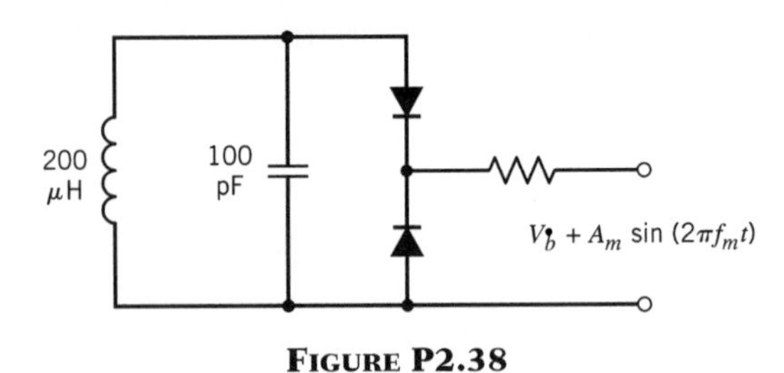


FIGURE P2.38

2.39 The FM signal

$$s(t) = A_c \cos\left[2\pi f_c t + 2\pi k_f \int_0^t m(\tau)\, d\tau\right]$$

is applied to the system shown in Figure P2.39 consisting of a high-pass RC filter and an envelope detector. Assume that (a) the resistance $R$ is small compared with the reactance of the capacitor C for all significant frequency components of $s(t)$, and (b) the envelope detector does not load the filter. Determine the resulting signal at the envelope detector output, assuming that $k_f|m(t)| < f_c$ for all $t$.

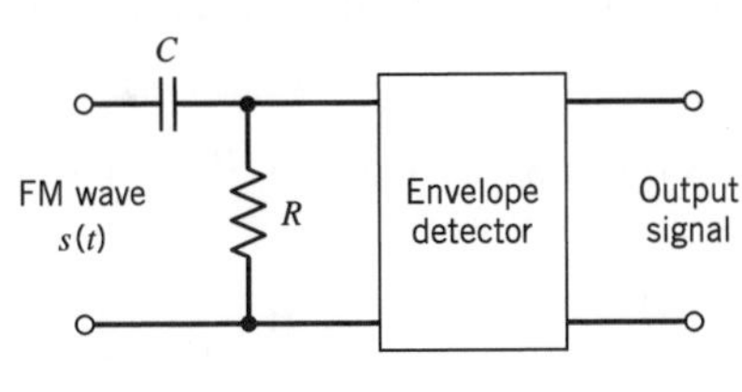


**FIGURE P2.39**

2.40 Consider the frequency demodulation scheme shown in Figure P2.40 in which the incoming FM signal $s(t)$ is passed through a delay line that produces a phase-shift of $\pi/2$ radians at the carrier frequency $f_c$. The delay-line output is subtracted from the incoming FM signal, and the resulting composite signal is then envelope-detected. This demodulator finds application in demodulating microwave FM signals. Assuming that

$$s(t) = A_c \cos[2\pi f_c t + \beta \sin(2\pi f_m t)]$$

analyze the operation of this demodulator when the modulation index $\beta$ is less than unity and the delay $T$ produced by the delay line is sufficiently small to justify making the approximations

$$\cos(2\pi f_m T) \simeq 1$$

and

$$\sin(2\pi f_m T) \simeq 2\pi f_m T$$

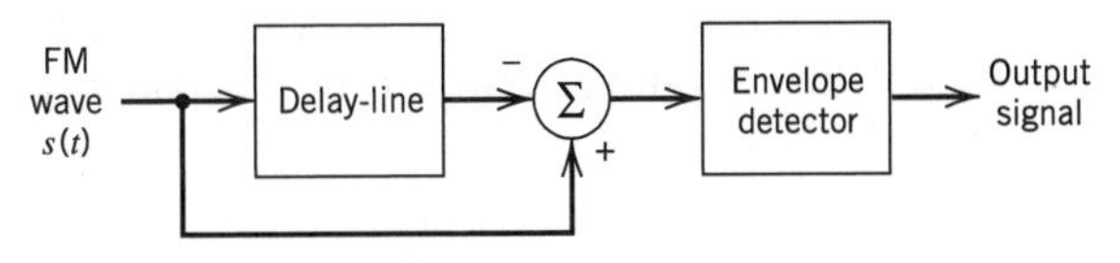


**FIGURE P2.40**

2.41 Figure 2.41 shows the block diagram of a *zero-crossing detector* for demodulating an FM signal. It consists of a limiter, a pulse generator for producing a short pulse at each zero-crossing of the input, and a low-pass filter for extracting the modulating wave.

(a) Show that the instantaneous frequency of the input FM signal is proportional to the number of zero crossings in the time interval $t - (T_1/2)$ to $t + (T_1/2)$, divided by $T_1$. Assume that the modulating signal is essentially constant during this time interval.

(b) Illustrate the operation of this demodulator, using the sawtooth wave of Figure P2.24 as the modulating wave.

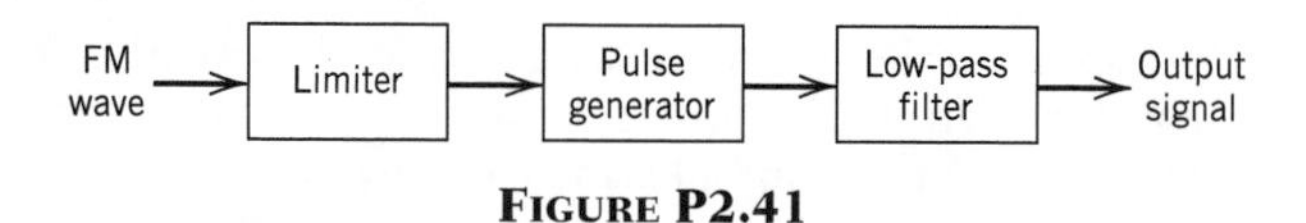


FIGURE P2.41

2.42 Suppose that the received signal in an FM system contains some residual amplitude modulation of positive amplitude $a(t)$, as shown by

$$s(t) = a(t) \cos[2\pi f_c t + \phi(t)]$$

where $f_c$ is the carrier frequency. The phase $\phi(t)$ is related to the modulating signal $m(t)$ by

$$\phi(t) = 2\pi k_f \int_0^t m(\tau)\, d\tau$$

where $k_f$ is a constant. Assume that the signal $s(t)$ is restricted to a frequency band of width $B_T$, centered at $f_c$, where $B_T$ is the transmission bandwidth of the FM signal in the absence of amplitude modulation, and that the amplitude modulation is slowly varying compared with $\phi(t)$. Show that the output of an ideal frequency discriminator produced by $s(t)$ is proportional to $a(t)m(t)$. *Hint:* Use the complex notation described in Appendix 2 to represent the modulated wave $s(t)$.

2.43 (a) Let the modulated wave $s(t)$ in Problem 2.42 be applied to a *hard limiter*, whose output $z(t)$ is defined by

$$z(t) = \text{sgn}[s(t)] = \begin{cases} +1, & s(t) > 0 \\ -1, & s(t) < 0 \end{cases}$$

Show that the limiter output may be expressed in the form of a Fourier series as follows:

$$z(t) = \frac{4}{\pi} \sum_{n=0}^{\infty} \frac{(-1)^n}{2n+1} \cos[2\pi f_c t(2n+1) + (2n+1)\phi(t)]$$

(b) Suppose that the limiter output is applied to a band-pass filter with a passband magnitude response of one and bandwidth $B_T$ centered about the carrier frequency $f_c$, where $B_T$ is the transmission bandwidth of the FM signal in the absence of amplitude modulation. Assuming that $f_c$ is much greater than $B_T$, show that the resulting filter output equals

$$y(t) = \frac{4}{\pi} \cos[2\pi f_c t + \phi(t)]$$

By comparing this output with the original modulated signal $s(t)$ defined in Problem 2.42, comment on the practical usefulness of the result.

2.44 (a) Consider an FM signal of carrier frequency $f_c$, which is produced by a modulating signal $m(t)$. Assume that $f_c$ is large enough to justify treating this FM signal as a narrowband signal. Find an approximate expression for its Hilbert transform.

(b) For the special case of a sinusoidal modulating wave $m(t) = A_m \cos(2\pi f_m t)$, find the exact expression for the Hilbert transform of the resulting FM signal. For this case, what is the error in the approximation used in part (a)?

2.45 The *single sideband version of angle modulation* is defined by

$$s(t) = \exp[-\hat{\phi}(t)]\cos[2\pi f_c t + \phi(t)]$$

where $\hat{\phi}(t)$ is the Hilbert transform of the phase function $\phi(t)$, and $f_c$ is the carrier frequency.

(a) Show that the spectrum of the modulated signal $s(t)$ contains no frequency components in the interval $-f_c < f < f_c$, and is of infinite extent.

(b) Given that the phase function

$$\phi(t) = \beta \sin(2\pi f_m t)$$

where $\beta$ is the modulation index and $f_m$ is the modulation frequency, derive the corresponding expression for the modulated wave $s(t)$.

*Note:* For Problems 2.44 and 2.45, you need to refer to Appendix 2 for a treatment of the Hilbert transform.

## Noise in CW Modulation Systems

2.46 A DSB-SC modulated signal is transmitted over a noisy channel, with the power spectral density of the noise being as shown in Figure P2.46. The message bandwidth is 4 kHz and the carrier frequency is 200 kHz. Assuming that the average power of the modulated wave is 10 watts, determine the output signal-to-noise ratio of the receiver.

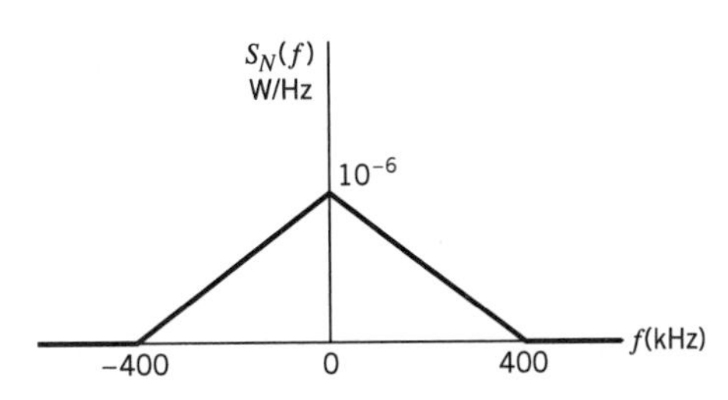


**FIGURE P2.46**

2.47 Evaluate the autocorrelation functions and cross-correlation functions of the in-phase and quadrature components of the narrowband noise at the coherent detector input for (a) the DSB-SC system, (b) an SSB system using the lower sideband, and (c) an SSB system using the upper sideband.

2.48 In a receiver using coherent detection, the sinusoidal wave generated by the local oscillator suffers from a phase error $\theta(t)$ with respect to the carrier wave $\cos(2\pi f_c t)$. Assuming that $\theta(t)$ is a sample function of a zero-mean Gaussian process of variance $\sigma_\Theta^2$, and that most of the time the maximum value of $\theta(t)$ is small compared with unity, find the mean-square error of the receiver output for DSB-SC modulation. The mean-square error is defined as the expected value of the squared difference between the receiver output and the message signal component of the receiver output.

2.49 Following a procedure similar to that described in Section 2.11 for the DSB-SC receiver, extend this noise analysis to a SSB receiver. Specifically, evaluate the following:

(a) The output signal-to-noise ratio.

(b) The channel signal-to-noise ratio.

Hence, show that the figure of merit for the SSB receiver is exactly the same as that for the DSB-SC receiver. Note that unlike the DSB-SC receiver, the midband frequency of the spectral density function of the narrowband-filtered noise at the front end of the SSB receiver is offset from the carrier frequency $f_c$ by an amount equal to $W/2$, where $W$ is the message bandwidth.

2.50 Let a message signal $m(t)$ be transmitted using single-sideband modulation. The power spectral density of $m(t)$ is

$$S_M(f) = \begin{cases} a\dfrac{|f|}{W}, & |f| \le W \\ 0, & \text{otherwise} \end{cases}$$

where $a$ and $W$ are constants. White Gaussian noise of zero mean and power spectral density $N_0/2$ is added to the SSB modulated wave at the receiver input. Find an expression for the output signal-to-noise ratio of the receiver.

2.51 Consider the output of an envelope detector defined by Equation (2.92), which is reproduced here for convenience

$$y(t) = \{[A_c + A_c k_a m(t) + n_I(t)]^2 + n_Q^2(t)\}^{1/2}$$

(a) Assume that the probability of the event

$$|n_Q(t)| > \varepsilon A_c |1 + k_a m(t)|$$

is equal to or less than $\delta_1$, where $\varepsilon \ll 1$. What is the probability that the effect of the quadrature component $n_Q(t)$ is negligible?

(b) Suppose that $k_a$ is adjusted relative to the message signal $m(t)$ such that the probability of the event

$$A_c[1 + k_a m(t)] + n_I(t) < 0$$

is equal to $\delta_2$. What is the probability that the approximation

$$y(t) \simeq A_c[1 + k_a m(t)] + n_I(t)$$

is valid?

(c) Comment on the significance of the result in part (b) for the case when $\delta_1$ and $\delta_2$ are both small compared with unity.

2.52 An unmodulated carrier of amplitude $A_c$ and frequency $f_c$ and band-limited white noise are summed and then passed through an ideal envelope detector. Assume the noise spectral density to be of height $N_0/2$ and bandwidth $2W$, centered about the carrier frequency $f_c$. Determine the output signal-to-noise ratio for the case when the carrier-to-noise ratio is high.

2.53 Let $R$ denote the random variable obtained by observing the output of an envelope detector at some fixed time. Intuitively, the envelope detector is expected to be operating well into the threshold region if the probability that the random variable $R$ exceeds the carrier amplitude $A_c$ is 0.5. On the other hand, if this same probability is only 0.01, the envelope detector is expected to be relatively free of loss of message and the threshold effect.

(a) Assuming that the narrowband noise at the detector input is white, zero-mean, Gaussian with spectral density $N_0/2$ and the message bandwidth is $W$, show that the probability of the event $R \ge A_c$ is

$$P(R \ge A_c) = \exp(-\rho)$$

where $\rho$ is the carrier-to-noise ratio:

$$\rho = \frac{A_c^2}{4WN_0}$$

(b) Using the formula for this probability, calculate the carrier-to-noise ratio when (1) the envelope detector is expected to be well into the threshold region, and (2) it is expected to be operating satisfactorily.

2.54 Consider a phase modulation (PM) system, with the modulated wave defined by

$$s(t) = A_c \cos[2\pi f_c t + k_p m(t)]$$

where $k_p$ is a constant and $m(t)$ is the message signal. The additive noise $n(t)$ at the phase detector input is

$$n(t) = n_I(t) \cos(2\pi f_c t) - n_Q(t) \sin(2\pi f_c t)$$

Assuming that the carrier-to-noise ratio at the detector input is high compared with unity, determine (a) the output signal-to-noise ratio and (b) the figure of merit of the system. Compare your results with the FM system for the case of sinusoidal modulation.

2.55 An FDM system uses single-sideband modulation to combine 12 independent voice signals and then uses frequency modulation to transmit the composite baseband signal. Each voice signal has an average power $P$ and occupies the frequency band 0.3 to 3.4 kHz; the system allocates it a bandwidth of 4 kHz. For each voice signal, only the lower sideband is transmitted. The subcarrier waves used for the first stage of modulation are defined by

$$c_k(t) = A_k \cos(2\pi k f_0 t), \qquad 0 \le k \le 11$$

The received signal consists of the transmitted FM signal plus white Gaussian noise of zero mean and power spectral density $N_0/2$.

(a) Sketch the power spectral density of the signal produced at the frequency discriminator output, showing both the signal and noise components.

(b) Find the relationship between the subcarrier amplitudes $A_k$ so that the modulated voice signals have equal signal-to-noise ratios.

2.56 In the discussion on FM threshold effect presented in Section 2.13, we described the conditions for positive-going and negative-going clicks in terms of the envelope $r(t)$ and phase $\psi(t)$ of the narrowband noise $n(t)$. Reformulate these conditions in terms of the in-phase component $n_I(t)$ and quadrature component $n_Q(t)$ of $n(t)$.

2.57 By using the pre-emphasis filter shown in Figure 2.50$a$ and with a voice signal as the modulating wave, an FM transmitter produces a signal that is essentially frequency-modulated by the lower audio frequencies and phase-modulated by the higher audio frequencies. Explain the reasons for this phenomenon.

2.58 Suppose that the transfer functions of the pre-emphasis and de-emphasis filters of an FM system are scaled as follows:

$$H_{\text{pe}}(f) = k\left(1 + \frac{jf}{f_0}\right)$$

and

$$H_{\text{de}}(f) = \frac{1}{k}\left(\frac{1}{1 + jf/f_0}\right)$$

The scaling factor $k$ is to be chosen so that the average power of the emphasized message signal is the same as that of the original message signal $m(t)$.

(a) Find the value of $k$ that satisfies this requirement for the case when the power spectral density of the message signal $m(t)$ is

$$S_M(f) = \begin{cases} \dfrac{S_0}{1 + (f/f_0)^2}, & -W \le f \le W \\ 0, & \text{elsewhere} \end{cases}$$

(b) What is the corresponding value of the improvement factor $I$ produced by using this pair of pre-emphasis and de-emphasis filters? Compare this ratio with that obtained in Example 2.6. The improvement factor $I$ is defined by Equation (2.160).

**2.59** A phase modulation (PM) system uses a pair of pre-emphasis and de-emphasis filters defined by the transfer functions

$$H_{\text{pe}}(f) = 1 + \frac{jf}{f_0}$$

and

$$H_{\text{de}}(f) = \frac{1}{1 + (jf/f_0)}$$

Show that the improvement in output signal-to-noise ratio produced by using this pair of filters is

$$I = \frac{W/f_0}{\tan^{-1}(W/f_0)}$$

where $W$ is the message bandwidth. Evaluate this improvement for the case when $W = 15$ kHz and $f_0 = 2.1$ kHz, and compare your result with the corresponding value for an FM system.

## Computer Experiments

**2.60** In this experiment we study the behavior of the envelope detector shown in Figure P2.5 for the following specifications:

Source resistance, $R_s = 75\ \Omega$
Load resistance, $R_l = 10\ \text{k}\Omega$
Capacitance, $C = 0.01\mu\text{F}$

The diode has a resistance of 25 Ω when it is forward-biased and infinite resistance when reverse-biased.

Compute the waveform of the envelope detector output, assuming an input sinusoidal AM wave with 50 percent modulation. The modulation frequency is 1 kHz, and the carrier frequency is 20 kHz.

**2.61** In this experiment we continue the study of the phase-locked loop considered in Section 2.14:

(a) Compute variations in the instantaneous frequency of the voltage-controlled oscillator in the loop for the following loop parameters:

$$\text{Loop-gain parameter, } K_0 = \frac{50}{2\pi}\ \text{Hz}$$

$$\text{Natural frequency, } f_n = \frac{1}{2\pi}\ \text{Hz}$$

$$\text{Damping factor, } \zeta = 0.707$$

Perform the computations for the following values of frequency step: $\Delta f = 0.125, 0.5, \frac{7}{12}, \frac{2}{3}$ Hz.

(b) For the parameters of the phase-locked loop as specified in Experiment 2 in Section 2.14, compute how variations in the relative frequency deviation $\Delta f \cdot f_m/f_n^2$ affect the peak phase error of the phase-locked loop.

CHAPTER 3

# PULSE MODULATION

This chapter, representing the transition from analog to digital communications, covers the following topics:

- *Sampling, which is basic to all forms of pulse modulation.*
- *Pulse-amplitude modulation, which is the simplest form of pulse modulation.*
- *Quantization, which, when combined with sampling, permits the representation of an analog signal in discrete form in both amplitude and time.*
- *Pulse-code modulation, which is the standard method for the transmission of an analog message signal by digital means.*
- *Time-division multiplexing, which provides for the time sharing of a common channel by a plurality of users by means of pulse modulation.*
- *Digital multiplexers, which combine many slow bit streams into a single faster stream.*
- *Other forms of digital pulse modulation, namely, delta modulation and differential pulse-code modulation.*
- *Linear prediction, which is basic to the encoding of analog message signals at reduced bit rates as in differential pulse-code modulation.*
- *Adaptive forms of differential pulse-code modulation and delta modulation.*
- *The MPEG-1/audio coding standard, which is a transparent, perceptually lossless compression system.*

## 3.1 *Introduction*

In *continuous-wave (CW) modulation*, which we studied in Chapter 2, some parameter of a sinusoidal carrier wave is varied continuously in accordance with the message signal. This is in direct contrast to pulse modulation, which we study in the present chapter. In *pulse modulation*, some parameter of a pulse train is varied in accordance with the message signal. We may distinguish two families of pulse modulation: *analog pulse modulation* and *digital pulse modulation*. In analog pulse modulation, a periodic pulse train is used as the carrier wave, and some characteristic feature of each pulse (e.g., amplitude, duration, or position) is varied in a continuous manner in accordance with the corresponding *sample* value of the message signal. Thus in analog pulse modulation, information is transmitted basically in analog form, but the transmission takes place at discrete times. In digital pulse modulation, on the other hand, the message signal is represented in a form that is discrete in both time and amplitude, thereby permitting its transmission in digital form as a sequence of *coded pulses*; this form of signal transmission has *no* CW counterpart.

The use of coded pulses for the transmission of analog information-bearing signals represents a basic ingredient in the application of digital communications. This chapter may therefore be viewed as a transition from analog to digital communications in our study of the principles of communication systems. We begin the discussion by describing the sampling process, which is basic to all pulse modulation systems, whether they are analog or digital.

## 3.2 *Sampling Process*

The *sampling process* is usually described in the time domain. As such, it is an operation that is basic to digital signal processing and digital communications. Through use of the sampling process, an analog signal is converted into a corresponding sequence of samples that are usually spaced uniformly in time. Clearly, for such a procedure to have practical utility, it is necessary that we choose the sampling rate properly, so that the sequence of samples uniquely defines the original analog signal. This is the essence of the sampling theorem, which is derived in what follows.

Consider an arbitrary signal $g(t)$ of finite energy, which is specified for all time. A segment of the signal $g(t)$ is shown in Figure 3.1*a*. Suppose that we sample the signal $g(t)$ instantaneously and at a uniform rate, once every $T_s$ seconds. Consequently, we obtain an infinite sequence of samples spaced $T_s$ seconds apart and denoted by $\{g(nT_s)\}$, where $n$ takes on all possible integer values. We refer to $T_s$ as the *sampling period*, and to its reciprocal $f_s = 1/T_s$ as the *sampling rate*. This ideal form of sampling is called *instantaneous sampling*.

Let $g_\delta(t)$ denote the signal obtained by individually weighting the elements of a periodic sequence of delta functions spaced $T_s$ seconds apart by the sequence of numbers $\{g(nT_s)\}$, as shown by (see Figure 3.1*b*)

$$g_\delta(t) = \sum_{n=-\infty}^{\infty} g(nT_s)\,\delta(t - nT_s) \tag{3.1}$$

We refer to $g_\delta(t)$ as the *ideal sampled signal*. The term $\delta(t - nT_s)$ represents a delta function positioned at time $t = nT_s$. From the definition of the delta function, we recall that such an idealized function has unit area; see Appendix 2. We may therefore view the multiplying factor $g(nT_s)$ in Equation (3.1) as a "mass" assigned to the delta function $\delta(t - nT_s)$. A delta function weighted in this manner is closely approximated by a rectangular pulse of

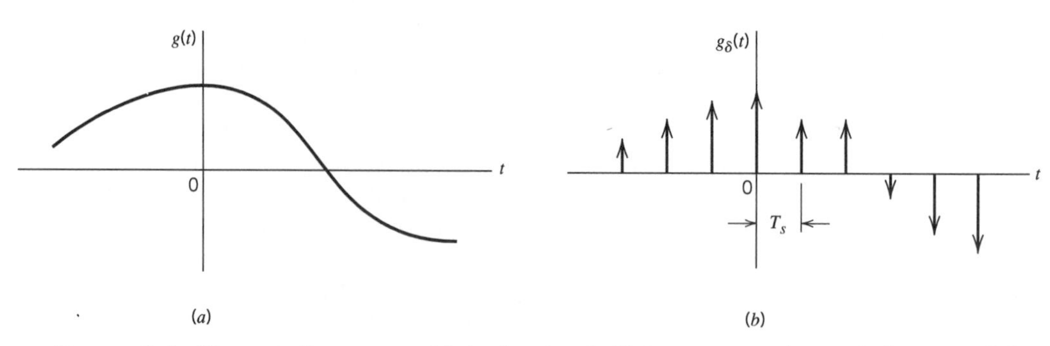


**FIGURE 3.1** The sampling process. (*a*) Analog signal. (*b*) Instantaneously sampled version of the analog signal.

duration $\Delta t$ and amplitude $g(nT_s)/\Delta t$; *the smaller we make* $\Delta t$ *the better will be the approximation.*

Using the table of Fourier-transform pairs, we may write (see the last item of Table A6.3)

$$g_\delta(t) \rightleftharpoons f_s \sum_{m=-\infty}^{\infty} G(f - mf_s) \tag{3.2}$$

where $G(f)$ is the Fourier transform of the original signal $g(t)$, and $f_s$ is the sampling rate. Equation (3.2) states that *the process of uniformly sampling a continuous-time signal of finite energy results in a periodic spectrum with a period equal to the sampling rate.*

Another useful expression for the Fourier transform of the ideal sampled signal $g_\delta(t)$ may be obtained by taking the Fourier transform of both sides of Equation (3.1) and noting that the Fourier transform of the delta function $\delta(t - nT_s)$ is equal to $\exp(-j2\pi nfT_s)$. Let $G_\delta(f)$ denote the Fourier transform of $g_\delta(t)$. We may therefore write

$$G_\delta(f) = \sum_{n=-\infty}^{\infty} g(nT_s)\exp(-j2\pi nfT_s) \tag{3.3}$$

This relation is called the *discrete-time Fourier transform.* It may be viewed as a complex Fourier series representation of the periodic frequency function $G_\delta(f)$, with the sequence of samples $\{g(nT_s)\}$ defining the coefficients of the expansion.

The relations, as derived here, apply to any continuous-time signal $g(t)$ of finite energy and infinite duration. Suppose, however, that the signal $g(t)$ is *strictly band-limited,* with no frequency components higher than $W$ Hertz. That is, the Fourier transform $G(f)$ of the signal $g(t)$ has the property that $G(f)$ is zero for $|f| \geq W$, as illustrated in Figure 3.2*a*; the shape of the spectrum shown in this figure is intended for the purpose of illustration only. Suppose also that we choose the sampling period $T_s = 1/2W$. Then the corresponding spectrum $G_\delta(f)$ of the sampled signal $g_\delta(t)$ is as shown in Figure 3.2*b*. Putting $T_s = 1/2W$ in Equation (3.3) yields

$$G_\delta(f) = \sum_{n=-\infty}^{\infty} g\left(\frac{n}{2W}\right)\exp\left(-\frac{j\pi nf}{W}\right) \tag{3.4}$$

From Equation (3.2), we readily see that the Fourier transform of $g_\delta(t)$ may also be expressed as

$$G_\delta(f) = f_sG(f) + f_s \sum_{\substack{m=-\infty \\ m\neq 0}}^{\infty} G(f - mf_s) \tag{3.5}$$

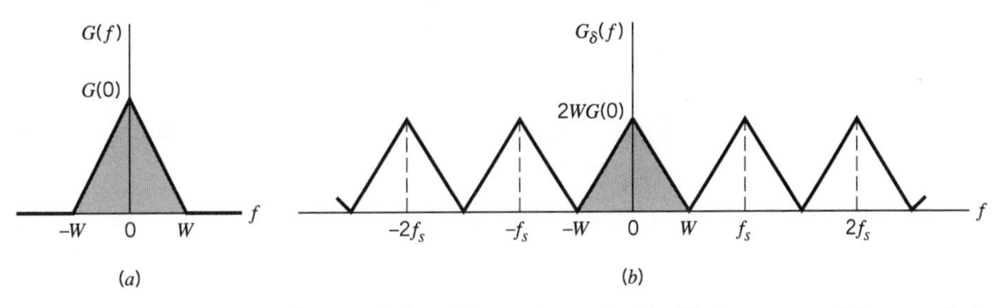


**FIGURE 3.2** (*a*) Spectrum of a strictly band-limited signal $g(t)$. (*b*) Spectrum of the sampled version of $g(t)$ for a sampling period $T_s = 1/2W$.

Hence, under the following two conditions:

1. $G(f) = 0$ for $|f| \geq W$
2. $f_s = 2W$

we find from Equation (3.5) that

$$G(f) = \frac{1}{2W} G_\delta(f), \qquad -W < f < W \tag{3.6}$$

Substituting Equation (3.4) into (3.6), we may also write

$$G(f) = \frac{1}{2W} \sum_{n=-\infty}^{\infty} g\left(\frac{n}{2W}\right) \exp\left(-\frac{j\pi n f}{W}\right), \qquad -W < f < W \tag{3.7}$$

Therefore, if the sample values $g(n/2W)$ of a signal $g(t)$ are specified for all $n$, then the Fourier transform $G(f)$ of the signal is uniquely determined by using the discrete-time Fourier transform of Equation (3.7). Because $g(t)$ is related to $G(f)$ by the inverse Fourier transform, it follows that the signal $g(t)$ is itself uniquely determined by the sample values $g(n/2W)$ for $-\infty < n < \infty$. In other words, the sequence $\{g(n/2W)\}$ has all the information contained in $g(t)$.

Consider next the problem of reconstructing the signal $g(t)$ from the sequence of sample values $\{g(n/2W)\}$. Substituting Equation (3.7) in the formula for the inverse Fourier transform defining $g(t)$ in terms of $G(f)$, we get

$$\begin{aligned} g(t) &= \int_{-\infty}^{\infty} G(f) \exp(j2\pi f t)\, df \\ &= \int_{-W}^{W} \frac{1}{2W} \sum_{n=-\infty}^{\infty} g\left(\frac{n}{2W}\right) \exp\left(-\frac{j\pi n f}{W}\right) \exp(j2\pi f t)\, df \end{aligned}$$

Interchanging the order of summation and integration:

$$g(t) = \sum_{n=-\infty}^{\infty} g\left(\frac{n}{2W}\right) \frac{1}{2W} \int_{-W}^{W} \exp\left[j2\pi f\left(t - \frac{n}{2W}\right)\right] df \tag{3.8}$$

The integral term in Equation (3.8) is readily evaluated, yielding the final result

$$\begin{aligned} g(t) &= \sum_{n=-\infty}^{\infty} g\left(\frac{n}{2W}\right) \frac{\sin(2\pi W t - n\pi)}{(2\pi W t - n\pi)} \\ &= \sum_{n=-\infty}^{\infty} g\left(\frac{n}{2W}\right) \operatorname{sinc}(2Wt - n), \qquad -\infty < t < \infty \end{aligned} \tag{3.9}$$

Equation (3.9) provides an *interpolation formula* for reconstructing the original signal $g(t)$ from the sequence of sample values $\{g(n/2W)\}$, with the sinc function $\operatorname{sinc}(2Wt)$ playing the role of an *interpolation function.* Each sample is multiplied by a delayed version of the interpolation function, and all the resulting waveforms are added to obtain $g(t)$.

We may now state the *sampling theorem* for strictly band-limited signals of finite energy in two equivalent parts, which apply to the transmitter and receiver of a pulse-modulation system, respectively:

1. *A band-limited signal of finite energy, which has no frequency components higher than W Hertz, is completely described by specifying the values of the signal at instants of time separated by 1/2W seconds.*
2. *A band-limited signal of finite energy, which has no frequency components higher than W Hertz, may be completely recovered from a knowledge of its samples taken at the rate of 2W samples per second.*

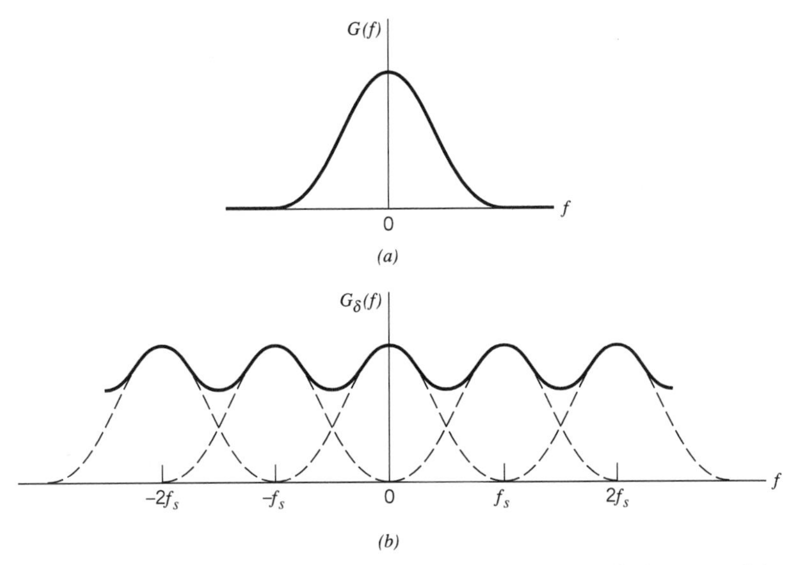


**FIGURE 3.3** (*a*) Spectrum of a signal. (*b*) Spectrum of an undersampled version of the signal exhibiting the aliasing phenomenon.

The sampling rate of $2W$ samples per second, for a signal bandwidth of $W$ Hertz, is called the *Nyquist rate*; its reciprocal $1/2W$ (measured in seconds) is called the *Nyquist interval.*

The derivation of the sampling theorem, as described herein, is based on the assumption that the signal $g(t)$ is strictly band limited. In practice, however, an information-bearing signal is *not* strictly band limited, with the result that some degree of undersampling is encountered. Consequently, some *aliasing* is produced by the sampling process. Aliasing refers to the phenomenon of a high-frequency component in the spectrum of the signal seemingly taking on the identity of a lower frequency in the spectrum of its sampled version, as illustrated in Figure 3.3. The aliased spectrum, shown by the solid curve in Figure 3.3*b*, pertains to an "undersampled" version of the message signal represented by the spectrum of Figure 3.3*a*.

To combat the effects of aliasing in practice, we may use two corrective measures, as described here:

1. Prior to sampling, a low-pass *anti-aliasing filter* is used to attenuate those high-frequency components of the signal that are not essential to the information being conveyed by the signal.
2. The filtered signal is sampled at a rate slightly higher than the Nyquist rate.

The use of a sampling rate higher than the Nyquist rate also has the beneficial effect of easing the design of the *reconstruction filter* used to recover the original signal from its sampled version. Consider the example of a message signal that has been anti-alias (low-pass) filtered, resulting in the spectrum shown in Figure 3.4*a*. The corresponding spectrum of the instantaneously sampled version of the signal is shown in Figure 3.4*b*, assuming a sampling rate higher than the Nyquist rate. According to Figure 3.4*b*, we readily see that the design of the reconstruction filter may be specified as follows (see Figure 3.4*c*):

- The reconstruction filter is low-pass with a passband extending from $-W$ to $W$, which is itself determined by the anti-aliasing filter.
- The filter has a transition band extending (for positive frequencies) from $W$ to $f_s - W$, where $f_s$ is the sampling rate.

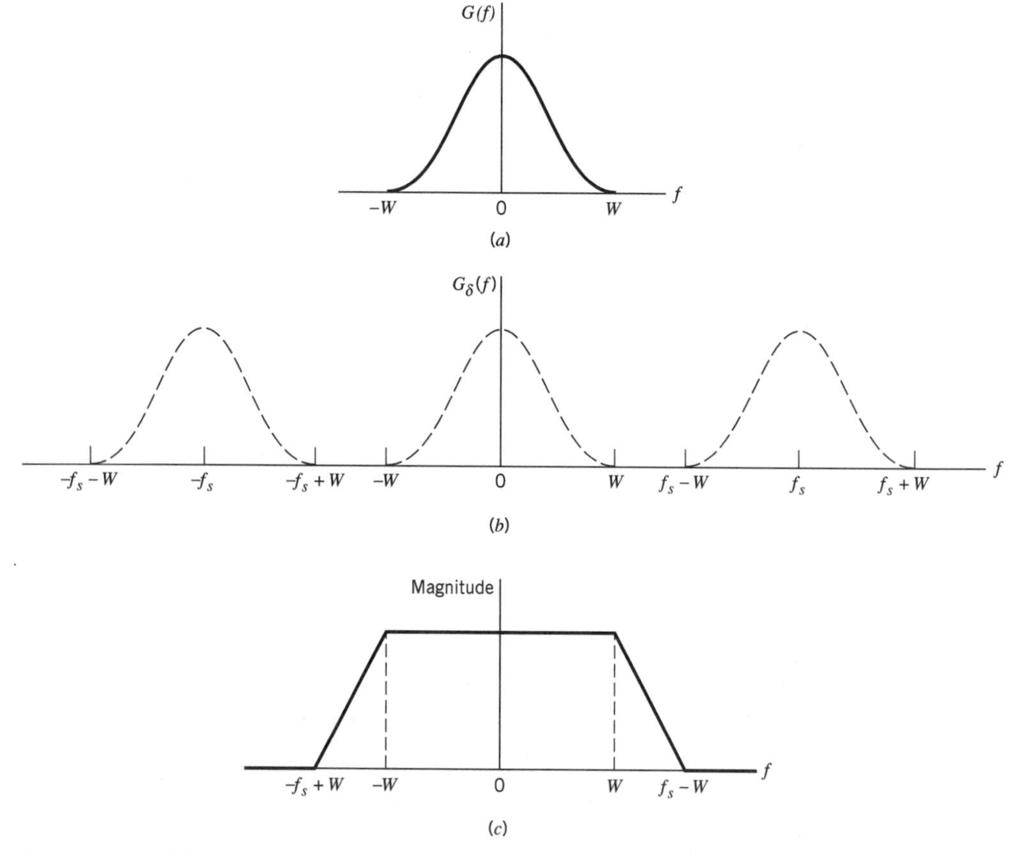


**FIGURE 3.4** (*a*) Anti-alias filtered spectrum of an information-bearing signal. (*b*) Spectrum of instantaneously sampled version of the signal, assuming the use of a sampling rate greater than the Nyquist rate. (*c*) Magnitude response of reconstruction filter.

The fact that the reconstruction filter has a well-defined transition band means that it is physically realizable.

## 3.3 *Pulse-Amplitude Modulation*

Now that we understand the essence of the sampling process, we are ready to formally define pulse-amplitude modulation, which is the simplest and most basic form of analog pulse modulation. In *pulse-amplitude modulation (PAM), the amplitudes of regularly spaced pulses are varied in proportion to the corresponding sample values of a continuous message signal*; the pulses can be of a rectangular form or some other appropriate shape. Pulse-amplitude modulation as defined here is somewhat similar to natural sampling, where the message signal is multiplied by a periodic train of rectangular pulses. However, in natural sampling the top of each modulated rectangular pulse varies with the message signal, whereas in PAM it is maintained flat; natural sampling is explored further in Problem 3.2.

The waveform of a PAM signal is illustrated in Figure 3.5. The dashed curve in this figure depicts the waveform of a message signal $m(t)$, and the sequence of amplitude-

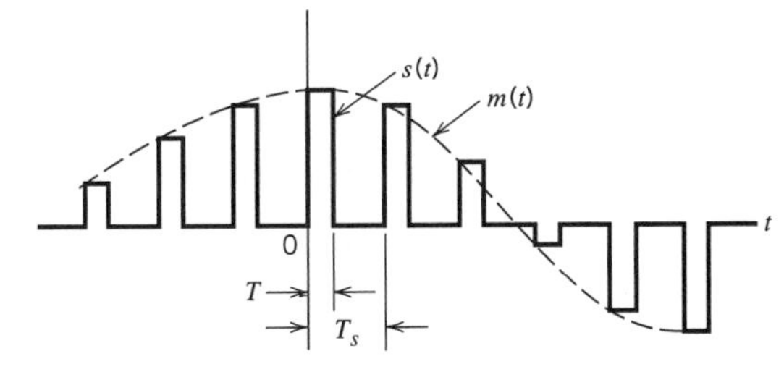


**FIGURE 3.5** Flat-top samples, representing an analog signal.

modulated rectangular pulses shown as solid lines represents the corresponding PAM signal $s(t)$. There are two operations involved in the generation of the PAM signal:

1. *Instantaneous sampling* of the message signal $m(t)$ every $T_s$ seconds, where the sampling rate $f_s = 1/T_s$ is chosen in accordance with the sampling theorem.
2. *Lengthening* the duration of each sample so obtained to some constant value $T$.

In digital circuit technology, these two operations are jointly referred to as "sample and hold." One important reason for intentionally lengthening the duration of each sample is to avoid the use of an excessive channel bandwidth, since bandwidth is inversely proportional to pulse duration. However, care has to be exercised in how long we make the sample duration $T$, as the following analysis reveals.

Let $s(t)$ denote the sequence of flat-top pulses generated in the manner described in Figure 3.5. We may express the PAM signal as

$$s(t) = \sum_{n=-\infty}^{\infty} m(nT_s)h(t - nT_s) \tag{3.10}$$

where $T_s$ is the *sampling period* and $m(nT_s)$ is the sample value of $m(t)$ obtained at time $t = nT_s$. The $h(t)$ is a standard rectangular pulse of unit amplitude and duration $T$, defined as follows (see Figure 3.6*a*):

$$h(t) = \begin{cases} 1, & 0 < t < T \\ \frac{1}{2}, & t = 0,\ t = T \\ 0, & \text{otherwise} \end{cases} \tag{3.11}$$

By definition, the instantaneously sampled version of $m(t)$ is given by

$$m_\delta(t) = \sum_{n=-\infty}^{\infty} m(nT_s)\ \delta(t - nT_s) \tag{3.12}$$

where $\delta(t - nT_s)$ is a time-shifted delta function. Therefore, convolving $m_\delta(t)$ with the pulse $h(t)$, we get

$$\begin{aligned} m_\delta(t) \star h(t) &= \int_{-\infty}^{\infty} m_\delta(\tau)h(t - \tau)\, d\tau \\ &= \int_{-\infty}^{\infty} \sum_{n=-\infty}^{\infty} m(nT_s)\ \delta(\tau - nT_s)h(t - \tau)\, d\tau \\ &= \sum_{n=-\infty}^{\infty} m(nT_s) \int_{-\infty}^{\infty} \delta(\tau - nT_s)h(t - \tau)\, d\tau \end{aligned} \tag{3.13}$$

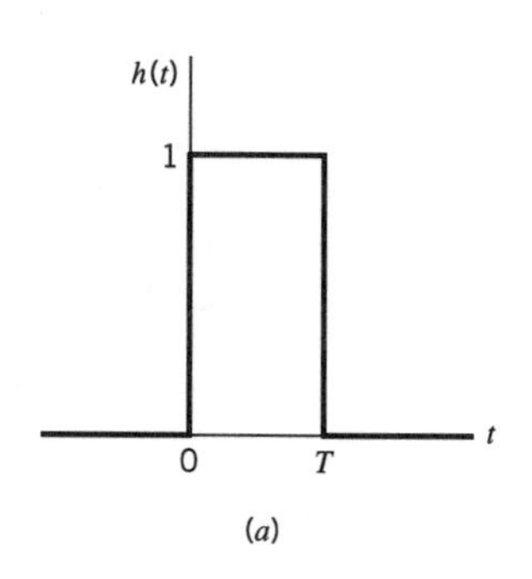


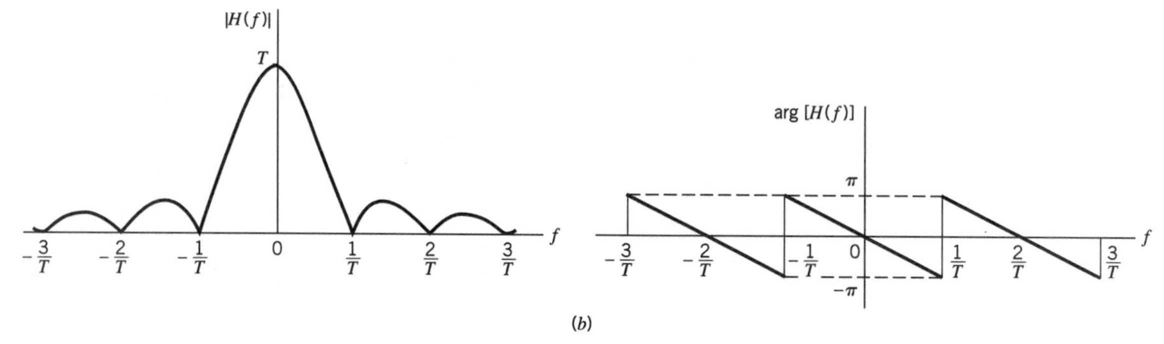


**FIGURE 3.6** (*a*) Rectangular pulse $h(t)$. (*b*) Spectrum $H(f)$, made up of the magnitude $|H(f)|$, and phase $\arg[H(f)]$.

Using the sifting property of the delta function (see Appendix 2), we thus obtain

$$m_\delta(t) \star h(t) = \sum_{n=-\infty}^{\infty} m(nT_s)h(t - nT_s) \tag{3.14}$$

From Equations (3.10) and (3.14) it follows that the PAM signal $s(t)$ is mathematically equivalent to the convolution of $m_\delta(t)$, the instantaneously sampled version of $m(t)$, and the pulse $h(t)$, as shown by

$$s(t) = m_\delta(t) \star h(t) \tag{3.15}$$

Taking the Fourier transform of both sides of Equation (3.15) and recognizing that the convolution of two time functions is transformed into the multiplication of their respective Fourier transforms, we get

$$S(f) = M_\delta(f)H(f) \tag{3.16}$$

where $S(f) = F[s(t)]$, $M_\delta(f) = F[m_\delta(t)]$, and $H(f) = F[h(t)]$. Adapting Equation (3.2) to the problem at hand, we note that the Fourier transform $M_\delta(f)$ is related to the Fourier transform $M(f)$ of the original message signal $m(t)$ as follows:

$$M_\delta(f) = f_s \sum_{k=-\infty}^{\infty} M(f - kf_s) \tag{3.17}$$

where $f_s$ is the sampling rate. Therefore, substitution of Equation (3.17) into (3.16) yields

$$S(f) = f_s \sum_{k=-\infty}^{\infty} M(f - kf_s)H(f) \tag{3.18}$$

Given a PAM signal $s(t)$ whose Fourier transform $S(f)$ is as defined in Equation (3.18), how do we recover the original message signal $m(t)$? As a first step in this recon-

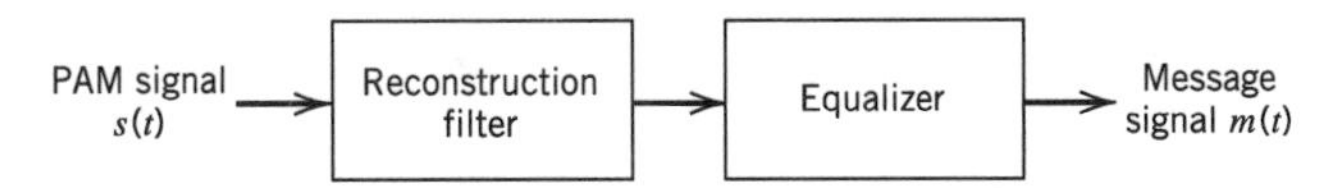


**FIGURE 3.7** System for recovering message signal $m(t)$ from PAM signal $s(t)$.

struction, we may pass $s(t)$ through a low-pass filter whose frequency response is defined in Figure 3.4*c*; here it is assumed that the message is limited to bandwidth $W$ and the sampling rate $f_s$ is larger than the Nyquist rate $2W$. Then, from Equation (3.18) we find that the spectrum of the resulting filter output is equal to $M(f)H(f)$. This output is equivalent to passing the original message signal $m(t)$ through another low-pass filter of frequency response $H(f)$.

From Equation (3.11) we note that the Fourier transform of the rectangular pulse $h(t)$ is given by

$$H(f) = T\,\text{sinc}(fT)\exp(-j\pi fT) \tag{3.19}$$

which is plotted in Figure 3.6*b*. We see therefore that by using flat-top samples to generate a PAM signal, we have introduced *amplitude distortion* as well as a *delay* of $T/2$. This effect is rather similar to the variation in transmission with frequency that is caused by the finite size of the scanning aperture in television. Accordingly, the distortion caused by the use of pulse-amplitude modulation to transmit an analog information-bearing signal is referred to as the *aperture effect.*

This distortion may be corrected by connecting an *equalizer* in cascade with the low-pass reconstruction filter, as shown in Figure 3.7. The equalizer has the effect of decreasing the in-band loss of the reconstruction filter as the frequency increases in such a manner as to compensate for the aperture effect. Ideally, the magnitude response of the equalizer is given by

$$\frac{1}{|H(f)|} = \frac{1}{T\,\text{sinc}(fT)} = \frac{\pi f}{\sin(\pi fT)} \tag{3.20}$$

The amount of equalization needed in practice is usually small. Indeed, for a duty cycle $T/T_s \leq 0.1$, the amplitude distortion is less than 0.5 percent, in which case the need for equalization may be omitted altogether.

The transmission of a PAM signal imposes rather stringent requirements on the magnitude and phase responses of the channel, because of the relatively short duration of the transmitted pulses. Furthermore, the noise performance of a PAM system can never be better than baseband-signal transmission. Accordingly, we find that for transmission over long distances, PAM would be used only as a means of message processing for time-division multiplexing, from which conversion to some other form of pulse modulation is subsequently made; time-division multiplexing is discussed in Section 3.9.

## 3.4 *Other Forms of Pulse Modulation*

In a pulse modulation system we may use the increased bandwidth consumed by the pulses to improve the noise performance of the system. This can be achieved by representing the sample values of the message signals by some property of the pulse other than amplitude:

- *Pulse-duration modulation* (PDM), also referred to as *pulse-width modulation*, where samples of the message signal are used to vary the duration of the individual pulses in the carrier.

- *Pulse-position modulation* (PPM), where the position of a pulse relative to its unmodulated time of occurrence is varied in accordance with the message signal.

These two other forms of pulse modulation are illustrated in Figure 3.8 for the case of a sinusoidal modulating wave.

In PDM, long pulses expend considerable power while bearing no additional information. If this unused power is subtracted from PDM so that only time transitions are preserved, we obtain PPM. Accordingly, PPM is a more efficient form of pulse modulation than PDM.

Since in a PPM system the transmitted information is contained in the relative positions of the modulated pulses, the presence of additive noise affects the performance of such a system by falsifying the time at which the modulated pulses are judged to occur. Immunity to noise can be established by making the pulse build up so rapidly that the time interval during which noise can exert any perturbation is very short. Indeed, additive noise would have no effect on the pulse positions if the received pulses were perfectly rectangular, because the presence of noise introduces only vertical perturbations. However, the reception of perfectly rectangular pulses would require an infinite channel bandwidth, which is of course impractical. Thus with a finite channel bandwidth in practice, we find that the received pulses have a finite rise time, so the performance of the PPM receiver is affected by noise, which is to be expected.

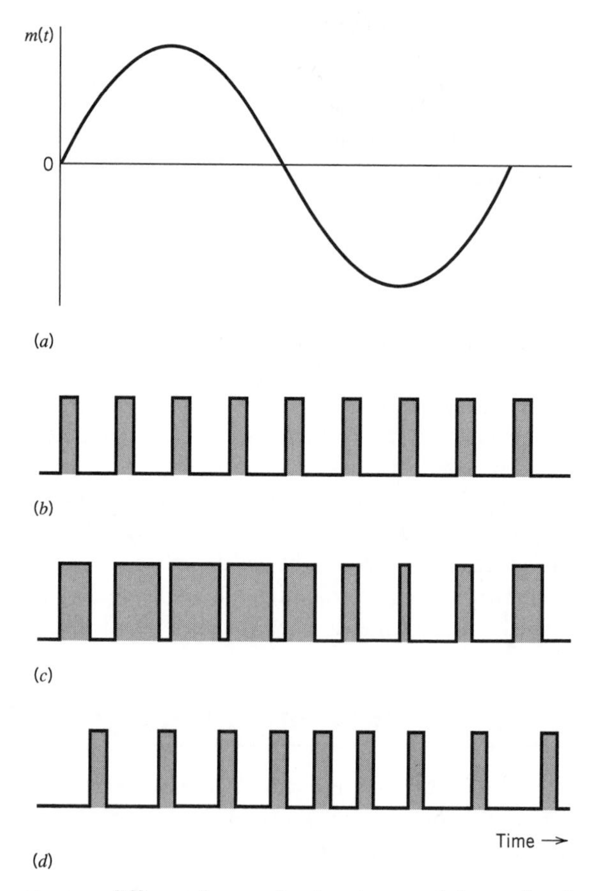


**FIGURE 3.8** Illustrating two different forms of pulse-time modulation for the case of a sinusoidal modulating wave. (*a*) Modulating wave. (*b*) Pulse carrier. (*c*) PDM wave. (*d*) PPM wave.

As in a CW modulation system, the noise performance of a PPM system may be described in terms of the output signal-to-noise ratio (SNR). Also, to find the noise improvement produced by PPM over baseband transmission of a message signal, we may use the figure of merit defined as the output signal-to-noise ratio of the PPM system divided by the channel signal-to-noise ratio; see Section 2.10. Assuming that the average power of the channel noise is small compared to the peak pulse power, the figure of merit of the PPM system is proportional to the square of the transmission bandwidth $B_T$ normalized with respect to the message bandwidth $W$. When, however, the input signal-to-noise ratio drops below a critical value, the system suffers a loss of the wanted message signal at the receiver output. That is, a PPM system suffers from a threshold effect of its own.

## 3.5 *Bandwidth–Noise Trade-Off*

In the context of noise performance, a PPM system is the optimum form of analog pulse modulation. The noise analysis of a PPM system reveals that pulse-position modulation (PPM) and frequency modulation (FM) systems exhibit a similar noise performance, as summarized here.[1]

1. Both systems have a figure of merit proportional to the square of the transmission bandwidth normalized with respect to the message bandwidth.
2. Both systems exhibit a threshold effect as the signal-to-noise ratio is reduced.

The practical implication of point 1 is that, in terms of a trade-off of increased transmission bandwidth for improved noise performance, the best that we can do with continuous-wave (CW) modulation and analog pulse modulation systems is to follow a *square law*. A question that arises at this point in the discussion is: Can we produce a trade-off better than a square law? The answer is an emphatic yes, and *digital pulse modulation* is the way to do it. The use of such a method is a radical departure from CW modulation.

Specifically, in a basic form of digital pulse modulation known as *pulse-code modulation* (PCM),[2] a message signal is represented in discrete form in both time and amplitude. This form of signal representation permits the transmission of the message signal as a sequence of *coded binary pulses*. Given such a sequence, the effect of channel noise at the receiver output can be reduced to a negligible level simply by making the average power of the transmitted binary PCM wave large enough compared to the average power of the noise.

Two fundamental processes are involved in the generation of a binary PCM wave: *sampling* and *quantization*. The sampling process takes care of the discrete-time representation of the message signal; for its proper application, we have to follow the sampling theorem described in Section 3.2. The quantization process takes care of the discrete-amplitude representation of the message signal; quantization is a new process, the details of which are described in the next section. For now it suffices to say that the combined use of sampling and quantization permits the transmission of a message signal in coded form. This, in turn, makes it possible to realize an *exponential law* for the bandwidth-noise trade-off, which is also demonstrated in the next section.

## 3.6 *Quantization Process*[3]

A continuous signal, such as voice, has a continuous range of amplitudes and therefore its samples have a continuous amplitude range. In other words, within the finite amplitude

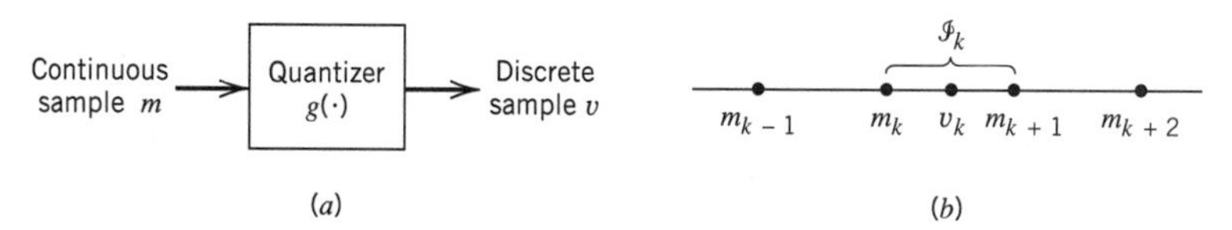


**FIGURE 3.9** Description of a memoryless quantizer.

range of the signal, we find an infinite number of amplitude levels. It is not necessary in fact to transmit the exact amplitudes of the samples. Any human sense (the ear or the eye), as ultimate receiver, can detect only finite intensity differences. This means that the original continuous signal may be *approximated* by a signal constructed of discrete amplitudes selected on a minimum error basis from an available set. The existence of a finite number of discrete amplitude levels is a basic condition of pulse-code modulation. Clearly, if we assign the discrete amplitude levels with sufficiently close spacing, we may make the approximated signal practically indistinguishable from the original continuous signal.

Amplitude *quantization* is defined as *the process of transforming the sample amplitude $m(nT_s)$ of a message signal $m(t)$ at time $t = nT_s$ into a discrete amplitude $v(nT_s)$ taken from a finite set of possible amplitudes.* We assume that the quantization process is *memoryless* and *instantaneous*, which means that the transformation at time $t = nT_s$ is not affected by earlier or later samples of the message signal. This simple form of scalar quantization, though not optimum, is commonly used in practice.

When dealing with a memoryless quantizer, we may simplify the notation by dropping the time index. We may thus use the symbol $m$ in place of $m(nT_s)$, as indicated in the block diagram of a quantizer shown in Figure 3.9*a*. Then, as shown in Figure. 3.9*b*, the signal amplitude $m$ is specified by the index $k$ if it lies inside the *partition cell*

$$\mathcal{I}_k: \{m_k < m \le m_{k+1}\}, \qquad k = 1, 2, \ldots, L \tag{3.21}$$

where $L$ is the total number of amplitude levels used in the quantizer. The discrete amplitudes $m_k$, $k = 1, 2, \ldots, L$, at the quantizer input are called *decision levels* or *decision thresholds*. At the quantizer output, the index $k$ is transformed into an amplitude $v_k$ that represents all amplitudes of the cell $\mathcal{I}_k$; the discrete amplitudes $v_k$, $k = 1, 2, \ldots, L$, are called *representation levels* or *reconstruction levels*, and the spacing between two adjacent representation levels is called a *quantum* or *step-size*. Thus, the quantizer output $v$ equals $v_k$ if the input signal sample $m$ belongs to the interval $\mathcal{I}_k$. The mapping (see Figure 3.9*a*)

$$v = g(m) \tag{3.22}$$

is the *quantizer characteristic*, which is a staircase function by definition.

Quantizers can be of a *uniform* or *nonuniform* type. In a uniform quantizer, the representation levels are uniformly spaced; otherwise, the quantizer is nonuniform. In this section, we consider only uniform quantizers; nonuniform quantizers are considered in Section 3.7. The quantizer characteristic can also be of *midtread* or *midrise type*. Figure 3.10*a* shows the input–output characteristic of a uniform quantizer of the midtread type, which is so called because the origin lies in the middle of a tread of the staircaselike graph. Figure 3.10*b* shows the corresponding input–output characteristic of a uniform quantizer of the midrise type, in which the origin lies in the middle of a rising part of the staircaselike graph. Note that both the midtread and midrise types of uniform quantizers illustrated in Figure 3.10 are *symmetric* about the origin.

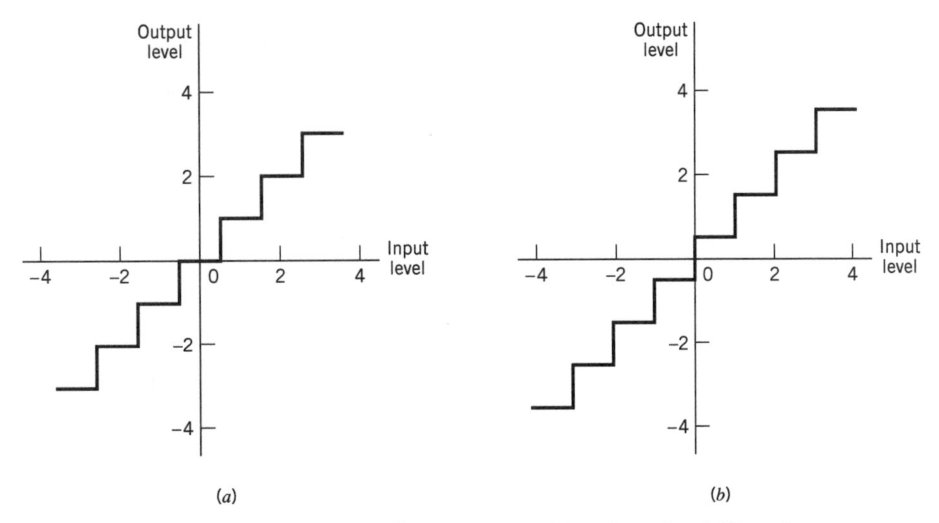


**FIGURE 3.10** Two types of quantization: (*a*) midtread and (*b*) midrise.

## ■ QUANTIZATION NOISE

The use of quantization introduces an error defined as the difference between the input signal $m$ and the output signal $v$. The error is called *quantization noise*. Figure 3.11 illustrates a typical variation of the quantization noise as a function of time, assuming the use of a uniform quantizer of the midtread type.

Let the quantizer input $m$ be the sample value of a zero-mean random variable $M$. (If the input has a nonzero mean, we can always remove it by subtracting the mean from the input and then adding it back after quantization.) A quantizer $g(\cdot)$ maps the input

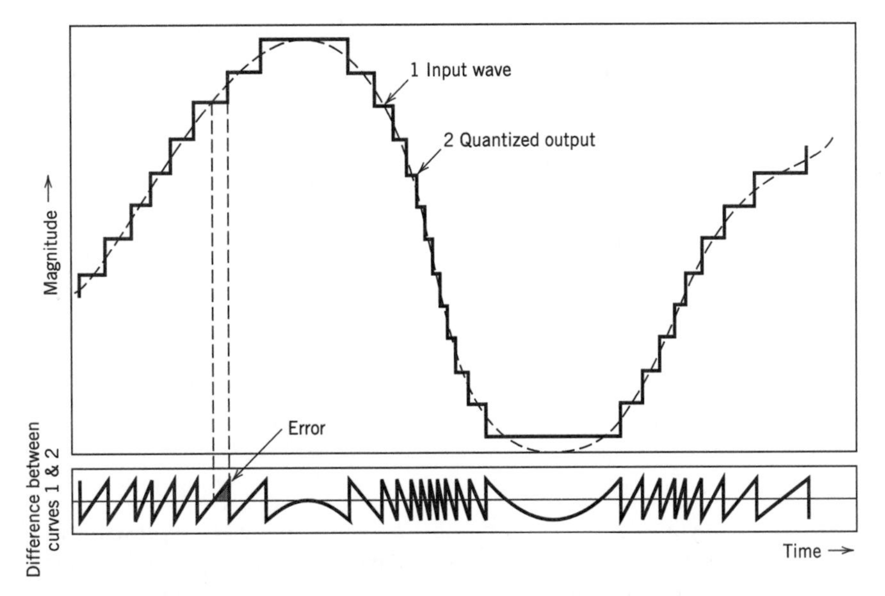


**FIGURE 3.11** Illustration of the quantization process. (Adapted from Bennett, 1948, with permission of AT&T.)

random variable $M$ of continuous amplitude into a discrete random variable $V$; their respective sample values $m$ and $v$ are related by Equation (3.22). Let the quantization error be denoted by the random variable $Q$ of sample value $q$. We may thus write

$$q = m - v \tag{3.23}$$

or, correspondingly,

$$Q = M - V \tag{3.24}$$

With the input $M$ having zero mean, and the quantizer assumed to be symmetric as in Figure 3.10, it follows that the quantizer output $V$ and therefore the quantization error $Q$, will also have zero mean. Thus for a partial statistical characterization of the quantizer in terms of output signal-to-(quantization) noise ratio, we need only find the mean-square value of the quantization error $Q$.

Consider then an input $m$ of continuous amplitude in the range $(-m_{\max}, m_{\max})$. Assuming a uniform quantizer of the midrise type illustrated in Figure 3.10*b*, we find that the step-size of the quantizer is given by

$$\Delta = \frac{2m_{\max}}{L} \tag{3.25}$$

where $L$ is the total number of representation levels. For a uniform quantizer, the quantization error $Q$ will have its sample values bounded by $-\Delta/2 \le q \le \Delta/2$. If the step-size $\Delta$ is sufficiently small (i.e., the number of representation levels $L$ is sufficiently large), it is reasonable to assume that the quantization error $Q$ is a *uniformly distributed* random variable, and the interfering effect of the quantization noise on the quantizer input is similar to that of thermal noise. We may thus express the probability density function of the quantization error $Q$ as follows:

$$f_Q(q) = \begin{cases} \dfrac{1}{\Delta}, & -\dfrac{\Delta}{2} < q \le \dfrac{\Delta}{2} \\ 0, & \text{otherwise} \end{cases} \tag{3.26}$$

For this to be true, however, we must ensure that the incoming signal does *not* overload the quantizer. Then, with the mean of the quantization error being zero, its variance $\sigma_Q^2$ is the same as the mean-square value:

$$\begin{aligned} \sigma_Q^2 &= E[Q^2] \\ &= \int_{-\Delta/2}^{\Delta/2} q^2 f_Q(q)\, dq \end{aligned} \tag{3.27}$$

Substituting Equation (3.26) into (3.27), we get

$$\begin{aligned} \sigma_Q^2 &= \frac{1}{\Delta}\int_{-\Delta/2}^{\Delta/2} q^2\, dq \\ &= \frac{\Delta^2}{12} \end{aligned} \tag{3.28}$$

Typically, the $L$-ary number $k$, denoting the $k$th representation level of the quantizer, is transmitted to the receiver in binary form. Let $R$ denote the number of *bits per sample* used in the construction of the binary code. We may then write

$$L = 2^R \tag{3.29}$$

or, equivalently,

$$R = \log_2 L \tag{3.30}$$

Hence, substituting Equation (3.29) into (3.25), we get the step size

$$\Delta = \frac{2m_{\max}}{2^R} \tag{3.31}$$

Thus the use of Equation (3.31) in (3.28) yields

$$\sigma_Q^2 = \tfrac{1}{3}m_{\max}^2 2^{-2R} \tag{3.32}$$

Let $P$ denote the average power of the message signal $m(t)$. We may then express the *output signal-to-noise ratio* of a uniform quantizer as

$$\begin{aligned}(\text{SNR})_O &= \frac{P}{\sigma_Q^2} \\ &= \left(\frac{3P}{m_{\max}^2}\right)2^{2R}\end{aligned} \tag{3.33}$$

Equation (3.33) shows that the output signal-to-noise ratio of the quantizer increases *exponentially* with increasing number of bits per sample, $R$. Recognizing that an increase in $R$ requires a proportionate increase in the channel (transmission) bandwidth $B_T$, we thus see that the use of a binary code for the representation of a message signal (as in pulse-code modulation) provides a more efficient method than either frequency modulation (FM) or pulse-position modulation (PPM) for the trade-off of increased channel bandwidth for improved noise performance. In making this statement, we presume that the FM and PPM systems are limited by receiver noise, whereas the binary-coded modulation system is limited by quantization noise. We have more to say on the latter issue in Section 3.8.

### ▶ EXAMPLE 3.1 Sinusoidal Modulating Signal

Consider the special case of a full-load sinusoidal modulating signal of amplitude $A_m$, which utilizes all the representation levels provided. The average signal power is (assuming a load of 1 ohm)

$$P = \frac{A_m^2}{2}$$

The total range of the quantizer input is $2A_m$, because the modulating signal swings between $-A_m$ and $A_m$. We may therefore set $m_{\max} = A_m$, in which case the use of Equation (3.32) yields the average power (variance) of the quantization noise as

$$\sigma_Q^2 = \tfrac{1}{3}A_m^2 2^{-2R}$$

Thus the output signal-to-noise ratio of a uniform quantizer, for a full-load test tone, is

$$(\text{SNR})_O = \frac{A_m^2/2}{A_m^2 2^{-2R}/3} = \frac{3}{2}\,(2^{2R}) \tag{3.34}$$

Expressing the signal-to-noise ratio in decibels, we get

$$10 \log_{10}(\text{SNR})_O = 1.8 + 6R \tag{3.35}$$

**TABLE 3.1** ***Signal-to-(quantization) noise ratio for varying number of representation levels for sinusoidal modulation***

| *Number of Representation Levels, L* | *Number of Bits per Sample, R* | *Signal-to-Noise Ratio (dB)* |
|---|---|---|
| 32 | 5 | 31.8 |
| 64 | 6 | 37.8 |
| 128 | 7 | 43.8 |
| 256 | 8 | 49.8 |

For various values of $L$ and $R$, the corresponding values of signal-to-noise ratio are as given in Table 3.1. From Table 3.1 we can make a quick estimate of the number of bits per sample required for a desired output signal-to-noise ratio, assuming sinusoidal modulation. ◀

Thus far in this section we have focused on how to characterize memoryless scalar quantizers and assess their performance. In so doing, however, we avoided the optimum design of quantizers, that is, the issue of selecting the representation levels and partition cells so as to minimize the average quantization power for a prescribed number of representation levels. Unfortunately, this optimization problem does not lend itself to a closed-form solution because of the highly *nonlinear* nature of the quantization process. Rather, we have effective algorithms for finding the optimum design in an iterative manner. A well-known algorithm that deserves to be mentioned in this context is the Lloyd-Max quantizer, which is discussed next.

## ■ CONDITIONS FOR OPTIMALITY OF SCALAR QUANTIZERS

In designing a scalar quantizer the challenge is how to select the representation levels and surrounding partition cells so as to minimize the average quantization power for a fixed number of representation levels.

To state the problem in mathematical terms, consider a message signal $m(t)$ drawn from a stationary process $M(t)$. Let $-A \le m \le A$ denote the dynamic range of $m(t)$, which is partitioned into a set of $L$ cells, as depicted in Figure 3.12. The boundaries of the partition cells are defined by a set of real numbers $m_1, m_2, \ldots, m_{L+1}$ that satisfy the following three conditions:

$$\begin{aligned} m_1 &= -A \\ m_{L+1} &= A \\ m_k &\le m_{k+1} \text{ for } k = 1, 2, \ldots, L \end{aligned}$$

The $k$th partition cell is defined by

$$\mathcal{I}_k\!: m_k < m \le m_{k+1} \text{ for } k = 1, 2, \ldots, L \tag{3.36}$$

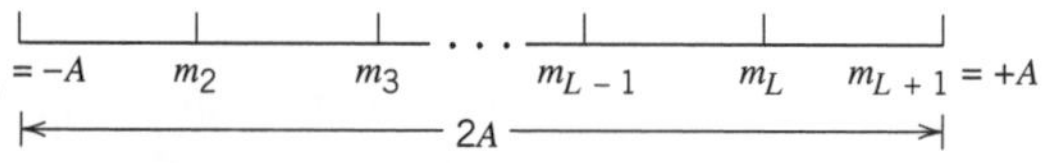

$m_1 = -A$ $m_2$ $m_3$ … $m_{L-1}$ $m_L$ $m_{L+1} = +A$; $2A$

**FIGURE 3.12** Illustrating the partitioning of the dynamic range $-A \le m \le A$ of a message signal $m(t)$ into a set of $L$ cells.

Let the representation levels (i.e., quantization values) be denoted by $v_k$, $k = 1, 2, \ldots, L$. Then, assuming that $d(m, v_k)$ denotes a *distortion measure* for using $v_k$ to represent all those values of the input $m$ that lie inside the partition cell $\mathcal{J}_k$, the goal is to find the two sets, $\{v_k\}_{k=1}^{L}$ and $\{\mathcal{J}_k\}_{k=1}^{L}$, that minimize the *average distortion*

$$D = \sum_{k=1}^{L} \int_{m \in \mathcal{J}_k} d(m, v_k) f_M(m)\, dm \tag{3.37}$$

where $f_M(m)$ is the probability density function of the random variable $M$ with sample value $m$.

A commonly used distortion measure is

$$d(m, v_k) = (m - v_k)^2 \tag{3.38}$$

in which case we speak of the *mean-square distortion*. In any event, the optimization problem stated herein is nonlinear, defying an explicit, closed-form solution. To get around this difficulty, we resort to an algorithmic approach for solving the problem in an iterative manner.

Structurally speaking, the quantizer consists of two components with interrelated design parameters:

- An encoder characterized by the set of partition cells $\{\mathcal{J}_k\}_{k=1}^{L}$; it is located in the transmitter.
- A decoder characterized by the set of representation levels $\{v_k\}_{k=1}^{L}$; it is located in the receiver.

Accordingly, we may identify two critically important conditions that provide the mathematical basis for all algorithmic solutions to the optimum quantization problem. One condition assumes that we are given a decoder and the problem is to find the optimum encoder in the transmitter. The other condition assumes that we are given an encoder and the problem is to find the optimum decoder in the receiver. Henceforth, these two conditions are referred to as condition I and condition II, respectively.

## Condition I. Optimality of the Encoder for a Given Decoder

The availability of a decoder means that we have a certain *codebook* in mind. Let the codebook be defined by

$$\mathscr{C}: \{v_k\}_{k=1}^{L} \tag{3.39}$$

Given the codebook $\mathscr{C}$, the problem is to find the set of partition cells $\{\mathcal{J}_k\}_{k=1}^{L}$ that minimizes the average distortion $D$. That is, we wish to find the encoder defined by the nonlinear mapping

$$g(m) = v_k, \qquad k = 1, 2, \ldots, L \tag{3.40}$$

such that we have

$$D = \int_{-A}^{A} d(m, g(m)) f_M(m)\, dM \geq \sum_{k=1}^{L} \int_{m \in \mathcal{J}_k} [\min_{v_k \in \mathscr{C}} d(m, v_k)] f_M(m)\, dm \tag{3.41}$$

For the lower bound specified in Equation (3.41) to be attained, we require that the nonlinear mapping of Equation (3.40) be satisfied only if the condition

$$d(m, v_k) \leq d(m, v_j) \qquad \text{holds for all } j \neq k \tag{3.42}$$

The necessary condition described in Equation (3.42) for optimality of the encoder for a specified codebook $\mathscr{C}$ is recognized as the *nearest neighbor condition*. In words, the nearest neighbor condition requires that the partition cell $\mathscr{J}_k$ should embody all those values of the input $m$ that are closer to $v_k$ than any other element of the codebook C. This optimality condition is indeed intuitively satisfying.

### Condition II. Optimality of the Decoder for a Given Encoder

Consider next the reverse situation to that described under condition I, which may be stated as follows: Optimize the codebook $\mathscr{C} = \{v_k\}_{k=1}^{L}$ for the decoder, given that the set of partition cells $\{\mathscr{J}_k\}_{k=1}^{L}$ characterizing the encoder is fixed. The criterion for optimization is the average (mean-square) distortion:

$$D = \sum_{k=1}^{L} \int_{m \in \mathscr{J}_k} (m - v_k)^2 f_M(m)\, dm \tag{3.43}$$

The probability density function $f_M(m)$ is clearly independent of the codebook $\mathscr{C}$. Hence, differentiating $D$ with respect to the representation level $v_k$, we readily obtain

$$\frac{\partial D}{\partial v_k} = -2 \sum_{k=1}^{L} \int_{m \in \mathscr{J}_k} (m - v_k) f_M(m)\, dm \tag{3.44}$$

Setting $\partial D / \partial v_k$ equal to zero and then solving for $v_k$, we obtain the optimum value

$$v_{k,\text{opt}} = \frac{\displaystyle\int_{m \in \mathscr{J}_k} m f_M(m)\, dm}{\displaystyle\int_{m \in \mathscr{J}_k} f_M(m)\, dm} \tag{3.45}$$

The denominator in Equation (3.45) is just the probability, $p_k$, that the random variable $M$ with sample value $m$ lies in the partition cell $\mathscr{J}_k$, as shown by

$$\begin{aligned} p_k &= P(m_k < M \leq m_k + 1) \\ &= \int_{m \in \mathscr{J}_k} f_M(m)\, dm \end{aligned} \tag{3.46}$$

Accordingly, we may interpret the optimality condition of Equation (3.45) as choosing the representation level $v_k$ to equal the *conditional mean* of the random variable $M$, given that $M$ lies in the partition cell $\mathscr{J}_k$. We can thus formally state the condition for optimality of the decoder for a given encoder as follows:

$$v_{k,\text{opt}} = E[M \mid m_k < M \leq m_{k+1}] \tag{3.47}$$

where $E$ is the expectation operator. Equation (3.47) is also intuitively satisfying.

Note that the nearest neighbor condition (condition I) for optimality of the encoder for a given decoder was proved for a generic average distortion. However, the conditional mean requirement (condition II) for optimality of the decoder for a given encoder was proved for the special case of a mean-square distortion. In any event, these two conditions are necessary for optimality of a scalar quantizer. Basically, the algorithm for designing the quantizer consists of alternately optimizing the encoder in accordance with condition I, then optimizing the decoder in accordance with condition II, and continuing in this manner until the average distortion $D$ reaches a minimum. An optimum quantizer designed in this manner is called a *Lloyd-Max quantizer*.[4]

# 3.7 *Pulse-Code Modulation*

With the sampling and quantization processes at our disposal, we are now ready to describe pulse-code modulation, which, as mentioned previously, is the most basic form of digital pulse modulation. In *pulse-code modulation (PCM), a message signal is represented by a sequence of coded pulses, which is accomplished by representing the signal in discrete form in both time and amplitude.* The basic operations performed in the transmitter of a PCM system are *sampling*, *quantizing*, and *encoding*, as shown in Figure 3.13*a*; the low-pass filter prior to sampling is included to prevent aliasing of the message signal. The quantizing and encoding operations are usually performed in the same circuit, which is called an *analog-to-digital converter*. The basic operations in the receiver are *regeneration* of impaired signals, *decoding*, and *reconstruction* of the train of quantized samples, as shown in Figure 3.13*c*. Regeneration also occurs at intermediate points along the transmission path as necessary, as indicated in Figure 3.13*b*. When time-division multiplexing is used, it becomes necessary to synchronize the receiver to the transmitter for the overall system to operate satisfactorily, as discussed in Section 3.9. In what follows, we describe the various operations that constitute a basic PCM system.

## Sampling

The incoming message signal is sampled with a train of narrow rectangular pulses so as to closely approximate the instantaneous sampling process. To ensure perfect reconstruction of the message signal at the receiver, the sampling rate must be greater than twice the highest frequency component $W$ of the message signal in accordance with the sampling theorem. In practice, a low-pass anti-aliasing filter is used at the front end of the sampler to exclude frequencies greater than $W$ before sampling. Thus the application of sampling

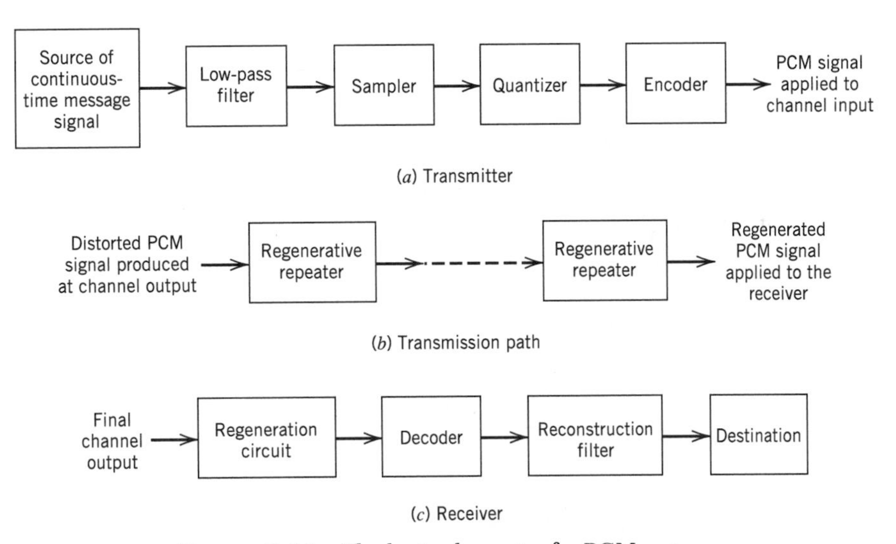


**Figure 3.13** The basic elements of a PCM system.

permits the reduction of the continuously varying message signal (of some finite duration) to a limited number of discrete values per second.

## ■ QUANTIZATION

The sampled version of the message signal is then quantized, thereby providing a new representation of the signal that is discrete in both time and amplitude. The quantization process may follow a uniform law as described in Section 3.6. In telephonic communication, however, it is preferable to use a variable separation between the representation levels. For example, the range of voltages covered by voice signals, from the peaks of loud talk to the weak passages of weak talk, is on the order of 1000 to 1. By using a *nonuniform quantizer* with the feature that the step-size increases as the separation from the origin of the input–output amplitude characteristic is increased, the large end steps of the quantizer can take care of possible excursions of the voice signal into the large amplitude ranges that occur relatively infrequently. In other words, the weak passages, which need more protection, are favored at the expense of the loud passages. In this way, a nearly uniform percentage precision is achieved throughout the greater part of the amplitude range of the input signal, with the result that fewer steps are needed than would be the case if a uniform quantizer were used.

The use of a nonuniform quantizer is equivalent to passing the baseband signal through a *compressor* and then applying the compressed signal to a uniform quantizer. A particular form of compression law that is used in practice is the so-called *$\mu$-law*,[5] which is defined by

$$|v| = \frac{\log(1 + \mu|m|)}{\log(1 + \mu)} \tag{3.48}$$

where $m$ and $v$ are the normalized input and output voltages, and $\mu$ is a positive constant. In Figure 3.14*a*, we have plotted the $\mu$-law for three different values of $\mu$. The case of uniform quantization corresponds to $\mu = 0$. For a given value of $\mu$, the reciprocal slope

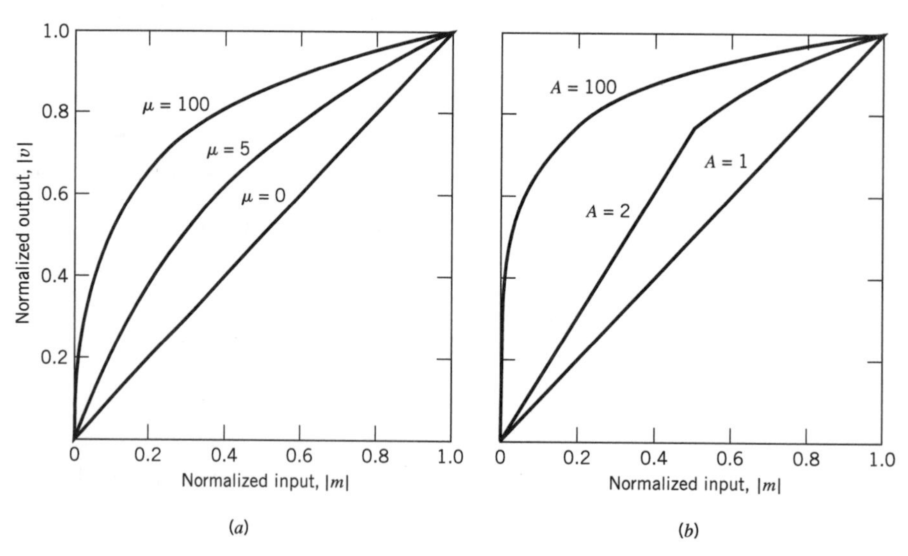


**FIGURE 3.14** Compression laws. (*a*) $\mu$-law. (*b*) A-law.

of the compression curve, which defines the quantum steps, is given by the derivative of $|m|$ with respect to $|v|$; that is,

$$\frac{d|m|}{d|v|} = \frac{\log(1+\mu)}{\mu}(1+\mu|m|) \tag{3.49}$$

We see therefore that the $\mu$-law is neither strictly linear not strictly logarithmic, but it is approximately linear at low input levels corresponding to $\mu|m| \ll 1$, and approximately logarithmic at high input levels corresponding to $\mu|m| \gg 1$.

Another compression law that is used in practice is the so-called *A-law* defined by

$$|v| = \begin{cases} \dfrac{A|m|}{1+\log A}, & 0 \le |m| \le \dfrac{1}{A} \\ \dfrac{1+\log(A|m|)}{1+\log A}, & \dfrac{1}{A} \le |m| \le 1 \end{cases} \tag{3.50}$$

which is plotted in Figure 3.14$b$ for varying $A$. The case of uniform quantization corresponds to $A = 1$. The reciprocal slope of this second compression curve is given by the derivative of $|m|$ with respect to $|v|$, as shown by (depending on the value assigned to the normalized input $|m|$)

$$\frac{d|m|}{d|v|} = \begin{cases} \dfrac{1+\log A}{A}, & 0 \le |m| \le \dfrac{1}{A} \\ (1+A)|m|, & \dfrac{1}{A} \le |m| \le 1 \end{cases} \tag{3.51}$$

To restore the signal samples to their correct relative level, we must, of course, use a device in the receiver with a characteristic complementary to the compressor. Such a device is called an *expander*. Ideally, the compression and expansion laws are exactly inverse so that, except for the effect of quantization, the expander output is equal to the compressor input. The combination of a *comp*ressor and an exp*ander* is called a *compander*.

For both the $\mu$-law and $A$-law, the dynamic range capability of the compander improves with increasing $\mu$ and $A$, respectively. The SNR for low-level signals increases at the expense of the SNR for high-level signals. To accommodate these two conflicting requirements (i.e., a reasonable SNR for both low- and high-level signals), a compromise is usually made in choosing the value of paramenter $\mu$ for the $\mu$-law and parameter $A$ for the $A$-law. The typical values used in practice are: $\mu = 255$ and $A = 87.6$.

It is also of interest to note that in actual PCM systems, the companding circuitry does not produce an exact replica of the nonlinear compression curves shown in Figure 3.14. Rather, it provides a *piecewise linear* approximation to the desired curve. By using a large enough number of linear segments, the approximation can approach the true compression curve very closely. This form of approximation is illustrated in Example 3.2.

## ENCODING

In combining the processes of sampling and quantization, the specification of a continuous message (baseband) signal becomes limited to a discrete set of values, but not in the form best suited to transmission over a telephone line or radio path. To exploit the advantages of sampling and quantizing for the purpose of making the transmitted signal more robust to noise, interference and other channel impairments, we require the use of an *encoding*

**TABLE 3.2 *Binary number system for R = 4 bits/sample***

| *Ordinal Number of Representation Level* | *Level Number Expressed as Sum of Powers of 2* | *Binary Number* |
|---|---|---|
| 0 | | 0000 |
| 1 | $2^0$ | 0001 |
| 2 | $2^1$ | 0010 |
| 3 | $2^1 + 2^0$ | 0011 |
| 4 | $2^2$ | 0100 |
| 5 | $2^2 + 2^0$ | 0101 |
| 6 | $2^2 + 2^1$ | 0110 |
| 7 | $2^2 + 2^1 + 2^0$ | 0111 |
| 8 | $2^3$ | 1000 |
| 9 | $2^3 + 2^0$ | 1001 |
| 10 | $2^3 + 2^1$ | 1010 |
| 11 | $2^3 + 2^1 + 2^0$ | 1011 |
| 12 | $2^3 + 2^2$ | 1100 |
| 13 | $2^3 + 2^2 + 2^0$ | 1101 |
| 14 | $2^3 + 2^2 + 2^1$ | 1110 |
| 15 | $2^3 + 2^2 + 2^1 + 2^0$ | 1111 |

*process* to translate the discrete set of sample values to a more appropriate form of signal. Any plan for representing each of this discrete set of values as a particular arrangement of discrete events is called a *code*. One of the discrete events in a code is called a *code element* or *symbol*. For example, the presence or absence of a pulse is a symbol. A particular arrangement of symbols used in a code to represent a single value of the discrete set is called a *code word* or *character*.

In a *binary code*, each symbol may be either of two distinct values or kinds, such as the presence or absence of a pulse. The two symbols of a binary code are customarily denoted as 0 and 1. In a *ternary code*, each symbol may be one of three distinct values or kinds, and so on for other codes. However, *the maximum advantage over the effects of noise in a transmission medium is obtained by using a binary code, because a binary symbol withstands a relatively high level of noise and is easy to regenerate*. Suppose that, in a binary code, each code word consists of $R$ bits: *bit* is an acronym for *binary digit*; thus $R$ denotes the number of *bits per sample*. Then, using such a code, we may represent a total of $2^R$ distinct numbers. For example, a sample quantized into one of 256 levels may be represented by an 8-bit code word.

There are several ways of establishing a one-to-one correspondence between representation levels and code words. A convenient method is to express the ordinal number of the representation level as a binary number. In the binary number system, each digit has a place-value that is a power of 2, as illustrated in Table 3.2 for the case of four bits per sample (i.e., $R = 4$).

### *Line Codes*

Any of several line codes can be used for the electrical representation of a binary data stream. Figure 3.15 displays the waveforms of five important line codes for the example data stream 01101001. Figure 3.16 displays their individual power spectra (for

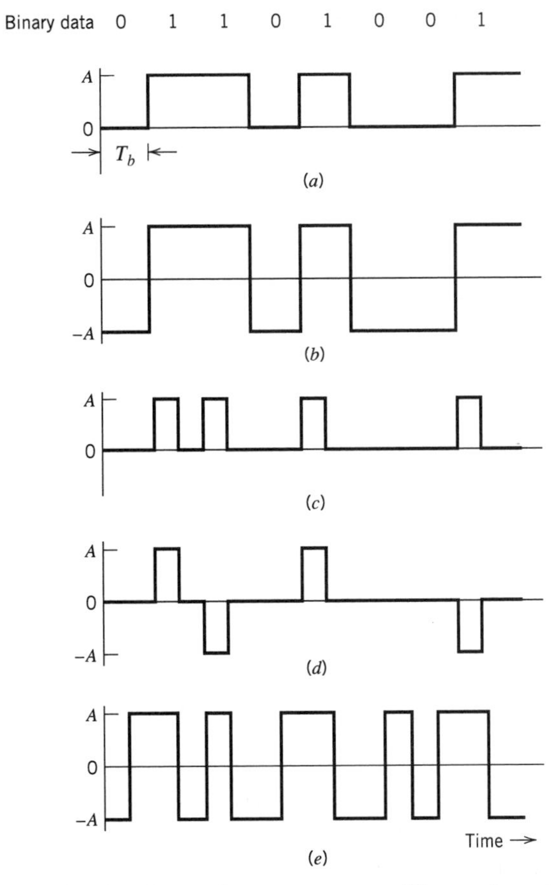


**FIGURE 3.15** Line codes for the electrical representations of binary data. (*a*) Unipolar NRZ signaling. (*b*) Polar NRZ signaling. (*c*) Unipolar RZ signaling. (*d*) Bipolar RZ signaling. (*e*) Split-phase or Manchester code.

positive frequencies) for randomly generated binary data, assuming that (1) symbols 0 and 1 are equiprobable, (2) the average power is normalized to unity, and (3) the frequency $f$ is normalized with respect to the bit rate $1/T_b$. (For the formulas used to plot the power spectra of Figure 3.16, the reader is referred to Problem 3.11.) The five line codes illustrated in Figure 3.15 are described here:

1. *Unipolar nonreturn-to-zero (NRZ) signaling*
In this line code, symbol 1 is represented by transmitting a pulse of amplitude $A$ for the duration of the symbol, and symbol 0 is represented by switching off the pulse, as in Figure 3.15*a*. This line code is also referred to as *on-off signaling*. Disadvantages of on-off signaling are the waste of power due to the transmitted DC level and the fact that the power spectrum of the transmitted signal does not approach zero at zero frequency.

2. *Polar nonreturn-to-zero (NRZ) signaling*
In this second line code, symbols 1 and 0 are represented by transmitting pulses of amplitudes $+A$ and $-A$, respectively, as illustrated in Figure 3.15*b*. This line code is relatively easy to generate but its disadvantage is that the power spectrum of the signal is large near zero frequency.

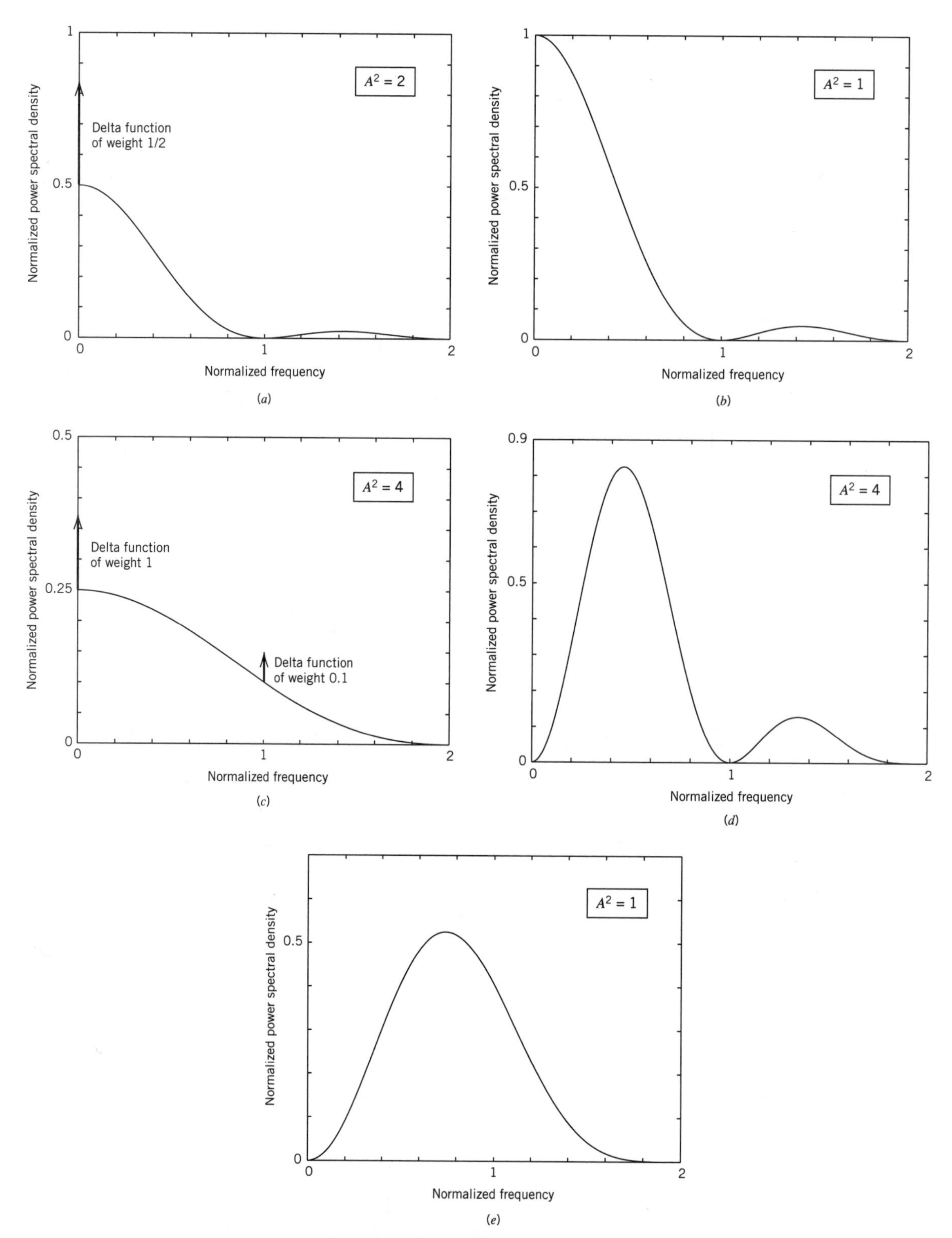


**FIGURE 3.16** Power spectra of line codes: (*a*) Unipolar NRZ signal. (*b*) Polar NRZ signal. (*c*) Unipolar RZ signal. (*d*) Bipolar RZ signal. (*e*) Manchester-encoded signal. The frequency is normalized with respect to the bit rate $1/T_b$, and the average power is normalized to unity.

3. *Unipolar return-to-zero (RZ) signaling*
In this other line code, symbol 1 is represented by a rectangular pulse of amplitude $A$ and half-symbol width, and symbol 0 is represented by transmitting *no* pulse, as illustrated in Figure 3.15*c*. An attractive feature of this line code is the presence of delta functions at $f = 0, \pm 1/T_b$ in the power spectrum of the transmitted signal, which can be used for bit-timing recovery at the receiver. However, its disadvantage is that it requires 3 dB more power than polar return-to-zero signaling for the same probability of symbol error; this issue is addressed in Chapter 4 under Problem 4.10.

4. *Bipolar return-to-zero (BRZ) signaling*
This line code uses three amplitude levels as indicated in Figure 3.15*d*. Specifically, positive and negative pulses of equal amplitude (i.e., $+A$ and $-A$) are used alternately for symbol 1, with each pulse having a half-symbol width; no pulse is always used for symbol 0. A useful property of the BRZ signaling is that the power spectrum of the transmitted signal has no DC component and relatively insignificant low-frequency components for the case when symbols 1 and 0 occur with equal probability. This line code is also called *alternate mark inversion* (AMI) signaling.

5. *Split-phase (Manchester code)*
In this method of signaling, illustrated in Figure 3.15*e*, symbol 1 is represented by a positive pulse of amplitude $A$ followed by a negative pulse of amplitude $-A$, with both pulses being half-symbol wide. For symbol 0, the polarities of these two pulses are reversed. The Manchester code suppresses the DC component and has relatively insignificant low-frequency components, regardless of the signal statistics. This property is essential in some applications.

### *Differential Encoding*

This method is used to encode information in terms of *signal transitions*. In particular, a transition is used to designate symbol 0 in the incoming binary data stream, while no transition is used to designate symbol 1, as illustrated in Figure 3.17. In Figure 3.17*b* we show the differentially encoded data stream for the example data specified in Figure 3.17*a*. The original binary data stream used here is the same as that used in Figure 3.15. The waveform of the differentially encoded data is shown in Figure 3.17*c*, assuming the use of unipolar nonreturn-to-zero signaling. From Figure 3.17 it is apparent that a differentially encoded signal may be inverted without affecting its interpretation. The original binary information is recovered simply by comparing the polarity of adjacent binary symbols to establish whether or not a transition has occurred. Note that differential encoding requires the use of a *reference bit* before initiating the encoding process. In Figure 3.17, symbol 1 is used as the reference bit.

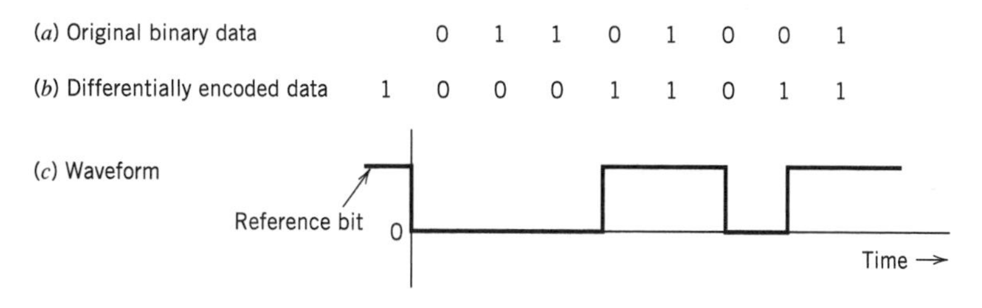


**FIGURE 3.17** (*a*) Original binary data. (*b*) Differentially encoded data, assuming reference bit 1. (*c*) Waveform of differentially encoded data using unipolar NRZ signaling.

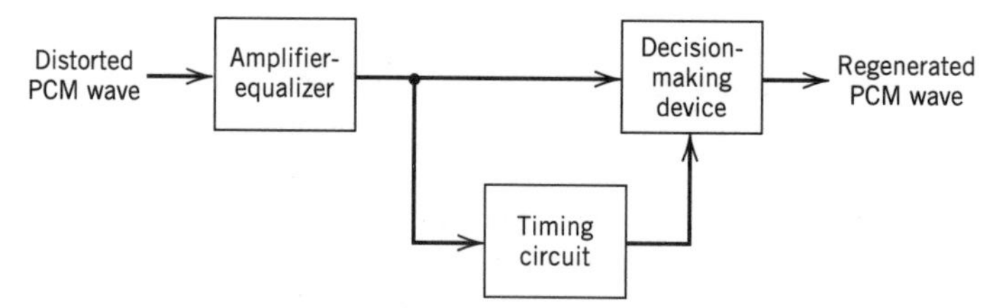


**FIGURE 3.18** Block diagram of regenerative repeater.

## ■ REGENERATION

The most important feature of PCM systems lies in the ability to control the effects of distortion and noise produced by transmitting a PCM signal through a channel. This capability is accomplished by reconstructing the PCM signal by means of a chain of *regenerative repeaters* located at sufficiently close spacing along the transmission route. As illustrated in Figure 3.18, three basic functions are performed by a regenerative repeater: *equalization*, *timing*, and *decision making*. The equalizer shapes the received pulses so as to compensate for the effects of amplitude and phase distortions produced by the nonideal transmission characteristics of the channel. The timing circuitry provides a periodic pulse train, derived from the received pulses, for sampling the equalized pulses at the instants of time where the signal-to-noise ratio is a maximum. Each sample so extracted is compared to a predetermined *threshold* in the decision-making device. In each bit interval, a decision is then made whether the received symbol is a 1 or a 0 on the basis of whether the threshold is exceeded or not. If the threshold is exceeded, a clean new pulse representing symbol 1 is transmitted to the next repeater. Otherwise, another clean new pulse representing symbol 0 is transmitted. In this way, the accumulation of distortion and noise in a repeater span is completely removed, provided that the disturbance is not too large to cause an error in the decision-making process. Ideally, except for delay, the regenerated signal is exactly the same as the signal originally transmitted. In practice, however, the regenerated signal departs from the original signal for two main reasons:

1. The unavoidable presence of channel noise and interference causes the repeater to make wrong decisions occasionally, thereby introducing *bit errors* into the regenerated signal.
2. If the spacing between received pulses deviates from its assigned value, a *jitter* is introduced into the regenerated pulse position, thereby causing distortion.

## ■ DECODING

The first operation in the receiver is to regenerate (i.e., reshape and clean up) the received pulses one last time. These clean pulses are then regrouped into code words and decoded (i.e., mapped back) into a quantized PAM signal. The *decoding* process involves generating a pulse the amplitude of which is the linear sum of all the pulses in the code word, with each pulse being weighted by its place value ($2^0, 2^1, 2^2, \ldots, 2^{R-1}$) in the code, where $R$ is the number of bits per sample.

## ■ FILTERING

The final operation in the receiver is to recover the message signal by passing the decoder output through a low-pass reconstruction filter whose cutoff frequency is equal to the message bandwidth $W$. Assuming that the transmission path is error free, the recovered signal includes no noise with the exception of the initial distortion introduced by the quantization process.

## 3.8 *Noise Considerations in PCM Systems*

The performance of a PCM system is influenced by two major sources of noise:

1. *Channel noise*, which is introduced anywhere between the transmitter output and the receiver input. Channel noise is always present, once the equipment is switched on.
2. *Quantization noise*, which is introduced in the transmitter and is carried all the way along to the receiver output. Unlike channel noise, quantization noise is *signal-dependent* in the sense that it disappears when the message signal is switched off.

Naturally, these two sources of noise appear simultaneously once the PCM system is in operation. However, the traditional practice is to consider them separately, so that we may develop insight into their individual effects on the system performance.

The main effect of channel noise is to introduce *bit errors* into the received signal. In the case of a binary PCM system, the presence of a bit error causes symbol 1 to be mistaken for symbol 0, or vice versa. Clearly, the more frequently bit errors occur, the more dissimilar the receiver output becomes compared to the original message signal. The fidelity of information transmission by PCM in the presence of channel noise may be measured in terms of the *average probability of symbol error*, which is defined as the probability that the reconstructed symbol at the receiver output differs from the transmitted binary symbol, on the average. The average probability of symbol error, also referred to as the *bit error rate* (BER), assumes that all the bits in the original binary wave are of equal importance. When, however, there is more interest in reconstructing the analog waveform of the original message signal, different symbol errors may need to be *weighted* differently; for example, an error in the most significant bit in a code word (representing a quantized sample of the message signal) is more harmful than an error in the least significant bit.

To optimize system performance in the presence of channel noise, we need to minimize the average probability of symbol error. For this evaluation, it is customary to model the channel noise as additive, white, and Gaussian. The effect of channel noise can be made practically negligible by ensuring the use of an adequate signal energy-to-noise density ratio through the provision of short-enough spacing between the regenerative repeaters in the PCM system. In such a situation, the performance of the PCM system is essentially limited by quantization noise acting alone.

From the discussion of quantization noise presented in Section 3.6, we recognize that quantization noise is essentially under the designer's control. It can be made negligibly small through the use of an adequate number of representation levels in the quantizer and the selection of a companding strategy matched to the characteristics of the type of message signal being transmitted. We thus find that the use of PCM offers the possibility of building a communication system that is *rugged* with respect to channel noise on a scale that is beyond the capability of any CW modulation or analog pulse modulation system.

### ■ Error Threshold

The underlying theory of bit error rate calculation in a PCM system is deferred until Chapter 4. For the present, it suffices to say that the average probability of symbol error in a binary encoded PCM receiver due to additive white Gaussian noise depends solely on

$E_b/N_0$, which is defined as *the ratio of the transmitted signal energy per bit, $E_b$, to the noise spectral density*, $N_0$. Note that the ratio $E_b/N_0$ is dimensionless even though the quantities $E_b$ and $N_0$ have different physical meaning. In Table 3.3 we present a summary of this dependence for the case of a binary PCM system using polar nonreturn-to-zero signaling. The results presented in the last column of the table assume a bit rate of $10^5$ b/s.

From Table 3.3 it is clear that there is an *error threshold* (at about 11 dB). For $E_b/N_0$ below the error threshold the receiver performance involves significant numbers of errors, and above it the effect of channel noise is practically negligible. In other words, provided that the ratio $E_b/N_0$ exceeds the error threshold, channel noise has virtually no effect on the receiver performance, which is precisely the goal of PCM. When, however, $E_b/N_0$ drops below the error threshold, there is a sharp increase in the rate at which errors occur in the receiver. Because decision errors result in the construction of incorrect code words, we find that when the errors are frequent, the reconstructed message at the receiver output bears little resemblance to the original message.

Comparing the figure of 11 dB for the error threshold in a PCM system using polar NRZ signaling with the 60–70 dB required for high-quality transmission of speech using amplitude modulation, we see that PCM requires much less power, even though the average noise power in the PCM system is increased by the $R$-fold increase in bandwidth, where $R$ is the number of bits in a code word (i.e., bits per sample).

In most transmission systems, the effects of noise and distortion from the individual links accumulate. For a given quality of overall transmission, the longer the physical separation between the transmitter and the receiver, the more severe are the requirements on each link in the system. In a PCM system, however, because the signal can be regenerated as often as necessary, the effects of amplitude, phase, and nonlinear distortions in one link (if not too severe) have practically no effect on the regenerated input signal to the next link. We have also seen that the effect of channel noise can be made practically negligible by using a ratio $E_b/N_0$ above threshold. For all practical purposes, then, the transmission requirements for a PCM link are almost independent of the physical length of the communication channel.

Another important characteristic of a PCM system is its *ruggedness to interference*, caused by stray impulses or cross-talk. The combined presence of channel noise and interference causes the error threshold necessary for satisfactory operation of the PCM system to increase. If an adequate margin over the error threshold is provided in the first place, however, the system can withstand the presence of relatively large amounts of interference. In other words, a PCM system is *robust* to channel noise and interference.

**TABLE 3.3** ***Influence of $E_b/N_0$ on the probability of error***

| $E_b/N_0$ | *Probability of Error $P_e$* | *For a Bit Rate of $10^5$ b/s, This Is About One Error Every* |
|---|---|---|
| 4.3 dB | $10^{-2}$ | $10^{-3}$ second |
| 8.4 | $10^{-4}$ | $10^{-1}$ second |
| 10.6 | $10^{-6}$ | 10 seconds |
| 12.0 | $10^{-8}$ | 20 minutes |
| 13.0 | $10^{-10}$ | 1 day |
| 14.0 | $10^{-12}$ | 3 months |

## 3.9 *Time-Division Multiplexing*

The sampling theorem provides the basis for transmitting the information contained in a band-limited message signal $m(t)$ as a sequence of samples of $m(t)$ taken uniformly at a rate that is usually slightly higher than the Nyquist rate. An important feature of the sampling process is a *conservation of time*. That is, the transmission of the message samples engages the communication channel for only a fraction of the sampling interval on a periodic basis, and in this way some of the time interval between adjacent samples is cleared for use by other independent message sources on a time-shared basis. We thereby obtain a *time-division multiplex* (TDM) *system*, which enables the joint utilization of a common communication channel by a plurality of independent message sources without mutual interference among them.

The concept of TDM is illustrated by the block diagram shown in Figure 3.19. Each input message signal is first restricted in bandwidth by a low-pass anti-aliasing filter to remove the frequencies that are nonessential to an adequate signal representation. The low-pass filter outputs are then applied to a *commutator*, which is usually implemented using electronic switching circuitry. The function of the commutator is twofold: (1) to take a narrow sample of each of the $N$ input messages at a rate $f_s$ that is slightly higher than $2W$, where $W$ is the cutoff frequency of the anti-aliasing filter, and (2) to sequentially interleave these $N$ samples inside the sampling interval $T_s$. Indeed, this latter function is the essence of the time-division multiplexing operation. Following the commutation process, the multiplexed signal is applied to a *pulse modulator*, the purpose of which is to transform the multiplexed signal into a form suitable for transmission over the common channel. It is clear that the use of time-division multiplexing introduces a bandwidth expansion factor $N$, because the scheme must squeeze $N$ samples derived from $N$ independent message sources into a time slot equal to one sampling interval. At the receiving end of the system, the received signal is applied to a *pulse demodulator*, which performs the reverse operation of the pulse modulator. The narrow samples produced at the pulse demodulator output are distributed to the appropriate low-pass reconstruction filters by means of a *decommutator*, which operates in *synchronism* with the commutator in the transmitter. This synchronization is essential for a satisfactory operation of the system. The way this synchronization is implemented depends naturally on the method of pulse modulation used to transmit the multiplexed sequence of samples.

The TDM system is highly sensitive to dispersion in the common channel, that is, to variations of amplitude with frequency or lack of proportionality of phase with frequency. Accordingly, accurate equalization of both magnitude and phase responses of the channel is necessary to ensure a satisfactory operation of the system; this issue is discussed in

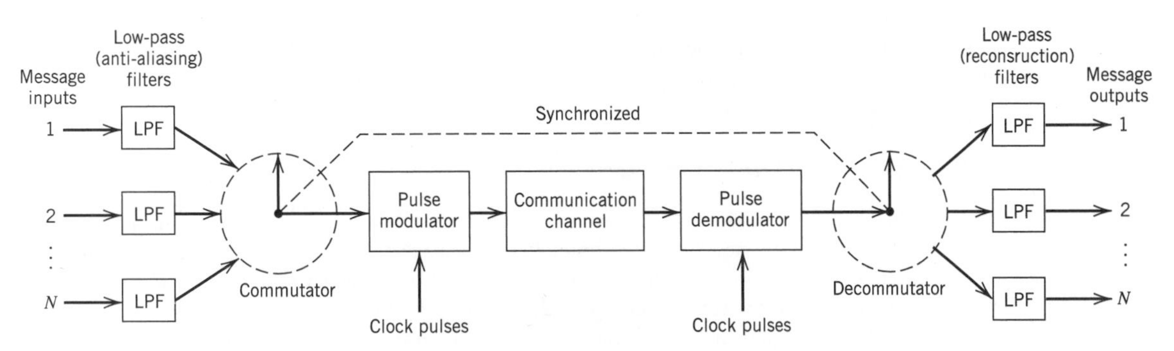


**FIGURE 3.19** Block diagram of TDM system.

Chapter 4. However, unlike FDM, to a first-order approximation TDM is immune to nonlinearities in the channel as a source of cross-talk. The reason for this behavior is that different message signals are not simultaneously applied to the channel.

## ■ SYNCHRONIZATION

In applications using PCM, it is natural to multiplex different messages sources by time division, whereby each source keeps its individuality throughout the journey from the transmitter to the receiver. This individuality accounts for the comparative ease with which message sources may be dropped or reinserted in a time-division multiplex system. As the number of independent message sources is increased, the time interval that may be allotted to each source has to be reduced, since all of them must be accommodated into a time interval equal to the reciprocal of the sampling rate. This, in turn, means that the allowable duration of a code word representing a single sample is reduced. However, pulses tend to become more difficult to generate and to transmit as their duration is reduced. Furthermore, if the pulses become too short, impairments in the transmission medium begin to interfere with the proper operation of the system. Accordingly, in practice, it is necessary to restrict the number of independent message sources that can be included within a time-division group.

In any event, for a PCM system with time-division multiplexing to operate satisfactorily, it is necessary that the timing operations at the receiver, except for the time lost in transmission and regenerative repeating, follow closely the corresponding operations at the transmitter. In a general way, this amounts to requiring a local clock at the receiver to keep the same time as a distant standard clock at the transmitter, except that the local clock is somewhat slower by an amount corresponding to the time required to transport the message signals from the transmitter to the receiver. One possible procedure to synchronize the transmitter and receiver clocks is to set aside a code element or pulse at the end of a *frame* (consisting of a code word derived from each of the independent message sources in succession) and to transmit this pulse every other frame only. In such a case, the receiver includes a circuit that would search for the pattern of 1s and 0s alternating at half the frame rate, and thereby establish synchronization between the transmitter and receiver.

When the transmission path is interrupted, it is highly unlikely that transmitter and receiver clocks will continue to indicate the same time for long. Accordingly, in carrying out a synchronization process, we must set up an orderly procedure for detecting the synchronizing pulse. The procedure consists of observing the code elements one by one until the synchronizing pulse is detected. That is, after observing a particular code element long enough to establish the absence of the synchronizing pulse, the receiver clock is set back by one code element and the next code element is observed. This *searching process* is repeated until the synchronizing pulse is detected. Clearly, the time required for synchronization depends on the epoch at which proper transmission is re-established.

### ▶ EXAMPLE 3.2 The T1 System

In this example, we describe the important characteristics of a PCM system known as the *T1 system*,[6] which carries 24 voice channels over separate pairs of wires with regenerative repeaters spaced at approximately 2-km intervals. The T1 carrier system is basic to the North American Digital Switching Hierarchy described in Section 3.10.

A voice signal (male or female) is essentially limited to a band from 300 to 3100 Hz in that frequencies outside this band do not contribute much to articulation efficiency. Indeed, telephone circuits that respond to this range of frequencies give quite satisfactory service. Accordingly, it is customary to pass the voice signal through a low-pass filter with a cutoff frequency of about 3.1 kHz prior to sampling. Hence, with $W = 3.1$ kHz, the nominal value of the Nyquist rate is 6.2 kHz. The filtered voice signal is usually sampled at a slightly higher rate, namely, 8 kHz, which is the *standard* sampling rate in telephone systems.

For companding, the T1 system uses a *piecewise-linear* characteristic (consisting of 15 linear segments) to approximate the logarithmic $\mu$-law of Equation (3.48) with the constant $\mu = 255$. This approximation is constructed in such a way that the segment end points lie on the compression curve computed from Equation (3.48), and their projections onto the vertical axis are spaced uniformly. Table 3.4 gives the projections of the segment end points onto the horizontal axis and the step-sizes of the individual segments. The table is normalized to 8159, so that all values are represented as integer numbers. Segment 0 of the approximation is a colinear segment, passing through the origin; it contains a total of 30 uniform decision levels. Linear segments $1a, 2a, \ldots, 7a$ lie above the horizontal axis, whereas linear segments $1b, 2b, \ldots, 7b$ lie below the horizontal axis; each of these 14 segments contains 16 uniform decision levels. For colinear segment 0 the decision levels at the quantizer input are $\pm 1, \pm 3, \ldots, \pm 31$, and the corresponding representation levels at the quantizer output are $0, \pm 1, \ldots, \pm 15$. For linear segments $1a$ and $1b$, the decision levels at the quantizer input are $\pm 31, \pm 35, \ldots, \pm 95$, and the corresponding representation levels at the quantizer output are $\pm 16, \pm 17, \ldots, \pm 31$, and so on for the other linear segments.

There are a total of $31 + (14 \times 16) = 255$ representation levels associated with the 15-segment companding characteristic described above. To accommodate this number of representation levels, each of the 24 voice channels uses a binary code with an 8-bit word. The first bit indicates whether the input voice sample is positive or negative; this bit is a 1 if positive and a 0 if negative. The next three bits of the code word identify the particular segment inside which the amplitude of the input voice sample lies, and the last four bits identify the actual representation level inside that segment.

With a sampling rate of 8 kHz, each frame of the multiplexed signal occupies a period of 125 $\mu$s. In particular, it consists of twenty-four 8-bit words, plus a single bit that is added at the end of the frame for the purpose of synchronization. Hence, each frame consists of a total of $(24 \times 8) + 1 = 193$ bits. Correspondingly, the duration of each bit equals 0.647 $\mu$s, and the resulting transmission rate is 1.544 megabits per second (Mb/s).

In addition to the voice signal, a telephone system must also pass special supervisory signals to the far end. This *signaling information* is needed to transmit dial pulses, as well as

**TABLE 3.4** ***The 15-segment companding characteristic ($\mu = 255$)***

| *Linear Segment Number* | *Step-Size* | *Projections of Segment End Points onto the Horizontal Axis* |
|---|---|---|
| 0 | 2 | ±31 |
| $1a, 1b$ | 4 | ±95 |
| $2a, 2b$ | 8 | ±223 |
| $3a, 3b$ | 16 | ±479 |
| $4a, 4b$ | 32 | ±991 |
| $5a, 5b$ | 64 | ±2015 |
| $6a, 6b$ | 128 | ±4063 |
| $7a, 7b$ | 256 | ±8159 |

telephone off-hook/on-hook signals. In the T1 system, this requirement is accomplished as follows. Every sixth frame, the least significant (that is, the eighth) bit of each voice channel is deleted and a *signaling bit* is inserted in its place, thereby yielding an average $7\frac{5}{6}$-bit operation for each voice input. The sequence of signaling bits is thus transmitted at a rate equal to sampling rate of 8 kHz divided by six, that is, 1.333 kb/s. This signaling rate applies to each of the 24 input channels. ◀

## 3.10 *Digital Multiplexers*

In Section 3.9 we introduced the idea of time-division multiplexing whereby a group of analog signals (e.g., voice signals) are sampled sequentially in time at a *common* sampling rate and then multiplexed for transmission over a common line. In this section we consider the multiplexing of digital signals at different bit rates. This enables us to combine several digital signals, such as computer outputs, digitized voice signals, digitized facsimile, and television signals, into a single data stream (at a considerably higher bit rate than any of the inputs). Figure 3.20 shows a conceptual diagram of the *digital multiplexing-demultiplexing* operation.

The multiplexing of digital signals is accomplished by using a *bit-by-bit interleaving procedure* with a selector switch that sequentially takes a bit from each incoming line and then applies it to the high-speed common line. At the receiving end of the system the output of this common line is separated out into its low-speed individual components and then delivered to their respective destinations.

Digital multiplexers are categorized into two major groups. One group of multiplexers is used to take relatively low bit-rate data streams originating from digital computers and multiplex them for TDM transmission over the public switched telephone network. The implementation of this first group of multiplexers requires the use of modems (modulators-demodulators), which are discussed in Chapter 6.

The second group of digital multiplexers forms part of the data transmission service provided by telecommunication carriers such as AT&T. In particular, these multiplexers constitute a *digital hierarchy* that time-division multiplexes low-rate bit streams into much higher-rate bit streams. The details of the bit rates that are accommodated in the hierarchy vary from one country to another. However, a worldwide feature of the hierarchy is that it starts at 64 kb/s, which corresponds to the standard PCM representation of a voice signal. An incoming bit stream at this rate, irrespective of its origin, is called a *digital signal zero* (*DS0*). In the *United States*, *Canada*, and *Japan*[7] the hierarchy follows the North American digital TDM hierarchy as described here:

- The *first-level hierarchy* combines twenty-four DS0 bit streams to obtain a *digital signal one* (*DS1*) at 1.544 Mb/s, which is carried on the T1 system described in

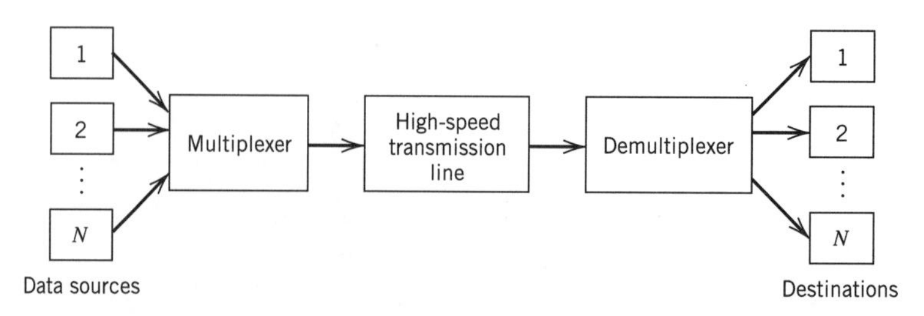


**FIGURE 3.20** Conceptual diagram of multiplexing-demultiplexing.

Example 3.2. These bit streams are called the *primary rate* in the digital hierarchy, because it is the lowest bit rate that exists outside a digital switch. The *digital switch* is a device consisting of memory and logic, the function of which is merely the switching of digital signals, hence the name.

- The *second-level multiplexer* combines four DS1 bit streams to obtain a *digital signal two* (*DS2*) at 6.312 Mb/s.
- The *third-level multiplexer* combines seven DS2 bit streams to obtain a *digital signal three* (*DS3*) at 44.736 Mb/s.
- The *fourth-level multiplexer* combines six DS3 bit streams to obtain a *digital signal four* (*DS4*) at 274.176 Mb/s.
- The *fifth-level multiplexer*, the final one in the hierarchy, combines two DS4 bit streams to obtain a *digital signal five* (*DS5*) at 560.160 Mb/s.

Note that the bit rate of a digital signal produced by any one of these multiplexers is slightly higher than the prescribed multiple of the incoming bit rate because of bit stuffing built into the design of each multiplexer; bit stuffing is discussed in the sequel.

Moreover, it is important to recognize that the functions of a digital transmission facility is merely to carry a bit stream without interpreting what the bits themselves mean. However, the digital switches at the two ends of the facility do have a common understanding of how to interpret the bits within the stream, such as whether the bits represent voice or data, framing format, signaling format, and so on.

There are some basic problems involved in the design of a digital multiplexer, irrespective of its grouping:

1. Digital signals cannot be directly interleaved into a format that allows for their eventual separation unless their bit rates are locked to a common clock. Rather, provision has to be made for *synchronization* of the incoming digital signals, so that they can be properly interleaved.
2. The multiplexed signal must include some form of *framing* so that its individual components can be identified at the receiver.
3. The multiplexer has to handle small variations in the bit rates of the incoming digital signals. For example, a 1000-km coaxial cable carrying $3 \times 10^8$ pulses per second will have about one million pulses in transit, with each pulse occupying about one meter of the cable. A 0.01 percent variation in the propagation delay, produced by a 1°F decrease in temperature, will result in 100 fewer pulses in the cable. Clearly, these pulses must be absorbed by the multiplexer.

To tailor the requirements of synchronization and rate adjustment to accommodate small variations in the input data rates, we may use a technique known as *bit stuffing*. The idea here is to have the outgoing bit rate of the multiplexer slightly higher than the sum of the maximum expected bit rates of the input channels by stuffing in additional non-information carrying pulses. All incoming digital signals are stuffed with a number of bits sufficient to raise each of their bit rates to equal that of a locally generated clock. To accomplish bit stuffing, each incoming digital signal or bit stream is fed into an *elastic store* at the multiplexer. The elastic store is a device that stores a bit stream in such a manner that the stream may be read out at a rate different from the rate at which it is read in. At the demultiplexer, the stuffed bits must obviously be removed from the multiplexed signal. This requires a method that can be used to identify the stuffed bits. To illustrate one such method, and also show one method of providing frame synchronization, we describe the signal format of the AT&T *M12 multiplexer*, which is designed to combine four DS1 bit streams into one DS2 bit stream. This is the second level of the digital hierarchy discussed earlier.

## ▶ EXAMPLE 3.3 Signal Format of the AT&T M12 Multiplexer

Figure 3.21 illustrates the signal format of the M12 multiplexer. Each frame is subdivided into four subframes. The first subframe (first line in Figure 3.21) is transmitted, then the second, the third, and the fourth, in that order.

Bit-by-bit interleaving of the incoming four DS1 bit streams is used to accumulate a total of 48 bits, 12 from each input. A *control bit* is then inserted by the multiplexer. Each frame contains a total of 24 control bits, separated by sequences of 48 data bits. Three types of control bits are used in the M12 multiplexer to provide synchronization and frame indication, and to identify which of the four input signals has been stuffed. These control bits are labeled $F$, $M$, and $C$ in Figure 3.21. Their functions are as follows:

1. The $F$-control bits, two per subframe, constitute the *main* framing pulses. The subscripts on the $F$-control bits denote the actual bit (0 or 1) transmitted. Thus the main framing sequence is $F_0F_1F_0F_1F_0F_1F_0F_1$ or 01010101.
2. The $M$-control bits, one per subframe, form *secondary* framing pulses to identify the four subframes. Here again the subscripts on the $M$-control bits denote the actual bit (0 or 1) transmitted. Thus the secondary framing sequence is $M_0M_1M_1M_1$ or 0111.
3. The $C$-control bits, three per subframe, are *stuffing indicators*. In particular, $C_{\text{I}}$ refers to input channel I, $C_{\text{II}}$ refers to input channel II, and so forth. For example, the three $C$-control bits in the first subframe following $M_0$ in the first subframe are stuffing indicators for the first DS1 bit stream. The insertion of a stuffed bit in this DS1 bit stream is indicated by setting all three $C$-control bits to 1. To indicate no stuffing, all three are set to 0. If the three $C$-control bits indicate stuffing, the stuffed bit is located in the position of the first information bit associated with the first DS1 bit stream that follows the $F_1$-control bit in the same subframe. In a similar way, the second, third, and fourth DS1 bit streams may be stuffed, as required. By using *majority logic decoding* in the receiver, a single error in any of the three $C$-control bits can be detected. This form of decoding means simply that the majority of the $C$-control bits determine whether an all-one or all-zero sequence was transmitted. Thus three 1s or combinations of two 1s and a 0 indicate that a stuffed bit is present in the information sequence, following the control bit $F_1$ in the pertinent subframe. On the other hand, three 0s or combinations of two 0s and a 1 indicate that no stuffing is used.

The demultiplexer at the receiving M12 unit first searches for the main framing sequence $F_0F_1F_0F_1F_0F_1F_0F_1$. This establishes identity for the four input DS1 bit streams and also for the $M$- and $C$-control bits. From the $M_0M_1M_1M_1$ sequence, the correct framing of the $C$-control bits is verified. Finally, the four DS1 bit streams are properly demultiplexed and destuffed.

The signal format described above has two safeguards:

1. It is possible, although unlikely, that with just the $F_0F_1F_0F_1F_0F_1F_0F_1$ sequence, one of the incoming DS1 bit streams may contain a similar sequence. This could then cause

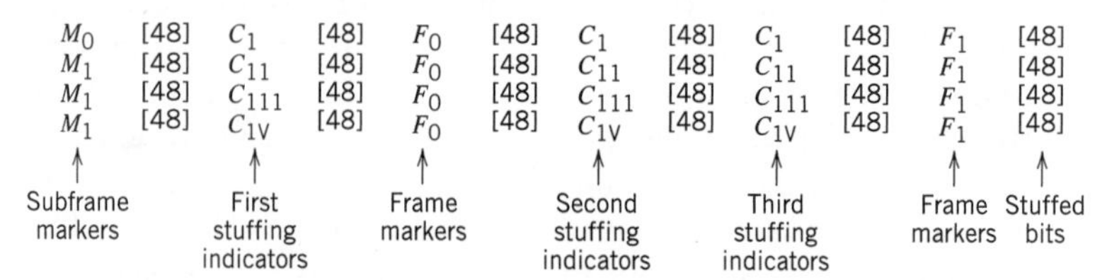


**FIGURE 3.21** Signal format of AT&T M12 multiplexer.

the receiver to lock onto the wrong sequence. The presence of the $M_0M_1M_1M_1$ sequence provides verification of the genuine $F_0F_1F_0F_1F_0F_1F_0F_1$ sequence, thereby ensuring that the four DS1 bit streams are properly demultiplexed.

2. The single-error correction capability built into the C-control bits ensures that the four DS1 bit streams are properly destuffed.

The capacity of the M12 multiplexer to accommodate small variations in the input data rates can be calculated from the format of Figure 3.21. In each $M$ frame, defined as the interval containing one cycle of $M_0M_1M_1M_1$ bits, one bit can be stuffed into each of four input DS1 bit streams. Each such signal has $12 \times 6 \times 4 = 288$ positions in each $M$ frame. Also, the T1 bit stream has a bit rate equal to 1.544 Mb/s. Hence, each input can be incremented by

$$1.544 \times 10^3 \times \frac{1}{288} = 5.4 \text{ kb/s}$$

This result is much larger than the expected change in the bit rate of the incoming DS1 bit stream. It follows therefore that the use of only one stuffed bit per input channel in each frame is sufficient to accommodate expected variations in the input signal rate.

The local clock that determines the outgoing bit rate also determines the nominal *stuffing rate* $S$, defined as the average number of bits stuffed per channel in any frame. The M12 multiplexer is designed for $S = 1/3$. Accordingly, the nominal bit rate of the DS2 bit stream is

$$1.544 \times 4 \times \frac{49}{48} \times \frac{288}{288\text{-}S} = 6.312 \text{ Mb/s}$$

This also ensures that the nominal DS2 clock frequency is a multiple of 8 kHz (the nominal sampling rate of a voice signal), which is a desirable feature. ◀

## 3.11 *Virtues, Limitations, and Modifications of PCM*

In a generic sense, pulse-code modulation (PCM) has emerged as the most favored modulation scheme for the transmission of analog information-bearing signals such as voice and video signals. The advantages of PCM may all be traced to the use of *coded pulses for the digital representation of analog signals*, a feature that distinguishes it from all other analog methods of modulation. We may summarize the important advantages of PCM as follows:

1. *Robustness* to channel noise and interference.
2. Efficient *regeneration* of the coded signal along the transmission path.
3. Efficient *exchange* of increased channel bandwidth for improved signal-to-noise ratio, obeying an exponential law.
4. A *uniform format* for the transmission of different kinds of baseband signals, hence their integration with other forms of digital data in a common network.
5. Comparative *ease* with which message sources may be dropped or reinserted in a time-division multiplex system.
6. *Secure* communication through the use of special modulation schemes or encryption; the encryption and decryption of data are discussed in Appendix 5.

These advantages, however, are attained at the cost of increased system complexity and increased channel bandwidth. These two issues are considered in the sequel in turn.

Although the use of PCM involves many complex operations, today they can all be implemented in a cost-effective fashion using commercially available and/or custom-made *very-large-scale integrated* (VLSI) chips. In other words, the requisite device technology for the implementation of a PCM system is already in place. Moreover, with continuing improvements in VLSI technology, we are likely to see an ever-expanding use of PCM for the digital transmission of analog signals.

If, however, the simplicity of implementation is a necessary requirement, then we may use delta modulation as an alternative to pulse-code modulation. In delta modulation, the baseband signal is intentionally "oversampled" to permit the use of a simple quantizing strategy for constructing the encoded signal; delta modulation is discussed in Section 3.12.

Turning next to the issue of bandwidth, we do recognize that the increased bandwidth requirement of PCM may have been a reason for justifiable concern in the past. Today, however, it is of no real concern for two different reasons. First, the increasing availability of *wideband communication channels* means that bandwidth is no longer a system constraint in the traditional way it used to be. Liberation from the bandwidth constraint has been made possible by the deployment of communication satellites for broadcasting and the ever-increasing use of fiber optics for networking; a discussion of these communication channel concepts was presented in the Background and Preview chapter.

The second reason is that through the use of sophisticated *data compression* techniques, it is indeed possible to remove the redundancy inherently present in a PCM signal and thereby reduce the bit rate of the transmitted data without serious degradation in system performance. In effect, increased processing complexity (and therefore increased cost of implementation) is traded off for a reduced bit rate and therefore reduced bandwidth requirement. A major motivation for bit-rate reduction is for secure communication over radio channels that are inherently of low capacity.

## 3.12 Delta Modulation

In *delta modulation*[8] (DM), an incoming message signal is oversampled (i.e., at a rate much higher than the Nyquist rate) to purposely increase the correlation between adjacent samples of the signal. This is done to permit the use of a simple quantizing strategy for constructing the encoded signal.

In its basic form, DM provides a *staircase approximation* to the oversampled version of the message signal, as illustrated in Figure 3.22*a*. The difference between the input and the approximation is quantized into only two levels, namely, $\pm\Delta$, corresponding to positive and negative differences. Thus if the approximation falls below the signal at any sampling epoch, it is increased by $\Delta$. If on the other hand, the approximation lies above the signal, it is diminished by $\Delta$. Provided that the signal does not change too rapidly from sample to sample, we find that the staircase approximation remains within $\pm\Delta$ of the input signal.

Let $m(t)$ denote the input (message) signal, and $m_q(t)$ denote its staircase approximation. For convenience of presentation, we adopt the following notation that is commonly used in the digital signal processing literature:

$$m[n] = m(nT_s), \qquad n = 0, \pm 1, \pm 2, \ldots$$

where $T_s$ is the sampling period and $m(nT_s)$ is a sample of the signal $m(t)$ taken at time $t = nT_s$, and likewise for the samples of other continuous-time signals. We may then

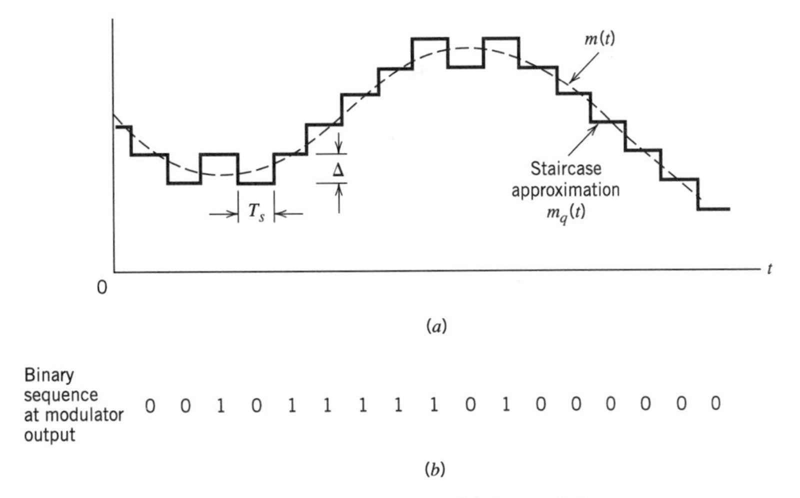


**FIGURE 3.22** Illustration of delta modulation.

formalize the basic principles of delta modulation in the following set of discrete-time relations:

$$e[n] = m[n] - m_q[n-1] \tag{3.52}$$

$$e_q = \Delta \operatorname{sgn}(e[n]) \tag{3.53}$$

$$m_q[n] = m_q[n-1] + e_q[n] \tag{3.54}$$

where $e[n]$ is an *error signal* representing the difference between the present sample $m[n]$ of the input signal and the latest approximation $m_q[n-1]$ to it, $e_q[n]$ is the quantized version of $e[n]$, and $\operatorname{sgn}(\cdot)$ is the signum function. Finally, the quantizer output $m_q[n]$ is coded to produce the DM signal.

Figure 3.22*a* illustrates the way in which the staircase approximation $m_q(t)$ follows variations in the input signal $m(t)$ in accordance with Equations (3.52)–(3.54), and Figure 3.22*b* displays the corresponding binary sequence at the delta modulator output. It is apparent that in a delta modulation system the rate of information transmission is simply equal to the sampling rate $f_s = 1/T_s$.

The principal virtue of delta modulation is its simplicity. It may be generated by applying the sampled version of the incoming message signal to a modulator that involves a *comparator*, *quantizer*, and *accumulator* interconnected as shown in Figure 3.23*a*. The block labeled $z^{-1}$ inside the accumulator represents a *unit delay*, that is, a delay equal to one sampling period. (The variable $z$ is commonly used in the $z$-transform, which is basic to the analysis of discrete-time signals and systems.) Details of the modulator follow directly from Equations (3.52)–(3.54). The comparator computes the difference between its two inputs. The quantizer consists of a *hard limiter* with an input-output relation that is a scaled version of the signum function. The quantizer output is then applied to an accumulator, producing the result

$$\begin{aligned} m_q[n] &= \Delta \sum_{i=1}^{n} \operatorname{sgn}(e[i]) \\ &= \sum_{i=1}^{n} e_q[i] \end{aligned} \tag{3.55}$$

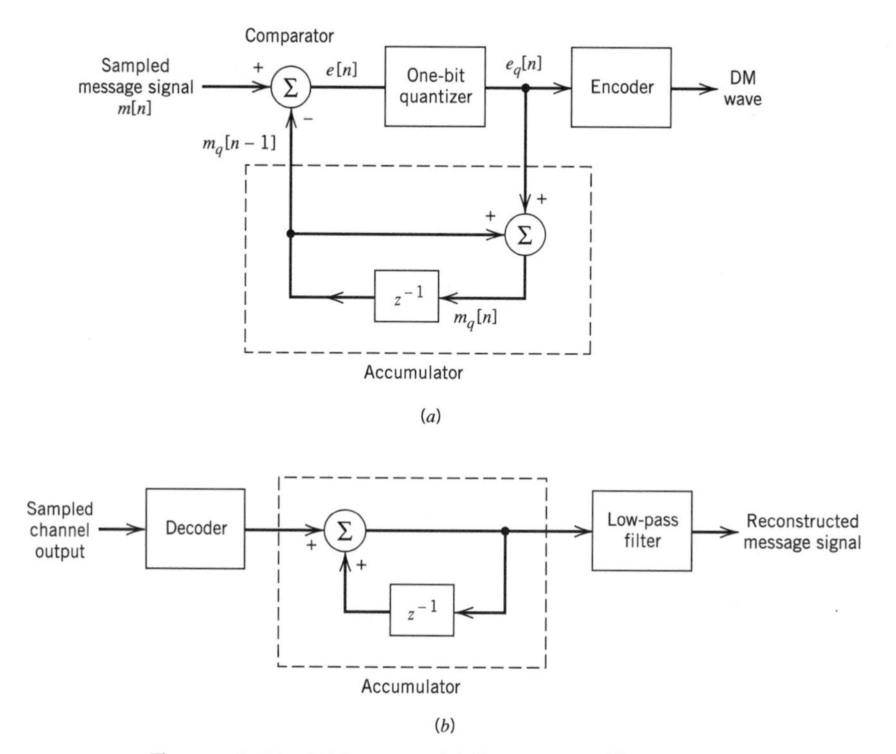


**FIGURE 3.23** DM system. (*a*) Transmitter. (*b*) Receiver.

which is obtained by solving Equations (3.53) and (3.54) for $m_q[n]$. Thus, at the sampling instant $nT_s$, the accumulator increments the approximation by a step $\Delta$ in a positive or negative direction, depending on the algebraic sign of the error sample $e[n]$. If the input sample $m[n]$ is greater than the most recent approximation $m_q[n]$, a positive increment $+\Delta$ is applied to the approximation. If, on the other hand, the input sample is smaller, a negative increment $-\Delta$ is applied to the approximation. In this way, the accumulator does the best it can to track the input samples by one step (of amplitude $+\Delta$ or $-\Delta$) at a time. In the receiver shown in Figure 3.23*b*, the staircase approximation $m_q(t)$ is reconstructed by passing the sequence of positive and negative pulses, produced at the decoder output, through an accumulator in a manner similar to that used in the transmitter. The out-of-band quantization noise in the high-frequency staircase waveform $m_q(t)$ is rejected by passing it through a low-pass filter, as in Figure 3.23*b*, with a bandwidth equal to the original message bandwidth.

Delta modulation is subject to two types of quantization error: slope overload distortion and granular noise. We will discuss the case of slope overload distortion first.

We observe that Equation (3.54) is the digital equivalent of integration in the sense that it represents the accumulation of positive and negative increments of magnitude $\Delta$. Also, denoting the quantization error by $q[n]$, as shown by

$$m_q[n] = m[n] + q[n] \tag{3.56}$$

we observe from Equation (3.52) that the input to the quantizer is

$$e[n] = m[n] - m[n-1] - q[n-1] \tag{3.57}$$

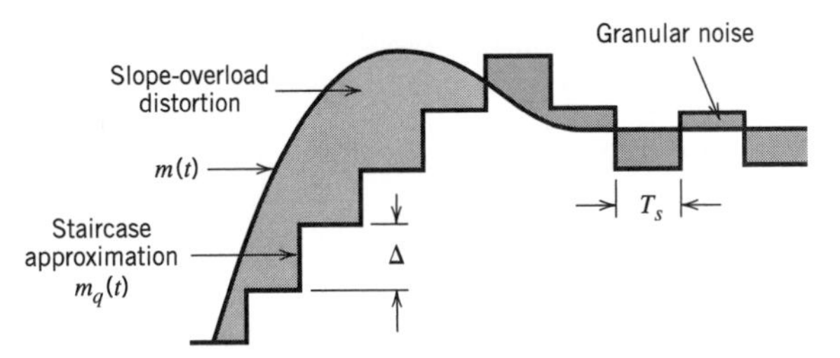


**Figure 3.24** Illustration of the two different forms of quantization error in delta modulation.

Thus except for the quantization error $q[n - 1]$, the quantizer input is a *first backward difference* of the input signal, which may be viewed as a digital approximation to the derivative of the input signal or, equivalently, as the inverse of the digital integration process. If we consider the maximum slope of the original input waveform $m(t)$, it is clear that in order for the sequence of samples $\{m_q[n]\}$ to increase as fast as the input sequence of samples $\{m[n]\}$ in a region of maximum slope of $m(t)$, we require that the condition

$$\frac{\Delta}{T_s} \geq \max\left|\frac{dm(t)}{dt}\right| \tag{3.58}$$

be satisfied. Otherwise, we find that the step-size $\Delta$ is too small for the staircase approximation $m_q(t)$ to follow a steep segment of the input waveform $m(t)$, with the result that $m_q(t)$ falls behind $m(t)$, as illustrated in Figure 3.24. This condition is called *slope overload*, and the resulting quantization error is called *slope-overload distortion (noise)*. Note that since the maximum slope of the staircase approximation $m_q(t)$ is fixed by the step size $\Delta$, increases and decreases in $m_q(t)$ tend to occur along straight lines. For this reason, a delta modulator using a fixed step size is often referred to as a *linear delta modulator*.

In contrast to slope-overload distortion, *granular noise* occurs when the step size $\Delta$ is too large relative to the local slope characteristics of the input waveform $m(t)$, thereby causing the staircase approximation $m_q(t)$ to hunt around a relatively flat segment of the input waveform; this phenomenon is also illustrated in Figure 3.24. Granular noise is analogous to quantization noise in a PCM system.

We thus see that there is a need to have a large step-size to accommodate a wide dynamic range, whereas a small step size is required for the accurate representation of relatively low-level signals. It is therefore clear that the choice of the optimum step size that minimizes the mean-square value of the quantization error in a linear delta modulator will be the result of a compromise between slope-overload distortion and granular noise. To satisfy such a requirement, we need to make the delta modulator "adaptive," in the sense that the step size is made to vary in accordance with the input signal; this issue is discussed further in a computer experiment presented in Section 3.16.

## ■ Delta-Sigma Modulation

As mentioned earlier, the quantizer input in the conventional form of delta modulation may be viewed as an approximation to the *derivative* of the incoming message signal. This behavior leads to a drawback of delta modulation in that transmission disturbances such as noise result in an accumulative error in the demodulated signal. This drawback can be overcome by *integrating* the message signal prior to delta modulation. The use of integration in the manner described here has also the following beneficial effects:

- The low-frequency content of the input signal is pre-emphasized.
- Correlation between adjacent samples of the delta modulator input is increased, which tends to improve overall system performance by reducing the variance of the error signal at the quantizer input.
- Design of the receiver is simplified.

A delta modulation scheme that incorporates integration at its input is called *delta-sigma modulation* (D-ΣM).[9] To be more precise, however, it should be called *sigma-delta modulation*, because the integration is in fact performed before the delta modulation. Nevertheless, the former terminology is the one commonly used in the literature.

Figure 3.25*a* shows the block diagram of a delta-sigma modulation system. In this diagram, the message signal $m(t)$ is defined in its continuous-time form, which means that the pulse modulator now consists of a hard-limiter followed by a multiplier; the latter component is also fed from an external pulse generator (clock) to produce a 1-bit encoded signal. The use of integration at the transmitter input clearly requires an inverse signal emphasis, namely, differentiation, at the receiver. The need for this differentiation is, however, eliminated because of its cancellation by integration in the conventional DM receiver.

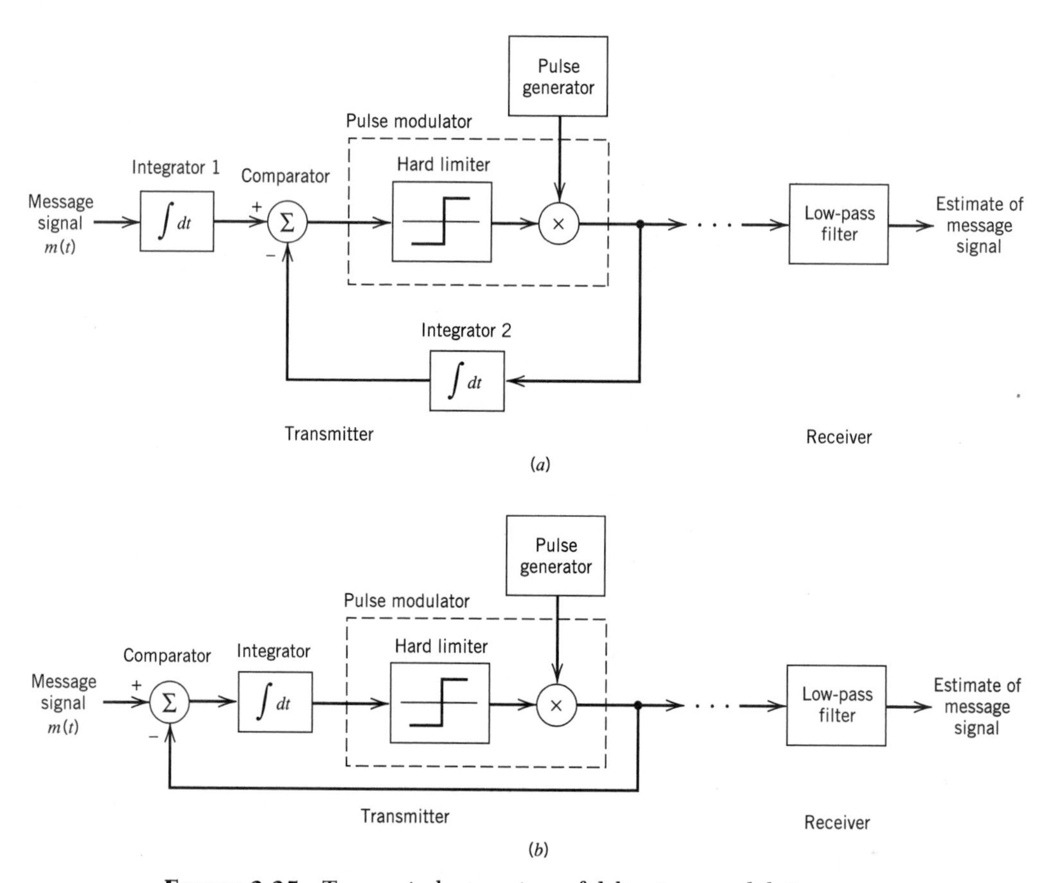


**FIGURE 3.25** Two equivalent versions of delta-sigma modulation system.

Thus the receiver of a delta-sigma modulation system consists simply of a low-pass filter, as indicated in Figure 3.25*a*.

Moreover, we note that integration is basically a linear operation. Accordingly, we may simplify the design of the transmitter by combining the two integrators 1 and 2 of Figure 3.25*a* into a single integrator placed after the comparator, as shown in Figure 3.25*b*. This latter form of the delta-sigma modulation system is not only simpler than that of Figure 3.25*a*, but it also provides an interesting interpretation of delta-sigma modulation as a "smoothed" version of 1-bit pulse-code modulation: The term *smoothness* refers to the fact that the comparator output is integrated prior to quantization, and the term *1-bit* merely restates that the quantizer consists of a hard-limiter with only two representation levels.

In delta modulation, simplicity of implementations of both the transmitter and receiver is attained by using a sampling rate far in excess of that needed for pulse-code modulation. The price paid for this benefit is a corresponding increase in the transmission and therefore channel bandwidth. There are, however, applications where channel bandwidth is at a premium, in which case we have the opposite requirement to that in delta modulation. Specifically, we may wish to trade increased system complexity for a reduced channel bandwidth. A signal-processing operation basic to the attainment of this latter design objective is prediction, the linear form of which is discussed next.

## 3.13 *Linear Prediction*

Consider a *finite-duration impulse response (FIR) discrete-time filter* configured as in Figure 3.26, which involves the use of three functional blocks:

1. Set of $p$ unit-delay elements, each of which is represented by $z^{-1}$.
2. Set of multipliers involving the filter coefficients $w_1, w_2, \ldots, w_p$.
3. Set of "adders" used to sum the scaled versions of the delayed inputs $x[n-1]$, $x[n-2], \ldots, x[n-p]$ to produce the output $\hat{x}[n]$. The filter output $\hat{x}[n]$ or more precisely, the *linear prediction* of the input, is thus defined by the *convolution sum*

$$\hat{x}[n] = \sum_{k=1}^{p} w_k x[n-k] \tag{3.59}$$

where $p$, the number of unit-delay elements, is called the *prediction order*.

The actual sample at time $nT_s$ is $x[n]$. The *prediction error*, denoted by $e[n]$, is defined as the difference between $x[n]$ and the prediction $\hat{x}[n]$, as shown by

$$e[n] = x[n] - \hat{x}[n] \tag{3.60}$$

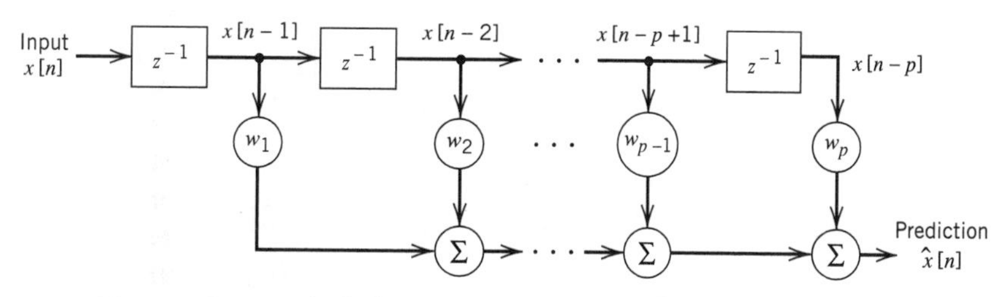


**FIGURE 3.26** Block diagram of a linear prediction filter of order $p$.

The design objective is to choose the filter coefficients $w_1, w_2, \ldots, w_p$ so as to minimize an *index of performance*, $J$, defined as the mean-square error:

$$J = E[e^2[n]] \tag{3.61}$$

Substituting Equations (3.59) and (3.60) into (3.61) and then expanding terms, we may reformulate the index of performance as

$$J = E[x^2[n]] - 2\sum_{k=1}^{p} w_k E[x[n]x[n-k]] + \sum_{j=1}^{p}\sum_{k=1}^{p} w_j w_k E[x[n-j]x[n-k]] \tag{3.62}$$

We assume that the input signal $x(t)$ is the sample function of a stationary process $X(t)$ of zero mean; that is, $E[x[n]]$ is zero for all $n$. Define

$$\begin{aligned}
\sigma_X^2 &= \text{variance of a sample of the process } X(t) \text{ at time } nT_s \\
&= E[x^2[n]] - (E[x[n]])^2 \\
&= E[x^2[n]] \\
R_X(kT_s) &= \text{autocorrelation of the process } X(t) \text{ for a lag of } kT_s \\
&= R_X[k] \\
&= E[x[n]x[n-k]]
\end{aligned}$$

Accordingly, we may rewrite Equation (3.62) in the simplified form

$$J = \sigma_X^2 - 2\sum_{k=1}^{p} w_k R_X[k] + \sum_{j=1}^{p}\sum_{k=1}^{p} w_j w_k R_X[k-j] \tag{3.63}$$

Hence differentiating the index of performance $J$ with respect to the filter coefficient $w_k$, setting the result equal to zero, and then rearranging terms, we obtain

$$\sum_{j=1}^{p} w_j R_X[k-j] = R_X[k] = R_X[-k], \qquad k = 1, 2, \ldots, p \tag{3.64}$$

The optimality equations (3.64) are called the *Wiener-Hopf equations* for linear prediction.

We find it convenient to reformulate the Wiener-Hopf equations (3.64) in matrix form. Let

$$\begin{aligned}
\mathbf{w}_o &= p\text{-by-1 optimum coefficient vector} \\
&= [w_1, w_2, \ldots, w_p]^T \\
\mathbf{r}_X &= p\text{-by-1 autocorrelation vector} \\
&= [R_X[1], R_X[2], \ldots, R_X[p]]^T \\
\mathbf{R}_X &= p\text{-by-}p \text{ autocorrelation matrix} \\
&= \begin{bmatrix}
R_X[0] & R_X[1] & \cdots & R_X[p-1] \\
R_X[1] & R_X[0] & \cdots & R_X[p-2] \\
\vdots & \vdots & & \vdots \\
R_X[p-1] & R_X[p-2] & \cdots & R_X[0]
\end{bmatrix}
\end{aligned}$$

We may thus simplify the set of equations (3.64) as

$$\mathbf{R}_X \mathbf{w}_o = \mathbf{r}_X \tag{3.65}$$

We assume that the autocorrelation matrix $\mathbf{R}_X$ is *nonsingular*, so that its inverse exists. We may then solve Equation (3.65) for the coefficient vector $\mathbf{w}_o$ by multiplying both sides of this equation by the *inverse matrix* $\mathbf{R}_X^{-1}$, obtaining the optimum solution

$$\mathbf{w}_o = \mathbf{R}_X^{-1}\mathbf{r}_X \tag{3.66}$$

Note that all the elements on the main diagonal of the autocorrelation matrix $\mathbf{R}_X$ are equal to $R_X[0] = \sigma_X^2$, and the elements on any other diagonal parallel to the main diagonal are also equal. A square matrix having this property is said to be *Toeplitz*, which is a direct consequence of the assumption that the input signal $x(t)$ is drawn from a stationary process. The practical significance of the Toeplitz property is that the correlation matrix $\mathbf{R}_X$ is uniquely defined by the set of autocorrelation values $R_X[0], R_X[1], \ldots, R_X[p-1]$. The autocorrelation vector $\mathbf{r}_X$ is defined by the set of autocorrelation values $R_X[1], R_X[2], \ldots, R_X[p]$. It follows therefore that the $p$ filter coefficients of the linear optimum predictor are uniquely defined by the variance $\sigma_X^2 = R_X[0]$ and $p$ values of the autocorrelation function of the process $X(t)$ for lags of $T_s, 2T_s, \ldots, pT_s$.

The minimum mean-square value of the prediction error is obtained by substituting Equation (3.64) into (3.63), which yields (after simplification)

$$J_{\min} = \sigma_X^2 - \mathbf{r}_X^T\mathbf{R}_X^{-1}\mathbf{r}_X \tag{3.67}$$

The quadratic term $\mathbf{r}_X^T\mathbf{R}_X^{-1}\mathbf{r}_X$ is always positive. Accordingly, the mean-square error $J_{\min}$ of the optimum linear predictor defined by Equation (3.67) is always less than the variance $\sigma_X^2$ of the input sample that is being predicted.

## Linear Adaptive Prediction

The use of Equation (3.66) for calculating the weight vector of a linear predictor requires knowledge of the autocorrelation function $R_X[k]$ of the input sequence $\{x[n]\}$ for lags $k = 0, 1, \ldots, p$, where $p$ is the prediction order. What if knowledge of $R_X[k]$ for varying $k$ is not available? In these situations, which occur frequently in practice, we may resort to the use of an *adaptive predictor*.

The predictor is adaptive in the following sense:

- Computation of the tap weights $w_k$, $k = 1, 2, \ldots, p$, proceeds in a "recursive" manner, starting from some arbitrary initial values of the tap weights.
- The algorithm used to adjust the tap weights (from one iteration to the next) is "self-designed," operating solely on the basis of available data.

The aim of the algorithm is to find the minimum point of the bowl-shaped *error surface* that describes the dependence of the cost function $J$ on the tap weights. It is therefore intuitively reasonable that successive adjustments to the tap-weights of the predictor be made in the direction of the steepest descent of the error surface, that is, in a direction opposite to the *gradient vector* whose elements are defined by

$$g_k = \frac{\partial J}{\partial w_k}, \qquad k = 1, 2, \ldots, p \tag{3.68}$$

This is indeed the idea behind the *method of steepest descent*. Let $w_k[n]$ denote the value of the $k$th tap-weight at iteration $n$. Then the updated value of this weight at iteration $n + 1$ is defined by

$$w_k[n+1] = w_k[n] - \frac{1}{2}\mu g_k, \qquad k = 1, 2, \ldots, p \tag{3.69}$$

where $\mu$ is a *step-size parameter* that controls the speed of adaptation, and the factor 1/2 is included for convenience of presentation. Differentiating the cost function $J$ of Equation (3.63) with respect to $w_k$, we readily find that

$$\begin{aligned} g_k &= -2R_X[k] + 2\sum_{j=1}^{p} w_j R_X[k-j] \\ &= -2E[x[n]x[n-k]] + 2\sum_{j=1}^{p} w_j E[x[n-j]x[n-k]], \qquad k = 1, 2, \ldots, p \end{aligned} \tag{3.70}$$

This formula for $g_k$ could do with further simplification, which is achieved by using *instantaneous values as estimates* of the autocorrelation functions $R_X[k]$ and $R_X[k-j]$. That is, we ignore the expectation operators in Equation (3.70) to facilitate the adaptive process on a step-by-step basis. We may thus express the corresponding *estimate* of $g_k$ at iteration $n$ as

$$\hat{g}_k[n] = -2x[n]x[n-k] + 2\sum_{j=1}^{p} w_j[n]x[n-j]x[n-k], \qquad k = 1, 2, \ldots, p \tag{3.71}$$

Note that for an input $x[n]$ drawn from a stationary process the gradient $g_k$ is a deterministic quantity, whereas the estimate $\hat{g}_k[n]$ is the sample value of a random variable.

In any event, substituting Equation (3.71) into (3.69) and factoring the common term $x[n-k]$, we may write

$$\begin{aligned} \hat{w}_k[n+1] &= \hat{w}_k[n] + \mu x[n-k]\left(x[n] - \sum_{j=1}^{p} \hat{w}_j[n]x[n-j]\right) \\ &= \hat{w}_k[n] + \mu x[n-k]e[n], \qquad k = 1, 2, \ldots, p \end{aligned} \tag{3.72}$$

where $e[n]$ is the *prediction error* defined as

$$e[n] = x[n] - \sum_{j=1}^{p} \hat{w}_j[n]x[n-j] \tag{3.73}$$

In Equations (3.72) and (3.73), we have used $\hat{w}_k$ as an *estimate* of the $k$th tap-weight to distinguish it from the actual value $w_k$. Note also that $x[n]$ plays the role of a "desired response" for computing the recursive adjustments applied to the tap-weights of the predictor.

Equations (3.72) and (3.73) constitute the popular *least-mean-square (LMS) algorithm* for linear adaptive prediction, the operation of which is depicted in Figure 3.27. The reason for popularity of this adaptive filtering algorithm is the simplicity of its implementation. In particular, the computational complexity of the algorithm, measured in terms of the number of additions and multiplications, is *linear* in the prediction order $p$.

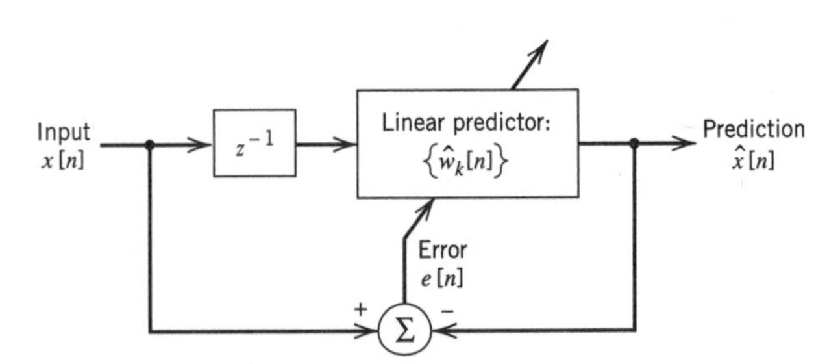


**FIGURE 3.27** Block diagram illustrating the linear adaptive prediction process.

The LMS algorithm is a *stochastic* adaptive filtering algorithm, stochastic in the sense that, starting from the *initial condition* defined by $\{w_k[0]\}_{k=1}^{p}$, it seeks to find the minimum point of the error surface by following a zig-zag path. Moreover, it never finds this minimum point exactly. Rather, it executes a random motion around the minimum point of the error surface, once steady-state conditions are established.

With this material on linear prediction at hand, we are ready to discuss practical improvements on the performance of pulse-code modulation.

## 3.14 *Differential Pulse-Code Modulation*

When a voice or video signal is sampled at a rate slightly higher than the Nyquist rate as usually done in pulse-code modulation, the resulting sampled signal is found to exhibit a high degree of correlation between adjacent samples. The meaning of this high correlation is that, in an average sense, the signal does not change rapidly from one sample to the next, and as a result, the difference between adjacent samples has a variance that is smaller than the variance of the signal itself. When these highly correlated samples are encoded, as in the standard PCM system, the resulting encoded signal contains *redundant information*. This means that symbols that are not absolutely essential to the transmission of information are generated as a result of the encoding process. By removing this redundancy before encoding, we obtain a more efficient coded signal, which is the basic idea behind differential pulse-code modulation.

Now if we know the past behavior of a signal up to a certain point in time, we may use prediction to make an estimate of a future value of the signal as described in Section 3.13. Suppose then a baseband signal $m(t)$ is sampled at the rate $f_s = 1/T_s$ to produce the sequence $\{m[n]\}$ whose samples are $T_s$ seconds apart. The fact that it is possible to predict future values of the signal $m(t)$ provides motivation for the *differential quantization* scheme shown in Figure 3.28*a*. In this scheme, the input signal to the quantizer is defined by

$$e[n] = m[n] - \hat{m}[n] \tag{3.74}$$

which is the difference between the unquantized input sample $m[n]$ and a prediction of it, denoted by $\hat{m}[n]$. This predicted value is produced by using a linear prediction filter whose input, as we will see, consists of a quantized version of the input sample $m[n]$. The difference signal $e[n]$ is the prediction error, since it is the amount by which the prediction filter fails to predict the input exactly. By encoding the quantizer output, as in Figure 3.28*a*, we obtain a variant of PCM known as *differential pulse-code modulation*[10] (DPCM).

The quantizer output may be expressed as

$$e_q[n] = e[n] + q[n] \tag{3.75}$$

where $q[n]$ is the quantization error. According to Figure 3.28*a*, the quantizer output $e_q[n]$ is added to the predicted value $\hat{m}[n]$ to produce the prediction-filter input

$$m_q[n] = \hat{m}[n] + e_q[n] \tag{3.76}$$

Substituting Equation (3.75) into (3.76), we get

$$m_q[n] = \hat{m}[n] + e[n] + q[n] \tag{3.77}$$

However, from Equation (3.74) we observe that the sum term $\hat{m}[n] + e[n]$ is equal to the input sample $m[n]$. Therefore, we may simplify Equation (3.77) as

$$m_q[n] = m[n] + q[n] \tag{3.78}$$

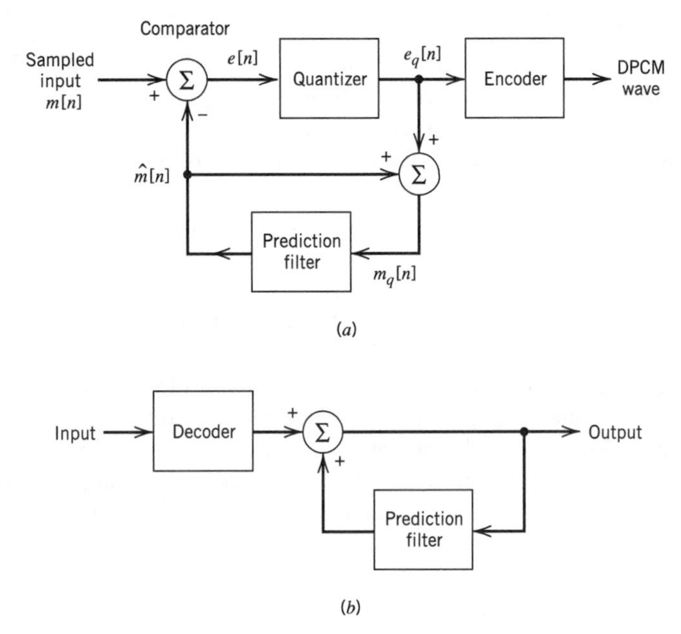


**FIGURE 3.28** DPCM system. (*a*) Transmitter. (*b*) Receiver.

which represents a quantized version of the input sample $m[n]$. That is, irrespective of the properties of the prediction filter, the quantized sample $m_q[n]$ at the prediction filter input differs from the original input sample $m[n]$ by the quantization error $q[n]$. Accordingly, if the prediction is good, the variance of the prediction error $e[n]$ will be smaller than the variance of $m[n]$, so that a quantizer with a given number of levels can be adjusted to produce a quantization error with a smaller variance than would be possible if the input sample $m[n]$ were quantized directly as in a standard PCM system.

The receiver for reconstructing the quantized version of the input is shown in Figure 3.28*b*. It consists of a decoder to reconstruct the quantized error signal. The quantized version of the original input is reconstructed from the decoder output using the same prediction filter used in the transmitter of Figure 3.28*a*. In the absence of channel noise, we find that the encoded signal at the receiver input is identical to the encoded signal at the transmitter output. Accordingly, the corresponding receiver output is equal to $m_q[n]$, which differs from the original input $m[n]$ only by the quantization error $q[n]$ incurred as a result of quantizing the prediction error $e[n]$.

From the foregoing analysis we observe that, in a noise-free environment, the prediction filters in the transmitter and receiver operate on the same sequence of samples, $m_q[n]$. It is with this purpose in mind that a feedback path is added to the quantizer in the transmitter, as shown in Figure 3.28*a*.

Differential pulse-code modulation includes delta modulation as a special case. In particular, comparing the DPCM system of Figure 3.28 with the DM system of Figure 3.23, we see that they are basically similar, except for two important differences: the use of a one-bit (two-level) quantizer in the delta modulator and the replacement of the prediction filter by a single delay element (i.e., zero prediction order). Simply put, DM is the 1-bit version of DPCM. Note that unlike a standard PCM system, the transmitters of both the DPCM and DM involve the use of *feedback*.

DPCM, like DM, is subject to slope-overload distortion whenever the input signal changes too rapidly for the prediction filter to track it. Also, like PCM, DPCM suffers from quantization noise.

### PROCESSING GAIN

The output signal-to-noise ratio of the DPCM system shown in Figure 3.28 is, by definition,

$$(\mathrm{SNR})_O = \frac{\sigma_M^2}{\sigma_Q^2} \tag{3.79}$$

where $\sigma_M^2$ is the variance of the original input sample $m[n]$, assumed to be of zero mean, and $\sigma_Q^2$ is the variance of the quantization error $q[n]$. We may rewrite Equation (3.79) as the product of two factors as follows:

$$\begin{aligned}(\mathrm{SNR})_O &= \left(\frac{\sigma_M^2}{\sigma_E^2}\right)\left(\frac{\sigma_E^2}{\sigma_Q^2}\right) \\ &= G_p(\mathrm{SNR})_Q\end{aligned} \tag{3.80}$$

where $\sigma_E^2$ is the variance of the prediction error. The factor $(\mathrm{SNR})_Q$ is the *signal-to-quantization noise ratio*, which is defined by

$$(\mathrm{SNR})_Q = \frac{\sigma_E^2}{\sigma_Q^2} \tag{3.81}$$

The other factor $G_p$ is the *processing gain* produced by the differential quantization scheme; it is defined by

$$G_p = \frac{\sigma_M^2}{\sigma_E^2} \tag{3.82}$$

The quantity $G_p$, when greater than unity, represents a gain in signal-to-noise ratio that is due to the differential quantization scheme of Figure 3.28. Now, for a given baseband (message) signal, the variance $\sigma_M^2$ is fixed, so that $G_p$ is maximized by minimizing the variance $\sigma_E^2$ of the prediction error $e[n]$. Accordingly, our objective should be to design the prediction filter so as to minimize $\sigma_E^2$.

In the case of voice signals, it is found that the optimum signal-to-quantization noise advantage of DPCM over standard PCM is in the neighborhood of 4 to 11 dB. The greatest improvement occurs in going from no prediction to first-order prediction, with some additional gain resulting from increasing the order of the prediction filter up to 4 or 5, after which little additional gain is obtained. Since 6 dB of quantization noise is equivalent to 1 bit per sample by virtue of Equation (3.35), the advantage of DPCM may also be expressed in terms of bit rate. For a constant signal-to-quantization noise ratio, and assuming a sampling rate of 8 kHz, the use of DPCM may provide a saving of about 8 to 16 kb/s (i.e., 1 to 2 bits per sample) compared to the standard PCM.

## 3.15 *Adaptive Differential Pulse-Code Modulation*

The use of PCM for speech coding at the standard rate of 64 kb/s demands a high channel bandwidth for its transmission. In certain applications, however, such as secure transmission over radio channels that are inherently of low capacity, channel bandwidth is at a premium. In applications of this kind, there is a definite need for *speech coding at low bit rates, while maintaining acceptable fidelity or quality of reproduction.*

For coding speech at low bit rates, a waveform coder of prescribed configuration is optimized by exploiting both *statistical characterization of speech waveforms* and *properties of hearing*. In particular, the design philosophy has two aims in mind:

1. To remove redundancies from the speech signal as far as possible.
2. To assign the available bits to code the nonredundant parts of the speech signal in a perceptually efficient manner.

As we strive to reduce the bit rate from 64 kb/s (used in standard PCM) to 32, 16, 8, and 4 kb/s, the schemes used for redundancy removal and bit assignment become increasingly more sophisticated. As a rule of thumb, in the 64 to 8 kb/s range, the computational complexity (measured in terms of multiply-add operations) required to code speech increases by an order of magnitude when the bit rate is halved, for approximately equal speech quality.

In this section, we describe *adaptive differential pulse-code modulation* (ADPCM),[11] which permits the coding of speech at 32 kb/s through the combined use of *adaptive quantization* and *adaptive prediction*; the number of eight bits per sample required in the standard PCM is thereby reduced to four. The term *adaptive* used herein means being responsive to changing level and spectrum of the input speech signal. The variation of performance with speakers and speech material, together with variations in signal level inherent in the speech communication process, make the combined use of adaptive quantization and adaptive prediction necessary to achieve best performance over a wide range of speakers and speaking situations.

Adaptive quantization refers to a quantizer that operates with a *time-varying* step-size $\Delta[n]$. At any given sampling instant identified by the index $n$, the adaptive quantizer is assumed to have a uniform transfer characteristic. The step-size $\Delta[n]$ is varied so as to match the variance $\sigma_M^2$ of the input sample $m[n]$. In particular, we write

$$\Delta[n] = \phi\hat{\sigma}_M[n] \tag{3.83}$$

where $\phi$ is a constant, and $\hat{\sigma}_M[n]$ is an *estimate* of the standard deviation $\sigma_M[n]$ (i.e., square root of the variance $\sigma_M^2$). For a nonstationary input, $\sigma_M[n]$ is time varying. The problem of adaptive quantization according to Equation (3.83) is, therefore, one of computing the estimate $\hat{\sigma}_M[n]$ continuously.

The implementation of Equation (3.83) may proceed in one of two ways:

1. *Adaptive quantization with forward estimation* (AQF), in which unquantized samples of the input signal are used to derive forward estimates of $\sigma_M[n]$.
2. *Adaptive quantization with backward estimation* (AQB), in which samples of the quantizer output are used to derive backward estimates of $\sigma_M[n]$.

The AQF scheme requires the use of a *buffer* to store unquantized samples of the input speech signal needed for the learning period. It also requires the explicit transmission of level information (typically, about 5 to 6 bits per step-size sample) to a remote decoder, thereby burdening the system with additional *side information* that has to be transmitted to the receiver. Moreover, a processing *delay* (on the order of 16 ms for speech) in the encoding operation results from the use of AQF, which is unacceptable in some applications. The problems of level transmission, buffering, and delay intrinsic to AQF are all avoided in AQB. In the latter scheme, the recent history of the quantizer output is used to extract information for the computation of the step size $\Delta[n]$. In practice, AQB is therefore usually preferred over AQF.

Figure 3.29 shows the block diagram of an adaptive quantizer with backward estimation. It represents a nonlinear feedback system; hence, it is not obvious that the system

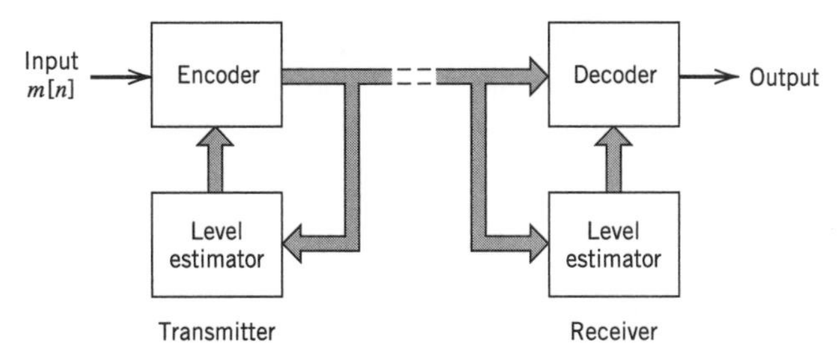


**FIGURE 3.29** Adaptive quantization with backward estimation (AQB).

will be stable. However, if the quantizer input $m[n]$ is *bounded*, then the backward estimate $\hat{\sigma}_M[n]$ and the corresponding step size $\Delta[n]$ are as well; under such a condition, the system is indeed *stable*.

The use of adaptive prediction in ADPCM is justified because speech signals are inherently *nonstationary*, a phenomenon that manifests itself in the fact that the autocorrelation function and power spectral density of speech signals are time-varying functions of their respective arguments. This implies that the design of predictors for such inputs should likewise be time varying, that is, adaptive. As with adaptive quantization, there are two schemes for performing adaptive prediction:

1. *Adaptive prediction with forward estimation* (APF), in which unquantized samples of the input signal are used to derive estimates of the predictor coefficients.
2. *Adaptive prediction with backward estimation* (APB), in which samples of the quantizer output and the prediction error are used to derive estimates of the predictor coefficients.

However, APF suffers from the same intrinsic disadvantages (side information, buffering, and delay) as AQF. These disadvantages are eliminated by using the APB scheme shown in Figure 3.30, where the box labeled "logic for adaptive prediction" represents the algorithm for updating the predictor coefficients. In the latter scheme, the optimum predictor coefficients are estimated on the basis of quantized and transmitted data; they can therefore be updated as frequently as desired, say, from sample to sample. Accordingly, APB is the preferred method of prediction for ADPCM.

The LMS algorithm for the predictor, described in Section 3.13, and an adaptive scheme for the quantizer, based on Equation (3.83), have been combined in a synchronous

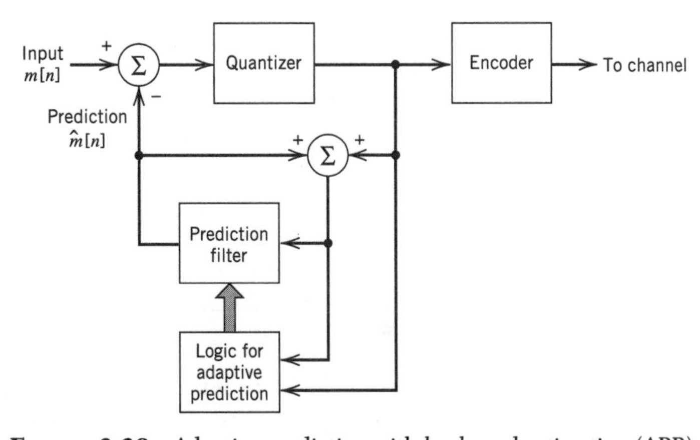


**FIGURE 3.30** Adaptive prediction with backward estimation (APB).

fashion for the design of both the encoder and decoder. The performance of this combination is so impressive at 32 kb/s that ADPCM is now accepted internationally as a standard coding technique for voice signals, along with 64 kb/s using standard PCM.

## 3.16 Computer Experiment: Adaptive Delta Modulation

A simple form of AQB is to be found in the modification of linear delta modulation (LDM) to form *adaptive delta modulation* (ADM). The principle underlying all ADM algorithms is two-fold:

1. If successive errors are of opposite polarity, then the delta modulator is operating in its granular mode; in this case, it may be advantageous to reduce the step size.
2. If, however, successive errors are of the same polarity, then the delta modulator is operating in its slope-overload mode; in this second case, the step size should be increased.

Thus by varying the step-size in accordance with this principle, the delta modulator is enabled to cope with changes in the input signal.

Figure 3.31 shows the block diagram of an ADM based on increasing or decreasing the step size by a factor of 50 percent at each iteration of the adaptive process. The algorithm for adaptation of the step size is defined by[12]

$$\Delta[n] = \begin{cases} \dfrac{|\Delta[n-1]|}{m_q[n]}(m_q[n] + 0.5m_q[n-1]) & \text{if } \Delta[n-1] \geq \Delta_{\min} \\ \Delta_{\min} & \text{if } \Delta[n-1] < \Delta_{\min} \end{cases} \tag{3.84}$$

where $\Delta[n]$ is the step size at iteration (time step) $n$ of the algorithm, and $m_q[n]$ is the 1-bit quantizer output that equals $\pm 1$.

In this experiment we use a sinusoidal input signal to demonstrate the reconstruction performance of the ADM algorithm based on Equation (3.84), and compare it to the performance of a corresponding linear delta modulator (LDM). Details of the experiment are as follows:

*Input signal:*

$$m(t) = A \sin(2\pi f_m t)$$

where amplitude $A = 10$, frequency $f_m = f_s/100$, and $f_s$ = sampling frequency.

*Linear delta modulation* (LDM):

$$\text{Step size } \Delta[n] = 1 \text{ for all } n$$

*Adaptive delta modulation* (ADM):

$$\Delta_{\min} = \frac{1}{8}$$

The results of the experiment are plotted in Figure 3.32. Part *a* of Figure 3.32 is for LDM, and part *b* of the figure is for ADM. From the waveforms presented here, we may make the following observations:

- ADM *tracks* changes in the sinusoidal input signal much better than LDM. This improvement in the performance of ADM is due to adaptation of the step size in

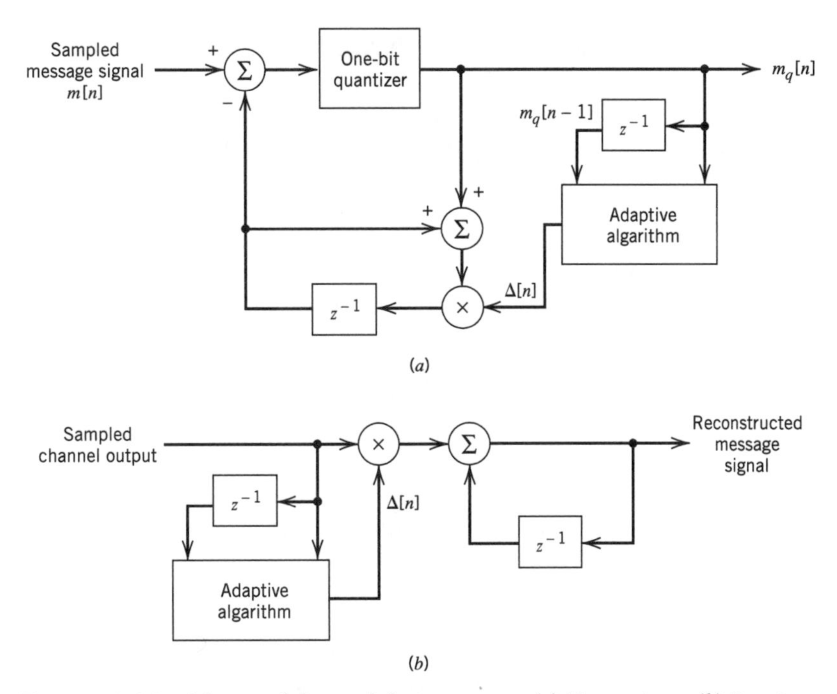


**FIGURE 3.31** Adaptive delta modulation system: (*a*) Transmitter. (*b*) Receiver.

successive iterations of the algorithm. In particular, the reduced step size of the ADM results in smaller quantization errors near the extremities of the input signal than the LDM. However, both modulation schemes produce comparable quantization errors in regions of the input signal where the slope is moderately high.

- The improved tracking performance of the ADM results in an output signal with a much lower bit rate, on the average, than the LDM.

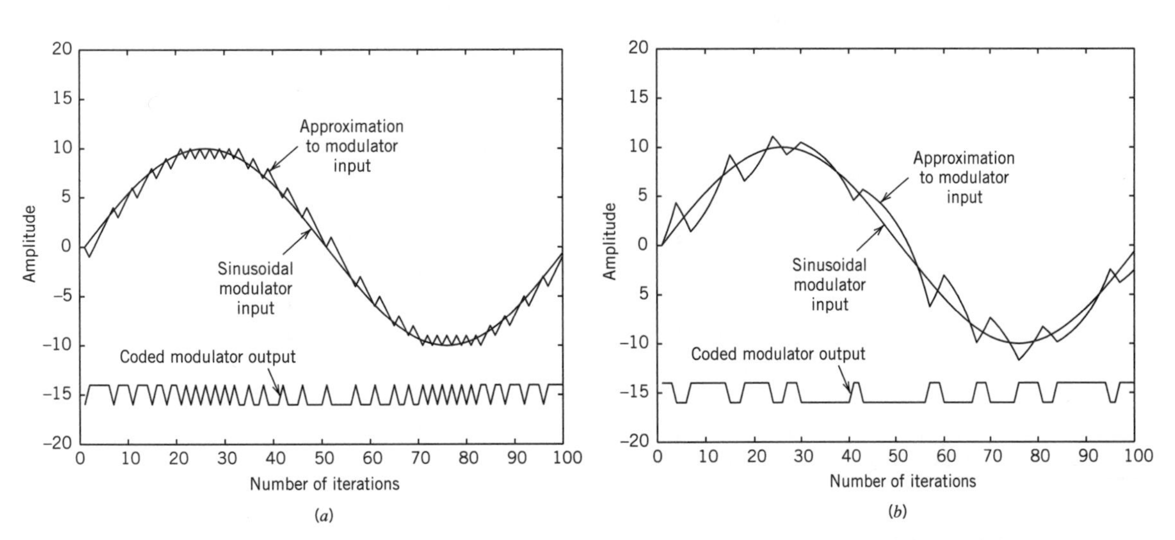


**FIGURE 3.32** Waveforms resulting from the computer experiment on delta modulation: (*a*) Linear delta modulation. (*b*) Adaptive delta modulation.

## 3.17 *MPEG Audio Coding Standard*

Speech (voice) and audio signals are similar in that, in both cases, the quality of a coding scheme is based on the properties of human auditory perception. In the case of speech signals, we have efficient coding schemes (e.g., ADPCM) because a speech production model is available. Unfortunately, nothing similar exists for audio signals.

In this section, we revisit the *MPEG-1/audio coding standard* briefly described in the Background and Preview chapter; MPEG stands for Motion Picture Experts Group, and the suffix 1 is intended to mean it is the first in a series of several standards.[13] Like ADPCM, the MPEG-1/audio coding standard is a lossy compression system, but it differs from ADPCM in an important practical respect: The MPEG-1 standard is capable of achieving *transparent, perceptually lossless compression* of stereophonic audio signals at high sampling rates. In particular, subjective listening tests performed by the MPEG/audio committee, under very difficult listening conditions, have shown that even with a 6-to-1 compression ratio, the coded and original audio signals are perceptually indistinguishable.

The MPEG-1/audio coding standard achieves this remarkable performance by exploiting two psychoacoustic characteristics of the human auditory system:

1. *Critical bands.*
The inner ear[14] of the auditory system represents the power spectra of incoming signals on a nonlinear scale in the form of limited frequency bands called the *critical bands*. The audible frequency band, extending up to 20 kHz, is covered by 25 critical bands, whose individual bandwidths increase with frequency. Loosely speaking, the auditory system may be modeled as a *band-pass filter bank*, consisting of 25 overlapping band-pass filters with bandwidths less than 100 Hz for the lowest audible frequencies and up to 5 kHz for the highest audible frequencies.

2. *Auditory masking.*
Auditory masking or noise masking is a frequency-domain phenomenon that arises when a low-level signal (the maskee) and a high-level signal (the masker) occur simultaneously and are close enough to each other in frequency. If the low-level signal lies below a *masking threshold*, it is made inaudible (i.e., masked) by the stronger signal. The auditory-masking phenomenon is most pronounced when both signals lie in the same critical band, and less effective when they lie in neighboring bands.

Figure 3.33 illustrates the definition of masking threshold and related parameters for a pair of adjacent frequency bands; it is assumed that the masker (i.e., the high-level signal) lies inside the dark-shaded critical band. The low-level signals lying inside this dark area and below the masking threshold are masked by the stronger signal. From Figure 3.33 we see that the masking threshold varies with frequency across the critical band. Accordingly, we may define a *minimum masking threshold* for a critical band, below which all low-level signals that lie inside that band are made inaudible by the stronger signal. The power difference, expressed in decibels, between the masker and the minimum masking threshold, is termed the *signal-to-mask ratio* (SMR). Figure 3.33 also includes the *signal-to-noise ratio* (SNR) for an $R$-bit quantizer. The difference between SMR and SNR is the *noise-to-mask ratio (NMR)* for an $R$-bit quantizer as shown by

$$\text{NMR} = \text{SMR} - \text{SNR} \tag{3.85}$$

where all three terms are expressed in dBs. Within a critical band, the quantization noise is inaudible as long as the NMR for the pertinent quantizer is negative.

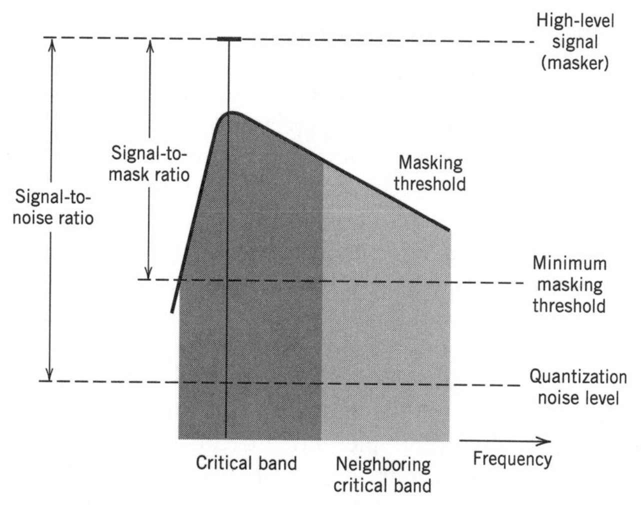


**FIGURE 3.33** Illustrating the definitions of masking threshold and related parameters. The high-level signal (masker) lies inside the darker-shaded critical band, hence the masking is more effective in this band than in the neighboring band shown in lighter shading. (Adapted from Noll (1998) with permission of the CRC Press.)

With this background on the psychoacoustics of the auditory system, we are now ready to describe the operation of the MPEG-1/audio coding standard. Figure 3.34 shows the basic block diagrams of the encoder and decoder. The encoder consists of four functional units: time-to-frequency mapping network, psychoacoustic model, quantizer and coder, and frame-packing unit. The decoder consists of three functional units: frame-unpacking unit, frequency-sample reconstruction network, and frequency-to-time mapping network. The psychoacoustic model is thus only necessary in the encoder.

Starting with a description of the encoder first, the function of the *time-to-frequency mapping network* is to decompose the input audio signal into multiple subbands for coding. The mapping is performed in three layers, labeled I, II, and III, which are of increasing complexity, delay, and subjective perceptual performance. The algorithm in layer I uses a

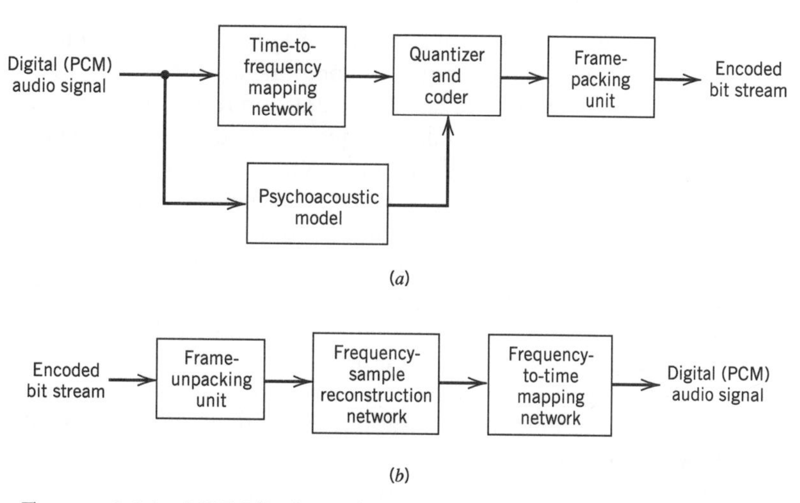


**FIGURE 3.34** MPEG/Audio coding system. (*a*) Transmitter. (*b*) Receiver.

band-pass filter bank that divides the audio signal into 32 *constant-width subbands*; this filter bank is also found in layers II and III. In light of our previous remarks on the non-uniformly spaced critical bands, the design of this filter bank is a compromise between computational efficiency and perceptual performance. The algorithm in layer II is a simple enhancement of layer I; it improves the compression performance by coding the data in larger groups. Finally, the layer III algorithm is much more refined in that it is designed to achieve frequency resolutions closer to the partitions between the critical bands.

The *psychoacoustic model* is the key component in the encoder. Its function is to analyze the spectral content of the input audio signal and thereby compute the signal-to-mask ratio for each subband in each of the three layers. This information is, in turn, used by the quantizer-coder to decide how to apportion the available number of bits for the quantization of the subband signals. This dynamic allocation of bits is performed so as to minimize the audibility of the quantization noise. Finally, the *frame-packing unit* assembles the quantized audio samples into a decodable bit stream.

The decoder simply reverses the signal-processing operations performed in the encoder, converting the received stream of encoded bits into a time-domain audio signal.

To sum up, the MPEG-1/audio coding standard represents the state of the art in the coding of audio signals. Layer I achieves a compression ratio of 4 at an approximate stereo bit rate of 384 kb/s for transparent quality of performance. The corresponding compression ratios for layers II and III are 8 and 12 at approximate stereo bit rates of 192 kb/s and 128 kb/s, respectively. The subjective quality of the MPEG-1/audio coding standard is equivalent to compact disc quality (16-bit PCM) for many types of music; the compact disc (CD) is today's de facto standard of digital audio representation.

## 3.18 *Summary and Discussion*

In this chapter we introduced two fundamental and complementary processes:

- *Sampling*, which operates in the time domain; the sampling process is the link between an analog waveform and its discrete-time representation.
- *Quantization*, which operates in the amplitude domain; the quantization process is the link between an analog waveform and its discrete-amplitude representation.

The sampling process builds on the *sampling theorem*, which states that a strictly band-limited signal with no frequency components higher than $W$ Hz is represented uniquely by a sequence of samples taken at a uniform rate equal to or greater than the Nyquist rate of $2W$ samples per second. The quantization process exploits the fact that any human sense, as ultimate receiver, can only detect finite intensity differences.

The sampling process is basic to the operation of all pulse modulation systems, which may be classified into analog pulse modulation and digital pulse modulation. The distinguishing feature between them is that analog pulse modulation systems maintain a continuous amplitude representation of the message signal, whereas digital pulse modulation systems also employ quantization to provide a representation of the message signal that is discrete in both time and amplitude.

Analog pulse modulation results from varying some parameter of the transmitted pulses, such as amplitude, duration, or position, in which case we speak of pulse-amplitude modulation (PAM), pulse-duration modulation (PDM), or pulse-position modulation (PPM), respectively. In time-division multiplexing (TDM) of several channels, signal processing usually begins with PAM. To use PDM or PPM in such an application, we have to ensure that full-scale modulation will not cause a pulse from one message signal to enter a time slot belonging to another message signal. This restriction results in a wasteful use of time space in telephone systems that are characterized by high peak factors, which is one reason for not using PDM or PPM in telephony. Also, despite the fact that PPM is more efficient than PDM, they both fall short of the ideal system for exchanging transmission bandwidth for improved noise performance.

Digital pulse modulation systems transmit analog message signals as a sequence of coded pulses, which is made possible through the combined use of sampling and quantization. Pulse-code modulation is an important form of digital pulse modulation that is endowed with some unique system advantages, which, in turn, have made it the standard method of modulation for the transmission of such analog signals as voice and video signals. The advantages of pulse-code modulation include robustness to noise and interference, efficient regeneration of the coded pulses along the transmission path, and a uniform format for different kinds of baseband signals.

Indeed, it is because of this list of advantages unique to pulse-code modulation that it has become the method of choice for the construction of public switched telephone networks (PSTNs). In this context, the reader should carefully note that the telephone channel viewed from the PSTN to an Internet service provider, for example, is *nonlinear* due to the use of companding and, most importantly, it is *entirely digital.* This observation has a significant impact on the design of high-speed modems for communication between a computer user and server, as discussed in Chapter 6.

Delta modulation and differential pulse-code modulation are two other useful forms of digital pulse modulation. The principal advantage of delta modulation is the simplicity of its circuitry. In contrast, differential pulse-code modulation employs increased circuit complexity to reduce channel bandwidth. The improvement is achieved by using the idea of prediction to remove redundant symbols from an incoming data stream. A further improvement in the operation of differential pulse-code modulation can be made through the use of adaptivity to account for statistical variations in the input data. By so doing, bandwidth requirement is reduced significantly without serious degradation in system performance.

Unlike adaptive differential pulse-code modulation, the MPEG audio coding standard achieves the compression of stereophonic audio signals in a transparent, perceptually lossless manner. This impressive performance is realized by exploiting certain psychoacoustic properties of the auditory system.

At this point in the discussion, it is informative to take a critical look at the different forms of pulse modulation that we have described in this chapter. In a strict sense, the term *pulse modulation* is a misnomer in that all of its different forms, be they analog or digital, are in fact *source coding* techniques. We say this for the simple reason that a message signal remains a baseband signal after undergoing all the changes involved in a pulse modulation process. The baseband nature of a pulse-modulated signal is exemplified by the fact that, irrespective of its exact description, it can be transmitted over a baseband channel of adequate bandwidth. Indeed, the material presented in the next chapter is devoted to the baseband transmission of data represented by a sequence of pulses.

It is also important to recognize that pulse modulation techniques are *lossy* in the sense that some information is lost as a result of the signal representation that they perform. For example, in pulse-amplitude modulation, the customary practice is to use low-pass anti-alias filtering prior to sampling; in so doing, information is lost by virtue of the fact that high-frequency components considered to be unessential are removed by the filter. The lossy nature of pulse modulation is most vividly seen in pulse-code modulation that is characterized by the generation of quantization noise (i.e., distortion); the transmitted sequence of encoded pulses does not have the infinite precision needed to represent continuous samples exactly. Nevertheless, the loss of information incurred by the use of a pulse-modulation process is *under the designer's control* in that it can be made small enough for it to be nondiscernible by the end user.

The material presented in this chapter on pulse modulation has been from a signal processing perspective. We will revisit pulse-code modulation in Chapter 9, which is devoted to information-theoretic considerations of communication systems. In so doing we will develop deeper insight into its operation as a source coding technique.

## NOTES AND REFERENCES

1. The classic book on pulse modulation is Black (1953). A more detailed treatment of the subject is presented in the book by Rowe (1965). For the noise analysis of a PPM system, see the third edition of the book by Haykin (1994).

2. Pulse-code modulation was invented by Reeves in 1937. For a historical account of this invention, see the paper by Reeves (1975). The book by Jayant and Noll (1984) presents detailed treatment of pulse-code modulation, differential pulse-code modulation, delta modulation, and their variants. The book edited by Jayant (1976) provides a collection of early papers written on waveform quantization and coding.

3. For a detailed discussion of quantization noise in PCM systems, see the paper by Bennett (1948) and also the book by Rowe (1965, pp. 311–321).

4. The two necessary conditions of Equations (3.42) and (3.47) for optimality of a scalar quantizer were reported independently by Lloyd (1957) and Max (1960), hence the name "Lloyd-Max quantizer." The derivation of these two optimality conditions presented in this chapter follows the book by Gersho and Gray (1992).

5. The $\mu$-law used for signal compression is described in Smith (1957).The $\mu$-law is used in the United States, Canada, and Japan. In Europe, the $A$-law is used for signal compression; this compression law is described in Cattermode (1969, pp. 133–140). For a discussion of the $\mu$-law and $A$-law, see also the paper by Kaneko (1970).

6. For a description of the original version of the T1-carrier system, see the paper by Fultz and Penick (1965). The description given in Example 3.2 is based on an updated version of this system; see Henning and Pan (1972).

7. The North America/Japan standards for digital multiplexers were originally adopted by AT&T. Another set of standards has been adopted by CCITT for the rest of the world. The CCITT digital hierarchy is similar to that described in Section 3.10, except for certain changes in the specifications of the number of channel inputs to the five digital multiplexers and their individual bit rates. For details of the CCITT digital hierarchy, see Couch (1997).

8. For the original papers on delta modulation, see Schouten, DeJager, and Greefkes (1952) and DeJager (1952). For a review paper on delta modulation, see the paper by Schindler (1970).

9. Delta-sigma modulation is described in the book by Jayant and Noll (1984, pp. 399–400); see also the paper by Inose, Yasuda, and Murakami (1962).

10. Differential pulse-code modulation was invented by Cutler; the invention is described in a patent issued in 1952. For a comparison of the noise performances of PCM and DPCM, see the paper by Jayant (1974); see also Rabiner and Schafer (1978, Chapter 5).

11. For a discussion of adaptive differential pulse-code modulation, see Jayant and Noll (1984).

12. The adaptive delta modulation algorithm (ADM) of Equation (3.84) is the corrected version of an algorithm presented in Sklar (1988, p. 641). Sklar's algorithm was adapted from an earlier paper by Song et al. (1971), where an optimum ADM system is derived; the highly nonlinear equations characterizing the optimum system are approximated in the latter paper by piecewise-linear equations for the purpose of implementation.

13. The MPEG-1/audio coding standard is described in the papers by Brandenburg and Stoll (1994), Pan (1993), and the article by Peter Noll in the handbook on Digital Signal Processing edited by Madisetti and Williams (1998); the latter article also discusses the follow-up standards to MPEG-1. In particular, the MPEG–2 offers stereophonic audio coding at sampling rates lower than MPEG–1.

14. The ear, the organ of hearing, responds to incoming acoustical waves. It has three main parts, with their functions as summarized here:

- The *outer ear* aids in the collection of sounds.
- The *middle ear* provides an acoustic impedance match between the air and the cochlea fluids, thereby conveying the vibrations of the tympanic membrane (eardrum) due to the incoming sounds to the inner ear in an efficient manner.
- The *inner ear* converts the mechanical vibrations from the middle ear to an electrochemical or neural signal for transmission to the brain for processing.

## PROBLEMS

### Sampling Process

3.1 A narrowband signal has a bandwidth of 10 kHz centered on a carrier frequency of 100 kHz. It is proposed to represent this signal in discrete-time form by sampling its in-phase and quadrature components individually. What is the minimum sampling rate that can be used for this representation? Justify your answer. How would you reconstruct the original narrowband signal from the sampled versions of its in-phase and quadrature components?

3.2 In *natural sampling*, an analog signal $g(t)$ is multiplied by a periodic train of rectangular pulses $c(t)$. Given that the pulse repetition frequency of this periodic train is $f_s$ and the duration of each rectangular pulse is $T$ (with $f_sT \ll 1$), do the following:

(a) Find the spectrum of the signal $s(t)$ that results from the use of natural sampling; you may assume that time $t = 0$ corresponds to the midpoint of a rectangular pulse in $c(t)$.

(b) Show that the original signal $m(t)$ may be recovered exactly from its naturally sampled version, provided that the conditions embodied in the sampling theorem are satisfied.

3.3 Specify the Nyquist rate and the Nyquist interval for each of the following signals:

(a) $g(t) = \text{sinc}(200t)$

(b) $g(t) = \text{sinc}^2(200t)$

(c) $g(t) = \text{sinc}(200t) + \text{sinc}^2(200t)$

3.4 (a) Plot the spectrum of a PAM wave produced by the modulating signal

$$m(t) = A_m \cos(2\pi f_m t)$$

assuming a modulation frequency $f_m = 0.25$ Hz, sampling period $T_s = 1$ s, and pulse duration $T = 0.45$ s.

(b) Using an ideal reconstruction filter, plot the spectrum of the filter output. Compare this result with the output that would be obtained if there were no aperture effect.

### Pulse-Amplitude Modulation

**3.5** Figure P3.5 shows the idealized spectrum of a message signal $m(t)$. The signal is sampled at a rate equal to 1 kHz using flat-top pulses, with each pulse being of unit amplitude and duration 0.1 ms. Determine and sketch the spectrum of the resulting PAM signal.

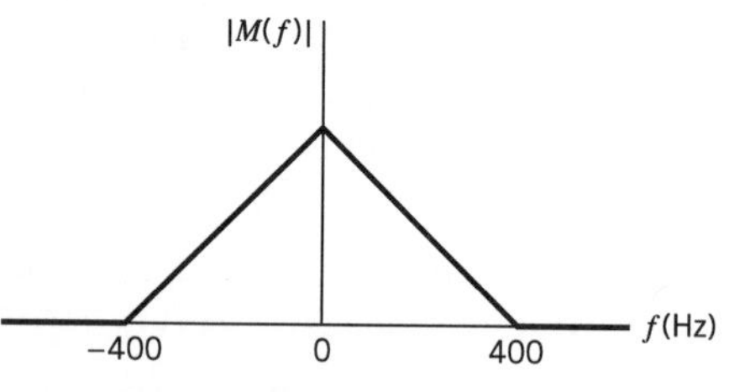


**FIGURE P3.5**

**3.6** In this problem, we evaluate the equalization needed for the aperture effect in a PAM system. The operating frequency $f = f_s/2$, which corresponds to the highest frequency component of the message signal for a sampling rate equal to the Nyquist rate. Plot $1/\text{sinc}\,(0.5T/T_s)$ versus $T/T_s$, and hence find the equalization needed when $T/T_s = 0.1$.

**3.7** Consider a PAM wave transmitted through a channel with white Gaussian noise and minimum bandwidth $B_T = 1/2T_s$, where $T_s$ is the sampling period. The noise is of zero mean and power spectral density $N_0/2$. The PAM signal uses a standard pulse $g(t)$ with its Fourier transform defined by

$$G(f) = \begin{cases} \dfrac{1}{2B_T}, & |f| < B_T \\ 0, & |f| > B_T \end{cases}$$

By considering a full-load sinusoidal modulating wave, show that PAM and baseband-signal transmission have equal signal-to-noise ratios for the same average transmitted power.

**3.8** Twenty-four voice signals are sampled uniformly and then time-division multiplexed. The sampling operation uses flat-top samples with 1 $\mu$s duration. The multiplexing operation includes provision for synchronization by adding an extra pulse of sufficient amplitude and also 1 $\mu$s duration. The highest frequency component of each voice signal is 3.4 kHz.

(a) Assuming a sampling rate of 8 kHz, calculate the spacing between successive pulses of the multiplexed signal.

(b) Repeat your calculation assuming the use of Nyquist rate sampling.

**3.9** Twelve different message signals, each with a bandwidth of 10 kHz, are to be multiplexed and transmitted. Determine the minimum bandwidth required for each method if the multiplexing/modulation method used is

(a) FDM, SSB.

(b) TDM, PAM.

**3.10** A PAM telemetry system involves the multiplexing of four input signals: $s_i(t)$, $i = 1, 2, 3, 4$. Two of the signals $s_1(t)$ and $s_2(t)$ have bandwidths of 80 Hz each, whereas the remaining two signals $s_3(t)$ and $s_4(t)$ have bandwidths of 1 kHz each. The signals $s_3(t)$ and $s_4(t)$ are each sampled at the rate of 2400 samples per second. This sampling rate is divided by $2^R$ (i.e., an integer power of 2) to derive the sampling rate for $s_1(t)$ and $s_2(t)$.

(a) Find the maximum value of $R$.

(b) Using the value of $R$ found in part (a), design a multiplexing system that first multiplexes $s_1(t)$ and $s_2(t)$ into a new sequence, $s_5(t)$, and then multiplexes $s_3(t)$, $s_4(t)$, and $s_5(t)$.

## Line Codes

**3.11** In this problem we derive the formulas used to compute the power spectra of Figure 3.16 for the five line codes described in Section 3.7. In the case of each line code, the bit duration is $T_b$ and the pulse amplitude $A$ is conditioned to normalize the average power of the line code to unity as indicated in Figure 3.16. Assume that the data stream is randomly generated, and symbols 0 and 1 are equally likely.

Derive the power spectral densities of these line codes as summarized here:

(a) Unipolar nonreturn-to-zero signals:

$$S(f) = \frac{A^2 T_b}{4} \operatorname{sinc}^2(fT_b)\left(1 + \frac{1}{T_b}\,\delta(f)\right)$$

(b) Polar nonreturn-to-zero signals:

$$S(f) = A^2 T_b \operatorname{sinc}^2(fT_b)$$

(c) Unipolar return-to-zero signals:

$$S(f) = \frac{A^2 T_b}{16} \operatorname{sinc}^2\left(\frac{fT_b}{2}\right)\left[1 + \frac{1}{T_b} \sum_{n=-\infty}^{\infty} \delta\left(f - \frac{n}{T_b}\right)\right]$$

(d) Bipolar return-to-zero signals:

$$S(f) = \frac{A^2 T_b}{4} \operatorname{sinc}^2\left(\frac{fT_b}{2}\right) \sin^2(\pi f T_b)$$

(e) Manchester-encoded signals:

$$S(f) = A^2 T_b \operatorname{sinc}^2\left(\frac{fT_b}{2}\right) \sin^2\left(\frac{\pi f T_b}{2}\right)$$

Hence, confirm the spectral plots displayed in Figure 3.16.

**3.12** Suppose a random binary data stream (with equiprobable symbols) is differentially encoded and then transmitted using one of the five line codes described in Problem 3.11. How is the power spectral density of the transmitted data affected by the use of differential encoding? Justify your answer.

**3.13** A randomly generated data stream consists of equiprobable binary symbols 0 and 1. It is encoded into a polar nonreturn-to-zero waveform with each binary symbol being defined as follows:

$$s(t) = \begin{cases} \cos\left(\dfrac{\pi t}{T_b}\right), & -\dfrac{T_b}{2} < t \le \dfrac{T_b}{2} \\ 0, & \text{otherwise} \end{cases}$$

(a) Sketch the waveform so generated, assuming that the data stream is 00101110.

(b) Derive an expression for the power spectral density of this signal, and sketch it. *Hint:* use Equation (1.52).

(c) Compare the power spectral density of this random waveform with that defined in part (b) of Problem 3.11.

**3.14** Given the data stream 1110010100, sketch the transmitted sequence of pulses for each of the following line codes:

(a) Unipolar nonreturn-to-zero

(b) Polar nonreturn-to-zero

(c) Unipolar return-to-zero

(d) Bipolar return-to-zero

(e) Manchester code

**3.15** Suppose the binary data stream considered in Problem 3.14 is differentially encoded and then transmitted using one of the five line codes considered therein. Sketch each of the transmitted data streams, assuming the use of symbol 1 for the reference bit. How is the result affected if symbol 0 is used for the reference bit?

## Pulse-Code Modulation

**3.16** A speech signal has a total duration of 10 s. It is sampled at the rate of 8 kHz and then encoded. The signal-to-(quantization) noise ratio is required to be 40 dB. Calculate the minimum storage capacity needed to accommodate this digitized speech signal.

**3.17** Consider a uniform quantizer characterized by the input–output relation illustrated in Figure 3.10*a*. Assume that a Gaussian-distributed random variable with zero mean and unit variance is applied to this quantizer input.

(a) What is the probability that the amplitude of the input lies outside the range $-4$ to $+4$?

(b) Using the result of part (a), show that the output signal-to-noise ratio of the quantizer is given by

$$(\text{SNR})_O = 6R - 7.2 \text{ dB}$$

where $R$ is the number of bits per sample. Specifically, you may assume that the quantizer input extends from $-4$ to $+4$. Compare the result of part (b) with that obtained in Example 3.1.

**3.18** A PCM system uses a uniform quantizer followed by a 7-bit binary encoder. The bit rate of the system is equal to $50 \times 10^6$ b/s.

(a) What is the maximum message bandwidth for which the system operates satisfactorily?

(b) Determine the output signal-to-(quantization) noise ratio when a full-load sinusoidal modulating wave of frequency 1 MHz is applied to the input.

**3.19** Show that, with a nonuniform quantizer, the mean-square value of the quantization error is approximately equal to $(1/12)\Sigma_i \Delta_i^2 p_i$, where $\Delta_i$ is the $i$th step size and $p_i$ is the probability that the input signal amplitude lies within the $i$th interval. Assume that the step size $\Delta_i$ is small compared with the excursion of the input signal.

**3.20** (a) A sinusoidal signal, with an amplitude of 3.25 volts, is applied to a uniform quantizer of the midtread type whose output takes on the values 0, $\pm 1$, $\pm 2$, $\pm 3$ volts. Sketch the waveform of the resulting quantizer output for one complete cycle of the input.

(b) Repeat this evaluation for the case when the quantizer is of the midrise type whose output takes on the values $=0.5$, $\pm 1.5$, $\pm 2.5$, $\pm 3.5$ volts.

**3.21** The signal

$$m(t) = 6 \sin(2\pi t) \text{ volts}$$

is transmitted using a 4-bit binary PCM system. The quantizer is of the midrise type, with a step size of 1 volt. Sketch the resulting PCM wave for one complete cycle of the input. Assume a sampling rate of four samples per second, with samples taken at $t = \pm 1/8$, $\pm 3/8$, $\pm 5/8$, . . . , seconds.

**3.22** Figure P3.22 shows a PCM signal in which the amplitude levels of $+1$ volt and $-1$ volt are used to represent binary symbols 1 and 0, respectively. The code word used consists of three bits. Find the sampled version of an analog signal from which this PCM signal is derived.

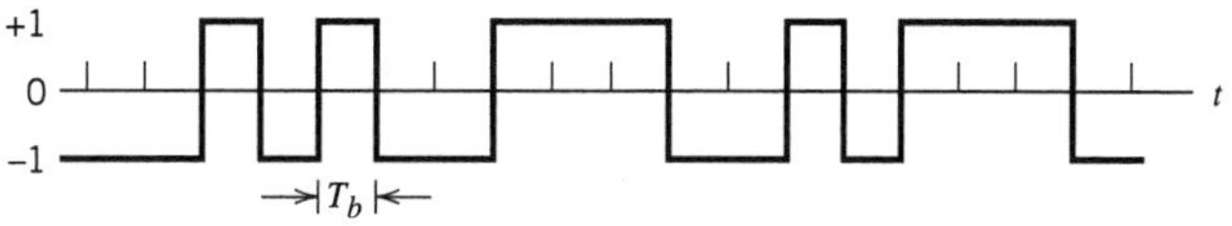

FIGURE P3.22

3.23 Consider a chain of $(n - 1)$ regenerative repeaters, with a total of $n$ sequential decisions made on a binary PCM wave, including the final decision made at the receiver. Assume that any binary symbol transmitted through the system has an independent probability $p_1$ of being inverted by any repeater. Let $p_n$ represent the probability that a binary symbol is in error after transmission through the complete system.

(a) Show that

$$p_n = \frac{1}{2}[1 - (1 - 2p_1)^n]$$

(b) If $p_1$ is very small and $n$ is not too large, what is the corresponding value of $p_n$?

3.24 Discuss the basic issues involved in the design of a regenerative repeater for pulse-code modulation.

## Delta Modulation

3.25 Consider a test signal $m(t)$ defined by a hyperbolic tangent function:

$$m(t) = A \tanh(\beta t)$$

where $A$ and $\beta$ are constants. Determine the minimum step size $\Delta$ for delta modulation of this signal, which is required to avoid slope overload.

3.26 Consider a sine wave of frequency $f_m$ and amplitude $A_m$, which is applied to a delta modulator of step size $\Delta$. Show that slope-overload distortion will occur if

$$A_m > \frac{\Delta}{2\pi f_m T_s}$$

where $T_s$ is the sampling period. What is the maximum power that may be transmitted without slope-overload distortion?

3.27 A linear delta modulator is designed to operate on speech signals limited to 3.4 kHz. The specifications of the modulator are as follows:

- Sampling rate $= 10 f_{\text{Nyquist}}$, where $f_{\text{Nyquist}}$ is the Nyquist rate of the speech signal.
- Step size $\Delta = 100$ mV.

The modulator is tested with a 1-kHz sinusoidal signal. Determine the maximum amplitude of this test signal required to avoid slope overload.

3.28 In this problem, we derive an empirical formula for the average signal-to-(quantization) noise ratio of a DM system with a sinusoidal signal of amplitude $A$ and frequency $f_m$ as the test signal. Assume that the power spectral density of the granular noise generated by the system is governed by the formula

$$S_N(f) = \frac{\Delta^2}{6f_s}$$

where $f_s$ is the sampling rate and $\Delta$ is the step size. (Note that this formula is basically the same as that for the power spectral density of quantization noise in a PCM system with $\Delta/2$ for PCM being replaced by $\Delta$ for DM.) The DM system is designed to handle analog message signals limited to bandwidth $W$.

(a) Show that the average quantization noise power produced by the system is

$$N = \frac{4\pi^2 A^2 f_m^2 W}{3f_s^3}$$

where it is assumed that the step size $\Delta$ has been chosen in accordance with the formula used in Problem 3.27 so as to avoid slope overload.

(b) Hence determine the signal-to-(quantization) noise ratio of the DM system for a sinusoidal input.

**3.29** Consider a DM system designed to accommodate analog message signals limited to bandwidth $W = 5$ kHz. A sinusoidal test signal of amplitude $A = 1$ volt and frequency $f_m = 1$ kHz is applied to the system. The sampling rate of the system is 50 kHz.

(a) Calculate the step size $\Delta$ required to minimize slope overload.

(b) Calculate the signal-to-(quantization) noise ratio of the system for the specified sinusoidal test signal.

For these calculations, use the formulas derived in Problems 3.27 and 3.28.

**3.30** Consider a low-pass signal with a bandwidth of 3 kHz. A linear delta modulation system, with step size $\Delta = 0.1$V, is used to process this signal at a sampling rate ten times the Nyquist rate.

(a) Evaluate the maximum amplitude of a test sinusoidal signal of frequency 1 kHz, which can be processed by the system without slope-overload distortion.

(b) For the specifications given in part (a), evaluate the output signal-to-noise ratio under (i) prefiltered, and (ii) postfiltered conditions.

### Linear Prediction

**3.31** A one-step linear predictor operates on the sampled version of a sinusoidal signal. The sampling rate is equal to $10f_0$ where $f_0$ is the frequency of the sinusoid. The predictor has a single coefficient denoted by $w_1$.

(a) Determine the optimum value of $w_1$ required to minimize the prediction error variance.

(b) Determine the minimum value of the prediction error variance.

**3.32** A stationary process $X(t)$ has the following values for its autocorrelation function:

$$R_X(0) = 1$$
$$R_X(1) = 0.8$$
$$R_X(2) = 0.6$$
$$R_X(3) = 0.4$$

(a) Calculate the coefficients of an optimum linear predictor involving the use of three unit-delays.

(b) Calculate the variance of the resulting prediction error.

**3.33** Repeat the calculations of Problem 3.32, but this time use a linear predictor with two unit-delays. Compare the performance of this second optimum linear predictor with that considered in Problem 3.32.

### Differential Pulse-Code Modulation

**3.34** A DPCM system uses a linear predictor with a single tap. The normalized autocorrelation function of the input signal for a lag of one sampling interval is 0.75. The predictor is designed to minimize the prediction error variance. Determine the processing gain attained by the use of this predictor.

3.35 Calculate the improvement in processing gain of a DPCM system using the optimized three-tap linear predictor of Problem 3.32 over that of the optimized two-tap linear predictor of Problem 3.33. For this calculation, use the autocorrelation function values of the input signal specified in Problem 3.32.

3.36 In this problem, we compare the performance of a DPCM system with that of an ordinary PCM system using companding.

For a sufficiently large number of representation levels, the signal-to-(quantization) noise ratio of PCM systems, in general, is defined by

$$10 \log_{10}(\text{SNR})_O = \alpha + 6n \text{ dB}$$

where $2^n$ is the number of representation levels. For a companded PCM system using the $\mu$-law, the constant $\alpha$ is itself defined by

$$\alpha \simeq 4.77 - 20 \log_{10}(\log(1 + \mu)) \text{ dB}$$

For a DPCM system, on the other hand, the constant $\alpha$ lies in the range $-3 < \alpha < 15$ dBs. The formulas quoted herein apply to telephone-quality speech signals.

Compare the performance of the DPCM system against that of the $\mu$-companded PCM system with $\mu = 255$ for each of the following scenarios:

(a) The improvement in $(\text{SNR})_O$ realized by DPCM over companded PCM for the same number of bits per sample.

(b) The reduction in the number of bits per sample required by DPCM, compared to the companded PCM for the same $(\text{SNR})_O$.

3.37 In the DPCM system depicted in Figure P3.37, show that in the absence of channel noise, the transmitting and receiving prediction filters operate on slightly different input signals.

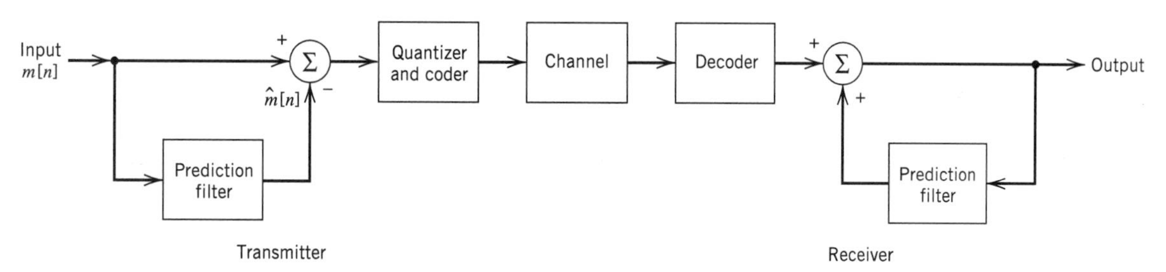


**FIGURE P3.37**

## Computer Experiments

3.38 A sinusoidal signal of frequency $f_0 = 10^4/2\pi$ Hz is sampled at the rate of 8 kHz and then applied to a sample-and-hold circuit to produce a flat-topped PAM signal $s(t)$ with pulse duration $T = 500\ \mu$s.

(a) Compute the waveform of the PAM signal $s(t)$.

(b) Compute $|S(f)|$, denoting the magnitude spectrum of the PAM signal $s(t)$.

(c) Compute the envelope of $|S(f)|$. Hence confirm that the frequency at which this envelope goes through zero for the first time is equal to $(1/T) = 20$ kHz.

3.39 In this problem, we use computer simulation to compare the performance of a companded PCM system using the $\mu$-law against that of the corresponding system using a uniform quantizer. The simulation is to be performed for a sinusoidal input signal of varying amplitude.

(a) Using the $\mu$-law described in Table 3.4, plot the output signal-to-noise ratio as a function of the input signal-to-noise ratio, both ratios being expressed in decibels.

(b) Compare the results of your computation in part (a) with a uniform quantizer having 256 representation levels.

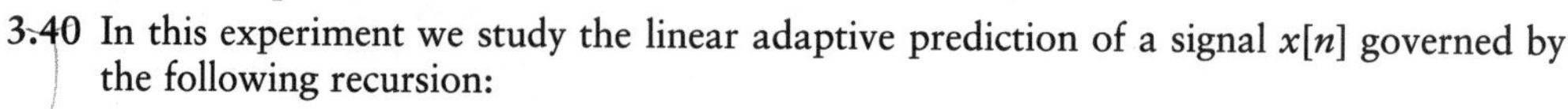

3.40 In this experiment we study the linear adaptive prediction of a signal $x[n]$ governed by the following recursion:

$$x[n] = 0.8x[n-1] - 0.1x[n-2] + 0.1v[n]$$

where $v[n]$ is drawn from a discrete–time white noise process of zero mean and unit variance. (A process generated in this manner is referred to as an *autoregressive process of order two.*) Specifically, the adaptive prediction is performed using the *normalized LMS algorithm* defined by

$$\hat{x}[n] = \sum_{k=1}^{p} w_k[n]x[n-k]$$

$$e[n] = x[n] - \hat{x}[n]$$

$$w_k[n+1] = w_k[n] + \left(\frac{\mu}{\sum_{k=1}^{p} x^2[n-k]}\right)x[n-k]e[n], \qquad k = 1, 2, \ldots, p$$

where $p$ is the prediction order and $\mu$ is the normalized step-size parameter. The important point to note here is that $\mu$ is dimensionless and stability of the algorithm is assured by choosing it in accordance with the formula

$$0 < \mu < 2$$

The algorithm is initiated by setting

$$w_k[0] = 0 \qquad \text{for all } k$$

The *learning curve* of the algorithm is defined as a plot of the mean-square error versus the number of iterations $n$ for specified parameter values, which is obtained by averaging the plot of $e^2[n]$ versus $n$ over a large number of different realizations of the algorithm.

(a) Plot the learning curves for the adaptive prediction of $x[n]$ for a fixed prediction order $p = 5$ and three different values of step-size parameter: $\mu = 0.0075, 0.05$, and $0.5$.

(b) What observations can you make from the learning curves of part (a)?

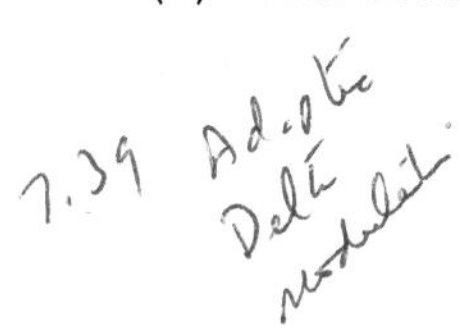

CHAPTER 4

# BASEBAND PULSE TRANSMISSION

This chapter discusses the transmission of digital data over a baseband channel, with emphasis on the following topics:

- *The matched filter, which is the optimum system for detecting a known signal in additive white Gaussian noise.*
- *Calculation of the bit error rate due to the presence of channel noise.*
- *Intersymbol interference, which arises when the channel is dispersive as is commonly the case in practice.*
- *Nyquist's criterion for distortionless baseband data transmission.*
- *Correlative-level coding or partial-response signaling for combatting the effects of intersymbol interference.*
- *Digital subscriber lines.*
- *Equalization of a dispersive baseband channel.*
- *The eye pattern for displaying the combined effects of intersymbol interference and channel noise in data transmission.*

## 4.1 *Introduction*

In Chapter 3 we described techniques for converting an analog information-bearing signal into digital form. There is another way in which digital data can arise in practice: The data may represent the output of a source of information that is inherently discrete in nature (e.g., a digital computer). In this chapter we study the transmission of digital data (of whatever origin) over a *baseband channel.*[1] Data transmission over a band-pass channel using modulation is covered in Chapter 6.

Digital data have a broad spectrum with a significant low-frequency content. Baseband transmission of digital data therefore requires the use of a low-pass channel with a bandwidth large enough to accommodate the essential frequency content of the data stream. Typically, however, the channel is *dispersive* in that its frequency response deviates from that of an ideal low-pass filter. The result of data transmission over such a channel is that each received pulse is affected somewhat by adjacent pulses, thereby giving rise to a common form of interference called *intersymbol interference* (ISI). Intersymbol interference is a major source of bit errors in the reconstructed data stream at the receiver output. To correct for it, control has to be exercised over the pulse shape in the overall system. Thus much of the material covered in this chapter is devoted to *pulse shaping* in one form or another.

Another source of bit errors in a baseband data transmission system is the ubiquitous *channel noise*. Naturally, noise and ISI arise in the system simultaneously. However, to understand how they affect the performance of the system, we first consider them separately; later on in the chapter, we study their combined effects.

We thus begin the chapter by describing a fundamental result in communication theory, which deals with the *detection* of a pulse signal of known waveform that is immersed in additive white noise. The device for the optimum detection of such a pulse involves the use of a linear-time-invariant filter known as a *matched filter*,[2] which is so called because its impulse response is matched to the pulse signal.

## 4.2 Matched Filter

A basic problem that often arises in the study of communication systems is that of *detecting* a pulse transmitted over a channel that is corrupted by channel noise (i.e., additive noise at the front end of the receiver). For the purpose of the discussion presented in this section, we assume that the major source of system limitation is the channel noise.

Consider then the receiver model shown in Figure 4.1, involving a linear time-invariant filter of impulse response $h(t)$. The filter input $x(t)$ consists of a pulse signal $g(t)$ corrupted by additive channel noise $w(t)$, as shown by

$$x(t) = g(t) + w(t), \qquad 0 \le t \le T \tag{4.1}$$

where $T$ is an arbitrary observation interval. The pulse signal $g(t)$ may represent a binary symbol 1 or 0 in a digital communication system. The $w(t)$ is the sample function of a white noise process of zero mean and power spectral density $N_0/2$. It is assumed that the receiver has knowledge of the waveform of the pulse signal $g(t)$. The source of uncertainty lies in the noise $w(t)$. The function of the receiver is to detect the pulse signal $g(t)$ in an optimum manner, given the received signal $x(t)$. To satisfy this requirement, we have to optimize the design of the filter so as to minimize the effects of noise at the filter output in some statistical sense, and thereby enhance the detection of the pulse signal $g(t)$.

Since the filter is linear, the resulting output $y(t)$ may be expressed as

$$y(t) = g_o(t) + n(t) \tag{4.2}$$

where $g_o(t)$ and $n(t)$ are produced by the signal and noise components of the input $x(t)$, respectively. A simple way of describing the requirement that the output signal component $g_o(t)$ be considerably greater than the output noise component $n(t)$ is to have the filter make the instantaneous power in the output signal $g_o(t)$, measured at time $t = T$, as large as possible compared with the average power of the output noise $n(t)$. This is equivalent to maximizing the *peak pulse signal-to-noise ratio*, defined as

$$\eta = \frac{|g_o(T)|^2}{E[n^2(t)]} \tag{4.3}$$

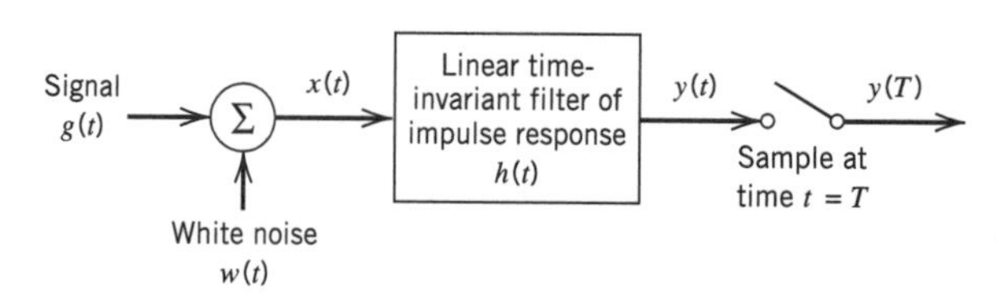


**FIGURE 4.1** Linear receiver.

where $|g_o(T)|^2$ is the instantaneous power in the output signal, $E$ is the statistical expectation operator, and $E[n^2(t)]$ is a measure of the average output noise power. The requirement is to specify the impulse response $h(t)$ of the filter such that the output signal-to-noise ratio in Equation (4.3) is maximized.

Let $G(f)$ denote the Fourier transform of the known signal $g(t)$, and $H(f)$ denote the frequency response of the filter. Then the Fourier transform of the output signal $g_o(t)$ is equal to $H(f)G(f)$, and $g_o(t)$ is itself given by the inverse Fourier transform

$$g_o(t) = \int_{-\infty}^{\infty} H(f)G(f) \exp(j2\pi ft)\, df \tag{4.4}$$

Hence, when the filter output is sampled at time $t = T$, we have (in the absence of channel noise)

$$|g_o(T)|^2 = \left| \int_{-\infty}^{\infty} H(f)G(f) \exp(j2\pi fT)\, df \right|^2 \tag{4.5}$$

Consider next the effect on the filter output due to the noise $w(t)$ acting alone. The power spectral density $S_N(f)$ of the output noise $n(t)$ is equal to the power spectral density of the input noise $w(t)$ times the squared magnitude response $|H(f)|^2$ (see Section 1.7). Since $w(t)$ is white with constant power spectral density $N_0/2$, it follows that

$$S_N(f) = \frac{N_0}{2} |H(f)|^2 \tag{4.6}$$

The average power of the output noise $n(t)$ is therefore

$$\begin{aligned} E[n^2(t)] &= \int_{-\infty}^{\infty} S_N(f)\, df \\ &= \frac{N_0}{2} \int_{-\infty}^{\infty} |H(f)|^2\, df \end{aligned} \tag{4.7}$$

Thus substituting Equations (4.5) and (4.7) into (4.3), we may rewrite the expression for the peak pulse signal-to-noise ratio as

$$\eta = \frac{\left| \int_{-\infty}^{\infty} H(f)G(f) \exp(j2\pi fT)\, df \right|^2}{\frac{N_0}{2} \int_{-\infty}^{\infty} |H(f)|^2\, df} \tag{4.8}$$

Our problem is to find, for a given $G(f)$, the particular form of the frequency response $H(f)$ of the filter that makes $\eta$ a maximum. To find the solution to this optimization problem, we apply a mathematical result known as Schwarz's inequality to the numerator of Equation (4.8).

A derivation of *Schwarz's inequality* is given in Chapter 5. For now it suffices to say that if we have two complex functions $\phi_1(x)$ and $\phi_2(x)$ in the real variable $x$, satisfying the conditions

$$\int_{-\infty}^{\infty} |\phi_1(x)|^2\, dx < \infty$$

and

$$\int_{-\infty}^{\infty} |\phi_2(x)|^2\, dx < \infty$$

then we may write

$$\left| \int_{-\infty}^{\infty} \phi_1(x)\phi_2(x)\, dx \right|^2 \leq \int_{-\infty}^{\infty} |\phi_1(x)|^2\, dx \int_{-\infty}^{\infty} |\phi_2(x)|^2\, dx \tag{4.9}$$

The equality in (4.9) holds if, and only if, we have

$$\phi_1(x) = k\phi_2^*(x) \tag{4.10}$$

where $k$ is an arbitrary constant, and the asterisk denotes complex conjugation.

Returning to the problem at hand, we readily see that by invoking Schwarz's inequality (4.9), and setting $\phi_1(x) = H(f)$ and $\phi_2(x) = G(f)\exp(j\pi fT)$, the numerator in Equation (4.8) may be rewritten as

$$\left| \int_{-\infty}^{\infty} H(f)G(f)\exp(j2\pi fT)\, df \right|^2 \leq \int_{-\infty}^{\infty} |H(f)|^2\, df \int_{-\infty}^{\infty} |G(f)|^2\, df \tag{4.11}$$

Using this relation in Equation (4.8), we may redefine the peak pulse signal-to-noise ratio as

$$\eta \leq \frac{2}{N_0} \int_{-\infty}^{\infty} |G(f)|^2\, df \tag{4.12}$$

The right-hand side of this relation does not depend on the frequency response $H(f)$ of the filter but only on the signal energy and the noise power spectral density. Consequently, the peak pulse signal-to-noise ratio $\eta$ will be a maximum when $H(f)$ is chosen so that the equality holds; that is,

$$\eta_{\max} = \frac{2}{N_0} \int_{-\infty}^{\infty} |G(f)|^2\, df \tag{4.13}$$

Correspondingly, $H(f)$ assumes its optimum value denoted by $H_{\text{opt}}(f)$. To find this optimum value we use Equation (4.10), which, for the situation at hand, yields

$$H_{\text{opt}}(f) = kG^*(f)\exp(-j2\pi fT) \tag{4.14}$$

where $G^*(f)$ is the complex conjugate of the Fourier transform of the input signal $g(t)$, and $k$ is a scaling factor of appropriate dimensions. This relation states that, except for the factor $k\exp(-j2\pi fT)$, the frequency response of the optimum filter is the same as the complex conjugate of the Fourier transform of the input signal.

Equation (4.14) specifies the optimum filter in the frequency domain. To characterize it in the time domain, we take the inverse Fourier transform of $H_{\text{opt}}(f)$ in Equation (4.14) to obtain the impulse response of the optimum filter as

$$h_{\text{opt}}(t) = k \int_{-\infty}^{\infty} G^*(f)\exp[-j2\pi f(T - t)]\, df \tag{4.15}$$

Since for a real signal $g(t)$ we have $G^*(f) = G(-f)$, we may rewrite Equation (4.15) as

$$\begin{aligned} h_{\text{opt}}(t) &= k \int_{-\infty}^{\infty} G(-f)\exp[-j2\pi f(T - t)]\, df \\ &= k \int_{-\infty}^{\infty} G(f)\exp[j2\pi f(T - t)]\, df \\ &= kg(T - t) \end{aligned} \tag{4.16}$$

Equation (4.16) shows that the impulse response of the optimum filter, except for the scaling factor $k$, is a time-reversed and delayed version of the input signal $g(t)$; that is, it is "matched" to the input signal. A linear time-invariant filter defined in this way is called a matched filter. Note that in deriving the matched filter the only assumption we have made about the input noise $w(t)$ is that it is stationary and white with zero mean and power spectral density $N_0/2$. In other words, no assumption was made on the statistics of the channel noise $w(t)$.

## Properties of Matched Filters

We note that a filter, which is matched to a pulse signal $g(t)$ of duration $T$, is characterized by an impulse response that is a time-reversed and delayed version of the input $g(t)$, as shown by

$$h_{\text{opt}}(t) = kg(T - t)$$

In other words, the impulse response $h_{\text{opt}}(t)$ is uniquely defined, except for the delay $T$ and the scaling factor $k$, by the waveform of the pulse signal $g(t)$ to which the filter is matched. In the frequency domain, the matched filter is characterized by a frequency response that is, except for a delay factor, the complex conjugate of the Fourier transform of the input $g(t)$, as shown by

$$H_{\text{opt}}(f) = kG^*(f)\exp(-j2\pi fT)$$

The most important result in the calculation of the performance of signal processing systems using matched filters is perhaps the following:

*The peak pulse signal-to-noise ratio of a matched filter depends only on the ratio of the signal energy to the power spectral density of the white noise at the filter input.*

To demonstrate this property, consider a filter matched to a known signal $g(t)$. The Fourier transform of the resulting matched filter output $g_o(t)$ is

$$\begin{aligned} G_o(f) &= H_{\text{opt}}(f)G(f) \\ &= kG^*(f)G(f)\exp(-j2\pi fT) \\ &= k|G(f)|^2\exp(-j2\pi fT) \end{aligned} \tag{4.17}$$

Using Equation (4.17) in the formula for the inverse Fourier transform, we find that the matched filter output at time $t = T$ is

$$\begin{aligned} g_o(T) &= \int_{-\infty}^{\infty} G_o(f)\exp(j2\pi fT)\,df \\ &= k\int_{-\infty}^{\infty} |G(f)|^2\,df \end{aligned}$$

According to *Rayleigh's energy theorem, the integral of the squared magnitude spectrum of a pulse signal with respect to frequency is equal to the signal energy E*:

$$E = \int_{-\infty}^{\infty} g^2(t)dt = \int_{-\infty}^{\infty} |G(f)|^2\,df$$

Hence

$$g_o(T) = kE \tag{4.18}$$

Substituting Equation (4.14) into (4.7), we find that the average output noise power is

$$\begin{aligned} E[n^2(t)] &= \frac{k^2N_0}{2}\int_{-\infty}^{\infty} |G(f)|^2\,df \\ &= k^2N_0E/2 \end{aligned} \tag{4.19}$$

where again we have made use of Rayleigh's energy theorem. Therefore, the peak pulse signal-to-noise ratio has the maximum value

$$\eta_{\max} = \frac{(kE)^2}{(k^2 N_0 E/2)} = \frac{2E}{N_0} \tag{4.20}$$

From Equation (4.20) we see that dependence on the waveform of the input $g(t)$ has been completely removed by the matched filter. Accordingly, in evaluating the ability of a matched-filter receiver to combat additive white noise, we find that all signals that have the same energy are equally effective. Note that the signal energy $E$ is in joules and the noise spectral density $N_0/2$ is in watts per Hertz, so that the ratio $2E/N_0$ is dimensionless; however, the two quantities have different physical meaning. We refer to $E/N_0$ as the *signal energy-to-noise spectral density ratio.*

### ▶ EXAMPLE 4.1 Matched Filter for Rectangular Pulse

Consider a signal $g(t)$ in the form of a rectangular pulse of amplitude $A$ and duration $T$, as shown in Figure 4.2*a*. In this example, the impulse response $h(t)$ of the matched filter has exactly the same waveform as the signal itself. The output signal $g_o(t)$ of the matched filter produced in response to the input signal $g(t)$ has a triangular waveform, as shown in Figure 4.2*b*.

The maximum value of the output signal $g_o(t)$ is equal to $kA^2T$, which is the energy of the input signal $g(t)$ scaled by the factor $k$; this maximum value occurs at $t = T$, as indicated in Figure 4.2*b*.

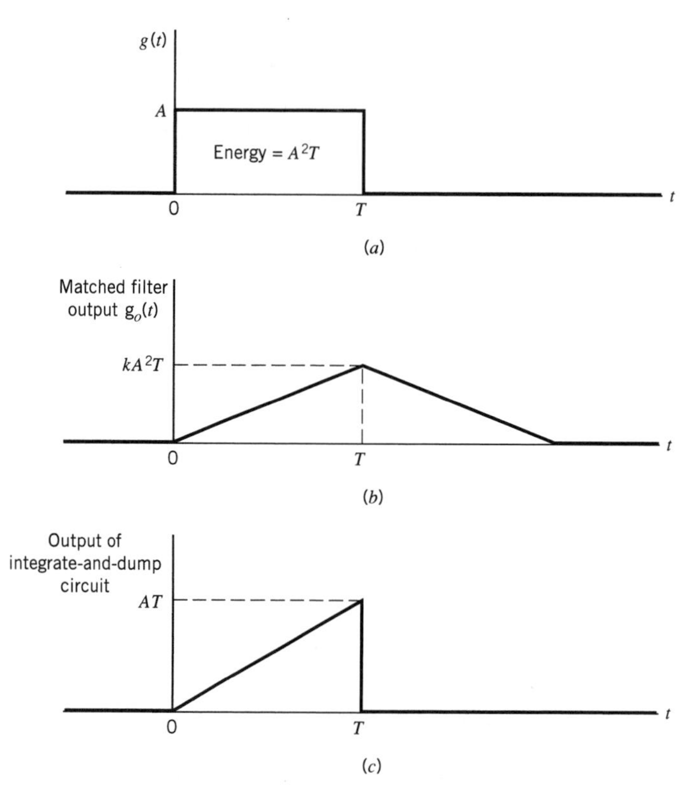


**FIGURE 4.2** (*a*) Rectangular pulse. (*b*) Matched filter output. (*c*) Integrator output.

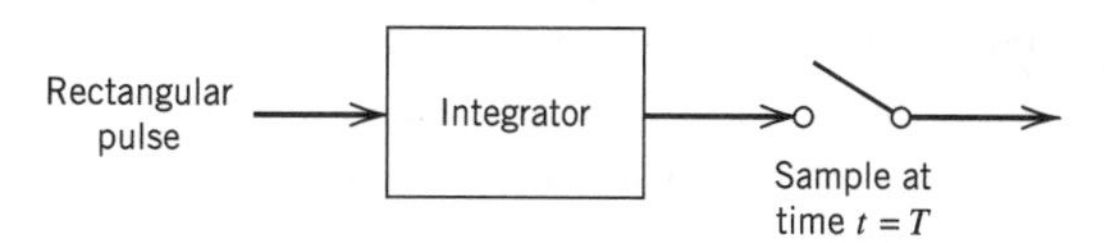


**FIGURE 4.3** Integrate-and-dump circuit.

For the special case of a rectangular pulse, the matched filter may be implemented using a circuit known as the *integrate-and-dump circuit*, a block diagram of which is shown in Figure 4.3. The integrator computes the area under the rectangular pulse, and the resulting output is then sampled at time $t = T$, where $T$ is the duration of the pulse. Immediately after $t = T$, the integrator is restored to its initial condition; hence the name of the circuit. Figure 4.2*c* shows the output waveform of the integrate-and-dump circuit for the rectangular pulse of Figure 4.2*a*. We see that for $0 \leq t \leq T$, the output of this circuit has the *same waveform* as that appearing at the output of the matched filter; the difference in the notations used to describe their peak values is of no practical significance. ◀

## 4.3 *Error Rate Due to Noise*

In Section 3.8 we presented a qualitative discussion of the effect of channel noise on the performance of a binary PCM system. Now that we are equipped with the matched filter as the optimum detector of a known pulse in additive white noise, we are ready to derive a formula for the error rate in such a system due to noise.

To proceed with the analysis, consider a binary PCM system based on *polar non-return-to-zero (NRZ) signaling*. In this form of signaling, symbols 1 and 0 are represented by positive and negative rectangular pulses of equal amplitude and equal duration. The channel noise is modeled as *additive white Gaussian noise* $w(t)$ of zero mean and power spectral density $N_0/2$; the Gaussian assumption is needed for later calculations. In the signaling interval $0 \leq t \leq T_b$, the received signal is thus written as follows:

$$x(t) = \begin{cases} +A + w(t), & \text{symbol 1 was sent} \\ -A + w(t), & \text{symbol 0 was sent} \end{cases} \tag{4.21}$$

where $T_b$ is the *bit duration*, and $A$ is the *transmitted pulse amplitude*. It is assumed that the receiver has acquired knowledge of the starting and ending times of each transmitted pulse; in other words, the receiver has prior knowledge of the pulse shape, but not its polarity. Given the noisy signal $x(t)$, the receiver is required to make a decision in each signaling interval as to whether the transmitted symbol is a 1 or a 0.

The structure of the receiver used to perform this decision-making process is shown in Figure 4.4. It consists of a matched filter followed by a sampler, and then finally a

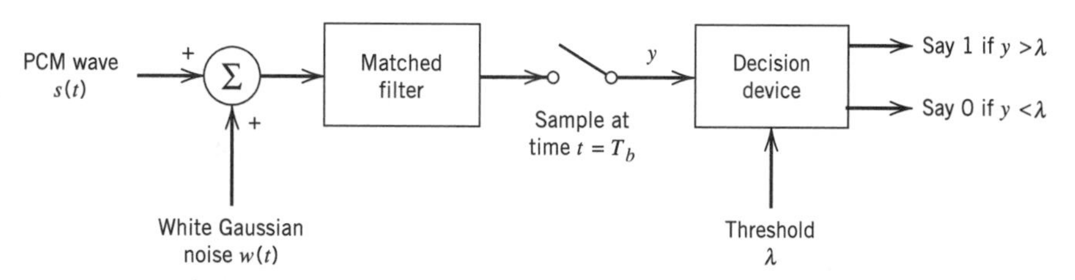


**FIGURE 4.4** Receiver for baseband transmission of binary-encoded PCM wave using polar NRZ signaling.

decision device. The filter is matched to a rectangular pulse of amplitude $A$ and duration $T_b$, exploiting the bit-timing information available to the receiver. The resulting matched filter output is sampled at the end of each signaling interval. The presence of channel noise $w(t)$ adds randomness to the matched filter output.

Let $y$ denote the sample value obtained at the end of a signaling interval. The sample value $y$ is compared to a preset *threshold* $\lambda$ in the decision device. If the threshold is exceeded, the receiver makes a decision in favor of symbol 1; if not, a decision is made in favor of symbol 0. We adopt the convention that when the sample value $y$ is exactly equal to the threshold $\lambda$, the receiver just makes a guess as to which symbol was transmitted; such a decision is the same as that obtained by flipping a fair coin, the outcome of which will not alter the average probability of error.

There are two possible kinds of error to be considered:

1. Symbol 1 is chosen when a 0 was actually transmitted; we refer to this error as an *error of the first kind.*
2. Symbol 0 is chosen when a 1 was actually transmitted; we refer to this error as an *error of the second kind.*

To determine the average probability of error, we consider these two situations separately.

Suppose that symbol 0 was sent. Then, according to Equation (4.21), the received signal is

$$x(t) = -A + w(t), \qquad 0 \le t \le T_b \tag{4.22}$$

Correspondingly, the matched filter output, sampled at time $t = T_b$, is given by (in light of Example 4.1 with $kAT_b$ set equal to unity for convenience of presentation)

$$\begin{aligned} y &= \int_0^{T_b} x(t)\,dt \\ &= -A + \frac{1}{T_b}\int_0^{T_b} w(t)\,dt \end{aligned} \tag{4.23}$$

which represents the sample value of a random variable $Y$. By virtue of the fact that the noise $w(t)$ is white and Gaussian, we may characterize the random variable $Y$ as follows:

- The random variable $Y$ is Gaussian distributed with a mean of $-A$.
- The variance of the random variable $Y$ is

$$\begin{aligned} \sigma_Y^2 &= E[(Y + A)^2] \\ &= \frac{1}{T_b^2}E\left[\int_0^{T_b}\int_0^{T_b} w(t)w(u)\,dt\,du\right] \\ &= \frac{1}{T_b^2}\int_0^{T_b}\int_0^{T_b} E[w(t)w(u)]\,dt\,du \\ &= \frac{1}{T_b^2}\int_0^{T_b}\int_0^{T_b} R_W(t, u)\,dt\,du \end{aligned} \tag{4.24}$$

where $R_W(t, u)$ is the autocorrelation function of the white noise $w(t)$. Since $w(t)$ is white with a power spectral density $N_0/2$, we have

$$R_W(t, u) = \frac{N_0}{2}\,\delta(t - u) \tag{4.25}$$

where $\delta(t - u)$ is a time-shifted delta function. Hence, substituting Equation (4.25) into (4.24) yields

$$\begin{aligned}\sigma_Y^2 &= \frac{1}{T_b^2}\int_0^{T_b}\int_0^{T_b}\frac{N_0}{2}\,\delta(t-u)\,dt\,du \\ &= \frac{N_0}{2T_b}\end{aligned} \tag{4.26}$$

where we have used the sifting property of the delta function and the fact that its area is unity. The conditional probability density function of the random variable $Y$, given that symbol 0 was sent, is therefore

$$f_Y(y\,|\,0) = \frac{1}{\sqrt{\pi N_0/T_b}}\exp\left(-\frac{(y+A)^2}{N_0/T_b}\right) \tag{4.27}$$

This function is plotted in Figure 4.5($a$). Let $p_{10}$ denote the *conditional probability of error, given that symbol 0 was sent.* This probability is defined by the shaded area under the curve of $f_Y(y\,|\,0)$ from the threshold $\lambda$ to infinity, which corresponds to the range of values assumed by $y$ for a decision in favor of symbol 1. In the absence of noise, the matched filter output $y$ sampled at time $t = T_b$ is equal to $-A$. When noise is present, $y$ occasionally assumes a value greater than $\lambda$, in which case an error is made. The probability of this error, conditional on sending symbol 0, is defined by

$$\begin{aligned}p_{10} &= P(y > \lambda\,|\,\text{symbol 0 was sent}) \\ &= \int_\lambda^\infty f_Y(y\,|\,0)\,dy \\ &= \frac{1}{\sqrt{\pi N_0/T_b}}\int_\lambda^\infty \exp\left(-\frac{(y+A)^2}{N_0/T_b}\right)dy\end{aligned} \tag{4.28}$$

At this point in the discussion we digress briefly and introduce the definition of the so-called *complementary error function:*[3]

$$\operatorname{erfc}(u) = \frac{2}{\sqrt{\pi}}\int_u^\infty \exp(-z^2)\,dz \tag{4.29}$$

which is closely related to the Gaussian distribution. For large positive values of $u$, we have the following *upper bound* on the complementary error function:

$$\operatorname{erfc}(u) < \frac{\exp(-u^2)}{\sqrt{\pi}u} \tag{4.30}$$

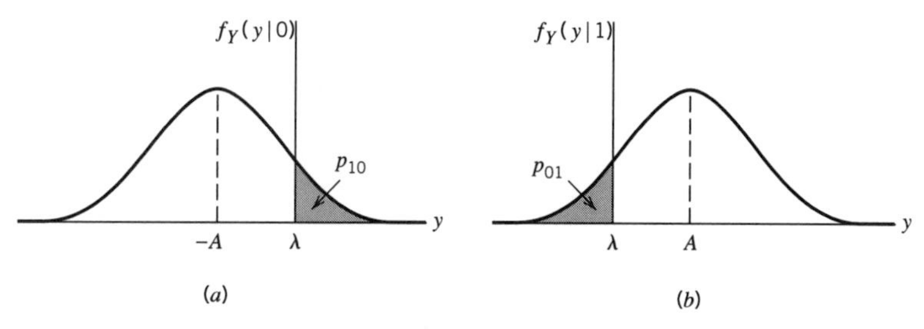


**FIGURE 4.5** Noise analysis of PCM system. ($a$) Probability density function of random variable $Y$ at matched filter output when 0 is transmitted. ($b$) Probability density function of $Y$ when 1 is transmitted.

To reformulate the conditional probability of error $p_{10}$ in terms of the complementary error function, we first define a new variable

$$z = \frac{y + A}{\sqrt{N_0/T_b}}$$

Accordingly, we may rewrite Equation (4.28) in the compact form

$$\begin{aligned} p_{10} &= \frac{1}{\sqrt{\pi}} \int_{(A+\lambda)/\sqrt{N_0/T_b}}^{\infty} \exp(-z^2)\, dz \\ &= \frac{1}{2} \operatorname{erfc}\left( \frac{A + \lambda}{\sqrt{N_0/T_b}} \right) \end{aligned} \tag{4.31}$$

Assume next that symbol 1 was transmitted. This time the Gaussian random variable $Y$ represented by the sample value $y$ of the matched filter output has a mean $+A$ and variance $N_0/2T_b$. Note that, compared to the situation when symbol 0 was sent, the mean of the random variable $Y$ has changed, but its variance is exactly the same as before. The conditional probability density function of $Y$, given that symbol 1 was sent, is therefore

$$f_Y(y|1) = \frac{1}{\sqrt{\pi N_0/T_b}} \exp\left( -\frac{(y - A)^2}{N_0/T_b} \right) \tag{4.32}$$

which is plotted in Figure 4.5*b*. Let $p_{01}$ denote the *conditional probability of error, given that symbol 1 was sent.* This probability is defined by the shaded area under the curve of $f_Y(y|1)$ extending from $-\infty$ to the threshold $\lambda$, which corresponds to the range of values assumed by $y$ for a decision in favor of symbol 0. In the absence of noise, the matched filter output $y$ sampled at time $t = T_b$ is equal to $+A$. When noise is present, $y$ occasionally assumes a value less than $\lambda$, and an error is then made. The probability of this error, conditional on sending symbol 1, is defined by

$$\begin{aligned} p_{01} &= P(y < \lambda | \text{symbol 1 was sent}) \\ &= \int_{-\infty}^{\lambda} f_Y(y|1)\, dy \\ &= \frac{1}{\sqrt{\pi N_0/T_b}} \int_{-\infty}^{\lambda} \exp\left( -\frac{(y - A)^2}{N_0/T_b} \right) dy \end{aligned} \tag{4.33}$$

To express $p_{01}$ in terms of the complementary error function, this time we define a new variable

$$z = \frac{A - y}{\sqrt{N_0/T_b}}$$

Accordingly, we may reformulate Equation (4.33) in the compact form

$$\begin{aligned} p_{01} &= \frac{1}{\sqrt{\pi}} \int_{(A-\lambda)/\sqrt{N_0/T_b}}^{\infty} \exp(-z^2)\, dz \\ &= \frac{1}{2} \operatorname{erfc}\left( \frac{A - \lambda}{\sqrt{N_0/T_b}} \right) \end{aligned} \tag{4.34}$$

Having determined the conditional probabilities of error, $p_{10}$ and $p_{01}$, our next task is to derive the formula for the *average probability of symbol error*, denoted by $P_e$. Here we note that these two possible kinds of error are mutually exclusive events in that if the receiver, at a particular sampling instant, chooses symbol 1, then symbol 0 is excluded from appearing, and vice versa. Let $p_0$ and $p_1$ denote the *a priori* probabilities of transmitting symbols 0 and 1, respectively. Hence, the *average probability of symbol error* $P_e$ in the receiver is given by

$$\begin{aligned} P_e &= p_0 p_{10} + p_1 p_{01} \\ &= \frac{p_0}{2}\,\text{erfc}\left(\frac{A+\lambda}{\sqrt{N_0/T_b}}\right) + \frac{p_1}{2}\,\text{erfc}\left(\frac{A-\lambda}{\sqrt{N_0/T_b}}\right) \end{aligned} \tag{4.35}$$

From Equation (4.35) we see that $P_e$ is in fact a function of the threshold $\lambda$, which immediately suggests the need for formulating an *optimum threshold* that minimizes $P_e$. For this optimization we use *Leibniz's rule.*

Consider the integral

$$\int_{a(u)}^{b(u)} f(z, u)\, dz$$

*Leibniz's rule* states that the derivative of this integral with respect to $u$ is

$$\frac{d}{du}\int_{a(u)}^{b(u)} f(z, u)\, dz = f(b(u), u)\frac{db(u)}{du} - f(a(u), u)\frac{d(a(u)}{du} + \int_{a(u)}^{b(u)} \frac{\partial f(z, u)}{\partial u}\, dz$$

For the problem at hand, we note from the definition of the complementary error function in Equation (4.29) that

$$\begin{aligned} f(z, u) &= \frac{2}{\sqrt{\pi}}\exp(-z^2) \\ a(u) &= u \\ b(u) &= \infty \end{aligned}$$

The application of Leibniz's rule to the complementary error function thus yields

$$\frac{d}{du}\,\text{erfc}(u) = -\frac{1}{\sqrt{\pi}}\exp(-u^2) \tag{4.36}$$

Hence, differentiating Equation (4.35) with respect to $\lambda$ by making use of the formula in Equation (4.36), then setting the result equal to zero and simplifying terms, we obtain the optimum threshold as

$$\lambda_{\text{opt}} = \frac{N_0}{4AT_b}\log\left(\frac{p_0}{p_1}\right) \tag{4.37}$$

For the special case when symbols 1 and 0 are equiprobable, we have

$$p_1 = p_0 = \frac{1}{2}$$

in which case Equation (4.37) reduces to

$$\lambda_{\text{opt}} = 0$$

This result is intuitively satisfying as it states that, for the transmission of equiprobable binary symbols, we should choose the threshold at the midpoint between the pulse heights $-A$ and $+A$ representing the two symbols 0 and 1. Note that for this special case we also have

$$p_{01} = p_{10}$$

A channel for which the conditional probabilities of error $p_{01}$ and $p_{10}$ are equal is said to be *binary symmetric*. Correspondingly, the average probability of symbol error in Equation (4.35) reduces to

$$P_e = \frac{1}{2}\,\text{erfc}\left(\frac{A}{\sqrt{N_0/T_b}}\right) \tag{4.38}$$

Now the *transmitted signal energy per bit* is defined by

$$E_b = A^2 T_b \tag{4.39}$$

Accordingly, we may finally formulate the average probability of symbol error for the receiver in Figure 4.4 as

$$P_e = \frac{1}{2}\,\text{erfc}\left(\sqrt{\frac{E_b}{N_0}}\right) \tag{4.40}$$

which shows that *the average probability of symbol error in a binary symmetric channel depends solely on $E_b/N_0$, the ratio of the transmitted signal energy per bit to the noise spectral density.*

Using the upper bound of Equation (4.30) on the complementary error function, we may correspondingly bound the average probability of symbol error for the PCM receiver as

$$P_e < \frac{\exp(-E_b/N_0)}{2\sqrt{\pi E_b/N_0}} \tag{4.41}$$

The PCM receiver of Figure 4.4 therefore exhibits an *exponential* improvement in the average probability of symbol error with increase in $E_b/N_0$.

This important result is further illustrated in Figure 4.6 where the average probability of symbol error $P_e$ is plotted versus the dimensionless ratio $E_b/N_0$. In particular, we see that $P_e$ decreases very rapidly as the ratio $E_b/N_0$ is increased, so that eventually a very "small increase" in transmitted signal energy will make the reception of binary pulses almost error free, as discussed previously in Section 3.8. Note, however, that in practical terms the increase in signal energy has to be viewed in the context of the bias; for example,

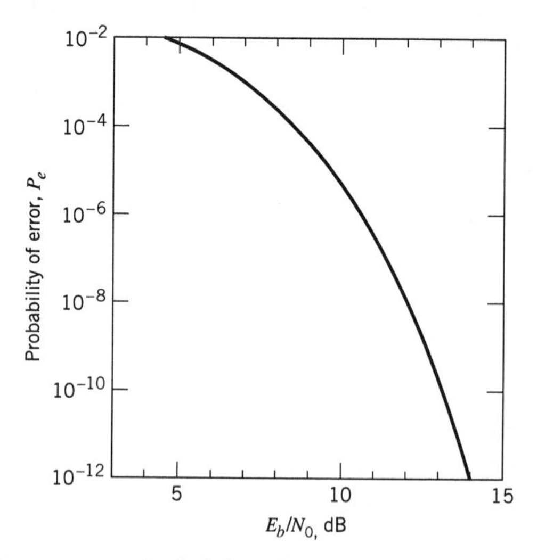


**FIGURE 4.6** Probability of error in a PCM receiver.

a 3-dB increase in $E_b/N_0$ is much easier to implement when $E_b$ has a small value than when its value is orders of magnitude larger.

## 4.4 *Intersymbol Interference*

The next source of bit errors in a baseband-pulse transmission system that we wish to study is intersymbol interference (ISI), which arises when the communication channel is *dispersive*. First of all, however, we need to address a key question: Given a pulse shape of interest, how do we use it to transmit data in $M$-ary form? The answer lies in the use of *discrete pulse modulation*, in which the amplitude, duration, or position of the transmitted pulses is varied in a discrete manner in accordance with the given data stream. However, for the baseband transmission of digital data, the use of *discrete pulse-amplitude modulation* (PAM) is one of the most efficient schemes in terms of power and bandwidth utilization. Accordingly, we confine our attention to discrete PAM systems. We begin the study by first considering the case of binary data; later in the chapter, we consider the more general case of $M$-ary data.

Consider then a *baseband binary PAM system*, a generic form of which is shown in Figure 4.7. The incoming binary sequence $\{b_k\}$ consists of symbols 1 and 0, each of duration $T_b$. The *pulse-amplitude modulator* modifies this binary sequence into a new sequence of short pulses (approximating a unit impulse), whose amplitude $a_k$ is represented in the polar form

$$a_k = \begin{cases} +1 & \text{if symbol } b_k \text{ is } 1 \\ -1 & \text{if symbol } b_k \text{ is } 0 \end{cases} \tag{4.42}$$

The sequence of short pulses so produced is applied to a *transmit filter* of impulse response $g(t)$, producing the transmitted signal

$$s(t) = \sum_k a_k g(t - kT_b) \tag{4.43}$$

The signal $s(t)$ is modified as a result of transmission through the *channel* of impulse response $h(t)$. In addition, the channel adds random noise to the signal at the receiver input. The noisy signal $x(t)$ is then passed through a *receive filter* of impulse response $c(t)$. The resulting filter output $y(t)$ is sampled *synchronously* with the transmitter, with the sampling instants being determined by a *clock* or *timing signal* that is usually extracted from the receive filter output. Finally, the sequence of samples thus obtained is used to reconstruct the original data sequence by means of a *decision device*. Specifically, the amplitude of each sample is compared to a *threshold* $\lambda$. If the threshold $\lambda$ is exceeded, a decision is made in favor of symbol 1. If the threshold $\lambda$ is not exceeded, a decision is made in favor of symbol 0. If the sample amplitude equals the threshold exactly, the flip of a

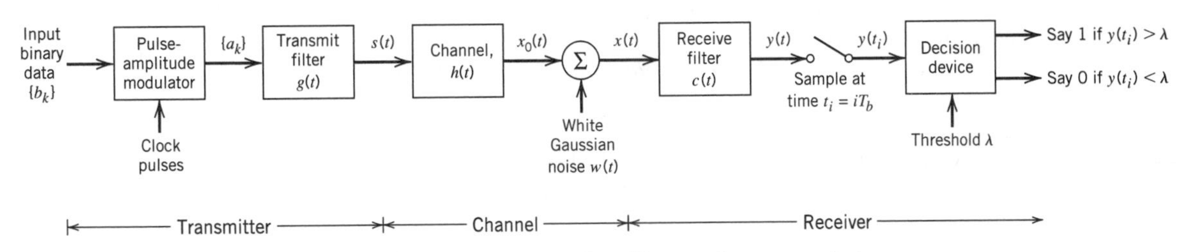


**FIGURE 4.7** Baseband binary data transmission system.

fair coin will determine which symbol was transmitted (i.e., the receiver simply makes a random guess).

The receive filter output is written as

$$y(t) = \mu \sum_{k} a_k p(t - kT_b) + n(t) \tag{4.44}$$

where $\mu$ is a scaling factor, and the pulse $p(t)$ is to be defined. To be precise, an arbitrary time delay $t_0$ should be included in the argument of the pulse $p(t - kT_b)$ in Equation (4.44) to represent the effect of transmission delay through the system. To simplify the exposition, we have put this delay equal to zero in Equation (4.44) without loss of generality.

The scaled pulse $\mu p(t)$ is obtained by a double convolution involving the impulse response $g(t)$ of the transmit filter, the impulse response $h(t)$ of the channel, and the impulse response $c(t)$ of the receive filter, as shown by

$$\mu p(t) = g(t) \star h(t) \star c(t) \tag{4.45}$$

where the star denotes convolution. We assume that the pulse $p(t)$ is *normalized* by setting

$$p(0) = 1 \tag{4.46}$$

which justifies the use of $\mu$ as a scaling factor to account for amplitude changes incurred in the course of signal transmission through the system.

Since convolution in the time domain is transformed into multiplication in the frequency domain, we may use the Fourier transform to change Equation (4.45) into the equivalent form

$$\mu P(f) = G(f)H(f)C(f) \tag{4.47}$$

where $P(f)$, $G(f)$, $H(f)$, and $C(f)$ are the Fourier transforms of $p(t)$, $g(t)$, $h(t)$, and $c(t)$, respectively.

Finally, the term $n(t)$ in Equation (4.44) is the noise produced at the output of the receive filter due to the channel noise $w(t)$. It is customary to model $w(t)$ as a white Gaussian noise of zero mean.

The receive filter output $y(t)$ is sampled at time $t_i = iT_b$ (with $i$ taking on integer values), yielding [in light of Equation (4.46)]

$$\begin{aligned} y(t_i) &= \mu \sum_{k=-\infty}^{\infty} a_k p[(i - k)T_b] + n(t_i) \\ &= \mu a_i + \mu \sum_{\substack{k=-\infty \\ k \neq i}}^{\infty} a_k p[(i - k)T_b] + n(t_i) \end{aligned} \tag{4.48}$$

In Equation (4.48), the first term $\mu a_i$ represents the contribution of the $i$th transmitted bit. The second term represents the residual effect of all other transmitted bits on the decoding of the $i$th bit; this residual effect due to the occurrence of pulses before and after the sampling instant $t_i$ is called intersymbol interference (ISI). The last term $n(t_i)$ represents the noise sample at time $t_i$.

In the absence of both ISI and noise, we observe from Equation (4.48) that

$$y(t_i) = \mu a_i$$

which shows that, under these ideal conditions, the $i$th transmitted bit is decoded correctly. The unavoidable presence of ISI and noise in the system, however, introduces errors in the decision device at the receiver output. Therefore, in the design of the transmit and receive filters, the objective is to minimize the effects of noise and ISI and thereby deliver the digital data to their destination with the smallest error rate possible.

When the signal-to-noise ratio is high, as is the case in a telephone system, for example, the operation of the system is largely limited by ISI rather than noise; in other words, we may ignore $n(t_i)$. In the next couple of sections, we assume that this condition holds so that we may focus our attention on ISI and the techniques for its control. In particular, the issue we wish to consider is to determine the pulse waveform $p(t)$ for which the ISI is completely eliminated.

## 4.5 Nyquist's Criterion for Distortionless Baseband Binary Transmission

Typically, the frequency response of the channel and the transmitted pulse shape are specified, and the problem is to determine the frequency responses of the transmit and receive filters so as to reconstruct the original binary data sequence $\{b_k\}$. The receiver does this by *extracting* and then *decoding* the corresponding sequence of coefficients, $\{a_k\}$, from the output $y(t)$. The *extraction* involves sampling the output $y(t)$ at time $t = iT_b$. The *decoding* requires that the weighted pulse contribution $a_k p(iT_b - kT_b)$ for $k = i$ be *free* from ISI due to the overlapping tails of all other weighted pulse contributions represented by $k \neq i$. This, in turn, requires that we *control* the overall pulse $p(t)$, as shown by

$$p(iT_b - kT_b) = \begin{cases} 1, & i = k \\ 0, & i \neq k \end{cases} \tag{4.49}$$

where $p(0) = 1$, by normalization. If $p(t)$ satisfies the conditions of Equation (4.49), the receiver output $y(t_i)$ given in Equation (4.48) simplifies to (ignoring the noise term)

$$y(t_i) = \mu a_i \qquad \text{for all } i$$

which implies zero intersymbol interference. Hence, the two conditions of Equation (4.49) ensure *perfect reception in the absence of noise.*

From a design point of view, it is informative to transform the conditions of Equation (4.49) into the frequency domain. Consider then the sequence of samples $\{p(nT_b)\}$, where $n = 0, \pm 1, \pm 2, \cdots$. From the discussion presented in Chapter 3 on the sampling process, we recall that sampling in the time domain produces periodicity in the frequency domain. In particular, we may write

$$P_\delta(f) = R_b \sum_{n=-\infty}^{\infty} P(f - nR_b) \tag{4.50}$$

where $R_b = 1/T_b$ is the *bit rate* in bits per second (b/s); $P_\delta(f)$ is the Fourier transform of an infinite periodic sequence of delta functions of period $T_b$, whose individual areas are weighted by the respective sample values of $p(t)$. That is, $P_\delta(f)$ is given by

$$P_\delta(f) = \int_{-\infty}^{\infty} \sum_{m=-\infty}^{\infty} [p(mT_b)\, \delta(t - mT_b)] \exp(-j2\pi f t)\, dt \tag{4.51}$$

Let the integer $m = i - k$. Then, $i = k$ corresponds to $m = 0$, and likewise $i \neq k$ corresponds to $m \neq 0$. Accordingly, imposing the conditions of Equation (4.49) on the sample values of $p(t)$ in the integral of Equation (4.51), we get

$$\begin{aligned} P_\delta(f) &= \int_{-\infty}^{\infty} p(0)\, \delta(t) \exp(-j2\pi f t)\, dt \\ &= p(0) \end{aligned} \tag{4.52}$$

where we have made use of the sifting property of the delta function. Since from Equation (4.46) we have $p(0) = 1$, it follows from Equations (4.50) and (4.52) that the condition for zero intersymbol interference is satisfied if

$$\sum_{n=-\infty}^{\infty} P(f - nR_b) = T_b \tag{4.53}$$

We may now state the *Nyquist criterion*[4] *for distortionless baseband transmission* in the absence of noise: *The frequency function $P(f)$ eliminates intersymbol interference for samples taken at intervals $T_b$ provided that it satisfies Equation (4.53).* Note that $P(f)$ refers to the overall system, incorporating the transmit filter, the channel, and the receive filter in accordance with Equation (4.47).

## ■ IDEAL NYQUIST CHANNEL

The simplest way of satisfying Equation (4.53) is to specify the frequency function $P(f)$ to be in the form of a *rectangular function*, as shown by

$$\begin{aligned} P(f) &= \begin{cases} \dfrac{1}{2W}, & -W < f < W \\ 0, & |f| > W \end{cases} \\ &= \frac{1}{2W} \operatorname{rect}\left(\frac{f}{2W}\right) \end{aligned} \tag{4.54}$$

where rect$(f)$ stands for a *rectangular function* of unit amplitude and unit support centered on $f = 0$, and the overall system bandwidth $W$ is defined by

$$W = \frac{R_b}{2} = \frac{1}{2T_b} \tag{4.55}$$

According to the solution described by Equations (4.54) and (4.55), no frequencies of absolute value exceeding half the bit rate are needed. Hence, from Fourier-transform pair 2 of Table A6.3 we find that a signal waveform that produces zero intersymbol interference is defined by the *sinc function*:

$$\begin{aligned} p(t) &= \frac{\sin(2\pi Wt)}{2\pi Wt} \\ &= \operatorname{sinc}(2Wt) \end{aligned} \tag{4.56}$$

The special value of the bit rate $R_b = 2W$ is called the *Nyquist rate*, and $W$ is itself called the *Nyquist bandwidth*. Correspondingly, the ideal baseband pulse transmission system described by Equation (4.54) in the frequency domain or, equivalently, Equation (4.56) in the time domain, is called the *ideal Nyquist channel*.

Figures 4.8*a* and 4.8*b* show plots of $P(f)$ and $p(t)$, respectively. In Figure 4.8*a*, the normalized form of the frequency function $P(f)$ is plotted for positive and negative frequencies. In Figure 4.8*b*, we have also included the signaling intervals and the corresponding centered sampling instants. The function $p(t)$ can be regarded as the impulse response of an ideal low-pass filter with passband magnitude response $1/2W$ and bandwidth $W$. The function $p(t)$ has its peak value at the origin and goes through zero at integer multiples of the bit duration $T_b$. It is apparent that if the received waveform $y(t)$ is sampled at the

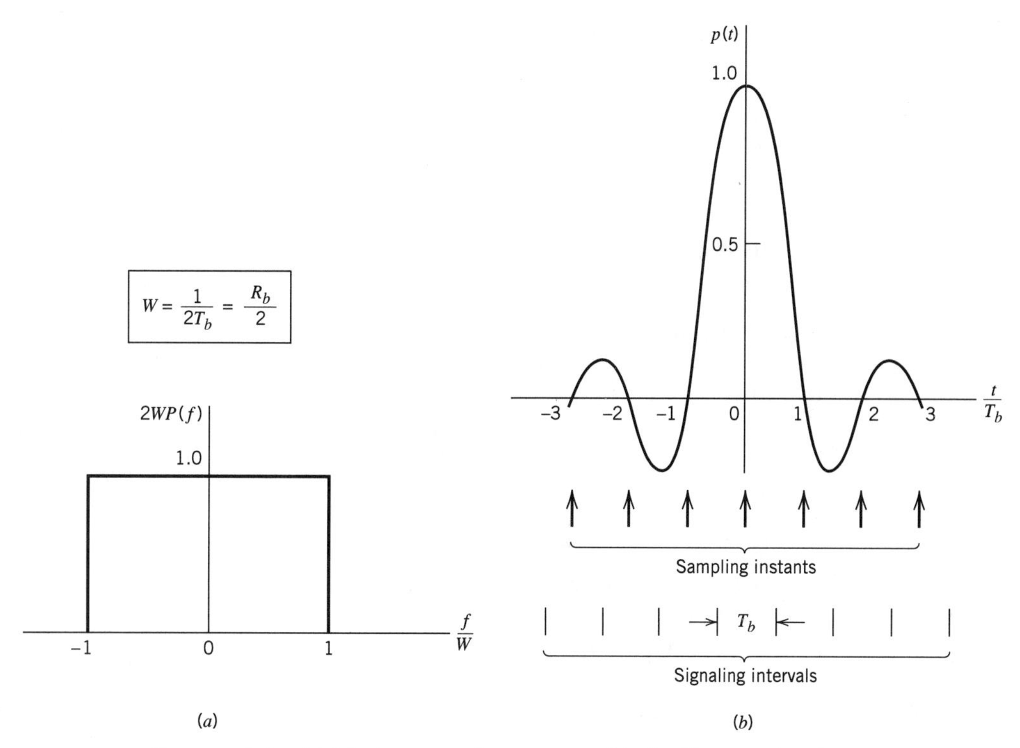


**FIGURE 4.8** (*a*) Ideal magnitude response. (*b*) Ideal basic pulse shape.

instants of time $t = 0, \pm T_b, \pm 2T_b, \cdots$, then the pulses defined by $\mu p(t - iT_b)$ with arbitrary amplitude $\mu$ and index $i = 0, \pm 1, \pm 2, \cdots$, will not interfere with each other. This condition is illustrated in Figure 4.9 for the binary sequence 1011010.

Although the use of the ideal Nyquist channel does indeed achieve economy in bandwidth in that it solves the problem of zero intersymbol interference with the minimum

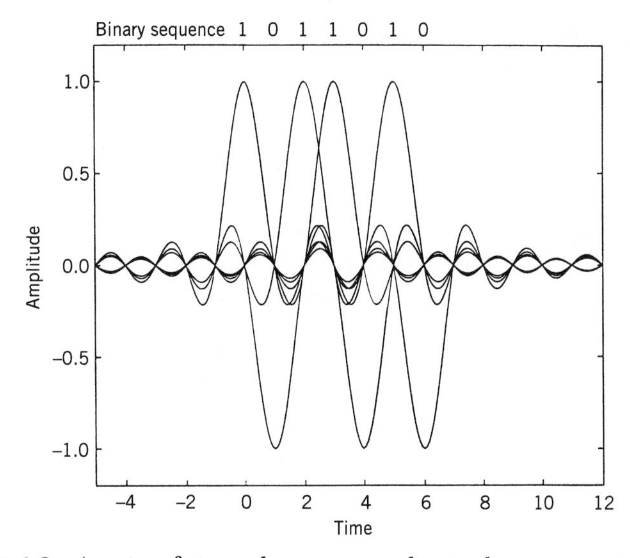


**FIGURE 4.9** A series of sinc pulses corresponding to the sequence 1011010.

bandwidth possible, there are two practical difficulties that make it an undesirable objective for system design:

1. It requires that the magnitude characteristic of $P(f)$ be flat from $-W$ to $W$, and zero elsewhere. This is physically unrealizable because of the abrupt transitions at the band edges $\pm W$.
2. The function $p(t)$ decreases as $1/|t|$ for large $|t|$, resulting in a slow rate of decay. This is also caused by the discontinuity of $P(f)$ at $\pm W$. Accordingly, there is practically no margin of error in sampling times in the receiver.

To evaluate the effect of this *timing error*, consider the sample of $y(t)$ at $t = \Delta t$, where $\Delta t$ is the timing error. To simplify the exposition, we may put the correct sampling time $t_i$ equal to zero. In the absence of noise, we thus have (from Equation (4.48))

$$\begin{aligned} y(\Delta t) &= \mu \sum_k a_k p(\Delta t - kT_b) \\ &= \mu \sum_k a_k \frac{\sin[2\pi W(\Delta t - kT_b)]}{2\pi W(\Delta t - kT_b)} \end{aligned} \tag{4.57}$$

Since $2WT_b = 1$, by definition, we may rewrite Equation (4.57) as

$$y(\Delta t) = \mu a_0 \operatorname{sinc}(2W\,\Delta t) + \frac{\mu \sin(2\pi W\,\Delta t)}{\pi} \sum_{\substack{k \\ k \neq 0}} \frac{(-1)^k a_k}{(2W\,\Delta t - k)} \tag{4.58}$$

The first term on the right-hand side of Equation (4.58) defines the desired symbol, whereas the remaining series represents the intersymbol interference caused by the timing error $\Delta t$ in sampling the output $y(t)$. Unfortunately, it is possible for this series to diverge, thereby causing erroneous decisions in the receiver.

## ■ RAISED COSINE SPECTRUM

We may overcome the practical difficulties encountered with the ideal Nyquist channel by extending the bandwidth from the minimum value $W = R_b/2$ to an adjustable value between $W$ and $2W$. We now specify the overall frequency response $P(f)$ to satisfy a condition more elaborate than that for the ideal Nyquist channel; specifically, we retain three terms of Equation (4.53) and restrict the frequency band of interest to $[-W, W]$, as shown by

$$P(f) + P(f - 2W) + P(f + 2W) = \frac{1}{2W}, \qquad -W \leq f \leq W \tag{4.59}$$

We may devise several band-limited functions that satisfy Equation (4.59). A particular form of $P(f)$ that embodies many desirable features is provided by a *raised cosine spectrum*. This frequency response consists of a *flat* portion and a *rolloff* portion that has a sinusoidal form, as follows:

$$P(f) = \begin{cases} \dfrac{1}{2W}, & 0 \leq |f| < f_1 \\ \dfrac{1}{4W}\left\{1 - \sin\left[\dfrac{\pi(|f| - W)}{2W - 2f_1}\right]\right\}, & f_1 \leq |f| < 2W - f_1 \\ 0, & |f| \geq 2W - f_1 \end{cases} \tag{4.60}$$

The frequency parameter $f_1$ and bandwidth $W$ are related by

$$\alpha = 1 - \frac{f_1}{W} \tag{4.61}$$

The parameter $\alpha$ is called the *rolloff factor*; it indicates the *excess bandwidth* over the ideal solution, $W$. Specifically, the transmission bandwidth $B_T$ is defined by

$$\begin{aligned} B_T &= 2W - f_1 \\ &= W(1 + \alpha) \end{aligned}$$

The frequency response $P(f)$, normalized by multiplying it by $2W$, is plotted in Figure 4.10*a* for three values of $\alpha$, namely, 0, 0.5, and 1. We see that for $\alpha = 0.5$ or 1, the

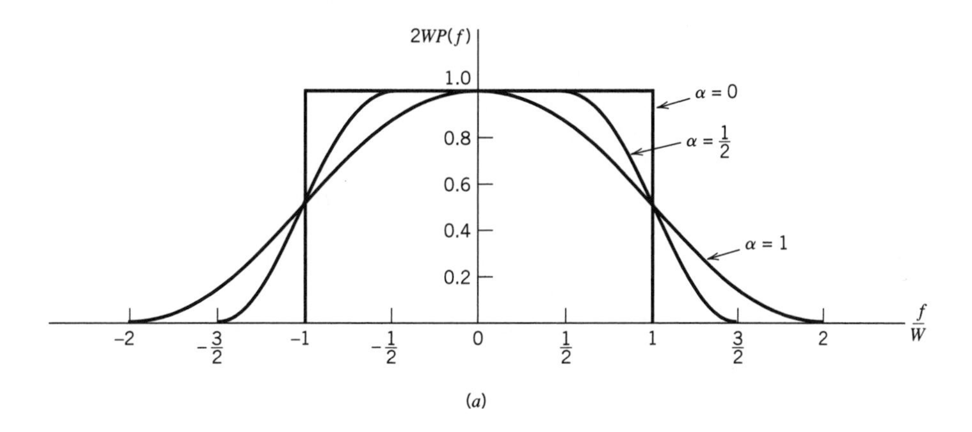


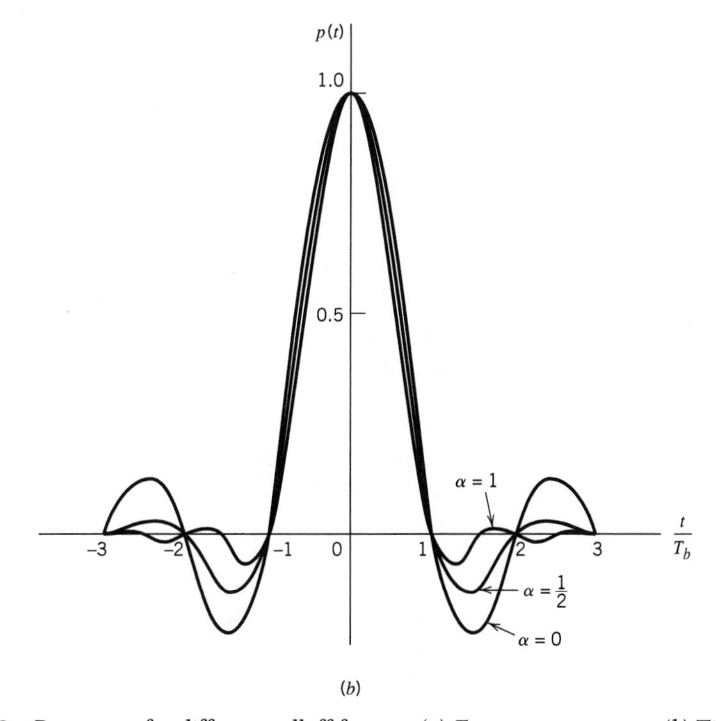


**FIGURE 4.10** Responses for different rolloff factors. (*a*) Frequency response. (*b*) Time response.

function $P(f)$ cuts off gradually as compared with the ideal Nyquist channel (i.e., $\alpha = 0$) and is therefore easier to implement in practice. Also the function $P(f)$ exhibits odd symmetry with respect to the Nyquist bandwidth $W$, making it possible to satisfy the condition of Equation (4.59).

The time response $p(t)$ is the inverse Fourier transform of the frequency response $P(f)$. Hence, using the $P(f)$ defined in Equation (4.60), we obtain the result (see Problem 4.13)

$$p(t) = (\text{sinc}(2Wt))\left(\frac{\cos(2\pi\alpha Wt)}{1 - 16\alpha^2W^2t^2}\right) \tag{4.62}$$

which is plotted in Figure 4.10*b* for $\alpha = 0$, 0.5, and 1.

The time response $p(t)$ consists of the product of two factors: the factor $\text{sinc}(2Wt)$ characterizing the ideal Nyquist channel and a second factor that decreases as $1/|t|^2$ for large $|t|$. The first factor ensures zero crossings of $p(t)$ at the desired sampling instants of time $t = iT$ with $i$ an integer (positive and negative). The second factor reduces the tails of the pulse considerably below that obtained from the ideal Nyquist channel, so that the transmission of binary waves using such pulses is relatively insensitive to sampling time errors. In fact, for $\alpha = 1$ we have the most gradual rolloff in that the amplitudes of the oscillatory tails of $p(t)$ are smallest. Thus the amount of intersymbol interference resulting from timing error decreases as the rolloff factor $\alpha$ is increased from zero to unity.

The special case with $\alpha = 1$ (i.e., $f_1 = 0$) is known as the *full-cosine rolloff* characteristic, for which the frequency response of Equation (4.60) simplifies to

$$P(f) = \begin{cases} \dfrac{1}{4W}\left[1 + \cos\left(\dfrac{\pi f}{2W}\right)\right], & 0 < |f| < 2W \\ 0, & |f| \geq 2W \end{cases} \tag{4.63}$$

Correspondingly, the time response $p(t)$ simplifies to

$$p(t) = \frac{\text{sinc}(4Wt)}{1 - 16W^2t^2} \tag{4.64}$$

This time response exhibits two interesting properties:

1. At $t = \pm T_b/2 = \pm 1/4W$, we have $p(t) = 0.5$; that is, the pulse width measured at half amplitude is exactly equal to the bit duration $T_b$.
2. There are zero crossings at $t = \pm 3T_b/2, \pm 5T_b/2, \cdots$ in addition to the usual zero crossings at the sampling times $t = \pm T_b, \pm 2T_b, \cdots$.

These two properties are extremely useful in extracting a timing signal from the received signal for the purpose of synchronization. However, the price paid for this desirable property is the use of a channel bandwidth double that required for the ideal Nyquist channel corresponding to $\alpha = 0$.

### ▶ EXAMPLE 4.2 Bandwidth Requirement of the T1 System

In Example 3.2 of Chapter 3, we described the signal format for the T1 carrier system that is used to multiplex 24 independent voice inputs, based on an 8-bit PCM word. It was shown that the bit duration of the resulting time-division multiplexed signal (including a framing bit) is

$$T_b = 0.647\ \mu\text{s}$$

Assuming the use of an ideal Nyquist channel, it follows that the minimum transmission bandwidth $B_T$ of the T1 system is (for $\alpha = 0$)

$$B_T = W = \frac{1}{2T_b} = 772\ \text{kHz}$$

However, a more realistic value for the necessary transmission bandwidth is obtained by using a full-cosine rolloff characteristic with $\alpha = 1$. In this case, we find that

$$B_T = W(1 + \alpha) = 2W = \frac{1}{T_b} = 1.544\ \text{MHz}$$

◀

## 4.6 *Correlative-Level Coding*

Thus far we have treated intersymbol interference as an undesirable phenomenon that produces a degradation in system performance. Indeed, its very name connotes a nuisance effect. Nevertheless, by adding intersymbol interference to the transmitted signal in a controlled manner, it is possible to achieve a signaling rate equal to the Nyquist rate of $2W$ symbols per second in a channel of bandwidth $W$ Hertz. Such schemes are called *correlative-level coding* or *partial-response signaling* schemes.[5] The design of these schemes is based on the following premise: Since intersymbol interference introduced into the transmitted signal is known, its effect can be interpreted at the receiver in a deterministic way. Thus correlative-level coding may be regarded as a practical method of achieving the theoretical maximum signaling rate of $2W$ symbols per second in a bandwidth of $W$ Hertz, as postulated by Nyquist, using realizable and perturbation-tolerant filters.

### ■ Duobinary Signaling

The basic idea of correlative-level coding will now be illustrated by considering the specific example of *duobinary signaling*, where "duo" implies doubling of the transmission capacity of a straight binary system. This particular form of correlative-level coding is also called *class I partial response*.

Consider a binary input sequence $\{b_k\}$ consisting of uncorrelated binary symbols 1 and 0, each having duration $T_b$. As before, this sequence is applied to a pulse-amplitude modulator producing a two-level sequence of short pulses (approximating a unit impulse), whose amplitude $a_k$ is defined by

$$a_k = \begin{cases} +1 & \text{if symbol } b_k \text{ is } 1 \\ -1 & \text{if symbol } b_k \text{ is } 0 \end{cases} \tag{4.65}$$

When this sequence is applied to a *duobinary encoder*, it is converted into a *three-level output*, namely, $-2$, 0, and $+2$. To produce this transformation, we may use the scheme shown in Figure 4.11. The two-level sequence $\{a_k\}$ is first passed through a simple filter involving a single delay element and summer. For every unit impulse applied to the

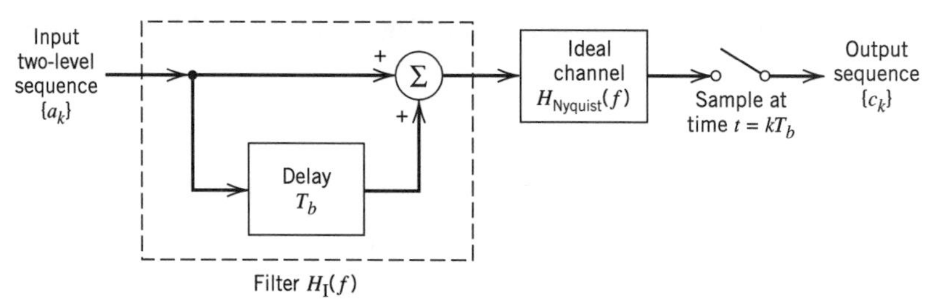


**FIGURE 4.11** Duobinary signaling scheme.

input of this filter, we get two unit impulses spaced $T_b$ seconds apart at the filter output. We may therefore express the duobinary coder output $c_k$ as the sum of the present input pulse $a_k$ and its previous value $a_{k-1}$, as shown by

$$c_k = a_k + a_{k-1} \tag{4.66}$$

One of the effects of the transformation described by Equation (4.66) is to change the input sequence $\{a_k\}$ of uncorrelated two-level pulses into a sequence $\{c_k\}$ of correlated three-level pulses. This correlation between the adjacent pulses may be viewed as introducing intersymbol interference into the transmitted signal in an artificial manner. However, the intersymbol interference so introduced is under the designer's control, which is the basis of correlative coding.

An ideal delay element, producing a delay of $T_b$ seconds, has the frequency response $\exp(-j2\pi fT_b)$, so that the frequency response of the simple delay-line filter in Figure 4.11 is $1 + \exp(-j2\pi fT_b)$. Hence, the overall frequency response of this filter connected in cascade with an ideal Nyquist channel is

$$\begin{aligned} H_I(f) &= H_{\text{Nyquist}}(f)[1 + \exp(-j2\pi fT_b)] \\ &= H_{\text{Nyquist}}(f)[\exp(j\pi fT_b) + \exp(-j\pi fT_b)] \exp(-j\pi fT_b) \\ &= 2H_{\text{Nyquist}}(f) \cos(\pi fT_b) \exp(-j\pi fT_b) \end{aligned} \tag{4.67}$$

where the subscript I in $H_I(f)$ indicates the pertinent class of partial response. For an ideal Nyquist channel of bandwidth $W = 1/2T_b$, we have (ignoring the scaling factor $T_b$)

$$H_{\text{Nyquist}}(f) = \begin{cases} 1, & |f| \le 1/2T_b \\ 0, & \text{otherwise} \end{cases} \tag{4.68}$$

Thus the overall frequency response of the duobinary signaling scheme has the form of a half-cycle cosine function, as shown by

$$H_I(f) = \begin{cases} 2\cos(\pi fT_b)\exp(-j\pi fT_b), & |f| \le 1/2T_b \\ 0, & \text{otherwise} \end{cases} \tag{4.69}$$

for which the magnitude response and phase response are as shown in Figures 4.12*a* and 4.12*b*, respectively. An advantage of this frequency response is that it can be easily approximated, in practice, by virtue of the fact that there is continuity at the band edges.

From the first line in Equation (4.67) and the definition of $H_{\text{Nyquist}}(f)$ in Equation (4.68), we find that the impulse response corresponding to the frequency response $H_I(f)$

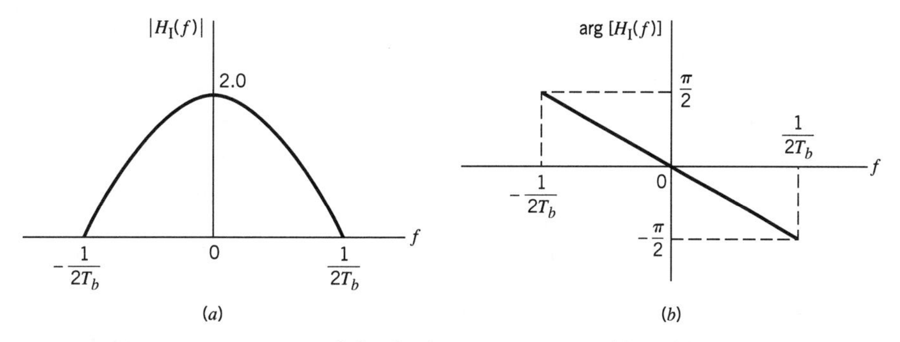


**FIGURE 4.12** Frequency response of the duobinary conversion filter. (*a*) Magnitude response. (*b*) Phase response.

consists of two sinc (Nyquist) pulses that are time-displaced by $T_b$ seconds with respect to each other, as shown by (except for a scaling factor)

$$
\begin{aligned}
h_{\mathrm{I}}(t) &= \frac{\sin(\pi t/T_b)}{\pi t/T_b} + \frac{\sin[\pi(t - T_b)/T_b]}{\pi(t - T_b)/T_b} \\
&= \frac{\sin(\pi t/T_b)}{\pi t/T_b} - \frac{\sin(\pi t/T_b)}{\pi(t - T_b)/T_b} \\
&= \frac{T_b^2 \sin(\pi t/T_b)}{\pi t(T_b - t)}
\end{aligned}
\tag{4.70}
$$

The impulse response $h_{\mathrm{I}}(t)$ is plotted in Figure 4.13, where we see that it has only *two* distinguishable values at the sampling instants. The form of $h_{\mathrm{I}}(t)$ shown here explains why we also refer to this type of correlative coding as partial-response signaling. The response to an input pulse is spread over more than one signaling interval; stated in another way, the response in any signaling interval is "partial." Note also that the tails of $h_{\mathrm{I}}(t)$ decay as $1/|t|^2$, which is a faster rate of decay than the $1/|t|$ encountered in the ideal Nyquist channel.

The original two-level sequence $\{a_k\}$ may be detected from the duobinary-coded sequence $\{c_k\}$ by invoking the use of Equation (4.66). Specifically, let $\hat{a}_k$ represent the *estimate* of the original pulse $a_k$ as conceived by the receiver at time $t = kt_b$. Then, subtracting the previous estimate $\hat{a}_{k-1}$ from $c_k$, we get

$$
\hat{a}_k = c_k - \hat{a}_{k-1} \tag{4.71}
$$

It is apparent that if $c_k$ is received without error and if also the previous estimate $\hat{a}_{k-1}$ at time $t = (k - 1)T_b$ corresponds to a correct decision, then the current estimate $\hat{a}_k$ will be

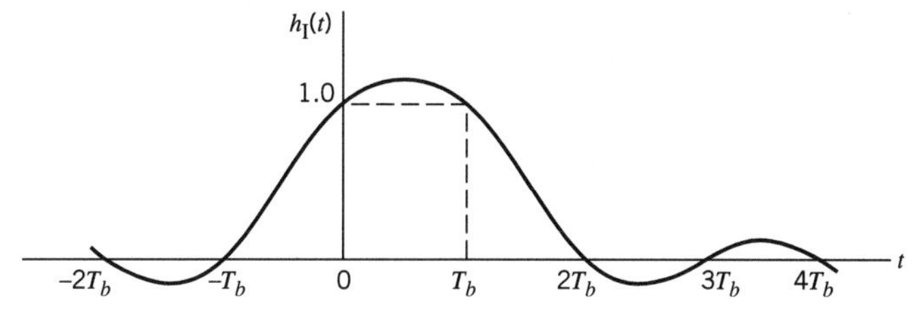


**FIGURE 4.13** Impulse response of the duobinary conversion filter.

correct too. The technique of using a stored estimate of the previous symbol is called *decision feedback*.

We observe that the detection procedure just described is essentially an inverse of the operation of the simple delay-line filter at the transmitter. However, a major drawback of this detection procedure is that once errors are made, they tend to *propagate* through the output because a decision on the current input $a_k$ depends on the correctness of the decision made on the previous input $a_{k-1}$.

A practical means of avoiding the error-propagation phenomenon is to use *precoding* before the duobinary coding, as shown in Figure 4.14. The precoding operation performed on the binary data sequence $\{b_k\}$ converts it into another binary sequence $\{d_k\}$ defined by

$$d_k = b_k \oplus d_{k-1} \tag{4.72}$$

where the symbol $\oplus$ denotes *modulo-two addition* of the binary digits $b_k$ and $d_{k-1}$. This addition is equivalent to a two-input EXCLUSIVE OR operation, which is performed as follows:

$$d_k = \begin{cases} \text{symbol 1} & \text{if either symbol } b_k \text{ or symbol } d_{k-1} \text{ (but not both) is 1} \\ \text{symbol 0} & \text{otherwise} \end{cases} \tag{4.73}$$

The precoded binary sequence $\{d_k\}$ is applied to a pulse-amplitude modulator, producing a corresponding two-level sequence of short pulses $\{a_k\}$, where $a_k = \pm 1$ as before. This sequence of short pulses is next applied to the duobinary coder, thereby producing the sequence $\{c_k\}$ that is related to $\{a_k\}$ as follows:

$$c_k = a_k + a_{k-1} \tag{4.74}$$

Note that unlike the linear operation of duobinary coding, the precoding described by Equation (4.72) is a *nonlinear* operation.

The combined use of Equations (4.72) and (4.74) yields

$$c_k = \begin{cases} 0 & \text{if data symbol } b_k \text{ is 1} \\ \pm 2 & \text{if data symbol } b_k \text{ is 0} \end{cases} \tag{4.75}$$

which is illustrated in Example 4.3. From Equation (4.75) we deduce the following decision rule for detecting the original binary sequence $\{b_k\}$ from $\{c_k\}$:

$$\begin{aligned} &\text{If } |c_k| < 1, \quad \text{say symbol } b_k \text{ is 1} \\ &\text{If } |c_k| > 1, \quad \text{say symbol } b_k \text{ is 0} \end{aligned} \tag{4.76}$$

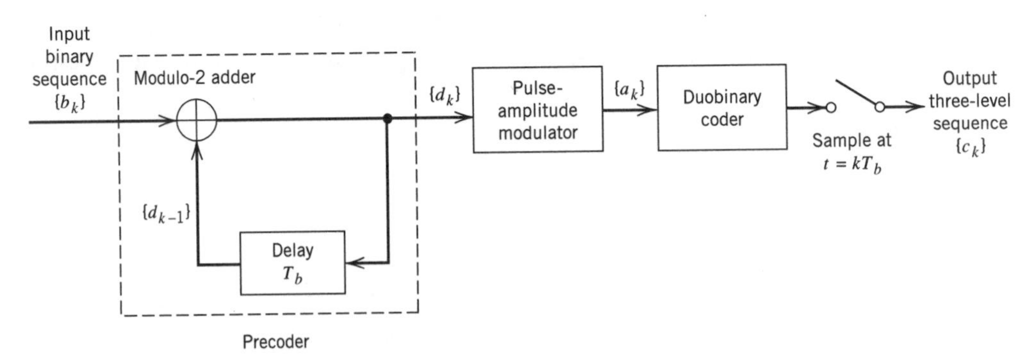


**FIGURE 4.14** A precoded duobinary scheme; details of the duobinary coder are given in Figure 4.11.

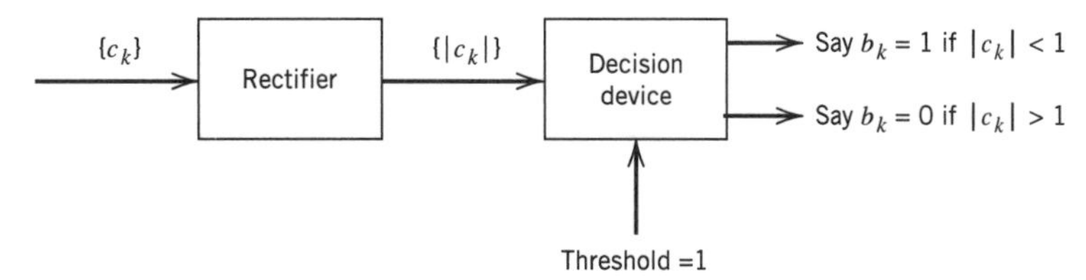


**FIGURE 4.15** Detector for recovering original binary sequence from the precoded duobinary coder output.

When $|c_k| = 1$, the receiver simply makes a random guess in favor of symbol 1 or 0. According to this decision rule, the detector consists of a rectifier, the output of which is compared in a decision device to a threshold of 1. A block diagram of the detector is shown in Figure 4.15. A useful feature of this detector is that no knowledge of any input sample other than the present one is required. Hence, error propagation cannot occur in the detector of Figure 4.15.

### ▶ EXAMPLE 4.3 Duobinary Coding with Precoding

Consider the binary data sequence 0010110. To proceed with the precoding of this sequence, which involves feeding the precoder output back to the input, we add an extra bit to the precoder output. This extra bit is chosen arbitrarily to be 1. Hence, using Equation (4.73), we find that the sequence $\{d_k\}$ at the precoder output is as shown in row 2 of Table 4.1. The polar representation of the precoded sequence $\{d_k\}$ is shown in row 3 of Table 4.1. Finally, using Equation (4.74), we find that the duobinary coder output has the amplitude levels given in row 4 of Table 4.1.

To detect the original binary sequence, we apply the decision rule of Equation (4.76), and so obtain the binary sequence given in row 5 of Table 4.1. This latter result shows that, in the absence of noise, the original binary sequence is detected correctly. ◀

## ■ MODIFIED DUOBINARY SIGNALING

In the duobinary signaling technique the frequency response $H(f)$, and consequently the power spectral density of the transmitted pulse, is nonzero at the origin. This is considered to be an undesirable feature in some applications, since many communications channels cannot transmit a DC component. We may correct for this deficiency by using the *class IV partial response* or *modified duobinary* technique, which involves a correlation span of two binary digits. This special form of correlation is achieved by subtracting amplitude-modulated pulses spaced $2T_b$ seconds apart, as indicated in the block diagram of Figure

**TABLE 4.1** *Illustrating Example 4.3 on duobinary coding*

| | | | | | | | | |
|---|---|---|---|---|---|---|---|---|
| Binary sequence $\{b_k\}$ | | 0 | 0 | 1 | 0 | 1 | 1 | 0 |
| Precoded sequence $\{d_k\}$ | 1 | 1 | 1 | 0 | 0 | 1 | 0 | 0 |
| Two-level sequence $\{a_k\}$ | +1 | +1 | +1 | −1 | −1 | +1 | −1 | −1 |
| Duobinary coder output $\{c_k\}$ | | +2 | +2 | 0 | −2 | 0 | 0 | −2 |
| Binary sequence obtained by applying decision rule of Eq. (4.76) | | 0 | 0 | 1 | 0 | 1 | 1 | 0 |

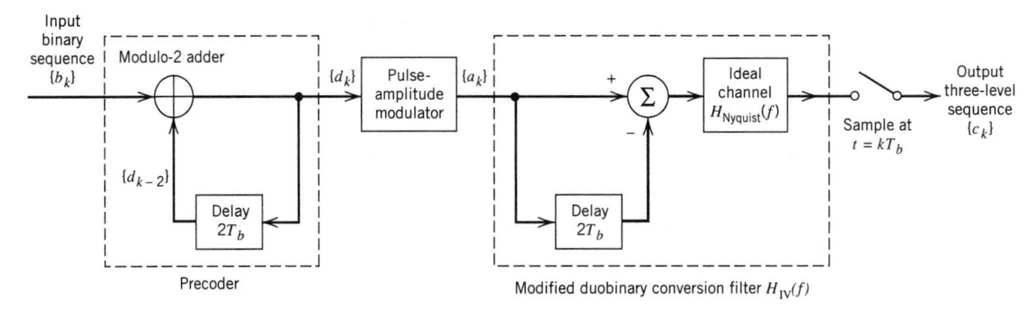


**FIGURE 4.16** Modified duobinary signaling scheme.

4.16. The precoder involves a delay of $2T_b$ seconds. The output of the modified duobinary conversion filter is related to the input two-level sequence $\{a_k\}$ at the pulse-amplitude modulator output as follows:

$$c_k = a_k - a_{k-2} \tag{4.77}$$

Here, again, we find that a three-level signal is generated. With $a_k = \pm 1$, we find that $c_k$ takes on one of three values: +2, 0, and −2.

The overall frequency response of the delay-line filter connected in cascade with an ideal Nyquist channel, as in Figure 4.16, is given by

$$\begin{aligned} H_{IV}(f) &= H_{Nyquist}(f)[1 - \exp(-j4\pi f T_b)] \\ &= 2jH_{Nyquist}(f)\sin(2\pi f T_b)\exp(-j2\pi f T_b) \end{aligned} \tag{4.78}$$

where the subscript IV in $H_{IV}(f)$ indicates the pertinent class of partial response and $H_{Nyquist}(f)$ is as defined in Equation (4.68). We therefore have an overall frequency response in the form of a half-cycle sine function, as shown by

$$H_{IV}(f) = \begin{cases} 2j\sin(2\pi f T_b)\exp(-j2\pi f T_b), & |f| \le 1/2T_b \\ 0, & \text{elsewhere} \end{cases} \tag{4.79}$$

The corresponding magnitude response and phase response of the modified duobinary coder are shown in Figures 4.17*a* and 4.17*b*, respectively. A useful feature of the modified duobinary coder is the fact that its output has no DC component. Note also that this

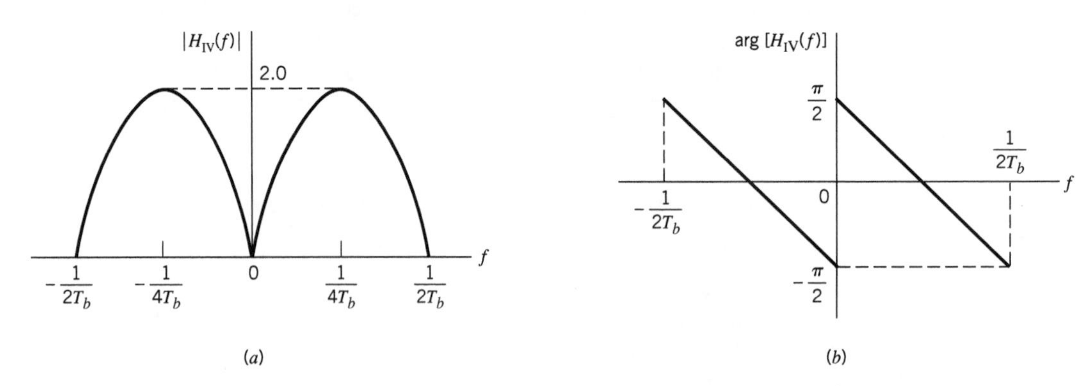


**FIGURE 4.17** Frequency response of the modified duobinary conversion filter. (*a*) Magnitude response. (*b*) Phase response.

second form of correlative-level coding exhibits the same continuity at the band edges as in duobinary signaling.

From the first line of Equation (4.78) and the definition of $H_{\text{Nyquist}}(f)$ in Equation (4.68), we find that the impulse response of the modified duobinary coder consists of two sinc (Nyquist) pulses that are time-displaced by $2T_b$ seconds with respect to each other, as shown by (except for a scaling factor)

$$\begin{aligned} h_{\text{IV}}(t) &= \frac{\sin(\pi t/T_b)}{\pi t/T_b} - \frac{\sin[\pi(t-2T_b)/T_b)]}{\pi(t-2T_b)/T_b} \\ &= \frac{\sin(\pi t/T_b)}{\pi t/T_b} - \frac{\sin(\pi t/T_b)}{\pi(t-2T_b)/T_b} \\ &= \frac{2T_b^2 \sin(\pi t/T_b)}{\pi t(2T_b - t)} \end{aligned} \tag{4.80}$$

This impulse response is plotted in Figure 4.18, which shows that it has *three* distinguishable levels at the sampling instants. Note also that, as with duobinary signaling, the tails of $h_{\text{IV}}(t)$ for the modified duobinary signaling decay as $1/|t|^2$.

To eliminate the possibility of error propagation in the modified duobinary system, we use a precoding procedure similar to that used for the duobinary case. Specifically, prior to the generation of the modified duobinary signal, a modulo-two logical addition is used on signals $2T_b$ seconds apart, as shown by (see the front end of Figure 4.16)

$$\begin{aligned} d_k &= b_k \oplus d_{k-2} \\ &= \begin{cases} \text{symbol 1} & \text{if either symbol } b_k \text{ or symbol } d_{k-2} \text{ (but not both) is 1} \\ \text{symbol 0} & \text{otherwise} \end{cases} \end{aligned} \tag{4.81}$$

where $\{b_k\}$ is the incoming binary data sequence and $\{d_k\}$ is the sequence at the precoder output. The precoded sequence $\{d_k\}$ thus produced is then applied to a pulse-amplitude modulator and then to the modified duobinary conversion filter.

In Figure 4.16, the output digit $c_k$ equals $-2$, 0, or $+2$, assuming that the pulse-amplitude modulator uses a polar representation for the precoded sequence $\{d_k\}$. Also we find that the detected digit $\hat{b}_k$ at the receiver output may be extracted from $c_k$ by disregarding the polarity of $c_k$. Specifically, we may formulate the following decision rule:

$$\begin{aligned} &\text{If } |c_k| > 1, \quad \text{say symbol } b_k \text{ is 1} \\ &\text{If } |c_k| < 1, \quad \text{say symbol } b_k \text{ is 0} \end{aligned} \tag{4.82}$$

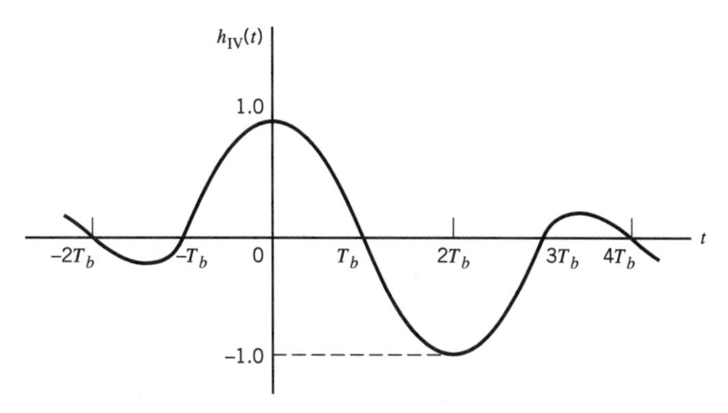


**FIGURE 4.18** Impulse response of the modified duobinary conversion filter.

When $|c_k| = 1$, the receiver makes a random guess in favor of symbol 1 or 0. As with the duobinary signaling, we may note the following:

- In the absence of channel noise, the detected binary sequence $\{\hat{b}_k\}$ is exactly the same as the original binary sequence $\{b_k\}$ at the transmitter input.
- The use of Equation (4.81) requires the addition of two extra bits to the precoded sequence $\{a_k\}$. The composition of the decoded sequence $\{\hat{b}_k\}$ using Equation (4.82) is invariant to the selection made for these two bits.

## ■ GENERALIZED FORM OF CORRELATIVE-LEVEL CODING (PARTIAL-RESPONSE SIGNALING)

The duobinary and modified duobinary techniques have correlation spans of 1 binary digit and 2 binary digits, respectively. It is a straightforward matter to generalize these two techniques to other schemes, which are known collectively as *correlative-level coding* or *partial-response signaling* schemes. This generalization is shown in Figure 4.19, where $H_{\text{Nyquist}}(f)$ is defined in Equation (4.68). It involves the use of a tapped-delay-line filter with tap-weights $w_0, w_1, \cdots, w_{N-1}$. Specifically, different classes of partial-response sig-

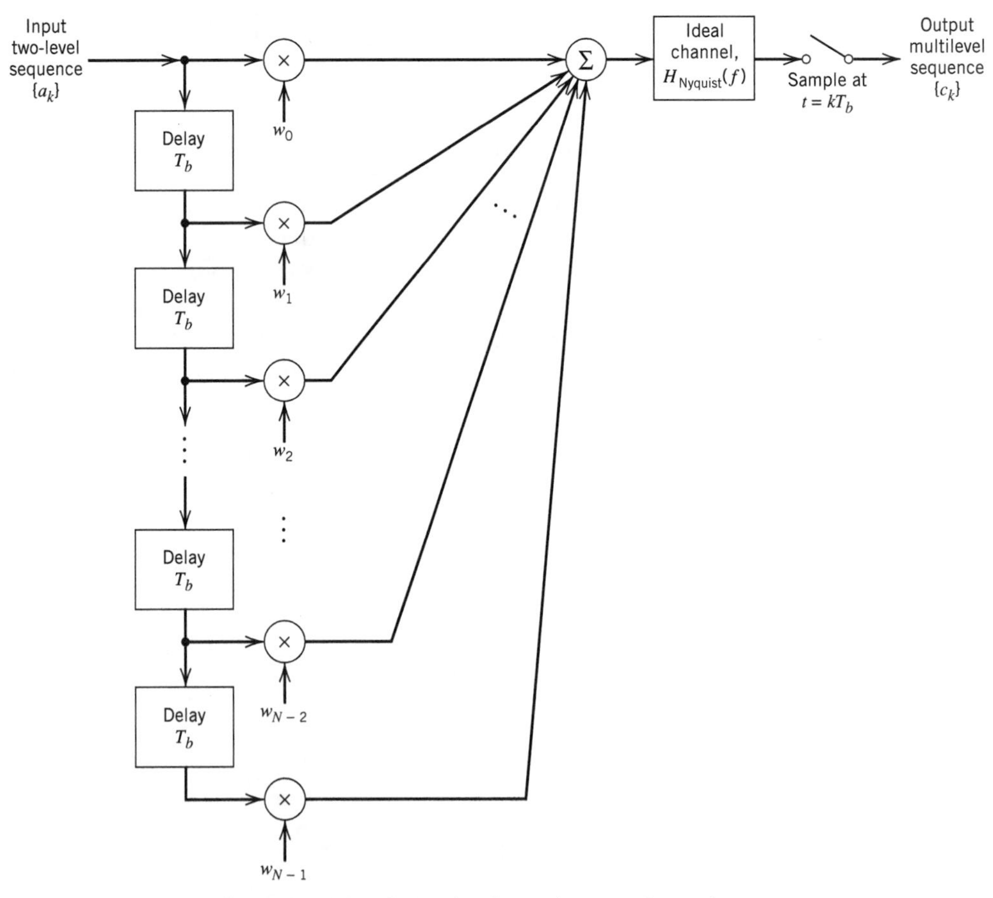


**FIGURE 4.19** Generalized correlative coding scheme.

**TABLE 4.2** ***Different classes of partial-response signaling schemes referring to Figure 4.19***

| *Type of Class* | N | $w_0$ | $w_1$ | $w_2$ | $w_3$ | $w_4$ | *Comments* |
|---|---|---|---|---|---|---|---|
| I | 2 | 1 | 1 | | | | Duobinary coding |
| II | 3 | 1 | 2 | 1 | | | |
| III | 3 | 2 | 1 | −1 | | | |
| IV | 3 | 1 | 0 | −1 | | | Modified duobinary coding |
| V | 5 | −1 | 0 | 2 | 0 | −1 | |

naling schemes may be achieved by using a weighted linear combination of $N$ ideal Nyquist (sinc) pulses, as shown by

$$h(t) = \sum_{n=0}^{N-1} w_n \operatorname{sinc}\left(\frac{t}{T_b} - n\right) \tag{4.83}$$

An appropriate choice of the tap-weights in Equation (4.83) results in a variety of spectral shapes designed to suit individual applications. Table 4.2 presents the specific details of five different classes of partial-response signaling schemes. For example, in the duobinary case (class I partial response), we have

$$w_0 = +1$$
$$w_1 = +1$$

and $w_n = 0$ for $n \geq 2$. In the modified duobinary case (class IV partial response), we have

$$w_0 = +1$$
$$w_1 = 0$$
$$w_2 = -1$$

and $w_n = 0$ for $n \geq 3$.

The useful characteristics of partial-response signaling schemes may now be summarized as follows:

- Binary data transmission over a physical baseband channel can be accomplished at a rate close to the Nyquist rate, using realizable filters with gradual cutoff characteristics.
- Different spectral shapes can be produced, appropriate for the application at hand.

However, these desirable characteristics are achieved at a price: A larger signal-to-noise ratio is required to yield the same average probability of symbol error in the presence of noise as in the corresponding binary PAM systems because of an increase in the number of signal levels used.

## 4.7 *Baseband M-ary PAM Transmission*

In the baseband binary PAM system of Figure 4.7, the pulse-amplitude modulator produces binary pulses, that is, pulses with one of two possible amplitude levels. On the other hand, in a *baseband M-ary PAM system*, the pulse-amplitude modulator produces one of $M$ possible amplitude levels with $M > 2$. This form of pulse modulation is illustrated in Figure 4.20*a* for the case of a *quaternary* ($M = 4$) system and the binary data sequence 0010110111. The waveform shown in Figure 4.20*a* is based on the electrical representation for each of the four possible *dibits* (pairs of bits) given in Figure 4.20*b*. Note that this representation is *Gray encoded*, which means that any dibit in the quaternary alphabet differs from an adjacent dibit in a single bit position.

In an $M$-ary system, the information source emits a sequence of symbols from an alphabet that consists of $M$ symbols. Each amplitude level at the pulse-amplitude modulator output corresponds to a distinct symbol, so that there are $M$ distinct amplitude levels to be transmitted. Consider then an $M$-ary PAM system with a signal alphabet that contains $M$ equally likely and statistically independent symbols, with the symbol duration denoted by $T$ seconds. We refer to $1/T$ as the *signaling rate* of the system, which is expressed in *symbols per second*, or *bauds*. It is informative to relate the signaling rate of this system to that of an equivalent binary PAM system for which the value of $M$ is 2 and the successive binary symbols 1 and 0 are equally likely and statistically independent, with the duration of either symbol denoted by $T_b$ seconds. Under the conditions described here, the binary PAM system produces information at the rate of $1/T_b$ bits per seconds. We also observe that in the case of a quaternary PAM system, for example, the four possible symbols may be identified with the dibits 00, 01, 10, and 11. We thus see that each symbol represents 2 bits of information, and 1 baud is equal to 2 bits per second. We may generalize this result by stating that in an $M$-ary PAM system, 1 baud is equal to $\log_2 M$ bits per second, and the symbol duration $T$ of the $M$-ary PAM system is related to the bit duration $T_b$ of the equivalent binary PAM system as

$$T = T_b \log_2 M \tag{4.84}$$

Therefore, in a given channel bandwidth, we find that by using an $M$-ary PAM system, we are able to transmit information at a rate that is $\log_2 M$ faster than the corresponding binary PAM system. However, to realize the same average probability of symbol error, an $M$-ary PAM system requires more transmitted power. Specifically, we find that for $M$ much larger than 2 and an average probability of symbol error small compared to 1, the trans-

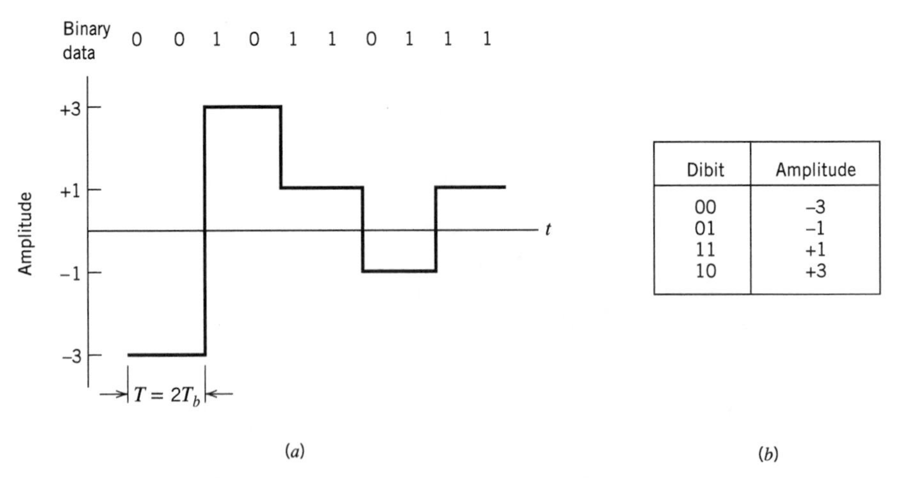


| Dibit | Amplitude |
|---|---|
| 00 | −3 |
| 01 | −1 |
| 11 | +1 |
| 10 | +3 |

**FIGURE 4.20** Output of a quaternary system. (*a*) Waveform. (*b*) Representation of the 4 possible dibits, based on Gray encoding.

mitted power must be increased by the factor $M^2/\log_2 M$, compared to a binary PAM system.

In a baseband $M$-ary system, first of all, the sequence of symbols emitted by the information source is converted into an $M$-level PAM pulse train by a pulse-amplitude modulator at the transmitter input. Next, as with the binary PAM system, this pulse train is shaped by a transmit filter and then transmitted over the communication channel, which corrupts the signal waveform with both noise and distortion. The received signal is passed through a receive filter and then sampled at an appropriate rate in synchronism with the transmitter. Each sample is compared with preset *threshold* values (also called *slicing* levels), and a decision is made as to which symbol was transmitted. We therefore find that the designs of the pulse-amplitude modulator and the decision-making device in an $M$-ary PAM are more complex than those in a binary PAM system. Intersymbol interference, noise, and imperfect synchronization cause errors to appear at the receiver output. The transmit and receive filters are designed to minimize these errors. Procedures used for the design of these filters are similar to those discussed in Sections 4.5 and 4.6 for baseband binary PAM systems.

## 4.8 *Digital Subscriber Lines*

At this point in our study of baseband data transmission it is rather appropriate that we digress from theoretical aspects of the study and consider a fast-growing application: *digital subscriber lines.*[6] A digital subscriber line (DSL) operates over a local loop (less than 1.5 km) that provides a direct connection between a user terminal (e.g., computer) and a telephone company's *central office* (CO), as illustrated in Figure 4.21. Through the CO, a DSL user is connected to a broadband backbone data network, which is based on technologies such as the asynchronous transfer mode (ATM) and Internet protocol (IP); these technologies and related network resources (i.e., optical fibers, SONET) are discussed in the Background and Preview chapter. Accordingly, the information-bearing signal is kept in the digital domain all the way from the user terminal to an Internet service provider, with the signal being switched or routed at regular intervals in the course of its transmission through the data network.

In the interest of an inexpensive implementation, digital subscriber lines use twisted pairs configured to provide a high data-rate, full duplex, digital transmission capability. (Twisted pairs are also used for ordinary telephonic communication, as discussed in the Background and Preview chapter.) To achieve full-duplex, two-wire transmission, we may use one of two possible modes of operation:

1. *Time compression* (*TC*) *multiplexing*, where data transmission in the two opposite directions on the common line are separated in time. Specifically, blocks of bits of data are sent in bursts in each direction on an alternate basis, as illustrated in

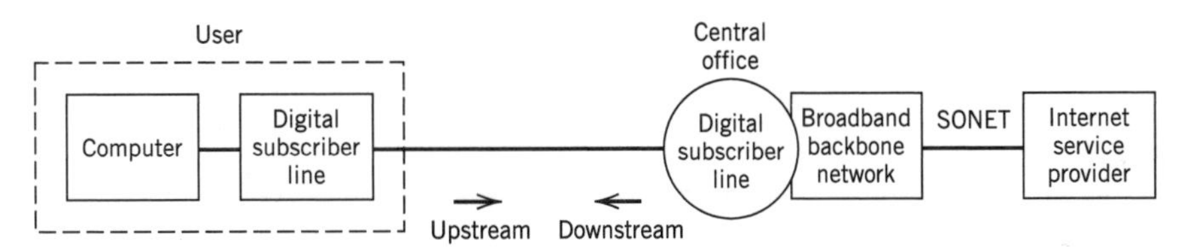


**Figure 4.21** Block diagram depicting the operational environment of digital subscriber lines.

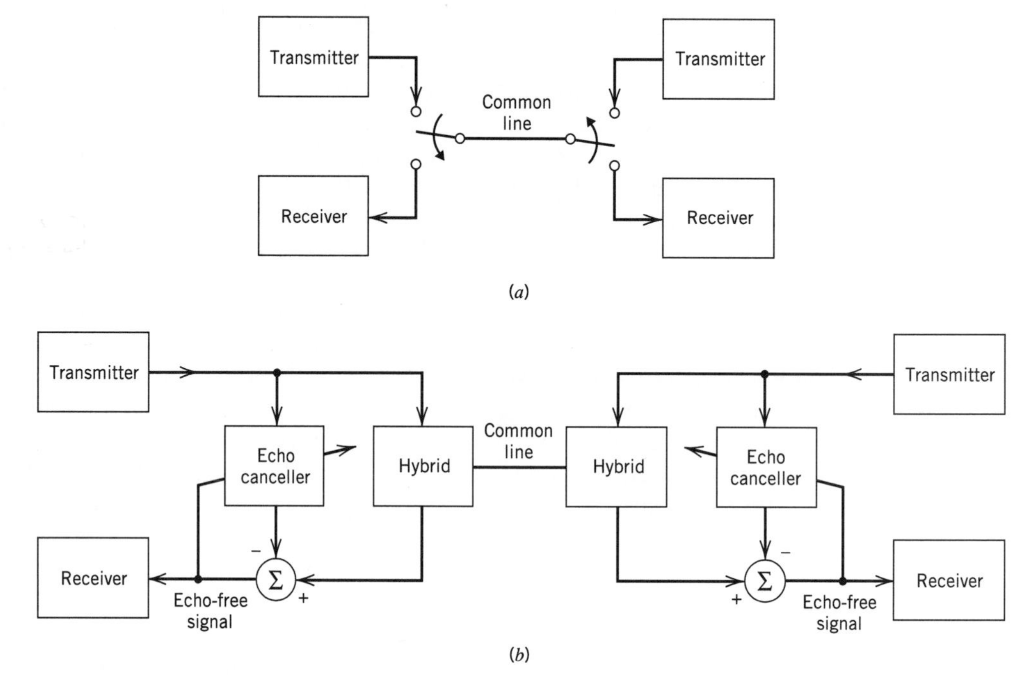


**FIGURE 4.22** Full-duplex operation using (*a*) time compression multiplexing, and (*b*) echo-cancellation.

Figure 4.22*a*. To account for propagation time across the line, a guard time is inserted between individual bursts of data. Accordingly, the line rate is slightly greater than twice the data rate.

2. *Echo-cancellation mode*, which supports the simultaneous flow of data along the common line in both directions. For this form of transmission to be feasible, each *transceiver* (transmitter/receiver) includes a hybrid for two purposes: the separation of the transmitted signal from the received signal and the two-to-four-wire conversion, as shown in Figure 4.22*b*. The *hybrid*, or more precisely, the *hybrid transformer*, is basically a bridge circuit with three ports (terminal pairs), as depicted in Figure 4.23. If the bridge is not perfectly balanced, the transmitter port of the hybrid

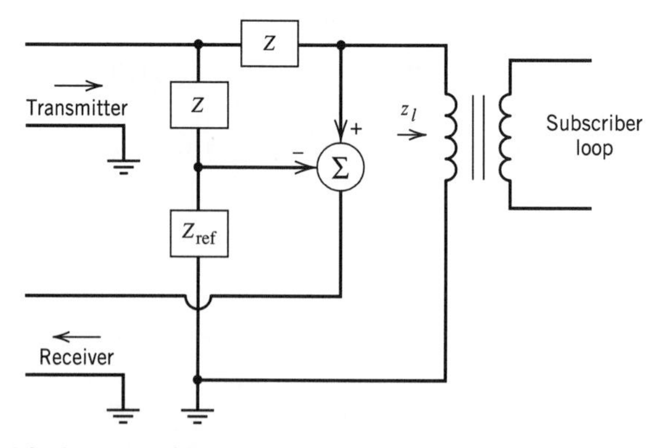


**FIGURE 4.23** Simplified circuit of hybrid transformer. For the bridge to be balanced, the reference impedance $Z_{ref}$ should equal the line impedance $z_l$.

becomes coupled to the receiver port, thereby giving rise to an *echo* due to leakage of the near-end (local) transmitted signal to the near-end (local) receiver. To cancel the unwanted echo, each transceiver includes an *echo canceller*, as shown in Figure 4.22*b*. Since data can flow through the line simultaneously in both directions, the line rate is the same as the data rate.

From this discussion, it is apparent that the echo-cancellation scheme offers a much better data-transmission performance than the time compression multiplexing scheme, but at the expense of increased complexity. However, by implementing the entire transceiver in a single very large-scale integrated (VLSI) chip, the cost is made affordable despite the increased complexity. In North America, the echo-cancellation scheme has been adopted as the basis for designing the transceivers. An adaptive implementation of the echo canceller is discussed in Problem 4.31.

In addition to echo, there are other impairments of the transmission medium that need to be considered. The two dominant impairments are intersymbol interference and crosstalk, which are discussed in what follows in that order.

To a first-order approximation, the squared magnitude response of a twisted pair is given by

$$|H(f)|^2 = \exp(-\alpha\sqrt{f}) \tag{4.85}$$

where

$$\alpha = k\frac{l}{l_0} \tag{4.86}$$

In Equation (4.85), the frequency $f$ is measured in kHz, $k$ is a physical constant of the twisted pair, $l_0$ is a reference length (e.g., kilometers), and $l$ is the actual length of the twisted pair. Equation (4.85) points to a major impairment in the use of a twisted pair for baseband data transmission: the gradual falloff in the frequency response, which, in turn, gives rise to intersymbol interference.

Turning next to crosstalk, the primary cause for its occurrence is the capacitive coupling that exists between adjacent twisted pairs in a cable. Typically, the nearest five to seven twisted pairs in the cable cause most of the crosstalk. In any event, two kinds of crosstalk can be observed in a receiver of interest:

1. *Near-end crosstalk (NEXT)*, which is generated by transmitters located at the same end of the cable as the receiver, as illustrated in Figure 4.24*a*.

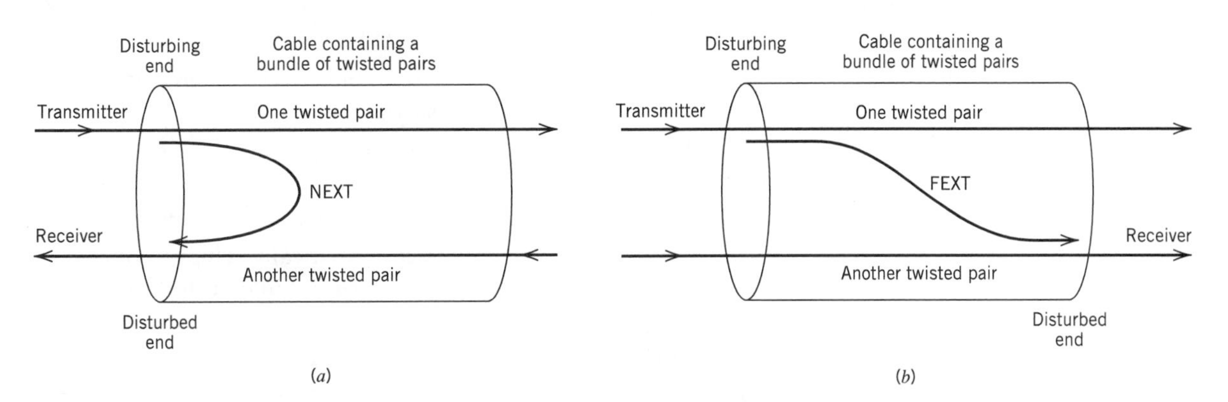


**FIGURE 4.24** (*a*) Near-end crosstalk (NEXT). (*b*) Far-end crosstalk (FEXT).

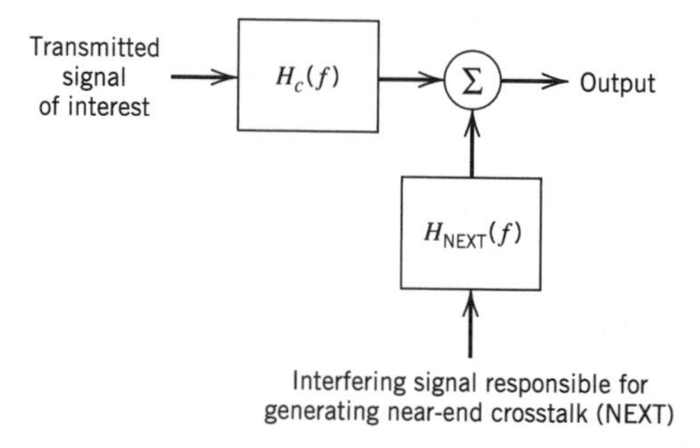


**FIGURE 4.25** Model of twisted-pair channel.

2. *Far-end crosstalk (FEXT)*, which is generated by transmitters located further away from the receiver, as illustrated in Figure 4.24*b*.

FEXT naturally suffers the same line loss as the signal, whereas NEXT does not. Accordingly, in the echo-cancellation scheme of Figure 4.22*b* where signals travel in both directions in the cable, NEXT will be much stronger than FEXT. Henceforth, we ignore the effect of FEXT.

Indeed, near-end crosstalk and intersymbol interference are the two most important factors in determining the performance of a digital subscriber loop. Figure 4.25 shows the model of a twisted-pair channel dominated by these two impairments. Since all twisted pairs are usually transmitting similar signals, we may model the NEXT as a signal with the same power spectral density as the transmitted signal passing through a *crosstalk* frequency response $H_{\text{NEXT}}(f)$, which is approximated by

$$H_{\text{NEXT}}(f) = \beta f^{3/2} \tag{4.87}$$

where $\beta$ is a constant of the cable. The interesting point to note from Figure 4.25 is that both the transmitted signal and the interfering signal have the same power spectral density; they differ from each other merely in their associated frequency responses, as shown in Equations (4.85) and (4.87), respectively. When the model described herein is used for simulation study, the transmitted signal is represented by a random data sequence, while the interference is represented by a Gaussian noise sequence.

## ■ LINE CODES FOR DIGITAL SUBSCRIBER LINES

Now that we have identified the major transmission impairments, we may describe the desirable features the spectrum of a transmitted signal should exhibit:

1. The power spectral density of the transmitted signal should be zero at zero frequency, since no DC transmission through a hybrid transformer is possible.
2. The power spectral density of the transmitted signal should be low at high frequencies for the following reasons:
   - Transmission attenuation in a twisted pair is most severe at high frequencies.
   - Crosstalk between adjacent twisted pairs increases dramatically at high frequencies because of increased capacitive coupling. In this regard, recall that the impedance of a capacitor is inversely proportional to frequency.

To satisfy these desirable properties, we have to be careful in choosing the *line code* that maps the incoming stream of data bits into electrical pulses for transmission on the line. Various possibilities, each with its own advantages and disadvantages, exist for such a choice. The list of potential candidates for line codes includes the following:

- *Manchester code*, which is simple and has zero DC component. Its disadvantage is the occupation of a large spectrum, which makes it vulnerable to near-end crosstalk and intersymbol interference. (The Manchester code was discussed in Section 3.7.)
- *Modified duobinary code*, which has zero DC, is moderately spectrally efficient, and causes minimal intersymbol interference. However, simulation studies of the crosstalk performance of the modified duobinary code have shown that its immunity to near-end crosstalk and intersymbol interference is about 2 to 3 dB poorer than that of block codes on worst-case subscriber lines. (The modified duobinary code was discussed in Section 4.6.)
- *Bipolar code*, in which successive 1s are represented alternately by positive and negative but equal levels, and symbol 0 is represented by a zero level. Bipolar signaling has zero DC. Computer simulations have shown that its near-end crosstalk and intersymbol interference performance is slightly inferior to the modified duobinary code on all digital subscriber loops. (The bipolar code, also known as the *alternate mark inversion (AMI) codes*, was discussed in Section 3.7.)
- *2B1Q code*, which stands for two binary digits encoded into one quaternary symbol. This code is a block code representing a four-level PAM signal, as illustrated in Figure 4.20. Assuming that symbols 1 and 0 are equiprobable, the 2B1Q code has zero DC on the average. Moreover, among all the line codes considered herein, it offers the greatest baud reduction, and the best performance with respect to near-end crosstalk and intersymbol interference.

It is because of the desirable properties of the 2BIQ code compared to the Manchester code, modified duobinary code, the bipolar code, and other line codes not mentioned here,[7] that the 2B1Q code has been adopted as the North American standard for digital subscriber loops.

Using the 2B1Q as the line code and VLSI implementation of a transceiver that incorporates adaptive equalizers and echo cancellers, it is possible to achieve a bit error rate of $10^{-7}$ operating full duplex at 160 kb/s on the vast majority of twisted-pair subscriber lines. A bit error rate of $10^{-7}$ with 12 dB noise margin, when 1 percent worst-case NEXT is present, is an accepted performance criterion for digital subscriber lines. *Noise margin* is the amount of receiver noise (including uncancelled echo) that can be tolerated without exceeding the $10^{-7}$ error rate.

## Asymmetric Digital Subscriber Lines

Another important type of DSL is the *asymmetric digital subscriber line* (ADSL), which is a local transmission system designed to simultaneously support three services on a single twisted-wire pair:

1. Data transmission *downstream* (toward the subscriber) at bit rates of up to 9 Mb/s.
2. Data transmission *upstream* (away from the subscriber) at bit rates of up to 1 Mb/s.
3. Plain old telephone service (POTS).

The downstream and upstream bit rates depend on the length of the twisted pair used to do the transmission. The DSL is said to be "asymmetric" because the downstream bit rate is much higher than the upstream bit rate. Analog voice is transmitted at baseband frequencies and combined with the passband transmissions of downstream and upstream

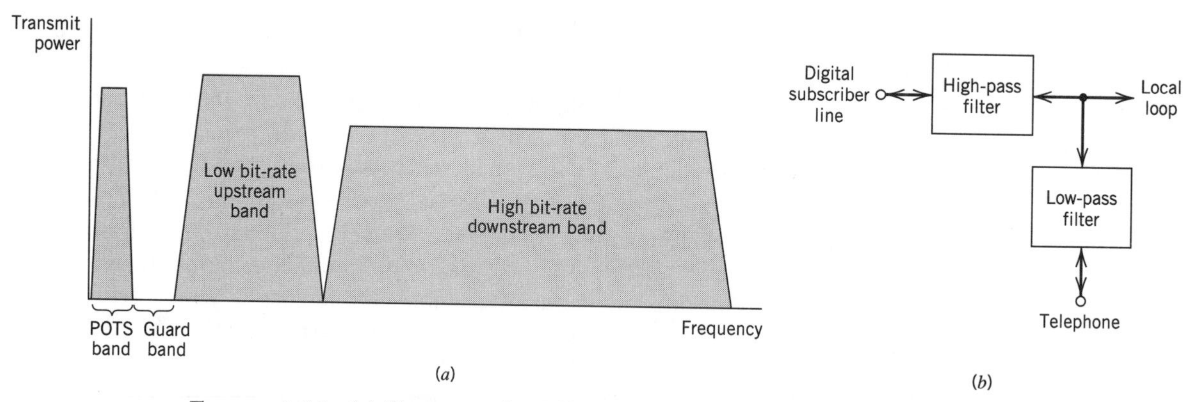


**FIGURE 4.26** (*a*) Illustrating the different band allocations for an FDM-based ADSL system. (*b*) Block diagram of splitter performing the function of a multiplexer or demultiplexer. Note: both filters in the splitter are bidirectional filters.

data using frequency-division multiplexing (FDM). As illustrated in Figure 4.26*a*, the upstream data transmission is placed in a frequency band different from the downstream data transmission to avoid crosstalk. Moreover, a guard band is inserted between the POTS band and the upstream transmission band. The coexistence of ADSL and POTS signals on the local loop is made possible through the use of a pair of *splitters*; one splitter is placed at the CO end of the local loop and the other one is placed at the user end. In functional terms, a splitter is divided into two *bidirectional* filters, as shown in Figure 4.26*b*:

- A low-pass filter for the baseband transmission or extraction of voice signals.
- A high-pass filter for the passband transmission or extraction of ADSL data.

In effect, the splitter performs the role of a frequency-division multiplexer or demultiplexer, depending on the direction of signal transmission.

The motivation for making DSL asymmetric is to accommodate "video-on-demand." In such applications, a subscriber needs a high-throughput channel to download high-bandwidth video data from a central office on demand. In the reverse direction, a much lower throughput channel is adequate to send order information as well as real-time control commands. For example, an ADSL for Internet providing downstream transmission at the DS1 rate of 1.544 Mb/s and upstream transmission of about 160 kb/s would meet the requirements of this application. The approximate 10:1 *asymmetry ratio* realized by such a system prevents the flow of acknowledgment packets in the IP from becoming a bottleneck to the faster direction of data transmission.

It is very difficult to transmit data over a twisted pair at the DS1 rate and higher, and doing so requires the use of sophisticated modulation techniques. The treatment of this subject is deferred to Chapter 6.

## 4.9 *Optimum Linear Receiver*

Resuming our study of the baseband data transmission system depicted in Figure 4.7, we have thus far treated the following two channel conditions separately:

- *Channel noise acting alone*, which led to formulation of the matched filter receiver.
- *Intersymbol interference acting alone*, which led to formulation of the pulse-shaping transmit filter so as to realize the Nyquist channel.

In a real-life situation, however, channel noise and intersymbol interference act together, affecting the behavior of a data transmission system in a combined manner. In this section, we formulate the basis for designing a linear receiver optimized for the general case of a linear channel that is both dispersive and noisy.

In one approach to the design of a linear receiver, the receiver is viewed as a *zero-forcing equalizer* followed by a decision-making device. The objective of this form of equalization is to have the "intersymbol interference forced to zero" at all the instants $t = kT$ at which the channel output is sampled, except for $k = 0$ where the symbol of interest is assumed to occur. Under this condition, symbol-to-symbol detection is assured to be optimal in accordance with the Nyquist criterion, provided that the channel noise $w(t)$ is zero.

The zero-forcing equalizer is relatively easy to design because it ignores the effect of the channel noise $w(t)$. A serious consequence of this oversight, however, is that it leads to overall performance degradation due to *noise enhancement*, a phenomenon that is an inherent feature of zero-forcing equalization; see Problem 4.32. A more refined approach for the receiver design is to use the *mean-square error criterion*, which provides a balanced solution to the problem of reducing the effects of both channel noise and intersymbol interference. Indeed, for a prescribed computational complexity, an equalizer designed on this latter basis always performs as well as, and often better than, its zero-forcing counterpart. Henceforth, we concentrate on the mean-square error criterion for receiver design.

Referring back to the baseband binary data transmission system of Figure 4.7, the receive filter characterized by the impulse response $c(t)$ produces the following response due to the channel output $x(t)$:

$$y(t) = \int_{-\infty}^{\infty} c(\tau)x(t - \tau)\, d\tau \tag{4.88}$$

The channel output $x(t)$ is itself defined by

$$x(t) = \sum_k a_k q(t - kT_b) + w(t) \tag{4.89}$$

where $a_k$ is the symbol transmitted at time $t = kT_b$ and $w(t)$ is the channel noise. The time function $q(t)$ is the convolution of two impulse responses: $g(t)$ pertaining to the pulse-shaping transmit filter, and $h(t)$ pertaining to the channel. Substituting Equation (4.89) into (4.88) and sampling the resulting output $y(t)$ at time $t = iT_b$, we may write

$$y(iT_b) = \xi_i + n_i \tag{4.90}$$

where $\xi_i$ is the signal component defined by

$$\xi_i = \sum_k a_k \int_{-\infty}^{\infty} c(\tau)q(iT_b - kT_b - \tau)\, d\tau \tag{4.91}$$

and $n_i$ is the noise component defined by

$$n_i = \int_{-\infty}^{\infty} c(\tau)w(iT_b - \tau)\, d\tau \tag{4.92}$$

The condition for perfect operation of the receiver is to have $y(iT_b) = a_i$, where $a_i$ is the transmitted symbol. Deviation from this condition results in the error signal

$$\begin{aligned} e_i &= y(iT_b) - a_i \\ &= \xi_i + n_i - a_i \end{aligned} \tag{4.93}$$

Accordingly, we may formally define the mean-square error as

$$J = \frac{1}{2} E[e_i^2] \tag{4.94}$$

where $E$ is the statistical expectation operator, and the factor 1/2 is introduced for convenience of presentation. Substituting Equation (4.93) into (4.94) and then expanding terms, we get

$$J = \frac{1}{2} E[\xi_i^2] + \frac{1}{2} E[n_i^2] + \frac{1}{2} E[a_i^2] + E[\xi_i n_i] - E[n_i a_i] - E[\xi_i a_i] \tag{4.95}$$

We now evaluate the six expectation terms in this equation in the order they appear here:

1. In a *stationary* environment the mean-square term $E[\xi_i^2]$ is independent of the instant of time $t = iT_b$ at which the receive filter output is sampled. Hence, we may simplify the expression of this term by writing

$$E[\xi_i^2] = \sum_l \sum_k E[a_l a_k] \int_{-\infty}^{\infty} \int_{-\infty}^{\infty} c(\tau_1)c(\tau_2)q(lT_b - \tau_1)q(kT_b - \tau_2)\, d\tau_1\, d\tau_2$$

Assuming that, first, the binary symbols $a_k = \pm 1$ as in Equation (4.42) and, second, the transmitted symbols are statistically independent, that is,

$$E[a_l a_k] = \begin{cases} 1 & \text{for } l = k \\ 0 & \text{otherwise} \end{cases} \tag{4.96}$$

we may further reduce the expression for the mean-square term $E[\xi_i^2]$ to

$$E[\xi_i^2] = \int_{-\infty}^{\infty} \int_{-\infty}^{\infty} R_q(\tau_1, \tau_2)c(\tau_1)c(\tau_2)\, d\tau_1\, d\tau_2 \tag{4.97}$$

where

$$R_q(\tau_1, \tau_2) = \sum_k q(kT_b - \tau_1)q(kT_b - \tau_2) \tag{4.98}$$

The factor $R_q(\tau_1, \tau_2)$ is the *temporal autocorrelation function* of the sequence $\{q(kT_b)\}$. Stationarity of this sequence means that (see Section 1.5)

$$R_q(\tau_1, \tau_2) = R_q(\tau_2 - \tau_1) = R_q(\tau_1 - \tau_2)$$

2. The mean-square term $E[n_i^2]$ due to channel noise is given by (using Equation (4.92))

$$\begin{aligned} E[n_i^2] &= \int_{-\infty}^{\infty} \int_{-\infty}^{\infty} c(\tau_1)c(\tau_2)E[w(iT_b - \tau_1)w(iT_b - \tau_2)]\, d\tau_1\, d\tau_2 \\ &= \int_{-\infty}^{\infty} \int_{-\infty}^{\infty} c(\tau_1)c(\tau_2)R_W(\tau_2 - \tau_1)\, d\tau_1\, d\tau_2 \end{aligned} \tag{4.99}$$

where $R_W(\tau_2 - \tau_1)$ is the *ensemble-averaged autocorrelation function* of the channel noise $w(t)$. With $w(t)$ assumed to be white with power spectral density $N_0/2$, we have

$$R_W(\tau_2 - \tau_1) = \frac{N_0}{2} \delta(\tau_2 - \tau_1) \tag{4.100}$$

Hence, the expression for $E[n_i^2]$ simplifies to

$$E[n_i^2] = \frac{N_0}{2} \int_{-\infty}^{\infty} \int_{-\infty}^{\infty} c(\tau_1)c(\tau_2)\, \delta(\tau_2 - \tau_1)\, d\tau_1\, d\tau_2 \tag{4.101}$$

3. The mean-square term $E[a_i^2]$ due to the transmitted symbol $a_i$ is unity by virtue of Equation (4.96); that is,

$$E[a_i^2] = 1 \qquad \text{for all } i \tag{4.102}$$

4. The expectation of the cross-product term $\xi_i n_i$ is zero for two reasons: first $\xi_i$ and $n_i$ are independent and, second, the channel noise $w(t)$, and therefore $n_i$, has zero mean; that is,

$$E[\xi_i n_i] = 0 \qquad \text{for all } i \tag{4.103}$$

5. For similar reasons, the expectation of the cross-product term $n_i a_i$ is also zero; that is,

$$E[n_i a_i] = 0 \qquad \text{for all } i \tag{4.104}$$

6. Finally, the expectation of the cross-product term $\xi_i a_i$ is given by (using Equation (4.91))

$$E[\xi_i a_i] = \sum_k E[a_k a_i] \int_{-\infty}^{\infty} c(\tau) q(iT_b - kT_b - \tau)\, d\tau \tag{4.105}$$

By virtue of the statistical independence of the transmitted symbols described in Equation (4.96), this expectation reduces to

$$E[\xi_i a_i] = \int_{-\infty}^{\infty} c(\tau) q(-\tau)\, d\tau \tag{4.106}$$

Thus substituting Equations (4.97), (4.101) to (4.104) and (4.106) into (4.95), we may express the mean-square error $J$ for the binary data transmission system of Figure 4.7 as

$$J = \frac{1}{2} + \frac{1}{2} \int_{-\infty}^{\infty} \int_{-\infty}^{\infty} \left( R_q(t - \tau) + \frac{N_0}{2}\, \delta(t - \tau) \right) c(t) c(\tau)\, dt\, d\tau - \int_{-\infty}^{\infty} c(t) q(-t)\, dt \tag{4.107}$$

For convenience of presentation, we have made the following changes in variables: $\tau_1$ and $\tau_2$ in the first integral are replaced by $t$ and $\tau$, respectively, and $\tau$ is replaced with $t$ in the second integral. Note also that this expression for the mean-square error $J$ is in actual fact normalized with respect to the variance of the transmitted symbols $a_k$ by virtue of the assumption made in Equation (4.96).

With the formula of Equation (4.107) for the mean-square error $J$ at hand, we are now ready to specify the design of the receive filter in Figure 4.7. Differentiating Equation (4.107) with respect to the impulse response $c(t)$ of the receive filter, and then setting the result equal to zero, we get

$$\int_{-\infty}^{\infty} \left( R_q(t - \tau) + \frac{N_0}{2}\, \delta(t - \tau) \right) c(\tau)\, d\tau = q(-t) \tag{4.108}$$

Equation (4.108) is the formula for finding the impulse response $c(t)$ of the equalizer optimized in the mean-square error sense. An equalizer so designed is referred to as the *minimum-mean square error (mmse) equalizer*.

Taking the Fourier transform of both sides of Equation (4.108), we obtain

$$\left( S_q(f) + \frac{N_0}{2} \right) C(f) = Q^*(f) \tag{4.109}$$

where $c(t) \rightleftharpoons C(f)$, $q(t) \rightleftharpoons Q(f)$, and $R_q \rightleftharpoons S_q(f)$. Solving Equation (4.109) for $C(f)$, we get

$$C(f) = \frac{Q^*(f)}{S_q(f) + \dfrac{N_0}{2}} \tag{4.110}$$

In Problem 4.33 it is shown that the power spectral density of the sequence $\{q(kT_b)\}$ can be expressed as

$$S_q(f) = \frac{1}{T_b} \sum_k \left| Q\left(f + \frac{k}{T_b}\right) \right|^2 \tag{4.111}$$

which means that the frequency response $C(f)$ of the optimum linear receiver is *periodic* with period $1/T_b$. Equation (4.110) suggests the interpretation of the optimum linear receiver as the cascade connection of two basic components:[8]

- A *matched filter* whose impulse response is $q(-t)$, where $q(t) = g(t) \star h(t)$.
- A *transversal (tapped-delay-line) equalizer* whose frequency response is the inverse of the periodic function $S_q(f) + (N_0/2)$.

To implement Equation (4.110) exactly we need an equalizer of infinite length. In practice, we may approximate the optimum solution by using an equalizer with a finite set of coefficients $\{c_k\}_{k=-N}^{N}$, provided $N$ is large enough. Thus the receiver takes the form shown in Figure 4.27. Note that the block labeled $z^{-1}$ in Figure 4.27 introduces a delay equal to $T_b$, which means that the tap spacing of the equalizer is exactly the same as the bit duration $T_b$. An equalizer so configured is said to be *synchronous* with the transmitter.

## ■ PRACTICAL CONSIDERATIONS

The mmse receiver of Figure 4.27 works well in the laboratory, where we have access to the system to be equalized, in which case we may determine a transversal equalizer characterized by the set of coefficients $\{c_k\}_{k=-N}^{N}$, which provides an adequate approximation to the frequency response $C(f)$ of Equation (4.110). In a real-life telecommunications environment, however, the channel is usually time varying. For example, in a public

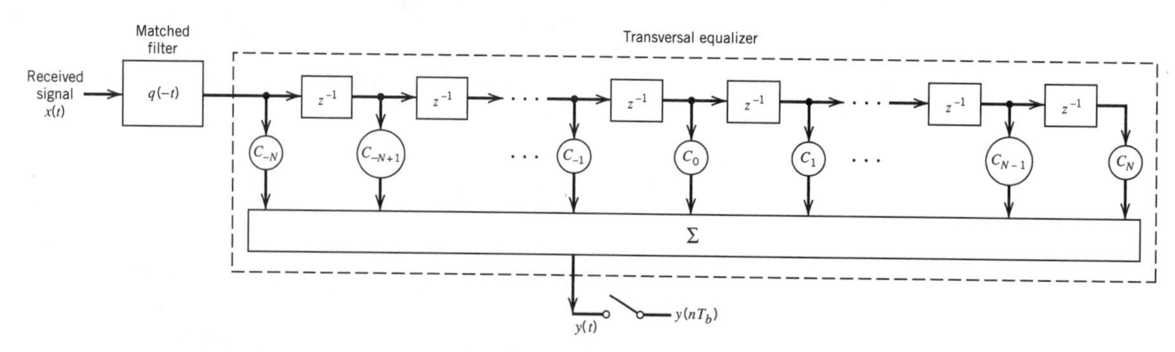


**FIGURE 4.27** Optimum linear receiver consisting of the cascade connection of matched filter and transversal equalizer.

switched telephone network, we find that two factors contribute to the distribution of pulse distortion on different link connections:

- Differences in the transmission characteristics of the individual links that may be switched together.
- Differences in the number of links in a connection.

The result is that the telephone channel is random in the sense of being one of an ensemble of possible physical realizations. Consequently, the use of a fixed pair of matched filter and equalizer designed on the basis of average channel characteristics may not adequately reduce the effects of intersymbol interference and channel noise. To realize the full transmission capability of the telephone channel, we need an *adaptive receiver*[9] that provides for the *adaptive implementation of both the matched filter and the equalizer in a combined manner*. The receiver is adaptive in the sense that the equalizer coefficients are adjusted automatically in accordance with a built-in algorithm.

Another point of interest is that it may be desirable to have the taps of the equalizer spaced by an amount closer than the symbol period; typically, the spacing between adjacent taps is set equal to $T/2$. The resulting structure is known as a *fractionally spaced equalizer (FSE)*. An FSE has the capability of compensating for delay distortion much more effectively than a conventional synchronous equalizer. Another advantage of the FSE is the fact that data transmission may begin with an arbitrary sampling phase. However, mathematical analysis of the FSE is more complicated than for a synchronous equalizer and will therefore not be pursued here.[10]

## 4.10 *Adaptive Equalization*

In this section we develop a simple and yet effective algorithm for the adaptive equalization of a linear channel of unknown characteristics. Figure 4.28 shows the structure of an adaptive synchronous equalizer, which incorporates the matched filtering action. The algorithm used to adjust the equalizer coefficients assumes the availability of a desired response. One's first reaction to the availability of a replica of the transmitted signal is: If such a signal is available at the receiver, why do we need adaptive equalization? To answer this question, we first note that a typical telephone channel changes little during an average data call. Accordingly, prior to data transmission, the equalizer is adjusted under the guid-

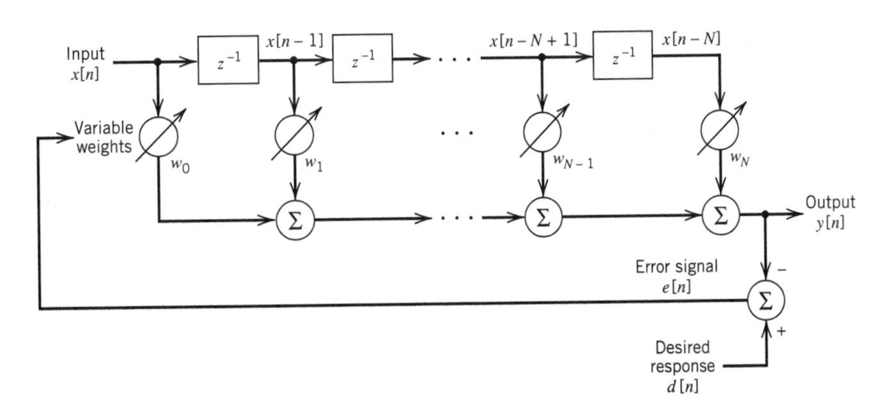


**FIGURE 4.28** Block diagram of adaptive equalizer.

ance of a *training sequence* transmitted through the channel. A synchronized version of this training sequence is generated at the receiver, where (after a time shift equal to the transmission delay through the channel) it is applied to the equalizer as the desired response. A training sequence commonly used in practice is the *pseudonoise (PN) sequence*, which consists of a deterministic periodic sequence with noise-like characteristics. Two identical PN sequence generators are used, one at the transmitter and the other at the receiver. When the training process is completed, the PN sequence generator is switched off, and the adaptive equalizer is ready for normal data transmission. Detailed description of PN sequence generators is presented in Chapter 7.

## ■ LEAST-MEAN-SQUARE ALGORITHM (REVISITED)

To simplify notational matters, we let

$$x[n] = x(nT)$$
$$y[n] = y(nT)$$

Then, the output $y[n]$ of the tapped-delay-line equalizer in response to the input sequence $\{x[n]\}$ is defined by the discrete convolution sum (see Figure 4.28)

$$y[n] = \sum_{k=0}^{N} w_k x[n-k] \tag{4.112}$$

where $w_k$ is the weight at the $k$th tap, and $N + 1$ is the total number of taps. The tap-weights constitute the adaptive filter coefficients. We assume that the input sequence $\{x[n]\}$ has finite energy. We have used a notation for the equalizer weights in Figure 4.28 that is different from the corresponding notation in Figure 4.27 to emphasize the fact that the equalizer in Figure 4.28 also incorporates matched filtering.

The adaptation may be achieved by observing the error between the desired pulse shape and the actual pulse shape at the filter output, measured at the sampling instants, and then using this error to estimate the direction in which the tap-weights of the filter should be changed so as to approach an optimum set of values. For the adaptation, we may use a criterion based on minimizing the *peak distortion*, defined as the worst-case intersymbol interference at the output of the equalizer. The development of an adaptive equalizer using such a criterion builds on the zero-forcing concept described briefly in Section 4.9. However, the equalizer is optimum only when the peak distortion at its input is less than 100 percent (i.e., the intersymbol interference is not too severe). A better approach is to use a mean-square error criterion, which is more general in application; also an adaptive equalizer based on the mean-square error criterion appears to be less sensitive to timing perturbations than one based on the peak distortion criterion. Accordingly, in what follows we use the mean-square error criterion to derive the adaptive equalization algorithm.

Let $a[n]$ denote the *desired response* defined as the polar representation of the $n$th transmitted binary symbol. Let $e[n]$ denote the *error signal* defined as the difference between the desired response $a[n]$ and the actual response $y[n]$ of the equalizer, as shown by

$$e[n] = a[n] - y[n] \tag{4.113}$$

In the *least-mean-square (LMS) algorithm*[11] for adaptive equalization, the error signal $e[n]$ actuates the adjustments applied to the individual tap weights of the equalizer as the algorithm proceeds from one iteration to the next. A derivation of the LMS algorithm for adaptive prediction was presented in Section 3.13. Recasting Equation (3.72) into its most general form, we may state the formula for the LMS algorithm in words as follows:

$$\begin{pmatrix}\text{Updated value}\\ \text{of } k\text{th tap-}\\ \text{weight}\end{pmatrix} = \begin{pmatrix}\text{Old value}\\ \text{of } k\text{th tap-}\\ \text{weight}\end{pmatrix} + \begin{pmatrix}\text{Step-size}\\ \text{parameter}\end{pmatrix} \cdot \begin{pmatrix}\text{Input signal}\\ \text{applied to}\\ k\text{th tap-}\\ \text{weight}\end{pmatrix}\begin{pmatrix}\text{Error}\\ \text{signal}\end{pmatrix} \quad (4.114)$$

Let $\mu$ denote the step-size parameter. From Figure 4.28 we see that the input signal applied to the $k$th tap-weight at time step $n$ is $x[n-k]$. Hence, using $\hat{w}_k(n)$ as the old value of the $k$th tap-weight at time step $n$, the updated value of this tap-weight at time step $n+1$ is, in light of Equation (4.114), defined by

$$\hat{w}_k[n+1] = \hat{w}_k[n] + \mu x[n-k]e[n], \qquad k = 0, 1, \ldots, N \quad (4.115)$$

where

$$e[n] = a[n] - \sum_{k=0}^{N} \hat{w}_k[n]x[n-k] \quad (4.116)$$

These two equations constitute the *LMS algorithm for adaptive equalization.* Note that the length of the adaptive equalizer in Figure 4.28 is not to be confused with the length of the equalizer in Figure 4.27.

We may simplify the formulation of the LMS algorithm using matrix notation. Let the $(N+1)$-by-1 vector $\mathbf{x}[n]$ denote the tap-inputs of the equalizer:

$$\mathbf{x}[n] = [x[n], \ldots, x[n-N+1], x[n-N]]^T \quad (4.117)$$

where the superscript $T$ denotes matrix transposition. Correspondingly, let the $(N+1)$-by-1 vector $\hat{\mathbf{w}}[n]$ denote the tap-weights of the equalizer:

$$\hat{\mathbf{w}}[n] = [\hat{w}_0[n], \hat{w}_1[n], \ldots, \hat{w}_N[n]]^T \quad (4.118)$$

We may then use matrix notation to recast the convolution sum of Equation (4.112) in the compact form

$$y[n] = \mathbf{x}^T[n]\hat{\mathbf{w}}[n] \quad (4.119)$$

where $\mathbf{x}^T[n]\hat{\mathbf{w}}[n]$ is referred to as the *inner product* of the vectors $\mathbf{x}[n]$ and $\hat{\mathbf{w}}[n]$. We may now summarize the LMS algorithm for adaptive equalization as follows:

1. Initialize the algorithm by setting $\hat{\mathbf{w}}[1] = \mathbf{0}$ (i.e., set all the tap-weights of the equalizer to zero at $n = 1$, which corresponds to time $t = T$).
2. For $n = 1, 2, \ldots,$ compute

$$\begin{aligned} y[n] &= \mathbf{x}^T[n]\hat{\mathbf{w}}[n] \\ e[n] &= a[n] - y[n] \\ \hat{\mathbf{w}}[n+1] &= \hat{\mathbf{w}}[n] + \mu e[n]\mathbf{x}[n] \end{aligned}$$

   where $\mu$ is the step-size parameter.
3. Continue the iterative computation until the equalizer reaches a "steady state," by which we mean that the actual mean-square error of the equalizer essentially reaches a constant value.

The LMS algorithm is an example of a feedback system, as illustrated in the block diagram of Figure 4.29, which pertains to the $k$th filter coefficient. It is therefore possible

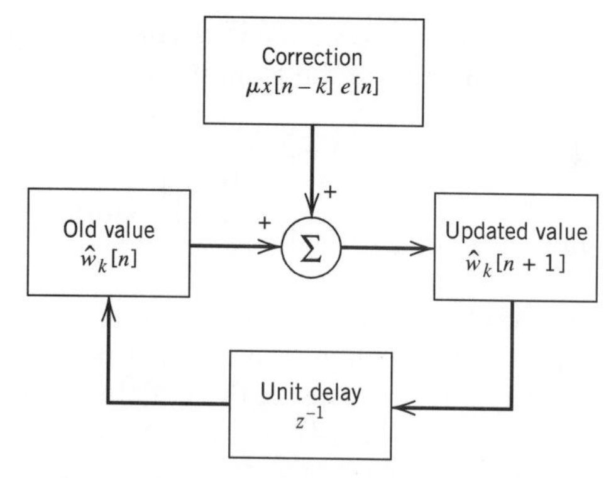


**FIGURE 4.29** Signal-flow graph representation of the LMS algorithm involving the $k$th tap weight.

for the algorithm to diverge (i.e., for the adaptive equalizer to become unstable). Unfortunately, the convergence behavior of the LMS algorithm is difficult to analyze. Nevertheless, provided that the step-size parameter $\mu$ is assigned a small value, we find that after a large number of iterations the behavior of the LMS algorithm is roughly similar to that of the *steepest-descent algorithm*, which uses the actual gradient rather than a noisy estimate for the computation of the tap-weights. (The steepest-descent algorithm was discussed in Section 3.13.)

## ■ OPERATION OF THE EQUALIZER

There are two modes of operation for an adaptive equalizer, namely, the training mode and decision-directed mode, as shown in Figure 4.30. During the *training mode*, as explained previously, a known PN sequence is transmitted and a synchronized version of it is generated in the receiver, where (after a time shift equal to the transmission delay) it is applied to the adaptive equalizer as the desired response; the tap-weights of the equalizer are thereby adjusted in accordance with the LMS algorithm.

When the training process is completed, the adaptive equalizer is switched to its second mode of operation: the *decision-directed mode*. In this mode of operation, the error signal is defined by

$$e[n] = \hat{a}[n] - y[n] \tag{4.120}$$

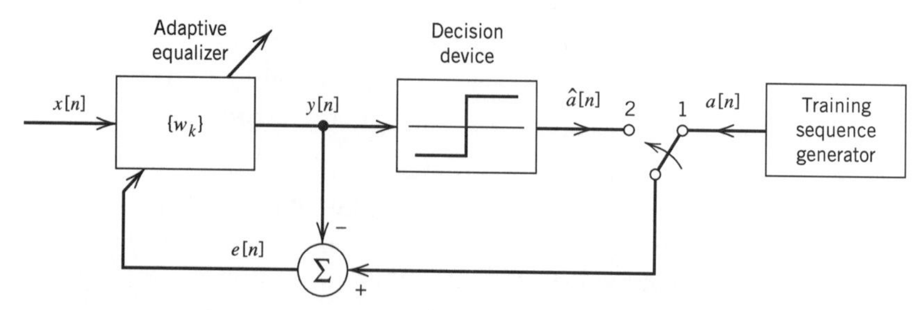


**FIGURE 4.30** Illustrating the two operating modes of an adaptive equalizer: For the training mode, the switch is in position 1; and for the tracking mode, it is moved to position 2.

where $y[n]$ is the equalizer output at time $t = nT$, and $\hat{a}[n]$ is the final (not necessarily) correct estimate of the transmitted symbol $a[n]$. Now, in normal operation the decisions made by the receiver are correct with high probability. This means that the error estimates are correct most of the time, thereby permitting the adaptive equalizer to operate satisfactorily. Furthermore, an adaptive equalizer operating in a decision-directed mode is able to *track* relatively slow variations in channel characteristics.

It turns out that the larger the step-size parameter $\mu$, the faster the tracking capability of the adaptive equalizer. However, a large step-size parameter $\mu$ may result in an unacceptably high *excess mean-square error*, defined as that part of the mean-square value of the error signal in excess of the minimum attainable value $J_{\min}$ (which results when the tap-weights are at their optimum settings). We therefore find that in practice the choice of a suitable value for the step-size parameter $\mu$ involves making a compromise between fast tracking and reducing the excess mean-square error.

## Decision-Feedback Equalization

To develop further insight into adaptive equalization, consider a baseband channel with impulse response denoted in its sampled form by the sequence $\{h[n]\}$ where $h[n] = h(nT)$. The response of this channel to an input sequence $\{x[n]\}$, in the absence of noise, is given by the discrete convolution sum

$$\begin{aligned} y[n] &= \sum_k h[k]x[n-k] \\ &= h[0]x[n] + \sum_{k<0} h[k]x[n-k] + \sum_{k>0} h[k]x[n-k] \end{aligned} \tag{4.121}$$

The first term of Equation (4.121) represents the desired data symbol. The second term is due to the *precursors* of the channel impulse response that occur before the main sample $h[0]$ associated with the desired data symbol. The third term is due to the *postcursors* of the channel impulse response that occur after the main sample $h[0]$. The precursors and postcursors of a channel impluse response are illustrated in Figure 4.31. The idea of *decision-feedback equalization*[12] is to use data decisions made on the basis of precursors of the channel impulse response to take care of the postcursors; for the idea to work, however, the decisions would obviously have to be correct. Provided that this condition is satisfied,

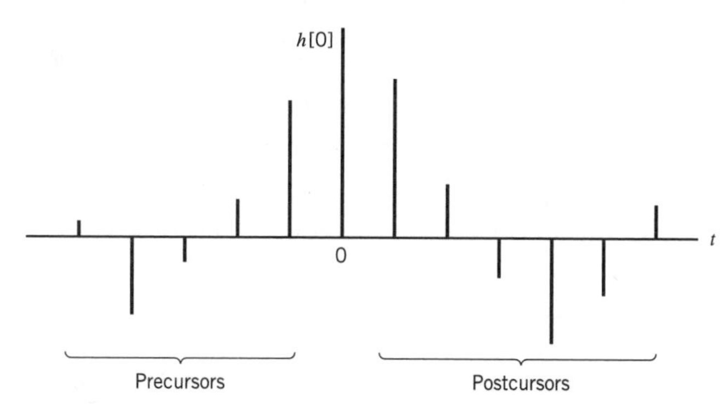


**FIGURE 4.31** Impulse response of a discrete-time channel, depicting the precursors and postcursors.

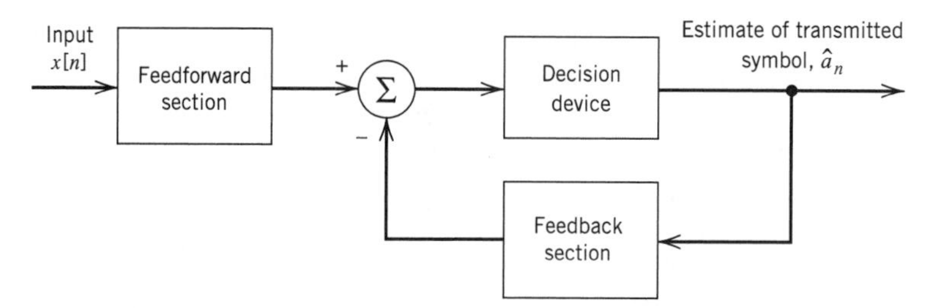


**FIGURE 4.32** Block diagram of decision-feedback equalizer.

a decision-feedback equalizer is able to provide an improvement over the performance of the tapped-delay-line equalizer.

A *decision-feedback equalizer* (DFE) consists of a feedforward section, a feedback section, and a decision device connected together as shown in Figure 4.32. The feedforward section consists of a tapped-delay-line filter whose taps are spaced at the reciprocal of the signaling rate. The data sequence to be equalized is applied to this section. The feedback section consists of another tapped-delay-line filter whose taps are also spaced at the reciprocal of the signaling rate. The input applied to the feedback section consists of the decisions made on previously detected symbols of the input sequence. The function of the feedback section is to subtract out that portion of the intersymbol interference produced by previously detected symbols from the estimates of future samples.

Note that the inclusion of the decision device in the feedback loop makes the equalizer intrinsically *nonlinear* and therefore more difficult to analyze than an ordinary tapped-delay-line equalizer. Nevertheless, the mean-square error criterion can be used to obtain a mathematically tractable optimization of a decision-feedback equalizer. Indeed, the LMS algorithm can be used to jointly adapt both the feedforward tap-weights and the feedback tap-weights based on a *common* error signal; see Problem 4.37.

On the basis of extensive comparative evaluations of a linear equalizer and decision-feedback equalizer reported in the literature,[13] we may report that when the frequency response of a linear channel is characterized by severe amplitude distortion or relatively sharp amplitude cutoff, the decision-feedback equalizer offers a significant improvement in performance over a linear equalizer for an equal number of taps. It is presupposed here that the feedback decisions in the DFE are all correct. For an example of sharp amplitude cutoff, see the frequency response of a telephone channel depicted in Figure 8 in the Background and Preview chapter.

Unlike a linear equalizer, a decision-feedback equalizer suffers from *error propagation*. However, despite the fact that the DFE is a feedback system, error propagation will not persist indefinitely. Rather, decision errors tend to occur in *bursts*. To justify this kind of behavior, we offer the following intuitive reasoning:[14]

- Let $L$ denote the number of taps in the feedback section of a DFE. After a sequence of $L$ consecutive correct decisions, all decision errors in the feedback section will be flushed out. This points to an error propagation of finite duration.
- When a decision error is made, the probability of the next decision being erroneous too is clearly no worse than 1/2.
- Let $K$ denote the duration of error propagation, that is, the number of symbols needed to make $L$ consecutive correct decisions. Then the average error rate is $(K/2)P_0$, where $K/2$ is the average number of errors produced by a single decision error, and $P_0$ is the probability of error given that the past $L$ decisions are all correct.

- In a fair-coin tossing experiment, the average number of coin tosses, $K$, needed to get $L$ successive heads (representing no errors) turns out to be $2(2^L - 1)$.

It follows therefore that the effect of error propagation in a decision-feedback equalizer is to increase the average error rate by a factor approximately equal to $2^L$, compared to the probability of making the first error. For example, for $L = 3$ the average error rate is increased by less than an order of magnitude due to error propagation.

## 4.11 *Computer Experiments: Eye Patterns*

In previous sections of this chapter we have discussed various techniques for dealing with the effects of channel noise and intersymbol interference on the performance of a baseband pulse-transmission system. In the final analysis, what really matters is how to evaluate the combined effect of these impairments on overall system performance in an operational environment. An experimental tool for such an evaluation in an insightful manner is the so-called *eye pattern*, which is defined as the synchronized superposition of all possible realizations of the signal of interest (e.g., received signal, receiver output) viewed within a particular signaling interval. The eye pattern derives its name from the fact that it resembles the human eye for binary waves. The interior region of the eye pattern is called the *eye opening*.

An eye pattern provides a great deal of useful information about the performance of a data transmission system, as described in Figure 4.33. Specifically, we make the following statements:

- The width of the eye opening defines the *time interval over which the received signal can be sampled without error from intersymbol interference*; it is apparent that the preferred time for sampling is the instant of time at which the eye is open the widest.
- The *sensitivity of the system to timing errors* is determined by the rate of closure of the eye as the sampling time is varied.
- The height of the eye opening, at a specified sampling time, defines the *noise margin* of the system.

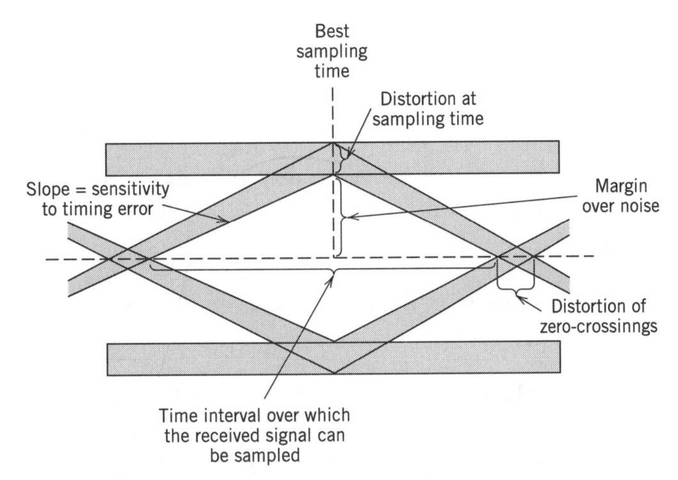


**FIGURE 4.33** Interpretation of the eye pattern.

When the effect of intersymbol interference is severe, traces from the upper portion of the eye pattern cross traces from the lower portion, with the result that the eye is completely closed. In such a situation, it is impossible to avoid errors due to the combined presence of intersymbol interference and noise in the system.

In the case of an $M$-ary system, the eye pattern contains $(M - 1)$ eye openings stacked up vertically one on the other, where $M$ is the number of discrete amplitude levels used to construct the transmitted signal. In a strictly linear system with truly random data, all these eye openings would be identical.

In the next two experiments, we use computer simulations to study the eye patterns for a quaternary ($M = 4$) baseband PAM transmission system under noiseless, noisy, and band-limited conditions. The effect of channel nonlinearity on eye patterns is discussed in Problem 4.38.

### Experiment 1: Effect of Channel Noise

Figure 4.34*a* shows the eye diagram of the system under idealized conditions: no channel noise and no bandwidth limitation. The source symbols used are randomly generated on a computer, with raised cosine pulse-shaping. The system parameters used for the generation of the eye diagram are as follows: Nyquist bandwidth $W = 0.5$ Hz, rolloff factor $\alpha = 0.5$, and symbol duration $T = T_b \log_2 M = 2T_b$. The openings in Figure 4.34 are perfect, indicating reliable operation of the system. Note that this figure has $M - 1 = 3$ openings.

Figures 4.34*b* and 4.34*c* show the eye diagrams for the system, but this time with channel noise corrupting the received signal. These two figures were simulated for signal-to-noise ratio SNR = 20 dB and 10 dB, respectively, with the SNR being measured at the channel output. When SNR = 20 dB the effect of channel noise is hardly discernible in Figure 4.34*b*, but when SNR = 10 dB the openings of the eye diagram in Figure 4.34*c* are barely visible.

### Experiment 2: Effect of Bandwidth Limitation

Figures 4.35*a* and 4.35*b* show the eye diagrams for the quaternary system using the same parameters as before, but this time under a bandwidth-limited condition and a noiseless channel. Specifically, the channel is now modeled by a low-pass *Butterworth filter*, whose squared magnitude response is defined by

$$|H(f)|^2 = \frac{1}{1 + (f/f_0)^{2N}}$$

where $N$ is the *order* of the filter, and $f_0$ is its 3-dB cutoff frequency. For the computer experiment described in Figure 4.35*a*, the following values are used:

$$N = 25 \text{ and } f_0 = 0.975 \text{ Hz}$$

The bandwidth required by the PAM trasmission system is computed to be

$$B_T = W(1 + \alpha) = 0.75 \text{ Hz}$$

Although the channel bandwidth (i.e., cutoff frequency) is greater than absolutely necessary, its effect on the passband is observed as a decrease in the size of the eye openings compared to those in Figure 4.34*a*. Instead of the distinct values at time $t = 1$ s (as shown in Figure 4.34*a*), now there is a blurred region.

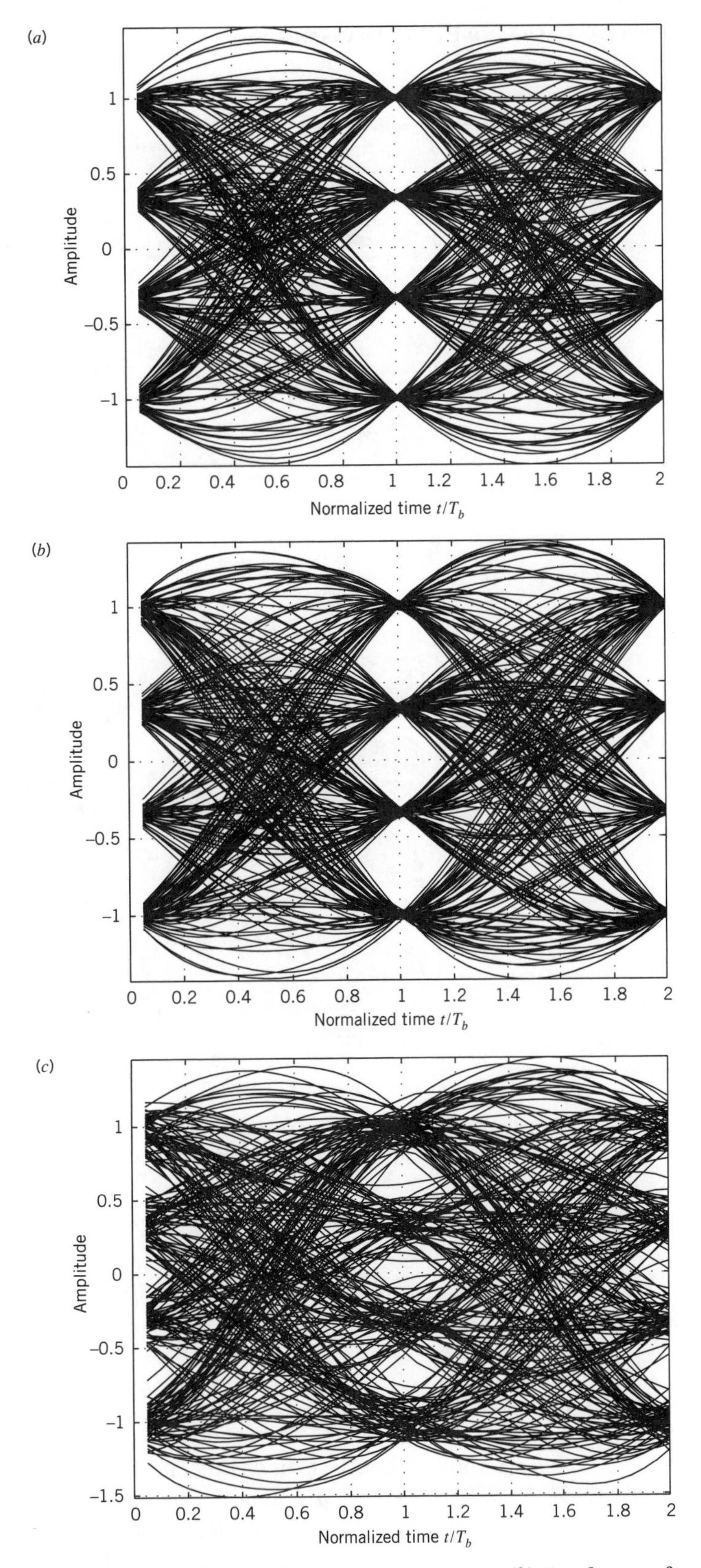


**FIGURE 4.34** (*a*) Eye diagram for noiseless quaternary system. (*b*) Eye diagram for quaternary system with SNR = 20 dB. (*c*) Eye diagram for quaternary system with SNR = 10 dB.

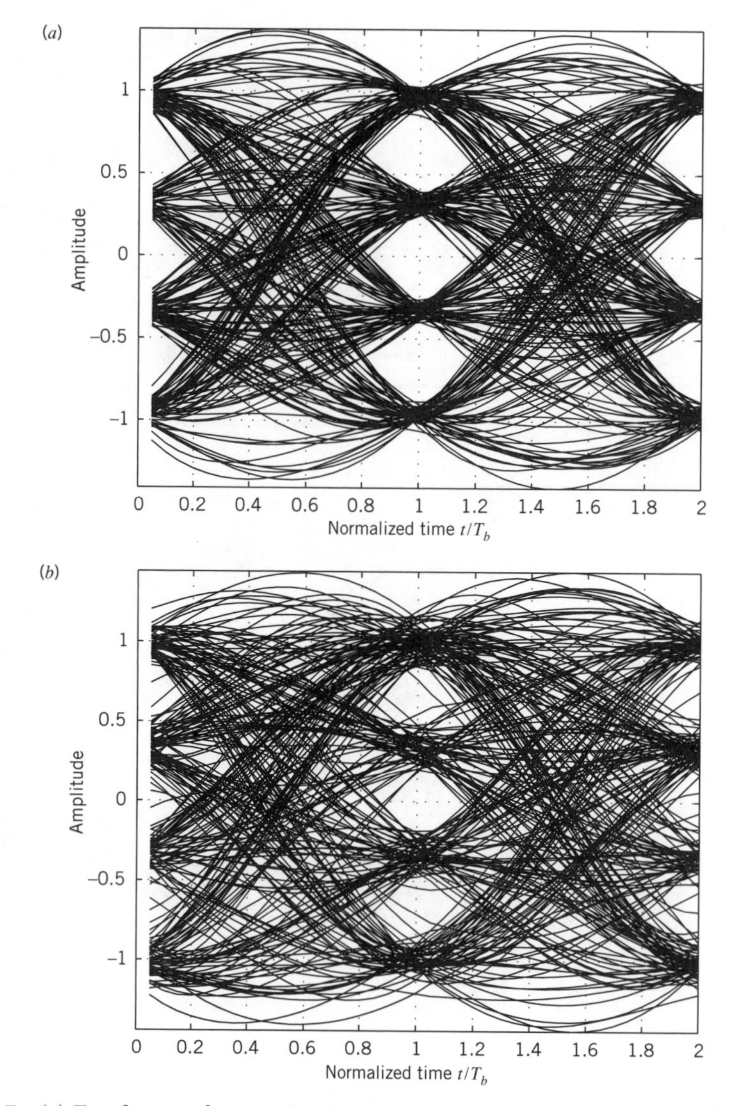


**FIGURE 4.35** (*a*) Eye diagram for noiseless band-limited quaternary system: cutoff frequency $f_0 = 0.975$ Hz. (*b*) Eye diagram for noiseless band-limited quaternary system: cutoff frequency $f_0 = 0.5$ Hz.

In Figure 4.35*b* the channel bandwidth is reduced further by modeling the channel as a low-pass Butterworth filter with $N = 25$ and $f_0 = 0.5$ Hz. The effect of reduced channel bandwidth is to further reduce the extent to which the eyes are open.

## 4.12 *Summary and Discussion*

In this chapter, we studied the effects of channel noise and intersymbol interference on the performance of baseband-pulse transmission systems. Intersymbol interference (ISI) is different from noise in that it is a *signal-dependent* form of interference that arises because of deviations in the frequency response of a channel from the ideal low-pass filter (Nyquist channel); it disappears when the transmitted signal is switched off. The result of these deviations is that the received pulse corresponding to a particular data symbol is affected by the tail ends of the pulses representing the previous symbols and the front ends of the pulses representing the subsequent symbols.

Depending on the received signal-to-noise ratio, we may distinguish three different situations that can arise in baseband-pulse transmission systems for channels with fixed characteristics:

1. *The effect of ISI is negligible in comparison to that of channel noise.*
The proper procedure in this case is to use a matched filter, which is the optimum linear time-invariant filter for maximizing the peak pulse signal-to-noise ratio.

2. *The received signal-to-noise ratio is high enough to ignore the effect of channel noise.*
In this case, we need to guard against the effects of ISI on the reconstruction of the transmitted data at the receiver. In particular, control must be exercised over the shape of the received pulse. This design objective can be achieved in one of two different ways:

   - Using a raised cosine spectrum for the overall frequency response of the baseband-pulse transmission system.
   - Using correlative-level coding or partial-response signaling that adds ISI to the transmitted signal in a controlled manner.

3. *The ISI and noise are both significant.*
For a mathematically tractable solution to this more difficult situation, we may use the mean-square error criterion. The resulting optimum linear receiver is called the minimum mean-square error (mmse) receiver. It consists of the cascade connection of a matched filter and linear transversal (tapped-delay-line) equalizer.

When, however, the channel is random in the sense of being one of an ensemble of possible physical realizations, which is frequently the case in a telecommunications environment, the use of fixed filter designs based on average channel characteristics may not be adequate. In situations of this kind, the preferred approach is to use an adaptive equalizer, the purpose of which is to compensate for variations in the frequency response of the channel automatically during the course of data transmission. The combined use of a tapped-delay-line filter and the least-mean-square (LMS) algorithm for adjusting the tap-weights provides the basis of a simple and yet highly effective method for implementing the adaptive equalizer. Such a device is capable of dealing with the combined effects of ISI and receiver noise in a nonstationary environment. Its practical value lies in the fact that almost every modem (modulator-demodulator) in commercial use today for the transmission of digital data over a voice-grade telephone channel uses an adaptive equalizer as an integral part.

Another important application of adaptive filtering is in the design of echo cancellers that constitute a critical component of transceivers for digital subscriber lines. Typically, a digital subscriber line uses a twisted pair as the transmission medium, the very same one used in ordinary telephone channels. However, unlike telephone channels, digital subscriber lines are designed to provide a high data-rate digital transmission capability between a digital network and subscriber plants, with a data rate of 64 kb/s and up.

## Notes and References

1. The classic books on baseband-pulse transmission are Lucky, Salz, and Weldon (1968) and Sunde (1969). For detailed treatment of different aspects of the subject, see Gitlin, Hayes, and Weinstein (1992). Proakis (1995), and Benedetto, Biglieri, and Castellani (1987).

2. The characterization of a matched filter was first derived by North in a classified report (RCA Laboratories Report PTR-6C, June 1943), which was published 20 years later: see the paper by North (1963). A similar result was obtained independently by Van Vleck and Middleton, who coined the term *matched filter*: see the paper by Van Vleck and Middleton (1946). For review material on the matched filter and its properties, see the papers by Turin (1960, 1976).

3. The *error function* denoted by erf($u$), is defined in a number of different ways in the literature. We shall use the following definition:

$$\text{erf}(u) = \frac{2}{\sqrt{\pi}} \int_0^u \exp(-z^2)\, dz$$

The error function has two useful properties:

(i) $\text{erf}(-u) = -\text{erf}(u)$
This is known as the *symmetry relation.*

(ii) As $u$ approaches infinity, erf($u$) approaches unity; that is,

$$\frac{2}{\sqrt{\pi}} \int_0^\infty \exp(-z^2)\, dz = 1$$

The *complementary error function* is defined by

$$\text{erfc}(u) = \frac{2}{\sqrt{\pi}} \int_u^\infty \exp(-z^2)\, dz$$

which is related to the error function as follows:

$$\text{erfc}(u) = 1 - \text{erf}(u)$$

Table A6.6 gives values of the error function erf($u$) for $u$ in the range 0 to 3.3.

For large positive values of $u$, we have two simple bounds on erfc($u$), one lower and the other upper, as shown by

$$\frac{\exp(-u^2)}{\sqrt{\pi}u}\left(1 - \frac{1}{2u^2}\right) < \text{erfc}(u) < \frac{\exp(-u^2)}{\sqrt{\pi}u}$$

The complementary error function provides the basis for a compact formulation of the probability of symbol error, as explained in Section 4.3. Another function that is also commonly used in the literature for this purpose is the Q-function. Consider a *standardized* Gaussian random variable $X$ of zero mean and unit variance. The probability that an observed value of the random variable $X$ will be greater than $v$ is given by the *Q-function*:

$$Q(v) = \frac{1}{\sqrt{2\pi}} \int_v^\infty \exp\left(-\frac{x^2}{2}\right) dx$$

*The Q-function defines the area under the standardized Gaussian tail.* The Q-function is related to the complementary error function as

$$Q(v) = \frac{1}{2}\,\text{erfc}\left(\frac{v}{\sqrt{2}}\right)$$

Conversely, putting $u = v/\sqrt{2}$, we have

$$\text{erfc}(u) = 2Q(\sqrt{2}u)$$

4. The criterion described in Equation (4.49) or Equation (4.53) was first formulated by Nyquist in the study of telegraph transmission theory; the 1928 paper by Nyquist is a classic. In the literature, this criterion is referred to as *Nyquist's first criterion.* In his 1928 paper, Nyquist described another method, referred to in the literature as *Nyquist's second*

*criterion*. The second method makes use of the instants of transition between unlike symbols in the received signal rather than centered samples. A discussion of the first and second criteria is presented in Bennett (1970, pp. 78–92) and in the paper by Gibby and Smith (1965). A third criterion attributed to Nyquist is discussed in Sunde (1969); see also the papers by Pasupathy (1974) and Sayar and Pasupathy (1987).

5. Correlative-level coding and partial-response signaling are synonymous; both terms are used in the literature. The idea of correlative coding was originated by Lender (1963). Lender's work was generalized for binary data transmission by Kretzmer (1966). For further details on correlative coding techniques, see the book by Gitlin, Hayes, and Weinstein (1992); see also the papers by Pasupathy (1977), Kabal and Pasupathy (1975), and Sousa and Pasupathy (1983).

6. The material on digital subscriber lines presented in Section 4.8 is based on the two papers by Lin and Tzeng (1988), and Lechleider (1989), and the books by Starr, Cioffi, and Silverman (1999) and Chen (1998).

7. For a discussion of line codes for digital subscriber loops, see Gitlin et al. (1992).

8. In Ericson (1971) it is shown that for every "reasonable" performance criterion, the optimum receiver can be realized as a matched filter followed by a tapped-delay-line equalizer, as shown in Figure 4.27. In addition to the mean-square error criterion considered in Section 4.9, reasonable performance criteria of interest include the following:

   (i) Minimization of the probability of symbol error.
   (ii) Zero-forcing equalization (to reduce the intersymbol interference to zero), followed by minimization of the probability of symbol error subject to this constraint.
   (iii) Minimization of signal-to-noise ratio at the sampling instants.

   Criterion (i) is the most natural approach to the optimization of a linear receiver; this approach, pursued in Aaron and Tufts (1966), is, unfortunately, complicated. Criterion (ii), due to Lucky et al. (1968), is a much simpler approach. Criterion (iii) is due to George (1965).

9. Adaptive equalization of telephone channels was pioneered by Lucky (1965, 1966). Since that time, numerous adaptive equalization schemes have been published in the literature, which provide equalization for specific synchronous data-transmission systems. For review papers on adaptive equalization, see Proakis (1975) and Qureshi (1982, 1985). Adaptive equalization is also discussed in detail in the books by Gitlin, Hayes, and Weinstein (1992, Chapter 8) and Proakis (1995, Chapter 6).

10. It appears that early work on fractionally spaced equalizers was initiated by Brady (1970). Other contributions to the subject include subsequent work by Ungerboeck (1976) and Gitlin and Weinstein (1981). A detailed discussion of fractionally spaced equalizers is also presented in Gitlin et al. (1992).

11. The LMS algorithm was originated by Widrow and Hoff, Jr. (1960). For a detailed convergence analysis of the LMS algorithm, see Haykin (1996, Chapter 9), and Widrow and Stearns (1985, Chapter 6).

12. Decision-feedback equalization was first described in a report by Austin (1967). The optimization of the decision-feedback equalizer for minimum mean-square error was first accomplished by Monsen (1971). A readable account of decision-feedback equalization is presented in the book by Gitlin, Hayes, and Weinstein (1992, pp. 500–510).

    Tomlinson (1971) and Harashima and Miyakawa (1972) describe a device for eliminating error propagation in a decision-feedback equalizer. The device, known as the *Tomlinson-Harashima precoder*, appears in the transmitter as a preprocessor to the modulator. The basic idea of this precoder is to move the feedback section in the decision feedback equalizer to the transmitter where it is impossible to make decision errors. However, this modification may result in a significant increase in transmit power; modulo arithmetic is used to overcome most of this power increase.

13. For performance comparison between linear equalizers and decision-feedback equalizers, see Gitlin et al. (1992) and Proakis (1995).

14. The intuitive discussion on error propagation in decision-feedback equalizers presented in Section 4.10 follows Gitlin et al. (1992).

    For a rigorous evaluation of the probability of symbol error $P_e$ in a decision-feedback equalizer with error propagation, see Duttweiler et al. (1974). In this paper it is shown that in the worst-case intersymbol interference, $P_e$ is multiplied by a factor of $2^L$ relative to the probability of error that results in the absence of decision errors at high signal-to-noise ratios, where $L$ is the number of taps in the feedback section. The result derived by Duttweiler et al. provides theoretical justification for the intuitive arguments presented in Section 4.10.

## PROBLEMS

### Matched Filters

**4.1** Consider the signal $s(t)$ shown in Figure P4.1.

(a) Determine the impulse response of a filter matched to this signal and sketch it as a function of time.

(b) Plot the matched filter output as a function of time.

(c) What is the peak value of the output?

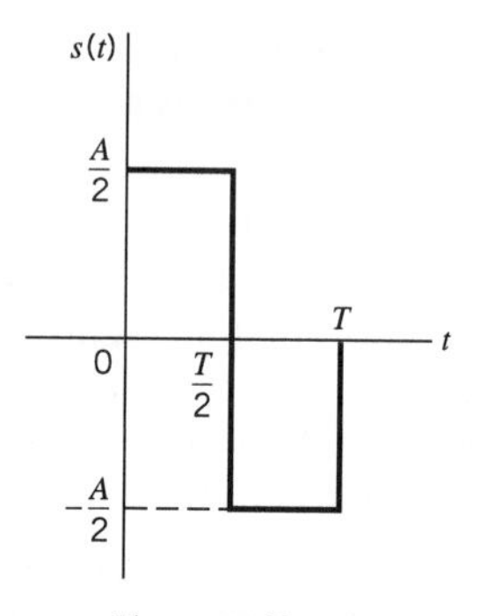


**FIGURE P4.1**

**4.2** Figure P4.2*a* shows a pair of pulses that are orthogonal to each other over the interval [0, $T$]. In this problem we investigate the use of this pulse-pair to study a *two-dimensional matched filter*.

(a) Determine the matched filters for the pulses $s_1(t)$ and $s_2(t)$ considered individually; for $s_1(t)$ the filter is the same as that considered in Problem 4.1.

(b) Form a two-dimensional matched filter by connecting the two matched filters of Part (a) in parallel, as shown in Figure P4.2*b*. Hence, demonstrate the following:

  (i) When the pulse $s_1(t)$ is applied to this two-dimensional filter, the response of the lower matched filter is zero.

  (ii) When the pulse $s_2(t)$ is applied to the two-dimensional filter, the response of the upper matched filter is zero.

Generalize the results of your investigation.

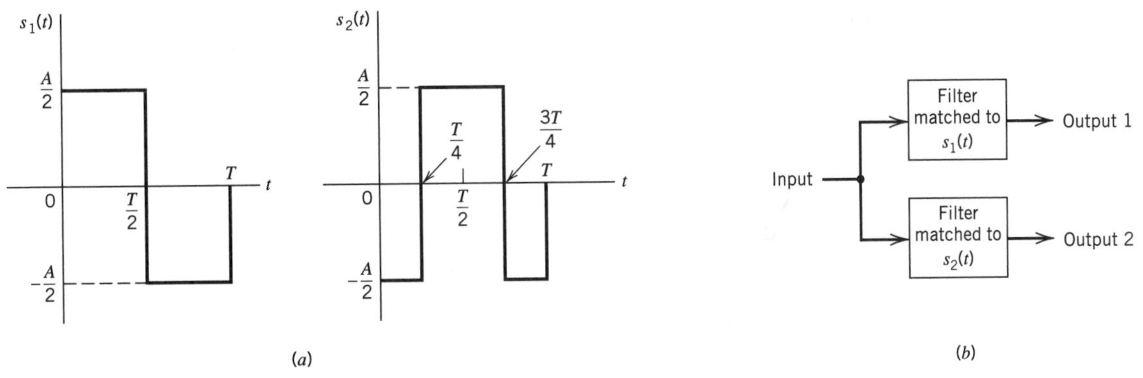


**FIGURE P4.2**

4.3 Consider a rectangular pulse defined by

$$g(t) = \begin{cases} A, & 0 \leq t \leq T \\ 0, & \text{otherwise} \end{cases}$$

It is proposed to approximate the matched filter for $g(t)$ by an ideal low-pass filter of bandwidth $B$; maximization of the peak pulse signal-to-noise ratio is the primary objective.

(a) Determine the optimum value of $B$ for which the ideal low-pass filter provides the best approximation to the matched filter.

(b) By how many decibels is the ideal low-pass filter worse off than the matched filter?

4.4 In this problem we explore another method for the approximate realization of a matched filter, this time using the simple resistance-capacitance (RC) low-pass filter shown in Figure P4.4. The frequency resonse of this filter is

$$H(f) = \frac{1}{1 + jf/f_0}$$

where $f_0 = 1/2\pi RC$. The input signal $g(t)$ is a rectangular pulse of amplitude $A$ and duration $T$. The requirement is to optimize the selection of the 3-dB cutoff frequency $f_0$ of the filter so that the peak pulse signal-to-noise ratio at the filter output is maximized. With this objective in mind, show that the optimum value of $f_0$ is $0.2/T$, for which the loss in signal-to-noise ratio compared to the matched filter is about 1 dB.

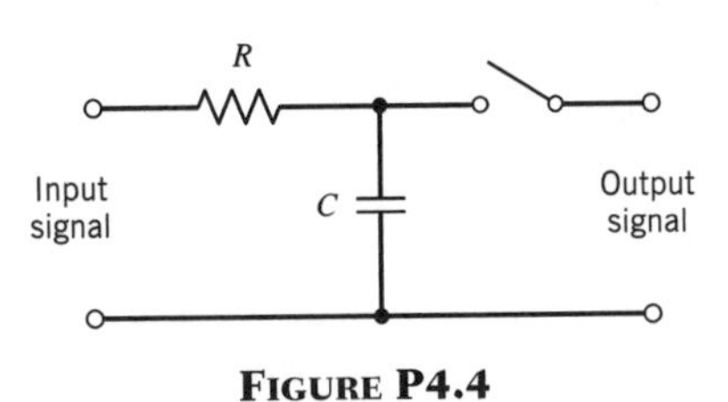


**FIGURE P4.4**

## Probability of Error Calculation

4.5 The formula for the optimum threshold in the receiver of Figure 4.4 is, in general, given by Equation (4.37). Discuss, in graphical terms, how this optimum choice affects the contributions of the two terms in Equation (4.35) for the average probability of symbol error $P_e$ by considering the following two cases:

(a) $p_0 > p_1$

(b) $p_1 < p_0$

where $p_0$ and $p_1$ are the *a priori* probabilities of symbols 0 and 1, respectively.

4.6 In a binary PCM system, symbols 0 and 1 have *a priori probabilities* $p_0$ and $p_1$, respectively. The conditional probability density function of the random variable $Y$ (with sample value $y$) obtained by sampling the matched filter output in the receiver of Figure 4.4 at the end of a signaling interval, given that symbol 0 was transmitted, is denoted by $f_Y(y|0)$. Similarly, $f_Y(y|1)$ denotes the conditional probability density function of $Y$, given that symbol 1 was transmitted. Let $\lambda$ denote the threshold used in the receiver, so that if the sample value $y$ exceeds $\lambda$, the receiver decides in favor of symbol 1; otherwise, it decides in favor of symbol 0. Show that the optimum threshold $\lambda_{\text{opt}}$, for which the average probability of error is a minimum, is given by the solution of

$$\frac{f_Y(\lambda_{\text{opt}}|1)}{f_Y(\lambda_{\text{opt}}|0)} = \frac{p_0}{p_1}$$

4.7 A binary PCM system using polar NRZ signaling operates just above the error threshold with an average probability of error equal to $10^{-6}$. Suppose that the signaling rate is doubled. Find the new value of the average probability of error. You may use Table A6.6 to evaluate the complementary error function.

4.8 A continuous-time signal is sampled and then transmitted as a PCM signal. The random variable at the input of the decision device in the receiver has a variance of 0.01 volts$^2$.

(a) Assuming the use of polar NRZ signaling, determine the pulse amplitude that must be transmitted for the average error rate not to exceed 1 bit in $10^8$ bits.

(b) If the added presence of interference causes the error rate to increase to 1 bit in $10^6$ bits, what is the variance of the interference?

4.9 A binary PCM wave uses unipolar NRZ signaling to transmit symbols 1 and 0; symbol 1 is represented by a rectangular pulse of amplitude $A$ and duration $T_b$. The channel noise is modeled as additive, white and Gaussian, with zero mean and power spectral density $N_0/2$. Assuming that symbols 1 and 0 occur with equal probability, find an expression for the average probability of error at the receiver output, using a matched filter as described in Section 4.3.

4.10 Repeat Problem 4.9 for the case of unipolar return-to-zero signaling, in which case symbol 1 is represented by a pulse of amplitude $A$ and duration $T_b/2$ and symbol 0 is represented by transmitting no pulse.

Hence show that this unipolar type of signaling requires twice the average power of unipolar nonreturn-to-zero (i.e., on-off) signaling for the same average probability of symbol error.

4.11 In this problem, we revisit the PCM receiver of Figure 4.4, but this time we consider the use of bipolar nonreturn-to-zero signaling, in which case the transmitted signal $s(t)$ is defined by

Binary symbol 1: $s(t) = \pm A$ for $0 < t \le T$

Binary symbol 0: $s(t) = 0, 0 < t \le T$

Determine the average probability of symbol error $P_e$ for this receiver assuming that the binary symbols 0 and 1 are equiprobable.

## Raised Cosine Spectrum

4.12 The nonreturn-to-zero pulse of Figure P4.12 may be viewed as a very crude form of a Nyquist pulse. Compare the spectral characteristics of these two pulses.

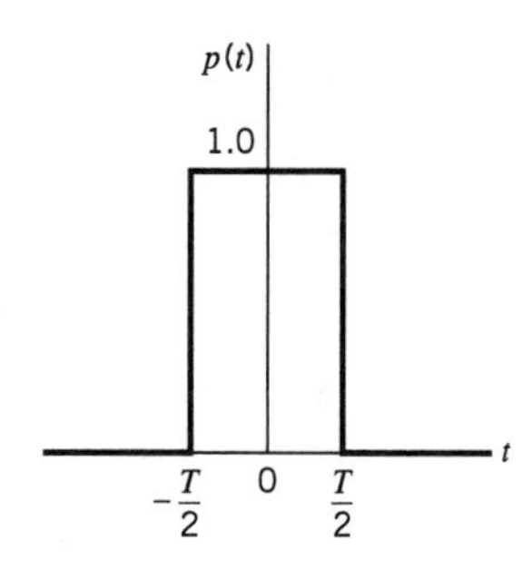


FIGURE P4.12

4.13 Determine the inverse Fourier transform of the frequency function $P(f)$ defined in Equation (4.60).

4.14 An analog signal is sampled, quantized, and encoded into a binary PCM wave. The specifications of the PCM system include the following:

Sampling rate = 8 kHz
Number of representation levels = 64

The PCM wave is transmitted over a baseband channel using discrete pulse-amplitude modulation. Determine the minimum bandwidth required for transmitting the PCM wave if each pulse is allowed to take on the following number of amplitude levels: 2, 4, or 8.

4.15 Consider a baseband binary PAM system that is designed to have a raised-cosine spectrum $P(f)$. The resulting pulse $p(t)$ is defined in Equation (4.62). How would this pulse be modified if the system was designed to have a linear phase response?

4.16 A computer puts out binary data at the rate of 56 kb/s. The computer output is transmitted using a baseband binary PAM system that is designed to have a raised-cosine spectrum. Determine the transmission bandwidth required for each of the following rolloff factors: $\alpha = 0.25, 0.5, 0.75, 1.0$.

4.17 Repeat Problem 4.16, given that each set of three successive binary digits in the computer output are coded into one of eight possible amplitude levels, and the resulting signal is transmitted using an eight-level PAM system designed to have a raised-cosine spectrum.

4.18 An analog signal is sampled, quantized, and encoded into a binary PCM wave. The number of representation levels used is 128. A synchronizing pulse is added at the end of each code word representing a sample of the analog signal. The resulting PCM wave is transmitted over a channel of bandwidth 12 kHz using a quaternary PAM system with raised-cosine spectrum. The rolloff factor is unity.

(a) Find the rate (b/s) at which information is transmitted through the channel.

(b) Find the rate at which the analog signal is sampled. What is the maximum possible value for the highest frequency component of the analog signal?

4.19 A binary PAM wave is to be transmitted over a baseband channel with an absolute maximum bandwidth of 75 kHz. The bit duration is 10 $\mu$s. Find a raised-cosine spectrum that satisfies these requirements.

## Correlative-Level Coding

4.20 The duobinary, ternary, and bipolar signaling techniques have one common feature: They all employ three amplitude levels. In what way does the duobinary technique differ from the other two?

4.21 The binary data stream 001101001 is applied to the input of a duobinary system.

(a) Construct the duobinary coder output and corresponding receiver output, without a precoder.

(b) Suppose that owing to error during transmission, the level at the receiver input produced by the second digit is reduced to zero. Construct the new receiver output.

4.22 Repeat Problem 4.21, assuming the use of a precoder in the transmitter.

4.23 The scheme shown in Figure P4.23 may be viewed as a differential encoder (consisting of the modulo-2 adder and the 1-unit delay element) connected in cascade with a special form of correlative coder (consisting of the 1-unit delay element and summer). A single-delay element is shown in Figure P4.23 since it is common to both the differential encoder and the correlative coder. In this differential encoder, a transition is represented by symbol 0 and no transition by symbol 1.

(a) Find the frequency response and impulse response of the correlative coder part of the scheme shown in Figure P4.23.

(b) Show that this scheme may be used to convert the on–off representation of a binary sequence (applied to the input) into the bipolar representation of the sequence at the output. You may illustrate this conversion by considering the sequence 010001101.

For descriptions of on–off, bipolar, and differential encoding of binary sequences, see Section 3.7.

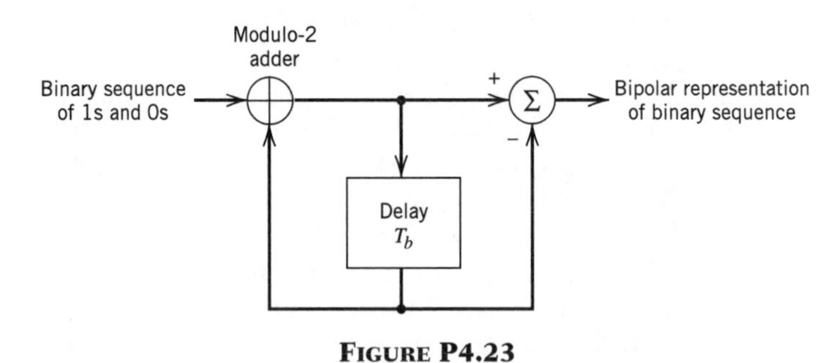


**FIGURE P4.23**

4.24 Consider a random binary wave $x(t)$ in which the 1s and 0s occur with equal probability, the symbols in adjacent time slots are statistically independent, and symbol 1 is represented by $A$ volts and symbol 0 by zero volts. This on–off binary wave is applied to the circuit of Figure P4.23.

(a) Using the result of Problem 4.23, show that the power spectral density of the bipolar wave $y(t)$ appearing at the output of the circuit equals

$$S_X(f) = T_b A^2 \sin^2(\pi f T_b) \operatorname{sinc}^2(f T_b)$$

(b) Plot the power spectral densities of the on–off and bipolar binary waves, and compare them.

4.25 The binary data stream 011100101 is applied to the input of a modified duobinary system.

(a) Construct the modified duobinary coder output and corresponding receiver output, without a precoder.

(b) Suppose that due to error during transmission, the level produced by the third digit is reduced to zero. Construct the new receiver output.

4.26 Repeat Problem 4.25 assuming the use of a precoder in the transmitter.

### *M*-ary PAM Systems

4.27 Consider a baseband $M$-ary system using $M$ discrete amplitude levels. The receiver model is as shown in Figure P4.27, the operation of which is governed by the following assumptions:

(a) The signal component in the received wave is

$$m(t) = \sum_n a_n \operatorname{sinc}\left(\frac{t}{T} - n\right)$$

where $1/T$ is the signaling rate in bauds.

(b) The amplitude levels are $a_n = \pm A/2, \pm 3A/2, \ldots, \pm(M-1)A/2$ if $M$ is even, and $a_n = 0, \pm A, \ldots, \pm(M-1)A/2$ if $M$ is odd.

(c) The $M$ levels are equiprobable, and the symbols transmitted in adjacent time slots are statistically independent.

(d) The channel noise $w(t)$ is white and Gaussian with zero mean and power spectral density $N_0/2$.

(e) The low-pass filter is ideal with bandwidth $B = 1/2T$.

(f) The threshold levels used in the decision device are $0, \pm A, \ldots, \pm(M-2)A/2$ if $M$ is even, and $\pm A/2, \pm 3A/2, \ldots, \pm(M-2)A/2$ if $M$ is odd.

The average probability of symbol error in this system is defined by

$$P_e = \left(1 - \frac{1}{M}\right) \operatorname{erfc}\left(\frac{A}{2\sqrt{2}\sigma}\right)$$

where $\sigma$ is the standard deviation of the noise at the input of the decision device. Demonstrate the validity of this general formula by determining $P_e$ for the following three cases: $M = 2, 3, 4$.

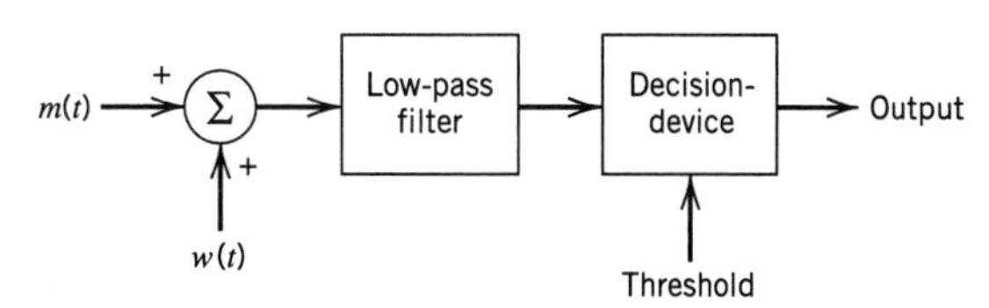


**FIGURE P4.27**

4.28 Suppose that in a baseband $M$-ary PAM system with $M$ equally likely amplitude levels, as described in Problem 4.27, the average probability of symbol error $P_e$ is less than $10^{-6}$ so as to make the occurrence of decoding errors negligible. Show that the minimum value of received signal-to-noise ratio in such a system is approximately given by

$$(\text{SNR})_{\min} \simeq 7.8(M^2 - 1)$$

### Digital Subscriber Lines

4.29 The amplitude distribution of cross-talk in a digital subscriber line may be modeled as Gaussian. Justify the validity of such a model. *Hint:* Typically, a cable contains many twisted pairs.

4.30 (a) Derive the formula for the power spectral density of a transmitted signal using the 2B1Q line code.

(b) Plot the power spectrum of the following line codes:

- Manchester code
- Modified duobinary code
- Bipolar return-to-zero code
- 2B1Q code

Hence compare the relative merits of these line codes for their suitability in a digital subscriber loop.

4.31 In this problem we use the LMS algorithm to formulate an adaptive echo canceller for use in a digital subscriber line. The basic principle of adaptive echo cancellation is to synthesize a replica of the echo and subtract it from the returned signal in an adaptive manner, as illustrated in Figure P4.31. The synthesized echo, denoted by $\hat{r}[n]$, is generated by passing the transmitted signal through an adaptive filter that ideally matches the transfer function of the echo path. The returned signal, consisting of the sum of actual echo $r[n]$ and the received signal $x[n]$, may be viewed as the desired response for the adaptive filtering process.

Using the LMS algorithm, formulate the equations that define the operation of the adaptive echo canceller in Figure P4.31.

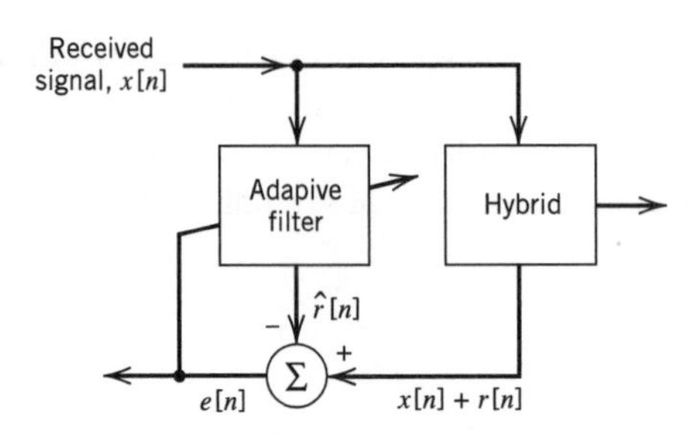


**FIGURE P4.31**

## Equalization

4.32 Figure P4.32 shows the cascade connection of a linear channel and a synchronous tapped-delay-line equalizer. The impulse response of the channel is denoted by $c(t)$, and that of the equalizer is denoted by $h(t)$. The $h(t)$ is defined by

$$h(t) = \sum_{k=-N}^{N} w_k \, \delta(t - kT)$$

where $T$ is the spacing between adjacent taps of the equalizer, and the $w_k$ are its tap-weights (coefficients). The impulse response of the cascaded system of Figure P4.32 is denoted by $p(t)$. The $p(t)$ is sampled uniformly at the rate $1/T$. To eliminate intersymbol interference, we require that the Nyquist criterion for distortionless transmission be satisfied, as shown by

$$p(nT) = \begin{cases} 1, & n = 0 \\ 0, & n \neq 0 \end{cases}$$

(a) By imposing this condition, show that the $(2N + 1)$ tap-weights of the resulting zero-forcing equalizer satisfy the following set of $(2N + 1)$ simultaneous equations:

$$\sum_{k=-N}^{N} w_k c_{n-k} = \begin{cases} 1, & n = 0 \\ 0, & n \neq \pm 1, \pm 2, \ldots, \pm N \end{cases}$$

where $c_n = c(nT)$. Hence, show that the zero-forcing equalizer is an *inverse filter* in that its transfer function is equal to the reciprocal of the transfer function of the channel.

(b) A shortcoming of the zero-forcing equalizer is *noise enhancement* that can result in poor performance in the presence of channel noise. To explore this phenomenon, consider a low-pass channel with a notch at the Nyquist frequency, that is, $H(f)$ is zero at $f = 1/2T$. Assuming that the channel noise is additive and white, show that the power spectral density of the noise at the equalizer output approaches infinity at $f = 1/2T$.

Even if the channel has no notch in its frequency response, the power spectral density of the noise at the equalizer output can assume high values. Justify the validity of this general statement.

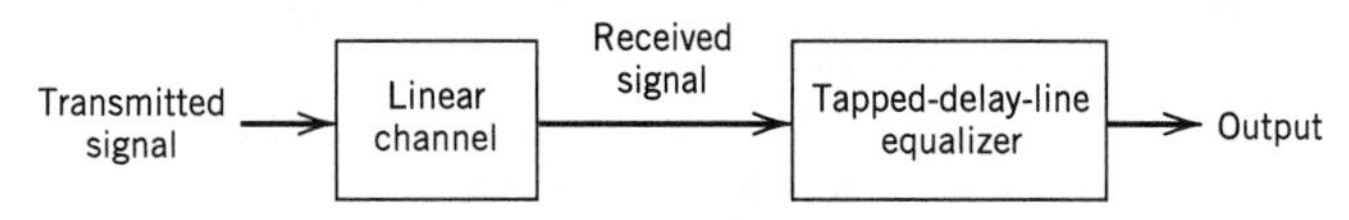


**FIGURE P4.32**

4.33 Consider Equation (4.108), which defines the impulse response of a minimum mean-square error receiver.

(a) Justify the validity of Equation (4.109) that is the Fourier-transformed version of Equation (4.108).

(b) The power spectral density $S_q(f)$ in Equation (4.109) is the Fourier transform of the autocorrelation $R_q(\tau_1, \tau_2)$ of the time function $q(t)$. The $R_q(\tau_1, \tau_2)$ is defined by Equation (4.98). Starting with Equation (4.98), derive the formula of Equation (4.111).

4.34 Some radio systems suffer from *multipath distortion*, which is caused by the existence of more than one propagation path between the transmitter and the receiver. Consider a channel the output of which, in response to a signal $s(t)$, is defined by (in the absence of noise)

$$x(t) = a_1 s(t - t_{01}) + a_2 s(t - t_{02})$$

where $a_1$ and $a_2$ are constant, and $t_{01}$ and $t_{02}$ represent transmission delays. It is proposed to use the three-tap delay-line-filter of Figure P4.34 to equalize the multipath distortion produced by this channel.

(a) Evaluate the transfer function of the channel.

(b) Evaluate the parameters of the tapped-delay-line filter in terms of $a_1$, $a_2$, $t_{01}$, and $t_{02}$, assuming that $a_2 \ll a_1$ and $t_{02} > t_{01}$.

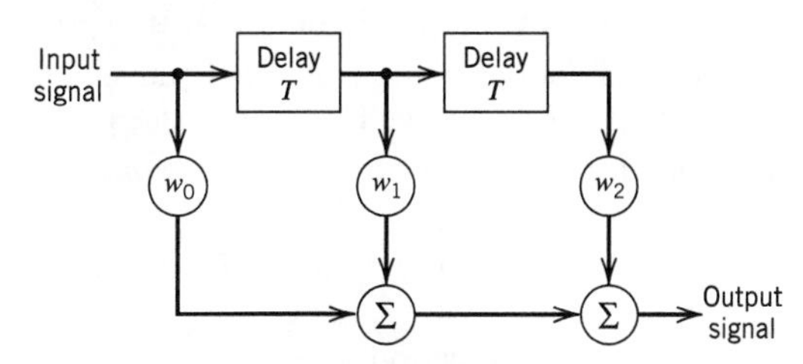


**FIGURE P4.34**

4.35 Let the sequence $\{x(nT)\}$ denote the input applied to a tapped-delay-line equalizer. Show that intersymbol interference is eliminated completely by the equalizer provided that its frequency response satisfies the condition

$$H(f) = \frac{T}{\sum_{k} X(f - k/T)}$$

where $T$ is the symbol duration.

As the number of taps in the equalizer approaches infinity, the frequency response of the equalizer becomes a Fourier series with real coefficients and can therefore approximate any function in the interval $(-1/2T, 1/2T)$. Demonstrate this property of the equalizer.

4.36 The step-size parameter $\mu$ plays a critical role in the operation of the LMS algorithm. In this context, discuss the following two issues:

(a) *Stability.* If $\mu$ exceeds a certain critical value, the algorithm diverges (i.e., the system becomes unstable).

(b) *Memory.* The reciprocal of $\mu$ may be viewed as a measure of the algorithm's memory: As we make $\mu$ smaller, more of the past samples of the input signal influence operation of the algorithm.

4.37 Let the vectors $\mathbf{w}^{(1)}[n]$ and $\mathbf{w}^{(2)}[n]$ denote the tap-weights of the feed-forward and feedback sections of the decision-feedback equalizer in Figure 4.32. Formulate the LMS algorithm for adjusting the tap-weights of this equalizer.

## Computer Experiments

4.38 In Section 4.11 we studied the eye diagram of a quaternary ($M = 4$) PAM baseband transmission system under both noisy and band-limited conditions. In that experiment, the channel was assumed linear. In a strictly linear system with truly random data, all the eye openings would be identical. In practice, however, it is often possible to discern asymmetries in the eye pattern, which are caused by nonlinearities in the communication channel.

In this experiment, we study the effect of a nonlinear channel on the openings of an eye pattern. Specifically, we repeat the computer experiment pertaining to the noiseless eye pattern of Figure 4.34*a* for $M = 4$, but this time assume that the channel is nonlinear with the following input-output relation:

$$x(t) = s(t) + as^2(t)$$

where $s(t)$ is the channel input and $x(t)$ is the channel output, and $a$ is a constant.

(a) Do the experiment for $a = 0, 0.05, 0.1, 0.2$.

(b) Hence, discuss how varying $a$ affects the shape of the eye pattern.

4.39 In this experiment we study the *root raised-cosine pulse* due to Chennakeshu and Saulnier (1993). This pulse, denoted by $p(t)$, has the following properties:

- The pulse $p(t)$ is symmetric in time, that is, $p(-t) = p(t)$.
- The squared Fourier transform of $p(t)$, namely, $P^2(f)$, satisfies the raised cosine spectrum of Equation (4.60), but the Fourier transform $P(f)$ itself does not.
- The pulse $p(t)$ satisfies the *orthogonality constraint*:

$$\int_{-\infty}^{\infty} p(t)p(t - nT)\,dt = 0, \qquad n = \pm 1, \pm 2, \ldots$$

where $T$ is the symbol period.

(a) Compute the baseband waveform of the binary data stream 101100 for rolloff factor $\alpha = 0.3$.

(b) Compare the waveform computed in part (a) with that obtained using the ordinary raised-cosine spectrum.

# CHAPTER 5

# SIGNAL-SPACE ANALYSIS

This chapter discusses some basic issues that pertain to the transmission of signals over an additive white Gaussian noise (AWGN) channel. Specifically, it addresses the following topics:

- *Geometric representation of signals with finite energy, which provides a mathematically elegant and highly insightful tool for the study of data transmission.*
- *Maximum likelihood procedure for the detection of a signal in AWGN channel.*
- *Derivation of the correlation receiver that is equivalent to the matched filter receiver discussed in the previous chapter.*
- *Probability of symbol error and the union bound for its approximate calculation.*

The material presented herein naturally leads to the study of passband data transmission covered in Chapter 6.

## 5.1 *Introduction*

Consider the most basic form of a digital communication system depicted in Figure 5.1. A *message source* emits one *symbol* every $T$ seconds, with the symbols belonging to an alphabet of $M$ symbols denoted by $m_1, m_2, \ldots, m_M$. Consider, for example, the remote connection of two digital computers, with one computer acting as an information source that calculates digital outputs based on observations and inputs fed into it. The resulting computer output is expressed as a sequence of 0s and 1s, which are transmitted to a second computer over a communication channel. In this case, the alphabet consists simply of two binary symbols: 0 and 1. A second example is that of a quaternary PCM encoder with an alphabet consisting of four possible symbols: 00, 01, 10, and 11. In any event, the *a priori* probabilities $p_1, p_2, \ldots, p_M$ specify the message source output. In the absence of prior information, it is customary to assume that the $M$ symbols of the alphabet are *equally likely*. Then we may express the probability that symbol $m_i$ is emitted by the source as

$$\begin{aligned} p_i &= P(m_i) \\ &= \frac{1}{M} \text{ for } i = 1, 2, \ldots, M \end{aligned} \tag{5.1}$$

The transmitter takes the message source output $m_i$ and codes it into a *distinct* signal $s_i(t)$ suitable for transmission over the channel. The signal $s_i(t)$ occupies the full duration $T$ allotted to symbol $m_i$. Most important, $s_i(t)$ is a real-valued *energy signal* (i.e., a signal with finite energy), as shown by

$$E_i = \int_0^T s_i^2(t)\, dt, \qquad i = 1, 2, \ldots, M \tag{5.2}$$

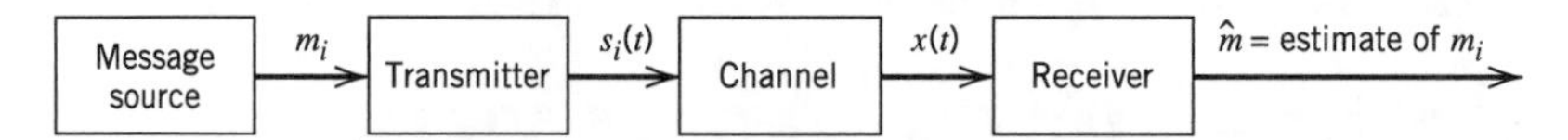


**FIGURE 5.1** Block diagram of a generic digital communication system.

The channel is assumed to have two characteristics:

1. The channel is *linear*, with a bandwidth that is wide enough to accommodate the transmission of signal $s_i(t)$ with negligible or no distortion.
2. The channel noise, $w(t)$, is the sample function of a *zero-mean white Gaussian noise process*. The reasons for this second assumption are that it makes receiver calculations tractable, and it is a reasonable description of the type of noise present in many practical communication systems.

We refer to such a channel as an *additive white Gaussian noise* (AWGN) *channel*. Accordingly, we may express the *received signal* $x(t)$ as

$$x(t) = s_i(t) + w(t), \qquad \begin{cases} 0 \le t \le T \\ i = 1, 2, \ldots, M \end{cases} \tag{5.3}$$

and thus model the channel as in Figure 5.2.

The receiver has the task of observing the received signal $x(t)$ for a duration of $T$ seconds and making a best *estimate* of the transmitted signal $s_i(t)$ or, equivalently, the symbol $m_i$. However, owing to the presence of channel noise, this decision-making process is statistical in nature, with the result that the receiver will make occasional errors. The requirement is therefore to design the receiver so as to minimize the *average probability of symbol error*, defined as

$$P_e = \sum_{i=1}^{M} p_i \, P(\hat{m} \neq m_i \,|m_i) \tag{5.4}$$

where $m_i$ is the transmitted symbol, $\hat{m}$ is the estimate produced by the receiver, and $P(\hat{m} \neq m_i \,|m_i)$ is the conditional error probability given that the $i$th symbol was sent. The resulting receiver is said to be *optimum in the minimum probability of error* sense.

This model provides a basis for the design of the optimum receiver, for which we will use geometric representation of the known set of transmitted signals, $\{s_i(t)\}$. This method, discussed in Section 5.2, provides a great deal of insight, with considerable simplification of detail.

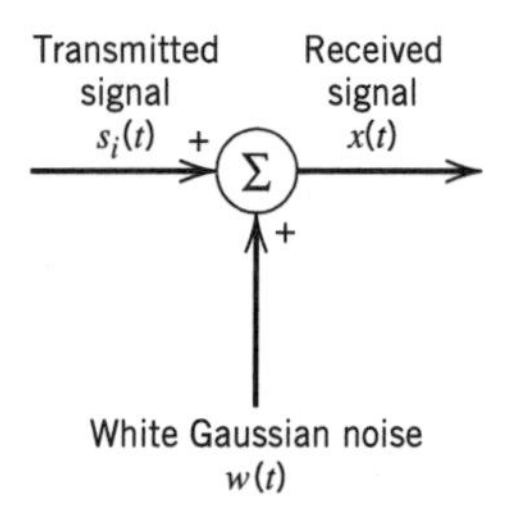


**FIGURE 5.2** Additive white Gaussian noise (AWGN) model of a channel.

## 5.2 Geometric Representation of Signals

The essence of *geometric representation of signals*[1] is to represent any set of $M$ energy signals $\{s_i(t)\}$ as linear combinations of $N$ *orthonormal basis functions*, where $N \le M$. That is to say, given a set of real-valued energy signals $s_1(t), s_2(t), \ldots, s_M(t)$, each of duration $T$ seconds, we write

$$s_i(t) = \sum_{j=1}^{N} s_{ij}\phi_j(t), \qquad \begin{cases} 0 \le t \le T \\ i = 1, 2, \ldots, M \end{cases} \tag{5.5}$$

where the coefficients of the expansion are defined by

$$s_{ij} = \int_0^T s_i(t)\phi_j(t)\,dt, \qquad \begin{cases} i = 1, 2, \ldots, M \\ j = 1, 2, \ldots, N \end{cases} \tag{5.6}$$

The real-valued basis functions $\phi_1(t), \phi_2(t), \ldots, \phi_N(t)$ are *orthonormal*, by which we mean

$$\int_0^T \phi_i(t)\phi_j(t)\,dt = \delta_{ij} = \begin{cases} 1 \text{ if } i = j \\ 0 \text{ if } i \ne j \end{cases} \tag{5.7}$$

where $\delta_{ij}$ is the *Kronecker delta*. The first condition of Equation (5.7) states that each basis function is *normalized* to have unit energy. The second condition states that the basis functions $\phi_1(t), \phi_2(t), \ldots, \phi_N(t)$ are *orthogonal* with respect to each other over the interval $0 \le t \le T$.

The set of coefficients $\{s_{ij}\}_{j=1}^{N}$ may naturally be viewed as an *N-dimensional vector*, denoted by $\mathbf{s}_i$. The important point to note here is that the vector $\mathbf{s}_i$ bears a *one-to-one* relationship with the transmitted signal $s_i(t)$:

- Given the $N$ elements of the vectors $\mathbf{s}_i$ (i.e., $s_{i1}, s_{i2}, \ldots, s_{iN}$) operating as input, we may use the scheme shown in Figure 5.3*a* to generate the signal $s_i(t)$, which follows

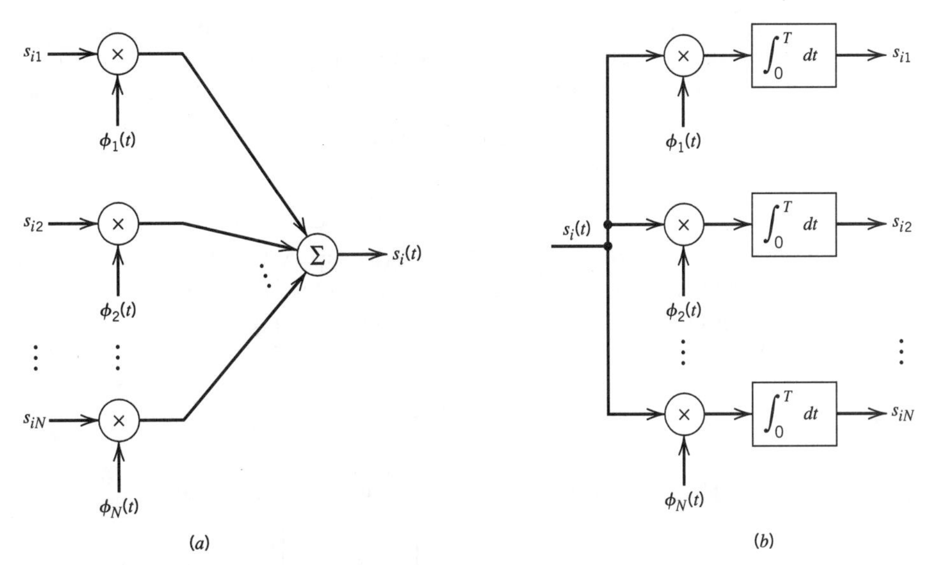


**FIGURE 5.3** (*a*) Synthesizer for generating the signal $s_i(t)$. (*b*) Analyzer for generating the set of signal vectors $\{\mathbf{s}_i\}$.

directly from Equation (5.5). It consists of a bank of $N$ multipliers, with each multiplier having its own basis function, followed by a summer. This scheme may be viewed as a *synthesizer*.

- Conversely, given the signals $s_i(t)$, $i = 1, 2, \ldots, M$, operating as input, we may use the scheme shown in Figure 5.3*b* to calculate the coefficients $s_{i1}, s_{i2}, \ldots, s_{iN}$ which follows directly from Equation (5.6). This second scheme consists of a bank of $N$ *product-integrators* or *correlators* with a common input, and with each one of them supplied with its own basis function. The scheme of Figure 5.3*b* may be viewed as an *analyzer*.

Accordingly, we may state that each signal in the set $\{s_i(t)\}$ is completely determined by the *vector* of its coefficients

$$\mathbf{s}_i = \begin{bmatrix} s_{i1} \\ s_{i2} \\ \vdots \\ s_{iN} \end{bmatrix}, \qquad i = 1, 2, \ldots, M \tag{5.8}$$

The vector $\mathbf{s}_i$ is called a *signal vector*. Furthermore, if we conceptually extend our conventional notion of two- and three-dimensional Euclidean spaces to an *N-dimensional Euclidean space*, we may visualize the set of signal vectors $\{\mathbf{s}_i | i = 1, 2, \ldots, M\}$ as defining a corresponding set of $M$ points in an $N$-dimensional Euclidean space, with $N$ mutually perpendicular axes labeled $\phi_1, \phi_2, \ldots, \phi_N$. This $N$-dimensional Euclidean space is called the *signal space*.

The idea of visualizing a set of energy signals geometrically, as just described, is of profound importance. It provides the mathematical basis for the geometric representation of energy signals, thereby paving the way for the noise analysis of digital communication systems in a conceptually satisfying manner. This form of representation is illustrated in Figure 5.4 for the case of a two-dimensional signal space with three signals, that is, $N = 2$ and $M = 3$.

In an $N$-dimensional Euclidean space, we may define *lengths* of vectors and *angles* between vectors. It is customary to denote the length (also called the *absolute value* or *norm*) of a signal vector $\mathbf{s}_i$ by the symbol $\| \mathbf{s}_i \|$. The squared-length of any signal vector $\mathbf{s}_i$ is defined to be the *inner product* or *dot product* of $\mathbf{s}_i$ with itself, as shown by

$$\begin{aligned} \| \mathbf{s}_i \|^2 &= \mathbf{s}_i^T \mathbf{s}_i \\ &= \sum_{j=1}^{N} s_{ij}^2, \qquad i = 1, 2, \ldots, M \end{aligned} \tag{5.9}$$

where $s_{ij}$ is the $j$th element of $\mathbf{s}_i$, and the superscript $T$ denotes matrix transposition.

There is an interesting relationship between the energy content of a signal and its representation as a vector. By definition, the energy of a signal $s_i(t)$ of duration $T$ seconds is

$$E_i = \int_0^T s_i^2(t)\, dt \tag{5.10}$$

Therefore, substituting Equation (5.5) into (5.10), we get

$$E_i = \int_0^T \left[ \sum_{j=1}^{N} s_{ij}\phi_j(t) \right] \left[ \sum_{k=1}^{N} s_{ik}\phi_k(t) \right] dt$$

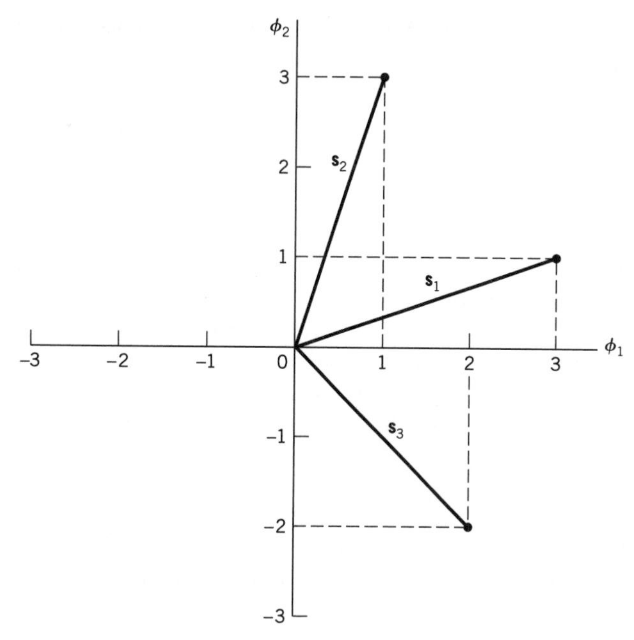


**FIGURE 5.4** Illustrating the geometric representation of signals for the case when $N = 2$ and $M = 3$.

Interchanging the order of summation and integration, and then rearranging terms, we get

$$E_i = \sum_{j=1}^{N} \sum_{k=1}^{N} s_{ij} s_{ik} \int_0^T \phi_j(t)\phi_k(t)dt \tag{5.11}$$

But since the $\phi_j(t)$ form an orthonormal set, in accordance with the two conditions of Equation (5.7), we find that Equation (5.11) reduces simply to

$$\begin{aligned} E_i &= \sum_{j=1}^{N} s_{ij}^2 \\ &= \| \mathbf{s}_i \|^2 \end{aligned} \tag{5.12}$$

Thus Equations (5.9) and (5.12) show that the energy of a signal $s_i(t)$ is equal to the squared length of the signal vector $\mathbf{s}_i(t)$ representing it.

In the case of a pair of signals $s_i(t)$ and $s_k(t)$, represented by the signal vectors $\mathbf{s}_i$ and $\mathbf{s}_k$, respectively, we may also show that

$$\int_0^T s_i(t)s_k(t)\, dt = \mathbf{s}_i^T \mathbf{s}_k \tag{5.13}$$

Equation (5.13) states that the *inner product* of the signals $s_i(t)$ and $s_k(t)$ over the interval $[0, T]$, using their time-domain representations, is equal to the inner product of their respective vector representations $\mathbf{s}_i$ and $\mathbf{s}_k$. Note that the inner product of $s_i(t)$ and $s_k(t)$ is *invariant* to the choice of basis functions $\{\phi_j(t)\}_{j=1}^{N}$ in that it only depends on the components of the signals $s_i(t)$ and $s_k(t)$ projected onto each of the basis functions.

Yet another useful relation involving the vector representations of the signals $s_i(t)$ and $s_k(t)$ is described by

$$\begin{aligned} \| \mathbf{s}_i - \mathbf{s}_k \|^2 &= \sum_{j=1}^{N} (s_{ij} - s_{kj})^2 \\ &= \int_0^T (s_i(t) - s_k(t))^2 dt \end{aligned} \tag{5.14}$$

where $\| \mathbf{s}_i - \mathbf{s}_k \|$ is the *Euclidean distance*, $d_{ik}$, between the points represented by the signal vectors $\mathbf{s}_i$ and $\mathbf{s}_k$.

To complete the geometric representation of energy signals, we need to have a representation for the angle $\theta_{ik}$ subtended between two signal vectors $\mathbf{s}_i$ and $\mathbf{s}_k$. By definition, the *cosine of the angle* $\theta_{ik}$ is equal to the inner product of these two vectors divided by the product of their individual norms, as shown by

$$\cos \theta_{ik} = \frac{\mathbf{s}_i^T \mathbf{s}_k}{\| \mathbf{s}_i \| \, \| \mathbf{s}_k \|} \tag{5.15}$$

The two vectors $\mathbf{s}_i$ and $\mathbf{s}_k$ are thus *orthogonal* or *perpendicular* to each other if their inner product $\mathbf{s}_i^T \mathbf{s}_k$ is zero, in which case $\theta_{ik} = 90$ degrees; this condition is intuitively satisfying.

### ▶ EXAMPLE 5.1 Schwarz Inequality

Consider any pair of energy signals $s_1(t)$ and $s_2(t)$. The *Schwarz inequality* states that

$$\left( \int_{-\infty}^{\infty} s_1(t) s_2(t) dt \right)^2 \leq \left( \int_{-\infty}^{\infty} s_1^2(t) dt \right) \left( \int_{-\infty}^{\infty} s_2^2(t) dt \right) \tag{5.16}$$

The equality holds if and only if $s_2(t) = c s_1(t)$, where $c$ is any constant.

To prove this important inequality, let $s_1(t)$ and $s_2(t)$ be expressed in terms of the pair of orthonormal basis functions $\phi_1(t)$ and $\phi_2(t)$ as follows:

$$s_1(t) = s_{11} \phi_1(t) + s_{12} \phi_2(t)$$
$$s_2(t) = s_{21} \phi_1(t) + s_{22} \phi_2(t)$$

where $\phi_1(t)$ and $\phi_2(t)$ satisfy the orthonormality conditions over the entire time interval $(-\infty, \infty)$:

$$\int_{-\infty}^{\infty} \phi_i(t) \phi_j(t) dt = \delta_{ij} = \begin{cases} 1 & \text{for } j = i \\ 0 & \text{otherwise} \end{cases}$$

On this basis, we may represent the signals $s_1(t)$ and $s_2(t)$ by the following respective pair of vectors, as illustrated in Figure 5.5:

$$\mathbf{s}_1 = \begin{bmatrix} s_{11} \\ s_{12} \end{bmatrix}$$
$$\mathbf{s}_2 = \begin{bmatrix} s_{21} \\ s_{22} \end{bmatrix}$$

From Figure 5.5 we readily see that angle $\theta$ subtended between the vectors $\mathbf{s}_1$ and $\mathbf{s}_2$ is

$$\begin{aligned} \cos \theta &= \frac{\mathbf{s}_1^T \mathbf{s}_2}{\| \mathbf{s}_1 \| \, \| \mathbf{s}_2 \|} \\ &= \frac{\int_{-\infty}^{\infty} s_1(t) s_2(t) dt}{\left( \int_{-\infty}^{\infty} s_1^2(t) dt \right)^{1/2} \left( \int_{-\infty}^{\infty} s_2^2(t) dt \right)^{1/2}} \end{aligned} \tag{5.17}$$

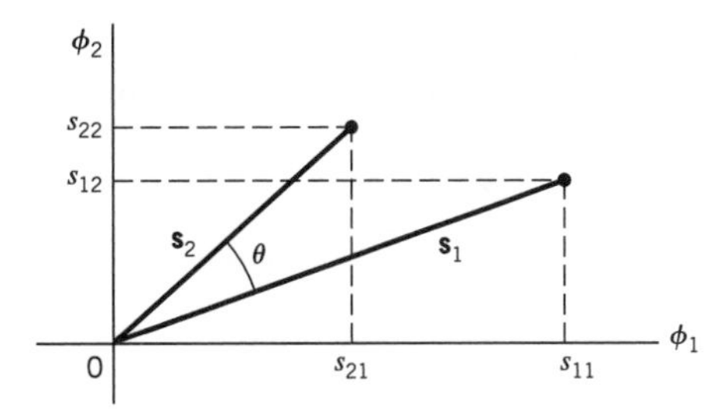


**FIGURE 5.5** Vector representations of signals $s_1(t)$ and $s_2(t)$, providing the background picture for proving the Schwarz inequality.

where we have made use of Equations (5.15), (5.13) and (5.9). Recognizing that $|\cos\theta| \leq 1$, the Schwarz inequality of Equation (5.16) immediately follows from Equation (5.17). Moreover, from the first line of Equation (5.17) we note that $|\cos\theta| = 1$ if and only if $\mathbf{s}_2 = c\mathbf{s}_1$, that is, $s_2(t) = cs_1(t)$, where $c$ is an arbitrary constant.

The proof of the Schwarz inequality, as presented here, applies to real-valued signals. It may be readily extended to complex-valued signals, in which case Equation (5.16) is reformulated as

$$\left|\int_{-\infty}^{\infty} s_1(t)s_2^*(t)dt\right|^2 \leq \left(\int_{-\infty}^{\infty} |s_1(t)|^2 dt\right)\left(\int_{-\infty}^{\infty} |s_2(t)|^2 dt\right) \tag{5.18}$$

where the equality holds if and only if $s_2(t) = cs_1(t)$, where $c$ is a constant; see Problem 5.9. It is the complex form of the Schwarz inequality that was used in Chapter 4 to derive the matched filter. ◀

## ■ Gram-Schmidt Orthogonalization Procedure

Having demonstrated the elegance of the geometric representation of energy signals, how do we justify it in mathematical terms? The answer lies in the *Gram-Schmidt orthogonalization procedure*, for which we need a *complete orthonormal set of basis functions*. To proceed with the formulation of this procedure, suppose we have a set of $M$ energy signals denoted by $s_1(t), s_2(t), \ldots, s_M(t)$. Starting with $s_1(t)$ chosen from this set arbitrarily, the first basis function is defined by

$$\phi_1(t) = \frac{s_1(t)}{\sqrt{E_1}} \tag{5.19}$$

where $E_1$ is the energy of the signal $s_1(t)$. Then, clearly, we have

$$\begin{aligned} s_1(t) &= \sqrt{E_1}\phi_1(t) \\ &= s_{11}\phi_1(t) \end{aligned} \tag{5.20}$$

where the coefficient $s_{11} = \sqrt{E_1}$ and $\phi_1(t)$ has unit energy, as required.

Next, using the signal $s_2(t)$, we define the coefficient $s_{21}$ as

$$s_{21} = \int_0^T s_2(t)\phi_1(t)dt \tag{5.21}$$

We may thus introduce a new intermediate function

$$g_2(t) = s_2(t) - s_{21}\phi_1(t) \tag{5.22}$$

which is orthogonal to $\phi_1(t)$ over the interval $0 \le t \le T$ by virtue of Equation (5.21) and the fact that the basis function $\phi_1(t)$ has unit energy. Now, we are ready to define the second basis function as

$$\phi_2(t) = \frac{g_2(t)}{\sqrt{\int_0^T g_2^2(t)dt}} \tag{5.23}$$

Substituting Equation (5.22) into (5.23) and simplifying, we get the desired result

$$\phi_2(t) = \frac{s_2(t) - s_{21}\phi_1(t)}{\sqrt{E_2 - s_{21}^2}} \tag{5.24}$$

where $E_2$ is the energy of the signal $s_2(t)$. It is clear from Equation (5.23) that

$$\int_0^T \phi_2^2(t)dt = 1$$

and from Equation (5.24) that

$$\int_0^T \phi_1(t)\phi_2(t)dt = 0$$

That is to say, $\phi_1(t)$ and $\phi_2(t)$ form an orthonormal pair, as required.

Continuing in this fashion, we may in general define

$$g_i(t) = s_i(t) - \sum_{j=1}^{i-1} s_{ij}\phi_j(t) \tag{5.25}$$

where the coefficients $s_{ij}$ are themselves defined by

$$s_{ij} = \int_0^T s_i(t)\phi_j(t)dt, \qquad j = 1, 2, \ldots, i-1 \tag{5.26}$$

Equation (5.22) is a special case of Equation (5.25) with $i = 2$. Note also that for $i = 1$, the function $g_i(t)$ reduces to $s_i(t)$.

Given the $g_i(t)$, we may now define the set of basis functions

$$\phi_i(t) = \frac{g_i(t)}{\sqrt{\int_0^T g_i^2(t)dt}}, \qquad i = 1, 2, \ldots, N \tag{5.27}$$

which form an orthonormal set. The dimension $N$ is less than or equal to the number of given signals, $M$, depending on one of two possibilities:

- The signals $s_1(t), s_2(t), \ldots, s_M(t)$ form a *linearly independent set*, in which case $N = M$.
- The signals $s_1(t), s_2(t), \ldots, s_M(t)$ are *not* linearly independent, in which case $N < M$, and the intermediate function $g_i(t)$ is zero for $i > N$.

**TABLE 5.1** *Amplitude Levels of the 2B1Q Code*

| *Signal* | | |
|---|---|---|
| *Symbol* | *Amplitude* | *Gray code* |
| $s_1(t)$ | $-3$ | 00 |
| $s_2(t)$ | $-1$ | 01 |
| $s_3(t)$ | $+1$ | 11 |
| $s_4(t)$ | $+3$ | 10 |

Note that the conventional Fourier series expansion of a periodic signal is an example of a particular expansion of the type described herein. Also, the representation of a band-limited signal in terms of its samples taken at the Nyquist rate may be viewed as another sample of a particular expansion of this type. However, two important distinctions should be made:

1. The form of the basis functions $\phi_1(t)$, $\phi_2(t)$, . . ., $\phi_N(t)$ has not been specified. That is to say, unlike the Fourier series expansion of a periodic signal or the sampled representation of a band-limited signal, we have not restricted the Gram-Schmidt orthogonalization procedure to be in terms of sinusoidal functions or sinc functions of time.
2. The expansion of the signal $s_i(t)$ in terms of a finite number of terms is not an approximation wherein only the first $N$ terms are significant but rather an *exact* expression where $N$ and only $N$ terms are significant.

### ▶ EXAMPLE 5.2 2B1Q Code

The 2B1Q code was described in Chapter 4 as the North American line code for digital subscriber lines. It represents a quaternary PAM signal as shown in the Gray-encoded alphabet of Table 5.1. The four possible signals, $s_1(t)$, $s_2(t)$, $s_3(t)$, and $s_4(t)$, are amplitude-scaled versions of a Nyquist pulse. Each signal represents a dibit. We wish to find the vector representation of the 2B1Q code.

This example is simple enough for us to solve it by inspection. Let $\phi_1(t)$ denote the Nyquist pulse, normalized to have unit energy. The $\phi_1(t)$ so defined is the only basis function for the vector representation of the 2B1Q code. Accordingly, the signal-space representation of this code is as shown in Figure 5.6. It consists of four signal vectors $\mathbf{s}_1$, $\mathbf{s}_2$, $\mathbf{s}_3$, and $\mathbf{s}_4$, which are located on the $\phi_1$-axis in a symmetric manner about the origin. In this example, we thus have $M = 4$ and $N = 1$.

We may generalize the result depicted in Figure 5.6 for the 2B1Q code as follows. The signal-space diagram of an $M$-ary pulse-amplitude modulated signal, in general, is one-dimensional with $M$ signal points uniformly positioned on the only axis of the diagram. ◀

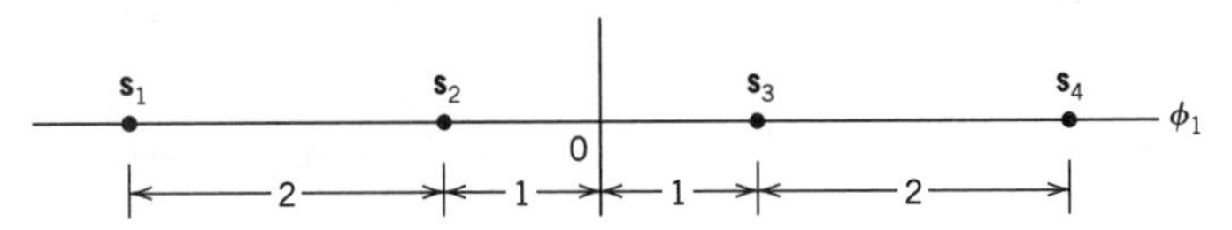


**FIGURE 5.6** Signal-space representation of the 2B1Q code.

## 5.3 *Conversion of the Continuous AWGN Channel into a Vector Channel*

Suppose that the input to the bank of $N$ product integrators or correlators in Figure 5.3*b* is not the transmitted signal $s_i(t)$ but rather the received signal $x(t)$ defined in accordance with the idealized AWGN channel of Figure 5.2. That is to say,

$$x(t) = s_i(t) + w(t), \qquad \begin{cases} 0 \le t \le T \\ i = 1, 2, \ldots, M \end{cases} \tag{5.28}$$

where $w(t)$ is a sample function of a white Gaussian noise process $W(t)$ of zero mean and power spectral density $N_0/2$. Correspondingly, we find that the output of correlator $j$, say, is the sample value of a random variable $X_j$, as shown by

$$\begin{aligned} x_j &= \int_0^T x(t)\phi_j(t)dt \\ &= s_{ij} + w_j, \qquad j = 1, 2, \ldots, N \end{aligned} \tag{5.29}$$

The first component, $s_{ij}$, is a deterministic quantity contributed by the transmitted signal $s_i(t)$; it is defined by

$$s_{ij} = \int_0^T s_i(t)\phi_j(t)dt \tag{5.30}$$

The second component, $w_j$, is the sample value of a random variable $W_j$ that arises because of the presence of the channel noise $w(t)$; it is defined by

$$w_j = \int_0^T w(t)\phi_j(t)dt \tag{5.31}$$

Consider next a new random process $X'(t)$ whose sample function $x'(t)$ is related to the received signal $x(t)$ as follows:

$$x'(t) = x(t) - \sum_{j=1}^{N} x_j\phi_j(t) \tag{5.32}$$

Substituting Equations (5.28) and (5.29) into (5.32), and then using the expansion of Equation (5.5), we get

$$\begin{aligned} x'(t) &= s_i(t) + w(t) - \sum_{j=1}^{N} (s_{ij} + w_j)\phi_j(t) \\ &= w(t) - \sum_{j=1}^{N} w_j\phi_j(t) \\ &= w'(t) \end{aligned} \tag{5.33}$$

The sample function $x'(t)$ therefore depends solely on the channel noise $w(t)$. On the basis of Equations (5.32) and (5.33), we may thus express the received signal as

$$\begin{aligned} x(t) &= \sum_{j=1}^{N} x_j\phi_j(t) + x'(t) \\ &= \sum_{j=1}^{N} x_j\phi_j(t) + w'(t) \end{aligned} \tag{5.34}$$

Accordingly, we may view $w'(t)$ as a sort of *remainder* term that must be included on the right to preserve the equality in Equation (5.34). It is informative to contrast the expansion of the received signal $x(t)$ given in Equation (5.34) with the corresponding expansion of the transmitted signal $s_i(t)$ given in Equation (5.5). The latter expansion is entirely deterministic, whereas that of Equation (5.34) is random (stochastic), which is to be expected.

## Statistical Characterization of the Correlator Outputs

We now wish to develop a statistical characterization of the set of $N$ correlator outputs. Let $X(t)$ denote the random process, a sample function of which is represented by the received signal $x(t)$. Correspondingly, let $X_j$ denote the random variable whose sample value is represented by the correlator output $x_j$, $j = 1, 2, \ldots, N$. According to the AWGN model of Figure 5.2, the random process $X(t)$ is a Gaussian process. It follows therefore that $X_j$ is a Gaussian random variable for all $j$ (see Property 1 of a Gaussian process, Section 1.8). Hence, $X_j$ is characterized completely by its mean and variance, which are determined next.

Let $W_j$ denote the random variable represented by the sample value $w_j$ produced by the $j$th correlator in response to the white Gaussian noise component $w(t)$. The random variable $W_j$ has zero mean, because the noise process $W(t)$ represented by $w(t)$ in the AWGN model of Figure 5.2 has zero mean by definition. Consequently, the mean of $X_j$ depends only on $s_{ij}$, as shown by

$$\begin{aligned}\mu_{X_j} &= E[X_j] \\ &= E[s_{ij} + W_j] \\ &= s_{ij} + E[W_j] \\ &= s_{ij}\end{aligned} \tag{5.35}$$

To find the variance of $X_j$, we note that

$$\begin{aligned}\sigma^2_{X_j} &= \text{var}[X_j] \\ &= E[(X_j - s_{ij})^2] \\ &= E[W_j^2]\end{aligned} \tag{5.36}$$

where the last line follows from Equation (5.29) with $x_j$ and $w_j$ replaced by $X_j$ and $W_j$, respectively. According to Equation (5.31), the random variable $W_j$ is defined by

$$W_j = \int_0^T W(t)\phi_j(t)dt$$

We may therefore expand Equation (5.36) as follows:

$$\begin{aligned}\sigma^2_{X_j} &= E\left[\int_0^T W(t)\phi_j(t)dt \int_0^T W(u)\phi_j(u)du\right] \\ &= E\left[\int_0^T \int_0^T \phi_j(t)\phi_j(u)W(t)W(u)dtdu\right]\end{aligned} \tag{5.37}$$

Interchanging the order of integration and expectation:

$$\begin{aligned}\sigma^2_{X_j} &= \int_0^T \int_0^T \phi_j(t)\phi_j(u)E[W(t)W(u)]dtdu \\ &= \int_0^T \int_0^T \phi_j(t)\phi_j(u)R_W(t, u)dtdu\end{aligned} \tag{5.38}$$

where $R_W(t, u)$ is the autocorrelation function of the noise process $W(t)$. Since this noise is stationary, $R_W(t, u)$ depends only on the time difference $t - u$. Furthermore, since the noise $W(t)$ is white with a constant power spectral density $N_0/2$, we may express $R_W(t, u)$ as follows [see Equation (1.95)]:

$$R_W(t, u) = \frac{N_0}{2} \delta(t - u) \tag{5.39}$$

Therefore, substituting Equation (5.39) into (5.38), and then using the sifting property of the delta function $\delta(t)$, we get

$$\begin{aligned} \sigma_{X_j}^2 &= \frac{N_0}{2} \int_0^T \int_0^T \phi_j(t)\phi_j(u)\, \delta(t - u)dtdu \\ &= \frac{N_0}{2} \int_0^T \phi_j^2(t)dt \end{aligned} \tag{5.40}$$

Since the $\phi_j(t)$ have unit energy, by definition, we finally get the simple result

$$\sigma_{X_j}^2 = \frac{N_0}{2} \qquad \text{for all } j \tag{5.41}$$

This important result shows that all the correlator outputs denoted by $X_j$ with $j = 1, 2, \ldots, N$, have a variance equal to the power spectral density $N_0/2$ of the noise process $W(t)$.

Moreover, since the $\phi_j(t)$ form an orthogonal set, we find that the $X_j$ are mutually uncorrelated, as shown by

$$\begin{aligned} \text{cov}[X_j X_k] &= E[(X_j - \mu_{X_j})(X_k - \mu_{X_k})] \\ &= E[(X_j - s_{ij})(X_k - s_{ik})] \\ &= E[W_j W_k] \\ &= E\left[\int_0^T W(t)\phi_j(t)dt \int_0^T W(u)\phi_k(u)du\right] \\ &= \int_0^T \int_0^T \phi_j(t)\phi_k(u)R_W(t, u)dtdu \\ &= \frac{N_0}{2} \int_0^T \int_0^T \phi_j(t)\phi_k(u)\, \delta(t - u)dtdu \\ &= \frac{N_0}{2} \int_0^T \phi_j(t)\phi_k(t)dt \\ &= 0, \qquad j \neq k \end{aligned} \tag{5.42}$$

Since the $X_j$ are Gaussian random variables, Equation (5.42) implies that they are also statistically independent (see Property 4 of a Gaussian Process, Section 1.8).

Define the vector of $N$ random variables

$$\mathbf{X} = \begin{bmatrix} X_1 \\ X_2 \\ \vdots \\ X_N \end{bmatrix} \tag{5.43}$$

whose elements are independent Gaussian random variables with mean values equal to $s_{ij}$ and variances equal to $N_0/2$. Since the elements of the vector $\mathbf{X}$ are statistically independent, we may express the conditional probability density function of the vector **X**, given that the signal $s_i(t)$ or correspondingly the symbol $m_i$ was transmitted, as the product of the conditional probability density functions of its individual elements as shown by

$$f_{\mathbf{X}}(\mathbf{x}|m_i) = \prod_{j=1}^{N} f_{X_j}(x_j|m_i), \qquad i = 1, 2, \ldots, M \tag{5.44}$$

where the vector **x** and scalar $x_j$ are sample values of the random vector **X** and random variable $X_j$, respectively. The vector **x** is called the *observation vector*; correspondingly, $x_i$ is called an *observable element*. Any channel that satisfies Equation (5.44) is called a *memoryless channel*.

Since each $X_j$ is a Gaussian random variable with mean $s_{ij}$ and variance $N_0/2$, we have

$$f_{X_j}(x_j|m_i) = \frac{1}{\sqrt{\pi N_0}} \exp\left[-\frac{1}{N_0}(x_j - s_{ij})^2\right], \qquad \begin{array}{l} j = 1, 2, \ldots, N \\ i = 1, 2, \ldots, M \end{array} \tag{5.45}$$

Therefore, substituting Equation (5.45) into (5.44) yields

$$f_{\mathbf{X}}(\mathbf{x}|m_i) = (\pi N_0)^{-N/2} \exp\left[-\frac{1}{N_0}\sum_{j=1}^{N}(x_j - s_{ij})^2\right], \qquad i = 1, 2, \ldots, M \tag{5.46}$$

It is now clear that the elements of the random vector **X** completely characterize the summation term $\Sigma_j X_j \phi_j(t)$, whose sample value is represented by the first term in Equation (5.34). However, there remains the noise term $w'(t)$ in this equation, which depends only on the channel noise $w(t)$. Since the noise process $W(t)$ represented by $w(t)$ is Gaussian with zero mean, it follows that the noise process $W'(t)$ represented by the sample function $w'(t)$ is also a zero-mean Gaussian process. Finally, we note that any random variable $W'(t_k)$, say, derived from the noise process $W'(t)$ by sampling it at time $t_k$, is in fact statistically independent of the set of random variables $\{X_j\}$; that is to say (see Problem 5.10),

$$E[X_j W'(t_k)] = 0, \qquad \begin{cases} j = 1, 2, \ldots, N \\ 0 \le t_k \le T \end{cases} \tag{5.47}$$

Since any random variable based on the remainder noise process $W'(t)$ is independent of the set of random variables $\{X_j\}$ as well as the set of transmitted signals $\{s_i(t)\}$, Equation (5.47) states that the random variable $W'(t_k)$ is irrelevant to the decision as to which particular signal was actually transmitted. In other words, the correlator outputs determined by the received signal $x(t)$ are the only data that are useful for the decision-making process and, hence, represent *sufficient statistics* for the problem at hand. By definition, sufficient statistics summarize the whole of the relevant information supplied by an observation vector.

We may now summarize the results presented in this section by formulating the *theorem of irrelevance*:

> Insofar as signal detection in additive white Gaussian noise is concerned, only the projections of the noise onto the basis functions of the signal set $\{s_i(t)\}_{i=1}^{M}$ affects the sufficient statistics of the detection problem; the remainder of the noise is irrelevant.

As a corollary to this theorem, we may state that the AWGN channel of Figure 5.2 is equivalent to an $N$-dimensional *vector channel* described by the observation vector

$$\mathbf{x} = \mathbf{s}_i + \mathbf{w}, \qquad i = 1, 2, \ldots, M \tag{5.48}$$

where the dimension $N$ is the number of basis functions involved in formulating the signal vector $\mathbf{s}_i$. The individual components of the signal vector $\mathbf{s}_i$ and noise vector $\mathbf{w}$ are defined by Equations (5.6) and (5.31), respectively. The theorem of irrelevance and its corollary are indeed basic to the understanding of the signal detection problem as described next.

## 5.4 *Likelihood Functions*

The conditional probability density functions $f_{\mathbf{X}}(\mathbf{x}|m_i)$, $i = 1, 2, \ldots, M$, are the very characterization of an AWGN channel. Their derivation leads to a functional dependence on the observation vector $\mathbf{x}$, given the transmitted message symbol $m_i$. However, at the receiver we have the exact opposite situation: We are given the observation vector $\mathbf{x}$ and the requirement is to estimate the message symbol $m_i$ that is responsible for generating $\mathbf{x}$. To emphasize this latter viewpoint, we introduce the idea of a *likelihood function*, denoted by $L(m_i)$ and defined by

$$L(m_i) = f_{\mathbf{X}}(\mathbf{x}|m_i), \qquad i = 1, 2, \ldots, M \tag{5.49}$$

It is important however to recognize that although the $L(m_i)$ and $f_{\mathbf{X}}(\mathbf{x}|m_i)$ have exactly the same mathematical form, their individual meanings are different.

In practice, we find it more convenient to work with the *log-likelihood function*, denoted by $l(m_i)$ and defined by

$$l(m_i) = \log L(m_i), \qquad i = 1, 2, \ldots, M \tag{5.50}$$

The log-likelihood function bears a one-to-one relationship to the likelihood function for two reasons:

1. By definition, a probability density function is always nonnegative. It follows therefore that the likelihood function is likewise a nonnegative quantity.
2. The logarithmic function is a monotonically increasing function of its argument.

The use of Equation (5.46) in (5.50) yields the log-likelihood functions for an AWGN channel as

$$l(m_i) = -\frac{1}{N_0}\sum_{j=1}^{N}(x_j - s_{ij})^2, \qquad i = 1, 2, \ldots, M \tag{5.51}$$

where we have ignored the constant term $-(N/2)\log(\pi N_0)$ as it bears no relation whatsoever to the message symbol $m_i$. Note that the $s_{ij}$, $j = 1, 2, \ldots, N$, are the elements of the signal vector $\mathbf{s}_i$ representing the message symbol $m_i$. With Equation (5.51) at our disposal, we are now ready to address the basic receiver design problem.

## 5.5 *Coherent Detection of Signals in Noise: Maximum Likelihood Decoding*

Suppose that in each time slot of duration $T$ seconds, one of the $M$ possible signals $s_1(t)$, $s_2(t), \ldots, s_M(t)$ is transmitted with equal probability, $1/M$. For geometric signal representation, the signal $s_i(t)$, $i = 1, 2, \ldots, M$, is applied to a bank of correlators, with a common input and supplied with an appropriate set of $N$ orthonormal basis functions. The resulting correlator outputs define the *signal vector* $\mathbf{s}_i$. Since knowledge of the signal vector $\mathbf{s}_i$ is as good as knowing the transmitted signal $s_i(t)$ itself, and vice versa, we may represent $s_i(t)$ by a point in a Euclidean space of dimension $N \le M$. We refer to this point as the *trans-*

*mitted signal point* or *message point*. The set of message points corresponding to the set of transmitted signals $\{s_i(t)\}_{i=1}^{M}$ is called a *signal constellation*.

However, the representation of the received signal $x(t)$ is complicated by the presence of additive noise $w(t)$. We note that when the received signal $x(t)$ is applied to the bank of $N$ correlators, the correlator outputs define the observation vector $\mathbf{x}$. From Equation (5.48), the vector $\mathbf{x}$ differs from the signal vector $\mathbf{s}_i$ by the *noise vector* $\mathbf{w}$ whose orientation is completely random. The noise vector $\mathbf{w}$ is completely characterized by the noise $w(t)$; the converse of this statement, however, is not true. The noise vector $\mathbf{w}$ represents that portion of the noise $w(t)$ that will interfere with the detection process; the remaining portion of this noise, denoted by $w'(t)$, is tuned out by the bank of correlators.

Now, based on the observation vector $\mathbf{x}$, we may represent the received signal $x(t)$ by a point in the same Euclidean space used to represent the transmitted signal. We refer to this second point as the *received signal point*. The received signal point wanders about the message point in a completely random fashion, in the sense that it may lie anywhere inside a Gaussian-distributed "cloud" centered on the message point. This is illustrated in Figure 5.7*a* for the case of a three-dimensional signal space. For a particular realization of the noise vector $\mathbf{w}$ (i.e., a particular point inside the random cloud of Figure 5.7*a*), the relationship between the observation vector $\mathbf{x}$ and the signal vector $\mathbf{s}_i$ is as illustrated in Figure 5.7*b*.

We are now ready to state the signal detection problem:

> Given the observation vector $\mathbf{x}$, perform a mapping from $\mathbf{x}$ to an estimate $\hat{m}$ of the transmitted symbol, $m_i$, in a way that would minimize the probability of error in the decision-making process.

Suppose that, given the observation vector $\mathbf{x}$, we make the decision $\hat{m} = m_i$. The probability of error in this decision, which we denote by $P_e(m_i|\,\mathbf{x})$, is simply

$$\begin{aligned} P_e(m_i|\,\mathbf{x}) &= P(m_i \text{ not sent}\,|\,\mathbf{x}) \\ &= 1 - P(m_i \text{ sent}\,|\,\mathbf{x}) \end{aligned} \tag{5.52}$$

The decision-making criterion is to minimize the probability of error in mapping each given observation vector $\mathbf{x}$ into a decision. On the basis of Equation (5.52), we may therefore state the *optimum decision rule*:

$$\begin{gathered} \text{Set } \hat{m} = m_i \text{ if} \\ P(m_i \text{ sent}\,|\,\mathbf{x}) \geq P(m_k \text{ sent}\,|\,\mathbf{x}) \qquad \text{for all } k \neq i \end{gathered} \tag{5.53}$$

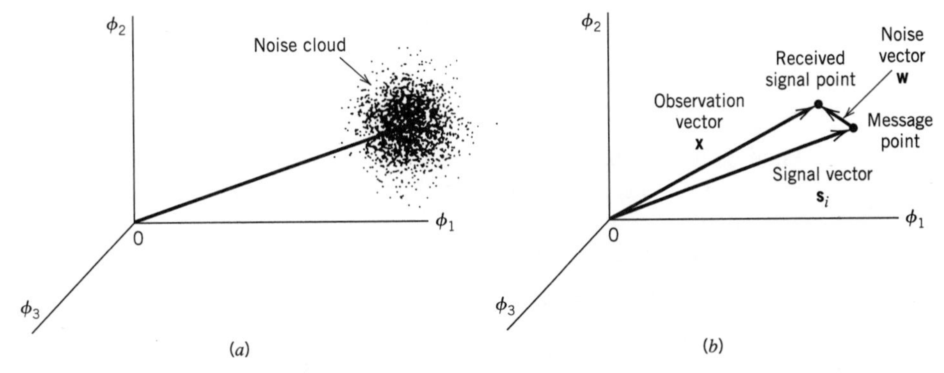


**FIGURE 5.7** Illustrating the effect of noise perturbation, depicted in (*a*), on the location of the received signal point, depicted in (*b*).

where $k = 1, 2, \ldots, M$. This decision rule is referred to as the *maximum a posteriori probability (MAP) rule.*

The condition of Equation (5.53) may be expressed more explicitly in terms of the *a priori* probabilities of the transmitted signals and in terms of the likelihood functions. Using Bayes' rule in Equation (5.53), and for the moment ignoring possible ties in the decision-making process, we may restate the MAP rule as follows:

$$\begin{gathered}\text{Set } \hat{m} = m_i \text{ if} \\ \frac{p_k f_{\mathbf{X}}(\mathbf{x}|m_k)}{f_{\mathbf{X}}(\mathbf{x})} \text{ is maximum for } k = i\end{gathered} \tag{5.54}$$

where $p_k$ is the *a priori* probability of transmitting symbol $m_k$, $f_{\mathbf{X}}(\mathbf{x}|m_k)$ is the conditional probability density function of the random observation vector $\mathbf{X}$ given the transmission of symbol $m_k$, and $f_{\mathbf{X}}(\mathbf{x})$ is the unconditional probability density function of $\mathbf{X}$. In Equation (5.54) we may note the following:

- The denominator term $f_{\mathbf{X}}(\mathbf{x})$ is independent of the transmitted symbol.
- The *a priori* probability $p_k = p_i$ when all the source symbols are transmitted with equal probability.
- The conditional probability density function $f_{\mathbf{X}}(\mathbf{x}|m_k)$ bears a one-to-one relationship to the log-likelihood function $l(m_k)$.

Accordingly, we may restate the decision rule of Equation (5.54) in terms of $l(m_k)$ simply as follows:

$$\begin{gathered}\text{Set } \hat{m} = m_i \text{ if} \\ l(m_k) \text{ is maximum for } k = i\end{gathered} \tag{5.55}$$

This decision rule is referred to as the *maximum likelihood rule*, and the device for its implementation is correspondingly referred to as the *maximum likelihood decoder*. According to Equation (5.55), a maximum likelihood decoder computes the log-likelihood functions as metrics for all the $M$ possible message symbols, compares them, and then decides in favor of the maximum. Thus the maximum likelihood decoder differs from the maximum *a posteriori* decoder in that it assumes equally likely message symbols.

It is useful to have a graphical interpretation of the maximum likelihood decision rule. Let $Z$ denote the $N$-dimensional space of all possible observation vectors $\mathbf{x}$. We refer to this space as the *observation space*. Because we have assumed that the decision rule must say $\hat{m} = m_i$, where $i = 1, 2, \ldots, M$, the total observation space $Z$ is correspondingly partitioned into *M-decision regions*, denoted by $Z_1, Z_2, \ldots, Z_M$. Accordingly, we may restate the decision rule of Equation (5.55) as follows:

$$\begin{gathered}\text{Observation vector } \mathbf{x} \text{ lies in region } Z_i \text{ if} \\ l(m_k) \text{ is maximum for } k = i\end{gathered} \tag{5.56}$$

Aside from the boundaries between the decision regions $Z_1, Z_2, \ldots, Z_M$, it is clear that this set of regions covers the entire space of possible observation vectors $\mathbf{x}$. We adopt the convention that all ties are resolved at random; that is, the receiver simply makes a guess. Specifically, if the observation vector $\mathbf{x}$ falls on the boundary between any two decision regions, $Z_i$ and $Z_k$, say, the choice between the two possible decisions $\hat{m} = m_i$ and $\hat{m} = m_k$ is resolved *a priori* by the flip of a fair coin. Clearly, the outcome of such an event does not affect the ultimate value of the probability of error since, on this boundary, the condition of Equation (5.53) is satisfied with the equality sign.

The maximum likelihood decision rule of Equation (5.55) or its geometric counterpart described in Equation (5.56) is of a generic kind, with the channel noise $w(t)$ being additive as the only restriction imposed on it. We next specialize this rule for the case when $w(t)$ is both white and Gaussian.

From the log-likelihood function defined in Equation (5.51) for an AWGN channel we note that $l(m_k)$ attains its maximum value when the summation term

$$\sum_{j=1}^{N} (x_j - s_{kj})^2$$

is minimized by the choice $k = i$. Accordingly, we may formulate the maximum likelihood decision rule for an AWGN channel as

$$\begin{gathered}\text{Observation vector } \mathbf{x} \text{ lies in region } Z_i \text{ if} \\ \sum_{j=1}^{N} (x_j - s_{kj})^2 \text{ is minimum for } k = i\end{gathered} \tag{5.57}$$

Next, we note from our earlier discussion that (see Equation (5.14) for comparison)

$$\sum_{j=1}^{N} (x_j - s_{kj})^2 = \| \mathbf{x} - \mathbf{s}_k \|^2 \tag{5.58}$$

where $\| \mathbf{x} - \mathbf{s}_k \|$ is the Euclidean distance between the received signal point and message point, represented by the vectors $\mathbf{x}$ and $\mathbf{s}_k$, respectively. Accordingly, we may restate the decision rule of Equation (5.57) as follows:

$$\begin{gathered}\text{Observation vector } \mathbf{x} \text{ lies in region } Z_i \text{ if} \\ \text{the Euclidean distance } \| \mathbf{x} - \mathbf{s}_k \| \text{ is minimum for } k = i\end{gathered} \tag{5.59}$$

Equation (5.59) states that *the maximum likelihood decision rule is simply to choose the message point closest to the received signal point*, which is intuitively satisfying.

In practice, the need for squarers in the decision rule of Equation (5.59) is avoided by recognizing that

$$\sum_{j=1}^{N} (x_j - s_{kj})^2 = \sum_{j=1}^{N} x_j^2 - 2 \sum_{j=1}^{N} x_j s_{kj} + \sum_{j=1}^{N} s_{kj}^2 \tag{5.60}$$

The first summation term of this expansion is independent of the index $k$ and may therefore be ignored. The second summation term is the inner product of the observation vector $\mathbf{x}$ and signal vector $\mathbf{s}_k$. The third summation term is the energy of the transmitted signal $s_k(t)$. Accordingly, we may formulate a decision rule equivalent to that of Equation (5.59) as follows:

$$\begin{gathered}\text{Observation vector } \mathbf{x} \text{ lies in region } Z_i \text{ if} \\ \sum_{j=1}^{N} x_j s_{kj} - \frac{1}{2} E_k \text{ is maximum for } k = i\end{gathered} \tag{5.61}$$

where $E_k$ is the energy of the transmitted signal $s_k(t)$:

$$E_k = \sum_{j=1}^{N} s_{kj}^2 \tag{5.62}$$

From Equation (5.61) we deduce that, for an AWGN channel, the decision regions are regions of the $N$-dimensional observation space $Z$, bounded by linear [$(N - 1)$-dimensional hyperplane] boundaries. Figure 5.8 shows the example of decision regions for

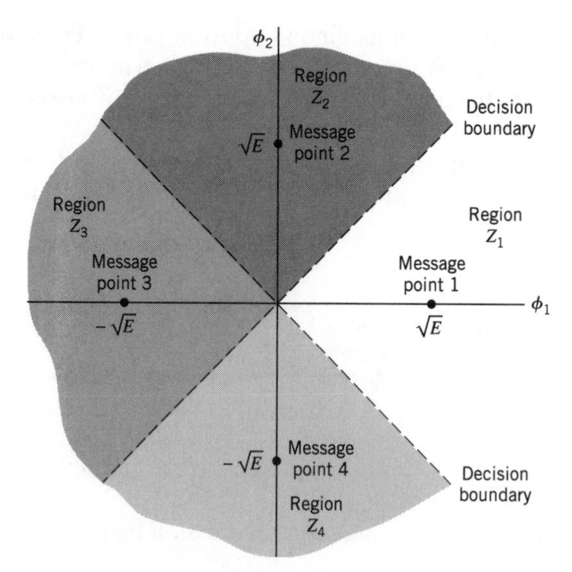


**FIGURE 5.8** Illustrating the partitioning of the observation space into decision regions for the case when $N = 2$ and $M = 4$; it is assumed that the $M$ transmitted symbols are equally likely.

$M = 4$ signals and $N = 2$ dimensions, assuming that the signals are transmitted with equal energy, $E$, and equal probability.

## 5.6 *Correlation Receiver*

From the material presented in the previous sections, we find that for an AWGN channel and for the case when the transmitted signals $s_1(t), s_2, \ldots, s_M(t)$ are equally likely, the optimum receiver consists of two subsystems, which are detailed in Figure 5.9 and described here:

1. The *detector* part of the receiver is shown in Figure 5.9*a*. It consists of a bank of $M$ *product-integrators* or *correlators*, supplied with a corresponding set of coherent reference signals or orthonormal basis functions $\phi_1(t), \phi_2(t), \ldots, \phi_N(t)$ that are generated locally. This bank of correlators operates on the received signal $x(t)$, $0 \le t \le T$, to produce the observation vector $\mathbf{x}$.
2. The second part of the receiver, namely, the *signal transmission decoder* is shown in Figure 5.9*b*. It is implemented in the form of a maximum-likelihood decoder that operates on the observation vector $\mathbf{x}$ to produce an estimate, $\hat{m}$, of the transmitted symbol $m_i$, $i = 1, 2, \ldots, M$, in a way that would minimize the average probability of symbol error. In accordance with Equation (5.61), the $N$ elements of the observation vector $\mathbf{x}$ are first multiplied by the corresponding $N$ elements of each of the $M$ signal vectors $\mathbf{s}_1, \mathbf{s}_2, \ldots, \mathbf{s}_M$, and the resulting products are successively summed in accumulators to form the corresponding set of inner products $\{\mathbf{x}^T\mathbf{s}_k \mid k = 1, 2, \ldots, M\}$. Next, the inner products are corrected for the fact that the transmitted signal energies may be unequal. Finally, the largest in the resulting set of numbers is selected, and an appropriate decision on the transmitted message is made.

The optimum receiver of Figure 5.9 is commonly referred to as a *correlation receiver*.

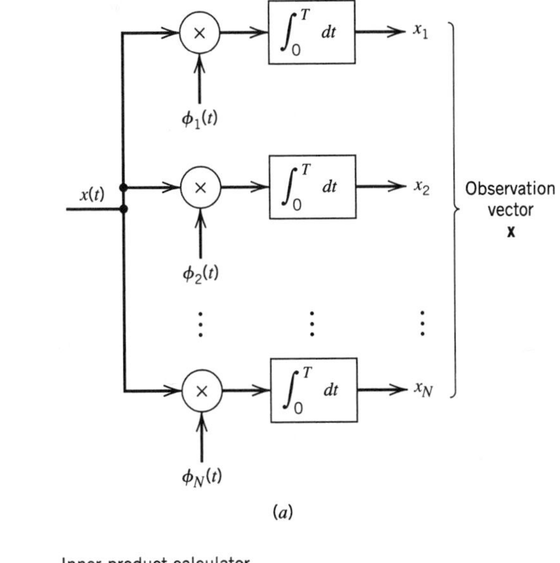


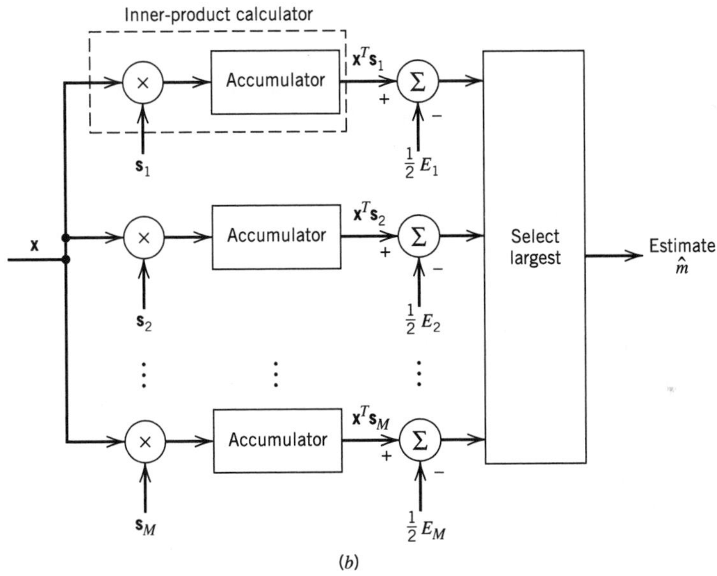


**FIGURE 5.9** (*a*) Detector or demodulator. (*b*) Signal transmission decoder.

## ■ EQUIVALENCE OF CORRELATION AND MATCHED FILTER RECEIVERS

The detector shown in Figure 5.9*a* involves a set of correlators. Alternatively, we may use a corresponding set of *matched filters* to build the detector; the matched filter and its properties were considered in Section 4.2. To demonstrate the equivalence of a correlator and a matched filter, consider a linear time-invariant filter with impulse response $h_j(t)$. With the received signal $x(t)$ used as the filter input, the resulting filter output, $y_j(t)$, is defined by the convolution integral:

$$y_j(t) = \int_{-\infty}^{\infty} x(\tau)h_j(t - \tau)d\tau \tag{5.63}$$

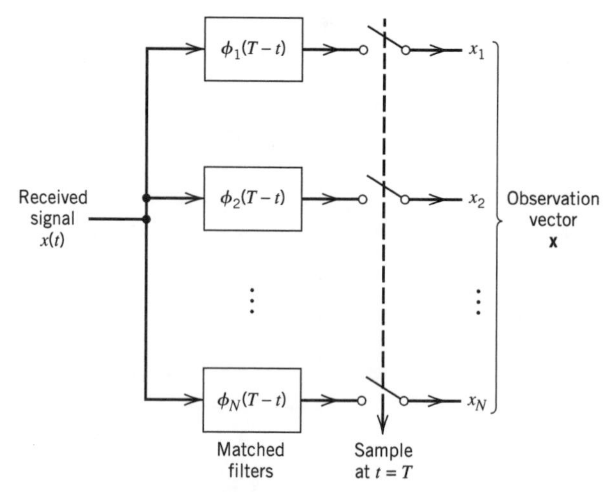


**FIGURE 5.10** Detector part of matched filter receiver; the signal transmission decoder is as shown in Fig. 5.9*b*.

From the definition of a matched filter presented in Section 4.2, we recall that the impulse response $h_j(t)$ of a linear time-invariant filter matched to an input signal $\phi_j(t)$ is a time-reversed and delayed version of the input $\phi_j(t)$. Suppose that we set

$$h_j(t) = \phi_j(T - t) \tag{5.64}$$

Then the resulting filter output is

$$y_j(t) = \int_{-\infty}^{\infty} x(\tau)\phi_j(T - t + \tau)d\tau \tag{5.65}$$

Sampling this output at time $t = T$, we get

$$y_j(T) = \int_{-\infty}^{\infty} x(\tau)\phi_j(\tau)d\tau$$

Since, by definition, $\phi_j(t)$ is zero outside the interval $0 \le t \le T$, we find that $y_j(T)$ is in actual fact the $j$th correlator output $x_j$ produced by the received signal $x(t)$ in Figure 5.9*a*, as shown by

$$y_j(T) = \int_{0}^{T} x(\tau)\phi_j(\tau)d\tau \tag{5.66}$$

Accordingly, the detector part of the optimum receiver may also be implemented using a bank of matched filters, as shown in Figure 5.10. It is important to note, however, that the output of each correlator in Figure 5.9*a* is equivalent to the output of a corresponding matched filter in Figure 5.10 only when that output is sampled at time $t = T$.

## 5.7 *Probability of Error*

To complete the statistical characterization of the correlation receiver depicted in Figure 5.9, we need to evaluate its noise performance. To do so, suppose that the observation space $Z$ is partitioned, in accordance with the maximum likelihood decision rule, into a set if $M$ regions $\{Z_i\}_{i=1}^{M}$. Suppose also that symbol $m_i$ (or, equivalently, signal vector $\mathbf{s}_i$) is transmitted, and an observation vector $\mathbf{x}$ is received. Then an error occurs whenever the received signal point represented by $\mathbf{x}$ does not fall inside region $Z_i$ associated with the message point represented by $\mathbf{s}_i$. Averaging over all possible transmitted symbols, we readily see that the *average probability of symbol error*, $P_e$ is

$$\begin{aligned} P_e &= \sum_{i=1}^{M} p_i \, P(\mathbf{x} \text{ does not lie in } Z_i | m_i \text{ sent}) \\ &= \frac{1}{M} \sum_{i=1}^{M} P(\mathbf{x} \text{ does not lie in } Z_i | m_i \text{ sent}) \\ &= 1 - \frac{1}{M} \sum_{i=1}^{M} P(\mathbf{x} \text{ lies in } Z_i | m_i \text{ sent}) \end{aligned} \tag{5.67}$$

where we have used standard notation to denote the probability of an event and the conditional probability of an event. Since $\mathbf{x}$ is the sample value of random vector $\mathbf{X}$, we may rewrite Equation (5.67) in terms of the likelihood function (when $m_i$ is sent) as follows:

$$P_e = 1 - \frac{1}{M} \sum_{i=1}^{M} \int_{Z_i} f_{\mathbf{X}}(\mathbf{x} | m_i) \, d\mathbf{x} \tag{5.68}$$

For an $N$-dimensional observation vector, the integral in Equation (5.68) is likewise $N$-dimensional.

## Invariance of the Probability of Error to Rotation and Translation

The way in which the observation space $Z$ is partitioned into the set of regions $Z_1, Z_2, \ldots, Z_M$, in the maximum likelihood detection of a signal in additive white Gaussian noise, is uniquely defined by the signal constellation under study. Accordingly, changes in the orientation of the signal constellation with respect to both the coordinate axes and origin of the signal space do *not* affect the probability of symbol error $P_e$ defined in Equation (5.68). This result is a consequence of two facts:

1. In maximum likelihood detection, the probability of symbol error $P_e$ depends solely on the relative Euclidean distances between the message points in the constellation.
2. The additive white Gaussian noise is *spherically symmetric* in all directions in the signal space.

Consider first the invariance of $P_e$ with respect to rotation. The effect of a rotation applied to all the message points in a constellation is equivalent to multiplying the $N$-dimensional signal vector $\mathbf{s}_i$ by an $N$-by-$N$ *orthonormal matrix* denoted by $\mathbf{Q}$ for all $i$. The matrix $\mathbf{Q}$ satisfies the condition

$$\mathbf{Q}\mathbf{Q}^T = \mathbf{I} \tag{5.69}$$

where $\mathbf{I}$ is the *identity matrix* whose diagonal elements are all unity and its off-diagonal elements are all zero. Note that according to Equation (5.69), the inverse of a real-valued orthonormal matrix is equal to its transposed form. Thus the signal vector $\mathbf{s}_i$ is replaced by its rotated version

$$\mathbf{s}_{i,\text{rotate}} = \mathbf{Q}\mathbf{s}_i, \qquad i = 1, 2, \ldots, M \tag{5.70}$$

Correspondingly, the $N$-by-1 noise vector $\mathbf{w}$ is replaced by its rotated version

$$\mathbf{w}_{\text{rotate}} = \mathbf{Q}\mathbf{w} \tag{5.71}$$

However, the statistical characteristics of the noise vector are unaffected by this rotation for the following reasons:

- From Chapter 1 we recall that a linear combination of Gaussian random variables is also Gaussian. Since the noise vector $\mathbf{w}$ is Gaussian, by assumption, it follows that the rotated noise vector $\mathbf{w}_{\text{rotate}}$ is also Gaussian.
- Since the noise vector $\mathbf{w}$ has zero mean, the rotated noise vector $\mathbf{w}_{\text{rotate}}$ also has zero mean, as shown by

$$\begin{aligned} E[\mathbf{w}_{\text{rotate}}] &= E[\mathbf{Q}\mathbf{w}] \\ &= \mathbf{Q}E[\mathbf{w}] \\ &= 0 \end{aligned} \tag{5.72}$$

- The covariance matrix of the noise vector $\mathbf{w}$ is equal to $(N_0/2)\mathbf{I}$, where $N_0/2$ is the power spectral density of the AWGN $w(t)$; that is,

$$E[\mathbf{w}\mathbf{w}^T] = \frac{N_0}{2}\mathbf{I} \tag{5.73}$$

  Hence, the covariance matrix of the rotated noise vector $\mathbf{w}_{\text{rotate}}$ is

$$\begin{aligned} E[\mathbf{w}_{\text{rotate}}\mathbf{w}_{\text{rotate}}^T] &= E[\mathbf{Q}\mathbf{w}(\mathbf{Q}\mathbf{w})^T] \\ &= E[\mathbf{Q}\mathbf{w}\mathbf{w}^T\mathbf{Q}^T] \\ &= \mathbf{Q}E[\mathbf{w}\mathbf{w}^T]\mathbf{Q}^T \\ &= \frac{N_0}{2}\mathbf{Q}\mathbf{Q}^T \\ &= \frac{N_0}{2}\mathbf{I} \end{aligned} \tag{5.74}$$

  where in the last two lines we have made use of Equations (5.73) and (5.69).

In light of these observations, we may express the observation vector for the rotated signal constellation as

$$\mathbf{x}_{\text{rotate}} = \mathbf{Q}\mathbf{s}_i + \mathbf{w}, \qquad i = 1, 2, \ldots, M \tag{5.75}$$

From Equation (5.59) we know that the decision rule for maximum likelihood detection is based on the Euclidean distance from the observation vector $\mathbf{x}_{\text{rotate}}$ to the rotated signal vector $\mathbf{s}_{i,\text{rotate}} = \mathbf{Q}\mathbf{s}_i$. Comparing Equation (5.75) to Equation (5.48), we readily see that

$$\| \mathbf{x}_{\text{rotate}} - \mathbf{s}_{i,\text{rotate}} \| = \| \mathbf{x} - \mathbf{s}_i \| \qquad \text{for all } i \tag{5.76}$$

We may therefore formally state the *principle of rotational invariance* as follows:

> If a signal constellation is rotated by an orthonormal transformation, that is,
>
> $$\mathbf{s}_{i,\text{rotate}} = \mathbf{Q}\mathbf{s}_i, \qquad i = 1, 2, \ldots, M$$
>
> where $\mathbf{Q}$ is an orthonormal matrix, then the probability of symbol error $P_e$ incurred in maximum likelihood signal detection over an AWGN channel is completely unchanged.

We illustrate this principle with an example. The signal constellation shown in Figure 5.11*b* is the same as that of Figure 5.11*a*, except that it has been rotated through 45 degrees. Although these two constellations do indeed look different, the principle of rotational invariance tells us immediately that the $P_e$ is the same for both of them.

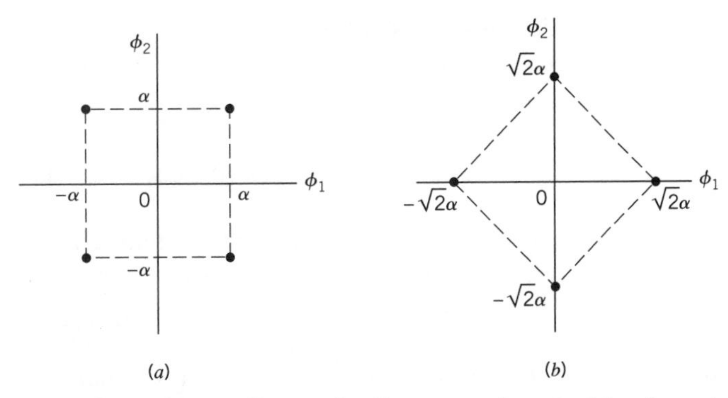


**FIGURE 5.11** A pair of signal constellations for illustrating the principle of rotational invariance.

Consider next the issue of invariance to translation. Suppose all the message points in a signal constellation are translated by a constant vector amount **a**, as shown by

$$\mathbf{s}_{i,\text{translate}} = \mathbf{s}_i - \mathbf{a}, \qquad i = 1, 2, \ldots, M \tag{5.77}$$

The observation vector is correspondingly translated by the same vector amount, as shown by

$$\mathbf{x}_{\text{translate}} = \mathbf{x} - \mathbf{a} \tag{5.78}$$

From Equations (5.77) and (5.78) we see that the translate **a** is common to both the translated signal vector $\mathbf{s}_i$ and translated observation vector **x**. We therefore immediately deduce that

$$\| \mathbf{x}_{\text{translate}} - \mathbf{s}_{i,\text{translate}} \| = \| \mathbf{x} - \mathbf{s}_i \| \qquad \text{for all } i \tag{5.79}$$

and thus formulate the *principle of translational invariance* as follows:

> If a signal constellation is translated by a constant vector amount, then the probability of symbol error $P_e$ incurred in maximum likelihood signal detection over an AWGN channel is completely unchanged.

As an example, consider the two signal constellations shown in Figure 5.12, which pertain to a pair of different 4-level PAM signals. The constellation of Figure 5.12*b* is the same as that of Figure 5.12*a*, except for a translation of $3\alpha/2$ to the right along the $\phi_1$-axis. The principle of translational invariance says that the $P_e$ is the same for both of these constellations.

## MINIMUM ENERGY SIGNALS

A useful application of the principle of translational invariance is in the translation of a given signal constellation in such a way that the average energy is minimized. To explore

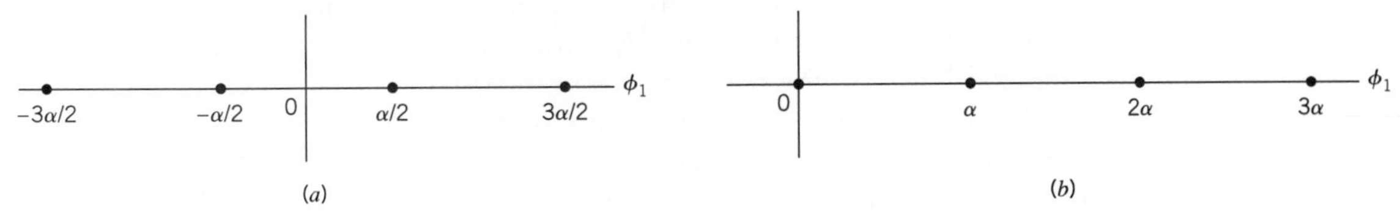


**FIGURE 5.12** A pair of signal constellations for illustrating the principle of translational invariance.

this issue, consider a set of symbols $m_1, m_2, \ldots, m_M$ represented by the signal vectors $\mathbf{s}_1, \mathbf{s}_2, \ldots, \mathbf{s}_M$, respectively. The average energy of this signal constellation translated by a vector amount $\mathbf{a}$ is

$$\mathscr{E}_{\text{translate}} = \sum_{i=1}^{M} \| \mathbf{s}_i - \mathbf{a} \|^2 p_i \tag{5.80}$$

where $p_i$ is the probability that symbol $m_i$ is emitted by the source of information. The squared Euclidean distance between $\mathbf{s}_i$ and $\mathbf{a}$ is expanded as

$$\| \mathbf{s}_i - \mathbf{a} \|^2 = \| \mathbf{s}_i \|^2 - 2\mathbf{a}^T\mathbf{s}_i + \| \mathbf{a} \|^2$$

We may therefore rewrite Equation (5.80) in the expanded form

$$\begin{aligned} \mathscr{E}_{\text{translate}} &= \sum_{i=1}^{M} \| \mathbf{s}_i \|^2 p_i - 2 \sum_{i=1}^{M} \mathbf{a}^T\mathbf{s}_i p_i + \| \mathbf{a} \|^2 \sum_{i=1}^{M} p_i \\ &= \mathscr{E} - 2\mathbf{a}^T E[\mathbf{s}] + \| \mathbf{a} \|^2 \end{aligned} \tag{5.81}$$

where $\mathscr{E}$ is the average energy of the original signal constellation, and

$$E[\mathbf{s}] = \sum_{i=1}^{M} \mathbf{s}_i p_i \tag{5.82}$$

Differentiating Equation (5.81) with respect to the vector $\mathbf{a}$ and then setting the result equal to zero, we readily find that the minimizing translate is

$$\mathbf{a}_{\text{min}} = E[\mathbf{s}] \tag{5.83}$$

The minimum average energy of the signal constellation translated in this way is

$$\mathscr{E}_{\text{translate,min}} = \mathscr{E} - \| \mathbf{a}_{\text{min}} \|^2 \tag{5.84}$$

We may now state the procedure for finding the *minimum energy translate*:

> Given a signal constellation $\{\mathbf{s}_i\}_{i=1}^{M}$, the corresponding signal constellation with minimum average energy is obtained by subtracting from each signal vector $\mathbf{s}_i$ in the given constellation an amount equal to the constant vector $E[\mathbf{s}]$, where $E[\mathbf{s}]$ is defined by Equation (5.82).

Recalling that the energy (or power) needed for signal transmission is a primary resource, the minimum energy translate provides a principled method for translating a signal constellation of interest so as to minimize the energy requirement. For example, the constellation of Figure 5.12*a* has minimum average energy, whereas that of Figure 5.12*b* does not.

## ■ UNION BOUND ON THE PROBABILITY OF ERROR[2]

For AWGN channels, the formulation of the average probability of symbol error, $P_e$, is conceptually straightforward. We simply write $P_e$ in integral form by substituting Equation (5.46) into Equation (5.68). Unfortunately, however, numerical computation of the integral is impractical, except in a few simple (but important) cases. To overcome this computational difficulty, we may resort to the use of *bounds*, which are usually adequate to predict the signal-to-noise ratio (within a decibel or so) required to maintain a prescribed error rate. The approximation to the integral defining $P_e$ is made by simplifying the integral or simplifying the region of integration. In the sequel, we use the latter procedure to develop a simple yet useful upper bound called the *union bound* as an approximation to the average probability of symbol error for a set of $M$ equally likely signals (symbols) in an AWGN channel.

Let $A_{ik}$, with $(i, k) = 1, 2, \ldots, M$, denote the event that the observation vector $\mathbf{x}$ is closer to the signal vector $\mathbf{s}_k$ than to $\mathbf{s}_i$, when the symbol $m_i$ (vector $\mathbf{s}_i$) is sent. The conditional probability of symbol error when symbol $m_i$ is sent, $P_e(m_i)$, is equal to the probability of the *union of events*, $A_{i1}, A_{i2}, \ldots, A_{i,i-1}, A_{i,i+1}, \ldots, A_{i,M}$. From probability theory we know that *the probability of a finite union of events is overbounded by the sum of the probabilities of the constituent events*. We may therefore write

$$P_e(m_i) \leq \sum_{\substack{k=1 \\ k \neq i}}^{M} P(A_{ik}), \qquad i = 1, 2, \ldots, M \tag{5.85}$$

This relationship is illustrated in Figure 5.13 for the case of $M = 4$. In Figure 5.13*a*, we show the four message points and associated decision regions, with the point $\mathbf{s}_1$ assumed to represent the transmitted symbol. In Figure 5.13*b*, we show the three constituent signal-space descriptions where, in each case, the transmitted message point $\mathbf{s}_1$ and one other message point are retained. According to Figure 5.13*a* the conditional probability of symbol error, $P_e(m_i)$, is equal to the probability that the observation vector $\mathbf{x}$ lies in the shaded region of the two-dimensional signal-space diagram. Clearly, this probability is less than the sum of the probabilities of the three individual events that $\mathbf{x}$ lies in the shaded regions of the three constituent signal spaces depicted in Figure 5.13*b*.

It is important to note that, in general, the probability $P(A_{ik})$ is different from the probability $P(\hat{m} = m_k | m_i)$. The latter is the probability that the observation vector $\mathbf{x}$ is

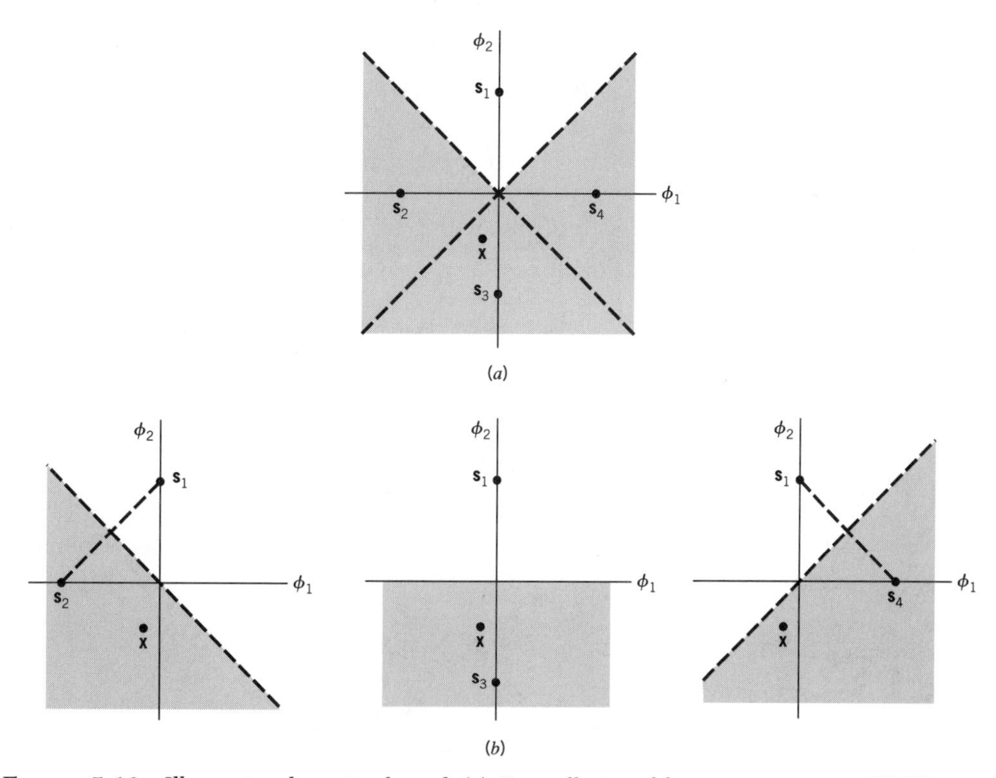


**FIGURE 5.13** Illustrating the union bound. (*a*) Constellation of four message points. (*b*) Three constellations with a common message point and one other message point retained from the original constellation.

closer to the signal vector $\mathbf{s}_k$ than every other, when $\mathbf{s}_i$ (or $m_i$) is sent. On the other hand, the probability $P(A_{ik})$ depends on only two signal vectors, $\mathbf{s}_i$ and $\mathbf{s}_k$. To emphasize this difference, we rewrite Equation (5.85) by adopting $P_2(\mathbf{s}_i, \mathbf{s}_k)$ in place of $P(A_{ik})$. We thus write

$$P_e(m_i) \leq \sum_{\substack{k=1 \\ k \neq i}}^{M} P_2(\mathbf{s}_i, \mathbf{s}_k), \qquad i = 1, 2, \ldots, M \tag{5.86}$$

The probability $P_2(\mathbf{s}_i, \mathbf{s}_k)$ is called the *pairwise error probability* in that if a data transmission system uses only a pair of signals, $\mathbf{s}_i$ and $\mathbf{s}_k$, then $P_2(\mathbf{s}_i, \mathbf{s}_k)$ is the probability of the receiver mistaking $\mathbf{s}_k$ for $\mathbf{s}_i$.

Consider then a simplified digital communication system that involves the use of two equally likely messages represented by the vectors $\mathbf{s}_i$ and $\mathbf{s}_k$. Since white Gaussian noise is identically distributed along any set of orthogonal axes, we may temporarily choose the first axis in such a set as one that passes through the points $\mathbf{s}_i$ and $\mathbf{s}_k$; for three examples, see Figure 5.13*b*. The corresponding decision boundary is represented by the bisector that is perpendicular to the line joining the points $\mathbf{s}_i$ and $\mathbf{s}_k$. Accordingly, when the symbol $m_i$ (vector $\mathbf{s}_i$) is sent, and if the observation vector $\mathbf{x}$ lies on the side of the bisector where $\mathbf{s}_k$ lies, an error is made. The probability of this event is given by

$$\begin{aligned} P_2(\mathbf{s}_i, \mathbf{s}_k) &= P(\mathbf{x} \text{ is closer to } \mathbf{s}_k \text{ than } \mathbf{s}_i, \text{ when } \mathbf{s}_i \text{ is sent}) \\ &= \int_{d_{ik}/2}^{\infty} \frac{1}{\sqrt{\pi N_0}} \exp\left(-\frac{v^2}{N_0}\right) dv \end{aligned} \tag{5.87}$$

where $d_{ik}$ is the Euclidean distance between $\mathbf{s}_i$ and $\mathbf{s}_k$; that is,

$$d_{ik} = \| \mathbf{s}_i - \mathbf{s}_k \| \tag{5.88}$$

From the definition of the complementary error function, we have

$$\operatorname{erfc}(u) = \frac{2}{\sqrt{\pi}} \int_u^{\infty} \exp(-z^2)\, dz$$

Thus, in terms of this function, with $z$ set equal to $v/\sqrt{N_0}$, we find that Equation (5.87) takes on the compact form

$$P_2(\mathbf{s}_i, \mathbf{s}_k) = \frac{1}{2} \operatorname{erfc}\left(\frac{d_{ik}}{2\sqrt{N_0}}\right) \tag{5.89}$$

Substituting Equation (5.89) into Equation (5.86), we get

$$P_e(m_i) \leq \frac{1}{2} \sum_{\substack{k=1 \\ k \neq i}}^{M} \operatorname{erfc}\left(\frac{d_{ik}}{2\sqrt{N_0}}\right), \qquad i = 1, 2, \ldots, M \tag{5.90}$$

The probability of symbol error, averaged over all the $M$ symbols, is therefore overbounded as follows:

$$\begin{aligned} P_e &= \sum_{i=1}^{M} p_i P_e(m_i) \\ &\leq \frac{1}{2} \sum_{i=1}^{M} \sum_{\substack{k=1 \\ k \neq i}}^{M} p_i \operatorname{erfc}\left(\frac{d_{ik}}{2\sqrt{N_0}}\right) \end{aligned} \tag{5.91}$$

where $p_i$ is the probability of transmitting symbol $m_i$.

There are two special forms of Equation (5.91) that we will find useful in Chapter 6 on passband data transmission:

1. Suppose that the signal constellation is *circularly symmetric* about the origin. Then the conditional probability of error $P_e(m_i)$ is the same for all $i$, in which case Equation (5.91) reduces to

$$P_e \leq \frac{1}{2} \sum_{\substack{k=1 \\ k \neq i}}^{M} \text{erfc}\left(\frac{d_{ik}}{2\sqrt{N_0}}\right) \quad \text{for all } i \tag{5.92}$$

2. Define the *minimum distance* of a signal constellation, $d_{\min}$, as the smallest Euclidean distance between any two transmitted signal points in the constellation, as shown by

$$d_{\min} = \min_{k \neq i} d_{ik} \qquad \text{for all } i \text{ and } k \tag{5.93}$$

Then, recognizing that the complementary error function erfc($u$) is a monotonically decreasing function of its argument $u$, we may write

$$\text{erfc}\left(\frac{d_{ik}}{2\sqrt{N_0}}\right) \leq \text{erfc}\left(\frac{d_{\min}}{2\sqrt{N_0}}\right) \text{ for all } i \text{ and } k \tag{5.94}$$

We may therefore, in general, simplify the bound on the average probability of symbol error in Equation (5.91) as

$$P_e \leq \frac{(M-1)}{2} \text{erfc}\left(\frac{d_{\min}}{2\sqrt{N_0}}\right) \tag{5.95}$$

The complementary error function is itself bounded as[3]

$$\text{erfc}\left(\frac{d_{\min}}{2\sqrt{N_0}}\right) \leq \frac{1}{\sqrt{\pi}} \exp\left(-\frac{d_{\min}^2}{4N_0}\right) \tag{5.96}$$

Accordingly, we may further simplify the union bound on $P_e$ given in Equation (5.95) as

$$P_e \leq \frac{(M-1)}{2\sqrt{\pi}} \exp\left(-\frac{d_{\min}^2}{4N_0}\right) \tag{5.97}$$

Equation (5.97) shows that for a prescribed AWGN channel, the average probability of symbol error $P_e$ decreases exponentially as the squared minimum distance, $d_{\min}^2$.

## ■ Bit versus Symbol Error Probabilities

Thus far, the only figure of merit we have used to assess the noise performance of a digital passband transmission system has been the average probability of symbol error. This figure of merit is the natural choice when messages of length $m = \log_2 M$ are transmitted, such as alphanumeric symbols. However, when the requirement is to transmit binary data such as digital computer data, it is often more meaningful to use another figure of merit called the *bit error rate* (BER). Although, in general, there are no unique relationships between these two figures of merit, it is fortunate that such relationships can be derived for two cases of practical interest, as discussed next.

### *Case 1*

In the first case, we assume that it is possible to perform the mapping from binary to $M$-ary symbols in such a way that the two binary $M$-tuples corresponding to any pair of adjacent symbols in the $M$-ary modulation scheme differ in only one bit position. This mapping constraint is satisfied by using a *Gray code.* When the probability of symbol error $P_e$ is acceptably small, we find that the probability of mistaking one symbol for either one of the two "nearest" symbols is much greater than any other kind of symbol error. Moreover, given a symbol error, the most probable number of bit errors is one, subject to the aforementioned mapping constraint. Since there are $\log_2 M$ bits per symbol, it follows that the average probability of symbol error is related to the bit error rate as follows:

$$\begin{aligned} P_e &= P\left( \bigcup_{i=1}^{\log_2 M} \{i\text{th bit is in error}\} \right) \\ &\le \sum_{i=1}^{\log_2 M} P(i\text{th bit is in error}) \\ &= \log_2 M \cdot (\text{BER}) \end{aligned} \tag{5.97}$$

We also note that

$$P_e \ge P(i\text{th bit is in error}) = \text{BER} \tag{5.98}$$

It follows therefore that the bit error rate is bounded as follows:

$$\frac{P_e}{\log_2 M} \le \text{BER} \le P_e \tag{5.99}$$

### *Case 2*

Let $M = 2^K$, where $K$ is an integer. We assume that all symbol errors are equally likely and occur with probability

$$\frac{P_e}{M-1} = \frac{P_e}{2^K - 1}$$

where $P_e$ is the average probability of symbol error. What is the probability that the $i$th bit in a symbol is in error? Well, there are $2^{K-1}$ cases of symbol error in which this particular bit is changed, and there are $2^{K-1}$ cases in which it is not changed. Hence, the bit error rate is

$$\text{BER} = \left( \frac{2^{K-1}}{2^K - 1} \right) P_e \tag{5.100}$$

or, equivalently,

$$\text{BER} = \left( \frac{M/2}{M-1} \right) P_e \tag{5.101}$$

Note that for large $M$, the bit error rate approaches the limiting value of $P_e/2$. The same idea described here also shows that bit errors are not independent, since we have

$$P(i\text{th and } j\text{th bits are in error}) = \frac{2^{K-2}}{2^K - 1} P_e \neq (\text{BER})^2$$

## 5.8 Summary and Discussion

The primary goal of the material presented in this chapter is the formulation of a systematic procedure for the analysis and design of a digital communication receiver in the presence of *additive white Gaussian noise* (AWGN). The procedure, known as *maximum likelihood detection*, decides which particular transmitted symbol is the most likely cause of the noisy signal observed at the channel output. The approach that led to the formulation of the maximum likelihood detector (receiver) is called *signal-space analysis*. The basic idea of the approach is to represent each member of a set of transmitted signals by an $N$-dimensional vector, where $N$ is the number of orthonormal basis functions needed for a unique geometric representation of the transmitted signals. The set of signal vectors so formed defines a *signal constellation* in an $N$-dimensional *signal space*.

For a given signal constellation, the (average) probability of symbol error $P_e$ incurred in maximum likelihood signal detection over an AWGN channel is invariant to rotation of the signal constellation as well as its translation. However, except for a few simple (but important) cases, the numerical calculation of $P_e$ is an impractical proposition. To overcome this difficulty, the customary practice is to resort to the use of bounds that lend themselves to computation in a straightforward manner. In this context, we described the *union bound* that follows directly from the signal-space diagram. The union bound is based on an intuitively satisfying idea: The probability of symbol error $P_e$ is dominated by the nearest neighbors to the transmitted signal. The results obtained using the union bound are usually fairly accurate when the signal-to-noise ratio is high.

With the material on signal-space analysis and related issues on hand, we are well-equipped to study passband data transmission systems, which we do in Chapter 6.

## Notes and References

1. The geometric representation of signals was first developed by Kotel'nikov in 1947: V. A. Kotel'nikov, *The Theory of Optimum Noise Immunity* (Dover Publications, 1960), which is a translation of the original doctoral dissertation presented in January 1947 before the Academic Council of the Molotov Energy Institute in Moscow. In particular, see Part II of the book. This method was subsequently brought to fuller fruition in the classic book by Wozencraft and Jacobs (1965). Signal-space analysis is also discussed in Cioffi (1998), Anderson (1999), and Proakis (1995).

2. In Section 5.7, we derived the union bound on the average probability of symbol error; the classic reference for this bound is Wozencraft and Jacobs (1965). For the derivation of tighter bounds, see Viterbi and Omura (1979, pp. 58–59).

3. In Chapter 4, we used the following upper bound on the complementary error function

$$\operatorname{erfc}(u) < \frac{\exp(-u^2)}{\sqrt{\pi}u}$$

For large positive $u$, a second bound on the complementary error function is obtained by omitting the multiplying factor $1/u$ in the above upper bound, as shown by

$$\operatorname{erfc}(u) < \frac{\exp(-u^2)}{\sqrt{\pi}}$$

It is this second upper bound that is used in Equation (5.97).

## PROBLEMS

### Representation of Signals

5.1 In Section 3.7 we described line codes for pulse-code modulation. Referring to the material presented therein, formulate the signal constellations for the following line codes:

(a) Unipolar nonreturn-to-zero code

(b) Polar nonreturn-to-zero code

(c) Unipolar return-to-zero code

(d) Manchester code

5.2 An 8-level PAM signal is defined by

$$s_i(t) = A_i \, \text{rect}\left(\frac{t}{T} - \frac{1}{2}\right)$$

where $A_i = \pm 1, \pm 3, \pm 5, \pm 7$. Formulate the signal constellation of $\{s_i(t)\}_{i=1}^{8}$.

5.3 Figure P5.3 displays the waveforms of four signals $s_1(t)$, $s_2(t)$, $s_3(t)$, and $s_4(t)$.

(a) Using the Gram-Schmidt orthogonalization procedure, find an orthonormal basis for this set of signals.

(b) Construct the corresponding signal-space diagram.

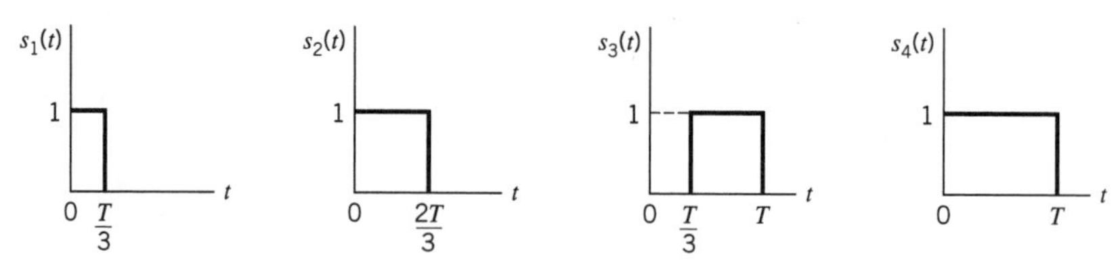


FIGURE P5.3

5.4 (a) Using the Gram-Schmidt orthogonalization procedure, find a set of orthonormal basis functions to represent the three signals $s_1(t)$, $s_2(t)$, and $s_3(t)$ shown in Figure P5.4.

(b) Express each of these signals in terms of the set of basis functions found in part (a).

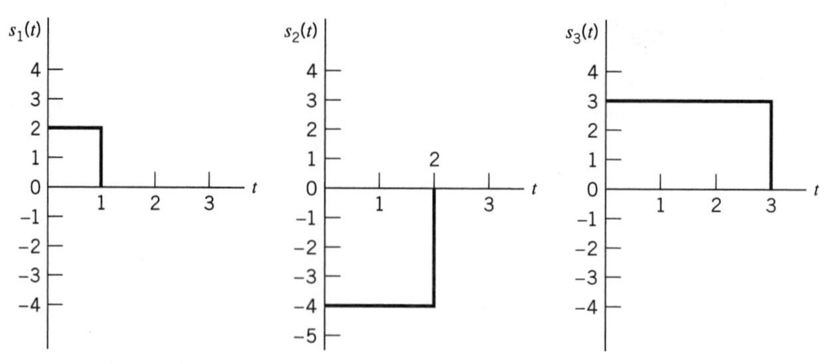


FIGURE P5.4

5.5 An *orthogonal set of signals* is characterized by the property that the inner product of any pair of signals in the set is zero. Figure P5.5 shows a pair of signals $s_1(t)$ and $s_2(t)$ that satisfy this condition. Construct the signal constellation for $s_1(t)$ and $s_2(t)$.

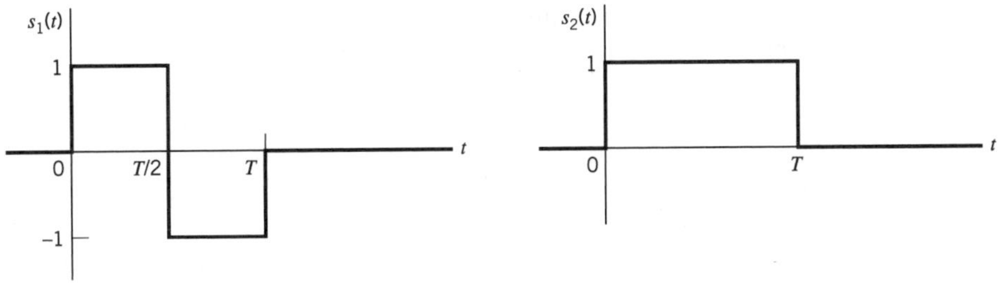

**FIGURE P5.5**

5.6 A source of information emits a set of symbols denoted by $\{m_i\}_{i=1}^{M}$. Two candidate modulation schemes, namely, pulse-duration modulation (PDM) and pulse-position modulation (PPM), are considered for the electrical representation of this set of symbols. In PDM, the $i$th symbol is represented by a pulse of unit amplitude and duration $(i/M)T$. On the other hand, in PPM, the $i$th symbol is represented by a short pulse of unit amplitude and fixed duration, which is transmitted at time $t = (i/M)T$. Show that PPM is the only one of the two that can produce an orthogonal set of signals over the interval $0 \leq t \leq T$.

5.7 A set of $2M$ *biorthogonal signals* is obtained from a set of $M$ orthogonal signals by augmenting it with the negative of each signal in the set.

(a) The extension of orthogonal to biorthogonal signals leaves the dimensionality of the signal space unchanged. Why?

(b) Construct the signal constellation for the biorthogonal signals corresponding to the pair of orthogonal signals shown in Figure P5.5.

5.8 (a) A pair of signals $s_i(t)$ and $s_k(t)$ have a common duration $T$. Show that the inner product of this pair of signals is given by

$$\int_0^T s_i(t)s_k(t)dt = \mathbf{s}_i^T\mathbf{s}_k$$

where $\mathbf{s}_i$ and $\mathbf{s}_k$ are the vector representations of $s_i(t)$ and $s_k(t)$, respectively.

(b) As a followup to part (a), show that

$$\int_0^T (s_i(t) - s_k(t))^2 dt = \| \mathbf{s}_i - \mathbf{s}_k \|^2$$

5.9 Consider a pair of complex-valued signals $s_1(t)$ and $s_2(t)$ that are respectively represented by

$$s_1(t) = a_{11}\phi_1(t) + a_{12}\phi_2(t), \qquad -\infty < t < \infty$$
$$s_2(t) = a_{21}\phi_1(t) + a_{22}\phi_2(t), \qquad -\infty < t < \infty$$

where the basis functions $\phi_1(t)$ and $\phi_2(t)$ are both real valued, but the coefficients $a_{11}$, $a_{12}$, $a_{21}$, and $a_{22}$ are complex valued. Prove the complex form of the Schwarz inequality:

$$\left| \int_{-\infty}^{\infty} s_1(t)s_2^*(t)dt \right|^2 \leq \int_{-\infty}^{\infty} |s_1(t)|^2 dt \int_{-\infty}^{\infty} |s_2(t)|^2 dt$$

where the asterisk denotes complex conjugation. When is this relation satisfied with the equality sign?

## Random Processes

5.10 Consider a random process $X(t)$ expanded in the form

$$X(t) = \sum_{j=1}^{N} X_j\phi_j(t) + W'(t), \qquad 0 \leq t \leq T$$

where $W'(t)$ is a remainder noise term. The $\{\phi_j(t)\}_{j=1}^N$ form an orthonormal set over the interval $0 \le t \le T$, and the $X_j$ are defined by

$$X_j = \int_0^T X(t)\phi_j(t)dt$$

Let $W'(t_k)$ denote a random variable obtained by observing $W'(t)$ at time $t = t_k$. Show that

$$E[X_j W'(t_k)] = 0, \qquad \begin{cases} j = 1, 2, \ldots, N \\ 0 \le t_k \le T \end{cases}$$

**5.11** Consider the optimum detection of the sinusoidal signal

$$s(t) = \sin\left(\frac{8\pi t}{T}\right), \qquad 0 \le t \le T$$

in additive white Gaussian noise.

(a) Determine the correlator output assuming a noiseless input.

(b) Determine the corresponding matched filter output, assuming that the filter includes a delay $T$ to make it causal.

(c) Hence show that these two outputs are the same only at time instant $t = T$.

## Probability of Error

**5.12** Figure P5.12 shows a pair of signals $s_1(t)$ and $s_2(t)$ that are orthogonal to each other over the observation interval $0 \le t \le 3T$. The received signal is defined by

$$x(t) = s_k(t) + w(t), \qquad \begin{matrix} 0 \le t \le 3T \\ k = 1, 2 \end{matrix}$$

where $w(t)$ is white Gaussian noise of zero mean and power spectral density $N_0/2$.

(a) Design a receiver that decides in favor of signals $s_1(t)$ or $s_2(t)$, assuming that these two signals are equiprobable.

(b) Calculate the average probability of symbol error incurred by this receiver for $E/N_0 = 4$, where $E$ is the signal energy.

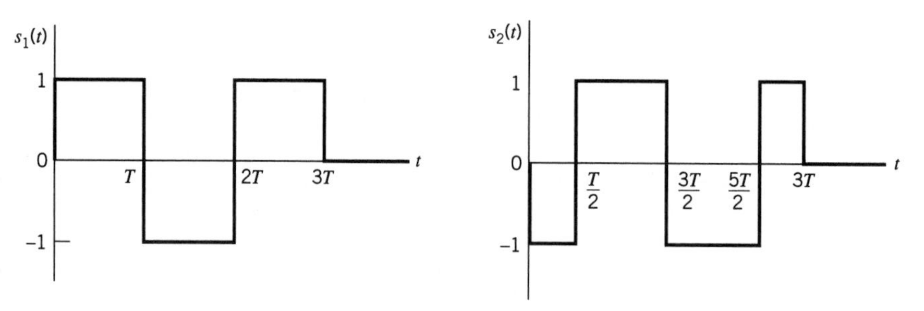


FIGURE P5.12

**5.13** In the Manchester code, binary symbol 1 is represented by the doublet pulse $s(t)$ shown in Figure P5.13, and binary symbol 0 is represented by the negative of this pulse. Derive the formula for the probability of error incurred by the maximum likelihood detection procedure applied to this form of signaling over an AWGN channel.

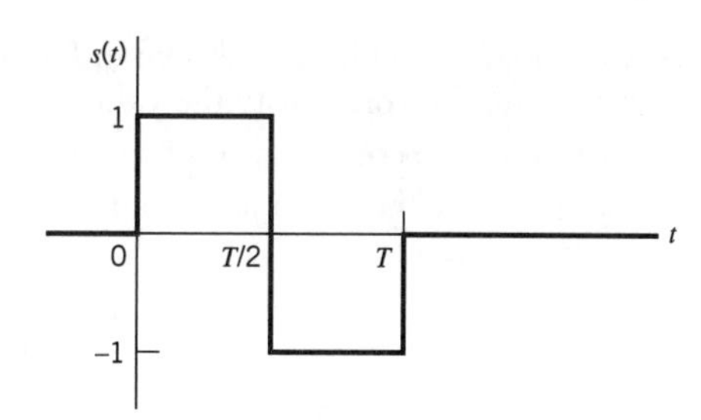


FIGURE P5.13

5.14 In the *Bayes test*, applied to a binary hypothesis testing problem where we have to choose one of two possible hypotheses $H_0$ and $H_1$, we minimize the *risk* $R$ defined by

$$\begin{aligned} R = {} & C_{00}p_0P(\text{say } H_0 \mid H_0 \text{ is true}) \\ & + C_{10}p_0P(\text{say } H_1 \mid H_0 \text{ is true}) \\ & + C_{11}p_1P(\text{say } H_1 \mid H_1 \text{ is true}) \\ & + C_{01}p_1P(\text{say } H_0 \mid H_1 \text{ is true}) \end{aligned}$$

The terms $C_{00}$, $C_{10}$, $C_{11}$, and $C_{01}$ denote the costs assigned to the four possible outcomes of the experiment: The first subscript indicates the hypothesis chosen, and the second the hypothesis that is true. Assume that $C_{10} > C_{00}$ and $C_{01} > C_{11}$. The $p_0$ and $p_1$ denote the *a priori* probabilities of hypotheses $H_0$ and $H_1$, respectively.

(a) Given the observation vector $\mathbf{x}$, show that the partitioning of the observation space so as to minimize the risk $R$ leads to the *likelihood ratio test*:

$$\begin{aligned} &\text{say } H_0 \text{ if } \Lambda(\mathbf{x}) < \lambda \\ &\text{say } H_1 \text{ if } \Lambda(\mathbf{x}) > \lambda \end{aligned}$$

where $\Lambda(\mathbf{x})$ is the *likelihood ratio*

$$\Lambda(\mathbf{x}) = \frac{f_{\mathbf{X}}(\mathbf{x} \mid H_1)}{f_{\mathbf{X}}(\mathbf{x} \mid H_0)}$$

and $\lambda$ is the *threshold* of the test defined by

$$\lambda = \frac{p_0(C_{10} - C_{00})}{p_1(C_{01} - C_{11})}$$

(b) What are the cost values for which the Bayes' criterion reduces to the minimum probability of error criterion?

## Principles of Rotational and Translational Invariance

5.15 Continuing with the four line codes considered in Problem 5.1, identify the line codes that have minimum average energy and those that do not. Compare your answers with the observations made on these line codes in Section 3.7.

5.16 Consider the two constellations shown in Figure 5.11. Determine the orthonormal matrix $\mathbf{Q}$ that transforms the constellation shown in Figure 5.11*a* into the one shown in Figure 5.11*b*.

5.17 (a) The two signal constellations shown in Figure P5.17 exhibit the same average probability of symbol error. Justify the validity of this statement.

(b) Which of these two constellations has minimum average energy? Justify your answer.

You may assume that the symbols pertaining to the message points displayed in Figure P5.17 are equally likely.

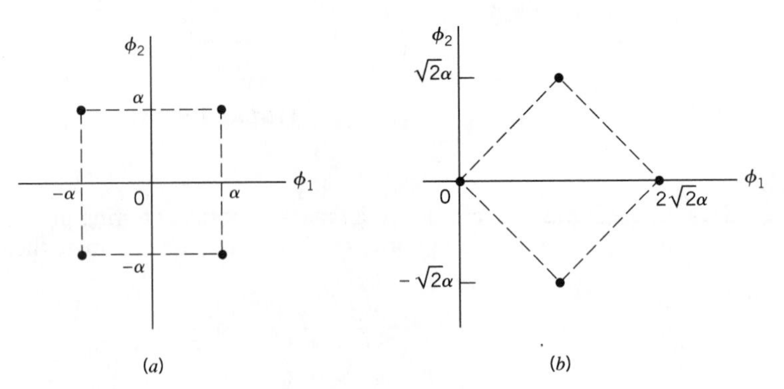


FIGURE P5.17

5.18 *Simplex (transorthogonal) signals* are equally likely highly-correlated signals with the most negative correlation that can be achieved with a set of $M$ orthogonal signals. That is, the correlation coefficient between any pair of signals in the set is defined by

$$\rho_{ij} = \begin{cases} 1 & \text{for } i = j \\ \dfrac{-1}{M-1} & \text{for } i \neq j \end{cases}$$

One method of constructing simplex signals is to start with a set of $M$ orthogonal signals, each with energy $E$, and then apply the minimum energy translate.

Consider a set of three equally likely symbols whose signal constellation consists of the vertices of an equilateral triangle. Show that these three symbols constitute a simplex code.

## Bounds on Probability of Error

5.19 In this problem we explore the approximations to the probability of an error, $P_e$, for the pair of antipodal signals shown in Figure P5.19 in the presence of additive white Gaussian noise of power spectral density $N_0/2$. The exact formula for $P_e$ is

$$P_e = \frac{1}{2}\,\text{erfc}\left(\sqrt{\frac{E_b}{N_0}}\right)$$

(This formula is derived in Section 6.3)

(a) Using the two upper bounds for the complementary error function given in Note 3, derive the corresponding approximations to $P_e$.

(b) Compare the approximations derived in part (a) for $P_e$ to the exact formula for $E_b/N_0 = 9$. For the exact calculation of $P_e$, you may use Table A6.6 on the error function.

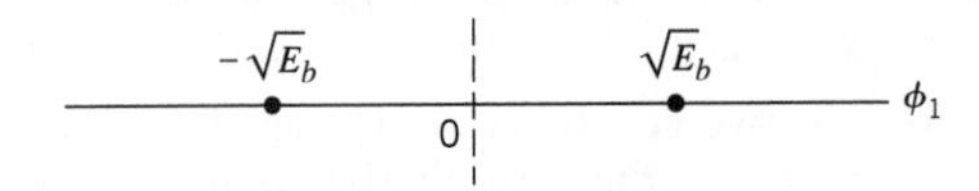


FIGURE P5.19

5.20 Consider the special case of a signal constellation that has a *symmetric geometry* with respect to the origin. Assume that the $M$ message points of the constellation, pertaining to symbols $m_1, m_2, \cdots, m_M$, are equally likely. Using the upper bound on the complementary error function given in Equation (5.94), show that the average probability of symbol error for the constellation is bounded as

$$P_e \leq \frac{M_{\min}}{2\sqrt{\pi}} \exp\left(-\min_{\substack{i,\,k \\ i \neq k}} \left(\frac{d_{ik}^2}{4N_0}\right)\right)$$

where $d_{ik}$ is the Euclidean distance between message points $i$ and $k$, and $M_{\min}$ is the number of transmitted signals that attain the minimum Euclidean distance for each $m_i$.

# CHAPTER 6

# PASSBAND DATA TRANSMISSION

This chapter builds on the material developed in Chapter 5 on signal-space analysis. It discusses the subject of digital data transmission over a band-pass channel that can be linear or nonlinear. As with analog communications, this mode of data transmission relies on the use of a sinusoidal carrier wave modulated by the data stream.

Specifically, the following topics are covered:

- *Different methods of digital modulation, namely, phase-shift keying, quadrature-amplitude modulation, and frequency-shift keying, and their individual variants.*
- *Coherent detection of modulated signals in additive white Gaussian noise, which requires the receiver to be synchronized to the transmitter with respect to both carrier phase and bit timing.*
- *Noncoherent detection of modulated signals in additive white Gaussian noise, disregarding phase information in the received signal.*
- *Modems for the transmission and reception of digital data over the public switched telephone network.*
- *Sophisticated modulation techniques, namely, carrierless amplitude/phase modulation and discrete multitone, for data transmission over a wideband channel with medium to severe intersymbol interference.*
- *Techniques for synchronizing the receiver to the transmitter.*

## 6.1 *Introduction*

In *baseband pulse transmission*, which we studied in Chapter 4, a data stream represented in the form of a discrete pulse-amplitude modulated (PAM) signal is transmitted directly over a low-pass channel. In *digital passband transmission*, on the other hand, the incoming data stream is modulated onto a carrier (usually sinusoidal) with fixed frequency limits imposed by a band-pass channel of interest; passband data transmission is studied in this chapter.

The communication channel used for passband data transmission may be a microwave radio link, a satellite channel, or the like. Yet other applications of passband data transmission are in the design of passband line codes for use on digital subscriber loops and orthogonal frequency-division multiplexing techniques for broadcasting. In any event, the modulation process making the transmission possible involves switching (keying) the amplitude, frequency, or phase of a sinusoidal carrier in some fashion in accordance with the incoming data. Thus there are three basic signaling schemes, and they are known as

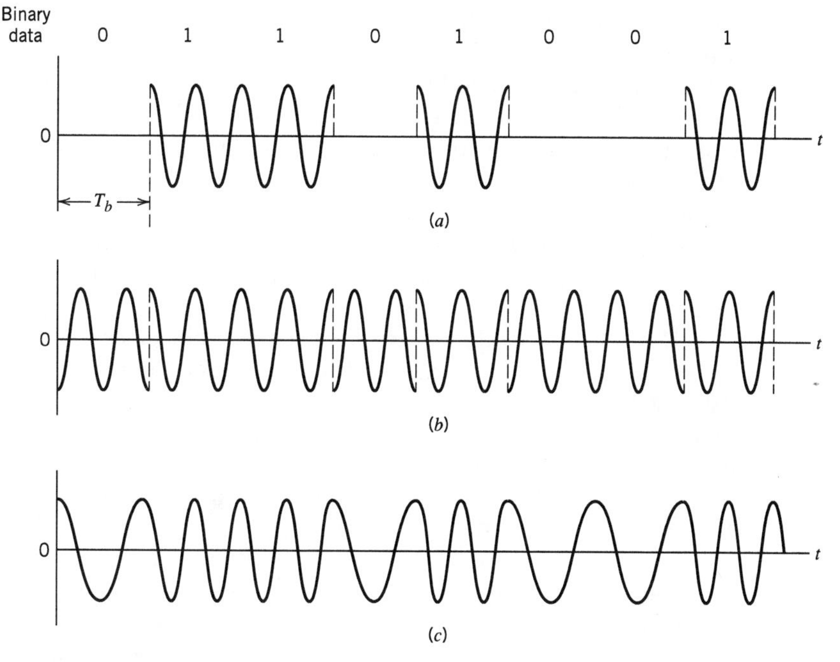


**FIGURE 6.1** Illustrative waveforms for the three basic forms of signaling binary information. (*a*) Amplitude-shift keying. (*b*) Phase-shift keying. (*c*) Frequency-shift keying with continuous phase.

*amplitude-shift keying* (ASK), *frequency-shift keying* (FSK), and *phase-shift keying* (PSK). They may be viewed as special cases of amplitude modulation, frequency modulation, and phase modulation, respectively.

Figure 6.1 illustrates these three methods of modulation for the case of a source supplying binary data. The following points are noteworthy from Figure 6.1:

- Although in continuous-wave modulation it is usually difficult to distinguish between phase-modulated and frequency-modulated signals by merely looking at their waveforms, this is not true for PSK and FSK signals.
- Unlike ASK signals, both PSK and FSK signals have a constant envelope.

This latter property makes PSK and FSK signals impervious to amplitude nonlinearities, commonly encountered in microwave radio and satellite channels. It is for this reason, in practice, we find that PSK and FSK signals are preferred to ASK signals for passband data transmission over nonlinear channels.

## Hierarchy of Digital Modulation Techniques[1]

Digital modulation techniques may be classified into *coherent* and *noncoherent* techniques, depending on whether the receiver is equipped with a phase-recovery circuit or not. The phase-recovery circuit ensures that the oscillator supplying the locally generated carrier wave in the receiver is synchronized (in both frequency and phase) to the oscillator supplying the carrier wave used to originally modulate the incoming data stream in the transmitter.

As discussed in Chapter 4, in an *M-ary signaling scheme*, we may send any one of $M$ possible signals $s_1(t), s_2(t), \dots, s_M(t)$, during each signaling interval of duration $T$. For almost all applications, the number of possible signals $M = 2^n$, where $n$ is an integer. The symbol duration $T = nT_b$, where $T_b$ is the bit duration. In passband data transmission, these signals are generated by changing the amplitude, phase, or frequency of a sinusoidal carrier in $M$ discrete steps. Thus we have *M-ary ASK*, *M-ary PSK*, and *M-ary FSK* digital modulation schemes. Another way of generating $M$-ary signals is to combine different methods of modulation into a hybrid form. For example, we may combine discrete changes in both the amplitude and phase of a carrier to produce *M-ary amplitude-phase keying* (APK). A special form of this hybrid modulation is *M-ary quadrature-amplitude modulation* (QAM), which has some attractive properties. $M$-ary ASK is a special case of $M$-ary QAM.

$M$-ary signaling schemes are preferred over binary signaling schemes for transmitting digital information over band-pass channels when the requirement is to conserve bandwidth at the expense of increased power. In practice, we rarely find a communication channel that has the exact bandwidth required for transmitting the output of an information source by means of binary signaling schemes. Thus when the bandwidth of the channel is less than the required value, we may use $M$-ary signaling schemes for maximum efficiency. To illustrate the bandwidth-conservation capability of $M$-ary signaling schemes, consider the transmission of information consisting of a binary sequence with bit duration $T_b$. If we were to transmit this information by means of binary PSK, for example, we would require a bandwidth that is inversely proportional to $T_b$. However, if we take blocks of $n$ bits and use an $M$-ary PSK scheme with $M = 2^n$ and symbol duration $T = nT_b$, the bandwidth required is proportional to $1/nT_b$. This shows that the use of $M$-ary PSK enables a reduction in transmission bandwidth by the factor $n = \log_2 M$ over binary PSK.

$M$-ary PSK and $M$-ary QAM are examples of *linear modulation*. However, they differ from each other in one important respect: An $M$-ary PSK signal has a constant envelope, whereas an $M$-ary QAM signal involves changes in the carrier amplitude. Accordingly, $M$-ary PSK can be used to transmit digital data over a nonlinear band-pass channel, whereas $M$-ary QAM requires the use of a linear channel.

$M$-ary PSK, $M$-ary QAM, and $M$-ary FSK are commonly used in coherent systems. Amplitude-shift keying and frequency-shift keying lend themselves naturally to use in noncoherent systems whenever it is impractical to maintain carrier phase synchronization. But in the case of phase-shift keying, we cannot have "noncoherent PSK" because the term noncoherent means doing without carrier phase information. Instead, we employ a "pseudo PSK" technique known as *differential phase-shift keying* (DPSK), which (in a loose sense) may be viewed as the noncoherent form of PSK. In practice, $M$-ary FSK and $M$-ary DPSK are the commonly used forms of digital modulation in noncoherent systems.

## ■ PROBABILITY OF ERROR

A major goal of passband data transmission systems is the optimum design of the receiver so as to minimize the average probability of symbol error in the presence of *additive white Gaussian noise* (AWGN). With this goal in mind, much of the material presented in this chapter builds on the signal-space analysis tools presented in Chapter 5. Specifically, in the study of each system we begin with the formulation of a signal constellation and the construction of decision regions in accordance with maximum likelihood signal detection over an AWGN channel. These formulations set the stage for evaluating the probability of symbol error $P_e$. Depending on the method of digital modulation under study, the evaluation of $P_e$ proceeds in one of two ways:

- In the case of certain simple methods such as coherent binary PSK and coherent binary FSK, exact formulas are derived for $P_e$.

- In the case of more elaborate methods such as coherent $M$-ary PSK and coherent $M$-ary FSK, we resort to the use of the union bound for deriving an approximate formula for $P_e$.

## Power Spectra

To fully appreciate the practical virtues of different methods of digital modulation, we also need to study the power spectra of the resulting modulated signals. This latter issue is particularly important in two contexts: occupancy of the channel bandwidth and co-channel interference in multiplexed systems.

Given a modulated signal $s(t)$, we may describe it in terms of its in-phase and quadrature components as

$$\begin{aligned} s(t) &= s_I(t)\cos(2\pi f_c T) - s_Q(t)\sin(2\pi f_c t) \\ &= \text{Re}[\tilde{s}(t)\exp(j2\pi f_c t)] \end{aligned} \tag{6.1}$$

where Re[·] is the real part of the expression contained inside the square brackets. We also have

$$\tilde{s}(t) = s_I(t) + js_Q(t) \tag{6.2}$$

and

$$\exp(j2\pi f_c T) = \cos(2\pi f_c t) + j\sin(2\pi f_c t) \tag{6.3}$$

The signal $\tilde{s}(t)$ is the *complex envelope* (i.e., baseband version) of the modulated (band-pass) signal $s(t)$. The components $s_I(t)$ and $s_Q(t)$ and therefore $\tilde{s}(t)$ are all low-pass signals. They are uniquely defined in terms of the band-pass signal $s(t)$ and the carrier frequency $f_c$, provided that the half-bandwidth of $s(t)$ is less than the carrier frequency $f_c$.

Let $S_B(f)$ denote the power spectral density of the complex envelope $\tilde{s}(t)$. We refer to $S_B(f)$ as the *baseband power spectral density*. The power spectral density, $S_S(f)$, of the original band-pass signal $s(t)$ is a frequency-shifted version of $S_B(f)$, except for a scaling factor, as shown by

$$S_S(t) = \frac{1}{4}[S_B(f - f_c) + S_B(f + f_c)] \tag{6.4}$$

It is therefore sufficient to evaluate the baseband power spectral density $S_B(f)$. Since $\tilde{s}(t)$ is a low-pass signal, the calculation of $S_B(f)$ should be simpler than the calculation of $S_S(f)$. (See Example 1.7.)

## Bandwidth Efficiency

Throughout this book we have emphasized that *channel bandwidth* and *transmitted power* constitute two primary "communication resources," the efficient utilization of which provides the motivation for the search for *spectrally efficient* schemes. The primary objective of spectrally efficient modulation is to maximize the bandwidth efficiency defined as *the ratio of the data rate in bits per second to the effectively utilized channel bandwidth*. A secondary objective is to achieve this bandwidth efficiency at a minimum practical expenditure of average signal power or, equivalently, in a channel perturbed by additive white Gaussian noise, a minimum practical expenditure of average signal-to-noise ratio.

With the data rate denoted by $R_b$ and the effectively used channel bandwidth by $B$, we may express the bandwidth efficiency, $\rho$, as

$$\rho = \frac{R_b}{B} \text{ bits/s/Hz} \tag{6.5}$$

Recall from Chapter 4 that bandwidth efficiency is the product of two independent factors: one due to the possible use of *multilevel encoding* and the other due to *spectral shaping*. In multilevel encoding, information transmission through the channel is carried out on the basis of blocks of bits rather than single bits. With efficient spectral shaping, bandwidth requirement on the channel is reduced by the use of pulse-shaping filters that smooth out the sharp transitions in the transmitted waveform. These two factors are therefore important in their own individual ways in determining the bandwidth efficiency of a passband data transmission system of interest.

## 6.2 Passband Transmission Model

In a functional sense, we may model a passband data transmission system as shown in Figure 6.2. First, there is assumed to exist a *message source* that emits one *symbol* every $T$ seconds, with the symbols belonging to an alphabet of $M$ symbols, which we denote by $m_1, m_2, \ldots, m_M$. The *a priori probabilities* $P(m_1), P(m_2), \ldots, P(m_M)$ specify the message source output. When the $M$ symbols of the alphabet are *equally likely*, we write

$$\begin{aligned} p_i &= P(m_i) \\ &= \frac{1}{M} \quad \text{for all } i \end{aligned} \tag{6.6}$$

The $M$-ary output of the message source is presented to a *signal transmission encoder*, producing a corresponding vector $\mathbf{s}_i$ made up of $N$ real elements, one such set for each of the $M$ symbols of the source alphabet; the dimension $N$ is less than or equal to $M$. With the vector $\mathbf{s}_i$ as input, the *modulator* then constructs a *distinct* signal $s_i(t)$ of duration $T$ seconds as the representation of the symbol $m_i$ generated by the message source. The signal $s_i(t)$ is necessarily an energy signal, as shown by

$$E_i = \int_0^T s_i^2(t)\, dt, \qquad i = 1, 2, \ldots, M \tag{6.7}$$

Note that $s_i(t)$ is real valued. One such signal is transmitted every $T$ seconds. The particular signal chosen for transmission depends in some fashion on the incoming message and possibly on the signals transmitted in preceding time slots. With a sinusoidal carrier, the feature that is used by the modulator to distinguish one signal from another is a *step* change in the amplitude, frequency, or phase of the carrier. (Sometimes, a hybrid form of modulation that combines changes in both amplitude and phase or amplitude and frequency is used.)

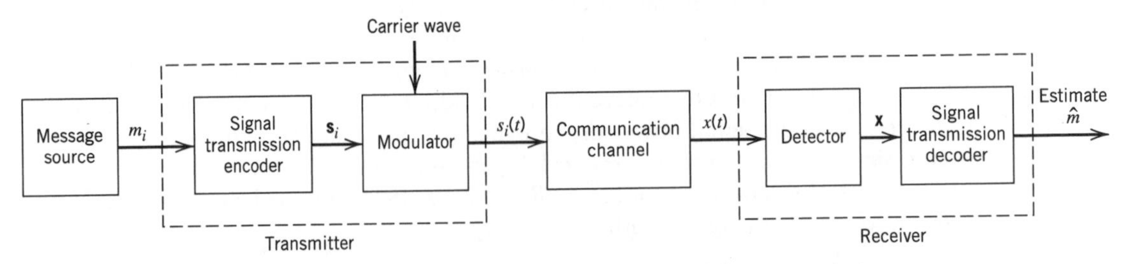


**FIGURE 6.2** Functional model of passband data transmission system.

Returning to the functional model of Figure 6.2, the bandpass communication channel, coupling the transmitter to the receiver, is assumed to have two characteristics:

1. The channel is linear, with a bandwidth that is wide enough to accommodate the transmission of the modulated signal $s_i(t)$ with negligible or no distortion.
2. The channel noise $w(t)$ is the sample function of a white Gaussian noise process of zero mean and power spectral density $N_0/2$.

The assumptions made herein are basically the same as those invoked in Chapter 5 dealing with signal-space analysis.

The receiver, which consists of a *detector* followed by a *signal transmission decoder*, performs two functions:

1. It reverses the operations performed in the transmitter.
2. It minimizes the effect of channel noise on the estimate $\hat{m}$ computed for the transmitted symbol $m_i$.

## 6.3 *Coherent Phase-Shift Keying*

With the background material on the coherent detection of signals in additive white Gaussian noise that was presented in Chapter 5 at our disposal, we are now ready to study specific passband data transmission systems. In this section we focus on coherent phase-shift keying (PSK) by considering binary PSK, QPSK and its variants, and finish up with $M$-ary PSK.

### Binary Phase-Shift Keying

In a coherent binary PSK system, the pair of signals $s_1(t)$ and $s_2(t)$ used to represent binary symbols 1 and 0, respectively, is defined by

$$s_1(t) = \sqrt{\frac{2E_b}{T_b}} \cos(2\pi f_c t) \tag{6.8}$$

$$s_2(t) = \sqrt{\frac{2E_b}{T_b}} \cos(2\pi f_c t + \pi) = -\sqrt{\frac{2E_b}{T_b}} \cos(2\pi f_c t) \tag{6.9}$$

where $0 \le t \le T_b$, and $E_b$ is the *transmitted signal energy per bit.* To ensure that each transmitted bit contains an integral number of cycles of the carrier wave, the carrier frequency $f_c$ is chosen equal to $n_c/T_b$ for some fixed integer $n_c$. A pair of sinusoidal waves that differ only in a relative phase-shift of 180 degrees, as defined in Eqautions (6.8) and (6.9), are referred to as *antipodal signals.*

From this pair of equations it is clear that, in the case of binary PSK, there is only one basis function of unit energy, namely,

$$\phi_1(t) = \sqrt{\frac{2}{T_b}} \cos(2\pi f_c t), \qquad 0 \le t < T_b \tag{6.10}$$

Then we may express the transmitted signals $s_1(t)$ and $s_2(t)$ in terms of $\phi_1(t)$ as follows:

$$s_1(t) = \sqrt{E_b}\phi_1(t), \qquad 0 \le t < T_b \tag{6.11}$$

and

$$s_2(t) = -\sqrt{E_b}\phi_1(t), \qquad 0 \le t < T_b \tag{6.12}$$

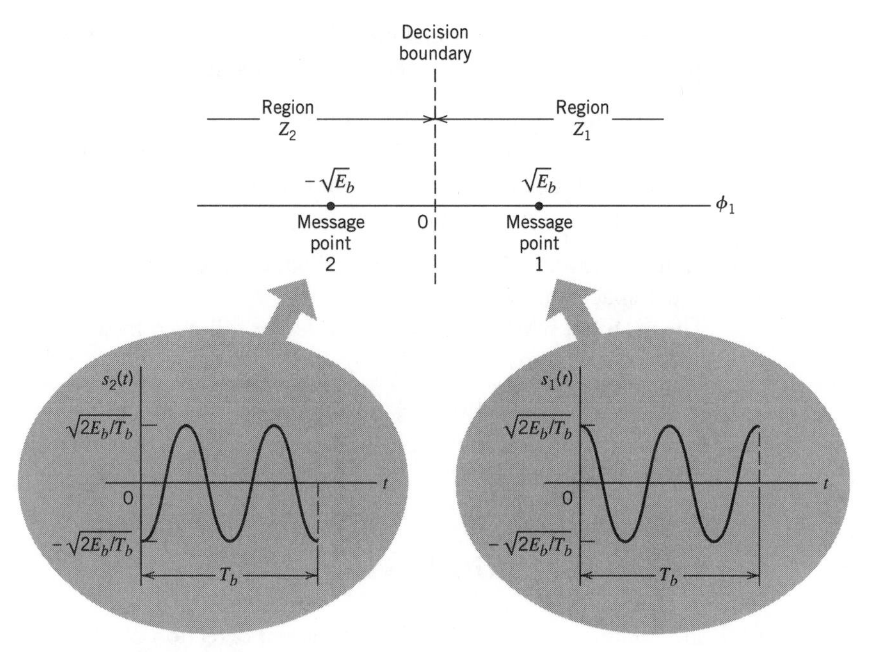


**FIGURE 6.3** Signal-space diagram for coherent binary PSK system. The waveforms depicting the transmitted signals $s_1(t)$ and $s_2(t)$, displayed in the inserts, assume $n_c = 2$.

A coherent binary PSK system is therefore characterized by having a signal space that is one-dimensional (i.e., $N = 1$), with a signal constellation consisting of two message points (i.e., $M = 2$). The coordinates of the message points are

$$\begin{aligned} s_{11} &= \int_0^{T_b} s_1(t)\phi_1(t)\, dt \\ &= +\sqrt{E_b} \end{aligned} \tag{6.13}$$

and

$$\begin{aligned} s_{21} &= \int_0^{T_b} s_2(t)\phi_1(t)\, dt \\ &= -\sqrt{E_b} \end{aligned} \tag{6.14}$$

The message point corresponding to $s_1(t)$ is located at $s_{11} = +\sqrt{E_b}$, and the message point corresponding to $s_2(t)$ is located at $s_{21} = -\sqrt{E_b}$. Figure 6.3 displays the signal-space diagram for binary PSK. This figure also includes two inserts, showing example waveforms of antipodal signals representing $s_1(t)$ and $s_2(t)$. Note that the constellation of Figure 6.3 has minimum average energy.

### *Error Probability of Binary PSK*

To realize a *rule for making a decision* in favor of symbol 1 or symbol 0, we apply Equation (5.59) of Chapter 5. Specifically, we partition the signal space of Figure 6.3 into two regions:

- The set of points closest to message point 1 at $+\sqrt{E_b}$.
- The set of points closest to message point 2 at $-\sqrt{E_b}$.

This is accomplished by constructing the midpoint of the line joining these two message points, and then marking off the appropriate decision regions. In Figure 6.3 these decision regions are marked $Z_1$ and $Z_2$, according to the message point around which they are constructed.

The decision rule is now simply to decide that signal $s_1(t)$ (i.e., binary symbol 1) was transmitted if the received signal point falls in region $Z_1$, and decide that signal $s_2(t)$ (i.e., binary symbol 0) was transmitted if the received signal point falls in region $Z_2$. Two kinds of erroneous decisions may, however, be made. Signal $s_2(t)$ is transmitted, but the noise is such that the received signal point falls inside region $Z_1$ and so the receiver decides in favor of signal $s_1(t)$. Alternatively, signal $s_1(t)$ is transmitted, but the noise is such that the received signal point falls inside region $Z_2$ and so the receiver decides in favor of signal $s_2(t)$.

To calculate the probability of making an error of the first kind, we note from Figure 6.3 that the decision region associated with symbol 1 or signal $s_1(t)$ is described by

$$Z_1\colon 0 < x_1 < \infty$$

where the observable element $x_1$ is related to the received signal $x(t)$ by

$$x_1 = \int_0^{T_b} x(t)\phi_1(t)\, dt \tag{6.15}$$

The conditional probability density function of random variable $X_1$, given that symbol 0 [i.e., signal $s_2(t)$] was transmitted, is defined by

$$\begin{aligned} f_{X_1}(x_1|0) &= \frac{1}{\sqrt{\pi N_0}} \exp\left[-\frac{1}{N_0}(x_1 - s_{21})^2\right] \\ &= \frac{1}{\sqrt{\pi N_0}} \exp\left[-\frac{1}{N_0}(x_1 + \sqrt{E_b})^2\right] \end{aligned} \tag{6.16}$$

The conditional probability of the receiver deciding in favor of symbol 1, given that symbol 0 was transmitted, is therefore

$$\begin{aligned} p_{10} &= \int_0^{\infty} f_{X_1}(x_1|0)\, dx_1 \\ &= \frac{1}{\sqrt{\pi N_0}} \int_0^{\infty} \exp\left[-\frac{1}{N_0}(x_1 + \sqrt{E_b})^2\right] dx_1 \end{aligned} \tag{6.17}$$

Putting

$$z = \frac{1}{\sqrt{N_0}}(x_1 + \sqrt{E_b}) \tag{6.18}$$

and changing the variable of integration from $x_1$ to $z$, we may rewrite Equation (6.17) in the compact form

$$\begin{aligned} p_{10} &= \frac{1}{\sqrt{\pi}} \int_{\sqrt{E_b/N_0}}^{\infty} \exp(-z^2)\, dz \\ &= \frac{1}{2}\,\mathrm{erfc}\left(\sqrt{\frac{E_b}{N_0}}\right) \end{aligned} \tag{6.19}$$

where erfc(·) is the complementary error function.

Consider next an error of the second kind. We note that the signal space of Figure 6.3 is symmetric with respect to the origin. It follows therefore that $p_{01}$, the conditional probability of the receiver deciding in favor of symbol 0, given that symbol 1 was transmitted, also has the same value as in Equation (6.19).

Thus, averaging the conditional error probabilities $p_{10}$ and $p_{01}$, we find that the *average probability of symbol error* or, equivalently, the *bit error rate for coherent binary PSK* is (assuming equiprobable symbols)

$$P_e = \frac{1}{2}\,\mathrm{erfc}\left(\sqrt{\frac{E_b}{N_0}}\right) \tag{6.20}$$

As we increase the transmitted signal energy per bit, $E_b$, for a specified noise spectral density $N_0$, the message points corresponding to symbols 1 and 0 move further apart, and the average probability of error $P_e$ is correspondingly reduced in accordance with Equation (6.20), which is intuitively satisfying.

### *Generation and Detection of Coherent Binary PSK Signals*

To generate a binary PSK signal, we see from Equations (6.8)–(6.10) that we have to represent the input binary sequence in polar form with symbols 1 and 0 represented by constant amplitude levels of $+\sqrt{E_b}$ and $-\sqrt{E_b}$, respectively. This signal transmission encoding is performed by a polar nonreturn-to-zero (NRZ) level encoder. The resulting binary wave and a sinusoidal carrier $\phi_1(t)$, whose frequency $f_c = (n_c/T_b)$ for some fixed integer $n_c$, are applied to a product modulator, as in Figure 6.4*a*. The carrier and the timing pulses used to generate the binary wave are usually extracted from a common master clock. The desired PSK wave is obtained at the modulator output.

To detect the original binary sequence of 1s and 0s, we apply the noisy PSK signal $x(t)$ (at the channel output) to a correlator, which is also supplied with a locally generated coherent reference signal $\phi_1(t)$, as in Figure 6.4*b*. The correlator output, $x_1$, is compared with a threshold of zero volts. If $x_1 > 0$, the receiver decides in favor of symbol 1. On the

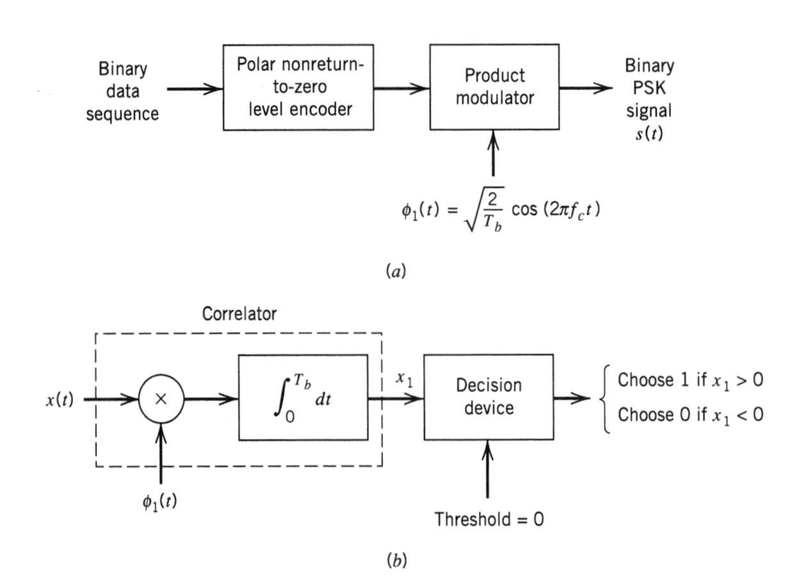


**FIGURE 6.4** Block diagrams for (*a*) binary PSK transmitter and (*b*) coherent binary PSK receiver.

other hand, if $x_1 < 0$, it decides in favor of symbol 0. If $x_1$ is exactly zero, the receiver makes a random guess in favor of 0 or 1.

### Power Spectra of Binary PSK Signals

From the modulator of Figure 6.4*a*, we see that the complex envelope of a binary PSK wave consists of an in-phase component only. Furthermore, depending on whether we have symbol 1 or symbol 0 at the modulator input during the signaling interval $0 \le t \le T_b$, we find that this in-phase component equals $+g(t)$ or $-g(t)$, respectively, where $g(t)$ is the *symbol shaping function* defined by

$$g(t) = \begin{cases} \sqrt{\dfrac{2E_b}{T_b}}, & 0 \le t \le T_b \\ 0, & \text{otherwise} \end{cases} \tag{6.21}$$

We assume that the input binary wave is random, with symbols 1 and 0 equally likely and the symbols transmitted during the different time slots being statistically independent. In Example 1.6 of Chapter 1 it is shown that the power spectral density of a random binary wave so described is equal to the energy spectral density of the symbol shaping function divided by the symbol duration. The energy spectral density of a Fourier transformable signal $g(t)$ is defined as the squared magnitude of the signal's Fourier transform. Hence, the baseband power spectral density of a binary PSK signal equals

$$\begin{aligned} S_B(f) &= \frac{2E_b \sin^2(\pi T_b f)}{(\pi T_b f)^2} \\ &= 2E_b \operatorname{sinc}^2(T_b f) \end{aligned} \tag{6.22}$$

This power spectrum falls off as the inverse square of frequency, as shown in Figure 6.5.

Figure 6.5 also includes a plot of the baseband power spectral density of a binary FSK signal, details of which are presented in Section 6.5. Comparison of these two spectra is deferred to that section.

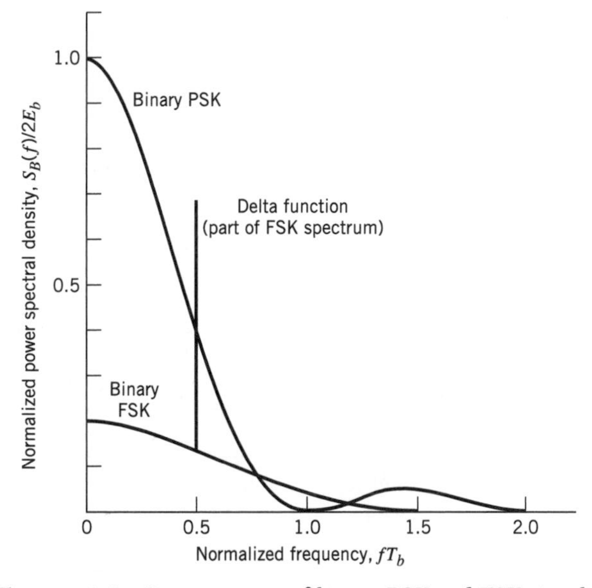


**Figure 6.5** Power spectra of binary PSK and FSK signals.

## ■ QUADRIPHASE-SHIFT KEYING

The provision of reliable performance, exemplified by a very low probability of error, is one important goal in the design of a digital communication system. Another important goal is the efficient utilization of channel bandwidth. In this subsection, we study a bandwidth-conserving modulation scheme known as coherent quadriphase-shift keying, which is an example of *quadrature-carrier multiplexing*.

In *quadriphase-shift keying* (QPSK), as with binary PSK, information carried by the transmitted signal is contained in the phase. In particular, the phase of the carrier takes on one of four equally spaced values, such as $\pi/4$, $3\pi/4$, $5\pi/4$, and $7\pi/4$. For this set of values we may define the transmitted signal as

$$s_i(t) = \begin{cases} \sqrt{\dfrac{2E}{T}} \cos\left[2\pi f_c t + (2i - 1)\dfrac{\pi}{4}\right], & 0 \leq t \leq T \\ 0, & \text{elsewhere} \end{cases} \tag{6.23}$$

where $i = 1, 2, 3, 4$; $E$ is the transmitted signal energy per symbol, and $T$ is the symbol duration. The carrier frequency $f_c$ equals $n_c/T$ for some fixed integer $n_c$. Each possible value of the phase corresponds to a unique dibit. Thus, for example, we may choose the foregoing set of phase values to represent the *Gray-encoded* set of dibits: 10, 00, 01, and 11, where only a single bit is changed from one dibit to the next.

### *Signal-Space Diagram of QPSK*

Using a well-known trigonometric identity, we may use Equation (6.23) to redefine the transmitted signal $s_i(t)$ for the interval $0 \leq t \leq T$ in the equivalent form:

$$s_i(t) = \sqrt{\frac{2E}{T}} \cos\left[(2i - 1)\frac{\pi}{4}\right] \cos(2\pi f_c t) - \sqrt{\frac{2E}{T}} \sin\left[(2i - 1)\frac{\pi}{4}\right] \sin(2\pi f_c t) \tag{6.24}$$

where $i = 1, 2, 3, 4$. Based on this representation, we can make the following observations:

- There are two orthonormal basis functions, $\phi_1(t)$ and $\phi_2(t)$, contained in the expansion of $s_i(t)$. Specifically, $\phi_1(t)$ and $\phi_2(t)$ are defined by a pair of *quadrature carriers*:

$$\phi_1(t) = \sqrt{\frac{2}{T}} \cos(2\pi f_c t), \quad 0 \leq t \leq T \tag{6.25}$$

$$\phi_2(t) = \sqrt{\frac{2}{T}} \sin(2\pi f_c t), \quad 0 \leq t \leq T \tag{6.26}$$

**TABLE 6.1** ***Signal-space characterization of QPSK***

| Gray-encoded Input Dibit | Phase of QPSK Signal (radians) | Coordinates of Message Points $s_{i1}$ | Coordinates of Message Points $s_{i2}$ |
|---|---|---|---|
| 10 | $\pi/4$ | $+\sqrt{E/2}$ | $-\sqrt{E/2}$ |
| 00 | $3\pi/4$ | $-\sqrt{E/2}$ | $-\sqrt{E/2}$ |
| 01 | $5\pi/4$ | $-\sqrt{E/2}$ | $+\sqrt{E/2}$ |
| 11 | $7\pi/4$ | $+\sqrt{E/2}$ | $+\sqrt{E/2}$ |

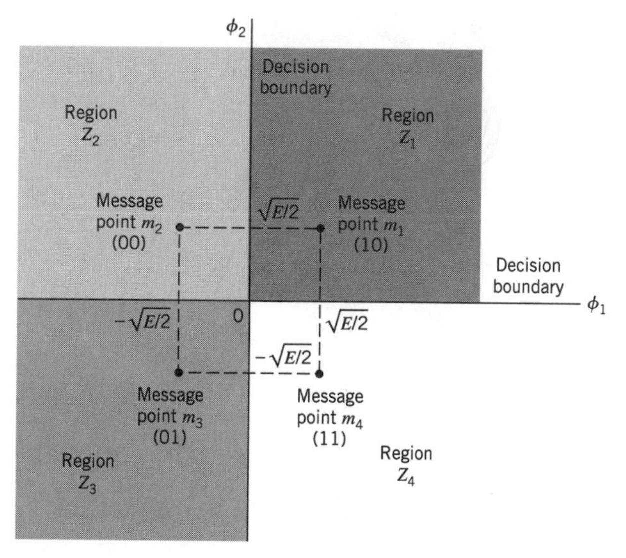


**FIGURE 6.6** Signal-space diagram of coherent QPSK system.

▶ There are four message points, and the associated signal vectors are defined by

$$\mathbf{s}_i = \begin{bmatrix} \sqrt{E}\cos\left((2i-1)\dfrac{\pi}{4}\right) \\ -\sqrt{E}\sin\left((2i-1)\dfrac{\pi}{4}\right) \end{bmatrix}, \qquad i = 1, 2, 3, 4 \tag{6.27}$$

The elements of the signal vectors, namely, $s_{i1}$ and $s_{i2}$, have their values summarized in Table 6.1. The first two columns of this table give the associated dibit and phase of the QPSK signal.

Accordingly, a QPSK signal has a two-dimensional signal constellation (i.e., $N = 2$) and four message points (i.e., $M = 4$) whose phase angles increase in a counterclockwise direction, as illustrated in Figure 6.6. As with binary PSK, the QPSK signal has minimum average energy.

## ▶ EXAMPLE 6.1

Figure 6.7 illustrates the sequences and waveforms involved in the generation of a QPSK signal. The input binary sequence 01101000 is shown in Figure 6.7*a*. This sequence is divided into two other sequences, consisting of odd- and even-numbered bits of the input sequence. These two sequences are shown in the top lines of Figures 6.7*b* and 6.7*c*. The waveforms representing the two components of the QPSK signal, namely, $s_{i1}\phi_1(t)$ and $s_{i2}\phi_2(t)$, are also shown in Figures 6.7*b* and 6.7*c*, respectively. These two waveforms may individually be viewed as examples of a binary PSK signal. Adding them, we get the QPSK waveform shown in Figure 6.7*d*.

To define the decision rule for the detection of the transmitted data sequence, we partition the signal space into four regions, in accordance with Equation (5.59) of Chapter 5. The individual regions are defined by the set of points closest to the message point represented by signal vectors $\mathbf{s}_1$, $\mathbf{s}_2$, $\mathbf{s}_3$, and $\mathbf{s}_4$. This is readily accomplished by constructing the perpendicular bisectors of the square formed by joining the four message points and then marking

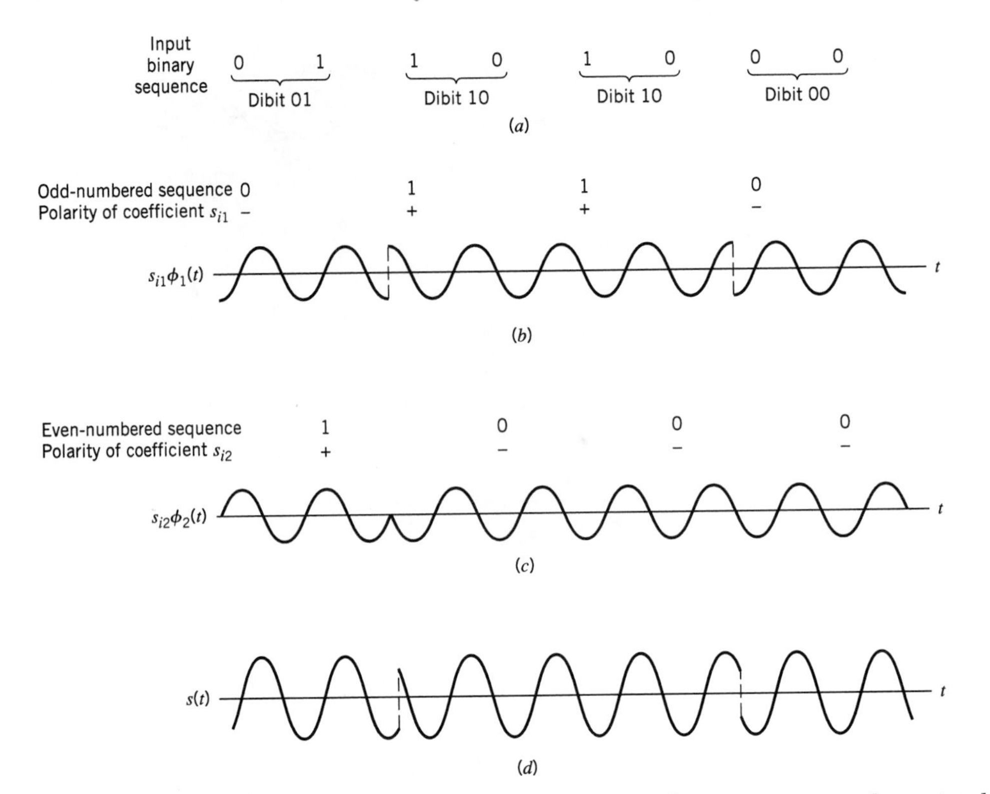


**FIGURE 6.7** (*a*) Input binary sequence. (*b*) Odd-numbered bits of input sequence and associated binary PSK wave. (*c*) Even-numbered bits of input sequence and associated binary PSK wave. (*d*) QPSK waveform defined as $s(t) = s_{i1}\phi_1(t) + s_{i2}\phi_2(t)$.

off the appropriate regions. We thus find that the decision regions are quadrants whose vertices coincide with the origin. These regions are marked $Z_1$, $Z_2$, $Z_3$, and $Z_4$, in Figure 6.6, according to the message point around which they are constructed. ◀

### *Error Probability of QPSK*

In a coherent QPSK system, the received signal $x(t)$ is defined by

$$x(t) = s_i(t) + w(t), \qquad \begin{cases} 0 \le t \le T \\ i = 1, 2, 3, 4 \end{cases} \tag{6.28}$$

where $w(t)$ is the sample function of a white Gaussian noise process of zero mean and power spectral density $N_0/2$. Correspondingly, the observation vector **x** has two elements, $x_1$ and $x_2$, defined by

$$\begin{aligned} x_1 &= \int_0^T x(t)\phi_1(t)\,dt \\ &= \sqrt{E}\cos\left[(2i-1)\frac{\pi}{4}\right] + w_1 \\ &= \pm\sqrt{\frac{E}{2}} + w_1 \end{aligned} \tag{6.29}$$

and

$$\begin{aligned} x_2 &= \int_0^T x(t)\phi_2(t)\,dt \\ &= -\sqrt{E}\sin\left[(2i-1)\frac{\pi}{4}\right] + w_2 \\ &= \mp\sqrt{\frac{E}{2}} + w_2 \end{aligned} \tag{6.30}$$

Thus the observable elements $x_1$ and $x_2$ are sample values of independent Gaussian random variables with mean values equal to $\pm\sqrt{E/2}$ and $\mp\sqrt{E/2}$, respectively, and with a common variance equal to $N_0/2$.

The decision rule is now simply to decide that $s_1(t)$ was transmitted if the received signal point associated with the observation vector $\mathbf{x}$ falls inside region $Z_1$, decide that $s_2(t)$ was transmitted if the received signal point falls inside region $Z_2$, and so on. An erroneous decision will be made if, for example, signal $s_4(t)$ is transmitted but the noise $w(t)$ is such that the received signal point falls outside region $Z_4$.

To calculate the average probability of symbol error, we note from Equation (6.24) that a coherent QPSK system is in fact equivalent to two coherent binary PSK systems working in parallel and using two carriers that are in phase quadrature; this is merely a statement of the quadrature-carrier multiplexing property of coherent QPSK. The in-phase channel output $x_1$ and the quadrature channel output $x_2$ (i.e., the two elements of the observation vector $\mathbf{x}$) may be viewed as the individual outputs of the two coherent binary PSK systems. Thus, according to Equations (6.29) and (6.30), these two binary PSK systems may be characterized as follows:

- The signal energy per bit is $E/2$.
- The noise spectral density is $N_0/2$.

Hence, using Equation (6.20) for the average probability of bit error of a coherent binary PSK system, we may now state that the average probability of bit error in *each* channel of the coherent QPSK system is

$$\begin{aligned} P' &= \frac{1}{2}\,\text{erfc}\left(\sqrt{\frac{E/2}{N_0}}\right) \\ &= \frac{1}{2}\,\text{erfc}\left(\sqrt{\frac{E}{2N_0}}\right) \end{aligned} \tag{6.31}$$

Another important point to note is that the bit errors in the in-phase and quadrature channels of the coherent QPSK system are statistically independent. The in-phase channel makes a decision on one of the two bits constituting a symbol (dibit) of the QPSK signal, and the quadrature channel takes care of the other bit. Accordingly, the *average probability of a correct decision* resulting from the combined action of the two channels working together is

$$\begin{aligned} P_c &= (1-P')^2 \\ &= \left[1 - \frac{1}{2}\,\text{erfc}\left(\sqrt{\frac{E}{2N_0}}\right)\right]^2 \\ &= 1 - \text{erfc}\left(\sqrt{\frac{E}{2N_0}}\right) + \frac{1}{4}\,\text{erfc}^2\left(\sqrt{\frac{E}{2N_0}}\right) \end{aligned} \tag{6.32}$$

The average probability of symbol error for coherent QPSK is therefore

$$\begin{aligned} P_e &= 1 - P_c \\ &= \operatorname{erfc}\left(\sqrt{\frac{E}{2N_0}}\right) - \frac{1}{4}\operatorname{erfc}^2\left(\sqrt{\frac{E}{2N_0}}\right) \end{aligned} \tag{6.33}$$

In the region where $(E/2N_0) \gg 1$, we may ignore the quadratic term on the right-hand side of Equation (6.33), so we approximate the formula for the average probability of symbol error for coherent QPSK as

$$P_e \simeq \operatorname{erfc}\left(\sqrt{\frac{E}{2N_0}}\right) \tag{6.34}$$

The formula of Equation (6.34) may also be derived in another insightful way, using the signal-space diagram of Figure 6.6. Since the four message points of this diagram are circularly symmetric with respect to the origin, we may apply Equation (5.92), reproduced here in the form

$$P_e \leq \frac{1}{2}\sum_{\substack{k=1 \\ k\neq i}}^{4} \operatorname{erfc}\left(\frac{d_{ik}}{2\sqrt{N_0}}\right) \qquad \text{for all } i \tag{6.35}$$

Consider, for example, message point $m_1$ (corresponding to dibit 10) chosen as the transmitted message point. The message points $m_2$ and $m_4$ (corresponding to dibits 00 and 11) are the *closest* to $m_1$. From Figure 6.6 we readily find that $m_1$ is equidistant from $m_2$ and $m_4$ in a Euclidean sense, as shown by

$$d_{12} = d_{14} = \sqrt{2E}$$

Assuming that $E/N_0$ is large enough to ignore the contribution of the most distant message point $m_3$ (corresponding to dibit 01) relative to $m_1$, we find that the use of Equation (6.35) yields an approximate expression for $P_e$ that is the same as Equation (6.34). Note that in mistaking either $m_2$ or $m_4$ for $m_1$, a single bit error is made; on the other hand, in mistaking $m_3$ for $m_1$, two bit errors are made. For a high enough $E/N_0$, the likelihood of both bits of a symbol being in error is much less than a single bit, which is a further justification for ignoring $m_3$ in calculating $P_e$ when $m_1$ is sent.

In a QPSK system, we note that since there are two bits per symbol, the transmitted signal energy per symbol is twice the signal energy per bit, as shown by

$$E = 2E_b \tag{6.36}$$

Thus expressing the average probability of symbol error in terms of the ratio $E_b/N_0$, we may write

$$P_e \simeq \operatorname{erfc}\left(\sqrt{\frac{E_b}{N_0}}\right) \tag{6.37}$$

With Gray encoding used for the incoming symbols, we find from Equations (6.31) and (6.36) that the *bit error rate* of QPSK is exactly

$$\text{BER} = \frac{1}{2}\operatorname{erfc}\left(\sqrt{\frac{E_b}{N_0}}\right) \tag{6.38}$$

We may therefore state that a coherent QPSK system achieves the same average probability of bit error as a coherent binary PSK system for the same bit rate and the same $E_b/N_0$, but uses only half the channel bandwidth. Stated in a different way, for the same $E_b/N_0$ and therefore the same average probability of bit error, a coherent QPSK system transmits information at twice the bit rate of a coherent binary PSK system for the same channel bandwidth. For a prescribed performance, QPSK uses channel bandwidth better than binary PSK, which explains the preferred use of QPSK over binary PSK in practice.

### *Generation and Detection of Coherent QPSK Signals*

Consider next the generation and detection of QPSK signals. Figure 6.8*a* shows a block diagram of a typical QPSK transmitter. The incoming binary data sequence is first transformed into polar form by a *nonreturn-to-zero level* encoder. Thus, symbols 1 and 0 are represented by $+\sqrt{E_b}$ and $-\sqrt{E_b}$, respectively. This binary wave is next divided by means of a *demultiplexer* into two separate binary waves consisting of the odd- and even-numbered input bits. These two binary waves are denoted by $a_1(t)$ and $a_2(t)$. We note that in any signaling interval, the amplitudes of $a_1(t)$ and $a_2(t)$ equal $s_{i1}$ and $s_{i2}$, respectively, depending on the particular dibit that is being transmitted. The two binary waves $a_1(t)$ and $a_2(t)$ are used to modulate a pair of quadrature carriers or orthonormal basis functions: $\phi_1(t)$ equal to $\sqrt{2/T}\cos(2\pi f_c t)$ and $\phi_2(t)$ equal to $\sqrt{2/T}\sin(2\pi f_c t)$. The result is a pair of

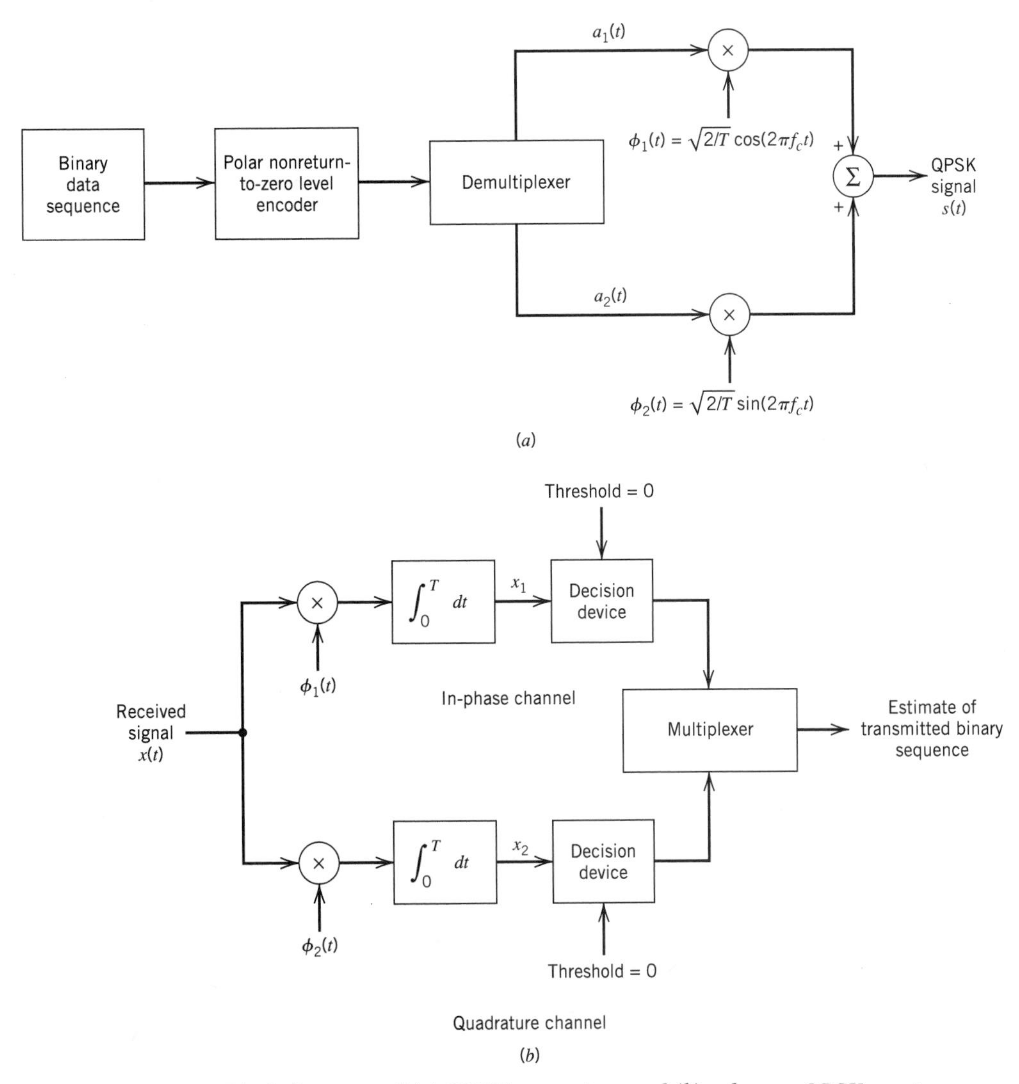


**FIGURE 6.8** Block diagrams of (*a*) QPSK transmitter and (*b*) coherent QPSK receiver.

binary PSK signals, which may be detected independently due to the orthogonality of $\phi_1(t)$ and $\phi_2(t)$. Finally, the two binary PSK signals are added to produce the desired QPSK signal.

The QPSK receiver consists of a pair of correlators with a common input and supplied with a locally generated pair of coherent reference signals $\phi_1(t)$ and $\phi_2(t)$, as in Figure 6.8*b*. The correlator outputs $x_1$ and $x_2$, produced in response to the received signal $x(t)$, are each compared with a threshold of zero. If $x_1 > 0$, a decision is made in favor of symbol 1 for the in-phase channel output, but if $x_1 < 0$, a decision is made in favor of symbol 0. Similarly, if $x_2 > 0$, a decision is made in favor of symbol 1 for the quadrature channel output, but if $x_2 < 0$, a decision is made in favor of symbol 0. Finally, these two binary sequences at the in-phase and quadrature channel outputs are combined in a *multiplexer* to reproduce the original binary sequence at the transmitter input with the minimum probability of symbol error in an AWGN channel.

### *Power Spectra of QPSK Signals*

Assume that the binary wave at the modulator input is random, with symbols 1 and 0 being equally likely, and with the symbols transmitted during adjacent time slots being statistically independent. We make the following observations pertaining to the in-phase and quadrature components of a QPSK signal:

1. Depending on the dibit sent during the signaling interval $-T_b \leq t \leq T_b$, the in-phase component equals $+g(t)$ or $-g(t)$, and similarly for the quadrature component. The $g(t)$ denotes the symbol shaping function, defined by

$$g(t) = \begin{cases} \sqrt{\dfrac{E}{T}}, & 0 \leq t \leq T \\ 0, & \text{otherwise} \end{cases} \tag{6.39}$$

Hence, the in-phase and quadrature components have a common power spectral density, namely, $E \operatorname{sinc}^2(Tf)$.

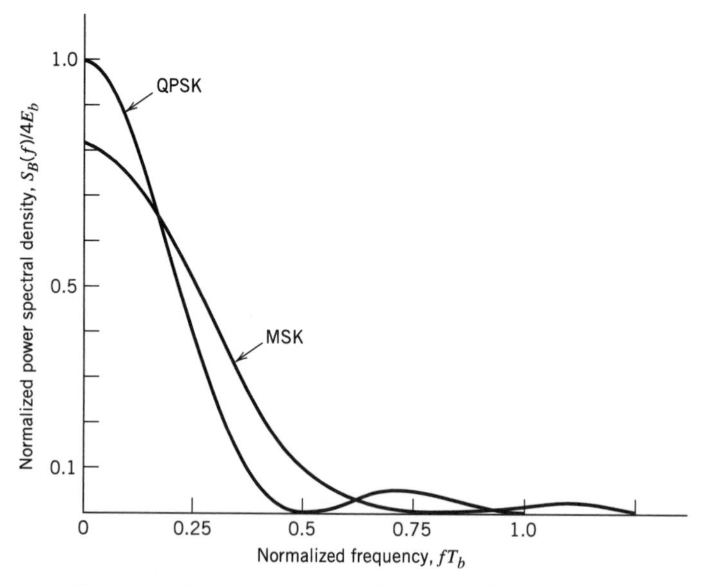


FIGURE 6.9 Power spectra of QPSK and MSK signals.

2. The in-phase and quadrature components are statistically independent. Accordingly, the baseband power spectral density of the QPSK signal equals the sum of the individual power spectral densities of the in-phase and quadrature components, so we may write

$$\begin{aligned} S_B(f) &= 2E\ \mathrm{sinc}^2(Tf) \\ &= 4E_b\ \mathrm{sinc}^2(2T_b f) \end{aligned} \tag{6.40}$$

Figure 6.9 plots $S_B(f)$, normalized with respect to $4E_b$, versus the normalized frequency $fT_b$. This figure also includes a plot of the baseband power spectral density of a certain form of binary FSK called minimum shift keying, the evaluation of which is presented in Section 6.5. Comparison of these two spectra is deferred to that section.

## ■ OFFSET QPSK

The signal space diagram of Figure 6.10*a* embodies all the possible phase transitions that can arise in the generation of a QPSK signal. More specifically, examining the QPSK waveform illustrated in Figure 6.7 for Example 6.1, we may make the following observations:

1. The carrier phase changes by $\pm 180$ degrees whenever both the in-phase and quadrature components of the QPSK signal changes sign. An example of this situation is illustrated in Figure 6.7 when the input binary sequence switches from dibit 01 to dibit 10.
2. The carrier phase changes by $\pm 90$ degrees whenever the in-phase or quadrature component changes sign. An example of this second situation is illustrated in Figure 6.7 when the input binary sequence switches from dibit 10 to dibit 00, during which the in-phase component changes sign, whereas the quadrature component is unchanged.
3. The carrier phase is unchanged when neither the in-phase component nor the quadrature component changes sign. This last situation is illustrated in Figure 6.7 when dibit 10 is transmitted in two successive symbol intervals.

Situation 1 and, to a much lesser extent, situation 2 can be of a particular concern when the QPSK signal is filtered during the course of transmission, prior to detection. Specifically, the 180- and 90-degree shifts in carrier phase can result in changes in the carrier amplitude (i.e., envelope of the QPSK signal), thereby causing additional symbol errors on detection.

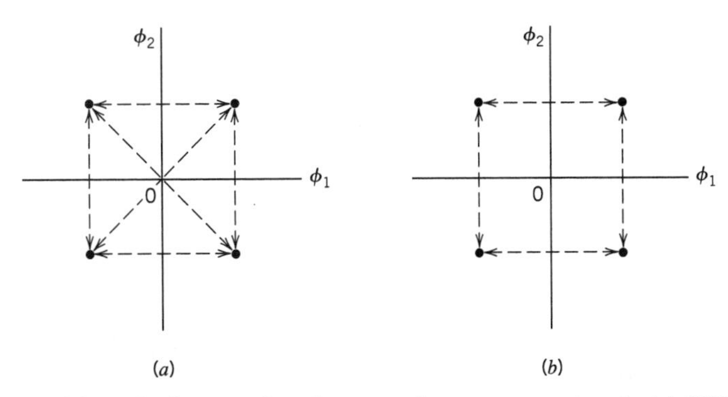


**FIGURE 6.10** Possible paths for switching between the message points in (*a*) QPSK and (*b*) offset QPSK.

The extent of amplitude fluctuations exhibited by QPSK signals may be reduced by using *offset QPSK*.[2] In this variant of QPSK, the bit stream responsible for generating the quadrature component is delayed (i.e., offset) by half a symbol interval with respect to the bit stream responsible for generating the in-phase component. Specifically, the two basis functions of offset QPSK are defined by

$$\phi_1(t) = \sqrt{\frac{2}{T}} \cos(2\pi f_c t), \qquad 0 \le t \le T \tag{6.41}$$

$$\phi_2(t) = \sqrt{\frac{2}{T}} \sin(2\pi f_c t), \qquad \frac{T}{2} \le t \le \frac{3T}{2} \tag{6.42}$$

Accordingly, unlike QPSK, the phase transitions likely to occur in offset QPSK are confined to ±90 degrees, as indicated in the signal space diagram of Figure 6.10*b*. However, ±90 degree phase transitions in offset QPSK occur twice as frequently but with half the intensity encountered in QPSK. Since, in addition to ±90-degree phase transitions, ±180-degree phase transitions also occur in QPSK, we find that amplitude fluctuations in offset QPSK due to filtering have a smaller amplitude than in the case of QPSK.

Despite the delay $T/2$ applied to the basis function $\phi_2(t)$ in Equation (6.42) compared to that in Equation (6.26), the offset QPSK has exactly the same probability of symbol error in an AWGN channel as QPSK. The equivalence in noise performance between these phase-shift keying schemes assumes the use of coherent detection. The reason for the equivalence is that the statistical independence of the in-phase and quadrature components applies to both QPSK and offset QPSK. We may therefore say that the error probability in the in-phase or quadrature channel of a coherent offset QPSK receiver is still equal to $(1/2)\,\text{erfc}(\sqrt{E/2N_0})$. Hence the formula of Equation (6.34) applies equally well to the offset QPSK.

## ■ $\pi$/4-SHIFTED QPSK

An ordinary QPSK signal may reside in either one of the two commonly used constellations shown in Figures 6.11*a* and 6.11*b*, which are shifted by $\pi/4$ radians with respect to each other. In another variant of QPSK known as *$\pi$/4-shifted QPSK*,[3] the carrier phase used for the transmission of successive symbols (i.e., dibits) is alternately picked from one of the two QPSK constellations in Figure 6.11 and then the other. It follows therefore that a $\pi/4$-shifted QPSK signal may reside in any one of eight possible phase states, as indicated

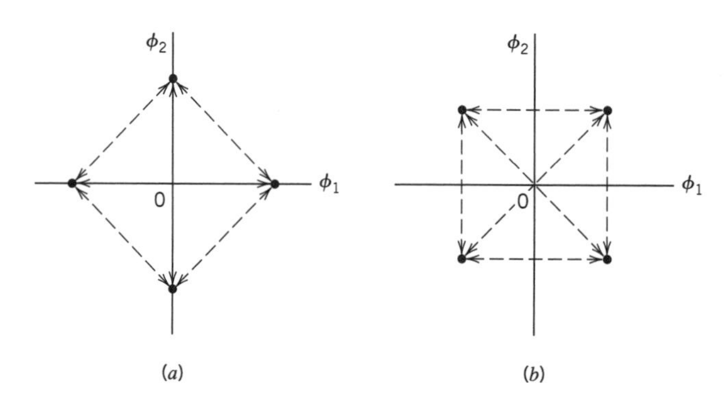


**FIGURE 6.11** Two commonly used signal constellations for QPSK; the arrows indicate the paths along which the QPSK modulator can change its state.

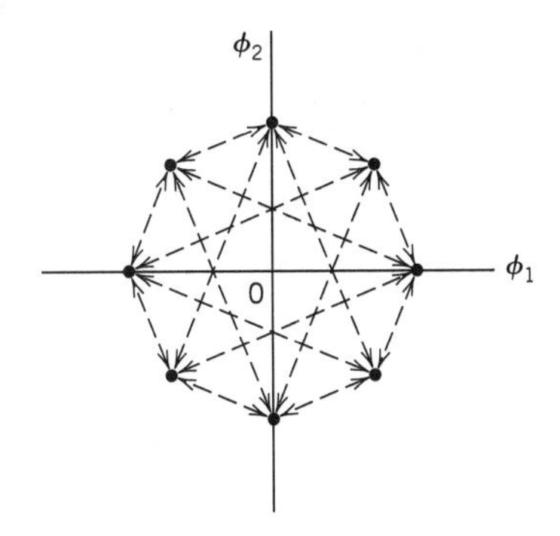


**FIGURE 6.12** Eight possible phase states for the $\pi/4$-shifted QPSK modulator.

in Figure 6.12. The four dashed lines emanating from each possible message point in Figure 6.12 define the phase transitions that are feasible in $\pi/4$-shifted QPSK.

Table 6.2 summarizes a possible set of relationships between the phase transitions in this new digital modulation scheme and the incoming Gray-encoded dibits. For example, if the modulator is in one of the phase states portrayed in Figure 6.11*b*, then on receiving the dibit 00 it shifts into a phase state portrayed in Figure 6.11*a* by rotating through $\pi/4$ radians in a counterclockwise direction.

Attractive features of the $\pi/4$-shifted QPSK scheme include the following:

- The phase transitions from one symbol to the next are restricted to $\pm\pi/4$ and $\pm 3\pi/4$ radians, which is to be contrasted with the $\pm\pi/2$ and $\pm\pi$ phase transitions in QPSK. Consequently, envelope variations of $\pi/4$-shifted QPSK signals due to filtering are significantly reduced, compared to those in QPSK.
- Unlike offset QPSK signals, $\pi/4$-shifted QPSK signals can be noncoherently detected, thereby considerably simplifying the receiver design. Moreover, like QPSK signals, $\pi/4$-shifted QPSK can be differently encoded, in which case we should really speak of *$\pi/4$-shifted DQPSK*.

The generation of $\pi/4$-shifted DQPSK symbols, represented by the symbol pair $(I, Q)$, is described by the following pair of relationships (see Problem 6.13):

$$\begin{aligned} I_k &= \cos(\theta_{k-1} + \Delta\theta_k) \\ &= \cos\,\theta_k \end{aligned} \tag{6.43}$$

$$\begin{aligned} Q_k &= \sin(\theta_{k-1} + \Delta\theta_k) \\ &= \sin\,\theta_k \end{aligned} \tag{6.44}$$

**TABLE 6.2** ***Correspondence between input dibit and phase change for $\pi/4$-shifted DQPSK***

| *Gray-Encoded Input Dibit* | *Phase Change, $\Delta\theta$ (radians)* |
|---|---|
| 00 | $\pi/4$ |
| 01 | $3\pi/4$ |
| 11 | $-3\pi/4$ |
| 10 | $-\pi/4$ |

**TABLE 6.3** *$\pi/4$-shifted DQPSK results for Example 6.2*

| *Step* k | *Phase* $\theta_{k-1}$ *(radians)* | *Input Dibit* | *Phase Change* $\Delta\theta_k$ *(radians)* | *Transmitted Phase* $\theta_k$ *(radians)* |
|---|---|---|---|---|
| 1 | $\pi/4$ | 00 | $\pi/4$ | $\pi/2$ |
| 2 | $\pi/2$ | 10 | $-\pi/4$ | $\pi/4$ |
| 3 | $\pi/4$ | 10 | $-\pi/4$ | 0 |
| 4 | 0 | 01 | $3\pi/4$ | $3\pi/4$ |

where $\theta_{k-1}$ is the absolute phase angle of symbol $k - 1$, and $\Delta\theta_k$ is the differentially encoded phase change defined in accordance with Table 6.2.

### ▶ EXAMPLE 6.2

Continuing with the input binary sequence of Example 6.1, namely, 01101000, suppose that the phase angle $\theta_0 = \pi/4$ in the constellation of Figure 6.11*b* is assigned as the *initial* phase state of the $\pi/4$-shifted DQPSK modulator. Then, arranging the input binary sequence as a sequence of dibits and following the convention of Table 6.2, we get the results presented in Table 6.3 for the example at hand. ◀

#### *Detection of $\pi/4$-Shifted DQPSK Signals*

Having familiarized ourselves with the generation of $\pi/4$-shifted DQPSK signals, we go on to consider their differential detection. Given the noisy channel output $x(t)$, the receiver first computes the projections of $x(t)$ onto the basis functions $\phi_1(t)$ and $\phi_2(t)$. The resulting outputs, denoted by $I$ and $Q$, respectively, are applied to a *differential detector* that consists of the following components, as indicated in Figure 6.13:

- *Arctangent computer* for extracting the phase angle $\theta$ of the channel output (received signal).
- *Phase-difference computer* for determining the change in the phase $\theta$ occurring over one symbol interval.
- *Modulo-$2\pi$ correction logic* for correcting errors due to the possibility of phase angles wrapping around the real axis.

Elaborating further on the latter point, let $\Delta\theta_k$ denote the computed phase difference between $\theta_k$ and $\theta_{k-1}$ representing the phase angles of the channel output for symbols $k$ and $k - 1$, respectively. Then the modulo-$2\pi$ correction logic operates as follows:

$$\begin{aligned}&\text{IF } \Delta\theta_k < -180 \text{ degrees THEN } \Delta\theta_k = \Delta\theta_k + 360 \text{ degrees}\\&\text{IF } \Delta\theta_k > 180 \text{ degrees THEN } \Delta\theta_k = \Delta\theta_k - 360 \text{ degrees}\end{aligned} \tag{6.45}$$

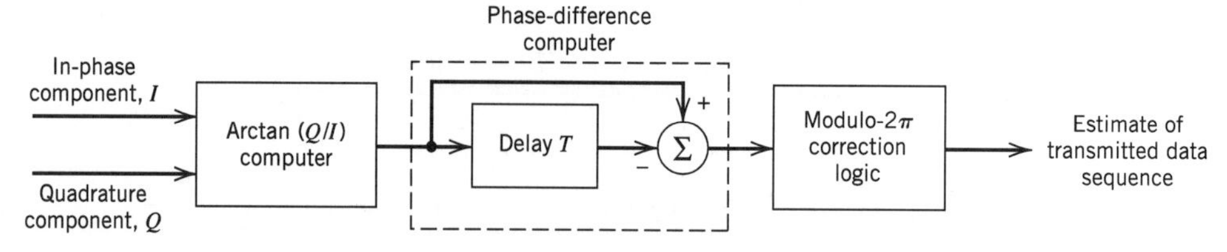


**FIGURE 6.13** Block diagram of the $\pi/4$-shifted DQPSK detector.

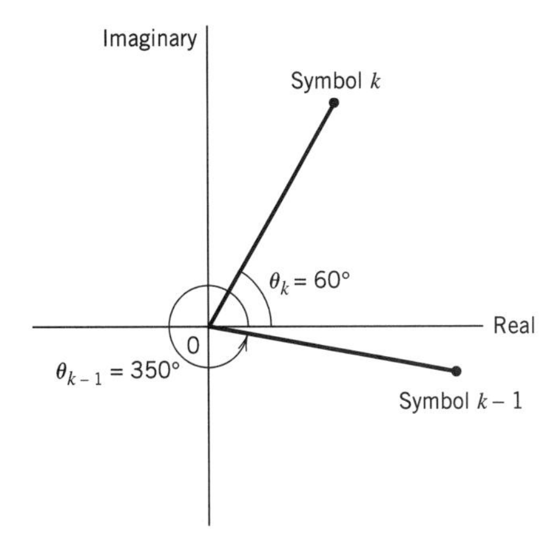


**FIGURE 6.14** Illustrating the possibility of phase angles wrapping around the positive real axis.

To illustrate the need for this phase correction, consider the situation depicted in Figure 6.14, where $\theta_{k-1} = 350$ degrees and $\theta_k = 60$ degrees, both phase angles being measured in a counterclockwise direction. From this figure we readily see that the phase change $\Delta\theta_k$, measured in a counterclockwise direction, is 70 degrees. However, without correction the phase change $\Delta\theta_k$ is computed as 60 degrees $-$ 350 degrees $= -290$ degrees. Applying the first line of Equation (6.45), the modulo-$2\pi$ correction logic compensates for the wraparound the positive real axis, yielding the corrected result

$$\Delta\theta_k = -290 \text{ degrees} + 360 \text{ degrees} = 70 \text{ degrees}$$

The tangent type differential detector of Figure 6.13 for the demodulation of $\pi/4$-shifted DQPSK signals is relatively simple to implement. It offers a satisfactory performance in a Rayleigh fading channel as in a static multipath environment. However, when the multipath environment is time varying as experienced in a commercial digital wireless communication system, computer simulation results appear to show that the receiver performance degrades very rapidly.[4]

## *M*-ARY PSK

QPSK is a special case of *M-ary PSK*, where the phase of the carrier takes on one of $M$ possible values, namely, $\theta_i = 2(i-1)\pi/M$, where $i = 1, 2, \ldots, M$. Accordingly, during each signaling interval of duration $T$, one of the $M$ possible signals

$$s_i(t) = \sqrt{\frac{2E}{T}} \cos\left(2\pi f_c t + \frac{2\pi}{M}(i-1)\right), \qquad i = 1, 2, \ldots, M \tag{6.46}$$

is sent, where $E$ is the signal energy per symbol. The carrier frequency $f_c = n_c/T$ for some fixed integer $n_c$.

Each $s_i(t)$ may be expanded in terms of the same two basis functions $\phi_1(t)$ and $\phi_2(t)$ defined in Equations (6.25) and (6.26), respectively. The signal constellation of $M$-ary PSK is therefore two-dimensional. The $M$ message points are equally spaced on a circle of radius $\sqrt{E}$ and center at the origin, as illustrated in Figure 6.15*a*, for the case of *octaphase-shift-keying* (i.e., $M = 8$).

From Figure 6.15*a* we note that the signal-space diagram is circularly symmetric. We may therefore apply Equation (5.92), based on the union bound, to develop an approximate formula for the average probability of symbol error for $M$-ary PSK. Suppose that the

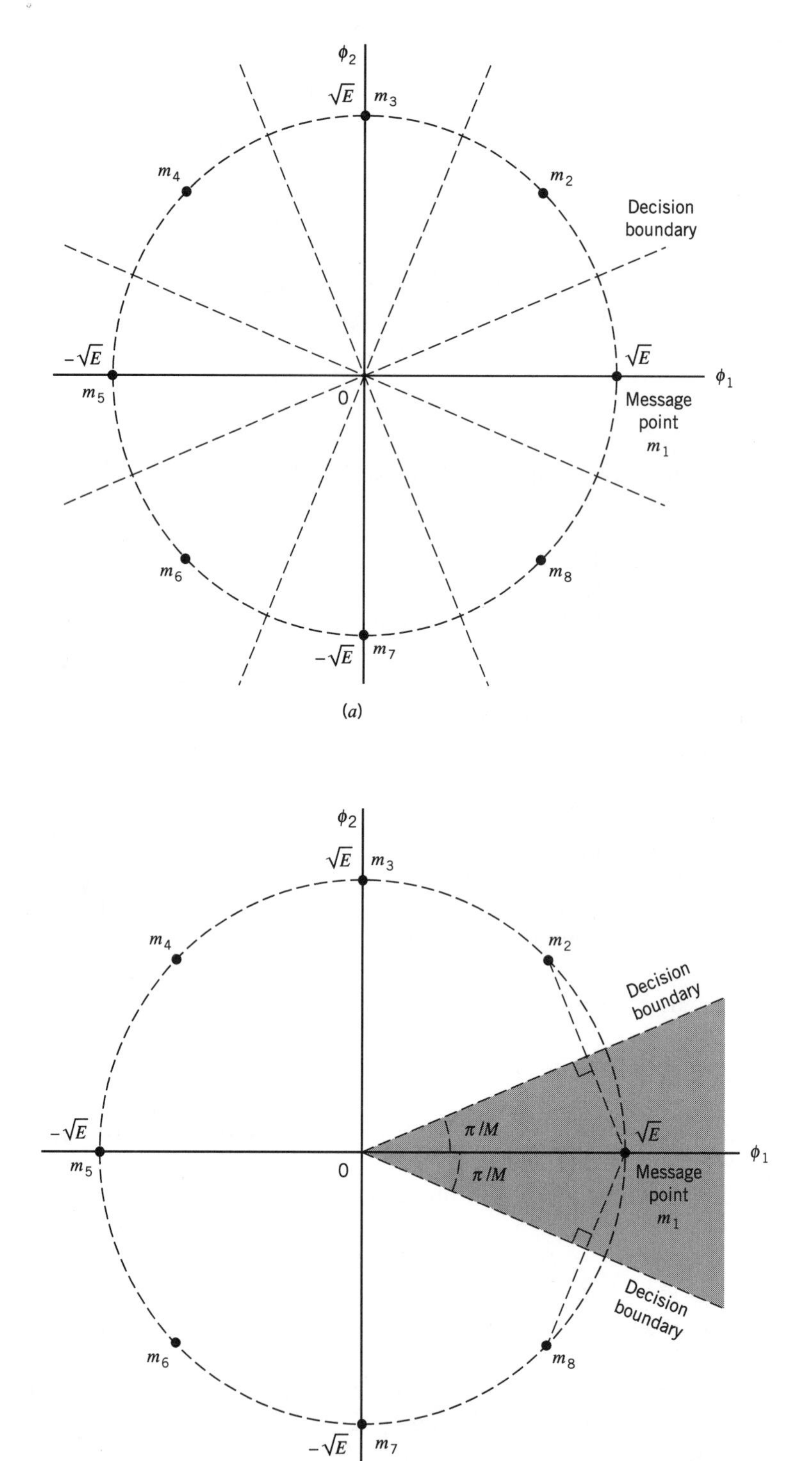


**FIGURE 6.15** (*a*) Signal-space diagram for octaphase-shift keying (i.e., $M = 8$). The decision boundaries are shown as dashed lines. (*b*) Signal-space diagram illustrating the application of the union bound for octaphase-shift keying.

transmitted signal corresponds to the message point $m_1$, whose coordinates along the $\phi_1$- and $\phi_2$-axes are $+\sqrt{E}$ and 0, respectively. Suppose that the ratio $E/N_0$ is large enough to consider the nearest two message points, one on either side of $m_1$, as potential candidates for being mistaken for $m_1$ due to channel noise. This is illustrated in Figure 6.15*b* for the case of $M = 8$. The Euclidean distance of each of these two points from $m_1$ is (for $M = 8$)

$$d_{12} = d_{18} = 2\sqrt{E}\,\sin\left(\frac{\pi}{M}\right)$$

Hence, the use of Equation (5.92) of Chapter 5 yields the average probability of symbol error for coherent $M$-ary PSK as

$$P_e \simeq \text{erfc}\left(\sqrt{\frac{E}{N_0}}\,\sin\left(\frac{\pi}{M}\right)\right) \tag{6.47}$$

where it is assumed that $M \geq 4$. The approximation becomes extremely tight, for fixed $M$, as $E/N_0$ is increased. For $M = 4$, Equation (6.47) reduces to the same form given in Equation (6.34) for QPSK.

### *Power Spectra of M-ary PSK Signals*

The symbol duration of $M$-ary PSK is defined by

$$T = T_b \log_2 M \tag{6.48}$$

where $T_b$ is the bit duration. Proceeding in a manner similar to that described for a QPSK signal, we may show that the baseband power spectral density of an $M$-ary PSK signal is given by

$$\begin{aligned} S_B(f) &= 2E\,\text{sinc}^2(Tf) \\ &= 2E_b \log_2 M\,\text{sinc}^2(T_b f \log_2 M) \end{aligned} \tag{6.49}$$

In Figure 6.16, we show the normalized power spectral density $S_B(f)/2E_b$ plotted versus the normalized frequency $fT_b$ for three different values of $M$, namely, $M = 2, 4, 8$.

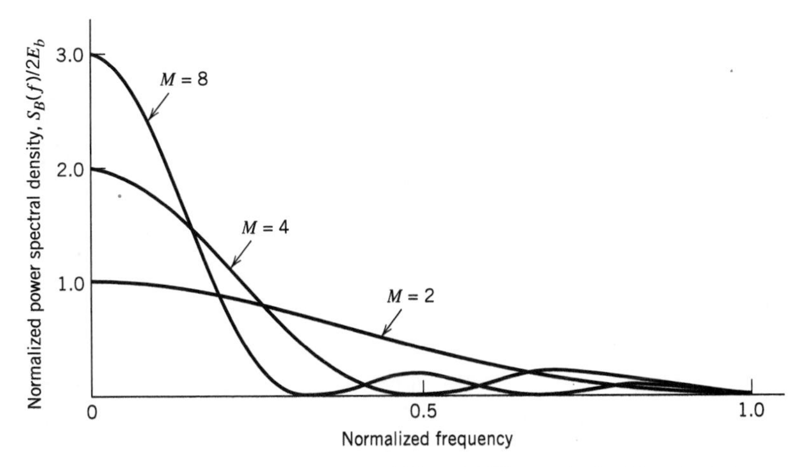


**FIGURE 6.16** Power spectra of $M$-ary PSK signals for $M = 2, 4, 8$.

**TABLE 6.4** *Bandwidth efficiency of M-ary PSK signals*

| M | 2 | 4 | 8 | 16 | 32 | 64 |
|---|---|---|---|---|---|---|
| $\rho$ (bits/s/Hz) | 0.5 | 1 | 1.5 | 2 | 2.5 | 3 |

### ■ BANDWIDTH EFFICIENCY OF $M$-ARY PSK SIGNALS

The power spectra of $M$-ary PSK signals possess a *main lobe* bounded by well-defined *spectral nulls* (i.e., frequencies at which the power spectral density is zero). Accordingly, the spectral width of the main lobe provides a simple and popular measure for the bandwidth of $M$-ary PSK signals. This definition is referred to as the *null-to-null bandwidth.* With the null-to-null bandwidth encompassing the main lobe of the power spectrum of an $M$-ary signal, we find that it contains most of the signal power. This is readily seen by looking at the power spectral plots of Figure 6.16.

For the passband basis functions defined in Equations (6.25) and (6.26), the channel bandwidth required to pass $M$-ary PSK signals (more precisely, the main spectral lobe of $M$-ary signals) is given by

$$B = \frac{2}{T} \tag{6.50}$$

where $T$ is the symbol duration. But the symbol duration $T$ is related to the bit duration $T_b$ by Equation (6.48). Moreover, the bit rate $R_b = 1/T_b$. Hence, we may redefine the channel bandwidth of Equation (6.50) in terms of the bit rate $R_b$ as

$$B = \frac{2R_b}{\log_2 M} \tag{6.51}$$

Based on this formula, the bandwidth efficiency of $M$-ary PSK signals is given by

$$\begin{aligned} \rho &= \frac{R_b}{B} \\ &= \frac{\log_2 M}{2} \end{aligned} \tag{6.52}$$

Table 6.4 gives the values of $\rho$ calculated from Equation (6.52) for varying $M$.

In light of Equation (6.47) and Table 6.4, we make the following observation in the context of $M$-ary PSK: As the number of states, $M$, is increased, the bandwidth efficiency is improved at the expense of error performance. To ensure that there is no degradation in error performance, we have to increase $E_b/N_0$ to compensate for the increase in $M$.

## 6.4 *Hybrid Amplitude/Phase Modulation Schemes*

In an $M$-ary PSK system, the in-phase and quadrature components of the modulated signal are interrelated in such a way that the envelope is constrained to remain constant. This constraint manifests itself in a circular constellation for the message points. However, if this constraint is removed, and the in-phase and quadrature components are thereby permitted to be independent, we get a new modulation scheme called *M-ary quadrature amplitude modulation* (QAM). This latter modulation scheme is *hybrid* in nature in that the carrier experiences amplitude as well as phase modulation.

The passband basis functions in $M$-ary QAM may not be periodic for an arbitrary choice of the carrier frequency $f_c$ with respect to the symbol rate $1/T$. Ordinarily, this aperiodicity is of no real concern. By reformulating the expression for the transmitted signal in a certain way, it is possible to eliminate the time variation of the basis functions on successive symbol transmissions, and thereby simplify implementation of the transmitter. In particular, the transmitter is made to appear "carrierless," while fully retaining the essence of the hybridized amplitude and phase modulation process. This is indeed the idea behind the carrierless amplitude/phase modulation (CAP).

Despite the differences between QAM and CAP in their implementation details, they have exactly the same signal constellations. Accordingly, they are fundamentally equivalent in performance for a prescribed receiver complexity. In what follows we first discuss QAM and then CAP.

## *M*-ARY QUADRATURE AMPLITUDE MODULATION

In Chapters 4 and 5, we studied $M$-ary pulse amplitude modulation (PAM), which is one-dimensional. $M$-ary QAM is a two-dimensional generalization of $M$-ary PAM in that its formulation involves two orthogonal passband basis functions, as shown by

$$\phi_1(t) = \sqrt{\frac{2}{T}}\cos(2\pi f_c t), \qquad 0 \le t \le T \tag{6.53}$$

$$\phi_2(t) = \sqrt{\frac{2}{T}}\sin(2\pi f_c t), \qquad 0 \le t \le T \tag{6.54}$$

Let the $i$th message point $\mathbf{s}_i$ in the $(\phi_1, \phi_2)$ plane be denoted by $(a_i d_{\min}/2, b_i d_{\min}/2)$, where $d_{\min}$ is the minimum distance between any two message points in the constellation, $a_i$ and $b_i$ are integers, and $i = 1, 2, \ldots, M$. Let $(d_{\min}/2) = \sqrt{E_0}$, where $E_0$ is the energy of the signal with the lowest amplitude. The transmitted $M$-ary QAM signal for symbol $k$, say, is then defined by

$$s_k(t) = \sqrt{\frac{2E_0}{T}}\, a_k \cos(2\pi f_c t) - \sqrt{\frac{2E_0}{T}}\, b_k \sin(2\pi f_c t), \qquad \begin{array}{l} 0 \le t \le T \\ k = 0, \pm 1, \pm 2, \ldots \end{array} \tag{6.55}$$

The signal $s_k(t)$ consists of two phase-quadrature carriers with each one being modulated by a set of discrete amplitudes, hence the name *quadrature amplitude modulation.*

Depending on the number of possible symbols $M$, we may distinguish two distinct QAM constellations: square constellations for which the number of bits per symbol is even, and cross constellations for which the number of bits per symbol is odd. These two cases are considered in the sequel in that order.

### *QAM Square Constellations*

With an *even* number of bits per symbol, we may write

$$L = \sqrt{M} \tag{6.56}$$

where $L$ is a positive integer. Under this condition, an $M$-ary QAM square constellation can always be viewed as the *Cartesian product* of a one-dimensional $L$-ary PAM constellation with itself. By definition, the Cartesian product of two sets of coordinates (representing a pair of one-dimensional constellations) is made up of the set of all possible ordered pairs of coordinates with the first coordinate in each such pair taken from the first set involved in the product and the second coordinate taken from the second set in the product.

In the case of a QAM square constellation, the ordered pairs of coordinates naturally form a square matrix, as shown by

$$\{a_i, b_i\} = \begin{bmatrix} (-L+1, L-1) & (-L+3, L-1) & \cdots & (L-1, L-1) \\ (-L+1, L-3) & (-L+3, L-3) & \cdots & (L-1, L-3) \\ \vdots & \vdots & & \vdots \\ (-L+1, -L+1) & (-L+3, -L+1) & \cdots & (L-1, -L+1) \end{bmatrix} \tag{6.57}$$

## ▶ EXAMPLE 6.3

Consider a 16-QAM whose signal constellation is depicted in Figure 6.17*a*. The encoding of the message points shown in this figure is as follows:

- ▶ Two of the four bits, namely, the left-most two bits, specify the quadrant in the $(\phi_1, \phi_2)$-plane in which a message point lies. Thus, starting from the first quadrant and proceeding counterclockwise, the four quadrants are represented by the dibits 11, 10, 00, and 01.
- ▶ The remaining two bits are used to represent one of the four possible symbols lying within each quadrant of the $(\phi_1, \phi_2)$-plane.

Note that the encoding of the four quadrants and also the encoding of the symbols in each quadrant follow the Gray coding rule.

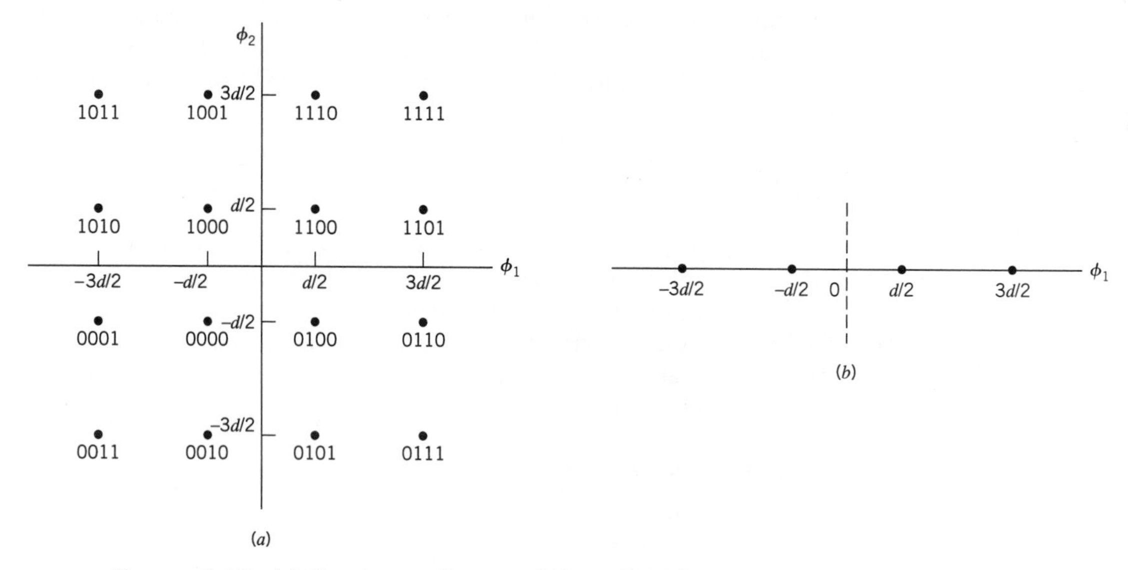


FIGURE 6.17 (*a*) Signal-space diagram of *M*-ary QAM for $M = 16$; the message points in each quadrant are identified with Gray-encoded quadbits. (*b*) Signal-space diagram of the corresponding 4-PAM signal.

For the example at hand, we have $L = 4$. Thus the square constellation of Figure 6.17*a* is the Cartesian product of the 4-PAM constellation shown in Figure 6.17*b* with itself. Moreover, the matrix of Equation (6.57) has the value

$$\{a_i, b_i\} = \begin{bmatrix} (-3, 3) & (-1, 3) & (1, 3) & (3, 3) \\ (-3, 1) & (-1, 1) & (1, 1) & (3, 1) \\ (-3, -1) & (-1, -1) & (1, -1) & (3, -1) \\ (-3, -3) & (-1, -3) & (1, -3) & (3, -3) \end{bmatrix}$$

◀

To calculate the probability of symbol error for $M$-ary QAM, we exploit the property that a QAM square constellation can be factored into the product of the corresponding PAM constellation with itself. We may thus proceed as follows:

1. The probability of correct detection for $M$-ary QAM may be written as

$$P_c = (1 - P_e')^2 \tag{6.58}$$

where $P_e'$ is the probability of symbol error for the corresponding $L$-ary PAM with $L = \sqrt{M}$.

2. The probability of symbol error $P_e'$ is defined by

$$P_e' = \left(1 - \frac{1}{\sqrt{M}}\right) \operatorname{erfc}\left(\sqrt{\frac{E_0}{N_0}}\right) \tag{6.59}$$

(Note that $L = \sqrt{M}$ in the $M$-ary QAM corresponds to $M$ in the $M$-ary PAM considered in Problem 4.27.)

3. The probability of symbol error for $M$-ary QAM is given by

$$\begin{aligned} P_e &= 1 - P_c \\ &= 1 - (1 - P_e')^2 \\ &\simeq 2P_e' \end{aligned} \tag{6.60}$$

where it is assumed that $P_e'$ is small enough compared to unity to justify ignoring the quadratic term.

Hence, using Equations (6.58) and (6.59) in Equation (6.60), we find that the probability of symbol error for $M$-ary QAM is approximately given by

$$P_e \simeq 2\left(1 - \frac{1}{\sqrt{M}}\right) \operatorname{erfc}\left(\sqrt{\frac{E_0}{N_0}}\right) \tag{6.61}$$

The transmitted energy in $M$-ary QAM is variable in that its instantaneous value depends on the particular symbol transmitted. It is therefore more logical to express $P_e$ in terms of the *average* value of the transmitted energy rather than $E_0$. Assuming that the $L$ amplitude levels of the in-phase or quadrature component are equally likely, we have

$$E_{\text{av}} = 2\left[\frac{2E_0}{L} \sum_{i=1}^{L/2} (2i - 1)^2\right] \tag{6.62}$$

where the multiplying factor of 2 outside the square brackets accounts for the equal contributions made by the in-phase and quadrature components. The limits of the summation and the multiplying factor of 2 inside the square brackets take account of the symmetric nature of the pertinent amplitude levels around zero. Summing the series in Equation (6.62), we get

$$\begin{aligned} E_{\text{av}} &= \frac{2(L^2 - 1)E_0}{3} \\ &= \frac{2(M - 1)E_0}{3} \end{aligned} \tag{6.63}$$

Accordingly, we may rewrite Equation (6.61) in terms of $E_{\text{av}}$ as

$$P_e \simeq 2\left(1 - \frac{1}{\sqrt{M}}\right) \operatorname{erfc}\left(\sqrt{\frac{3E_{\text{av}}}{2(M - 1)N_0}}\right) \tag{6.64}$$

which is the desired result.

The case of $M = 4$ is of special interest. The signal constellation for this value of $M$ is the same as that for QPSK. Indeed, putting $M = 4$ in Equation (6.64) and noting that for this special case $E_{\text{av}}$ equals $E$, where $E$ is the energy per symbol, we find that the resulting formula for the probability of symbol error becomes identical to that in Equation (6.34), and so it should.

### *QAM Cross Constellation*

To generate an $M$-ary QAM signal with an odd number of bits per symbol, we require the use of a *cross constellation.* As illustrated in Figure 6.18, we may construct such a signal constellation with $n$ bits per symbol by proceeding as follows:

- Start with a QAM square constellation with $n$-1 bits per symbol.
- Extend each side of the QAM square constellation by adding $2^{n-3}$ symbols.
- Ignore the corners in the extension.

The inner square represents $2^{n-1}$ symbols. The four side extensions add $4 \times 2^{n-3} = 2^{n-1}$ symbols. The total number of symbols in the cross constellation is therefore $2^{n-1} + 2^{n-1}$, which equals $2^n$ and therefore represents $n$ bits per symbol as desired.

Unlike QAM square constellation, it is *not* possible to express a QAM cross constellation as the product of a PAM constellation with itself. The absence of such a factor-

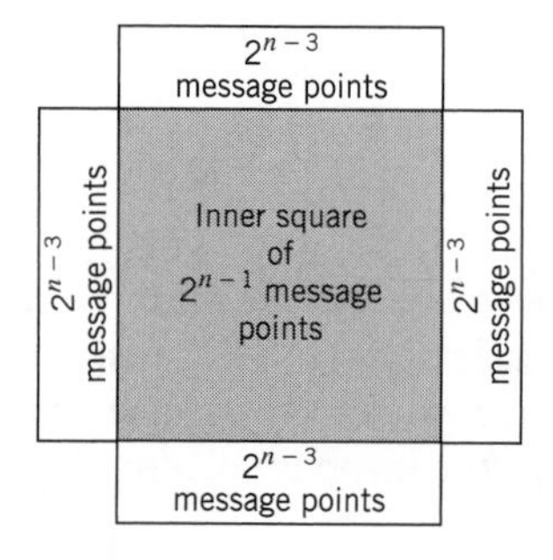


**FIGURE 6.18** Illustrating how a square QAM constellation can be expanded to form a QAM cross-constellation.

ization complicates the determination of the probability of symbol error $P_e$ incurred in the use of $M$-ary QAM characterized by a cross constellation. We therefore simply state the formula for $P_e$ without proof, as shown here

$$P_e \simeq 2\left(1 - \frac{1}{\sqrt{2M}}\right) \operatorname{erfc}\left(\sqrt{\frac{E_0}{N_0}}\right) \quad \text{for high } E_0/N_0 \tag{6.65}$$

which agrees with the formula of Equation (6.61) for a square constellation, except for the inclusion of an extra 0.5 bit per dimension in the constellation.[5] Note also that it is not possible to perfectly Gray code a QAM cross constellation.

## Carrierless Amplitude/Phase Modulation

The passband basis functions of Equations (6.53) and (6.54) assume the use of a rectangular pulse for the pulse-shaping function. For reasons that will become apparent, we redefine the transmitted $M$-ary QAM signal of Equation (6.55) in terms of a general pulse-shaping function $g(t)$ as

$$s_k(t) = a_k g(t - kT)\cos(2\pi f_c t) - b_k g(t - kT)\sin(2\pi f_c t), \quad \begin{array}{l} 0 \le t \le T \\ k = 0, \pm 1, \pm 2, \ldots \end{array} \tag{6.66}$$

It is assumed that carrier frequency $f_c$ has an arbitrary value with respect to the symbol rate $1/T$. On the basis of Equation (6.66), we may express the transmitted $M$-ary QAM signal $s(t)$ for an infinite succession of symbols as

$$\begin{aligned} s(t) &= \sum_{k=-\infty}^{\infty} s_k(t) \\ &= \sum_{k=-\infty}^{\infty} [a_k g(t - kT)\cos(2\pi f_c t) - b_k g(t - kT)\sin(2\pi f_c t)] \end{aligned} \tag{6.67}$$

This equation shows that for an arbitrary $f_c$, the passband functions $g(t - kT)\cos(2\pi f_c t)$ and $g(t - kT)\sin(2\pi f_c t)$ are *aperiodic* in that they vary from one symbol to another.

How can we eliminate the time variations of these passband basis functions from symbol to symbol? To answer this question, we find it convenient to change our formalism from real to complex notation. Specifically, we rewrite Equation (6.67) in the equivalent form

$$\begin{aligned} s(t) &= \operatorname{Re}\left\{\sum_{k=-\infty}^{\infty} (a_k + jb_k)g(t - kT)\exp(j2\pi f_c t)\right\} \\ &= \operatorname{Re}\left\{\sum_{k=-\infty}^{\infty} A_k g(t - kT)\exp(j2\pi f_c t)\right\} \end{aligned} \tag{6.68}$$

where $A_k$ is a complex number defined by

$$A_k = a_k + jb_k \tag{6.69}$$

and $\operatorname{Re}\{\cdot\}$ denotes the real part of the complex quantity enclosed inside the braces. Clearly, Equation (6.68) is unchanged by multiplying the summand in this equation by unity expressed as the product of the complex exponential $\exp(-j2\pi f_c kT)$ and its complex conjugate $\exp(j2\pi f_c kT)$. We may thus rewrite Equation (6.68) in the new form

$$\begin{aligned} s(t) &= \text{Re}\left\{ \sum_{k=-\infty}^{\infty} A_k g(t - kT) \exp(j2\pi f_c t) \exp(-j2\pi f_c kT) \exp(j2\pi f_c kT) \right\} \\ &= \text{Re}\left\{ \sum_{k=-\infty}^{\infty} A_k \exp(j2\pi f_c kT) g(t - kT) \exp(j2\pi f_c (t - kT)) \right\} \end{aligned} \tag{6.70}$$

Define

$$\tilde{A}_k = A_k \exp(j2\pi f_c kT) \tag{6.71}$$

$$g_+(t) = g(t) \exp(j2\pi f_c t) \tag{6.72}$$

The scalar $\tilde{A}_k$ is simply a *rotated* version of the complex representation of the coordinates of the $k$th transmitted symbol in the $(\phi_1, \phi_2)$-plane. Before presenting an interpretation of the complex-valued signal $g_+(t)$, we assume that the pulse-shaping function $g(t)$ is a low-pass signal whose highest frequency component is smaller than the carrier frequency $f_c$. Then following the material presented in Appendix 2, we recognize $g_+(t)$ as the *analytic signal*, or *pre-envelope*, representation of the band-pass signal $g(t) \cos(2\pi f_c t)$. To be more specific, we expand $g_+(t)$ as

$$\begin{aligned} g_+(t) &= g(t) \cos(2\pi f_c t) + jg(t) \sin(2\pi f_c t) \\ &= p(t) + j\hat{p}(t) \end{aligned} \tag{6.73}$$

where $p(t)$ and $\hat{p}(t)$ are defined by

$$p(t) = g(t) \cos(2\pi f_c t) \tag{6.74}$$

and

$$\hat{p}(t) = g(t) \sin(2\pi f_c t) \tag{6.75}$$

We may then say that the quadrature (imaginary) component $\hat{p}(t)$ of the analytic signal $g_+(t)$ is the *Hilbert transform* of the in-phase (real) component $p(t)$. Note that whereas the pulse-shaping function $g(t)$ is a baseband function, the in-phase and quadrature components of the corresponding analytic signal $g_+(t)$ are both passband functions. Henceforth, $g(t)$ is referred to as the *baseband pulse*, and $p(t)$ and $\hat{p}(t)$ are referred to as *passband in-phase* and *passband quadrature pulses*, respectively.

With the definitions of $\tilde{A}_k$ and $g_+(t)$ at hand, we are ready to finally redefine the transmitted signal of Equation (6.70) simply as

$$s(t) = \text{Re}\left\{ \sum_{k=-\infty}^{\infty} \tilde{A}_k g_+(t - kT) \right\} \tag{6.76}$$

Three important observations are noteworthy from this new formulation of the transmitted signal:

- The transmitted signal $s(t)$ appears to be carrierless.
- Since the formulation of $s(t)$ in Equation (6.76) is indeed equivalent to that of the $M$-ary QAM signal presented in Equation (6.67), the new formulation of $s(t)$ fully retains the hybridized amplitude and phase modulation characterizing the original $M$-ary QAM signal.
- The transmitted signal $s(t)$ represents a symbol-time-invariant realization of this hybrid modulation process.

For a prescribed carrier frequency $f_c$, the sequence of rotations described in Equation (6.71) is known. Hence, the receiver need only detect $\tilde{A}_k$, in which case we may compute the corresponding value of $A_k$ by applying the reverse rotations as described here:

$$A_k = \tilde{A}_k \exp(-j2\pi f_c kT), \qquad k = 0, \pm 1, \pm 2, \ldots$$

In practice, however, the rotations are ignored because they do *not* have any bearing on operation or performance of the hybrid modulation system; application of the rotations is in fact necessary only when its equivalence to QAM is an issue of interest. Accordingly, we may ignore Equation (6.71) and redefine the transmitted signal of Equation (6.76) simply as

$$\begin{aligned} s(t) &= \text{Re}\left\{ \sum_{k=-\infty}^{\infty} A_k g_+(t - kT) \right\} \\ &= \text{Re}\left\{ \sum_{k=-\infty}^{\infty} (a_k + jb_k)(p(t - kT) + j\hat{p}(t - kT)) \right\} \\ &= \sum_{k=-\infty}^{\infty} [a_k p(t - kT) - b_k \hat{p}(t - kT)] \end{aligned} \tag{6.77}$$

where, as mentioned previously, $\hat{p}(t)$ is the Hilbert transform of $p(t)$. For obvious reasons, the transmitted signal of Equation (6.77) is referred to as *carrierless amplitude/phase modulation* (CAP).[6]

### *Properties of the Passband In-phase and Quadrature Pulses*

From the definitions of the passband in-phase and quadrature pulses given in Equations (6.74) and (6.75), we deduce the following properties:

**Property 1**

*The passband in-phase pulse* p(t) *and quadrature pulse* p̂(t) *are even and odd functions of time* t, *respectively, given that the baseband pulse* g(t) *is an even function of time* t.

This property follows directly from Equations (6.74) and (6.75).

**Property 2**

*The passband pulses* p(t) *and* p̂(t) *form an orthogonal set over the entire interval* $(-\infty, \infty)$ *as shown by*

$$\int_{-\infty}^{\infty} p(t)\hat{p}(t)\, dt = 0 \tag{6.78}$$

The reason for not restricting the integration interval in Equation (6.78) to a symbol period $T$ is that the pulses $p(t)$ and $\hat{p}(t)$ are bandwidth-efficient for high-performing CAP systems. To prove Property 2, we transform Equation (6.78) into the frequency domain by using the Fourier transform to write

$$\int_{-\infty}^{\infty} p(t)\hat{p}(t)\, dt = \int_{-\infty}^{-\infty} P(f)\hat{P}^*(f)\, df \tag{6.79}$$

where $p(t) \rightleftharpoons P(f)$ and $\hat{p}(t) \rightleftharpoons \hat{P}(f)$. The asterisk in $\hat{P}^*(f)$ denotes complex conjugation. (The frequency function $\hat{P}(f)$ should *not* be viewed as the Hilbert transform of $P(f)$;

rather, following the usual terminology in Fourier analysis, $\hat{P}(f)$ is simply the Fourier transform of $\hat{p}(t)$.) The Fourier transform $\hat{P}(f)$ is related to the Fourier transform $P(f)$ by (see Appendix 2)

$$\hat{P}(f) = -j \operatorname{sgn}(f) P(f) \tag{6.80}$$

where $\operatorname{sgn}(f)$ is the signum function. The Fourier transforms $P(f)$ and $\hat{P}(f)$ have the same magnitude spectrum, but their phase spectra differ by $+90$ degrees for negative frequencies and $-90$ degrees for positive frequencies. We may therefore rewrite Equation (6.79) as

$$\begin{aligned} \int_{-\infty}^{\infty} p(t)\hat{p}(t)\, dt &= j \int_{-\infty}^{\infty} P(f)P^*(f) \operatorname{sgn}(f)\, df \\ &= j \int_{-\infty}^{\infty} |P(f)|^2 \operatorname{sgn}(f)\, df \end{aligned} \tag{6.81}$$

Recognizing that the magnitude response $|P(f)|$ is an even function of frequency $f$ and the signum function $\operatorname{sgn}(f)$ is an odd function of frequency $f$, the integral of Equation (6.81) is zero, proving Property 2.

Next, let the passband pulses $p(t)$ and $\hat{p}(t)$ be passed through a linear time-invariant channel of impulse response $h(t)$, yielding the following passband outputs:

$$u(t) = p(t) \star h(t) \tag{6.82}$$

and

$$\hat{u}(t) = \hat{p}(t) \star h(t) \tag{6.83}$$

where the symbol $\star$ denotes convolution. We may then formulate the third property.

**Property 3**

*The passband pulses* u(t) *and* û(t), *defined in Equations (6.82) and (6.83), form a Hilbert-transform pair and are therefore orthogonal over the entire interval* $(-\infty, \infty)$ *for any* h(t).

It is this important property that makes it possible for the CAP receiver to separate the transmitted real and imaginary symbols, $a_k$ and $b_k$, given the channel output. To prove Property 3, we again use the Fourier transform (in a manner similar to Equations (6.79) and (6.81)) to write

$$\begin{aligned} \int_{-\infty}^{\infty} u(t)\hat{u}(t)\, dt &= \int_{-\infty}^{\infty} U(f)\hat{U}^*(f)\, df \\ &= \int_{-\infty}^{\infty} (P(f)H(f))(-j \operatorname{sgn}(f)P(f)H(f))^* df \\ &= \int_{-\infty}^{\infty} j|P(f)|^2|H(f)|^2 \operatorname{sgn}(f)\, df \\ &= 0 \end{aligned}$$

### ▶ EXAMPLE 6.4 Bandwidth-Efficient Spectral Shaping

The CAP signal of Equation (6.77) uses multilevel encoding via the complex scalar $A_k$ for spectral efficiency. We may further improve the bandwidth efficiency of CAP by using a spectrally efficient formulation of the baseband pulse $g(t)$. For the selection of $g(t)$, we may draw upon the *raised-cosine family* of pulse-shaping functions discussed in Chapter 4.

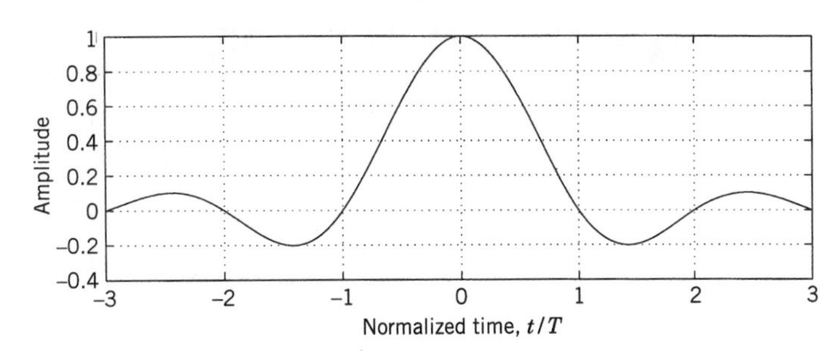


**FIGURE 6.19** The baseband pulse $g(t)$ for rolloff factor $\alpha = 0.2$.

Consider, for example, a baseband raised-cosine shaping filter with a rolloff factor $\alpha = 0.2$ (i.e., excess bandwidth of 20 percent). The formula for the baseband pulse $g(t)$ is*

$$g(t) = (\text{sinc}(t))\,\frac{\cos(\pi\alpha t)}{1 - 4\alpha^2 t^2} \tag{6.84}$$

where time $t$ is normalized with respect to the symbol duration $T$. Hence putting $\alpha = 0.2$ in Equation (6.84), we get the plot of Figure 6.19. Using Equations (6.74) and (6.75), we may compute the passband in-phase pulse $p(t)$ and quadrature pulse $\hat{p}(t)$ plotted in Figures 6.20*a* and 6.20*b* for a normalized carrier frequency $f_c T = 0.5(1 + \alpha) = 0.6$. The waveforms of Figure 6.20 show that $p(t)$ and $\hat{p}(t)$ are even and odd function of time $t$, respectively, in accordance with Property 1 and they are orthogonal over the interval $(-\infty, \infty)$ in accordance with Property 2. ◀

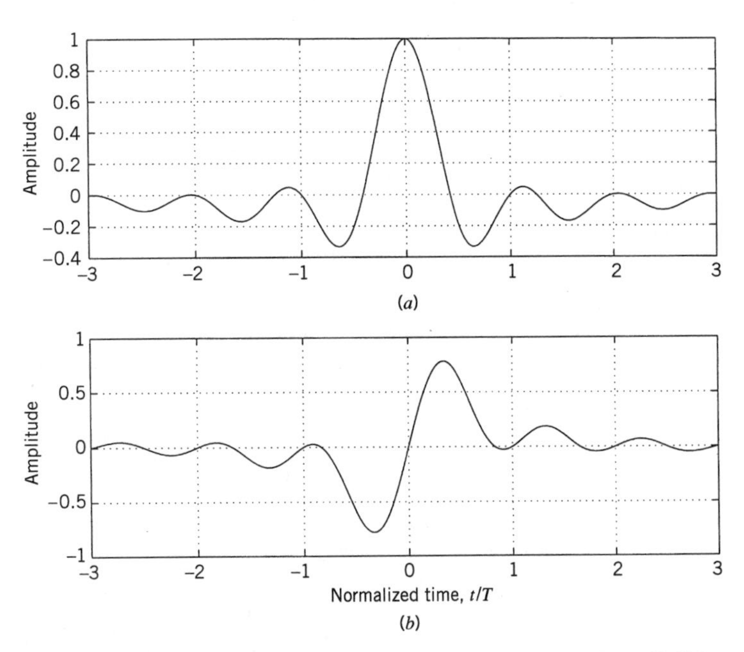


**FIGURE 6.20** (*a*) In-phase pulse $p(t)$, and (*b*) quadrature pulse $\hat{p}(t)$ for rolloff factor $\alpha = 0.2$.

*Equation (6.84) is obtained by using the normalized time $t$ in place of $2Wt$ in Eq. (4.62) of Chapter 4. Recall that the Nyquist bandwidth $W$ is equal to 0.5 times the symbol (bit) rate. Note also that the $p(t)$ in Chapter 4 is not to be confused with the $p(t)$ in this chapter.

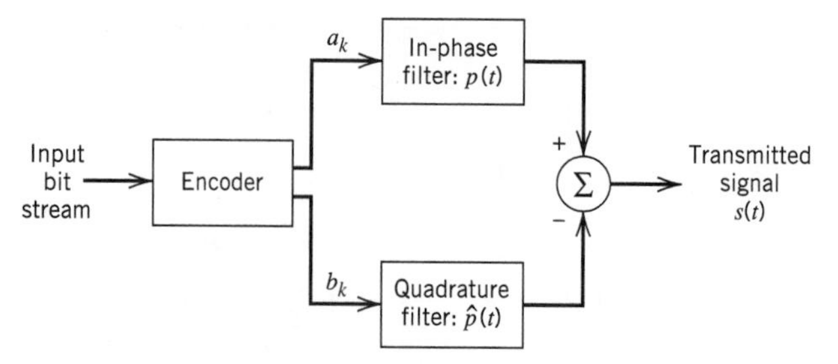


**FIGURE 6.21** Block diagram of CAP transmitter.

### *Basic Structure of the CAP System*

Figure 6.21 shows a functional block diagram of the CAP transmitter, which builds on the material embodied in Equations (6.74), (6.75), and (6.77). The transmitter consists of a multilevel encoder and a pair of passband filters. The *multilevel* encoder partitions the incoming serial data stream into successive blocks of $n$ bits each; these blocks are, in turn, mapped into multilevel symbols $a_k$ and $b_k$, where $k$ refers to the $k$th symbol period. The *passband in-phase and quadrature filters* process the symbol streams $\{a_k\}$ and $\{b_k\}$ in parallel, respectively. The impulse responses of these two filters, namely, $p(t)$ and $\hat{p}(t)$, are designed in accordance with Equations (6.74) and (6.75), for a chosen baseband pulse $g(t)$. The resulting outputs of these two filters are subtracted to produce the transmitted CAP signal $s(t)$ in accordance with Equation (6.77).

The transmitted signal $s(t)$ propagates through a channel characterized by impulse response $h(t)$ and additive noise $w(t)$. The resulting channel output is defined by

$$\begin{aligned} x(t) &= s(t) \star h(t) + w(t) \\ &= \sum_{k=-\infty}^{\infty} [a_k p(t - kT) \star h(t) - b_k \hat{p}(t - kT) \star h(t)] + w(t) \\ &= \sum_{k=-\infty}^{\infty} [a_k u(t - kT) - b_k \hat{u}(t - kT)] + w(t) \end{aligned} \tag{6.85}$$

where $u(t)$ and $\hat{u}(t)$ are defined by Equations (6.82) and (6.83), respectively. Given $x(t)$ as the input signal, the function of the receiver is to recover the transmitted symbols $a_k$ and $b_k$ in an optimum fashion and on a symbol-by-symbol basis.

The receiver consists of a two-dimensional optimum receiver, the details of which depend on the channel impairments. Specifically, we may mention two different situations:

- *Additive white Gaussian noise is the only impairment.* In this idealized situation, the optimum CAP receiver consists of a *two-dimensional matched filter*, as indicated in

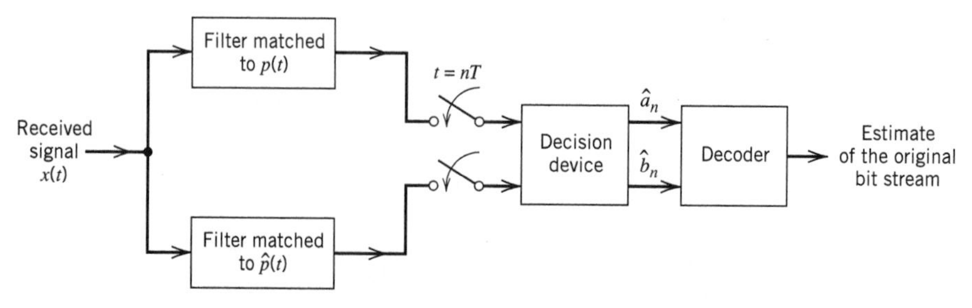


**FIGURE 6.22** Block diagram of CAP receiver using a two-dimensional matched filter for the case of an ideal white Gaussian noise channel.

Figure 6.22. The optimum in-phase filter consists of a filter matched to the passband in-phase pulse $p(t)$. The optimum quadrature filter consists of a filter matched to the passband quadrature pulse $\hat{p}(t)$.

- *Intersymbol interference and additive white Gaussian noise are the channel impairments.* In this more realistic situation in practical terms, the structure of the optimum CAP receiver follows from the optimum linear receiver theory presented in Section 4.9. In particular, the CAP receiver consists of a two-dimensional matched filter, followed by a pair of identical equalizers and a pair of synchronous samplers, as shown in Figure 6.23. The equalizers (implemented in tapped-delay-line form) compensate for dispersion in the channel.

From Property 3, we deduce that the output of the in-phase matched filter due to the passband quadrature pulse is zero, and vice versa. Accordingly, in both of these situations, the transmitted symbols $a_k$ and $b_k$ can be detected separately by the two-dimensional optimum CAP receiver.

### Digital Implementation of the CAP Receiver

The receiver structure of Figure 6.23 can be further simplified by first reformulating it in the form shown in Figure 6.24*a*, where we have introduced an analog-to-digital (A/D) converter at the receiver input to facilitate the use of digital signal processing. Next, we recognize that the matched filters and equalizers in Figure 6.24*a* are all linear systems. Accordingly, the samples at points A and B in Figure 6.24*a* are *linear combinations* of the input samples $x[n] = x(nT_s)$, where $T_s$ is the sampling period. It follows therefore that we may replace the combination of matched filter and equalizer in each receiver path in Figure 6.24*a* by a single *finite-duration impulse response* (FIR) *filter* operating at the sampling rate $1/T_s$, as shown in Figure 6.24*b*. With a common input, it is natural for the two FIR filters in Figure 6.24*b* to share a common set of unit-delay elements but different sets of coefficients of their own. Note also that the receiver structure of Figure 6.24*b* is the mirror image of the transmitter structure in Figure 6.21.

One last comment is in order. In practice, the two FIR filters in the receiver of Figure 6.24*b* are made *adaptive* so as to accommodate operation of the CAP receiver in an unknown environment. With adaptive equalization (filtering) as the method of choice, we have two options in light of the material presented in Chapter 4: linear equalization and decision feedback equalization (DFE). When the frequency response of the channel is approximately flat, the use of linear equalization is adequate for the task at hand. However, when the frequency response of the channel and/or the noise power spectrum are not approximately flat, performance of the CAP receiver can be improved significantly by the use of DFE.

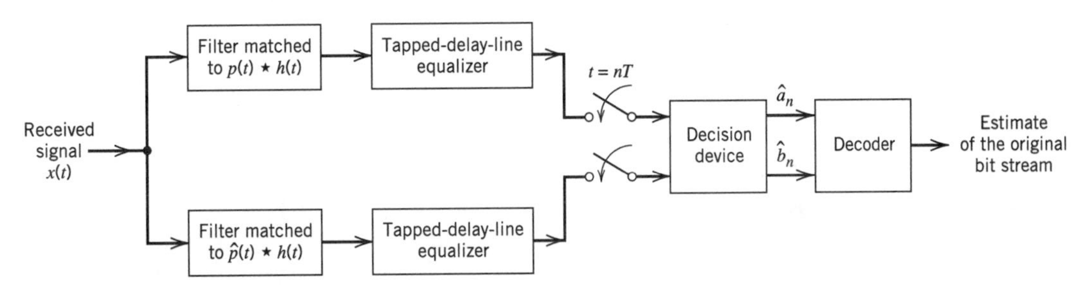


**FIGURE 6.23** Block diagram of CAP receiver using a pair of optimum linear receivers for the case of a noisy, dispersive channel.

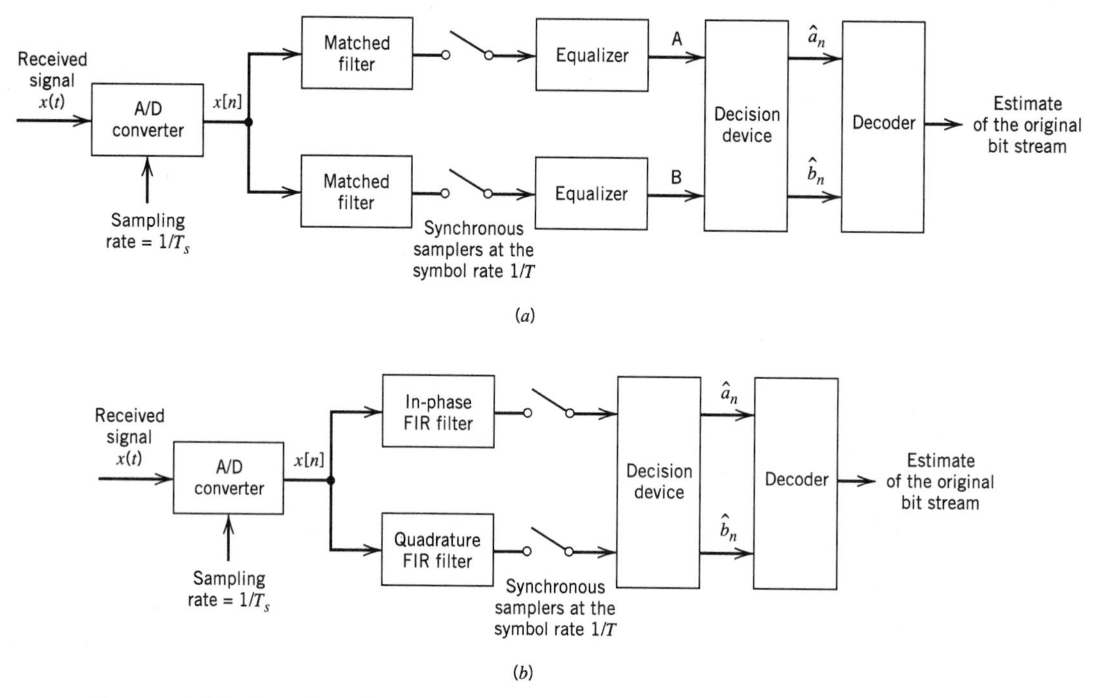


**FIGURE 6.24** Digital implementation of the CAP receiver. (*a*) Reformulation of the receiver in Figure 6.23 using an A/D converter and associated circuitry. (*b*) Replacement of the matched filter/equalizer pairs with equivalent (digitally implemented) FIR filters.

### *An Application of CAP*

An important application of CAP (using 32 or 64 constellation points) is in the passband transmission of digital data over twisted-pair wiring of lengths less than 100 m, as in local area networks (LANs) suitable for premises' distribution systems. (All modern LAN standards limit the length of the wiring to a maximum of 100 m; the CAP systems can actually operate over longer loops, if so required.) The data rates may range from 51 up to 155 Mb/s, with the usable channel bandwidth being strictly limited to 30 MHz.

The two major impairments for transceivers providing duplex operation over twisted-pairs cables are propagation loss and near-end crosstalk (NEXT); these impairments were discussed in Section 4.8, which dealt with digital subscriber lines.

In the environment described herein, a practical issue is how to adapt the receiver to wide variations in twisted-pair cables. To cater to this requirement, adaptive filters are used to implement the optimum in-phase and quadrature filters in the receiver, as remarked earlier. The wider the range of twisted-pair cables that have to be accommodated, the more complex these adaptive filters must be.

## 6.5 *Coherent Frequency-Shift Keying*

*M*-ary PSK and *M*-ary QAM share a common property: Both are examples of linear modulation. In this section we study a nonlinear method of passband data transmission, namely, *coherent frequency-shift keying* (FSK). We begin the study by considering the simple case of binary FSK.

## ■ Binary FSK

In a *binary FSK system*, symbols 1 and 0 are distinguished from each other by transmitting one of two sinusoidal waves that differ in frequency by a fixed amount. A typical pair of sinusoidal waves is described by

$$s_i(t) = \begin{cases} \sqrt{\dfrac{2E_b}{T_b}} \cos(2\pi f_i t), & 0 \le t \le T_b \\ 0, & \text{elsewhere} \end{cases} \tag{6.86}$$

where $i = 1, 2$, and $E_b$ is the transmitted signal energy per bit; the transmitted frequency is

$$f_i = \frac{n_c + i}{T_b} \qquad \text{for some fixed integer } n_c \text{ and } i = 1, 2 \tag{6.87}$$

Thus symbol 1 is represented by $s_1(t)$, and symbol 0 by $s_2(t)$. The FSK signal described here is known as *Sunde's FSK*. It is a *continuous-phase signal* in the sense that phase continuity is always maintained, including the inter-bit switching times. This form of digital modulation is an example of *continuous-phase frequency-shift keying* (CPFSK), on which we have more to say later on in the section.

From Equations (6.86) and (6.87), we observe directly that the signals $s_1(t)$ and $s_2(t)$ are orthogonal, but not normalized to have unit energy. We therefore deduce that the most useful form for the set of orthonormal basis functions is

$$\phi_i(t) = \begin{cases} \sqrt{\dfrac{2}{T_b}} \cos(2\pi f_i t), & 0 \le t \le T_b \\ 0, & \text{elsewhere} \end{cases} \tag{6.88}$$

where $i = 1, 2$. Correspondingly, the coefficient $s_{ij}$ for $i = 1, 2$, and $j = 1, 2$ is defined by

$$\begin{aligned} s_{ij} &= \int_0^{T_b} s_i(t)\phi_j(t)\, dt \\ &= \int_0^{T_b} \sqrt{\frac{2E_b}{T_b}} \cos(2\pi f_i t) \sqrt{\frac{2}{T_b}} \cos(2\pi f_j t)\, dt \\ &= \begin{cases} \sqrt{E_b}, & i = j \\ 0, & i \ne j \end{cases} \end{aligned} \tag{6.89}$$

Thus, unlike coherent binary PSK, a coherent binary FSK system is characterized by having a signal space that is two-dimensional (i.e., $N = 2$) with two message points (i.e., $M = 2$), as shown in Figure 6.25. The two message points are defined by the

$$\mathbf{s}_1 = \begin{bmatrix} \sqrt{E_b} \\ 0 \end{bmatrix} \tag{6.90}$$

and

$$\mathbf{s}_2 = \begin{bmatrix} 0 \\ \sqrt{E_b} \end{bmatrix} \tag{6.91}$$

with the Euclidean distance between them equal to $\sqrt{2E_b}$. Figure 6.25 also includes a couple of inserts, which show waveforms representative of signals $s_1(t)$ and $s_2(t)$.

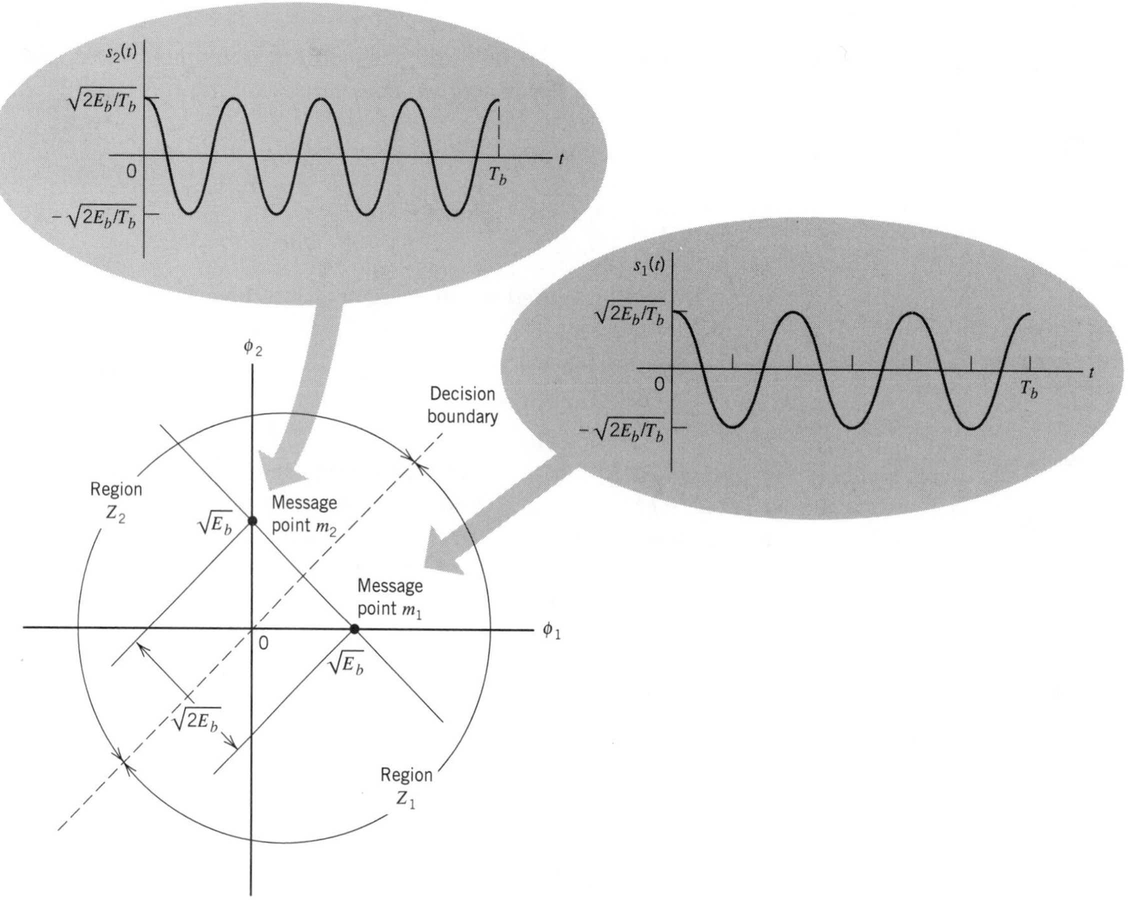


**FIGURE 6.25** Signal-space diagram for binary FSK system. The diagram also includes two inserts showing example waveforms of the two modulated signals $s_1(t)$ and $s_2(t)$.

### *Error Probability of Binary FSK*

The observation vector **x** has two elements $x_1$ and $x_2$ that are defined by, respectively,

$$x_1 = \int_0^{T_b} x(t)\phi_1(t)\, dt \tag{6.92}$$

and

$$x_2 = \int_0^{T_b} x(t)\phi_2(t)\, dt \tag{6.93}$$

where $x(t)$ is the received signal, the form of which depends on which symbol was transmitted. Given that symbol 1 was transmitted, $x(t)$ equals $s_1(t) + w(t)$, where $w(t)$ is the sample function of a white Gaussian noise process of zero mean and power spectral density $N_0/2$. If, on the other hand, symbol 0 was transmitted, $x(t)$ equals $s_2(t) + w(t)$.

Now, applying the decision rule of Equation (5.59), we find that the observation space is partitioned into two decision regions, labeled $Z_1$ and $Z_2$ in Figure 6.25. The decision boundary, separating region $Z_1$ from region $Z_2$ is the perpendicular bisector of the line joining the two message points. The receiver decides in favor of symbol 1 if the received signal point represented by the observation vector $\mathbf{x}$ falls inside region $Z_1$. This occurs when $x_1 > x_2$. If, on the other hand, we have $x_1 < x_2$, the received signal point falls inside region $Z_2$, and the receiver decides in favor of symbol 0. On the decision boundary, we have $x_1 = x_2$, in which case the receiver makes a random guess in favor of symbol 1 or 0.

Define a new Gaussian random variable $Y$ whose sample value $y$ is equal to the difference between $x_1$ and $x_2$; that is,

$$y = x_1 - x_2 \tag{6.94}$$

The mean value of the random variable $Y$ depends on which binary symbol was transmitted. Given that symbol 1 was transmitted, the Gaussian random variables $X_1$ and $X_2$, whose sample values are denoted by $x_1$ and $x_2$, have mean values equal to $\sqrt{E_b}$ and zero, respectively. Correspondingly, the conditional mean of the random variable $Y$, given that symbol 1 was transmitted, is

$$\begin{aligned} E[Y|1] &= E[X_1|1] - E[X_2|1] \\ &= +\sqrt{E_b} \end{aligned} \tag{6.95}$$

On the other hand, given that symbol 0 was transmitted, the random variables $X_1$ and $X_2$ have mean values equal to zero and $\sqrt{E_b}$, respectively. Correspondingly, the conditional mean of the random variable $Y$, given that symbol 0 was transmitted, is

$$\begin{aligned} E[Y|0] &= E[X_1|0] - E[X_2|0] \\ &= -\sqrt{E_b} \end{aligned} \tag{6.96}$$

The variance of the random variable $Y$ is independent of which binary symbol was transmitted. Since the random variables $X_1$ and $X_2$ are statistically independent, each with a variance equal to $N_0/2$, it follows that

$$\begin{aligned} \text{var}[Y] &= \text{var}[X_1] + \text{var}[X_2] \\ &= N_0 \end{aligned} \tag{6.97}$$

Suppose we know that symbol 0 was transmitted. The conditional probability density function of the random variable $Y$ is then given by

$$f_Y(y|0) = \frac{1}{\sqrt{2\pi N_0}} \exp\left[-\frac{(y + \sqrt{E_b})^2}{2N_0}\right] \tag{6.98}$$

Since the condition $x_1 > x_2$, or equivalently, $y > 0$, corresponds to the receiver making a decision in favor of symbol 1, we deduce that the conditional probability of error, given that symbol 0 was transmitted, is

$$\begin{aligned} p_{10} &= P(y > 0|\text{symbol 0 was sent}) \\ &= \int_0^\infty f_Y(y|0)\, dy \\ &= \frac{1}{\sqrt{2\pi N_0}} \int_0^\infty \exp\left[-\frac{(y + \sqrt{E_b})^2}{2N_0}\right] dy \end{aligned} \tag{6.99}$$

Put

$$\frac{y + \sqrt{E_b}}{\sqrt{2N_0}} = z \tag{6.100}$$

Then, changing the variable of integration from $y$ to $z$, we may rewrite Equation (6.99) as follows:

$$
\begin{aligned}
p_{10} &= \frac{1}{\sqrt{\pi}} \int_{\sqrt{E_b/2N_0}}^{\infty} \exp(-z^2)\, dz \\
&= \frac{1}{2}\operatorname{erfc}\left(\sqrt{\frac{E_b}{2N_0}}\right)
\end{aligned}
\tag{6.101}
$$

Similarly, we may show the $p_{01}$, the conditional probability of error given that symbol 1 was transmitted, has the same value as in Equation (6.101). Accordingly, averaging $p_{10}$ and $p_{01}$, we find that the *average probability of bit error* or, equivalently, the *bit error rate for coherent binary FSK* is (assuming equiprobable symbols)

$$
P_e = \frac{1}{2}\operatorname{erfc}\left(\sqrt{\frac{E_b}{2N_0}}\right) \tag{6.102}
$$

Comparing Equations (6.20) and (6.102), we see that, in a coherent binary FSK system, we have to double the *bit energy-to-noise density ratio*, $E_b/N_0$, to maintain the same bit error rate as in a coherent binary PSK system. This result is in perfect accord with the signal-space diagrams of Figures 6.3 and 6.25, where we see that in a binary PSK system the Euclidean distance between the two message points is equal to $2\sqrt{E_b}$, whereas in a binary FSK system the corresponding distance is $\sqrt{2E_b}$. For a prescribed $E_b$, the minimum distance $d_{\min}$ in binary PSK is therefore $\sqrt{2}$ times that in binary FSK. Recall from Chapter 5 that the probability of error decreases exponentially as $d_{\min}^2$, hence the difference between the formulas of Equations (6.20) and (6.102).

### *Generation and Detection of Coherent Binary FSK Signals*

To generate a binary FSK signal, we may use the scheme shown in Figure 6.26*a*. The incoming binary data sequence is first applied to an *on–off level encoder*, at the output of which symbol 1 is represented by a constant amplitude of $\sqrt{E_b}$ volts and symbol 0 is represented by zero volts. By using an *inverter* in the lower channel in Figure 6.26*a*, we in effect make sure that when we have symbol 1 at the input, the oscillator with frequency $f_1$ in the upper channel is switched on while the oscillator with frequency $f_2$ in the lower channel is switched off, with the result that frequency $f_1$ is transmitted. Conversely, when we have symbol 0 at the input, the oscillator in the upper channel is switched off and the oscillator in the lower channel is switched on, with the result that frequency $f_2$ is transmitted. The two frequencies $f_1$ and $f_2$ are chosen to equal different integer multiples of the bit rate $1/T_b$, as in Equation (6.87).

In the transmitter of Figure 6.26*a*, we assume that the two oscillators are synchronized, so that their outputs satisfy the requirements of the two orthonormal basis functions $\phi_1(t)$ and $\phi_2(t)$, as in Equation (6.88). Alternatively, we may use a single keyed (voltage-controlled) oscillator. In either case, the frequency of the modulated wave is shifted with a continuous phase, in accordance with the input binary wave.

To detect the original binary sequence given the noisy received signal $x(t)$, we may use the receiver shown in Figure 6.26*b*. It consists of two correlators with a common input, which are supplied with locally generated coherent reference signals $\phi_1(t)$ and $\phi_2(t)$. The correlator outputs are then subtracted, one from the other, and the resulting difference, $y$, is compared with a threshold of zero volts. If $y > 0$, the receiver decides in favor of 1. On the other hand, if $y < 0$, it decides in favor of 0. If $y$ is exactly zero, the receiver makes a random guess in favor of 1 or 0.

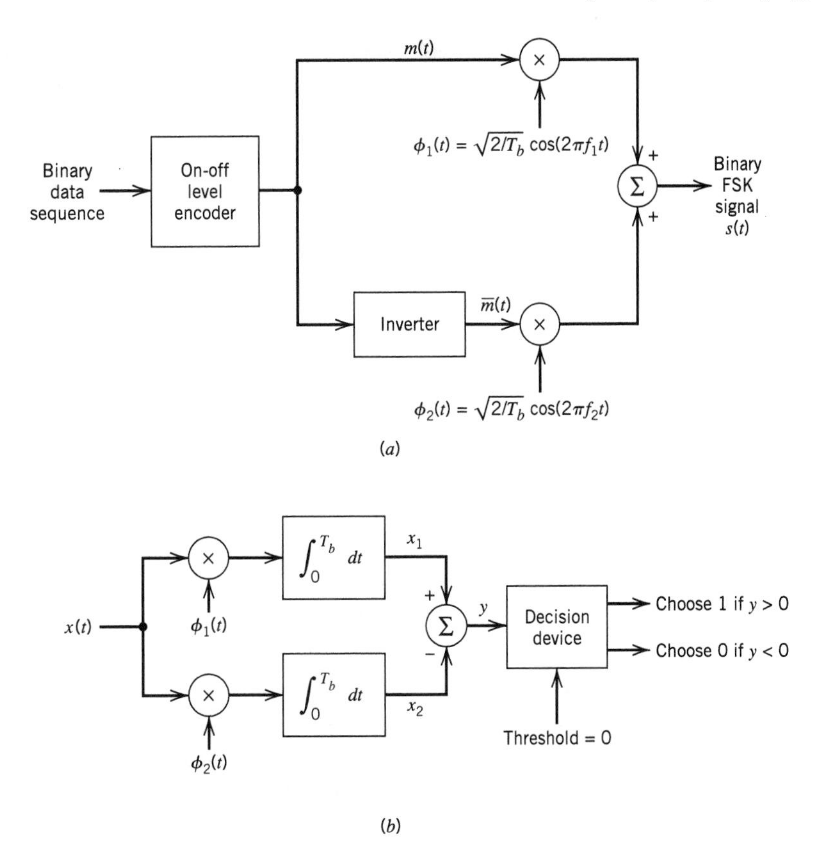


**FIGURE 6.26** Block diagrams for (*a*) binary FSK transmitter and (*b*) coherent binary FSK receiver.

### *Power Spectra of Binary FSK Signals*

Consider the case of Sunde's FSK, for which the two transmitted frequencies $f_1$ and $f_2$ differ by an amount equal to the bit rate $1/T_b$, and their arithmetic mean equals the nominal carrier frequency $f_c$; phase continuity is always maintained, including inter-bit switching times. We may express this special binary FSK signal as follows:

$$s(t) = \sqrt{\frac{2E_b}{T_b}} \cos\left(2\pi f_c t \pm \frac{\pi t}{T_b}\right), \qquad 0 \le t \le T_b \tag{6.103}$$

and using a well-known trigonometric identity, we get

$$\begin{aligned} s(t) &= \sqrt{\frac{2E_b}{T_b}} \cos\left(\pm \frac{\pi t}{T_b}\right) \cos(2\pi f_c t) - \sqrt{\frac{2E_b}{T_b}} \sin\left(\pm \frac{\pi t}{T_b}\right) \sin(2\pi f_c t) \\ &= \sqrt{\frac{2E_b}{T_b}} \cos\left(\frac{\pi t}{T_b}\right) \cos(2\pi f_c t) \mp \sqrt{\frac{2E_b}{T_b}} \sin\left(\frac{\pi t}{T_b}\right) \sin(2\pi f_c t) \end{aligned} \tag{6.104}$$

In the last line of Equation (6.104), the plus sign corresponds to transmitting symbol 0, and the minus sign corresponds to transmitting symbol 1. As before, we assume that the symbols 1 and 0 in the random binary wave at the modulator input are equally likely, and that the symbols transmitted in adjacent time slots are statistically independent. Then, based on the representation of Equation (6.104), we may make the following observations pertaining to the in-phase and quadrature components of a binary FSK signal with continuous phase:

1. The in-phase component is completely independent of the input binary wave. It equals $\sqrt{2E_b/T_b}\cos(\pi t/T_b)$ for all values of time $t$. The power spectral density of this component therefore consists of two delta functions, weighted by the factor $E_b/2T_b$, and occurring at $f = \pm 1/2T_b$.
2. The quadrature component is directly related to the input binary wave. During the signaling interval $0 \le t \le T_b$, it equals $-g(t)$ when we have symbol 1, and $+g(t)$ when we have symbol 0. The symbol shaping function $g(t)$ is defined by

$$g(t) = \begin{cases} \sqrt{\dfrac{2E_b}{T_b}}\sin\left(\dfrac{\pi t}{T_b}\right), & 0 \le t \le T_b \\ 0, & \text{elsewhere} \end{cases} \tag{6.105}$$

The energy spectral density of this symbol shaping function equals

$$\Psi_g(f) = \frac{8E_b T_b \cos^2(\pi T_b f)}{\pi^2(4T_b^2 f^2 - 1)^2} \tag{6.106}$$

The power spectral density of the quadrature component equals $\Psi_g(f)/T_b$. It is also apparent that the in-phase and quadrature components of the binary FSK signal are independent of each other. Accordingly, the baseband power spectral density of Sunde's FSK signal equals the sum of the power spectral densities of these two components, as shown by

$$S_B(f) = \frac{E_b}{2T_b}\left[\delta\left(f - \frac{1}{2T_b}\right) + \delta\left(f + \frac{1}{2T_b}\right)\right] + \frac{8E_b\cos^2(\pi T_b f)}{\pi^2(4T_b^2 f^2 - 1)^2} \tag{6.107}$$

Substituting Equation (6.107) in Equation (6.4), we find that the power spectrum of the binary FSK signal contains two discrete frequency components located at $(f_c + 1/2T_b) = f_1$ and $(f_c - 1/2T_b) = f_2$, with their average powers adding up to one-half the total power of the binary FSK signal. The presence of these two discrete frequency components provides a means of synchronizing the receiver with the transmitter.

Note also that the baseband power spectral density of a binary FSK signal with continuous phase ultimately falls off as the inverse fourth power of frequency. This is readily established by taking the limit in Equation (6.107) as $f$ approaches infinity. If, however, the FSK signal exhibits phase discontinuity at the inter-bit switching instants (this arises when the two oscillators applying frequencies $f_1$ and $f_2$ operate independently of each other), the power spectral density ultimately falls off as the inverse square of frequency; see Problem 6.23. Accordingly, an FSK signal with continuous phase does not produce as much interference outside the signal band of interest as an FSK signal with discontinuous phase.

In Figure 6.5, we have plotted the baseband power spectra of Equations (6.22) and (6.107). (To simplify matters, we have only plotted the results for positive frequency.) In both cases, $S_B(f)$ is shown normalized with respect to $2E_b$, and the frequency is normalized with respect to the bit rate $R_b = 1/T_b$. The difference in the falloff rates of these spectra can be explained on the basis of the pulse shape $g(t)$. The smoother the pulse, the faster the drop of spectral tails to zero. Thus, since binary FSK (with continuous phase) has a smoother pulse shape, it has lower sidelobes than binary PSK.

## MINIMUM SHIFT KEYING

In the coherent detection of binary FSK signal, the phase information contained in the received signal is not fully exploited, other than to provide for synchronization of the receiver to the transmitter. We now show that by proper use of the phase when performing detection, it is possible to improve the noise performance of the receiver significantly. This improvement is, however, achieved at the expense of increased receiver complexity.

Consider a continuous-phase frequency-shift keying (CPFSK) signal, which is defined for the interval $0 \le t \le T_b$ as follows:

$$s(t) = \begin{cases} \sqrt{\dfrac{2E_b}{T_b}} \cos[2\pi f_1 t + \theta(0)] & \text{for symbol 1} \\ \sqrt{\dfrac{2E_b}{T_b}} \cos[2\pi f_2 t + \theta(0)] & \text{for symbol 0} \end{cases} \tag{6.108}$$

where $E_b$ is the transmitted signal energy per bit, and $T_b$ is the bit duration. The phase $\theta(0)$, denoting the value of the phase at time $t = 0$, sums up the past history of the modulation process up to time $t = 0$. The frequencies $f_1$ and $f_2$ are sent in response to binary symbols 1 and 0 appearing at the modulator input, respectively.

Another useful way of representing the CPFSK signal $s(t)$ is to express it in the conventional form of an *angle-modulated signal* as follows:

$$s(t) = \sqrt{\frac{2E_b}{T_b}} \cos[2\pi f_c t + \theta(t)] \tag{6.109}$$

where $\theta(t)$ is the phase of $s(t)$. When the phase $\theta(t)$ is a continuous function of time, we find that the modulated signal $s(t)$ itself is also continuous at all times, including the inter-bit switching times. The phase $\theta(t)$ of a CPFSK signal increases or decreases linearly with time during each bit duration of $T_b$ seconds, as shown by

$$\theta(t) = \theta(0) \pm \frac{\pi h}{T_b} t, \qquad 0 \le t \le T_b \tag{6.110}$$

where the plus sign corresponds to sending symbol 1, and the minus sign corresponds to sending symbol 0; the parameter $h$ is to be defined. Substituting Equation (6.110) into (6.109), and then comparing the angle of the cosine function with that of Equation (6.108), we deduce the following pair of relations:

$$f_c + \frac{h}{2T_b} = f_1 \tag{6.111}$$

$$f_c - \frac{h}{2T_b} = f_2 \tag{6.112}$$

Solving Equations (6.111) and (6.112) for $f_c$ and $h$, we thus get

$$f_c = \frac{1}{2}(f_1 + f_2) \tag{6.113}$$

and

$$h = T_b(f_1 - f_2) \tag{6.114}$$

The nominal carrier frequency $f_c$ is therefore the arithmetic mean of the frequencies $f_1$ and $f_2$. The difference between the frequencies $f_1$ and $f_2$, normalized with respect to the bit rate $1/T_b$, defines the dimensionless parameter $h$, which is referred to as the *deviation ratio*.

### *Phase Trellis*

From Equation (6.110) we find that at time $t = T_b$,

$$\theta(T_b) - \theta(0) = \begin{cases} \pi h & \text{for symbol 1} \\ -\pi h & \text{for symbol 0} \end{cases} \tag{6.115}$$

That is to say, the sending of symbol 1 increases the phase of a CPFSK signal $s(t)$ by $\pi h$ radians, whereas the sending of symbol 0 reduces it by an equal amount.

The variation of phase $\theta(t)$ with time $t$ follows a path consisting of a sequence of straight lines, the slopes of which represent frequency changes. Figure 6.27 depicts possible paths starting from time $t = 0$. A plot like that shown in Figure 6.27 is called a *phase tree.* The tree makes clear the transitions of phase across interval boundaries of the incoming sequence of data bits. Moreover, it is evident from Figure 6.27 that the phase of a CPFSK signal is an odd or even multiple of $\pi h$ radians at odd or even multiples of the bit duration $T_b$, respectively.

The phase tree described in Figure 6.27 is a manifestation of phase continuity, which is an inherent characteristic of a CPFSK signal. To appreciate the notion of phase continuity, let us go back for a moment to Sunde's FSK, which is a CPFSK scheme as previously described. In this case, the deviation ratio $h$ is exactly unity. Hence, according to Figure 6.27 the phase change over one bit interval is $\pm\pi$ radians. But, a change of $+\pi$ radians is exactly the same as a change of $-\pi$ radians, modulo $2\pi$. It follows therefore that in the case of Sunde's FSK there is *no* memory; that is, knowing which particular change occurred in the *previous* bit interval provides no help in the *current* bit interval.

In contrast, we have a completely different situation when the deviation ratio $h$ is assigned the special value of 1/2. We now find that the phase can take on only the two values $\pm\pi/2$ at odd multiples of $T_b$, and only the two values 0 and $\pi$ at even multiples of $T_b$, as in Figure 6.28. This second graph is called a *phase trellis*, since a "trellis" is a treelike structure with remerging branches. Each path from left to right through the trellis of Figure 6.28 corresponds to a specific binary sequence input. For example, the path shown in boldface in Figure 6.28 corresponds to the binary sequence 1101000 with $\theta(0) = 0$. Henceforth, we assume that $h = 1/2$.

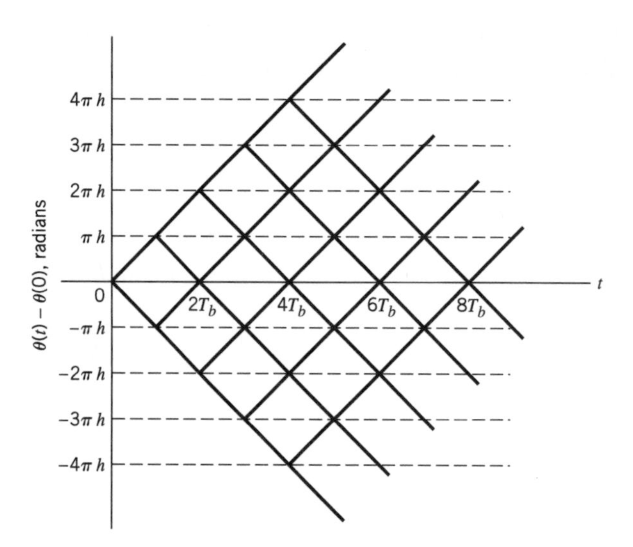


**FIGURE 6.27** Phase tree.

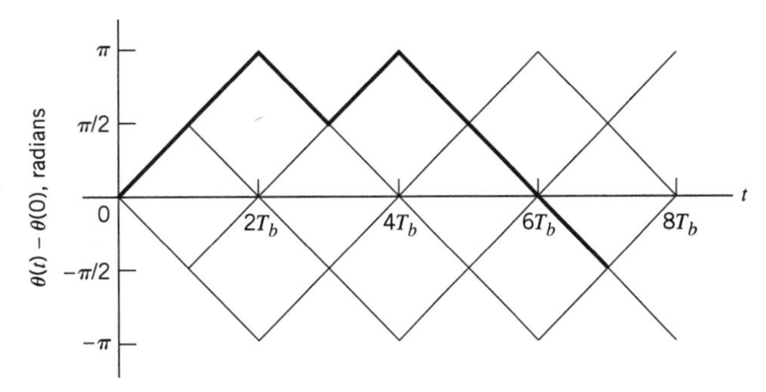


**FIGURE 6.28** Phase trellis; boldfaced path represents the sequence 1101000.

With $h = 1/2$, we find from Equation (6.114) that the frequency deviation (i.e., the difference between the two signaling frequencies $f_1$ and $f_2$) equals half the bit rate. This is the minimum frequency spacing that allows the two FSK signals representing symbols 1 and 0, as in Equation (6.108), to be coherently orthogonal in the sense that they do not interfere with one another in the process of detection. It is for this reason that a CPFSK signal with a deviation ratio of one half is commonly referred to as *minimum shift keying* (MSK).[7]

### *Signal-Space Diagram of MSK*

Using a well-known trigonometric identity in Equation (6.109), we may express the CPFSK signal $s(t)$ in terms of its in-phase and quadrature components as follows:

$$s(t) = \sqrt{\frac{2E_b}{T_b}} \cos[\theta(t)] \cos(2\pi f_c t) - \sqrt{\frac{2E_b}{T_b}} \sin[\theta(t)] \sin(2\pi f_c t) \qquad (6.116)$$

Consider first the in-phase component $\sqrt{2E_b/T_b}\cos[\theta(t)]$. With the deviation ratio $h = 1/2$, we have from Equation (6.110) that

$$\theta(t) = \theta(0) \pm \frac{\pi}{2T_b} t, \qquad 0 \le t \le T_b \qquad (6.117)$$

where the plus sign corresponds to symbol 1 and the minus sign corresponds to symbol 0. A similar result holds for $\theta(t)$ in the interval $-T_b \le t \le 0$, except that the algebraic sign is not necessarily the same in both intervals. Since the phase $\theta(0)$ is 0 or $\pi$, depending on the past history of the modulation process, we find that, in the interval $-T_b \le t \le T_b$, the polarity of $\cos[\theta(t)]$ depends only on $\theta(0)$, regardless of the sequence of 1s and 0s transmitted before or after $t = 0$. Thus, for this time interval, the in-phase component $s_I(t)$ consists of a *half-cycle cosine pulse* defined as follows:

$$\begin{aligned} s_I(t) &= \sqrt{\frac{2E_b}{T_b}} \cos[\theta(t)] \\ &= \sqrt{\frac{2E_b}{T_b}} \cos[\theta(0)] \cos\left(\frac{\pi}{2T_b} t\right) \\ &= \pm\sqrt{\frac{2E_b}{T_b}} \cos\left(\frac{\pi}{2T_b} t\right), \qquad -T_b \le t \le T_b \end{aligned} \qquad (6.118)$$

where the plus sign corresponds to $\theta(0) = 0$ and the minus sign corresponds to $\theta(0) = \pi$. In a similar way, we may show that, in the interval $0 \le t \le 2T_b$, the quadrature component

$s_Q(t)$ consists of a *half-cycle sine pulse*, whose polarity depends only on $\theta(T_b)$, as shown by

$$
\begin{aligned}
s_Q(t) &= \sqrt{\frac{2E_b}{T_b}} \sin[\theta(t)] \\
&= \sqrt{\frac{2E_b}{T_b}} \sin[\theta(T_b)] \sin\left(\frac{\pi}{2T_b} t\right) \\
&= \pm\sqrt{\frac{2E_b}{T_b}} \sin\left(\frac{\pi}{2T_b} t\right), \qquad 0 \le t \le 2T_b
\end{aligned}
\tag{6.119}
$$

where the plus sign corresponds to $\theta(T_b) = \pi/2$ and the minus sign corresponds to $\theta(T_b) = -\pi/2$.

From the foregoing discussion we see that since the phase states $\theta(0)$ and $\theta(T_b)$ can each assume one of two possible values, any one of four possibilities can arise, as described here:

- The phase $\theta(0) = 0$ and $\theta(T_b) = \pi/2$, corresponding to the transmission of symbol 1.
- The phase $\theta(0) = \pi$ and $\theta(T_b) = \pi/2$, corresponding to the transmission of symbol 0.
- The phase $\theta(0) = \pi$ and $\theta(T_b) = -\pi/2$ (or, equivalently, $3\pi/2$ modulo $2\pi$), corresponding to the transmission of symbol 1.
- The phase $\theta(0) = 0$ and $\theta(T_b) = -\pi/2$, corresponding to the transmission of symbol 0.

This, in turn, means that the MSK signal itself may assume any one of four possible forms, depending on the values of $\theta(0)$ and $\theta(T_b)$.

From the expansion of Equation (6.116), we deduce that the orthonormal basis functions $\phi_1(t)$ and $\phi_2(t)$ for MSK are defined by a pair of sinusoidally modulated quadrature carriers:

$$
\phi_1(t) = \sqrt{\frac{2}{T_b}} \cos\left(\frac{\pi}{2T_b} t\right) \cos(2\pi f_c t), \qquad 0 \le t \le T_b \tag{6.120}
$$

$$
\phi_2(t) = \sqrt{\frac{2}{T_b}} \sin\left(\frac{\pi}{2T_b} t\right) \sin(2\pi f_c t), \qquad 0 \le t \le T_b \tag{6.121}
$$

Correspondingly, we may express the MSK signal in the expanded form

$$
s(t) = s_1\phi_1(t) + s_2\phi_2(t), \qquad 0 \le t \le T_b \tag{6.122}
$$

where the coefficients $s_1$ and $s_2$ are related to the phase states $\theta(0)$ and $\theta(T_b)$, respectively. To evaluate $s_1$, we integrate the product $s(t)\phi_1(t)$ between the limits $-T_b$ and $T_b$, as shown by

$$
\begin{aligned}
s_1 &= \int_{-T_b}^{T_b} s(t)\phi_1(t)\, dt \\
&= \sqrt{E_b} \cos[\theta(0)], \qquad -T_b \le t \le T_b
\end{aligned}
\tag{6.123}
$$

Similarly, to evaluate $s_2$ we integrate the product $s(t)\phi_2(t)$ between the limits 0 and $2T_b$, as shown by

$$
\begin{aligned}
s_2 &= \int_{0}^{2T_b} s(t)\phi_2(t)\, dt \\
&= -\sqrt{E_b} \sin[\theta(T_b)], \qquad 0 \le t \le 2T_b
\end{aligned}
\tag{6.124}
$$

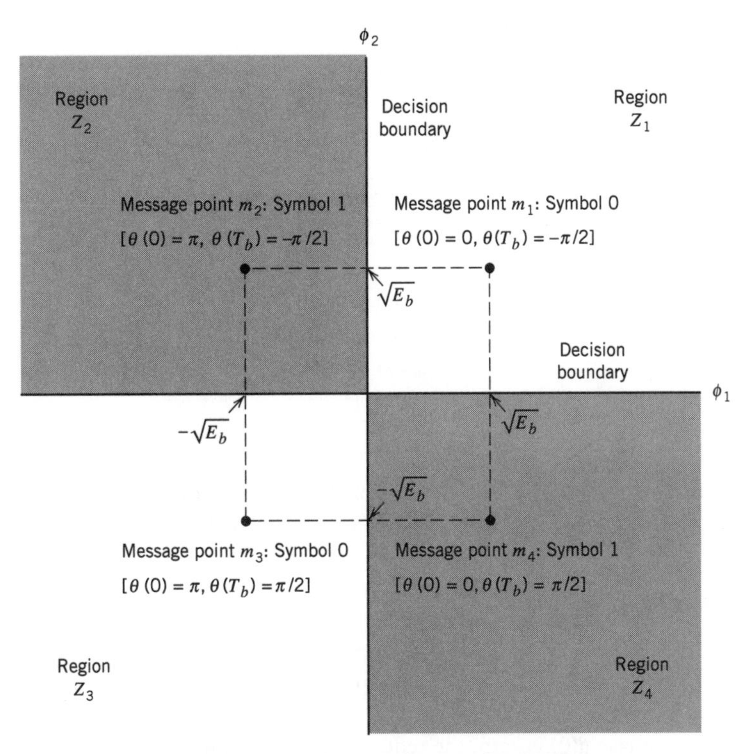


**FIGURE 6.29** Signal-space diagram for MSK system.

Note that in Equations (6.123) and (6.124):

- Both integrals are evaluated for a time interval equal to twice the bit duration.
- Both the lower and upper limits of the product integration used to evaluate the coefficient $s_1$ are shifted by the bit duration $T_b$ with respect to those used to evaluate the coefficient $s_2$.
- The time interval $0 \leq t \leq T_b$, for which the phase states $\theta(0)$ and $\theta(T_b)$ are defined, is common to both integrals.

Accordingly, the signal constellation for an MSK signal is two-dimensional (i.e., $N = 2$), with four possible message points (i.e., $M = 4$), as illustrated in Figure 6.29. The coordinates of the message points are as follows in a counterclockwise direction: $(+\sqrt{E_b}, +\sqrt{E_b})$, $(-\sqrt{E_b}, +\sqrt{E_b})$, $(-\sqrt{E_b}, -\sqrt{E_b})$, and $(+\sqrt{E_b}, -\sqrt{E_b})$. The possible values of $\theta(0)$ and $\theta(T_b)$, corresponding to these four message points, are also included in Figure 6.29. The signal-space diagram of MSK is thus similar to that of QPSK in that both of them have four message points. However, they differ in a subtle way that should be carefully noted: In QPSK the transmitted symbol is represented by any one of the four message points, whereas in MSK one of two message points is used to represent the transmitted symbol at any one time, depending on the value of $\theta(0)$.

Table 6.5 presents a summary of the values of $\theta(0)$ and $\theta(T_b)$, as well as the corresponding values of $s_1$ and $s_2$ that are calculated for the time intervals $-T_b \leq t \leq T_b$ and $0 \leq t \leq 2T_b$, respectively. The first column of this table indicates whether symbol 1 or symbol 0 was sent in the interval $0 \leq t \leq T_b$. Note that the coordinates of the message points, $s_1$ and $s_2$, have opposite signs when symbol 1 is sent in this interval, but the same sign when symbol 0 is sent. Accordingly, for a given input data sequence, we may use the

**TABLE 6.5** ***Signal-space characterization of MSK***

| *Transmitted Binary Symbol,* $0 \le t \le T_b$ | *Phase States (radians)* | | *Coordinates of Message Points* | |
|---|---|---|---|---|
| | $\theta(0)$ | $\theta(T_b)$ | $s_1$ | $s_2$ |
| 0 | 0 | $-\pi/2$ | $+\sqrt{E_b}$ | $+\sqrt{E_b}$ |
| 1 | $\pi$ | $-\pi/2$ | $-\sqrt{E_b}$ | $+\sqrt{E_b}$ |
| 0 | $\pi$ | $+\pi/2$ | $-\sqrt{E_b}$ | $-\sqrt{E_b}$ |
| 1 | 0 | $+\pi/2$ | $+\sqrt{E_b}$ | $-\sqrt{E_b}$ |

entries of Table 6.5 to derive, on a bit-by-bit basis, the two sequences of coefficients required to scale $\phi_1(t)$ and $\phi_2(t)$, and thereby determine the MSK signal $s(t)$.

### ▶ EXAMPLE 6.5

Figure 6.30 shows the sequences and waveforms involved in the generation of an MSK signal for the binary sequence 1101000. The input binary sequence is shown in Figure 6.30*a*. The two modulation frequences are: $f_1 = 5/4T_b$ and $f_2 = 3/4T_b$. Assuming that, at time $t = 0$ the phase $\theta(0)$ is zero, the sequence of phase states is as shown in Figure 6.30, modulo $2\pi$. The polarities of the two sequences of factors used to scale the time functions $\phi_1(t)$ and $\phi_2(t)$ are shown in the top lines of Figures 6.30*b* and 6.30*c*. Note that these two sequences are *offset* relative to each other by an interval equal to the bit duration $T_b$. The waveforms of the resulting two components of $s(t)$, namely, $s_1\phi_1(t)$ and $s_2\phi_2(t)$, are also shown in Figures 6.30*b* and 6.30*c*. Adding these two modulated waveforms, we get the desired MSK signal $s(t)$ shown in Figure 6.30*d*. ◀

### *Error Probability of MSK*

In the case of an AWGN channel, the received signal is given by

$$x(t) = s(t) + w(t)$$

where $s(t)$ is the transmitted MSK signal, and $w(t)$ is the sample function of a white Gaussian noise process of zero mean and power spectral density $N_0/2$. To decide whether symbol 1 or symbol 0 was transmitted in the interval $0 \le t \le T_b$, say, we have to establish a procedure for the use of $x(t)$ to detect the phase states $\theta(0)$ and $\theta(T_b)$. For the optimum detection of $\theta(0)$, we first determine the projection of the received signal $x(t)$ onto the reference signal $\phi_1(t)$ over the interval $-T_b \le t \le T_b$, obtaining

$$\begin{aligned} x_1 &= \int_{-T_b}^{T_b} x(t)\phi_1(t)\,dt \\ &= s_1 + w_1, \qquad -T_b \le t \le T_b \end{aligned} \tag{6.125}$$

where $s_1$ is as defined by Equation (6.123) and $w_1$ is the sample value of a Gaussian random variable of zero mean and variance $N_0/2$. From the signal-space diagram of Figure 6.29, we observe that if $x_1 > 0$, the receiver chooses the estimate $\hat{\theta}(0) = 0$. On the other hand, if $x_1 < 0$, it chooses the estimate $\hat{\theta}(0) = \pi$.

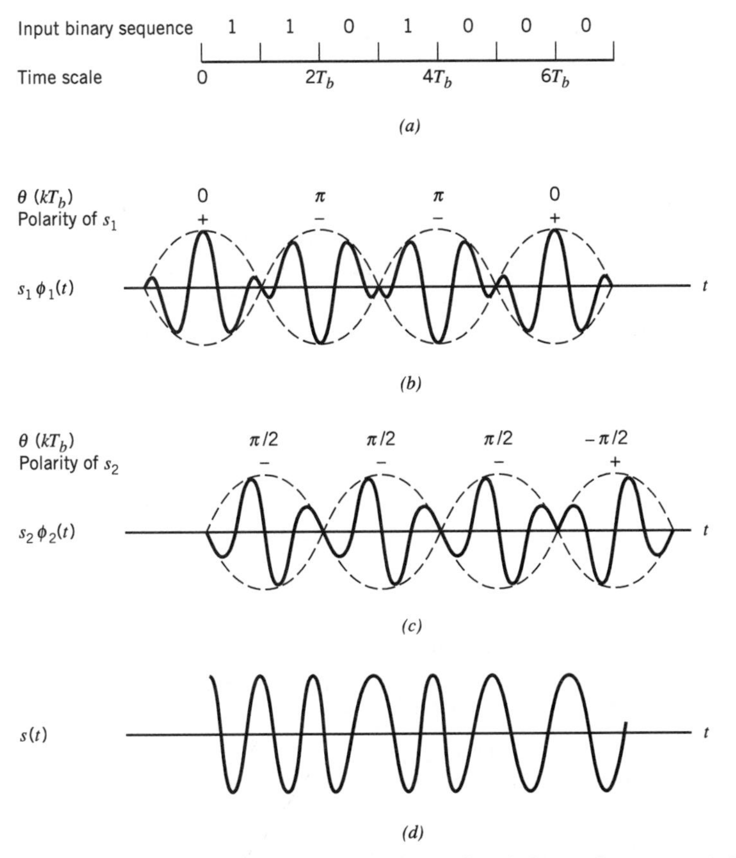


**FIGURE 6.30** (*a*) Input binary sequence. (*b*) Waveform of scaled time function $s_1\phi_1(t)$. (*c*) Waveform of scaled time function $s_2\phi_2(t)$. (*d*) Waveform of the MSK signal $s(t)$ obtained by adding $s_1\phi_1(t)$ and $s_2\phi_2(t)$ on a bit-by-bit basis.

Similarly, for the optimum detection of $\theta(T_b)$, we determine the projection of the received signal $x(t)$ onto the second reference signal $\phi_2(t)$ over the interval $0 \le t \le 2T_b$, obtaining

$$\begin{aligned} x_2 &= \int_0^{2T_b} x(t)\phi_2(t)\, dt \\ &= s_2 + w_2, \qquad 0 \le t \le 2T_b \end{aligned} \tag{6.126}$$

where $s_2$ is as defined by Equation (6.124) and $w_2$ is the sample value of another independent Gaussian random variable of zero mean and variance $N_0/2$. Referring again to the signal space diagram of Figure 6.29, we observe that if $x_2 > 0$, the receiver chooses the estimate $\hat{\theta}(T_b) = -\pi/2$. If, on the other hand, $x_2 < 0$, it chooses the estimate $\hat{\theta}(T_b) = \pi/2$.

To reconstruct the original binary sequence, we interleave the above two sets of phase decisions, as described next (see Table 6.5):

- If we have the estimates $\hat{\theta}(0) = 0$ and $\hat{\theta}(T_b) = -\pi/2$, or alternatively if we have the estimates $\hat{\theta}(0) = \pi$ and $\hat{\theta}(T_b) = \pi/2$, the receiver makes a decision in favor of symbol 0.

- If we have the estimates $\hat{\theta}(0) = \pi$ and $\hat{\theta}(T_b) = -\pi/2$, or alternatively if we have the estimates $\hat{\theta}(0) = 0$ and $\hat{\theta}(T_b) = \pi/2$, the receiver makes a decision in favor of symbol 1.

Thus referring to the signal-space diagram of Figure 6.29, we see that the decision made by the receiver is between the message points $m_1$ and $m_3$ for symbol 1, or between the message points $m_2$ and $m_4$ for symbol 0. The corresponding decisions whether $\theta(0)$ is 0 or $\pi$ and whether $\theta(T_b)$ is $-\pi/2$ or $+\pi/2$ (i.e., the bit decisions) are made *alternately* in the $I$- and $Q$-channels of the receiver, with each channel looking at the input signal for $2T_b$ seconds. The signal from *other bits* does not interfere with the receiver's decision for a given bit in either channel. The receiver makes an error when the $I$-channel assigns the wrong value to $\theta(0)$ or the $Q$-channel assigns the wrong value to $\theta(T_b)$. Accordingly, using the statistical characterizations of the product-integrator outputs $x_1$ and $x_2$ of these two channels, defined by Equations (6.125) and (6.126), respectively, we readily find that the bit error rate for coherent MSK is given by

$$P_e = \frac{1}{2}\,\text{erfc}\left(\sqrt{\frac{E_b}{N_0}}\right) \tag{6.127}$$

which is exactly the same as that for binary PSK and QPSK. It is important to note, however, that this good performance is the result of the detection of the MSK signal being performed in the receiver on the basis of observations over $2T_b$ seconds.

### *Generation and Detection of MSK Signals*

Consider next the generation and demodulation of MSK. Figure 6.31*a* shows the block diagram of a typical MSK transmitter. The advantage of this method of generating MSK signals is that the signal coherence and deviation ratio are largely unaffected by variations in the input data rate. Two input sinusoidal waves, one of frequency $f_c = n_c/4T_b$ for some fixed integer $n_c$, and the other of frequency $1/4T_b$, are first applied to a product modulator. This produces two phase-coherent sinusoidal waves at frequencies $f_1$ and $f_2$, which are related to the carrier frequency $f_c$ and the bit rate $1/T_b$ by Equations (6.111) and (6.112) for $h = 1/2$. These two sinusoidal waves are separated from each other by two narrowband filters, one centered at $f_1$ and the other at $f_2$. The resulting filter outputs are next linearly combined to produce the pair of quadrature carriers or orthonormal basis functions $\phi_1(t)$ and $\phi_2(t)$. Finally, $\phi_1(t)$ and $\phi_2(t)$ are multiplied with two binary waves $a_1(t)$ and $a_2(t)$, both of which have a bit rate equal to $1/2T_b$. These two binary waves are extracted from the incoming binary sequence in the manner described in Example 6.5.

Figure 6.31*b* shows the block diagram of a typical MSK receiver. The received signal $x(t)$ is correlated with locally generated replicas of the coherent reference signals $\phi_1(t)$ and $\phi_2(t)$. Note that in both cases the integration interval is $2T_b$ seconds, and that the integration in the quadrature channel is delayed by $T_b$ seconds with respect to that in the in-phase channel. The resulting in-phase and quadrature channel correlator outputs, $x_1$ and $x_2$, are each compared with a threshold of zero, and estimates of the phase $\theta(0)$ and $\theta(T_b)$ are derived in the manner described previously. Finally, these phase decisions are interleaved so as to reconstruct the original input binary sequence with the minimum average probability of symbol error in an AWGN channel.

### *Power Spectra of MSK Signals*

As with the binary FSK signal, we assume that the input binary wave is random, with symbols 1 and 0 equally likely, and the symbols transmitted during different time slots being statistically independent. In this case, we make the following observations:

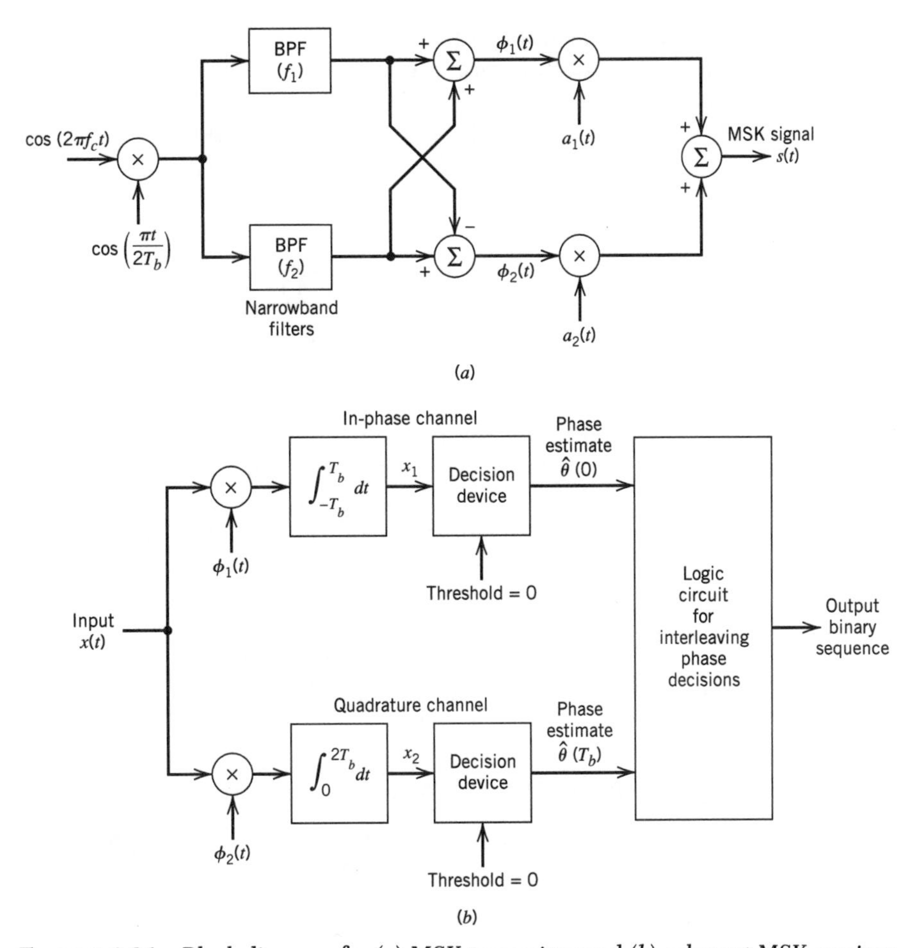


**FIGURE 6.31** Block diagrams for (*a*) MSK transmitter and (*b*) coherent MSK receiver.

1. Depending on the value of phase state $\theta(0)$, the in-phase component equals $+g(t)$ or $-g(t)$, where

$$g(t) = \begin{cases} \sqrt{\dfrac{2E_b}{T_b}} \cos\left(\dfrac{\pi t}{2T_b}\right), & -T_b \le t \le T_b \\ 0, & \text{otherwise} \end{cases} \tag{6.128}$$

The energy spectral density of this symbol-shaping function is

$$\psi_g(f) = \frac{32E_bT_b}{\pi^2}\left[\frac{\cos(2\pi T_b f)}{16T_b^2 f^2 - 1}\right]^2 \tag{6.129}$$

Hence, the power spectral density of the in-phase component equals $\psi_g(f)/2T_b$.

2. Depending on the value of the phase state $\theta(T_b)$, the quadrature component equals $+g(t)$ or $-g(t)$, where we now have

$$g(t) = \begin{cases} \sqrt{\dfrac{2E_b}{T_b}} \sin\left(\dfrac{\pi t}{2T_b}\right), & 0 \le t \le 2T_b \\ 0, & \text{otherwise} \end{cases} \tag{6.130}$$

The energy spectral density of this second symbol-shaping function is also given by Equation (6.129). Hence, the in-phase and quadrature components have the same power spectral density.

3. The in-phase and quadrature component of the MSK signal are also statistically independent. Hence, the baseband power spectral density of the MSK signal is given by

$$\begin{aligned} S_B(f) &= 2\left[\frac{\psi_g(f)}{2T_b}\right] \\ &= \frac{32E_b}{\pi^2}\left[\frac{\cos(2\pi T_b f)}{16T_b^2 f^2 - 1}\right]^2 \end{aligned} \tag{6.131}$$

The baseband power spectrum of Equation (6.131) is plotted in Figure 6.9, where the power spectrum is normalized with respect to $4E_b$ and the frequency $f$ is normalized with respect to the bit rate $1/T_b$. This figure also includes the corresponding plot of Equation (6.40) for the QPSK signal, which was considered earlier. For $f \gg 1/T_b$, the baseband power spectral density of the MSK signal falls off as the inverse fourth power of frequency, whereas in the case of the QPSK signal it falls off as the inverse square of frequency. Accordingly, MSK does not produce as much interference outside the signal band of interest as QPSK. This is a desirable characteristic of MSK, especially when the digital communication system operates with a bandwidth limitation.

## ■ GAUSSIAN-FILTERED MSK

From the detailed study of minimum shift keying (MSK) just presented, we may summarize the desirable properties of the MSK signal as follows:

- Constant envelope
- Relatively narrow bandwidth
- Coherent detection performance equivalent to that of QPSK

However, the out-of-band spectral characteristics of MSK signals, as good as they are, still do not satisfy the stringent requirements of certain applications such as wireless communications. To illustrate this limitation, we find from Equation (6.131) that at $fT_b = 0.5$, the baseband power spectral density of the MSK signal drops by only $10 \log_{10} 9 = 9.54$ dB below its midband value. Hence, when the MSK signal is assigned a transmission bandwidth of $1/T_b$, the adjacent channel interference of a wireless communication system using MSK is not low enough to satisfy the practical requirements of such a multiuser communications environment.

Recognizing that the MSK signal can be generated by direct frequency modulation of a voltage-controlled oscillator, we may overcome this serious limitation of MSK by modifying its power spectrum into a *compact* form, while maintaining the constant-envelope property of the MSK signal. This modification can be achieved through the use of a premodulation low-pass filter, hereafter referred to as a baseband *pulse-shaping filter*. Desirably, the pulse-shaping filter should satisfy the following properties:

1. Frequency response with narrow bandwidth and sharp cutoff characteristics.
2. Impulse response with relatively low overshoot.

3. Evolution of a phase trellis where the carrier phase of the modulated signal assumes the two values $\pm\pi/2$ at odd multiples of $T_b$ and the two values 0 and $\pi$ at even multiples of $T_b$ as in MSK.

Condition 1 is needed to suppress the high-frequency components of the transmitted signal. Condition 2 avoids excessive deviations in the instantaneous frequency of the FM signal. Finally, condition 3 ensures that the modified FM signal can be coherently detected in the same way as the MSK signal, or it can be noncoherently detected as a simple binary FSK signal.

These desirable properties can be achieved by passing a nonreturn-to-zero (NRZ) binary data stream through a baseband pulse-shaping filter whose impulse response (and likewise its frequency response) is defined by a *Gaussian* function. The resulting method of binary frequency modulation is naturally referred to as *Gaussian-filtered MSK* or just *GMSK*.[8]

Let $W$ denote the *3 dB baseband bandwidth* of the pulse-shaping filter. We may then define the transfer function $H(f)$ and impulse response $h(t)$ of the pulse-shaping filter as follows, respectively:

$$H(f) = \exp\left(-\frac{\log 2}{2}\left(\frac{f}{W}\right)^2\right) \tag{6.132}$$

and

$$h(t) = \sqrt{\frac{2\pi}{\log 2}}\, W \exp\left(-\frac{2\pi^2}{\log 2} W^2 t^2\right) \tag{6.133}$$

The response of this Gaussian filter to a rectangular pulse of unit amplitude and duration $T_b$ (centered on the origin) is given by (see Problem 6.28)

$$\begin{aligned} g(t) &= \int_{-T_b/2}^{T_b/2} h(t-\tau)\, d\tau \\ &= \sqrt{\frac{2\pi}{\log 2}}\, W \int_{-T_b/2}^{T_b/2} \exp\left(-\frac{2\pi^2}{\log 2} W^2 (t-\tau)^2\right) d\tau \end{aligned} \tag{6.134}$$

which may be expressed as the difference between two complementary error functions, as shown by

$$g(t) = \frac{1}{2}\left[\operatorname{erfc}\left(\pi\sqrt{\frac{2}{\log 2}}\, WT_b\left(\frac{t}{T_b} - \frac{1}{2}\right)\right) - \operatorname{erfc}\left(\pi\sqrt{\frac{2}{\log 2}}\, WT_b\left(\frac{t}{T_b} + \frac{1}{2}\right)\right)\right] \tag{6.135}$$

The pulse response $g(t)$ constitutes the *frequency shaping pulse* of the GMSK modulator, with the dimensionless *time-bandwidth product* $WT_b$ playing the role of a design parameter.

The frequency-shaping pulse $g(t)$, as defined in Equation (6.135), is noncausal in that it is nonzero for $t < -T_b/2$, where $t = -T_b/2$ is the time at which the input rectangular pulse (symmetrically positioned around the origin) is applied to the Gaussian filter. For a causal response, $g(t)$ must be truncated and shifted in time. Figure 6.32 presents plots of $g(t)$, which has been truncated at $t = \pm 2.5T_b$ and then shifted in time by $2.5T_b$. The plots shown here are for $WT_b = 0.2$, 0.25, and 0.3. Note that as $WT_b$ is reduced, the time spread of the frequency-shaping pulse is correspondingly increased.

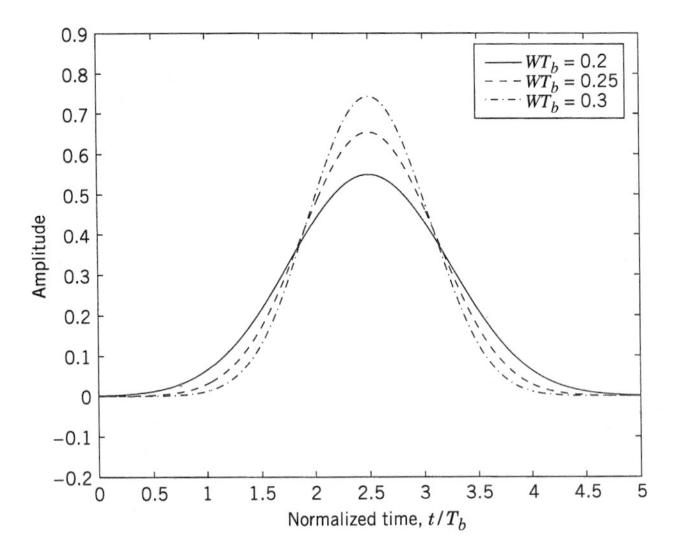


**FIGURE 6.32** Frequency-shaping pulse $g(t)$ of Equation (6.135) shifted in time by $2.5T_b$ and truncated at $\pm 2.5T_b$ for varying time-bandwidth product $WT_b$.

Figure 6.33 shows the machine-computed power spectra of MSK signals (expressed in decibels) versus the normalized frequency difference $(f - f_c)T_b$, where $f_c$ is the mid-band frequency and $T_b$ is the bit duration.[9] The results plotted in Figure 6.33 are for varying values of the time-bandwidth product $WT_b$. From this figure we may make the following observations:

- The curve for the limiting condition $WT_b = \infty$ corresponds to the case of ordinary MSK.
- When $WT_b$ is less than unity, increasingly more of the transmit power is concentrated inside the passband of the GMSK signal.

An undesirable feature of GMSK is that the processing of NRZ binary data by a Gaussian filter generates a modulating signal that is no longer confined to a single bit interval as in ordinary MSK, which is readily apparent from Figure 6.32. Stated in another way, the tails of the Gaussian impulse response of the pulse-shaping filter cause the modulating signal to spread out to adjacent symbol intervals. The net result is the generation of *intersymbol interference*, the extent of which increases with decreasing $WT_b$. In light of this observation and the observation we made on the basis of Figure 6.33 on the power spectra of GMSK signals, we may say that the choice of the time-bandwidth product $WT_b$ offers a trade-off between spectral compactness and performance loss.

To explore the issue of performance degradation, consider the probability of error $P_e$ of GMSK using coherent detection in the presence of additive white Gaussian noise. Recognizing that GMSK is a special kind of binary frequency modulation, we may express $P_e$ by the empirical formula

$$P_e = \frac{1}{2}\,\text{erfc}\left(\sqrt{\frac{\alpha E_b}{2N_0}}\right) \tag{6.136}$$

where, as before, $E_b$ is the signal energy per bit and $N_0/2$ is the noise spectral density. The factor $\alpha$ is a constant whose value depends on the time-bandwidth product $WT_b$. Comparing the formula of Equation (6.136) for GMSK with that of Equation (6.127) for

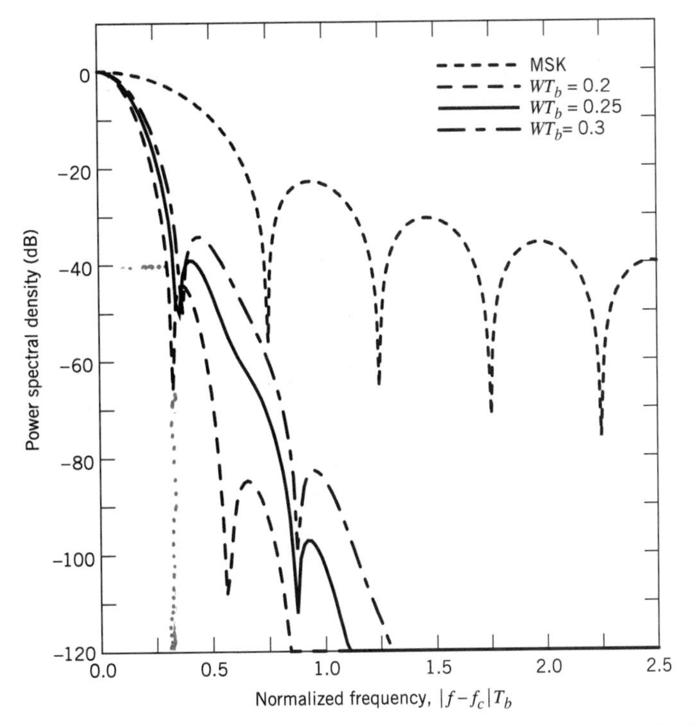


**FIGURE 6.33** Power spectra of MSK and GMSK signals for varying time-bandwidth product. (Reproduced with permission from Dr. Gordon Stüber, Georgia Tech.)

ordinary MSK, we may view $10\log_{10}(\alpha/2)$, expressed in decibels, as a measure of performance degradation of GMSK (with prescribed $WT_b$) compared to ordinary MSK. Figure 6.34 shows the machine-computed value of $10\log_{10}(\alpha/2)$ versus $WT_b$. For ordinary MSK we have $WT_b = \infty$, in which case Equation (6.136) with $\alpha = 2$ assumes exactly the same form as Equation (6.127) and there is no degradation in performance, which is confirmed by Figure 6.34. For GMSK with $WT_b = 0.3$ we find from Figure 6.34 that there is a

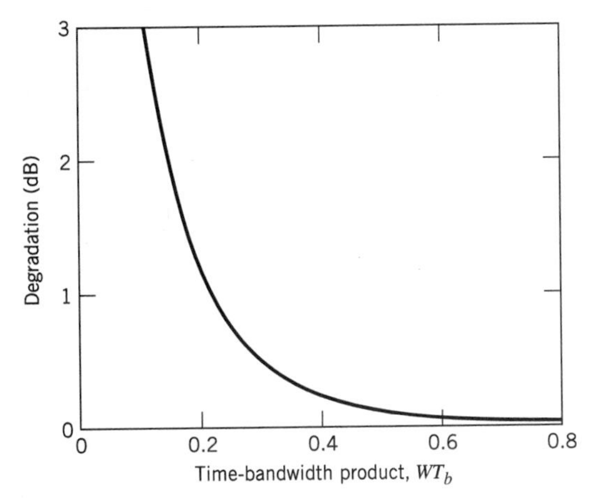


**FIGURE 6.34** Theoretical $E_b/N_0$ degradation of GMSK for varying time-bandwidth product. (Taken from Murata and Hirade, 1981, with permission of the IEEE.)

degradation in performance of about 0.46 dB, which corresponds to $(\alpha/2) = 0.9$. This degradation in performance is a small price to pay for the highly desirable spectral compactness of the GMSK signal.

### ▶ EXAMPLE 6.6 GMSK for GSM Wireless Communications

An important application of GMSK is in a standardized wireless communication system widely known as GSM, which is a time-division multiple-access system that is discussed in Chapter 8. For this application, the time-bandwidth product $WT_b$ of GMSK is standardized at 0.3, which provides the best compromise between increased bandwidth occupancy and resistance to co-channel interference. Ninety-nine percent of the radio frequency (RF) power of GMSK signals so specified is confined to a bandwidth of 250 kHz, which means that, for all practical purposes, the sidelobes are virtually zero outside this frequency band.

The available spectrum is divided into 200 kHz-wide subchannels. Each subchannel is assigned to a GSM system transmitting data at 271 kb/s. Figure 6.35 depicts the power spectrum of a subchannel in relation to its two adjacent subchannels; this plot is the passband version of the baseband power spectrum of Figure 6.33 corresponding to $WT_b = 0.3$. From Figure 6.35 we may make the following important observation: The RF power spectrum of the subchannel shown shaded is down by an amount larger than 40 dB at the carrier frequencies of both adjacent subchannels, which means that the effect of co-channel interference is practically negligible. ◀

## ■ M-ARY FSK

Consider next the $M$-ary version of FSK, for which the transmitted signals are defined by

$$s_i(t) = \sqrt{\frac{2E}{T}} \cos\left[\frac{\pi}{T}(n_c + i)t\right], \qquad 0 \le t \le T \tag{6.137}$$

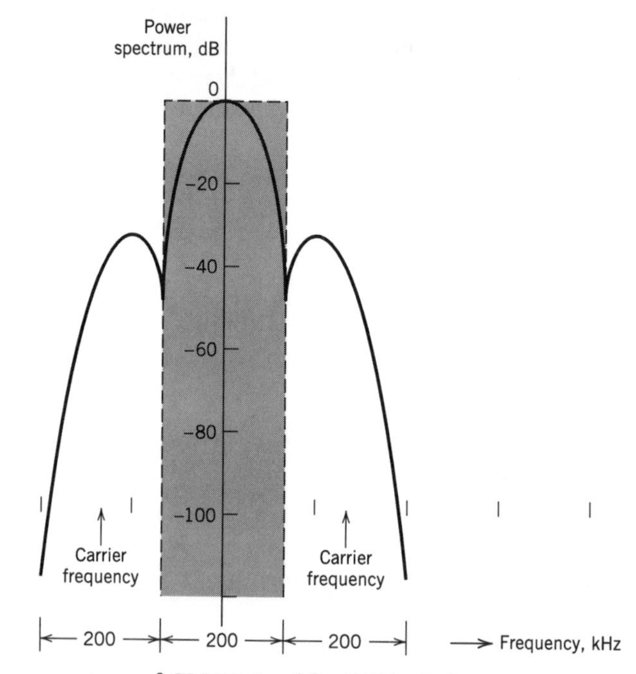


FIGURE 6.35 Power spectrum of GMSK signal for GSM wireless communications.

where $i = 1, 2, \ldots, M$, and the carrier frequency $f_c = n_c/2T$ for some fixed integer $n_c$. The transmitted symbols are of equal duration $T$ and have equal energy $E$. Since the individual signal frequencies are separated by $1/2T$ Hz, the signals in Equation (6.137) are orthogonal; that is

$$\int_0^T s_i(t)s_j(t)\,dt = 0, \qquad i \neq j \tag{6.138}$$

This property of $M$-ary FSK suggests that we may use the transmitted signals $s_i(t)$ themselves, except for energy normalization, as a complete orthonormal set of basis functions, as shown by

$$\phi_i(t) = \frac{1}{\sqrt{E}}\, s_i(t), \qquad \begin{array}{l} 0 \le t \le T \\ i = 1, 2, \ldots, M \end{array} \tag{6.139}$$

Accordingly, the $M$-ary FSK is described by an $M$-dimensional signal-space diagram.

For coherent $M$-ary FSK, the optimum receiver consists of a bank of $M$ correlators or matched filters, with the $\phi_i(t)$ of Equation (6.139) providing the pertinent reference signals. At the sampling times $t = kT$, the receiver makes decisions based on the largest matched filter output in accordance with the maximum likelihood decoding rule. An exact formula for the probability of symbol error is however difficult to derive for a coherent $M$-ary FSK system. Nevertheless, we may use the union bound of Equation (5.96) of Chapter 5 to place an upper bound on the average probability of symbol error for $M$-ary FSK. Specifically, noting that the minimum distance $d_{\min}$ in $M$-ary FSK is $\sqrt{2E}$, the use of Equation (5.96) yields (assuming equiprobable symbols)

$$P_e \le \frac{1}{2}(M-1)\,\text{erfc}\left(\sqrt{\frac{E}{2N_0}}\right) \tag{6.140}$$

For fixed $M$, this bound becomes increasingly tight as $E/N_0$ is increased. Indeed, it becomes a good approximation to $P_e$ for values of $P_e \le 10^{-3}$. Moreover, for $M = 2$ (i.e., binary FSK), the bound of Equation (6.140) becomes an equality.

### *Power Spectra of* M*-ary FSK Signals*

The spectral analysis of $M$-ary FSK signals[10] is much more complicated than that of $M$-ary PSK signals. A case of particular interest occurs when the frequencies assigned to the multilevels make the frequency spacing uniform and the frequency deviation $k = 0.5$. That is, the $M$ signal frequencies are separated by $1/2T$, where $T$ is the symbol duration. For $k = 0.5$, the baseband power spectral density of $M$-ary FSK signals is plotted in Figure 6.36 for $M = 2, 4, 8$.

### *Bandwidth Efficiency of* M*-ary FSK Signals*

When the orthogonal signals of an $M$-ary FSK signal are detected coherently, the adjacent signals need only be separated from each other by a frequency difference $1/2T$ so as to maintain orthogonality. Hence, we may define the channel bandwidth required to transmit $M$-ary FSK signals as

$$B = \frac{M}{2T} \tag{6.141}$$

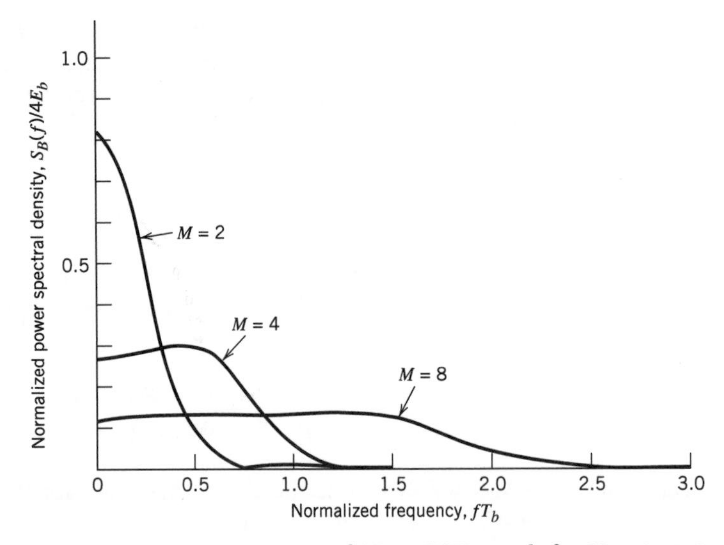


**FIGURE 6.36** Power spectra of $M$-ary PSK signals for $M = 2, 4, 8$.

For multilevels with frequency assignments that make the frequency spacing uniform and equal to $1/2T$, the bandwidth $B$ of Equation (6.141) contains a large fraction of the signal power. This is readily confirmed by looking at the baseband power spectral plots shown in Figure 6.36. From Equation (6.48) we recall that the symbol period $T$ is equal to $T_b \log_2 M$. Hence, using $R_b = 1/T_b$, we may redefine the channel bandwidth $B$ for $M$-ary FSK signals as

$$B = \frac{R_b M}{2 \log_2 M} \tag{6.142}$$

The bandwidth efficiency of $M$-ary signals is therefore

$$\begin{aligned} \rho &= \frac{R_b}{B} \\ &= \frac{2 \log_2 M}{M} \end{aligned} \tag{6.143}$$

Table 6.6 gives the values of $\rho$ calculated from Equation (6.143) for varying $M$.

Comparing Tables 6.4 and 6.6, we see that increasing the number of levels $M$ tends to increase the bandwidth efficiency of $M$-ary PSK signals, but it also tends to decrease the bandwidth efficiency of $M$-ary FSK signals. In other words, $M$-ary PSK signals are spectrally efficient, whereas $M$-ary FSK signals are spectrally inefficient.

**TABLE 6.6** ***Bandwidth efficiency of M-ary FSK signals***

| $M$ | 2 | 4 | 8 | 16 | 32 | 64 |
|---|---|---|---|---|---|---|
| $\rho$ (bits/s/Hz) | 1 | 1 | 0.75 | 0.5 | 0.3125 | 0.1875 |

## 6.6 Detection of Signals with Unknown Phase

Up to this point in our discussion, we have assumed that the receiver is perfectly synchronized to the transmitter, and the only channel impairment is noise. In practice, however, it is often found that in addition to the uncertainty due to channel noise, there is also uncertainty due to the randomness of certain signal parameters. The usual cause of this uncertainty is distortion in the transmission medium. Perhaps the most common random signal parameter is the carrier phase, which is especially true for narrowband signals. For example, transmission over a multiplicity of paths of different and variable lengths, or rapidly varying delays in the propagating medium from transmitter to receiver, may cause the phase of the received signal to change in a way that the receiver cannot follow. Synchronization with the phase of the transmitted carrier may then be too costly, and the designer may simply choose to disregard the phase information in the received signal at the expense of some degradation in noise performance. A digital communication receiver with no provision made for carrier phase recovery is said to be *noncoherent*.

### OPTIMUM QUADRATIC RECEIVER

Consider a binary digital communication system in which the transmitted signal is

$$s_i(t) = \sqrt{\frac{2E}{T}} \cos(2\pi f_i t), \qquad \begin{array}{l} 0 \le t \le T \\ i = 1,2 \end{array} \tag{6.144}$$

where $E$ is the signal energy, $T$ is the duration of the signaling interval, and the carrier frequency $f_i$ for symbol $i$ is an integral multiple of $1/2T$. The system is assumed to be noncoherent, in which case the received signal for an AWGN channel may be written in the form

$$x(t) = \sqrt{\frac{2E}{T}} \cos(2\pi f_i t + \theta) + w(t), \qquad \begin{array}{l} 0 \le t \le T \\ i = 1, 2, \end{array} \tag{6.145}$$

where $\theta$ is the unknown carrier phase, and $w(t)$ is the sample function of a white Gaussian noise process of zero mean and power spectral density $N_0/2$. In a real-life situation it is realistic to assume complete lack of prior information about $\theta$ and to treat it as a random variable with *uniform distribution*:

$$f_\Theta(\theta) = \begin{cases} \dfrac{1}{2\pi}, & -\pi < \theta \le \pi \\ 0, & \text{otherwise} \end{cases} \tag{6.146}$$

The binary detection problem to be solved may now be stated as follows:

> Given the received signal $x(t)$ and confronted with the unknown carrier phase $\theta$, design an optimum receiver for detecting symbol $s_i$ represented by the signal component $\sqrt{E/2T}\cos(2\pi f_i t + \theta)$ that is contained in $x(t)$.

Proceeding in a manner similar to that described in Sections 5.3–5.6, we may formulate the *conditional likelihood function* of symbol $s_i$, given the carrier phase $\theta$, as

$$L(s_i(\theta)) = \exp\left(\sqrt{\frac{E}{N_0 T}} \int_0^T x(t) \cos(2\pi f_i t + \theta)\, dt\right) \tag{6.147}$$

To proceed further, we have to remove dependence of $L(s_i(\theta))$ on phase $\theta$, which is achieved by integrating it over all possible values of $\theta$. We may thus write

$$\begin{aligned} L(s_i) &= \int_{-\pi}^{\pi} L(s_i(\theta)) f_\Theta(\theta)\, d\theta \\ &= \frac{1}{2\pi} \int_{-\pi}^{\pi} \exp\left( \sqrt{\frac{E}{N_0 T}} \int_0^T x(t) \cos(2\pi f_i t + \theta)\, dt \right) d\theta \end{aligned} \tag{6.148}$$

Note that the dependence on $\theta$ was removed by integrating the likelihood function and *not* the log-likelihood function.

Using a well-known trigonometric formula, we may expand $\cos(2\pi f_i t + \theta)$ as

$$\cos(2\pi f_i t + \theta) = \cos(2\pi f_i t) \cos\theta - \sin(2\pi f_i t) \sin\theta$$

Correspondingly, we may rewrite the integral in the exponent of Equation (6.148) as

$$\int_0^T x(t) \cos(2\pi f_i t + \theta)\, dt = \cos\theta \int_0^T x(t) \cos(2\pi f_i t)\, dt - \sin\theta \int_0^T x(t) \sin(2\pi f_i t)\, dt \tag{6.149}$$

Define

$$l_i = \left[ \left( \int_0^T x(t) \cos(2\pi f_i t)\, dt \right)^2 + \left( \int_0^T x(t) \sin(2\pi f_i t)\, dt \right)^2 \right]^{1/2} \tag{6.150}$$

$$\beta_i = \tan^{-1} \left( \frac{\int_0^T x(t) \sin(2\pi f_i t)\, dt}{\int_0^T x(t) \cos(2\pi f_i t)\, dt} \right) \tag{6.151}$$

Hence, we may go one step further and simplify Equation (6.149) to

$$\begin{aligned} \int_0^T x(t) \cos(2\pi f_i t + \theta)\, dt &= l_i(\cos\theta \cos\beta_i - \sin\theta \sin\beta_i) \\ &= l_i \cos(\theta + \beta_i) \end{aligned} \tag{6.152}$$

Accordingly, using Equation (6.152) in Equation (6.148), we obtain

$$\begin{aligned} L(s_i) &= \frac{1}{2\pi} \int_{-\pi}^{\pi} \exp\left( \sqrt{\frac{E}{N_0 T}}\, l_i \cos(\theta + \beta_i) \right) d\theta \\ &= \frac{1}{2\pi} \int_{-\pi+\beta_i}^{\pi+\beta_i} \exp\left( \sqrt{\frac{E}{N_0 T}}\, l_i \cos\theta \right) d\theta \\ &= \frac{1}{2\pi} \int_{-\pi}^{\pi} \exp\left( \sqrt{\frac{E}{N_0 T}}\, l_i \cos\theta \right) d\theta \end{aligned} \tag{6.153}$$

From Appendix 3 on Bessel functions, we immediately recognize the integral of Equation (6.153) as the *modified Bessel function of zero order*:

$$I_0\left( \sqrt{\frac{E}{N_0 T}}\, l_i \right) = \frac{1}{2\pi} \int_{-\pi}^{\pi} \exp\left( \sqrt{\frac{E}{N_0 T}}\, l_i \cos\theta \right) d\theta \tag{6.154}$$

Hence, we may express the likelihood function for the signal detection problem described herein in the compact form

$$L(s_i) = I_0\left(\sqrt{\frac{E}{N_0T}}\, l_i\right) \tag{6.155}$$

The *binary hypothesis test* (i.e., the hypothesis that signal $s_1(t)$ or signal $s_2(t)$ was transmitted) can now be written as

$$I_0\left(\sqrt{\frac{E}{N_0T}}\, l_1\right) \underset{H_2}{\overset{H_1}{\gtrless}} I_0\left(\sqrt{\frac{E}{N_0T}}\, l_2\right) \tag{6.156}$$

where hypothesis $H_1$ and $H_2$ correspond to signals $s_1(t)$ and $s_2(t)$, respectively. From Appendix 3 we note that the modified Bessel function $I(\cdot)$ is a monotonically increasing function of its argument. Hence the hypothesis test can be carried out in terms of either $I_0(\sqrt{E/N_0T}l_i)$ or simply $l_i$. For convenience of implementation, however, the hypothesis test is carried out in terms of $l_i^2$ instead of $l_i$, as shown by

$$l_1^2 \underset{H_2}{\overset{H_1}{\gtrless}} l_2^2 \tag{6.157}$$

A receiver based on Equation (6.157) is known as the *quadratic receiver*. In light of the definition of $l_i$ given in Equation (6.150), the receiver structure for computing $l_i$ is as shown in Figure 6.37*a*. Note that the test described in Equation (6.157) is independent of the symbol energy $E$. It is for this reason that this hypothesis test is said to be *uniformly most powerful* with respect to the symbol energy $E$.

## Two Equivalent Forms of the Quadratic Receiver

We next derive two equivalent forms of the quadrature receiver shown in Figure 6.37*a*. The first form is obtained easily by replacing each correlator in Figure 6.37*a* with a corresponding equivalent matched filter. We thus obtain the alternative form of quadrature receiver shown in Figure 6.37*b*. In one branch of this receiver, we have a filter matched to the signal $\cos(2\pi f_i t)$, and in the other branch we have a filter matched to $\sin(2\pi f_i t)$, both of which are defined for the time interval $0 \le t \le T$. The filter outputs are sampled at time $t = T$, squared, and then added together.

To obtain the second equivalent form of the quadrature receiver, suppose we have a filter that is matched to $s(t) = \cos(2\pi f_i t + \theta)$ for $0 \le t \le T$. The envelope of the matched filter output is obviously unaffected by the value of phase $\theta$. Therefore, for convenience, we may simply choose a matched filter with impulse response $\cos[2\pi f_i(T - t)]$, corresponding to $\theta = 0$. The output of such a filter in response to the received signal $x(t)$ is given by

$$\begin{aligned} y(t) &= \int_0^T x(\tau)\cos[2\pi f_i(T - t + \tau)]\, d\tau \\ &= \cos[2\pi f_i(T - t)]\int_0^T x(\tau)\cos(2\pi f_i\tau)\, d\tau - \sin[2\pi f_i(T - t)]\int_0^T x(\tau)\sin(2\pi f_i\tau)\, d\tau \end{aligned} \tag{6.158}$$

The envelope of the matched filter output is proportional to the square root of the sum of the squares of the integrals in Equation (6.158). The envelope, evaluated at time $t = T$, is therefore

$$l_i = \left\{\left[\int_0^T x(\tau)\cos(2\pi f_i\tau)\, d\tau\right]^2 + \left[\int_0^T x(\tau)\sin(2\pi f_i\tau)\, d\tau\right]^2\right\}^{1/2} \tag{6.159}$$

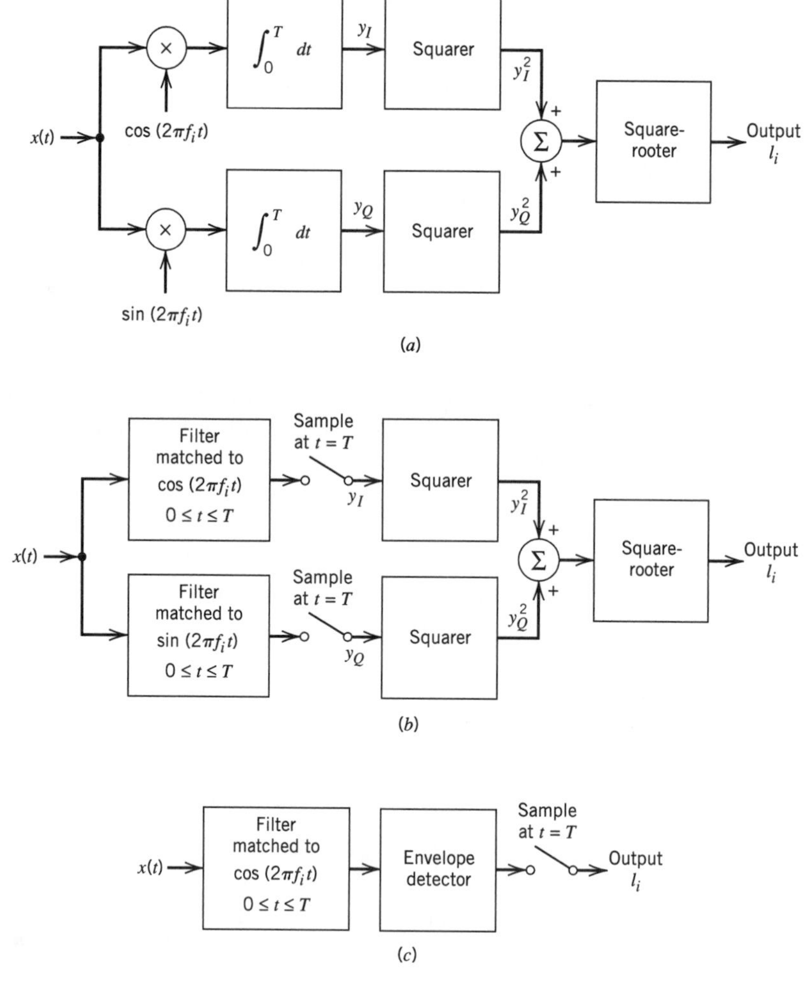


**FIGURE 6.37** Noncoherent receivers. (*a*) Quadrature receiver using correlators. (*b*) Quadrature receiver using matched filters. (*c*) Noncoherent matched filter.

But this is just the output of the quadrature receiver. Therefore, the output (at time $T$) of a filter matched to the signal $\cos(2\pi f_i t + \theta)$, of arbitrary phase $\theta$, followed by an envelope detector is the same as the corresponding output of the quadrature receiver of Figure 6.37*a*. This form of receiver is shown in Figure 6.37*c*. The combination of matched filter and envelope detector shown in Figure 6.37*c* is called a *noncoherent matched filter*.

The need for an envelope detector following the matched filter in Figure 6.37*c* may also be justified intuitively as follows. The output of a filter matched to a rectangular RF wave reaches a positive peak at the sampling instant $t = T$. If, however, the phase of the filter is not matched to that of the signal, the peak may occur at a time different from the sampling instant. In actual fact, if the phases differ by 180 degrees, we get a negative peak at the sampling instant. Figure 6.38 illustrates the matched filter output for the two limiting conditions: $\theta = 0$ and $\theta = 180$ degrees. To avoid poor sampling that arises in the absence of prior information about the phase $\theta$, it is reasonable to retain only the envelope of the matched filter output, since it is completely independent of the phase mismatch $\theta$.

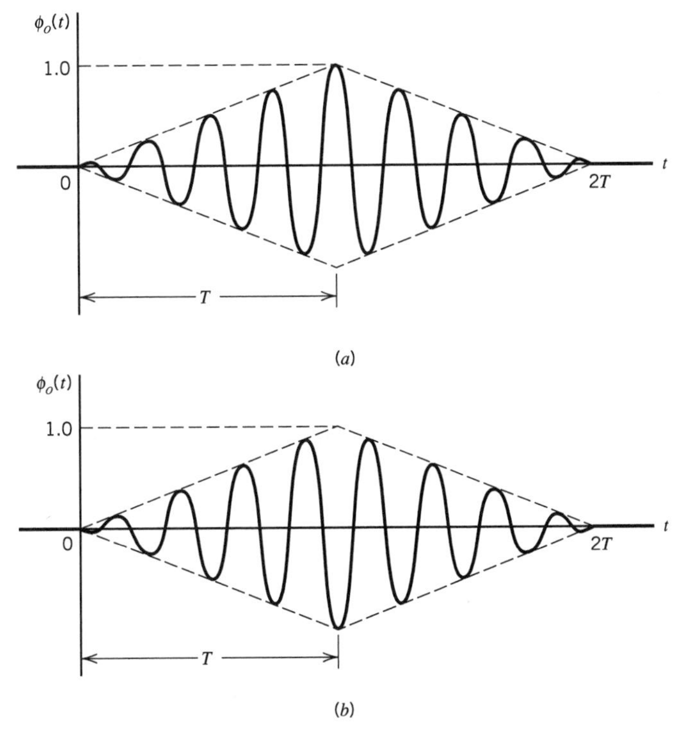


**FIGURE 6.38** Output of matched filter for a rectangular RF wave: (*a*) $\theta = 0$, and (*b*) $\theta = 180$ degrees.

## 6.7 Noncoherent Orthogonal Modulation

With the noncoherent receiver structures of Figure 6.37 at our disposal, we may now proceed to study the noise performance of *noncoherent orthogonal modulation* that includes two noncoherent receivers as special cases: noncoherent binary frequency-shift keying and differential phase-shift keying.

Consider a binary signaling scheme that involves the use of two orthogonal signals $s_1(t)$ and $s_2(t)$, which have equal energy. During the interval $0 \le t \le T$, one of these two signals is sent over an imperfect channel that shifts the carrier phase by an unknown amount. Let $g_1(t)$ and $g_2(t)$ denote the phase-shifted versions of $s_1(t)$ and $s_2(t)$, respectively. It is assumed that the signals $g_1(t)$ and $g_2(t)$ remain orthogonal and have the same energy $E$, regardless of the unknown carrier phase. We refer to such a signaling scheme as *noncoherent orthogonal modulation*. Depending on how we define the orthogonal pair of signals $s_1(t)$ and $s_2(t)$, noncoherent binary FSK and DPSK may be treated as special cases of this modulation scheme.

The channel also introduces an additive white Gaussian noise $w(t)$ of zero mean and power spectral density $N_0/2$. We may thus express the received signal $x(t)$ as

$$x(t) = \begin{cases} g_1(t) + w(t), & s_1(t) \text{ sent}, \ 0 \le t \le T \\ g_2(t) + w(t), & s_2(t) \text{ sent}, \ 0 \le t \le T \end{cases} \tag{6.160}$$

The requirement is to use $x(t)$ to discriminate between $s_1(t)$ and $s_2(t)$, regardless of the carrier phase.

For this purpose, we employ the receiver shown in Figure 6.39*a*. The receiver consists of a pair of filters matched to the transmitted signals $s_1(t)$ and $s_2(t)$. Because the carrier phase is unknown, the receiver relies on amplitude as the only possible discriminant. Accordingly, the matched filter outputs are envelope detected, sampled, and then compared with each other. If the upper path in Figure 6.39*a* has an output amplitude $l_1$ greater than the output amplitude $l_2$ of the lower path, the receiver makes a decision in favor of $s_1(t)$. If the converse is true, it decides in favor of $s_2(t)$. When they are equal, the decision may be made by flipping a fair coin. In any event, a decision error occurs when the matched filter that rejects the signal component of the received signal $x(t)$ has a larger output amplitude (due to noise alone) than the matched filter that passes it.

From the discussion presented in Section 6.6, we note that a noncoherent matched filter (constituting the upper or lower path in the receiver of Figure 6.39*a*) may be viewed as being equivalent to a *quadrature receiver*. The quadrature receiver itself has two channels. One version of the quadrature receiver is shown in Figure 6.39*b*. In the upper channel, called the *in-phase channel*, the received signal $x(t)$ is correlated with the function $\psi_i(t)$, which represents a scaled version of the transmitted signal $s_1(t)$ or $s_2(t)$ with zero carrier phase. In the lower channel, called the *quadrature channel*, on the other hand, $x(t)$ is

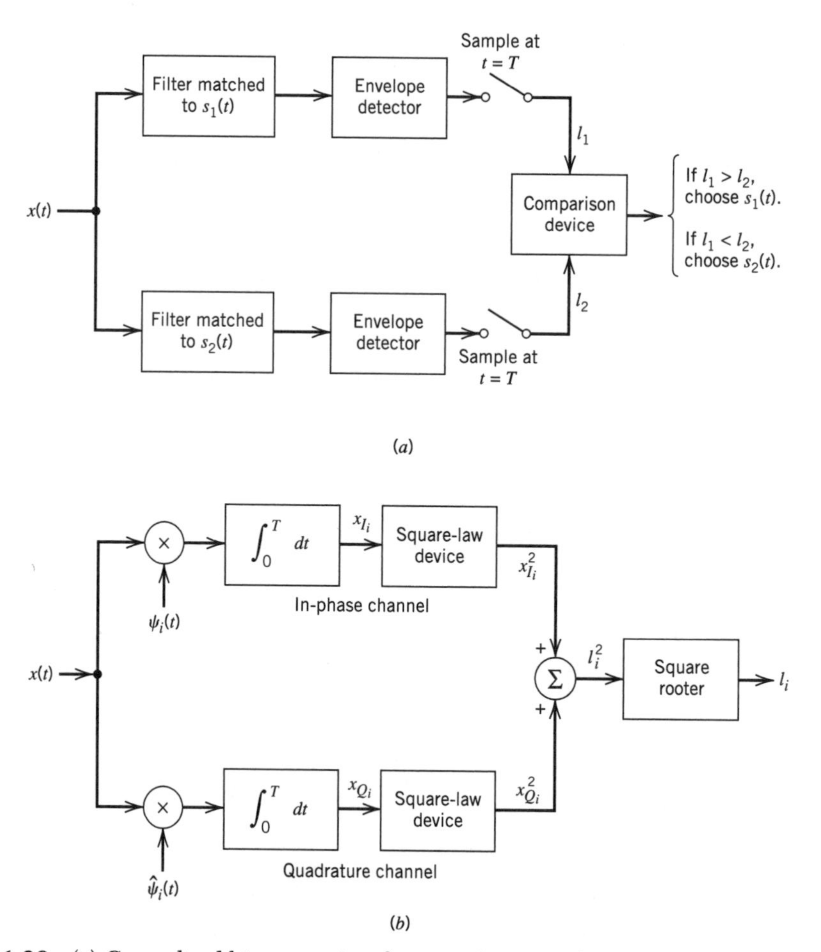


**FIGURE 6.39** (*a*) Generalized binary receiver for noncoherent orthogonal modulation. (*b*) Quadrature receiver equivalent to either one of the two matched filters in part (*a*); the index $i = 1, 2$.

correlated with another function $\hat{\psi}_i(t)$, which represents the version of $\psi_i(t)$ that results from shifting the carrier phase by $-90$ degrees. Naturally, $\psi_i(t)$ and $\hat{\psi}_i(t)$ are orthogonal to each other.

The signal $\hat{\psi}_i(t)$ is in fact the *Hilbert transform of* $\psi_i(t)$; the Hilbert transform is discussed in Appendix 2. To illustrate the nature of this relationship, let

$$\psi_i(t) = m(t)\cos(2\pi f_i t) \tag{6.161}$$

where $m(t)$ is a band-limited message signal. Typically, the carrier frequency $f_i$ is greater than the highest frequency component of $m(t)$. Then (in a manner similar to the discussion on CAP presented in Section 6.4) the Hilbert transform of $\psi_i(t)$ is defined by

$$\hat{\psi}_i(t) = m(t)\sin(2\pi f_i t) \tag{6.162}$$

Since

$$\cos\left(2\pi f_i t - \frac{\pi}{2}\right) = \sin(2\pi f_i t)$$

we see that $\hat{\psi}_i(t)$ is indeed obtained from $\psi_i(t)$ by shifting the carrier $\cos(2\pi f_i t)$ by $-90$ degrees. An important property of Hilbert transformation is that a signal and its Hilbert transform are orthogonal to each other. Thus, $\psi_i(t)$ and $\hat{\psi}_i(t)$ are orthogonal to each other, as already stated.

The average probability of error for the noncoherent receiver of Figure 6.39*a* is given by the simple formula

$$P_e = \frac{1}{2}\exp\left(-\frac{E}{2N_0}\right) \tag{6.163}$$

where $E$ is the signal energy per symbol, and $N_0/2$ is the noise spectral density.

## Derivation of Equation (6.163)*

To derive the formula of Equation (6.163), we make use of the equivalence depicted in Figure 6.39. In particular, we observe that since the carrier phase is unknown, noise at the output of each matched filter in Figure 6.39*a* has *two degrees of freedom*, namely, in-phase and quadrature. Accordingly, the noncoherent receiver of Figure 6.39*a* has a total of four noisy parameters that are *conditionally independent* given the phase $\theta$, and also *identically distributed*. These four noisy parameters have sample values denoted by $x_{I1}$, $x_{Q1}$, $x_{I2}$, and $x_{Q2}$; the first two account for degrees of freedom associated with the upper path of Figure 6.39*a*, and the latter two account for degrees of freedom associated with the lower path.

The receiver of Figure 6.39*a* has a *symmetric* structure. Hence, the probability of choosing $s_2(t)$, given that $s_1(t)$ was transmitted, is the same as the probability of choosing $s_1(t)$, given that $s_2(t)$ was transmitted. This means that the average probability of error may be obtained by transmitting $s_1(t)$ and calculating the probability of choosing $s_2(t)$, or vice versa, assuming that $s_1(t)$ and $s_2(t)$ are equiprobable.

Suppose that signal $s_1(t)$ is transmitted for the interval $0 \le t \le T$. An error occurs if the channel noise $w(t)$ is such that the output $l_2$ of the lower path in Figure 6.39*a* is greater than the output $l_1$ of the upper path. Then the receiver makes a decision in favor of $s_2(t)$

---

*Readers who are not interested in the formal derivation of Eq. (6.163) may at this point wish to move on to the treatment of noncoherent binary frequency-shift keying (in Section 6.7) and differential phase-shift keying (in Section 6.8) as special cases of noncoherent orthogonal modulation, without loss of continuity.

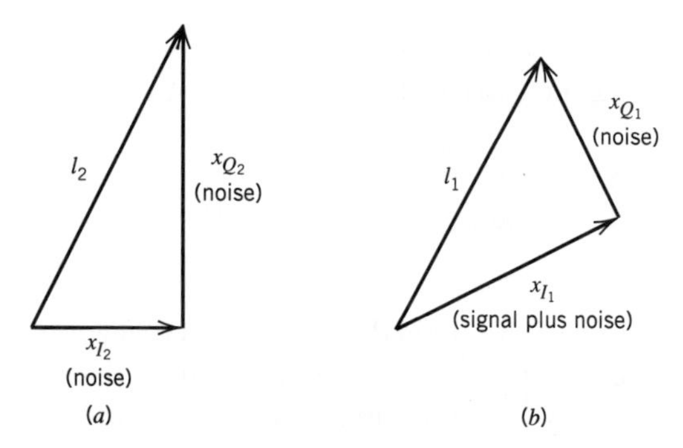


**FIGURE 6.40** Geometric interpretations of the two path outputs $l_1$ and $l_2$ in the generalized noncoherent receiver.

rather than $s_1(t)$. To calculate the probability of error so made, we must have the probability density function of the random variable $L_2$ (represented by sample value $l_2$). Since the filter in the lower path is matched to $s_2(t)$, and $s_2(t)$ is orthogonal to the transmitted signal $s_1(t)$, it follows that the output of this matched filter is due to *noise alone*. Let $x_{I2}$ and $x_{Q2}$ denote the in-phase and quadrature components of the matched filter output in the lower path of Figure 6.39*a*. Then, from the equivalent structure depicted in Figure 6.39*b*, we see that (for $i = 2$)

$$l_2 = \sqrt{x_{I2}^2 + x_{Q2}^2} \tag{6.164}$$

Figure 6.40*a* shows a geometric interpretation of this relation. The channel noise $w(t)$ is both white (with power spectral density $N_0/2$) and Gaussian (with zero mean). Correspondingly, we find that the random variables $X_{I2}$ and $X_{Q2}$ (represented by sample values $x_{I2}$ and $x_{Q2}$) are both Gaussian-distributed with zero mean and variance $N_0/2$, given the phase $\theta$. Hence, we may write

$$f_{X_{I2}}(x_{I2}) = \frac{1}{\sqrt{\pi N_0}} \exp\left(-\frac{x_{I2}^2}{N_0}\right) \tag{6.165}$$

and

$$f_{X_{Q2}}(x_{Q2}) = \frac{1}{\sqrt{\pi N_0}} \exp\left(-\frac{x_{Q2}^2}{N_0}\right) \tag{6.166}$$

Next, we use a well-known result in probability theory, namely, the fact that the envelope of a Gaussian process is *Rayleigh-distributed* and independent of the phase $\theta$ (see Section 1.12). Specifically, for the situation at hand, we may state that the random variable $L_2$ [whose sample value $l_2$ is related to $x_{I2}$ and $x_{Q2}$ by Equation (6.164)] has the following probability density function:

$$f_{L_2}(l_2) = \begin{cases} \dfrac{2l_2}{N_0} \exp\left(-\dfrac{l_2^2}{N_0}\right), & l_2 \geq 0 \\ 0, & \text{elsewhere} \end{cases} \tag{6.167}$$

Figure 6.41 shows a plot of this probability density function. The conditional probability that $l_2 > l_1$, given the sample value $l_1$, is defined by the shaded area in Figure 6.41. Hence, we have

$$P(l_2 > l_1 | l_1) = \int_{l_1}^{\infty} f_{L_2}(l_2)\, dl_2 \tag{6.168}$$

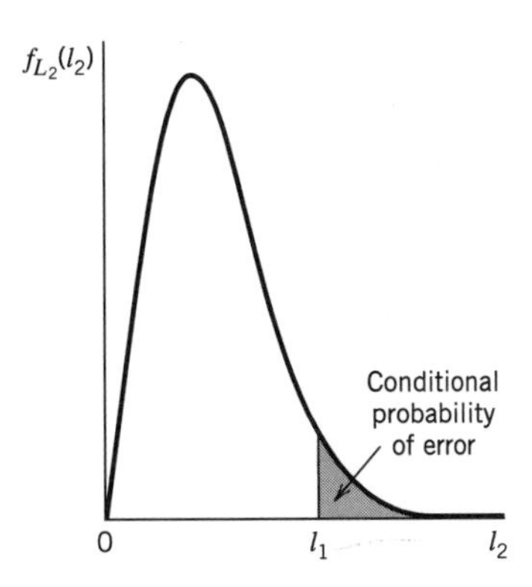


**FIGURE 6.41** Calculation of the conditional probability that $l_2 > l_1$, given $l_1$.

Substituting Equation (6.167) into Equation (6.168) and integrating, we get

$$P(l_2 > l_1 | l_1) = \exp\left(-\frac{l_1^2}{N_0}\right) \tag{6.169}$$

Consider next the output amplitude $l_1$, pertaining to the upper path in Figure 6.39*a*. Since the filter in this path is matched to $s_1(t)$, and it is assumed that $s_1(t)$ is transmitted, it follows that $l_1$ is due to *signal plus noise*. Let $x_{I1}$ and $x_{Q1}$ denote the components at the output of the matched filter (in the upper path of Figure 6.39*a*) that are in phase and in quadrature with respect to the received signal, respectively. Then from the equivalent structure depicted in Figure 6.39*b*, we see that (for $i = 1$)

$$l_1 = \sqrt{x_{I1}^2 + x_{Q1}^2} \tag{6.170}$$

Figure 6.40*b* presents a geometric interpretation of this relation. Since a Fourier-transformable signal and its Hilbert transform form an orthogonal pair, it follows that $x_{I1}$ is due to signal plus noise, whereas $x_{Q1}$ is due to noise alone. This means that (1) the random variable $X_{I1}$ represented by the sample value $x_{I1}$ is Gaussian distributed with mean $\sqrt{E}$ and variance $N_0/2$, where $E$ is the signal energy per symbol, and (2) the random variable $X_{Q1}$ represented by the sample value $x_{Q1}$ is Gaussian distributed with zero mean and variance $N_0/2$. Hence, we may express the probability density functions of these two independent random variables as follows:

$$f_{X_{I1}}(x_{I1}) = \frac{1}{\sqrt{\pi N_0}} \exp\left(-\frac{(x_{I1} - \sqrt{E})^2}{N_0}\right) \tag{6.171}$$

and

$$f_{X_{Q1}}(x_{Q1}) = \frac{1}{\sqrt{\pi N_0}} \exp\left(-\frac{x_{Q1}^2}{N_0}\right) \tag{6.172}$$

Since the two random variables $X_{I1}$ and $X_{Q1}$ are independent, their joint probability density function is simply the product of the probability density functions given in Equations (6.171) and (6.172).

To find the average probability of error, we have to average the conditional probability of error given in Equation (6.169) over all possible values of $l_1$. Naturally, this calculation requires knowledge of the probability density function of random variables $L_1$ represented by sample value $l_1$. The standard method is now to combine Equations (6.171) and (6.172) to find the probability density function of $L_1$ due to signal plus noise. However, this leads to rather complicated calculations involving the use of Bessel functions. This analytic difficulty may be circumvented by the following approach. Given $x_{I1}$ and $x_{Q1}$, an error occurs when, in Figure 6.39*a*, the lower path's output amplitude $l_2$ due to noise alone exceeds $l_1$ due to signal plus noise; from Equation (6.170) we have

$$l_1^2 = x_{I1}^2 + x_{Q1}^2 \tag{6.173}$$

The probability of such an occurrence is obtained by substituting Equation (6.173) into Equation (6.169), as shown by

$$P(\text{error} \mid x_{I1}, x_{Q1}) = \exp\left(-\frac{x_{I1}^2 + x_{Q1}^2}{N_0}\right) \tag{6.174}$$

This is now a conditional probability of error, conditional on the output of the matched filter in the upper path taking on values $X_{I1}$ and $X_{Q1}$. This conditional probability multiplied by the joint probability density function of $X_{I1}$ and $X_{Q1}$ is then the *error-density, given $x_{I1}$ and $x_{Q1}$*. Since $X_{I1}$ and $X_{Q1}$ are statistically independent, their joint probability density function equals the product of their individual probability density functions. The resulting error-density is a complicated expression in $x_{I1}$ and $x_{Q1}$. However, the average probability of error, which is the issue of interest, may be obtained in a relatively simple manner. We first use Equations (6.171), (6.172), and (6.174) to evaluate the desired error-density as

$$\begin{aligned} &P(\text{error} \mid x_{I1}, x_{Q1}) f_{X_{I1}}(x_{I1}) f_{X_{Q1}}(x_{Q1}) \\ &= \frac{1}{\pi N_0} \exp\left\{-\frac{1}{N_0}\left[x_{I1}^2 + x_{Q1}^2 + (x_{I1} - \sqrt{E})^2 + x_{Q1}^2\right]\right\} \end{aligned} \tag{6.175}$$

Completing the square in the exponent of Equation (6.175), we may rewrite the exponent except for $-1/N_0$ as

$$x_{I1}^2 + x_{Q1}^2 + (x_{I1} - \sqrt{E})^2 + x_{Q1}^2 = 2\left(x_{I1} - \frac{\sqrt{E}}{2}\right)^2 + 2x_{Q1}^2 + \frac{E}{2} \tag{6.176}$$

Next, we substitute Equation (6.176) into Equation (6.175) and integrate the error-density over all $x_{I1}$ and $x_{Q1}$. We thus evaluate the average probability of error as

$$\begin{aligned} P_e &= \int_{-\infty}^{\infty}\int_{-\infty}^{\infty} P(\text{error} \mid x_{I1}, x_{Q1}) f_{X_{I1}}(x_{I1}) f_{X_{Q1}}(x_{Q1})\, dx_{I1}\, dx_{Q1} \\ &= \frac{1}{\pi N_0} \exp\left(-\frac{E}{2N_0}\right) \int_{-\infty}^{\infty} \exp\left[-\frac{2}{N_0}\left(x_{I1} - \frac{\sqrt{E}}{2}\right)^2\right] dx_{I1} \\ &\quad \cdot \int_{-\infty}^{\infty} \exp\left(-\frac{2x_{Q1}^2}{N_0}\right) dx_{Q1} \end{aligned} \tag{6.177}$$

We now use the following two identities:

$$\int_{-\infty}^{\infty} \exp\left[-\frac{2}{N_0}\left(x_{I1} - \frac{\sqrt{E}}{2}\right)^2\right] dx_{I1} = \sqrt{\frac{N_0 \pi}{2}} \tag{6.178}$$

and

$$\int_{-\infty}^{\infty} \exp\left(-\frac{2x_{Q1}^2}{N_0}\right) dx_{Q1} = \sqrt{\frac{N_0 \pi}{2}} \tag{6.179}$$

The identity of Equation (6.178) is obtained by considering a Gaussian-distributed variable with mean $\sqrt{E}/2$ and variance $N_0/4$, and recognizing that the total area under the curve of a random variable's probability density function equals unity; the identity of Equation

(6.179) follows as a special case of Equation (6.178). Thus, in light of these two identities, Equation (6.177) simplifies as follows:

$$P_e = \frac{1}{2}\exp\left(-\frac{E}{2N_0}\right)$$

which is the desired result presented previously as Equation (6.163).

With this formula at our disposal, we are ready to consider noncoherent binary FSK and DPSK as special cases, which we do in the next two sections, respectively.[11]

## 6.8 *Noncoherent Binary Frequency-Shift Keying*

In the binary FSK case, the transmitted signal is defined by

$$s_i(t) = \begin{cases} \sqrt{\dfrac{2E_b}{T_b}}\cos(2\pi f_i t), & 0 \le t \le T_b \\ 0, & \text{elsewhere} \end{cases} \tag{6.180}$$

where the carrier frequency $f_i$ equals one of two possible values, $f_1$ and $f_2$; to ensure that the signals representing these two frequencies are orthogonal, we choose $f_i = n_i/T_b$, where $n_i$ is an integer. The transmission of frequency $f_1$ represents symbol 1, and the transmission of frequency $f_2$ represents symbol 0. For the noncoherent detection of this frequency-modulated wave, the receiver consists of a pair of matched filters followed by envelope detectors, as in Figure 6.42. The filter in the upper path of the receiver is matched to $\cos(2\pi f_1 t)$, and the filter in the lower path is matched to $\cos(2\pi f_2 t)$, and in both cases $0 \le t \le T_b$. The resulting envelope detector outputs are sampled at $t = T_b$, and their values are compared. The envelope samples of the upper and lower paths in Figure 6.42 are shown as $l_1$ and $l_2$, respectively. Then, if $l_1 > l_2$, the receiver decides in favor of symbol 1, and if $l_1 < l_2$, it decides in favor of symbols 0. If $l_1 = l_2$, the receiver simply makes a guess in favor of symbol 1 or 0.

The noncoherent binary FSK described herein is a special case of noncoherent orthogonal modulation with $T = T_b$ and $E = E_b$, where $T_b$ is the bit duration and $E_b$ is the

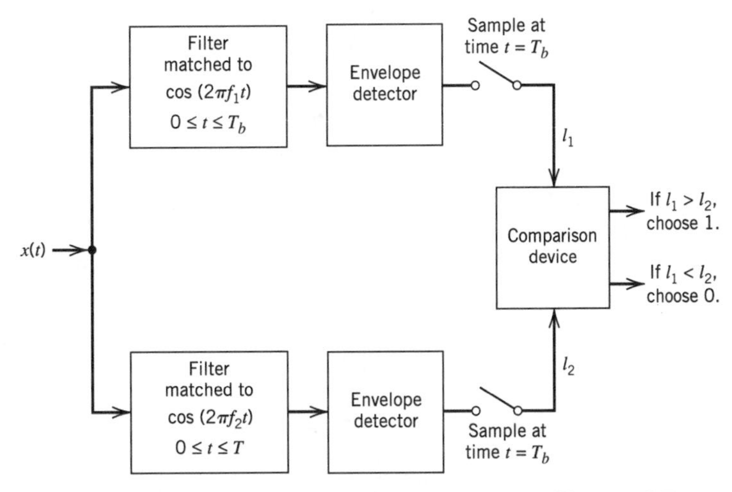


**FIGURE 6.42** Noncoherent receiver for the detection of binary FSK signals.

signal energy per bit. Hence, using (Equation (6.163), we find that the *bit error rate for noncoherent binary FSK* is

$$P_e = \frac{1}{2}\exp\left(-\frac{E_b}{2N_0}\right) \tag{6.181}$$

The formula of Equation (6.181) is derived as a special case of noncoherent orthogonal modulation. In Problem 6.31 we address the same issue using a direct approach that invokes the application of Rayleigh and Rician distributions; these distributions pertain respectively to the random variables $L_2$ and $L_1$ whose sample values are defined by Equations (6.164) and (6.170), respectively.

## 6.9 Differential Phase-Shift Keying

As remarked earlier in Section 6.1, we may view *differential phase-shift keying* (DPSK) as the noncoherent version of PSK. It eliminates the need for a coherent reference signal at the receiver by combining two basic operations at the transmitter: (1) *differential encoding* of the input binary wave and (2) *phase-shift keying*—hence, the name, *differential phase-shift keying* (DPSK). In effect, to send symbol 0, we phase advance the current signal waveform by 180 degrees, and to send symbol 1 we leave the phase of the current signal waveform unchanged. The receiver is equipped with a *storage* capability, so that it can measure the *relative phase difference* between the waveforms received during two successive bit intervals. Provided that the unknown phase $\theta$ contained in the received wave varies slowly (that is, slow enough for it to be considered essentially constant over two bit intervals), the phase difference between waveforms received in two successive bit intervals will be independent of $\theta$.

DPSK is another example of noncoherent orthogonal modulation, when it is considered over two bit intervals. Suppose the transmitted DPSK signal equals $\sqrt{E_b/2T_b}\cos(2\pi f_c t)$ for $0 \le t \le T_b$, where $T_b$ is the bit duration and $E_b$ is the signal energy per bit. Let $s_1(t)$ denote the transmitted DPSK signal for $0 \le t \le 2T_b$ for the case when we have binary symbol 1 at the transmitter input for the second part of this interval, namely, $T_b \le t \le 2T_b$. The transmission of symbol 1 leaves the carrier phase unchanged over the interval $0 \le t \le 2T_b$, and so we define $s_1(t)$ as

$$s_1(t) = \begin{cases} \sqrt{\dfrac{E_b}{2T_b}}\cos(2\pi f_c t), & 0 \le t \le T_b \\ \sqrt{\dfrac{E_b}{2T_b}}\cos(2\pi f_c t), & T_b \le t \le 2T_b \end{cases} \tag{6.182}$$

Let $s_2(t)$ denote the transmitted DPSK signal for $0 \le t \le 2T_b$ for the case when we have binary symbol 0 at the transmitter input for $T_b \le t \le 2T_b$. The transmission of 0 advances the carrier phase by 180 degrees, and so we define $s_2(t)$ as

$$s_2(t) = \begin{cases} \sqrt{\dfrac{E_b}{2T_b}}\cos(2\pi f_c t), & 0 \le t \le T_b \\ \sqrt{\dfrac{E_b}{2T_b}}\cos(2\pi f_c t + \pi), & T_b \le t \le 2T_b \end{cases} \tag{6.183}$$

We readily see from Equations (6.182) and (6.183) that $s_1(t)$ and $s_2(t)$ are indeed orthogonal over the two-bit interval $0 \leq t \leq 2T_b$. In other words, DPSK is a special case of noncoherent orthogonal modulation with $T = 2T_b$ and $E = 2E_b$. Hence, using Equation (6.163), we find that the *bit error rate for DPSK* is given by

$$P_e = \frac{1}{2}\exp\left(-\frac{E_b}{N_0}\right) \tag{6.184}$$

which provides a gain of 3 dB over noncoherent FSK for the same $E_b/N_0$.

### Generation and Detection of DPSK

The next issue to be considered is the generation of DPSK signals. The differential encoding process at the transmitter input starts with an arbitrary first bit, serving as reference. Let $\{d_k\}$ denote the differentially encoded sequence with this added reference bit. We now introduce the following definitions in the generation of this sequence:

- If the incoming binary symbol $b_k$ is 1, leave the symbol $d_k$ unchanged with respect to the previous bit.
- If the incoming binary symbol $b_k$ is 0, change the symbol $d_k$ with respect to the previous bit.

The differentially encoded sequence $\{d_k\}$ thus generated is used to phase-shift a carrier with phase angles 0 and $\pi$ radians representing symbols 1 and 0, respectively. The differential-phase encoding process is illustrated in Table 6.7. Note that $d_k$ is the complement of the modulo-2 sum of $b_k$ and $d_{k-1}$.

The block diagram of a DPSK transmitter is shown in Figure 6.43*a*. It consists, in part, of a logic network and a one-bit delay element interconnected so as to convert the raw binary sequence $\{b_k\}$ into a differentially encoded sequence $\{d_k\}$. This sequence is amplitude-level encoded and then used to modulate a carrier wave of frequency $f_c$, thereby producing the desired DPSK signal.

Suppose next, in differentially coherent detection of binary DPSK, the carrier phase is unknown. Then, in light of the receiver being equipped with an in-phase and a quadrature channel, we have a signal space diagram where the received signal points are $(A \cos\theta, A \sin\theta)$ and $(-A\cos\theta, -A\sin\theta)$, with $\theta$ denoting the unknown phase and $A$ denoting the amplitude. This geometry of possible signals is illustrated in Figure 6.44. The receiver measures the coordinates $(x_{I_0}, x_{Q_0})$ at time $t = T_b$ and $(x_{I_1}, x_{Q_1})$ at time $t = 2T_b$. The issue to be resolved is whether these two points map to the same signal point or different ones. Recognizing that the two vectors $\mathbf{x}_0$ and $\mathbf{x}_1$, with end points $(x_{I_0}, x_{Q_0})$ and $(x_{I_1}, x_{Q_1})$ are pointed roughly in the same direction if their inner product is positive, we may formulate the hypothesis test as follows:

Is the inner product $\mathbf{x}_0^T\mathbf{x}_1$ positive or negative?

**TABLE 6.7** *Illustrating the generation of DPSK signal*

| | | | | | | | | | |
|---|---|---|---|---|---|---|---|---|---|
| $\{b_k\}$ | | 1 | 0 | 0 | 1 | 0 | 0 | 1 | 1 |
| $\{d_{k-1}\}$ | | 1 | 1 | 0 | 1 | 1 | 0 | 1 | 1 |
| Differentially encoded sequence $\{d_k\}$ | 1 | 1 | 0 | 1 | 1 | 0 | 1 | 1 | 1 |
| Transmitted phase (radians) | 0 | 0 | $\pi$ | 0 | 0 | $\pi$ | 0 | 0 | 0 |

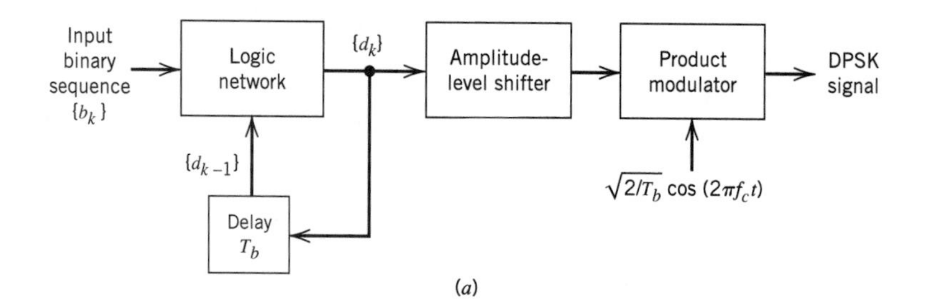


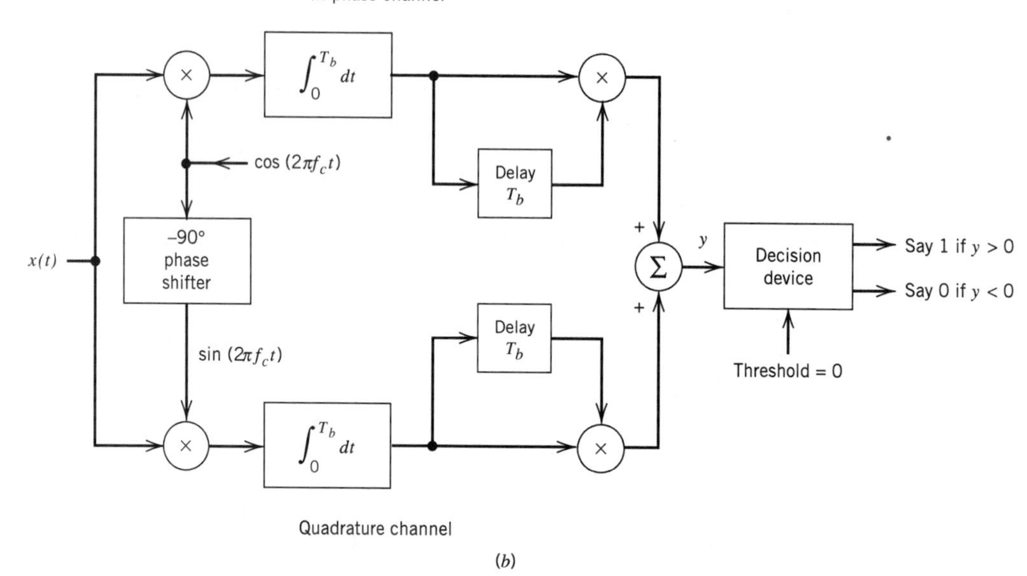


**FIGURE 6.43** Block diagrams of (*a*) DPSK transmitter and (*b*) DPSK receiver.

Accordingly, we may write

$$x_{I_0}x_{I_1} + x_{Q_0}x_{Q_1} \underset{\text{say } 0}{\overset{\text{say } 1}{\gtrless}} 0 \tag{6.185}$$

We now note the following identity:

$$x_{I_0}x_{I_1} + x_{Q_0}x_{Q_1} = \frac{1}{4}\left[(x_{I_0} + x_{I_1})^2 - (x_{I_0} - x_{I_1})^2 + (x_{Q_0} + x_{Q_1})^2 - (x_{Q_0} - x_{Q_1})^2\right]$$

Hence substituting this identity into Equation (6.185) and multiplying both sides of the test by 4, we get the equivalent test:

$$(x_{I_0} + x_{I_1})^2 + (x_{Q_0} + x_{Q_1})^2 - (x_{I_0} - x_{I_1})^2 - (x_{Q_0} - x_{I_1})^2 \underset{\text{say } 0}{\overset{\text{say } 1}{\gtrless}} 0 \tag{6.186}$$

The decision-making process may therefore be thought of as testing whether the point $(x_{I_0}, x_{Q_0})$ is closer to $(x_{I_1}, x_{Q_1})$ or its image $(-x_{I_1}, -x_{Q_1})$.

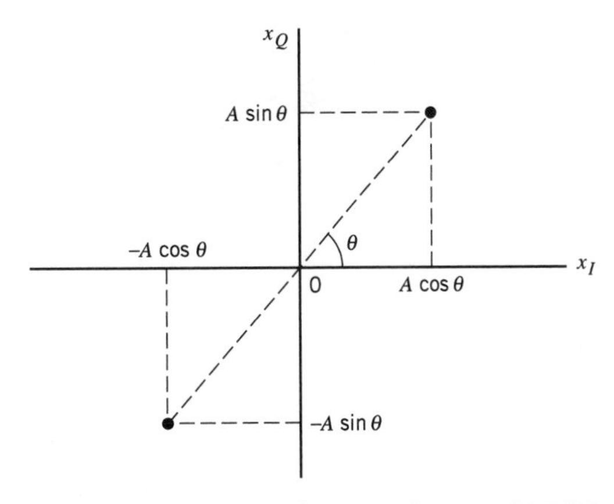


**FIGURE 6.44** Signal-space diagram of received DPSK signal.

Thus, the *optimum receiver*[12] for differentially coherent detection of binary DPSK is as shown in Figure 6.43*b*, which follows directly from Equation (6.185). This implementation merely requires that *sample* values be stored, thereby avoiding the need for fancy delay lines that may be needed otherwise. The equivalent receiver implementation that tests squared elements as in Equation (6.186) is more complicated, but its use makes the *analysis* easier to handle in that the two signals to be considered are orthogonal over the interval $(0, 2T_b)$; hence, the noncoherent orthogonal demodulation analysis applies.

## 6.10 Comparison of Digital Modulation Schemes Using a Single Carrier

### ■ PROBABILITY OF ERROR

In Table 6.8 we have summarized the expressions for the bit error rate (BER) for coherent binary PSK, conventional coherent binary FSK with one-bit decoding, DPSK, noncoherent binary FSK, coherent QPSK, and coherent MSK, when operating over an AWGN channel. In Figure 6.45 we have used the expressions summarized in Table 6.8 to plot the BER as a function of the signal energy per bit-to-noise spectral density ratio, $E_b/N_0$.

**TABLE 6.8** *Summary of formulas for the bit error rate of different digital modulation schemes*

| *Signaling Scheme* | *Bit Error Rate* |
|---|---|
| (a) Coherent binary PSK<br>Coherent QPSK<br>Coherent MSK | $\frac{1}{2}\,\text{erfc}(\sqrt{E_b/N_0})$ |
| (b) Coherent binary FSK | $\frac{1}{2}\,\text{erfc}(\sqrt{E_b/2N_0})$ |
| (c) DPSK | $\frac{1}{2}\exp(-E_b/N_0)$ |
| (d) Noncoherent binary FSK | $\frac{1}{2}\exp(-E_b/2N_0)$ |

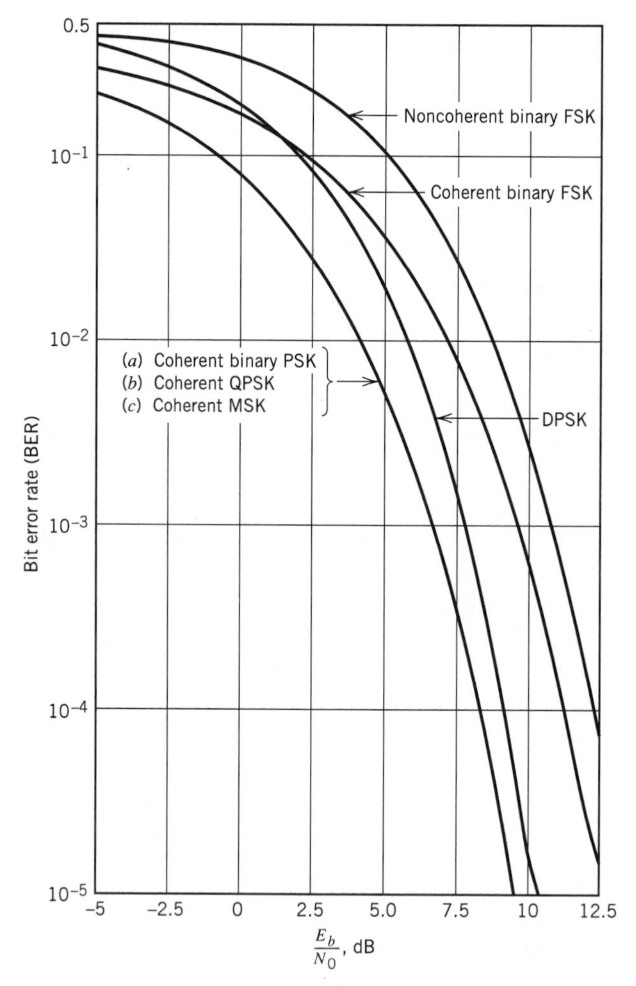


**FIGURE 6.45** Comparison of the noise performance of different PSK and FSK schemes.

Based on the performance curves shown in Figure 6.45, the summary of formulas given in Table 6.8, and the defining equations for the pertinent modulation formats, we can make the following statements:

1. The bit error rates for all the systems decrease monotonically with increasing values of $E_b/N_0$; the defining curves have a similar shape in the form of a *waterfall.*
2. For any value of $E_b/N_0$, coherent binary PSK, QPSK, and MSK produce a smaller bit error rate than any of the other modulation schemes.
3. Coherent binary PSK and DPSK require an $E_b/N_0$ that is 3 dB less than the corresponding values for conventional coherent binary FSK and noncoherent binary FSK, respectively, to realize the same bit error rate.
4. At high values of $E_b/N_0$, DPSK and noncoherent binary FSK perform almost as well (to within about 1 dB) as coherent binary PSK and conventional coherent binary FSK, respectively, for the same bit rate and signal energy per bit.
5. In coherent QPSK, two orthogonal carriers $\sqrt{2/T}\cos(2\pi f_c t)$ and $\sqrt{2/T}\sin(2\pi f_c t)$ are used, where the carrier frequency $f_c$ is an integer multiple of the symbol rate

$1/T$, with the result that two independent bit streams can be transmitted simultaneously and subsequently detected in the receiver.

6. In the case of coherent MSK, there are two orthogonal carriers, namely, $\sqrt{2/T_b}\cos(2\pi f_c t)$ and $\sqrt{2/T_b}\sin(2\pi f_c t)$, which are modulated by the two antipodal symbol shaping pulses $\cos(\pi t/2T_b)$ and $\sin(\pi t/2T_b)$, respectively, over $2T_b$ intervals, where $T_b$ is the bit duration. Correspondingly, the receiver uses a coherent phase decoding process over two successive bit intervals to recover the original bit stream.
7. The MSK scheme differs from its counterpart, the QPSK, in that its receiver has *memory*. In particular, the MSK receiver makes decisions based on observations over two successive bit intervals. Thus, although the transmitted signal has a binary format represented by the transmission of two distinct frequencies, the presence of memory in the receiver makes it assume a two-dimensional signal space diagram. There are four message points, depending on which binary symbol (0 or 1) was sent and the past phase history of the FSK signal.

## Bandwidth Efficiency of *M*-ary Digital Modulation Techniques

In Table 6.9, we have summarized typical values of power-bandwidth requirements for coherent binary and $M$-ary PSK schemes, assuming an average probability of symbol error equal to $10^{-4}$ and the systems operating in identical noise environments. This table shows that, among the family of $M$-ary PSK signals, QPSK (corresponding to $M = 4$) offers the best trade-off between power and bandwidth requirements. For this reason, we find that QPSK is widely used in practice. For $M > 8$, power requirements become excessive; accordingly, $M$-ary PSK schemes with $M > 8$ are not as widely used in practice. Also, coherent $M$-ary PSK schemes require considerably more complex equipment than coherent binary PSK schemes for signal generation or detection, especially when $M > 8$. (Coherent 8-PSK is used in digital satellite communications.)

Basically, $M$-ary PSK and $M$-ary QAM have similar spectral and bandwidth characteristics. For $M > 4$, however, the two schemes have different signal constellations. For $M$-ary PSK the signal constellation is circular, whereas for $M$-ary QAM it is rectangular. Moreover, a comparison of these two constellations reveals that the distance between the message points of $M$-ary PSK is smaller than the distance between the message points of $M$-ary QAM, for a fixed peak transmitted power. This basic difference between the two schemes is illustrated in Figure 6.46 for $M = 16$. Accordingly, in an AWGN channel, $M$-ary QAM outperforms the corresponding $M$-ary PSK in error performance for $M > 4$.

**TABLE 6.9** ***Comparison of power-bandwidth requirements for M-ary PSK with binary PSK. Probability of symbol error = $10^{-4}$***

| *Value of* M | $\frac{(\text{Bandwidth})_{M\text{-ary}}}{(\text{Bandwidth})_{\text{Binary}}}$ | $\frac{(\text{Average power})_{M\text{-ary}}}{(\text{Average power})_{\text{Binary}}}$ |
|---|---|---|
| 4 | 0.5 | 0.34 dB |
| 8 | 0.333 | 3.91 dB |
| 16 | 0.25 | 8.52 dB |
| 32 | 0.2 | 13.52 dB |

From Shanmugan (1979, p. 424).

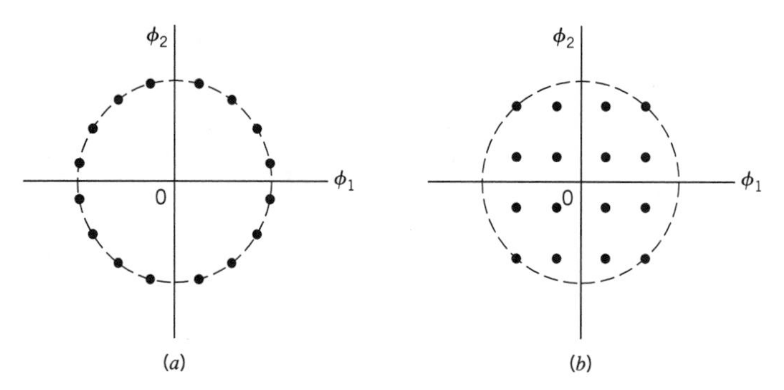


**FIGURE 6.46** Signal constellations for (*a*) $M$-ary PSK and (*b*) corresponding $M$-ary QAM, for $M = 16$.

However, the superior performance of $M$-ary QAM can be realized only if the channel is free of nonlinearities.

As for $M$-ary FSK, we find that for a fixed probability of error, increasing $M$ results in a reduced power requirement. However, this reduction in transmitted power is achieved at the cost of increased channel bandwidth. In other words, $M$-ary FSK behaves in an opposite manner to that of $M$-ary PSK. We will revisit this issue in an information-theoretical context in Chapter 9, and thereby develop further insight into the contrasting behaviors of $M$-ary PSK and $M$-ary FSK.

## 6.11 *Voiceband Modems*

The "modem," a contraction of the term *modulator-demodulator*, is a conversion device that facilitates the transmission and reception of data over the *public switched telephone network* (PSTN).[13] The data of interest may be digital signals generated by computers or service providers. In such an application, the modulator portion of the modem converts the incoming digital signal into a standard form suitable for transmission over a telephone channel in the PSTN. The demodulator portion of the modem receives the channel output and reconverts it into the original digital signal format. In yet another application, namely, *fax modems*, or more precisely modems with facsimile capability, the data may represent text, graphics, pictures, or combinations thereof. In this latter application, the document of interest is coded into a series of compressed picture elements (pixels), which are then transmitted over the telephone channel by modulating their values according to a predefined modulation standard. When the fax modem is in a receiving mode of operation, the demodulator portion of the modem operates on the received analog signal and decompresses the corresponding binary data representation of the demodulated signal into a near or actual duplicate of the original transmitted image. In what follows, we focus our attention on modems that provide communication between a *user* and an *Internet Service Provider* (ISP) over the PSTN.

Traditionally, the PSTN has been viewed as an analog network. In reality, however, the PSTN as we presently know it has become an almost entirely digital network. In most cases, the only part of the PSTN that has remained analog (and will likely remain so for many years to come) is the local loop, which represents the relatively short connection from a home to the central office. Thus, depending on how the PSTN is used, we may identify two distinct classes of modem configurations, symmetric and asymmetric, as described next.

## SYMMETRIC MODEM CONFIGURATIONS

The simplest approach to the design of modems is to treat the entire PSTN as a linear analog network, as indicated in Figure 6.47*a*. (Recall from Chapter 3 that the PSTN is almost entirely digital due to the use of pulse-code modulation (PCM) for the transmission of voice signals.) In such a setting, analog-to-digital and digital-to-analog conversions are needed whenever the modems send signals to and receive signals from the PSTN. The modem configuration depicted in Figure 6.47*a* exhibits "symmetry" in that both modems are identical and the data rate *downstream* (from the ISP to the user) is exactly the same as the data rate *upstream* (from the user to the ISP).

The symmetric modem configuration of Figure 6.47*a* embodies a large number of modem types, ranging in data rate from 300 b/s to 36,600 b/s, as summarized in Table A6.7 on a selection of standard modems. The design of modems began with frequency-shift keying, which catered to relatively low data rates. As the demand for data transmission over telephone channels increased, increasingly more sophisticated modulation techniques were employed to better use the information capacity of the telephone channel.

Consider, for example, the popular *V.32 modem standard* that has the following characteristics:

Carrier frequency = 1,800 Hz
Modulation rate = 2,400 bauds
Data rate = 9,600 b/s

The signaling data rate of 9,600 b/s assumes a high signal-to-noise ratio. The V.32 standard specifies two alternative modulation schemes:

*Nonredundant coding.* Under this scheme, the incoming data stream is divided into quadbits (i.e., groups of four successive bits) and then transmitted over the telephone channel as 16-QAM. In each quadbit, the most significant input dibit undergoes phase modulation, whereas the least significant input dibit undergoes amplitude modulation. Discussing the phase modulation first, practical considerations favor the use of differential phase modulation for the receiver need only be concerned with the detection of phase charges. This matter is taken care of by using a *differential encoder*, which consists of a read-only memory and a couple of delay units, as shown in Figure 6.48*a*. Let $Q_{1,n}Q_{2,n}$ denote the current value of the most significant

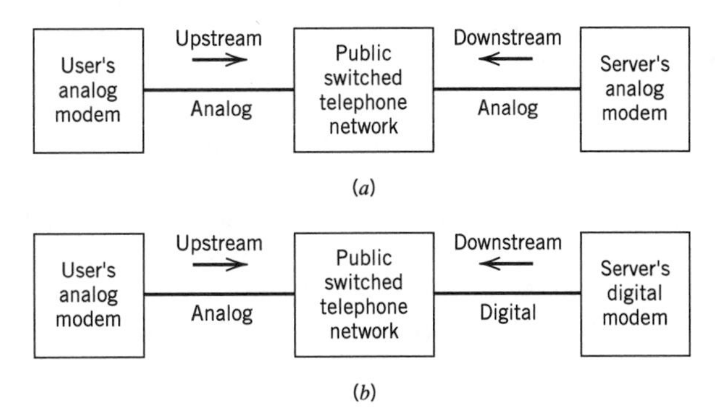


**FIGURE 6.47** (*a*) Environmental overview of symmetric modem configuration: the upstream and downstream data rates are equal. (*b*) Environmental overview of "asymmetric" modem configuration: data rate downstream is higher than upstream.

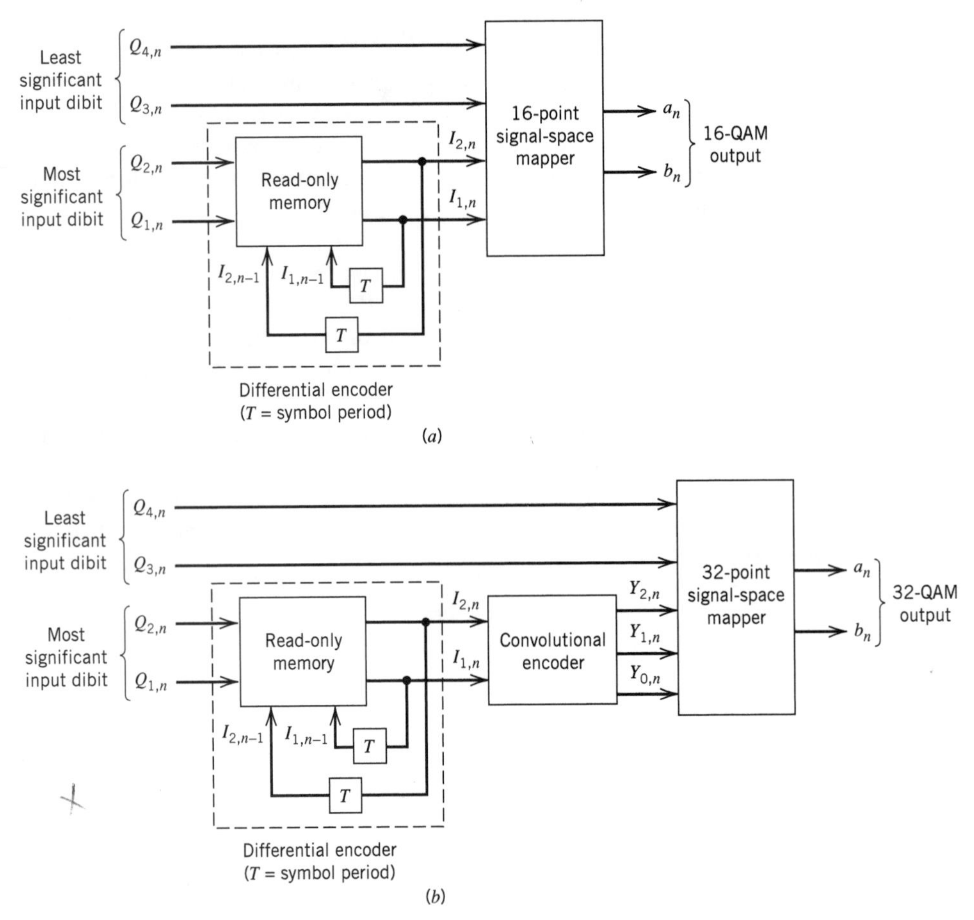


**FIGURE 6.48** Block diagrams of V.32 modem. (*a*) Nonredundant coding. (*b*) Trellis coding.

input dibit, and let $I_{1,n-1}I_{2,n-1}$ denote the previous value of the corresponding dibit output by the encoder. Then, in response to the dibits $Q_{1,n}Q_{2,n}$ and $I_{1,n-1}I_{2,n-1}$, the differential encoder produces the dibit $I_{1,n}I_{2,n}$, which, in turn, induces a phase change in the modulated signal. This phase change, measured in the counterclockwise direction, is governed by the Gray coding scheme of Table 6.10. Note that the phase change is determined entirely by the input dibit $Q_{1,n}Q_{2,n}$. Insofar as the differential phase modulation is concerned, there is one other matter that needs to be addressed: a code for identifying the four quadrants of the two-dimensional signal space. This second matter is resolved by adopting the Gray coding scheme included in Figure 6.49.

Turning next to the amplitude modulation, a code has to be specified for the four possible values which the least significant input dibit, denoted by $Q_{3,n}Q_{4,n}$, can assume in, say, the first quadrant. This matter is taken care of by adopting the Gray code for the four signal points in the first quadrant shown lightly shaded in Figure 6.49.

The final issue that needs to be resolved is the *90° rotational invariance*, which is mandated by the use of differential encoding. This form of invariance means that the overall $M$-ary QAM constellation looks exactly the same when it is rotated

**TABLE 6.10** ***Phase changes induced by differential encoding in the V.32 modem due to varying input dibits***

| *Current input dibit* | | *Phase change* |
|---|---|---|
| $Q_{1,n}$ | $Q_{2,n}$ | *(degrees)* |
| 0 | 0 | 90 |
| 0 | 1 | 0 |
| 1 | 0 | 180 |
| 1 | 1 | 270 |

through an integer multiple of 90 degrees, regardless of whether it is coded or uncoded; then the receiver can correctly decode the transmitted message sequence when the local oscillator phase differs from the carrier phase by an integer multiple of 90 degrees. This final requirement is satisfied by filling in the Gray codes for the signal points in the remaining three quadrants in the manner shown in Figure 6.49. Dashed arrows are included in Figure 6.49 to illustrate the 90° rotational invariance.

Putting all of these matters together for the combined amplitude and phase modulation, we get the 16-QAM constellation shown previously in Figure 6.17*a*, which is reproduced here as Figure 6.50*a*. Correspondingly, the encoding system consists of a differential encoder followed by a 16-point signal-space mapper, as shown in Figure 6.48*a*. The V.32 modem so configured is said to be nonredundant because, with 16 constellation points, the transmitted 4-bit code word has *no* redundant bits.

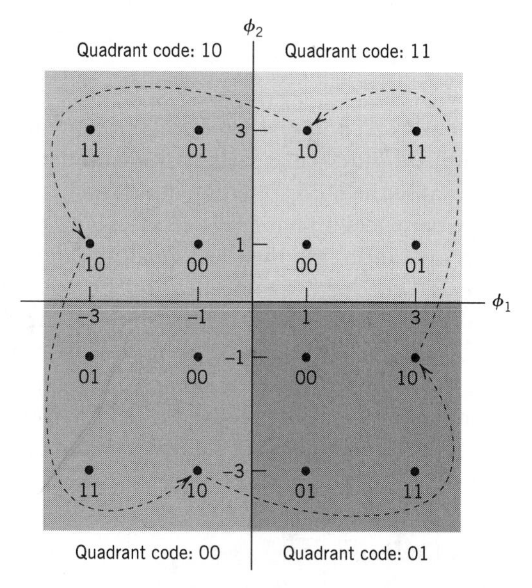


**FIGURE 6.49** Illustrating the Gray encoding of the four quadrants and dibits in each quadrant for the V.32 modem. The dashed arrows illustrate the 90° rotational invariance.

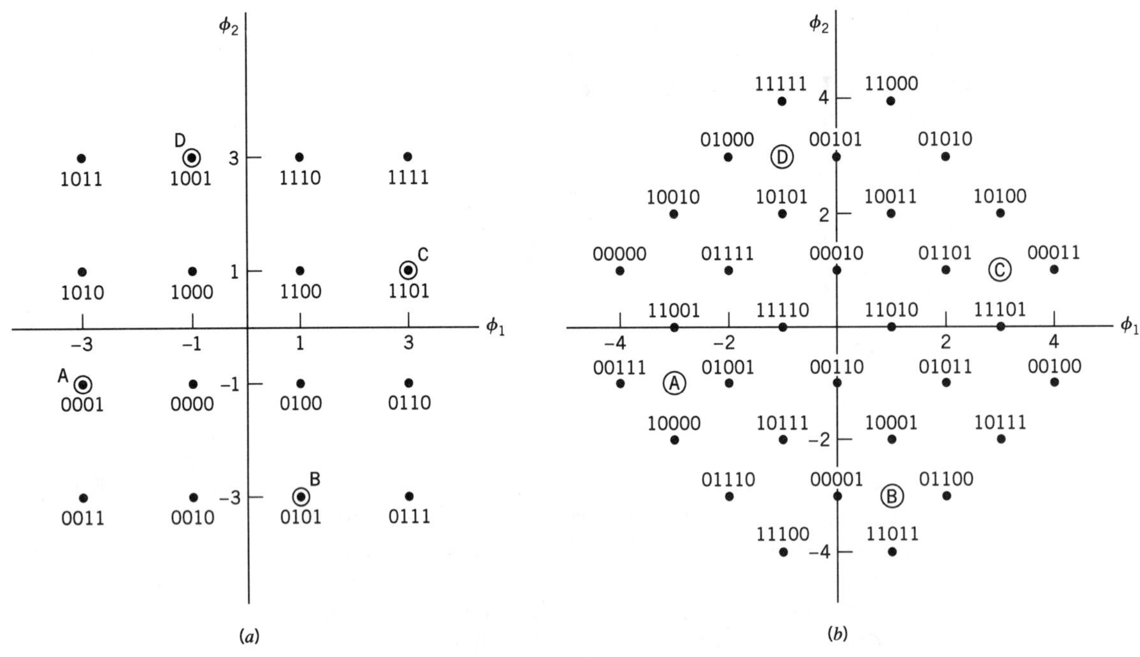


**FIGURE 6.50** (*a*) Signal constellation of V.32 modem using nonredundant coding. (*b*) Signal constellation of V.32 modem using trellis coding.

As an illustrative example of how this particular V.32 modem operates, let the current group of four input bits be 1001 and the dibit previously output by the modem be 11. For this example, we thus have

$$\begin{aligned} Q_{1,n}Q_{2,n} &= 10 \\ Q_{3,n}Q_{4,n} &= 01 \\ I_{1,n-1}I_{2,n-1} &= 11 \end{aligned}$$

Then in light of the coding scheme for the four quadrants specified in Figure 6.49, the previous output dibit 11 means that the modulator was previously residing in the first quadrant. Because the corresponding input dibit is 10, it follows from Table 6.10 that the modulator experiences a phase change of 180° in the counterclockwise direction, thereby switching its operation into the third quadrant identified by the dibit 00. Finally, with the current value of the least significant dibit $Q_{3,n}Q_{4,n}$ being 01, the modulator outputs a QAM signal whose coordinates are $a_n = -3$ (along the $\phi_1$-axis) and $b_n = -1$ (along the $\phi_2$-axis). This output corresponds to the code word 0001.

When the signal-to-noise ratio is not high enough, the V.32 modem switches to its QPSK mode, operating at the reduced rate of 4,800 b/s. In this latter mode of operation, the four states of the modem are signified by the points labeled A, B, C, and D in Figure 6.50*a*.

### *Trellis Coding*

Trellis coding is a forward-error correction scheme where coding and modulation are treated as a combined entity rather than as two separate operations. Figure 6.48*b* shows the encoding system of the V.32 modem with trellis coding. The incoming data stream is divided into quadbits, but unlike the case of nonredundant coding, they are transmitted over the channel as a 32-QAM signal.

As indicated in Figure 6.48*b*, the trellis encoder involves the use of a *convolutional encoder*, which operates on the output of the differential encoder. (Convolutional encoders are discussed in Chapter 10.) However, the choice of convolutional encoding is restricted in the V.32 modem to accommodate the use of differential encoding (i.e., 90 degrees rotational invariance). Indeed, this requirement cannot be satisfied by a linear convolutional encoder. Rather, the convolutional encoder must be *nonlinear*;[14] see Problem 10.30.

The data-encoding process in the V.32 modem with trellis coding proceeds in three stages:

1. The differential encoder in Figure 6.48*b*, in response to the current input dibit $Q_{1,n}Q_{2,n}$ and the previous differentially encoded dibit $I_{1,n-1}I_{2,n-1}$, produces the dibit $I_{1,n}I_{2,n}$.
2. The differentially encoded current dibit $I_{1,n}I_{2,n}$ is input to the convolutional encoder in Figure 6.48*b*, which produces a three-bit output. One of these bits is a *parity-check bit*, denoted by $Y_{0,n}$. The value of $Y_{0,n}$ depends on the other two bits, $Y_{1,n}$ and $Y_{2,n}$, produced by the convolutional encoder.
3. The bits $Y_{0,n}$, $Y_{1,n}$ and $Y_{2,n}$ produced by the convolutional encoder, together with the least significant input dibit $Q_{3,n}Q_{4,n}$ are applied to the signal-space mapper in Figure 6.48*b*, which selects one of the states in the 32-point constellation shown in Figure 6.50*b* as the modem output.

The parity-check bit $Y_{0,n}$ provides a modem with trellis coding better immunity to channel impairments than a V.32 modem with nonredundant coding, an advantage that is gained without an increase in bandwidth requirements. In quantitative terms, trellis coding provides an effective coding gain of 4 dB compared to 16-QAM. *Coding gain* expresses how much more signal energy per data bit is needed by the uncoded modem for the same level of noise performance.

However, for this advantage of trellis coding to be realized in practice, the signal-to-noise ratio must be high enough. Otherwise, the V.32 modem is switched to its QPSK mode of operation, which is signified by the four states labeled A, B, C, and D in Figure 6.50*b*. In this latter mode of operation, the data rate of the modem is reduced to 4,800 b/s.

## Asymmetric Modem Configurations

For a more efficient use of the PSTN, we should treat it as what it really is: an *almost entirely digital network that is nonlinear*. In particular, since the ISP is digitally implemented, the need for analog-to-digital conversion at the ISP modem is eliminated. This means that the communication between the ISP and the PSTN can be entirely digital, as portrayed in Figure 6.47*b*. However, the user's modem has to remain analog because the local loop is analog. This, in turn, requires the use of analog-to-digital and digital-to-analog conversions each time the user's modem sends signals to and receives signals from the PSTN. The modem configuration depicted in Figure 6.47*b* is "asymmetric" in that it is possible for the downstream signaling data rate to be much higher than the upstream signaling data rate, as explained next.

As mentioned earlier, a digital PSTN is based on the use of PCM for the transmission of voice signals. Features of the system relevant to the present discussion are as follows (see Chapter 3):

- Data signaling rate of 64 kb/s, which is made up of a sampling rate of 8 kHz and the representation of each voice sample by an 8-bit code word.
- Fifteen-segment companding law (e.g., a logarithmic $\mu$-law with $\mu = 255$) for compressing the voice signal at the transmitter and expanding it at the receiver.

From the discussion on PCM presented in Chapter 3 we also recall that quantization only affects analog-to-digital conversion but *not* digital-to-analog conversion. These observations have a profound impact on the optimum strategy for the design of asymmetric modems.

Suppose there is no analog-to-digital conversion between a digital modem at the ISP and the digital portion of the PSTN, and the digitally connected transmitter of the modem is designed to properly use the nonuniformly spaced 256 (discrete) threshold levels of the digital PSTN. Then, since digital-to-analog conversion is completely unaffected by quantization noise, it follows that the information transmitted by the ISP's digital modem reaches the user's analog modem with no loss whatsoever. On the basis of these arguments, in theory, it should be possible to transmit data from the ISP to the user at a rate equal to the 64 kb/s data rate of the digital PSTN. But system limitations inherent to the PSTN reduce the attainable data rate down to 56 kb/s, as explained in the sequel.

### *Digital Modem*

From the description of a PCM voiceband channel presented in Chapter 3, we find that the design of the digital modem is constrained by three factors *not* under our control. The design constraints are:

1. A sampling rate $f_s = 8$ kHz.
2. A set of $M = 256$ allowable threshold levels built into the construction of the compressor (i.e., transmitter portion of the compander).
3. A baseband (antialiasing) filter of about 3.5 kHz bandwidth, built into the front end of the PCM transmitter.

In light of these constraints, we may now state the fundamental philosophy underlying the design of the digital modem as follows:

> Design a signal $s(t)$ at the digital modem's input such that each of its samples taken at the rate $f_s = 8$ kHz matches one of the $M = 256$ threshold levels of the compressor, and the transmitted signal satisfies Nyquist's criterion for zero intersymbol interference.

(Nyquist's criterion for zero intersymbol interference was discussed in Chapter 4.)

### *One Realization of the Digital Modem*

A solution to this signal design problem is made particularly difficult by the fact that the PCM transmit filter has a bandwidth of about 3.5 kHz and not 4 kHz (half the sampling rate $f_s$). The immediate implication of this constraint is that instead of the desired set of 8,000 samples, we can only generate $2 \times 3{,}500 = 7{,}000$ *independent* samples every second in accordance with Nyquist's criterion for zero intersymbol interference. How then do we

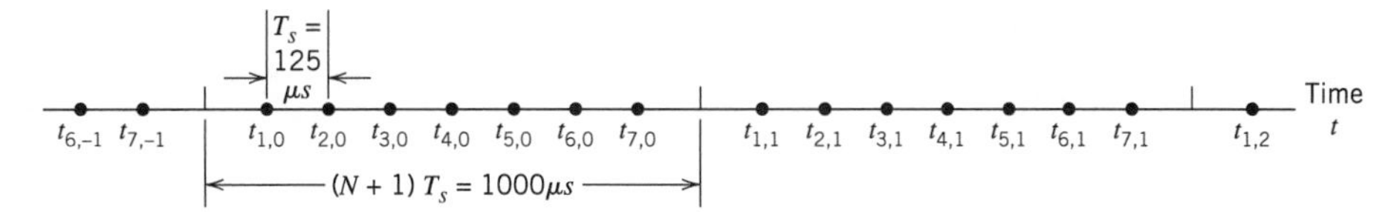


**FIGURE 6.51** Group of $N$ uniformly spaced samples, repeating every $(N + 1)T_s$ seconds.

fit 7,000 independent samples per second within the prescribed framework of 8,000 samples per second?

To answer this fundamental question, we make use of the *recurrent nonuniform equivalent* form of the sampling theorem. To be more specific, consider the situation depicted in Figure 6.51, where the samples are divided into groups, with each group containing $N$ uniformly spaced samples, and the groups having a recurrent period of $(N + 1)T_s$ seconds, where $T_s = 1/f_s$. The illustration presented in Figure 6.51 is for the problem at hand: $T_s = 125\ \mu s$ and $N = 7$. The sampling instants in the nonuniform distribution of Figure 6.51 are written as

$$
\begin{aligned}
t_{k,l} &= t_k + (N + 1)lT_s \\
&= (k - 1)T_s + (N + 1)lT_s, \qquad \begin{array}{l} k = 1, 2, \ldots, N \\ l = 0, \pm 1, \pm 2, \ldots \end{array}
\end{aligned} \tag{6.187}
$$

The stage is now set for us to define the band-limited signal $s(t)$ as follows:[15]

$$
s(t) = \sum_{l=-\infty}^{\infty} \sum_{k=1}^{N} s(t_{k,l})\psi_k(t - (N + 1)lT_s) \tag{6.188}
$$

where the *interpolation function* $\psi_k(t)$ is itself defined by

$$
\psi_k(t) = \operatorname{sinc}\left(\frac{t - t_k}{(N + 1)T_s}\right) \prod_{\substack{q=1 \\ q \neq k}}^{N} \frac{\sin\left(\dfrac{\pi}{(N + 1)T_s}(t - t_q)\right)}{\sin\left(\dfrac{\pi}{(N + 1)T_s}(t_k - t_q)\right)} \tag{6.189}
$$

Computing Equation (6.189) for $N = 7$, we obtain the seven *standard pulses* plotted in Figure 6.52, where time is normalized with respect to the sampling period $T_s$. These pulses exhibit the following properties:

- Each standard pulse is normalized so that we have

$$
\psi_k\left(\frac{t_k}{T_s}\right) = \psi_k(k - 1) = 1 \qquad \text{for } k = 1, 2, \ldots, 7
$$

  Note, however, that the peak of the $k$th pulse does *not* occur at time $t_k = (k - 1)T_s$.
- For $k = 1, 2, \ldots, 7$ the pulse $\psi_k(t/T_s)$ goes through zero at times $t \neq (k - 1)T_s$ modulo $(N + 1)$, except at those times that are congruent to $t = (-1)$ modulo $(N + 1)$.

Accordingly, the signaling scheme for the digital modem consists of a recurrent nonuniform pulse amplitude modulation scheme. The amplitudes of seven uniformly spaced samples in each group of eight samples are determined by the incoming data stream and in conformity to the threshold levels of the compressor in the PCM transmitter. In effect, these seven samples are the independent samples that are responsible for carrying the

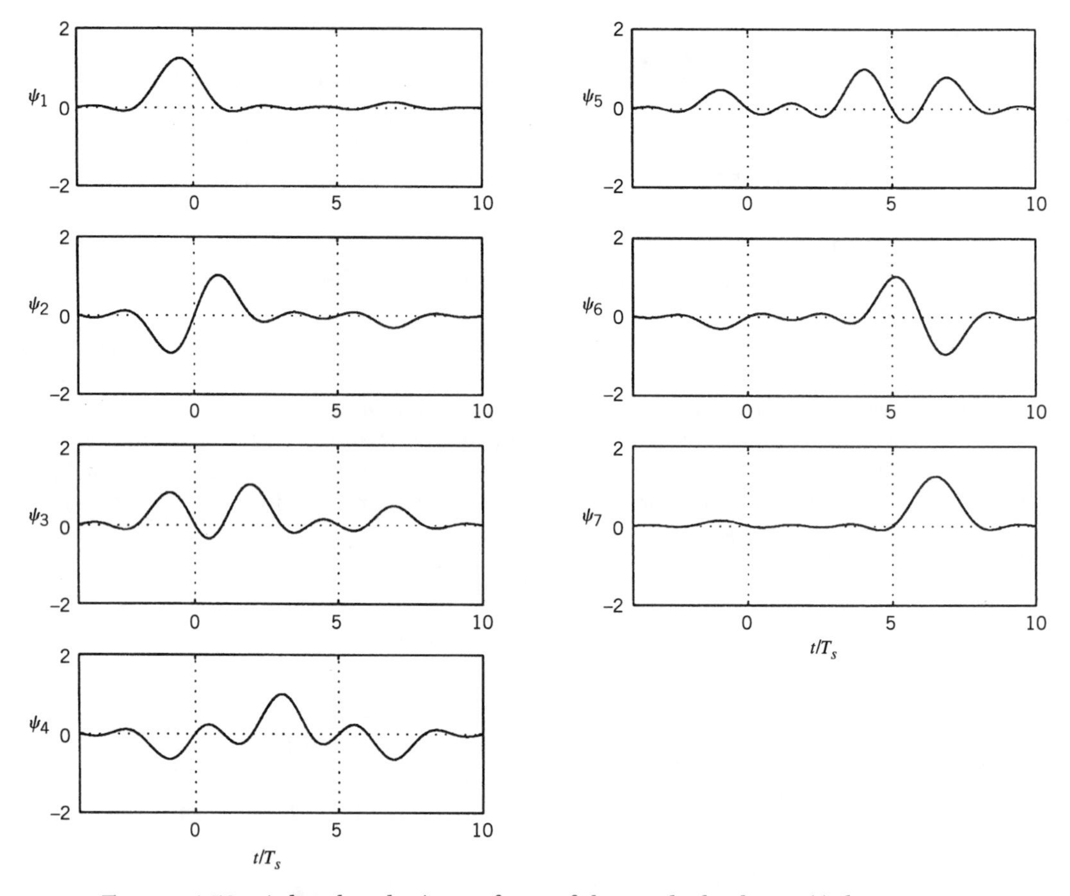


**FIGURE 6.52** A digital modem's waveforms of the standard pulses $\psi_k(t)$, $k = 1, 2, \ldots, 7$.

incoming data stream across the PSTN every 1,000 $\mu$s. Moreover, they deliver the data to the receiver with *zero* intersymbol interference. The remaining "eighth" samples are completely determined by the independent samples and known beforehand to the system; they do not carry information and are therefore discarded at the receiver. Thus the digital modem is capable of transmitting digital data across the PSTN almost errorless at a rate equal to 56 kb/s, which is calculated as follows:

$$7 \times 1{,}000 \times \log_2 256 = 56{,}000 \text{ b/s}$$

One last comment is in order. The standard pulses $\psi_k(t)$ can be constructed so as to decay at a rate faster than $1/t$. To do so, we simply replace the sinc function in Equation (6.189) by a Nyquist pulse with a rolloff in a manner similar to that described in Chapter 4.

### *Another Realization of the Digital Modem*

The kind of digital modem just described is *bidirectional*, assuming that both ends of the data link are analog. However, a simpler solution to the digital modem design problem ensues when one end of the link is digital and asymmetric data rates are possible.[16]

Consider what happens when a data sequence consisting of *octets* (i.e., 8-bit code words) arrives at the PSTN. There they will be treated as octets representing speech encoded in accordance with the $\mu$-law or A-law, depending on the part of the world where the PSTN is located. Consequently, the D/A converter, which drives the analog modem, produces a continuous-time signal defined by

$$s(t) = \sum_k a(c_k)g(t - kT_s) \tag{6.190}$$

where $c_k$ is the $k$th octet in the data sequence, $a(c_k)$ is the representation level specified by the pertinent companding law, $T_s$ is the sampling interval (equal to 125 $\mu$s), and $g(t)$ is an interpolation function bandlimited to a frequency below $1/2T_s$, or about 4 kHz, to satisfy the reconstruction part of the sampling theorem; see Section 3.2.

In the normal operation of the PSTN, the signal $s(t)$ represents a reconstructed speech signal. However, in the case of input data, $s(t)$ appears like noise. In any event, from a communication theoretical perspective, the signal $s(t)$ in Equation (6.190) may be viewed as a *pulse-amplitude modulated* signal. Herein lies the theoretical basis for the design of the digital modem. Specifically, the design is based on a signal constellation as in an analog modem, except that the constellation is constructed from *one-dimensional PCM symbols* rather than two-dimensional QAM symbols.

Ordinarily, the data rate achievable by a digital modem is limited to about 56 kb/s because of the following factors:

1. The inner levels of the compander in the PSTN are very closely spaced, as shown in Table 3.4; hence they are susceptible to residual intersymbol interference and noise following the modem's equalizer.
2. Least significant bits (LSBs) are robbed from the data stream for various purposes internal to the PSTN; this "bit-robbing" can be as much as (but usually less than) 8 kb/s and always in a periodic pattern.

### *Analog Modem*

Unlike the digital modem, the noise performance of the analog modem is limited essentially by quantization noise in the $\mu$-law or $A$-law governing the operation of the PCM compander. Typically, the signal-to-noise ratio on a good PCM voiceband channel is on the order of 34 to 38 dB. The other channel impairment that limits the operation of the analog modem is the effect of bandlimiting imposed by the antialiasing and interpolation filters, which, as already mentioned, is typically about 3.5 kHz.

A sophisticated choice for the analog modem is the standard *V.34 modem*, which operates at rates extending up to 33.6 kb/s. The fundamental design philosophy of this modem embodies five distinctive features.[17]

1. *960-QAM super-constellation.*
The signal constellation is said to be a super- or nested-constellation in that it consists of four constellations: the QAM constellation shown in Figure 6.53 with 240 message points, and its rotated versions through 90, 180, and 270 degrees.

2. *Adaptive bandwidth.*
The transmitter probes the channel by sending a set of tones, which permits measurement of the signal-to-noise ratio at the channel output as a function of frequency. The modem is thereby enabled to select the appropriate carrier frequency and bandwidth according to the probing results and available symbol rates.

3. *Adaptive bit rates.*
During the training of the receiver, the bit rate is selected according to the receiver's estimate of the maximum bit rate, which the modem can support at bit error rates as low as $10^{-6}$ to $10^{-5}$.

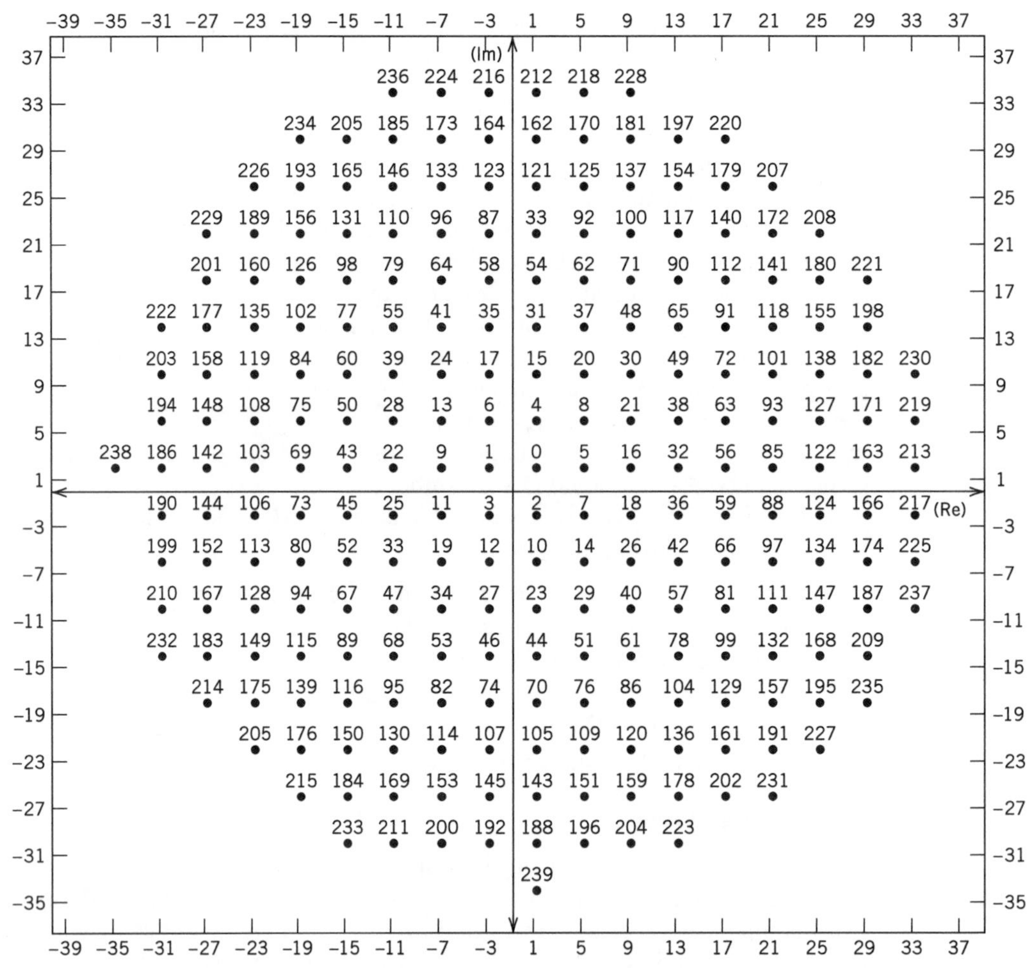


**FIGURE 6.53** Quarter-superconstellation of V.34 modem with 240 signal points. The full superconstellation is obtained by combining the rotated versions of these points by 0, 90, 180, and 270 degrees. (Taken from Forney et al., 1996, with permission of the IEEE.)

4. *Trellis coding.*
This error-control coding technique is used to provide an effective coding gain of about 3.6 dB; there is an optional more powerful trellis code with an effective coding gain of about 4.7 dB.

5. *Decision feedback equalization.*
To make full use of the available telephone channel bandwidth, including frequencies near the band edges where there can be attenuation as much as 10 to 20 dB, a decision feedback equalizer (DFE) is used. (The DFE is discussed in Chapter 4.) However, it is not a straightforward matter to combine coding with DFE because decision feedback requires immediate decisions, whereas coding inherently involves decoding delay. The overcome this problem, the feedback section of the DFE is moved to the transmitter, which is made possible through the use of the Tomlinson-Harashima precoding. (This form of equalization via precoding is discussed briefly in Note 12 of Chapter 4.)

### V.90 Modem

The V.90 modem standard embodies digital and analog modems. The digital modem at the ISP end is based on the second realization described earlier; it sends data downstream at the rate of 56 kb/s. The analog modem at the user's end is a V.34 modem standard, transmitting data upstream at the rate of 33.6 kb/s. These two highly different rates confirm the asymmetric nature of the V.90 modem.

The outstanding feature of the V.90 modem, namely, the downstream data rate of 56 kb/s, makes it suitable for use on the Internet for downloading graphics in intensive Web pages, audio, and video at near-ISDN speeds.

## 6.12 Multichannel Modulation

The *asymmetric digital subscriber line (ADSL)*, described in Section 4.8, is a data transmission system capable of realizing megabit rates over existing twisted-pair telephone lines. Specifically, ADSL runs at a downstream data rate up to 9 Mb/s and an upstream data rate up to 1 Mb/s. These data signaling rates fit the access requirements of the Internet perfectly. (As mentioned in Section 4.8, the upstream bit rate should be about 10 percent of the downstream bit rate for efficient operation of the Internet protocol.) The challenge in designing ADSL is to develop a line code that exploits the information capacity of the channel as fully as possible. The carrierless amplitude phase modulation (CAP), discussed in Section 6.4, provides one approach for solving this difficult passband data transmission problem. Another approach is to use an equally elegant modulation technique called discrete multitone. This latter approach is a form of *multichannel modulation*[18] that allows the modulator characteristics to be a function of measured channel characteristics. It is fitting that we begin the discussion by describing multichannel modulation, which we do in this section, followed by discrete multitone in the next section.

The basic idea of *multichannel modulation* is rooted in a commonly used engineering principle: *divide and conquer*. According to this principle, a difficult problem is solved by dividing it into a number of simpler problems, and then combining the solutions to those simple problems. In the context of our present discussion, the difficult problem is that of data transmission over a wideband channel with severe intersymbol interference, and the simpler problems are exemplified by data transmission over AWGN channels. We may thus summarize the essence of multichannel modulation as follows:

> Data transmission over a difficult channel is transformed through the use of advanced signal processing techniques into the parallel transmission of the given data stream over a large number of subchannels, such that each subchannel may be viewed effectively as an AWGN channel.

Naturally, the overall data rate is the sum of the individual data rates over the subchannels operating in parallel.

### ■ Capacity of AWGN Channel

From the Background and Preview material presented in the opening chapter, we recall that, according to *Shannon's information capacity theorem*, the capacity of an AWGN channel (that is free from intersymbol interference) is defined by

$$C = B \log_2(1 + \text{SNR}) \text{ b/s} \tag{6.191}$$

where $B$ is the channel bandwidth, and SNR denotes the signal-to-noise ratio measured at the channel output. A proof of this important theorem is formally presented in Chapter 9. For now it suffices to say that for a given SNR, we can transmit data over an AWGN channel of bandwidth $B$ at the maximum rate of $C$ bits per second with arbitrarily small probability of error, provided that we employ an encoding system of sufficiently high complexity. Equivalently, we may express the capacity $C$ in bits per transmission or channel use as

$$C = \frac{1}{2}\log_2(1 + \text{SNR}) \qquad \text{bits/transmission} \tag{6.192}$$

In practice, we usually find that a physically realizable encoding system must transmit data at a rate $R$ less than the maximum possible rate $C$ for it to be reliable. For an implementable system operating at low enough probability of symbol error, we thus need to introduce a *signal-to-noise ratio gap* or just *gap*, denoted by $\Gamma$. The gap is a function of the permissible probability of symbol error $P_e$ and the encoding system of interest. It provides a measure of the "efficiency" of an encoding system with respect to the ideal transmission system of Equation (6.192). With $C$ denoting the capacity of the ideal encoding system and $R$ denoting the capacity of the corresponding implementable encoding system, the gap is defined by

$$\begin{aligned}\Gamma &= \frac{2^{2C} - 1}{2^{2R} - 1} \\ &= \frac{\text{SNR}}{2^{2R} - 1}\end{aligned} \tag{6.193}$$

Equivalently, we may write

$$R = \frac{1}{2}\log_2\left(1 + \frac{\text{SNR}}{\Gamma}\right) \qquad \text{bits/transmission} \tag{6.194}$$

For encoded PAM or QAM operating at $P_e = 10^{-6}$, for example, the gap $\Gamma$ is constant at 8.8 dB. Through the use of codes (e.g., trellis codes discussed in Chapter 10), the gap $\Gamma$ may be reduced to as low as 1 dB.

Let $P$ denote the transmitted signal power, and $\sigma^2$ denote the channel noise variance measured over the bandwidth $B$. The signal-to-noise ratio is therefore

$$\text{SNR} = \frac{P}{\sigma^2}$$

where

$$\sigma^2 = N_0 B$$

We may thus finally define the attainable data rate as

$$R = \frac{1}{2}\log_2\left(1 + \frac{P}{\Gamma\sigma^2}\right) \qquad \text{bits/transmission} \tag{6.195}$$

With this formula at hand, we are ready to describe multichannel modulation in quantitative terms.

## ▪ CONTINUOUS-TIME CHANNEL PARTITIONING

Consider a linear wideband channel (e.g., twisted pair) with an arbitrary frequency response $H(f)$. Let the squared magnitude response $|H(f)|$ be approximated by a staircase

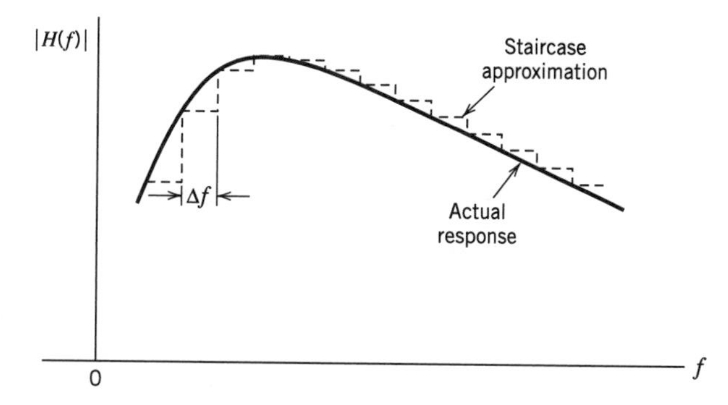


**FIGURE 6.54** Staircase approximation of an arbitrary magnitude response $|H(f)|$; only positive-frequency portion of the response is shown.

function as illustrated in Figure 6.54, with $\Delta f$ denoting the width of each step. In the limit, as the frequency increment $\Delta f$ approaches zero, the staircase approximation of the channel approaches the actual $H(f)$. Along each step of the approximation, the channel may be assumed to operate as an AWGN channel free from intersymbol interference. The problem of transmitting a single wideband signal is thereby transformed into the transmission of a set of narrowband orthogonal signals. Each narrowband orthogonal signal, with its own carrier, is generated using a spectrally efficient modulation technique such as $M$-ary QAM, with additive white Gaussian noise being essentially the only primary source of transmission impairment. This, in turn, means that data transmission over each subchannel of bandwidth $\Delta f$ can be optimized by invoking Shannon's information capacity theorem, with the optimization of each subchannel being performed independently of all the others. Thus, in practical signal-processing terms, *the need for complicated equalization of a wideband channel is replaced by the need for multiplexing and demultiplexing the transmission of the incoming data stream over a large number of narrowband subchannels that are contiguous and disjoint.* Although the resulting complexity of a multicarrier system is indeed high for a large number of subchannels, implementation of the entire system can be accomplished in a cost-effective manner through the use of VLSI technology.

Figure 6.55 shows a block diagram of the multichannel data transmission system in its most basic form. The system is configured here using quadrature-amplitude modulation whose choice is justified by virtue of its spectral efficiency. The incoming binary data stream is first applied to a demultiplexer (not shown in the figure), thereby producing a set of $N$ substreams. Each substream represents a sequence of *two-element subsymbols*, which, for the symbol interval $0 \le t \le T$, is denoted by

$$(a_n, b_n), \qquad n = 1, 2, \ldots, N$$

where $a_n$ and $b_n$ are element values along the two coordinates of subchannel $n$.

Correspondingly, the passband basis functions of the quadrature-amplitude modulators are defined by the function pairs

$$\{\phi(t)\cos(2\pi f_n t),\ \phi(t)\sin(2\pi f_n t)\}, \qquad n = 1, 2, \ldots, N \tag{6.196}$$

where the carrier frequency $f_n$ of the $n$th modulator is an integer multiple of the symbol rate $1/T$, as shown by

$$f_n = \frac{n}{T}, \qquad n = 1, 2, \ldots, N$$

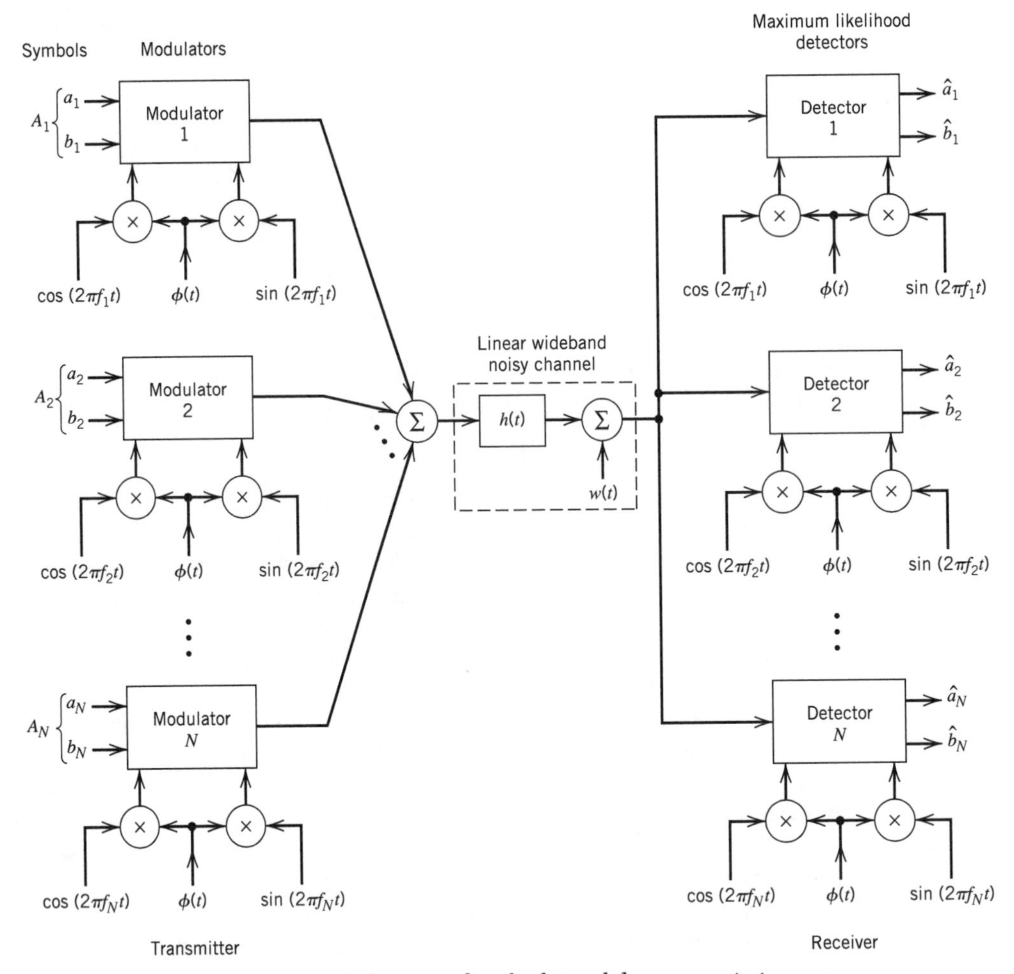


**FIGURE 6.55** Block diagram of multichannel data transmission system.

and the low-pass function $\phi(t)$ is the sinc function:

$$\phi(t) = \sqrt{\frac{2}{T}}\,\text{sinc}\left(\frac{t}{T}\right), \qquad -\infty < t < \infty \tag{6.197}$$

The passband basis functions defined here have the following desirable properties (see Problem 6.41 for their proofs):

**Property 1**

*For each* n, *the two quadrature-modulated sinc functions form an orthogonal pair as shown by*

$$\int_{-\infty}^{\infty} (\phi(t)\cos(2\pi f_n t))(\phi(t)\sin(2\pi f_n t))\,dt = 0 \qquad \text{for all } n \tag{6.198}$$

This orthogonal relationship provides the basis for formulating the signal constellation for each of the $N$ modulators in the form of a squared lattice.

**Property 2**

*Recognizing that*

$$\exp(j2\pi f_n t) = \cos(2\pi f_n t) + j \sin(2\pi f_n t)$$

*we may completely redefine the passband basis functions in the complex form*

$$\left\{\frac{1}{\sqrt{2}}\,\phi(t)\exp(j2\pi f_n t)\right\}, \qquad n = 1, 2, \ldots, N \tag{6.199}$$

*where the factor* $1/\sqrt{2}$ *has been introduced to ensure that the scaled function* $\phi(t)/\sqrt{2}$ *has unit energy. Hence, these passband basis functions form an orthonormal set, as shown by*

$$\int_{-\infty}^{\infty}\left(\frac{1}{\sqrt{2}}\,\phi(t)\exp(j2\pi f_n t)\right)\left(\frac{1}{\sqrt{2}}\,\phi(t)\exp(j2\pi f_k t)\right)^{*} dt = \begin{cases} 1, & k = n \\ 0, & k \neq n \end{cases} \tag{6.200}$$

*where the asterisk denotes complex conjugation.*

Equation (6.200) provides the mathematical basis for ensuring that the $N$ modulator-demodulator pairs operate independently of each other.

**Property 3**

*The set of channel-output functions* $\{h(t) \star \phi(t)\}$ *remains orthogonal for a linear channel with arbitrary impulse response* $h(t)$, *where* $\star$ *denotes convolution.*

The channel is thus partitioned into a set of independent subchannels operating in continuous time.

Figure 6.55 also includes the structure of the receiver. It consists of a bank of $N$ coherent detectors, with the channel output being simultaneously applied to the detector inputs. Each detector is supplied with a locally generated pair of quadrature modulated sinc functions operating in synchrony with the pair of passband basis function applied to the corresponding modulator in the transmitter.

Each subchannel may have some residual intersymbol interference (ISI). However, as the number of subchannels $N$ approaches infinity, the ISI disappears. Thus, for a sufficiently large $N$, the bank of coherent detectors in Figure 6.55 operates as *maximum likelihood detectors*, independently of each other and on a subsymbol-by-subsymbol basis.

To define the detector outputs in response to the input subsymbols, we find it convenient to use complex notation. Let $A_n$ denote the subsymbol applied to the $n$th modulator during the symbol interval $0 \leq t \leq T$:

$$A_n = a_n + jb_n, \qquad n = 1, 2, \ldots, N \tag{6.201}$$

The corresponding detector output is

$$Y_n = H_n A_n + W_n, \qquad n = 1, 2, \ldots, N \tag{6.202}$$

where $H_n$ is the complex-valued frequency response of the channel evaluated at the subchannel carrier frequency $f = f_n$:

$$H_n = H(f_n), \qquad n = 1, 2, \ldots, N \tag{6.203}$$

The $W_n$ is a complex-valued random variable due to the channel noise $w(t)$; the real and imaginary parts of $W_n$ have zero mean and variance $N_0/2$. With knowledge of the measured frequency response $H(f)$ available, we may therefore use Equation (6.202) to compute a maximum likelihood estimate of the transmitted subsymbol $A_n$. The estimates $\hat{A}_1, \hat{A}_2, \ldots, \hat{A}_N$ so obtained are finally multiplexed to produce the corresponding estimate of the original binary data transmitted during the interval $0 \le t \le T$.

To summarize, for a sufficiently large $N$, we may implement the receiver as an optimum maximum likelihood detector, operating as $N$ subsymbol-by-subsymbol detectors. The reason why it is possible to build a maximum likelihood receiver in such a simple way is the fact that the passband basis functions constitute an orthonormal set, and their orthogonality is maintained for any channel impulse response $h(t)$.

## ■ GEOMETRIC SIGNAL-TO-NOISE RATIO

In the multichannel transmission system of Figure 6.55, each subchannel is characterized by a SNR of its own. It would be highly desirable to derive a single measure for the performance of the entire system of Figure 6.55.

To simplify the derivation of such a measure, we assume that all of the subchannels in Figure 6.55 are represented by one-dimensional constellations. Then the channel capacity of the entire system in bits per transmission is given by

$$\begin{aligned} R &= \frac{1}{N}\sum_{n=1}^{N} R_n \\ &= \frac{1}{2N}\sum_{n=1}^{N} \log_2\left(1 + \frac{P_n}{\Gamma\sigma_n^2}\right) \\ &= \frac{1}{2N}\log_2 \prod_{n=1}^{N}\left(1 + \frac{P_n}{\Gamma\sigma_n^2}\right) \\ &= \frac{1}{2}\log_2\left[\prod_{n=1}^{N}\left(1 + \frac{P_n}{\Gamma\sigma_n^2}\right)\right]^{1/N} \end{aligned} \tag{6.204}$$

Let $(\text{SNR})_{\text{overall}}$ denote the overall signal-to-noisé ratio of the entire system. We may then express $R$ in bits per transmission as

$$R = \frac{1}{2}\log_2\left(1 + \frac{(\text{SNR})_{\text{overall}}}{\Gamma}\right) \tag{6.205}$$

Comparing Equations (6.205) with (6.204), we may thus write

$$(\text{SNR})_{\text{overall}} = \Gamma\left(\prod_{n=1}^{N}\left(1 + \frac{P_n}{\Gamma\sigma_n^2}\right)^{1/N} - 1\right) \tag{6.206}$$

Assuming that $P_n/\Gamma\sigma_n^2$ is high enough to ignore the two unity terms in Equation (6.206), we may approximate the overall SNR as

$$(\text{SNR}) = \prod_{n=1}^{N}\left(\frac{P_n}{\sigma_n^2}\right)^{1/N} \tag{6.207}$$

We may thus characterize the overall system by a SNR that is the *geometric mean* of the SNRs of the individual subchannels.

The geometric SNR of Equation (6.207) can be improved considerably by distributing the available transmit power among the $N$ subchannels on a nonuniform basis. This objective is attained through the use of loading as discussed next.

## ■ LOADING OF THE MULTICHANNEL TRANSMISSION SYSTEM

Equation (6.204) for the bit rate of the entire multichannel system ignores the effect of the channel on system performance. To account for this effect, define

$$g_n = |H(f_n)|, \qquad n = 1, 2, \ldots, N \tag{6.208}$$

Then assuming that the number of subchannels $N$ is large enough, we may assume that $g_n$ is constant over the entire bandwidth $\Delta f$ assigned to subchannel $n$ for all $n$. In such a case, we may modify the second line of Equation (6.204) for the overall SNR of the system as

$$R = \frac{1}{2N} \sum_{n=1}^{N} \log_2\left(1 + \frac{g_n^2 P_n}{\Gamma \sigma_n^2}\right) \tag{6.209}$$

The $g_n^2$ and $\Gamma$ are usually fixed. The noise variance $\sigma_n^2$ is $\Delta f N_0$ for all $n$, where $\Delta f$ is the bandwidth of each subchannel and $N_0/2$ is the noise power spectral density. We may therefore optimize the overall bit rate $R$ through a proper allocation of the total transmit power among the various channels. However, for this optimization to be of practical value, we must maintain the total transmit power at some constant value $P$, say, as shown by

$$\sum_{n=1}^{N} P_n = P = \text{constant} \tag{6.210}$$

The optimization we therefore have to deal with is a *constrained optimization problem*, which may be stated as follows:

> Maximize the bit rate $R$ for the entire multichannel transmission system through an optimal sharing of the total transmit power $P$ between the $N$ subchannels, subject to the constraint that $P$ is maintained constant.

To solve this optimization problem, we first use the *method of Lagrange multipliers*[19] to set up an objective function that incorporates the constraint of Equation (6.210), as shown by

$$\begin{aligned} J &= \frac{1}{2N} \sum_{n=1}^{N} \log_2\left(1 + \frac{g_n^2 P_n}{\Gamma \sigma_n^2}\right) + \lambda\left(P - \sum_{n=1}^{N} P_n\right) \\ &= \frac{1}{2N} \log_2 e \sum_{n=1}^{N} \log_e\left(1 + \frac{g_n^2 P_n}{\Gamma \sigma_n^2}\right) + \lambda\left(P - \sum_{n=1}^{N} P_n\right) \end{aligned} \tag{6.211}$$

where $\lambda$ is the *Lagrange multiplier*. Hence, differentiating $J$ with respect to $P_n$, then setting the result equal to zero and finally rearranging terms, we get

$$\frac{\dfrac{1}{2N} \log_2 e}{P_n + \dfrac{\Gamma \sigma_n^2}{g_n^2}} = \lambda \tag{6.212}$$

This result indicates that the solution to our constrained optimization problem is to have

$$P_n + \frac{\Gamma \sigma_n^2}{g_n^2} = K \qquad \text{for } n = 1, 2, \ldots, N \tag{6.213}$$

where $K$ is a prescribed constant under the designer's control. That is, the sum of the transmit power and the noise variance (power) scaled by the ratio $\Gamma/g_n^2$ must be maintained constant for each subchannel. The process of allocating the transmit power $P$ to the individual subchannels so as to maximize the bit rate of the entire multichannel transmission system is called *loading*.

## ■ WATER-FILLING INTERPRETATION OF THE OPTIMIZATION PROBLEM

In solving the constrained optimization problem just described, two conditions must be satisfied, namely, Equations (6.210) and (6.213). The optimum solution so defined has an interesting interpretation as illustrated in Figure 6.56 for $N = 6$, assuming that the gap $\Gamma$ is constant over all the subchannels. To simplify the illustration in Figure 6.56 we have set $\sigma_n^2 = N_0 \Delta f = 1$, that is, the average noise power is unity for all $N$ subchannels. Referring to this figure, we may now make the following observations:

- The sum of power $P_n$ allocated to channel $n$ and the scaled noise power $\Gamma/g_n^2$ satisfies the constraint of Equation (6.213) for four of the subchannels for a prescribed value of the constant $K$.
- The sum of power allocations to these four subchannels consumes all the available transmit power, maintained at the constant value $P$.
- The remaining two subchannels have been eliminated from consideration because they would each require negative power to satisfy Equation (6.213) for the prescribed value of the constant $K$; this condition is clearly unacceptable.

The interpretation illustrated in Figure 6.56 prompts us to refer to the optimum solution of Equation (6.213), subject to the constraint of Equation (6.210), as the *water-filling solution*. This terminology follows from analogy of our optimization problem with a fixed amount of water (standing for transmit power) being poured into a container with a number of connected regions, each having a different depth (standing for noise power). The water distributes itself in such a way that a constant water level is attained across the whole container. We have more to say on the water-filling interpretation of information capacity in Chapter 9.

Returning to the task of how to allocate the fixed transmit power $P$ among the various subchannels of a multichannel transmission system so as to optimize the bit rate

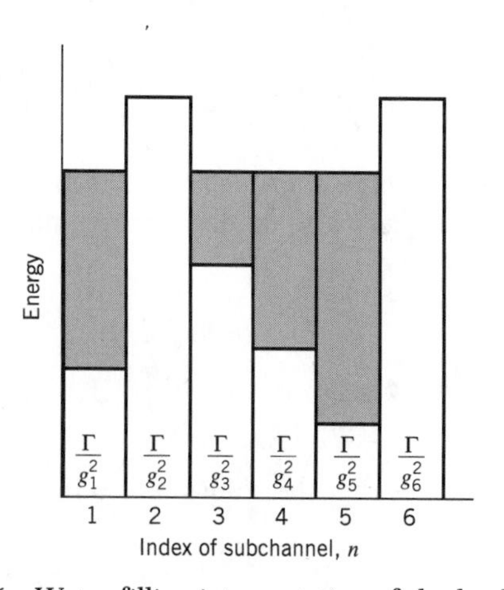


**FIGURE 6.56** Water-filling interpretation of the loading problem.

of the entire system, we may proceed as follows. Let the total transmit power be fixed at the constant value $P$ as in Equation (6.210). Let $K$ denote the constant value prescribed for the sum $P_n + \Gamma\sigma_n^2/g_n^2$ for all $n$ as in Equation (6.213). We may then use this pair of equations to set up the following system of simultaneous equations:

$$\begin{aligned} P_1 + P_2 + \cdots P_N &= P \\ P_1 - K \qquad\quad &= -\Gamma\sigma^2/g_1^2 \\ P_2 - K \qquad\quad &= -\Gamma\sigma^2/g_2^2 \\ \vdots \qquad\qquad &\qquad \vdots \\ P_N - K \qquad\quad &= -\Gamma\sigma^2/g_N^2 \end{aligned} \tag{6.214}$$

where we have a total of $(N + 1)$ unknowns and $(N + 1)$ equations to solve for them. We may rewrite this set of simultaneous equations in matrix form as

$$\begin{bmatrix} 1 & 1 & \cdots & 1 & 0 \\ 1 & 0 & \cdots & 0 & -1 \\ 0 & 1 & \cdots & 0 & -1 \\ \vdots & \vdots & & \vdots & \vdots \\ 0 & 0 & \cdots & 1 & -1 \end{bmatrix} \begin{bmatrix} P_1 \\ P_2 \\ \vdots \\ P_N \\ K \end{bmatrix} = \begin{bmatrix} P \\ -\Gamma\sigma^2/g_1^2 \\ -\Gamma\sigma^2/g_2^2 \\ \vdots \\ -\Gamma\sigma^2/g_N^2 \end{bmatrix} \tag{6.215}$$

Premultiplying both sides of Equation (6.215) by the inverse of the $(N + 1)$-by-$(N + 1)$ matrix on the left-hand side of the equation, we obtain solutions for the unknowns $P_1$, $P_2, \ldots, P_N$, and $K$. We should always find that $K$ is positive, but it is possible for some of the $P$s to be negative. The negative $P$s are discarded as power cannot be negative.

### ▶ EXAMPLE 6.7

Consider a linear channel whose squared magnitude response $|H(f)|^2$ has the piecewise-linear form shown in Figure 6.57. To simplify the example, we set the gap $\Gamma = 1$ and the noise variance $\sigma^2 = 1$. In the situation so described, the application of Equation (6.214) yields

$$\begin{aligned} P_1 + P_2 &= P \\ P_1 - K &= -1 \\ P_2 - K &= -1/l \end{aligned}$$

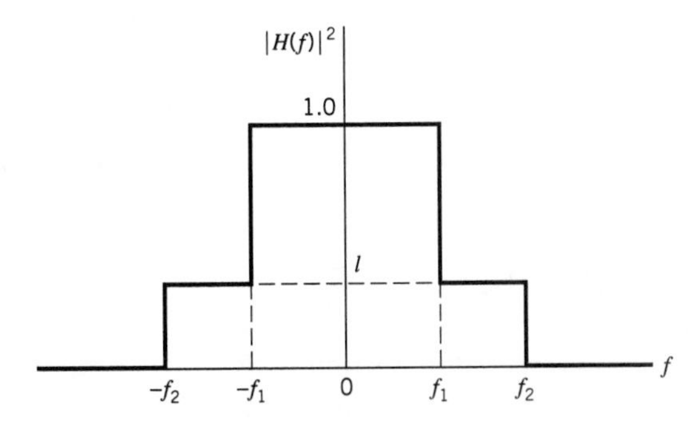


**FIGURE 6.57** Squared magnitude response for Example 6.7.

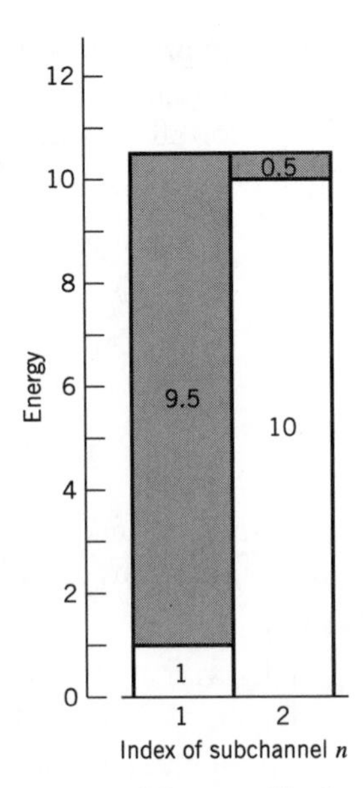


**FIGURE 6.58** Water-filling profile for Example 6.7.

where the total transmit power $P$ is normalized with respect to the noise variance. Solving these three simultaneous equations for $P_1$, $P_2$, and $K$, we get

$$P_1 = \frac{1}{2}\left(P - 1 + \frac{1}{l}\right)$$

$$P_2 = \frac{1}{2}\left(P + 1 - \frac{1}{l}\right)$$

$$K = \frac{1}{2}\left(P + 1 + \frac{1}{l}\right)$$

Since $0 < l < 1$, it follows that $P_1 > 0$, but it is possible for $P_2$ to be negative. This latter condition can arise if

$$l < \frac{1}{P + 1}$$

But then $P_1$ exceeds the prescribed value of transmit power $P$. It follows therefore that in this example the only acceptable solution is to have $1/(P + 1) < l < 1$. Suppose then we have $P = 10$ and $l = 0.1$, for which the solution is

$$K = 10.5$$
$$P_1 = 9.5$$
$$P_2 = 0.5$$

The corresponding water-filling picture is portrayed in Figure 6.58. ◀

## 6.13 *Discrete Multitone*

The material presented in Section 6.12 provides an insightful introduction to the notion of multichannel modulation. In particular, the continuous-time channel partitioning induced by the passband basis functions of Equation (6.196) or equivalently (6.199) exhibits a highly desirable property: Orthogonality of the basis functions (and therefore the channel partitioning) is preserved despite their convolution with the impulse response of the channel. However, the system has two shortcomings:

1. The passband basis functions use a sinc function that is nonzero for an infinite time interval, whereas practical considerations favor a finite observation interval.
2. For a finite number of subchannels, $N$, the system is suboptimal; optimality of the system is assured only when $N$ approaches infinity.

We may overcome these shortcomings by using *discrete multitone* (DMT), the basic idea of which is to transform a wideband channel into a set of $N$ subchannels operating in parallel. What makes DMT distinctive is the fact that the transformation is performed in discrete time as well as discrete frequency. Consequently, the transmitter input-output behavior of the entire communication system admits a linear matrix representation, which lends itself to implementation using the discrete Fourier transform.

To explore this new approach, we first recognize that in a realistic situation the channel has its nonzero impulse response, $h(t)$, essentially confined to a finite interval $[0, T_b]$. So, let the sequence $h_0, h_1, \ldots, h_\nu$ denote the baseband equivalent impulse response of the channel sampled at the rate $1/T_s$, with

$$T_b = (1 + \nu)T_s \tag{6.216}$$

The sampling rate $1/T_s$ is chosen to be greater than twice the higher frequency component of interest in accordance with the sampling theorem. To continue with the discrete-time description of the system, let $s[n] = s(nT_s)$ denote a sample of the transmitted symbol $s(t)$, $w[n] = w(nT_s)$ denote a sample of the channel noise $w(t)$, and $x[n] = x(nT_s)$ denote the corresponding sample of the channel output (received signal). The channel performs linear convolution on the incoming symbol sequence $\{s[n]\}$ of length $N$, producing a channel output sequence $\{x[n]\}$ of length $N + \nu$. Extension of the channel output sequence by $\nu$ samples compared to the channel input sequence is due to the intersymbol interference produced by the channel.

To overcome the effect of intersymbol interference, we create a cyclically extended *guard interval* whereby each symbol sequence is preceded by a periodic extension of the sequence itself. Specifically, the last $\nu$ samples of the symbol sequence are repeated at the beginning of the sequence being transmitted, as shown by

$$s[-k] = s[N - k] \qquad \text{for } k = 1, 2, \ldots, \nu \tag{6.217}$$

This condition is called a *cyclic prefix*. The *excess bandwidth factor* due to the inclusion of the cyclic prefix is therefore $\nu/N$, where $N$ is the number of transmitted samples after the guard interval.

With the cyclic prefix in place, the matrix description of the channel takes the form

$$\begin{bmatrix} x[N-1] \\ x(N-2] \\ \vdots \\ x[N-\nu-1] \\ x[N-\nu-2] \\ \vdots \\ x[0] \end{bmatrix} = \begin{bmatrix} h_0 & h_1 & h_2 & \cdots & h_{\nu-1} & h_\nu & 0 & \cdots & 0 \\ 0 & h_0 & h_1 & \cdots & h_{\nu-2} & h_{\nu-1} & h_\nu & \cdots & 0 \\ \vdots & \vdots & \vdots & & \vdots & \vdots & \vdots & & \vdots \\ 0 & 0 & 0 & \cdots & 0 & h_0 & h_1 & \cdots & h_\nu \\ h_\nu & 0 & 0 & \cdots & 0 & 0 & h_0 & \cdots & h_{\nu-1} \\ \vdots & \vdots & \vdots & & \vdots & \vdots & \vdots & & \vdots \\ h_1 & h_2 & h_3 & \cdots & h_\nu & 0 & 0 & \cdots & h_0 \end{bmatrix} \begin{bmatrix} s(N-1) \\ s(N-2) \\ \vdots \\ s[N-\nu-1] \\ s[N-\nu-2] \\ \vdots \\ s[0] \end{bmatrix} + \begin{bmatrix} w[N-1] \\ w[N-2] \\ \vdots \\ w[N-\nu-1] \\ w[N-\nu-2] \\ \vdots \\ w[0] \end{bmatrix} \tag{6.218}$$

Equivalently, we may describe the discrete-time representation of the channel in the compact matrix form

$$\mathbf{x} = \mathbf{H}\mathbf{s} + \mathbf{w} \tag{6.219}$$

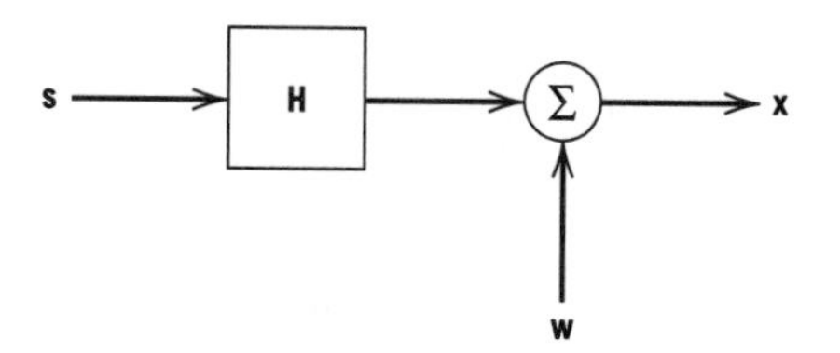


**FIGURE 6.59** Discrete-time representation of multichannel data transmission system.

where the transmitted symbol vector **s**, the channel noise vector **w**, and the received signal vector **x** are all $N$-by-1 vectors which are respectively defined by

$$\mathbf{s} = [s[N-1], s[N-2], \ldots, s[0]]^T \tag{6.220}$$

$$\mathbf{w} = [w[N-1], w[N-2], \ldots, w[0]]^T \tag{6.221}$$

and

$$\mathbf{x} = [x[N-1], x[N-2], \ldots, x[0]]^T \tag{6.222}$$

We may thus depict the discrete-time representation of the channel as in Figure 6.59. The $N$-by-$N$ channel matrix **H** is defined by

$$\mathbf{H} = \begin{bmatrix} h_0 & h_1 & h_2 & \cdots & h_{\nu-1} & h_\nu & 0 & \cdots & 0 \\ 0 & h_0 & h_1 & \cdots & h_{\nu-2} & h_{\nu-1} & h_\nu & \cdots & 0 \\ \vdots & \vdots & \vdots & & \vdots & \vdots & \vdots & & \vdots \\ 0 & 0 & 0 & \cdots & 0 & h_0 & h_1 & \cdots & h_\nu \\ h_\nu & 0 & 0 & \cdots & 0 & 0 & h_0 & \cdots & h_{\nu-1} \\ \vdots & \vdots & \vdots & & \vdots & \vdots & \vdots & & \vdots \\ h_1 & h_2 & h_3 & \cdots & h_\nu & 0 & 0 & \cdots & h_0 \end{bmatrix} \tag{6.223}$$

From this definition, we readily see that the matrix **H** has the following structural composition: Every row of the matrix is obtained by applying a right-shift to the previous row by one position, with the added proviso that the rightmost element of the previous row spills over in the shifting process to be "circulated" back to the leftmost element of the new row. Accordingly, the matrix **H** is referred to as a *circulant matrix*.

Before proceeding further, it is befitting that we briefly review the discrete Fourier transform and its role in the spectral decomposition of the circulant matrix **H**.

## ■ DISCRETE FOURIER TRANSFORM

Consider the $N$-by-1 vector **x** of Equation (6.222). The *discrete Fourier transform* (DFT) of the vector **x** is defined by the $N$-by-1 vector

$$\mathbf{X} = [X[N-1], X[N-2], \ldots, X[0]]^T \tag{6.224}$$

where

$$X[k] = \frac{1}{\sqrt{N}} \sum_{n=1}^{N-1} x[n] \exp\left(-j\frac{2\pi}{N}kn\right), \qquad k = 0, 1, \ldots, N-1 \tag{6.225}$$

The exponential term $\exp(-j2\pi kn/N)$ is referred to as the *kernel* of the DFT. Correspondingly, the *inverse discrete Fourier transform (IDFT)* of the $N$-by-1 vector **X** is defined by

$$x[n] = \frac{1}{\sqrt{N}} \sum_{k=0}^{N-1} X[k] \exp\left(j\frac{2\pi}{N}kn\right), \qquad n = 0, 1, \ldots, N-1 \tag{6.226}$$

Although Equations (6.225) and (6.226) appear to be similar, they have different interpretations. Given the signal vector **x**, Equation (6.225) provides a spectral representation of the signal computed at a set of discrete frequencies: $f_k = k/N$, which are normalized with respect to the sampling rate. Given the transformed vector **X**, Equation (6.226) recovers the original signal vector **x**. We may therefore view Equation (6.225) as the *analysis equation* and Equation (6.226) as the *synthesis equation*.

An important property of a circulant matrix, exemplified by the channel matrix **H** of Equation (6.223), is that it permits *spectral decomposition* as shown by

$$\mathbf{H} = \mathbf{Q}^{\dagger}\mathbf{\Lambda}\mathbf{Q} \tag{6.227}$$

where the superscript $\dagger$ denotes *Hermitian transposition* (i.e., the combination of complex conjugation and ordinary matrix transposition). Descriptions of the matrices **Q** and **Λ** are presented in the sequel in that order.

The matrix **Q** is a square matrix defined in terms of the kernel of the $N$-point DFT as follows:

$$\mathbf{Q} = \frac{1}{\sqrt{N}}\begin{bmatrix} \exp\left(-j\frac{2\pi}{N}(N-1)(N-1)\right) & \cdots & \exp\left(-j\frac{2\pi}{N}2(N-1)\right) & \exp\left(-j\frac{2\pi}{N}(N-1)\right) & 1 \\ \exp\left(-j\frac{2\pi}{N}(N-1)(N-2)\right) & \cdots & \exp\left(-j\frac{2\pi}{N}2(N-2)\right) & \exp\left(-j\frac{2\pi}{N}(N-2)\right) & 1 \\ \vdots & & \vdots & \vdots & \vdots \\ \exp\left(-j\frac{2\pi}{N}(N-1)\right) & \cdots & \exp\left(-j\frac{2\pi}{N}2\right) & \exp\left(-j\frac{2\pi}{N}\right) & 1 \\ 1 & \cdots & 1 & 1 & 1 \end{bmatrix} \tag{6.228}$$

From this definition, we readily see that the $kl$th element of the $N$-by-$N$ matrix, **Q**, starting from the *bottom right* at $k = 0$ and $l = 0$ and counting up step-by-step, is

$$q_{kl} = \frac{1}{\sqrt{N}}\exp\left(-j\frac{2\pi}{N}kl\right), \qquad (k, l) = 0, 1, \ldots, N-1 \tag{6.229}$$

The matrix **Q** is an *orthonormal matrix* or *unitary matrix* in that it satisfies the condition

$$\mathbf{Q}^{\dagger}\mathbf{Q} = \mathbf{I} \tag{6.230}$$

where **I** is the identity matrix. That is, the inverse matrix of **Q** is equal to the Hermitian transpose of **Q**.

The matrix **Λ** is a *diagonal matrix* that contains the $N$ discrete Fourier transform values of the sequence $h_0, h_1, \ldots, h_\nu$ characterizing the channel. Denoting these transform values by $\lambda_{N-1}, \ldots, \lambda_1, \lambda_0$, we may express **Λ** as

$$\mathbf{\Lambda} = \begin{bmatrix} \lambda_{N-1} & 0 & \cdots & 0 \\ 0 & \lambda_{N-2} & \cdots & 0 \\ \vdots & \vdots & & \vdots \\ 0 & 0 & \cdots & \lambda_0 \end{bmatrix} \tag{6.231}$$

(The $\lambda$s here are not to be confused with the Lagrange multipliers in Section 6.12.)

The DFT has established itself as one of the principal tools of digital signal processing by virtue of its efficient computation using the *fast Fourier transform (FFT) algorithm*.[20]

Specifically, the FFT algorithm requires on the order of $N \log_2 N$ operations rather than the $N^2$ operations for direct computation of the DFT. For efficient implementation of the FFT algorithm, we should choose the block length $N$ an integer power of two. The computational savings obtained by using the FFT algorithm are made possible by exploiting the special structure of the DFT defined in Equation (6.225). Moreover, these savings become more substantial as we increase the data length $N$.

## ■ FREQUENCY-DOMAIN DESCRIPTION OF THE CHANNEL

With this brief description of the DFT on hand, we are ready to resume our discussion of discrete multitone. First, we define

$$\mathbf{s} = \mathbf{Q}^{\dagger}\mathbf{S} \tag{6.232}$$

where $\mathbf{S}$ is the frequency-domain vector representation of the transmitter input. Each element of the $N$-by-1 vector $\mathbf{S}$ may be viewed as a complex-valued point in a two-dimensional QAM signal constellation. Given the channel output vector $\mathbf{x}$, we define its corresponding frequency-domain representation as

$$\mathbf{X} = \mathbf{Q}\mathbf{x} \tag{6.233}$$

Using Equations (6.227), (6.232) and (6.233), we may rewrite Equation (6.219) in the equivalent form

$$\mathbf{X} = \mathbf{Q}(\mathbf{Q}^{\dagger}\mathbf{\Lambda}\mathbf{Q}\mathbf{Q}^{\dagger}\mathbf{S} + \mathbf{W}) \tag{6.234}$$

Hence, using the relation of Equation (6.230), we simply get

$$\mathbf{X} = \mathbf{\Lambda}\mathbf{S} + \mathbf{W} \tag{6.235}$$

where

$$\mathbf{W} = \mathbf{Q}\mathbf{w} \tag{6.236}$$

In expanded form, Equation (6.235) reads as

$$X_k = \lambda_k S_k + W_k, \qquad k = 0, 1, \ldots, N-1 \tag{6.237}$$

where the set of frequency-domain values $\{\lambda_k\}_{k=0}^{N-1}$ is known for a prescribed channel.

For a channel with additive white noise, Equation (6.237) implies that the receiver is composed of a set of independent processors operating in parallel. With the $\lambda_k$ all known, we may thus use the block of frequency-domain values $\{X_k\}_{k=0}^{N-1}$ to compute estimates of the corresponding transmitted block of frequency domain-values $\{S_k\}_{k=0}^{N-1}$.

## ■ DFT-BASED DMT SYSTEM

Equations (6.235), (6.225), (6.226), and (6.237) provide the mathematical basis for the implementation of DMT using the DFT. Figure 6.60 illustrates the block diagram of the system derived from these equations and their practical implications.

The transmitter consists of the following functional blocks:

- *Demultiplexer*, which converts the incoming serial data stream into parallel form.
- *Constellation encoder*, which maps the parallel data into $N/2$ multibit subchannels with each subchannel being represented by a QAM signal constellation. Bit allocation among the subchannels is also performed here in accordance with a loading algorithm.

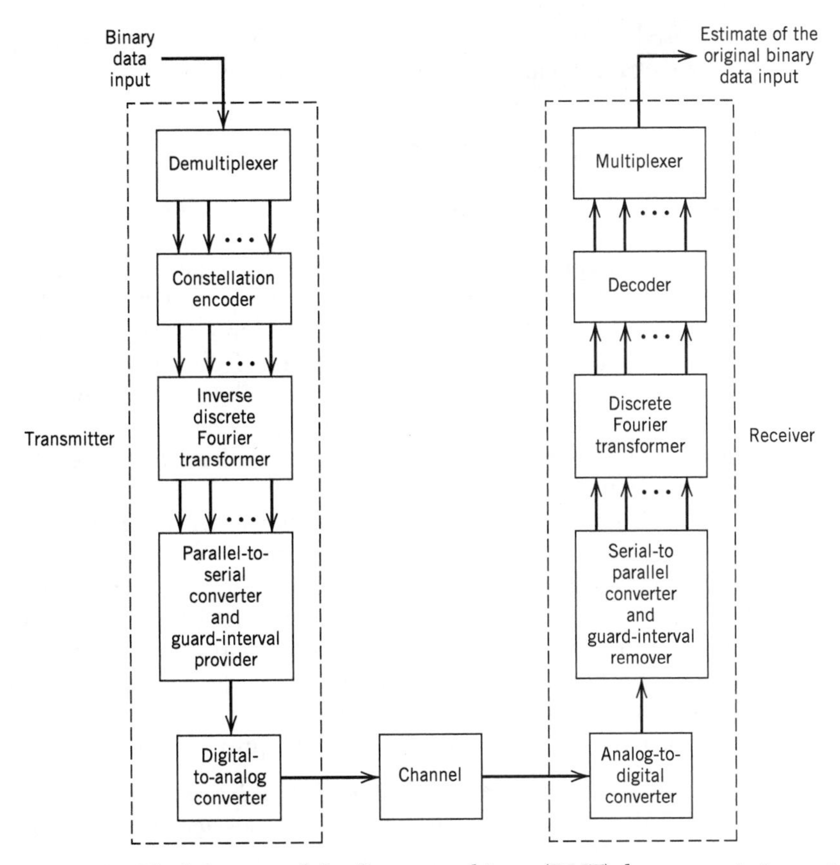


**FIGURE 6.60** Block diagram of the discrete-multitone (DMT) data-transmission system.

- *Inverse discrete Fourier transformer (IDFT)*, which transforms the frequency-domain parallel data at the constellation encoder output into parallel time-domain data. For efficient implementation of the IDFT using the fast Fourier transform (FFT) algorithm, we need to choose $N = 2^k$ where $k$ is a positive integer.
- *Parallel-to-serial converter*, which converts the parallel time-domain data into serial form. Guard intervals stuffed with cyclic prefixes are inserted into the serial data on a periodic basis before conversion into analog form.
- *Digital-to-analog converter (DAC)*, which converts the digital data into analog form ready for transmission over the channel.

Typically, the DAC includes a transmit filter. Accordingly, the time function $h(t)$ should be redefined as the combined impulse response of the cascade connection of the transmit filter and the channel.

The receiver performs the inverse operations of the transmitter, as described here:

- *Analog-to-digital converter (ADC)*, which converts the analog channel output into digital form.
- *Serial-to-parallel converter*, which converts the resulting bit stream into parallel form. Before this conversion takes place, the guard intervals (cyclic prefixes) are removed.
- *Discrete Fourier transformer (DFT)*, which transforms the time-domain parallel data into frequency-domain parallel data; as with the IDFT, the FFT algorithm is used to implement the DFT.

- *Decoder*, which uses the DFT output to compute estimates of the original multi-bit subchannel data supplied to the transmitter.
- *Multiplexer*, which combines the estimates so computed to produce a reconstruction of the transmitted serial data stream.

## ■ APPLICATIONS OF DMT

An important application of DMT is in the transmission of data over two-way channels. Indeed, DMT has been standardized for use on *asymmetric digital subscriber lines* (ADSLs) using twisted pairs. The ADSL was described in Chapter 4. For example, DMT provides for the transmission of data downstream (i.e., from an Internet service provider to a subscriber) at the DS1 rate of 1.544 Mb/s and the simultaneous transmission of data upstream (i.e., from the subscriber to the Internet service provider) at 160 kb/s. This kind of data transmission capability is well suited for handling data-intensive applications such as video-on-demand.

DMT is also a core technology in implementing the asymmetric *very-high-rate digital subscriber lines*[21] (VDSL), which differs from all other DSL transmission techniques because of its ability to deliver extremely high data rates. For example, VDSL can provide data rates of 13 to 26 Mb/s downstream and 2 to 3 MB/s upstream over twisted pairs that emanate from an optical network unit and connect to the subscriber over distances less than about 1 km. These high data rates allow the delivery of digital TV, super-fast Web surfing and file transfer, and virtual offices at home.

The use of DMT for ADSL and VDSL provides a number of advantages:

- *The ability to maximize the transmitted bit rate*, which is provided by tailoring the distribution of information-bearing signals across the channel according to channel attenuation and noise conditions.
- *Adaptivity to changing line conditions*, which is realized by virtue of the fact that the channel is partitioned into a number of subchannels.
- *Reduced sensitivity to impulse noise*, which is achieved by spreading its energy over the many subchannels of the receiver. As the name implies, *impulse noise* is characterized by long, quiet intervals followed by narrow pulses of randomly varying amplitude. In an ADSL or VDSL environment, impulse noise arises due to switching transients coupled to wire pairs in the central office and to various electrical devices on the user's premises.

## ■ COMPARISON OF DIGITAL SUBSCRIBER LINES AND VOICEBAND MODEMS

In Section 6.11 we discussed voiceband modems that are already close to operating at their theoretical limits of 33.6 kb/s upstream and 56 kb/s downstream. In this section we have discussed the application of DMT to VDSLs that can operate at data rates of about 2 to 3 Mb/s upstream and 13 to 26 Mb/s downstream. These two vastly different sets of upstream/downstream data rates prompt the following question: How is it possible for VDSL to operate at rates about three orders of magnitude faster than voiceband modems over the same twisted pairs (i.e., phone lines)? The reason for this vast difference in operating data rates between voiceband modems and VDSLs is not the twisted pairs; rather, it is the *digital switches* built into a public switched telephone network that prevent the transport of broadband data to subscribers (users) via voiceband modems. Simply put, the digital switches treat digital data in the same way as voice signals for which they are primarily designed.

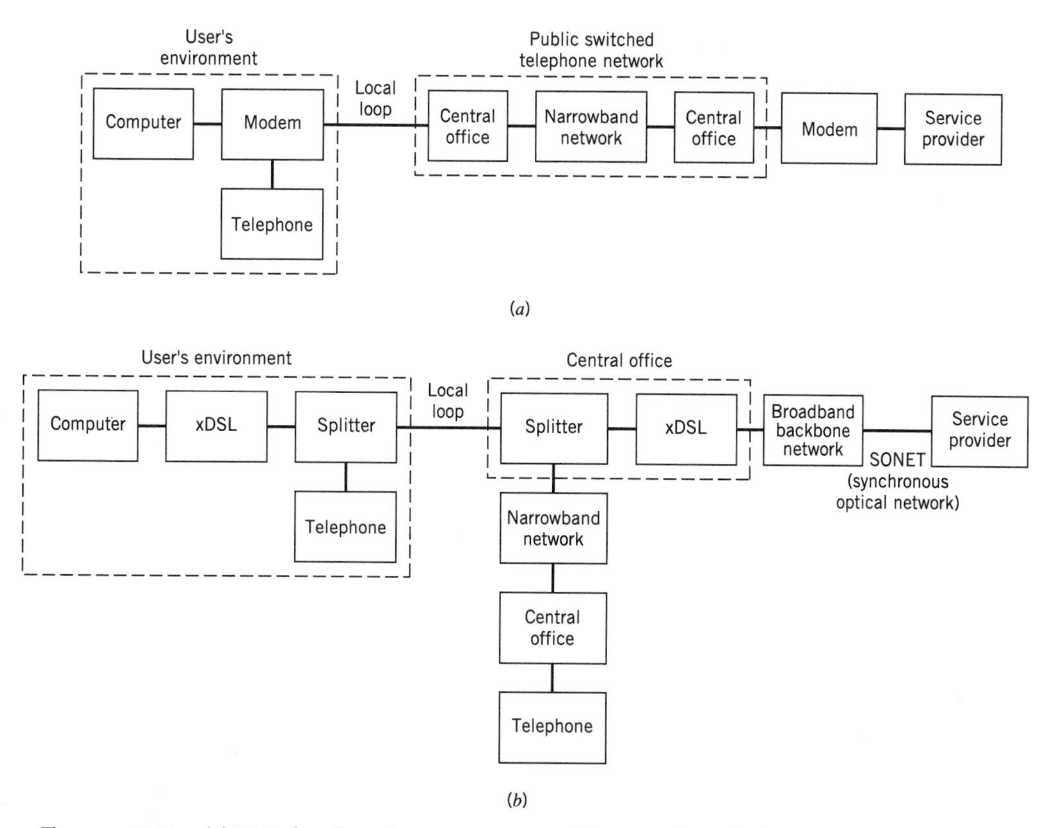


**FIGURE 6.61** (*a*) Voiceband modem environment. (*b*) xDSL (digital subscriber line) environment, where x stands for "asymmetric" or "very high-rate."

Figure 6.61 highlights the operational environments of voiceband modems and xDSLs, where x stands for A in ADSL and V in VDSL. In the model of Figure 6.61*a* pertaining to a voiceband modem, we have a relatively long transmission path between an Internet service provider (ISP) and a subscriber. Most importantly, the transmission path traverses through a narrowband public switched telephone network (PSTN), which limits the available channel bandwidth to about 3.5 kHz. In contrast, in the model of Figure 6.61*b* pertaining to xDSL, the transmission path accommodates the transport of broadband data between the ISP and subscriber via a broadband integrated services digital network and a relatively short local loop consisting of a twisted pair. The system permits the coexistence of POTS and xDSL signals on the same local loop, which is made possible through the use of a pair of splitters, as indicated in Figure 6.61*b*; splitters, consisting of bidirectional low-pass and high-pass filters, are discussed in Section 4.8.

## ■ ORTHOGONAL FREQUENCY DIVISION MULTIPLEXING[22]

Discrete multitone is one particular discrete form of multichannel modulation. Another closely related form of this method of modulation is *orthogonal frequency-division multiplexing* (OFDM) that differs from DMT in areas of application and some aspects of its design.

OFDM is used for data transmission over radio broadcast channels and wireless communication channels. This domain of application requires some changes to the design of the OFDM system. Unlike DMT that uses loading for bit allocation, OFDM uses a fixed number of bits per subchannel. This restriction is made necessary by the fact that a broadcast channel involves one-way transmission, and in a wireless communications environment the channel is varying too rapidly. Accordingly, in both cases it is not feasible for the transmitter to know the channel and how to "load" it.

Thus, the block diagram of Figure 6.60 applies equally to OFDM except for the fact that the signal constellation encoder does not include a loading algorithm for bit allocation. In addition, two other changes have to be made to the design of the system:

- In the transmitter, an *upconverter* is included after the digital-to-analog converter to translate the transmitted frequency, thereby facilitating the propagation of the transmitted signal over a radio channel.
- In the receiver, a *downconverter* is included before the analog-to-digital converter to undo the frequency translation that was performed by the upconverter in the transmitter.

Applications of OFDM include the following:

1. *Wireless communications.*
OFDM, combined with coding and interleaving, provides an effective technique to combat multipath fading that is a characteristic feature of wireless communication channels.

2. *Digital audio broadcasting.*
OFDM has been adopted as the standard for digital audio broadcasting in Europe. Here again the system involves the combined use of coding and interleaving.

(Error-control coding and related issues are discussed in Chapter 10.)

## 6.14 Synchronization

The coherent reception of a digitally modulated signal, irrespective of its form, requires that the receiver be synchronous to the transmitter. We say that two sequences of events (representing a transmitter and a receiver) are *synchronous* relative to each other when the events in one sequence and the corresponding events in the other occur simultaneously. The process of making a situation synchronous, and maintaining it in this condition, is called *synchronization.*[23]

From the discussion presented on the operation of digital modulation techniques, we recognize the need for two basic modes of synchronization:

1. When coherent detection is used, knowledge of both the frequency and phase of the carrier is necessary. The estimation of carrier phase and frequency is called *carrier recovery* or *carrier synchronization.*
2. To perform demodulation, the receiver has to know the instants of time at which the modulation can change its state. That is, it has to know the starting and finishing times of the individual symbols, so that it may determine when to sample and when to quench the product-integrators. The estimation of these times is called *clock recovery* or *symbol synchronization.*

These two modes of synchronization can be coincident with each other, or they can occur sequentially one after the other. Naturally, in a noncoherent system, carrier synchronization is of no concern.

Synchronization can be implemented in one of two fundamentally different ways:

1. *Data-aided synchronization.*
In data-aided synchronization systems, a preamble is transmitted along with the data-bearing signal in a time-multiplexed manner on a periodic basis. The preamble contains information about the carrier and symbol timing, which is extracted by appropriate processing of the channel output at the receiver. Such an approach is commonly used in digital satellite and wireless communications, where the motivation is to minimize the time required to synchronize the receiver to the transmitter. Its limitations are two-fold: (1) reduced data-throughput efficiency that is incurred by assigning a certain portion of each transmitted frame to the preamble, and (2) reduced power efficiency by allocating a certain fraction of the transmitted power to the transmission of the preamble.

2. *Nondata-aided synchronization.*
In this second approach, the use of a preamble is avoided, and the receiver has the task of establishing synchronization by extracting the necessary information from the modulated signal. Both throughput and power efficiency are thereby improved but at the expense of an increase in the time taken to establish synchronization.

In any event, synchronization is basically a statistical parameter estimation problem. A principled approach for solving such a problem is *maximum likelihood estimation* (see Section 5.5), which proceeds by first formulating a log-likelihood function of the parameter of interest given the received signal. This formulation is relatively straightforward by treating the channel noise as a Gaussian process. Most important, it requires no prior information about the modulated signal.

In this section we confine our attention to nondata-aided forms of carrier and timing synchronization systems. In this context, we may identify two approaches for solving the synchronization problem, given a modulated signal with suppressed carrier to conserve power:

1. *Classical approach.*
An essential building block in the classical approach to synchronization is the *phase-locked loop*. (The phase-locked loop was discussed in Chapter 2.) Specifically, for carrier recovery the receiver requires the use of a *suppressed-carrier tracking loop* for providing a coherent secondary carrier (subcarrier) reference. For example, we may use a variant of the Costas loop or the *Mth power loop* for $M$-ary PSK. The standard Costas loop for double sideband-suppressed carrier (DSB-SC) modulation was discussed in Chapter 2. As for the $M$th power loop, it consists of the cascade connection of an $M$th power-law device, band-pass filter, phase-locked loop, and frequency divider by $M$. The objective here is to exploit the acquisition and tracking properties of the phase-locked loop. For further discussion of the $M$th power loop, the reader is referred to Problem 6.47.

2. *Algorithmic (modern) approach.*
In the modern approach, the solution to maximum likelihood estimation is formulated in algorithmic form using discrete-time signal processing. Specifically, implementation of the synchronizer is built on an algorithm that provides an estimate of carrier phase or symbol timing on an iteration-by-iteration basis. The processing is performed in the baseband domain to pave the way for the use of discrete-time (digital) signal processing.

In this section we describe the algorithmic approach to synchronization for $M$-ary PSK systems for both carrier recovery and symbol-timing recovery.

The approach taken in the exposition is sequential in that timing recovery is performed *before* phase recovery. The reason for so doing is that if we know the group delay incurred by transmission through the channel, then one sample per symbol at the matched filter output in the receiver is sufficient for estimating the unknown carrier phase. Moreover, the computational complexity of the receiver is minimized by using synchronization algorithms that operate at the symbol rate $1/T$.

## ■ DECISION-DIRECTED RECURSIVE ALGORITHM FOR PHASE RECOVERY

As remarked earlier, the first important step in solving the synchronization problem is to formulate the log-likelihood function for the carrier phase $\theta$, given the Gaussian noise-contaminated received signal. Let $l(\theta)$ denote this log-likelihood function, which serves as the objective function for estimating $\theta$. The next step is to determine the derivative of $l(\theta)$ with respect to $\theta$. The final step is to formulate a recursive (iterative) algorithm for computing a maximum likelihood estimate of the unknown $\theta$ in a step-by-step manner.

### *Evaluation of $\partial l(\theta)/\partial \theta$**

Let $s_k(t)$ denote the transmitted signal for symbol $k = 0, 1, \ldots, M-1$:

$$s_k(t) = \sqrt{\frac{2E}{T}}\cos(2\pi f_c t + \alpha_k), \qquad 0 \le t \le T \tag{6.238}$$

where $E$ is the symbol energy, $T$ is the symbol period, and

$$\alpha_k = 0, \frac{2\pi}{M}, \ldots, (M-1)\frac{2\pi}{M} \tag{6.239}$$

Equivalently, we may write

$$s_k(t) = \sqrt{\frac{2E}{T}}\cos(2\pi f_c t + \alpha_k)g(t) \tag{6.240}$$

where $g(t)$ is the shaping pulse, namely, a rectangular pulse of unit amplitude and duration $T$. Let $\tau_c$ denote the carrier (phase) delay, and $\tau_g$ denote the envelope (group) delay, both of which are introduced by the channel. By definition, $\tau_c$ affects the carrier and $\tau_g$ affects the envelope. Then the received signal is

$$\begin{aligned} x(t) &= \sqrt{\frac{2E}{T}}\cos(2\pi f_c(t-\tau_c) + \alpha_k)g(t-\tau_g) + w(t) \\ &= \sqrt{\frac{2E}{T}}\cos(2\pi f_c t + \theta + \alpha_k)g(t-\tau_g) + w(t) \end{aligned} \tag{6.241}$$

where $w(t)$ is the channel noise and $\theta$ is defined as $-2\pi f_c \tau_c$ to be consistent with the notation in Section 6.6. Both the carrier phase $\theta$ and group delay $\tau_g$ are unknown. However, it is assumed that they remain constant over the observation interval $0 \le t \le T_0$ or through the transmission of $L_0 = T_0/T$ symbols. Equivalently, we may write (using $\tau$ in place of $\tau_g$ to simplify matters)

$$x(t) = \sqrt{\frac{2E}{T}}\cos(2\pi f_c t + \theta + \alpha_k) + w(t), \qquad \tau \le t \le T + \tau \tag{6.242}$$

---

*A reader who is not interested in the formal derivation of $\partial l(\theta)/\partial\theta$ may omit this subsection and move onto the next subsection without loss of continuity.

At the receiver the basis functions are defined by

$$\phi_1(t) = \sqrt{\frac{2}{T}}\cos(2\pi f_c t), \qquad \tau \le t \le T + \tau \tag{6.243}$$

$$\phi_2(t) = \sqrt{\frac{2}{T}}\sin(2\pi f_c t), \qquad \tau \le t \le T + \tau \tag{6.244}$$

Here it is assumed that the receiver has perfect knowledge of the carrier frequency $f_c$; otherwise, a carrier frequency offset has to be included, which complicates the analysis. Accordingly, we may represent the received signal $x(t)$ by the vector

$$\mathbf{x}(\tau) = \begin{bmatrix} x_1(\tau) \\ x_2(\tau) \end{bmatrix} \tag{6.245}$$

where

$$x_i(\tau) = \int_{\tau}^{T+\tau} x(t)\phi_i(t)\,dt, \qquad i = 1, 2 \tag{6.246}$$

In a corresponding fashion, we may express the signal component of $\mathbf{x}(t)$ by the vector

$$\mathbf{s}(a_k, \theta, \tau) = \begin{bmatrix} s_1(a_k, \theta, \tau) \\ s_2(a_k, \theta, \tau) \end{bmatrix} \tag{6.247}$$

where $a_k$ is the transmitted symbol and

$$s_i(a_k, \theta, \tau) = \int_{\tau}^{T+\tau} \sqrt{\frac{2E}{T}}\cos(2\pi f_c t + \theta + \alpha_k)\phi_i(t)\,dt \text{ for } i = 1, 2 \tag{6.248}$$

Assuming that $f_c$ is an integer multiple of the symbol rate $1/T$, we have

$$s_1(a_k, \theta, \tau) = \sqrt{E}\cos(\theta + \alpha_k) \tag{6.249}$$

$$s_2(a_k, \theta, \tau) = -\sqrt{E}\sin(\theta + \alpha_k) \tag{6.250}$$

We may thus write

$$\mathbf{x}_k(\tau) = \mathbf{s}(a_k, \theta, \tau) + \mathbf{w} \tag{6.251}$$

where $\mathbf{w}$ is the noise vector

$$\mathbf{w} = \begin{bmatrix} w_1 \\ w_2 \end{bmatrix} \tag{6.252}$$

with

$$w_i = \int_{\tau}^{T+\tau} w(t)\phi_i(t)\,dt, \qquad i = 1, 2 \tag{6.253}$$

The $w_i$ is the sample value of a Gaussian random variable $W$ of zero mean and variance $N_0/2$, where $N_0/2$ is the (two-sided) power spectral density of the channel noise $w(t)$.

The conditional probability density function of the random vector $\mathbf{X}$, given the transmission of symbol $a_k$ and the occurrence of carrier phase $\theta$ and group delay $\tau$, is

$$f_{\mathbf{X}}(\mathbf{x} \mid a_k, \theta, \tau) = \frac{1}{\pi N_0}\exp\left(-\frac{1}{N_0}\,\| \mathbf{x}_k(\tau) - \mathbf{s}(a_k, \theta, \tau) \|^2\right) \tag{6.254}$$

For $a_k = 0$ the received signal $x(t)$ equals the channel noise $w(t)$, so

$$f_{\mathbf{X}}(\mathbf{x}|a_k = 0) = \frac{1}{\pi N_0} \exp\left(-\frac{1}{N_0} \| \mathbf{x}_k(\tau) \|^2\right) \tag{6.255}$$

Hence we may define the likelihood function for $M$-ary PSK at the receiver as

$$\begin{aligned} L(a_k, \theta, \tau) &= \frac{f_{\mathbf{X}}(\mathbf{x}|a_k, \theta, \tau)}{f_{\mathbf{X}}(\mathbf{x}|a_k = 0)} \\ &= \exp\left(\frac{2}{N_0} \mathbf{x}_k^T(\tau)\mathbf{s}(a_k, \theta, \tau) - \frac{1}{N_0} \| \mathbf{s}(a_k, \theta, \tau) \|^2\right) \end{aligned} \tag{6.256}$$

In $M$-ary PSK,

$$\| \mathbf{s}(a_k, \theta, \tau) \| = \text{constant}$$

as the message points lie on a circle of radius $\sqrt{E}$. Hence, ignoring the second term in the exponent, we may simplify the likelihood function as

$$L(a_k, \theta, \tau) = \exp\left(\frac{2}{N_0} \mathbf{x}_k^T(\tau)\mathbf{s}(a_k, \theta, \tau)\right) \tag{6.257}$$

Assuming that we transmit a sequence of $L_0$ statistically independent symbols, namely,

$$\mathbf{a} = [a_0, a_1, \ldots, a_{L_0-1}]^T \tag{6.258}$$

the resulting likelihood function is

$$L(\mathbf{a}, \theta, \tau) = \prod_{k=0}^{L_0-1} \exp\left(\frac{2}{N_0} \mathbf{x}_k^T(\tau)\mathbf{s}(a_k, \theta, \tau)\right) \tag{6.259}$$

The log-likelihood function is therefore

$$\begin{aligned} l(a, \theta, \tau) &= \log L(a, \theta, \tau) \\ &= \frac{2}{N_0} \sum_{k=0}^{L_0-1} \mathbf{x}_k^T(\tau)\mathbf{s}(a_k, \theta, \tau) \end{aligned} \tag{6.260}$$

From Equations (6.249) and (6.250) we deduce

$$\begin{aligned} \hat{\mathbf{s}}_k(\theta) &= \mathbf{s}(\hat{a}_k, \theta, \tau) \\ &= \sqrt{E} \begin{bmatrix} \cos(\hat{\alpha}_k + \theta) \\ -\sin(\hat{\alpha}_k + \theta) \end{bmatrix}, \qquad k = 0, 1, \ldots, L_0 - 1 \end{aligned} \tag{6.261}$$

where $\hat{\alpha}_k$ is an estimate of the actual $\alpha_k$ produced at the detector output for the symbol $a_k$. Correspondingly, we may express the matched filter output as

$$\mathbf{x}_k = \begin{bmatrix} x_{1,k} \\ -x_{2,k} \end{bmatrix}$$

Hence, using this definition and Equation (6.261) in Equation (6.260), we get

$$\begin{aligned} l(\theta) &= \frac{2\sqrt{E}}{N_0} \sum_{k=0}^{L_0-1} [x_{1,k} \cos(\hat{\alpha}_k + \theta) + x_{2,k} \sin(\hat{\alpha}_k + \theta)] \\ &= \frac{2\sqrt{E}}{N_0} \sum_{k=0}^{L_0-1} [(x_{1,k} \cos \hat{\alpha}_k + x_{2,k} \sin \hat{\alpha}_k) \cos \theta \\ &\qquad - (x_{1,k} \sin \hat{\alpha}_k - x_{2,k} \cos \hat{\alpha}_k) \sin \theta] \end{aligned} \tag{6.262}$$

Differentiating $l(\theta)$ with respect to $\theta$, we obtain

$$\begin{aligned}\frac{\partial l(\theta)}{\partial \theta} = -\frac{2\sqrt{E}}{N_0}\sum_{k=0}^{L_0-1} &[(x_{1,k}\cos\hat{\alpha}_k + x_{2,k}\sin\hat{\alpha}_k)\sin\theta \\ &+ (x_{1,k}\sin\hat{\alpha}_k - x_{2,k}\cos\hat{\alpha}_k)\cos\theta]\end{aligned} \tag{6.263}$$

We may simplify Equation (6.263) by introducing the following notations:

$$\tilde{x}_k = x_{1,k} + jx_{2,k} \tag{6.264}$$

and

$$\begin{aligned}a_k &= e^{j\alpha_k} \\ &= \cos\alpha_k + j\sin\alpha_k\end{aligned} \tag{6.265}$$

where $\tilde{x}_k$ is the complex envelope (i.e., baseband value) of the matched filter output due to the $k$th transmitted symbol, and $a_k$ is a symbol indicator in the message constellation of the $M$-ary PSK. We may thus write

$$\begin{aligned}\text{Re}[\hat{a}_k^*\tilde{x}_k] &= \text{Re}[(\cos\hat{\alpha}_k - j\sin\hat{\alpha}_k)(x_{1,k} + jx_{2,k})] \\ &= x_{1,k}\cos\hat{\alpha}_k + x_{2,k}\sin\hat{\alpha}_k\end{aligned} \tag{6.266}$$

$$\begin{aligned}\text{Im}[\hat{a}_k^*\tilde{x}_k] &= \text{Im}[(\cos\hat{\alpha}_k - j\sin\hat{\alpha}_k)(x_{1,k} + jx_{2,k})] \\ &= -x_{1,k}\sin\hat{\alpha}_k + x_{2,k}\cos\hat{\alpha}_k\end{aligned} \tag{6.267}$$

We may also note from Euler's formula:

$$e^{-j\theta} = \cos\theta - j\sin\theta \tag{6.268}$$

Accordingly, we may rewrite Equation (6.263) in the compact form:

$$\begin{aligned}\frac{\partial l(\theta)}{\partial \theta} &= \frac{2\sqrt{E}}{N_0}\sum_{k=0}^{L_0-1}\{(\text{Re}[\hat{a}_k^*\tilde{x}_k])(\text{Im}[e^{-j\theta}]) + (\text{Im}[\hat{a}_k^*\tilde{x}_k])(\text{Re}[e^{-j\theta}])\} \\ &= \frac{2\sqrt{E}}{N_0}\sum_{k=0}^{L_0-1}\text{Im}[\hat{a}_k^*\tilde{x}_k e^{-j\theta}]\end{aligned} \tag{6.269}$$

where $\hat{a}_k$ is an estimate of $a_k$, and the asterisk denotes complex conjugation.

## Recursive Algorithm for Maximum Likelihood Estimation of the Carrier Phase

With the formula of Equation (6.269) for the derivative of the log-likelihood function $l(\theta)$ with respect to the carrier phase $\theta$ at hand, we are now ready to formulate an algorithm that seeks to maximize $l(\theta)$. We would like to perform the maximization in an iterative fashion so that the receiver is enabled to respond to the received signal on a symbol-by-symbol basis. To that end, we may build on the following algorithmic idea borrowed from adaptive filtering (see the discussions on the LMS algorithm presented in Chapters 3 and 4):

$$\begin{pmatrix}\text{Updated} \\ \text{estimate}\end{pmatrix} = \begin{pmatrix}\text{Old} \\ \text{estimate}\end{pmatrix} + \begin{pmatrix}\text{Step-size} \\ \text{parameter}\end{pmatrix}\begin{pmatrix}\text{Error} \\ \text{signal}\end{pmatrix} \tag{6.270}$$

where the *error signal*, or the adjustment signal to be more precise, is defined as the instantaneous value of the gradient of the log-likelihood function $l(\theta)$ with respect to $\theta$. Note that the parameter adjustment applied to the old estimate in Equation (6.270) is

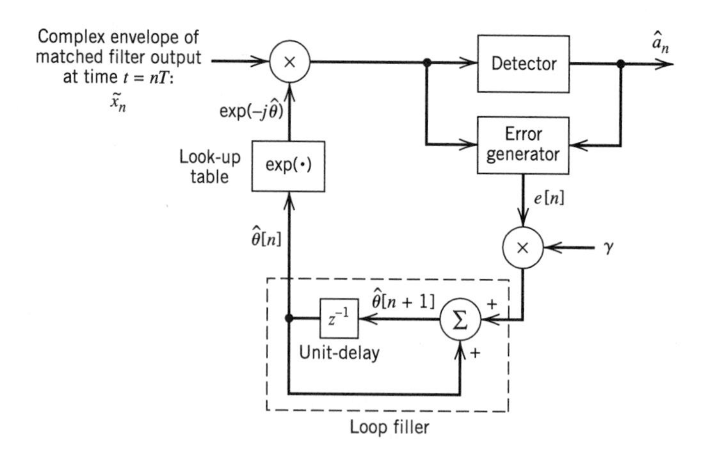


**FIGURE 6.62** Recursive Costas loop.

positive as the objective here is to perform *gradient ascent*. From Equation (6.269) we readily see that the error signal (i.e., the instantaneous value of $\partial l(\theta)/\partial\theta$ due to the transmission of a single symbol) is given by

$$e[n] = \text{Im}[\hat{a}_n^* \tilde{x}_n e^{-j\hat{\theta}}] \tag{6.271}$$

where the scaling factor $2\sqrt{E}/N_0$ is accounted for in what follows. Also, we have used $n$ in place of $k$ to denote a time step or iteration of the algorithm. Accordingly, we use Equation (6.270) to write

$$\hat{\theta}[n+1] = \hat{\theta}[n] + \gamma e[n] \tag{6.272}$$

where $\hat{\theta}[n]$ is the *old estimate* of the carrier phase $\theta$, $\hat{\theta}[n+1]$ is the *updated estimate* of $\theta$, and $\gamma$ is the *step-size parameter*; the scaling factor $2\sqrt{E}/N_0$ is absorbed in $\gamma$.

Equations (6.271) and (6.272) define the recursive algorithm for phase recovery. This algorithm is implemented using the system shown in Figure 6.62, which may be viewed as a recursive generalization of the Costas loop. We may therefore refer to it as the *recursive Costas loop* for phase synchronization.

The following points should be noted in Figure 6.62:

- The detector supplies an estimate of the transmitted symbol $a_n$, given the matched filter output.
- The look-up table supplies the value of $\exp(-j\hat{\theta}[n]) = \cos\hat{\theta}[n] - \sin\hat{\theta}[n]$ for an input $\hat{\theta}[n]$.
- The output of the error generator is the error signal $e[n]$.
- The block labeled $z^{-1}$ is a unit-delay element with the delay equal to the symbol period $T$.

The recursive Costas loop of Figure 6.62 uses a first-order digital filter. To improve the tracking performance of this synchronization system we may use a second-order digital filter. Figure 6.63 shows an example of a second-order digital filter made up of a cascade of two first-order sections, with $\rho$ as an adjustable *loop parameter*. An important property of a second-order filter used in the Costas loop for phase recovery is that it will eventually lock onto the incoming carrier with no static error, provided that the frequency error between the receiver and transmitter is initially small.

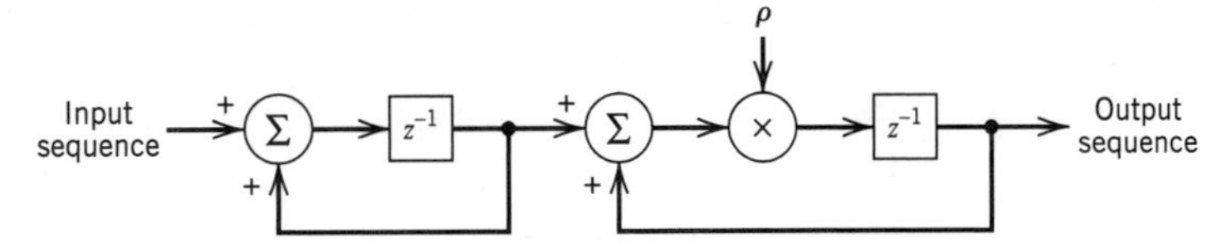


**FIGURE 6.63** Second-order digital filter.

## ■ NONDATA-AIDED RECURSIVE ALGORITHM FOR SYMBOL TIMING

For timing synchronization the only assumption made is that the receiver has knowledge of the carrier frequency $f_c$. The requirement is to develop an algorithm for recursive estimation of the group delay $\tau$ incurred in the course of transmitting the modulated signal through the channel.

Let $L(a_k, \theta, \tau)$ denote the likelihood function of $\tau$, which is also a function of transmitted symbol $a_k$ and carrier phase $\theta$. The likelihood function is defined by Equation (6.257). To proceed further we must remove the dependencies of $L(a_k, \theta, \tau)$ on the transmitted data sequence $\{a_k\}$ and carrier phase $\theta$, as described next.

To remove the dependence on $\theta$ we average the likelihood function $L(a_k, \theta, \tau)$, but *not* its logarithm, over all possible values of $\theta$ inside the range $[0, 2\pi]$. Assuming that $\theta$ is uniformly distributed inside this range, which is usually justifiable, we may write

$$\begin{aligned} L_{\text{av}}(a_k, \tau) &= \int_0^{2\pi} L(a_k, \theta, \tau) f_\Theta(\theta)\, d\theta \\ &= \frac{1}{2\pi}\int_0^{2\pi} \exp\left(\frac{2}{N_0}\,\mathbf{x}_k^T(\tau)\mathbf{s}(a_k, \theta, \tau)\right) d\theta \end{aligned}$$

The exponent in $L(a_k, \theta, \tau)$ is expressed by (see Problem 6.49)

$$\begin{aligned} \frac{2}{N_0}\mathbf{x}_k^T(\tau)\mathbf{s}(a_k, \theta, \tau) &= \frac{2\sqrt{E}}{N_0}\,\text{Re}[a_k^*\tilde{x}_k(\tau)e^{-j\theta}] \\ &= \frac{2\sqrt{E}}{N_0}\,\text{Re}[|a_k\tilde{x}_k(\tau)|\exp(j(\arg[\tilde{x}_k(\tau)] - \arg[a_k] - \theta))] \\ &= \frac{2\sqrt{E}}{N_0}\,|a_k\tilde{x}_k(\tau)|\cos(\arg[\tilde{x}_k(\tau)] - \arg[a_k] - \theta) \end{aligned} \tag{6.273}$$

Hence,

$$\begin{aligned} L_{\text{av}}(a_k, \tau) &= \frac{1}{2\pi}\int_0^{2\pi} \exp\left(\frac{2\sqrt{E}}{N_0}\,|a_k\tilde{x}_k(\tau)|\cos(\arg[\tilde{x}_k(\tau)] - \arg[a_k] - \theta)\right) d\theta \\ &= \frac{1}{2\pi}\int_{-\arg[\tilde{x}_k(\tau)]+\arg[a_k]}^{2\pi-\arg[\tilde{x}_k(\tau)]+\arg[a_k]} \exp\left(\frac{2\sqrt{E}}{N_0}\,|a_k\tilde{x}_k(\tau)|\cos(\varphi)\, d\varphi\right) \end{aligned} \tag{6.274}$$

where, in the last line, we have made the substitution

$$\varphi = \arg[\tilde{x}_k(\tau)] - \arg[a_k] - \theta$$

We now invoke the definition of the modified Bessel function of zero order, as shown by (see Appendix 3)

$$I_0(x) = \frac{1}{2\pi}\int_0^{2\pi} e^{x\cos\varphi}\, d\varphi \tag{6.275}$$

Hence, we may express the average likelihood function $L_{av}(a_k, \tau)$ as

$$L_{av}(a_k, \tau) = I_0\left(\frac{2\sqrt{E}}{N_0}|a_k\tilde{x}_k(\tau)|\right) \tag{6.276}$$

where $\tilde{x}_k(\tau)$ is the complex envelope of the matched filter output in the receiver due to the $k$th transmitted symbol $a_k$. For $M$-ary PSK, we have

$$|a_k| = 1 \qquad \text{for all } k$$

Hence, Equation (6.276) reduces to

$$L_{av}(a_k, \tau) = I_0\left(\frac{2\sqrt{E}}{N_0}|\tilde{x}_k(\tau)|\right) \tag{6.277}$$

We thus see that averaging the likelihood function over the carrier phase $\theta$ has also removed dependence on the transmitted symbol $a_k$ for $M$-ary PSK.

Finally, taking account of the transmission of $L_0$ independent symbols $a_0, a_1, \ldots, a_{L_0-1}$, we may express the overall likelihood function of $\tau$ as

$$\begin{aligned} L_{av}(\tau) &= \prod_{k=0}^{L_0-1} L_{av}(a_k, \tau) \\ &= \prod_{k=0}^{L_0-1} I_0\left(\frac{2\sqrt{E}}{N_0}|\tilde{x}_k(\tau)|\right) \end{aligned} \tag{6.278}$$

Now we can take the logarithm of $L_{av}(\tau)$ to obtain the log-likelihood function of $\tau$ as

$$\begin{aligned} l_{av}(\tau) &= \log L_{av}(\tau) \\ &= \sum_{k=0}^{L_0-1} \log I_0\left(\frac{2\sqrt{E}}{N_0}|\tilde{x}_k(\tau)|\right) \end{aligned} \tag{6.279}$$

To proceed further, we need to approximate $l_{av}(\tau)$. To that end we first note that the modified Bessel function $I_0(x)$ may be expanded in a power series as (see Appendix 3)

$$I_0(x) = \sum_{m=0}^{\infty} \frac{\left(\frac{1}{2}x\right)^{2m}}{(m!)^2}$$

For small values of $x$ we may thus approximate $I_0(x)$ as

$$I_0(x) \simeq 1 + \frac{x^2}{4}$$

We may further simplify matters by using the approximation

$$\begin{aligned} \log I_0(x) &\simeq \log\left(1 + \frac{x^2}{4}\right) \\ &\simeq \frac{x^2}{4} \qquad \text{for small } x \end{aligned}$$

For the problem at hand, small $x$ corresponds to small signal-to-noise ratio. Under this condition, we may approximate Equation (6.279) as

$$l_{av}(\tau) \simeq \frac{E}{N_0^2}\sum_{k=0}^{L_0-1}|\tilde{x}_k(\tau)|^2 \tag{6.280}$$

where, as mentioned earlier, $\tilde{x}_k(\tau)$ is the complex envelope of the matched filter output due to the $k$th transmitted symbol.

Differentiating $l_{\text{av}}(\tau)$ with respect to the group delay $\tau$, we obtain

$$\begin{aligned}\frac{\partial l_{\text{av}}(\tau)}{\partial \tau} &= \frac{E}{N_0^2}\sum_{k=0}^{L_0-1}\frac{\partial}{\partial \tau}|\tilde{x}_k(\tau)|^2 \\ &= \frac{2E}{N_0^2}\sum_{k=0}^{L_0-1}\text{Re}[\tilde{x}_k^*(\tau)\tilde{x}_k'(\tau)]\end{aligned} \tag{6.281}$$

where $\tilde{x}_k^*(\tau)$ is the complex conjugate of $\tilde{x}_k(\tau)$ and $\tilde{x}_k'(\tau)$ is its derivative with respect to $\tau$. Accordingly, we may define the error signal for timing recovery as (accounting for the scaling factor $2E/N_0^2$ in what follows)

$$e[n] = \text{Re}[\hat{x}_n^*(\tau)\hat{x}_n'(\tau)]$$

where we have used $n$ in place of $k$ to be consistent with the notation in Figure 6.62. Let $\hat{\tau}_n$ denote the estimate of the unknown delay $\tau$ at time $t = nT$. Then, introducing the definitions

$$\tilde{x}_n(\tau) = \tilde{x}(nT + \hat{\tau}_n)$$

and

$$\tilde{x}_n'(\tau) = \tilde{x}'(nT + \hat{\tau}_n)$$

we may reformulate the error signal $e(n)$ as

$$e[n] = \text{Re}[\tilde{x}^*(nT + \hat{\tau}_n)\tilde{x}'(nT + \hat{\tau}_n)] \tag{6.282}$$

Calculation of the error signal $e[n]$ requires the use of two filters:

1. The complex matched filter for generating $\tilde{x}_n(\tau)$.
2. The derivative matched filter for generating $\tilde{x}_n'(\tau)$.

The receiver is already equipped with the first filter. The second one is new. In practice, the additional computational complexity due to the derivative matched filter is objectionable. We may dispense with the need for it by using a finite difference to approximate the derivative $\tilde{x}_n'(\tau)$ as

$$\tilde{x}'(nT + \hat{\tau}_n) \simeq \frac{1}{T}\left[\tilde{x}\left(nT + \frac{T}{2} + \hat{\tau}_{n+1/2}\right) - \tilde{x}\left(nT - \frac{T}{2} + \hat{\tau}_{n-1/2}\right)\right] \tag{6.283}$$

where $\hat{\tau}_{n\pm 1/2}$ are the timing estimates computed at $nT \pm T/2$. It is desirable to make one further modification to account for the fact that timing estimates are updated at multiples of the symbol period $T$ and the only available quantities are $\hat{\tau}_n$. Consequently, we replace $\hat{\tau}_{n+1/2}$ by $\hat{\tau}_n$ (which represents the latest estimate of $\tau$) and replace $\hat{\tau}_{n-1/2}$ by $\hat{\tau}_{n-1}$ (which is the estimate of $\tau$ before the last one). We may thus rewrite Equation (6.283) as

$$\tilde{x}'(nT + \hat{\tau}_n) \simeq \frac{1}{T}\left[\tilde{x}\left(nT + \frac{T}{2} + \hat{\tau}_n\right) - \tilde{x}\left(nT - \frac{T}{2} + \hat{\tau}_{n-1}\right)\right] \tag{6.284}$$

and so finally redefine the error signal as

$$e[n] = \text{Re}\left\{\tilde{x}^*(nT + \hat{\tau}_n)\left[\tilde{x}\left(nT + \frac{T}{2} + \hat{\tau}_n\right) - \tilde{x}\left(nT - \frac{T}{2} + \hat{\tau}_{n-1}\right)\right]\right\} \tag{6.285}$$

where the scaling factor $1/T$ is also accounted for in what follows.

We are now ready to formulate the recursive algorithm for timing recovery:

$$c[n + 1] = c[n] + \gamma e[n] \tag{6.286}$$

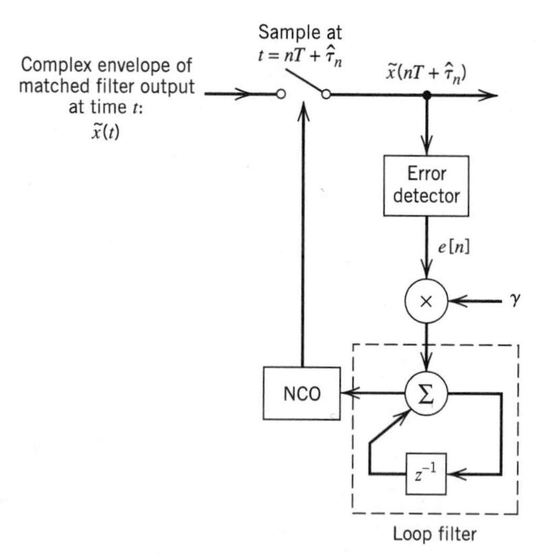


**FIGURE 6.64** Nondata-aided early-late delay synchronizer.

where $\gamma$ is the step-size parameter in which $2E/N_0^2$ and $1/T$ are absorbed, and the error signal $e[n]$ is defined by Equation (6.285). The $c[n]$ is a real number employed as the control for the frequency of an oscillator, referred to as a *number-controlled oscillator* (NCO). The scheme for implementing the timing recovery algorithm of Equations (6.285) and (6.286) is shown in Figure 6.64. This scheme is analogous to the continuous-time version of the early-late gate synchronizer widely used for timing recovery. It is thus referred to as a *nondata-aided early-late delay (NDA-ELD) synchronizer*. At every iteration, it works on three successive samples of the matched filter output, namely, $\tilde{x}\left(nT + \frac{T}{2} + \hat{\tau}_n\right)$, $\tilde{x}\left(nT + \hat{\tau}_n\right)$ and $\tilde{x}\left(nT + \frac{T}{2} - \hat{\tau}_{n-1}\right)$. The first sample is early and the last one is late, both with respect to the middle one.

Note that we could have simplified the derivations presented in this section by using the band-pass to complex low-pass transformation described in Appendix 2. We did not do so merely for the sake of simplifying the understanding of the material presented here.

## 6.15 *Computer Experiments: Carrier Recovery and Symbol Timing*

In this section we illustrate the operations of the recursive Costas loop and nondata-aided early-late delay synchronizer by considering a coherent QPSK system with the following specifications:

(i) Channel response: raised cosine (Nyquist) with rolloff factor $\alpha = 0.5$.

(ii) Loop filter: first-order digital filter with its transfer function defined by

$$H(z) = \frac{1}{z - (1 - \gamma A)} \tag{6.287}$$

where $\gamma$ is the step-size parameter and $A$ is a parameter to be defined.

(iii) Loop bandwidth, $B_L$ = 2% of the symbol rate $1/T$; that is, $B_L T = 0.02$.

### Experiment 1: Carrier Phase Recovery

In order to investigate the phase-acquisition behavior of the recursive Costas loop, we need to have the so-called *S-curve* of the phase-error generator. This is defined as the expectation of the adjustment signal $e[n]$, conditioned on a fixed value of the phase error

$$\varphi = \theta - \hat{\theta}$$

where $\theta$ is the actual value of the carrier phase and $\hat{\theta}$ is its estimate. That is,

$$S(\varphi) = E[e[n] \mid \varphi] \tag{6.288}$$

Experimentally, $S(\varphi)$ is measured by opening the recursive Costas loop of Figure 6.62 and measuring the average of the adjustment signal $e[n]$, as indicated in Figure 6.65.

The implementation procedure consists of the following steps. First, the complex envelope of the received signal is generated, which is given by

$$\tilde{x}_k(t) = \sqrt{\frac{2E}{T}} \exp(-j(2\pi f_c \tau_c - \alpha_k)) g(t - \tau_g) + \tilde{w}(t) \tag{6.289}$$

where $\alpha_k = 0, \pi/2, \pi, 3\pi/4$; $\tau_c$ is the carrier delay and $\tau_g$ is the group delay; and $\tilde{w}(t)$ is the complex-valued channel noise. The overall channel response $g(t)$ is given by the Nyquist pulse (see Section 4.5)

$$g(t) = \frac{\sin(\pi t/T)}{(\pi t/T)} \cdot \frac{\cos(\pi \alpha t/T)}{1 - 4\alpha^2 t^2/T^2} \tag{6.290}$$

where $\alpha = 0.5$. As pointed out earlier, we assume that the symbol timing (i.e., group delay $\tau_g$) is known, and the problem is to estimate the carrier phase $\theta = -2\pi f_c \tau_c$. The effect of $\theta$ is to shift an element of the signal constellation in the manner indicated in Figure 6.66.

Using the experimental procedure described in Figure 6.65, the S-curve of the QPSK system may now be measured. Figure 6.67*a* shows the ideal S-curve, assuming an infinitely large signal-to-noise ratio. This curve displays discontinuities at $\varphi = \pm m\pi/4$, where $m = 0, 1, 3, \ldots$, because of ambiguity encountered in the detection of the transmitted

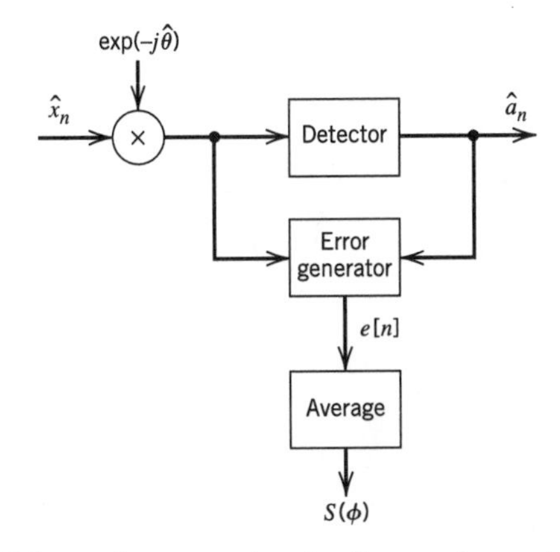


**FIGURE 6.65** Scheme for measuring the S-curve for carrier phase recovery.

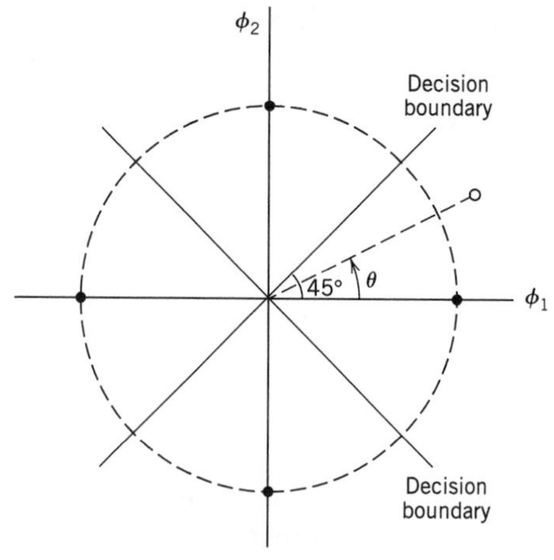


**FIGURE 6.66** Illustrating the effect of carrier phase $\theta$ on a state of the QPSK signal.

symbol $a_k$. The presence of channel noise tends to roundoff the discontinuities, as shown in the experimentally measured $S$-curve of Figure 6.67$b$. The results presented in Figure 6.67$b$ were obtained for $E/N_0 = 10$ dB. Recall that the in-phase and quadrature components of the narrowband noise have an identical Gaussian distribution with zero mean and the same variance as the original narrowband noise; these two components define $\tilde{w}(t)$.

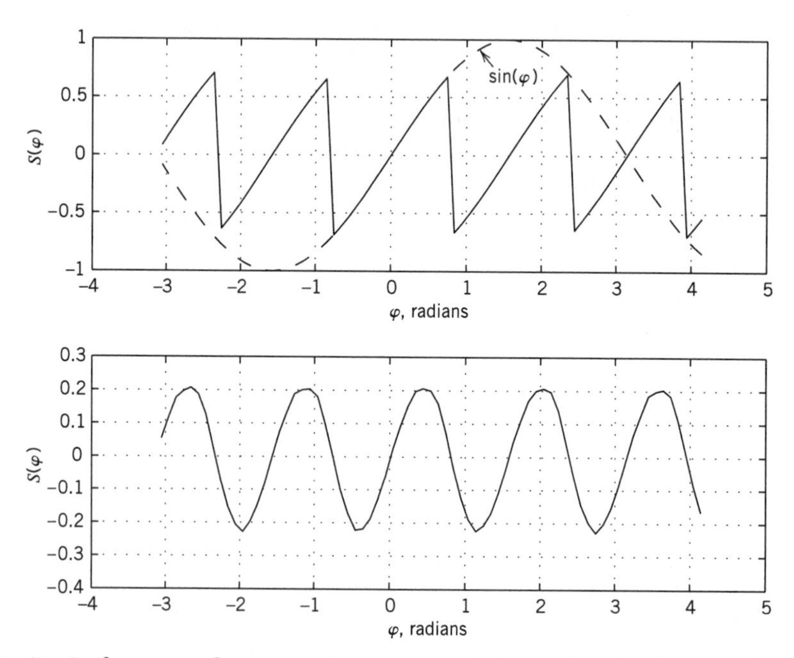


**FIGURE 6.67** Performance of recursive Costas loop. ($a$) $S$-curve for $(E/N_0) = \infty$. ($b$) $S$-curve for $(E/N_0) = 10$ dB.

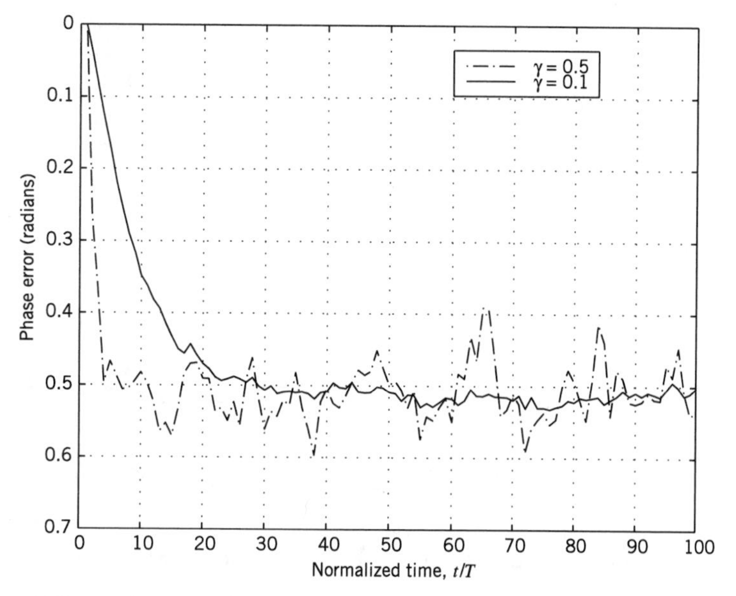


**FIGURE 6.68** Effects of varying the step-size parameter on convergence behavior of the recursive Costas loop.

When steady-state conditions have been established, the estimated phase $\hat{\theta}$ will fluctuate around the true value $\theta$. The extent of these fluctuations depends on the step-size parameter $\gamma$ and the received signal-to-noise ratio:

(i) Figure 6.68 plots the phase error $\varphi$ versus the normalized time $t/T$ for two different values of step-size parameter $\gamma$, namely, 0.1 and 0.5, and fixed $E/N_0 = 20$ dB. This figure clearly shows that the smaller we make $\gamma$ the smaller the steady-state fluctuations in the phase error $\varphi$ will be. However, this improvement is attained at the expense of a slower rate of convergence of the algorithm. The number of iterations needed by the algorithm to reach steady-state is approximately given by

$$L_0 \simeq \frac{1}{2B_L T} \tag{6.291}$$

The normalized bandwidth $B_L T$ is itself approximately given by

$$B_L T \simeq \frac{\gamma A}{4} \tag{6.292}$$

where $A$ is the slope of the $S$-curve measured at the origin. For $\gamma = 0.1$, and $B_L T = 0.02$, Equation (6.291) yields $L_0 = 25$ iterations, which checks with the solid curve plotted in Figure 6.68. Moreover, from Equations (6.291) and (6.292) we see that $L_0$ is inversely proportional to $\gamma$, which again checks with the results presented in Figure 6.68.

(ii) Figure 6.69 plots the phase error $\varphi$ versus the normalized time $t/T$ for three different values of $E/N_0$, namely, 5, 10, and 30 dB, and fixed $\gamma = 0.08$. We now see that the larger we make the signal-to-noise ratio, the smaller the steady-state fluctuations in the phase error $\varphi$ will be. Moreover, the rate of convergence of the algorithm also improves with increased signal-to-noise ratio, which is intuitively satisfying.

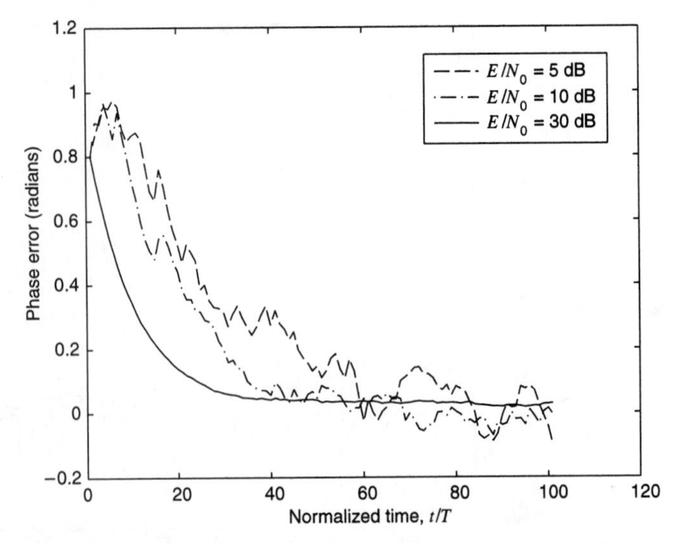


**FIGURE 6.69** Convergence behavior of the recursive Costas loop for varying $E/N_0$.

Figure 6.70 plots the variance of the phase error (averaged over 100 trials of the experiment) versus $E/N_0$ (measured in decibels) for $B_LT = 0.02$ and $\gamma = 0.08$. This figure also includes a plot of the *modified Cramér-Rao bound* defined by[24]

$$\text{MCRB}(\theta) = \frac{1}{2L_0(E/N_0)} \tag{6.293}$$

This bound is a modification of the ordinary Cramér-Rao bound, which is a lower bound on the variance of any *unbiased* estimator. The modification to this bound is made to overcome computational difficulties encountered in practical synchronization problems.

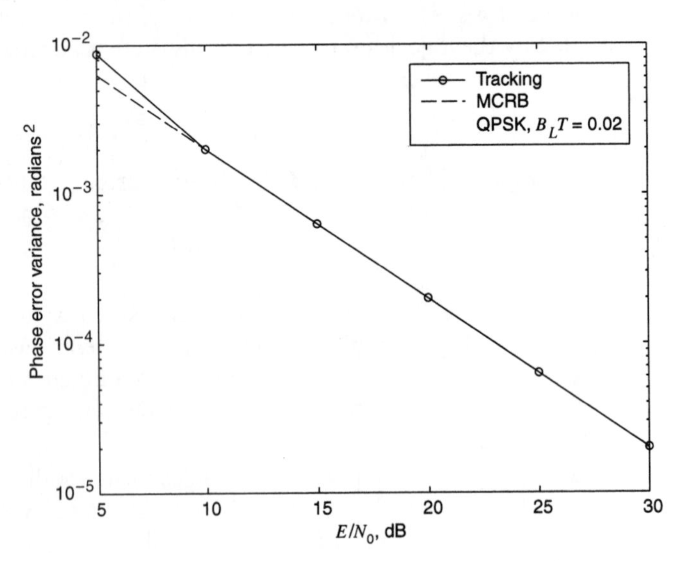


**FIGURE 6.70** Comparison of the measured tracking-error variance of the recursive Costas loop against theory for varying $E/N_0$.

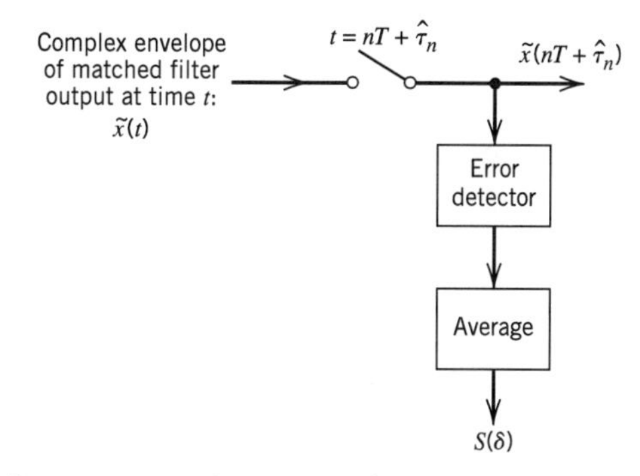


**FIGURE 6.71** Scheme for measuring the $S$-curve for the recursive early-late-delay synchronizer.

In any event, the experimental and theoretical results presented in Figure 6.70 are in very close agreement for $(E/N_0) \geq 10$ dB.

### Experiment 2: Symbol Timing Recovery

To measure the $S$-curve for the nondata-aided early-late delay synchronizer for symbol timing recovery, we may use the experimental set-up shown ion Figure 6.71, where the $\delta$ in $S(\delta)$ refers to the timing offset. The $S$-curve so measured is plotted in Figure 6.72 for $E/N_0 = 10$ dB and $E/N_0 = \infty$.

Figure 6.73 plots the normalized value of the experimentally measured symbol timing error versus $E/N_0$ for two different values of step-size parameter $\gamma$, namely, $T/20$ and $T/200$. This figure also includes theoretical plots of the corresponding modified Cramér-Rao bound of Equation (6.293) adapted for symbol-timing error. From the results presented here, we observe that as the step-size parameter $\gamma$ is reduced, the normalized timing

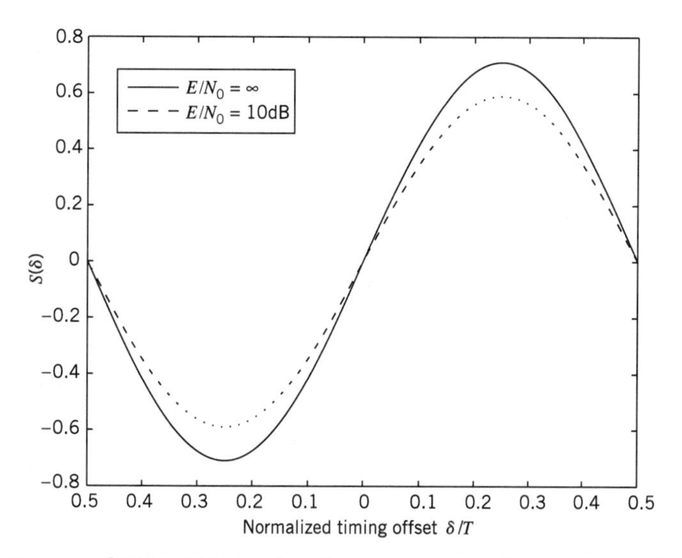


**FIGURE 6.72** S-curve of NDA-ELD synchronizer measured under noiseless and noisy conditions.

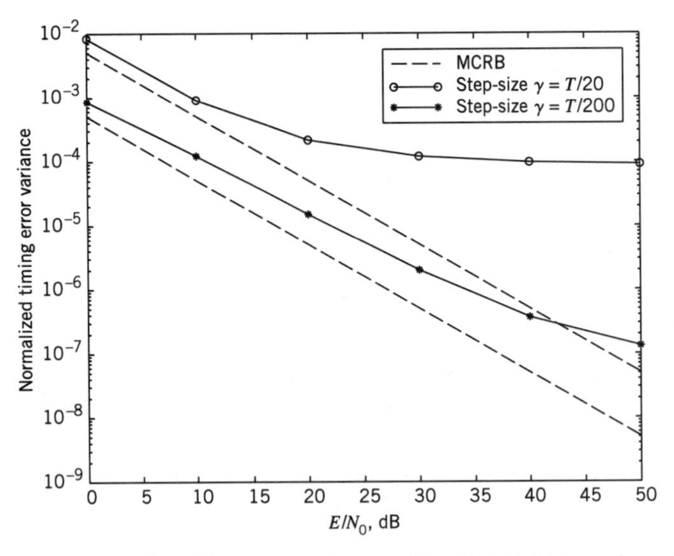


**FIGURE 6.73** Comparison of tracking-error variance of the NDA-ELD synchronizer against theory for varying $E/N_0$ and two step-size parameters.

error is reduced and the range of $E/N_0$ for which the modified Cramér-Rao bound holds (albeit in an approximate fashion) is enlarged.

## 6.16 *Summary and Discussion*

With the basic background theory on optimum receivers of Chapter 5 at our disposal, in this chapter we derived formulas for, or bounds on, the bit error rate for some important digital modulation techniques in an AWGN channel:

1. Phase-shift keying (PSK), represented by
   - Coherent binary phase-shift keying (BPSK).
   - Coherent quadriphase shift keying (QPSK) and its variants, namely, the offset QPSK and $\pi/4$-shifted QPSK.
   - Coherent $M$-ary PSK, which includes BPSK and QPSK as special cases with $M = 2$ and $M = 4$, respectively. Coherent $M$-ary PSK is used in digital satellite communications.
   - Differential phase-shift keying (DPSK), which may be viewed as the pseudo-noncoherent form of PSK.
2. Coherent $M$-ary quadrature amplitude modulation (QAM), which is a hybrid form of modulation that combines amplitude and phase-shift keying. For $M = 4$ it includes QPSK as a special case. $M$-ary QAM is basic to the construction of high-speed voiceband modems.
3. Frequency-shift keying (FSK), represented by
   - Coherent binary frequency-shift keying.
   - Coherent forms of minimum shift keying (MSK) and Gaussian minimum shift keying (GMSK); GMSK is basic to the construction of GSM wireless communications.

- Coherent $M$-ary FSK.
- Noncoherent binary FSK.

In this chapter we also studied two alternative techniques for passband data transmission: carrierless amplitude/phase modulation (CAP) and discrete multitone (DMT). In the case of an AWGN channel, the performance of CAP and DMT are equivalent because the DMT may be viewed as a linear reversible transformation of a single-carrier modulated signal. However, they perform quite differently in a practical setting that deviates from this idealized model.[25] DMT has been standardized for use on asymmetric digital subscriber lines (ADSLs) using twisted pairs. CAP, with the use of decision feedback equalization, provides another approach for solving the ADSL problem. CAP is also used for data transmission in local area networks for premises' distribution systems.

DMT is a form of multichannel modulation, and so is orthogonal frequency-division multiplexing (OFDM). The basic difference between DMT and OFDM is that DMT permits the use of loading to optimize information capacity, whereas OFDM does not. This difference arises because of their different domains of application. DMT applies to two-wire channels such as ADSLs, whereas OFDM applies to broadcasting and wireless channels.

Irrespective of the digital modulation system of interest, synchronization of the receiver to the transmitter is essential to the operation of the system. Symbol timing recovery is required whether the receiver is coherent or not. If the receiver is coherent, we also require provision for carrier recovery. In the latter part of the chapter, we discussed non-data-aided synchronizers to cater to these two requirements with emphasis on $M$-ary phase-shift keying signals in which the carrier is suppressed. The presentation focused on iterative synchronization techniques that are naturally suited for the use of digital signal processing.

## Notes and References

1. For an early tutorial paper reviewing different digital modulation techniques (ASK, FSK, and PSK) based on a geometric viewpoint, see Arthurs and Dym, (1962). See also the following list of books:
   - Anderson (1998, Chapter 3)
   - Benedetto and Biglieri (1999, Chapters 4 and 5)
   - Lee and Messerschmitt (1994, Part II)
   - Proakis (1995, Chapter 5)
   - Sklar (1988, Chapter 3)
   - Viterbi and Omura (1979, pp. 47–127)
2. For an early paper on the offset QPSK, see Gitlin and Ho (1975).
3. The $\pi/4$-shifted QPSK was first described in the open literature in Akaiwa and Negata (1987).
4. Chennakeshu and Sauliner (1993) use computer simulations to study the performance of $\pi/4$-shifted QPSK in a digital wireless communications environment. The pulse-shaping signal used in the generation of the $\pi/4$-shifted QPSK signal is based on the square root raised cosine spectrum (see Problem 4.38). In this latter paper, it is shown that the performance of $\pi/4$-shifted QPSK may degrade rapidly in such an environment. The differential detector of Figure 6.13 follows Chennakeshu and Sauliner (1993).

5. For a derivation of Equation (6.65), see Cioffi (1998).

6. The derivation of Equation (6.76) was first reported in a 1975 internal Bell Laboratories memorandum authored by Werner. A little later, Falconer (1975) issued another Bell Laboratories memorandum, in which it was pointed out the symbol rotation is not really needed if we did not want to be compatible with existing QAM or other bandpass signals, thereby simplifying the mathematical representation of CAP signals (and therefore their implementation), as shown in Equation (6.77). However, the terminology "CAP" was not coined until 1987 when carrierless amplitude/phase modulation was replaced by CAP by the standards representative, Garry Smith, of Bell Laboratories. The first detailed discussion of CAP in the context of digital subscriber lines was presented in a two-part report by Werner (1992, 1993). In a separate report by Chen, Im, and Werner (1992), the feasibility of CAP for use on digital subscriber lines was studied; see also the book by Chen (1998), pp. 461–473. The application of CAP to local area networks, involving the use of twisted pairs for lengths less than 100 m, is discussed in the paper by Im and Werner (1995); the maximum length of 100 m is specified by a standard for the wiring of premises.

   The digital implementation of a baseband equalizer similar to the CAP receiver of Figure 6.24 is discussed in Mueller and Werner (1982).

7. The MSK signal was first described in Doelz and Heald (1961). For a tutorial review of MSK and comparison with QPSK, see Pasupathy (1979). Since the frequency spacing is only half as much as the conventional spacing of $1/T_b$ that is used in the coherent detection of binary FSK signals, this signaling scheme is also referred to as *fast* FSK; see deBuda (1972).

8. For early discussions of Gaussian MSK, see Murota and Hirade (1981) and Ishizuka and Hirade (1980).

9. The analytical specification of the power spectral density of digital FM is difficult to handle, except for the case of a rectangular shaped modulating pulse. The paper by Garrison (1975) presents a procedure based on the selection of an appropriate duration-limited/level-quantized approximation for the modulating pulse. The equations developed therein are particularly suitable for machine computation of the power spectra of digital FM signals; see the book by Stüber (1996).

10. A detailed analysis of the spectra of $M$-ary FSK for an arbitrary value of frequency deviation is presented in the paper by Anderson and Salz (1965). The results shown plotted in Figure 6.36 represent a special case of a formula derived in that paper for a frequency deviation of $k = 0.5$.

11. The standard method of deriving the bit error rate for noncoherent binary FSK, presented in McDonough and Whalen (1995) and that for differential phase-shift keying presented in Arthurs and Dym (1962), involves the use of the Rician distribution. This distribution arises when the envelope of a sine wave plus additive Gaussian noise is of interest; see Chapter 1 for a discussion of the Rician distribution. The derivations presented in Section 6.6 avoid the complications encountered in the standard method.

12. The optimum receiver for differential phase-shift keying is discussed in Simon and Divsalar (1992).

13. For a technical discussion of various kinds of modems, with emphasis on their operational characteristics, see the books of Lewart (1988) and Hold (1997).

14. In a two-part paper by Wei (1984), differential encoding is applied to convolutional channel coding. Several eight-state convolutional encoders are described therein, which result in codes that are transparent to signal element rotations. In particular, in part II of the paper, Wei describes design rules and procedures for a 90-degree rotationally invariant convolutional code that has been adopted for use in the V.32 modem with trellis coding.

15. Nonuniform sampling of band-limited signals is discussed in the paper by Yen (1956). The main results derived in that paper are contained in four generalized theorems. Equation (6.188) is based on Theorem III of Yen's paper.

    In the paper by Kalet, Mazo, and Saltzberg (1993), this particular theorem due to Yen is used to formulate the fundamental philosophy underlying the design of the bidirectional digital modem; see also the patent by Ayanoglu et al. (1995).

16. For a discussion of the second realization of a digital modem, see the article by Humblet and Troulis (1996).

17. For a detailed description of the V.34 high-speed modem standard, see Forney et al. (1996). The trellis codes used in the V.34 modem are due to Wei (1984, 1987).

18. The idea of multichannel modulation may be traced to the early work of Chang (1966), Saltzberg (1967), and Weinstein and Ebert (1971). A mathematical treatment of the optimality of multitone modulation for a linear channel with severe intersymbol interference is presented in Kalet (1989). However, it was the work done by Cioffi and co-workers that led to the standardization of discrete multitone (DMT) for asymmetric digital subscriber lines; for details, see Ruiz et al. (1992), Chow and Cioffi (1995), Section 7.2 of the book by Starr et al. (1999), and Chapter 11 of Cioffi (1998). Problem 6.44 is adapted from Cioffi (1998).

19. The *method of Lagrange multipliers* for determining the extreme values of the function

    $$y = f(x)$$

    subject to the constraint

    $$\varphi(x) = 0$$

    follows from the following theorem: A necessary and sufficient condition for an extremum of a continuously differentiable function $f(x)$ is that its differential with respect to $x$ vanishes at the critical (i.e., maximum and minimum) points of the function. Accordingly, at the critical points of $f(x)$ we have

    $$\frac{\partial f}{\partial x}\,dx = 0 \tag{1}$$

    Moreover, since $\varphi(x) = 0$, its differential also vanishes as shown by

    $$\frac{\partial \varphi}{\partial x}\,dx = 0 \tag{2}$$

    Hence multiplying (2) by some parameter $\lambda$ and then adding the result to (1), we get

    $$\left(\frac{\partial f}{\partial x} + \lambda\,\frac{\partial \varphi}{\partial x}\right) dx = 0$$

    Since $dx$ is an independent increment, we immediately deduce that

    $$\frac{\partial}{\partial x}\,(f(x) + \lambda\,\varphi(x)) = 0$$

    This equation is a mathematical statement of the method of Lagrange multipliers. The parameter $\lambda$ is called the *Lagrange multiplier*. The material presented in this note follows Sokolnikoff and Redheffer (1966, pp. 341–344).

20. In the discrete Fourier transform (DFT), both the input and the output consist of sequences of numbers defined at uniformly spaced points in time and frequency, respectively. This feature makes the DFT ideally suited for numerical computation using the *fast Fourier transform (FFT) algorithm*. FFT algorithms are efficient because they use a greatly reduced number of arithmetic operations compared to the brute-force computation of the DFT. Basically, an FFT algorithm attains its computational efficiency by following a "divide and conquer" strategy, whereby the original DFT computation is decomposed successively into smaller DFT computations. For the case of an $N$-point DFT and $N = 2^L$, the FFT algorithm

requires $L = \log_2 N$ stages of computation, with each stage of the computation involving complex multiplications and additions of order $N$. For detailed discussion of the FFT algorithm, see Oppenheim and Schafer (1989, Chapter 9).

21. An overview of very-high-rate digital subscriber lines (VDSL) is presented in the paper by Cioffi et al. (1999); this paper also includes a comparative discussion on VDSLs and Voiceband modems.

22. For discussion of OFDM and its applications, see Casas and Leung (1991), LeFloch et al. (1989), and Zou and Wu (1995). For tutorial notes on OFDM and an extensive list of references, see Cimini and Li (1999).

23. For detailed descriptions of phase recovery and symbol-timing recovery using classical synchronization systems, see Stiffler (1971), Lindsey (1972), and Lindsey and Simon (1973, Chapters 2 and 9).

    For a modern treatment of synchronization systems with emphasis on the use of discrete-time signal processing algorithms, see Mengali and D'Andrea (1997), Meyr, Moeneclaey, and Fechtel (1998).

24. Equation (6.293) on the modified Cramér-Rao bound for phase recovery is derived in Mengali and D'Andrea (1997).

25. Saltzberg (1998) discusses how the performances of CAP and DMT are affected by channel impairments and system imperfections in the context of ADSL application. The impairments/imperfections considered therein include impulse noise, narrowband interference (e.g., RF ingress from an over-the-air AM radio transmission), timing jitter caused by imperfect synchronization, and system nonlinearities.

## PROBLEMS

### Amplitude-Shift Keying

**6.1** In the on-off keying version of an ASK system, symbol 1 is represented by transmitting a sinusoidal carrier of amplitude $\sqrt{2E_b/T_b}$, where $E_b$ is the signal energy per bit and $T_b$ is the bit duration. Symbol 0 is represented by switching off the carrier. Assume that symbols 1 and 0 occur with equal probability.

For an AWGN channel, determine the average probability of error for this ASK system under the following scenarios:

**(a)** Coherent reception.

**(b)** Noncoherent reception, operating with a large value of bit energy-to-noise spectral density ratio $E_b/N_0$.

*Note:* When $x$ is large, the modified Bessel function of the first kind of zero order may be approximated as follows (see Appendix 3):

$$I_0(x) \simeq \frac{\exp(x)}{\sqrt{2\pi x}}$$

### Phase-Shift Keying

**6.2** A PSK signal is applied to a correlator supplied with a phase reference that lies within $\varphi$ radians of the exact carrier phase. Determine the effect of the phase error $\varphi$ on the average probability of error of the system.

**6.3** Consider a phase-locked loop consisting of a multiplier, loop filter, and voltage-controlled oscillator (VCO). Let the signal applied to the multiplier input be a PSK signal defined by

$$s(t) = A_c \cos[2\pi f_c t + k_p m(t)]$$

where $k_p$ is the phase sensitivity, and the data signal $m(t)$ takes on the value $+1$ for binary symbol 1 and $-1$ for binary symbol 0. The VCO output is

$$r(t) = A_c \sin[2\pi f_c t + \theta(t)]$$

(a) Evaluate the loop filter output, assuming that this filter removes only modulated components with carrier frequency $2f_c$.

(b) Show that this output is proportional to the data signal $m(t)$ when the loop is phase locked, that is, $\theta(t) = 0$.

**6.4** The signal component of a coherent PSK system is defined by

$$s(t) = A_c k \sin(2\pi f_c t) \pm A_c\sqrt{1 - k^2} \cos(2\pi f_c t)$$

where $0 \le t \le T_b$, and the plus sign corresponds to symbol 1 and the minus sign corresponds to symbol 0. The first term represents a carrier component included for the purpose of synchronizing the receiver to the transmitter.

(a) Draw a signal-space diagram for the scheme described here; what observations can you make about this diagram?

(b) Show that, in the presence of additive white Gaussian noise of zero mean and power spectral density $N_0/2$, the average probability of error is

$$P_e = \frac{1}{2} \operatorname{erfc}\left(\sqrt{\frac{E_b}{N_0}(1 - k^2)}\right)$$

where

$$E_b = \frac{1}{2} A_c^2 T_b$$

(c) Suppose that 10 percent of the transmitted signal power is allocated to the carrier component. Determine the $E_b/N_0$ required to realize a probability of error equal to $10^{-4}$.

(d) Compare this value of $E_b/N_0$ with that required for a conventional PSK system with the same probability of error.

**6.5** (a) Given the input binary sequence 1100100010, sketch the waveforms of the in-phase and quadrature components of a modulated wave obtained by using the QPSK based on the signal set of Figure 6.6.

(b) Sketch the QPSK waveform itself for the input binary sequence specified in part (a).

**6.6** Let $P_{eI}$ and $P_{eQ}$ denote the probabilities of symbol error for the in-phase and quadrature channels of a narrowband digital communication system. Show that the average probability of symbol error for the overall system is given by

$$P_e = P_{eI} + P_{eQ} - P_{eI}P_{eQ}$$

**6.7** Equation (6.47) is an approximate formula for the average probability of symbol error for coherent $M$-ary PSK. This formula was derived using the union bound in light of the signal-space diagram of Figure 6.15*b*. Given that message point $m_1$ was transmitted, show that the approximate formula of Equation (6.47) may be derived directly from Figure 6.15*b*.

**6.8** Find the power spectral density of an offset QPSK signal produced by a random binary sequence in which symbols 1 and 0 (represented by $\pm 1$) are equally likely, and the symbols in different time slots are statistically independent and identically distributed.

**6.9** Vestigial sideband modulation (VSB), discussed in Chapter 2, offers another modulation method for passband data transmission.

(a) In particular, a digital VSB transmission system may be viewed as a time-varying one-dimensional system operating at a rate of $2/T$ dimensions per second, where $T$ is the symbol period. Justify the validity of this statement.

(b) Show that digital VSB is indeed equivalent in performance to the offset QPSK.

**6.10** The binary data stream 01101000 is applied to a $\pi/4$-shifted DQPSK modulator that is initially in the state ($\phi_1 = \sqrt{E}$, $\phi_2 = 0$) in Figure 6.11*a*. Using the relationships between input dibits and carrier-phase shifts summarized in Table 6.2, determine the phase states occupied by the modulator in response to the specified data stream.

**6.11** Just as in an ordinary QPSK modulator, the output of a $\pi/4$-shifted DQPSK modulator may be expressed in terms of its in-phase and quadrature components as follows:

$$s(t) = s_I(t)\cos(2\pi f_c t) - s_Q(t)\sin(2\pi f_c t)$$

Formulate the in-phase component $s_I(t)$ and quadrature component $s_Q(t)$ of the $\pi/4$-shifted DQPSK signal. Hence, outline a scheme for the generation of $\pi/4$-shifted DQPSK signals.

**6.12** An interesting property of $\pi/4$-shifted DQPSK signals is that they can be demodulated using an FM discriminator. Demonstrate the validity of this property. The FM discriminator is discussed in Chapter 2.

**6.13** Let $\Delta\theta_k$ denote the differentially encoded phase in the $\pi/4$-shifted DQPSK. The symbol pairs $(I, Q)$ generated by this scheme may be defined as

$$I_k = I_{k-1}\cos(\Delta\theta_k) - Q_{k-1}\sin(\Delta\theta_k)$$
$$Q_k = I_{k-1}\sin(\Delta\theta_k) + Q_{k-1}\cos(\Delta\theta_k)$$

where $I_k$ and $Q_k$ are the in-phase and quadrature components corresponding to the $k$th symbol. Show that this pair of relations can be expressed simply as

$$I_k = \cos\theta_k$$
$$Q_k = \sin\theta_k$$

where $\theta_k$ is the absolute phase angle for the $k$th symbol.

### Quadrature-Amplitude Modulation

**6.14** Figure 6.53 shows a 240-QAM signal constellation, which may be viewed as an extended form of QAM cross constellation.

(a) Identify the portion of Figure 6.53 that is a QAM square constellation.

(b) Build on part (a) to identify the portion of Figure 6.53 that is a QAM cross constellation.

(c) Hence, identify the portion of Figure 6.53 that is an extension to QAM cross constellation.

**6.15** Determine the transmission bandwidth reduction and average signal energy of 256-QAM, compared to 64-QAM.

**6.16** Two passband data transmission systems are to be compared. One system uses 16-PSK, and the other uses 16-QAM. Both systems are required to produce an average probability of symbol error equal to $10^{-3}$. Compare the signal-to-noise ratio requirements of these two systems.

### Carrierless Amplitude/Phase Modulation (CAP)

**6.17** The two-dimensional CAP and $M$-ary QAM schemes are closely related. Do the following:

(a) Given a QAM system, with a prescribed number of amplitude levels, derive the equivalent CAP system.

(b) Perform the reverse of part (a).

**6.18** Show that the power spectral density of a CAP signal with a total of $L$ amplitude levels is defined by

$$S(f) = \frac{\sigma_A^2}{T}|P(f)|^2$$

where $|P(f)|$ is the magnitude spectrum of the passband in-phase pulse $p(t)$; the $\sigma_A^2$ is the variance of the complex symbols $A_i = a_i + jb_i$, which is defined by

$$\sigma_A^2 = \frac{1}{L}\sum_{i=1}^{L}(a_i^2 + b_i^2)$$

**6.19** You are given the baseband raised-cosine spectrum $G(f)$ pertaining to a certain rolloff factor $\alpha$. Describe a frequency-domain procedure for evaluating the passband in-phase pulse $p(t)$ and quadrature pulse $\hat{p}(t)$ that characterize the corresponding CAP signal.

### Frequency-Shift Keying

**6.20** The signal vectors $\mathbf{s}_1$ and $\mathbf{s}_2$ are used to represent binary symbols 1 and 0, respectively, in a coherent binary FSK system. The receiver decides in favor of symbol 1 when

$$\mathbf{x}^T\mathbf{s}_1 > \mathbf{x}^T\mathbf{s}_2$$

where $\mathbf{x}^T\mathbf{s}_i$ is the inner product of the observation vector $\mathbf{x}$ and the signal vector $\mathbf{s}_i$, where $i = 1, 2$. Show that this decision rule is equivalent to the condition $x_1 > x_2$, where $x_1$ and $x_2$ are the elements of the observation vector $\mathbf{x}$. Assume that the signal vectors $\mathbf{s}_1$ and $\mathbf{s}_2$ have equal energy.

**6.21** An FSK system transmits binary data at the rate of $2.5 \times 10^6$ bits per second. During the course of transmission, white Gaussian noise of zero mean and power spectral density $10^{-20}$ W/Hz is added to the signal. In the absence of noise, the amplitude of the received sinusoidal wave for digit 1 or 0 is 1 mV. Determine the average probability of symbol error for the following system configurations:

(a) Coherent binary FSK
(b) Coherent MSK
(c) Noncoherent binary FSK

**6.22** (a) In a coherent FSK system, the signals $s_1(t)$ and $s_2(t)$ representing symbols 1 and 0, respectively, are defined by

$$s_1(t),\, s_2(t) = A_c \cos\left[2\pi\left(f_c \pm \frac{\Delta f}{2}\right)t\right], \qquad 0 \le t \le T_b$$

Assuming that $f_c > \Delta f$, show that the correlation coefficient of the signals $s_1(t)$ and $s_2(t)$ is approximately given by

$$\rho = \frac{\int_0^{T_b} s_1(t)s_2(t)\,dt}{\int_0^{T_b} s_1^2(t)\,dt} \simeq \text{sinc}(2\Delta f T_b)$$

(b) What is the minimum value of frequency shift $\Delta f$ for which the signals $s_1(t)$ and $s_2(t)$ are orthogonal?

(c) What is the value of $\Delta f$ that minimizes the average probability of symbol error?

(d) For the value of $\Delta f$ obtained in part (c), determine the increase in $E_b/N_0$ required so that this coherent FSK system has the same noise performance as a coherent binary PSK system.

**6.23** A binary FSK signal with *discontinuous phase* is defined by

$$s(t) = \begin{cases} \sqrt{\dfrac{2E_b}{T_b}}\cos\left[2\pi\left(f_c + \dfrac{\Delta f}{2}\right)t + \theta_1\right] & \text{for symbol 1} \\ \sqrt{\dfrac{2E_b}{T_b}}\cos\left[2\pi\left(f_c - \dfrac{\Delta f}{2}\right)t + \theta_2\right] & \text{for symbol 0} \end{cases}$$

where $E_b$ is the signal energy per bit, $T_b$ is the bit duration, and $\theta_1$ and $\theta_2$ are sample values of uniformly distributed random variables over the interval 0 to $2\pi$. In effect, the two oscillators supplying the transmitted frequencies $f_c \pm \Delta f/2$ operate independently of each other. Assume that $f_c >> \Delta f$.

**(a)** Evaluate the power spectral density of the FSK signal.

**(b)** Show that for frequencies far removed from the carrier frequency $f_c$, the power spectral density falls off as the inverse square of frequency.

**6.24** Set up a block diagram for the generation of Sunde's FSK signal $s(t)$ with continuous phase by using the representation given in Equation (6.104), which is reproduced here:

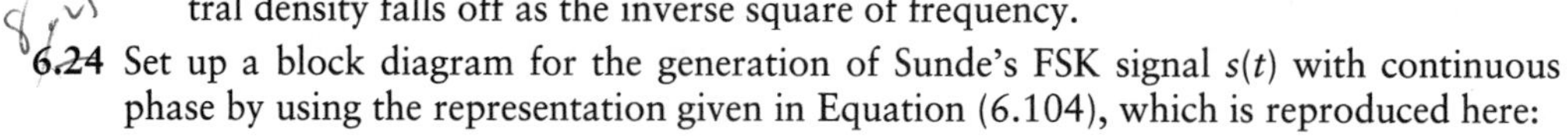

$$s(t) = \sqrt{\frac{2E_b}{T_b}} \cos\left(\frac{\pi t}{T_b}\right) \cos(2\pi f_c t) \mp \sqrt{\frac{2E_b}{T_b}} \sin\left(\frac{\pi t}{T_b}\right) \sin(2\pi f_c t)$$

**6.25** Discuss the similarities between MSK and offset QPSK, and the features that distinguish them.

**6.26** There are two ways of detecting an MSK signal. One way is to use a coherent receiver to take full account of the phase information content of the MSK signal. Another way is to use a noncoherent receiver and disregard the phase information. The second method offers the advantage of simplicity of implementation, at the expense of a degraded noise performance. By how many decibels do we have to increase the bit energy-to-noise density ratio $E_b/N_0$ in the second case so as to realize an average probability of symbol error equal to $10^{-5}$ in both cases?

**6.27** **(a)** Sketch the waveforms of the in-phase and quadrature components of the MSK signal in response to the input binary sequence 1100100010.

**(b)** Sketch the MSK waveform itself for the binary sequence specified in part (a).

**6.28** A nonreturn-to-zero data stream (of amplitude levels $\pm 1$) is passed through a low-pass filter whose impulse response is defined by the Gaussian function

$$h(t) = \frac{\sqrt{\pi}}{\alpha} \exp\left(-\frac{\pi^2 t^2}{\alpha^2}\right)$$

where $\alpha$ is a design parameter defined in terms of the filter's 3-dB bandwidth by

$$\alpha = \sqrt{\frac{\log 2}{2}} \frac{1}{W}$$

**(a)** Show that the transfer function of the filter is defined by

$$H(f) = \exp(-\alpha^2 f^2)$$

Hence demonstrate that the 3-dB bandwidth of the filter is indeed equal to $W$. You may use Table A6.3 on Fourier-transform pairs.

**(b)** Show that the response of the filter to a rectangular pulse of unit amplitude and duration $T$ centered on the origin is defined by Equation (6.135).

**6.29** Plot the waveform of a GMSK modulator produced in response to the binary sequence 1101000, assuming the use of a gain-bandwidth product $WT_b = 0.3$. Compare your result with that of Example 6.5.

**6.30** Summarize the similarities and differences between the standard MSK and Gaussian-filtered MSK signals.

### Noncoherent Receivers

**6.31** In Section 6.8 we derived the formula for the bit error rate of noncoherent binary FSK as a special case of noncoherent orthogonal modulation. In this problem we revisit this issue. As before, we assume that binary symbol 1 represented by signal $s_1(t)$ is transmitted. According to the material presented in Section 6.8, we note the following:

- The random variable $L_2$ represented by the sample value $l_2$ of Equation (6.164) is Rayleigh distributed.
- The random variable $L_1$ represented by the sample value $l_1$ of Equation (6.170) is Rician-distributed.

The Rayleigh and Rician distributions are discussed in Chapter 1. Using the probability distributions defined in that chapter, derive the formula of Equation (6.181) for the BER of noncoherent binary FSK.

**6.32** Figure P6.32*a* shows a noncoherent receiver using a matched filter for the detection of a sinusoidal signal of known frequency but random phase, in the presence of additive white Gaussian noise. An alternative implementation of this receiver is its mechanization in the frequency domain as a *spectrum analyzer receiver*, as in Figure P6.32*b*, where the correlator computes the finite time autocorrelation function $R_x(\tau)$ defined by

$$R_x(\tau) = \int_0^{T-\tau} x(t)x(t+\tau)\,dt, \qquad 0 \le \tau \le T$$

Show that the square-law envelope detector output sampled at time $t = T$ in Figure P6.32*a* is twice the spectral output of the Fourier transformer sampled at frequency $f = f_c$ in Figure P6.32*b*.

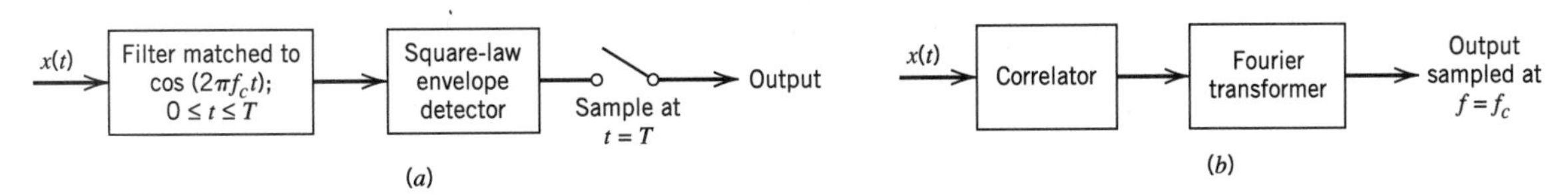


**FIGURE P6.32**

**6.33** The binary sequence 1100100010 is applied to the DPSK transmitter of Figure 6.43*a*.

(a) Sketch the resulting waveform at the transmitter output.

(b) Applying this waveform to the DPSK receiver of Figure 6.43*b*, show that, in the absence of noise, the original binary sequence is reconstructed at the receiver output.

**6.34** *Differential M-ary PSK* is the *M*-ary extension of binary DPSK. The present phase angle $\theta_n$ of the modulator at symbol time $n$ is determined recursively by the relation

$$\theta_n = \theta_{n-1} + \left(\frac{2\pi}{M}\right)m_n, \quad \text{modulo } 2\pi$$

where $\theta_{n-1}$ is the previous phase angle and $m_n \in \{0, 1, \ldots, M-1\}$ is the present modulator input. The probability of symbol error for this *M*-ary modulation scheme is approximately given by

$$P_e \simeq \operatorname{erfc}\left(\sqrt{\frac{2E}{N_0}}\sin\left(\frac{\pi}{2M}\right)\right), \qquad M \ge 4$$

where it is assumed that $E/N_0$ is large.

(a) Determine the factor by which the transmitted energy per symbol would have to be increased for the differential *M*-ary PSK to attain the same probability of symbol error as coherent *M*-ary PSK for $M \ge 4$.

(b) For $M = 4$, by how many decibels is differential QPSK poorer in performance than coherent QPSK?

## Comparison of Digital Modulation Schemes Using a Single Carrier

**6.35** Binary data are transmitted over a microwave link at the rate of $10^6$ b/s, and the power spectral density of the noise at the receiver input is $10^{-10}$ W/Hz. Find the average carrier power required to maintain an average probability of error $P_e \leq 10^{-4}$ for (a) coherent binary PSK, and (b) DPSK.

**6.36** The values of $E_b/N_0$ required to realize an average probability of symbol error $P_e = 10^{-4}$ using coherent binary PSK and coherent FSK (conventional) systems are equal to 7.2 and 13.5, respectively. Using the approximation

$$\text{erfc}(u) \simeq \frac{1}{\sqrt{\pi}u} \exp(-u^2)$$

determine the separation in the values of $E_b/N_0$ for $P_e = 10^{-4}$, using

(a) Coherent binary PSK and DPSK.
(b) Coherent binary PSK and QPSK.
(c) Coherent binary FSK (conventional) and noncoherent binary FSK.
(d) Coherent binary FSK (conventional) and coherent MSK.

**6.37** In Section 6.10 we compared the noise performances of coherent binary PSK, coherent binary FSK, QPSK, MSK, DPSK, and noncoherent FSK by using the bit error rate as the basis of comparison. In this problem we take a different viewpoint and use the average probability of symbol error, $P_e$, to do the comparison. Plot $P_e$ versus $E_b/N_0$ for each of these schemes and comment on your results.

**6.38** The *noise equivalent bandwidth* of a bandpass signal is defined as the value of bandwidth that satisfies the relation

$$2BS(f_c) = P/2$$

where $2B$ is the noise equivalent bandwidth centered around the midband frequency $f_c$, $S(f_c)$ is the maximum value of the power spectral density of the signal at $f = f_c$, and $P$ is the average power of the signal. Show that the noise equivalent bandwidths of binary PSK, QPSK, and MSK are as follows:

| *Type of Modulation* | *Noise Bandwidth/Bit Rate* |
|---|---|
| Binary PSK | 1.0 |
| QPSK | 0.5 |
| MSK | 0.62 |

*Note:* You may use the definite integrals in Table A6.10. A discussion of noise equivalent bandwidth is presented in Appendix 2.

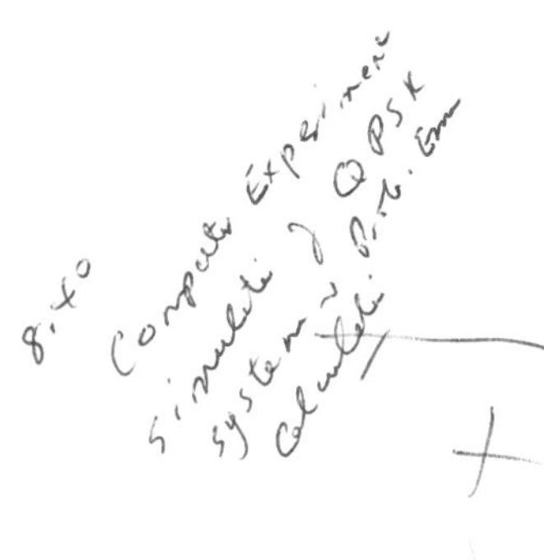

## Voiceband Modems

**6.39** (a) Refer to the differential encoder used in Figure 6.48*a*. Table 6.10 defines the phase changes induced in the V.32 modem by varying input dibits. Expand this table by including the corresponding previous and current values of the differential encoder's output. Note that for every input dibit $Q_{1,n}Q_{2,n}$, there are four possible values for the differentially encoded dibit $I_{1,n}I_{2,n}$ and likewise for its previous value $I_{1,n-1}I_{2,n-1}$.

(b) The current quadbit applied to the V.32 modem with nonredundant coding is 0001. The previous output of the modem is 01. Find the code word output produced by the modem and its coordinates.

6.40 The V.32 modem standard with nonredundant coding uses a rectangular 16-QAM constellation. The model specifications are as follows:

Carrier frequency = 1,800 Hz
Symbol rate = 2,400 bauds
Data rate = 9,600 b/s

Calculate (a) the average signal-to-noise ratio, and (b) the average probability of symbol error for this modem, assuming that $E_{av}/N_0 = 20\text{dB}$.

## Multichannel Line Codes

6.41 Consider the passband basis functions defined in Equation (6.196), where $\phi(t)$ is itself defined by Equation (6.197). Demonstrate the validity of Properties 1, 2, and 3 of these passband basis functions mentioned on pages 434 and 435.

6.42 The water-filling solution for the loading problem is defined by Equation (6.213) subject to the constraint of Equation (6.210). Using this pair of relations, formulate a recursive algorithm for computing the allocation of the transmit power $P$ among the $N$ subchannels. The algorithm should start with (a) an initial total or sum *noise-to-signal ratio* $\text{NSR}(i) = 0$ for iteration $i = 0$, and (b) the subchannels sorted in terms of those with the smallest power allocation to the largest.

6.43 The squared magnitude response of a linear channel, denoted by $|H(f)|^2$, is shown in Figure P6.43. Assume that the gap $\Gamma = 1$ and the noise variance $\sigma_n^2 = 1$ for all subchannels.

(a) Derive the formulas for the optimum powers $P_1$, $P_2$, and $P_3$ allocated to the three subchannels of frequency bands $(0, W_1)$, $(W_1, W_2)$, and $(W_2, W)$.

(b) Given that the total transmit power $P = 10$, $l_1 = 2/3$ and $l_2 = 1/3$, calculate the corresponding values of $P_1$, $P_2$, and $P_3$.

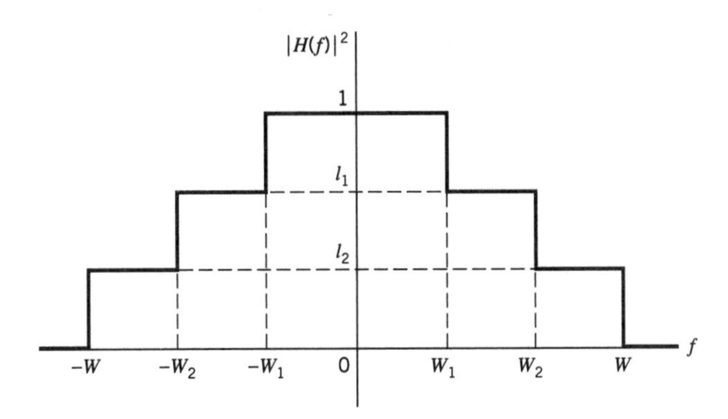


**FIGURE P6.43**

6.44 In this problem we explore the use of *singular value decomposition* (SVD) as an alternative to the discrete Fourier transform for vector coding. This approach avoids the need for a cyclic prefix, with the channel matrix being formulated as

$$\mathbf{H} = \begin{bmatrix} h_0 & h_1 & h_2 & \cdots & h_\nu & 0 & \cdots & 0 \\ 0 & h_0 & h_1 & \cdots & h_{\nu-1} & h_\nu & \cdots & 0 \\ \vdots & \vdots & \vdots & & \vdots & \vdots & & \vdots \\ 0 & 0 & 0 & \cdots & h_0 & h_1 & \cdots & h_\nu \end{bmatrix}$$

where the sequence $h_0, h_1, \ldots, h_\nu$ denotes the sampled impulse response of the channel. The SVD of the matrix $\mathbf{H}$ is defined by

$$\mathbf{H} = \mathbf{U}[\mathbf{\Lambda} \,\vdots\, \mathbf{O}_{N,\nu}]\mathbf{V}^{\dagger}$$

where $\mathbf{U}$ is an $N$-by-$N$ unitary matrix and $\mathbf{V}$ is an $(N+\nu)$-by-$(N+\nu)$ unitary matrix; that is,

$$\mathbf{U}\mathbf{U}^{\dagger} = \mathbf{I}$$
$$\mathbf{V}\mathbf{V}^{\dagger} = \mathbf{I}$$

where $\mathbf{I}$ is the identity matrix and the superscript $\dagger$ denotes Hermitian transposition. The $\mathbf{\Lambda}$ is an $N$-by-$N$ diagonal matrix with singular values $\lambda_n$, $n = 1, 2, \ldots, N$. The $\mathbf{O}_{N,\nu}$ is an $N$-by-$\nu$ matrix of zeros.

(a) Using this decomposition, show that the $N$ subchannels resulting from the use of vector coding are mathematically described by

$$X_n = \lambda_n A_n + W_n$$

The $X_n$ is an element of the matrix product $\mathbf{U}^{\dagger}\mathbf{x}$, where $\mathbf{x}$ is the received signal (channel output) vector. The $A_n$ is the $n$th symbol $a_n + jb_n$ and $W_n$ is a random variable due to channel noise.

(b) Show that the signal-to-noise ratio for vector coding as described herein is given by

$$(\text{SNR})_{\text{vector coding}} = \Gamma\left(\prod_{n=1}^{N^*}\left(1 + \frac{(\text{SNR})_n}{\Gamma}\right)\right)^{1/(N+\nu)} - \Gamma$$

where $N^*$ is the number of channels for each of which the allocated transmit power is nonnegative, $(\text{SNR})_n$ is the signal-to-noise ratio of subchannel $n$, and $\Gamma$ is a prescribed gap.

(c) As the block length $N$ approaches infinity, the singular values approach the magnitudes of the channel Fourier transform. Using this result, comment on the relationship between vector coding and discrete multitone.

6.45 Compare the performance of DMT and CAP with respect to the following channel impairments:

(a) Impulse noise.

(b) Narrowband interference.

Assume that (1) the DMT has a large number of subchannels, and (2) the CAP system is uncoded and its receiver uses a pair of adaptive filters for implementation.

6.46 Orthogonal frequency-division multiplexing may be viewed as a generalization of $M$-ary FSK. Validate the rationale of this statement.

## Synchronization

6.47 Figure P6.47 shows the block diagram of a continuous-time *Mth power loop* for phase recovery in an $M$-ary PSK receiver.

(a) Show that the output of the $M$th power-law device contains a tone of frequency $Mf_c$, where $f_c$ is the original carrier.

(b) The oscillator in the phase-locked loop is set to a frequency equal to $Mf_c$. Justify this choice.

(c) The $M$th power loop suffers from a phase ambiguity problem in that it exhibits $M$ phase ambiguities in the interval $[0, 2\pi]$. Explain how this problem arises in the $M$th power loop. How would you overcome the problem?

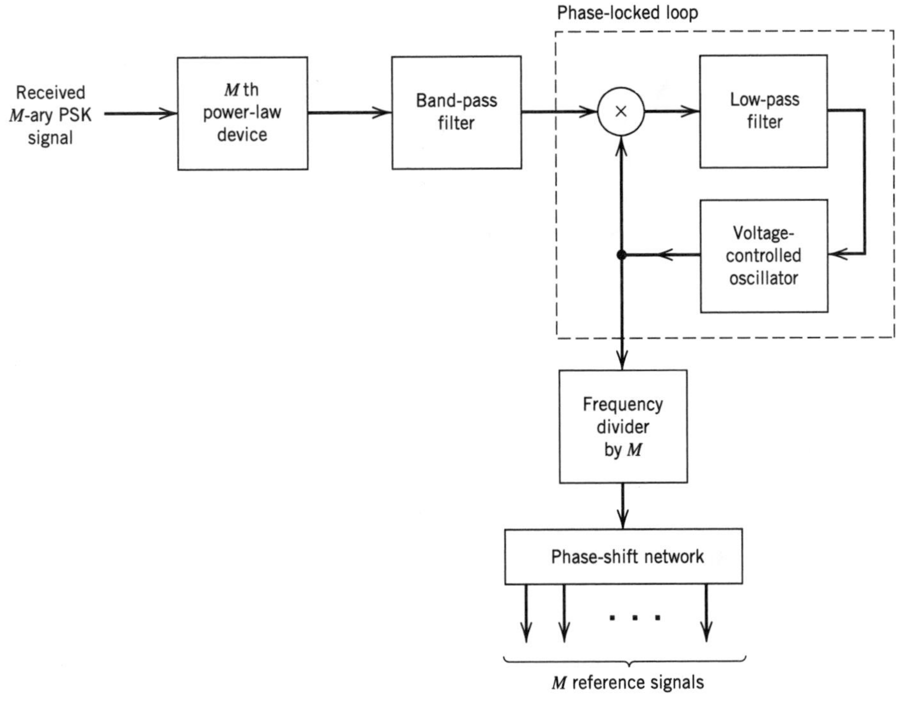


**FIGURE P6.47**

**6.48** (a) In the recursive algorithm of Equation (6.272) for phase recovery, the old estimate $\hat{\theta}[n]$ and the updated estimate $\hat{\theta}[n+1]$ of the carrier phase $\theta$ are both measured in radians. Discuss the units in which the error signal $e[n]$ and step-size parameter $\gamma$ are measured.

(b) In the recursive algorithm of Equation (6.286) for symbol timing recovery, the control signals $c[n]$ and $c[n+1]$ are both dimensionless. Discuss the units in which the error signal $e[n]$ and step-size parameter $\gamma$ are measured.

**6.49** Using the definitions of Equations (6.264) and (6.265) for $\tilde{x}_k$ and $a_k$, respectively, show that the exponent in the likelihood function $L(a_k, \theta, \tau)$ can be expressed as in Equation (6.273).

**6.50** In Section 6.14 we studied a non–data-aided scheme for carrier phase recovery, based on the log-likelihood function of Equation (6.260). In this problem we explore the use of this equation for *data-aid carrier phase recovery*.

(a) Consider a receiver designed for a linear modulation system. Given that the receiver has knowledge of a preamble of length $L_0$, show that the maximum likelihood estimate of the carrier phase is defined by

$$\hat{\theta} = \arg\left\{\sum_{k=0}^{L_0-1} a_k^* \tilde{x}(k)\right\}$$

where the preamble $\{a_k\}_{k=0}^{L_0-1}$ is a known sequence of complex symbols, and $\{\tilde{x}(k)\}_{k=0}^{L_0-1}$ is the complex envelope of the corresponding received signal.

(b) Using the result derived in part (a), construct a block diagram for the maximum likelihood phase estimator.

## Computer Experiment

6.51 The purpose of this computer experiment is to compare the effect of a dispersive channel on the waveforms generated by the following passband modulation techniques:

(a) Binary phase-shift keying (BPSK)
(b) Quadriphase-shift keying (QPSK)
(c) Minimum shift keying (MSK)
(d) Gaussian MSK with time-bandwidth product $WT_b = 0.3$

The channel consists of a band-pass *Butterworth filter* of order $2N = 10$ and 3-dB bandwidth $2B$ centered on the midband frequency $f_c$. The low-pass equivalent of the channel has the squared magnitude response

$$|H(f)|^2 = \frac{1}{1 + (f/B)^{2N}}$$

The channel bandwidth is variable so as to illustrate its effect on the filtered modulated wave.

Assuming the use of a coherent receiver, plot the waveforms of the modulated signals under (a), (b), (c) and (d) for the following channel bandwidths:

(i) $2B = 12$ kHz
(ii) $2B = 16$ kHz
(iii) $2B = 20$ kHz
(iv) $2B = 24$ kHz
(v) $2B = 30$ kHz

Comment on your results.

*Hint.* To perform the computations needed for this experiment, it is advisable to perform the computations in baseband by performing the band-pass to low-pass transformation described in Appendix 2.

CHAPTER 7

# SPREAD-SPECTRUM MODULATION

This chapter introduces a modulation technique called spread-spectrum modulation, which is radically different from the modulation techniques that are covered in preceding chapters. In spread-spectrum modulation, channel bandwidth and transmit power are sacrificed for the sake of secure communications.

Specifically, we cover the following topics:

- *Spreading sequences in the form of pseudo-noise sequences, their properties, and methods of generation.*
- *The basic notion of spread-spectrum modulation.*
- *The two commonly used types of spread-spectrum modulation: direct sequence and frequency hopping.*

The material presented in this chapter is basic to wireless communications using code-division multiple access, which is covered in Chapter 8.

## 7.1 *Introduction*

A major issue of concern in the study of digital communications as considered in Chapters 4, 5, and 6 is that of providing for the efficient use of bandwidth and power. Notwithstanding the importance of these two primary communication resources, there are situations where it is necessary to sacrifice this efficiency in order to meet certain other design objectives. For example, the system may be required to provide a form of *secure* communication in a *hostile* environment such that the transmitted signal is not easily detected or recognized by unwanted listeners. This requirement is catered to by a class of signaling techniques known collectively as *spread-spectrum modulation.*

The primary advantage of a spread-spectrum communication system is its ability to reject *interference* whether it be the *unintentional* interference by another user simultaneously attempting to transmit through the channel, or the *intentional* interference by a hostile transmitter attempting to jam the transmission.

The definition of spread-spectrum modulation[1] may be stated in two parts:

1. Spread spectrum is a means of transmission in which the data sequence occupies a bandwidth in excess of the minimum bandwidth necessary to send it.
2. The spectrum spreading is accomplished before transmission through the use of a code that is independent of the data sequence. The same code is used in the receiver

(operating in synchronism with the transmitter) to despread the received signal so that the original data sequence may be recovered.

Although standard modulation techniques such as frequency modulation and pulse-code modulation do satisfy part 1 of this definition, they are not spread-spectrum techniques because they do not satisfy part 2 of the definition.

Spread-spectrum modulation was originally developed for military applications, where resistance to jamming (interference) is of major concern. However, there are civilian applications that also benefit from the unique characteristics of spread-spectrum modulation. For example, it can be used to provide *multipath rejection* in a ground-based mobile radio environment. Yet another application is in *multiple-access* communications in which a number of independent users are required to share a common channel without an external synchronizing mechanism; here, for example, we may mention a ground-based radio environment involving mobile vehicles that must communicate with a central station. More is said about this latter application in Chapter 8.

In this chapter, we discuss principles of spread-spectrum modulation, with emphasis on direct-sequence and frequency-hopping techniques. In a *direct-sequence spread-spectrum* technique, two stages of modulation are used. First, the incoming data sequence is used to modulate a wideband code. This code transforms the narrowband data sequence into a noiselike wideband signal. The resulting wideband signal undergoes a second modulation using a phase-shift keying technique. In a *frequency-hop spread-spectrum* technique, on the other hand, the spectrum of a data-modulated carrier is widened by changing the carrier frequency in a pseudo-random manner. For their operation, both of these techniques rely on the availability of a noiselike spreading code called a *pseudo-random* or *pseudo-noise sequence*. Since such a sequence is basic to the operation of spread-spectrum modulation, it is logical that we begin our study by describing the generation and properties of pseudo-noise sequences.

## 7.2 *Pseudo-Noise Sequences*

A *pseudo-noise (PN) sequence* is a periodic binary sequence with a noiselike waveform that is usually generated by means of a *feedback shift register*, a general block diagram of which is shown in Figure 7.1. A feedback shift register consists of an ordinary *shift register* made up of $m$ flip-flops (two-state memory stages) and a *logic circuit* that are interconnected to form a multiloop *feedback* circuit. The flip-flops in the shift register are regulated by a single timing *clock*. At each pulse (tick) of the clock, the *state* of each flip-flop is shifted to the next one down the line. With each clock pulse the logic circuit computes a

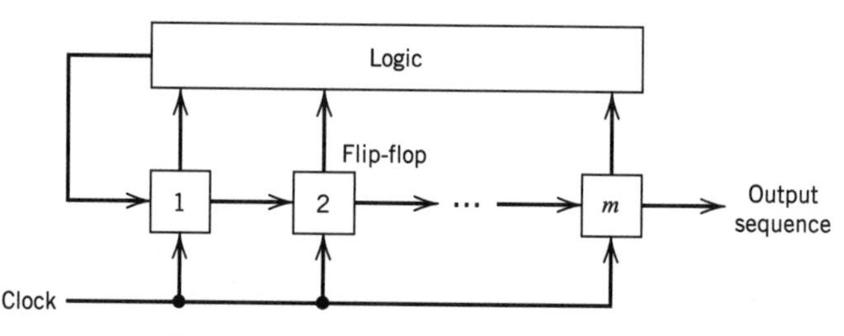


**FIGURE 7.1** Feedback shift register.

Boolean function of the states of the flip-flops. The result is then fed back as the input to the first flip-flop, thereby preventing the shift register from emptying. The PN sequence so generated is determined by the length $m$ of the shift register, its initial state, and the feedback logic.

Let $s_j(k)$ denote the state of the $j$th flip-flop after the $k$th clock pulse; this state may be represented by symbol 0 or 1. The state of the shift register after the $k$th clock pulse is then defined by the set $\{s_1(k), s_2(k), \ldots, s_m(k)\}$, where $k \geq 0$. For the initial state, $k$ is zero. From the definition of a shift register, we have

$$s_j(k+1) = s_{j-1}(k), \qquad \begin{cases} k \geq 0 \\ 1 \leq j \leq m \end{cases} \tag{7.1}$$

where $s_0(k)$ is the input applied to the first flip-flop after the $k$th clock pulse. According to the configuration described in Figure 7.1, $s_0(k)$ is a Boolean function of the individual states $s_1(k), s_2(k), \ldots, s_m(k)$. For a specified length $m$, this Boolean function uniquely determines the subsequent sequence of states and therefore the PN sequence produced at the output of the final flip-flop in the shift register. With a total number of $m$ flip-flops, the number of possible states of the shift register is at most $2^m$. It follows therefore that the PN sequence generated by a feedback shift register must eventually become *periodic* with a period of at most $2^m$.

A feedback shift register is said to be *linear* when the feedback logic consists entirely of *modulo-2 adders*. In such a case, the *zero state* (e.g., the state for which all the flip-flops are in state 0) is *not* permitted. We say so because for a zero state, the input $s_0(k)$ produced by the feedback logic would be 0, the shift register would then continue to remain in the zero state, and the output would therefore consist entirely of 0s. Consequently, the period of a PN sequence produced by a linear feedback shift register with $m$ flip-flops cannot exceed $2^m - 1$. When the period is exactly $2^m - 1$, the PN sequence is called a *maximal-length-sequence* or simply *m-sequence*.

### ▶ EXAMPLE 7.1

Consider the linear feedback shift register shown in Figure 7.2, involving three flip-flops. The input $s_0$ applied to the first flip-flop is equal to the modulo-2 sum of $s_1$ and $s_3$. It is assumed that the initial state of the shift register is 100 (reading the contents of the three flip-flops from left to right). Then, the succession of states will be as follows:

100, 110, 111, 011, 101, 010, 001, 100, . . . .

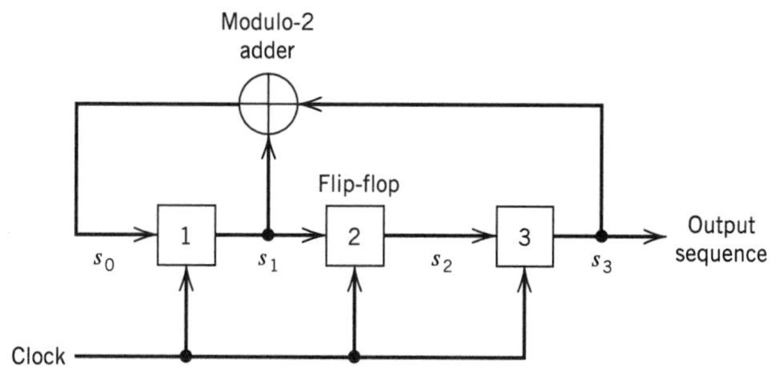


**FIGURE 7.2** Maximal-length sequence generator for $m = 3$.

The output sequence (the last position of each state of the shift register) is therefore

$$00111010\ldots$$

which repeats itself with period $2^3 - 1 = 7$.

Note that the choice of 100 as the initial state is arbitrary. Any of the other six permissible states could serve equally well as an initial state. The resulting output sequence would then simply experience a cyclic shift. ◀

## ■ PROPERTIES OF MAXIMAL-LENGTH SEQUENCES[2]

Maximal-length sequences have many of the properties possessed by a truly *random binary sequence*. A random binary sequence is a sequence in which the presence of binary symbol 1 or 0 is equally probable. Some properties of maximal-length sequences are as follows:

**1.** In each period of a maximal-length sequence, the number of 1s is always one more than the number of 0s. This property is called the *balance property*.

**2.** Among the runs of 1s and of 0s in each period of a maximal-length sequence, one-half the runs of each kind are of length one, one-fourth are of length two, one-eighth are of length three, and so on as long as these fractions represent meaningful numbers of runs. This property is called the *run property*. By a "run" we mean a subsequence of identical symbols (1s or 0s) within one period of the sequence. The length of this subsequence is the length of the run. For a maximal-length sequence generated by a linear feedback shift register of length $m$, the total number of runs is $(N + 1)/2$, where $N = 2^m - 1$.

**3.** The autocorrelation function of a maximal-length sequence is periodic and binary-valued. This property is called the *correlation property*.

The period of a maximum-length sequence is defined by

$$N = 2^m - 1 \tag{7.2}$$

where $m$ is the length of the shift register. Let binary symbols 0 and 1 of the sequence be denoted by the levels $-1$ and $+1$, respectively. Let $c(t)$ denote the resulting waveform of the maximal-length sequence, as illustrated in Figure 7.3*a* for $N = 7$. The period of the waveform $c(t)$ is (based on terminology used in subsequent sections)

$$T_b = NT_c \tag{7.3}$$

where $T_c$ is the duration assigned to symbol 1 or 0 in the maximal-length sequence. By definition, the autocorrelation function of a periodic signal $c(t)$ of period $T_b$ is

$$R_c(\tau) = \frac{1}{T_b} \int_{-T_b/2}^{T_b/2} c(t)c(t - \tau)\, dt \tag{7.4}$$

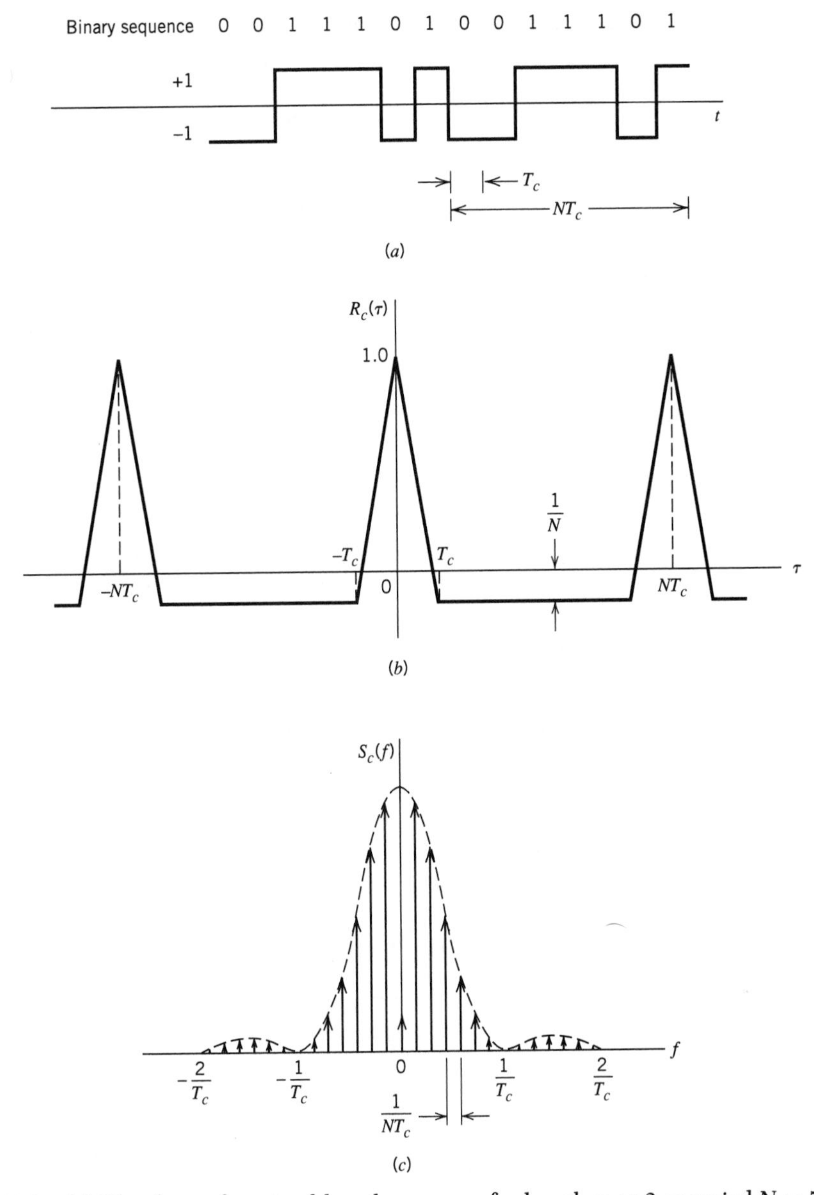


**FIGURE 7.3** (*a*) Waveform of maximal-length sequence for length $m = 3$ or period $N = 7$. (*b*) Autocorrelation function. (*c*) Power spectral density. All three parts refer to the output of the feedback shift register of Figure 7.2.

where the lag $\tau$ lies in the interval $(-T_b/2, T_b/2)$; Equation (7.4) is a special case of Equation (1.26). Applying this formula to a maximal-length sequence represented by $c(t)$, we get

$$R_c(\tau) = \begin{cases} 1 - \dfrac{N+1}{NT_c}|\tau|, & |\tau| \leq T_c \\ -\dfrac{1}{N}, & \text{for the remainder of the period} \end{cases} \tag{7.5}$$

This result is plotted in Figure 7.3*b* for the case of $m = 3$ or $N = 7$.

From Fourier transform theory we know that periodicity in the time domain is transformed into uniform sampling in the frequency domain. This interplay between the time and frequency domains is borne out by the power spectral density of the maximal-length wave $c(t)$. Specifically, taking the Fourier transform of Equation (7.5), we get the sampled spectrum

$$S_c(f) = \frac{1}{N^2}\,\delta(f) + \frac{1+N}{N^2}\sum_{\substack{n=-\infty \\ n\neq 0}}^{\infty} \text{sinc}^2\left(\frac{n}{N}\right)\delta\left(f - \frac{n}{NT_c}\right) \tag{7.6}$$

which is plotted in Figure 7.3$c$ for $m = 3$ or $N = 7$.

Comparing the results of Figure 7.3 for a maximal-length sequence with the corresponding results shown in Figure 1.11 for a random binary sequence, we may make the following observations:

- For a period of the maximal-length sequence, the autocorrelation function $R_c(\tau)$ is somewhat similar to that of a random binary wave.
- The waveforms of both sequences have the same envelope, $\text{sinc}^2(fT)$, for their power spectral densities. The fundamental difference between them is that whereas the random binary sequence has a continuous spectral density characteristic, the corresponding characteristic of a maximal-length sequence consists of delta functions spaced $1/NT_c$ Hz apart.

As the shift-register length $m$, or equivalently, the period $N$ of the maximal-length sequence is increased, the maximal-length sequence becomes increasingly similar to the random binary sequence. Indeed, in the limit, the two sequences become identical when $N$ is made infinitely large. However, the price paid for making $N$ large is an increasing storage requirement, which imposes a practical limit on how large $N$ can actually be made.

## ■ CHOOSING A MAXIMAL-LENGTH SEQUENCE

Now that we understand the properties of a maximal-length sequence and the fact that we can generate it using a linear feedback shift register, the key question that we need to address is: How do we find the feedback logic for a desired period $N$? The answer to this

**TABLE 7.1** ***Maximal-length sequences of shift-register lengths 2–8***

| *Shift-Register Length,* m | *Feedback Taps* |
|---|---|
| 2* | [2, 1] |
| 3* | [3, 1] |
| 4 | [4, 1] |
| 5* | [5, 2], [5, 4, 3, 2], [5, 4, 2, 1] |
| 6 | [6, 1], [6, 5, 2, 1], [6, 5, 3, 2] |
| 7* | [7, 1], [7, 3], [7, 3, 2, 1], [7, 4, 3, 2], [7, 6, 4, 2], [7, 6, 3, 1], [7, 6, 5, 2], [7, 6, 5, 4, 2, 1], [7, 5, 4, 3, 2, 1] |
| 8 | [8, 4, 3, 2], [8, 6, 5, 3], [8, 6, 5, 2], [8, 5, 3, 1], [8, 6, 5, 1], [8, 7, 6, 1], [8, 7, 6, 5, 2, 1], [8, 6, 4, 3, 2, 1] |

question is to be found in the theory of error-control codes, which is covered in Chapter 10. The task of finding the required feedback logic is made particularly easy for us by virtue of the extensive tables of the necessary feedback connections for varying shift-register lengths that have been compiled in the literature. In Table 7.1, we present the sets of maximal (feedback) taps pertaining to shift-register lengths $m = 2, 3, \ldots, 8$.[3] Note that as $m$ increases, the number of alternative schemes (codes) is enlarged. Also, for every set of feedback connections shown in this table, there is an "image" set that generates an identical maximal-length code, reversed in time sequence.

The particular sets identified with an asterisk in Table 7.1 correspond to *Mersenne prime length sequences*, for which the period $N$ is a prime number.

## ▶ EXAMPLE 7.2

Consider a maximal-length sequence requiring the use of a linear feedback-shift register of length $m = 5$. For feedback taps, we select the set [5, 2] from Table 7.1. The corresponding configuration of the code generator is shown in Figure 7.4*a*. Assuming that the initial state is 10000, the evolution of one period of the maximal-length sequence generated by this scheme is shown in Table 7.2*a*, where we see that the generator returns to the initial 10000 after 31 iterations; that is, the period is 31, which agrees with the value obtained from Equation (7.2).

Suppose next we select another set of feedback taps from Table 7.1, namely, [5, 4, 2, 1]. The corresponding code generator is thus as shown in Figure 7.4*b*. For the initial state 10000, we now find that the evolution of the maximal-length sequence is as shown in Table 7.2*b*. Here again, the generator returns to the initial state 10000 after 31 iterations, and so it should. But the maximal-length sequence generated is different from that shown in Table 7.2*a*.

Clearly, the code generator of Figure 7.4*a* has an advantage over that of Figure 7.4*b*, as it requires fewer feedback connections. ◀

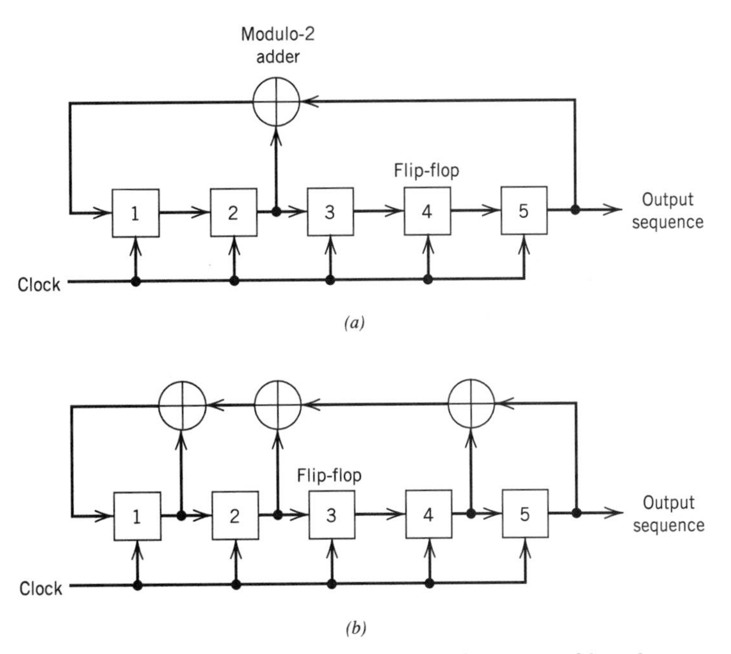


**FIGURE 7.4** Two different configurations of feedback shift register of length $m = 5$. (*a*) Feedback connections [5, 2]. (*b*) Feedback connections [5, 4, 2, 1].

**TABLE 7.2a** ***Evolution of the maximal-length sequence generated by the feedback-shift register of Fig. 7.4a***

| *Feedback Symbol* | *State of Shift Register* | | | | | *Output Symbol* |
|---|---|---|---|---|---|---|
| | 1 | 0 | 0 | 0 | 0 | |
| 0 | 0 | 1 | 0 | 0 | 0 | 0 |
| 1 | 1 | 0 | 1 | 0 | 0 | 0 |
| 0 | 0 | 1 | 0 | 1 | 0 | 0 |
| 1 | 1 | 0 | 1 | 0 | 1 | 0 |
| 1 | 1 | 1 | 0 | 1 | 0 | 1 |
| 1 | 1 | 1 | 1 | 0 | 1 | 0 |
| 0 | 0 | 1 | 1 | 1 | 0 | 1 |
| 1 | 1 | 0 | 1 | 1 | 1 | 0 |
| 1 | 1 | 1 | 0 | 1 | 1 | 1 |
| 0 | 0 | 1 | 1 | 0 | 1 | 1 |
| 0 | 0 | 0 | 1 | 1 | 0 | 1 |
| 0 | 0 | 0 | 0 | 1 | 1 | 0 |
| 1 | 1 | 0 | 0 | 0 | 1 | 1 |
| 1 | 1 | 1 | 0 | 0 | 0 | 1 |
| 1 | 1 | 1 | 1 | 0 | 0 | 0 |
| 1 | 1 | 1 | 1 | 1 | 0 | 0 |
| 1 | 1 | 1 | 1 | 1 | 1 | 0 |
| 0 | 0 | 1 | 1 | 1 | 1 | 1 |
| 0 | 0 | 0 | 1 | 1 | 1 | 1 |
| 1 | 1 | 0 | 0 | 1 | 1 | 1 |
| 1 | 1 | 1 | 0 | 0 | 1 | 1 |
| 0 | 0 | 1 | 1 | 0 | 0 | 1 |
| 1 | 1 | 0 | 1 | 1 | 0 | 0 |
| 0 | 0 | 1 | 0 | 1 | 1 | 0 |
| 0 | 0 | 0 | 1 | 0 | 1 | 1 |
| 1 | 1 | 0 | 0 | 1 | 0 | 1 |
| 0 | 0 | 1 | 0 | 0 | 1 | 0 |
| 0 | 0 | 0 | 1 | 0 | 0 | 1 |
| 0 | 0 | 0 | 0 | 1 | 0 | 0 |
| 0 | 0 | 0 | 0 | 0 | 1 | 0 |
| 1 | 1 | 0 | 0 | 0 | 0 | 1 |

Code: 0000101011101100011111001101001

**TABLE 7.2*b*** ***Evolution of the maximal-length sequence generated by the feedback-shift register of Fig. 7.4b***

| Feedback Symbol | State of Shift Register | | | | | Output Symbol |
|---|---|---|---|---|---|---|
| | 1 | 0 | 0 | 0 | 0 | |
| 1 | 1 | 1 | 0 | 0 | 0 | 0 |
| 0 | 0 | 1 | 1 | 0 | 0 | 0 |
| 1 | 1 | 0 | 1 | 1 | 0 | 0 |
| 0 | 0 | 1 | 0 | 1 | 1 | 0 |
| 1 | 1 | 0 | 1 | 0 | 1 | 1 |
| 0 | 0 | 1 | 0 | 1 | 0 | 1 |
| 0 | 0 | 0 | 1 | 0 | 1 | 0 |
| 1 | 1 | 0 | 0 | 1 | 0 | 1 |
| 0 | 0 | 1 | 0 | 0 | 1 | 0 |
| 0 | 0 | 0 | 1 | 0 | 0 | 1 |
| 0 | 0 | 0 | 0 | 1 | 0 | 0 |
| 1 | 1 | 0 | 0 | 0 | 1 | 0 |
| 0 | 0 | 1 | 0 | 0 | 0 | 1 |
| 1 | 1 | 0 | 1 | 0 | 0 | 0 |
| 1 | 1 | 1 | 0 | 1 | 0 | 0 |
| 1 | 1 | 1 | 1 | 0 | 1 | 0 |
| 1 | 1 | 1 | 1 | 1 | 0 | 1 |
| 1 | 1 | 1 | 1 | 1 | 1 | 0 |
| 0 | 0 | 1 | 1 | 1 | 1 | 1 |
| 1 | 1 | 0 | 1 | 1 | 1 | 1 |
| 1 | 1 | 1 | 0 | 1 | 1 | 1 |
| 0 | 0 | 1 | 1 | 0 | 1 | 1 |
| 0 | 0 | 0 | 1 | 1 | 0 | 1 |
| 1 | 1 | 0 | 0 | 1 | 1 | 0 |
| 1 | 1 | 1 | 0 | 0 | 1 | 1 |
| 1 | 1 | 1 | 1 | 0 | 0 | 1 |
| 0 | 0 | 1 | 1 | 1 | 0 | 0 |
| 0 | 0 | 0 | 1 | 1 | 1 | 0 |
| 0 | 0 | 0 | 0 | 1 | 1 | 1 |
| 0 | 0 | 0 | 0 | 0 | 1 | 1 |
| 1 | 1 | 0 | 0 | 0 | 0 | 1 |

Code: 0000110101001000101111101100111

## 7.3 A Notion of Spread Spectrum

An important attribute of spread-spectrum modulation is that it can provide protection against externally generated interfering (jamming) signals with finite power. The jamming signal may consist of a fairly powerful broadband noise or multitone waveform that is directed at the receiver for the purpose of disrupting communications. Protection against jamming waveforms is provided by purposely making the information-bearing signal occupy a bandwidth far in excess of the minimum bandwidth necessary to transmit it. This has the effect of making the transmitted signal assume a noiselike appearance so as to blend into the background. The transmitted signal is thus enabled to propagate through the channel undetected by anyone who may be listening. We may therefore think of spread spectrum as a method of "camouflaging" the information-bearing signal.

One method of widening the bandwidth of an information-bearing (data) sequence involves the use of *modulation.* Let $\{b_k\}$ denote a binary data sequence, and $\{c_k\}$ denote a pseudo-noise (PN) sequence. Let the waveforms $b(t)$ and $c(t)$ denote their respective polar nonreturn-to-zero representations in terms of two levels equal in amplitude and opposite in polarity, namely, $\pm 1$. We will refer to $b(t)$ as the information-bearing (data) signal, and to $c(t)$ as the PN signal. The desired modulation is achieved by applying the data signal $b(t)$ and the PN signal $c(t)$ to a product modulator or multiplier, as in Figure 7.5*a*. We know from Fourier transform theory that multiplication of two signals produces a signal whose spectrum equals the convolution of the spectra of the two component signals. Thus, if the message signal $b(t)$ is narrowband and the PN signal $c(t)$ is wideband, *the product (modulated) signal $m(t)$ will have a spectrum that is nearly the same as the wideband PN signal.* In other words, in the context of our present application, the PN sequence performs the role of a *spreading code.*

By multiplying the information-bearing signal $b(t)$ by the PN signal $c(t)$, each information bit is "chopped" up into a number of small time increments, as illustrated in the waveforms of Figure 7.6. These small time increments are commonly referred to as *chips.*

For *baseband* transmission, the product signal $m(t)$ represents the *transmitted signal.* We may thus express the transmitted signal as

$$m(t) = c(t)b(t) \tag{7.7}$$

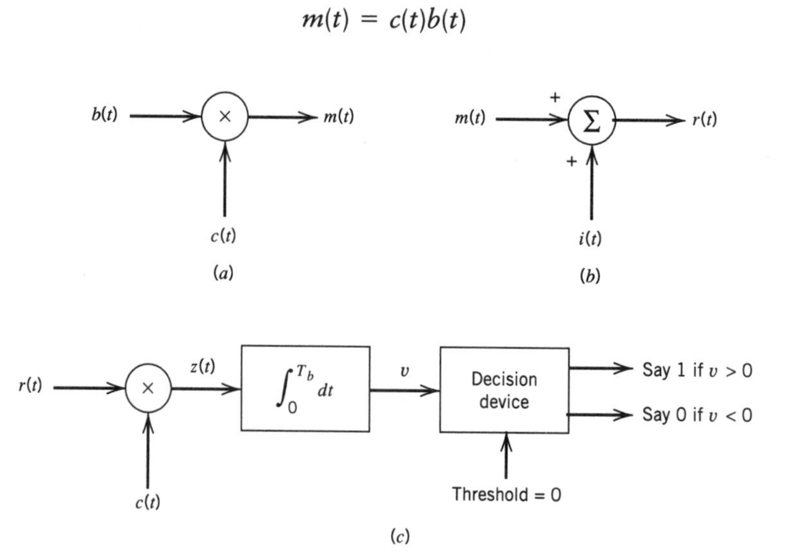


**FIGURE 7.5** Idealized model of baseband spread-spectrum system. (*a*) Transmitter. (*b*) Channel. (*c*) Receiver.

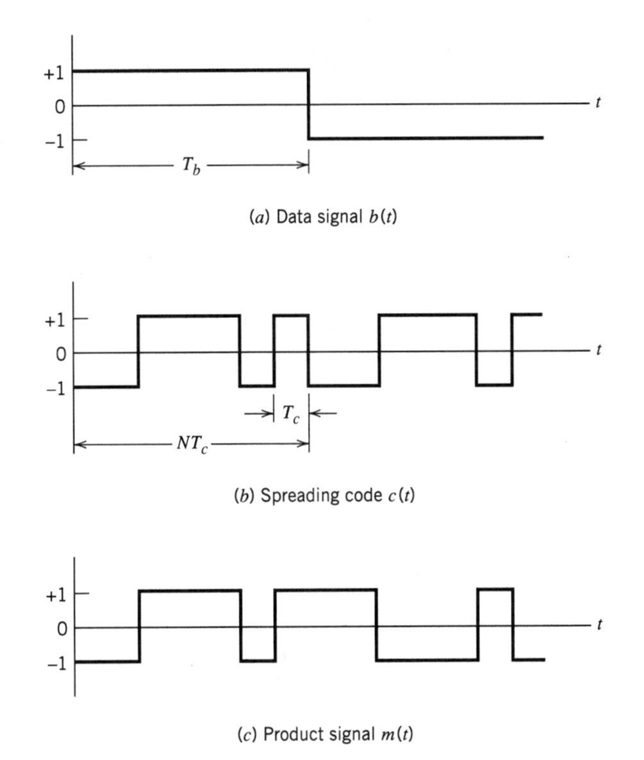


**FIGURE 7.6** Illustrating the waveforms in the transmitter of Figure 7.5*a*.

The received signal $r(t)$ consists of the transmitted signal $m(t)$ plus an additive *interference* denoted by $i(t)$, as shown in the channel model of Figure 7.5*b*. Hence,

$$\begin{aligned} r(t) &= m(t) + i(t) \\ &= c(t)b(t) + i(t) \end{aligned} \tag{7.8}$$

To recover the original message signal $b(t)$, the received signal $r(t)$ is applied to a *demodulator* that consists of a multiplier followed by an integrator, and a decision device, as in Figure 7.5*c*. The multiplier is supplied with a locally generated PN sequence that is an exact *replica* of that used in the transmitter. Moreover, we assume that the receiver operates in perfect *synchronism* with the transmitter, which means that the PN sequence in the receiver is lined up exactly with that in the transmitter. The multiplier output in the receiver is therefore given by

$$\begin{aligned} z(t) &= c(t)r(t) \\ &= c^2(t)b(t) + c(t)i(t) \end{aligned} \tag{7.9}$$

Equation (7.9) shows that the data signal $b(t)$ is multiplied *twice* by the PN signal $c(t)$, whereas the unwanted signal $i(t)$ is multiplied only *once*. The PN signal $c(t)$ alternates between the levels $-1$ and $+1$, and the alternation is destroyed when it is squared; hence,

$$c^2(t) = 1 \qquad \text{for all } t \tag{7.10}$$

Accordingly, we may simplify Equation (7.9) as

$$z(t) = b(t) + c(t)i(t) \tag{7.11}$$

We thus see from Equation (7.11) that the data signal $b(t)$ is reproduced at the multiplier output in the receiver, except for the effect of the interference represented by the additive term $c(t)i(t)$. Multiplication of the interference $i(t)$ by the locally generated PN signal $c(t)$ means that the spreading code will affect the interference just as it did the original signal at the transmitter. We now observe that the data component $b(t)$ is narrowband, whereas the spurious component $c(t)i(t)$ is wideband. Hence, by applying the multiplier output to a baseband (low-pass) filter with a bandwidth just large enough to accommodate the recovery of the data signal $b(t)$, most of the power in the spurious component $c(t)i(t)$ is filtered out. The effect of the interference $i(t)$ is thus significantly reduced at the receiver output.

In the receiver shown in Figure 7.5*c*, the low-pass filtering action is actually performed by the integrator that evaluates the area under the signal produced at the multiplier output. The integration is carried out for the bit interval $0 \leq t \leq T_b$, providing the sample value $v$. Finally, a decision is made by the receiver: If $v$ is greater than the threshold of zero, the receiver says that binary symbol 1 of the original data sequence was sent in the interval $0 \leq t \leq T_b$, and if $v$ is less than zero, the receiver says that symbol 0 was sent; if $v$ is exactly zero the receiver makes a random guess in favor of 1 or 0.

In summary, the use of a spreading code (with pseudo-random properties) in the transmitter produces a wideband transmitted signal that appears *noiselike* to a receiver that has *no* knowledge of the spreading code. From the discussion presented in Section 7.2, we recall that (for a prescribed data rate) the longer we make the period of the spreading code, the closer will the transmitted signal be to a truly random binary wave, and the harder it is to detect. Naturally, the price we have to pay for the improved protection against interference is increased transmission bandwidth, system complexity, and processing delay. However, when our primary concern is the security of transmission, these are not unreasonable costs to pay.

## 7.4 Direct-Sequence Spread Spectrum with Coherent Binary Phase-Shift Keying

The spread-spectrum technique described in the previous section is referred to as *direct-sequence spread spectrum*. The discussion presented there was in the context of baseband transmission. To provide for the use of this technique in passband transmission over a satellite channel, for example, we may incorporate *coherent binary phase-shift keying* (PSK) into the transmitter and receiver, as shown in Figure 7.7. The transmitter of Figure 7.7*a* first converts the incoming binary data sequence $\{b_k\}$ into a polar NRZ waveform $b(t)$, which is followed by two stages of modulation. The first stage consists of a product modulator or multiplier with the data signal $b(t)$ (representing a data sequence) and the PN signal $c(t)$ (representing the PN sequence) as inputs. The second stage consists of a binary PSK modulator. The transmitted signal $x(t)$ is thus a *direct-sequence spread binary phase-shift-keyed* (DS/BPSK) *signal*. The phase modulation $\theta(t)$ of $x(t)$ has one of two values, 0 and $\pi$, depending on the polarities of the message signal $b(t)$ and PN signal $c(t)$ at time $t$ in accordance with the truth table of Table 7.3.

Figure 7.8 illustrates the waveforms for the second stage of modulation. Part of the modulated waveform shown in Figure 7.6*c* is reproduced in Figure 7.8*a*; the waveform shown here corresponds to one period of the PN sequence. Figure 7.8*b* shows the waveform of a sinusoidal carrier, and Figure 7.8*c* shows the DS/BPSK waveform that results from the second stage of modulation.

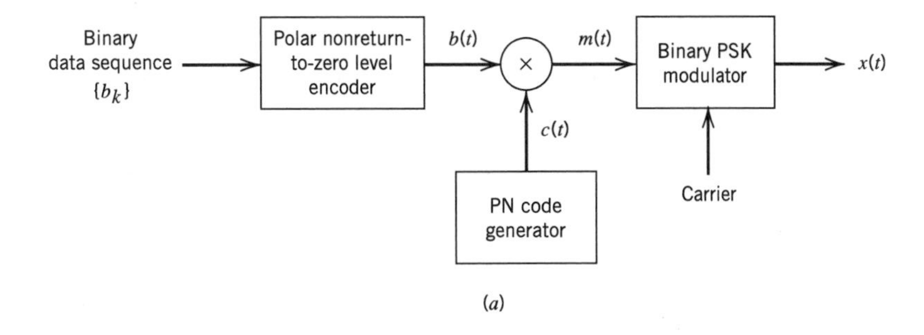


(a)

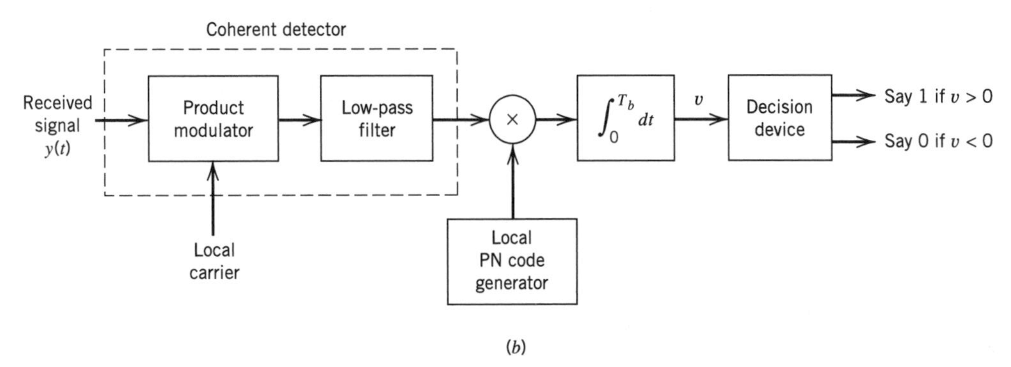


(b)

**FIGURE 7.7** Direct-sequence spread coherent phase-shift keying. (*a*) Transmitter. (*b*) Receiver.

The receiver, shown in Figure 7.7*b*, consists of two stages of demodulation. In the first stage, the received signal $y(t)$ and a locally generated carrier are applied to a product modulator followed by a low-pass filter whose bandwidth is equal to that of the original message signal $m(t)$. This stage of the demodulation process reverses the phase-shift keying applied to the transmitted signal. The second stage of demodulation performs spectrum despreading by multiplying the low-pass filter output by a locally generated replica of the PN signal $c(t)$, followed by integration over a bit interval $0 \leq t \leq T_b$, and finally decision-making in the manner described in Section 7.3.

## ■ MODEL FOR ANALYSIS

In the normal form of the transmitter, shown in Figure 7.7*a*, the spectrum spreading is performed prior to phase modulation. For the purpose of analysis, however, we find it more convenient to interchange the order of these operations, as shown in the model of

**TABLE 7.3** ***Truth table for phase modulation $\theta(t)$, radians***

| | | *Polarity of Data Sequence* b(t) *at Time* t | |
|---|---|---|---|
| | | + | − |
| *Polarity of PN sequence* c(t) *at time* t | + | 0 | $\pi$ |
| | − | $\pi$ | 0 |

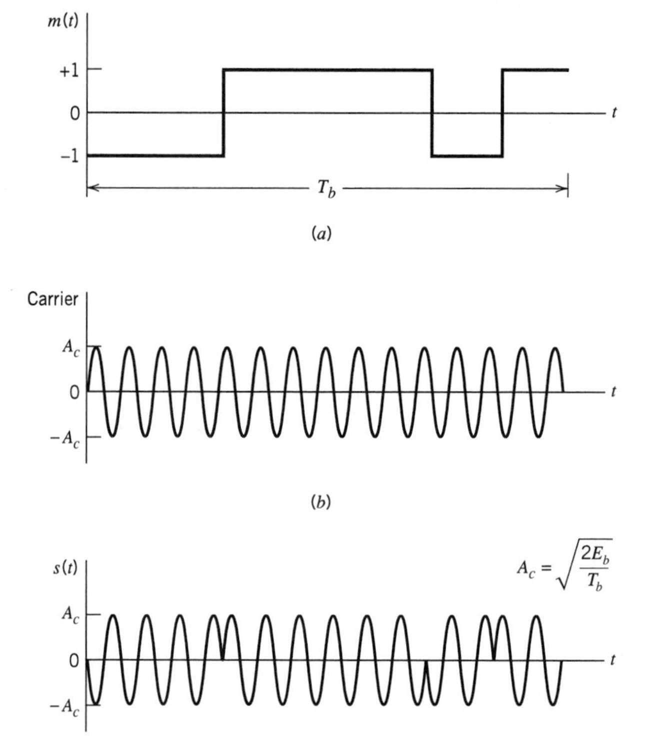


**FIGURE 7.8** (*a*) Product signal $m(t) = c(t)b(t)$. (*b*) Sinusoidal carrier. (*c*) DS/BPSK signal.

Figure 7.9. We are permitted to do this because the spectrum spreading and the binary phase-shift keying are both linear operations; likewise for the phase demodulation and spectrum despreading. But for the interchange of operations to be feasible, it is important to synchronize the incoming data sequence and the PN sequence. The model of Figure 7.9 also includes representations of the channel and the receiver. In this model, it is assumed that the interference $j(t)$ limits performance, so that the effect of channel noise may be ignored. Accordingly, the channel output is given by

$$\begin{aligned} y(t) &= x(t) + j(t) \\ &= c(t)s(t) + j(t) \end{aligned} \tag{7.12}$$

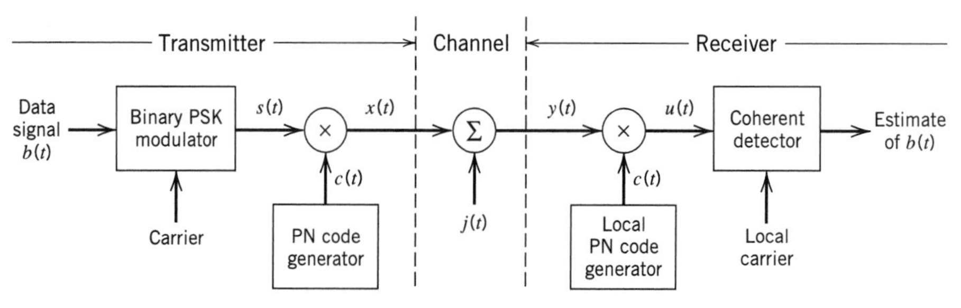


**FIGURE 7.9** Model of direct-sequence spread binary PSK system.

where $s(t)$ is the binary PSK signal, and $c(t)$ is the PN signal. In the channel model included in Figure 7.9, the interfering signal is denoted by $j(t)$. This notation is chosen purposely to be different from that used for the interference in Figure 7.5*b*. The channel model in Figure 7.9 is passband in spectral content, whereas that in Figure 7.5*b* is in baseband form.

In the receiver, the received signal $y(t)$ is first multiplied by the PN signal $c(t)$ yielding an output that equals the coherent detector input $u(t)$. Thus,

$$\begin{aligned} u(t) &= c(t)y(t) \\ &= c^2(t)s(t) + c(t)j(t) \\ &= s(t) + c(t)j(t) \end{aligned} \tag{7.13}$$

In the last line of Equation (7.13), we have noted that, by design, the PN signal $c(t)$ satisfies the property described in Equation (7.10), reproduced here for convenience:

$$c^2(t) = 1 \qquad \text{for all } t$$

Equation (7.13) shows that the coherent detector input $u(t)$ consists of a binary PSK signal $s(t)$ embedded in additive code-modulated interference denoted by $c(t)j(t)$. The modulated nature of the latter component forces the interference signal (jammer) to spread its spectrum such that the detection of information bits at the receiver output is afforded increased reliability.

### ■ SYNCHRONIZATION

For its proper operation, a spread-spectrum communication system requires that the locally generated PN sequence used in the receiver to despread the received signal be *synchronized* to the PN sequence used to spread the transmitted signal in the transmitter.[4] A solution to the synchronization problem consists of two parts: *acquisition* and *tracking*. In acquisition, or *coarse* synchronization, the two PN codes are aligned to within a fraction of the chip in as short a time as possible. Once the incoming PN code has been acquired, tracking, or *fine* synchronization, takes place. Typically, PN acquisition proceeds in two steps. First, the received signal is multiplied by a locally generated PN code to produce a measure of *correlation* between it and the PN code used in the transmitter. Next, an appropriate *decision-rule and search strategy* is used to process the measure of correlation so obtained to determine whether the two codes are in synchronism and what to do if they are not. As for tracking, it is accomplished using phase-lock techniques very similar to those used for the local generation of coherent carrier references. The principal difference between them lies in the way in which phase discrimination is implemented.

## 7.5 *Signal-Space Dimensionality and Processing Gain*

Having developed a conceptual understanding of spread-spectrum modulation and a method for its implementation, we are ready to undertake a detailed mathematical analysis of the technique. The approach we have in mind is based on the signal-space theoretic ideas of Chapter 5. In particular, we develop signal-space representations of the transmitted signal and the interfering signal (jammer).

In this context, consider the set of orthonormal basis functions:

$$\phi_k(t) = \begin{cases} \sqrt{\dfrac{2}{T_c}} \cos(2\pi f_c t), & kT_c \leq t \leq (k+1)T_c \\ 0, & \text{otherwise} \end{cases} \tag{7.14}$$

$$\tilde{\phi}_k(t) = \begin{cases} \sqrt{\dfrac{2}{T_c}} \sin(2\pi f_c t), & kT_c \leq t \leq (k+1)T_c \\ 0, & \text{otherwise} \end{cases} \tag{7.15}$$

$$k = 0, 1, \ldots, N-1$$

where $T_c$ is the *chip duration*, and $N$ is the number of chips per bit. Accordingly, we may describe the transmitted signal $x(t)$ for the interval of an information bit as follows:

$$\begin{aligned} x(t) &= c(t)s(t) \\ &= \pm\sqrt{\frac{2E_b}{T_b}}\, c(t) \cos(2\pi f_c t) \\ &= \pm\sqrt{\frac{E_b}{N}} \sum_{k=0}^{N-1} c_k \phi_k(t), \qquad 0 \leq t \leq T_b \end{aligned} \tag{7.16}$$

where $E_b$ is the signal energy per bit; the plus sign corresponds to information bit 1, and the minus sign corresponds to information bit 0. The code sequence $\{c_0, c_1, \ldots, c_{N-1}\}$ denotes the PN sequence, with $c_k = \pm 1$. The transmitted signal $x(t)$ is therefore $N$-dimensional in that it requires a minimum of $N$ orthonormal functions for its representation.

Consider next the representation of the interfering signal (jammer), $j(t)$. Ideally, the jammer likes to place all of its available energy in exactly the same $N$-dimensional signal space as the transmitted signal $x(t)$; otherwise, part of its energy goes to waste. However, the best that the jammer can hope to know is the transmitted signal bandwidth. Moreover, there is no way that the jammer can have knowledge of the signal phase. Accordingly, we may represent the jammer by the general form

$$j(t) = \sum_{k=0}^{N-1} j_k \phi_k(t) + \sum_{k=0}^{N-1} \tilde{j}_k \tilde{\phi}_k(t), \qquad 0 \leq t \leq T_b \tag{7.17}$$

where

$$j_k = \int_0^{T_b} j(t)\phi_k(t)\, dt, \qquad k = 0, 1, \ldots, N-1 \tag{7.18}$$

and

$$\tilde{j}_k = \int_0^{T_b} j(t)\tilde{\phi}_k(t)\, dt, \qquad k = 0, 1, \ldots, N-1 \tag{7.19}$$

Thus the interference $j(t)$ is $2N$-dimensional; that is, it has twice the number of dimensions required for representing the transmitted DS/BPSK signal $x(t)$. In terms of the representation given in Equation (7.17), we may express the average power of the interference $j(t)$ as follows:

$$
\begin{aligned}
J &= \frac{1}{T_b} \int_0^{T_b} j^2(t)\, dt \\
&= \frac{1}{T_b} \sum_{k=0}^{N-1} j_k^2 + \frac{1}{T_b} \sum_{k=0}^{N-1} \tilde{j}_k^2
\end{aligned}
\tag{7.20}
$$

Moreover, due to lack of knowledge of signal phase, the best strategy a jammer can apply is to place equal energy in the cosine and sine coordinates defined in Equations (7.18) and (7.19); hence, we may safely assume

$$
\sum_{k=0}^{N-1} j_k^2 = \sum_{k=0}^{N-1} \tilde{j}_k^2 \tag{7.21}
$$

Correspondingly, we may simplify Equation (7.20) as

$$
J = \frac{2}{T_b} \sum_{k=0}^{N-1} j_k^2 \tag{7.22}
$$

Our aim is to tie these results together by finding the signal-to-noise ratios measured at the input and output of the DS/BPSK receiver in Figure 7.9. To that end, we use Equation (7.13) to express the coherent detector output as

$$
\begin{aligned}
v &= \sqrt{\frac{2}{T_b}} \int_0^{T_b} u(t) \cos(2\pi f_c t)\, dt \\
&= v_s + v_{cj}
\end{aligned}
\tag{7.23}
$$

where the components $v_s$ and $v_{cj}$ are due to the despread binary PSK signal, $s(t)$, and the spread interference, $c(t)j(t)$, respectively. These two components are defined as follows:

$$
v_s = \sqrt{\frac{2}{T_b}} \int_0^{T_b} s(t) \cos(2\pi f_c t)\, dt \tag{7.24}
$$

and

$$
v_{cj} = \sqrt{\frac{2}{T_b}} \int_0^{T_b} c(t) j(t) \cos(2\pi f_c t)\, dt \tag{7.25}
$$

Consider first the component $v_s$ due to the signal. The despread binary PSK signal $s(t)$ equals

$$
s(t) = \pm\sqrt{\frac{2E_b}{T_b}} \cos(2\pi f_c t), \qquad 0 \le t \le T_b \tag{7.26}
$$

where the plus sign corresponds to information bit 1, and the minus sign corresponds to information bit 0. Hence, assuming that the carrier frequency $f_c$ is an integer multiple of $1/T_b$, we have

$$
v_s = \pm\sqrt{E_b} \tag{7.27}
$$

Consider next the component $v_{cj}$ due to interference. Expressing the PN signal $c(t)$ in the explicit form of a sequence, $\{c_0, c_1, \ldots, c_{N-1}\}$, we may rewrite Equation (7.25) in the corresponding form

$$v_{cj} = \sqrt{\frac{2}{T_b}} \sum_{k=0}^{N-1} c_k \int_{kT_c}^{(k+1)T_c} j(t) \cos(2\pi f_c t)\, dt \tag{7.28}$$

Using Equation (7.14) for $\phi_k(t)$, and then Equation (7.18) for the coefficient $j_k$, we may redefine $v_{cj}$ as

$$\begin{aligned} v_{cj} &= \sqrt{\frac{T_c}{T_b}} \sum_{k=0}^{N-1} c_k \int_0^{T_b} j(t)\phi_k(t)\, dt \\ &= \sqrt{\frac{T_c}{T_b}} \sum_{k=0}^{N-1} c_k j_k \end{aligned} \tag{7.29}$$

We next approximate the PN sequence as an *independent and identically distributed (i.i.d.) binary sequence*. We emphasize the implication of this approximation by recasting Equation (7.29) in the form

$$V_{cj} = \sqrt{\frac{T_c}{T_b}} \sum_{k=0}^{N-1} C_k j_k \tag{7.30}$$

where $V_{cj}$ and $C_k$ are random variables with sample values $v_{cj}$ and $c_k$, respectively. In Equation (7.30), the jammer is assumed to be fixed. With the $C_k$ treated as i.i.d. random variables, we find that the probability of the event $C_k = \pm 1$ is

$$P(C_k = 1) = P(C_k = -1) = \tfrac{1}{2} \tag{7.31}$$

Accordingly, the mean of the random variable $V_{cj}$ is zero since, for fixed $k$, we have

$$\begin{aligned} E[C_k j_k \mid j_k] &= j_k P(C_k = 1) - j_k P(C_k = -1) \\ &= \tfrac{1}{2} j_k - \tfrac{1}{2} j_k \\ &= 0 \end{aligned} \tag{7.32}$$

For a fixed vector $\mathbf{j}$, representing the set of coefficients $j_0, j_1, \ldots, j_{N-1}$, the variance of $V_{cj}$ is given by

$$\text{var}[V_{cj} \mid \mathbf{j}] = \frac{1}{N} \sum_{k=0}^{N-1} j_k^2 \tag{7.33}$$

Since the *spread factor* $N = T_b/T_c$, we may use Equation (7.22) to express this variance in terms of the average interference power $J$ as

$$\text{var}[V_{cj} \mid \mathbf{j}] = \frac{JT_c}{2} \tag{7.34}$$

Thus the random variable $V_{cj}$ has zero mean and variance $JT_c/2$.

From Equation (7.27), we note that the signal component at the coherent detector output (during each bit interval) equals $\pm\sqrt{E_b}$, where $E_b$ is the signal energy per bit. Hence, the peak instantaneous power of the signal component is $E_b$. Accordingly, we may define the *output signal-to-noise ratio* as the instantaneous peak power $E_b$ divided by the variance of the equivalent noise component in Equation (7.34). We thus write

$$(\text{SNR})_O = \frac{2E_b}{JT_c} \tag{7.35}$$

The average signal power at the receiver input equals $E_b/T_b$. We thus define an *input signal-to-noise* ratio as

$$(\text{SNR})_I = \frac{E_b/T_b}{J} \tag{7.36}$$

Hence, eliminating $E_b/J$ between Equations (7.35) and (7.36), we may express the output signal-to-noise ratio in terms of the input signal-to-noise ratio as

$$(\text{SNR})_O = \frac{2T_b}{T_c}(\text{SNR})_I \tag{7.37}$$

It is customary practice to express signal-to-noise ratios in decibels. To that end, we introduce a term called the *processing gain* (PG), which is defined as *the gain in SNR obtained by the use of spread spectrum.* Specifically, we write

$$\text{PG} = \frac{T_b}{T_c} \tag{7.38}$$

which represents the gain achieved by processing a spread-spectrum signal over an unspread signal. We may thus write Equation (7.37) in the equivalent form:

$$10\log_{10}(\text{SNR})_O = 10\log_{10}(\text{SNR})_I + 3 + 10\log_{10}(\text{PG})\ \text{dB} \tag{7.39}$$

The 3-dB term on the right-hand side of Equation (7.39) accounts for the gain in SNR that is obtained through the use of coherent detection (which presumes exact knowledge of the signal phase by the receiver). This gain in SNR has nothing to do with the use of spread spectrum. Rather, it is the last term, $10\log_{10}(\text{PG})$, that accounts for the processing gain. Note that both the processing gain PG and the spread factor $N$ (i.e., PN sequence length) equal the ratio $T_b/T_c$. Thus, the longer we make the PN sequence (or, correspondingly, the smaller the chip time $T_c$ is), the larger will the processing gain be.

## 7.6 *Probability of Error*

Let the coherent detector output $v$ in the direct-sequence spread BPSK system of Figure 7.9 denote the sample value of a random variable $V$. Let the equivalent noise component $v_{cj}$ produced by external interference denote the sample value of a random variable $V_{cj}$. Then, from Equations (7.23) and (7.27) we deduce that

$$V = \pm\sqrt{E_b} + V_{cj} \tag{7.40}$$

where $E_b$ is the transmitted signal energy per bit. The plus sign refers to sending symbol (information bit) 1, and the minus sign refers to sending symbol 0. The decision rule used by the coherent detector of Figure 7.9 is to declare that the received bit in an interval $(0, T_b)$ is 1 if the detector output exceeds a threshold of zero, and that it is 0 if the detector output is less than the threshold; if the detector output is exactly zero, the receiver makes a random guess in favor of 1 or 0. With both information bits assumed equally likely, we find that (because of the symmetric nature of the problem) the average probability of error $P_e$ is the same as the conditional probability of (say) the receiver making a decision in favor of symbol 1, given that symbol 0 was sent. That is,

$$\begin{aligned} P_e &= P(V > 0 \mid \text{symbol 0 was sent}) \\ &= P(V_{cj} > \sqrt{E_b}) \end{aligned} \tag{7.41}$$

Naturally, the probability of error $P_e$ depends on the random variable $V_{cj}$ defined by Equation (7.30). According to this definition, $V_{cj}$ is the sum of $N$ identically distributed random variables. Hence, from the *central limit theorem*, we deduce that for large $N$, the random variable $V_{cj}$ assumes a Gaussian distribution. Indeed, the spread factor or PN sequence length $N$ is typically large in the direct-sequence spread-spectrum systems encountered in practice, under which condition the application of the central limit theorem is justified.

Earlier we evaluated the mean and variance of $V_{cj}$; see Equations (7.32) and (7.34). We may therefore state that the equivalent noise component $V_{cj}$ contained in the coherent detector output may be approximated as a Gaussian random variable with zero mean and variance $JT_c/2$, where $J$ is the average interference power and $T_c$ is the chip duration. With this approximation at hand, we may then proceed to calculate the probability of the event $V_{cj} > \sqrt{E_b}$, and thus express the average probability of error in accordance with Equation (7.41) as

$$P_e \simeq \frac{1}{2}\,\text{erfc}\left(\sqrt{\frac{E_b}{JT_c}}\right) \tag{7.42}$$

This simple formula, which invokes the Gaussian assumption, is appropriate for DS/BPSK binary systems with large spread factor $N$.

## ■ ANTIJAM CHARACTERISTICS

It is informative to compare Equation (7.42) with the formula for the average probability of error for a coherent binary PSK system reproduced here for convenience of presentation [see Equation (6.20)]

$$P_e = \frac{1}{2}\,\text{erfc}\left(\sqrt{\frac{E_b}{N_0}}\right) \tag{7.43}$$

Based on this comparison, we see that insofar as the calculation of bit error rate in a direct-sequence spread binary PSK system is concerned, the interference may be treated as wideband noise of power spectral density $N_0/2$, defined by

$$\frac{N_0}{2} = \frac{JT_c}{2} \tag{7.44}$$

This relation is simply a restatement of an earlier result given in Equation (7.34).

Since the signal energy per bit $E_b = PT_b$, where $P$ is the average signal power and $T_b$ is the bit duration, we may express the signal energy per bit-to-noise spectral density ratio as

$$\frac{E_b}{N_0} = \left(\frac{T_b}{T_c}\right)\left(\frac{P}{J}\right) \tag{7.45}$$

Using the definition of Equation (7.38) for the processing gain PG we may reformulate this result as

$$\frac{J}{P} = \frac{\text{PG}}{E_b/N_0} \tag{7.46}$$

The ratio $J/P$ is termed the *jamming margin*. Accordingly, the jamming margin and the processing gain, both expressed in decibels, are related by

$$(\text{Jamming margin})_{\text{dB}} = (\text{Procesing gain})_{\text{dB}} - 10 \log_{10}\left(\frac{E_b}{N_0}\right)_{\text{min}} \tag{7.47}$$

where $(E_b/N_0)_{\text{min}}$ is the minimum value needed to support a prescribed average probability of error.

▶ **EXAMPLE 7.3**

A spread-spectrum communication system has the following parameters:

Information bit duration, $T_b = 4.095$ ms

PN chip duration, $T_c = 1\ \mu\text{s}$

Hence, using Equation (7.38) we find that the processing gain is

$$\text{PG} = 4095$$

Correspondingly, the required period of the PN sequence is $N = 4095$, and the shift-register length is $m = 12$.

For a satisfactory reception, we may assume that the average probability of error is not to exceed $10^{-5}$. From the formula for a coherent binary PSK receiver, we find that $E_b/N_0 = 10$ yields an average probability of error equal to $0.387 \times 10^{-5}$. Hence, using this value for $E_b/N_0$, and the value calculated for the processing gain, we find from Equation (7.47) that the jamming margin is

$$\begin{aligned}(\text{Jamming margin})_{\text{dB}} &= 10 \log_{10} 4095 - 10 \log_{10}(10) \\ &= 36.1 - 10 \\ &= 26.1 \text{ dB}\end{aligned}$$

That is, information bits at the receiver output can be detected reliably even when the noise or interference at the receiver input is up to 409.5 times the received signal power. Clearly, this is a powerful advantage against interference (jamming), which is realized through the clever use of spread-spectrum modulation. ◀

## 7.7 *Frequency-Hop Spread Spectrum*

In the type of spread-spectrum systems discussed in Section 7.4, the use of a PN sequence to modulate a phase-shift-keyed signal achieves *instantaneous* spreading of the transmission bandwidth. The ability of such a system to combat the effects of jammers is determined by the processing gain of the system, which is a function of the PN sequence period. The processing gain can be made larger by employing a PN sequence with narrow chip duration, which, in turn, permits a greater transmission bandwidth and more chips per bit. However, the capabilities of physical devices used to generate the PN spread-spectrum signals impose a practical limit on the attainable processing gain. Indeed, it may turn out that the processing gain so attained is still not large enough to overcome the effects of some jammers of concern, in which case we have to resort to other methods. One such alternative method is to force the jammer to cover a wider spectrum by *randomly hopping* the data-modulated carrier from one frequency to the next. In effect, the spectrum of the transmitted signal is spread *sequentially* rather than instantaneously; the term "sequentially" refers to the pseudo-random-ordered sequence of frequency hops.

The type of spread spectrum in which the carrier hops randomly from one frequency to another is called *frequency-hop (FH) spread spectrum.* A common modulation format for FH systems is that of *M-ary frequency-shift keying* (MFSK). The combination of these two techniques is referred to simply as FH/MFSK. (A description of $M$-ary FSK is presented in Chapter 6.)

Since frequency hopping does not cover the entire spread spectrum instantaneously, we are led to consider the rate at which the hops occur. In this context, we may identify two basic (technology-independent) characterizations of frequency hopping:

1. *Slow-frequency hopping*, in which the *symbol rate* $R_s$ of the MFSK signal is an integer multiple of the *hop rate* $R_h$. That is, several symbols are transmitted on each frequency hop.
2. *Fast-frequency hopping*, in which the hop rate $R_h$ is an integer multiple of the MFSK symbol rate $R_s$. That is, the carrier frequency will change or hop several times during the transmission of one symbol.

Obviously, slow-frequency hopping and fast-frequency hopping are the converse of one another. In the following, these two characterizations of frequency hopping are considered in turn.

## ■ SLOW-FREQUENCY HOPPING

Figure 7.10*a* shows the block diagram of an FH/MFSK transmitter, which involves *frequency modulation* followed by *mixing*. First, the incoming binary data are applied to an $M$-ary FSK modulator. The resulting modulated wave and the output from a digital *frequency synthesizer* are then applied to a mixer that consists of a multiplier followed by a band-pass filter. The filter is designed to select the sum frequency component resulting from the multiplication process as the transmitted signal. In particular, successive $k$-bit segments of a PN sequence drive the frequency synthesizer, which enables the carrier frequency to hop over $2^k$ distinct values. On a single hop, the bandwidth of the transmitted signal is the same as that resulting from the use of a conventional MFSK with an alphabet of $M = 2^K$ orthogonal signals. However, for a complete range of $2^k$ frequency hops, the transmitted FH/MFSK signal occupies a much larger bandwidth. Indeed, with present-day technology, FH bandwidths on the order of several GHz are attainable, which is an order of magnitude larger than that achievable with direct-sequence spread spectra. An implication of these large FH bandwidths is that coherent detection is possible only within each hop, because frequency synthesizers are unable to maintain phase coherence over successive hops. Accordingly, most frequency-hop spread-spectrum communication systems use noncoherent $M$-ary modulation schemes.

In the receiver depicted in Figure 7.10*b*, the frequency hopping is first removed by *mixing* (down-converting) the received signal with the output of a local frequency synthesizer that is synchronously controlled in the same manner as that in the transmitter. The resulting output is then band-pass filtered, and subsequently processed by a *noncoherent* $M$-ary FSK detector. To implement this $M$-ary detector, we may use a bank of $M$ noncoherent matched filters, each of which is matched to one of the MFSK tones. (Noncoherent matched filters are described in Chapter 6.) An estimate of the original symbol transmitted is obtained by selecting the largest filter output.

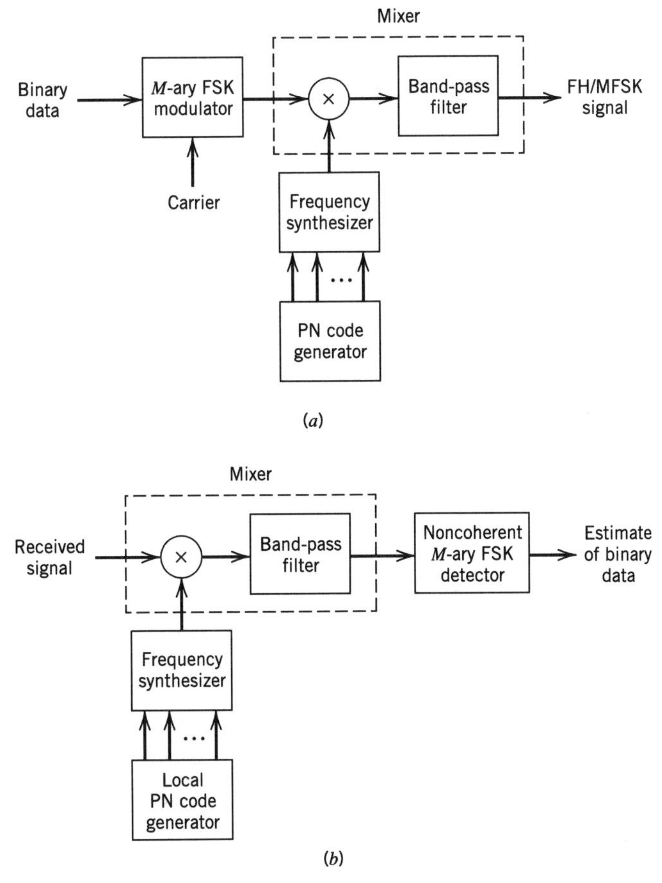


**FIGURE 7.10** Frequency-hop spread $M$-ary frequency-shift keying. (*a*) Transmitter. (*b*) Receiver.

An individual FH/MFSK tone of shortest duration is referred to as a *chip*; this terminology should not be confused with that used in Section 7.4 describing DS/BPSK. The *chip rate*, $R_c$, for an FH/MFSK system is defined by

$$R_c = \max(R_h, R_s) \tag{7.48}$$

where $R_h$ is the *hop rate*, and $R_s$ is the *symbol rate*.

A slow FH/MFSK signal is characterized by having multiple symbols transmitted per hop. Hence, each symbol of a slow FH/MFSK signal is a chip. Correspondingly, in a slow FH/MFSK system, the bit rate $R_b$ of the incoming binary data, the symbol rate $R_s$ of the MFSK signal, the chip rate $R_c$, and the hop rate $R_h$ are related by

$$R_c = R_s = \frac{R_b}{K} \geq R_h \tag{7.49}$$

where $K = \log_2 M$.

At each hop, the MFSK tones are separated in frequency by an integer multiple of the chip rate $R_c = R_s$, ensuring their orthogonality. The implication of this condition is that any transmitted symbol will not produce any crosstalk in the other $M - 1$ noncoherent matched filters constituting the MFSK detector of the receiver in Figure 7.10*b*. By "crosstalk" we mean the spillover from one filter output into an adjacent one. The resulting performance of the slow FH/MFSK system is the same as that for the noncoherent detection of conventional (unhopped) MFSK signals in additive white Gaussian noise. Thus the interfering (jamming) signal has an effect on the FH/MFSK receiver, in terms of average probability of symbol error, equivalent to that of additive white Gaussian noise on a conventional noncoherent $M$-ary FSK receiver experiencing no interference. On the basis of this equivalence, we may use Equation (6.140) for approximate evaluation of the probability of symbol error in the FH/MFSK system.

Assuming that the jammer decides to spread its average power $J$ over the entire frequency-hopped spectrum, the jammer's effect is equivalent to an AWGN with power spectral density $N_0/2$, where $N_0 = J/W_c$ and $W_c$ is the FH bandwidth. The spread-spectrum system is thus characterized by the *symbol energy-to-noise spectral density ratio*:

$$\frac{E}{N_0} = \frac{P/J}{W_c/R_s} \tag{7.50}$$

where the ratio $P/J$ is the reciprocal of the jamming margin. The other ratio in the denominator of Equation (7.50) is the processing gain of the slow FH/MFSK system, which is defined by

$$\begin{aligned} \text{PG} &= \frac{W_c}{R_s} \\ &= 2^k \end{aligned} \tag{7.51}$$

That is, the processing gain (expressed in decibels) is equal to $10 \log_{10} 2^k \simeq 3k$, where $k$ is the length of the PN segment employed to select a frequency hop.

This result assumes that the jammer spreads its power over the entire FH spectrum. However, if the jammer decides to concentrate on just a few of the hopped frequencies, then the processing gain realized by the receiver would be less than $3k$ decibels.

### ▶ EXAMPLE 7.4

Figure 7.11*a* illustrates the variation of the frequency of a slow FH/MFSK signal with time for one complete period of the PN sequence. The period of the PN sequence is $2^4 - 1 = 15$. The FH/MFSK signal has the following parameters:

| | |
|---|---|
| Number of bits per MFSK symbol | $K = 2$ |
| Number of MFSK tones | $M = 2^K = 4$ |
| Length of PN segment per hop | $k = 3$ |
| Total number of frequency hops | $2^k = 8$ |

In this example, the carrier is hopped to a new frequency after transmitting two symbols or equivalently, four information bits. Figure 7.11*a* also includes the input binary data, and the PN sequence controlling the selection of FH carrier frequency. It is noteworthy that although there are eight distinct frequencies available for hopping, only three of them are utilized by the PN sequence.

Figure 7.11*b* shows the variation of the dehopped frequency with time. This variation is recognized to be the same as that of a conventional MFSK signal produced by the given input data. ◀

## ■ FAST-FREQUENCY HOPPING

A fast FH/MFSK system differs from a slow FH/MFSK system in that there are multiple hops per $M$-ary symbol. Hence, in a fast FH/MFSK system, each hop is a chip. In general,

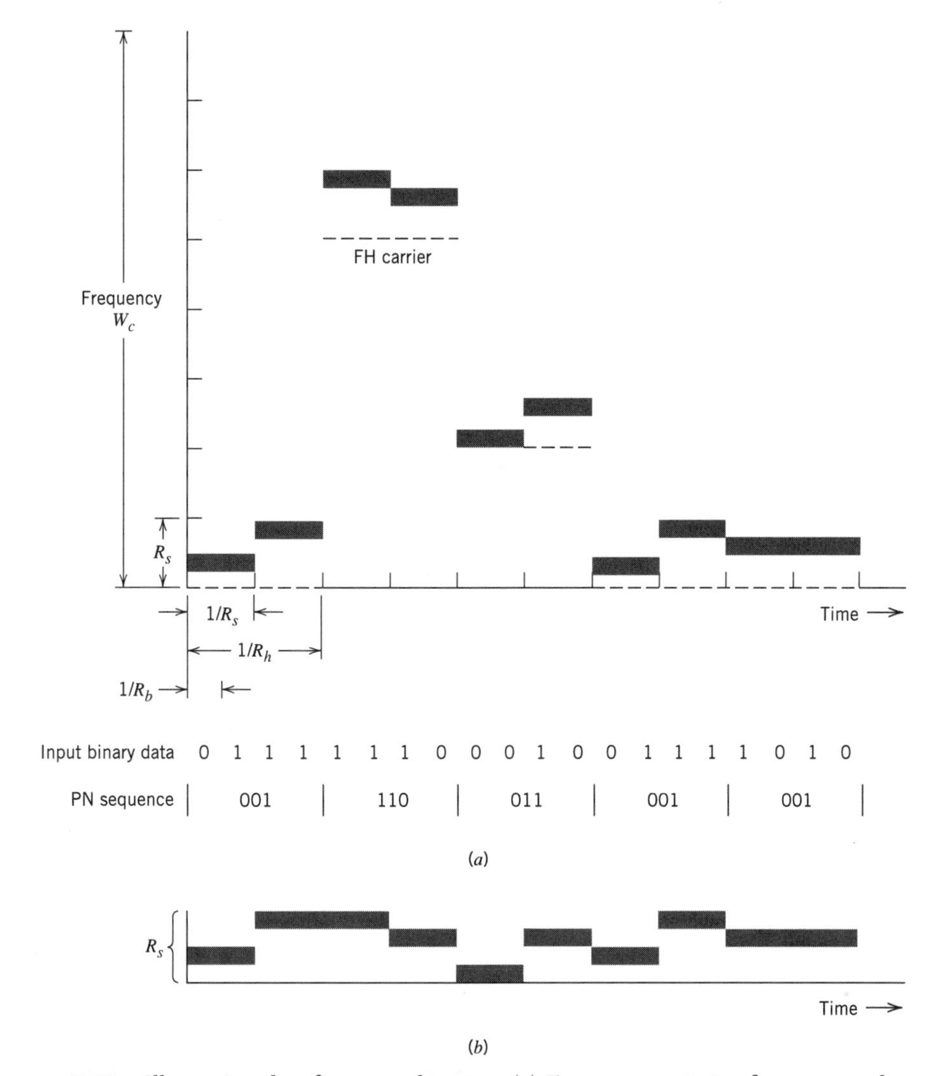


**FIGURE 7.11** Illustrating slow-frequency hopping. (*a*) Frequency variation for one complete period of the PN sequence. (*b*) Variation of the dehopped frequency with time.

fast-frequency hopping is used to defeat a smart jammer's tactic that involves two functions: measurements of the spectral content of the transmitted signal, and retuning of the interfering signal to that portion of the frequency band. Clearly, to overcome the jammer, the transmitted signal must be hopped to a new carrier frequency *before* the jammer is able to complete the processing of these two functions.

For data recovery at the receiver, noncoherent detection is used. However, the detection procedure is quite different from that used in a slow FH/MFSK receiver. In particular, two procedures may be considered:

1. For each FH/MFSK symbol, separate decisions are made on the $K$ frequency-hop chips received, and a simple rule based on *majority vote* is used to make an estimate of the dehopped MFSK symbol.
2. For each FH/MFSK symbol, likelihood functions are computed as functions of the total signal received over $K$ chips, and the largest one is selected.

A receiver based on the second procedure is optimum in the sense that it minimizes the average probability of symbol error for a given $E_b/N_0$.

### ▶ EXAMPLE 7.5

Figure 7.12*a* illustrates the variation of the transmitted frequency of a fast FH/MFSK signal with time. The signal has the following parameters:

| | |
|---|---|
| Number of bits per MFSK symbol | $K = 2$ |
| Number of MFSK tones | $M = 2^K = 4$ |
| Length of PN segment per hop | $k = 3$ |
| Total number of frequency hops | $2^k = 8$ |

In this example, each MFSK symbol has the same number of bits and chips; that is, the chip rate $R_c$ is the same as the bit rate $R_b$. After each chip, the carrier frequency of the transmitted MFSK signal is hopped to a different value, except for few occasions when the $k$-chip segment of the PN sequence repeats itself.

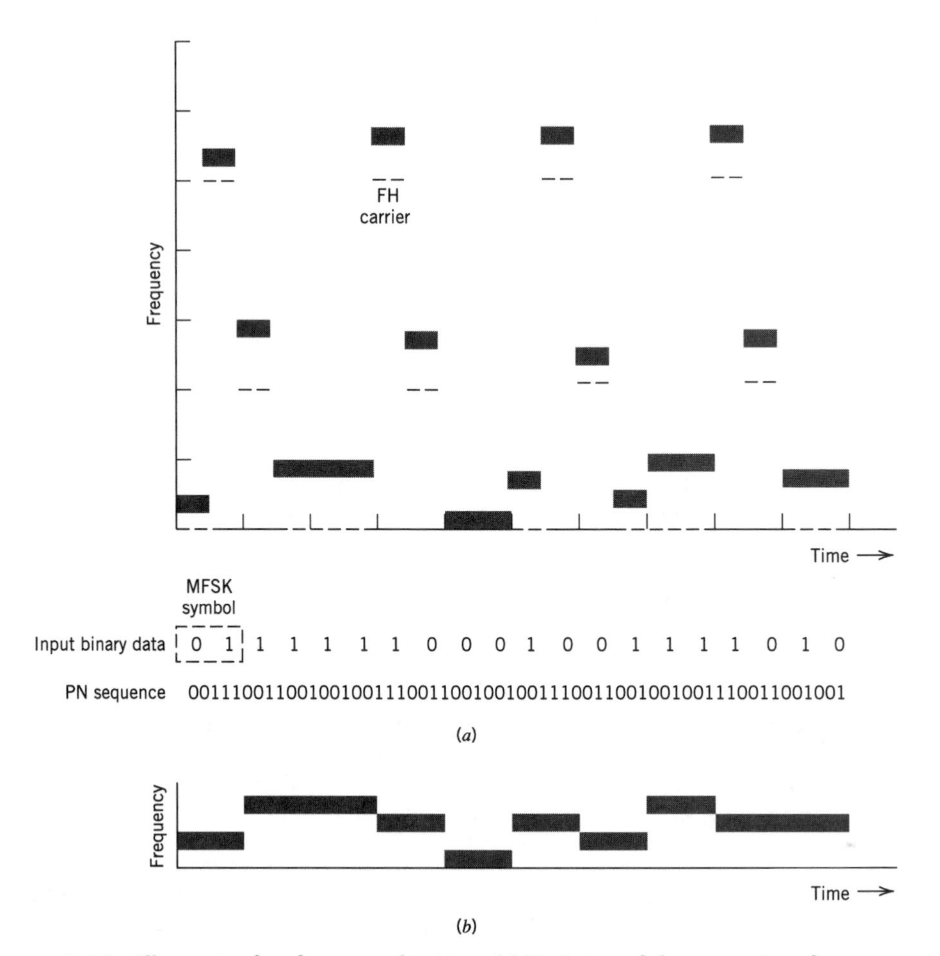


FIGURE 7.12 Illustrating fast-frequency hopping. (*a*) Variation of the transmitter frequency with time. (*b*) Variation of the dehopped frequency with time.

Figure 7.12*b* depicts the time variation of the frequency of the dehopped MFSK signal, which is the same as that in Example 7.4. ◀

## 7.8 Computer Experiments: Maximal-Length and Gold Codes

*Code-division multiplexing (CDM)* provides an alternative to the traditional methods of frequency-division multiplexing (FDM) and time-division multiplexing (TDM). It does not require the bandwidth allocation of FDM (discussed in Chapter 2) nor the time synchronization needed in TDM (discussed in Chapter 3). Rather, users of a common channel are permitted access to the channel through the assignment of a "spreading code" to each individual user under the umbrella of spread-spectrum modulation. The purpose of this computer experiment is to study a certain class of spreading codes for CDM systems that provide a satisfactory performance.

In an ideal CDM system, the cross-correlation between any two users of the system is zero. For this ideal condition to be realized, we require that the cross-correlation function between the spreading codes assigned to any two users of the system be zero for all cyclic shifts. Unfortunately, ordinary PN sequences do not satisfy this requirement because of their relatively poor cross-correlation properties.

As a remedy for this shortcoming of ordinary PN sequences, we may use a special class of PN sequences called *Gold sequences (codes)*,[5] the generation of which is embodied in the following theorem:

> Let $g_1(X)$ and $g_2(X)$ be a preferred pair of primitive polynomials of degree $n$ whose corresponding shift registers generate maximal-length sequences of period $2^n - 1$ and whose cross-correlation function has a magnitude less than or equal to
>
> $$2^{(n+1)/2} + 1 \qquad \text{for } n \text{ odd} \tag{7.52}$$
>
> or
>
> $$2^{(n+2)/2} + 1 \qquad \text{for } n \text{ even and } n \neq 0 \bmod 4 \tag{7.53}$$
>
> Then the shift register corresponding to the product polynomial $g_1(X) \cdot g_2(X)$ will generate $2^n + 1$ different sequences, with each sequence having a period of $2^n - 1$, and the cross-correlation between any pair of such sequences satisfying the preceding condition.

Hereafter, this theorem is referred to as *Gold's theorem*.

To understand Gold's theorem, we need to define what we mean by a primitive polynomial. Consider a polynomial $g(X)$ defined over a *binary field* (i.e., a finite set of two elements, 0 and 1, which is governed by the rules of binary arithmetic). The polynomial $g(X)$ is said to be an *irreducible polynomial* if it cannot be factored using any polynomials from the binary field. An irreducible polynomial $g(X)$ of degree $m$ is said to be a *primitive polynomial* if the smallest integer $m$ for which the polynomial $g(X)$ divides the factor $X^n + 1$ is $n = 2^m - 1$. Further discussion of this topic is deferred to Chapter 8; in particular, see Example 8.3.

### Experiment 1. Correlation Properties of PN Sequences

Consider a pair of shift registers for generating two PN sequences of period $2^7 - 1 = 127$. One feedback shift register has the feedback taps [7, 1] and the other one has the feedback taps [7, 6, 5, 4]. Both sequences have the same autocorrelation function shown in Figure 7.13*a*, which follows readily from the definition presented in Equation (7.5).

However, the calculation of the cross-correlation function between PN sequences is a more difficult proposition, particularly for large $n$. To perform this calculation, we resort to the use of computer simulation for varying cyclic shift $\tau$ inside the interval $0 < \tau \leq 2^n - 1$. The results of this computation are presented in Figure 7.13*b*. This figure confirms the poor cross-correlation property of PN sequences compared to their autocorrelation function. The magnitude of the cross-correlation function exceeds 40.

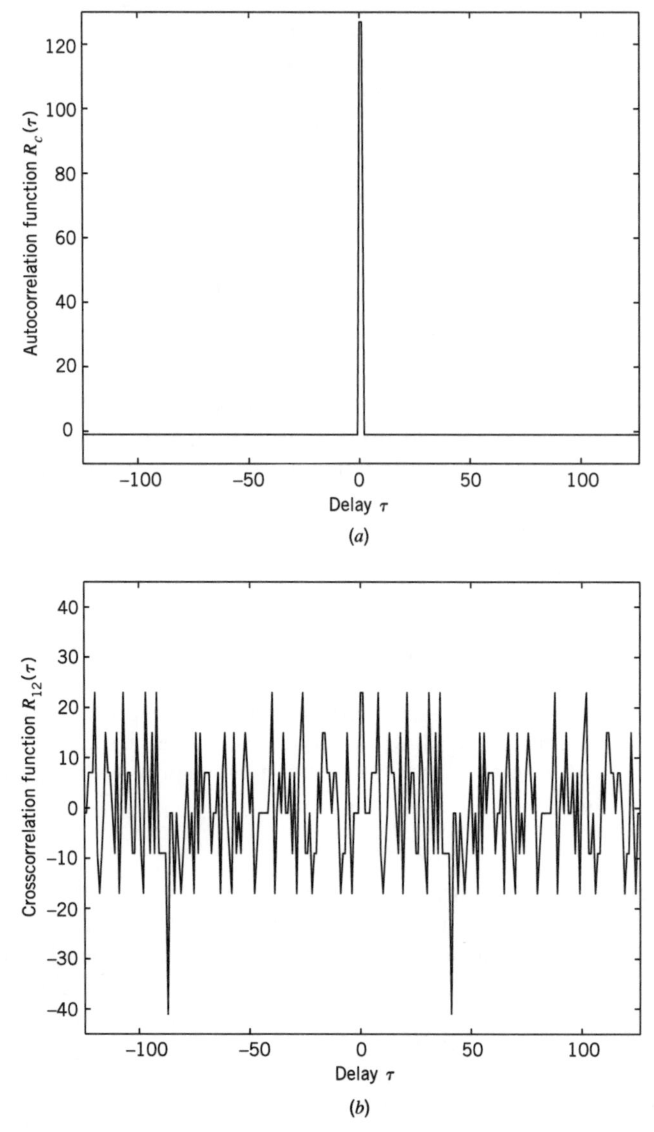


**FIGURE 7.13** (*a*) Autocorrelation function $R_c(\tau)$, and (*b*) cross-correlation function $R_{12}(\tau)$ of the two PN sequences [7, 1] and [7, 6, 5, 4].

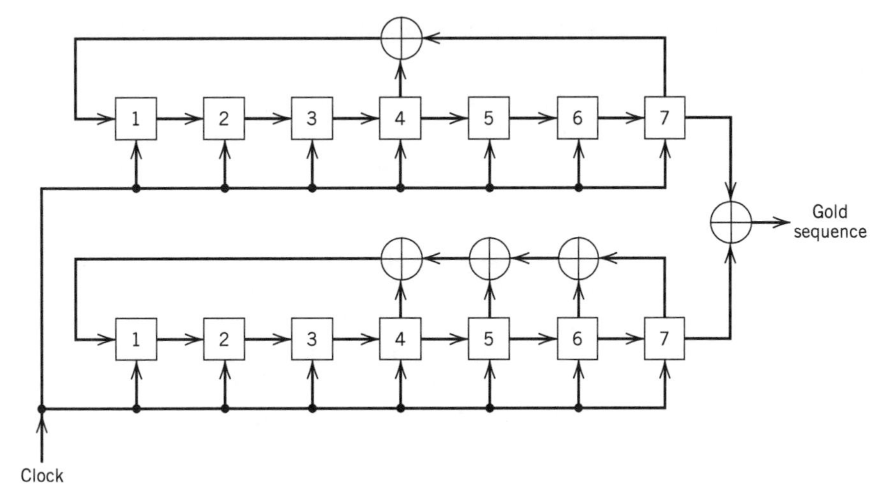


**FIGURE 7.14** Generator for a Gold sequence of period $2^7 - 1 = 127$.

### Experiment 2. Correlation Properties of Gold Sequences

For our next experiment, we consider Gold sequences with period $2^7 - 1 = 127$. To generate such a sequence for $n = 7$ we need a preferred pair of PN sequences that satisfy Equation (7.52) ($n$ odd), as shown by

$$2^{(n+1)/2} + 1 = 2^4 + 1 = 17$$

This requirement is satisfied by the PN sequences with feedback taps [7, 4] and [7, 6, 5, 4]. The Gold-sequence generator is shown in Figure 7.14 that involves the modulo-2 addition of these two sequences. According to Gold's theorem, there are a total of

$$2^n + 1 = 2^7 + 1 = 129$$

sequences that satisfy Equation (7.52). The cross-correlation between any pair of such sequences is shown in Figure 7.15, which is indeed in full accord with Gold's theorem. In particular, the magnitude of the cross-correlation is less than or equal to 17.

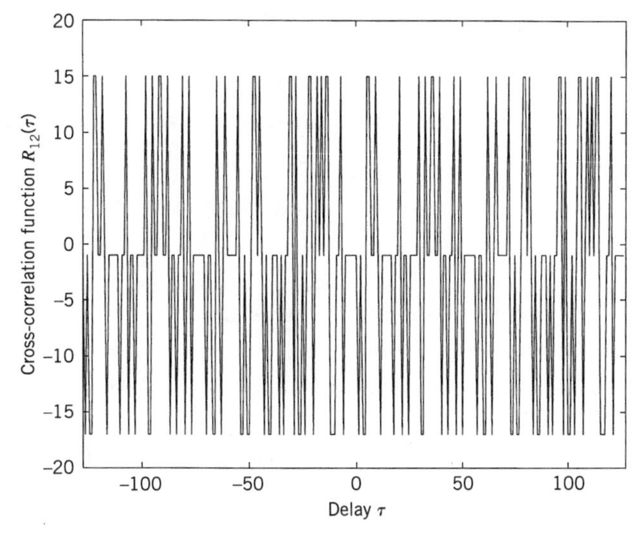


**FIGURE 7.15** Cross-correlation function $R_{12}(\tau)$ of a pair of Gold sequences based on the two PN sequences [7, 4] and [7, 6, 5, 4].

## 7.9 Summary and Discussion

*Direct-sequence M-ary phase shift keying* (DS/MPSK) and *frequency-hop M-ary frequency shift-keying* (FH/MFSK) represent two principal categories of spread-spectrum communications. Both of them rely on the use of a pseudo-noise (PN) sequence, which is applied differently in the two categories.

In a DS/MPSK system, the PN sequence makes the transmitted signal assume a noiselike appearance by spreading its spectrum over a broad range of frequencies simultaneously. For the phase-shift keying, we may use binary PSK (i.e., $M = 2$) with a single carrier. Alternatively, we may use QPSK (i.e., $M = 4$), in which case the data are transmitted using a pair of carriers in phase quadrature. (Both PSK and QPSK are discussed in Section 6.3.) The usual motivation for using QPSK is to provide for improved bandwidth efficiency. In a spread-spectrum system, bandwidth efficiency is usually not of prime concern. Rather, the use of QPSK is motivated by the fact that it is less sensitive to some types of interference (jamming).

In an FH/MFSK system, the PN sequence makes the carrier hop over a number of frequencies in a pseudo-random manner, with the result that the spectrum of the transmitted signal is spread in a sequential manner.

Naturally, the direct-sequence and frequency-hop spectrum-spreading techniques may be employed in a single system. The resulting system is referred to as *hybrid* DS/FH *spread-spectrum system*. The reason for seeking a hybrid approach is that advantages of both the direct-sequence and frequency-hop spectrum-spreading techniques are realized in the same system.

A discussion of spread-spectrum communications would be incomplete without some reference to jammer waveforms. The jammers encountered in practice include the following types:

1. *The barrage noise jammer*, which consists of band-limited white Gaussian noise of high average power. The barrage noise jammer is a brute-force jammer that does not exploit any knowledge of the antijam communication system except for its spread bandwidth.
2. *The partial-band noise jammer*, which consists of noise whose total power is evenly spread over some frequency band that is a subset of the total spread bandwidth. Owing to the smaller bandwidth, the partial-band noise jammer is easier to generate than the barrage noise jammer.
3. *The pulsed noise jammer*, which involves transmitting wideband noise of power

$$J_{\text{peak}} = \frac{J}{p}$$

   for a fraction $p$ of the time, and nothing for the remaining fraction $1 - p$ of the time. The average noise power equals $J$.
4. *The single-tone jammer*, which consists of a sinusoidal wave whose frequency lies inside the spread bandwidth; as such, it is the easiest of all jamming signals to generate.
5. *The multitone jammer*, which is the tone equivalent of the partial-band noise jammer.

In addition to these five, many other kinds of jamming waveforms occur in practice. In any event, there is no single jamming waveform that is worst for all spread-spectrum systems, and there is no single spread-spectrum system that is best against all possible jamming waveforms.

## Notes and References

1. The definition of spread-spectrum modulation presented in the Introduction is adapted from Pickholtz, Schilling, and Milstein (1982). This paper presents a tutorial review of the theory of spread-spectrum communications.

   For introductory papers on the subject, see Viterbi (1979), and Cook and Marsh (1983). For books on the subject, see Dixon (1984), Holmes (1982), Ziemer and Peterson (1985, pp. 327–649), Cooper and McGillem (1986, pp. 269–411), and Simon, Omura, Scholtz, and Levitt (1985, Volumes I, II, and III). The three-volume book by Simon et al. is the most exhaustive treatment of spread-spectrum communications available in the open literature. The development of spread-spectrum communications dates back to about the mid-1950s. For a historical account of these techniques, see Scholtz (1982). This latter paper traces the origins of spread-spectrum communications back to the 1920s. Much of the historical material presented in this paper is reproduced in Chapter 2, Volume I, of the book by Simon et al.

   The book edited by Tantaratana and Ahmed (1998) includes introductory and advanced papers on wireless applications of spread-spectrum modulation. The papers are grouped into the following categories: spread-spectrum technology, cellular mobile systems, satellite communications, wireless local area networks, and global positioning systems (GPS).

2. For further details on maximal-length sequences, see Golomb (1964, pp. 1–32), Simon, Omura, Scholtz, and Levitt (1985, pp. 283–295), and Peterson and Weldon (1972). The last reference includes an extensive list of polynomials for generating maximal-length sequences; see also Dixon (1984). For a tutorial paper on pseudo-noise sequences, see Sarwate and Pursley (1980).

3. Table 7.1 is extracted from the book by Dixon (1984, pp. 81–83), where feedback connections of maximal-length sequences are tabulated for shift-register length $m$ extending up to 89.

4. For detailed discussion of the synchronization problem in spread-spectrum communications, see Ziemer and Peterson (1985, Chapters 9 and 10) and Simon et al. (1985, Volume III).

5. The original papers on Gold sequences are Gold (1967, 1968). A detailed discussion of Gold sequences is presented in Holmes (1982).

## Problems

### Pseudo-Noise Sequences

7.1 A pseudo-noise (PN) sequence is generated using a feedback shift register of length $m = 4$. The chip rate is $10^7$ chips per second. Find the following parameters:

(a) PN sequence length.

(b) Chip duration of the PN sequence.

(c) PN sequence period.

7.2 Figure P7.2 shows a four-stage feedback shift register. The initial state of the register is 1000. Find the output sequence of the shift register.

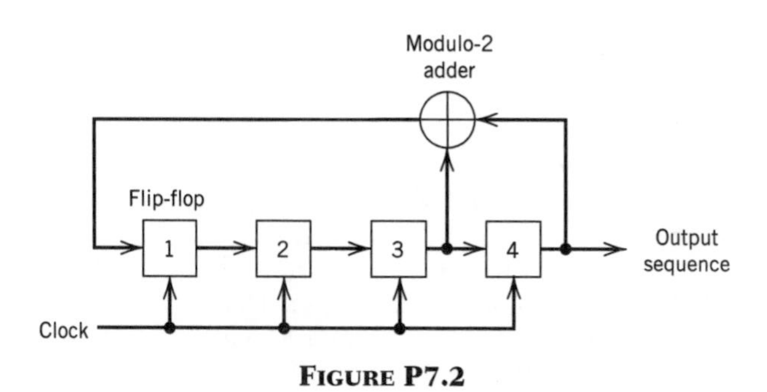


FIGURE P7.2

7.3 For the feedback shift register given in Problem 7.2, demonstrate the balance property and run property of a PN sequence. Also, calculate and plot the autocorrelation function of the PN sequence produced by this shift register.

7.4 Referring to Table 7.1, develop the maximal-length codes for the three feedback configurations [6, 1], [6, 5, 2, 1], and [6, 5, 3, 2], whose period is $N = 63$.

7.5 Figure P7.5 shows the modular multitap version of the linear feedback shift-register shown in Figure 7.4*b*. Demonstrate that the PN sequence generated by this scheme is exactly the same as that described in Table 7.2*b*.

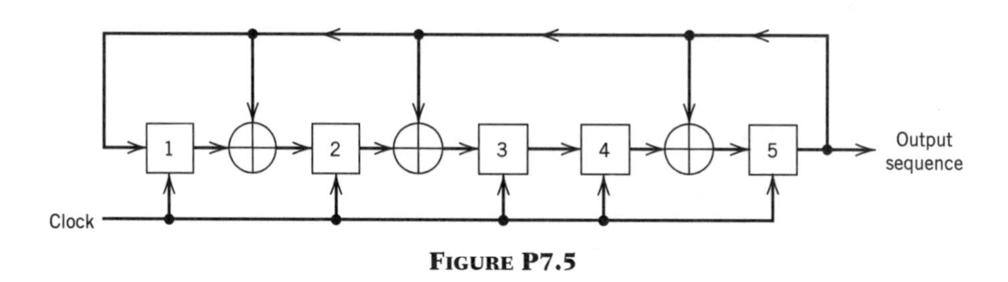


FIGURE P7.5

## Direct Sequence/Phase-Shift Keying System

7.6 Show that the truth table given in Table 7.3 can be constructed by combining the following two steps:

(a) The message signal $b(t)$ and PN signal $c(t)$ are added modulo-2.

(b) Symbols 0 and 1 at the modulo-2 adder output are represented by phase shifts of 0 and 180 degrees, respectively.

7.7 A single-tone jammer

$$j(t) = \sqrt{2J}\cos(2\pi f_c t + \theta)$$

is applied to a DS/BPSK system. The $N$-dimensional transmitted signal $x(t)$ is described by Equation (7.16). Find the $2N$ coordinates of the jammer $j(t)$.

7.8 The processing gain of a spread-spectrum system may be expressed as the ratio of the spread bandwidth of the transmitted signal to the despread bandwidth of the received signal. Justify this statement for the DS/BPSK system.

7.9 A direct-sequence spread binary phase-shift keying system uses a feedback shift register of length 19 for the generation of the PN sequence. Calculate the processing gain of the system.

7.10 In a DS/BPSK system, the feedback shift register used to generate the PN sequence has length $m = 19$. The system is required to have an average probability of symbol error due to externally generated interfering signals that does not exceed $10^{-5}$. Calculate the following system parameters in decibels:

(a) Processing gain.

(b) Antijam margin.

7.11 In Section 7.5, we presented an analysis on the signal-space dimensionality and processing gain of a direct sequence spread-spectrum system using binary phase-shift keying. Extend the analysis presented therein to the case of such a system using quadriphase-shift keying.

### Frequency-Hop Spread Spectrum

7.12 A slow FH/MFSK system has the following parameters:

Number of bits per MFSK symbol = 4

Number of MFSK symbols per hop = 5

Calculate the processing gain of the system.

7.13 A fast FH/MFSK system has the following parameters:

Number of bits per MFSK symbol = 4

Number of hops per MFSK symbol = 4

Calculate the processing gain of the system.

### Computer Experiments

7.14 Consider two PN sequences of period $N = 63$. One sequence has the feedback taps [6, 1] and the other sequence has the feedback taps [6, 5, 2, 1], which are picked in accordance with Table 7.1.

(a) Compute the autocorrelation function of these two sequences, and their cross-correlation function.

(b) Compare the cross-correlation function computed in part (a) with the cross-correlation function between the sequence [6, 5 2, 1] and its mirror image [6, 5, 4, 1].

Comment on your results.

7.15 (a) Compute the partial cross-correlation function of a PN sequence with feedback taps [5, 2] and its image sequence defined by the feedback taps [5, 3].

(b) Repeat the computation for the PN sequence with feedback taps [5, 2] and the PN sequence with feedback taps [5, 4, 2, 1].

(c) Repeat the computation for the PN sequence with feedback taps [5, 4, 3, 2] and the PN sequence with feedback taps [5, 4, 2, 1].

The feedback taps [5, 2], [5, 4, 3, 2], and [5, 4, 2, 1] are possible taps for a maximal-length sequence of period 31, in accordance with Table 7.1.

CHAPTER 8

# MULTIUSER RADIO COMMUNICATIONS

As its name implies, multiuser communications refers to the simultaneous use of a communication channel by a number of users. In this chapter, we discuss multiuser communication systems that rely on radio propagation for linking the receivers to the transmitters.

In particular, we focus on the following topics:

- *Multiple-access techniques, which are basic to multiuser communication systems.*
- *Satellite communications, offering global coverage.*
- *Radio link analysis, highlighting the roles of transmitting and receiving antennas and free-space propagation.*
- *Wireless communications with emphasis on mobility and the multipath phenomenon.*
- *Speech coding for wireless communications.*
- *Adaptive antennas for wireless communications.*

## 8.1 *Introduction*

Much of the material on communication theory presented in earlier chapters has been based on a particular idealization of the communication channel, namely, a *channel model limited in bandwidth and corrupted by additive white Gaussian noise* (AWGN). The *classical communication theory* so developed is mathematically elegant, providing a sound introduction to the ever-expanding field of communication systems. An example of a physical channel that is well represented by such a model is the satellite communications channel. It is therefore befitting that the first type of multiuser communications discussed in this chapter is *satellite communications*.

A satellite communication system in geostationary orbit relies on line-of-sight radio propagation for the operation of its uplink from an earth terminal to the transponder and the downlink from the transponder to another earth terminal. Thus the discussion of satellite communications naturally leads to the analysis of radio propagation in free space, linking a receiving antenna to a transmitting antenna.

The use of satellite communications offers *global coverage*. The other multiuser communication system studied in this chapter, namely, *wireless communications*, offers *mobility* which, in conjunction with existing telephone networks and satellite communication systems, permits a mobile unit to communicate with anyone, anywhere in the world. Another characteristic feature of wireless communication systems is that they are *tetherless* (i.e., total freedom of location is permitted), hence the interest in their use for local area networks (i.e., data networks confined to buildings up to a few kilometers in size) due to significant advantages over conventional cabling: elimination of wiring and rewiring, flexibility of creating new communication services, and mobility of users.

The radio propagation channel characterizing wireless communications deviates from the idealized AWGN channel model due to the presence of *multipath*, which is a non-Gaussian form of signal-dependent phenomenon that arises because of reflections of the transmitted signal from fixed and moving objects. The presence of multipath raises practical difficulties in the use of a radio propagation channel and complicates its mathematical analysis. Simply put, multipath is a physical phenomenon that is intrinsic to the operation of indoor and outdoor forms of wireless communications.

Before proceeding to discuss specific aspects of satellite communications and wireless communications, however, it is appropriate that we begin the discussion by describing multiple-access techniques, which enable different users to simultaneously (or nearly so) access a common channel.

## 8.2 *Multiple-Access Techniques*

*Multiple access* is a technique whereby many subscribers or local stations can share the use of a communication channel at the same time or nearly so, despite the fact that their individual transmissions may originate from widely different locations. Stated in another way, a multiple-access technique permits the communication resources of the channel to be shared by a large number of users seeking to communicate with each other.

There are subtle differences between multiple access and multiplexing that should be noted:

- Multiple access refers to the remote sharing of a communication channel such as a satellite or radio channel by users in highly dispersed locations. On the other hand, multiplexing refers to the sharing of a channel such as a telephone channel by users confined to a local site.
- In a multiplexed system, user requirements are ordinarily fixed. In contrast, in a multiple-access system user requirements can change dynamically with time, in which case provisions are necessary for dynamic channel allocation.

For obvious reasons it is desirable that in a multiple-access system the sharing of resources of the channel be accomplished without causing serious interference between users of the system. In this context, we may identify four basic types of multiple access:

1. *Frequency-division multiple access (FDMA).*
In this technique, disjoint subbands of frequencies are allocated to the different users on a continuous-time basis. In order to reduce interference between users allocated adjacent channel bands, *guard bands* are used to act as buffer zones, as illustrated in Figure 8.1*a*. These guard bands are necessary because of the impossibility of achieving ideal filtering for separating the different users.

2. *Time-division multiple access (TDMA).*
In this second technique, each user is allocated the full spectral occupancy of the channel, but only for a short duration of time called a *time slot*. As shown in Figure 8.1*b*, buffer zones in the form of *guard times* are inserted between the assigned time slots. This is done

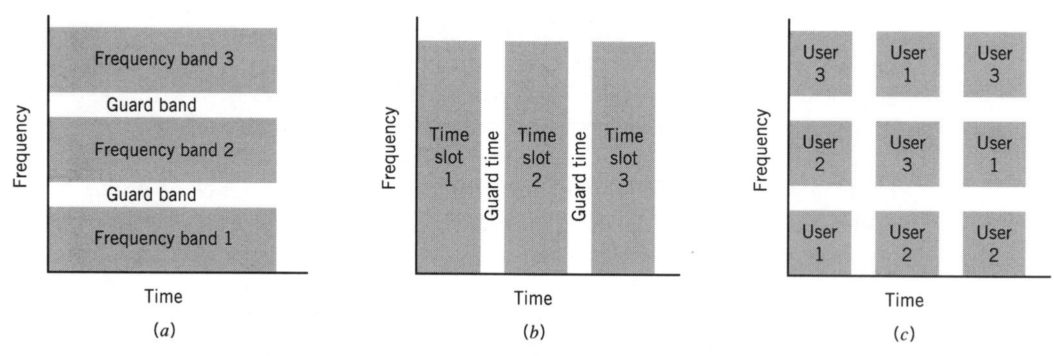


**FIGURE 8.1** Illustrating the ideas behind multiple-access techniques. (*a*) Frequency-division multiple access. (*b*) Time-division multiple access. (*c*) Frequency-hop multiple access.

to reduce interference between users by allowing for time uncertainty that arises due to system imperfections, especially in synchronization schemes.

3. *Code-division multiple access (CDMA).*
In FDMA, the resources of the channel are shared by dividing them along the frequency coordinate into disjoint frequency bands, as illustrated in Figure 8.1*a*. In TDMA, the resources are shared by dividing them along the time coordinate into disjoint time slots, as illustrated in Figure 8.1*b*. In Figure 8.1*c*, we illustrate another technique for sharing the channel resources by using a hybrid combination of FDMA and TDMA, which represents a specific form of code-division multiple access (CDMA). For example, *frequency hopping* may be employed to ensure that during each successive time slot, the frequency bands assigned to the users are reordered in an essentially random manner. To be specific, during time slot 1, user 1 occupies frequency band 1, user 2 occupies frequency band 2, user 3 occupies frequency band 3, and so on. During time slot 2, user 1 hops to frequency band 3, user 2 hops to frequency band 1, user 3 hops to frequency band 2, and so on. Such an arrangement has the appearance of the users playing a game of musical chairs. An important advantage of CDMA over both FDMA and TDMA is that it can provide for *secure* communications. In the type of CDMA illustrated in Figure 8.1*c*, the frequency hopping mechanism can be implemented through the use of a pseudo-noise (PN) sequence.

4. *Space-division multiple access (SDMA).*
In this multiple-access technique, resource allocation is achieved by exploiting the spatial separation of the individual users. In particular, *multibeam antennas* are used to separate radio signals by pointing them along different directions. Thus, different users are enabled to access the channel simultaneously on the same frequency or in the same time slot.

These multiple-access techniques share a common feature: allocating the communication resources of the channel through the use of disjointedness (or orthogonality in a loose sense) in time, frequency, or space.

With this background material at hand, we are now ready to discuss some important multiuser communication systems.

## 8.3 Satellite Communications

In a geostationary satellite communication system,[1] a message signal is transmitted from an earth station via an *uplink* to a satellite, amplified in a *transponder* (i.e., electronic

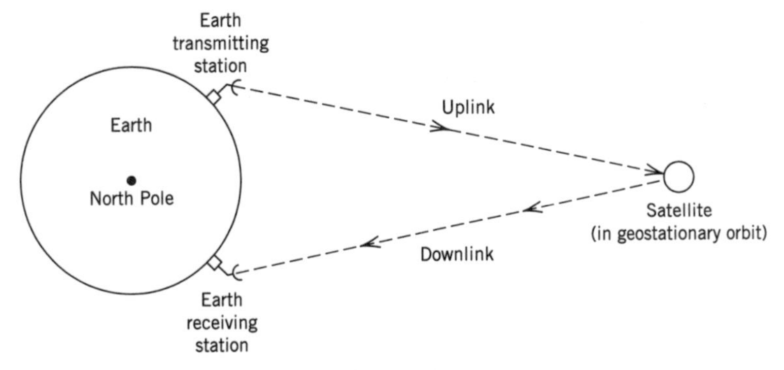


**FIGURE 8.2** Satellite communications system.

circuitry) on board the satellite, and then retransmitted from the satellite via a *downlink* to another earth station, as illustrated in Figure 8.2. The most popular frequency band for satellite communications is 6 GHz (C-band) for the uplink and 4 GHz for the downlink. The use of this frequency band offers the following advantages:

- Relatively inexpensive microwave equipment.
- Low attenuation due to rainfall; rainfall is the primary atmospheric cause of signal degradation.
- Insignificant sky background noise; the sky background noise (due to random noise emissions from galactic, solar, and terrestrial sources) reaches its lowest level between 1 and 10 GHz.

However, radio interference limits the applications of communication satellites operating in the 6/4 GHz band, because the transmission frequencies of this band coincide with those used for terrestrial microwave systems. This problem is eliminated in the more powerful "second-generation" communication satellites that operate in the 14/12 GHz band (i.e., Ku-band); moreover, the use of these higher frequencies makes it possible to build smaller and therefore less expensive antennas.

The block diagram of Figure 8.3 shows the basic components of a single transponder channel of a typical communication satellite. Specifically, the receiving antenna output of the uplink is applied to the cascade connection of the following components:

- *Band-pass filter*, designed to separate the received signal from among the different radio channels.
- *Low-noise amplifier.*

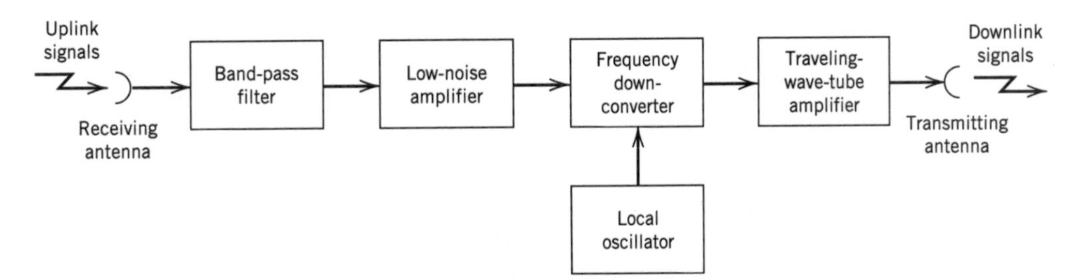


**FIGURE 8.3** Block diagram of transponder.

- *Frequency down-converter*, the purpose of which is to convert the received radio frequency (RF) signal to the desired downlink frequency.
- *Traveling-wave tube amplifier*, which provides high gain over a wide band of frequencies. In a traveling-wave tube (TWT), an electromagnetic signal travels along a helix (i.e., a spring-shaped coil of wire), while electrons in a high-voltage beam travel through the helix at a speed close to that of the signal wave; the net result is the transfer of power from the electrons to the wave, which grows rapidly as the signal wave travels down the helix.

The channel configuration shown in Figure 8.3 uses a single frequency translation. Other channel configurations do the frequency conversion from the uplink to the downlink frequency in two stages: down-conversion to an intermediate frequency, followed by amplification, and then up-conversion to the desired transmit frequency.

Propagation time delay becomes particularly pronounced in a satellite channel because of the large distances involved. Specifically, speech signals sent by satellite incur a transmission delay of approximately 270 ms. Hence, for speech signals, any impedance mismatch at the receiving end of a satellite link results in an *echo* of the speaker's voice, which is heard back at the transmitting end after a round-trip delay of approximately 540 ms. We may overcome this problem by using an *echo canceller*, which is a device that subtracts an estimate of the echo from the return path; elimination of the echo is performed by means of a special filter that adapts itself to the changing channel characteristics.

The satellite channel is closely represented by an *additive white Gaussian noise (AWGN) model*, which applies to both the uplink and downlink portions of the satellite communication system. Accordingly, much of the material presented in Chapter 6 on passband systems for the transmission of data, with particular reference to phase-shift keying and frequency-shift keying techniques, is directly applicable to digital satellite communications.

A satellite transponder differs from a conventional microwave line-of-sight repeater in that many earth stations can access the satellite from widely different locations on earth at the same time or nearly so. This capability is made possible by using one of the multiple-access techniques discussed in Section 8.2. In this context we may offer the following observations:

- In a satellite channel, nonlinearity of the transponder is the primary cause of interference between users. To contain this serious problem, the traveling-wave tube amplifier in the transponder is purposely operated below capacity. Consequently, we find that in an FDMA system the power efficiency of the system is reduced because of the necessary *power backoff* of the traveling-wave tube amplifier.
- In a TDMA system, the users access the satellite transponder one at a time. Accordingly, the satellite transponder is now able to operate close to full power efficiency by permitting the traveling-wave tube amplifier to run into saturation. This, in turn, means that TDMA uses the transponder more efficiently than FDMA, hence its wide use in the implementation of digital satellite communication systems.
- SDMA operates by exploiting the spatial locations of earth stations, which is achieved by means of *onboard switching*. Specifically, the transponder is equipped with multiple antennas, with the proper antenna beam being selected for radio transmission to the particular earth station demanding use of the transponder.

In addition to multiple access, another capability of a satellite channel is that of *broadcasting* with emphasis on broad area coverage. Here we mention broadcasting satellites, which are characterized by their high power transmission to inexpensive receivers.

This characteristic is exploited in the use of *direct broadcast satellites* (DBS), designed for home reception of television services on a very wide scale. By comparison with the large earth stations used for satellite communications, the earth stations for DBS are very simple and therefore inexpensive.

## 8.4 *Radio Link Analysis*

An important issue that arises in the design of satellite communication systems is that of link budget analysis.[2] As its name implies, a *link budget*, or more precisely "link power budget," is the totaling of all the gains and losses incurred in operating a communication link. In particular, the balance sheet constituting the link budget provides a detailed accounting of three broadly defined items:

1. Apportionment of the resources available to the transmitter and the receiver.
2. Sources responsible for the loss of signal power.
3. Sources of noise.

Putting all these items together into the link budget, we end up with an *estimation* procedure for evaluating the performance of a radio link, which could be the uplink or downlink of a satellite communication system. Needless to say, the essence of the communication link analysis presented in this section also applies to other radio links that rely on *line of sight* for their operation. It is for this reason the treatment of radio link analysis presented in this section is of a generic nature. The section finishes with an illustrative example on the budget analysis of a downlink of a digital satellite communication system.

From the material presented in Chapter 6 we learned that the performance of a digital communication system, in the presence of channel noise modeled as additive white Gaussian noise, is defined by a formula having the shape of a "waterfall" curve as shown in Figure 8.4. This figure portrays the probability of symbol error, $P_e$, plotted versus the bit

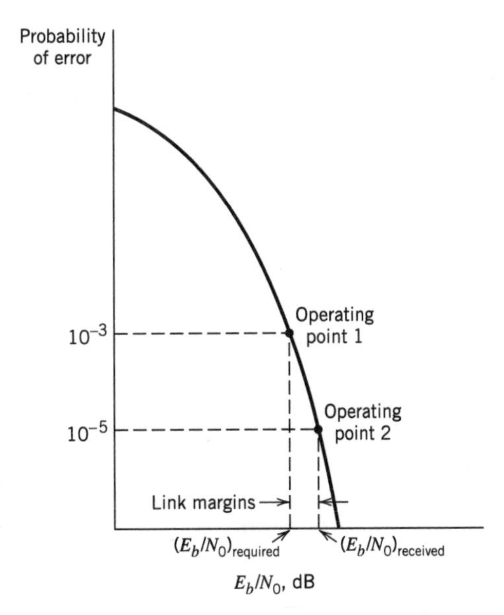


**FIGURE 8.4** "Waterfall" curve relating the probability of error to the $E_b/N_0$ ratio.

energy-to-noise spectral density ratio, $E_b/N_0$. Once a modulation scheme has been chosen, the first design task is to specify two particular values of $E_b/N_0$ as described here:

1. *Required* $E_b/N_0$.
Suppose for example, the prescribed probability of symbol error is $P_e = 10^{-3}$. Using the waterfall curve of Figure 8.4 pertaining to the modulation scheme of interest, the $E_b/N_0$ required to realize the prescribed $P_e$ is determined. Let $(E_b/N_0)_{\text{req}}$ denote the value of $E_b/N_0$ obtained from this calculation. The prescribed $P_e$ and the calculated $(E_b/N_0)_{\text{req}}$ define a point on the waterfall curve of Figure 8.4, which is designated as operating point 1.

2. *Received* $E_b/N_0$.
To assure *reliable* operation of the communication link, the link budget includes a safety measure called the *link margin.* The link margin provides protection against change and the unexpected. Thus the $(E_b/N_0)$ actually received by the system is somewhat larger than $(E_b/N_0)_{\text{req}}$. Let $(E_b/N_0)_{\text{rec}}$ denote the actual or received $E_b/N_0$, which defines a second point on the waterfall curve of Figure 8.4, designated as operating point 2. The $P_e$ corresponding to operating point 2 is shown as $10^{-5}$ in Figure 8.4 merely for the purpose of illustration. In any event, introducing the link margin denoted by $M$, we may write

$$\left(\frac{E_b}{N_0}\right)_{\text{rec}} = M\left(\frac{E_b}{N_0}\right)_{\text{req}} \tag{8.1}$$

Equivalently, expressing the two $E_b/N_0$ values of interest in decibels, we may define the link margin as

$$M(\text{dB}) = \left(\frac{E_b}{N_0}\right)_{\text{rec}}(\text{dB}) - \left(\frac{E_b}{N_0}\right)_{\text{req}}(\text{dB}) \tag{8.2}$$

Clearly, the larger we make the link margin $M$, the more reliable is the communication link. However, the increased reliability of the link is attained at the cost of a higher $E_b/N_0$.

## ■ FREE-SPACE PROPAGATION MODEL

The next step in formulating the link budget is to calculate the received signal power. Naturally, this calculation accounts for all the gains and losses incurred in the transmission and reception of the carrier.

In a radio communication system, the propagation of the modulated signal is accomplished by means of a *transmitting antenna*, the function of which is twofold:

- To convert the electrical modulated signal into an electromagnetic field. In this capacity, the transmitting antenna acts as an "impedance-transforming" transducer, matching the impedance of the antenna to that of free space.
- To radiate the electromagnetic energy in desired directions.

At the receiver, we have a *receiving antenna* whose function is the opposite of that of the transmitting antenna: It converts the electromagnetic field into an electrical signal from which the modulated signal is extracted. In addition, the receiving antenna may be required to suppress radiation originating from directions where it is not wanted.

Typically, the receiver is located in the farfield of the transmitting antenna, in which case, for all practical purposes, we may view the transmitting antenna as a fictitious volumeless emitter or *point source.* A complete description of the far field of the point source requires knowledge of the electromagnetic field as a function of both time and space.

However, insofar as link calculations are concerned, such a complete knowledge is not necessary. Rather, it is sufficient to merely specify the variation of the power density for the antenna.

By definition, the *Poynting vector* or *power density is the rate of energy flow per unit area*; it has the dimensions of watts per square meter. The treatment of the transmitting antenna as a point source greatly simplifies matters in that the power density of a point source has only a radial component; that is, the radiated energy streams from the source along radial lines.

It is useful to have a "reference" antenna against which the performance of the transmitting and receiving antennas can be compared. The customary practice is to assume that the reference antenna is an *isotropic source*, defined as an *omnidirectional (i.e., completely nondirectional) antenna that radiates uniformly in all directions*. An isotropic source is hypothetical because, in reality, all radio antennas have some directivity, however small. Nonetheless, the notion of an isotropic source is useful, especially for gain comparison purposes.

Consider then an isotropic source radiating a total power denoted by $P_t$, measured in watts. The radiated power passes uniformly through a sphere of surface area $4\pi d^2$, where $d$ is the distance (in meters) from the source. Hence, the power density, denoted by $\rho(d)$, at any point on the surface of the sphere is given by

$$\rho(d) = \frac{P_t}{4\pi d^2} \quad \text{watts/m}^2 \tag{8.3}$$

Equation (8.3) states that the power density varies inversely as the square of the distance from a point source. This statement is the familiar *inverse-square law* that governs the propagation of electromagnetic waves in free space.

Multiplying the power density $\rho(d)$ by the square of the distance $d$ at which it is measured, we get a quantity called *radiation intensity* denoted by $\Phi$. We may thus write

$$\Phi = d^2\rho(d) \tag{8.4}$$

Whereas the power density $\rho(d)$ is measured in watts per square meter, the radiation intensity $\Phi$ is measured in watts per unit solid angle (watts per steradian).

In the case of a typical transmitting or receiving radio antenna, the radiation intensity is a function of the spherical coordinates $\theta$ and $\phi$ defined in Figure 8.5. Thus, in general,

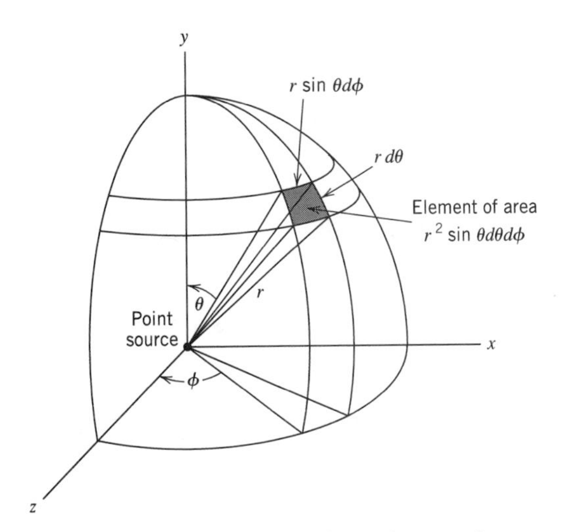


**FIGURE 8.5** Illustrating the spherical coordinates of a point source.

we may express the radiation intensity as $\Phi(\theta, \phi)$, and so speak of a *radiation-intensity pattern*. The power radiated inside an infinitesimal solid angle $d\Omega$ is given by $\Phi(\theta, \phi)\, d\Omega$, where (referring to Figure 8.5)

$$d\Omega = \sin\theta \, d\theta \, d\phi \quad \text{steradians} \tag{8.5}$$

The total power radiated is therefore

$$P = \int \Phi(\theta, \phi)\, d\Omega \quad \text{watts} \tag{8.6}$$

which is a mathematical statement of the *power theorem*. In words, the power theorem states that if the radiation-intensity pattern $\Phi(\theta, \phi)$ is known for all values of angle pair $(\theta, \phi)$, then the total power radiated is given by the integral of $\Phi(\theta, \phi)$ over a solid angle of $4\pi$ steradians. The *average* power radiated per unit solid angle is

$$\begin{aligned} P_{\text{av}} &= \frac{1}{4\pi} \int \Phi(\theta, \phi)\, d\Omega \\ &= \frac{P}{4\pi} \qquad \text{watts/steradian} \end{aligned} \tag{8.7}$$

which represents the radiation intensity that is produced by an isotropic source radiating the same total power $P$.

### *Directive Gain, Directivity, and Power Gain*[3]

Now the ability of an antenna to concentrate the radiated power in a given direction as in the case of the transmitting antenna or, conversely, to effectively absorb the incident power from that direction as in the case of the receiving antenna, is specified in terms of its directive gain or directivity. For a direction specified by the angle pair $(\theta, \phi)$, the *directive gain* of an antenna, denoted by $g(\theta, \phi)$ is defined as *the ratio of the radiation intensity in that direction to the average radiated power*, as shown by

$$\begin{aligned} g(\theta, \phi) &= \frac{\Phi(\theta, \phi)}{P_{\text{av}}} \\ &= \frac{\Phi(\theta, \phi)}{P/4\pi} \end{aligned} \tag{8.8}$$

The *directivity* of an antenna, denoted by $D$, is defined as *the ratio of the maximum radiation intensity from the antenna to the radiation intensity from an isotropic source*. That is, the directivity $D$ is the maximum value of the directive gain $g(\theta, \phi)$. Thus, whereas the directive gain of the antenna is a function of the angle pair $(\theta, \phi)$, the directivity $D$ is a constant that has been maximized for a particular direction.

The definition of directivity is based on the shape of the radiation-intensity pattern $\Phi(\theta, \phi)$; as such, it does not involve the effect of antenna imperfections due to dissipation loss and impedance mismatch. A quantity called *power gain* does involve the radiation efficiency of the antenna. Specifically, the power gain of an antenna, denoted by $G$, is defined as *the ratio of the maximum radiation intensity from the antenna to the radiation intensity from a lossless isotropic source, under the constraint that the same input power is applied to both antennas*. Specifically, using $\eta_{\text{radiation}}$ to denote the *radiation efficiency factor* of the antenna, we may relate the power gain $G$ to the directivity $D$ as

$$G = \eta_{\text{radiation}} D \tag{8.9}$$

Thus, the power gain of an antenna over a lossless isotropic source equals the directivity if the antenna is 100 percent efficient (i.e., $\eta_{\text{radiation}} = 1$), but it is less than the directivity if any losses are present in the antenna (i.e., $\eta_{\text{radiation}} < 1$). Henceforth, we assume that the antenna is 100 percent efficient and therefore refer only to the power gain of the antenna.

The concept of power gain, which is based on the transmitted power-pattern shape, can be extended to a receiving antenna by virtue of the *reciprocity principle*. An antenna is said to be reciprocal if the transmission medium is linear, passive and isotropic. For a given antenna structure, the power gains of transmitting and receiving antennas are then identical.

The power gain of an antenna is the result of concentrating the power density in a restricted region smaller than $4\pi$ steradians, as illustrated in Figure 8.6. In light of the picture portrayed in this figure, we may introduce the following two parameters:

1. *Effective radiated power referenced to an isotropic source (EIRP)*; the EIRP is defined as *the product of the transmitted power, $P_t$, and the power gain of the transmitting antenna, $G_t$*, as shown by

$$\text{EIRP} = P_t G_t \quad \text{watts} \tag{8.10}$$

2. *Antenna beamwidth*, representing a "planar" measure of the antenna's solid angle of view; the beamwidth, in degrees or radians, is defined as *the angle that subtends the two points on the mainlobe of the field-power pattern at which the peak field power is reduced by 3 dBs*. The higher the power gain of the antenna, the narrower is the antenna beamwidth.

Another matter of interest discernible from Figure 8.6 is the *sidelobes* of the field-power pattern. Unfortunately, every physical antenna has sidelobes, which are responsible for absorbing unwanted interfering radiations.

### *Effective Aperture*

A term that has a special significance for a receiving antenna is the *effective aperture* of the antenna, which is defined as *the ratio of the power available at the antenna terminals to the power per unit area of the appropriately polarized incident electromagnetic wave*. The effective aperture, denoted by $A$, is defined in terms of the antenna's power gain G as

$$A = \frac{\lambda^2}{4\pi} G \tag{8.11}$$

where $\lambda$ is the *wavelength* of the carrier. The wavelength $\lambda$ and frequency $f$ are reciprocally related as

$$\lambda = \frac{c}{f} \tag{8.12}$$

where $c$ is the speed of light (approximately equal to $3 \times 10^8$ m/s).

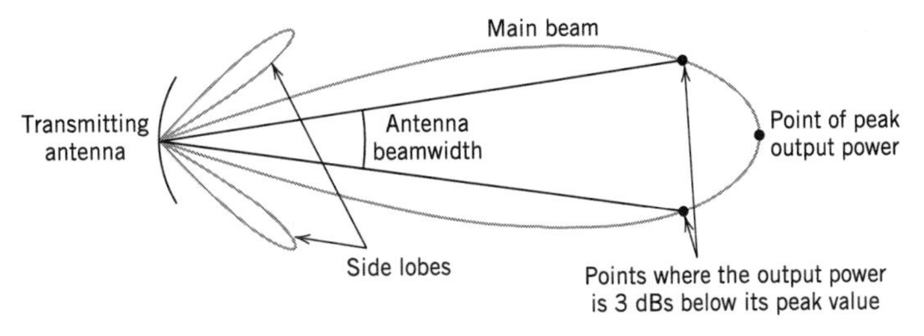


**FIGURE 8.6** Illustrating the concentration of power density of a transmitting antenna inside a region smaller than $4\pi$ steradians.

The term effective aperture has particular significance in the context of reflector antennas and electromagnetic horns that are characterized by a well-defined aperture. For these antennas, *the ratio of the antenna's effective aperture to its physical aperture is a direct measure of the antenna's aperture efficiency*, $\eta_{\text{aperture}}$, in radiating power to a desired direction or absorbing power from that direction. Nominal values for the efficiency $\eta_{\text{aperture}}$ of reflector antennas lie in the range of 45 to 75 percent.

### *Friis Free-Space Equation*

With this introductory material on antennas at hand, we are now ready to formulate the basic propagation equation for a radio communication link. Consider a transmitting antenna with an EIRP defined in Equation (8.10). Invoking the inverse-square law of Equation (8.3), we may express the power density of the transmitting antenna as $\text{EIRP}/4\pi d^2$, where $d$ is the distance between the receiving and transmitting antennas. The power $P_r$ absorbed by the receiving antenna is the product of this power density and the antenna's effective area denoted by $A_r$, as shown by

$$\begin{aligned} P_r &= \left(\frac{\text{EIRP}}{4\pi d^2}\right) A_r \\ &= \frac{P_t G_t A_r}{4\pi d^2} \quad \text{watts} \end{aligned} \tag{8.13}$$

According to the reciprocity principle, we may use Equation (8.11) to express the effective area of the receiving antenna as

$$A_r = \frac{\lambda^2}{4\pi} G_r$$

where $G_r$ is the power gain of the receiving antenna. Substituting this formula for $A_r$ into Equation (8.13), we may express the received signal power in the equivalent form

$$P_r = P_t G_t G_r \left(\frac{\lambda}{4\pi d}\right)^2 \tag{8.14}$$

Equation (8.14) is called the *Friis free-space equation*.[4]

The *path loss*, PL, representing signal "attenuation" in decibels across the entire communication link, is defined as *the difference (in decibels) between the transmitted signal power $P_t$ and received signal power $P_r$*, as shown by

$$\begin{aligned} \text{PL} &= 10 \log_{10}\left(\frac{P_t}{P_r}\right) \\ &= -10 \log_{10}(G_t G_r) + 10 \log_{10}\left(\frac{4\pi d}{\lambda}\right)^2 \end{aligned} \tag{8.15}$$

The minus sign associated with the first term in Equation (8.15) signifies the fact that this term represents a "gain." The second term, due to the collection of terms $(4\pi d/\lambda)^2$, is called the *free-space loss*, denoted by $L_{\text{free space}}$. Note that increasing the distance $d$ separating the receiving antenna from the transmitting antenna causes the free-space loss to increase, which, in turn, compels us to operate the radio communication link at lower frequencies so as to maintain the path loss at a manageable level.

The Friis free-space equation enables us to calculate the path loss PL for specified values of power gains $G_t$ and $G_r$, the carrier wavelength $\lambda$, and distance $d$. To complete the budget link analysis, we need to calculate the average noise power in the received signal, which is considered next.

## Noise Figure

To perform noise analysis at the receiver of a communication system, we need a convenient measure of the noise performance of a linear two-port device. One such measure is furnished by the so-called *noise figure*. Consider a linear two-port device connected to a signal source of internal impedance $Z(f) = R(f) + jX(f)$ at the input, as in Figure 8.7. The noise voltage $v(t)$ represents the thermal noise associated with the internal resistance $R(f)$ of the source. The output noise of the device is made up of two contributions, one due to the source and the other due to the device itself. We define *the available output noise power in a band of width $\Delta f$ centered at frequency $f$ as the maximum average noise power in this band, obtainable at the output of the device.* The maximum noise power that the two-port device can deliver to an external load is obtained when the load impedance is the complex conjugate of the output impedance of the device, that is, when the resistance is matched and the reactance is tuned out. We define *the noise figure of the two-port device as the ratio of the total available output noise power (due to the device and the source) per unit bandwidth to the portion thereof due solely to the source.*

Let the spectral density of the total available noise power of the device output be $S_{NO}(f)$, and the spectral density of the available noise power due to the source at the device input be $S_{NS}(f)$. Also let $G(f)$ denote the *available power gain* of the two-port device, defined as *the ratio of the available signal power at the output of the device to the available signal power of the source when the signal is a sinusoidal wave of frequency $f$*. Then we may express the noise figure $F$ of the device as

$$F = \frac{S_{NO}(f)}{G(f)S_{NS}(f)} \tag{8.16}$$

If the device were noise free, $S_{NO}(f) = G(f)S_{NS}(f)$, and the noise figure would then be unity. In a physical device, however, $S_{NO}(f)$ is larger than $G(f)S_{NS}(f)$, so that the noise figure is always larger than unity. The noise figure is commonly expressed in decibels, that is, as $10 \log_{10} F$.

The noise figure may also be expressed in an alternative form. Let $P_S(f)$ denote the available signal power from the source, which is the maximum average signal power that can be obtained. For the case of a source providing a single-frequency signal component

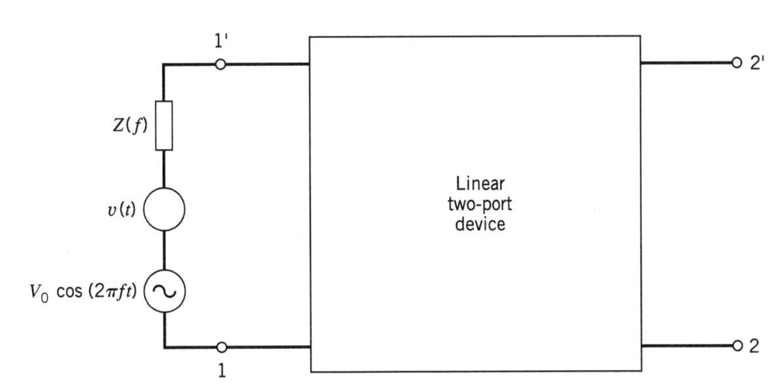


**FIGURE 8.7** Linear two-port device.

with open-circuit voltage $V_0 \cos(2\pi f t)$, the available signal power is obtained when the load connected to the source is

$$Z^*(f) = R(f) - jX(f)$$

where the asterisk denotes complex conjugation. Under this condition, we find that

$$\begin{aligned} P_S(f) &= \left[\frac{V_0}{2R(f)}\right]^2 R(f) \\ &= \frac{V_0^2}{4R(f)} \end{aligned} \tag{8.17}$$

The available signal power at the output of the device is therefore

$$P_O(f) = G(f)P_S(f) \tag{8.18}$$

Then, multiplying both the numerator and denominator of the right-hand side of Equation (8.16) by $P_S(f)\,\Delta(f)$, we obtain

$$\begin{aligned} F &= \frac{P_S(f)S_{NO}(f)\,\Delta f}{G(f)P_S(f)S_{NS}(f)\,\Delta f} \\ &= \frac{P_S(f)S_{NO}(f)\,\Delta f}{P_O(f)S_{NS}(f)\,\Delta f} \\ &= \frac{\rho_S(f)}{\rho_O(f)} \end{aligned} \tag{8.19}$$

where

$$\rho_S(f) = \frac{P_S(f)}{S_{NS}(f)\,\Delta f} \tag{8.20}$$

$$\rho_O(f) = \frac{P_O(f)}{S_{NO}(f)\,\Delta f} \tag{8.21}$$

We refer to $\rho_S(f)$ as the *available signal-to-noise ratio of the source* and to $\rho_O(f)$ as the *available signal-to-noise ratio at the device output*, both measured in a narrow band of width $\Delta f$ centered at $f$. Since the noise figure is always greater than unity, it follows from Equation (8.19) that the signal-to-noise ratio always decreases with amplification, which is a significant result.

The noise figure $F$ is a function of the operating frequency $f$; it is therefore referred to as the *spot noise figure*. In contrast, we may define an *average noise figure* $F_0$ of a two-port device as the ratio of the total noise power at the device output to the output noise power due solely to the source. That is,

$$F_0 = \frac{\displaystyle\int_{-\infty}^{\infty} S_{NO}(f)\,df}{\displaystyle\int_{-\infty}^{\infty} G(f)S_{NS}(f)\,df} \tag{8.22}$$

It is apparent that in the case of thermal noise in the input circuit with $R(f)$ constant and constant gain throughout a fixed band with zero gain at other frequencies, the spot noise figure $F$ and the average noise figure $F_0$ are identical.

### *Equivalent Noise Temperature*

A disadvantage of the noise figure $F$ is that when it is used to compare low-noise devices, the values obtained are all close to unity, which makes the comparison rather

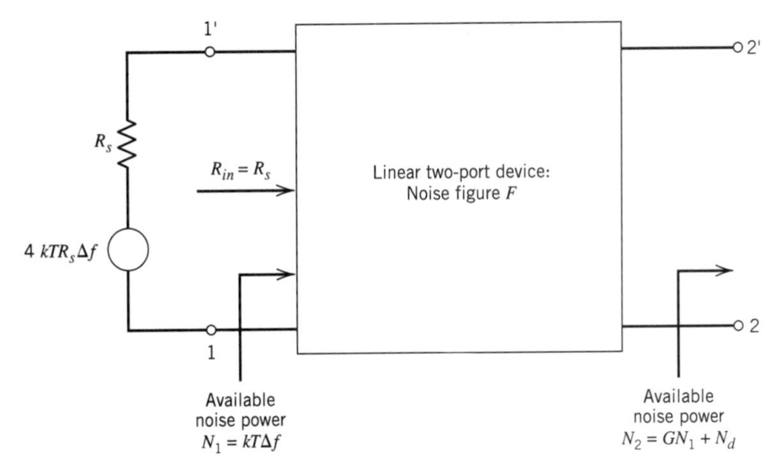


**FIGURE 8.8** Linear two-port device matched to the internal resistance of a source connected to the input.

difficult. In such cases, it is preferable to use the *equivalent noise temperature*. Consider a linear two-port device whose input resistance is matched to the internal resistance of the source as shown in Figure 8.8. In this diagram, we have also included the noise voltage generator associated with the internal resistance $R_s$ of the source. The mean-square value of this noise voltage is $4kTR_s\,\Delta f$, where $k$ is *Boltzmann's constant*. Hence, the available noise power at the device input is

$$N_1 = kT\,\Delta f \tag{8.23}$$

Let $N_d$ denote the noise power contributed by the two-port device to the total available output noise power $N_2$. We define $N_d$ as

$$N_d = GkT_e\,\Delta f \tag{8.24}$$

where $G$ is the available power gain of the device and $T_e$ is its equivalent noise temperature. Then it follows that the total output noise power is

$$\begin{aligned} N_2 &= GN_1 + N_d \\ &= Gk(T + T_e)\,\Delta f \end{aligned} \tag{8.25}$$

The noise figure of the device is therefore (see the output port of Figure 8.8)

$$F = \frac{N_2}{N_2 - N_d} = \frac{T + T_e}{T} \tag{8.26}$$

Solving for the equivalent noise temperature:

$$T_e = T(F - 1) \tag{8.27}$$

The noise figure $F$ is measured under matched input conditions, and with the noise source at temperature $T$. By convention the temperature $T$ is taken as "room temperature," namely 290 K, where K stands for "degree Kelvin."

### *Cascade Connection of Two-Port Networks*

It is often necessary to evaluate the noise figure of a cascade connection of two-port networks whose individual noise figures are known. Consider Figure 8.9, consisting of a

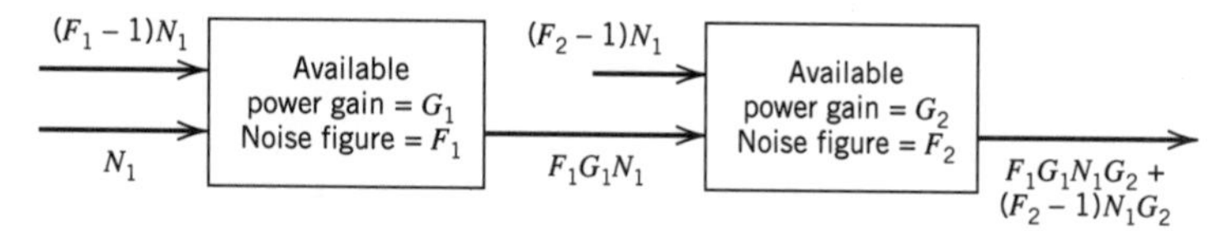


**FIGURE 8.9** A cascade of two noisy two-port networks.

pair of two-port networks of noise figures $F_1$ and $F_2$ and power gains $G_1$ and $G_2$, connected in cascade. It is assumed that the devices are matched, and that the noise figure $F_2$ of the second network is defined assuming an input noise power $N_1$.

At the input of the first network, we have a noise power $N_1$ contributed by the source, plus an equivalent noise power $(F_1 - 1)N_1$ contributed by the network itself. The output noise power from the first network is therefore $F_1N_1G_1$. Added to this noise power at the input of the second network, we have the equivalent extra power $(F_2 - 1)N_1$ contributed by the second network itself. The output noise power from this second network is therefore equal to $F_1G_1N_1G_2 + (F_2 - 1)N_1G_2$. We may consider the noise figure $F$ as the ratio of the actual output noise power to the output noise power assuming the networks to be noiseless. We may therefore express the overall noise figure of the cascade connection of Figure 8.9 as

$$
\begin{aligned}
F &= \frac{F_1G_1N_1G_2 + (F_2 - 1)N_1G_2}{N_1G_1G_2} \\
&= F_1 + \frac{F_2 - 1}{G_1}
\end{aligned}
\tag{8.28}
$$

The result may be readily extended to the cascade connection of any number of two-port networks, as shown by

$$
F = F_1 + \frac{F_2 - 1}{G_1} + \frac{F_3 - 1}{G_1G_2} + \frac{F_4 - 1}{G_1G_2G_3} + \cdots \tag{8.29}
$$

where $F_1, F_2, F_3, \ldots$ are the individual noise figures, and $G_1, G_2, G_3, \ldots$ are the available power gains, respectively. Equation (8.29) shows that if the first stage of the cascade connection in Figure 8.9 has a high gain, the overall noise figure $F$ is dominated by the noise figure of the first stage.

Correspondingly, we may express the overall equivalent noise temperature of the cascade connection of any number of noisy two-port networks as follows:

$$
T_e = T_1 + \frac{T_2}{G_1} + \frac{T_3}{G_1G_2} + \frac{T_4}{G_1G_2G_3} + \cdots \tag{8.30}
$$

where $T_1, T_2, T_3, \ldots$ are the equivalent noise temperatures of the individual networks, and $G_1, G_2, G_3, \ldots$ are the available power gains, respectively. Equation (8.30) is known as the *Friis formula*. Here again we note that if the gain $G_1$ of the first stage is high, the equivalent noise temperature $T_e$ is dominated by that of the first stage.

### ▶ EXAMPLE 8.1 Noise Temperature of Earth-Terminal Receiver

Figure 8.10 shows a typical earth-terminal receiver, consisting of a low-noise radio-frequency (RF) amplifier (LNA), frequency down-converter (mixer), and intermediate frequency (IF)

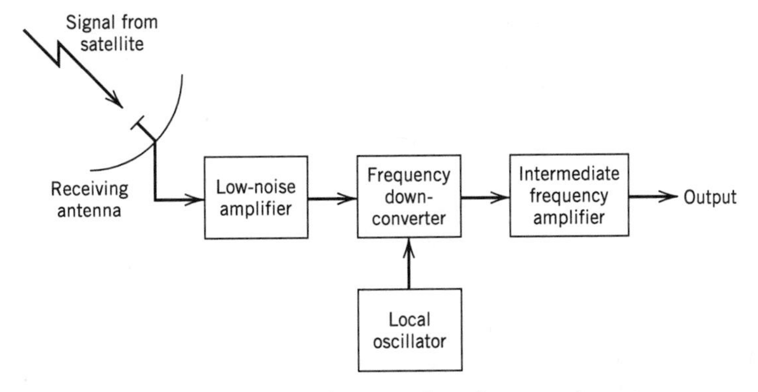


**FIGURE 8.10** Block diagram of earth terminal receiver.

amplifier. The equivalent noise temperatures of these components, including the receiving antenna, are

$$\begin{aligned} T_{\text{antenna}} &= 50 \text{ K} \\ T_{\text{RF}} &= 50 \text{ K} \\ T_{\text{mixer}} &= 500 \text{ K} \\ T_{\text{IF}} &= 1000 \text{ K} \end{aligned}$$

The available power gains of the two amplifiers are

$$\begin{aligned} G_{\text{RF}} &= 200 = 23 \text{ dB} \\ G_{\text{IF}} &= 1000 = 30 \text{ dB} \end{aligned}$$

To calculate the equivalent noise temperature of the receiver, we use Equation (8.30), obtaining

$$\begin{aligned} T_e &= T_{\text{antenna}} + T_{\text{RF}} + \frac{T_{\text{mixer}} + T_{\text{IF}}}{G_{\text{RF}}} \\ &= 50 + 50 + \frac{500 + 1000}{200} \\ &= 107.5 \text{ K} \end{aligned}$$

◀

## ▶ EXAMPLE 8.2 Downlink Budget Analysis of a Digital Satellite Communication System

In a digital satellite communication system, one of the key elements in the overall design and analysis of the system is the downlink power budget, which is usually more critical than the uplink power budget because of the practical constraints imposed on downlink power and satellite antenna size. The example presented here addresses a sample downlink budget analysis, assuming that any required uplink power (within limits) is available for satisfactory operation of the system.

The critical parameter to be calculated is the *ratio of received carrier power-to-noise spectral density*, denoted by $C/N_0$. According to the Friis free-space equation (8.14), the average power received at the earth terminal to the average power $P_t$ transmitted by the satellite is

$$P_r = P_t G_t G_r \left( \frac{\lambda}{4\pi d} \right)^2$$

where, in this example, $G_t$ is the power gain of the satellite antenna, $G_r$ is the power gain of the receiving earth-terminal antenna, $\lambda$ is the carrier wavelength for the downlink, and $d$ is the distance between the satellite and the earth terminal. Given that the equivalent noise temperature of the system is $T_e$, we may use Equation (1.94) of Chapter 1 to express the noise spectral density $N_0$ as $kT_e$, where $k$ is Boltzmann's constant. Moreover, from Equation (8.10) we note that $P_tG_t$ is equal to the EIRP of the satellite. Hence, dividing $P_r$ by $N_0$, we may express the $C/N_0$ ratio for the downlink as

$$\left(\frac{C}{N_0}\right)_{\text{downlink}} = (\text{EIRP})_{\text{satellite}}\left(\frac{G_r}{T_e}\right)_{\text{earth terminal}}\left(\frac{\lambda}{4\pi d}\right)^2\frac{1}{k} \tag{8.31}$$

For a given satellite system, the free-space loss $(4\pi d/\lambda)^2$ is a constant. Viewing the system from the earth terminal, we see from Equation (8.31) that the $(C/N_0)$ ratio is proportional to $G_r/T_e$. The ratio $G_r/T_e$ may therefore be used to assess the "quality" of an earth terminal; it is usually shortened to the *G/T ratio*, which is referred to as the *figure of merit* of the receiving earth terminal. Thus, rewriting the formula (8.31) for the $(C/N_0)$ ratio measured in decibels, we may express it as the sum of gains and losses as itemized here:

1. $(\text{EIRP})_{\text{satellite}}$, measured in dBW, where dBW denotes decibels referenced to 1 watt, that is, 0 dBW.
2. $(G/T)_{\text{earth terminal}}$, measured in dB/K, where K refers to degree Kelvin.
3. $L_{\text{free space}}$, denoting the free-space loss $10 \log_{10}(4\pi d/\lambda)^2$ in dB.
4. $-10 \log_{10} k$, representing the gain in dBW/K-Hz due to division by the Boltzmann constant $k = 1.38 \times 10^{-23}$ joule/K.

Table 8.1 presents the values of these four terms for the downlink of a typical domestic digital satellite communication system, based on the following:

1. The transponder is operated at its maximum output power (i.e., no power backoff is employed), yielding an EIRP of 46.5 dBW.
2. The receiving earth terminal uses a 2m-dish antenna with a power gain $G = 45$ dB, and the receiver is configured as in Example 8.1 with equivalent temperature $T = 107.5$ K. Hence

$$\begin{aligned}\frac{G}{T} &= 45 - 10\log_{10} 107.5\\ &= 45 - 20.3\\ &= 24.7 \text{ dB/K}\end{aligned}$$

3. The free-space loss is

$$L_{\text{free-space}} = 92.4 + 20\log_{10} f + 20\log_{10} d \text{ dB} \tag{8.32}$$

**TABLE 8.1** ***Downlink power budget for Example 8.2***

| *Variable* | *Value* |
|---|---|
| EIRP | +46.5 dBW |
| *G/T* ratio | +24.7 dB/K |
| Free-space loss | −206 dB |
| Boltzmann constant | +228.6 dBW/K-Hz |
| $C/N_0$ | 93.8 dB-Hz |

where the downlink carrier frequency $f$ is in GHz and the distance $d$ between the satellite and the earth terminal is in kilometers. For a geostationary satellite, the distance between the satellite and an earth terminal lies in the range of 36,000 to 41,000 km. Thus choosing $d = 40{,}000$ km and assuming $f = 12$ GHz, the use of Equation (8.32) yields

$$\begin{aligned} L_{\text{free-space}} &= 92.4 + 20\log_{10} 12 + 20\log_{10} 40{,}000 \\ &= 92.4 + 21.6 + 92.0 \\ &= 206 \text{ dB} \end{aligned}$$

4. With the Boltzmann constant $k = 1.39 \times 10^{-23}$ joule/K, its contribution to the $C/N_0$ ratio is

$$\begin{aligned} -10\log_{10} k &= 10\log_{10} 1.38 \times 10^{-23} \\ &= 228.6 \text{ dBW/K-Hz} \end{aligned}$$

Totaling the gains and losses, we thus get

$$\left(\frac{C}{N_0}\right)_{\text{downlink}} = 93.8 \text{ dB-Hz}$$

The "received" downlink value of the $(C/N_0)$ ratio may also be expressed in terms of the "required" value of the bit energy-to-noise spectral density ratio, $(E_b/N_0)_{\text{req}}$ dB, at the receiving earth terminal as (see Equation (8.2))

$$\left(\frac{C}{N_0}\right)_{\text{downlink}} = \left(\frac{E_b}{N_0}\right)_{\text{req}} + 10\log_{10} M + 10\log_{10} R \text{ dB} \tag{8.33}$$

where $10\log_{10} M$ is the link margin in decibels, and $R$ is the data rate in b/s. The link margin allows for excess rain losses in propagation and other power degradations. Typically, the link margin is selected as 4 dB for C-band, 6 dB for *Ku*-band, and higher for the higher *K*-band frequencies because of the higher rain losses. For operation at the *Ku*-band frequency of 12 GHz, we choose a link margin of 6 dB. Thus, using the value $C/N_0 = 93.8$ dB-Hz calculated from the link budget, the link margin $10\log_{10} M = 6$ dB, and assuming $(E_b/N_0)_{\text{req}} = 12.5$ dB, the use of Equation (8.33) yields

$$\begin{aligned} 10\log_{10} R &= 93.8 - 12.5 - 6 \\ &= 75.3 \end{aligned}$$

Hence,

$$R = 33.9 \text{ Mb/s}$$

Assuming the use of coherent 8-PSK for the transmission of digital data via the satellite, and substituting $(E_b/N_0) = 12.5$ dB in Equation (6.47) of Chapter 6, we find that the probability of symbol error $P_e = 0.6 \times 10^{-3}$.

To summarize, the digital satellite communication system analyzed in this example permits, under the worst operating conditions, data transmission on the downlink at a rate $R = 33.9$ Mb/s and with a probability of symbol error $P_e = 0.6 \times 10^{-3}$, assuming the use of 8-phase PSK. ◀

## 8.5 *Wireless Communications*

In this section we study the second type of multiuser radio communication system, namely, *wireless communications*, which is synonymous with *mobile radio*. The term mobile radio is usually meant to encompass indoor or outdoor forms of wireless communications where a radio transmitter or receiver is capable of being moved, regardless of whether it actually moves or not. Due to the stochastic nature of the mobile radio channel, its characterization mandates the use of practical measurements and statistical analysis. The aim of such an evaluation is to quantify two factors of primary concern:

1. *Median signal strength*, which enables us to predict the minimum power needed to radiate from the transmitter so as to provide an acceptable quality of coverage over a predetermined service area.
2. *Signal variability*, which characterizes the fading nature of the channel.

Our specific interest in wireless communications is in the context of *cellular radio*[5] that has the inherent capability of building mobility into the telephone network. With such a capability, a user can move freely within a service area and simultaneously communicate with any telephone subscriber in the world. An idealized model of the cellular radio system, illustrated in Figure 8.11, consists of an array of hexagonal *cells* with a *base station* located at the center of each cell; a typical cell has a radius of 1 to 12 miles. The function of the base stations is to act as an interface between *mobile subscribers* and the cellular radio system. The base stations are themselves connected to a *switching center* by dedicated wirelines.

The mobile switching center has two important roles. First, it acts as the interface between the cellular radio system and the public switched telephone network. Second, it performs overall supervision and control of the mobile communications. It performs the latter function by monitoring the signal-to-noise ratio of a call in progress, as measured at the base station in communication with the mobile subscriber involved in the call. When the SNR falls below a prescribed threshold, which happens when the mobile subscriber leaves its cell or when the radio channel fades, it is switched to another base station. This switching process, called a *handover* or *handoff*, is designed to move a mobile subscriber from one base station to another during a call in a transparent fashion, that is, without interruption of service.

The cellular concept relies on two essential features, as described here:

1. *Frequency reuse.* The term *frequency reuse* refers to the use of radio channels on the same carrier frequency to cover different areas, which are physically separated from

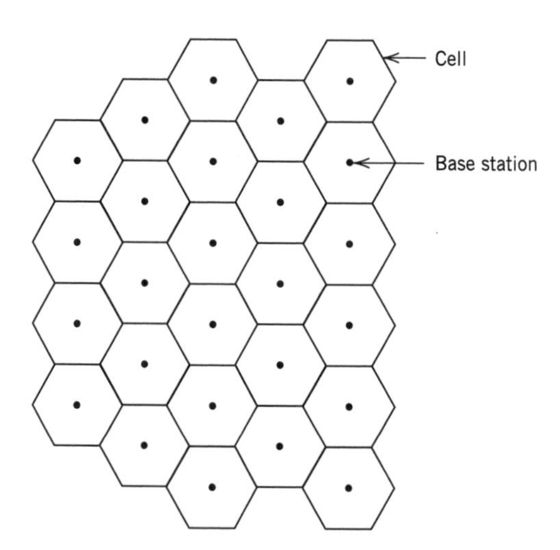


**FIGURE 8.11** Idealized model of cellular radio.

each other sufficiently to ensure that *co-channel interference* is not objectionable. Thus, instead of covering an entire local area from a single transmitter with high power at a high elevation, frequency reuse makes it possible to achieve two commonsense objectives: keep the transmitted power from each base station to a minimum, and position the antennas of the base stations just high enough to provide for the area coverage of the respective cells.

2. *Cell splitting.* When the demand for service exceeds the number of channels allocated to a particular cell, cell splitting is used to handle the additional growth in traffic within that particular cell. Specifically, cell splitting involves a revision of cell boundaries, so that the local area formerly regarded as a single cell can now contain a number of smaller cells and use the channel complements of these new cells. The new cells, which have a smaller radius than the original cells, are called *microcells.* The transmitter power and the antenna height of the new base stations are correspondingly reduced, and the same set of frequencies are reused in accordance with a new plan.

For a hexagonal model of the cellular radio system, we may exploit the basic properties of hexagonal cellular geometry to lay out a radio channel assignment plan that determines which channel set should be assigned to which cell. We begin with two integers $i$ and $j$ ($i \geq j$), called *shift parameters,* which are predetermined in some manner. We note that with a hexagonal cellular geometry there are six "chains" of hexagons that emanate from each hexagon and that extend in different directions. Thus, starting with any cell as a reference, we find the nearest *co-channel cells* by proceeding as follows:

- Move $i$ cells along any chain of hexagons, turn counterclockwise 60 degrees, and move $j$ cells along the chain that lies on this new direction. The $j$th cells so located and the reference cell constitute the set of co-channel cells.

This procedure is repeated for a different reference cell, until all the cells in the system are covered. Figure 8.12 illustrates the application of this procedure for a single reference cell and the example of $i = 2$ and $j = 2$.

In North America, the band of radio frequencies assigned to the cellular system is 800–900 MHz. The subband 824–849 MHz is used to receive signals from the mobile units, and the subband 869–894 MHz is used to transmit signals to the mobile units. The

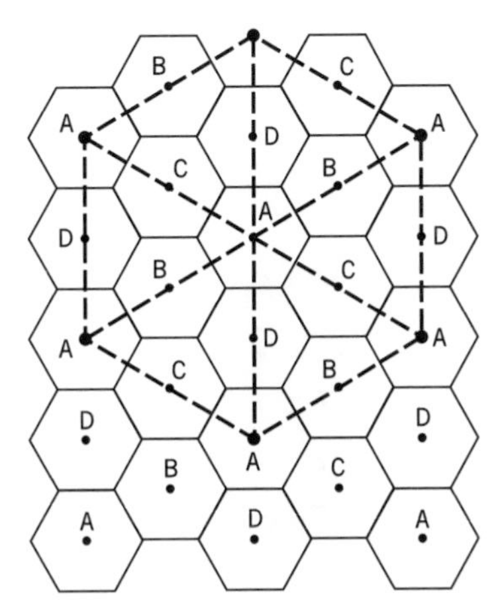


**FIGURE 8.12** Illustrating the determination of co-channel cells.

use of these relatively high frequencies has the beneficial feature of providing a good portable coverage by penetrating buildings. In Europe and elsewhere, the base–mobile and mobile–base subbands are reversed.

## ■ PROPAGATION EFFECTS[6]

The major propagation problems encountered in the use of cellular radio in built-up areas are due to the fact that the antenna of a mobile unit may lie well below the surrounding buildings. Simply put, there is no "line-of-sight" path to the base station. Instead, radio propagation takes place mainly by way of scattering from the surfaces of the surrounding buildings and by diffraction over and/or around them, as illustrated in Figure 8.13. The important point to note from Figure 8.13 is that energy reaches the receiving antenna via more than one path. Accordingly, we speak of a *multipath phenomenon* in that the various incoming radio waves reach their destination from different directions and with different time delays.

To understand the nature of the multipath phenomenon, consider first a "static" multipath environment involving a stationary receiver and a transmitted signal that consists of a narrowband signal (e.g., unmodulated sinusoidal carrier). Let it be assumed that two attenuated versions of the transmitted signal arrive sequentially at the receiver. The effect of the differential time delay is to introduce a relative phase shift between the two components of the received signal. We may then identify one of two extreme cases that can arise:

- The relative phase shift is zero, in which case the two components add constructively, as illustrated in Figure 8.14*a*.
- The relative phase shift is 180 degrees, in which case the two component add destructively, as illustrated in Figure 8.14*b*.

We may also use *phasors* to demonstrate the constructive and destructive effects of multipath, as shown in Figures 8.15*a* and 8.15*b*, respectively. Note that in the static multipath environment described herein, the amplitude of the received signal does not vary with time.

Consider next a "dynamic" multipath environment in which the receiver is in motion and two versions of the transmitted narrowband signal reach the receiver via paths of

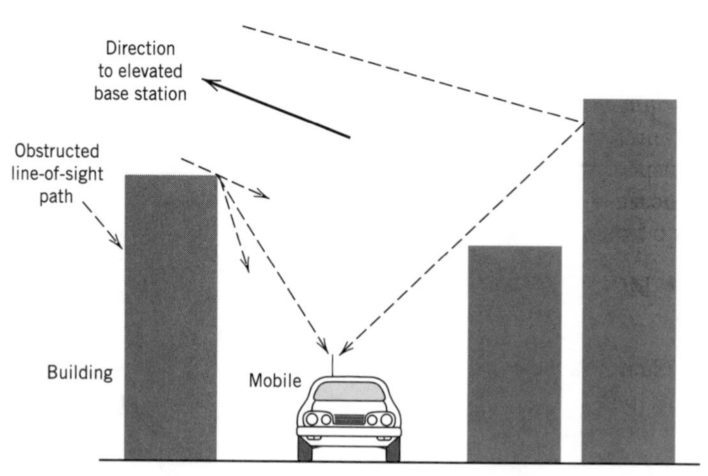


FIGURE 8.13 Illustrating the mechanism of radio propagation in urban areas. (From Parsons, 1992, with permission.)

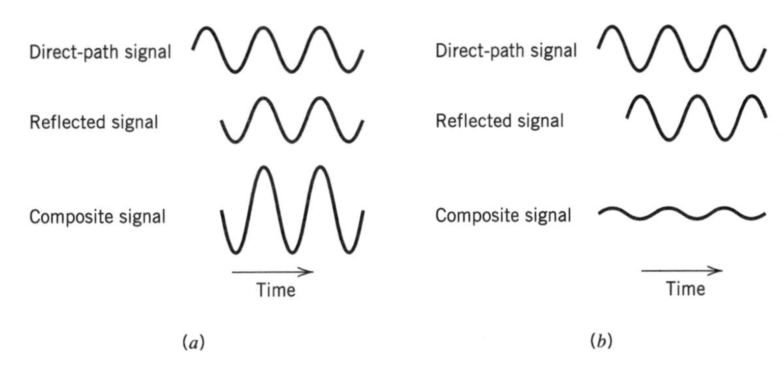


**FIGURE 8.14** (*a*) Constructive and (*b*) destructive forms of the multipath phenomenon for sinusoidal signals.

different lengths. Due to motion of the receiver, there is a continuous change in the length of each propagation path. Hence, the relative phase shift between the two components of the received signal is a function of spatial location of the receiver. As the receiver moves, we now find that the received amplitude (envelope) is no longer constant as was the case in a static environment; rather, it varies with distance, as illustrated in Figure 8.16. At the top of this figure, we have also included the phasor relationships for the two components of the received signal at various locations of the receiver. Figure 8.16 shows that there is constructive addition at some locations, and almost complete cancellation at some other locations. This phenomenon is referred to as *signal fading*.

In a mobile radio environment encountered in practice, there may of course be a multitude of propagation paths with different lengths, and their contributions to the re-

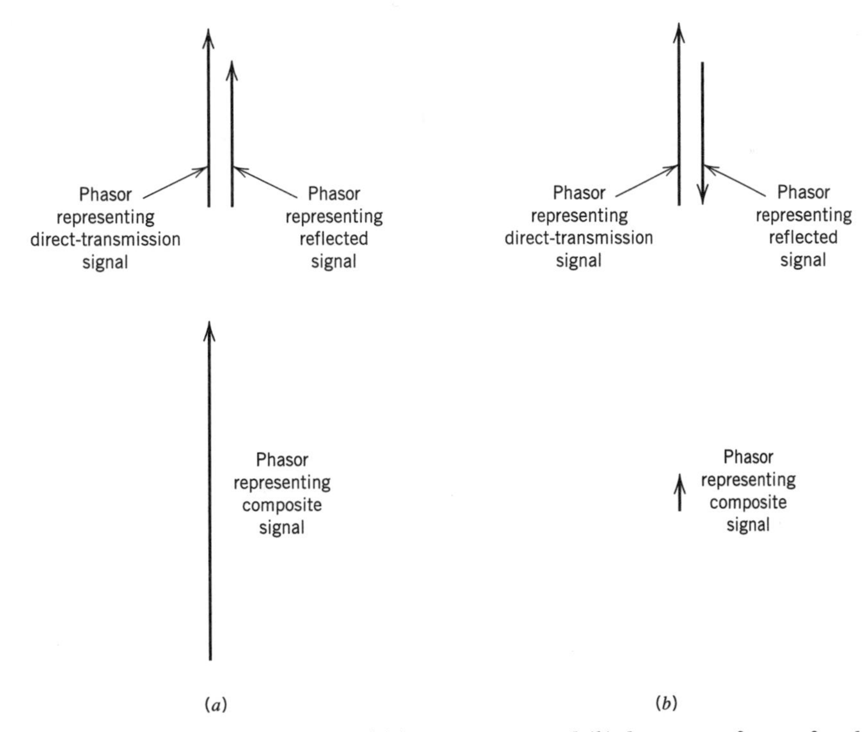


**FIGURE 8.15** Phasor representations of (*a*) constructive and (*b*) destructive forms of multipath.

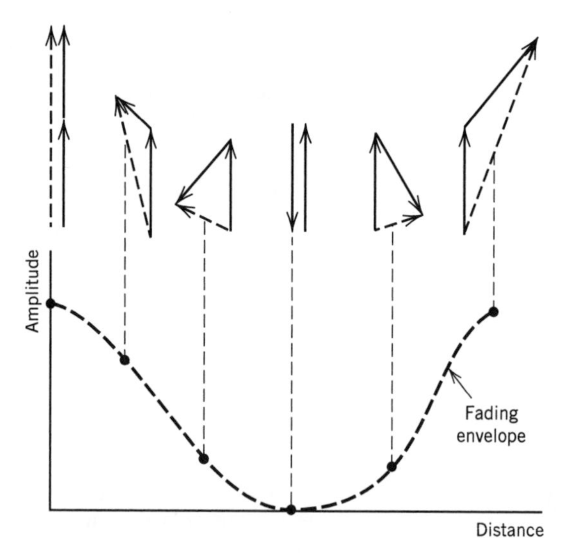


**FIGURE 8.16** Illustrating how the envelope fades as two incoming signals combine with different phases. (From Parsons, 1992, with permission.)

ceived signal could combine in a variety of ways. The net result is that the envelope of the received signal varies with location in a complicated fashion, as shown by the experimental record of received signal envelope in an urban area that is presented in Figure 8.17. This figure clearly displays the fading nature of the received signal. The received signal envelope in Figure 8.17 is measured in dBm. The unit dBm is defined as $10 \log_{10}(P/P_0)$, with $P$ denoting the power being measured and $P_0 = 1$ milliwatt. In the case of Figure 8.17, $P$ is the instantaneous power in the received signal envelope.

Signal fading is essentially a *spatial phenomenon* that manifests itself in the time domain as the receiver moves. These variations can be related to the motion of the receiver as follows. To be specific, consider the situation illustrated in Figure 8.18, where the receiver is assumed to be moving along the line $AA'$ with a constant velocity $v$. It is also

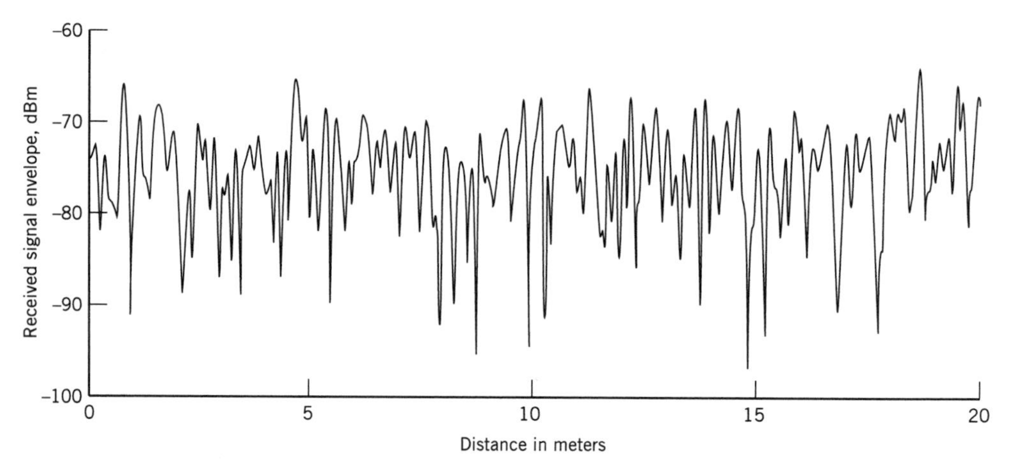


**FIGURE 8.17** Experimental record of received signal envelope in an urban area. (From Parsons, 1992, with permission.)

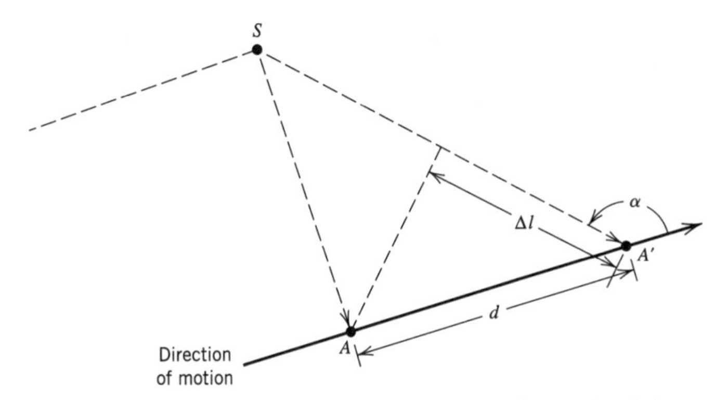


**FIGURE 8.18** Illustrating the calculation of Doppler shift.

assumed that the received signal is due to a radio wave from a scatterer labeled $S$. Let $\Delta t$ denote the time taken for the receiver to move from point $A$ to $A'$. Using the notation described in Figure 8.18, the incremental change in the path length of the radio wave is deduced to be

$$\begin{aligned}\Delta l &= d \cos \alpha \\ &= -\, v \, \Delta t \cos \alpha\end{aligned} \tag{8.34}$$

where $\alpha$ is the spatial angle between the incoming radio wave and the direction of motion of the receiver. Correspondingly, the change in the phase angle of the received signal at point $A'$ with respect to that at point $A$ is given by

$$\begin{aligned}\Delta\phi &= \frac{2\pi}{\lambda} \Delta l \\ &= -\frac{2\pi v \, \Delta t}{\lambda} \cos \alpha\end{aligned} \tag{8.35}$$

where $\lambda$ is the radio wavelength. The apparent change in frequency, or the *Doppler-shift*, is therefore

$$\begin{aligned}\nu &= -\frac{1}{2\pi} \frac{\Delta\phi}{\Delta t} \\ &= \frac{v}{\lambda} \cos \alpha\end{aligned} \tag{8.36}$$

The Doppler-shift $\nu$ is positive (resulting in an increase in frequency) when the radio waves arrive from ahead of the mobile unit, and it is negative when the radio waves arrive from behind the mobile unit.

## 8.6 Statistical Characterization of Multipath Channels

The narrowband characterization of the multipath environment described in Section 8.5 is appropriate for mobile radio transmissions where the signal bandwidth is very small compared to the reciprocal of the spread in propagation path delays. Multipath in such an environment results in two effects: rapid fading of the received signal envelope and a spread in Doppler shifts in the received spectrum. Real-life signals radiated in a mobile radio environment may, however, occupy a bandwidth wide enough to require more detailed considerations of the effects of multipath propagation on the received signal. In this section, we present a statistical characterization of a mobile radio channel.*

Consider a mobile radio channel with multiple propagation paths. In accordance with the complex notation described in Appendix 2, we may express the transmitted band-pass signal as

$$s(t) = \text{Re}[\tilde{s}(t) \exp(j2\pi f_c t)] \tag{8.37}$$

where $\tilde{s}(t)$ is the complex (low-pass) envelope of $s(t)$, and $f_c$ is a nominal carrier frequency. Since the channel is time varying due to multipath effects, the impulse response of the channel is delay dependent and therefore a time-varying function. Let the impulse response of the channel be expressed as

$$h(\tau;t) = \text{Re}[\tilde{h}(\tau;t) \exp(j2\pi f_c t)] \tag{8.38}$$

where $\tilde{h}(\tau;t)$ is the (low-pass) complex impulse response of the channel, and $\tau$ is a delay variable. The complex impulse response $\tilde{h}(\tau;t)$ is called the *input delay-spread function* of the channel. The (low-pass) complex envelope of the channel output is defined by the convolution integral

$$\tilde{s}_o(t) = \frac{1}{2}\int_{-\infty}^{\infty} \tilde{s}(t - \tau)\tilde{h}(\tau;t)\, d\tau \tag{8.39}$$

where the scaling factor $\frac{1}{2}$ is the result of using complex notation.

In general, the behavior of a mobile radio channel can be described only in statistical terms. For analytic purposes, the delay-spread function $\tilde{h}(\tau;t)$ may thus be modeled as a zero-mean complex-valued Gaussian process. Then, at any time $t$ the envelope $|\tilde{h}(\tau;t)|$ is Rayleigh distributed, and the channel is referred to as a *Rayleigh fading channel*. When, however, the mobile radio environment includes *fixed* scatterers, we are no longer justified in using a zero-mean model to describe the input delay-spread function $\tilde{h}(\tau;t)$. In such a case, it is more appropriate to use a Rician distribution to describe the envelope $|\tilde{h}(\tau;t)|$, and the channel is referred to as a *Rician fading channel*. The Rayleigh and Rician distributions for a real-valued random process were considered in Chapter 1. In the discussion presented in this chapter, we consider only a Rayleigh fading channel.

The *time-varying transfer function* of the channel is defined as the Fourier transform of the input delay-spread function $\tilde{h}(\tau;t)$ with respect to the delay variable $\tau$, as shown by

$$\tilde{H}(f;t) = \int_{-\infty}^{\infty} \tilde{h}(\tau;t) \exp(-j2\pi f\tau)\, d\tau \tag{8.40}$$

where $f$ denotes the frequency variable. The time-varying transfer function $\tilde{H}(f;t)$ may be viewed as a frequency transmission characteristic of the channel.

*Readers who are not interested in the mathematical details pertaining to the statistical characterization of fading multipath channels, may skip the material presented in this section, except for the subsection on the classification of multipath channels at the end of the section.

For a statistical characterization of the channel, we make the following assumptions:

- The input delay-spread function $\tilde{h}(\tau;t)$ is a *zero-mean, complex-valued Gaussian process*. Our interest is confined to short-term fading; it is therefore reasonable to assume that $\tilde{h}(\tau;t)$ is also *stationary*. Because Fourier transformation is linear, the time-varying transfer function $\tilde{H}(f;t)$ has similar statistics.
- The channel is an *uncorrelated scattering channel*, which means that contributions from scatterers with different propagation delays are uncorrelated.

Consider then the autocorrelation function of the input delay-spread function $\tilde{h}(\tau;t)$. Since $\tilde{h}(\tau;t)$ is complex valued, we use the following definition for the autocorrelation function:

$$R_{\tilde{h}}(\tau_1,t_1;\tau_2,t_2) = E[\tilde{h}^*(\tau_1;t_1)\tilde{h}(\tau_2;t_2)] \tag{8.41}$$

where $E$ is the statistical expectation operator, the asterisk denotes complex conjugation, $\tau_1$ and $\tau_2$ are the propagation delays of the two paths involved in the calculation, and $t_1$ and $t_2$ are the times at which the outputs of the two paths are observed. Invoking stationarity in the time variable $t$ and uncorrelated scattering in the time-delay variable $\tau$, we may reformulate the autocorrelation function of $\tilde{h}(\tau;t)$ as

$$\begin{aligned} R_{\tilde{h}}(\tau_1,\tau_2;\Delta t) &= E[\tilde{h}^*(\tau_1;t)\tilde{h}(\tau_2;t + \Delta t)] \\ &= r_{\tilde{h}}(\tau_1;\Delta t)\ \delta(\tau_1 - \tau_2) \end{aligned} \tag{8.42}$$

where $\Delta t$ is the difference between the observation times, and $\delta(\tau_1 - \tau_2)$ is a delta function. Using $\tau$ in place of $\tau_1$, the remaining function in Equation (8.42) is redefined as

$$r_{\tilde{h}}(\tau;\Delta t) = E[\tilde{h}(\tau;t)\tilde{h}^*(\tau;t + \Delta t)] \tag{8.43}$$

The function $r_{\tilde{h}}(\tau;\Delta t)$ is called the *multipath autocorrelation profile* of the channel.

Consider next a statistical characterization of the channel in terms of the complex-valued, time-varying transfer function $\tilde{H}(f;t)$. Following a formulation similar to that described in Equation (8.41), the autocorrelation function of $\tilde{H}(f;t)$ is defined by

$$R_{\tilde{H}}(f_1,t_1;f_2,t_2) = E[\tilde{H}^*(f_1;t_1)\tilde{H}(f_2;t_2)] \tag{8.44}$$

where $f_1$ and $f_2$ represent two frequencies in the spectrum of a transmitted signal. The autocorrelation function $R_{\tilde{H}}(f_1,t_1;f_2,t_2)$ provides a statistical measure of the extent to which the signal is distorted by transmission through the channel. From Equations (8.40), (8.41), and (8.44) we find that the autocorrelation functions $R_{\tilde{H}}(f_1,t_1;f_2,t_2)$ and $R_{\tilde{h}}(\tau_1,t_1;\tau_2,t_2)$ are related by a form of two-dimensional Fourier transformation as follows:

$$R_{\tilde{H}}(f_1,t_1;f_2,t_2) = \int_{-\infty}^{\infty}\int_{-\infty}^{\infty} R_{\tilde{h}}(\tau_1,t_1;\tau_2,t_2)\exp[j2\pi(f_1\tau_1 - f_2\tau_2)]\,d\tau_1\,d\tau_2 \tag{8.45}$$

Invoking stationarity in the time domain, we may reformulate Equation (8.44) as

$$R_{\tilde{H}}(f_1,f_2;\Delta t) = E[\tilde{H}^*(f_1;t)\tilde{H}(f_2;t + \Delta t)] \tag{8.46}$$

This definition suggests that the autocorrelation function $R_{\tilde{H}}(f_1,f_2;\Delta t)$ may be measured by pairs of spaced tones to carry out cross-correlation measurements on the resulting channel outputs. Such a measurement presumes stationarity in the time domain. If we also assume stationarity in the frequency domain, we may go one step further and write

$$\begin{aligned} R_{\tilde{H}}(f, f + \Delta f; \Delta t) &= r_{\tilde{H}}(\Delta f; \Delta t) \\ &= E[\tilde{H}^*(f;t)\tilde{H}(f + \Delta f; t + \Delta t)] \end{aligned} \tag{8.47}$$

This specialized form of the autocorrelation function of $\tilde{H}(f;t)$ is in fact the Fourier transform of the multipath autocorrelation profile $r_{\tilde{h}}(\tau;\Delta t)$ with respect to the delay-time variable $\tau$, as shown by

$$r_{\tilde{H}}(\Delta f;\Delta t) = \int_{-\infty}^{\infty} r_{\tilde{h}}(\tau;\Delta t) \exp(-j2\pi\tau\,\Delta f)\, d\tau \tag{8.48}$$

The function $r_{\tilde{H}}(\Delta f;\Delta t)$ is called the *spaced-frequency spaced-time correlation function* of the channel.

Finally, we introduce a function $S(\tau;\nu)$ that forms a Fourier-transform pair with the multipath autocorrelation profile $r_{\tilde{h}}(\tau;\Delta t)$ with respect to the variable $\Delta t$, as shown by

$$S(\tau;\nu) = \int_{-\infty}^{\infty} r_{\tilde{h}}(\tau;\Delta t) \exp(-j2\pi\nu\,\Delta t)\, d(\Delta t) \tag{8.49}$$

and

$$r_{\tilde{h}}(\tau;\Delta t) = \int_{-\infty}^{\infty} S(\tau;\nu) \exp(j2\pi\nu\,\Delta t)\, d\nu \tag{8.50}$$

The function $S(\tau;\nu)$ may also be defined in terms of $r_{\tilde{H}}(\Delta f;\Delta t)$ by applying a form of double Fourier transformation: a Fourier transform with respect to the time variable $\Delta t$ and an inverse Fourier transform with respect to the frequency variable $\Delta f$. That is to say,

$$S(\tau;\nu) = \int_{-\infty}^{\infty}\int_{-\infty}^{\infty} r_{\tilde{H}}(\Delta f;\Delta t) \exp(-j2\pi\nu\,\Delta t) \exp(j2\pi\tau\,\Delta f)\, d(\Delta t)\, d(\Delta f) \tag{8.51}$$

Figure 8.19 displays the functional relationships between $r_{\tilde{h}}(\tau;\Delta t)$, $r_{\tilde{H}}(\Delta f;\Delta t)$, and $S(\tau;\nu)$ in terms of the Fourier transform and its inverse.

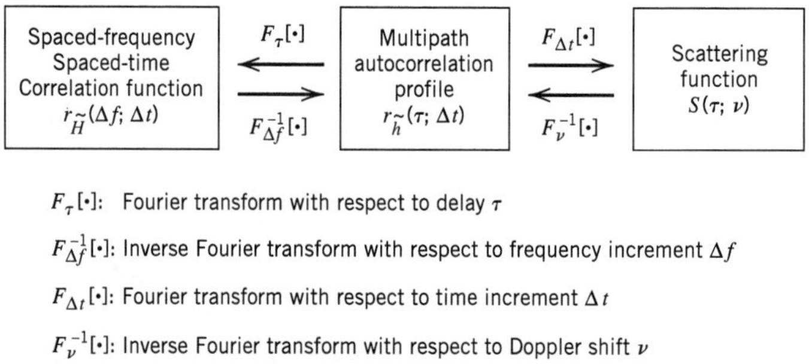

Spaced-frequency Spaced-time Correlation function $r_{\tilde{H}}(\Delta f;\Delta t)$

$F_\tau[\cdot]$

$F_{\Delta f}^{-1}[\cdot]$

Multipath autocorrelation profile $r_{\tilde{h}}(\tau;\Delta t)$

$F_{\Delta t}[\cdot]$

$F_\nu^{-1}[\cdot]$

Scattering function $S(\tau;\nu)$

$F_\tau[\cdot]$: Fourier transform with respect to delay $\tau$

$F_{\Delta f}^{-1}[\cdot]$: Inverse Fourier transform with respect to frequency increment $\Delta f$

$F_{\Delta t}[\cdot]$: Fourier transform with respect to time increment $\Delta t$

$F_\nu^{-1}[\cdot]$: Inverse Fourier transform with respect to Doppler shift $\nu$

**FIGURE 8.19** Functional relationships between the multipath autocorrelation profile $r_{\tilde{h}}(\tau;\Delta t)$, the spaced-frequency spaced-time correlation function $r_{\tilde{H}}(\Delta f;\Delta t)$, and the scattering function $S(\tau;\nu)$.

The function $S(\tau;\nu)$ is called the *scattering function* of the channel. For a physical interpretation of it, consider the transmission of a single tone of frequency $f'$ (relative to the carrier). The complex envelope of the resulting filter output is

$$\tilde{s}_o(t) = \exp(j2\pi f't)\tilde{H}(f';t) \tag{8.52}$$

The autocorrelation function of $\tilde{s}_o(t)$ is

$$\begin{aligned} E[\tilde{s}_o^*(t)\tilde{s}_o(t + \Delta t)] &= \exp(j2\pi f' \Delta t)E[\tilde{H}^*(f';t)\tilde{H}(f';t + \Delta t)] \\ &= \exp(j2\pi f' \Delta t)r_{\tilde{H}}(0;\Delta t) \end{aligned} \tag{8.53}$$

where, in the last line, we have made use of Equation (8.47). Putting $\Delta f = 0$ in Equation (8.48), and then using Equation (8.50), we may write

$$\begin{aligned} r_{\tilde{H}}(0;\Delta t) &= \int_{-\infty}^{\infty} r_{\tilde{h}}(\tau;\Delta t)\, d\tau \\ &= \int_{-\infty}^{\infty} \left[\int_{-\infty}^{\infty} S(\tau;\nu)\, d\tau\right] \exp(j2\pi\nu\, \Delta t)\, d\nu \end{aligned} \tag{8.54}$$

Hence, we may view the integral

$$\int_{-\infty}^{\infty} S(\tau;\nu)\, d\tau$$

as the power spectral density of the channel output relative to the frequency $f'$ of the transmitted tone, and with the Doppler shift $\nu$ acting as the frequency variable. Generalizing this result, we may state that the scattering function $S(\tau;\nu)$ provides a statistical measure of the output power of the channel, expressed as a function of the time delay $\tau$ and the Doppler shift $\nu$.

## Delay Spread and Doppler Spread

Putting $\Delta t = 0$ in Equation (8.43), we may write

$$\begin{aligned} P_{\tilde{h}}(\tau) &= r_{\tilde{h}}(\tau;0) \\ &= E[|\tilde{h}(\tau;t)|^2] \end{aligned} \tag{8.55}$$

The function $P_{\tilde{h}}(\tau)$ describes the intensity (averaged over the fading fluctuations) of the scattering process at propagation delay $\tau$. Accordingly, $P_{\tilde{h}}(\tau)$ is called the *delay power spectrum* or the *multipath intensity profile* of the channel. The delay power spectrum may also be defined in terms of the scattering function $S(\tau;\nu)$ by averaging it over all Doppler shifts. Specifically, putting $\Delta t = 0$ in Equation (8.50) and then using the first line of Equation (8.55), we may write

$$P_{\tilde{h}}(\tau) = \int_{-\infty}^{\infty} S(\tau;\nu)\, d\nu \tag{8.56}$$

Figure 8.20 shows an example of a delay power spectrum that depicts a typical plot of the power spectral density versus excess delay; the excess delay is measured with respect to the time delay for the shortest echo path. Note, as in Figure 8.17, the power is measured in dBm. The "threshold level" included in Figure 8.20 defines the power level below which the receiver fails to operate satisfactorily.

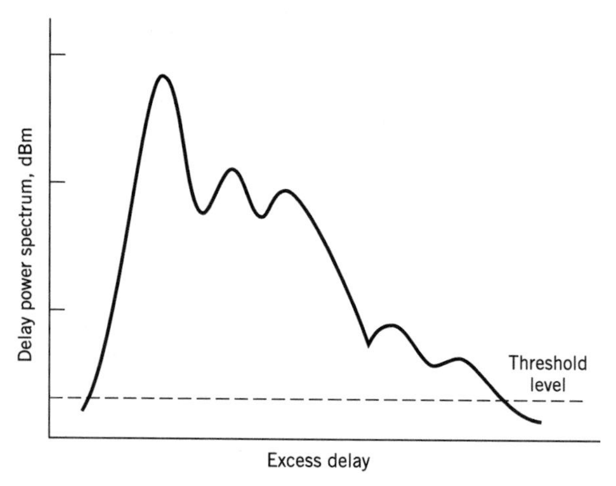


**FIGURE 8.20** Example of a power-delay profile for a mobile radio channel. (From Parsons, 1992, with permission.)

Two statistical moments of $P_{\tilde{h}}(\tau)$ of interest are the *average delay*, $\tau_{\text{av}}$, and the *delay spread*, $\sigma_\tau$. The average delay is defined as the first central moment (i.e., the mean) of $P_{\tilde{h}}(\tau)$, as shown by

$$\tau_{\text{av}} = \frac{\int_0^\infty \tau P_{\tilde{h}}(\tau)\, d\tau}{\int_0^\infty P_{\tilde{h}}(\tau)\, d\tau} \tag{8.57}$$

The delay spread is defined as the square root of the second central moment of $P_{\tilde{h}}(\tau)$, as shown by

$$\sigma_\tau = \left( \frac{\int_0^\infty (\tau - \tau_{\text{av}})^2 P_{\tilde{h}}(\tau)\, d\tau}{\int_0^\infty P_{\tilde{h}}(\tau)\, d\tau} \right)^{1/2} \tag{8.58}$$

The reciprocal of the delay spread $\sigma_\tau$ is a measure of the *coherence bandwidth* of the channel, which is denoted by $B_c$.

Consider next the issue of relating the Doppler effects to time variations of the channel. For this purpose, we first set $\Delta f = 0$, which corresponds to the transmission of a single tone (of some appropriate frequency) over the channel. The spaced-frequency spaced-time correlation function of the channel then reduces to $r_{\tilde{H}}(0;\Delta t)$. Hence, evaluating the Fourier transform of this function with respect to the time variable $\Delta t$, we may write

$$S_{\tilde{H}}(\nu) = \int_{-\infty}^{\infty} r_{\tilde{H}}(0;\Delta t) \exp(-j2\pi\nu\, \Delta t)\, d(\Delta t) \tag{8.59}$$

The function $S_{\tilde{H}}(\nu)$ defines the power spectrum of the channel output expressed as a function of the Doppler shift $\nu$; it is therefore called the *Doppler spectrum* of the channel. The

Doppler spectrum may also be defined in terms of the scattering function by averaging it over all possible propagation delays, as shown by

$$S_{\tilde{H}}(\nu) = \int_{-\infty}^{\infty} S(\tau;\nu)\, d\tau \tag{8.60}$$

The Doppler shift $\nu$ may assume positive and negative values with equal likelihood. The mean Doppler shift is therefore zero. The square root of the second moment of the Doppler spectrum is thus defined by

$$\sigma_\nu = \left( \frac{\int_{-\infty}^{\infty} \nu^2 S_{\tilde{H}}(\nu)\, d\nu}{\int_{-\infty}^{\infty} S_{\tilde{H}}(\nu)\, d\nu} \right)^{1/2} \tag{8.61}$$

The parameter $\sigma_\nu$ provides a measure of the width of the Doppler spectrum; it is therefore called the *Doppler spread* of the channel. The reciprocal of the Doppler spread is called the *coherence time* of the channel, which is denoted by $\tau_c$.

Another useful parameter that is often used in measurements is the *fade rate* of the channel. For a Rayleigh fading channel, the average fade rate is related to the Doppler spread $\sigma_\nu$ as

$$f_e = 1.475\sigma_\nu \text{ crossings per second} \tag{8.62}$$

As the name implies, the fade rate provides a measure of the rapidity of fading of the channel.

Some typical values encountered in a mobile radio environment are as follows:

- The delay spread, $\sigma_\tau$, amounts to about 20 $\mu$s.
- The Doppler spread, $\sigma_\nu$, due to the motion of a vehicle may extend up to 40–80 Hz.

## Classification of Multipath Channels

The particular form of fading experienced by a multipath channel depends on whether the channel characterization is viewed in the frequency domain or the time domain.

When the channel is viewed in the frequency domain, the parameter of concern is the channel's coherence bandwidth, $B_c$, which is a measure of the transmission bandwidth for which signal distortion across the channel becomes noticeable. A multipath channel is said to be *frequency selective* if the coherence bandwidth of the channel is small compared to the bandwidth of the transmitted signal. In such a situation, the channel has a filtering effect in that two sinusoidal components, with a frequency separation greater than the channel's coherence bandwidth, are treated differently. If, however, the coherence bandwidth of the channel is large compared to the message bandwidth, the fading is said to be *frequency nonselective*, or *frequency flat*.

When the channel is viewed in the time domain, the parameter of concern is the coherence time, $\tau_c$, which provides a measure of the transmitted signal duration for which distortion across the channel becomes noticeable. The fading is said to be *time selective* if the coherence time of the channel is small compared to the duration of the received signal (i.e., the time for which the signal is in flight). For digital transmission, the received signal's duration is taken as the symbol duration plus the channel's delay spread. If, however, the channel's coherence time is large compared to the received signal duration, the fading is

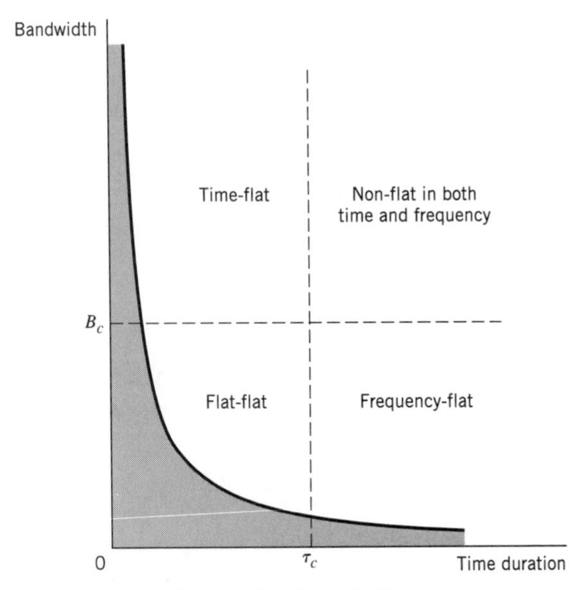


**FIGURE 8.21** Illustrating the four classes of multipath channels: $\tau_c$ = coherence time, $B_c$ = coherence bandwidth.

said to be *time nonselective*, or *time flat*, in the sense that the channel appears to the transmitted signal as time invariant.

In light of this discussion, we may classify multipath channels as follows:

- *Flat-flat channel*, which is flat in both frequency and time.
- *Frequency-flat channel*, which is flat in frequency only.
- *Time-flat channel*, which is flat in time only.
- *Nonflat channel*, which is flat neither in frequency nor in time; such a channel is sometimes referred to as a *doubly dispersive channel.*

The classification of multipath channels, based on this approach, is shown in Figure 8.21. The forbidden area, shown shaded in this figure, follows from the inverse relationship that exists between bandwidth and time duration.

## 8.7 Binary Signaling over a Rayleigh Fading Channel

In Chapter 6, we determined the average probability of symbol error for the transmission of binary data over a channel corrupted by additive white Gaussian noise. In a mobile radio environment, we have an additional effect to consider, namely, the fluctuations in the amplitude and phase of the received signal due to multipath effects. To be specific, consider the transmission of binary data over a Rayleigh fading channel, for which the (low-pass) complex envelope of the received signal is modeled as follows:

$$\tilde{x}(t) = \alpha \exp(-j\phi)\tilde{s}(t) + \tilde{w}(t) \tag{8.63}$$

where $\tilde{s}(t)$ is the complex envelope of the transmitted (band-pass) signal, $\alpha$ is a Rayleigh-distributed random variable describing the attenuation in transmission, $\phi$ is a uniformly distributed random variable describing the phase-shift in transmission, and $\tilde{w}(t)$ is a complex-valued white Gaussian noise process. It is assumed that the channel is flat in both time and frequency, so that we can estimate the phase-shift $\phi$ from the received signal without error. Suppose then that coherent binary phase-shift keying is used to do the data transmission. Under the condition that $\alpha$ is fixed or constant over a bit interval, we may adapt Equation (6.20) of Chapter 6 for the situation at hand by expressing the average probability of symbol error (i.e., bit error rate) due to the additive white Gaussian noise acting alone as follows:

$$P_e(\gamma) = \tfrac{1}{2}\,\mathrm{erfc}(\sqrt{\gamma}) \tag{8.64}$$

where $\gamma$ is an attenuated version of the transmitted signal energy per bit-to-noise spectral density ratio $E_b/N_0$, as shown by

$$\gamma = \frac{\alpha^2 E_b}{N_0} \tag{8.65}$$

Now, insofar as a mobile radio channel is concerned, we may view $P_e(\gamma)$ as a conditional probability given that $\alpha$ is fixed. Thus, to evaluate the average probability of symbol error in the combined presence of fading and noise, we must average $P_e(\gamma)$ over all possible values of $\gamma$, as shown by

$$P_e = \int_0^\infty P_e(\gamma) f(\gamma)\, d\gamma \tag{8.66}$$

where $f(\gamma)$ is the probability density function of $\gamma$. From Equation (8.65) we note that $\gamma$ depends on the squared value of $\alpha$. Since $\alpha$ is Rayleigh distributed, we find that $\gamma$ has a *chi-square distribution* with two degrees of freedom.[7] In particular, we may express the probability density function of $\gamma$ as

$$f(\gamma) = \frac{1}{\gamma_0}\exp\left(-\frac{\gamma}{\gamma_0}\right), \qquad \gamma \geq 0 \tag{8.67}$$

The term $\gamma_0$ is the *mean value of the received signal energy per bit-to-noise spectral density ratio*, which is defined by

$$\begin{aligned} \gamma_0 &= E[\gamma] \\ &= \frac{E_b}{N_0} E[\alpha^2] \end{aligned} \tag{8.68}$$

where $E[\alpha^2]$ is the mean-square value of the Rayleigh-distributed random variable $\alpha$. Substituting Equations (8.64) and (8.67) into (8.66), and carrying out the integration, we get the final result

$$P_e = \frac{1}{2}\left(1 - \sqrt{\frac{\gamma_0}{1 + \gamma_0}}\right) \tag{8.69}$$

Equation (8.69) defines the bit error rate for coherent binary phase-shift keying (PSK) over a flat-flat Rayleigh fading channel. Following a similar approach, we may derive the corresponding bit error rates for coherent binary frequency-shift keying (FSK), binary differential phase-shift keying (DPSK), and noncoherent binary FSK. The results of these evaluations are summarized in Table 8.2. In Figure 8.22, we have used the exact formulas of Table 8.2 to plot the bit error rate versus $\gamma_0$ expressed in decibels. For the sake of comparison, we have also included in Figure 8.22 plots for the bit error rates of coherent binary PSK and noncoherent binary FSK for a nonfading channel. We see that Rayleigh

**TABLE 8.2** ***Bit error rates for binary signaling over a flat-flat Rayleigh fading channel***

| *Type of Signaling* | *Exact Formula for the Bit Error Rate* $P_e$ | *Approximate Formula for the Bit Error Rate, Assuming Large* $\gamma_0$ |
|---|---|---|
| Coherent binary PSK | $\frac{1}{2}\left(1 - \sqrt{\frac{\gamma_0}{1+\gamma_0}}\right)$ | $\frac{1}{4\gamma_0}$ |
| Coherent binary FSK | $\frac{1}{2}\left(1 - \sqrt{\frac{\gamma_0}{2+\gamma_0}}\right)$ | $\frac{1}{2\gamma_0}$ |
| Binary DPSK | $\frac{1}{2(1+\gamma_0)}$ | $\frac{1}{2\gamma_0}$ |
| Noncoherent binary FSK | $\frac{1}{2+\gamma_0}$ | $\frac{1}{\gamma_0}$ |

fading results in a severe degradation in the noise performance of a digital passband transmission system, the degradation being measured in tens of decibels of additional mean signal-to-noise ratio compared to a nonfading channel for the same bit error rate. In particular, for large $\gamma_0$ we may derive the approximate formulas given in the last column of Table 8.2, according to which the asymptotic decrease in the bit error rate with the average signal energy per bit-to-noise spectral density ratio $\gamma_0$ follows an *inverse* law. This behavior is dramatically different from the case of a nonfading channel, for which the asymptotic decrease in the bit error rate with $\gamma_0$ follows an *exponential law*.

The practical implication of this difference is that in a mobile radio environment, we have to provide a large increase in mean signal-to-noise ratio (relative to a nonfading environment), so as to ensure a bit error rate that is low enough for practical use. To meet such a requirement, we have to increase the transmitted power, antenna size, and so on, which can be costly in terms of implementation. Alternatively, we may utilize special modulation and reception techniques that are less vulnerable to fading effects. Among these techniques, the best known and most widely used are the multiple-receiver combining techniques referred to collectively as *diversity*, a brief discussion of which is presented next.

## ■ DIVERSITY TECHNIQUES

Diversity may be viewed as a form of redundancy. In particular, if several replicas of the message signal can be transmitted simultaneously over independently fading channels, then there is a good likelihood that at least one of the received signals will not be severely degraded by fading. There are several methods for making such a provision. In the context of our present discussion, the following diversity techniques are of particular interest:

- Frequency diversity
- Time (signal-repetition) diversity
- Space diversity

In *frequency diversity*, the message signal is transmitted using several carriers that are spaced sufficiently apart form each other to provide independently fading versions of

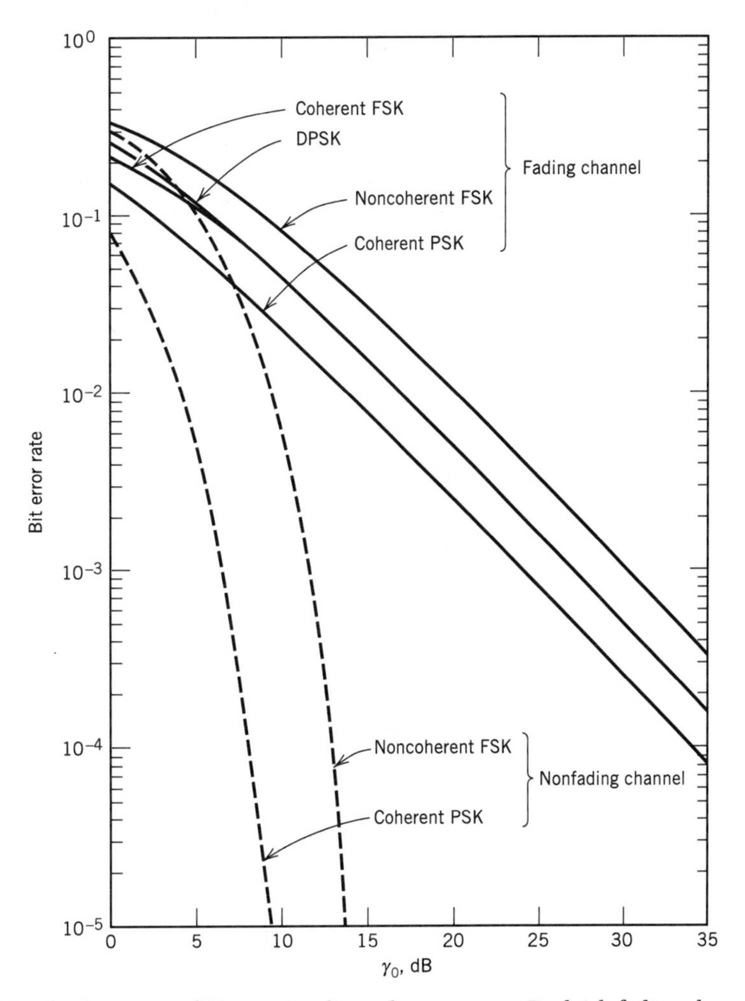


**FIGURE 8.22** Performance of binary signaling schemes over a Rayleigh fading channel, shown as continuous curves; the dashed curves pertain to a nonfading channel.

the signal. This may be accomplished by choosing a frequency spacing equal to or larger than the coherence bandwidth of the channel.

In *time diversity*, the same message signal is transmitted in different time slots, with the spacing between successive time slots being equal to or greater than the coherence time of the channel. Time diversity may be likened to the use of a repetition code for error-control coding. (Error-control coding is discussed in Chapter 10.)

In *space diversity*, multiple transmitting or receiving antennas (or both) are used, with the spacing between adjacent antennas being chosen so as to assure the independence of fading events; this may be satisfied by spacing the adjacent antennas by at least seven times the radio wavelength.

Given that by one of these means we create $L$ independently fading channels, we may then use a *linear diversity combining structure* involving $L$ separate receivers, as depicted in Figure 8.23. The system is designed to compensate only for *short-term effects* of a fading channel. Moreover, it is assumed that *noise-free estimates* of the channel attenuation factors $\{\alpha_\ell\}$ and the channel phase-shifts $\{\phi_\ell\}$ are available. Then, the linear combiner achieves optimum performance for binary data transmission (discussed here for

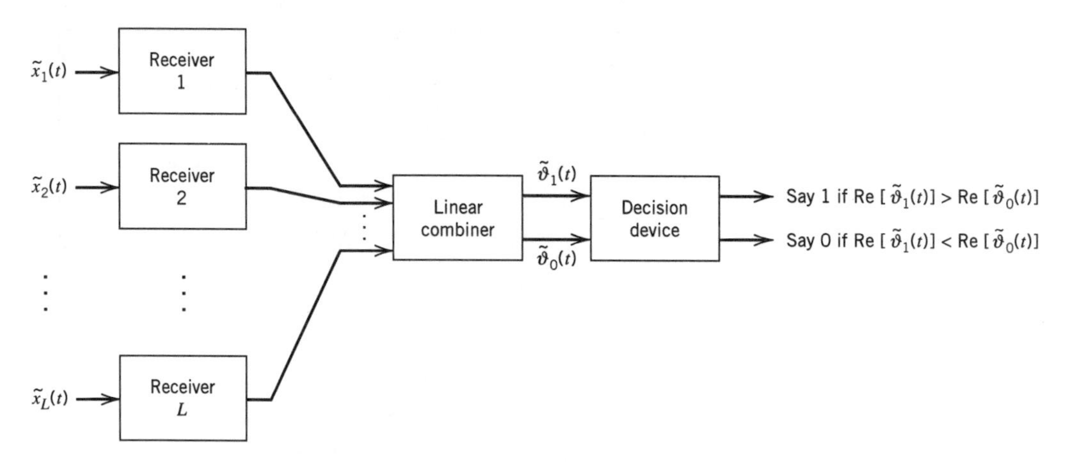


**FIGURE 8.23** Block diagram illustrating the space diversity technique.

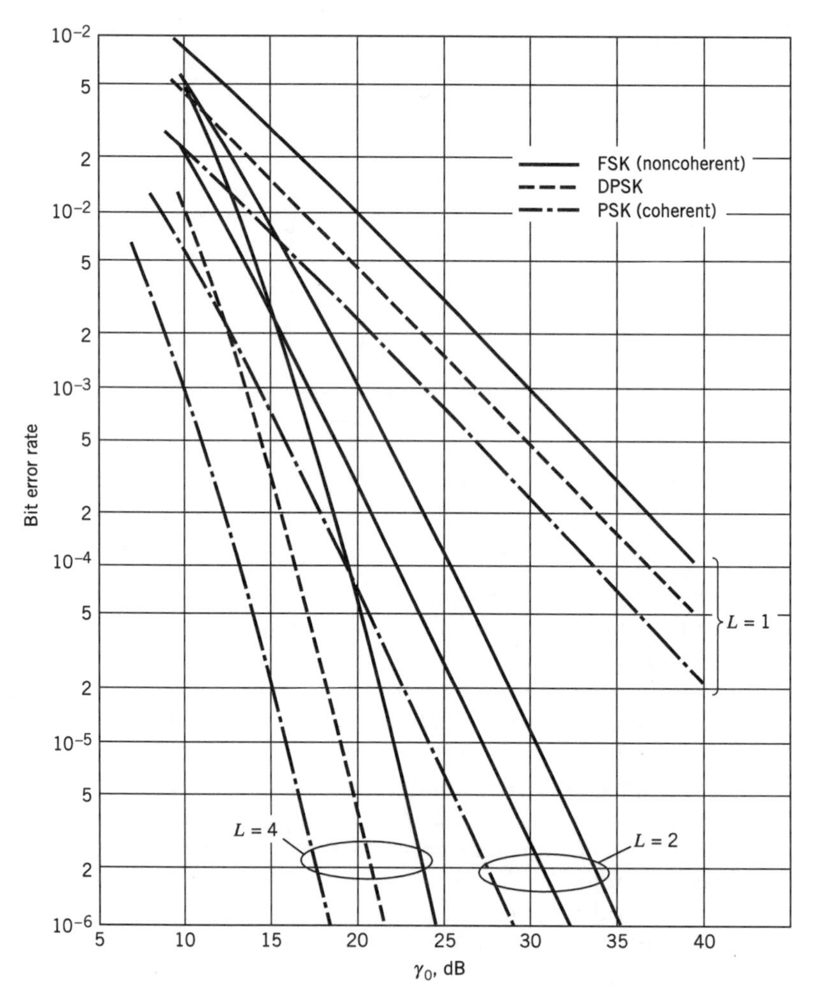


**FIGURE 8.24** Performance of binary signaling schemes with diversity. (From Proakis, 1995, with permission of McGraw-Hill.)

the purpose of illustration) by proceeding as follows: The output of the $k$th matched filter in the $\ell$th receiver, $\tilde{v}_{\ell k}(t)$, is multiplied by $\alpha_\ell \exp(j\phi_\ell)$ that represents the complex conjugate of the $\ell$th channel gain, where $\ell = 1, 2, \ldots, L$, and $k = 0, 1$. Thus, the linear combiner results in two output complex envelopes defined by

$$\tilde{v}_k(t) = \sum_{\ell=1}^{L} \alpha_\ell \exp(j\phi_\ell)\tilde{v}_{\ell k}(t), \qquad k = 0, 1 \tag{8.70}$$

according to which $\alpha_\ell \exp(j\phi_\ell)$ plays the role of a *weighting factor*. One output complex envelope $\tilde{v}_0(t)$ corresponds to the transmission of symbol 0, and the other $\tilde{v}_1(t)$ corresponds to the transmission of symbol 1. The real parts of $\tilde{v}_0(t)$ and $\tilde{v}_1(t)$ are then used in the decision-making process. The situation described here applies to binary FSK. In the case of binary PSK, only a single matched filter is needed, in which case the linear combiner produces a single output complex envelope. Here again, however, the real part of the combiner output is used in the decision-making process.

In the linear combiner described herein, the "instantaneous" output signal-to-noise ratio (SNR) is the sum of the instantaneous SNRs on the individual diversity branches (channels). This optimum form of a linear combiner is therefore referred to as a *maximal-ratio combiner;* see Problem 8.17.

Figure 8.24 shows the noise performance of coherent binary PSK, binary DPSK, and noncoherent binary FSK for $L = 2, 4$ independently fading channels. For the sake of comparison, we have also included in this figure the corresponding graphs for a fading channel with no diversity (i.e., $L = 1$). Figure 8.24 clearly illustrates the effectiveness of diversity as a means of mitigating the short-term effects of Rayleigh fading.

## 8.8 *TDMA and CDMA Wireless Communication Systems*[8]

In wireless communications, as with ordinary telephony, a user would like to talk and listen simultaneously. To cater to this natural desire, some form of duplexing is required. One way in which this requirement can be satisfied is to provide two frequency bands, one for the forward link from the base station to a mobile and the other for the reverse link from the mobile to the base station. As pointed out earlier, in North America the subband 869–894 MHz is used for the forward link, and the subband 824–849 MHz is used for the reverse link. This form of duplexing is called *frequency division duplexing* (FDD). Indeed, FDD is an integral part of the two widely used wireless communication systems summarized in Table 8.3.

The first of these systems, namely, GSM, uses TDMA. From Section 8.2 we recall that in a TDMA system each subscriber is permitted to access the radio channel during a set of predetermined time slots, during which time that particular subscriber will have full use of the channel. Consequently, data are transmitted over the channel in *bursts*, as shown in the *frame structure* of Figure 8.25. The basic frame of GSM is composed of eight 577 $\mu$s slots. The 1-bit *flag* preceding each data burst of 57 bits is used to identify whether the data bits are digitized speech or some other information-bearing signal. The 3 *tail* bits, all logical zeros, are used in convolutional decoding of the channel-encoded data bits. (Convolutional codes are discussed in Chapter 10.) The 26-bit training sequence in the middle of the time slot is used for channel equalization. Finally, the *guard time*, occupying 8.25 bits, is included at the end of each slot to prevent data bursts received at the base

**TABLE 8.3** ***Summary of two widely used wireless communication systems***

| *Item* | *GSM** | *IS-95†* | *Comments* |
|---|---|---|---|
| Number of duplex channels | 125 | 20 | CDMA assumes 12.5 MHz in each direction; see the next line |
| Channel bandwidth (kHz) | 200 | 1,250 | |
| Type of multiple access | TDMA | CDMA | |
| Access users per channel | 8 | 20 to 35 | A TDMA system is *deterministic* in that the number of access users per channel is defined by the number of available time slots. On the other hand, a CDMA system is *interference-limited* in that it has a *soft* limit on the number of access users per channel. |
| Modulation type | GMSK | BPSK/QPSK | In CDMA, data are modulated as BPSK, but the spreading is QPSK |
| Data rate (kb/s) | 270.833 | 9.6 or 14.4 | |
| Frame period (ms) | 4.615 | 20 | For CDMA, the frame period equals that of the speech codec (coder/decoder) |

*GSM stands for *Global System for Mobile Communications*; originally, it was introduced as an acronym for *Groupe de travail Spéciale pour les services Mobiles.*

†IS stands for *Interim Standard.*

station from mobiles from overlapping with each other; this is achieved by transmitting no signal at all during the guard time. With each slot consisting of 156.25 bits, of which 40.25 bits are overhead (ignoring the 2 flag bits), the *frame efficiency* of GSM is

$$\left(1 - \frac{40.25}{156.25}\right) \times 100 = 74.24\%$$

The second wireless communication system, IS-95, summarized in Table 8.3 uses CDMA. From Section 8.2 we recall that in CDMA, each subscriber is assigned a distinct spreading code (PN sequence), thereby permitting the subscriber full access to the channel all of the time. Consequently, in a CDMA system we have a new form of interference called *multiple-access interference* (MAI), which arises because of deviation of the spreading codes from perfect orthogonality. A related phenomenon that needs attention is the *near-far problem*, which occurs if the received signals from the mobile units do not have equal power at the base station. In such a situation, the strongest received signal from a mobile user captures the demodulation process at the base station to the detriment of the

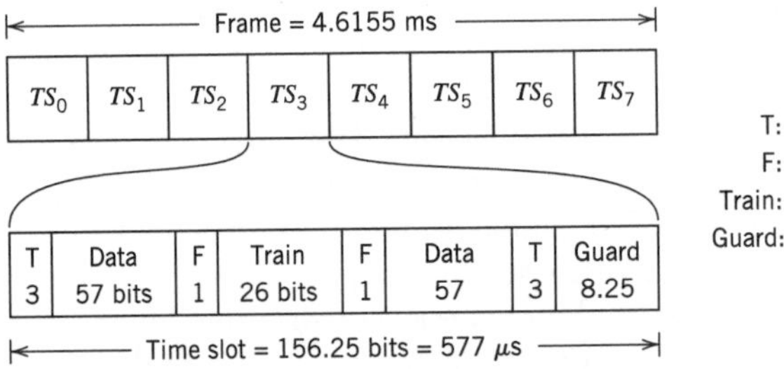


**FIGURE 8.25** Frame structure of the GSM wireless communication system.

other users. To overcome the near-far problem, it is customary to use *power control* at the base station, whereby the base station maintains control over the power level of the transmitted signal from every mobile being served by that base station. The use of power control is particularly important in CDMA systems for another reason. A goal of multiple-access systems is to maximize *system capacity*, which is defined as the largest possible number of users that can be reliably served by the system, given prescribed resources. Clearly, system capacity is compromised if each mobile is free to raise its transmitted power level regardless of other users, since that increase in transmitted power will, in turn, raise the level of multiple-access interference in the system. To maximize system capacity, it is therefore essential that each mobile's transmitter be under the control of the serving base station so that the signal-to-interference ratio is maintained at the minimum acceptable level needed for reliable service.

## RAKE Receiver

A discussion of wireless communications using CDMA would be incomplete without a description of the *RAKE receiver.*[9] The RAKE receiver was originally developed in the 1950s as a "diversity" receiver designed expressly to equalize the effect of multipath. First, and foremost, it is recognized that useful information about the transmitted signal is contained in the multipath component of the received signal. Thus, taking the viewpoint that multipath may be approximated as a linear combination of differently delayed echoes, the RAKE receiver seeks to combat the effect of multipath by using a correlation method to detect the echo signals individually and then adding them algebraically. In this way, intersymbol interference due to multipath is dealt with by reinserting different delays into the detected echoes so that they perform a constructive rather than destructive role.

Figure 8.26 shows the basic idea behind the RAKE receiver. The receiver consists of a number of *correlators* connected in parallel and operating in a synchronous fashion. Each correlator has two inputs: (1) a delayed version of the received signal and (2) a replica of the pseudo-noise (PN) sequence used as the spreading code to generate the spread-

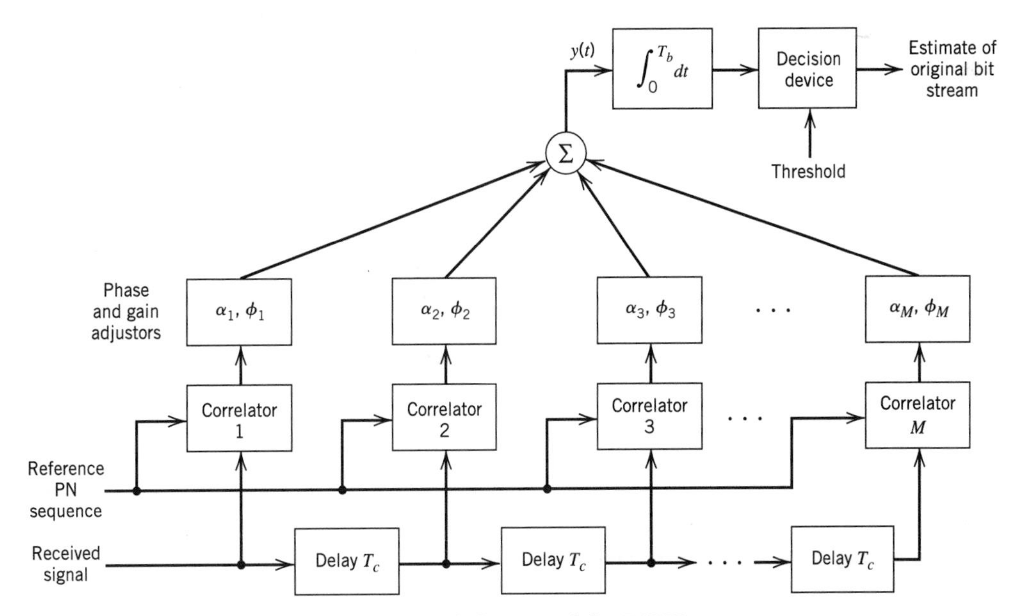


**FIGURE 8.26** Block diagram of the RAKE receiver.

spectrum modulated signal at the transmitter. In effect, the PN sequence acts as a "reference signal." Let the nominal bandwidth of the PN sequence be denoted as $W = 1/T_c$, where $T_c$ is the chip duration. From the discussion of spread-spectrum modulation presented in Chapter 7, we recall that the autocorrelation function of a PN sequence has a single peak of width $1/W$, and it disappears toward zero elsewhere inside one period of the PN sequence (i.e., one symbol period). Thus we need only make the bandwidth $W$ of the PN sequence sufficiently large to "identify" the significant echoes in the received signal. To be sure that the correlator outputs all add constructively, two other operations are performed in the receiver by the functional blocks labeled "phase and gain adjustors":

1. An appropriate delay is introduced into each correlator output so that the phase angles of the correlator outputs are in agreement with each other.
2. The correlator outputs are weighted so that the correlators responding to strong paths in the multipath environment have their contributions accentuated, while the correlators not synchronizing with any significant path are correspondingly suppressed.

The weighting coefficients, $\alpha_k$, are computed in accordance with the *maximal ratio combining principle*:[10]

> The signal-to-noise ratio of a weighted sum, where each element of the sum consists of a signal plus additive noise of fixed power, is maximized when the amplitude weighting is performed in proportion to the pertinent signal strength.

The linear combiner output is

$$y(t) = \sum_{k=1}^{M} \alpha_k z_k(t) \tag{8.71}$$

where $z_k(t)$ is the phase-compensated output of the $k$th correlator, and $M$ is the number of correlators in the receiver. Provided we use enough correlators in the receiver to span a region of delays sufficiently wide to encompass all the significant echoes that are likely to occur in the multipath environment, the output $y(t)$ behaves essentially as though there was a single propagation path between the transmitter and receiver rather than a series of multiple paths spread in time.

To simplify the presentation, the receiver of Figure 8.26 assumes the use of binary phase-shift keying in performing spread-spectrum modulation at the transmitter. Thus the final operation performed in Figure 8.26 is that of integrating the linear combiner output $y(t)$ over the bit interval $T_b$ and then determining whether binary symbol 1 or 0 was transmitted in that bit interval.

The RAKE receiver derives its name from the fact that the bank of parallel correlators has an appearance similar to the fingers of a rake. Because spread spectrum modulation is basic to the operation of CDMA wireless communications, it is natural for the RAKE receiver to be central to the design of the receiver used in this type of multiuser radio communication.[11]

## 8.9 *Source Coding of Speech for Wireless Communications*

For the efficient use of channel bandwidth, digital wireless communication systems, be they of the TDMA or CDMA type, rely on the use of *speech coding* to remove almost all of the natural redundancy in speech, while maintaining a high-quality speech on decoding. The common approach is to use source coding, which, in one form or another, exploits the *linear predictive coding (LPC)* of speech.

In this section, we describe two different techniques for speech coding: multi-pulse excited LPC and code-excited LPC, versions of which are used in GSM and IS-95, respectively. Our treatment of both of these speech coding techniques is in conceptual terms.[12]

## ■ MULTI-PULSE EXCITED LPC

This form of speech coding exploits the *principle of analysis by synthesis*, which means that the encoder includes a replica of the decoder in its design. Specifically, the encoder consists of three main parts as indicated in Figure 8.27*a*:

1. *Synthesis filter* for the predictive modeling of speech. It may consist of an all-pole filter (i.e., a filter whose transfer function has poles only), which is designed to model the *short-term* spectral envelope of speech; the term *short-term* refers to the fact that the filter parameters are computed on the basis of predicting the present sample of the speech signal using eight to sixteen previous samples. The synthesis filter may also include a *long-term* predictor for modeling the fine structure of the speech spectrum; in such a case, the long-term predictor is connected in cascade with the short-term predictor. In any event, the function of the synthesis filter is to produce a synthetic version of the original speech that is of high quality.
2. *Excitation generator* for producing the excitation applied to the synthesis filter. The excitation consists of a definite number of pulses every 5 to 15 ms. The amplitudes and positions of the individual pulses are adjustable.
3. *Error minimization* for optimizing the perceptually weighted error between the original speech and synthesized speech. The aim of this minimization is to optimize the amplitudes and positions of the pulses used in the excitation. Typically, a mean-square error criterion is used for the minimization.

Thus, as shown in Figure 8.27*a*, the three parts of the encoder form a *closed-loop optimization procedure*, which permits the encoder to operate at a bit rate below 16 kb/s, while maintaining high-quality speech.

The encoding procedure itself has two main steps:

- The free parameters of the synthesis filter are computed using the actual speech samples as input. This computation is performed outside the optimization loop over

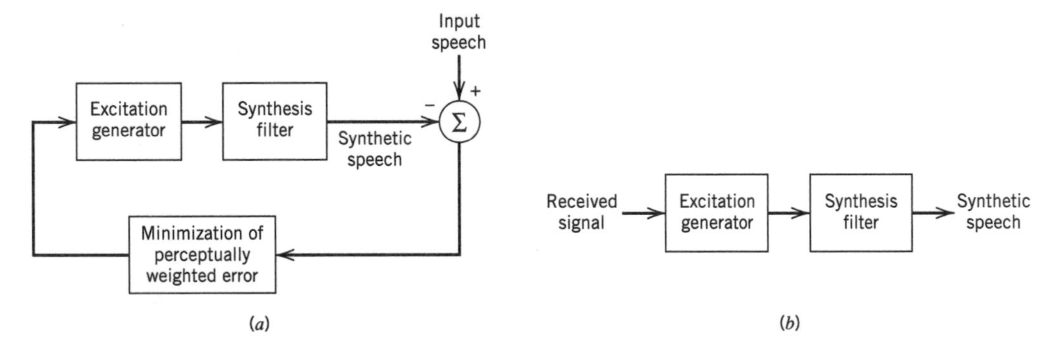


**FIGURE 8.27** Multi-pulse excited linear predictive codec. (*a*) Encoder. (*b*) Decoder whose input (the received signal) consists of quantized filter parameters and quantized excitation as produced by the encoder.

a period of 10 to 30 ms, during which the speech signal is treated as pseudo-stationary.

- The optimum excitation for the synthesis filter is computed by minimizing the perceptually weighted error with the loop closed as in Figure 8.27*a*.

Thus the speech samples are divided into *frames* (10 to 30 ms long) for computing the filter parameters, and each frame is divided further into *subframes* (5 to 15 ms) for optimizing the excitation. The quantized filter parameters and quantized excitation constitute the transmitted signal.

Note that by first permitting the filter parameters to vary from one frame to the next, and then permitting the excitation to vary from one subframe to the next, the encoder is enabled to track the nonstationary behavior of speech, albeit on a batch-by-batch basis.

The decoder, located in the receiver, consists simply of two parts: excitation generator and synthesis filter, as shown in Figure 8.27*b*. These two parts are identical to the corresponding ones in the encoder. The function of the decoder is to use the received signal to produce a synthetic version of the original speech signal. This is achieved by passing the decoded excitation through the synthesis filter whose parameters are set equal to those in the encoder.

To reduce the computational complexity of the codec (i.e., contraction of coder/decoder), the intervals between the individual pulses in the excitation are constrained to assume a common value. The resulting analysis-by-synthesis codec is said to have a *regular-pulse excitation.*

## ■ CODE-EXCITED LPC

Figure 8.28 shows the block diagram of the *code-excited LPC*, commonly referred to as CELP. The distinguishing feature of CELP is the use of a predetermined *codebook* of stochastic (zero-mean white Gaussian) vectors as the source of excitation for the synthesis filter. The synthesis filter itself consists of two all-pole filters connected in cascade, one of which performs short-term prediction and the other performs long-term prediction.

As with the multi-pulse excited LPC, the free parameters of the synthesis filter are computed first, using the actual speech samples as input. Next, the choice of a particular vector (code) stored in the excitation codebook and the gain factor G in Figure 8.28 is optimized by minimizing the average power of the perceptually weighted error between

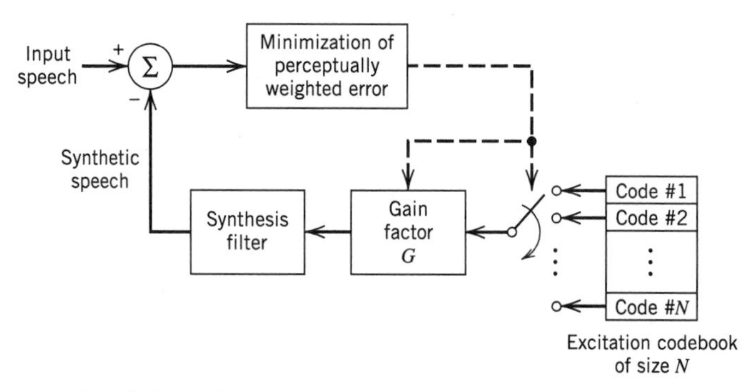


**FIGURE 8.28** Encoder of the code-excited linear predictive codec (CELP): the transmitted signal consists of the address of the code selected from the codebook, quantized G, and quantized filter parameters.

the original speech and synthesized speech (i.e., output of the synthesis filter). The address of the stochastic vector selected from the codebook and the corresponding quantized gain factor, together with the quantized filter parameters, constitute the transmitted signal.

An identical copy of the codebook is made available to the decoder, and likewise for the synthesis filter. Hence, given the received signal, the decoder is enabled to parameterize its own synthesis filter and determine the appropriate excitation for the synthesis filter, thereby producing a synthetic version of the original speech signal.

CELP is capable of producing good-quality speech at bit rates below 8 kb/s. However, its computational complexity is intensive because of the exhaustive search of the excitation codebook. In particular, the weighted synthesized speech in the encoder has to be computed for all the entries in the codebook and then compared with the weighted original speech. Nevertheless, real-time implementation of CELP codecs has been made possible by virtue of advances in digital signal processing and VLSI technology.

## 8.10 *Adaptive Antenna Arrays for Wireless Communications*[13]

The goal of wireless communications is to allow as many users as possible to communicate reliably without regard to location and mobility. From the discussion presented in Sections 8.5 and 8.6, we find that this goal is seriously impeded by three major channel impairments:

1. *Multipath* can cause severe fading due to phase cancellation between different propagation paths. Fading leads to a reduction in available signal power and therefore a degraded noise performance.
2. *Delay spread* results from differences in propagation delays among the multiple propagation paths. When the delay spread exceeds about 10 percent of the symbol duration, the intersymbol interference experienced by the received signal reaches a significant level, thereby causing a reduction in the attainable data rate.
3. *Co-channel interference* arises in cellular systems where the available frequency channels are divided into different sets, with each set being assigned to a specific cell and with several cells in the system using the same set of frequencies. Co-channel interference limits the *system capacity* (i.e., the largest possible number of users that can be reliably served by the system).

Typically, cellular systems use 120° sectorization at each base station, and only one user accesses a sector of a base station at a given frequency. We may combat the effects of multipath fading and co-channel interference at the base station by using three identical but separate *antenna arrays*, one for each section of the base station. The compensation of delay spread is considered later in the section. Figure 8.29 shows the block diagram of an *array signal processor*, where it is assumed that there are $N$ users whose signals are received at a particular sector of the base station, and the array for that sector consists of $M$ identical antenna elements. A particular user is treated as the one of interest, and the remaining $N$-1 users give rise to co-channel interference. In addition to the co-channel interference, each component of the array signal processor's input is corrupted by additive white Gaussian noise (AWGN). The analysis presented herein is for baseband signals, which, in general, are complex valued. This, in turn, means that both the channel and array signal processor require complex characterizations of their own. The structure depicted in Figure 8.29 is drawn for one output pertaining to the user of

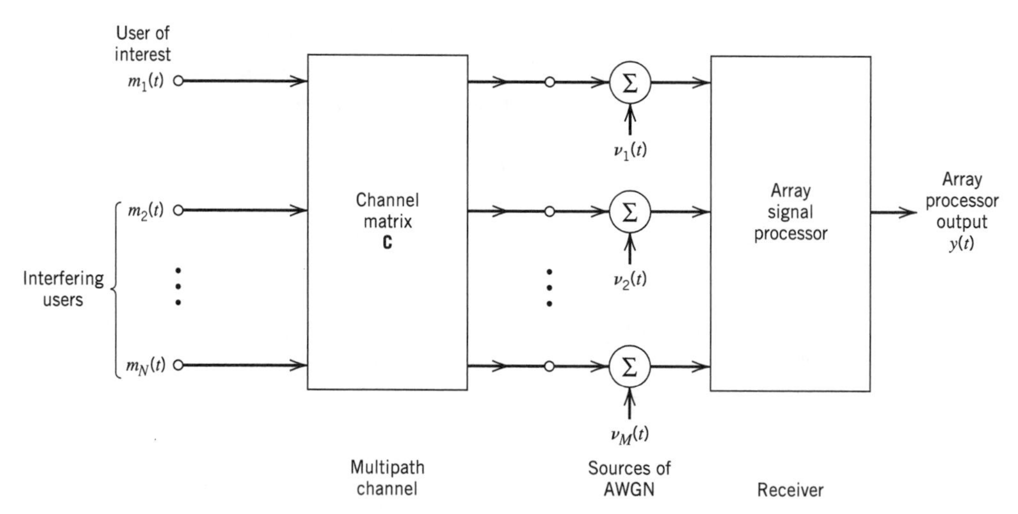


**FIGURE 8.29** Block diagram of array signal processor that involves $M$ antenna elements, and that is being driven by a multipath channel.

interest. The array signal processor is duplicated for users at other frequencies at the base station.

The multipath channel is characterized by the channel matrix, which is denoted by $\mathbf{C}$. The matrix $\mathbf{C}$ has dimensions $M$-by-$N$ and may therefore be expanded into $N$ column vectors, as shown by

$$\mathbf{C} = [\mathbf{c}_1, \mathbf{c}_2, \ldots, \mathbf{c}_N] \tag{8.72}$$

where each column vector is of dimension $M$.

Given the configuration described in Figure 8.29, the goal is to design a *linear array signal processor* for the receiver, which satisfies two requirements:

1. The co-channel interference produced by the $N$-1 interfering users is cancelled.
2. The output signal-to-noise ratio (SNR) for the user of interest is maximized.

Hereafter, these two requirements are referred to as design requirements 1 and 2.

To proceed with this design task, it is assumed that the multipath channel is described by flat Rayleigh fading. Then, in light of the material presented in Section 8.7, we find that the use of diversity permits the treatment of the column vectors $\mathbf{c}_1, \mathbf{c}_2, \ldots, \mathbf{c}_N$ as *linearly independent*, which is justified provided that the spacing between antenna elements of the array is large enough (e.g., seven times the wavelength) for independent fading. To simplify the presentation, we suppose that user 1 is the user of interest and the remaining $N - 1$ users are responsible for co-channel interference, as indicated in Figure 8.29. The key design issue is how to find the *weight vector* denoted by $\mathbf{w}$, which characterizes the array signal processor. To that end, we may proceed as follows:

1. We choose the $M$-dimensional weight vector $\mathbf{w}$ to be orthogonal to the vectors $\mathbf{c}_2, \ldots, \mathbf{c}_N$, which are associated with the interfering users. This choice fulfills design requirement 1 (i.e., cancellation of co-channel interference).
2. To satisfy design requirement 2 (i.e., maximization of the SNR), we will briefly digress from the issue at hand to introduce the notion of a subspace. Given a *vector*

*space*, or just *space*, formed by a set of linearly independent vectors, a *subspace* of the space is a subset that satisfies two conditions:[14]

(i) If we add any two vectors $\mathbf{z}_1$ and $\mathbf{z}_2$ in the subspace, their sum $\mathbf{z}_1$ and $\mathbf{z}_2$ is still in the subspace.

(ii) If we multiply any vector $\mathbf{z}$ in the subspace by any scalar $a$, the multiple $a\mathbf{z}$ is still in the subspace.

Returning to the issue of how to maximize the output SNR for user 1, we first construct a subspace denoted by $\mathcal{W}$, whose dimension is equal to the difference between the number of antenna elements and the number of interfering users, that is, $M - (N - 1) = M - N + 1$. Next, we project the complex conjugate of the channel vector $\mathbf{c}_1$ (pertaining to user 1) onto the subspace $\mathcal{W}$. The projection so computed defines the weight vector $\mathbf{w}$.

### ▶ EXAMPLE 8.3

To illustrate the two-step subspace method for determining the weight vector $\mathbf{w}$, consider the simple example of a system involving two users characterized by the channel vectors $\mathbf{c}_1$ and $\mathbf{c}_2$, and an antenna array consisting of three elements; that is, $N = 2$ and $M = 3$. Then, for this example, the subspace $\mathcal{W}$ is two-dimensional, as shown by

$$M - N + 1 = 3 - 2 + 1 = 2$$

With user 1 viewed as the user of interest and user 2 viewed as the interferer, we may construct the signal-space diagram shown in Figure 8.30. The subspace $\mathcal{W}$, shown shaded in this figure, is orthogonal to channel vector $\mathbf{c}_2$. The weight vector $\mathbf{w}$ of the array signal processor is determined by the projection of the complex-conjugated channel vector of user 1, that is, $\mathbf{c}_1^*$, onto the subspace $\mathcal{W}$, as depicted in Figure 8.30. ◀

The important conclusion drawn from this discussion is that a linear receiver using optimum combining with $M$ antenna elements and involving $N - 1$ interfering users has the same performance as a linear receiver with $M - N + 1$ antenna elements without interference, independent of the multipath environment. For this equivalence to be realized,

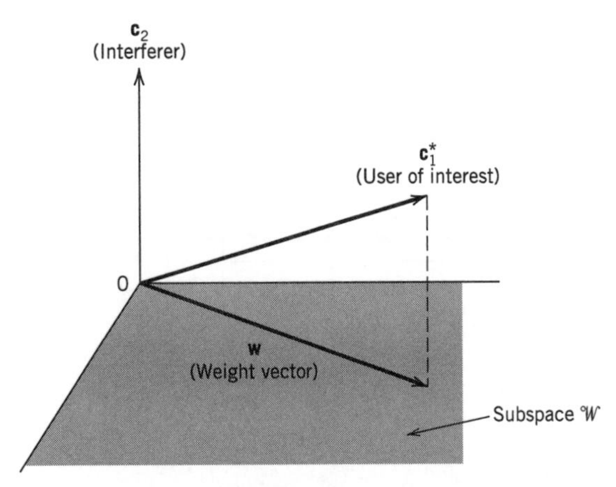


**FIGURE 8.30** Signal-space diagram for Example 8.3, involving a user of interest, a single interferer, and an antenna array of 3 elements. The subspace $\mathcal{W}$, shown shaded, is two-dimensional in this example.

we of course require that $M > N - 1$. Provided that this condition is satisfied, the receiver cancels the co-channel interference with a diversity improvement equal to $M - N + 1$, which represents an $N$-fold increase in system capacity.

The design of an array signal processor in accordance with the two-step subspace procedure described herein is of the *zero-forcing* kind. We say so because, given $M$ antenna elements, the array has enough degrees of freedom to *force* the output due to the $N - 1$ interfering users represented by the linearly independent channel vectors $\mathbf{c}_2, \ldots, \mathbf{c}_M$ to zero so long as $M$ is greater than $N - 1$. Note also that this procedure includes $N = 1$ (i.e., a single user with no interfering users) as a special case. In this case, the channel matrix consists of vector $\mathbf{c}_1$, which lies in the subspace $\mathcal{W}$, and the zero-forcing solution $\mathbf{w}$ equals $\mathbf{c}_1^*$.

The analysis presented thus far has been entirely of a *spatial* kind, which ignores the effect of delay spread. What if the delay spread is significant compared to the symbol duration and cannot therefore be ignored? Recognizing that delay spread is responsible for intersymbol interference, we may, in light of the material presented in Chapter 4 on the equalization of a telephone channel, incorporate a *linear* equalizer in each antenna branch of the array to compensate for delay spread. The resulting array signal processor takes the form shown in Figure 8.31, which combines temporal and spatial processing.

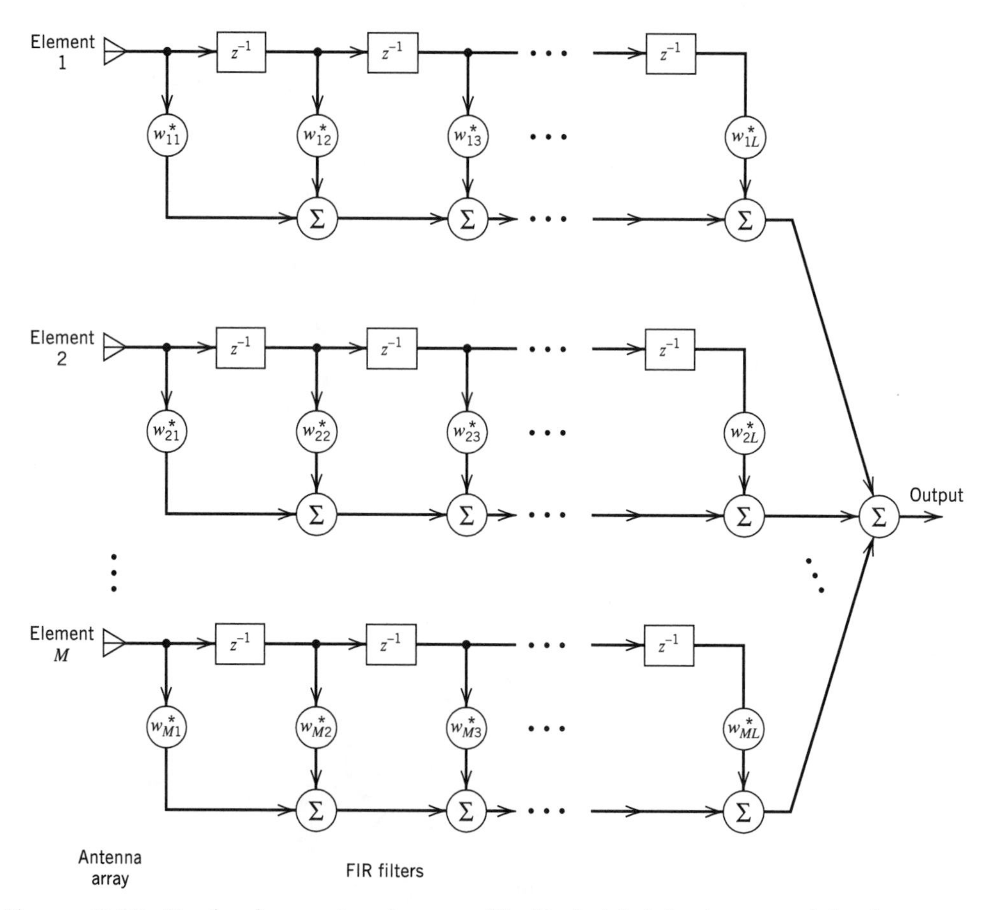


**FIGURE 8.31** Baseband space-time processor. The blocks labeled $z^{-1}$ are unit-delay elements with each delay being equal to the symbol period. The filter coefficients are complex valued. The FIR filters are all assumed to be of length $L$.

Spatial processing is provided by the antenna array, and the temporal processing is provided by a bank of finite-duration impulse response (FIR) filters. For obvious reasons, this structure is called a *space-time processor.*[15]

## ■ Adaptive Antenna Array

The subspace design procedure for the array signal processor in Figure 8.29 assumes that the channel impairments are stationary, and that we have knowledge of the channel matrix C. In reality, however, multipath fading, delay spread, and co-channel interference are all nonstationary in their own individual ways. Also, the channel characterization may be unknown. To deal with these practical issues, we need to make the receiving array signal processor in Figure 8.29 adaptive. Bearing in mind the scope of this book, we confine the discussion to adaptive spatial processing, assuming that the delay spread is negligible. We further assume that the multipath fading phenomenon is slow enough to justify the *least-mean-square (LMS) algorithm* to perform the adaptation.

Figure 8.32 shows the structure of an *adaptive antenna array*, where the output of each antenna element is multiplied by an adjustable (controllable) weight, and then the weighted elemental outputs of the array are summed to produce the array output signal. The adaptive antenna array does not require knowledge of the direction of arrival of the desired signal originating from a user of interest as long as the system is supplied with a *reference signal*, which is *correlated* with the desired signal. The output signal of the array is subtracted from the reference signal to generate an *error signal*, which is used to apply the appropriate adjustments to the elemental weights of the array. In this way, a feedback system to control the elemental weights is built into the operation of the antenna array, thereby making it adaptive to changes in the environment. Note that the block diagram of Figure 8.32 is drawn for baseband processing, hence the complex conjugation of the elemental weights. In a practical system, a quadrature hybrid is used for each antenna element of the array to split the complex-valued received signal at each element into two components: one real and the other imaginary. The use of a hybrid has been omitted in Figure 8.32 to simplify the diagram.

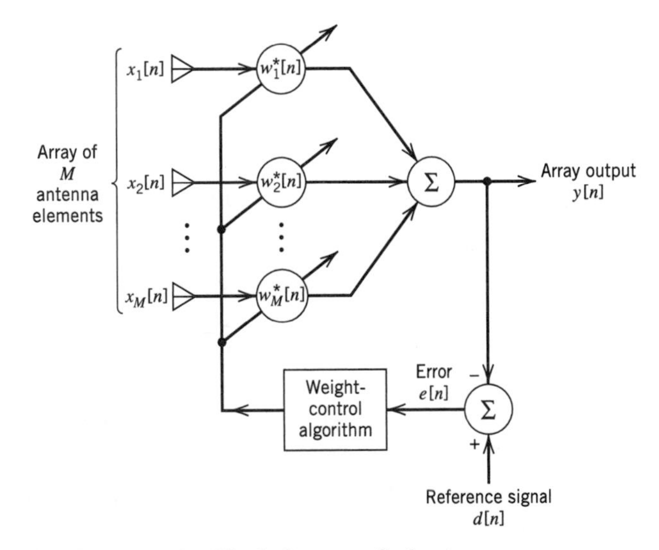


**Figure 8.32** Block diagram of adaptive antenna array.

To optimize the performance of the adaptive antenna array, it is customary to use the *mean-square error*

$$J = E[|e[n]|^2] \tag{8.73}$$

as the cost function to be *minimized*. The $e[n]$ is the error signal at time $t = nT$, where $T$ is the symbol period and $n$ is an integer serving as discrete time. Minimization of the cost function $J$ suppresses the interfering signals and enhances the desired signal in the array output. However, the LMS algorithm minimizes the instantaneous value of the cost function $J$ and, through successive iterations, it strives to reach the minimum mean-square error (MMSE) (i.e., optimum solution for the elemental weights). In light of the discussion presented in Chapter 4 on temporal equalizers, which carries over to the spatial domain, we may say that an adaptive antenna array based on the minimum mean-square error criterion is highly likely to provide a better solution than one based on the zero-forcing criterion embodied in the two-step subspace method.

Let $x_k[n]$ denote the output of the $k$th element in the array at discrete time $n$, and let $w_k[n]$ denote the corresponding value of the weight connected to this element. The output signal of the array (consisting of $M$ antenna elements) is therefore

$$y[n] = \sum_{k=1}^{M} w_k^*[n]x_k[n] \tag{8.74}$$

where $w_k^*[n]x_k[n]$ is the inner product of the complex-valued quantities $w_k[n]$ and $x_k[n]$. Denoting the reference signal as $d[n]$, we may evaluate the error signal as

$$e[n] = d[n] - y[n] \tag{8.75}$$

Hence, the adjustment applied to the $k$th elemental weight is

$$\Delta w_k[n] = \mu e^*[n]x_k[n], \qquad k = 1, 2, \ldots, M \tag{8.76}$$

where $\mu$ is the *step-size parameter*, and the updated value of this weight is

$$w_k[n+1] = w_k[n] + \Delta w_k[n], \qquad k = 1, 2, \ldots, M \tag{8.77}$$

Equations (8.74)–(8.77), in that order, constitute the *complex LMS algorithm*, which includes the LMS algorithm for real signals (studied in Chapters 3 and 4) as a special case. The algorithm is initiated by setting $w_k[0] = 0$ for all $k$. The derivation of the complex LMS algorithm is posed as Problem 8.19.

The advantages of an adaptive antenna array using the complex LMS algorithm are three-fold:

- Simplicity of implementation.
- Linear growth in complexity with the number of antenna elements.
- Robust performance with respect to disturbances.

However, the system suffers from the following drawbacks:

- Slow rate of convergence, which is typically ten times the number of weights. This limits the use of the complex LMS algorithm to a slow-fading environment, for which the Doppler spread is small compared to the reciprocal of the duration of the observation interval.
- Sensitivity of the convergence behavior to variations in the reference signal and co-channel interference powers.

These limitations of the complex LMS algorithm can be overcome by using an algorithm known as *direct matrix inversion* (DMI), which follows directly from the Wiener filter discussed in Chapter 4; see Problem 8.21. Unlike the LMS algorithm, the DMI algorithm operates in the *batch* mode in that the computation of the elemental weights is based on a batch of $K$ snapshots. The batch size $K$ is chosen as a compromise between two conflicting requirements:

- The size $K$ should be small enough for the batch of snapshots used in the computation to be justifiably treated as pseudo-stationary.
- The size $K$ should be large enough for the computed values of the elemental weights to approach the MMSE solution.

The DMI algorithm is the optimum combining technique for array antennas currently deployed in many base stations today. The DMI algorithm may be reformulated for recursive computation,[16] if so desired.

When the teletraffic is high, the base stations are ordinarily configured as microcells, which are small cells such as an office floor or a station deployed along a highway with directional antennas. In such a configuration, there are many inexpensive base stations in close proximity to each other. The use of adaptive antenna arrays provides the means for an alternative configuration where there are fewer (but more expensive) base stations and further apart from each other than in the corresponding microcellular system.

## 8.11 *Summary and Discussion*

In this chapter, we discussed two important types of multiuser communications: satellite communications and wireless communications. Satellite communication systems offer global coverage, whereas wireless communication systems offer mobility. The global coverage and mobility offered by these two communication systems have profoundly transformed the way we communicate, both locally and globally.

Although satellite communication and wireless communication systems function in entirely different ways, both rely on radio propagation to link the receiver to the transmitter. In satellite communications, we have an *uplink* from an earth terminal to the satellite transponder and a *downlink* from the satellite to another earth terminal. The satellite operates like a repeater in the sky. Moreover, with the satellite positioned in a geostationary orbit, the uplink and downlink operate as line-of-sight paths of fixed lengths. Accordingly, the satellite communication channel, encompassing both of these links, is closely modeled as an *additive white Gaussian noise (AWGN) channel.*

The wireless communication system also has two links of its own: an *uplink*, or *reverse link*, for the mobile-to-base station transmission, and a *downlink*, or *forward link*, for the base station-to-mobile transmission. The base station is fixed, being located at the center or on the edge of a coverage region; it consists of radio channels, and transmitter, and receiver antennas mounted on a tower. Three major sources of degradation in wireless communications, discussed in the chapter, are co-channel interference, fading, and delay spread; the latter two are byproducts of multipath. A common characteristic of these channel impairments is that they are all *signal-dependent phenomena.* Unlike the ubiquitous channel noise, the degrading effects of interference and multipath cannot therefore be combatted by simply increasing the transmitted signal power. Rather, both interference and multipath require the use of specialized techniques, tailor-made to their particular physical characteristics. These specialized techniques include diversity, adaptive array antennas, and the RAKE receiver.

We close the discussion with remarks contrasting wireless communications to wired communications. From Chapter 3 we recall that a major source of concern in wired communication systems is noise; these systems have sufficient channel bandwidth to permit the use of pulse-code modulation (PCM) as the standard method for converting speech into a 64 kb/s stream, which provides the basic data for an almost noise-free performance. In wireless communications, on the other hand, channel bandwidth is a precious resource, the conservation of which necessitates the use of spectrally efficient speech coding techniques to produce toll-quality digitized speech at rates that are a small fraction of the PCM rate. Unfortunately, the waveform coders exemplified by adaptive differential pulse-code modulation, discussed in Chapter 3, do not satisfy this stringent requirement. The preferred approach is to use the spectrally efficient source-coding techniques: multi-pulse excited linear predictive coding (LPC) or its regular-pulse excited variant, and code-excited LPC (CELP); these source coding techniques produce bit rates below 16 kb/s by removing almost all of the natural redundancy in speech, while maintaining high-quality speech, albeit of a synthetic kind. To provide protection against noise, channel coding is used whereby redundant bits are inserted into the transmitted data stream in a controlled manner. The use of channel coding also helps in other ways: It extends the range of low-power handsets as well as battery life. Channel coding is discussed in Chapter 10.

## NOTES AND REFERENCES

1. For detailed treatment of satellite communications and related issues, see the following books: Sklar (1988), Pratt and Bostian (1986), Wu (1984), Bhargava et al. (1981), and Spilker, Jr. (1977). The first, third, fourth, and fifth books emphasize the use of satellites for digital communications. The book by Pratt and Bostian presents a broad treatment of satellite communications, emphasizing such diverse topics as radio-wave propagation, antennas, orbital mechanics, signal processing, and radio electronics.

2. Link budget analysis is discussed in the books by Sklar (1988) and Anderson (1999); for satellite communications, it is discussed in Bhargava et al. (1981).

3. For the fundamentals of antennas, see the book by Kraus (1950) and Chapter 11 of the book by Jordan and Balmain (1973).

4. The free-space equation (Equation 8.14) is named in honor of Friis (1946). For the origin of the Friis formula of Equation (8.30), see Friis (1944).

5. For an original treatment of cellular radio, see the paper by MacDonald (1979).

6. For a comprehensive treatment of the mobile radio propagation channel, see the book by Parsons (1992). This book presents the fundamentals of VHF and UHF propagation, propagation over irregular terrain and in built-up areas, and a statistical characterization of the mobile radio channel. The statistical characterization of a mobile radio channel is also discussed in Proakis (1995). This book provides a readable account of the effect of fading on the error performance of Rayleigh fading channels and a good discussion of diversity techniques. For a full treatment of the subject, see Chapters 9–11 by Stein in the book edited by Schwartz, Bennett, and Stein (1966).

7. The chi-square distribution is a special case of the gamma distribution. The probability density function of a *gamma-distributed random variable* $X$ has two parameters: $\alpha > 0$ and $\lambda > 0$; it is defined by

$$f_X(x) = \frac{\lambda(\lambda x)^{\alpha-1}e^{-\lambda x}}{\Gamma(\alpha)}, \quad 0 < x < \infty$$

where $\Gamma(\alpha)$ is the *gamma function*, which is itself defined by

$$\Gamma(\alpha) = \int_0^\infty z^{\alpha-1}e^{-\alpha}dz, \quad \alpha > 0$$

The gamma function has the following properties:

$$\Gamma(1/2) = \sqrt{\pi}$$
$$\Gamma(\alpha + 1) = \alpha\Gamma(\alpha), \quad \alpha > 0$$

By letting $\lambda = 1/2$ and $\alpha = k/2$, where $k$ is a positive integer, we get the *chi-square distribution* with $2k$ degrees of freedom, as shown by

$$f_X(x) = \frac{x^{(k-2)/2}e^{-x/2}}{2^{k/2}\Gamma(k/2)}, \quad 0 < x < \infty$$

8. For a survey article on the evolution of wireless communications, see Oliphant (1999). For books on the fundamentals of wireless communication systems, see Steele and Hanzo (1999), Stüber (1996) and Rappaport (1996).
   For a detailed description of GSM, see Chapter 8 of the book by Steele and Hanzo (1999). For a detailed description of the IS-95 system, see the handbook by Lee and Miller (1998).

9. The classic paper on the RAKE receiver is due to Price and Green (1958).

10. For the original paper on how to maximize the signal-to-noise ratio realizable from the sum of several noisy signals, see the classic paper by Brennan (1955).

11. The application of the RAKE receiver in CDMA wireless communication systems is discussed in detail in the book by Viterbi (1995).

12. The idea of multi-pulse excitation for speech coding is due to Atal and Remde (1982). Code-excited linear prediction (CELP) of speech was first introduced by Atal and Schroeder (1984). For a detailed mathematical discussion of multi-pulse excited, regular-pulse excited, and code-excited types of speech coding, particularly as they relate to wireless communications, see Chapter 3 in the book edited by Steele and Hanzo (1999).

13. In the wireless communications literature, adaptive antenna arrays are often referred to as *smart antennas*. For an overview of the various issues involved in the use of adaptive antenna arrays for wireless communications, see the article by Winters (1998) and the course notes by Winters (1999). The two-step subspace procedure for designing the array signal processor in Figure 8.29 is based on material presented in Winters (1999). The book by Rappaport (1999) presents a collection of papers on adaptive antenna arrays, which are grouped into algorithms, architectures, hardware applications, channel models, and performance evaluation.

14. The idea of subspace is rooted in matrix algebra. For a discussion of this idea, see Strang (1980) and Stewart (1973). For a discussion of subspace decomposition in the context of statistical signal processing, see Scharf (1991).

15. For tutorial discussions of space-time processing for wireless communications, see the articles by Paulraj and Ng (1998), Paulraj and Papadias (1997), and Kohno (1998).

16. Recursive implementation of the DMI algorithm leads to a new algorithm commonly referred to as the *recursive least squares (RLS) algorithm*; for a derivation of the RLS algorithm and its variants, see Haykin (1996).

## PROBLEMS

### Free-Space Propagation

8.1 A radio link uses a pair of 2m dish antennas with an efficiency of 60 percent each, as transmitting and receiving antennas. Other specifications of the link are:

Transmitted power = 1 dBw
Carrier frequency = 4 GHz
Distance of the receiver
from the transmitter = 150 m

Calculate (a) the free-space loss, (b) the power gain of each antenna, and (c) the received power in dBW.

8.2 Repeat Problem 8.1 for a carrier frequency of 12 GHz.

8.3 Equation (8.14) is one formulation of the Friis free-space equation. Show that this equation can also be formulated in the following equivalent forms:

(a) $P_r = \dfrac{P_t A_t A_r}{\lambda^2 d^2}$

(b) $P_r = \dfrac{P_t A_t G_r}{4\pi d^2}$

where $P_t$ is the transmitted power, $A_t$ is the effective area of the transmitting antenna, $\lambda$ is the carrier wavelength, $d$ is the distance of the receiver from the transmitter, $G_r$ is the power gain of the receiving antenna, $A_r$ is the effective area of the receiving antenna, and $P_r$ is the received power.

Discuss the situations that favor the use of one of these equations over the other.

8.4 From the mathematical definition of the free-space loss

$$L_{\text{free space}} = \left(\frac{4\pi d}{\lambda}\right)^2$$

we see that it is dependent on the carrier wavelength $\lambda$ or frequency $f$. How can this dependence on wavelength or frequency be justified in physical terms?

8.5 In a satellite communication system, the carrier frequency used on the uplink is always higher than the carrier frequency used on the downlink. Justify the rationale for this choice.

8.6 A continuous-wave (CW) beacon transmitter is located on a satellite in geostationary orbit. The beacon's 12 GHz output is monitored by an earth station positioned 40,000 km from the satellite. The satellite transmitting antenna is a 1m dish with an aperture efficiency of 70 percent, and the earth station receiving antenna is a 10m dish with an aperture efficiency of 55 percent. Calculate the received power, given that the beacon's output power is 100 mW.

### Noise Figure

8.7 Consider a 75-Ω resistor maintained at "room temperature" of 290K. Assuming a bandwidth of 1 MHz, calculate the following:

(a) The root-mean-square (RMS) value of the voltage appearing across the terminals of this resistor due to thermal noise.

(b) The maximum available noise power delivered to a matched load.

8.8 In this problem, we revisit Example 8.1 based on the receiver configuration of Figure 8.10. Suppose that a lossy waveguide is inserted between the receiving antenna and the low-noise amplifier. The waveguide loss is 1 dB, and its physical temperature is 290K. Recalculate the effective noise temperature of the receiver.

8.9 Consider the receiver of Figure P8.9, which consists of a lossy waveguide, low-noise RF amplifier, frequency down-converter (mixer), and IF amplifier. The figure includes the noise figures and power gains of these four components. The antenna temperature is 50K.

(a) Calculate the equivalent noise temperature for each of the four components in Figure P8.9, assuming a room temperature $T = 290$K.

(b) Calculate the effective noise temperature of the whole receiver.

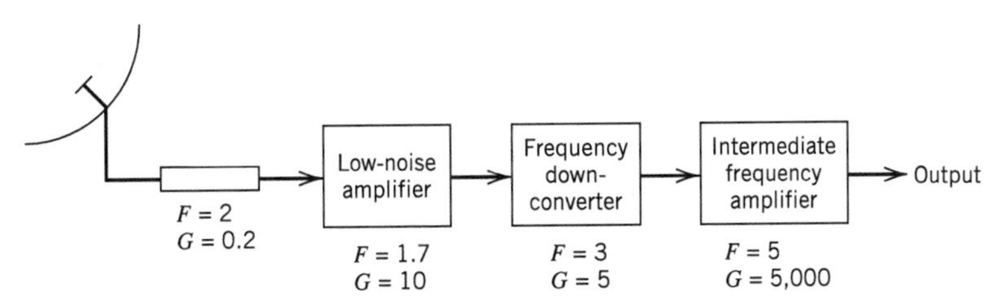


**FIGURE P8.9**

## Budget Link Calculations

8.10 In this problem we address the uplink power budget of the digital satellite communication system considered in Example 8.2. The parameters of the link are as follows:

| | |
|---|---|
| Carrier frequency | = 14 GHz |
| Power density at the TWT amplifier in saturation | = $-81$ dBW/m$^2$ |
| Satellite figure of merit, $G/T$ | = 1.9 dB/K |
| Distance of the satellite from the transmitting earth terminal | = 40,000 km |

(a) Assuming no power backoff of the TWT, calculate the $C/N_0$ ratio at the satellite.

(b) Given that the data rate in the uplink is the same as that calculated for the downlink in Example 8.2, calculate the probability of symbol error incurred in the uplink allowing for a link margin of 6 dB. Compare your result with that in Example 8.2.

8.11 The downlink $C/N_0$ ratio in a direct broadcast satellite (DBS) system is estimated to be 85 dB-Hz. The specifications of the link are:

| | |
|---|---|
| Satellite EIRP | = 57 dBW |
| Downlink carrier frequency | = 12.5 GHz |
| Data rate | = 10 Mb/s |
| Required $E_b/N_0$ at the receiving earth terminal | = 10 dB |
| Distance of the satellite from the receiving earth terminal | = 41,000 km |

Calculate the minimum diameter of the dish antenna needed to provide a satisfactory TV reception, assuming that the dish has an efficiency of 55 percent and it is located alongside the home where the temperature is 310K. For this calculation, assume that the operation of the DBS system is essentially downlink-limited.

## Wireless Communications

8.12 Both wireless communications and satellite communications rely on radio propagation for their operations. Summarize (**a**) the similarities of these two multiuser communication systems, and (**b**) the major differences that distinguish them from each other.

8.13 In wireless communication systems, the carrier frequency on the uplink (reverse link) is smaller than the carrier frequency on the downlink (forward link). Justify the rationale for this choice.

8.14 Figure P8.14 depicts the direct (line-of-sight) and indirect (reflected) paths of a radio link operating over a plane earth. The heights of the transmitting antenna at the base station and the receiving antenna of a mobile unit are $h_b$ and $h_m$, respectively. Assume the following:

- The reflection coefficient of the ground is $-1$.
- The distance $d$ between the two antennas is large enough to make the phase difference $\phi$ between the reflected and direct paths small compared to 1 radian, so that we may set $\sin\phi \simeq \phi$.

Hence, show that the received power $P_r$ is given by the approximation

$$P_r \simeq P_t G_b G_m \left(\frac{h_b h_m}{d^2}\right)^2$$

where $P_t$ is the transmitted power, and $G_b$ and $G_m$ are the power gains of the transmitting base and mobile antennas, respectively. Compare this result with the Friis free-space equation.

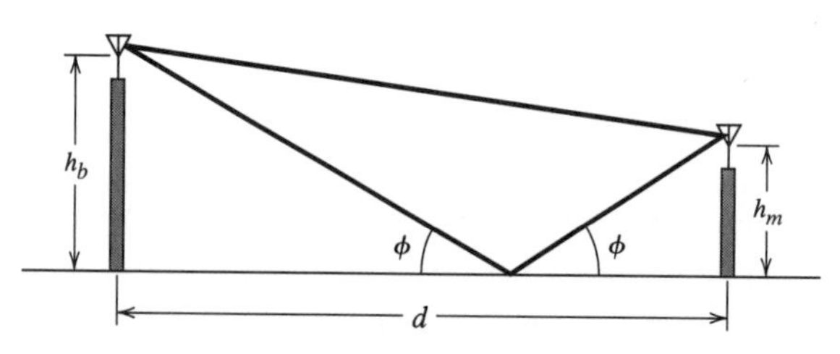


**FIGURE P8.14**

8.15 The *two-path model* defined by the impulse response

$$h(t) = a_1\,\delta(t - \tau_1) + a_2\exp(-j\theta)\,\delta(t - \tau_2)$$

is frequently used in the analytic treatment of wireless communication systems. The model parameters are the delay times $\tau_1$ and $\tau_2$, the uniformly distributed phase $\theta$, and the real coefficients $a_1$ and $a_2$.

(**a**) Determine (i) the transfer function of the model, and (ii) its power-delay profile.

(**b**) Show that the model exhibits frequency-selective fading due to variations in the coefficients $a_1$ and $a_2$.

8.16 In the RAKE receiver illustrated in Figure 8.26, each correlator is synchronized by inserting the right delay into the received signal.

(**a**) Show that, in theory, the same result is obtained by inserting the right delay into the reference signal (i.e., pseudo-noise sequence).

(**b**) In practice, the preferred method is to use the procedure described in Figure 8.26. What reason can you suggest for this preference?

**8.17** In this problem we study the maximal-ratio combining diversity scheme. To proceed, consider a set of noisy signals $\{x_j(t)\}_{j=1}^{N}$, where $x_j(t)$ is defined by

$$x_j(t) = s_j(t) + n_j(t), \qquad j = 1, 2, \ldots, N$$

Assume the following:

- The signal components $s_j(t)$ are locally coherent, that is,

$$s_j(t) = z_j m(t), \qquad j = 1, 2, \ldots, N$$

where the $z_j$ are positive real numbers, and $m(t)$ denotes a message signal with unit power.

- The noise components $n_j(t)$ have zero mean, and they are statistically independent, that is,

$$E[n_j(t)n_k(t)] = \begin{cases} \sigma_j^2 & \text{for } k = j \\ 0 & \text{otherwise} \end{cases}$$

The output of the linear combiner is defined by

$$x(t) = \sum_{j=1}^{N} \alpha_j x_j(t)$$

where the parameters $\alpha_j$ are to be determined.

(a) Show that the output signal-to-noise ratio is

$$(\text{SNR})_O = \frac{\left( \sum_{j=1}^{N} \alpha_j z_j \right)^2}{\sum_{j=1}^{N} \alpha_j^2 \sigma_j^2}$$

(b) Set

$$u_j = \alpha_j \sigma_j$$

$$v_j = \frac{z_j}{\sigma_j}$$

and reformulate the expression for $(\text{SNR})_O$. Hence, applying the Schwarz inequality to this reformulation, show that

(i) $(\text{SNR})_O \le \sum_{j=1}^{N} (\text{SNR})_j$

where $(\text{SNR})_j = z_j^2/\sigma_j^2$.

(ii) The optimum values of the combiner's coefficients are defined by

$$\alpha_j = \frac{z_j}{\sigma_j^2}$$

in which case the Schwarz inequality is satisfied with the equality sign.

The Schwarz inequality is discussed in Section 5.2.

## Adaptive Antenna Arrays

**8.18** Consider the array signal processor of Figure 8.29 where there are only two users ($N$ = 2) and the array consists of two elements ($M$ = 2). Construct the subspace $\mathcal{W}$ for this problem. Hence, using a signal-space diagram, illustrate the computation of the weight characterizing the array signal processor.

8.19 In this problem we derive the complex LMS algorithm. Referring to Figure 8.32 and starting with the instantaneous cost function

$$J = \frac{1}{2}|e[n]|^2$$

where $e[n]$ is the error signal and $M$ is the number of antenna elements, do the following:

(a) Determine the derivative of the cost function $J$ with respect to the $k$th elemental weight $w_k[n]$.

(b) Using the instantaneous derivative $\partial J/\partial w_k[n]$, denoted by $\hat{\nabla}J[k]$, determine the adjustment $\Delta w_k[n]$ made to the $k$th elemental weight in accordance with the rule

$$\Delta w_k[n] = -\mu\hat{\nabla}J[k]$$

(c) Verify the composition of the complex LMS algorithm described in Equations (8.75) to (8.77).

Note that $w_k[n]$ is complex valued, and you need to consider its real and imaginary parts separately.

8.20 A practical limitation of an adaptive antenna array using the LMS algorithm is the dynamic range over which the array can operate. This limitation is due to the fact that the speed of response of the weights in the LMS algorithm is proportional to the average signal power at the array input.

(a) Justify the assertion that the dynamic range of average signal power at the array input is proportional to $R_b/f_{max}$, where $R_b$ is the data rate in b/s and $f_{max}$ is the maximum fade rate in Hz.

(b) Assuming a proportionality factor of 0.2, by which the ratio $R_b/f_{max}$ is scaled, calculate the dynamic range of an adaptive antenna array using the LMS algorithm for $R_b = 32$ kb/s and $f_{max} = 70$ Hz. Comment on your result. (The proportionality factor of 0.2 is a reasonable choice for systems using PSK.)

8.21 In this problem we derive the direct matrix inversion algorithm for adjusting the weights of an adaptive antenna array. To do so, we revisit the derivation of the Wiener filter presented in Chapter 3.

(a) Show that

$$\hat{\mathbf{R}}_{\mathbf{x}}\mathbf{w} = \hat{\mathbf{r}}_{\mathbf{x}d}$$

where $\hat{\mathbf{R}}_{\mathbf{x}}$ is an estimate of the correlation matrix of the input vector $\mathbf{x}[k]$:

$$\hat{\mathbf{R}}_{\mathbf{x}} = \frac{1}{K}\sum_{k=1}^{K}\mathbf{x}[k]\mathbf{x}^H[k]$$

and $\hat{\mathbf{r}}_{\mathbf{x}d}$ is an estimate of the cross-correlation vector between $\mathbf{x}[k]$ and the reference signal $d[k]$:

$$\hat{\mathbf{r}}_{\mathbf{x}d} = \frac{1}{K}\sum_{k=1}^{K}\mathbf{x}[k]d^*[k]$$

The superscript $H$ in the formula for $\hat{\mathbf{R}}_{\mathbf{x}}$ denotes Hermitian transportation (i.e., transposition and complex conjugation), so $\mathbf{x}[k]\mathbf{x}^H[k]$ denotes the outer product of $\mathbf{x}[k]$ with itself. The summations for both $\hat{\mathbf{R}}_{\mathbf{x}}$ and $\hat{\mathbf{r}}_{\mathbf{x}d}$ are performed over a total of $K$ snapshots, with each snapshot being represented by the pair $\{\mathbf{x}[k], d[k]\}$.

(b) Using the formulas of part (a), describe an algorithm for computing the weight vector $\mathbf{w}$, given a data set consisting of $K$ snapshots. Hence demonstrate that the complexity of this algorithm grows as $M^3$ with the size of the weight vector $\mathbf{w}$ denoted by $M$.

CHAPTER 9

# FUNDAMENTAL LIMITS IN INFORMATION THEORY

Shannon's landmark paper on information theory in 1948, and its refinements by other researchers, were in direct response to the need of electrical engineers to design communication systems that are both efficient and reliable. Efficient communication from a source to a user destination is attained through source coding. Reliable communication over a noisy channel is attained through error-control coding. This chapter addresses these important issues as summarized here:

- *Entropy as the basic measure of information.*
- *Source coding theorem and data compaction algorithms.*
- *Mutual information and its relation to the capacity of a communication channel for information transmission.*
- *Channel coding theorem as the basis for reliable communication.*
- *Information capacity theorem as the basis for a tradeoff between channel bandwidth and signal-to-noise ratio.*
- *Rate-distortion theory for source coding with a fidelity criterion.*

## 9.1 *Introduction*

As mentioned in the Background and Preview chapter and reiterated along the way, the purpose of a communication system is to carry information-bearing baseband signals from one place to another over a communication channel. In preceding chapters of the book, we have described a variety of modulation schemes for accomplishing this objective. But what do we mean by the term *information*? To address this issue, we need to invoke *information theory*.[1] This broadly based mathematical discipline has made fundamental contributions, not only to communications, but also to computer science, statistical physics, statistical inference, and probability and statistics.

In the context of communications, information theory deals with mathematical modeling and analysis of a communication system rather than with physical sources and physical channels. In particular, it provides answers to two fundamental questions (among others):

- What is the irreducible complexity below which a signal cannot be compressed?
- What is the ultimate transmission rate for reliable communication over a noisy channel?

The answers to these questions lie in the entropy of a source and the capacity of a channel, respectively. *Entropy* is defined in terms of the probabilistic behavior of a source of information; it is so named in deference to the parallel use of this concept in thermodynamics. *Capacity* is defined as the intrinsic ability of a channel to convey information; it is naturally related to the noise characteristics of the channel. A remarkable result that emerges from information theory is that if the entropy of the source is less than the capacity of the channel, then error-free communication over the channel can be achieved. It is therefore befitting that we begin our study of information theory by discussing the relationships among uncertainty, information, and entropy.

## 9.2 *Uncertainty, Information, and Entropy*

Suppose that a *probabilistic experiment* involves the observation of the output emitted by a discrete source during every unit of time (signaling interval). The source output is modeled as a discrete random variable, $S$, which takes on symbols from a fixed finite *alphabet*

$$\mathscr{S} = \{s_0, s_1, \ldots, s_{K-1}\} \tag{9.1}$$

with probabilities

$$P(S = s_k) = p_k, \qquad k = 0, 1, \ldots, K-1 \tag{9.2}$$

Of course, this set of probabilities must satisfy the condition

$$\sum_{k=0}^{K-1} p_k = 1 \tag{9.3}$$

We assume that the symbols emitted by the source during successive signaling intervals are statistically independent. A source having the properties just described is called a *discrete memoryless source*, memoryless in the sense that the symbol emitted at any time is independent of previous choices.

Can we find a measure of how much information is produced by such a source? To answer this question, we note that the idea of information is closely related to that of uncertainty or surprise, as described next.

Consider the event $S = s_k$, describing the emission of symbol $s_k$ by the source with probability $p_k$, as defined in Equation (9.2). Clearly, if the probability $p_k = 1$ and $p_i = 0$ for all $i \neq k$, then there is no "surprise," and therefore no "information," when symbol $s_k$ is emitted, because we know what the message from the source must be. If, on the other hand, the source symbols occur with different probabilities, and the probability $p_k$ is low, then there is more surprise, and therefore information, when symbol $s_k$ is emitted by the source than when symbol $s_i$, $i \neq k$, with higher probability is emitted. Thus, the words *uncertainty*, *surprise*, and *information* are all related. Before the event $S = s_k$ occurs, there is an amount of uncertainty. When the event $S = s_k$ occurs there is an amount of surprise. After the occurrence of the event $S = s_k$, there is gain in the amount of information, the essence of which may be viewed as the *resolution of uncertainty*. Moreover, the amount of information is related to the *inverse* of the probability of occurrence.

We define the amount of information gained after observing the event $S = s_k$, which occurs with probability $p_k$, as the *logarithmic* function[2]

$$I(s_k) = \log\left(\frac{1}{p_k}\right) \tag{9.4}$$

This definition exhibits the following important properties that are intuitively satisfying:

1.
$$I(s_k) = 0 \qquad \text{for } p_k = 1 \tag{9.5}$$
Obviously, if we are absolutely *certain* of the outcome of an event, even before it occurs, there is *no* information gained.

2.
$$I(s_k) \geq 0 \qquad \text{for } 0 \leq p_k \leq 1 \tag{9.6}$$
That is to say, the occurrence of an event $S = s_k$ either provides some or no information, but never brings about a *loss* of information.

3.
$$I(s_k) > I(s_i) \qquad \text{for } p_k < p_i \tag{9.7}$$
That is, the less probable an event is, the more information we gain when it occurs.

4. $I(s_k s_l) = I(s_k) + I(s_l)$ if $s_k$ and $s_l$ are statistically independent.

The base of the logarithm in Equation (9.4) is quite arbitrary. Nevertheless, it is the standard practice today to use a logarithm to base 2. The resulting unit of information is called the *bit* (a contraction of *bi*nary digi*t*). We thus write

$$\begin{aligned} I(s_k) &= \log_2\left(\frac{1}{p_k}\right) \\ &= -\log_2 p_k \qquad \text{for } k = 0, 1, \ldots, K-1 \end{aligned} \tag{9.8}$$

When $p_k = 1/2$, we have $I(s_k) = 1$ bit. Hence, *one bit is the amount of information that we gain when one of two possible and equally likely (i.e., equiprobable) events occurs.* Note that the information $I(s_k)$ is positive, since the logarithm of a number less than one, such as a probability, is negative.

The amount of information $I(s_k)$ produced by the source during an arbitrary signaling interval depends on the symbol $s_k$ emitted by the source at that time. Indeed, $I(s_k)$ is a discrete random variable that takes on the values $I(s_0), I(s_1), \ldots, I(s_{K-1})$ with probabilities $p_0, p_1, \ldots, p_{K-1}$ respectively. The mean of $I(s_k)$ over the source alphabet $\mathcal{S}$ is given by

$$\begin{aligned} H(\mathcal{S}) &= E[I(s_k)] \\ &= \sum_{k=0}^{K-1} p_k I(s_k) \\ &= \sum_{k=0}^{K-1} p_k \log_2\left(\frac{1}{p_k}\right) \end{aligned} \tag{9.9}$$

The important quantity $H(\mathcal{S})$ is called the *entropy*[3] of a discrete memoryless source with source alphabet $\mathcal{S}$. It is a measure of the *average information content per source symbol.* Note that the entropy $H(\mathcal{S})$ depends only on the probabilities of the symbols in the alphabet $\mathcal{S}$ of the source. Thus the symbol $\mathcal{S}$ in $H(\mathcal{S})$ is not an argument of a function but rather a label for a source.

## ■ SOME PROPERTIES OF ENTROPY

Consider a discrete memoryless source whose mathematical model is defined by Equations (9.1) and (9.2). The entropy $H(\mathcal{S})$ of such a source is bounded as follows:

$$0 \leq H(\mathcal{S}) \leq \log_2 K \tag{9.10}$$

where $K$ is the *radix* (number of symbols) of the alphabet $\mathcal{S}$ of the source. Furthermore, we may make two statements:

1. $H(\mathcal{S}) = 0$, if and only if the probability $p_k = 1$ for some $k$, and the remaining probabilities in the set are all zero; this lower bound on entropy corresponds to *no uncertainty*.
2. $H(\mathcal{S}) = \log_2 K$, if and only if $p_k = 1/K$ for all $k$ (i.e., all the symbols in the alphabet $\mathcal{S}$ are *equiprobable*); this upper bound on entropy corresponds to *maximum uncertainty*.

To prove these properties of $H(\mathcal{S})$, we proceed as follows. First, since each probability $p_k$ is less than or equal to unity, it follows that each term $p_k \log_2(1/p_k)$ in Equation (9.9) is always nonnegative, and so $H(\mathcal{S}) \geq 0$. Next, we note that the product term $p_k \log_2(1/p_k)$ is zero if, and only if, $p_k = 0$ or 1. We therefore deduce that $H(\mathcal{S}) = 0$ if, and only if, $p_k = 0$ or 1, that is, $p_k = 1$ for some $k$ and all the rest are zero.

This completes the proofs of the lower bound in Equation (9.10) and statement (1).

To prove the upper bound in Equation (9.10) and statement (2), we make use of a property of the natural logarithm:

$$\log x \leq x - 1, \qquad x \geq 0 \tag{9.11}$$

This inequality can be readily verified by plotting the functions $\log x$ and $(x - 1)$ versus $x$, as shown in Figure 9.1. Here we see that the line $y = x - 1$ always lies above the curve $y = \log x$. The equality holds *only* at the point $x = 1$, where the line is tangential to the curve.

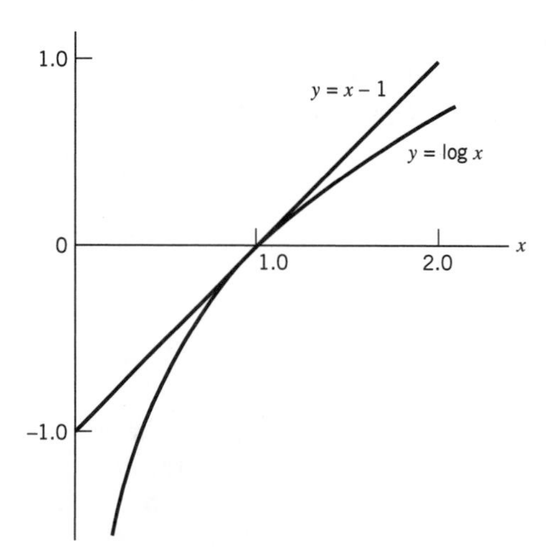


**FIGURE 9.1** Graphs of the functions $x - 1$ and $\log x$ versus $x$.

To proceed with the proof, consider first any two probability distributions $\{p_0, p_1, \ldots, p_{K-1}\}$ and $\{q_0, q_1, \ldots, q_{K-1}\}$ on the alphabet $\mathcal{S} = \{s_0, s_1, \ldots, s_{K-1}\}$ of a discrete memoryless source. Then, changing to the natural logarithm, we may write

$$\sum_{k=0}^{K-1} p_k \log_2\left(\frac{q_k}{p_k}\right) = \frac{1}{\log 2} \sum_{k=0}^{K-1} p_k \log\left(\frac{q_k}{p_k}\right)$$

Hence, using the inequality of Equation (9.11), we get

$$\begin{aligned}\sum_{k=0}^{K-1} p_k \log_2\left(\frac{q_k}{p_k}\right) &\le \frac{1}{\log 2} \sum_{k=0}^{K-1} p_k\left(\frac{q_k}{p_k} - 1\right) \\ &\le \frac{1}{\log 2} \sum_{k=0}^{K-1} (q_k - p_k) \\ &\le \frac{1}{\log 2}\left(\sum_{k=0}^{K-1} q_k - \sum_{k=0}^{K-1} p_k\right) = 0\end{aligned}$$

We thus have the *fundamental inequality*

$$\sum_{k=0}^{K-1} p_k \log_2\left(\frac{q_k}{p_k}\right) \le 0 \tag{9.12}$$

where the equality holds only if $q_k = p_k$ for all $k$.

Suppose we next put

$$q_k = \frac{1}{K}, \qquad k = 0, 1, \ldots, K - 1 \tag{9.13}$$

which corresponds to an alphabet $\mathcal{S}$ with *equiprobable* symbols. The entropy of a discrete memoryless source with such a characterization equals

$$\sum_{k=0}^{K-1} q_k \log_2\left(\frac{1}{q_k}\right) = \log_2 K \tag{9.14}$$

Also, the use of Equation (9.13) in Equation (9.12) yields

$$\sum_{k=0}^{K-1} p_k \log_2\left(\frac{1}{p_k}\right) \le \log_2 K$$

Equivalently, the entropy of a discrete memoryless source with an arbitrary probability distribution for the symbols of its alphabet $\mathcal{S}$ is bounded as

$$H(\mathcal{S}) \le \log_2 K$$

Thus $H(\mathcal{S})$ is always less than or equal to $\log_2 K$. The equality holds only if the symbols in the alphabet $\mathcal{S}$ are equiprobable, as in Equation (9.13). This completes the proof of Equation (9.10) and statements (1) and (2).

### ▶ Example 9.1 Entropy of Binary Memoryless Source

To illustrate the properties of $H(\mathcal{S})$, we consider a binary source for which symbol 0 occurs with probability $p_0$ and symbol 1 with probability $p_1 = 1 - p_0$. We assume that the source is memoryless so that successive symbols emitted by the source are statistically independent.

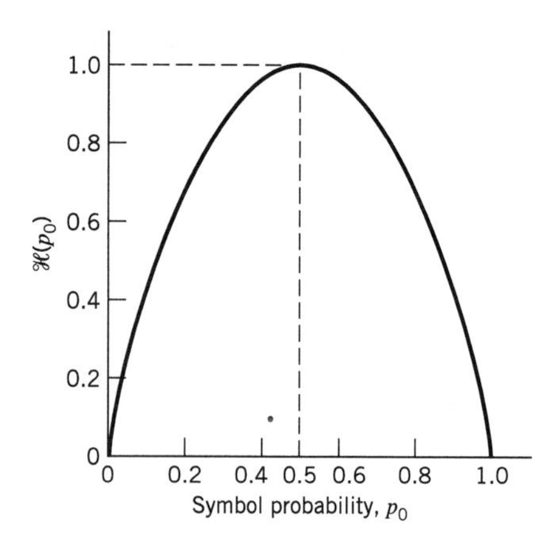


**FIGURE 9.2** Entropy function $\mathcal{H}(p_0)$.

The entropy of such a source equals

$$\begin{aligned} H(\mathcal{S}) &= -p_0 \log_2 p_0 - p_1 \log_2 p_1 \\ &= -p_0 \log_2 p_0 - (1 - p_0) \log_2(1 - p_0) \text{ bits} \end{aligned} \tag{9.15}$$

from which we observe the following:

1. When $p_0 = 0$, the entropy $H(\mathcal{S}) = 0$; this follows from the fact that $x \log x \to 0$ as $x \to 0$.
2. When $p_0 = 1$, the entropy $H(\mathcal{S}) = 0$.
3. The entropy $H(\mathcal{S})$ attains its maximum value, $H_{max} = 1$ bit, when $p_1 = p_0 = 1/2$, that is, symbols 1 and 0 are equally probable.

The function of $p_0$ given on the right-hand side of Equation (9.15) is frequently encountered in information-theoretic problems. It is therefore customary to assign a special symbol to this function. Specifically, we define

$$\mathcal{H}(p_0) = -p_0 \log_2 p_0 - (1 - p_0) \log_2(1 - p_0) \tag{9.16}$$

We refer to $\mathcal{H}(p_0)$ as the *entropy function*. The distinction between Equation (9.15) and Equation (9.16) should be carefully noted. The $H(\mathcal{S})$ of Equation (9.15) gives the entropy of a discrete memoryless source with source alphabet $\mathcal{S}$. The $\mathcal{H}(p_0)$ of Equation (9.16), on the other hand, is a function of the prior probability $p_0$ defined on the interval [0, 1]. Accordingly, we may plot the entropy function $\mathcal{H}(p_0)$ versus $p_0$, defined on the interval [0, 1], as in Figure 9.2. The curve in Figure 9.2 highlights the observations made under points 1, 2, and 3. ◀

## ■ EXTENSION OF A DISCRETE MEMORYLESS SOURCE

In discussing information-theoretic concepts, we often find it useful to consider *blocks* rather than individual symbols, with each block consisting of $n$ successive source symbols. We may view each such block as being produced by an *extended source* with a source alphabet $\mathcal{S}^n$ that has $K^n$ *distinct* blocks, where $K$ is the number of distinct symbols in the source alphabet $\mathcal{S}$ of the original source. In the case of a discrete memoryless source, the source symbols are statistically independent. Hence, the probability of a source symbol in $\mathcal{S}^n$ is equal to the product of the probabilities of the $n$ source symbols in $\mathcal{S}$ constituting the particular source symbol in $\mathcal{S}^n$. We may thus intuitively expect that $H(\mathcal{S}^n)$, the entropy of the extended source, is equal to $n$ times $H(\mathscr{S})$, the entropy of the original source. That is, we may write

$$H(\mathscr{S}^n) = nH(\mathscr{S}) \tag{9.17}$$

### ▶ EXAMPLE 9.2 Entropy of Extended Source

Consider a discrete memoryless source with source alphabet $\mathscr{S} = \{s_0, s_1, s_2\}$ with respective probabilities

$$p_0 = \tfrac{1}{4}$$
$$p_1 = \tfrac{1}{4}$$
$$p_2 = \tfrac{1}{2}$$

Hence, the use of Equation (9.9) yields the entropy of the source as

$$\begin{aligned} H(\mathscr{S}) &= p_0 \log_2\left(\frac{1}{p_0}\right) + p_1 \log_2\left(\frac{1}{p_1}\right) + p_2 \log_2\left(\frac{1}{p_2}\right) \\ &= \frac{1}{4}\log_2(4) + \frac{1}{4}\log_2(4) + \frac{1}{2}\log_2(2) \\ &= \frac{3}{2} \text{ bits} \end{aligned}$$

Consider next the second-order extension of the source. With the source alphabet $\mathscr{S}$ consisting of three symbols, it follows that the source alphabet $\mathscr{S}^2$ of the extended source has nine symbols. The first row of Table 9.1 presents the nine symbols of $\mathscr{S}^2$, denoted as $\sigma_0, \sigma_1, \ldots, \sigma_8$. The second row of the table presents the composition of these nine symbols in terms of the corresponding sequences of source symbols $s_0$, $s_1$, and $s_2$, taken two at a time. The probabilities of the nine source symbols of the extended source are presented in the last row of the table. Accordingly, the use of Equation (9.9) yields the entropy of the extended source as

$$\begin{aligned} H(\mathscr{S}^2) &= \sum_{i=0}^{8} p(\sigma_i) \log_2 \frac{1}{p(\sigma_i)} \\ &= \frac{1}{16}\log_2(16) + \frac{1}{16}\log_2(16) + \frac{1}{8}\log_2(8) + \frac{1}{16}\log_2(16) \\ &\quad + \frac{1}{16}\log_2(16) + \frac{1}{8}\log_2(8) + \frac{1}{8}\log_2(8) + \frac{1}{8}\log_2(8) + \frac{1}{4}\log_2(4) \\ &= 3 \text{ bits} \end{aligned}$$

We thus see that $H(\mathscr{S}^2) = 2H(\mathscr{S})$ in accordance with Equation (9.17). ◀

**TABLE 9.1** ***Alphabet particulars of second-order extension of a discrete memoryless source***

| Symbols of $\mathscr{S}^2$ | $\sigma_0$ | $\sigma_1$ | $\sigma_2$ | $\sigma_3$ | $\sigma_4$ | $\sigma_5$ | $\sigma_6$ | $\sigma_7$ | $\sigma_8$ |
|---|---|---|---|---|---|---|---|---|---|
| Corresponding sequences of symbols of $\mathscr{S}$ | $s_0s_0$ | $s_0s_1$ | $s_0s_2$ | $s_1s_0$ | $s_1s_1$ | $s_1s_2$ | $s_2s_0$ | $s_2s_1$ | $s_2s_2$ |
| Probability $p(\sigma_i)$, $i = 0, 1, \ldots, 8$ | $\frac{1}{16}$ | $\frac{1}{16}$ | $\frac{1}{8}$ | $\frac{1}{16}$ | $\frac{1}{16}$ | $\frac{1}{8}$ | $\frac{1}{8}$ | $\frac{1}{8}$ | $\frac{1}{4}$ |

## 9.3 Source-Coding Theorem

An important problem in communications is the *efficient* representation of data generated by a discrete source. The process by which this representation is accomplished is called *source encoding*. The device that performs the representation is called a *source encoder*. For the source encoder to be *efficient*, we require knowledge of the statistics of the source. In particular, if some source symbols are known to be more probable than others, then we may exploit this feature in the generation of a *source code* by assigning *short* code words to *frequent* source symbols, and *long* code words to *rare* source symbols. We refer to such a source code as a *variable-length code*. The *Morse code* is an example of a variable-length code. In the Morse code, the letters of the alphabet and numerals are encoded into streams of *marks* and *spaces*, denoted as dots "." and dashes "-", respectively. In the English language, the letter $E$ occurs more frequently than the letter $Q$, for example, so the Morse code encodes $E$ into a single dot ".", the shortest code word in the code, and it encodes $Q$ into "- - . -", the longest code word in the code.

Our primary interest is in the development of an efficient source encoder that satisfies two functional requirements:

1. The code words produced by the encoder are in *binary* form.
2. The source code is *uniquely decodable*, so that the original source sequence can be reconstructed perfectly from the encoded binary sequence.

Consider then the scheme shown in Figure 9.3, which depicts a discrete memoryless source whose output $s_k$ is converted by the source encoder into a block of 0s and 1s, denoted by $b_k$. We assume that the source has an alphabet with $K$ different symbols, and that the $k$th symbol $s_k$ occurs with probability $p_k$, $k = 0, 1, \ldots, K-1$. Let the binary code word assigned to symbol $s_k$ by the encoder have length $l_k$, measured in bits. We define the average code-word length, $\bar{L}$, of the source encoder as

$$\bar{L} = \sum_{k=0}^{K-1} p_k l_k \tag{9.18}$$

In physical terms, the parameter $\bar{L}$ represents the *average number of bits per source symbol* used in the source encoding process. Let $L_{\min}$ denote the *minimum* possible value of $\bar{L}$. We then define the *coding efficiency* of the source encoder as

$$\eta = \frac{L_{\min}}{\bar{L}} \tag{9.19}$$

With $\bar{L} \geq L_{\min}$, we clearly have $\eta \leq 1$. The source encoder is said to be *efficient* when $\eta$ approaches unity.

But how is the minimum value $L_{\min}$ determined? The answer to this fundamental question is embodied in Shannon's first theorem: the *source-coding theorem*,[4] which may be stated as follows:

> Given a discrete memoryless source of entropy $H(\mathcal{S})$, the average code-word length $\bar{L}$ for any distortionless source encoding scheme is bounded as
>
> $$\bar{L} \geq H(\mathcal{S}) \tag{9.20}$$

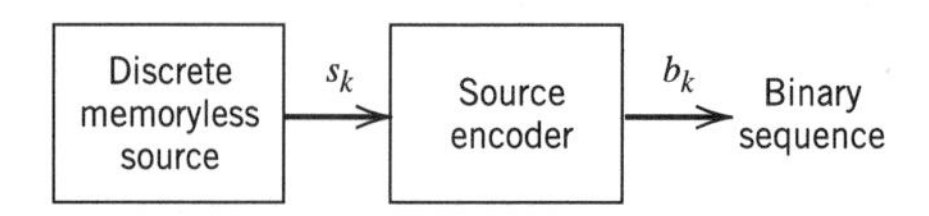


**FIGURE 9.3** Source encoding.

(A proof of this theorem for a particular class of source codes is presented in the next section.) According to the source-coding theorem, the entropy $H(\mathcal{S})$ represents a *fundamental limit* on the average number of bits per source symbol necessary to represent a discrete memoryless source in that it can be made as small as, but no smaller than, the entropy $H(\mathcal{S})$. Thus with $L_{\min} = H(\mathcal{S})$, we may rewrite the efficiency of a source encoder in terms of the entropy $H(\mathcal{S})$ as

$$\eta = \frac{H(\mathcal{S})}{\bar{L}} \tag{9.21}$$

## 9.4 *Data Compaction*

A common characteristic of signals generated by physical sources is that, in their natural form, they contain a significant amount of information that is *redundant*, the transmission of which is therefore wasteful of primary communication resources. For *efficient* signal transmission, the *redundant information should be removed from the signal prior to transmission*. This operation, with *no* loss of information, is ordinarily performed on a signal in digital form, in which case we refer to it as *data compaction* or *lossless data compression*. The code resulting from such an operation provides a representation of the source output that is not only efficient in terms of the average number of bits per symbol but also exact in the sense that the original data can be reconstructed with no loss of information. The entropy of the source establishes the fundamental limit on the removal of redundancy from the data. Basically, data compaction is achieved by assigning short descriptions to the most frequent outcomes of the source output and longer descriptions to the less frequent ones.

In this section, we discuss some source-coding schemes for data compaction. We begin the discussion by describing a type of source code known as a prefix code, which is not only decodable but also offers the possibility of realizing an average code-word length that can be made arbitrarily close to the source entropy.

### ■ Prefix Coding

Consider a discrete memoryless source of alphabet $\{s_0, s_1, \ldots, s_{K-1}\}$ and statistics $\{p_0, p_1, \ldots, p_{K-1}\}$. For a source code representing the output of this source to be of practical use, the code has to be uniquely decodable. This restriction ensures that for each finite sequence of symbols emitted by the source, the corresponding sequence of code words is different from the sequence of code words corresponding to any other source sequence. We are specifically interested in a special class of codes satisfying a restriction known as the *prefix condition*. To define the prefix condition, let the code word assigned to source symbol $s_k$ be denoted by $(m_{k_1}, m_{k_2}, \ldots, m_{k_n})$, where the individual elements $m_{k_1}, \ldots, m_{k_n}$ are 0s and 1s, and $n$ is the code-word length. The initial part of the code word is represented by the elements $m_{k_1}, \ldots, m_{k_i}$ for some $i \le n$. Any sequence made up of the initial part of the code word is called a *prefix* of the code word. A *prefix code* is defined as a code in which no code word is the prefix of any other code word.

To illustrate the meaning of a prefix code, consider the three source codes described in Table 9.2. Code I is not a prefix code since the bit 0, the code word for $s_0$, is a prefix of 00, the code word for $s_2$. Likewise, the bit 1, the code word for $s_1$, is a prefix of 11, the code word for $s_3$. Similarly, we may show that code III is not a prefix code, but code II is.

To decode a sequence of code words generated from a prefix source code, the *source decoder* simply starts at the beginning of the sequence and decodes one code word at a time. Specifically, it sets up what is equivalent to a *decision tree*, which is a graphical

**TABLE 9.2** ***Illustrating the definition of a prefix code***

| *Source Symbol* | *Probability of Occurrence* | *Code I* | *Code II* | *Code III* |
|---|---|---|---|---|
| $s_0$ | 0.5 | 0 | 0 | 0 |
| $s_1$ | 0.25 | 1 | 10 | 01 |
| $s_2$ | 0.125 | 00 | 110 | 011 |
| $s_3$ | 0.125 | 11 | 111 | 0111 |

portrayal of the code words in the particular source code. For example, Figure 9.4 depicts the decision tree corresponding to code II in Table 9.2. The tree has an *initial state* and four *terminal states* corresponding to source symbols $s_0$, $s_1$, $s_2$, and $s_3$. The decoder always starts at the initial state. The first received bit moves the decoder to the terminal state $s_0$ if it is 0, or else to a second decision point if it is 1. In the latter case, the second bit moves the decoder one step further down the tree, either to terminal state $s_1$ if it is 0, or else to a third decision point if it is 1, and so on. Once each terminal state emits its symbol, the decoder is reset to its initial state. Note also that each bit in the received encoded sequence is examined only once. For example, the encoded sequence 1011111000 . . . is readily decoded as the source sequence $s_1s_3s_2s_0s_0 \ldots$. The reader is invited to carry out this decoding.

A prefix code has the important property that it is *always* uniquely decodable. But the converse is not necessarily true. For example, code III in Table 9.2 does not satisfy the prefix condition, yet it is uniquely decodable since the bit 0 indicates the beginning of each code word in the code.

Moreover, if a prefix code has been constructed for a discrete memoryless source with source alphabet $\{s_0, s_1, \ldots, s_{K-1}\}$ and source statistics $\{p_0, p_1, \ldots, p_{K-1}\}$ and the code word for symbol $s_k$ has length $l_k$, $k = 0, 1, \ldots, K-1$, then the code-word lengths of the code always satisfy a certain inequality known as the *Kraft–McMillan Inequality*,[5] as shown by

$$\sum_{k=0}^{K-1} 2^{-l_k} \leq 1 \tag{9.22}$$

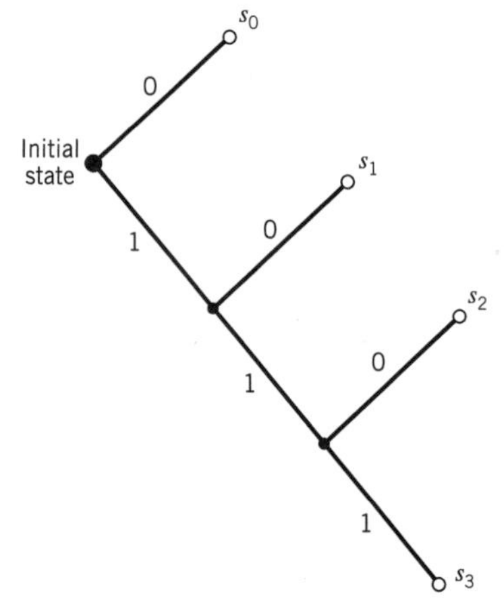


**FIGURE 9.4** Decision tree for code II of Table 9.2.

where the factor 2 refers to the radix (number of symbols) in the binary alphabet. It is important to note, however, that the Kraft–McMillan inequality does *not* tell us that a source code is a prefix code. Rather, it is merely a condition on the code-word lengths of the code and not on the code words themselves. For example, referring to the three codes listed in Table 9.2, we note the following:

- Code I violates the Kraft–McMillan inequality; it cannot therefore be a prefix code.
- The Kraft–McMillan inequality is satisfied by both codes II and III; but only code II is a prefix code.

Prefix codes are distinguished from other uniquely decodable codes by the fact that the end of a code word is always recognizable. Hence, the decoding of a prefix can be accomplished as soon as the binary sequence representing a source symbol is fully received. For this reason, prefix codes are also referred to as *instantaneous codes*.

Given a discrete memoryless source of entropy $H(\mathcal{S})$, a prefix code can be constructed with an average code-word length $\bar{L}$, which is bounded as follows:

$$H(\mathcal{S}) \leq \bar{L} < H(\mathcal{S}) + 1 \tag{9.23}$$

The left-hand bound of Equation (9.23) is satisfied with equality under the condition that symbol $s_k$ is emitted by the source with probability

$$p_k = 2^{-l_k} \tag{9.24}$$

where $l_k$ is the length of the code word assigned to source symbol $s_k$. We then have

$$\sum_{k=0}^{K-1} 2^{-l_k} = \sum_{k=0}^{K-1} p_k = 1$$

Under this condition, the Kraft–McMillan inequality of Equation (9.22) tells us that we can construct a prefix code, such that the length of the code word assigned to source symbol $s_k$ is $-\log_2 p_k$. For such a code, the average code-word length is

$$\bar{L} = \sum_{k=0}^{K-1} \frac{l_k}{2^{l_k}} \tag{9.25}$$

and the corresponding entropy of the source is

$$\begin{aligned} H(\mathcal{S}) &= \sum_{k=0}^{K-1} \left(\frac{1}{2^{l_k}}\right) \log_2(2^{l_k}) \\ &= \sum_{k=0}^{K-1} \frac{l_k}{2^{l_k}} \end{aligned} \tag{9.26}$$

Hence, in this special (rather meretricious) case, we find from Equations (9.25) and (9.26) that the prefix code is *matched* to the source in that $\bar{L} = H(\mathcal{S})$.

But how do we match the prefix code to an arbitrary discrete memoryless source? The answer to this problem lies in the use of an *extended code*. Let $\bar{L}_n$ denote the average code-word length of the extended prefix code. For a uniquely decodable code, $\bar{L}_n$ is the smallest possible. From Equation (9.23), we deduce that

$$H(\mathcal{S}^n) \leq \bar{L}_n < H(\mathcal{S}^n) + 1 \tag{9.27}$$

Substituting Equation (9.17) for an extended source into Equation (9.27), we get

$$nH(\mathcal{S}) \leq \bar{L}_n < nH(\mathcal{S}) + 1$$

or, equivalently,

$$H(\mathcal{S}) \le \frac{\overline{L}_n}{n} < H(\mathcal{S}) + \frac{1}{n} \tag{9.28}$$

In the limit, as $n$ approaches infinity, the lower and upper bounds in Equation (9.28) converge, as shown by

$$\lim_{n \to \infty} \frac{1}{n} \overline{L}_n = H(\mathcal{S}) \tag{9.29}$$

We may therefore state that by making the order $n$ of an extended prefix souce encoder large enough, we can make the code faithfully represent the discrete memoryless source $\mathcal{S}$ as closely as desired. In other words, the average code-word length of an extended prefix code can be made as small as the entropy of the source provided the extended code has a high enough order, in accordance with the source-coding theorem. However, the price we have to pay for decreasing the average code-word length is increased decoding complexity, which is brought about by the high order of the extended prefix code.

## ■ HUFFMAN CODING

We next describe an important class of prefix codes known as Huffman codes. The basic idea behind *Huffman coding*[6] is to assign to each symbol of an alphabet a sequence of bits roughly equal in length to the amount of information conveyed by the symbol in question. The end result is a source code whose average code-word length approaches the fundamental limit set by the entropy of a discrete memoryless source, namely, $H(\mathcal{S})$. The essence of the *algorithm* used to synthesize the Huffman code is to replace the prescribed set of source statistics of a discrete memoryless source with a simpler one. This *reduction* process is continued in a step-by-step manner until we are left with a final set of only two source statistics (symbols), for which (0, 1) is an optimal code. Starting from this trivial code, we then work backward and thereby construct the Huffman code for the given source.

Specifically, the Huffman *encoding algorithm* proceeds as follows:

1. The source symbols are listed in order of decreasing probability. The two source symbols of lowest probability are assigned a 0 and a 1. This part of the step is referred to as a *splitting* stage.
2. These two source symbols are regarded as being *combined* into a new source symbol with probability equal to the sum of the two original probabilities. (The list of source symbols, and therefore source statistics, is thereby *reduced* in size by one.) The probability of the new symbol is placed in the list in accordance with its value.
3. The procedure is repeated until we are left with a final list of source statistics (symbols) of only two for which a 0 and a 1 are assigned.

The code for each (original) source symbol is found by working backward and tracing the sequence of 0s and 1s assigned to that symbol as well as its successors.

### ▶ EXAMPLE 9.3 Huffman Tree

The five symbols of the alphabet of a discrete memoryless source and their probabilities are shown in the two leftmost columns of Figure 9.5*a*. Following through the Huffman algorithm, we reach the end of the computation in four steps, resulting in the *Huffman tree* shown in Figure 9.5*a*. The code words of the Huffman code for the source are tabulated in Figure 9.5*b*.

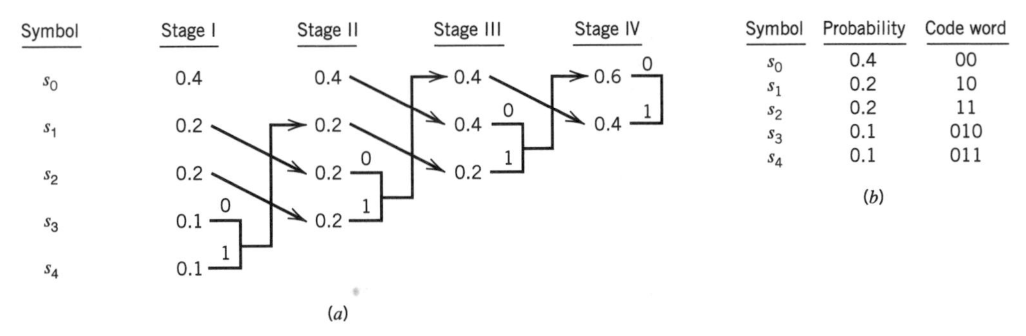


| Symbol | Probability | Code word |
|---|---|---|
| $s_0$ | 0.4 | 00 |
| $s_1$ | 0.2 | 10 |
| $s_2$ | 0.2 | 11 |
| $s_3$ | 0.1 | 010 |
| $s_4$ | 0.1 | 011 |

**FIGURE 9.5** (*a*) Example of the Huffman encoding algorithm. (*b*) Source code.

The average code-word length is therefore

$$\begin{aligned}\bar{L} &= 0.4(2) + 0.2(2) + 0.2(2) + 0.1(3) + 0.1(3) \\ &= 2.2\end{aligned}$$

The entropy of the specified discrete memoryless source is calculated as follows [see Equation (9.9)]:

$$\begin{aligned}H(\mathcal{S}) &= 0.4 \log_2\left(\frac{1}{0.4}\right) + 0.2 \log_2\left(\frac{1}{0.2}\right) + 0.2 \log_2\left(\frac{1}{0.2}\right) \\ &\quad + 0.1 \log_2\left(\frac{1}{0.1}\right) + 0.1 \log_2\left(\frac{1}{0.1}\right) \\ &= 0.52877 + 0.46439 + 0.46439 + 0.33219 + 0.33219 \\ &= 2.12193 \text{ bits}\end{aligned}$$

For the example at hand, we may make two observations:

1. The average code-word length $\bar{L}$ exceeds the entropy $H(\mathcal{S})$ by only 3.67 percent.
2. The average code-word length $\bar{L}$ does indeed satisfy Equation (9.23). ◀

It is noteworthy that the Huffman encoding process (i.e., the Huffman tree) is not unique. In particular, we may cite two variations in the process that are responsible for the nonuniqueness of the Huffman code. First, at each splitting stage in the construction of a Huffman code, there is arbitrariness in the way a 0 and a 1 are assigned to the last two source symbols. Whichever way the assignments are made, however, the resulting differences are trivial. Second, ambiguity arises when the probability of a *combined* symbol (obtained by adding the last two probabilities pertinent to a particular step) is found to equal another probability in the list. We may proceed by placing the probability of the new symbol as *high* as possible, as in Example 9.3. Alternatively, we may place it as *low* as possible. (It is presumed that whichever way the placement is made, high or low, it is consistently adhered to throughout the encoding process.) But this time, noticeable differences arise in that the code words in the resulting source code can have different lengths. Nevertheless, the average code-word length remains the same.

As a measure of the variability in code-word lengths of a source code, we define the *variance* of the average code-word length $\bar{L}$ over the ensemble of source symbols as

$$\sigma^2 = \sum_{k=0}^{K-1} p_k(l_k - \bar{L})^2 \tag{9.30}$$

where $p_0, p_1, \ldots, p_{K-1}$ are the source statistics, and $l_k$ is the length of the code word assigned to source symbol $s_k$. It is usually found that when a combined symbol is moved as high as possible, the resulting Huffman code has a significantly smaller variance $\sigma^2$ than when it is moved as low as possible. On this basis, it is reasonable to choose the former Huffman code over the latter.

## ■ LEMPEL–ZIV CODING

A drawback of the Huffman code is that it requires knowledge of a probabilistic model of the source; unfortunately, in practice, source statistics are not always known *a priori*. Moreover, in modeling text we find that storage requirements prevent the Huffman code from capturing the higher-order relationships between words and phrases, thereby compromising the efficiency of the code. To overcome these practical limitations, we may use the *Lempel–Ziv algorithm*,[7] which is intrinsically *adaptive* and simpler to implement than Huffman coding.

Basically, encoding in the Lempel–Ziv algorithm is accomplished by *parsing the source data stream into segments that are the shortest subsequences not encountered previously*. To illustrate this simple yet elegant idea, consider the example of an input binary sequence specified as follows:

$$000101110010100101 \ldots$$

It is assumed that the binary symbols 0 and 1 are already stored in that order in the code book. We thus write

Subsequences stored: 0, 1
Data to be parsed: 000101110010100101 ...

The encoding process begins at the left. With symbols 0 and 1 already stored, the *shortest subsequence* of the data stream encountered for the first time and not seen before is 00; so we write

Subsequences stored: 0, 1, 00
Data to be parsed: 0101110010100101 ...

The second shortest subsequence not seen before is 01; accordingly, we go on to write

Subsequences stored: 0, 1, 00, 01
Data to be parsed: 01110010100101 ...

The next shortest subsequence not encountered previously is 011; hence, we write

Subsequences stored: 0, 1, 00, 01, 011
Data to be parsed: 10010100101 ...

We continue in the manner described here until the given data stream has been completely parsed. Thus, for the example at hand, we get the *code book* of binary subsequences shown in the second row of Figure 9.6.

| Numerical positions: | 1 | 2 | 3 | 4 | 5 | 6 | 7 | 8 | 9 |
|---|---|---|---|---|---|---|---|---|---|
| Subsequences: | 0 | 1 | 00 | 01 | 011 | 10 | 010 | 100 | 101 |
| Numerical representations: | | | 11 | 12 | 42 | 21 | 41 | 61 | 62 |
| Binary encoded blocks: | | | 0010 | 0011 | 1001 | 0100 | 1000 | 1100 | 1101 |

**FIGURE 9.6** Illustrating the encoding process performed by the Lempel-Ziv algorithm on the binary sequence 000101110010100101. . . .

The first row shown in this figure merely indicates the numerical positions of the individual subsequences in the code book. We now recognize that the first subsequence of the data stream, 00, is made up of the concatenation of the *first* code book entry, 0, with itself; it is therefore represented by the number 11. The second subsequence of the data stream, 01, consists of the *first* code book entry, 0, concatenated with the *second* code book entry, 1; it is therefore represented by the number 12. The remaining subsequences are treated in a similar fashion. The complete set of numerical representations for the various subsequences in the code book is shown in the third row of Figure 9.6. As a further example illustrating the composition of this row, we note that the subsequence 010 consists of the concatenation of the subsequence 01 in position 4 and symbol 0 in position 1; hence, the numerical representation 41. The last row shown in Figure 9.6 is the binary encoded representation of the different subsequences of the data stream.

The last symbol of each subsequence in the code book (i.e., the second row of Figure 9.6) is an *innovation symbol*, which is so called in recognition of the fact that its appendage to a particular subsequence distinguishes it from all previous subsequences stored in the code book. Correspondingly, the last bit of each uniform block of bits in the binary encoded representation of the data stream (i.e., the fourth row in Figure 9.6) represents the innovation symbol for the particular subsequence under consideration. The remaining bits provide the equivalent binary representation of the "pointer" to the *root subsequence* that matches the one in question except for the innovation symbol.

The decoder is just as simple as the encoder. Specifically, it uses the pointer to identify the root subsequence and then appends the innovation symbol. Consider, for example, the binary encoded block 1101 in position 9. The last bit, 1, is the innovation symbol. The remaining bits, 110, point to the root subsequence 10 in position 6. Hence, the block 1101 is decoded into 101, which is correct.

From the example described here, we note that, in contrast to Huffman coding, the Lempel–Ziv algorithm uses fixed-length codes to represent a variable number of source symbols; this feature makes the Lempel–Ziv code suitable for synchronous transmission. In practice, fixed blocks of 12 bits long are used, which implies a code book of 4096 entries.

For a long time, Huffman coding was unchallenged as the algorithm of choice for data compaction. However, the Lempel–Ziv algorithm has taken over almost completely from the Huffman algorithm. The Lempel–Ziv algorithm is now the standard algorithm for file compression. When it is applied to ordinary English text, the Lempel–Ziv algorithm achieves a compaction of approximately 55 percent. This is to be contrasted with a compaction of approximately 43 percent achieved with Huffman coding. The reason for this behavior is that, as mentioned previously, Huffman coding does not take advantage of the intercharacter redundancies of the language. On the other hand, the Lempel–Ziv algorithm is able to do the best possible compaction of text (within certain limits) by working effectively at higher levels.

## 9.5 *Discrete Memoryless Channels*

Up to this point in the chapter, we have been preoccupied with discrete memoryless sources responsible for information generation. We next consider the issue of information transmission, with particular emphasis on reliability. We start the discussion by considering a discrete memoryless channel, the counterpart of a discrete memoryless source.

A *discrete memoryless channel* is a statistical model with an input $X$ and an output $Y$ that is a *noisy* version of $X$; both $X$ and $Y$ are random variables. Every unit of time, the

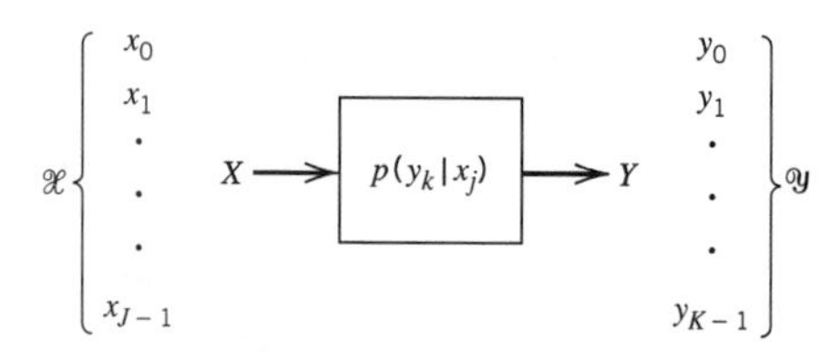


**FIGURE 9.7** Discrete memoryless channel.

channel accepts an input symbol $X$ selected from an alphabet $\mathscr{X}$ and, in response, it emits an output symbol $Y$ from an alphabet $\mathscr{Y}$. The channel is said to be "discrete" when both of the alphabets $\mathscr{X}$ and $\mathscr{Y}$ have *finite* sizes. It is said to be "memoryless" when the current output symbol depends *only* on the current input symbol and *not* any of the previous ones.

Figure 9.7 depicts a view of a discrete memoryless channel. The channel is described in terms of an *input alphabet*

$$\mathscr{X} = \{x_0, x_1, \ldots, x_{J-1}\}, \tag{9.31}$$

an *output alphabet*,

$$\mathscr{Y} = \{y_0, y_1, \ldots, y_{K-1}\}, \tag{9.32}$$

and a set of *transition probabilities*

$$p(y_k|x_j) = P(Y = y_k|X = x_j) \qquad \text{for all } j \text{ and } k \tag{9.33}$$

Naturally, we have

$$0 \leq p(y_k|x_j) \leq 1 \qquad \text{for all } j \text{ and } k \tag{9.34}$$

Also, the input alphabet $\mathscr{X}$ and output alphabet $\mathscr{Y}$ need not have the same size. For example, in channel coding, the size $K$ of the output alphabet $\mathscr{Y}$ may be larger than the size $J$ of the input alphabet $\mathscr{X}$; thus, $K \geq J$. On the other hand, we may have a situation in which the channel emits the same symbol when either one of two input symbols is sent, in which case we have $K \leq J$.

A convenient way of describing a discrete memoryless channel is to arrange the various transition probabilities of the channel in the form of a matrix as follows:

$$\mathbf{P} = \begin{bmatrix} p(y_0|x_0) & p(y_1|x_0) & \cdots & p(y_{K-1}|x_0) \\ p(y_0|x_1) & p(y_1|x_1) & \cdots & p(y_{K-1}|x_1) \\ \vdots & \vdots & & \vdots \\ p(y_0|x_{J-1}) & p(y_1|x_{J-1}) & \cdots & p(y_{K-1}|x_{J-1}) \end{bmatrix} \tag{9.35}$$

The $J$-by-$K$ matrix $\mathbf{P}$ is called the *channel matrix*, or *transition matrix*. Note that each *row* of the channel matrix $\mathbf{P}$ corresponds to a *fixed channel input*, whereas each column of the matrix corresponds to a *fixed channel output*. Note also that a fundamental property of the channel matrix $\mathbf{P}$, as defined here, is that the sum of the elements along any row of the matrix is always equal to one; that is,

$$\sum_{k=0}^{K-1} p(y_k|x_j) = 1 \qquad \text{for all } j \tag{9.36}$$

Suppose now that the inputs to a discrete memoryless channel are selected according to the *probability distribution* $\{p(x_j), j = 0, 1, \ldots, J-1\}$. In other words, the event that the channel input $X = x_j$ occurs with probability

$$p(x_j) = P(X = x_j) \qquad \text{for } j = 0, 1, \ldots, J-1 \tag{9.37}$$

Having specified the random variable $X$ denoting the channel input, we may now specify the second random variable $Y$ denoting the channel output. The *joint probability distribution* of the random variables $X$ and $Y$ is given by

$$\begin{aligned} p(x_j, y_k) &= P(X = x_j, Y = y_k) \\ &= P(Y = y_k | X = x_j)P(X = x_j) \\ &= p(y_k | x_j)p(x_j) \end{aligned} \tag{9.38}$$

The *marginal probability distribution* of the output random variable $Y$ is obtained by averaging out the dependence of $p(x_j, y_k)$ on $x_j$, as shown by

$$\begin{aligned} p(y_k) &= P(Y = y_k) \\ &= \sum_{j=0}^{J-1} P(Y = y_k | X = x_j)P(X = x_j) \\ &= \sum_{j=0}^{J-1} p(y_k | x_j)p(x_j) \qquad \text{for } k = 0, 1, \ldots, K-1 \end{aligned} \tag{9.39}$$

The probabilities $p(x_j)$ for $j = 0, 1, \ldots, J-1$, are known as the *a priori probabilities* of the various input symbols. Equation (9.39) states that if we are given the input *a priori* probabilities $p(x_j)$ and the channel matrix [i.e., the matrix of transition probabilities $p(y_k | x_j)$], then we may calculate the probabilities of the various output symbols, the $p(y_k)$.

### ▶ EXAMPLE 9.4 Binary Symmetric Channel

The *binary symmetric channel* is of great theoretical interest and practical importance. It is a special case of the discrete memoryless channel with $J = K = 2$. The channel has two input symbols ($x_0 = 0, x_1 = 1$) and two output symbols ($y_0 = 0, y_1 = 1$). The channel is symmetric because the probability of receiving a 1 if a 0 is sent is the same as the probability of receiving a 0 if a 1 is sent. This conditional probability of error is denoted by $p$. The *transition probability diagram* of a binary symmetric channel is as shown in Figure 9.8. ◀

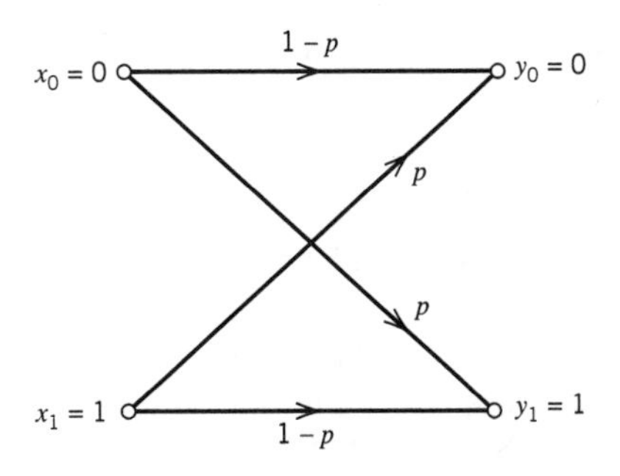


**FIGURE 9.8** Transition probability diagram of binary symmetric channel.

It is of interest to relate the transition probability diagram of Figure 9.8 to the conditional probabilities of error $p_{10}$ and $p_{01}$ that were determined for the PCM receiver in Section 3.3. For the case when the binary symbols 0 and 1 are equiprobable, we showed that the optimized values of these two error probabilities are equal. Indeed, recalling the following definitions (using the terminology of Figure 9.8):

$$p_{10} = P(y = 1 | x = 0)$$

and

$$p_{01} = P(y = 0 | x = 1)$$

we immediately see that for the PCM receiver of Figure 3.4:

$$p_{10} = p_{01} = p$$

## 9.6 *Mutual Information*

Given that we think of the channel output $Y$ (selected from alphabet $\mathcal{Y}$) as a noisy version of the channel input $X$ (selected from alphabet $\mathcal{X}$), and that the entropy $H(\mathcal{X})$ is a measure of the prior uncertainty about $X$, how can we measure the uncertainty about $X$ after observing $Y$? To answer this question, we extend the ideas developed in Section 9.2 by defining the *conditional entropy* of $X$ selected from alphabet $\mathcal{X}$, given that $Y = y_k$. Specifically, we write

$$H(\mathcal{X} | Y = y_k) = \sum_{j=0}^{J-1} p(x_j | y_k) \log_2 \left[ \frac{1}{p(x_j | y_k)} \right] \tag{9.40}$$

This quantity is itself a random variable that takes on the values $H(\mathcal{X} | Y = y_0), \ldots, H(\mathcal{X} | Y = y_{K-1})$ with probabilities $p(y_0), \ldots, p(y_{K-1})$, respectively. The mean of entropy $H(\mathcal{X} | Y = y_k)$ over the output alphabet $\mathcal{Y}$ is therefore given by

$$\begin{aligned} H(\mathcal{X} | \mathcal{Y}) &= \sum_{k=0}^{K-1} H(\mathcal{X} | Y = y_k) p(y_k) \\ &= \sum_{k=0}^{K-1} \sum_{j=0}^{J-1} p(x_j | y_k) p(y_k) \log_2 \left[ \frac{1}{p(x_j | y_k)} \right] \\ &= \sum_{k=0}^{K-1} \sum_{j=0}^{J-1} p(x_j, y_k) \log_2 \left[ \frac{1}{p(x_j | y_k)} \right] \end{aligned} \tag{9.41}$$

where, in the last line, we have made use of the relation

$$p(x_j, y_k) = p(x_j | y_k) p(y_k) \tag{9.42}$$

The quantity $H(\mathcal{X} | \mathcal{Y})$ is called a *conditional entropy*. It represents *the amount of uncertainty remaining about the channel input after the channel output has been observed.*

Since the entropy $H(\mathcal{X})$ represents our uncertainty about the channel input *before* observing the channel output, and the conditional entropy $H(\mathcal{X} | \mathcal{Y})$ represents our uncertainty about the channel input *after* observing the channel output, it follows that the difference $H(\mathcal{X}) - H(\mathcal{X} | \mathcal{Y})$ must represent our uncertainty about the channel input that is *resolved* by observing the channel output. This important quantity is called the *mutual information* of the channel. Denoting the mutual information by $I(\mathscr{X}; \mathscr{Y})$, we may thus write

$$I(\mathscr{X}; \mathscr{Y}) = H(\mathscr{X}) - H(\mathscr{X}|\mathscr{Y}) \tag{9.43}$$

Similarly, we may write

$$I(\mathscr{Y}; \mathscr{X}) = H(\mathscr{Y}) - H(\mathscr{Y}|\mathscr{X}) \tag{9.44}$$

where $H(\mathscr{Y})$ is the entropy of the channel output and $H(\mathscr{Y}|\mathscr{X})$ is the conditional entropy of the channel output given the channel input.

## ■ Properties of Mutual Information

The mutual inforamtion $I(\mathscr{X}; \mathscr{Y})$ has the following important properties.

### Property 1

*The mutual information of a channel is symmetric; that is*

$$I(\mathscr{X}; \mathscr{Y}) = I(\mathscr{Y}; \mathscr{X}) \tag{9.45}$$

where the mutual information $I(\mathscr{X}; \mathscr{Y})$ is a measure of the uncertainty about the channel input that is resolved by *observing* the channel output, and the mutual information $I(\mathscr{Y}; \mathscr{X})$ is a measure of the uncertainty about the channel output that is resolved by *sending* the channel input.

To prove this property, we first use the formula for entropy and then use Equations (9.36) and (9.38), in that order, to express $H(\mathscr{X})$ as

$$\begin{aligned}
H(\mathscr{X}) &= \sum_{j=0}^{J-1} p(x_j) \log_2\left[\frac{1}{p(x_j)}\right] \\
&= \sum_{j=0}^{J-1} p(x_j) \log_2\left[\frac{1}{p(x_j)}\right] \sum_{k=0}^{K-1} p(y_k|x_j) \\
&= \sum_{j=0}^{J-1} \sum_{k=0}^{K-1} p(y_k|x_j)p(x_j) \log_2\left[\frac{1}{p(x_j)}\right] \\
&= \sum_{j=0}^{J-1} \sum_{k=0}^{K-1} p(x_j, y_k) \log_2\left[\frac{1}{p(x_j)}\right]
\end{aligned} \tag{9.46}$$

Hence, substituting Equations (9.41) and (9.46) into Equation (9.43) and then combining terms, we obtain

$$I(\mathscr{X}; \mathscr{Y}) = \sum_{j=0}^{J-1} \sum_{k=0}^{K-1} p(x_j, y_k) \log_2\left[\frac{p(x_j|y_k)}{p(x_j)}\right] \tag{9.47}$$

From *Bayes' rule* for conditional probabilities, we have [see Equations (9.38) and (9.42)]

$$\frac{p(x_j|y_k)}{p(x_j)} = \frac{p(y_k|x_j)}{p(y_k)} \tag{9.48}$$

Hence, substituting Equation (9.48) into Equation (9.47) and interchanging the order of summation, we may write

$$\begin{aligned} I(\mathscr{X}; \mathscr{Y}) &= \sum_{k=0}^{K-1} \sum_{j=0}^{J-1} p(x_j, y_k) \log_2\left[\frac{p(y_k|x_j)}{p(y_k)}\right] \\ &= I(\mathscr{Y}; \mathscr{X}) \end{aligned} \tag{9.49}$$

which is the desired result.

**Property 2**

*The mutual information is always nonnegative; that is*

$$I(\mathscr{X}; \mathscr{Y}) \geq 0 \tag{9.50}$$

To prove this property, we first note from Equation (9.42) that

$$p(x_j|y_k) = \frac{p(x_j, y_k)}{p(y_k)} \tag{9.51}$$

Hence, substituting Equation (9.51) into Equation (9.47), we may express the mutual information of the channel as

$$I(\mathscr{X}; \mathscr{Y}) = \sum_{j=0}^{J-1} \sum_{k=0}^{K-1} p(x_j, y_k) \log_2\left(\frac{p(x_j, y_k)}{p(x_j)p(y_k)}\right) \tag{9.52}$$

Next, a direct application of the fundamental inequality [defined by Equation (9.12)] yields the desired result

$$I(\mathscr{X}; \mathscr{Y}) \geq 0$$

with equality if, and only if,

$$p(x_j, y_k) = p(x_j)p(y_k) \qquad \text{for all } j \text{ and } k \tag{9.53}$$

Property 2 states that *we cannot lose information, on the average, by observing the output of a channel*. Moreover, the mutual information is zero if, and only if, the input and output symbols of the channel are statistically independent, as in Equation (9.53).

**Property 3**

*The mutual information of a channel is related to the joint entropy of the channel input and channel output by*

$$I(\mathscr{X}; \mathscr{Y}) = H(\mathscr{X}) + H(\mathscr{Y}) - H(\mathscr{X}, \mathscr{Y}) \tag{9.54}$$

*where the joint entropy $H(\mathscr{X}, \mathscr{Y})$ is defined by*

$$H(\mathscr{X}, \mathscr{Y}) = \sum_{j=0}^{J-1} \sum_{k=0}^{K-1} p(x_j, y_k) \log_2\left(\frac{1}{p(x_j, y_k)}\right) \tag{9.55}$$

To prove Equation (9.54), we first rewrite the definition for the joint entropy $H(\mathscr{X}, \mathscr{Y})$ as

$$\begin{aligned} H(\mathscr{X}, \mathscr{Y}) &= \sum_{j=0}^{J-1} \sum_{k=0}^{K-1} p(x_j, y_k) \log_2\left[\frac{p(x_j)p(y_k)}{p(x_j, y_k)}\right] \\ &\quad + \sum_{j=0}^{J-1} \sum_{k=0}^{K-1} p(x_j, y_k) \log_2\left[\frac{1}{p(x_j)p(y_k)}\right] \end{aligned} \tag{9.56}$$

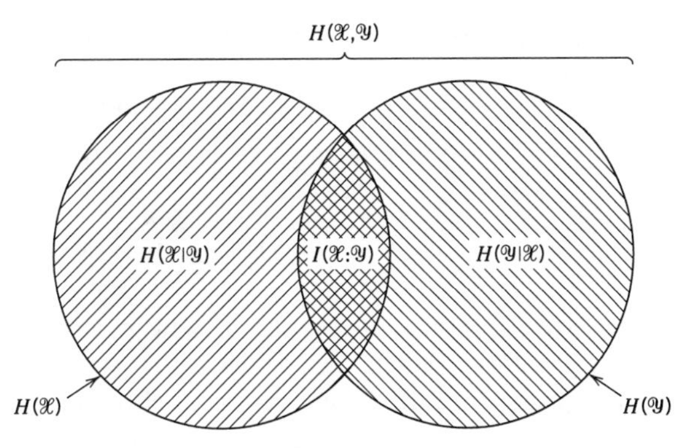


**FIGURE 9.9** Illustrating the relations among various channel entropies.

The first double summation term on the right-hand side of Equation (9.56) is recognized as the negative of the mutual information of the channel, $I(\mathscr{X}; \mathscr{Y})$, previously given in Equation (9.52). As for the second summation term, we manipulate it as follows:

$$
\begin{aligned}
\sum_{j=0}^{J-1} \sum_{k=0}^{K-1} p(x_j, y_k) \log_2\left[\frac{1}{p(x_j)p(y_k)}\right] &= \sum_{j=0}^{J-1} \log_2\left[\frac{1}{p(x_j)}\right] \sum_{k=0}^{K-1} p(x_j, y_k) \\
&\quad + \sum_{k=0}^{K-1} \log_2\left[\frac{1}{p(y_k)}\right] \sum_{j=0}^{J-1} p(x_j, y_k) \\
&= \sum_{j=0}^{J-1} p(x_j) \log_2\left[\frac{1}{p(x_j)}\right] \\
&\quad + \sum_{k=0}^{K-1} p(y_k) \log_2\left[\frac{1}{p(y_k)}\right] \\
&= H(\mathscr{X}) + H(\mathscr{Y})
\end{aligned}
\tag{9.57}
$$

Accordingly, using Equations (9.52) and (9.57) in Equation (9.56), we get the result

$$
H(\mathscr{X}, \mathscr{Y}) = -I(\mathscr{X}; \mathscr{Y}) + H(\mathscr{X}) + H(\mathscr{Y}) \tag{9.58}
$$

Rearranging terms in this equation, we get the result given in Equation (9.54), thereby confirming Property 3.

We conclude our discussion of the mutual information of a channel by providing a diagramatic interpretation of Equations (9.43), (9.44), and (9.54). The interpretation is given in Figure 9.9. The entropy of channel input $X$ is represented by the circle on the left. The entropy of channel output $Y$ is represented by the circle on the right. The mutual information of the channel is represented by the overlap between these two circles.

## 9.7 *Channel Capacity*

Consider a discrete memoryless channel with input alphabet $\mathscr{X}$, output alphabet $\mathscr{Y}$, and transition probabilities $p(y_k | x_j)$, where $j = 0, 1, \ldots, J - 1$ and $k = 0, 1, \ldots, K - 1$. The mutual information of the channel is defined by the first line of Equation (9.49), which is reproduced here for convenience:

$$
I(\mathscr{X}; \mathscr{Y}) = \sum_{k=0}^{K-1} \sum_{j=0}^{J-1} p(x_j, y_k) \log_2\left[\frac{p(y_k | x_j)}{p(y_k)}\right]
$$

Here we note that [see Equation (9.38)]

$$p(x_j, y_k) = p(y_k | x_j) p(x_j)$$

Also, from Equation (9.39), we have

$$p(y_k) = \sum_{j=0}^{J-1} p(y_k | x_j) p(x_j)$$

From these three equations we see that it is necessary for us to know the input probability distribution $\{p(x_j) | j = 0, 1, \ldots, J-1\}$ so that we may calculate the mutual information $I(\mathscr{X}; \mathscr{Y})$. The mutual information of a channel therefore depends not only on the channel but also on the way in which the channel is used.

The input probability distribution $\{p(x_j)\}$ is obviously independent of the channel. We can then maximize the mutual information $I(\mathscr{X}; \mathscr{Y})$ of the channel with respect to $\{p(x_j)\}$. Hence, *we define the channel capacity of a discrete memoryless channel as the maximum mutual information* $I(\mathscr{X}; \mathscr{Y})$ *in any single use of the channel (i.e., signaling interval), where the maximization is over all possible input probability distributions* $\{p(x_j)\}$ *on* $\mathscr{X}$. The channel capacity is commonly denoted by $C$. We thus write

$$C = \max_{\{p(x_j)\}} I(\mathscr{X}; \mathscr{Y}) \tag{9.59}$$

The channel capacity $C$ is measured in *bits per channel use*, or *bits per transmission.*

Note that the channel capacity $C$ is a function only of the transition probabilities $p(y_k | x_j)$, which define the channel. The calculation of $C$ involves maximization of the mutual information $I(\mathscr{X}; \mathscr{Y})$ over $J$ variables [i.e., the input probabilities $p(x_0), \ldots, p(x_{J-1})$] subject to two constraints:

$$p(x_j) \geq 0 \text{ for all } j$$

and

$$\sum_{j=0}^{J-1} p(x_j) = 1$$

In general, the variational problem of finding the channel capacity $C$ is a challenging task.

### ▶ EXAMPLE 9.5 Binary Symmetric Channel (Revisited)

Consider again the *binary symmetric channel*, which is described by the *transition probability diagram* of Figure 9.8. This diagram is uniquely defined by the conditional probability of error $p$.

The entropy $H(X)$ is maximized when the channel input probability $p(x_0) = p(x_1) = 1/2$, where $x_0$ and $x_1$ are each 0 or 1. The mutual information $I(\mathscr{X}; \mathscr{Y})$ is similarly maximized, so that we may write

$$C = I(\mathscr{X}; \mathscr{Y}) |_{p(x_0)=p(x_1)=1/2}$$

From Figure 9.8, we have

$$p(y_0 | x_1) = p(y_1 | x_0) = p$$

and

$$p(y_0 | x_0) = p(y_1 | x_1) = 1 - p$$

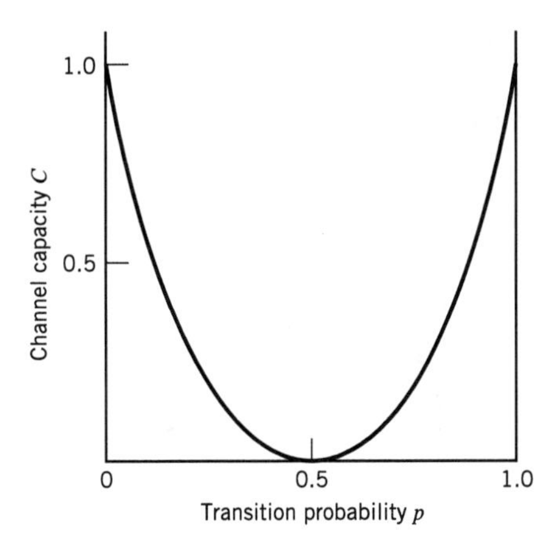


**FIGURE 9.10** Variation of channel capacity of a binary symmetric channel with transition probability $p$.

Therefore, substituting these channel transition probabilities into Equation (9.49) with $J = K = 2$, and then setting the input probability $p(x_0) = p(x_1)$ in accordance with Equation (9.59), we find that the capacity of the binary symmetric channel is

$$C = 1 + p \log_2 p + (1 - p) \log_2(1 - p) \tag{9.60}$$

Using the definition of the entropy function given in Equation (9.16), we may reduce Equation (9.60) to

$$C = 1 - H(p)$$

The channel capacity $C$ varies with the probability of error (transition probability) $p$ in a convex manner as shown in Figure 9.10, which is symmetric about $p = 1/2$. Comparing the curve in this figure with that in Figure 9.2, we may make the following observations:

1. When the channel is *noise free*, permitting us to set $p = 0$, the channel capacity $C$ attains its maximum value of one bit per channel use, which is exactly the information in each channel input. At this value of $p$, the entropy function $H(p)$ attains its minimum value of zero.
2. When the conditional probability of error $p = 1/2$ due to noise, the channel capacity $C$ attains its minimum value of zero, whereas the entropy function $H(p)$ attains its maximum value of unity; in such a case the channel is said to be *useless*. ◀

## 9.8 Channel-Coding Theorem

The inevitable presence of *noise* in a channel causes discrepancies (errors) between the output and input data sequences of a digital communication system. For a relatively noisy channel (e.g., wireless communication channel), the probability of error may reach a value as high as $10^{-1}$, which means that (on the average) only 9 out of 10 transmitted bits are received correctly. For many applications, this *level of reliability* is unacceptable. Indeed, a probability of error equal to $10^{-6}$ or even lower is often a necessary requirement. To achieve such a high level of performance, we resort to the use of channel coding.

The design goal of channel coding is to increase the resistance of a digital communication system to channel noise. Specifically, *channel coding* consists of *mapping* the incoming data sequence into a channel input sequence, and *inverse mapping* the channel output sequence into an output data sequence in such a way that the overall effect of channel noise on the system is minimized. The first mapping operation is performed in the transmitter by a *channel encoder*, whereas the inverse mapping operation is performed in the receiver by a *channel decoder*, as shown in the block diagram of Figure 9.11; to simplify the exposition, we have not included source encoding (before channel encoding) and source decoding (after channel decoding) in Figure 9.11.

The channel encoder and channel decoder in Figure 9.11 are both under the designer's control and should be designed to optimize the overall reliability of the communication system. The approach taken is to introduce *redundancy* in the channel encoder so as to reconstruct the original source sequence as accurately as possible. Thus, in a rather loose sense, we may view channel coding as the *dual* of source coding in that the former introduces controlled redundancy to improve reliability, whereas the latter reduces redundancy to improve efficiency.

The subject of channel coding is treated in detail in Chapter 10. For the purpose of our present discussion, it suffices to confine our attention to *block codes*. In this class of codes, the message sequence is subdivided into sequential blocks each $k$ bits long, and each $k$-bit block is *mapped* into an $n$-bit block, where $n > k$. The number of redundant bits added by the encoder to each transmitted block is $n - k$ bits. The ratio $k/n$ is called the *code rate*. Using $r$ to denote the code rate, we may thus write

$$r = \frac{k}{n}$$

where, of course, $r$ is less than unity. For a prescribed $k$, the code rate $r$ (and therefore the system's coding efficiency) approaches zero as the block length $n$ approaches infinity.

The accurate reconstruction of the original source sequence at the destination requires that the *average probability of symbol error* be arbitrarily low. This raises the following important question: Does there exist a channel coding scheme such that the probability that a message bit will be in error is less than any positive number $\epsilon$ (i.e., as small as we want it), and yet the channel coding scheme is efficient in that the code rate need not be too small? The answer to this fundamental question is an emphatic "yes." Indeed, the answer to the question is provided by Shannon's second theorem in terms of the channel capacity $C$, as described in what follows. Up until this point, *time* has not played an important role in our discussion of channel capacity. Suppose then the discrete memoryless source in Figure 9.11 has the source alphabet $\mathcal{S}$ and entropy $H(\mathcal{S})$ bits per source symbol. We assume that the source emits symbols once every $T_s$ seconds. Hence, the *average information rate* of the source is $H(\mathcal{S})/T_s$ bits per second. The decoder delivers decoded symbols to the destination from the source alphabet $\mathcal{S}$ and at the same source rate of one symbol every $T_s$ seconds. The discrete memoryless channel has a channel capacity equal to $C$ bits per use of the channel. We assume that the channel is capable of being used once every $T_c$ seconds. Hence, the *channel capacity per unit time is* $C/T_c$ bits per second, which represents the maximum rate of information transfer over the channel. We are now ready to state Shannon's second theorem, known as the channel coding theorem.

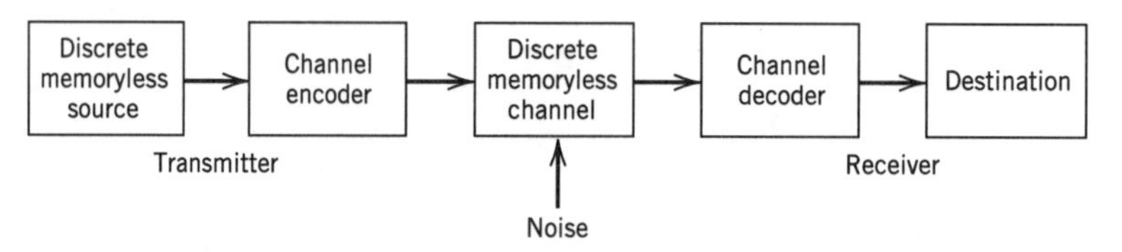


**FIGURE 9.11** Block diagram of digital communication system.

Specifically, the *channel coding theorem*[8] for a discrete memoryless channel is stated in two parts as follows.

(i) Let a discrete memoryless source with an alphabet $\mathcal{S}$ have entropy $H(\mathcal{S})$ and produce symbols once every $T_s$ seconds. Let a discrete memoryless channel have capacity $C$ and be used once every $T_c$ seconds. Then, if

$$\frac{H(\mathcal{S})}{T_s} \leq \frac{C}{T_c} \tag{9.61}$$

there exists a coding scheme for which the source output can be transmitted over the channel and be reconstructed with an arbitrarily small probability of error. The parameter $C/T_c$ is called the *critical rate*. When Equation (9.61) is satisfied with the equality sign, the system is said to be signaling at the critical rate.

(ii) Conversely, if

$$\frac{H(\mathcal{S})}{T_s} > \frac{C}{T_c}$$

it is not possible to transmit information over the channel and reconstruct it with an arbitrarily small probability of error.

The channel coding theorem is the single most important result of information theory. The theorem specifies the channel capacity $C$ as a *fundamental limit* on the rate at which the transmission of reliable error-free messages can take place over a discrete memoryless channel. However, it is important to note the following:

- The channel coding theorem does not show us how to construct a good code. Rather, the theorem should be viewed as an *existence proof* in the sense that it tells us that if the condition of Equation (9.61) is satisfied, then good codes do exist. (Later in Chapter 10 we describe several good codes for discrete memoryless channels.)
- The theorem does not have a precise result for the probability of symbol error after decoding the channel output. Rather, it tells us that the probability of symbol error tends to zero as the length of the code increases, again provided that the condition of Equation (9.61) is satisfied.

Note also that power and bandwidth constraints were hidden in the discussion presented here. Nevertheless, these two system constraints do actually show up in the channel matrix **P** of the discrete memoryless channel. This observation is readily confirmed by linking the results of Example 9.5 on the binary symmetric channel with the noise analysis for the PCM receiver presented in Section 5.3.

## Application of the Channel Coding Theorem to Binary Symmetric Channels

Consider a discrete memoryless source that emits equally likely binary symbols (0s and 1s) once every $T_s$ seconds. With the source entropy equal to one bit per source symbol (see Example 9.1), the information rate of the source is $(1/T_s)$ bits per second. The source sequence is applied to a channel encoder with code rate $r$. The channel encoder produces a symbol once every $T_c$ seconds. Hence, the encoded symbol transmission rate is $(1/T_c)$ symbols per second. The channel encoder engages a binary symmetric channel once every $T_c$ seconds. Hence, the channel capacity per unit time is $(C/T_c)$ bits per second, where $C$

is determined by the prescribed channel transition probability $p$ in accordance with Equation (9.60). Accordingly, the channel coding theorem [part (i)] implies that if

$$\frac{1}{T_s} \leq \frac{C}{T_c} \tag{9.62}$$

the probability of error can be made arbitrarily low by the use of a suitable channel encoding scheme. But the ratio $T_c/T_s$ equals the code rate of the channel encoder:

$$r = \frac{T_c}{T_s} \tag{9.63}$$

Hence, we may restate the condition of Equation (9.62) simply as

$$r \leq C \tag{9.64}$$

That is, for $r \leq C$, there exists a code (with code rate less than or equal to C) capable of achieving an arbitrarily low probability of error.

### ▶ EXAMPLE 9.6 Repetition Code

In this example, we present a graphical interpretation of the channel coding theorem. We also bring out a surprising aspect of the theorem by taking a look at a simple coding scheme.

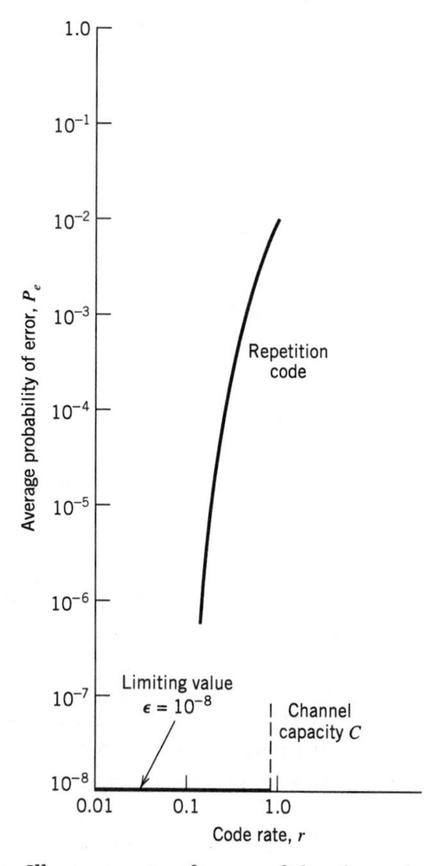


**FIGURE 9.12** Illustrating significance of the channel coding theorem.

**TABLE 9.3 Average probability of error for repetition code**

| Code Rate, r = 1/n | Average Probability of Error, $P_e$ |
|---|---|
| 1 | $10^{-2}$ |
| $\frac{1}{3}$ | $3 \times 10^{-4}$ |
| $\frac{1}{5}$ | $10^{-6}$ |
| $\frac{1}{7}$ | $4 \times 10^{-7}$ |
| $\frac{1}{9}$ | $10^{-8}$ |
| $\frac{1}{11}$ | $5 \times 10^{-10}$ |

Consider first a binary symmetric channel with transition probability $p = 10^{-2}$. For this value of $p$, we find from Equation (9.60) that the channel capacity $C = 0.9192$. Hence, from the channel coding theorem, we may state that for any $\epsilon > 0$ and $r \leq 0.9192$, there exists a code of large enouigh length $n$ and code rate $r$, and an appropriate decoding algorithm, such that when the coded bit stream is sent over the given channel, the average probability of channel decoding error is less than $\epsilon$. This result is depicted in Figure 9.12 for the limiting value $\epsilon = 10^{-8}$.

To put the significance of this result in perspective, consider next a simple coding scheme that involves the use of a *repetition code*, in which each bit of the message is repeated several times. Let each bit (0 or 1) be repeated $n$ times, where $n = 2m + 1$ is an odd integer. For example, for $n = 3$, we transmit 0 and 1 as 000 and 111, respectively. Intuitively, it would seem logical to use a *majority rule* for decoding, which operates as follows: *If in a block of n received bits (representing one bit of the message), the number of 0s exceeds the number of 1s, the decoder decides in favor of a 0. Otherwise, it decides in favor of a 1.* Hence, an error occurs when $m + 1$ or more bits out of $n = 2m + 1$ bits are received incorrectly. Because of the assumed symmetric nature of the channel, the *average probability of error* $P_e$ is independent of the *a priori* probabilities of 0 and 1. Accordingly, we find that $P_e$ is given by (see Problem 9.24)

$$P_e = \sum_{i=m+1}^{n} \binom{n}{i} p^i (1 - p)^{n-i} \tag{9.65}$$

where $p$ is the transition probability of the channel.

Table 9.3 gives the average probability of error $P_e$ for a repetition code, which is calculated by using Equation (9.65) for different values of the code rate $r$. The values given here assume the use of a binary symmetric channel with transition probability $p = 10^{-2}$. The improvement in reliability displayed in Table 9.3 is achieved at the cost of decreasing code rate. The results of this table are also shown plotted as the curve labeled "repetition code" in Figure 9.12. This curve illustrates the *exchange of code rate for message reliability*, which is a characteristic of repetition codes.

This example highlights the unexpected result presented to us by the channel coding theorem. The result is that it is not necessary to have the code rate $r$ approach zero (as in the case of repetition codes) so as to achieve more and more reliable operation of the communication link. The theorem merely requires that the code rate be less than the channel capacity $C$. ◀

## 9.9 Differential Entropy and Mutual Information for Continuous Ensembles

The sources and channels considered in our discussion of information-theoretic concepts thus far have involved ensembles of random variables that are *discrete* in amplitude. In this section, we extend some of these concepts to *continuous* random variables and random vectors. The motivation for doing so is to pave the way for the description of another fundamental limit in information theory, which we take up in Section 9.10.

Consider a continuous random variable $X$ with the *probability density function* $f_X(x)$. By analogy with the entropy of a discrete random variable, we introduce the following definition:

$$h(X) = \int_{-\infty}^{\infty} f_X(x) \log_2\left[\frac{1}{f_X(x)}\right] dx \tag{9.66}$$

We refer to $h(X)$ as the *differential entropy of* $X$ to distinguish it from the ordinary or *absolute entropy*. We do so in recognition of the fact that although $h(X)$ is a useful mathematical quantity to know, it is *not* in any sense a measure of the randomness of $X$. Nevertheless, we justify the use of Equation (9.66) in what follows. We begin by viewing the continuous random variable $X$ as the limiting form of a discrete random variable that assumes the value $x_k = k\,\Delta x$, where $k = 0, \pm 1, \pm 2, \ldots$, and $\Delta x$ approaches zero. By definition, the continuous random variable $X$ assumes a value in the interval $[x_k, x_k + \Delta x]$ with probability $f_X(x_k)\,\Delta x$. Hence, permitting $\Delta x$ to approach zero, the ordinary entropy of the continuous random variable $X$ may be written in the limit as follows:

$$\begin{aligned} H(X) &= \lim_{\Delta x \to 0} \sum_{k=-\infty}^{\infty} f_X(x_k)\,\Delta x \log_2\left(\frac{1}{f_X(x_k)\,\Delta x}\right) \\ &= \lim_{\Delta x \to 0} \left[\sum_{k=-\infty}^{\infty} f_X(x_k) \log_2\left(\frac{1}{f_X(x_k)}\right) \Delta x - \log_2 \Delta x \sum_{k=-\infty}^{\infty} f_X(x_k)\,\Delta x\right] \\ &= \int_{-\infty}^{\infty} f_X(x) \log_2\left(\frac{1}{f_X(x)}\right) dx - \lim_{\Delta x \to 0} \log_2 \Delta x \int_{-\infty}^{\infty} f_X(x)\,dx \\ &= h(X) - \lim_{\Delta x \to 0} \log_2 \Delta x \end{aligned} \tag{9.67}$$

where, in the last line, we have made use of Equation (9.66) and the fact that the total area under the curve of the probability density function $f_X(x)$ is unity. In the limit as $\Delta x$ approaches zero, $-\log_2 \Delta x$ approaches infinity. This means that the entropy of a continuous random variable is infinitely large. Intuitively, we would expect this to be true, because a continuous random variable may assume a value anywhere in the interval $(-\infty, \infty)$ and the uncertainty associated with the variable is on the order of infinity. We avoid the problem associated with the term $\log_2 \Delta x$ by adopting $h(X)$ as a differential entropy, with the term $-\log_2 \Delta x$ serving as reference. Moreover, since the information transmitted over a channel is actually the difference between two entropy terms that have a common reference, the information will be the same as the difference between the corresponding differential entropy terms. We are therefore perfectly justified in using the term $h(X)$, defined in Equation (9.66), as the differential entropy of the continuous random variable $X$.

When we have a continuous random vector $\mathbf{X}$ consisting of $n$ random variables $X_1, X_2, \ldots, X_n$, we define the differential entropy of $\mathbf{X}$ as the *n-fold integral*

$$h(\mathbf{X}) = \int_{-\infty}^{\infty} f_{\mathbf{X}}(\mathbf{x}) \log_2\left[\frac{1}{f_{\mathbf{X}}(\mathbf{x})}\right] d\mathbf{x} \tag{9.68}$$

where $f_{\mathbf{X}}(\mathbf{x})$ is the *joint probability density function* of $\mathbf{X}$.

### ▶ EXAMPLE 9.7 Uniform Distribution

Consider a random variable $X$ uniformly distributed over the interval $(0, a)$. The probability density function of $X$ is

$$f_X(x) = \begin{cases} \dfrac{1}{a}, & 0 < x < a \\ 0, & \text{otherwise} \end{cases}$$

Applying Equation (9.66) to this distribution, we get

$$\begin{aligned} h(X) &= \int_0^a \frac{1}{a} \log(a)\, dx \\ &= \log a \end{aligned} \tag{9.69}$$

Note that $\log a < 0$ for $a < 1$. Thus this example shows that, unlike a discrete random variable, the differential entropy of a continuous random variable can be negative. ◀

### ▶ EXAMPLE 9.8 Gaussian Distribution

Consider an arbitrary pair of random variables $X$ and $Y$, whose probability density functions are respectively denoted by $f_Y(x)$ and $f_X(x)$ where $x$ is merely a dummy variable. Adapting the fundamental inequality of Equation (9.12) to the situation at hand, we may write[9]

$$\int_{-\infty}^{\infty} f_Y(x) \log_2\left(\frac{f_X(x)}{f_Y(x)}\right) dx \le 0 \tag{9.70}$$

or, equivalently,

$$-\int_{-\infty}^{\infty} f_Y(x) \log_2 f_Y(x)\, dx \le -\int_{-\infty}^{\infty} f_Y(x) \log_2 f_X(x)\, dx \tag{9.71}$$

The quantity on the left-hand side of Equation (9.71) is the differential entropy of the random variable $Y$; hence,

$$h(Y) \le -\int_{-\infty}^{\infty} f_Y(x) \log_2 f_X(x)\, dx \tag{9.72}$$

Suppose now the random variables $X$ and $Y$ are described as follows:

- ▶ The random variables $X$ and $Y$ have the *same mean* $\mu$ and the *same variance* $\sigma^2$.
- ▶ The random variable $X$ is *Gaussian distributed* as shown by

$$f_X(x) = \frac{1}{\sqrt{2\pi}\sigma} \exp\left(-\frac{(x-\mu)^2}{2\sigma^2}\right) \tag{9.73}$$

Hence, substituting Equation (9.73) into Equation (9.72), and changing the base of the logarithm from 2 to $e = 2.7183$, we get

$$h(Y) \le -\log_2 e \int_{-\infty}^{\infty} f_Y(x)\left(-\frac{(x-\mu)^2}{2\sigma^2} - \log(\sqrt{2\pi}\sigma)\right) dx \tag{9.74}$$

We now recognize the following properties of the random variable $Y$ (given that its mean is $\mu$ and its variance is $\sigma^2$):

$$\int_{-\infty}^{\infty} f_Y(x)\, dx = 1$$

$$\int_{-\infty}^{\infty} (x-\mu)^2 f_Y(x)\, dx = \sigma^2$$

We may therefore simplify Equation (9.74) as

$$h(Y) \leq \tfrac{1}{2} \log_2(2\pi e \sigma^2) \tag{9.75}$$

The quantity on the right-hand side of Equation (9.75) is in fact the differential entropy of the Gaussian random variable $X$:

$$h(X) = \tfrac{1}{2} \log_2(2\pi e \sigma^2) \tag{9.76}$$

Finally, combining Equations (9.75) and (9.76), we may write

$$h(Y) \leq h(X), \qquad \begin{cases} X\text{: Gaussian random variable} \\ Y\text{: another random variable} \end{cases} \tag{9.77}$$

where equality holds if, and only if, $Y = X$.

We may now summarize the results of this important example as two entropic properties of a Gaussian random variable:

1. For a finite variance $\sigma^2$, the Gaussian random variable has the largest differential entropy attainable by any random variable.
2. The entropy of a Gaussian random variable $X$ is uniquely determined by the variance of $X$ (i.e., it is independent of the mean of $X$).

Indeed, it is because of Property 1 that the Gaussian channel model is so widely used as a conservative model in the study of digital communication systems. ◀

## ■ MUTUAL INFORMATION

Consider next a pair of continuous random variables $X$ and $Y$. By analogy with Equation (9.47), we define the *mutual information* between the random variables $X$ and $Y$ as follows:

$$I(X; Y) = \int_{-\infty}^{\infty} \int_{-\infty}^{\infty} f_{X,Y}(x,y) \log_2 \left[ \frac{f_X(x|y)}{f_X(x)} \right] dx\, dy \tag{9.78}$$

where $f_{X,Y}(x, y)$ is the joint probability density function of $X$ and $Y$, and $f_X(x|y)$ is the conditional probability density function of $X$, given that $Y = y$. Also, by analogy with Equations (9.45), (9.50), (9.43), and (9.44) we find that the mutual information $I(X; Y)$ has the following properties:

1. $I(X; Y) = I(Y; X)$ (9.79)
2. $I(X; Y) \geq 0$ (9.80)
3. $I(X; Y) = h(X) - h(X|Y)$
   $= h(Y) - h(Y|X)$ (9.81)

The parameter $h(X)$ is the differential entropy of $X$; likewise for $h(Y)$. The parameter $h(X|Y)$ is the *conditional differential entropy* of $X$, given $Y$; it is defined by the double integral (see Equation (9.41))

$$h(X|Y) = \int_{-\infty}^{\infty} \int_{-\infty}^{\infty} f_{X,Y}(x, y) \log_2 \left[ \frac{1}{f_X(x|y)} \right] dx\, dy \tag{9.82}$$

The parameter $h(Y|X)$ is the conditional differential entropy of $Y$, given $X$; it is defined in a manner similar to $h(X|Y)$.

## 9.10 *Information Capacity Theorem*

In this section, we use the idea of mutual information to formulate the information capacity theorem for *band-limited, power-limited Gaussian channels*. To be specific, consider a zero-mean stationary process $X(t)$ that is band-limited to $B$ hertz. Let $X_k$, $k = 1, 2, \dots, K$, denote the continuous random variables obtained by uniform sampling of the process $X(t)$ at the Nyquist rate of $2B$ samples per second. These samples are transmitted in $T$ seconds over a noisy channel, also band-limited to $B$ hertz. Hence, the number of samples, $K$, is given by

$$K = 2BT \tag{9.83}$$

We refer to $X_k$ as a sample of the *transmitted signal*. The channel output is perturbed by *additive white Gaussian noise* (AWGN) of zero mean and power spectral density $N_0/2$. The noise is band-limited to $B$ hertz. Let the continuous random variables $Y_k$, $k = 1, 2, \dots, K$ denote samples of the received signal, as shown by

$$Y_k = X_k + N_k, \qquad k = 1, 2, \dots, K \tag{9.84}$$

The noise sample $N_k$ is Gaussian with zero mean and variance given by

$$\sigma^2 = N_0 B \tag{9.85}$$

We assume that the samples $Y_k$, $k = 1, 2, \dots, K$ are statistically independent.

A channel for which the noise and the received signal are as described in Equations (9.84) and (9.85) is called a *discrete-time, memoryless Gaussian channel*. It is modeled as in Figure 9.13. To make meaningful statements about the channel, however, we have to assign a *cost* to each channel input. Typically, the transmitter is *power limited*; it is therefore reasonable to define the cost as

$$E[X_k^2] = P, \qquad k = 1, 2, \dots, K \tag{9.86}$$

where $P$ is the *average transmitted power*. The *power-limited Gaussian channel* described herein is of not only theoretical but also practical importance in that it models many communication channels, including line-of-sight radio and satellite links.

The *information capacity* of the channel is defined as the maximum of the mutual information between the channel input $X_k$ and the channel output $Y_k$ over all distributions on the input $X_k$ that satisfy the power constraint of Equation (9.86). Let $I(X_k; Y_k)$ denote

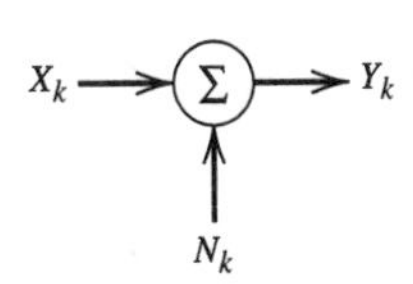


**FIGURE 9.13** Model of discrete-time, memoryless Gaussian channel.

the mutual information between $X_k$ and $Y_k$. We may then define the information capacity of the channel as

$$C = \max_{f_{X_k}(x)} \{I(X_k; Y_k) : E[X_k^2] = P\} \tag{9.87}$$

where the maximization is performed with respect to $f_{X_k}(x)$, the probability density function of $X_k$.

The mutual information $I(X_k; Y_k)$ can be expressed in one of the two equivalent forms shown in Equation (9.81). For the purpose at hand, we use the second line of this equation and so write

$$I(X_k; Y_k) = h(Y_k) - h(Y_k | X_k) \tag{9.88}$$

Since $X_k$ and $N_k$ are independent random variables, and their sum equals $Y_k$, as in Equation (9.84), we find that the conditional differential entropy of $Y_k$, given $X_k$, is equal to the differential entropy of $N_k$ (see Problem 9.28):

$$h(Y_k \mid X_k) = h(N_k) \tag{9.89}$$

Hence, we may rewrite Equation (9.88) as

$$I(X_k; Y_k) = h(Y_k) - h(N_k) \tag{9.90}$$

Since $h(N_k)$ is independent of the distribution of $X_k$, maximizing $I(X_k; Y_k)$ in accordance with Equation (9.87) requires maximizing $h(Y_k)$, the differential entropy of sample $Y_k$ of the received signal. For $h(Y_k)$ to be maximum, $Y_k$ has to be a Gaussian random variable (see Example 9.8). That is, the samples of the received signal represent a noiselike process. Next, we observe that since $N_k$ is Gaussian by assumption, the sample $X_k$ of the transmitted signal must be Gaussian too. We may therefore state that the maximization specified in Equation (9.87) is attained by choosing the samples of the transmitted signal from a noiselike process of average power $P$. Correspondingly, we may reformulate Equation (9.87) as

$$C = I(X_k; Y_k) : X_k \text{ Gaussian}, \quad E[X_k^2] = P \tag{9.91}$$

where the mutual information $I(X_k; Y_k)$ is defined in accordance with Equation (9.90).

For the evaluation of the information capacity $C$, we proceed in three stages:

1. The variance of sample $Y_k$ of the received signal equals $P + \sigma^2$. Hence, the use of Equation (9.76) yields the differential entropy of $Y_k$ as

$$h(Y_k) = \tfrac{1}{2} \log_2[2\pi e(P + \sigma^2)] \tag{9.92}$$

2. The variance of the noise sample $N_k$ equals $\sigma^2$. Hence, the use of Equation (9.76) yields the differential entropy of $N_k$ as

$$h(N_k) = \tfrac{1}{2} \log_2(2\pi e \sigma^2) \tag{9.93}$$

3. Substituting Equations (9.92) and (9.93) into Equation (9.90) and recognizing the definition of information capacity given in Equation (9.91), we get the desired result:

$$C = \frac{1}{2} \log_2\left(1 + \frac{P}{\sigma^2}\right) \text{ bits per transmission} \tag{9.94}$$

With the channel used $K$ times for the transmission of $K$ samples of the process $X(t)$ in $T$ seconds, we find that the information capacity per unit time is $(K/T)$ times the result given in Equation (9.94). The number $K$ equals $2BT$, as in Equation (9.83). Accordingly, we may express the information capacity in the equivalent form:

$$C = B \log_2\left(1 + \frac{P}{N_0 B}\right) \text{ bits per second} \tag{9.95}$$

where we have used Equation (9.85) for the noise variance $\sigma^2$.

Based on the formula of Equation (9.95), we may now state Shannon's third (and most famous) theorem, the *information capacity theorem*,[10] as follows:

The information capacity of a continuous channel of bandwidth $B$ hertz, perturbed by additive white Gaussian noise of power spectral density $N_0/2$ and limited in bandwidth to $B$, is given by

$$C = B \log_2\left(1 + \frac{P}{N_0 B}\right) \text{ bits per second}$$

where $P$ is the average transmitted power.

The information capacity theorem is one of the most remarkable results of information theory for, in a single formula, it highlights most vividly the interplay among three key system parameters: channel bandwidth, average transmitted power (or, equivalently, average received signal power), and noise power spectral density at the channel output. The dependence of information capacity $C$ on channel bandwidth $B$ is *linear*, whereas its dependence on signal-to-noise ratio $P/N_0B$ is *logarithmic*. Accordingly, *it is easier to increase the information capacity of a communication channel by expanding its bandwidth than increasing the transmitted power for a prescribed noise variance.*

The theorem implies that, for given average transmitted power $P$ and channel bandwidth $B$, we can transmit information at the rate of $C$ bits per second, as defined in Equation (9.95), with arbitrarily small probability of error by employing sufficiently complex encoding systems. It is not possible to transmit at a rate higher than $C$ bits per second by any encoding system without a definite probability of error. Hence, the channel capacity theorem defines the *fundamental limit* on the rate of error-free transmission for a power-limited, band-limited Gaussian channel. To approach this limit, however, the transmitted signal must have statistical properties approximating those of white Gaussian noise.

## ■ Sphere Packing[11]

To provide a plausible argument supporting the information capacity theorem, suppose that we use an encoding scheme that yields $K$ code words, one for each sample of the transmitted signal. Let $n$ denote the length (i.e., the number of bits) of each code word. It is presumed that the coding scheme is designed to produce an acceptably low probability of symbol error. Furthermore, the code words satisfy the power constraint; that is, the average power contained in the transmission of each code word with $n$ bits is $nP$, where $P$ is the average power per bit.

Suppose that any code word in the code is transmitted. The received vector of $n$ bits is Gaussian distributed with mean equal to the transmitted code word and variance equal to $n\sigma^2$, where $\sigma^2$ is the noise variance. With high probability, the received vector lies inside a sphere of radius $\sqrt{n\sigma^2}$, centered on the transmitted code word. This sphere is itself contained in a larger sphere of radius $\sqrt{n(P + \sigma^2)}$, where $n(P + \sigma^2)$ is the average power of the received vector.

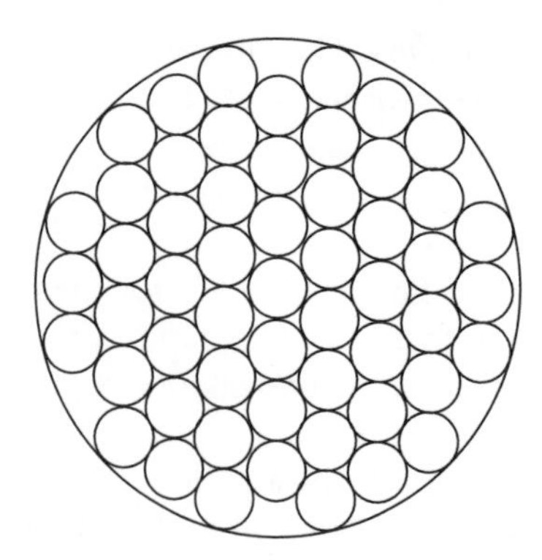

**FIGURE 9.14** The sphere-packing problem.

We may thus visualize the picture portrayed in Figure 9.14. With everything inside a small sphere of radius $\sqrt{n\sigma^2}$ assigned to the code word on which it is centered, it is reasonable to say that when this particular code word is transmitted, the probability that the received vector will lie inside the correct "decoding" sphere is high. The key question is: How many decoding spheres can be packed inside the larger sphere of received vectors? In other words, how many code words can we in fact choose? To answer this question, we first recognize that the volume of an $n$-dimensional sphere of radius $r$ may be written as $A_n r^n$, where $A_n$ is a scaling factor. We may therefore make the following statements:

- The volume of the sphere of received vectors is $A_n[n(P + \sigma^2)]^{n/2}$.
- The volume of the decoding sphere is $A_n(n\sigma^2)^{n/2}$.

Accordingly, it follows that the maximum number of *nonintersecting* decoding spheres that can be packed inside the sphere of possible received vectors is

$$\begin{aligned} \frac{A_n[n(P + \sigma^2)]^{n/2}}{A_n(n\sigma^2)^{n/2}} &= \left(1 + \frac{P}{\sigma^2}\right)^{n/2} \\ &= 2^{(n/2)\log_2(1+P/\sigma^2)} \end{aligned} \tag{9.96}$$

Taking the logarithm of this result to base 2, we readily see that the maximum number of bits per transmission for a low probability of error is indeed as defined previously in Equation (9.94).

### ▶ EXAMPLE 9.9 Reconfiguration of Constellation for Reduced Power

To illustrate the idea of sphere packing, consider the 64-QAM square constellation of Figure 9.15*a*. The figure depicts two-dimensional nonintersecting decoding spheres centered on the message points in the constellation. In trying to pack the decoding spheres as tightly as possible while maintaining the same Euclidean distance between the message points as before, we obtain the alternative constellation shown in Figure 9.15*b*. With a common Euclidean distance between the message points, the two constellations of Figure 9.15 produce approximately the same bit error rate, assuming the use of a high enough signal-to-noise ratio over an AWGN channel; see, for example, Equation (5.95). However, comparing these two constellations, we find that the sum of squared Euclidean distances from the message points to the origin in Figure 9.15*b* is smaller than that in Figure 9.15*a*. It follows therefore that the tightly packed constellation of Figure 9.15*b* has an advantage over the square constellation

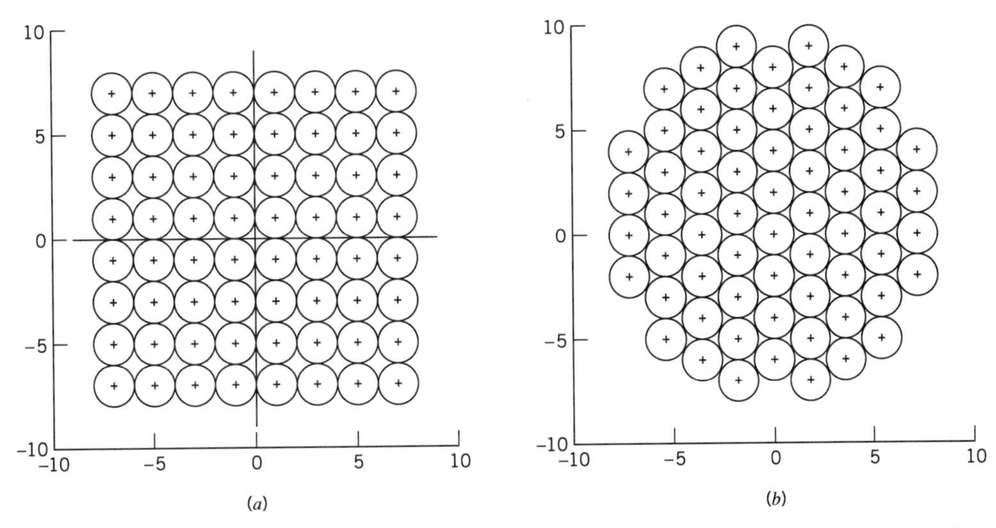


**FIGURE 9.15** (*a*) Square 64-QAM constellation. (*b*) The most tightly coupled alternative to that of part *a*.

of Figure 9.15*a*: a smaller transmitted average signal energy per symbol for the same bit error rate on an AWGN channel. ◀

## 9.11 *Implications of the Information Capacity Theorem*

Now that we have an intuitive feel for the information capacity theorem, we may go on to discuss its implications in the context of a Gaussian channel that is limited in both power and bandwidth. For the discussion to be useful, however, we need an ideal framework against which the performance of a practical communication system can be assessed. To this end, we introduce the notion of an *ideal system* defined as one that transmits data at a bit rate $R_b$ equal to the information capacity $C$. We may then express the average transmitted power as

$$P = E_b C \tag{9.97}$$

where $E_b$ is the transmitted energy per bit. Accordingly, the ideal system is defined by the equation

$$\frac{C}{B} = \log_2\left(1 + \frac{E_b}{N_0}\frac{C}{B}\right) \tag{9.98}$$

Equivalently, we may define the *signal energy-per-bit to noise power spectral density ratio* $E_b/N_0$ in terms of the ratio $C/B$ for the ideal system as

$$\frac{E_b}{N_0} = \frac{2^{C/B} - 1}{C/B} \tag{9.99}$$

A plot of bandwidth efficiency $R_b/B$ versus $E_b/N_0$ is called the *bandwidth-efficiency diagram*. A generic form of this diagram is displayed in Figure 9.16, where the curve labeled "capacity boundary" corresponds to the ideal system for which $R_b = C$. Based on Figure 9.16, we can make the following observations:

1. For *infinite bandwidth*, the ratio $E_b/N_0$ approaches the limiting value

$$\begin{aligned}\left(\frac{E_b}{N_0}\right)_\infty &= \lim_{B\to\infty}\left(\frac{E_b}{N_0}\right)\\ &= \log 2 = 0.693\end{aligned} \tag{9.100}$$

   This value is called the *Shannon limit* for an AWGN channel, assuming a code rate of zero. Expressed in decibels, it equals $-1.6$ dB. The corresponding limiting value of the channel capacity is obtained by letting the channel bandwidth $B$ in Equation (9.95) approach infinity; we thus find that

$$\begin{aligned}C_\infty &= \lim_{B\to\infty} C\\ &= \frac{P}{N_0}\log_2 e\end{aligned} \tag{9.101}$$

   where $e$ is the base of the natural logarithm.
2. The *capacity boundary*, defined by the curve for the critical bit rate $R_b = C$, separates combinations of system parameters that have the potential for supporting error-free transmission ($R_b < C$) from those for which error-free transmission is not possible ($R_b > C$). The latter region is shown shaded in Figure 9.16.
3. The diagram highlights potential *trade-offs* among $E_b/N_0$, $R_b/B$, and probability of symbol error $P_e$. In particular, we may view movement of the operating point along

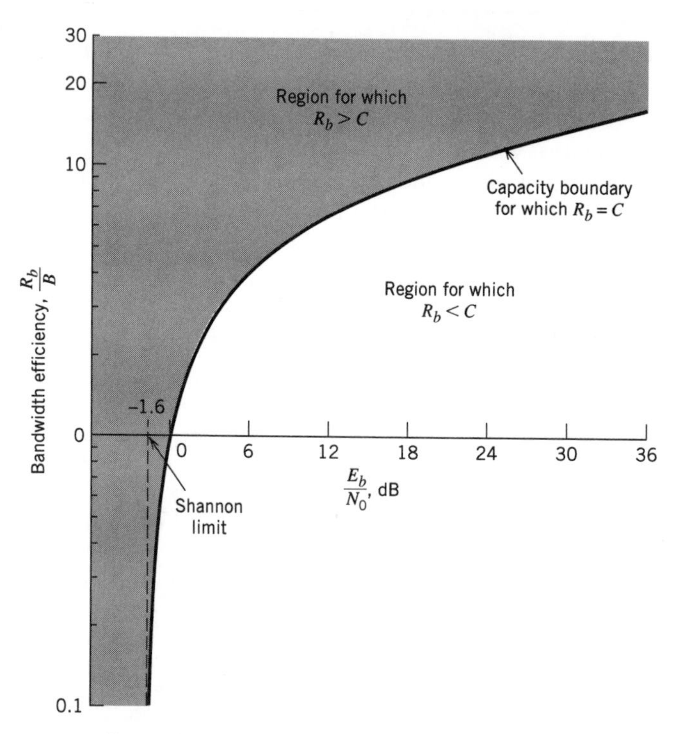


**FIGURE 9.16** Bandwidth-efficiency diagram.

a horizontal line as trading $P_e$ versus $E_b/N_0$ for a fixed $R_b/B$. On the other hand, we may view movement of the operating point along a vertical line as trading $P_e$ versus $R_b/B$ for a fixed $E_b/N_0$.

## ▶ EXAMPLE 9.10 *M*-ary PCM

In this example, we look at an $M$-ary PCM system in light of the channel capacity theorem under the assumption that the system operates above the error threshold. That is, the average probability of error due to channel noise is negligible.

We assume that the $M$-ary PCM system uses a code word consisting of $n$ code elements, each having one of $M$ possible discrete amplitude levels; hence the name "$M$-ary." From Chapter 3 we recall that for a PCM system to operate above the error threshold, there must be provision for a noise margin that is sufficiently large to maintain a negligible error rate due to channel noise. This, in turn, means there must be a certain separation between these $M$ discrete amplitude levels. Call this separation $k\sigma$, where $k$ is a constant and $\sigma^2 = N_0B$ is the noise variance measured in a channel bandwidth $B$. The number of amplitude levels $M$ is usually an integer power of 2. The average transmitted power will be least if the amplitude range is symmetrical about zero. Then the discrete amplitude levels, normalized with respect to the separation $k\sigma$, will have the values $\pm 1/2, +3/2, \ldots, \pm (M-1)/2$. We assume that these $M$ different amplitude levels are equally likely. Accordingly, we find that the average transmitted power is given by

$$\begin{aligned} P &= \frac{2}{M}\left[\left(\frac{1}{2}\right)^2 + \left(\frac{3}{2}\right)^2 + \cdots + \left(\frac{M-1}{2}\right)^2\right](k\sigma)^2 \\ &= k^2\sigma^2\left(\frac{M^2-1}{12}\right) \end{aligned} \tag{9.102}$$

Suppose that the $M$-ary PCM system described herein is used to transmit a message signal with its highest frequency component equal to $W$ hertz. The signal is sampled at the Nyquist rate of $2W$ samples per second. We assume that the system uses a quantizer of the midrise type, with $L$ equally likely representation levels. Hence, the probability of occurrence of any one of the $L$ representation levels is $1/L$. Correspondingly, the amount of information carried by a single sample of the signal is $\log_2 L$ bits. With a maximum sampling rate of $2W$ samples per second, the maximum rate of information transmission of the PCM system, measured in bits per second, is given by

$$R_b = 2W \log_2 L \text{ bits per second} \tag{9.103}$$

Since the PCM system uses a code word consisting of $n$ code elements, each having one of $M$ possible discrete amplitude values, we have $M^n$ different possible code words. For a unique encoding process, we require

$$L = M^n \tag{9.104}$$

Clearly, the rate of information transmission in the system is unaffected by the use of an encoding process. We may therefore eliminate $L$ between Equations (9.103) and (9.104) to obtain

$$R_b = 2Wn \log_2 M \text{ bits per second} \tag{9.105}$$

Equation (9.102) defines the average transmitted power required to maintain an $M$-ary PCM system operating above the error threshold. Hence, solving this equation for the number of discrete amplitude levels, $M$, we get

$$M = \left(1 + \frac{12P}{k^2N_0B}\right)^{1/2} \tag{9.106}$$

where $\sigma^2 = N_0 B$ is the variance of the channel noise measured in a bandwidth $B$. Therefore, substituting Equation (9.106) into Equation (9.105), we obtain

$$R_b = Wn \log_2\left(1 + \frac{12P}{k^2 N_0 B}\right) \tag{9.107}$$

The channel bandwidth $B$ required to transmit a rectangular pulse of duration $1/2nW$ (representing a code element in the code word) is given by (see Chapter 3)

$$B = \kappa n W$$

where $\kappa$ is a constant with a value lying between 1 and 2. Using the minimum possible value $\kappa = 1$, we find that the channel bandwidth $B = nW$. We may thus rewrite Equation (9.107) as

$$R_b = B \log_2\left(1 + \frac{12P}{k^2 N_0 B}\right) \tag{9.108}$$

The *ideal system* is described by Shannon's channel capacity theorem, given in Equation (9.95). Hence, comparing Equation (9.108) with Equation (9.95), we see that they are identical if the average transmitted power in the PCM system is increased by the factor $k^2/12$, compared with the ideal system. Perhaps the most interesting point to note about Equation (9.108) is that the form of the equation is right: *Power and bandwidth in a PCM system are exchanged on a logarithmic basis, and the information capacity C is proportional to the channel bandwidth B.* ◀

### ▶ EXAMPLE 9.11 *M*-ary PSK and *M*-ary FSK

In this example, we compare the bandwidth-power exchange capabilities of $M$-ary PSK and $M$-ary FSK signals in light of Shannon's information capacity theorem. Consider first a coherent $M$-ary PSK system that employs a *nonorthogonal* set of $M$ phase-shifted signals for the transmission of binary data. Each signal in the set represents a symbol with $\log_2 M$ bits. Using the definition of null-to-null bandwidth, we may express the bandwidth efficiency of $M$-ary PSK as follows [see Equation (6.51)]:

$$\frac{R_b}{B} = \frac{\log_2 M}{2}$$

In Figure 9.17*a*, we show the operating points for different numbers of phase levels $M = 2$, 4, 8, 16, 32, 64. Each point corresponds to an average probability of symbol error $P_e = 10^{-5}$. In the figure we have also included the capacity boundary for the ideal system. We observe from Figure 9.17*a* that as $M$ is increased, the bandwidth efficiency is improved, but the value of $E_b/N_0$ required for error-free transmission moves away from the Shannon limit.

Consider next a coherent $M$-ary FSK system that uses an *orthogonal* set of $M$ frequency-shifted signals for the transmission of binary data, with the separation between adjacent signal frequencies set at $1/2T$, where $T$ is the symbol period. As with the $M$-ary PSK, each signal in the set represents a symbol with $\log_2 M$ bits. The bandwidth efficiency of $M$-ary FSK is as follows [see Equation (6.143)]:

$$\frac{R_b}{B} = \frac{2 \log_2 M}{M}$$

In Figure 9.17*b*, we show the operating points for different numbers of frequency levels $M =$ 2, 4, 8, 16, 32, 64 for an average probability of symbol error $P_e = 10^{-5}$. In the figure, we have also included the capacity boundary for the ideal system. We see that increasing $M$ in (orthogonal) $M$-ary FSK has the opposite effect to that in (nonorthogonal) $M$-ary PSK. In particular, as $M$ is increased, which is equivalent to increased bandwidth requirement, the operating point moves closer to the Shannon limit. ◀

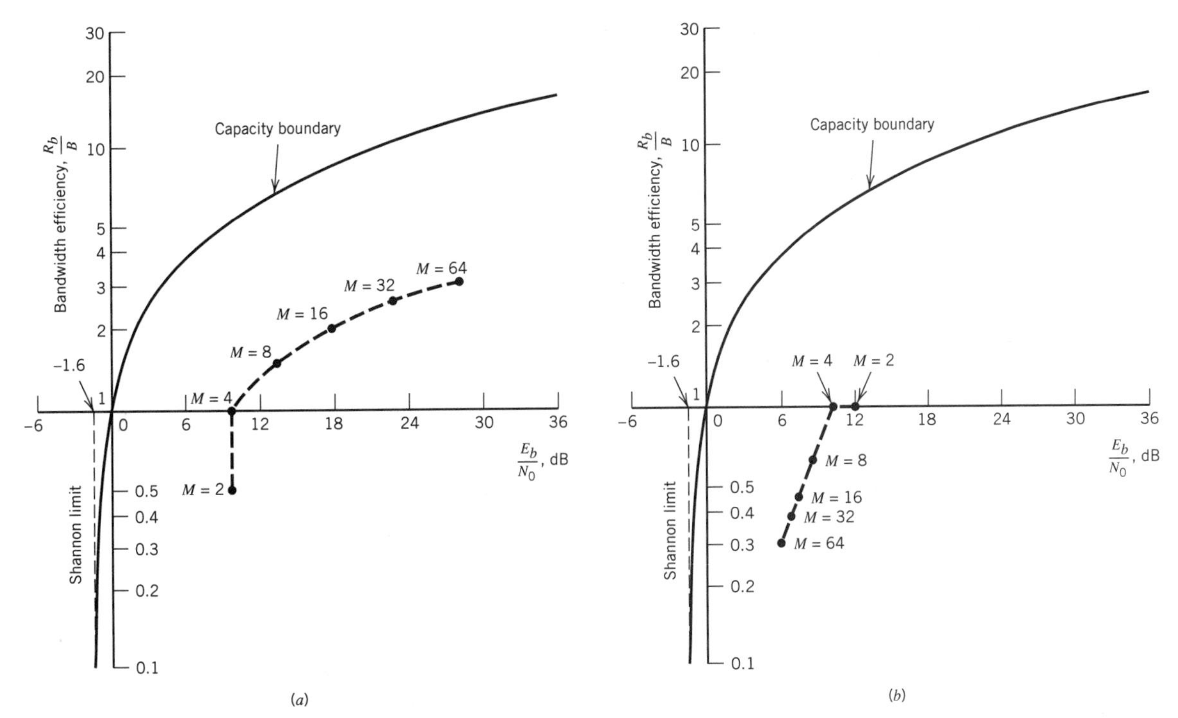


**FIGURE 9.17** (*a*) Comparison of *M*-ary PSK against the ideal system for $P_e = 10^{-5}$ and increasing *M*. (*b*) Comparison of *M*-ary FSK against the ideal system for $P_e = 10^{-5}$ and increasing *M*.

## ▶ EXAMPLE 9.12 Capacity of Binary-Input AWGN Channel

In this example, we investigate the capacity of an AWGN channel using *encoded* binary antipodal signaling (i.e., levels $-1$ and $+1$ for binary symbols 0 and 1, respectively). In particular, we address the issue of determining the minimum achievable bit error rate as a function of $E_b/N_0$ for varying code rate $r$. It is assumed that the binary symbols 0 and 1 are equiprobable.

Let the random variables $X$ and $Y$ denote the channel input and channel output, respectively; $X$ is a discrete variable, whereas $Y$ is a continuous variable. In light of the second line of Equation (9.81), we may express the mutual information between the channel input and channel output as

$$I(X; Y) = h(Y) - h(Y|X)$$

The second term, $h(Y|X)$, is the conditional differential entropy of the channel output $Y$, given the channel input $X$. By virtue of Equations (9.89) and (9.93), this term is just the entropy of a Gaussian distribution. Hence, using $\sigma^2$ to denote the variance of the channel noise, we may write

$$h(Y|X) = \frac{1}{2}\log_2(2\pi e\sigma^2)$$

Next, the first term, $h(Y)$, is the differential entropy of the channel output $Y$. With the use of binary antipodal signaling, the probability density function of $Y$, given $X = x$, is a mixture of two Gaussian distributions with common variance $\sigma^2$ and mean values $-1$ and $+1$, as shown by

$$f_Y(y_i|x) = \frac{1}{2}\left[\frac{\exp(-(y_i+1)^2/2\sigma^2)}{\sqrt{2\pi}\sigma} + \frac{\exp(-(y_i-1)^2/2\sigma^2)}{\sqrt{2\pi}\sigma}\right] \tag{9.109}$$

Hence, we may determine the differential entropy of $Y$ using the formula

$$h(Y) = -\int_{-\infty}^{\infty} f_Y(y_i|x) \log_2[f_Y(y_i|x)]\, dy_i$$

where $f_Y(y_i|x)$ is defined by Equation (9.109). From the formulas of $h(Y|X)$ and $h(Y)$, it is clear that the mutual information is solely a function of the noise variance $\sigma^2$. Using $M(\sigma^2)$ to denote this functional dependence, we may thus write

$$I(X;\, Y) = M(\sigma^2)$$

Unfortunately, there is no closed formula that we can derive for $M(\sigma^2)$ because of the difficulty of determining $h(Y)$. Nevertheless, the differential entropy $h(Y)$ can be well approximated using *Monte Carlo integration*, which is straightforward to program on a digital computer; see Problem 9.36.

Because symbols 0 and 1 are equiprobable, it follows that the channel capacity $C$ is equal to the mutual information between $X$ and $Y$. Hence, for error-free data transmission over the AWGN channel, the code rate $r$ must satisfy the condition

$$r < M(\sigma^2) \tag{9.110}$$

A robust measure of the ratio $E_b/N_0$ is

$$\frac{E_b}{N_0} = \frac{P}{N_0 r} = \frac{P}{2\sigma^2 r}$$

where $P$ is the average transmitted power, and $N_0/2$ is the two-sided power spectral density of the channel noise. Without loss of generality, we may set $P = 1$. We may then express the noise variance as

$$\sigma^2 = \frac{N_0}{2E_b r} \tag{9.111}$$

Substituting Equation (9.111) into (9.110) and rearranging terms, we get the desired relation:

$$\frac{E_b}{N_0} = \frac{1}{2rM^{-1}(r)} \tag{9.112}$$

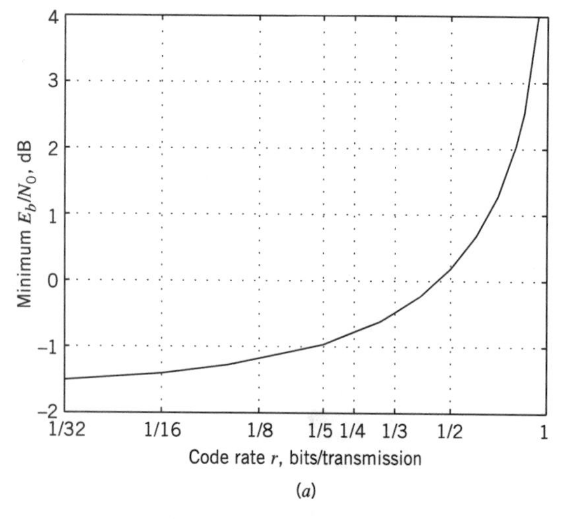


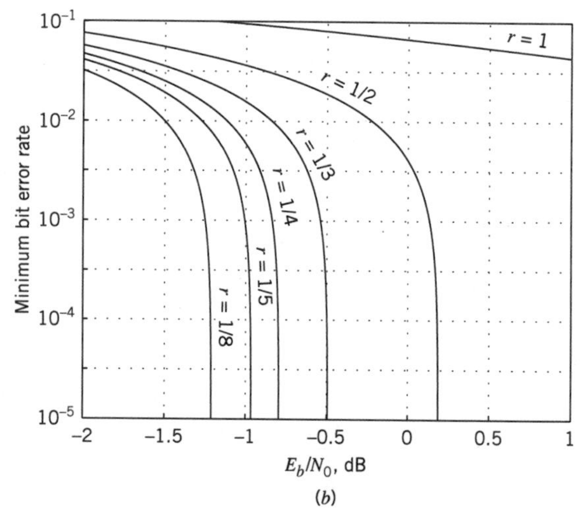


**FIGURE 9.18** Binary antipodal signaling over an AWGN channel. (*a*) Minimum $E_b/N_0$ versus the code rate $r$. (*b*) Minimum bit error rate (BER) versus $E_b/N_0$ for varying code rate $r$.

where $M^{-1}(r)$ is the *inverse* of the mutual information between the channel input and output, expressed as a function of the code rate $r$.

Using the Monte Carlo method to estimate the differential entropy $h(Y)$ and therefore $M^{-1}(r)$, the plots of Figure 9.18 are computed.[12] Figure 9.18*a* plots the minimum $E_b/N_0$ versus the code rate $r$ for error-free communication. Figure 9.18*b* plots the minimum achievable bit error rate versus $E_b/N_0$ with the code rate $r$ as a running parameter. From Figure 9.18 we may draw the following conclusions:

- For uncoded binary signaling (i.e., $r = 1$), an infinite $E_b/N_0$ is required for error-free communication, which agrees with what we know about uncoded data transmission over an AWGN channel.
- The minimum $E_b/N_0$ decreases with decreasing code rate $r$, which is intuitively satisfying. For example, for $r = 1/2$, the minimum value of $E_b/N_0$ is slightly less than 0.2 dB.
- As $r$ approaches zero, the minimum $E_b/N_0$ approaches the limiting value of $-1.6$ dB, which agrees with the Shannon limit derived earlier; see Equation (9.100). ◀

## 9.12 *Information Capacity of Colored Noise Channel*[13]

The information capacity theorem as formulated in Equation (9.95) applies to a band-limited white noise channel. In this section, we extend Shannon's information capacity theorem to the more general case of a *nonwhite*, or *colored, noise channel.* To be specific, consider the channel model shown in Figure 9.19*a* where the transfer function of the channel is denoted by $H(f)$. The channel noise $n(t)$, which appears additively at the channel output, is modeled as the sample function of a stationary Gaussian process of zero mean and power spectral density $S_N(f)$. The requirement is twofold:

1. Find the input ensemble, described by the power spectral density $S_X(f)$, that maximizes the mutual information between the channel output $y(t)$ and the channel input $x(t)$, subject to the constraint that the average power of $x(t)$ is fixed at a constant value $P$.
2. Hence, determine the optimum information capacity of the channel.

This problem is a constrained optimization problem. To solve it, we proceed as follows:

- Because the channel is linear, we may replace the model of Figure 9.19*a* with the equivalent model shown in Figure 9.19*b*. From the viewpoint of the spectral characteristics of the signal plus noise measured at the channel output, the two models of Figure 9.19 are equivalent, provided that the power spectral density of the noise

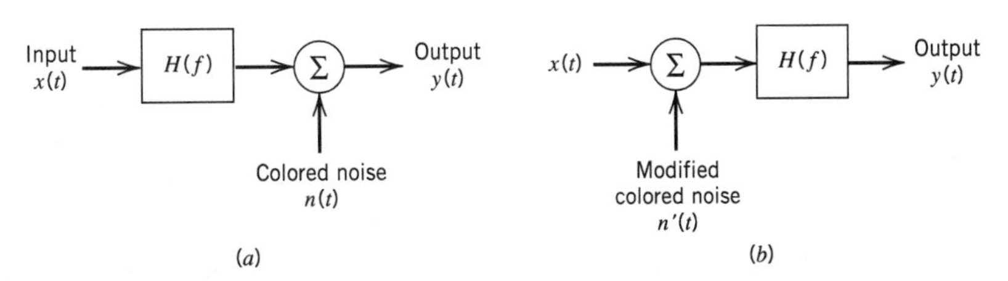


**FIGURE 9.19** (*a*) Model of band-limited, power-limited noisy channel. (*b*) Equivalent model of the channel.

$n'(t)$ in Figure 9.19*b* is defined in terms of the power spectral density of the noise $n(t)$ in Figure 9.19*a* as

$$S_{N'}(f) = \frac{S_N(f)}{|H(f)|^2} \tag{9.113}$$

where $|H(f)|$ is the magnitude response of the channel.

- To simplify the analysis, we use the "principle of divide and conquer" in a manner similar to that described in Section 6.12. Specifically, the channel is divided into a large number of adjoining frequency slots, as illustrated in Figure 9.20. The smaller we make the incremental frequency interval $\Delta f$ of each subchannel, the better is this approximation.

The net result of these two points is that the original model of Figure 9.19*a* is replaced by the parallel combination of a finite number of subchannels, $N$, each of which is corrupted essentially by "band-limited white Gaussian noise."

The $k$th subchannel in the approximation to the model of Figure 9.19*b* is described by

$$y_k(t) = x_k(t) + n_k(t), \qquad k = 1, 2, \ldots, N \tag{9.14}$$

The average power of the signal component $x_k(t)$ is

$$P_k = S_X(f_k)\,\Delta f, \qquad k = 1, 2, \ldots, N \tag{9.115}$$

where $S_X(f_k)$ is the power spectral density of the input signal evaluated at the frequency $f = f_k$. The variance of the noise component $n_k(t)$ is

$$\sigma_k^2 = \frac{S_N(f_k)}{|H(f_k)|^2}\,\Delta f, \qquad k = 1, 2, \ldots, N \tag{9.116}$$

where $S_N(f_k)$ and $|H(f_k)|$ are the noise spectral density and the channel's magnitude response evaluated at the frequency $f_k$, respectively. The information capacity of the $k$th subchannel is

$$C_k = \frac{1}{2}\,\Delta f \log_2\left(1 + \frac{P_k}{\sigma_k^2}\right), \qquad k = 1, 2, \ldots, N \tag{9.117}$$

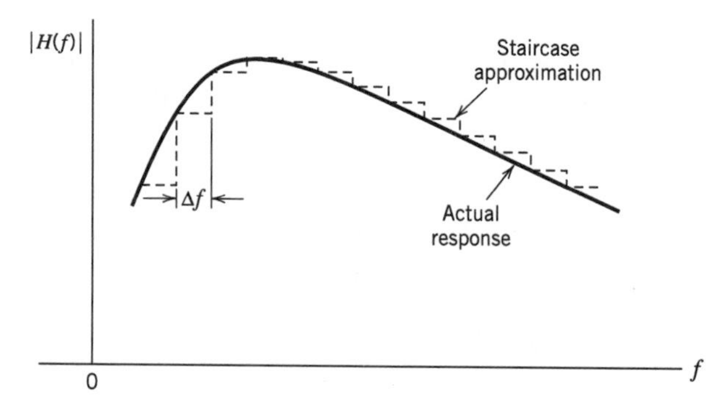


**FIGURE 9.20** Staircase approximation of an arbitrary magnitude response $|H(f)|$; only positive-frequency portion of the response is shown.

where the factor 1/2 accounts for the fact that $\Delta f$ applies to both positive and negative frequencies. All the $N$ subchannels are independent of one another. Hence the total capacity of the overall channel is approximately given by the summation

$$\begin{aligned} C &\simeq \sum_{k=1}^{N} C_k \\ &= \frac{1}{2} \sum_{k=1}^{N} \Delta f \log_2\left(1 + \frac{P_k}{\sigma_k^2}\right) \end{aligned} \tag{9.118}$$

The problem we have to address is to maximize the overall information capacity $C$ subject to the constraint:

$$\sum_{k=1}^{N} P_k = P = \text{constant} \tag{9.119}$$

The usual procedure to solve a constrained optimization problem is to use the *method of Lagrange multipliers*; see Note 19 in Chapter 6. To proceed with this optimization, we first define an objective function that incorporates both the information capacity $C$ and the constraint [i.e., Equations (9.118) and (9.119)], as shown by

$$J = \frac{1}{2} \sum_{k=1}^{N} \Delta f \log_2\left(1 + \frac{P_k}{\sigma_k^2}\right) + \lambda\left(P - \sum_{k=1}^{N} P_k\right) \tag{9.120}$$

where $\lambda$ is the Lagrange multiplier. Next, differentiating the objective function $J$ with respect to $P_k$ and setting the result equal to zero, we obtain

$$\frac{\Delta f \log_2 e}{P_k + \sigma_k^2} - \lambda = 0$$

To satisfy this optimizing solution, we impose the following requirement:

$$P_k + \sigma_k^2 = K\,\Delta f \qquad \text{for } k = 1, 2, \ldots, N \tag{9.121}$$

where $K$ is a constant that is the same for all $k$. The constant $K$ is chosen to satisfy the average power constraint.

Inserting the defining values of Equations (9.115) and (9.116) in the optimizing condition of Equation (9.121), simplifying, and rearranging terms, we get

$$S_X(f_k) = K - \frac{S_N(f_k)}{|H(f_k)|^2}, \qquad k = 1, 2, \ldots, N \tag{9.122}$$

Let $\mathcal{F}_A$ denote the frequency range for which the constant $K$ satisfies the condition

$$K \geq \frac{S_N(f)}{|H(f)|^2}$$

Then, as the incremental frequency interval $\Delta f$ is allowed to approach zero and the number of subchannels $N$ goes to infinity, we may use Equation (9.122) to formally state that the power spectral density of the input ensemble that achieves the optimum information capacity is a nonnegative quantity defined by

$$S_X(f) = \begin{cases} K - \dfrac{S_N(f)}{|H(f)|^2} & \text{for } f \in \mathcal{F}_A \\ 0 & \text{otherwise} \end{cases} \tag{9.123}$$

Since the average power of a random process is the total area under the curve of the power spectral density of the process, we may express the average power of the channel input $x(t)$ as

$$P = \int_{f \in \mathcal{F}_A} \left( K - \frac{S_N(f)}{|H(f)|^2} \right) df \tag{9.124}$$

For a prescribed $P$ and specified $S_N(f)$ and $H(f)$, the constant $K$ is the solution to Equation (9.124).

The only thing that remains for us to do is to find the optimum information capacity. Substituting the optimizing solution of Equation (9.121) into Equation (9.118) and then using the defining values of Equations (9.115) and (9.116), we obtain

$$C \simeq \frac{1}{2} \sum_{k=1}^{N} \Delta f \log_2 \left( K \frac{|H(f_k)|^2}{S_N(f_k)} \right)$$

When the incremental frequency interval $\Delta f$ is allowed to approach zero, this equation takes the limiting form:

$$C = \frac{1}{2} \int_{-\infty}^{\infty} \log_2 \left( K \frac{|H(f)|^2}{S_N(f)} \right) df \tag{9.125}$$

where the constant $K$ is chosen as the solution to Equation (9.124) for a prescribed input signal power $P$.

## ■ WATER-FILLING INTERPRETATION OF THE INFORMATION CAPACITY THEOREM

Equations (9.123) and (9.124) suggest the picture portrayed in Figure 9.21. Specifically, we make the following observations:

- The appropriate input power spectral density $S_X(f)$ is described as the bottom regions of the function $S_N(f)/|H(f)|^2$ that lie below the constant level $K$, which are shown shaded.
- The input power $P$ is defined by the total area of these shaded regions.

The spectral domain picture portrayed here is called the *water-filling (pouring) interpretation* in the sense that the process by which the input power is distributed across

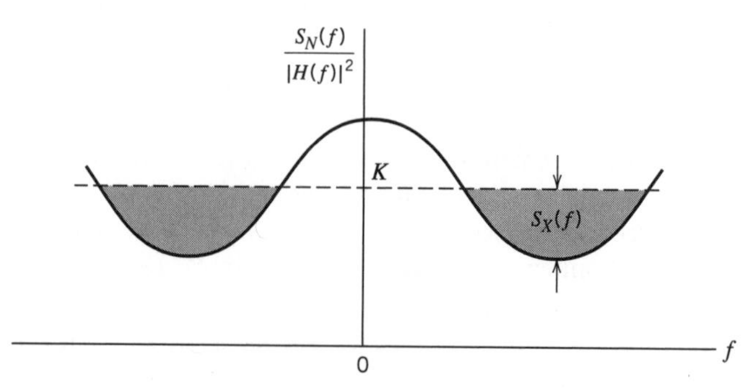


**FIGURE 9.21** Water-filling interpretation of information-capacity theorem for a colored noisy channel.

the function $S_N(f)/|H(f)|^2$ is identical to the way in which water distributes itself in a vessel.

Consider now the idealized case of a band-limited signal in additive white Gaussian noise of power spectral density $N(f) = N_0/2$. The transfer function $H(f)$ is that of an ideal band-pass filter defined by

$$H(f) = \begin{cases} 1, & 0 \le f_c - \dfrac{B}{2} \le |f| \le f_c + \dfrac{B}{2} \\ 0, & \text{otherwise} \end{cases}$$

where $f_c$ is the midband frequency and $B$ is the channel bandwidth. For this special case, Equations (9.124) and (9.125) reduce to, respectively,

$$P = 2B\left(K - \frac{N_0}{2}\right)$$

and

$$C = B \log_2\left(\frac{2K}{N_0}\right)$$

Hence, eliminating $K$ between these two equations, we get the standard form of Shannon's capacity theorem, defined by Equation (9.95).

### ▶ Example 9.13 Capacity of NEXT-Dominated Channel

From the discussion presented in Section 4.8, we recall that a major channel impairment in digital subscriber lines is near-end crosstalk (NEXT). The power spectral density of this crosstalk may be taken as

$$S_N(f) = |H_{\text{NEXT}}(f)|^2 S_X(f) \tag{9.126}$$

where $S_X(f)$ is the power spectral density of the transmitted signal and $H_{\text{NEXT}}(f)$ is the transfer function that couples adjacent twisted pairs. The only constraint we have to satisfy in this example is that the power spectral density function $S_X(f)$ be *nonnegative for all f*. Substituting Equation (9.126) into (9.123), we readily find that this condition is satisfied by solving for $K$ as

$$K = \left(1 + \frac{|H_{\text{NEXT}}(f)|^2}{|H(f)|^2}\right)S_X(f)$$

Finally, using this result in Equation (9.125), we find that the capacity of the NEXT-dominated digital subscriber channel is given by

$$C = \frac{1}{2}\int_{\mathcal{F}_A} \log_2\left(1 + \frac{|H(f)|^2}{|H_{\text{NEXT}}(f)|^2}\right) df$$

where $\mathcal{F}_A$ is the set of positive and negative frequencies for which $S_X(f) > 0$.

## 9.13 *Rate Distortion Theory*

In Section 9.3 we introduced the source coding theorem for a discrete memoryless source, according to which the average code-word length must be at least as large as the source entropy for perfect coding (i.e., perfect representation of the source). However, in many practical situations there are constraints that force the coding to be imperfect, thereby resulting in unavoidable *distortion*. For example, constraints imposed by a communication channel may place an upper limit on the permissible code rate and therefore average code-word length assigned to the information source. As another example, the information source may have a continuous amplitude as in the case of speech, and the requirement is to quantize the amplitude of each sample generated by the source to permit its representation by a code word of finite length as in pulse-code modulation. In such cases, the problem is referred to as *source coding with a fidelity criterion*, and the branch of information theory that deals with it is called *rate distortion theory*.[14] Rate distortion theory finds applications in two types of situations:

- Source coding where the permitted coding alphabet cannot exactly represent the information source, in which case we are forced to do lossy *data compression*.
- Information transmission at a rate greater than channel capacity.

Accordingly, rate distortion theory may be viewed as a natural extension of Shannon's coding theorems.

## RATE DISTORTION FUNCTION

Consider a discrete memoryless source defined by an $M$-ary alphabet $X$: $\{x_i \mid i = 1, 2, \ldots, M\}$, which consists of a set of statistically independent symbols together with the associated symbol probabilities $\{p_i \mid i = 1, 2, \ldots, M\}$. Let $R$ be the average code rate in bits per code word. The representation code words are taken from another alphabet $Y$: $\{y_j \mid j = 1, 2, \ldots, N\}$. The source coding theorem states that this second alphabet provides a perfect representation of the source provided that $R > H$, where $H$ is the source entropy. But if we are forced to have $R < H$, then there is unavoidable distortion and therefore loss of information.

Let $p(x_i, y_j)$ denote the joint probability of occurrence of source symbol $x_i$ and representation symbol $y_j$. From probability theory, we have

$$p(x_i, y_j) = p(y_j \mid x_i)p(x_i) \tag{9.127}$$

where $p(y_j \mid x_i)$ is a transition probability. Let $d(x_i, y_j)$ denote a measure of the cost incurred in representing the source symbol $x_i$ by the symbol $y_j$; the quantity $d(x_i, y_j)$ is referred to as a *single-letter distortion measure*. The statistical average of $d(x_i, y_j)$ over all possible source symbols and representation symbols is given by

$$\bar{d} = \sum_{i=1}^{M} \sum_{j=1}^{N} p(x_i)p(y_j \mid x_i)\, d(x_i, y_j) \tag{9.128}$$

Note that the average distortion $\bar{d}$ is a nonnegative continuous function of the transition probabilities $p(y_j \mid x_i)$ that are determined by the source encoder-decoder pair.

A conditional probability assignment $p(y_j \mid x_i)$ is said to be *D-admissible* if and only if the average distortion $\bar{d}$ is less than or equal to some acceptable value $D$. The set of all $D$-admissible conditional probability assignments is denoted by

$$P_D = \{p(y_j \mid x_i) : \bar{d} \le D\} \tag{9.129}$$

For each set of transition probabilities, we have a mutual information

$$I(X; Y) = \sum_{i=1}^{M} \sum_{j=1}^{N} p(x_i)p(y_j \mid x_i) \log\left(\frac{p(y_j \mid x_i)}{p(y_j)}\right) \tag{9.130}$$

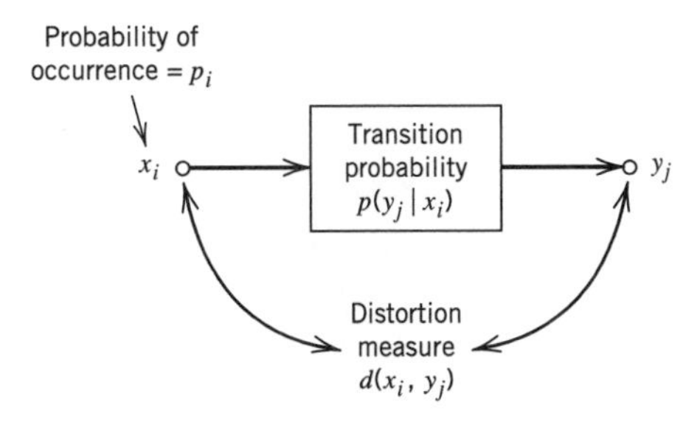


**FIGURE 9.22** Summary of rate distortion theory.

A *rate distortion function* $R(D)$ is defined as *the smallest coding rate possible for which the average distortion is guaranteed not to exceed D*. Let $P_D$ denote the set to which the conditional probability $p(y_j|x_i)$ belongs for a prescribed $D$. Then, for a fixed $D$ we write[15]

$$R(D) = \min_{p(y_j|x_i) \in P_D} I(X; Y) \tag{9.131}$$

subject to the constraint

$$\sum_{j=1}^{N} p(y_j|x_i) = 1 \qquad \text{for } i = 1, 2, \ldots, M \tag{9.132}$$

The rate distortion function $R(D)$ is measured in units of bits if the base-2 logarithm is used in Equation (9.130). Intuitively, we expect the distortion $D$ to decrease as the rate distortion function $R(D)$ is increased. We may say conversely that tolerating a large distortion $D$ permits the use of a smaller rate for coding and/or transmission of information.

Figure 9.22 summarizes the main parameters of rate distortion theory. In particular, given the source symbols $\{x_i\}$ and their probabilities $\{p_i\}$ and given a definition of the single-letter distortion measure $d(x_i, y_j)$, the calculation of the rate distortion function $R(D)$ involves finding the conditional probability assignment $p(y_j|x_i)$ subject to certain constraints imposed on $p(y_j|x_i)$. This is a variational problem, the solution of which is unfortunately not straightforward in general.

### ▶ EXAMPLE 9.14 Gaussian Source

Consider a discrete-time, memoryless Gaussian source with zero mean and variance $\sigma^2$. Let $x$ denote the value of a sample generated by such a source. Let $y$ denote a quantized version of $x$ that permits a finite representation of it. The *squared error distortion*

$$d(x, y) = (x - y)^2$$

provides a distortion measure that is widely used for continuous alphabets. The rate distortion function for the Gaussian source with squared error distortion, as described herein, is given by

$$R(D) = \begin{cases} \dfrac{1}{2} \log\left(\dfrac{\sigma^2}{D}\right), & 0 \le D \le \sigma^2 \\ 0, & D > \sigma^2 \end{cases} \tag{9.133}$$

In this case, we see that $R(D) \rightarrow \infty$ as $D \rightarrow 0$, and $R(D) = 0$ for $D = \sigma^2$. ◀

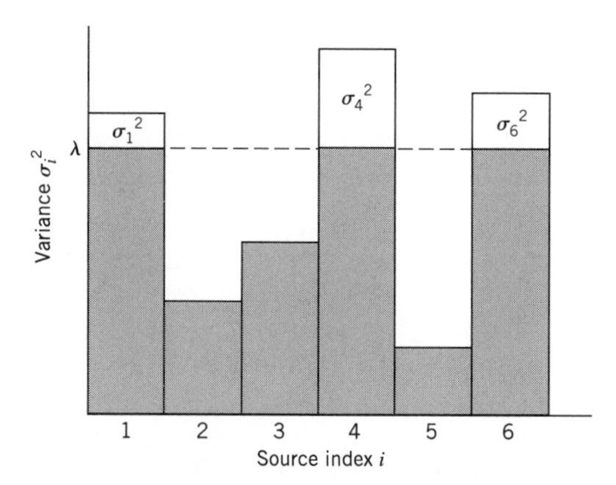


**FIGURE 9.23** Reverse water-filling picture for a set of parallel Gaussian processes.

### ▶ EXAMPLE 9.15 Set of Parallel Gaussian Sources

Consider next a set of $N$ independent Gaussian random variables $\{X_i\}_{i=1}^N$, where $X_i$ has zero mean and variance $\sigma_i^2$. Using the distortion measure

$$d = \sum_{i=1}^{N} (x_i - \hat{x}_i)^2$$

and building on the result of Example 9.14, we may express the rate distortion function for the set of parallel Gaussian sources described here as

$$R(D) = \sum_{i=1}^{N} \frac{1}{2} \log\left(\frac{\sigma_i^2}{D_i}\right) \tag{9.134}$$

where $D_i$ is itself defined by

$$D_i = \begin{cases} \lambda & \text{if } \lambda < \sigma_i \\ \sigma_i^2 & \text{if } \lambda \geq \sigma_i^2 \end{cases} \tag{9.135}$$

and the constant $\lambda$ is chosen so as to satisfy the condition

$$\sum_{i=1}^{N} D_i = D \tag{9.136}$$

Equations (9.135) and (9.136) may be interpreted as a kind of "water-filling in reverse," as illustrated in Figure 9.23. First, we choose a constant $\lambda$ and only the subset of random variables whose variances exceed the constant $\lambda$. No bits are used to describe the remaining subset of random variables whose variances are less than the constant $\lambda$. ◀

## 9.14 Data Compression

Rate distortion theory naturally leads us to consider the idea of *data compression* that involves a purposeful or unavoidable reduction in the information content of data from a continuous or discrete source. Specifically, we may think of a *data compressor*, or *signal compressor*, as a device that supplies a code with the least number of symbols for the representation of the source output, subject to a permissible or acceptable *distortion*. The data compressor thus retains the essential information content of the source output by blurring fine details in a deliberate but controlled manner. Accordingly, data compression is a *lossy* operation in the sense that the source entropy is reduced (i.e., information is lost), irrespective of the type of source being considered.

In the case of a discrete source, the reason for using data compression is to encode the source output at a rate smaller than the source entropy. By so doing, the source coding theorem is violated, which means that exact reproduction of the original data is *no longer* possible.

In the case of a continuous source, the entropy is infinite, and therefore a signal compression code must always be used to encode the source output at a finite rate. Consequently, it is impossible to digitally encode an analog signal with a finite number of bits without producing some distortion. This statement is in perfect accord with the idea of pulse-code modulation, which was studied in Chapter 3. There it was shown that quantization, which is basic to the analog-to-digital conversion process in pulse-code modulation, always introduces distortion (known as quantization noise) into the transmitted signal. A quantizer may therefore be viewed as a signal compressor.

The uniform and nonuniform quantizers considered in Chapter 3 are said to be *scalar quantizers* in the sense that they deal with samples of the analog signal (i.e., continuous source output) one at a time. Each sample is converted into a quantized value, with the conversion being independent from sample to sample. A scalar quantizer is a rather simple signal compressor, which makes it attractive for practical use. Yet it can provide a surprisingly good performance; this is especially so if nonuniform quantization is used.

There is another class of quantizers known as *vector quantizers* that use blocks of consecutive samples of the source output to form vectors, each of which is treated as a single entity. The essential operation in a vector quantizer is the quantization of a random vector[16] by encoding it as a binary code word. The vector is encoded by comparing it with a *codebook* consisting of a set of stored reference vectors known as *code vectors* or *patterns*. Each pattern in the codebook is used to represent input vectors that are identified by the encoder to be similar to the particular pattern, subject to the maximization of an appropriate fidelity criterion. The encoding process in a vector quantizer may thus be viewed as a *pattern matching operation*.

Let $N$ be the number of code vectors in the codebook, $k$ be the dimension of each vector (i.e., the number of samples in each pattern), and $r$ be the coded transmission rate in bits per sample. These three parameters are related as follows:

$$r = \frac{\log_2 N}{k} \tag{9.137}$$

Then, assuming that the size of the code book is sufficiently large, the signal-to-quantization noise ratio (SNR) for the vector quantizer is given by

$$10 \log_{10}(\text{SNR}) = 6\left(\frac{\log_2 N}{k}\right) + C_k \text{ dB} \tag{9.138}$$

where $C_k$ is a constant (expressed in dB) that depends on the dimensions $k$. According to Equation (9.138), the SNR for a vector quantizer increases approximately at the rate of $6/k$ dB for each doubling of the codebook size. Equivalently, we may state that the SNR increases by 6 dB per unit increase in rate (bits per sample) as in the standard PCM using a uniform scalar quantizer. The advantage of the vector quantizer over the scalar quantizer is that its constant term $C_k$ has a higher value, because the vector quantizer optimally exploits the correlations among the samples constituting a vector. Specifically, the constant $C_k$ increases with the dimension $k$, approaching the ultimate rate-distortion limit for a given source of information. However, the improvement in SNR is attained at the cost of increased encoding complexity, which grows exponentially with the dimension $k$ for a specified rate $r$. Unfortunately, this is the main obstacle to the wide use of vector quantization in practice. Nevertheless, in certain applications, the issue of computational complexity is mitigated by exploiting the capability of VLSI technology to concentrate a highly complex signal processor on a silicon chip. For example, that is precisely what is done in the use of code-excited linear predictive (CELP) modeling of speech in wireless communication systems of the CDMA type, namely, the IS-95 system. From the description of CELP presented in Section 8.9, it is clear that the CELP modeling of speech is an example of vector quantization.

## 9.15 *Summary and Discussion*

In this chapter we established four fundamental limits on different aspects of a communication system. The limits are embodied in the source coding theorem, the channel coding theorem, the information capacity theorem, and the rate distortion function.

The *source coding theorem*, Shannon's first theorem, provides the mathematical tool for assessing *data compaction*, that is, *lossless compression* of data generated by a discrete memoryless source. The theorem tells us that we can make the average number of binary code elements (bits) per source symbol as small as, but no smaller than, the entropy of the source measured in bits. The *entropy* of a source is a function of the probabilities of the source symbols that constitute the alphabet of the source. Since entropy is a measure of uncertainty, the entropy is maximum when the associated probability distribution generates maximum uncertainty.

The *channel coding theorem*, Shannon's second theorem, is both the most surprising and the single most important result of information theory. For a *binary symmetric channel*, the channel coding theorem tells us that for any *code rate* $r$ less than or equal to the *channel capacity* $C$, codes do exist such that the average probability of error is as small as we want it. A binary symmetric channel is the simplest form of a discrete memoryless channel. It is symmetric because the probability of receiving a 1 if a 0 is sent is the same as the probability of receiving a 0 if a 1 is sent. This probability, the probability that an error will occur, is termed a *transition probability*. The transition probability $p$ is determined not only by the additive noise at the channel output but also by the kind of receiver used. The value of $p$ uniquely defines the channel capacity $C$.

Shannon's third remarkable theorem, the *information capacity theorem*, tells us that there is a maximum to the rate at which any communication system can operate reliably (i.e., free of errors) when the system is constrained in power. This maximum rate is called the *information capacity*, measured in bits per second. When the system operates at a rate greater than the information capacity, it is condemned to a high probability of error, regardless of the choice of signal set used for transmission or the receiver used for processing the received signal.

Finally, the rate distortion function provides the mathematical tool for signal compression (i.e., solving the problem of source coding with a fidelity criterion): The rate distortion function can be applied to a discrete as well as continuous memoryless source.

When the output of a source of information is compressed in a lossless manner, the resulting data stream usually contains redundant bits. These redundant bits can be removed by using a lossless algorithm such as Huffman coding or the Lempel–Ziv algorithm for data compaction. We may thus speak of data compression followed by data compaction as two constituents of the *dissection of source coding*, which is so called because it refers exclusively to the sources of information. In some source coding applications, we have a third constituent, namely, *data encryption*, which follows data compaction. The purpose of data encryption is to disguise the data (bit) stream in such a way that it has no meaning to an unauthorized receiver. Some basic aspects of *cryptography*, which encompasses both encryption and decryption, follow quite naturally from information theory, as discussed in Appendix 5. Other issues relating to cryptography are also discussed in that appendix.

One last comment is in order. Shannon's information theory, as presented in this chapter, has been entirely in the context of memoryless sources and channels. The theory can be extended to deal with sources and channels with *memory*, in which case a symbol of interest depends on preceding symbols; however, the level of exposition needed to do this is beyond the scope of this book.[17]

## NOTES AND REFERENCES

1. According to Lucky (1989), the first mention of the term *information theory* by Shannon occurs in a 1945 memorandum entitled "A Mathematical Theory of Cryptography." It is rather curious that the term was never used in the classic 1948 paper by Shannon, which laid down the foundations of information theory. For an introductory treatment of information theory, see Chapter 2 of Lucky (1989) and the paper by Wyner (1981); see also the books of Adámek (1991), Hamming (1980), and Abramson (1963). For more advanced treatments of the subject, see the books of Cover and Thomas (1991), Blahut (1987), McEliece (1977), and Gallager (1968). For a collection of papers on the development of information theory (including the 1948 classic paper by Shannon), see Slepian (1974). For a collection of the papers published by Shannon, see Sloane and Wyner (1993).

2. The use of a logarithmic measure of information was first suggested by Hartley (1928); however, Hartley used logarithms to base 10.

3. In statistical physics, the entropy of a physical system is defined by (Reif, 1967, p. 147)

$$\mathcal{S} = k \log \Omega$$

   where $k$ is Boltzmann's constant, $\Omega$ is the number of states accessible to the system, and log denotes the natural logarithm. This entropy has the dimensions of energy because its definition involves the constant $k$. In particular, it provides a *quantitative measure of the degree of randomness of the system*. Comparing the entropy of statistical physics with that of information theory, we see that they have a similar form. For a detailed discussion of the relation between them, see Pierce (1961, pp. 184–207) and Brillouin (1962).

4. For the original proof of the source coding theorem, see Shannon (1948). A general proof of the source coding theorem is also given in the following books: Viterbi and Omura (1979, pp. 13–19), McEliece (1977, Chapter 3), and Gallager (1968, pp. 38–55). The source coding theorem is also referred to in the literature as the *noiseless coding theorem*, noiseless in the sense that it establishes the condition for error-free encoding to be possible.

5. For proof of the Kraft–McMillan inequality, see Cover and Thomas (1991, pp. 82–84), Blahut (1990, pp. 298–299), and McEliece (1977, pp. 239–240). For a proof of Equation (9.23), see Cover and Thomas (1991), pp. 87–88), Blahut (1990, pp. 300–301), and McEliece (1977, pp. 241–242).

6. The Huffman code is named after its inventor: D. A. Huffman (1952). For a readable account of Huffman coding and its use in data compaction, see Adámek (1991).

7. The original papers on the Lempel–Ziv algorithm are Ziv and Lempel (1977, 1978). For readable descriptions of the Lempel–Ziv algorithm, see Lucky (1989, pp. 118–122), Blahut

(1990, pp. 314–319), and Gitlin, Hayes, and Weinstein (1992, pp. 120–122). For the application of the Lempel–Ziv algorithm to the compaction of English text, see Lucky (1989, pp. 122–128) and the paper by Welch (1984); see also the review paper by Weiss and Shremp (1993).

8. The channel coding theorem is also known as the *noisy coding theorem*. The original proof of the theorem is given in Shannon (1948). A proof of the theorem is also presented in Hamming (1980, Chapters 9 and 10) in sufficient detail so that a general appreciation of relevant results is developed. The second part of the theorem is referred to in the literature as *the converse to the coding theorem*. A proof of this theorem is presented in the following references: Viterbi and Omura (1979, pp. 28–34) and Gallager (1968, pp. 76–82).

9. The quantity

$$\int_{-\infty}^{\infty} f_Y(x) \log_2\left(\frac{f_X(x)}{f_Y(x)}\right) dx$$

on the left-hand side of Equation (9.70) is called *relative entropy* or the *Kullback–Leibler divergence* between the probability density functions $f_X(x)$ and $f_Y(x)$; see Kullback (1968).

10. Shannon's information capacity theorem is also referred to in the literature as the *Shannon–Hartley law* in recognition of early work by Hartley on information transmission (Hartley, 1928). In particular, Hartley showed that the amount of information that can be transmitted over a given channel is proportional to the product of the channel bandwidth and the time of operation.

11. A lucid exposition of sphere packing is presented in Cover and Thomas (1991, pp. 242–243); see also Wozencraft and Jacobs (1965, pp. 323–341).

12. Parts *a* and *b* of Figure 9.18 follow the corresponding parts of Figure 6.2 in the book by Frey (1998).

13. For a rigorous treatment of the information capacity of a colored noisy channel, see Gallager (1968). The idea of replacing the channel model of Figure 9.19*a* with that of Figure 9.19*b* is discussed in Gitlin, Hayes, and Weinstein (1992).

14. For a complete treatment of rate distortion theory, see the book by Berger (1971); this subject is also treated in somewhat less detail in Cover and Thomas (1991), McEliece (1977), and Gallager (1968).

15. For the derivation of Equation (9.131), see Cover and Thomas (1991, p. 345). An algorithm for computation of the rate distortion function $R(D)$ defined in Equation (9.131) is described in Blahut (1987, pp. 220–221) and Cover and Thomas (1991, pp. 364–367).

16. For the early papers on vector quantization, see Gersho (1979) and Linde, Buzo, and Gray (1980). For a tutorial review of vector quantization, see Gray (1984). Equation (9.138), defining the SNR for a vector quantizer, is discussed in Gersho and Cuperman (1983). For a complete treatment of vector quantization, see the book by Gersho and Gray (1992).

17. For detailed discussion of discrete channels with memory, see Gallager (1968, pp. 97–112) and Ash (1965, pp. 211–229).

## PROBLEMS

### Entropy

9.1 Let $p$ denote the probability of some event. Plot the amount of information gained by the occurrence of this event for $0 \le p \le 1$.

**9.2** A source emits one of four possible symbols during each signaling interval. The symbols occur with the probabilities:

$$p_0 = 0.4$$
$$p_1 = 0.3$$
$$p_2 = 0.2$$
$$p_3 = 0.1$$

Find the amount of information gained by observing the source emitting each of these symbols.

**9.3** A source emits one of four symbols $s_0$, $s_1$, $s_2$, and $s_3$ with probabilities 1/3, 1/6, 1/4, and 1/4, respectively. The successive symbols emitted by the source are statistically independent. Calculate the entropy of the source.

**9.4** Let $X$ represent the outcome of a single roll of a fair die. What is the entropy of $X$?

**9.5** The sample function of a Gaussian process of zero mean and unit variance is uniformly sampled and then applied to a uniform quantizer having the input-output amplitude characteristic shown in Figure P9.5. Calculate the entropy of the quantizer output.

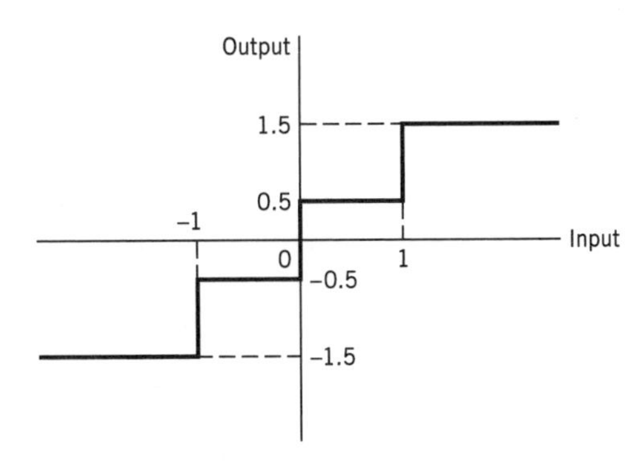


**FIGURE P9.5**

**9.6** Consider a discrete memoryless source with source alphabet $\mathcal{S} = \{s_0, s_1, \ldots, s_{K-1}\}$ and source statistics $\{p_0, p_1, \ldots, p_{K-1}\}$. The $n$th extension of this source is another discrete memoryless source with source alphabet $\mathcal{S}^n = \{\sigma_0, \sigma_1, \ldots, \sigma_{M-1}\}$, where $M = K^n$. Let $P(\sigma_i)$ denote the probability of $\sigma_i$.

(a) Show that

$$\sum_{i=0}^{M-1} P(\sigma_i) = 1$$

which is to be expected.

(b) Show that

$$\sum_{i=0}^{M-1} P(\sigma_i) \log_2\left(\frac{1}{p_{i_k}}\right) = H(\mathcal{S}), \qquad k = 1, 2, \ldots, n$$

where $p_{i_k}$ is the probability of symbol $s_{i_k}$, and $H(\mathcal{S})$ is the entropy of the original source.

(c) Hence, show that

$$H(\mathcal{S}^n) = \sum_{i=0}^{M-1} P(\sigma_i) \log_2 \frac{1}{P(\sigma_i)}$$
$$= nH(\mathcal{S})$$

9.7 Consider a discrete memoryless source with source alphabet $\mathcal{S} = \{s_0, s_1, s_2\}$ and source statistics $\{0.7, 0.15, 0.15\}$.

(a) Calculate the entropy of the source.

(b) Calculate the entropy of the second-order extension of the source.

9.8 It may come as a surprise, but the number of bits needed to store text is much less than that required to store its spoken equivalent. Can you explain the reason for it?

## Data Compaction

9.9 Consider a discrete memoryless source whose alphabet consists of $K$ equiprobable symbols.

(a) Explain why the use of a fixed-length code for the representation of such a source is about as efficient as any code can be.

(b) What conditions have to be satisfied by $K$ and the code-word length for the coding efficiency to be 100 percent?

9.10 Consider the four codes listed below:

| *Symbol* | *Code I* | *Code II* | *Code III* | *Code IV* |
|---|---|---|---|---|
| $s_0$ | 0 | 0 | 0 | 00 |
| $s_1$ | 10 | 01 | 01 | 01 |
| $s_2$ | 110 | 001 | 011 | 10 |
| $s_3$ | 1110 | 0010 | 110 | 110 |
| $s_4$ | 1111 | 0011 | 111 | 111 |

(a) Two of these four codes are prefix codes. Identify them, and construct their individual decision trees.

(b) Apply the Kraft–McMillan inequality to codes I, II, III, and IV. Discuss your results in light of those obtained in part (a).

9.11 Consider a sequence of leters of the English alphabet with their probabilities of occurrence as given here:

| Letter | *a* | *i* | *l* | *m* | *n* | *o* | *p* | *y* |
|---|---|---|---|---|---|---|---|---|
| Probability | 0.1 | 0.1 | 0.2 | 0.1 | 0.1 | 0.2 | 0.1 | 0.1 |

Compute two different Huffman codes for this alphabet. In one case, move a combined symbol in the coding procedure as high as possible, and in the second case, move it as low as possible. Hence, for each of the two codes, find the average code-word length and the variance of the average code-word length over the ensemble of letters.

9.12 A discrete memoryless source has an alphabet of seven symbols whose probabilities of occurrence are as described here:

| Symbol | $s_0$ | $s_1$ | $s_2$ | $s_3$ | $s_4$ | $s_5$ | $s_6$ |
|---|---|---|---|---|---|---|---|
| Probability | 0.25 | 0.25 | 0.125 | 0.125 | 0.125 | 0.0625 | 0.0625 |

Compute the Huffman code for this source, moving a "combined" symbol as high as possible. Explain why the computed source code has an efficiency of 100 percent.

9.13 Consider a discrete memoryless source with alphabet $\{s_0, s_1, s_2\}$ and statistics $\{0.7, 0.15, 0.15\}$ for its output.

(a) Apply the Huffman algorithm to this source. Hence, show that the average code-word length of the Huffman code equals 1.3 bits/symbol.

(b) Let the source be extended to order two. Apply the Huffman algorithm to the resulting extended source, and show that the average code-word length of the new code equals 1.1975 bits/symbol.

(c) Compare the average code-word length calculated in part (b) with the entropy of the original source.

**9.14** Figure P9.14 shows a Huffman tree. What is the code word for each of the symbols A, B, C, D, E, F, and G represented by this Huffman tree? What are their individual code-word lengths?

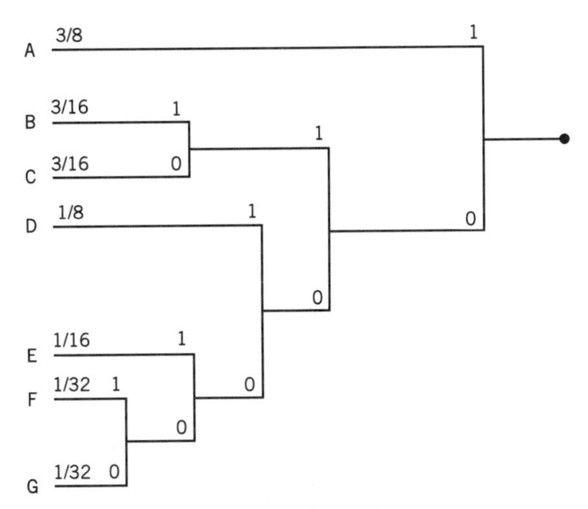


**FIGURE P9.14**

**9.15** A computer executes four instructions that are designated by the code words (00, 01, 10, 11). Assuming that the instructions are used independently with probabilities (1/2, 1/8, 1/8, 1/4), calculate the percentage by which the number of bits used for the instructions may be reduced by the use of an optimum source code. Construct a Huffman code to realize the reduction.

**9.16** Consider the following binary sequence

11101001100010110100 . . .

Use the Lempel–Ziv algorithm to encode this sequence. Assume that the binary symbols 0 and 1 are already in the codebook.

## Binary Symmetric Channel

**9.17** Consider the transition probability diagram of a binary symmetric channel shown in Figure 9.8. The input binary symbols 0 and 1 occur with equal probability. Find the probabilities of the binary symbols 0 and 1 appearing at the channel output.

**9.18** Repeat the calculation in Problem 9.17, assuming that the input binary symbols 0 and 1 occur with probabilities 1/4 and 3/4, respectively.

## Mutual Information and Channel Capacity

**9.19** Consider a binary symmetric channel characterized by the transition probability $p$. Plot the mutual information of the channel as a function of $p_1$, the *a priori* probability of symbol 1 at the channel input; do your calculations for the transition probability $p = 0$, 0.1, 0.2, 0.3, 0.5.

**9.20** Figure 9.10 depicts the variation of the channel capacity of a binary symmetric channel with the transition probability $p$. Use the results of Problem 9.19 to explain this variation.

**9.21** Consider the binary symmetric channel described in Figure 9.8. Let $p_0$ denote the probability of sending binary symbol $x_0 = 0$, and let $p_1 = 1 - p_0$ denote the probability of sending binary symbol $x_1 = 1$. Let $p$ denote the transition probability of the channel.

(a) Show that the mutual information between the channel input and channel output is given by

$$I(\mathscr{X}; \mathscr{Y}) = \mathscr{H}(z) - \mathscr{H}(p)$$

where

$$H(z) = z \log_2\left(\frac{1}{z}\right) + (1 - z) \log_2\left(\frac{1}{1 - z}\right)$$

$$z = p_0 p + (1 - p_0)(1 - p)$$

and

$$H(p) = p \log_2\left(\frac{1}{p}\right) + (1 - p) \log_2\left(\frac{1}{1 - p}\right)$$

(b) Show that the value of $p_0$ that maximizes $I(\mathscr{X}; \mathscr{Y})$ is equal to 1/2.

(c) Hence, show that the channel capacity equals

$$C = 1 - H(p)$$

**9.22** Two binary symmetric channels are connected in cascade, as shown in Figure P9.22. Find the overall channel capacity of the cascaded connection, assuming that both channels have the same transition probability diagram shown in Figure 9.8.

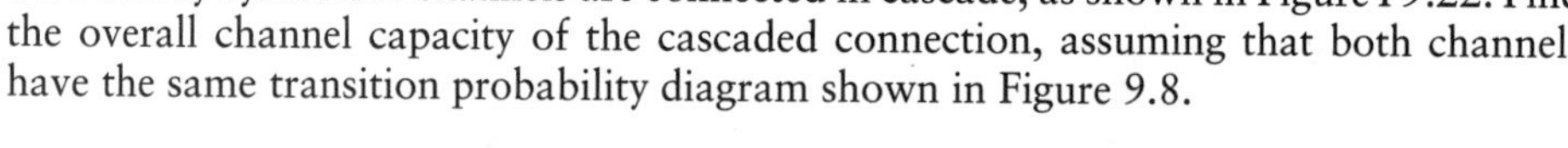

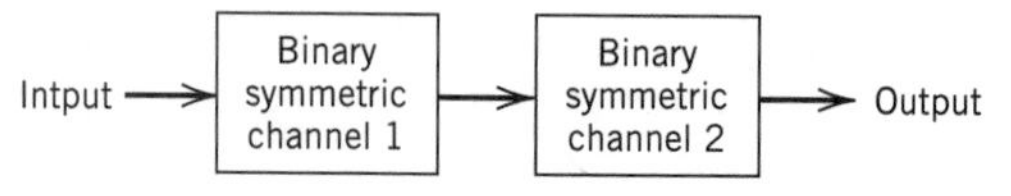


FIGURE P9.22

**9.23** The *binary erasure channel* has two inputs and three outputs as described in Figure P9.23. The inputs are labeled 0 and 1, and the outputs are labeled 0, 1, and $e$. A fraction $\alpha$ of the incoming bits are erased by the channel. Find the capacity of the channel.

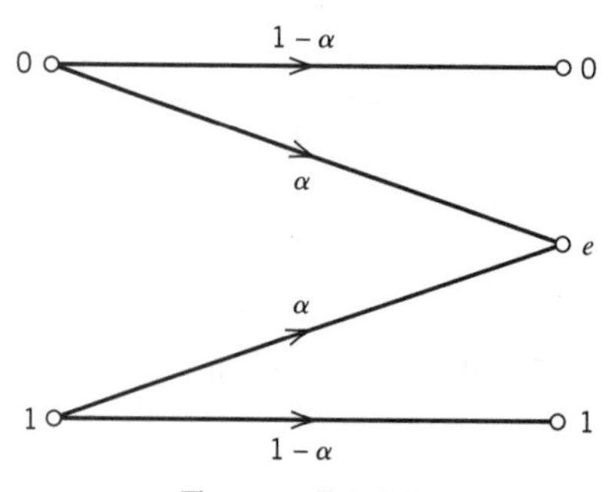


FIGURE P9.23

9.24 Consider a digital communication system that uses a *repetition code* for the channel encoding/decoding. In particular, each transmission is repeated $n$ times, where $n = 2m + 1$ is an odd integer. The decoder operates as follows. If in a block of $n$ received bits, the number of 0s exceeds the number of 1s, the decoder decides in favor of a 0. Otherwise, it decides in favor of a 1. An error occurs when $m + 1$ or more transmissions out of $n = 2m + 1$ are incorrect. Assume a binary symmetric channel.

(a) For $n = 3$, show that the average probability of error is given by

$$P_e = 3p^2(1 - p) + p^3$$

where $p$ is the transition probability of the channel.

(b) For $n = 5$, show that the average probability of error is given by

$$P_e = 10p^3(1 - p)^2 + 5p^4(1 - p) + p^5$$

(c) Hence, for the general case, deduce that the average probability of error is given by

$$P_e = \sum_{i=m+1}^{n} \binom{n}{i} p^i(1 - p)^{n-i}$$

## Differential Entropy

9.25 Let $X_1, X_2, \ldots, X_n$ denote the elements of a Gaussian vector $\mathbf{X}$. The $X_i$ are independent with mean $\mu_i$ and variance $\sigma_i^2$, $i = 1, 2, \ldots, n$. Show that the differential entropy of the vector $\mathbf{X}$ equals

$$h(\mathbf{X}) = \frac{n}{2} \log_2[2\pi e(\sigma_1^2\sigma_2^2 \ldots \sigma_n^2)^{1/n}]$$

What does $h(\mathbf{X})$ reduce to if the variances are equal?

9.26 A continuous random variable $X$ is constrained to a peak magnitude $M$; that is, $-M < X < M$.

(a) Show that the differential entropy of $X$ is maximum when it is uniformly distributed, as shown by

$$f_X(x) = \begin{cases} 1/2M, & -M < x \le M \\ 0, & \text{otherwise} \end{cases}$$

(b) Show that the maximum differential entropy of $X$ is $\log_2 2M$.

9.27 Prove the properties given in Equations (9.79) to (9.81) for the mutual information $I(X; Y)$.

9.28 Consider the continuous random variable $Y$ defined by

$$Y = X + N$$

where $X$ and $N$ are statistically independent. Show that the conditional differential entropy of $Y$, given $X$, equals

$$h(Y|X) = h(N)$$

where $h(N)$ is the differential entropy of $N$.

## Information Capacity

9.29 A voice-grade channel of the telephone network has a bandwidth of 3.4 kHz.

(a) Calculate the information capacity of the telephone channel for a signal-to-noise ratio of 30 dB.

(b) Calculate the minimum signal-to-noise ratio required to support information transmission through the telephone channel at the rate of 9,600 b/s.

**9.30** Alphanumeric data are entered into a computer from a remote terminal through a voice-grade telephone channel. The channel has a bandwidth of 3.4 kHz and output signal-to-noise ratio of 20 dB. The terminal has a total of 128 symbols. Assume that the symbols are equiprobable and the successive transmissions are statistically independent.

(a) Calculate the information capacity of the channel.

(b) Calculate the maximum symbol rate for which error-free transmission over the channel is possible.

**9.31** A black-and-white television picture may be viewed as consisting of approximately $3 \times 10^5$ elements, each of which may occupy one of 10 distinct brightness levels with equal probability. Assume that (1) the rate of transmission is 30 picture frames per second, and (2) the signal-to-noise ratio is 30 dB.

Using the information capacity theorem, calculate the minimum bandwidth required to support the transmission of the resulting video signal.

*Note:* As a matter of interest, commercial television transmissions actually employ a bandwidth of 4.2 MHz, which fits into an allocated bandwidth of 6 MHz.

**9.32** In this problem, we continue with Example 9.9. Suppose that the tightly packed constellation of Figure 9.15*b* is scaled upward so that the transmitted signal energy per symbol is maintained at the same average value as that consumed by the 64-QAM square constellation of Figure 9.15*a*. Construct the new constellation that results from this scaling. How does the bit error rate of this new constellation compare with that of Figure 9.15*a*? Justify your answer.

**9.33** The squared magnitude response of a twisted-pair channel can be modeled as

$$|H(f)|^2 = \exp(-\alpha\sqrt{f})$$

The constant $\alpha$ is defined by

$$\alpha = \frac{kl}{l_0}$$

where $k$ is a constant depending on wire gauge, $l_0$ is a reference line length, and $l$ is the actual length of the twisted pair under study. The squared magnitude response of the coupling responsible for NEXT has the form

$$|H_{\text{NEXT}}(f)|^2 = \beta f^{3/2}$$

where $\beta$ is a constant that depends on the type of cable used.

Formulate the expression for the information capacity of the NEXT-dominated channel described here.

## Data Compression

**9.34** Equation (9.138) for the signal-to-noise ratio (SNR) of a vector quantizer includes the SNR formula of Equation (3.33) for standard pulse-code modulation as a special case for which $k = 1$. Justify the validity of this inclusion.

**9.35** All practical data compression and data transmission schemes lie between two limits set by the rate distortion function and the channel capacity theorem. Both of these theorems involve the notion of mutual information, but in different ways. Elaborate on the issues raised by these two statements.

## Computer Experiment

**9.36** In this problem, we revisit Example 9.12, which deals with coded binary antipodal signaling over an additive white Gaussian noise (AWGN) channel. Starting with Equation (9.112) and the underlying theory, develop a software package for computing the minimum $E_b/N_0$ required for a given bit error rate, where $E_b$ is the signal energy per bit, and

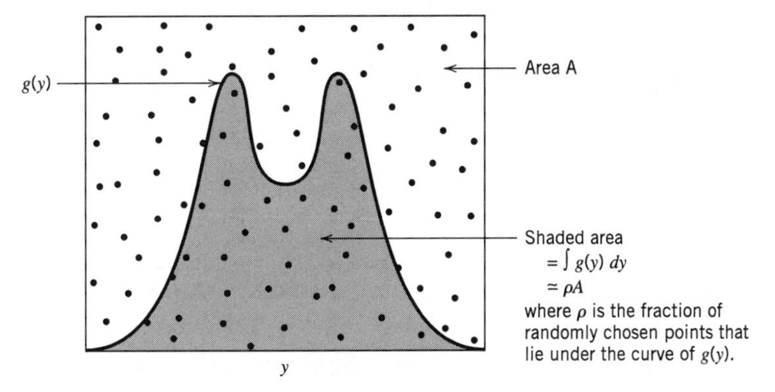


**FIGURE P9.36**

$N_0/2$ is the noise spectral density. Hence, compute the results plotted in parts *a* and *b* of Figure 9.18.

As mentioned in Example 9.12, the computation of the mutual information between the channel input and channel output is well approximated using Monte Carlo integration. To explain how this method works, consider a function $g(y)$ that is difficult to sample randomly, which is indeed the case for the problem at hand. (For our problem, the function $g(y)$ represents the complicated integrand in the formula for the differential entropy of the channel output.) For the computation, proceed as follows:

- Find an area $A$ that includes the region of interest and that is easily sampled.
- Choose $N$ points, uniformly randomly inside the area $A$.

Then the *Monte Carlo integration theorem* states that the integral of the function $g(y)$ with respect to $y$ is approximately equal to the area $A$ multiplied by the fraction of points that reside below the curve of $g$, as illustrated in Figure P9.36. The accuracy of the approximation improves with increasing $N$.

# CHAPTER 10

# ERROR-CONTROL CODING

This chapter is the natural sequel to the preceding chapter on Shannon's information theory. In particular, in this chapter we present error-control coding techniques that provide different ways of implementing Shannon's channel-coding theorem. Each error-control coding technique involves the use of a channel encoder in the transmitter and a decoding algorithm in the receiver.

The error-control coding techniques described herein include the following important classes of codes:

- *Linear block codes.*
- *Cyclic codes.*
- *Convolutional codes.*
- *Compound codes exemplified by turbo codes and low-density parity-check codes, and their irregular variants.*

## 10.1 *Introduction*

The task facing the designer of a digital communication system is that of providing a cost-effective facility for transmitting information from one end of the system at a rate and a level of reliability and quality that are acceptable to a user at the other end. The two key system parameters available to the designer are transmitted signal power and channel bandwidth. These two parameters, together with the power spectral density of receiver noise, determine the signal energy per bit-to-noise power spectral density ratio $E_b/N_0$. In Chapter 6, we showed that this ratio uniquely determines the bit error rate for a particular modulation scheme. Practical considerations usually place a limit on the value that we can assign to $E_b/N_0$. Accordingly, in practice, we often arrive at a modulation scheme and find that it is not possible to provide acceptable data quality (i.e., low enough error performance). For a fixed $E_b/N_0$, the only practical option available for changing data quality from problematic to acceptable is to use *error-control coding*.

Another practical motivation for the use of coding is to reduce the required $E_b/N_0$ for a fixed bit error rate. This reduction in $E_b/N_0$ may, in turn, be exploited to reduce the required transmitted power or reduce the hardware costs by requiring a smaller antenna size in the case of radio communications.

*Error control*[1] for data integrity may be exercised by means of *forward error correction* (FEC). Figure 10.1*a* shows the model of a digital communication system using such an approach. The discrete source generates information in the form of binary symbols. The *channel encoder* in the transmitter accepts message bits and adds *redundancy* according to a prescribed rule, thereby producing encoded data at a higher bit rate. The *channel*

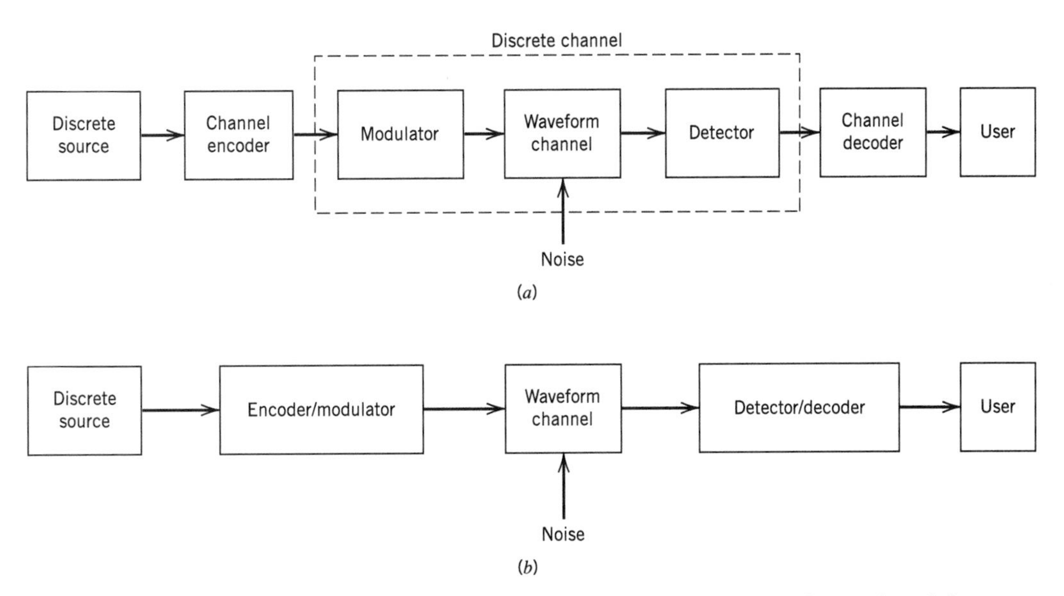


**FIGURE 10.1** Simplified models of digital communication system. (*a*) Coding and modulation performed separately. (*b*) Coding and modulation combined.

*decoder* in the receiver exploits the redundancy to decide which message bits were actually transmitted. The combined goal of the channel encoder and decoder is to minimize the effect of channel noise. That is, the number of errors between the channel encoder input (derived from the source) and the channel decoder output (delivered to the user) is minimized.

For a fixed modulation scheme, the addition of redundancy in the coded messages implies the need for increased transmission bandwidth. Moreover, the use of error-control coding adds *complexity* to the system, especially for the implementation of decoding operations in the receiver. Thus, the design trade-offs in the use of error-control coding to achieve acceptable error performance include considerations of bandwidth and system complexity.

There are many different error-correcting codes (with roots in diverse mathematical disciplines) that we can use. Historically, these codes have been classified into *block codes* and *convolutional codes*. The distinguishing feature for this particular classification is the presence or absence of *memory* in the encoders for the two codes.

To generate an $(n, k)$ block code, the channel encoder accepts information in successive $k$-bit *blocks*; for each block, it adds $n - k$ redundant bits that are algebraically related to the $k$ message bits, thereby producing an overall encoded block of $n$ bits, where $n > k$. The $n$-bit block is called a *code word*, and $n$ is called the *block length* of the code. The channel encoder produces bits at the rate $R_0 = (n/k)R_s$, where $R_s$ is the bit rate of the information source. The dimensionless ratio $r = k/n$ is called the *code rate*, where $0 < r < 1$. The bit rate $R_0$, coming out of the encoder, is called the *channel data rate*. Thus, the code rate is a dimensionless ratio, whereas the data rate produced by the source and the channel data rate are both measured in bits per second.

In a convolutional code, the encoding operation may be viewed as the *discrete-time convolution* of the input sequence with the impulse response of the encoder. The duration of the impulse response equals the memory of the encoder. Accordingly, the encoder for a convolutional code operates on the incoming message sequence, using a "sliding window" equal in duration to its own memory. This, in turn, means that in a convolutional code, unlike a block code, the channel encoder accepts message bits as a continuous sequence and thereby generates a continuous sequence of encoded bits at a higher rate.

In the model depicted in Figure 10.1*a*, the operations of channel coding and modulation are performed separately in the transmitter; likewise for the operations of detection and decoding in the receiver. When, however, bandwidth efficiency is of major concern, the most effective method of implementing forward error-control correction coding is to combine it with modulation as a single function, as shown in Figure 10.1*b*. In such an approach, coding is redefined as a process of imposing certain patterns on the transmitted signal.

## ■ AUTOMATIC-REPEAT REQUEST

Feed-forward error correction (FEC) relies on the controlled use of redundancy in the transmitted code word for both the *detection and correction* of errors incurred during the course of transmission over a noisy channel. Irrespective of whether the decoding of the received code word is successful, no further processing is performed at the receiver. Accordingly, channel coding techniques suitable for FEC require only a *one-way link* between the transmitter and receiver.

There is another approach known as *automatic-repeat request (ARQ)*[2] for solving the error-control problem. The underlying philosophy of ARQ is quite different from that of FEC. Specifically, ARQ uses redundancy merely for the purpose of *error detection*. Upon the detection of an error in a transmitted code word, the receiver requests a repeat transmission of the corrupted code word, which necessitates the use of a *return path* (i.e., a feedback channel). As such, ARQ can be used only on *half-duplex* or *full-duplex links*. In a half-duplex link, data transmission over the link can be made in either direction but *not* simultaneously. On the other hand, in a full-duplex link, it is possible for data transmission to proceed over the link in both directions simultaneously.

A half-duplex link uses the simplest ARQ scheme known as the *stop-and-wait strategy*. In this approach, a block of message bits is encoded into a code word and transmitted over the channel. The transmitter then stops and waits for feedback from the receiver. The feedback signal can be acknowledgment of a correct receipt of the code word or a request for transmission of the code word because of an error in its decoding. In the latter case, the transmitter resends the code word in question before moving onto the next block of message bits.

The idling problem in stop-and-wait ARQ results in reduced data throughput, which is alleviated in another type of ARQ known as *continuous ARQ with pullback*. This second strategy uses a full-duplex link, thereby permitting the receiver to send a feedback signal while the transmitter is engaged in sending code words over the forward channel. Specifically, the transmitter continues to send a succession of code words until it receives a request from the receiver (on the feedback channel) for a retransmission. At that point, the transmitter stops, pulls back to the particular code word that was not decoded correctly by the receiver, and retransmits the complete sequence of code words starting with the corrupted one.

In a refined version of continuous ARQ known as the *continuous ARQ with selective repeat*, data throughout is improved further by only retransmitting the code word that was received with detected errors. In other words, the need for retransmitting the successfully received code words following the corrupted code word is eliminated.

The three types of ARQ described here offer trade-offs of their own between the need for a half-duplex or full-duplex link and the requirement for efficient use of communication resources. In any event, they all rely on two premises:

- Error detection, which makes the design of the decoder relatively simple.
- Noiseless feedback channel, which is not a severe restriction because the rate of information flow over the feedback channel is typically quite low.

For these reasons, ARQ is widely used in computer-communication systems.

Nevertheless, the fact that FEC requires only one-way links for its operation makes the FEC much wider in application than ARQ. Moreover, the increased decoding complexity of FEC due to the combined need for error detection and correction is no longer a pressing practical issue because the decoder usually lends itself to microprocessor or VLSI implementation in a cost-effective manner.

## 10.2 *Discrete-Memoryless Channels*

Returning to the model of Figure 10.1*a*, the waveform channel is said to be memoryless if the detector output in a given interval depends only on the signal transmitted in that interval, and not on any previous transmission. Under this condition, we may model the combination of the modulator, the waveform channel, and the detector as a *discrete memoryless channel.* Such a channel is completely described by the set of transition probabilities $p(j|i)$, where $i$ denotes a modulator input symbol, $j$ denotes a demodulator output symbol, and $p(j|i)$ denotes the probability of receiving symbol $j$, given that symbol $i$ was sent. (Discrete memoryless channels were described previously at some length in Section 9.5.)

The simplest discrete memoryless channel results from the use of binary input and binary output symbols. When binary coding is used, the modulator has only the binary symbols 0 and 1 as inputs. Likewise, the decoder has only binary inputs if binary quantization of the demodulator output is used, that is, a *hard decision* is made on the demodulator output as to which symbol was actually transmitted. In this situation, we have a *binary symmetric channel* (BSC) with a *transition probability diagram* as shown in Figure 10.2. The binary symmetric channel, assuming a channel noise modeled as additive white Gaussian noise (AWGN) channel, is completely described by the *transition probability* $p$. The majority of coded digital communication systems employ binary coding with hard-decision decoding, due to the simplicity of implementation offered by such an approach. *Hard-decision decoders*, or *algebraic decoders*, take advantage of the special algebraic

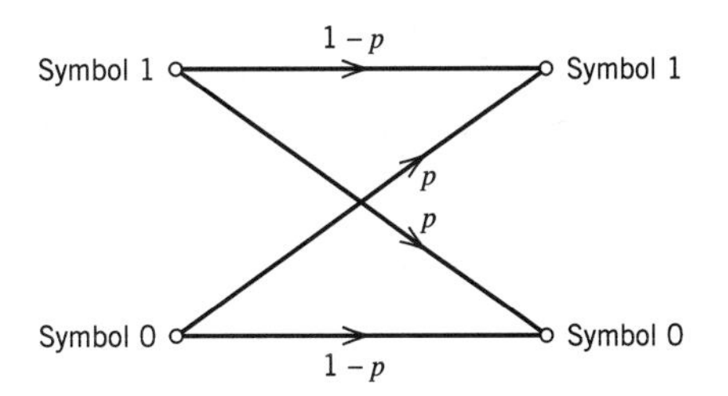


**FIGURE 10.2** Transition probability diagram of binary symmetric channel.

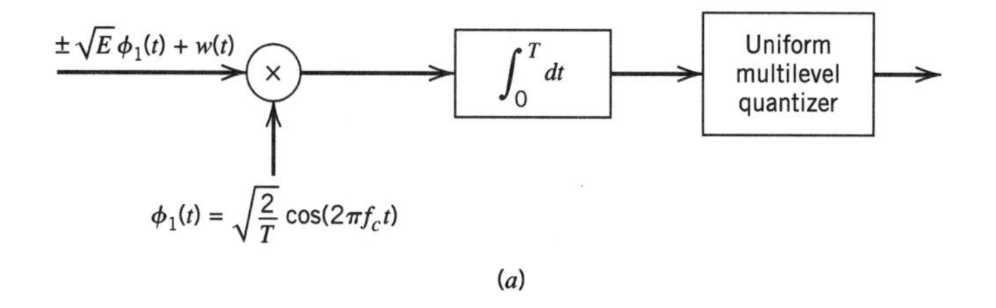


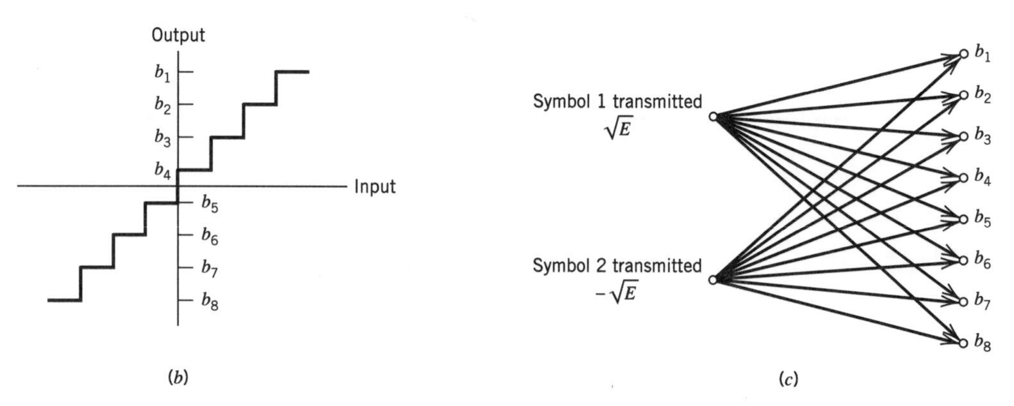


**FIGURE 10.3** Binary input $Q$-ary output discrete memoryless channel. (*a*) Receiver for binary phase-shift keying. (*b*) Transfer characteristic of multilevel quantizer. (*c*) Channel transition probability diagram. Parts (*b*) and (*c*) are illustrated for eight levels of quantization.

structure that is built into the design of channel codes to make the decoding relatively easy to perform.

The use of hard decisions prior to decoding causes an irreversible loss of information in the receiver. To reduce this loss, *soft-decision* coding is used. This is achieved by including a multilevel quantizer at the demodulator output, as illustrated in Figure 10.3*a* for the case of binary PSK signals. The input–output characteristic of the quantizer is shown in Figure 10.3*b*. The modulator has only the binary symbols 0 and 1 as inputs, but the demodulator output now has an alphabet with $Q$ symbols. Assuming the use of the quantizer as described in Figure 10.3*b*, we have $Q = 8$. Such a channel is called a *binary input Q-ary output discrete memoryless channel.* The corresponding channel transition probability diagram is shown in Figure 10.3*c*. The form of this distribution, and consequently the decoder performance, depends on the location of the representation levels of the quantizer, which, in turn, depends on the signal level and noise variance. Accordingly, the demodulator must incorporate automatic gain control if an effective multilevel quantizer is to be realized. Moreover, the use of soft decisions complicates the implementation of the decoder. Nevertheless, soft-decision decoding offers significant improvement in performance over hard-decision decoding by taking a probabilistic rather than an algebraic approach. It is for this reason that soft-decision decoders are also referred to as *probabilistic decoders.*

## ■ CHANNEL CODING THEOREM REVISITED

In Chapter 9, we established the concept of *channel capacity*, which, for a discrete memoryless channel, represents the maximum amount of information transmitted per channel use. The *channel coding theorem* states that if a discrete memoryless channel has capacity $C$ and a source generates information at a rate less than $C$, then there exists a coding technique such that the output of the source may be transmitted over the channel with an arbitrarily low probability of symbol error. For the special case of a binary symmetric channel, the theorem tells us that if the code rate $r$ is less than the channel capacity $C$, then it is possible to find a code that achieves error-free transmission over the channel. Conversely, it is not possible to find such a code if the code rate $r$ is greater than the channel capacity $C$.

The channel coding theorem thus specifies the channel capacity $C$ as a *fundamental limit* on the rate at which the transmission of reliable (error-free) messages can take place over a discrete memoryless channel. The issue that matters is not the signal-to-noise ratio, so long as it is large enough, but how the channel input is encoded.

The most unsatisfactory feature of the channel coding theorem, however, is its non-constructive nature. The theorem asserts the existence of good codes but does not tell us how to find them. By *good codes* we mean families of channel codes that are capable of providing reliable transmission of information (i.e., at arbitrarily small probability of symbol error) over a noisy channel of interest at bit rates up to a maximum value less than the capacity of that channel. The error-control coding techniques described in this chapter provide different methods of designing good codes.

## ■ NOTATION

The codes described in this chapter are *binary codes*, for which the alphabet consists only of symbols 0 and 1. In such a code, the encoding and decoding functions involve the binary arithmetic operations of *modulo-2 addition and multiplication* performed on code words in the code.

Throughout this chapter, we use an ordinary plus sign (+) to denote modulo-2 addition. The use of this terminology will not lead to confusion because the whole chapter relies on binary arithmetic. In so doing, we avoid the use of a special symbol $\oplus$, as we did in preceding chapters. Thus, according to the notation used in this chapter, the rules for modulo-2 addition are as follows:

$$0 + 0 = 0$$
$$1 + 0 = 1$$
$$0 + 1 = 1$$
$$1 + 1 = 0$$

Because $1 + 1 = 0$, it follows that $1 = -1$. Hence, in binary arithmetic, subtraction is the same as addition. The rules for modulo-2 multiplication are as follows:

$$0 \times 0 = 0$$
$$1 \times 0 = 0$$
$$0 \times 1 = 0$$
$$1 \times 1 = 1$$

Division is trivial in that we have

$$1 \div 1 = 1$$
$$0 \div 1 = 0$$

and division by 0 is not permitted. Modulo-2 addition is the EXCLUSIVE-OR operation in logic, and modulo-2 multiplication is the AND operation.

## 10.3 *Linear Block Codes*

A code is said to be *linear if any two code words in the code can be added in modulo-2 arithmetic to produce a third code word in the code.* Consider then an $(n, k)$ linear block code, in which $k$ bits of the $n$ code bits are always identical to the message sequence to be transmitted. The $n - k$ bits in the remaining portion are computed from the message bits in accordance with a prescribed encoding rule that determines the mathematical structure of the code. Accordingly, these $n - k$ bits are referred to as *generalized parity check bits* or simply *parity bits.* Block codes in which the message bits are transmitted in unaltered form are called *systematic codes.* For applications requiring *both* error detection and error correction, the use of systematic block codes simplifies implementation of the decoder.

Let $m_0, m_1, \ldots, m_{k-1}$ constitute a block of $k$ arbitrary message bits. Thus we have $2^k$ distinct message blocks. Let this sequence of message bits be applied to a linear block encoder, producing an $n$-bit code word whose elements are denoted by $c_0, c_1, \ldots, c_{n-1}$. Let $b_0, b_1, \ldots, b_{n-k-1}$ denote the $(n - k)$ parity bits in the code word. For the code to possess a systematic structure, a code word is divided into two parts, one of which is occupied by the message bits and the other by the parity bits. Clearly, we have the option of sending the message bits of a code word before the parity bits, or vice versa. The former option is illustrated in Figure 10.4, and its use is assumed in the sequel.

According to the representation of Figure 10.4, the $(n - k)$ left-most bits of a code word are identical to the corresponding parity bits, and the $k$ right-most bits of the code word are identical to the corresponding message bits. We may therefore write

$$c_i = \begin{cases} b_i, & i = 0, 1, \ldots, n-k-1 \\ m_{i+k-n}, & i = n-k, n-k+1, \ldots, n-1 \end{cases} \tag{10.1}$$

The $(n - k)$ parity bits are *linear sums* of the $k$ message bits, as shown by the generalized relation

$$b_i = p_{0i}m_0 + p_{1i}m_1 + \cdots + p_{k-1,i}m_{k-1} \tag{10.2}$$

where the coefficients are defined as follows:

$$p_{ij} = \begin{cases} 1 & \text{if } b_i \text{ depends on } m_j \\ 0 & \text{otherwise} \end{cases} \tag{10.3}$$

The coefficients $p_{ij}$ are chosen in such a way that the rows of the generator matrix are linearly independent and the parity equations are *unique.*

The system of Equations (10.1) and (10.2) defines the mathematical structure of the $(n, k)$ linear block code. This system of equations may be rewritten in a compact form

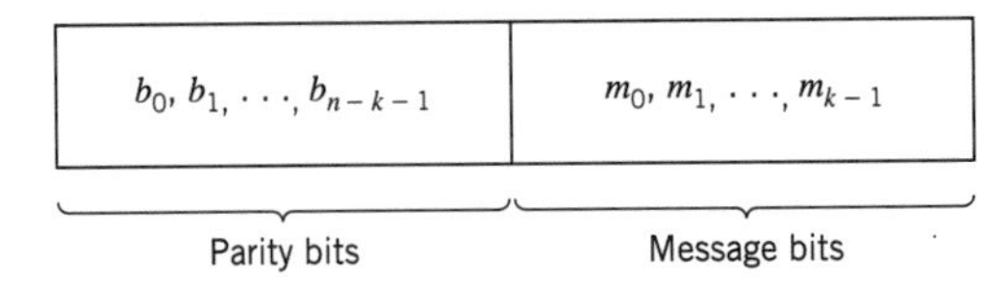


**FIGURE 10.4** Structure of systematic code word.

using matrix notation. To proceed with this reformulation, we define the 1-by-$k$ *message vector*, or *information vector*, **m**, the 1-by-$(n - k)$ parity vector **b**, and the 1-by-$n$ code vector **c** as follows:

$$\mathbf{m} = [m_0, m_1, \ldots, m_{k-1}] \tag{10.4}$$

$$\mathbf{b} = [b_0, b_1, \ldots, b_{n-k-1}] \tag{10.5}$$

$$\mathbf{c} = [c_0, c_1, \ldots, c_{n-1}] \tag{10.6}$$

Note that all three vectors are *row vectors*. The use of row vectors is adopted in this chapter for the sake of being consistent with the notation commonly used in the coding literature. We may thus rewrite the set of simultaneous equations defining the parity bits in the compact matrix form:

$$\mathbf{b} = \mathbf{mP} \tag{10.7}$$

where **P** is the $k$-by-$(n - k)$ *coefficient matrix* defined by

$$\mathbf{P} = \begin{bmatrix} p_{00} & p_{01} & \cdots & p_{0,n-k-1} \\ p_{10} & p_{11} & \cdots & p_{1,n-k-1} \\ \vdots & \vdots & & \vdots \\ p_{k-1,0} & p_{k-1,1} & \cdots & p_{k-1,n-k-1} \end{bmatrix} \tag{10.8}$$

where $p_{ij}$ is 0 or 1.

From the definitions given in Equations (10.4)–(10.6), we see that **c** may be expressed as a partitioned row vector in terms of the vectors **m** and **b** as follows:

$$\mathbf{c} = [\mathbf{b} \vdots \mathbf{m}] \tag{10.9}$$

Hence, substituting Equation (10.7) into Equation (10.9) and factoring out the common message vector **m**, we get

$$\mathbf{c} = \mathbf{m}[\mathbf{P} \vdots \mathbf{I}_k] \tag{10.10}$$

where $\mathbf{I}_k$ is the $k$-by-$k$ *identity matrix*:

$$\mathbf{I}_k = \begin{bmatrix} 1 & 0 & \cdots & 0 \\ 0 & 1 & \cdots & 0 \\ \vdots & \vdots & & \vdots \\ 0 & 0 & \cdots & 1 \end{bmatrix} \tag{10.11}$$

Define the $k$-by-$n$ *generator matrix*

$$\mathbf{G} = [\mathbf{P} \vdots \mathbf{I}_k] \tag{10.12}$$

The generator matrix **G** of Equation (10.12) is said to be in the *canonical form* in that its $k$ rows are linearly independent; that is, it is not possible to express any row of the matrix **G** as a linear combination of the remaining rows. Using the definition of the generator matrix **G**, we may simplify Equation (10.10) as

$$\mathbf{c} = \mathbf{mG} \tag{10.13}$$

The full set of code words, referred to simply as *the code*, is generated in accordance with Equation (10.13) by letting the message vector **m** range through the set of all $2^k$ binary $k$-tuples (1-by-$k$ vectors). Moreover, the sum of any two code words is another code word. This basic property of linear block codes is called *closure*. To prove its validity, consider a pair of code vectors $\mathbf{c}_i$ and $\mathbf{c}_j$ corresponding to a pair of message vectors $\mathbf{m}_i$ and $\mathbf{m}_j$, respectively. Using Equation (10.13) we may express the sum of $\mathbf{c}_i$ and $\mathbf{c}_j$ as

$$\begin{aligned}\mathbf{c}_i + \mathbf{c}_j &= \mathbf{m}_i\mathbf{G} + \mathbf{m}_j\mathbf{G} \\ &= (\mathbf{m}_i + \mathbf{m}_j)\mathbf{G}\end{aligned}$$

The modulo-2 sum of $\mathbf{m}_i$ and $\mathbf{m}_j$ represents a new message vector. Correspondingly, the modulo-2 sum of $\mathbf{c}_i$ and $\mathbf{c}_j$ represents a new code vector.

There is another way of expressing the relationship between the message bits and parity-check bits of a linear block code. Let $\mathbf{H}$ denote an $(n - k)$-by-$n$ matrix, defined as

$$\mathbf{H} = [\mathbf{I}_{n-k} \vdots \mathbf{P}^T] \tag{10.14}$$

where $\mathbf{P}^T$ is an $(n - k)$-by-$k$ matrix, representing the transpose of the coefficient matrix $\mathbf{P}$, and $\mathbf{I}_{n-k}$ is the $(n - k)$-by-$(n - k)$ identity matrix. Accordingly, we may perform the following multiplication of partitioned matrices:

$$\begin{aligned}\mathbf{H}\mathbf{G}^T &= [\mathbf{I}_{n-k} \vdots \mathbf{P}^T]\begin{bmatrix}\mathbf{P}^T \\ \cdots \\ \mathbf{I}_k\end{bmatrix} \\ &= \mathbf{P}^T + \mathbf{P}^T\end{aligned}$$

where we have used the fact that multiplication of a rectangular matrix by an identity matrix of compatible dimensions leaves the matrix unchanged. In modulo-2 arithmetic, we have $\mathbf{P}^T + \mathbf{P}^T = \mathbf{0}$, where $\mathbf{0}$ denotes an $(n - k)$-by-$k$ null matrix (i.e., a matrix that has zeros for all of its elements). Hence,

$$\mathbf{H}\mathbf{G}^T = \mathbf{0} \tag{10.15}$$

Equivalently, we have $\mathbf{G}\mathbf{H}^T = \mathbf{0}$, where $\mathbf{0}$ is a new null matrix. Postmultiplying both sides of Equation (10.13) by $\mathbf{H}^T$, the transpose of $\mathbf{H}$, and then using Equation (10.15), we get

$$\begin{aligned}\mathbf{c}\mathbf{H}^T &= \mathbf{m}\mathbf{G}\mathbf{H}^T \\ &= \mathbf{0}\end{aligned} \tag{10.16}$$

The matrix $\mathbf{H}$ is called the *parity-check matrix* of the code, and the set of equations specified by Equation (10.16) are called *parity-check equations*.

The generator equation (10.13) and the parity-check detector equation (10.16) are basic to the description and operation of a linear block code. These two equations are depicted in the form of block diagrams in Figure 10.5*a* and 10.5*b*, respectively.

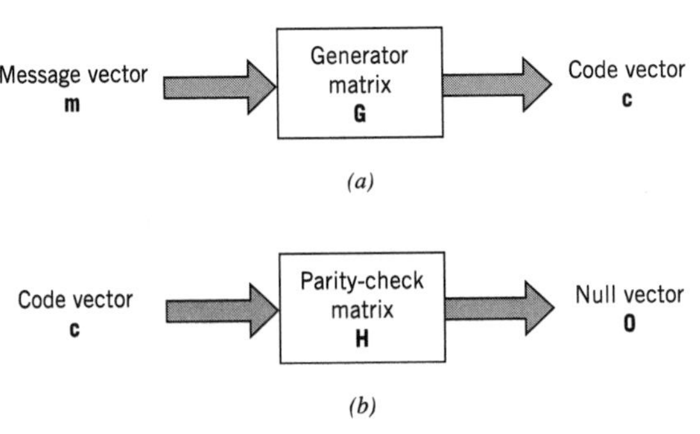


**FIGURE 10.5** Block diagram representations of the generator equation (10.13) and the parity-check equation (10.16).

### ▶ EXAMPLE 10.1 Repetition Codes

*Repetition codes* represent the simplest type of linear block codes. In particular, a single message bit is encoded into a block of $n$ identical bits, producing an $(n, 1)$ block code. Such a code allows provision for a variable amount of redundancy. There are only two code words in the code: an all-zero code word and an all-one code word.

Consider, for example, the case of a repetition code with $k = 1$ and $n = 5$. In this case, we have four parity bits that are the same as the message bit. Hence, the identity matrix $\mathbf{I}_k = 1$, and the coefficient matrix $\mathbf{P}$ consists of a 1-by-4 vector that has 1 for all of its elements. Correspondingly, the generator matrix equals a row vector of all 1s, as shown by

$$\mathbf{G} = [1 \quad 1 \quad 1 \quad 1 \,\vdots\, 1]$$

The transpose of the coefficient matrix $\mathbf{P}$, namely, matrix $\mathbf{P}^T$, consists of a 4-by-1 vector that has 1 for all of its elements. The identity matrix $\mathbf{I}_{n-k}$ consists of a 4-by-4 matrix. Hence, the parity-check matrix equals

$$\mathbf{H} = \begin{bmatrix} 1 & 0 & 0 & 0 & \vdots & 1 \\ 0 & 1 & 0 & 0 & \vdots & 1 \\ 0 & 0 & 1 & 0 & \vdots & 1 \\ 0 & 0 & 0 & 1 & \vdots & 1 \end{bmatrix}$$

Since the message vector consists of a single binary symbol, 0 or 1, it follows from Equation (10.13) that there are only two code words: 00000 and 11111 in the (5, 1) repetition code, as expected. Note also that $\mathbf{HG}^T = 0$, modulo-2, in accordance with Equation (10.15). ◀

## ■ SYNDROME: DEFINITION AND PROPERTIES

The generator matrix $\mathbf{G}$ is used in the encoding operation at the transmitter. On the other hand, the parity-check matrix $\mathbf{H}$ is used in the decoding operation at the receiver. In the context of the latter operation, let $\mathbf{r}$ denote the 1-by-$n$ *received vector* that results from sending the code vector $\mathbf{c}$ over a noisy channel. We express the vector $\mathbf{r}$ as the sum of the original code vector $\mathbf{c}$ and a vector $\mathbf{e}$, as shown by

$$\mathbf{r} = \mathbf{c} + \mathbf{e} \tag{10.17}$$

The vector $\mathbf{e}$ is called the *error vector* or *error pattern*. The $i$th element of $\mathbf{e}$ equals 0 if the corresponding element of $\mathbf{r}$ is the same as that of $\mathbf{c}$. On the other hand, the $i$th element of $\mathbf{e}$ equals 1 if the corresponding element of $\mathbf{r}$ is different from that of $\mathbf{c}$, in which case an error is said to have occurred in the $i$th location. That is, for $i = 1, 2, \dots, n$, we have

$$e_i = \begin{cases} 1 & \text{if an error has occurred in the } i\text{th location} \\ 0 & \text{otherwise} \end{cases} \tag{10.18}$$

The receiver has the task of decoding the code vector $\mathbf{c}$ from the received vector $\mathbf{r}$. The algorithm commonly used to perform this decoding operation starts with the computation of a 1-by-$(n - k)$ vector called the *error-syndrome vector* or simply the *syndrome*.[3] The importance of the syndrome lies in the fact that it depends only upon the error pattern.

Given a 1-by-$n$ received vector $\mathbf{r}$, the corresponding syndrome is formally defined as

$$\mathbf{s} = \mathbf{rH}^T \tag{10.19}$$

Accordingly, the syndrome has the following important properties.

**Property 1**

*The syndrome depends only on the error pattern, and not on the transmitted code word.*

To prove this property, we first use Equations (10.17) and (10.19), and then Equation (10.16) to obtain

$$\begin{aligned} \mathbf{s} &= (\mathbf{c} + \mathbf{e})\mathbf{H}^T \\ &= \mathbf{c}\mathbf{H}^T + \mathbf{e}\mathbf{H}^T \\ &= \mathbf{e}\mathbf{H}^T \end{aligned} \tag{10.20}$$

Hence, the parity-check matrix $\mathbf{H}$ of a code permits us to compute the syndrome $\mathbf{s}$, which depends only upon the error pattern $\mathbf{e}$.

**Property 2**

*All error patterns that differ by a code word have the same syndrome.*

For $k$ message bits, there are $2^k$ distinct code vectors denoted as $\mathbf{c}_i$, $i = 0, 1, \ldots, 2^k - 1$. Correspondingly, for any error pattern $\mathbf{e}$, we define the $2^k$ distinct vectors $\mathbf{e}_i$ as

$$\mathbf{e}_i = \mathbf{e} + \mathbf{c}_i, \qquad i = 0, 1, \ldots, 2^k - 1 \tag{10.21}$$

The set of vectors $\{\mathbf{e}_i, i = 0, 1, \ldots, 2^k - 1\}$ so defined is called a *coset* of the code. In other words, a coset has exactly $2^k$ elements that differ at most by a code vector. Thus, an $(n, k)$ linear block code has $2^{n-k}$ possible cosets. In any event, multiplying both sides of Equation (10.21) by the matrix $\mathbf{H}^T$, we get

$$\begin{aligned} \mathbf{e}_i\mathbf{H}^T &= \mathbf{e}\mathbf{H}^T + \mathbf{c}_i\mathbf{H}^T \\ &= \mathbf{e}\mathbf{H}^T \end{aligned} \tag{10.22}$$

which is independent of the index $i$. Accordingly, we may state that each coset of the code is characterized by a unique syndrome.

We may put Properties 1 and 2 in perspective by expanding Equation (10.20). Specifically, with the matrix $\mathbf{H}$ having the systematic form given in Equation (10.14), where the matrix $\mathbf{P}$ is itself defined by Equation (10.8), we find from Equation (10.20) that the $(n - k)$ elements of the syndrome $\mathbf{s}$ are linear combinations of the $n$ elements of the error pattern $\mathbf{e}$, as shown by

$$\begin{aligned} s_0 &= e_0 + e_{n-k}p_{00} + e_{n-k+1}p_{10} + \cdots + e_{n-1}p_{k-1,0} \\ s_1 &= e_1 + e_{n-k}p_{01} + e_{n-k+1}p_{11} + \cdots + e_{n-1}p_{k-1,1} \\ &\vdots \\ s_{n-k-1} &= e_{n-k-1} + e_{n-k}p_{0,n-k-1} + \cdots + e_{n-1}p_{k-1,n-k-1} \end{aligned} \tag{10.23}$$

This set of $(n - k)$ linear equations clearly shows that the syndrome contains information about the error pattern and may therefore be used for error detection. However, it should be noted that the set of equations is *underdetermined* in that we have more unknowns than equations. Accordingly, there is *no* unique solution for the error pattern. Rather, there are $2^n$ error patterns that satisfy Equation (10.23) and therefore result in the same syndrome, in accordance with Property 2 and Equation (10.22). In particular, with $2^{n-k}$ possible syndrome vectors, the information contained in the syndrome $\mathbf{s}$ about the error pattern $\mathbf{e}$ is *not* enough for the decoder to compute the exact value of the transmitted code vector. Nevertheless, knowledge of the syndrome $\mathbf{s}$ reduces the search for the true error pattern $\mathbf{e}$ from $2^n$ to $2^{n-k}$ possibilities. Given these possibilities, the decoder has the task of making the best selection from the cosets corresponding to $\mathbf{s}$.

## ■ Minimum Distance Considerations

Consider a pair of code vectors $\mathbf{c}_1$ and $\mathbf{c}_2$ that have the same number of elements. The *Hamming distance* $d(\mathbf{c}_1, \mathbf{c}_2)$ between such a pair of code vectors is defined as the number of locations in which their respective elements differ.

The *Hamming weight* $w(\mathbf{c})$ of a code vector $\mathbf{c}$ is defined as the number of nonzero elements in the code vector. Equivalently, we may state that the Hamming weight of a code vector is the distance between the code vector and the all-zero code vector.

The *minimum distance* $d_{\min}$ of a linear block code is defined as the smallest Hamming distance between any pair of code vectors in the code. That is, the minimum distance is the same as the smallest Hamming weight of the difference between any pair of code vectors. From the closure property of linear block codes, the sum (or difference) of two code vectors is another code vector. Accordingly, we may state that *the minimum distance of a linear block code is the smallest Hamming weight of the nonzero code vectors in the code.*

The minimum distance $d_{\min}$ is related to the structure of the parity-check matrix $\mathbf{H}$ of the code in a fundamental way. From Equation (10.16) we know that a linear block code is defined by the set of all code vectors for which $\mathbf{c}\mathbf{H}^T = \mathbf{0}$, where $\mathbf{H}^T$ is the transpose of the parity-check matrix $\mathbf{H}$. Let the matrix $\mathbf{H}$ be expressed in terms of its columns as follows:

$$\mathbf{H} = [\mathbf{h}_1, \mathbf{h}_2, \ldots, \mathbf{h}_n] \tag{10.24}$$

Then, for a code vector $\mathbf{c}$ to satisfy the condition $\mathbf{c}\mathbf{H}^T = \mathbf{0}$, the vector $\mathbf{c}$ must have 1s in such positions that the corresponding rows of $\mathbf{H}^T$ sum to the zero vector $\mathbf{0}$. However, by definition, the number of 1s in a code vector is the Hamming weight of the code vector. Moreover, the smallest Hamming weight of the nonzero code vectors in a linear block code equals the minimum distance of the code. Hence, *the minimum distance of a linear block code is defined by the minimum number of rows of the matrix* $\mathbf{H}^T$ whose sum is equal to the zero vector.

The minimum distance of a linear block code, $d_{\min}$, is an important parameter of the code. Specifically, it determines the error-correcting capability of the code. Suppose an $(n, k)$ linear block code is required to detect and correct all error patterns (over a binary symmetric channel), and whose Hamming weight is less than or equal to $t$. That is, if a code vector $\mathbf{c}_i$ in the code is transmitted and the received vector is $\mathbf{r} = \mathbf{c}_i + \mathbf{e}$, we require that the decoder output $\hat{\mathbf{c}} = \mathbf{c}_i$, whenever the error pattern $\mathbf{e}$ has a Hamming weight $w(\mathbf{e}) \le t$. We assume that the $2^k$ code vectors in the code are transmitted with equal probability. The best strategy for the decoder then is to pick the code vector closest to the received vector $\mathbf{r}$, that is, the one for which the Hamming distance $d(\mathbf{c}_i, \mathbf{r})$ is the smallest. With such a strategy, the decoder will be able to detect and correct all error patterns of Hamming weight $w(\mathbf{e}) \le t$, provided that the minimum distance of the code is equal to or greater than $2t + 1$. We may demonstrate the validity of this requirement by adopting a geometric interpretation of the problem. In particular, the 1-by-$n$ code vectors and the 1-by-$n$ received vector are represented as points in an $n$-dimensional space. Suppose that we construct two spheres, each of radius $t$, around the points that represent code vectors $\mathbf{c}_i$ and $\mathbf{c}_j$. Let these two spheres be disjoint, as depicted in Figure 10.6*a*. For this condition to be satisfied, we require that $d(\mathbf{c}_i, \mathbf{c}_j) \ge 2t + 1$. If then the code vector $\mathbf{c}_i$ is transmitted and the Hamming distance $d(\mathbf{c}_i, \mathbf{r}) \le t$, it is clear that the decoder will pick $\mathbf{c}_i$ as it is the

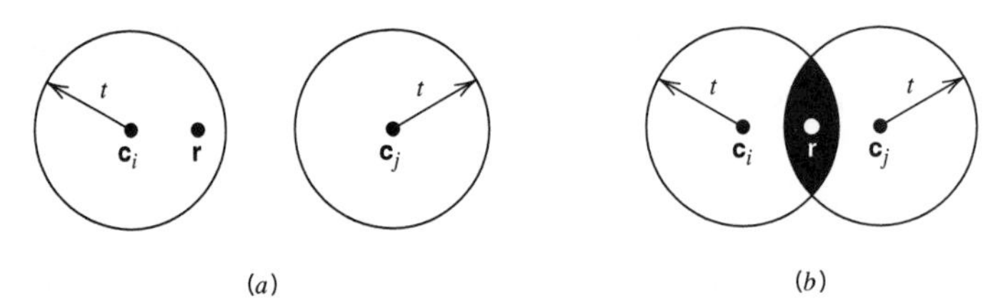


**FIGURE 10.6** (*a*) Hamming distance $d(\mathbf{c}_i, \mathbf{c}_j) \geq 2t + 1$. (*b*) Hamming distance $d(\mathbf{c}_i, \mathbf{c}_j) < 2t$. The received vector is denoted by **r**.

code vector closest to the received vector **r**. If, on the other hand, the Hamming distance $d(\mathbf{c}_i, \mathbf{c}_j) \leq 2t$, the two spheres around $\mathbf{c}_i$ and $\mathbf{c}_j$ intersect, as depicted in Figure 10.6*b*. Here we see that if $\mathbf{c}_i$ is transmitted, there exists a received vector **r** such that the Hamming distance $d(\mathbf{c}_i, \mathbf{r}) \leq t$, and yet **r** is as close to $\mathbf{c}_j$ as it is to $\mathbf{c}_i$. Clearly, there is now the possibility of the decoder picking the vector $\mathbf{c}_j$, which is wrong. We thus conclude that *an $(n, k)$ linear block code has the power to correct all error patterns of weight t or less if, and only if,*

$$d(\mathbf{c}_i, \mathbf{c}_j) \geq 2t + 1 \qquad \text{for all } \mathbf{c}_i \text{ and } \mathbf{c}_j$$

By definition, however, the smallest distance between any pair of code vectors in a code is the minimum distance of the code, $d_{\min}$. We may therefore state that *an $(n, k)$ linear block code of minimum distance $d_{\min}$ can correct up to t errors if, and only if,*

$$t \leq \lfloor \tfrac{1}{2}(d_{\min} - 1) \rfloor \tag{10.25}$$

where $\lfloor \ \rfloor$ *denotes the largest integer* less than or equal to the enclosed quantity. Equation (10.25) gives the error-correcting capability of a linear block code a quantitative meaning.

## ■ SYNDROME DECODING

We are now ready to describe a syndrome-based decoding scheme for linear block codes. Let $\mathbf{c}_1, \mathbf{c}_2, \ldots, \mathbf{c}_{2^k}$ denote the $2^k$ code vectors of an $(n, k)$ linear block code. Let **r** denote the received vector, which may have one of $2^n$ possible values. The receiver has the task of partitioning the $2^n$ possible received vectors into $2^k$ disjoint subsets $\mathcal{D}_1, \mathcal{D}_2, \ldots, \mathcal{D}_{2^k}$ in such a way that the *i*th subset $\mathcal{D}_i$ corresponds to code vector $\mathbf{c}_i$ for $1 \leq i \leq 2^k$. The received vector **r** is decoded into $\mathbf{c}_i$ if it is in the *i*th subset. For the decoding to be correct, **r** must be in the subset that belongs to the code vector $\mathbf{c}_i$ that was actually sent.

The $2^k$ subsets described herein constitute a *standard array* of the linear block code. To construct it, we may exploit the linear structure of the code by proceeding as follows:

1. The $2^k$ code vectors are placed in a row with the all-zero code vector $\mathbf{c}_1$ as the left-most element.
2. An error pattern $\mathbf{e}_2$ is picked and placed under $\mathbf{c}_1$, and a second row is formed by adding $\mathbf{e}_2$ to each of the remaining code vectors in the first row; it is important that the error pattern chosen as the first element in a row not have previously appeared in the standard array.
3. Step 2 is repeated until all the possible error patterns have been accounted for.

Figure 10.7 illustrates the structure of the standard array so constructed. The $2^k$ columns of this array represent the disjoint subsets $\mathcal{D}_1, \mathcal{D}_2, \ldots, \mathcal{D}_{2^k}$. The $2^{n-k}$ rows of the array

$$
\begin{array}{cccccccc}
\mathbf{c}_1 = \mathbf{0} & \mathbf{c}_2 & \mathbf{c}_3 & \cdots & \mathbf{c}_i & \cdots & \mathbf{c}_{2^k} \\
\mathbf{e}_2 & \mathbf{c}_2 + \mathbf{e}_2 & \mathbf{c}_3 + \mathbf{e}_2 & \cdots & \mathbf{c}_i + \mathbf{e}_2 & \cdots & \mathbf{c}_{2^k} + \mathbf{e}_2 \\
\mathbf{e}_3 & \mathbf{c}_2 + \mathbf{e}_3 & \mathbf{c}_3 + \mathbf{e}_3 & \cdots & \mathbf{c}_i + \mathbf{e}_3 & \cdots & \mathbf{c}_{2^k} + \mathbf{e}_3 \\
\vdots & \vdots & \vdots & & \vdots & & \vdots \\
\mathbf{e}_j & \mathbf{c}_2 + \mathbf{e}_j & \mathbf{c}_3 + \mathbf{e}_j & \cdots & \mathbf{c}_i + \mathbf{e}_j & \cdots & \mathbf{c}_{2^k} + \mathbf{e}_j \\
\vdots & \vdots & \vdots & & \vdots & & \vdots \\
\mathbf{e}_{2^{n-k}} & \mathbf{c}_2 + \mathbf{e}_{2^{n-k}} & \mathbf{c}_3 + \mathbf{e}_{2^{n-k}} & & \mathbf{c}_i + \mathbf{e}_{2^{n-k}} & & \mathbf{c}_{2^k} + \mathbf{e}_{2^{n-k}}
\end{array}
$$

**FIGURE 10.7** Standard array for an $(n, k)$ block code.

represent the cosets of the code, and their first elements $\mathbf{e}_2, \ldots, \mathbf{e}_{2^{n-k}}$ are called *coset leaders*.

For a given channel, the probability of decoding error is minimized when the most likely error patterns (i.e., those with the largest probability of occurrence) are chosen as the coset leaders. In the case of a binary symmetric channel, the smaller the Hamming weight of an error pattern the more likely it is to occur. Accordingly, the standard array should be constructed with each coset leader having the minimum Hamming weight in its coset.

We may now describe a decoding procedure for a linear block code:

1. For the received vector $\mathbf{r}$, compute the syndrome $\mathbf{s} = \mathbf{r}\mathbf{H}^T$.
2. Within the coset characterized by the syndrome $\mathbf{s}$, identify the coset leader (i.e., the error pattern with the largest probability of occurrence); call it $\mathbf{e}_0$.
3. Compute the code vector

$$\mathbf{c} = \mathbf{r} + \mathbf{e}_0 \tag{10.26}$$

as the decoded version of the received vector $\mathbf{r}$.

This procedure is called *syndrome decoding*.

## ▶ EXAMPLE 10.2 Hamming Codes[4]

Consider a family of $(n, k)$ linear block codes that have the following parameters:

| | |
|---|---|
| Block length: | $n = 2^m - 1$ |
| Number of message bits: | $k = 2^m - m - 1$ |
| Number of parity bits: | $n - k = m$ |

where $m \geq 3$. These are the so-called Hamming codes.

Consider, for example, the (7, 4) Hamming code with $n = 7$ and $k = 4$, corresponding to $m = 3$. The generator matrix of the code must have a structure that conforms to Equation (10.12). The following matrix represents an appropriate generator matrix for the (7, 4) Hamming code:

$$
\mathbf{G} = \left[\begin{array}{ccc|cccc}
1 & 1 & 0 & 1 & 0 & 0 & 0 \\
0 & 1 & 1 & 0 & 1 & 0 & 0 \\
1 & 1 & 1 & 0 & 0 & 1 & 0 \\
1 & 0 & 1 & 0 & 0 & 0 & 1
\end{array}\right]
$$
$$\underbrace{\qquad\quad}_{\mathbf{P}} \qquad \underbrace{\qquad\qquad}_{\mathbf{I}_k}$$

**TABLE 10.1** *Code words of a (7, 4) Hamming code*

| *Message Word* | *Code Word* | *Weight of Code Word* | *Message Word* | *Code Word* | *Weight of Code Word* |
|---|---|---|---|---|---|
| 0000 | 0000000 | 0 | 1000 | 1101000 | 3 |
| 0001 | 1010001 | 3 | 1001 | 0111001 | 4 |
| 0010 | 1110010 | 4 | 1010 | 0011010 | 3 |
| 0011 | 0100011 | 3 | 1011 | 1001011 | 4 |
| 0100 | 0110100 | 3 | 1100 | 1011100 | 4 |
| 0101 | 1100101 | 4 | 1101 | 0001101 | 3 |
| 0110 | 1000110 | 3 | 1110 | 0101110 | 4 |
| 0111 | 0010111 | 4 | 1111 | 1111111 | 7 |

The corresponding parity-check matrix is given by

$$\mathbf{H} = \left[\begin{array}{ccc:cccc} 1 & 0 & 0 & 1 & 0 & 1 & 1 \\ 0 & 1 & 0 & 1 & 1 & 1 & 0 \\ \underbrace{0 \quad 0 \quad 1}_{\mathbf{I}_{n-k}} & & & \underbrace{0 \quad 1 \quad 1 \quad 1}_{\mathbf{P}^T} \end{array}\right]$$

With $k = 4$, there are $2^k = 16$ distinct message words, which are listed in Table 10.1. For a given message word, the corresponding code word is obtained by using Equation (10.13). Thus, the application of this equation results in the 16 code words listed in Table 10.1.

In Table 10.1, we have also listed the Hamming weights of the individual code words in the (7, 4) Hamming code. Since the smallest of the Hamming weights for the nonzero code words is 3, it follows that the minimum distance of the code is 3. Indeed, Hamming codes have the property that the minimum distance $d_{\min} = 3$, independent of the value assigned to the number of parity bits $m$.

To illustrate the relation between the minimum distance $d_{\min}$ and the structure of the parity-check matrix $\mathbf{H}$, consider the code word 0110100. In the matrix multiplication defined by Equation (10.16), the nonzero elements of this code word "sift" out the second, third, and fifth columns of the matrix $\mathbf{H}$ yielding

$$\begin{bmatrix} 0 \\ 1 \\ 0 \end{bmatrix} + \begin{bmatrix} 0 \\ 0 \\ 1 \end{bmatrix} + \begin{bmatrix} 0 \\ 1 \\ 1 \end{bmatrix} = \begin{bmatrix} 0 \\ 0 \\ 0 \end{bmatrix}$$

We may perform similar calculations for the remaining 14 nonzero code words. We thus find that the smallest number of columns in $\mathbf{H}$ that sums to zero is 3, confirming the earlier statement that $d_{\min} = 3$.

An important property of Hamming codes is that they satisfy the condition of Equation (10.25) with the equality sign, assuming that $t = 1$. This means that Hamming codes are *single-error correcting binary perfect codes.*

Assuming single-error patterns, we may formulate the seven coset leaders listed in the right-hand column of Table 10.2. The corresponding $2^3$ syndromes, listed in the left-hand column, are calculated in accordance with Equation (10.20). The zero syndrome signifies no transmission errors.

Suppose, for example, the code vector [1110010] is sent, and the received vector is

**TABLE 10.2** ***Decoding table for the (7, 4) Hamming code defined in Table 10.1***

| *Syndrome* | *Error Pattern* |
|---|---|
| 0 0 0 | 0 0 0 0 0 0 0 |
| 1 0 0 | 1 0 0 0 0 0 0 |
| 0 1 0 | 0 1 0 0 0 0 0 |
| 0 0 1 | 0 0 1 0 0 0 0 |
| 1 1 0 | 0 0 0 1 0 0 0 |
| 0 1 1 | 0 0 0 0 1 0 0 |
| 1 1 1 | 0 0 0 0 0 1 0 |
| 1 0 1 | 0 0 0 0 0 0 1 |

[1100010] with an error in the third bit. Using Equation (10.19), the syndrome is calculated to be

$$
\begin{aligned}
\mathbf{s} &= [1100010]\begin{bmatrix} 1 & 0 & 0 \\ 0 & 1 & 0 \\ 0 & 0 & 1 \\ 1 & 1 & 0 \\ 0 & 1 & 1 \\ 1 & 1 & 1 \\ 1 & 0 & 1 \end{bmatrix} \\
&= [0 \quad 0 \quad 1]
\end{aligned}
$$

From Table 10.2 the corresponding coset leader (i.e., error pattern with the highest probability of occurrence) is found to be [0010000], indicating correctly that the third bit of the received vector is erroneous. Thus, adding this error pattern to the received vector, in accordance with Equation (10.26), yields the correct code vector actually sent. ◀

### ■ DUAL CODE

Given a linear block code, we may define its *dual* as follows. Taking the transpose of both sides of Equation (10.15), we have

$$\mathbf{G}\mathbf{H}^T = \mathbf{0}$$

where $\mathbf{H}^T$ is the transpose of the parity-check matrix of the code, and $\mathbf{0}$ is a new zero matrix. This equation suggests that every $(n, k)$ linear block code with generator matrix $\mathbf{G}$ and parity-check matrix $\mathbf{H}$ has a *dual code* with parameters $(n, n - k)$, generator matrix $\mathbf{H}$ and parity-check matrix $\mathbf{G}$.

## 10.4 *Cyclic Codes*

Cyclic codes form a subclass of linear block codes. Indeed, many of the important linear block codes discovered to date are either cyclic codes or closely related to cyclic codes. An advantage of cyclic codes over most other types of codes is that they are easy to encode. Furthermore, cyclic codes possess a well-defined mathematical structure, which has led to the development of very efficient decoding schemes for them.

A binary code is said to be a *cyclic code* if it exhibits two fundamental properties:

1. *Linearity property:* The sum of any two code words in the code is also a code word.
2. *Cyclic property:* Any cyclic shift of a code word in the code is also a code word.

Property 1 restates the fact that a cyclic code is a linear block code (i.e., it can be described as a parity-check code). To restate Property 2 in mathematical terms, let the $n$-tuple $(c_0, c_1, \dots, c_{n-1})$ denote a code word of an $(n, k)$ linear block code. The code is a cyclic code if the $n$-tuples

$$\begin{gathered}(c_{n-1}, c_0, \dots, c_{n-2}),\\ (c_{n-2}, c_{n-1}, \dots, c_{n-3}),\\ \vdots\\ (c_1, c_2, \dots, c_{n-1}, c_0)\end{gathered}$$

are all code words in the code.

To develop the algebraic properties of cyclic codes, we use the elements $c_0, c_1, \dots, c_{n-1}$ of a code word to define the *code polynomial*

$$c(X) = c_0 + c_1X + c_2X^2 + \cdots + c_{n-1}X^{n-1} \tag{10.27}$$

where $X$ is an indeterminate. Naturally, for binary codes, the coefficients are 1s and 0s. Each power of $X$ in the polynomial $c(X)$ represents a one-bit *shift* in time. Hence, multiplication of the polynomial $c(X)$ by $X$ may be viewed as a shift to the right. The key question is: How do we make such a shift *cyclic*? The answer to this question is addressed next.

Let the code polynomial $c(X)$ be multiplied by $X^i$, yielding

$$\begin{aligned}X^ic(X) &= X^i(c_0 + c_1X + \cdots + c_{n-i-1}X^{n-i-1} + c_{n-i}X^{n-i}\\ &\quad + \cdots + c_{n-1}X^{n-1})\\ &= c_0X^i + c_1X^{i+1} + \cdots + c_{n-i-1}X^{n-1} + c_{n-i}X^n\\ &\quad + \cdots + c_{n-1}X^{n+i-1}\\ &= c_{n-i}X^n + \cdots + c_{n-1}X^{n+i-1} + c_0X^i + c_1X^{i+1}\\ &\quad + \cdots + c_{n-i-1}X^{n-1}\end{aligned} \tag{10.28}$$

where, in the last line, we have merely rearranged terms. Recognizing, for example, that $c_{n-i} + c_{n-i} = 0$ in modulo-2 addition, we may manipulate the first $i$ terms of Equation (10.28) as follows:

$$\begin{aligned}X^ic(X) &= c_{n-i} + \cdots + c_{n-1}X^{i-1} + c_0X^i + c_1X^{i+1} + \cdots + c_{n-i-1}X^{n-1}\\ &\quad + c_{n-i}(X^n + 1) + \cdots + c_{n-1}X^{i-1}(X^n + 1)\end{aligned} \tag{10.29}$$

Next, we introduce the following definitions:

$$\begin{aligned}c^{(i)}(X) &= c_{n-i} + \cdots + c_{n-1}X^{i-1} + c_0X^i + c_1X^{i+1}\\ &\quad + \cdots + c_{n-i-1}X^{n-1}\end{aligned} \tag{10.30}$$

$$q(X) = c_{n-i} + c_{n-i+1}X + \cdots + c_{n-1}X^{i-1} \tag{10.31}$$

Accordingly, Equation (10.29) is reformulated in the compact form

$$X^i c(X) = q(X)(X^n + 1) + c^{(i)}(X) \tag{10.32}$$

The polynomial $c^{(i)}(X)$ is recognized as the code polynomial of the code word $(c_{n-i}, \ldots, c_{n-1}, c_0, c_1, \ldots, c_{n-i-1})$ obtained by applying $i$ cyclic shifts to the code word $(c_0, c_1, \ldots, c_{n-i-1}, c_{n-i}, \ldots, c_{n-1})$. Moreover, from Equation (10.32) we readily see that $c^{(i)}(X)$ is the remainder that results from dividing $X^i c(X)$ by $(X^n + 1)$. We may thus formally state the cyclic property in polynomial notation as follows: *If $c(X)$ is a code polynomial, then the polynomial*

$$c^{(i)}(X) = X^i c(X) \bmod (X^n + 1) \tag{10.33}$$

*is also a code polynomial for any cyclic shift $i$*; the term *mod* is the abbreviation for *modulo*. The special form of polynomial multiplication described in Equation (10.33) is referred to as *multiplication modulo* $X^n + 1$. In effect, the multiplication is subject to the constraint $X^n = 1$, the application of which restores the polynomial $X^i c(X)$ to order $n - 1$ for all $i < n$. (Note that in modulo-2 arithmetic, $X^n + 1$ has the same value as $X^n - 1$.)

## Generator Polynomial

The polynomial $X^n + 1$ and its factors play a major role in the generation of cyclic codes. Let $g(X)$ be a polynomial of degree $n - k$ that is a factor of $X^n + 1$; as such, $g(X)$ is *the polynomial of least degree in the code*. In general, $g(X)$ may be expanded as follows:

$$g(X) = 1 + \sum_{i=1}^{n-k-1} g_i X^i + X^{n-k} \tag{10.34}$$

where the coefficient $g_i$ is equal to 0 or 1. According to this expansion, the polynomial $g(X)$ has two terms with coefficient 1 separated by $n - k - 1$ terms. The polynomial $g(X)$ is called the *generator polynomial* of a cyclic code. A cyclic code is uniquely determined by the generator polynomial $g(X)$ in that each code polynomial in the code can be expressed in the form of a polynomial product as follows:

$$c(X) = a(X)g(X) \tag{10.35}$$

where $a(X)$ is a polynomial in $X$ with degree $k - 1$. The $c(X)$ so formed satisfies the condition of Equation (10.33) since $g(X)$ is a factor of $X^n + 1$.

Suppose we are given the generator polynomial $g(X)$ and the requirement is to encode the message sequence $(m_0, m_1, \ldots, m_{k-1})$ into an $(n, k)$ *systematic* cyclic code. That is, the message bits are transmitted in unaltered form, as shown by the following structure for a code word (see Figure 10.4):

$$(\underbrace{b_0, b_1, \ldots, b_{n-k-1}}_{n - k \text{ parity bits}}, \underbrace{m_0, m_1, \ldots, m_{k-1}}_{k \text{ message bits}})$$

Let the *message polynomial* be defined by

$$m(X) = m_0 + m_1 X + \cdots + m_{k-1} X^{k-1} \tag{10.36}$$

and let

$$b(X) = b_0 + b_1 X + \cdots + b_{n-k-1} X^{n-k-1} \tag{10.37}$$

According to Equation (10.1), we want the code polynomial to be in the form

$$c(X) = b(X) + X^{n-k}m(X) \tag{10.38}$$

Hence, the use of Equations (10.35) and (10.38) yields

$$a(X)g(X) = b(X) + X^{n-k}m(X)$$

Equivalently, in light of modulo-2 addition, we may write

$$\frac{X^{n-k}m(X)}{g(X)} = a(X) + \frac{b(X)}{g(X)} \tag{10.39}$$

Equation (10.39) states that the polynomial $b(X)$ is the *remainder* left over after dividing $X^{n-k}m(X)$ by $g(X)$.

We may now summarize the steps involved in the encoding procedure for an $(n, k)$ cyclic code assured of a systematic structure. Specifically, we proceed as follows:

1. Multiply the message polynomial $m(X)$ by $X^{n-k}$.
2. Divide $X^{n-k}m(X)$ by the generator polynomial $g(X)$, obtaining the remainder $b(X)$.
3. Add $b(X)$ to $X^{n-k}m(X)$, obtaining the code polynomial $c(X)$.

## ■ PARITY-CHECK POLYNOMIAL

An $(n, k)$ cyclic code is uniquely specified by its generator polynomial $g(X)$ of order $(n - k)$. Such a code is also uniquely specified by another polynomial of degree $k$, which is called the *parity-check polynomial*, defined by

$$h(X) = 1 + \sum_{i=1}^{k-1} h_i X^i + X^k \tag{10.40}$$

where the coefficients $h_i$ are 0 or 1. The parity-check polynomial $h(X)$ has a form similar to the generator polynomial in that there are two terms with coefficient 1, but separated by $k - 1$ terms.

The generator polynomial $g(X)$ is equivalent to the generator matrix $\mathbf{G}$ as a description of the code. Correspondingly, the parity-check polynomial, denoted by $h(X)$, is an equivalent representation of the parity-check matrix $\mathbf{H}$. We thus find that the matrix relation $\mathbf{HG}^T = \mathbf{0}$ presented in Equation (10.15) for linear block codes corresponds to the relationship

$$g(X)h(X) \bmod(X^n + 1) = 0 \tag{10.41}$$

Accordingly, we may state that *the generator polynomial $g(X)$ and the parity-check polynomial $h(X)$ are factors of the polynomial $X^n + 1$*, as shown by

$$g(X)h(X) = X^n + 1 \tag{10.42}$$

This property provides the basis for selecting the generator or parity-check polynomial of a cyclic code. In particular, we may state that if $g(X)$ is a polynomial of degree $(n - k)$ and it is also a factor of $X^n + 1$, then $g(X)$ is the generator polynomial of an $(n, k)$ cyclic code. Equivalently, we may state that if $h(X)$ is a polynomial of degree $k$ and it is also a factor of $X^n + 1$, then $h(X)$ is the parity-check polynomial of an $(n, k)$ cyclic code.

A final comment is in order. Any factor of $X^n + 1$ with degree $(n - k)$, the number of parity bits, can be used as a generator polynomial. For large values of $n$, the polynomial $X^n + 1$ may have many factors of degree $n - k$. Some of these polynomial factors generate good cyclic codes, whereas some of them generate bad cyclic codes. The issue of how to select generator polynomials that produce good cyclic codes is very difficult to resolve. Indeed, coding theorists have expended much effort in the search for good cyclic codes.

## ■ Generator and Parity-Check Matrices

Given the generator polynomial $g(X)$ of an $(n, k)$ cyclic code, we may construct the generator matrix $\mathbf{G}$ of the code by noting that the $k$ polynomials $g(X), Xg(X), \ldots, X^{k-1}g(X)$ span the code. Hence, the $n$-tuples corresponding to these polynomials may be used as rows of the $k$-by-$n$ generator matrix $\mathbf{G}$.

However, the construction of the parity-check matrix $\mathbf{H}$ of the cyclic code from the parity-check polynomial $h(X)$ requires special attention, as described here. Multiplying Equation (10.42) by $a(x)$ and then using Equation (10.35), we obtain

$$c(X)h(X) = a(X) + X^n a(X) \tag{10.43}$$

The polynomials $c(X)$ and $h(X)$ are themselves defined by Equations (10.27) and (10.40), respectively, which means that their product on the left-hand side of Equation (10.43) contains terms with powers extending up to $n + k - 1$. On the other hand, the polynomial $a(X)$ has degree $k - 1$ or less, the implication of which is that the powers of $X^k, X^{k+1}, \ldots, X^{n-1}$ do *not* appear in the polynomial on the right-hand side of Equation (10.43). Thus, setting the coefficients of $X^k, X^{k-1}, \ldots, X^{n-1}$ in the expansion of the product polynomial $c(X)h(X)$ equal to zero, we obtain the following set of $n - k$ equations:

$$\sum_{i=j}^{j+k} c_i h_{k+j-i} = 0 \qquad \text{for } 0 \le j \le n - k - 1 \tag{10.44}$$

Comparing Equation (10.44) with the corresponding relation of Equation (10.16), we may make the following important observation: The coefficients of the parity-check polynomial $h(X)$ involved in the polynomial multiplication described in Equation (10.44) are arranged in *reversed* order with respect to the coefficients of the parity-check matrix $\mathbf{H}$ involved in forming the inner product of vectors described in Equation (10.16). This observation suggests that we define the *reciprocal of the parity-check polynomial* as follows:

$$\begin{aligned} X^k h(X^{-1}) &= X^k\left(1 + \sum_{i=1}^{k-1} h_i X^{-i} + X^{-k}\right) \\ &= 1 + \sum_{i=1}^{k-1} h_{k-i} X^i + X^k \end{aligned} \tag{10.45}$$

which is also a factor of $X^n + 1$. The $n$-tuples pertaining to the $(n - k)$ polynomials $X^k h(X^{-1}), X^{k+1}h(X^{-1}), \ldots, X^{n-1}h(X^{-1})$ may now be used in rows of the $(n - k)$-by-$n$ parity-check matrix $\mathbf{H}$.

In general, the generator matrix $\mathbf{G}$ and the parity-check matrix $\mathbf{H}$ constructed in the manner described here are not in their systematic forms. They can be put into their systematic forms by performing simple operations on their respective rows, as illustrated in Example 10.3.

## ■ Encoder for Cyclic Codes

Earlier we showed that the encoding procedure for an $(n, k)$ cyclic code in systematic form involves three steps: (1) multiplication of the message polynomial $m(X)$ by $X^{n-k}$, (2) di-

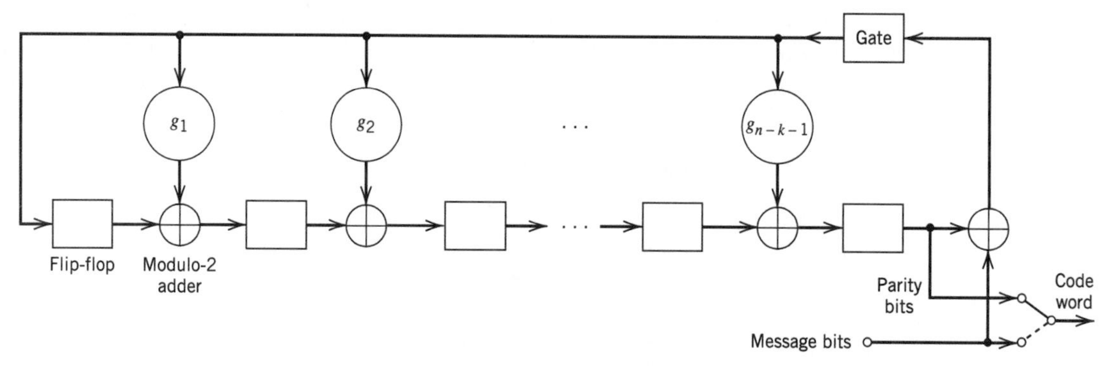


**FIGURE 10.8** Encoder for an $(n, k)$ cyclic code.

vision of $X^{n-k}m(X)$ by the generator polynomial $g(X)$ to obtain the remainder $b(X)$, and (3) addition of $b(X)$ to $X^{n-k}m(X)$ to form the desired code polynomial. These three steps can be implemented by means of the encoder shown in Figure 10.8, consisting of a *linear feedback shift register* with $(n - k)$ stages.

The boxes in Figure 10.8 represent *flip-flops*, or *unit-delay elements*. The flip-flop is a device that resides in one of two possible states denoted by 0 and 1. An *external clock* (not shown in Figure 10.8) controls the operation of all the flip-flops. Every time the clock ticks, the contents of the flip-flops (initially set to the state 0) are shifted out in the direction of the arrows. In addition to the flip-flops, the encoder of Figure 10.8 includes a second set of logic elements, namely, *adders*, which compute the modulo-2 sums of their respective inputs. Finally, the *multipliers* multiply their respective inputs by the associated coefficients. In particular, if the coefficient $g_i = 1$, the multiplier is just a direct "connection." If, on the other hand, the coefficient $g_i = 0$, the multiplier is "no connection."

The operation of the encoder shown in Figure 10.8 proceeds as follows:

1. The gate is switched on. Hence, the $k$ message bits are shifted into the channel. As soon as the $k$ message bits have entered the shift register, the resulting $(n - k)$ bits in the register form the parity bits [recall that the parity bits are the same as the coefficients of the remainder $b(X)$].
2. The gate is switched off, thereby breaking the feedback connections.
3. The contents of the shift register are read out into the channel.

## ■ CALCULATION OF THE SYNDROME

Suppose the code word $(c_0, c_1, \dots, c_{n-1})$ is transmitted over a noisy channel, resulting in the received word $(r_0, r_1, \dots, r_{n-1})$. From Section 10.3, we recall that the first step in the decoding of a linear block code is to calculate the syndrome for the received word. If the syndrome is zero, there are no transmission errors in the received word. If, on the other hand, the syndrome is nonzero, the received word contains transmission errors that require correction.

In the case of a cyclic code in systematic form, the syndrome can be calculated easily. Let the received word be represented by a polynomial of degree $n - 1$ or less, as shown by

$$r(X) = r_0 + r_1X + \cdots + r_{n-1}X^{n-1} \tag{10.46}$$

Let $q(X)$ denote the quotient and $s(X)$ denote the remainder, which are the results of dividing $r(X)$ by the generator polynomial $g(X)$. We may therefore express $r(X)$ as follows:

$$r(X) = q(X)g(X) + s(X) \tag{10.47}$$

The remainder $s(X)$ is a polynomial of degree $n - k - 1$ or less, which is the result of interest. It is called the *syndrome polynomial* because its coefficients make up the $(n - k)$-by-1 syndrome $\mathbf{s}$.

Figure 10.9 shows a *syndrome calculator* that is identical to the encoder of Figure 10.8 except for the fact that the received bits are fed into the $(n - k)$ stages of the feedback shift register from the left. As soon as all the received bits have been shifted into the shift register, its contents define the syndrome $\mathbf{s}$.

The syndrome polynomial $s(X)$ has the following useful properties that follow from the definition given in Equation (10.47).

**1.** *The syndrome of a received word polynomial is also the syndrome of the corresponding error polynomial.*

Given that a cyclic code with polynomial $c(X)$ is sent over a noisy channel, the received word polynomial is defined by

$$r(X) = c(X) + e(X) \tag{10.48}$$

where $e(X)$ is the *error polynomial*. Equivalently, we may write

$$e(X) = r(X) + c(X) \tag{10.49}$$

Hence, substituting Equations (10.35) and (10.47) into (10.49), we get

$$e(X) = u(X)g(X) + s(X) \tag{10.50}$$

where the quotient is $u(X) = a(X) + q(X)$. Equation (10.50) shows that $s(X)$ is also the syndrome of the error polynomial $e(X)$. The implication of this property is that when the syndrome polynomial $s(X)$ is nonzero, the presence of transmission errors in the received word is detected.

**2.** *Let $s(X)$ be the syndrome of a received word polynomial $r(X)$. Then, the syndrome of $Xr(X)$, a cyclic shift of $r(X)$, is $Xs(X)$.*

Applying a cyclic shift to both sides of Equation (10.47), we get

$$Xr(X) = Xq(X)g(X) + Xs(X) \tag{10.51}$$

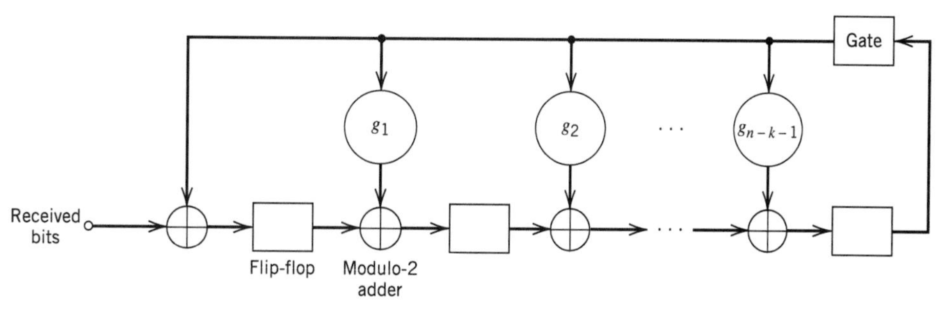


**FIGURE 10.9** Syndrome calculator for $(n, k)$ cyclic code.

from which we readily see that $Xs(X)$ is the remainder of the division of $Xr(X)$ by $g(X)$. Hence, the syndrome of $Xr(X)$ is $Xs(X)$ as stated. We may generalize this result by stating that if $s(X)$ is the syndrome of $r(X)$, then $X^i s(X)$ is the syndrome of $X^i r(X)$.

3. *The syndrome polynomial $s(X)$ is identical to the error polynomial $e(X)$, assuming that the errors are confined to the $(n - k)$ parity-check bits of the received word polynomial $r(X)$.*

The assumption made here is another way of saying that the degree of the error polynomial $e(X)$ is less than or equal to $(n - k - 1)$. Since the generator polynomial $g(X)$ is of degree $(n - k)$, by definition, it follows that Equation (10.50) can only be satisfied if the quotient $u(X)$ is zero. In other words, the error polynomial $e(X)$ and the syndrome polynomial $s(X)$ are one and the same. The implication of Property 3 is that, under the aforementioned conditions, error correction can be accomplished simply by adding the syndrome polynomial $s(X)$ to the received word polynomial $r(X)$.

### ▶ EXAMPLE 10.3 Hamming Codes Revisited

To illustrate the issues relating to the polynomial representation of cyclic codes, we consider the generation of a (7, 4) cyclic code. With the block length $n = 7$, we start by factorizing $X^7 + 1$ into three *irreducible polynomials*:

$$X^7 + 1 = (1 + X)(1 + X^2 + X^3)(1 + X + X^3)$$

By an "irreducible polynomial" we mean a polynomial that cannot be factored using only polynomials with coefficients from the binary field. An irreducible polynomial of degree $m$ is said to be *primitive* if the smallest positive integer $n$ for which the polynomial divides $X^n + 1$ is $n = 2^m - 1$. For the example at hand, the two polynomials $(1 + X^2 + X^3)$ and $(1 + X + X^3)$ are primitive. Let us take

$$g(X) = 1 + X + X^3$$

as the generator polynomial, whose degree equals the number of parity bits. This means that the parity-check polynomial is given by

$$\begin{aligned} h(X) &= (1 + X)(1 + X^2 + X^3) \\ &= 1 + X + X^2 + X^4 \end{aligned}$$

whose degree equals the number of message bits $k = 4$.

Next, we illustrate the procedure for the construction of a code word by using this generator polynomial to encode the message sequence 1001. The corresponding message polynomial is given by

$$m(X) = 1 + X^3$$

Hence, multiplying $m(X)$ by $X^{n-k} = X^3$, we get

$$X^{n-k}m(X) = X^3 + X^6$$

The second step is to divide $X^{n-k}m(X)$ by $g(X)$, the details of which (for the example at hand) are given below:

$$\begin{array}{r l}
 & X^3 + X \\
X^3 + X + 1 \,) & X^6 \qquad\qquad + X^3 \\
 & X^6 \quad + X^4 + X^3 \\
\hline
 & \qquad X^4 \\
 & \qquad X^4 \qquad + X^2 + X \\
\hline
 & \qquad\qquad\qquad X^2 + X
\end{array}$$

Note that in this long division we have treated subtraction the same as addition, since we are operating in modulo-2 arithmetic. We may thus write

$$\frac{X^3 + X^6}{1 + X + X^3} = X + X^3 + \frac{X + X^2}{1 + X + X^3}$$

That is, the quotient $a(X)$ and remainder $b(X)$ are as follows, respectively:

$$a(X) = X + X^3$$
$$b(X) = X + X^2$$

Hence, from Equation (10.38) we find that the desired code polynomial is

$$\begin{aligned} c(X) &= b(X) + X^{n-k}m(X) \\ &= X + X^2 + X^3 + X^6 \end{aligned}$$

The code word is therefore 0111001. The four right-most bits, 1001, are the specified message bits. The three left-most bits, 011, are the parity-check bits. The code word thus generated is exactly the same as the corresponding one shown in Table 10.1 for a (7, 4) Hamming code.

We may generalize this result by stating that *any cyclic code generated by a primitive polynomial is a Hamming code of minimum distance 3.*

We next show that the generator polynomial $g(X)$ and the parity-check polynomial $h(X)$ uniquely specify the generator matrix **G** and the parity-check matrix **H**, respectively.

To construct the 4-by-7 generator matrix **G**, we start with four polynomials represented by $g(X)$ and three cyclic-shifted versions of it, as shown by

$$\begin{aligned} g(X) &= 1 + X + X^3 \\ Xg(X) &= X + X^2 + X^4 \\ X^2g(X) &= X^2 + X^3 + X^5 \\ X^3g(X) &= X^3 + X^4 + X^6 \end{aligned}$$

The polynomials $g(X)$, $Xg(X)$, $X^2g(X)$, and $X^3g(X)$ represent code polynomials in the (7, 4) Hamming code. If the coefficients of these polynomials are used as the elements of the rows of a 4-by-7 matrix, we get the following generator matrix:

$$\mathbf{G}' = \begin{bmatrix} 1 & 1 & 0 & 1 & 0 & 0 & 0 \\ 0 & 1 & 1 & 0 & 1 & 0 & 0 \\ 0 & 0 & 1 & 1 & 0 & 1 & 0 \\ 0 & 0 & 0 & 1 & 1 & 0 & 1 \end{bmatrix}$$

Clearly, the generator matrix $\mathbf{G}'$ so constructed is not in systematic form. We can put it into a systematic form by adding the first row to the third row, and adding the sum of the first two rows to the fourth row. These manipulations result in the desired generator matrix:

$$\mathbf{G} = \begin{bmatrix} 1 & 1 & 0 & 1 & 0 & 0 & 0 \\ 0 & 1 & 1 & 0 & 1 & 0 & 0 \\ 1 & 1 & 1 & 0 & 0 & 1 & 0 \\ 1 & 0 & 1 & 0 & 0 & 0 & 1 \end{bmatrix}$$

which is exactly the same as that in Example 10.2.

We next show how to construct the 3-by-7 parity-check matrix **H** from the parity-check polynomial $h(X)$. To do this, we first take the *reciprocal* of $h(X)$, namely, $X^4h(X^{-1})$. For the problem at hand, we form three polynomials represented by $X^4h(X^{-1})$ and two shifted versions of it, as shown by

$$\begin{aligned} X^4h(X^{-1}) &= 1 + X^2 + X^3 + X^4 \\ X^5h(X^{-1}) &= X + X^3 + X^4 + X^5 \\ X^6h(X^{-1}) &= X^2 + X^4 + X^5 + X^6 \end{aligned}$$

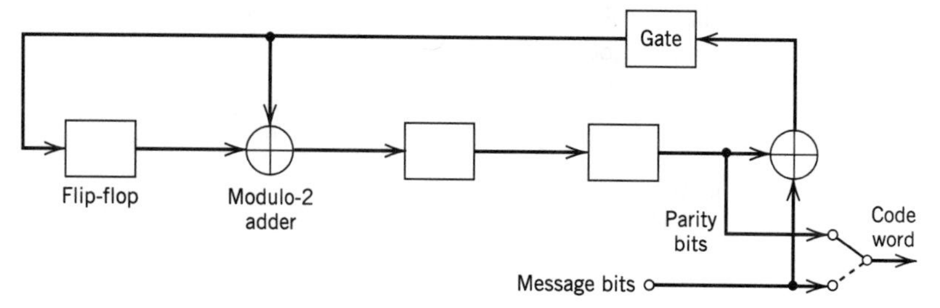


**FIGURE 10.10** Encoder for the (7, 4) cyclic code generated by $g(X) = 1 + X + X^3$.

Using the coefficients of these three polynomials as the elements of the rows of the 3-by-7 parity-check matrix, we get

$$\mathbf{H}' = \begin{bmatrix} 1 & 0 & 1 & 1 & 1 & 0 & 0 \\ 0 & 1 & 0 & 1 & 1 & 1 & 0 \\ 0 & 0 & 1 & 0 & 1 & 1 & 1 \end{bmatrix}$$

Here again we see that the matrix $\mathbf{H}'$ is not in systematic form. To put it into a systematic form, we add the third row to the first row to obtain

$$\mathbf{H} = \begin{bmatrix} 1 & 0 & 0 & 1 & 0 & 1 & 1 \\ 0 & 1 & 0 & 1 & 1 & 1 & 0 \\ 0 & 0 & 1 & 0 & 1 & 1 & 1 \end{bmatrix}$$

which is exactly the same as that of Example 10.2.

Figure 10.10 shows the encoder for the (7, 4) cyclic Hamming code generated by the polynomial $g(X) = 1 + X + X^3$. To illustrate the operation of this encoder, consider the message sequence (1001). The contents of the shift register are modified by the incoming message bits as in Table 10.3. After four shifts, the contents of the shift register, and therefore the parity bits, are (011). Accordingly, appending these parity bits to the message bits (1001), we get the code word (0111001); this result is exactly the same as that determined earlier in the example.

Figure 10.11 shows the corresponding syndrome calculator for the (7, 4) Hamming code. Let the transmitted code word be (0111001) and the received word be (0110001); that is, the middle bit is in error. As the received bits are fed into the shift register, initially set to zero, its contents are modified as in Table 10.4. At the end of the seventh shift, the syndrome is identified from the contents of the shift register as 110. Since the syndrome is nonzero, the received word is in error. Moreover, from Table 10.2, we see that the error pattern corresponding to this syndrome is 0001000. This indicates that the error is in the middle bit of the received word, which is indeed the case. ◀

**TABLE 10.3** ***Contents of the shift register in the encoder of Figure 10.10 for message sequence (1001)***

| *Shift* | *Input* | *Register Contents* |
|---|---|---|
| | | 0 0 0 (initial state) |
| 1 | 1 | 1 1 0 |
| 2 | 0 | 0 1 1 |
| 3 | 0 | 1 1 1 |
| 4 | 1 | 0 1 1 |

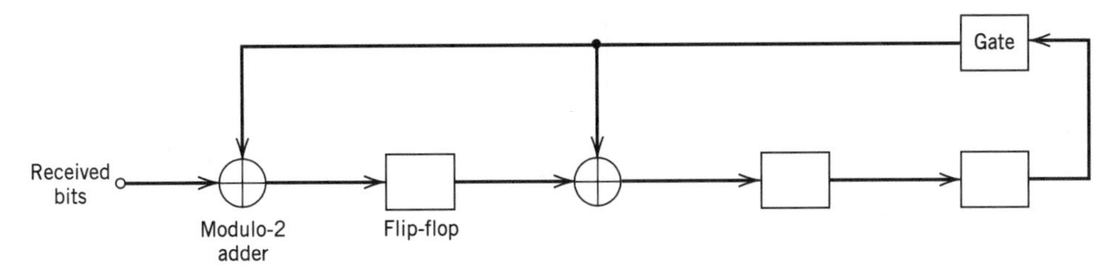


**FIGURE 10.11** Syndrome calculator for the (7, 4) cyclic code generated by the polynomial $g(X) = 1 + X + X^3$.

## ▶ EXAMPLE 10.4 Maximal-Length Codes

For any positive integer $m \geq 3$, there exists a *maximal-length code* with the following parameters:

| | |
|---|---|
| Block length: | $n = 2^m - 1$ |
| Number of message bits: | $k = m$ |
| Minimum distance: | $d_{\min} = 2^{m-1}$ |

Maximal-length codes are generated by polynomials of the form

$$g(X) = \frac{1 + X^n}{h(X)} \tag{10.52}$$

where $h(X)$ is any primitive polynomial of degree $m$. Earlier we stated that any cyclic code generated by a primitive polynomial is a Hamming code of minimum distance 3 (see Example 10.3). It follows therefore that maximal-length codes are the *dual* of Hamming codes.

The polynomial $h(X)$ defines the feedback connections of the encoder. The generator polynomial $g(X)$ defines one period of the maximal-length code, assuming that the encoder is in the initial state 00 . . . 01. To illustrate these points, consider the example of a (7, 3) maximal-length code, which is the dual of the (7, 4) Hamming code described in Example 10.3. Thus, choosing

$$h(X) = 1 + X + X^3$$

we find that the generator polynomial of the (7, 3) maximal-length code is

$$g(X) = 1 + X + X^2 + X^4$$

**TABLE 10.4** ***Contents of the syndrome calculator in Figure 10.11 for the received word 0110001***

| *Shift* | *Input Bit* | *Contents of Shift Register* |
|---|---|---|
| | | 0 0 0 (initial state) |
| 1 | 1 | 1 0 0 |
| 2 | 0 | 0 1 0 |
| 3 | 0 | 0 0 1 |
| 4 | 0 | 1 1 0 |
| 5 | 1 | 1 1 1 |
| 6 | 1 | 0 0 1 |
| 7 | 0 | 1 1 0 |

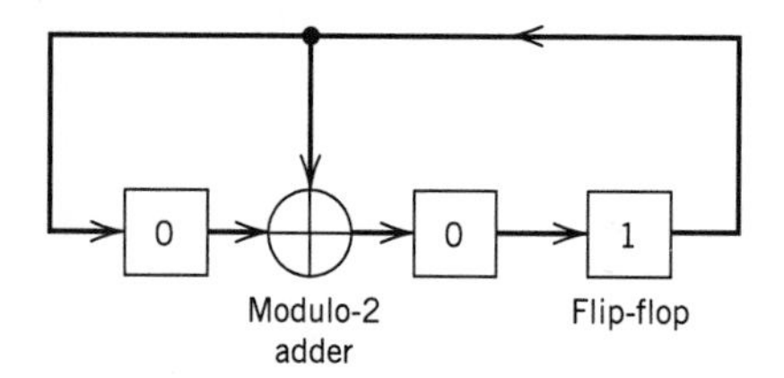


**FIGURE 10.12** Encoder for the (7, 3) maximal-length code; the initial state of the encoder is shown in the figure.

Figure 10.12 shows the encoder for the (7, 3) maximal-length code, the feedback connections of which are exactly the same as those shown in Figure 8.2 in Chapter 8. The period of the code is $n = 7$. Thus, assuming that the encoder is in the initial state 001, as indicated in Figure 10.12, we find the output sequence is described by

$$\underbrace{1\ 0\ 0}_{\text{initial state}} \qquad \underbrace{1\ 1\ 1\ 0\ 1\ 0\ 0}_{g(X) = 1 + X + X^2 + X^4}$$

This result may be readily validated by cycling through the encoder of Figure 10.12.

Note that if we were to choose the other primitive polynomial

$$h(X) = 1 + X^2 + X^3$$

for the (7, 3) maximal-length code, we would simply get the "image" of the code described above, and the output sequence would be "reversed" in time. ◀

## ■ OTHER CYCLIC CODES

We conclude the discussion of cyclic codes by presenting the characteristics of three other important classes of cyclic codes.

### *Cyclic Redundancy Check Codes*

Cyclic codes are extremely well-suited for *error detection*. We make this statement for two reasons. First, they can be designed to detect many combinations of likely errors. Second, the implementation of both encoding and error-detecting circuits is practical. It is for these reasons that many of the error-detecting codes used in practice are of the cyclic-code type. A cyclic code used for error-detection is referred to as *cyclic redundancy check (CRC) code.*

We define an *error burst* of length $B$ in an $n$-bit received word as a contiguous sequence of $B$ bits in which the first and last bits or any number of intermediate bits are received in error. Binary $(n, k)$ CRC codes are capable of detecting the following error patterns:

1. All error bursts of length $n - k$ or less.
2. A fraction of error bursts of length equal to $n - k + 1$; the fraction equals $1 - 2^{-(n-k-1)}$.
3. A fraction of error bursts of length greater than $n - k + 1$; the fraction equals $1 - 2^{-(n-k-1)}$.
4. All combinations of $d_{\min} - 1$ (or fewer) errors.
5. All error patterns with an odd number of errors if the generator polynomial $g(X)$ for the code has an even number of nonzero coefficients.

**TABLE 10.5** *CRC codes*

| *Code* | *Generator Polynomial, g(X)* | n − k |
|---|---|---|
| CRC-12 code | $1 + X + X^2 + X^3 + X^{11} + X^{12}$ | 12 |
| CRC-16 code (USA) | $1 + X^2 + X^{15} + X^{16}$ | 16 |
| CRC-ITU code | $1 + X^5 + X^{12} + X^{16}$ | 16 |

Table 10.5 presents the generator polynomials of three CRC codes that have become international standards. All three codes contain $1 + X$ as a prime factor. The CRC-12 code is used for 6-bit characters, and the other two codes are used for 8-bit characters. CRC codes provide a powerful method of error detection for use in automatic-repeat request (ARQ) strategies discussed in Section 10.1, and digital subscriber lines discussed in Chapter 4.

### *Bose–Chaudhuri–Hocquenghem (BCH) Codes*[5]

One of the most important and powerful classes of linear-block codes are *BCH codes*, which are cyclic codes with a wide variety of parameters. The most common binary BCH codes, known as *primitive BCH codes*, are characterized for any positive integers $m$ (equal to or greater than 3) and $t$ [less than $(2^m - 1)/2$] by the following parameters:

Block length: $n = 2^m - 1$

Number of message bits: $k \geq n - mt$

Minimum distance: $d_{\min} \geq 2t + 1$

Each BCH code is a *t-error correcting code* in that it can detect and correct up to $t$ random errors per code word. The Hamming single-error correcting codes can be described as BCH codes. The BCH codes offer flexibility in the choice of code parameters, namely, block length and code rate. Furthermore, for block lengths of a few hundred bits or less, the BCH codes are among the best known codes of the same block length and code rate.

A detailed treatment of the construction of BCH codes is beyond the scope of our present discussion. To provide a feel for their capability, we present in Table 10.6, the code parameters and generator polynomials for binary block BCH codes of length up to $2^5 - 1$. For example, suppose we wish to construct the generator polynomial for (15, 7)

**TABLE 10.6** *Binary BCH codes of length up to $2^5 - 1$*

| n | k | t | *Generator Polynomial* | | | | | | | | |
|---|---|---|---|---|---|---|---|---|---|---|---|
| 7 | 4 | 1 | | | | | | | | 1 | 011 |
| 15 | 11 | 1 | | | | | | | | 10 | 011 |
| 15 | 7 | 2 | | | | | | | 111 | 010 | 001 |
| 15 | 5 | 3 | | | | | | 10 | 100 | 110 | 111 |
| 31 | 26 | 1 | | | | | | | | 100 | 101 |
| 31 | 21 | 2 | | | | | | 11 | 101 | 101 | 001 |
| 31 | 16 | 3 | | | | 1 | 000 | 111 | 110 | 101 | 111 |
| 31 | 11 | 5 | | | 101 | 100 | 010 | 011 | 011 | 010 | 101 |
| 31 | 6 | 7 | 11 | 001 | 011 | 011 | 110 | 101 | 000 | 100 | 111 |

*Notation:* $n$ = block length
$k$ = number of message bits
$t$ = maximum number of detectable errors

The high-order coefficients of the generator polynomial $g(X)$ are at the left.

BCH code. From Table 10.6 we have (111 010 001) for the coefficients of the generator polynomial; hence, we write

$$g(X) = X^8 + X^7 + X^6 + X^4 + 1$$

### *Reed–Solomon Codes*[6]

The *Reed–Solomon codes* are an important subclass of *nonbinary* BCH codes; they are often abbreviated as RS codes. The encoder for an RS code differs from a binary encoder in that it operates on multiple bits rather than individual bits. Specifically, an RS $(n, k)$ code is used to encode $m$-bit symbols into blocks consisting of $n = 2^m - 1$ symbols, that is, $m(2^m - 1)$ bits, where $m \geq 1$. Thus, the encoding algorithm expands a block of $k$ symbols to $n$ symbols by adding $n - k$ redundant symbols. When $m$ is an integer power of two, the $m$-bit symbols are called *bytes*. A popular value of $m$ is 8; indeed, 8-bit RS codes are extremely powerful.

A $t$-error-correcting RS code has the following parameters:

| | |
|---|---|
| Block length: | $n = 2^m - 1$ symbols |
| Message size: | $k$ symbols |
| Parity-check size: | $n - k = 2t$ symbols |
| Minimum distance: | $d_{\min} = 2t + 1$ symbols |

The block length of the RS code is one less than the size of a code symbol, and the minimum distance is one greater than the number of parity-check symbols. The RS codes make highly efficient use of redundancy, and block lengths and symbol sizes can be adjusted readily to accommodate a wide range of message sizes. Moreover, the RS codes provide a wide range of code rates that can be chosen to optimize performance. Finally, efficient decoding techniques are available for use with RS codes, which is one more reason for their wide application (e.g., compact disc digital audio systems).

## 10.5 *Convolutional Codes*[7]

In block coding, the encoder accepts a $k$-bit message block and generates an $n$-bit code word. Thus, code words are produced on a block-by-block basis. Clearly, provision must be made in the encoder to buffer an entire message block before generating the associated code word. There are applications, however, where the message bits come in *serially* rather than in large blocks, in which case the use of a buffer may be undesirable. In such situations, the use of *convolutional coding* may be the preferred method. A convolutional coder generates redundant bits by using *modulo-2 convolutions*, hence the name.

The encoder of a binary convolutional code with rate $1/n$, measured in bits per symbol, may be viewed as a *finite-state machine* that consists of an $M$-stage shift register with prescribed connections to $n$ modulo-2 adders, and a multiplexer that serializes the outputs of the adders. An $L$-bit message sequence produces a coded output sequence of length $n(L + M)$ bits. The *code rate* is therefore given by

$$r = \frac{L}{n(L + M)} \quad \text{bits/symbol} \tag{10.53}$$

Typically, we have $L \gg M$. Hence, the code rate simplifies to

$$r \simeq \frac{1}{n} \quad \text{bits/symbol} \tag{10.54}$$

The *constraint length* of a convolutional code, expressed in terms of message bits, is defined as the number of shifts over which a single message bit can influence the encoder output. In an encoder with an $M$-stage shift register, the *memory* of the encoder equals $M$ message bits, and $K = M + 1$ shifts are required for a message bit to enter the shift register and finally come out. Hence, the constraint length of the encoder is $K$.

Figure 10.13*a* shows a convolutional encoder with $n = 2$ and $K = 3$. Hence, the code rate of this encoder is 1/2. The encoder of Figure 10.13*a* operates on the incoming message sequence, one bit at a time.

We may generate a binary convolutional code with rate $k/n$ by using $k$ separate shift registers with prescribed connections to $n$ modulo-2 adders, an input multiplexer and

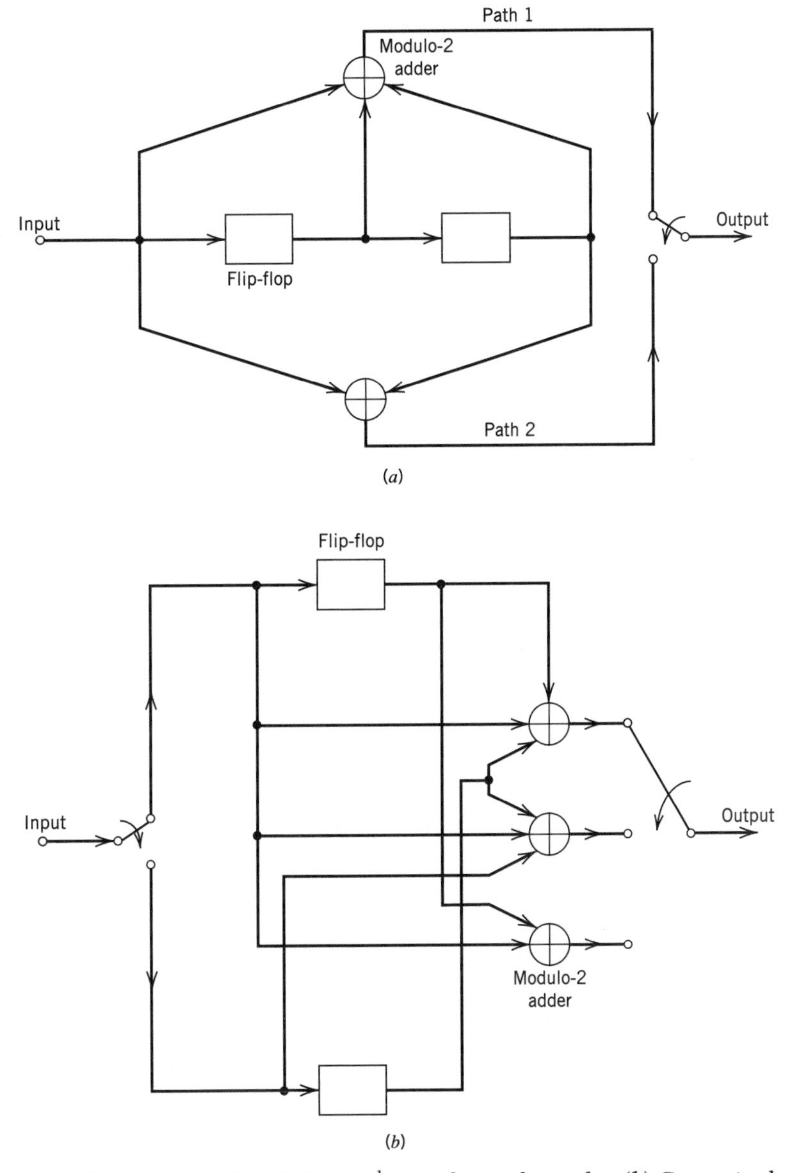


**FIGURE 10.13** (*a*) Constraint length-3, rate-$\frac{1}{2}$ convolutional encoder. (*b*) Constraint length-2, rate-$\frac{2}{3}$ convolutional encoder.

an output multiplexer. An example of such an encoder is shown in Figure 10.13*b*, where $k = 2$, $n = 3$, and the two shift registers have $K = 2$ each. The code rate is 2/3. In this second example, the encoder processes the incoming message sequence two bits at a time.

The convolutional codes generated by the encoders of Figure 10.13 are *nonsystematic* codes. Unlike block coding, the use of nonsystematic codes is ordinarily preferred over systematic codes in convolutional coding.

Each path connecting the output to the input of a convolutional encoder may be characterized in terms of its *impulse response*, defined as the response of that path to a symbol 1 applied to its input, with each flip-flop in the encoder set initially in the zero state. Equivalently, we may characterize each path in terms of a *generator polynomial*, defined as the *unit-delay transform* of the impulse response. To be specific, let the *generator sequence* $(g_0^{(i)}, g_1^{(i)}, g_2^{(i)}, \ldots, g_M^{(i)})$ denote the impulse response of the $i$th path, where the coefficients $g_0^{(i)}, g_1^{(i)}, g_2^{(i)}, \ldots, g_M^{(i)}$ equal 0 or 1. Correspondingly, the *generator polynomial* of the $i$th path is defined by

$$g^{(i)}(D) = g_0^{(i)} + g_1^{(i)}D + g_2^{(i)}D^2 + \cdots + g_M^{(i)}D^M \tag{10.55}$$

where $D$ denotes the unit-delay variable. The complete convolutional encoder is described by the set of generator polynomials $\{g^{(1)}(D), g^{(2)}(D), \ldots, g^{(n)}(D)\}$. Traditionally, different variables are used for the description of convolutional and cyclic codes, with $D$ being commonly used for convolutional codes and $X$ for cyclic codes.

### ▶ EXAMPLE 10.5

Consider the convolutional encoder of Figure 10.13*a*, which has two paths numbered 1 and 2 for convenience of reference. The impulse response of path 1 (i.e., upper path) is (1, 1, 1). Hence, the corresponding generator polynomial is given by

$$g^{(1)}(D) = 1 + D + D^2$$

The impulse response of path 2 (i.e., lower path) is (1, 0, 1). Hence, the corresponding generator polynomial is given by

$$g^{(2)}(D) = 1 + D^2$$

For the message sequence (10011), say, we have the polynomial representation

$$m(D) = 1 + D^3 + D^4$$

As with Fourier transformation, convolution in the time domain is transformed into multiplication in the $D$-domain. Hence, the output polynomial of path 1 is given by

$$\begin{aligned} c^{(1)}(D) &= g^{(1)}(D)m(D) \\ &= (1 + D + D^2)(1 + D^3 + D^4) \\ &= 1 + D + D^2 + D^3 + D^6 \end{aligned}$$

From this we immediately deduce that the output sequence of path 1 is (1111001). Similarly, the output polynomial of path 2 is given by

$$\begin{aligned} c^{(2)}(D) &= g^{(2)}(D)m(D) \\ &= (1 + D^2)(1 + D^3 + D^4) \\ &= 1 + D^2 + D^3 + D^4 + D^5 + D^6 \end{aligned}$$

The output sequence of path 2 is therefore (1011111). Finally, multiplexing the two output sequences of paths 1 and 2, we get the encoded sequence

$$\mathbf{c} = (11, 10, 11, 11, 01, 01, 11)$$

Note that the message sequence of length $L = 5$ bits produces an encoded sequence of length $n(L + K - 1) = 14$ bits. Note also that for the shift register to be restored to its zero initial state, a terminating sequence of $K - 1 = 2$ zeros is appended to the last input bit of the message sequence. The terminating sequence of $K - 1$ zeros is called the *tail of the message*. ◀

## ■ Code Tree, Trellis, and State Diagram

Traditionally, the structural properties of a convolutional encoder are portrayed in graphical form by using any one of three equivalent diagrams: code tree, trellis, and state diagram. We will use the convolutional encoder of Figure 10.13*a* as a running example to illustrate the insights that each one of these three diagrams can provide.

We begin the discussion with the *code tree* of Figure 10.14. Each branch of the tree represents an input symbol, with the corresponding pair of output binary symbols indicated on the branch. The convention used to distinguish the input binary symbols 0 and 1 is as follows. An input 0 specifies the upper branch of a bifurcation, whereas input 1 specifies the lower branch. A specific *path* in the tree is traced from left to right in accordance with the input (message) sequence. The corresponding coded symbols on the branches of that path constitute the input (message) sequence. Consider, for example, the message sequence (10011) applied to the input of the encoder of Figure 10.13*a*. Following the procedure just described, we find that the corresponding encoded sequence is (11, 10, 11, 11, 01), which agrees with the first 5 pairs of bits in the encoded sequence $\{c_i\}$ derived in Example 10.5.

From the diagram of Figure 10.14, we observe that the tree becomes *repetitive* after the first three branches. Indeed, beyond the third branch, the two nodes labeled *a* are identical, and so are all the other node pairs that are identically labeled. We may establish this repetitive property of the tree by examining the associated encoder of Figure 10.13*a*. The encoder has memory $M = K - 1 = 2$ message bits. Hence, when the third message bit enters the encoder, the first message bit is shifted out of the register. Consequently, after the third branch, the message sequences $(100\, m_3 m_4 \ldots)$ and $(000\, m_3 m_4 \ldots)$ generate the same code symbols, and the pair of nodes labeled *a* may be joined together. The same reasoning applies to other nodes. Accordingly, we may collapse the code tree of Figure 10.14 into the new form shown in Figure 10.15, which is called a *trellis*.[8] It is so called since a trellis is a treelike structure with remerging branches. The convention used in Figure 10.15 to distinguish between input symbols 0 and 1 is as follows. A code branch produced by an input 0 is drawn as a solid line, whereas a code branch produced by an input 1 is drawn as a dashed line. As before, each input (message) sequence corresponds to a specific path through the trellis. For example, we readily see from Figure 10.15 that the message sequence (10011) produces the encoded output sequence (11, 10, 11, 11, 01), which agrees with our previous result.

A trellis is more instructive than a tree in that it brings out explicitly the fact that the associated convolutional encoder is a *finite-state machine*. We define the *state* of a convolutional encoder of rate $1/n$ as the $(K - 1)$ message bits stored in the encoder's shift register. At time $j$, the portion of the message sequence containing the most recent $K$ bits is written as $(m_{j-K+1}, \ldots, m_{j-1}, m_j)$, where $m_j$ is the *current* bit. The $(K - 1)$-bit state of the encoder at time $j$ is therefore written simply as $(m_{j-1}, \ldots, m_{j-K+2}, m_{j-K+1})$. In the

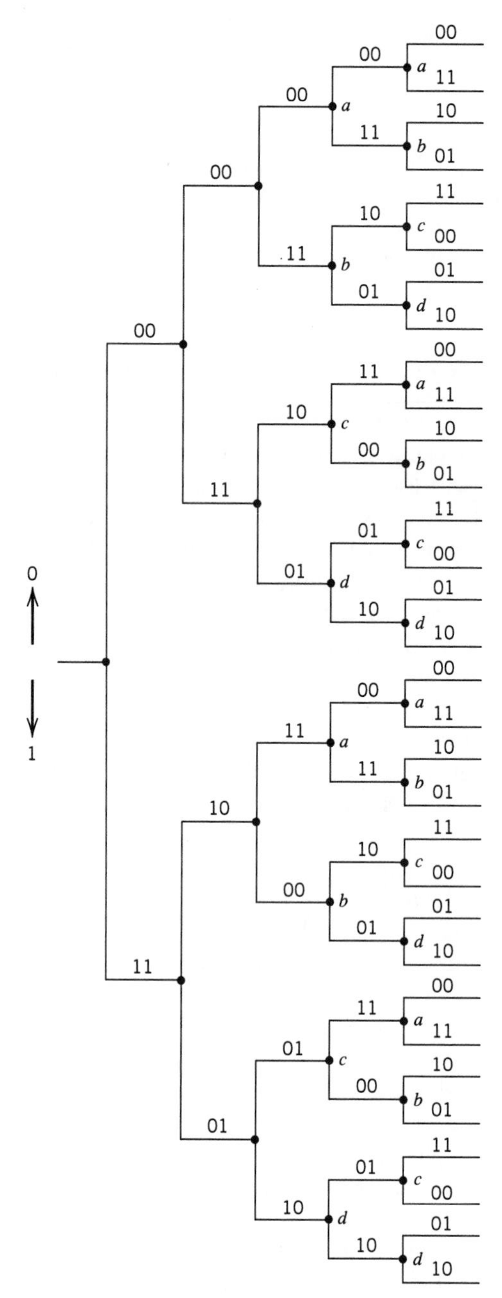


**FIGURE 10.14** Code tree for the convolutional encoder of Figure 10.13*a*.

case of the simple convolutional encoder of Figure 10.13*a* we have $(K - 1) = 2$. Hence, the state of this encoder can assume any one of four possible values, as described in Table 10.7. The trellis contains $(L + K)$ *levels*, where $L$ is the length of the incoming message sequence, and $K$ is the constraint length of the code. The levels of the trellis are labeled as $j = 0, 1, \ldots, L + K - 1$ in Figure 10.15 for $K = 3$. Level $j$ is also referred to as *depth* $j$; both terms are used interchangeably. The first $(K - 1)$ levels correspond to the encoder's departure from the initial state $a$, and the last $(K - 1)$ levels correspond to the encoder's

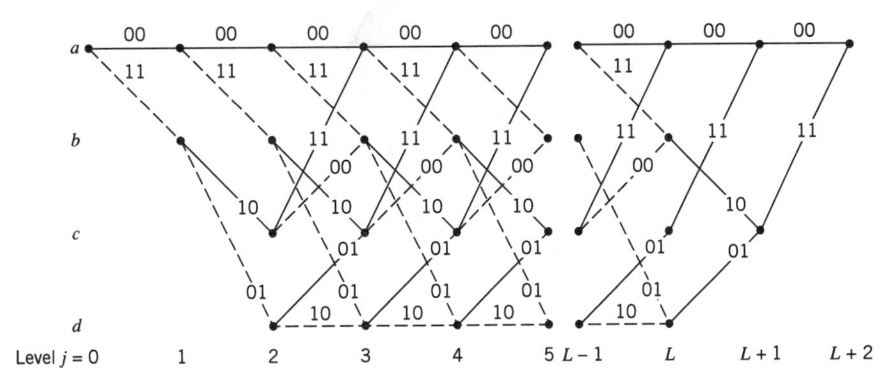


**FIGURE 10.15** Trellis for the convolutional encoder of Figure 10.13*a*.

return to the state $a$. Clearly, not all the states can be reached in these two portions of the trellis. However, in the central portion of the trellis, for which the level $j$ lies in the range $K - 1 \leq j \leq L$, all the states of the encoder are reachable. Note also that the central portion of the trellis exhibits a fixed periodic structure.

Consider next a portion of the trellis corresponding to times $j$ and $j + 1$. We assume that $j \geq 2$ for the example at hand, so that it is possible for the current state of the encoder to be $a$, $b$, $c$, or $d$. For convenience of presentation, we have reproduced this portion of the trellis in Figure 10.16*a*. The left nodes represent the four possible current states of the encoder, whereas the right nodes represent the next states. Clearly, we may coalesce the left and right nodes. By so doing, we obtain the *state diagram* of the encoder, shown in Figure 10.16*b*. The nodes of the figure represent the four possible states of the encoder, with each node having two incoming branches and two outgoing branches. A transition from one state to another in response to input 0 is represented by a solid branch, whereas a transition in response to input 1 is represented by a dashed branch. The binary label on each branch represents the encoder's output as it moves from one state to another. Suppose, for example, the current state of the encoder is (01), which is represented by node $c$. The application of input 1 to the encoder of Figure 10.13*a* results in the state (10) and the encoded output (00). Accordingly, with the help of this state diagram, we may readily determine the output of the encoder of Figure 10.13*a* for any incoming message sequence. We simply start at state $a$, the all-zero initial state, and walk through the state diagram in accordance with the message sequence. We follow a solid branch if the input is a 0 and a dashed branch if it is a 1. As each branch is traversed, we output the corresponding binary label on the branch. Consider, for example, the message sequence (10011). For this input we follow the path *abcabd*, and therefore output the sequence (11, 10, 11, 11, 01), which

**TABLE 10.7** ***State table for the convolutional encoder of Figure 10.13a***

| *State* | *Binary Description* |
|---|---|
| *a* | 00 |
| *b* | 10 |
| *c* | 01 |
| *d* | 11 |

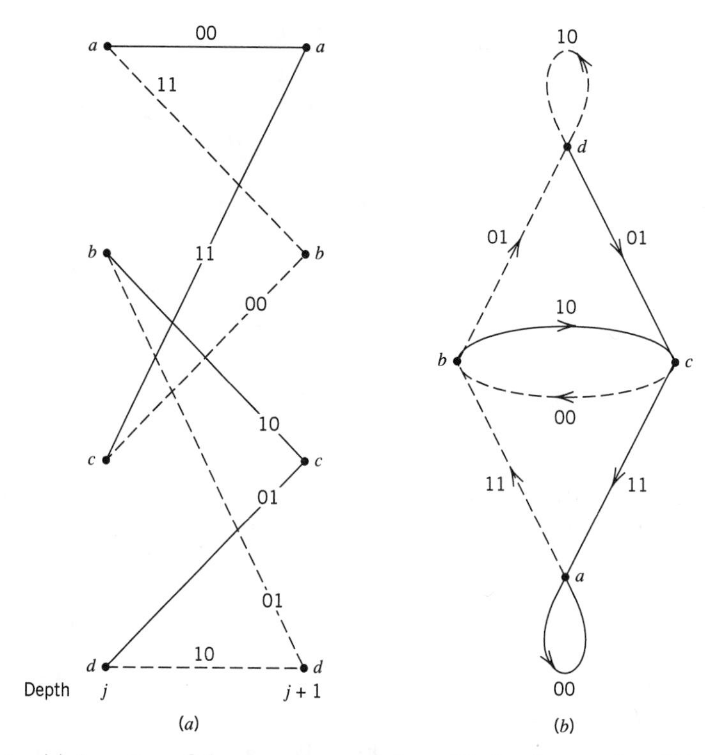


**FIGURE 10.16** (*a*) A portion of the central part of the trellis for the encoder of Figure 10.13*a*. (*b*) State diagram of the convolutional encoder of Figure 10.13*a*.

agrees exactly with our previous result. Thus, the input–output relation of a convolutional encoder is also completely described by its state diagram.

## 10.6 *Maximum Likelihood Decoding of Convolutional Codes*

Now that we understand the operation of a convolutional encoder, the next issue to be considered is the decoding of a convolutional code. In this section we first describe the underlying theory of maximum likelihood decoding, and then present an efficient algorithm for its practical implementation.

Let $\mathbf{m}$ denote a *message vector*, and $\mathbf{c}$ denote the corresponding *code vector* applied by the encoder to the input of a discrete memoryless channel. Let $\mathbf{r}$ denote the *received vector*, which may differ from the transmitted code vector due to channel noise. Given the received vector $\mathbf{r}$, the decoder is required to make an *estimate* $\hat{\mathbf{m}}$ of the message vector. Since there is a one-to-one correspondence between the message vector $\mathbf{m}$ and the code vector $\mathbf{c}$, the decoder may equivalently produce an estimate $\hat{\mathbf{c}}$ of the code vector. We may then put $\hat{\mathbf{m}} = \mathbf{m}$ if and only if $\hat{\mathbf{c}} = \mathbf{c}$. Otherwise, a *decoding error* is committed in the receiver. The *decoding rule* for choosing the estimate $\hat{\mathbf{c}}$, given the received vector $\mathbf{r}$, is said to be optimum when the *probability of decoding error* is minimized. From the material presented in Chapter 6, we may state that for equiprobable messages, the probability of decoding error is minimized if the estimate $\hat{\mathbf{c}}$ is chosen to maximize the *log-likelihood function*. Let $p(\mathbf{r}|\mathbf{c})$ denote the conditional probability of receiving $\mathbf{r}$, given that $\mathbf{c}$ was sent.

The log-likelihood function equals $\log p(\mathbf{r}|\mathbf{c})$. The *maximum likelihood decoder* or decision rule is described as follows:

$$\begin{array}{c}\text{Choose the estimate } \hat{\mathbf{c}} \text{ for which the} \\ \text{log-likelihood function } \log p(\mathbf{r}|\mathbf{c}) \text{ is maximum.}\end{array} \tag{10.56}$$

Consider now the special case of a binary symmetric channel. In this case, both the transmitted code vector $\mathbf{c}$ and the received vector $\mathbf{r}$ represent binary sequences of length $N$, say. Naturally, these two sequences may differ from each other in some locations because of errors due to channel noise. Let $c_i$ and $r_i$ denote the $i$th elements of $\mathbf{c}$ and $\mathbf{r}$, respectively. We then have

$$p(\mathbf{r}|\mathbf{c}) = \prod_{i=1}^{N} p(r_i|c_i) \tag{10.57}$$

Correspondingly, the log-likelihood is

$$\log p(\mathbf{r}|\mathbf{c}) = \sum_{i=1}^{N} \log p(r_i|c_i) \tag{10.58}$$

Let the transition probability $p(r_i|c_i)$ be defined as

$$p(r_i|c_i) = \begin{cases} p, & \text{if } r_i \neq c_i \\ 1-p, & \text{if } r_i = c_i \end{cases} \tag{10.59}$$

Suppose also that the received vector $\mathbf{r}$ differs from the transmitted code vector $\mathbf{c}$ in exactly $d$ positions. The number $d$ is the *Hamming distance* between vectors $\mathbf{r}$ and $\mathbf{c}$. Then, we may rewrite the log-likelihood function in Equation (10.58) as

$$\begin{aligned}\log p(\mathbf{r}|\mathbf{c}) &= d \log p + (N-d)\log(1-p) \\ &= d \log\left(\frac{p}{1-p}\right) + N\log(1-p)\end{aligned} \tag{10.60}$$

In general, the probability of an error occurring is low enough for us to assume $p < 1/2$. We also recognize that $N \log(1-p)$ is a constant for all $\mathbf{c}$. Accordingly, we may restate the maximum-likelihood decoding rule for the binary symmetric channel as follows:

$$\begin{array}{c}\text{Choose the estimate } \hat{\mathbf{c}} \text{ that minimizes the Hamming distance} \\ \text{between the received vector } \mathbf{r} \text{ and the transmitted vector } \mathbf{c}.\end{array} \tag{10.61}$$

That is, for the binary symmetric channel, the maximum-likelihood decoder reduces to a *minimum distance decoder*. In such a decoder, the received vector $\mathbf{r}$ is compared with each possible transmitted code vector $\mathbf{c}$, and the particular one closest to $\mathbf{r}$ is chosen as the correct transmitted code vector. The term "closest" is used in the sense of minimum number of differing binary symbols (i.e., Hamming distance) between the code vectors under investigation.

## ■ THE VITERBI ALGORITHM[9]

The equivalence between maximum likelihood decoding and minimum distance decoding for a binary symmetric channel implies that we may decode a convolutional code by choosing a path in the code tree whose coded sequence differs from the received sequence in the fewest number of places. Since a code tree is equivalent to a trellis, we may equally limit our choice to the possible paths in the trellis representation of the code. The reason for preferring the trellis over the tree is that the number of nodes at any level of the trellis does not continue to grow as the number of incoming message bits increases; rather, it remains constant at $2^{K-1}$, where $K$ is the constraint length of the code.

Consider, for example, the trellis diagram of Figure 10.15 for a convolutional code with rate $r = 1/2$ and constraint length $K = 3$. We observe that at level $j = 3$, there are two paths entering any of the four nodes in the trellis. Moreover, these two paths will be identical onward from that point. Clearly, a minimum distance decoder may make a decision at that point as to which of those two paths to retain, without any loss of performance. A similar decision may be made at level $j = 4$, and so on. This sequence of decisions is exactly what the *Viterbi algorithm* does as it walks through the trellis. The algorithm operates by computing a *metric* or discrepancy for every possible path in the trellis. The metric for a particular path is defined as the Hamming distance between the coded sequence represented by that path and the received sequence. Thus, for each node (state) in the trellis of Figure 10.15 the algorithm compares the two paths entering the node. The path with the lower metric is retained, and the other path is discarded. This computation is repeated for every level $j$ of the trellis in the range $M \leq j \leq L$, where $M = K - 1$ is the encoder's memory and $L$ is the length of the incoming message sequence. The paths that are retained by the algorithm are called *survivor* or *active paths*. For a convolutional code of constraint length $K= 3$, for example, no more than $2^{K-1} = 4$ survivor paths and their metrics will ever be stored. This list of $2^{K-1}$ paths is always guaranteed to contain the maximum-likelihood choice.

A difficulty that may arise in the application of the Viterbi algorithm is the possibility that when the paths entering a state are compared, their metrics are found to be identical. In such a situation, we make the choice by flipping a fair coin (i.e., simply make a guess).

In summary, the Viterbi algorithm is a maximum-likelihood decoder, which is optimum for an AWGN channel. It proceeds in a step-by-step fashion as follows:

#### *Initialization*

Label the left-most state of the trellis (i.e., the all-zero state at level 0) as 0, since there is no discrepancy at this point in the computation.

#### *Computation step* $j + 1$

Let $j = 0, 1, 2, \ldots$, and suppose that at the previous step $j$ we have done two things:

- All survivor paths are identified.
- The survivor path and its metric for each state of the trellis are stored.

Then, at level (clock time) $j + 1$, compute the metric for all the paths entering each state of the trellis by adding the metric of the incoming branches to the metric of the connecting survivor path from level $j$. Hence, for each state, identify the path with the lowest metric as the survivor of step $j + 1$, thereby updating the computation.

#### *Final Step*

Continue the computation until the algorithm completes its forward search through the trellis and therefore reaches the termination node (i.e., all-zero state), at which time it makes a decision on the maximum likelihood path. Then, like a block decoder, the sequence of symbols associated with that path is released to the destination as the decoded version of the received sequence. In this sense, it is therefore more correct to refer to the Viterbi algorithm as a *maximum likelihood sequence estimator*.

However, when the received sequence is very long (near infinite), the storage requirement of the Viterbi algorithm becomes too high, and some compromises must be made.

The approach usually taken is to "truncate" the path memory of the decoder as described here. A *decoding window* of length $\ell$ is specified, and the algorithm operates on a corresponding frame of the received sequence, always stopping after $\ell$ steps. A decision is then made on the "best" path and the symbol associated with the first branch on that path is released to the user. The symbol associated with the last branch of the path is dropped. Next, the decoding window is moved forward one time interval, and a decision on the next code frame is made, and so on. The decoding decisions made in this way are no longer truly maximum likelihood, but they can be made almost as good provided that the decoding window is long enough. Experience and analysis have shown that satisfactory results are obtained if the decoding window length $\ell$ is on the order of 5 times the constraint length $K$ of the convolutional code or more.

### ▶ EXAMPLE 10.6 Correct Decoding of Received All-Zero Sequence

Suppose that the encoder of Figure 10.13*a* generates an all-zero sequence that is sent over a binary symmetric channel, and that the received sequence is (0100010000 . . .). There are two errors in the received sequence due to noise in the channel: one in the second bit and the other in the sixth bit. We wish to show that this double-error pattern is correctable through the application of the Viterbi decoding algorithm.

In Figure 10.17, we show the results of applying the algorithm for level $j = 1, 2, 3, 4, 5$. We see that for $j = 2$ there are (for the first time) four paths, one for each of the four states of the encoder. The figure also includes the metric of each path for each level in the computation.

In the left side of Figure 10.17, for $j = 3$ we show the paths entering each of the states, together with their individual metrics. In the right side of the figure, we show the four survivors that result from application of the algorithm for level $j = 3, 4, 5$.

Examining the four survivors in Figure 10.17 for $j = 5$, we see that the all-zero path has the smallest metric and will remain the path of smallest metric from this point forward. This clearly shows that the all-zero sequence is the maximum likelihood choice of the Viterbi decoding algorithm, which agrees exactly with the transmitted sequence. ◀

### ▶ EXAMPLE 10.7 Incorrect Decoding of Received All-Zero Sequence

Suppose next that the received sequence is (1100010000 . . .), which contains three errors compared to the transmitted all-zero sequence.

In Figure 10.18, we show the results of applying the Viterbi decoding algorithm for $j = 1, 2, 3, 4$. We see that in this example the correct path has been eliminated by level $j = 3$. Clearly, a triple-error pattern is uncorrectable by the Viterbi algorithm when applied to a convolutional code of rate 1/2 and constraint length $K = 3$. The exception to this rule is a triple-error pattern spread over a time span longer than one constraint length, in which case it is very likely to be correctable. ◀

## ■ FREE DISTANCE OF A CONVOLUTIONAL CODE

The performance of a convolutional code depends not only on the decoding algorithm used but also on the distance properties of the code. In this context, the most important single measure of a convolutional code's ability to combat channel noise is the free distance, denoted by $d_{\text{free}}$. The *free distance* of a convolutional code is defined as *the minimum Hamming distance between any two code words in the code*. A convolutional code with free distance $d_{\text{free}}$ can correct $t$ errors if and only if $d_{\text{free}}$ is greater than $2t$.

The free distance can be obtained quite simply from the state diagram of the convolutional encoder. Consider, for example, Figure 10.16*b*, which shows the state diagram

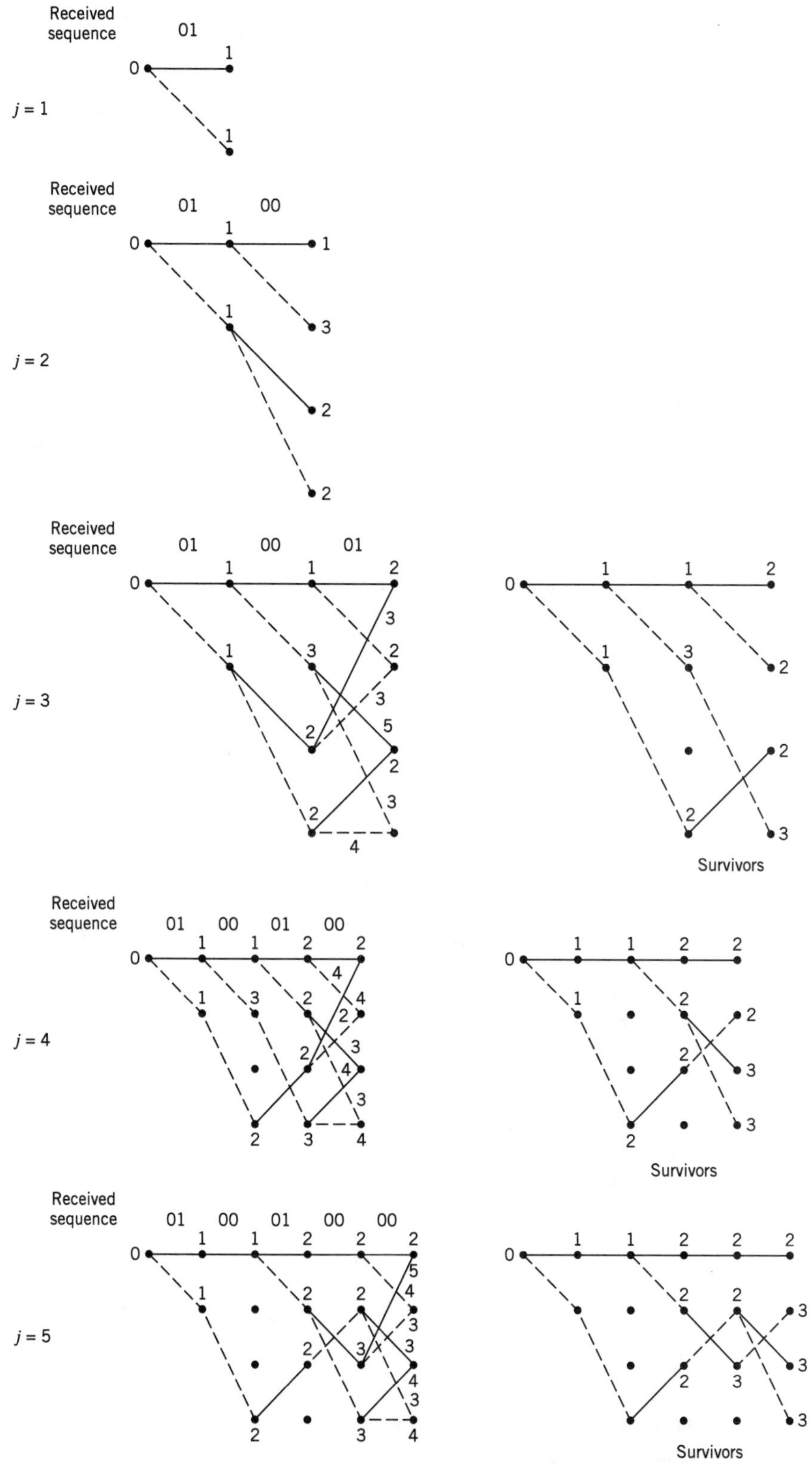


**FIGURE 10.17** Illustrating steps in the Viterbi algorithm for Example 10.6.

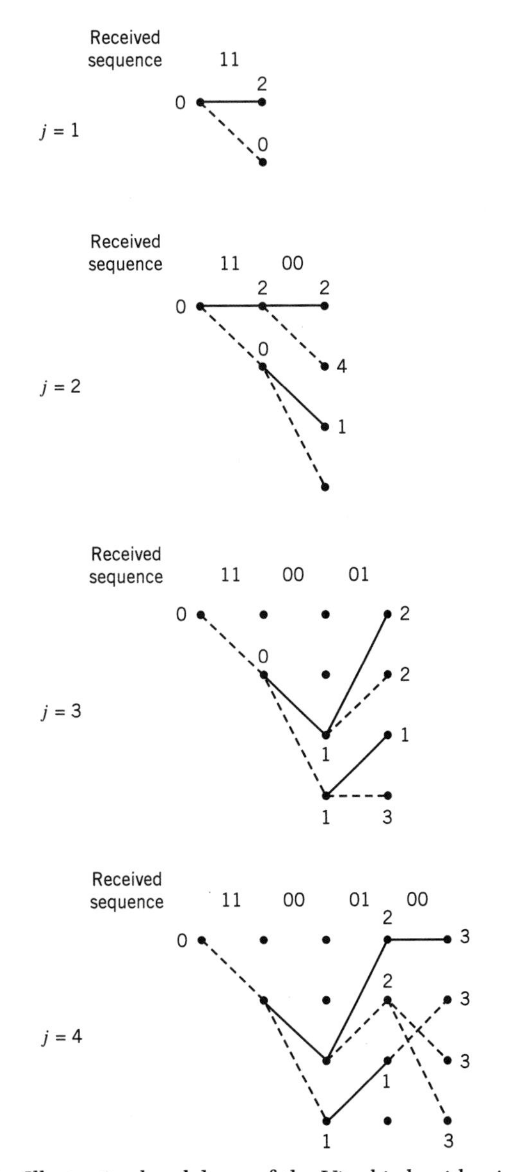


**FIGURE 10.18** Illustrating breakdown of the Viterbi algorithm in Example 10.7.

of the encoder of Figure 10.13*a*. Any nonzero code sequence corresponds to a complete path beginning and ending at the 00 state (i.e., node *a*). We thus find it useful to split this node in the manner shown in the modified state diagram of Figure 10.19, which may be viewed as a *signal-flow graph* with a single input and a single output. A signal-flow graph consists of *nodes* and directed *branches*; it operates by the following rules:

1. A branch multiplies the signal at its input node by the *transmittance* characterizing that branch.
2. A node with incoming branches *sums* the signals produced by all of those branches.
3. The signal at a node is applied equally to all the branches outgoing from that node.
4. The *transfer function* of the graph is the ratio of the output signal to the input signal.

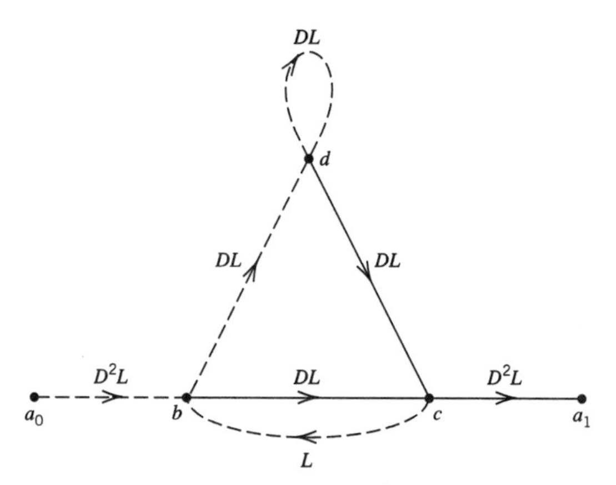


**FIGURE 10.19** Modified state diagram of convolutional encoder.

Returning to the signal-flow graph of Figure 10.19, we note that the exponent of $D$ on a branch in this graph describes the Hamming weight of the encoder output corresponding to that branch. The exponent of $L$ is always equal to one, since the length of each branch is one. Let $T(D, L)$ denote the transfer function of the signal-flow graph, with $D$ and $L$ playing the role of dummy variables. For the example of Figure 10.19, we may readily use rules 1, 2, and 3 to obtain the following input-output relations:

$$\left.\begin{aligned} b &= D^2La_0 + Lc \\ c &= DLb + DLd \\ d &= DLb + DLd \\ a_1 &= D^2Lc \end{aligned}\right\} \tag{10.62}$$

where $a_0$, $b$, $c$, $d$, and $a_1$ denote the node signals of the graph. Solving the set of Equations (10.62) for the ratio $a_1/a_0$, we find that the transfer function of the graph in Figure 10.19 is given by

$$T(D, L) = \frac{D^5L^3}{1 - DL(1 + L)} \tag{10.63}$$

Using the binomial expansion, we may equivalently write

$$T(D, L) = D^5L^3 \sum_{i=0}^{\infty} (DL(1 + L))^i \tag{10.64}$$

Setting $L = 1$ in Equation (10.64), we thus get the *distance transfer function* expressed in the form of a power series:

$$T(D, 1) = D^5 + 2D^6 + 4D^7 + \cdots \tag{10.65}$$

Since the free distance is the minimum Hamming distance between any two code words in the code and the distance transfer function $T(D, 1)$ enumerates the number of code words that are a given distance apart, it follows that the exponent of the first term in the expansion of $T(D, 1)$ defines the free distance. Thus, on the basis of Equation (10.65), the convolutional code of Figure 10.13*a* has a free distance $d_{\text{free}} = 5$.

This result indicates that up to two errors in the received sequence are correctable, for two or fewer transmission errors will cause the received sequence to be at most at a Hamming distance of 2 from the transmitted sequence but at least at a Hamming distance of 3 from any other code sequence in the code. In other words, in spite of the presence of

**TABLE 10.8** *Maximum free distances attainable with systematic and nonsystematic convolutional codes of rate 1/2*

| *Constraint Length* K | *Systematic* | *Nonsystematic* |
|---|---|---|
| 2 | 3 | 3 |
| 3 | 4 | 5 |
| 4 | 4 | 6 |
| 5 | 5 | 7 |
| 6 | 6 | 8 |
| 7 | 6 | 10 |
| 8 | 7 | 10 |

any pair of transmission errors, the received sequence remains closer to the transmitted sequence than any other possible code sequence. However, this statement is no longer true if there are three or more *closely spaced* transmission errors in the received sequence. These observations confirm the results reported earlier in Examples 10.6 and 10.7.

In using the distance transfer function $T(D, 1)$ to calculate the free distance of a convolutional code, it is assumed that the power series in the unit-delay variable $D$ representing $T(D, 1)$ is *convergent* (i.e., its sum has a "finite" value). This assumption is required to justify the expansion given in Equation (10.65) for the convolutional code of Figure 10.13*a*. However, there is no guarantee that $T(D, 1)$ is always convergent. When $T(D, 1)$ is nonconvergent, an infinite number of decoding errors are caused by a finite number of transmission errors; the convolutional code is then subject to catastrophic error propagation, and the code is called a *catastrophic code.*[10] In this context it is noteworthy that a *systematic* convolutional code cannot be catastrophic. Unfortunately, for a prescribed constraint length $K$, the free distances that can be attained with systematic convolutional codes using schemes such as those shown in Figure 10.13 are usually smaller than for the case of nonsystematic convolutional codes, as indicated in Table 10.8.

## ASYMPTOTIC CODING GAIN[11]

The transfer function of the encoder state diagram, modified in a manner similar to that illustrated in Figure 10.19, may be used to evaluate a *bound on the bit error rate* for a given decoding scheme; details of this evaluation are, however, beyond the scope of our present discussion. Here we simply summarize the results for two special channels, namely, the binary symmetric channel and the binary-input additive white Gaussian noise (AWGN) channel, assuming the use of binary phase-shift keying (PSK) with coherent detection.

1. *Binary symmetric channel.* The binary symmetric channel may be modeled as an additive white Gaussian noise channel with binary phase-shift keying (PSK) as the modulation and with hard-decision demodulation. The transition probability $p$ of the binary symmetric channel is then equal to the bit error rate (BER) for the uncoded binary PSK system. From Chapter 6 we recall that for large values of $E_b/N_0$, the ratio of signal energy per bit-to-noise power spectral density, the bit error rate for binary PSK without coding is dominated by the exponential factor $\exp(-E_b/N_0)$. On the other hand, the bit error rate for the same modulation scheme with convolutional coding is dominated by the exponential

factor $\exp(-d_{\text{free}}rE_b/2N_0)$, where $r$ is the code rate and $d_{\text{free}}$ is the free distance of the convolutional code. Therefore, as a figure of merit for measuring the improvement in error performance made by the use of coding with hard-decision decoding, we may use the exponents to define the *asymptotic coding gain* (in decibels) as follows:

$$G_a = 10 \log_{10}\left(\frac{d_{\text{free}}r}{2}\right) \text{ dB} \tag{10.66}$$

2. *Binary-input AWGN channel.* Consider next the case of a memoryless binary-input AWGN channel with no output quantization [i.e., the output amplitude lies in the interval $(-\infty, \infty)$]. For this channel, theory shows that for large values of $E_b/N_0$ the bit error rate for binary PSK with convolutional coding is dominated by the exponential factor $\exp(-d_{\text{free}}rE_b/N_0)$, where the parameters are as previously defined. Accordingly, in this case, we find that the asymptotic coding gain is defined by

$$G_a = 10 \log_{10}(d_{\text{free}}r) \text{ dB} \tag{10.67}$$

From Equations (10.66) and (10.67) we see that the asymptotic coding gain for the binary-input AWGN channel is greater than that for the binary symmetric channel by 3 dB. In other words, for large $E_b/N_0$, the transmitter for a binary symmetric channel must generate an additional 3 dB of signal energy (or power) over that for a binary-input AWGN channel if we are to achieve the same error performance. Clearly, there is an advantage to be gained by permitting an unquantized demodulator output instead of making hard decisions. This improvement in performance, however, is attained at the cost of increased decoder complexity due to the requirement for accepting analog inputs.

The asymptotic coding gain for a binary-input AWGN channel is approximated to within about 0.25 dB by a binary input $Q$-ary output discrete memoryless channel with the number of representation levels $Q = 8$. This means that we may avoid the need for an analog decoder by using a soft-decision decoder that performs finite output quantization (typically, $Q = 8$), and yet realize a performance close to the optimum.

## 10.7 *Trellis-Coded Modulation*[12]

In the traditional approach to channel coding described in the preceding sections of the chapter, encoding is performed separately from modulation in the transmitter; likewise for decoding and detection in the receiver. Moreover, error control is provided by transmitting additional redundant bits in the code, which has the effect of lowering the information bit rate per channel bandwidth. That is, bandwidth efficiency is traded for increased power efficiency.

To attain a more effective utilization of the available bandwidth and power, coding and modulation have to be treated as a single entity. We may deal with this new situation by redefining coding as *the process of imposing certain patterns on the transmitted signal*. Indeed, this definition includes the traditional idea of parity coding.

*Trellis codes* for band-limited channels result from the treatment of modulation and coding as a *combined* entity rather than as two separate operations. The combination itself is referred to as *trellis-coded modulation* (TCM). This form of signaling has three basic features:

1. The number of signal points in the constellation used is larger than what is required for the modulation format of interest with the same data rate; the additional points allow redundancy for forward error-control coding without sacrificing bandwidth.

2. Convolutional coding is used to introduce a certain dependency between successive signal points, such that only certain *patterns* or *sequences of signal points* are permitted.
3. Soft-decision decoding is performed in the receiver, in which the permissible sequence of signals is modeled as a trellis structure; hence, the name "trellis codes."

This latter requirement is the result of using an enlarged signal constellation. By increasing the size of the constellation, the probability of symbol error increases for a fixed signal-to-noise ratio. Hence, with hard-decision demodulation we would face a performance loss before we begin. Performing soft-decision decoding on the combined code and modulation trellis ameliorates this problem.

In the presence of AWGN, maximum likelihood decoding of trellis codes consists of finding that particular path through the trellis with *minimum squared Euclidean distance* to the received sequence. Thus, in the design of trellis codes, the emphasis is on maximizing the Euclidean distance between code vectors (or, equivalently, code words) rather than maximizing the Hamming distance of an error-correcting code. The reason for this approach is that, except for conventional coding with binary PSK and QPSK, maximizing the Hamming distance is not the same as maximizing the squared Euclidean distance. Accordingly, in what follows, the Euclidean distance is adopted as the distance measure of interest. Moreover, while a more general treatment is possible, the discussion is (by choice) confined to the case of *two-dimensional constellations of signal points*. The implication of such a choice is to restrict the development of trellis codes to multilevel amplitude and/or phase modulation schemes such as $M$-ary PSK and $M$-ary QAM.

The approach used to design this type of trellis codes involves partitioning an $M$-ary constellation of interest successively into 2, 4, 8, . . . subsets with size $M/2, M/4, M/8, \ldots$, and having progressively larger increasing minimum Euclidean distance between their respective signal points. Such a design approach by *set partitioning* represents the "key idea" in the construction of efficient coded modulation techniques for band-limited channels.

In Figure 10.20, we illustrate the partitioning procedure by considering a circular constellation that corresponds to 8-PSK. The figure depicts the constellation itself and the 2 and 4 subsets resulting from two levels of partitioning. These subsets share the common

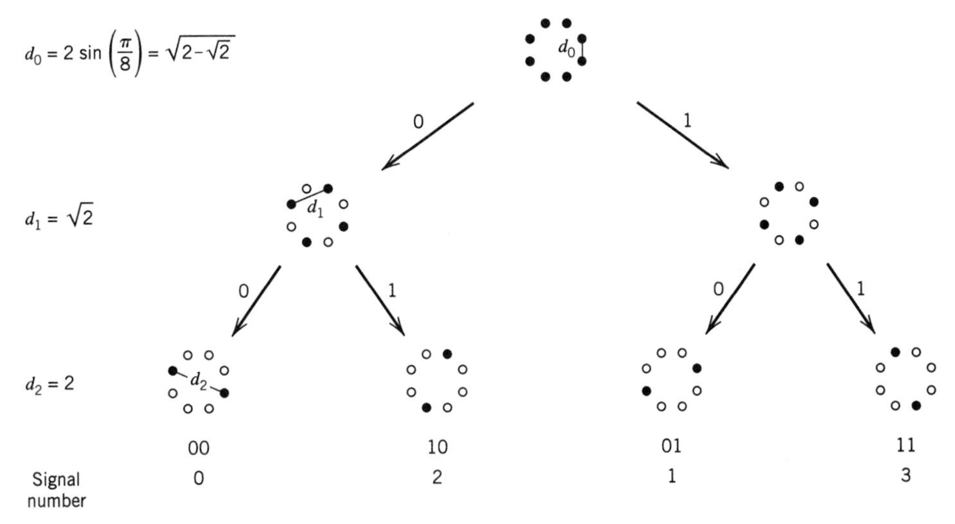


**FIGURE 10.20** Partitioning of 8-PSK constellation, which shows that $d_0 < d_1 < d_2$.

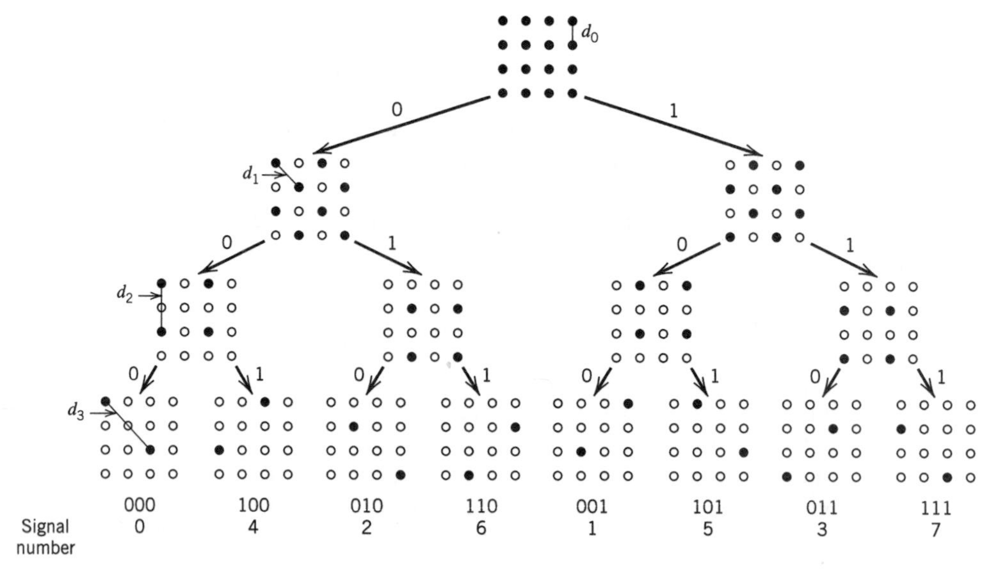


**FIGURE 10.21** Partitioning of 16-QAM constellation, which shows that $d_0 < d_1 < d_2 < d_3$.

property that the minimum Euclidean distances between their individual points follow an increasing pattern: $d_0 < d_1 < d_2$.

Figure 10.21 illustrates the partitioning of a rectangular constellation corresponding to 16-QAM. Here again we see that the subsets have increasing within-subset Euclidean distances: $d_0 < d_1 < d_2 < d_3$.

Based on the subsets resulting from successive partitioning of a two-dimensional constellation, we may devise relatively simple and yet highly effective coding schemes. Specifically, to send $n$ bits/symbol with *quadrature modulation* (i.e., one that has in-phase and quadrature components), we start with a two-dimensional constellation of $2^{n+1}$ signal points appropriate for the modulation format of interest; a circular grid is used for $M$-ary PSK, and a rectangular one for $M$-ary QAM. In any event, the constellation is partitioned into 4 or 8 subsets. One or two incoming bits per symbol enter a rate-1/2 or rate-2/3 binary convolutional encoder, respectively; the resulting two or three coded bits per symbol determine the selection of a particular subset. The remaining uncoded data bits determine which particular point from the selected subset is to be signaled. This class of trellis codes is known as *Ungerboeck codes.*

Since the modulator has memory, we may use the Viterbi algorithm to perform maximum likelihood sequence estimation at the receiver. Each branch in the trellis of the Ungerboeck code corresponds to a subset rather than an individual signal point. The first step in the detection is to determine the signal point within each subset that is closest to the received signal point in the Euclidean sense. The signal point so determined and its metric (i.e., the squared Euclidean distance between it and the received point) may be used thereafter for the branch in question, and the Viterbi algorithm may then proceed in the usual manner.

## ■ UNGERBOECK CODES FOR 8-PSK

The scheme of Figure 10.22*a* depicts the simplest Ungerboeck 8-PSK code for the transmission of 2 bits/symbol. The scheme uses a rate-1/2 convolutional encoder; the corre-

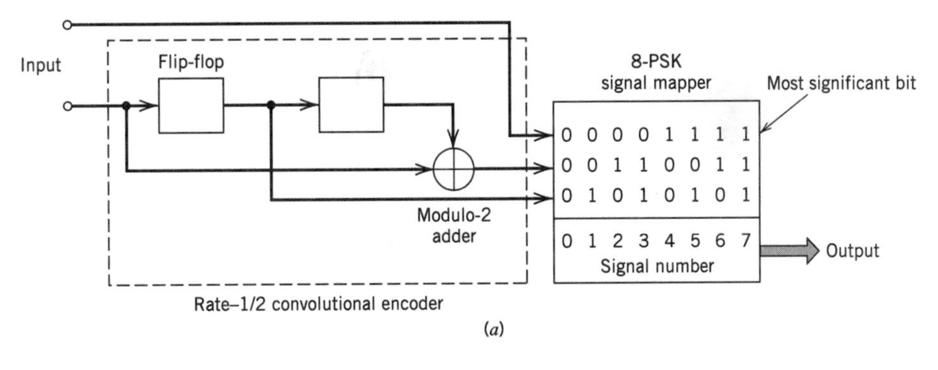


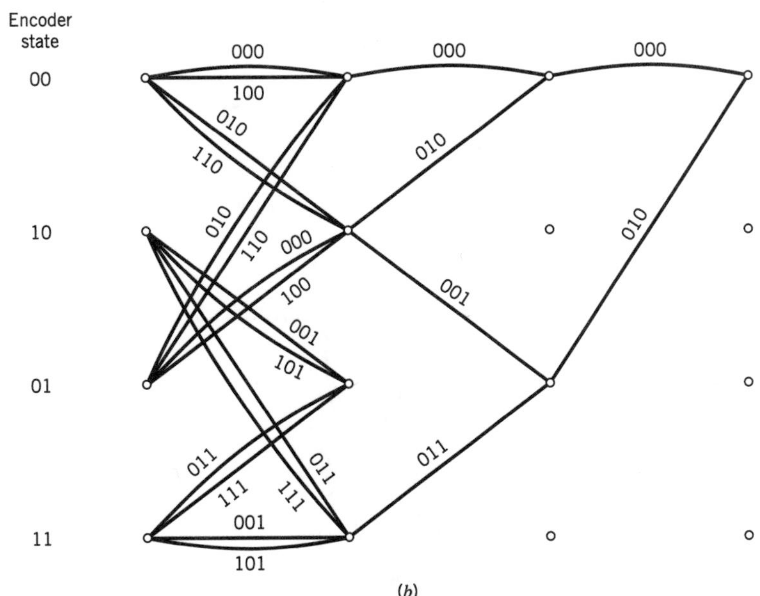


**FIGURE 10.22** (*a*) Four-state Ungerboeck code for 8-PSK; the mapper follows Figure 10.20. (*b*) Trellis of the code.

sponding trellis of the code is shown in Figure 10.22*b*, which has four states. Note that the most significant bit of the incoming binary word is left uncoded. Therefore, each branch of the trellis may correspond to two different output values of the 8-PSK modulator or, equivalently, to one of the four 2-point subsets shown in Figure 10.20. The trellis of Figure 10.22*b* also includes the minimum distance path.

The scheme of Figure 10.23*a* depicts another Ungerboeck 8-PSK code for transmitting 2 bits/sample; it is next in the level of complexity. This second scheme uses a rate-2/3 convolutional encoder. Therefore, the corresponding trellis of the code has eight states, as shown in Figure 10.23*b*. In this case, both bits of the incoming binary word are encoded. Hence, each branch of the trellis corresponds to a specific output value of the 8-PSK modulator. The trellis of Figure 10.23*b* also includes the minimum distance path.

Figures 10.22*b* and 10.23*b* also include the encoder states. In Figure 10.22, the state of the encoder is defined by the contents of the two-stage shift register. On the other hand, in Figure 10.23, it is defined by the content of the single-stage (top) shift register followed by that of the two-stage (bottom) shift register.

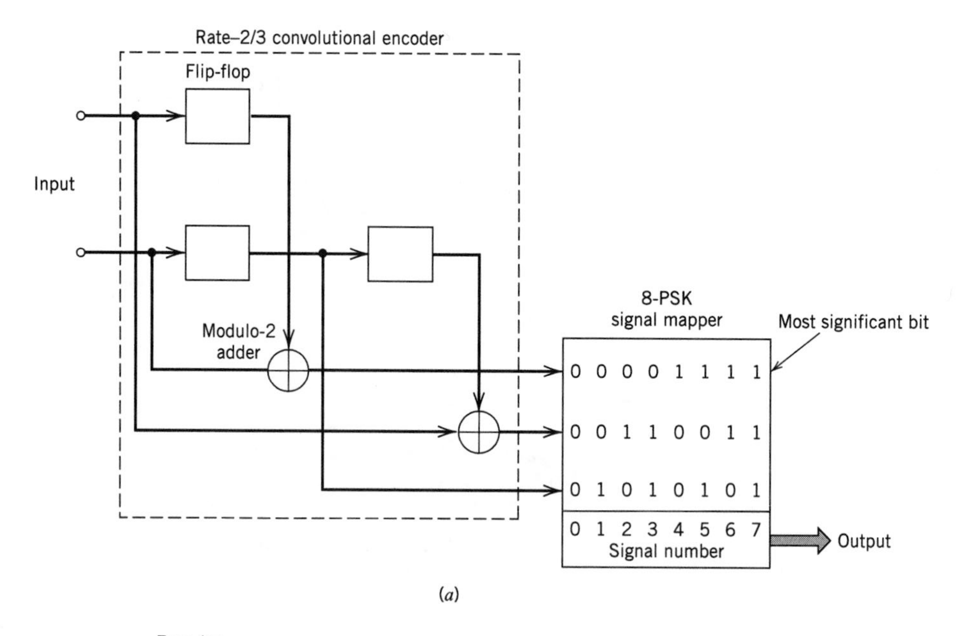


(*a*)

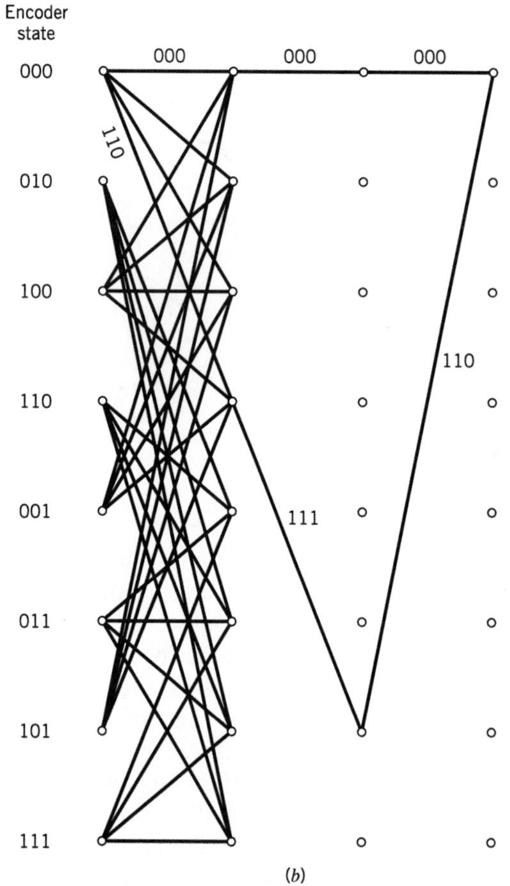


(*b*)

**FIGURE 10.23** (*a*) Eight-state Ungerboeck code for 8-PSK; the mapper follows Figure 10.20. (*b*) Trellis of the code with only some of the branches shown.

## ■ ASYMPTOTIC CODING GAIN

Following the discussion in Section 10.6, we define the *asymptotic coding gain* of Ungerboeck codes as

$$G_a = 10 \log_{10}\left(\frac{d_{\text{free}}^2}{d_{\text{ref}}^2}\right) \tag{10.68}$$

where $d_{\text{free}}$ is the *free Euclidean distance* of the code and $d_{\text{ref}}$ is the minimum Euclidean distance of an uncoded modulation scheme operating with the same signal energy per bit. For example, by using the Ungerboeck 8-PSK code of Figure 10.22*a*, the signal constellation has 8 message points, and we send 2 message bits per point. Hence, uncoded transmission requires a signal constellation with 4 message points. We may therefore regard uncoded 4-PSK as the reference for the Ungerboeck 8-PSK code of Figure 10.22*a*.

The Ungerboeck 8-PSK code of Figure 10.22*a* achieves an asymptotic coding gain of 3 dB, calculated as follows:

1. Each branch of the trellis in Figure 10.22*b* corresponds to a subset of two antipodal signal points. Hence, the free Euclidean distance $d_{\text{free}}$ of the code can be no larger than the Euclidean distance $d_2$ between the antipodal signal points of such a subset. We may therefore write

$$d_{\text{free}} = d_2 = 2$$

   where the distance $d_2$ is defined in Figure 10.24*a*; see also Figure 10.20.
2. The minimum Euclidean distance of an uncoded QPSK, viewed as a reference operating with the same signal energy per bit, equals (see Figure 10.24*b*)

$$d_{\text{ref}} = \sqrt{2}$$

Hence, as previously stated, the use of Equation (10.68) yields an asymptotic coding gain of $10 \log_{10} 2 = 3$ dB.

The asymptotic coding gain achievable with Ungerboeck codes increases with the number of states in the convolutional encoder. Table 10.9 presents the asymptotic coding gain (in dB) for Ungerboeck 8-PSK codes for increasing number of states, expressed with

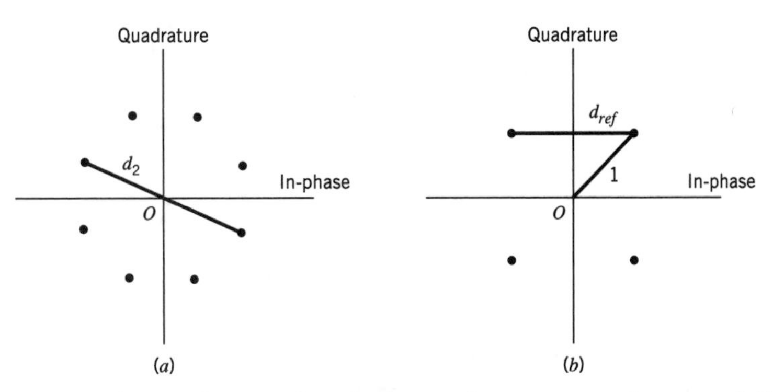


**FIGURE 10.24** Signal-space diagrams for calculation of asymptotic coding gain of Ungerboeck 8-PSK code. (*a*) Definition of distance $d_2$. (*b*) Definition of reference distance $d_{\text{ref}}$.

**TABLE 10.9** ***Asymptotic coding gain of Ungerboeck 8-PSK codes, with respect to uncoded 4-PSK***

| Number of states | 4 | 8 | 16 | 32 | 64 | 128 | 256 | 512 |
|---|---|---|---|---|---|---|---|---|
| Coding gain (dB) | 3 | 3.6 | 4.1 | 4.6 | 4.8 | 5 | 5.4 | 5.7 |

respect to uncoded 4-PSK. Note that improvements on the order of 6 dB require codes with a very large number of states.

## 10.8 *Turbo Codes*[13]

Traditionally, the design of good codes has been tackled by constructing codes with a great deal of algebraic structure, for which there are feasible decoding schemes. Such an approach is exemplified by the linear block codes and convolutional codes discussed in preceding sections. The difficulty with these traditional codes is that, in an effort to approach the theoretical limit for Shannon's channel capacity, we need to increase the code-word length of a linear block code or the constraint length of a convolutional code, which, in turn, causes the computational complexity of a maximum likelihood decoder to increase exponentially. Ultimately, we reach a point where complexity of the decoder is so high that it becomes physically unrealizable.

Various approaches have been proposed for the construction of powerful codes with large "equivalent" block lengths structured in such a way that the decoding can be split into a number of manageable steps. Building on these previous approaches, the development of *turbo codes* and *low-density parity-check codes* has been by far most successful. Indeed, this development has opened a brand new and exciting way of constructing good codes and decoding them with feasible complexity. Turbo codes are discussed in this section and low-density parity-check codes are discussed in Section 10.10.

### ■ TURBO CODING

In its most basic form, the encoder of a turbo code consists of two *constituent* systematic encoders joined together by means of an interleaver, as illustrated in Figure 10.25.

An *interleaver* is an input-output mapping device that *permutes* the ordering of a sequence of symbols from a fixed alphabet in a completely deterministic manner; that is, it takes the symbols at the input and produces identical symbols at the output but in a different temporal order. The interleaver can be of many types, of which the periodic and pseudo-random are two. Turbo codes use a pseudo-random interleaver, which operates

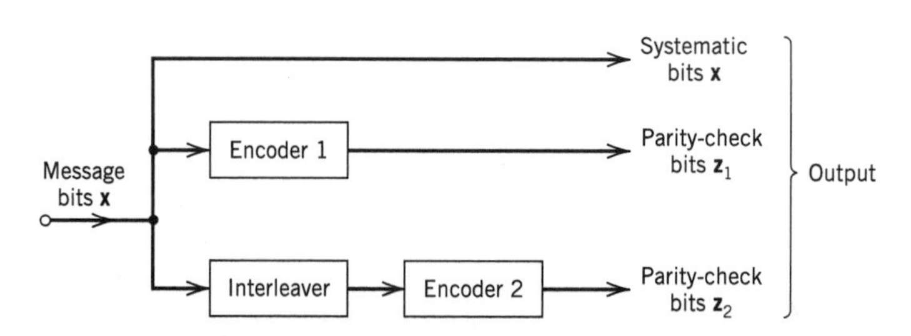


**FIGURE 10.25** Block diagram of turbo encoder.

only on the systematic bits. There are two reasons for the use of an interleaver in a turbo code:

- To tie together errors that are easily made in one half of the turbo code to errors that are exceptionally unlikely to occur in the other half. This is indeed the main reason why the turbo code performs better than a traditional code.
- To provide robust performance with respect to mismatched decoding, which is a problem that arises when the channel statistics are not known or have been incorrectly specified.

Typically, but not necessarily, the same code is used for both constituent encoders in Figure 10.25. The constituent codes recommended for turbo codes are short constraint-length *recursive systematic convolutional (RSC) codes*. The reason for making the convolutional codes recursive (i.e., feeding one or more of the tap outputs in the shift register back to the input) is to make the internal state of the shift register depend on past outputs. This affects the behavior of the error patterns (a single error in the systematic bits produces an infinite number of parity errors), with the result that a better performance of the overall coding strategy is attained.

## EXAMPLE 10.8 Eight-state RSC Encoder

Figure 10.26 shows an example eight-state RSC encoder. The generator matrix for this recursive convolutional code is

$$\mathbf{g}(D) = \left[1, \frac{1 + D + D^2 + D^3}{1 + D + D^3}\right] \tag{10.69}$$

where $D$ is the delay variable. The second entry of the matrix $\mathbf{g}(D)$ is the transfer function of the feedback shift register, defined as the transform of the output divided by the transform of the input. Let $M(D)$ denote the transform of the message sequence $\{m_i\}_{i=1}^{k}$ and $B(D)$ denote the transform of the parity sequence $\{b_i\}_{i=1}^{n-k}$. By definition, we have

$$\frac{B(D)}{M(D)} = \frac{1 + D + D^2 + D^3}{1 + D + D^3}$$

Cross-multiplying, we get:

$$(1 + D + D^2 + D^3)M(D) = (1 + D + D^3)B(D)$$

which, on inversion into the time domain, yields

$$m_i + m_{i-1} + m_{i-2} + m_{i-3} + b_i + b_{i-1} + b_{i-3} = 0 \tag{10.70}$$

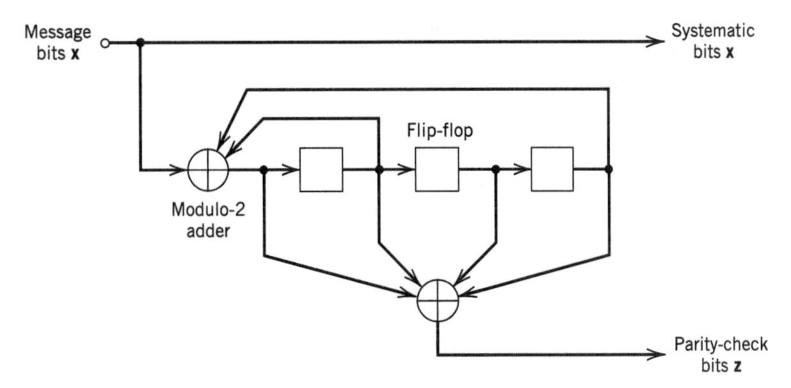


FIGURE 10.26 Example eight-state recursive systematic convolutional (RSC) encoder.

where the addition is modulo-2. Equation (10.70) is the parity-check equation, which the convolutional encoder of Figure 10.26 satisfies at each time step $i$. ◀

In Figure 10.25 the input data stream is applied directly to encoder 1, and the pseudo-randomly reordered version of the same data stream is applied to encoder 2. The systematic bits (i.e., original message bits) and the two sets of parity-check bits generated by the two encoders constitute the output of the turbo encoder. Although the constituent codes are convolutional, in reality turbo codes are block codes with the block size being determined by the size of the interleaver. Moreover, since both RSC encoders in Figure 10.25 are linear, we may describe turbo codes as *linear block codes*.

The block nature of the turbo code raises a practical issue: How do we know the beginning and the end of a code word? The common practice is to initialize the encoder to the *all-zero state* and then encode the data. After encoding a certain number of data bits a number of tail bits are added so as to make the encoder return to the all-zero state at the end of each block; thereafter the cycle is repeated. The termination approaches of turbo codes include the following:

- A simple approach is to terminate the first RSC code in the encoder and leave the second one unterminated. A drawback of this approach is that the bits at the end of the block due to the second RSC code are more vulnerable to noise than the other bits. Experimental work has shown that turbo codes exhibit a leveling off in performance as the SNR increases. This behavior is not like an error floor, but it has the appearance of an error floor compared to the steep drop in error performance at low SNR. This *error floor* is affected by a number of factors, the dominant one of which is the choice of interleaver.
- A more refined approach[14] is to terminate both constituent codes in the encoder in a symmetric manner. Through the combined use of a good interleaver and dual termination, the error floor can be reduced by an order of magnitude compared to the simple termination approach.

In the original version of the turbo encoder, the parity-check bits generated by the two encoders in Figure 10.25 were punctured prior to data transmission over the channel to maintain the rate at 1/2. A *punctured code* is constructed by deleting certain parity check bits, thereby increasing the data rate. Puncturing is the inverse of extending a code. It should, however, be emphasized that the use of a puncture map is not a necessary requirement for the generation of turbo codes.

The novelty of the parallel encoding scheme of Figure 10.25 is in the use of recursive systematic convolutional (RSC) codes and the introduction of a pseudo-random interleaver between the two encoders. Thus a turbo code appears essentially *random* to the channel by virtue of the pseudo-random interleaver, yet it possesses sufficient structure for the decoding to be physically realizable. Coding theory asserts that a code chosen at random is capable of approaching Shannon's channel capacity, provided that the block size is sufficiently large.[15] This is indeed the reason behind the impressive performance of turbo codes, as discussed next.

## ■ PERFORMANCE OF TURBO CODES

Figure 10.27 shows the error performance of a 1/2 rate, turbo code with a large block size for binary data transmission over an AWGN channel.[16] The code uses an interleaver of

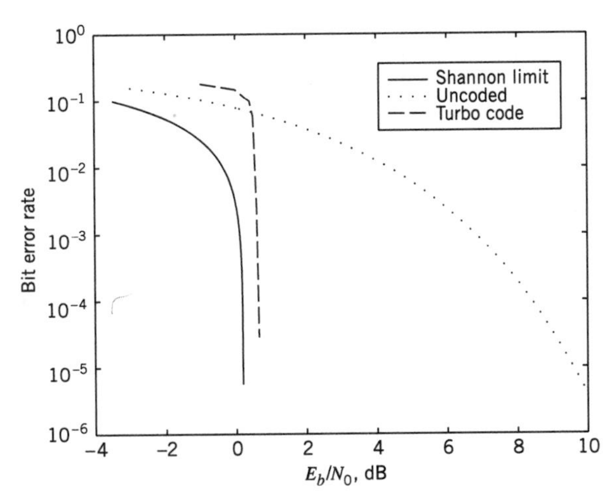


**FIGURE 10.27** Noise performances of 1/2 rate, turbo code and uncoded transmission for AWGN channel; the figure also includes Shannon's theoretical limit on channel capacity for code rate $r = 1/2$.

size 65,536 and a BCJR-based decoder; details of this decoder are presented later in the section. Eighteen iterations of turbo decoding were used in the computation.

For the purpose of comparison, Figure 10.27 also includes two other curves for the same AWGN channel:

- Uncoded transmission (i.e., code rate $r = 1$).
- Shannon's theoretical limit for code rate 1/2, which follows from Figure 9.18*b*.

From Figure 10.27, we may draw two important conclusions:

1. Although the bit error rate for the turbo-coded transmission is significantly higher than that for uncoded transmission at low $E_b/N_0$, the bit error rate for the turbo-coded transmission drops very rapidly once a critical value of $E_b/N_0$ has been reached.
2. At a bit error rate of $10^{-5}$, the turbo code is less than 0.5 dB from Shannon's theoretical limit.

Note, however, attaining this highly impressive performance requires that the size of the interleaver, or, equivalently, the block length of the turbo code, be large. Also, the large number of iterations needed to improve performance increases the decoder latency. This drawback is due to the fact that the digital processing of information does not lend itself readily to the application of feedback, which is a distinctive feature of the turbo decoder.

Now that we have an appreciation for the impressive performance of turbo codes, the stage is set for a discussion of how turbo decoding is actually performed.

## TURBO DECODING

Turbo codes derive their distinctive name from analogy of the decoding algorithm to the "turbo engine" principle. Figure 10.28*a* shows the basic structure of the turbo decoder. It operates on noisy versions of the systematic bits and the two sets of parity-check bits in two decoding stages to produce an estimate of the original message bits.

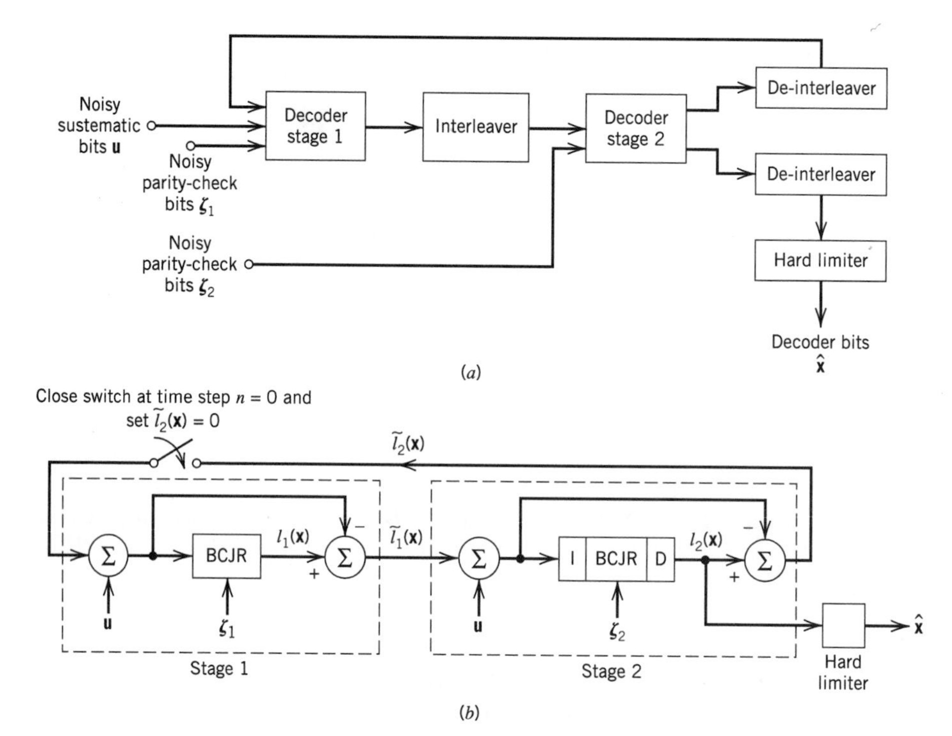


**FIGURE 10.28** (*a*) Block diagram of turbo decoder. (*b*) Extrinsic form of turbo decoder, where I stands for interleaver, D for de-interleaver, and BCJR for BCJR algorithm for log-MAP decoding.

Each of the two decoding stages uses a *BCJR algorithm*,[17] which was originally invented by Bahl, Cocke, Jelinek, and Raviv (hence the name) to solve a *maximum a posteriori probability (MAP) detection* problem. The BCJR algorithm differs from the Viterbi algorithm in two fundamental respects:

1. The BCJR algorithm is a *soft input–soft output* decoding algorithm with two recursions, one forward and the other backward, both of which involve soft decisions. In contrast, the Viterbi algorithm is a *soft input–hard output* decoding algorithm, with a single forward recursion involving soft decisions; the recursion ends with a hard decision, whereby a particular survivor path among several ones is retained. In computational terms, the BCJR algorithm is therefore more complex than the Viterbi algorithm because of the backward recursion.
2. The BCJR algorithm is a MAP decoder in that it minimizes the bit errors by estimating the *a posteriori* probabilities of the individual bits in a code word; to reconstruct the original data sequence, the soft outputs of the BCJR algorithm are hard-limited. On the other hand, the Viterbi algorithm is a maximum likelihood sequence estimator in that it maximizes the likelihood function for the whole sequence, not each bit. As such, the average bit error rate of the BCJR algorithm can be slightly better than the Viterbi algorithm; it is never worse.

Most important, formulation of the BCJR algorithm rests on the fundamental assumptions that (1) the channel encoding, namely, the convolutional encoding performed in the transmitter, is modeled as a *Markov process*, and (2) the channel is memoryless. In the context of our present discussion, the Markovian assumption means that if a code can be represented as a trellis, then the present state of the trellis depends only on the past state and the input bit. (A mathematical treatment of the BCJR algorithm is given later in this section.)

Before proceeding to describe the operation of the two-stage turbo decoder in Figure 10.28*a*, we find it desirable to introduce the notion of extrinsic information. The most convenient representation for this concept is as a log-likelihood ratio, in which case extrinsic information is computed as the difference between two log-likelihood ratios as depicted in Figure 10.29. Formally, *extrinsic information*, generated by a decoding stage for a set of systematic (message) bits, is defined as the difference between the log-likelihood ratio computed at the output of that decoding stage and the *intrinsic information* represented by a log-likelihood ratio fed back to the input of the decoding stage. In effect, extrinsic information is the incremental information gained by exploiting the dependencies that exist between a message bit of interest and incoming raw data bits processed by the decoder.

On this basis, we may depict the flow of information in the two-stage turbo decoder of Figure 10.28*a* in a *symmetric extrinsic* manner as shown in Figure 10.28*b*. The first decoding stage uses the BCJR algorithm to produce a soft estimate of systematic bit $x_j$, expressed as the log-likelihood ratio

$$l_1(x_j) = \log\left(\frac{P(x_j = 1 \,|\, \mathbf{u}, \boldsymbol{\zeta}_1, \tilde{l}_2(\mathbf{x}))}{P(x_j = 0 \,|\, \mathbf{u}, \boldsymbol{\zeta}_1, \tilde{l}_2(\mathbf{x}))}\right), \qquad j = 1, 2, \ldots, k \tag{10.71}$$

where $\mathbf{u}$ is the set of noisy systematic bits, $\boldsymbol{\zeta}_1$ is the set of noisy parity-check bits generated by encoder 1, and $\tilde{l}_2(\mathbf{x})$ is the extrinsic information about the set of message bits $\mathbf{x}$ derived from the second decoding stage and fed back to the first stage. Assuming that the $k$ message bits are statistically independent, the total log-likelihood ratio at the output of the first decoding stage is therefore

$$l_1(\mathbf{x}) = \sum_{j=1}^{k} l_1(x_j) \tag{10.72}$$

Hence, the extrinsic information about the message bits derived from the first decoding stage is

$$\tilde{l}_1(\mathbf{x}) = l_1(\mathbf{x}) - \tilde{l}_2(\mathbf{x}) \tag{10.73}$$

where $\tilde{l}_2(\mathbf{x})$ is to be defined.

Before application to the second decoding stage, the extrinsic information $\tilde{l}_1(\mathbf{x})$ is reordered to compensate for the psuedo-random interleaving introduced in the turbo encoder. In addition, the noisy parity-check bits $\boldsymbol{\zeta}_2$ generated by encoder 2 are used as input. Thus by using the BCJR algorithm, the second decoding stage produces a more refined

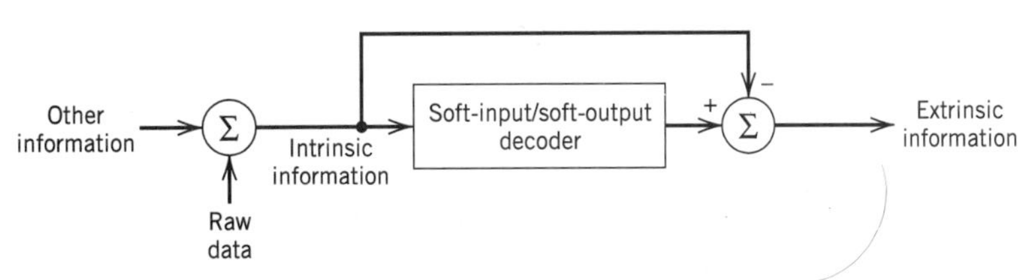


**FIGURE 10.29** Illustrating the concept of extrinsic information.

soft estimate of the message bits $\mathbf{x}$. This estimate is re-interleaved to produce the total log-likelihood ratio $l_2(\mathbf{x})$. The extrinsic information $\tilde{l}_2(\mathbf{x})$ fed back to the first decoding stage is therefore

$$\tilde{l}_2(\mathbf{x}) = l_2(\mathbf{x}) - \tilde{l}_1(\mathbf{x}) \tag{10.74}$$

where $\tilde{l}_1(\mathbf{x})$ is itself defined by Equation (10.73), and $l_2(\mathbf{x})$ is the log-likelihood ratio computed by the second stage. Specifically, for the $j$th element of the vector $\mathbf{x}$, we have

$$l_2(x_j) = \log_2\left(\frac{P(x_j = 1 \mid \mathbf{u}, \boldsymbol{\zeta}_2, \tilde{l}_1(\mathbf{x}))}{P(x_j = 0 \mid \mathbf{u}, \boldsymbol{\zeta}_2, \tilde{l}_1(\mathbf{x}))}\right), \qquad j = 1, 2, \ldots, k \tag{10.75}$$

Through the application of $\tilde{l}_2(\mathbf{x})$ to the first stage, the feedback loop around the pair of decoding stages is thereby closed. Note that although in actual fact the set of noisy systematic bits $\mathbf{u}$ is only applied to the first decoding stage as in Figure 10.28*a*, by formulating the information flow in the symmetric extrinsic manner depicted in Figure 10.28*b* we find that $\mathbf{u}$ is, in effect, also applied to the second decoding stage.

An estimate of the message bits $\mathbf{x}$ is computed by hard-limiting the log-likelihood ratio $l_2(\mathbf{x})$ at the output of the second stage, as shown by

$$\hat{\mathbf{x}} = \operatorname{sgn}(l_2(\mathbf{x})) \tag{10.76}$$

where the signum function operates on each element of $l_2(\mathbf{x})$ individually.

To initiate the turbo decoding algorithm, we simply set $\tilde{l}_2(\mathbf{x}) = 0$ on the first iteration of the algorithm; see Figure 10.28*b*.

The motivation for feeding only extrinsic information from one stage to the next in the turbo decoder of Figure 10.28 is to maintain as much statistical independence between the bits as possible from one iteration to the next. The feedback decoding strategy described herein implicitly relies on this assumption. If this assumption of statistical independence is strictly true, it can be shown that the estimate $\hat{\mathbf{x}}$ defined in Equation (10.76) approaches the MAP solution as the number of iterations approaches infinity.[18] The assumption of statistical independence appears to be close to the truth in the vast majority of cases encountered in practice.

## ■ THE BCJR ALGORITHM

For a discussion of turbo decoding to be complete, a mathematical exposition of the BCJR algorithm for MAP estimation is in order.

Let $x(t)$ be the input to a trellis encoder at time $t$. Let $y(t)$ be the corresponding output observed at the receiver. Note that $y(t)$ may include more than one observation; for example, a rate $1/n$ code produces $n$ bits for each input bit, in which case we have an $n$-dimensional observation vector. Let the observation vector be denoted by

$$\mathbf{y}_{(1,t)} = [y(1), y(2), \ldots, y(t)]$$

Let $\lambda_m(t)$ denote the probability that a state $\mathrm{s}(t)$ of the trellis encoder equals $m$, where $m = 1, 2, \ldots, M$. We may then write

$$\boldsymbol{\lambda}(t) = P[\mathrm{s}(t) \mid \mathbf{y}] \tag{10.77}$$

where $\mathbf{s}(t)$ and $\boldsymbol{\lambda}(t)$ are both $M$-by-1 vectors. Then, for a rate $1/n$ linear convolutional code with feedback as in the RSC code, the probability that a symbol "1" was the message bit is given by

$$P(x(t) = 1 \,|\, \mathbf{y}) = \sum_{s \in \mathcal{F}_A} \lambda_s(t) \tag{10.78}$$

where $\mathcal{F}_A$ is the set of transitions that correspond to a symbol "1" at the input, and $\lambda_s(t)$ is the $s$-component of $\boldsymbol{\lambda}(t)$.

Define the *forward estimation* of state probabilities as the $M$-by-1 vector

$$\boldsymbol{\alpha}(t) = P(\mathbf{s}(t) \,|\, \mathbf{y}_{(1,t)}) \tag{10.79}$$

where the observation vector $\mathbf{y}_{(1,t)}$ is defined above. Also define the *backward estimation* of state probabilities as the $M$-by-1 vector

$$\boldsymbol{\beta}(t) = P(\mathbf{s}(t) \,|\, \mathbf{y}_{(t,k)}) \tag{10.80}$$

where

$$\mathbf{y}_{(t,k)} = [y(t), y(t+1), \ldots, y(k)]$$

The vectors $\boldsymbol{\alpha}(t)$ and $\boldsymbol{\beta}(t)$ are estimates of the state probabilities at time $t$ based on the past and future data, respectively. We may then formulate the *separability theorem* as follows:

> The state probabilities at time $t$ are related to the forward estimator $\boldsymbol{\alpha}(t)$ and backward estimator $\boldsymbol{\beta}(t)$ by the vector
>
> $$\boldsymbol{\lambda}(t) = \frac{\boldsymbol{\alpha}(t) \cdot \boldsymbol{\beta}(t)}{\| \boldsymbol{\alpha}(t) \cdot \boldsymbol{\beta}(t) \|_1} \tag{10.81}$$
>
> where $\boldsymbol{\alpha}(t) \cdot \boldsymbol{\beta}(t)$ is the vector product of $\boldsymbol{\alpha}(t)$ and $\boldsymbol{\beta}(t)$, and $\| \boldsymbol{\alpha}(t) \cdot \boldsymbol{\beta}(t) \|_1$ is the $L_1$ norm of this vector product.

The *vector product* $\boldsymbol{\alpha}(t) \cdot \boldsymbol{\beta}(t)$ (not to be confused with the inner product) is defined in terms of the individual elements of $\boldsymbol{\alpha}(t)$ and $\boldsymbol{\beta}(t)$ by

$$\boldsymbol{\alpha}(t) \cdot \boldsymbol{\beta}(t) = \begin{bmatrix} \alpha_1(t)\beta_1(t) \\ \alpha_2(t)\beta_2(t) \\ \vdots \\ \alpha_M(t)\beta_M(t) \end{bmatrix} \tag{10.82}$$

and the $L_1$ *norm* of $\boldsymbol{\alpha}(t) \cdot \boldsymbol{\beta}(t)$ is defined by

$$\| \boldsymbol{\alpha}(t) \cdot \boldsymbol{\beta}(t) \|_1 = \sum_{m=1}^{M} \alpha_m(t)\beta_m(t) \tag{10.83}$$

The separability theorem says that the state distribution at time $t$ given the past is independent of the state distribution at time $t$ given the future, which is intuitively satisfying recalling the Markovian assumption for channel encoding, which is basic to the BCJR algorithm. Moreover, this theorem provides the basis of a simple way of combining the forward and backward estimates to obtain a complete description of the state probabilities.

To proceed further, let the state transition probability at time $t$ be defined by

$$\gamma_{m',m}(t) = P(s(t) = m, \mathbf{y}(t) \,|\, s(t-1) = m') \tag{10.84}$$

and denote the $M$-by-$M$ matrix of transition probabilities as

$$\mathbf{\Gamma}(t) = \{\gamma_{m',m}(t)\} \tag{10.85}$$

We may then formulate the *recursion theorem* as follows:

> The forward estimate $\boldsymbol{\alpha}(t)$ and backward estimate $\boldsymbol{\beta}(t)$ are computed recursively as
>
> $$\boldsymbol{\alpha}^T(t) = \frac{\boldsymbol{\alpha}^T(t-1)\mathbf{\Gamma}(t)}{\| \boldsymbol{\alpha}^T(t-1)\mathbf{\Gamma}(t) \|_1} \tag{10.86}$$
>
> and
>
> $$\boldsymbol{\beta}(t) = \frac{\mathbf{\Gamma}(t+1)\boldsymbol{\beta}(t+1)}{\| \mathbf{\Gamma}(t+1)\boldsymbol{\beta}(t+1) \|_1} \tag{10.87}$$
>
> where the superscript $T$ denotes matrix transposition.

The separability and recursion theorems together define the BCJR algorithm for the computation of *a posteriori* probabilities of the states and transitions of a code trellis, given the observation vector. Using these estimates, the likelihood ratios needed for turbo decoding may then be computed by performing summations over selected subsets of states as required.

## 10.9 *Computer Experiment: Turbo Decoding*

Two properties constitute the hallmark of turbo codes:

**Property 1:**

*The error performance of the turbo decoder improves with the number of iterations of the decoding algorithm. This is achieved by feeding extrinsic information from the output of the first decoding stage to the input of the second decoding stage in the forward path and feeding extrinsic information from the output of the second stage to the input of the first stage in the backward path, and then permitting the iterative decoding process to take its natural course in response to the received noisy message and parity bits.*

**Property 2**

*The turbo decoder is capable of approaching the Shannon theoretical limit of channel capacity in a computationally feasible manner; this property has been demonstrated experimentally but not yet proven theoretically.*

Property 2 requires that the block length of the turbo code be large. Unfortunately, a demonstration of this property requires the use of sophisticated implementations of the turbo decoding algorithm that are beyond the scope of this book. Accordingly, we focus our attention on a demonstration of Property 1 in this computer experiment.

So, as the primary objective of this computer experiment, we wish to use the log-MAP implementation of the BCJR algorithm to demonstrate Property 1 of turbo decoding.

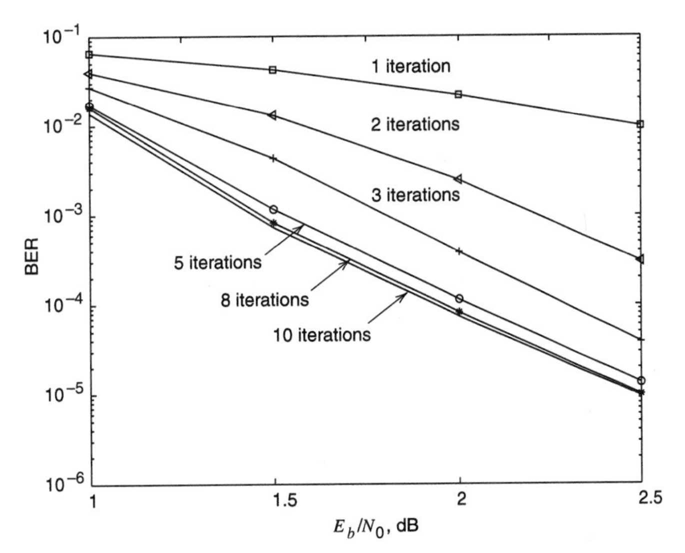


**FIGURE 10.30** Results of the computer experiment on turbo decoding, for increasing number of iterations.

The only channel impairment assumed in the experiment is additive white Gaussian noise. Details of the turbo encoder and decoder are as follows:

Turbo Encoder (described in Figure 10.25):

Encoder 1: convolutional encoder [1, 1, 1]
Encoder 2: convolutional encoder [1, 0, 1]
Block (i.e., interleaver) length: 1,200 bits

Turbo Decoder (described in Figure 10.28):

The BCJR algorithm for log-MAP decoding.

The experiment was carried out for $E_b/N_0 = 1$, 1.5, 2, and 2.5 dB, with varying number of iterations at each $E_b/N_0$. For each trial of the experiment, the number of bit errors was calculated after accumulating a total of 20 blocks of data (each 1,200 bits long) that were noise-corrupted. The probability of error was then evaluated as the ratio of bit errors to the total number of encoded bits. Note that in this calculation, many of the blocks of encoded bits were correctly decoded.

The results of the experiment are plotted in Figure 10.30. The following observations can be made from this figure:

1. For a given $E_b/N_0$, the probability of error decreases with increasing number of iterations, confirming Property 1 of turbo decoding.
2. After eight iterations, there is no significant improvement in decoding performance.
3. For a fixed number of iterations, the probability of error decreases with increasing $E_b/N_0$, which is to be expected.

## 10.10 *Low-Density Parity-Check Codes*[19]

Turbo codes, discussed in Section 10.8, and low-density parity-check (LDPC) codes, discussed in this section, belong to a broad family of error-control coding techniques called *compound codes*. The two most important advantages of LDPC codes over turbo codes are:

- Absence of low-weight code words.
- Iterative decoding of lower complexity.

With regard to the issue of low-weight code words, we usually find that a small number of code words in a turbo code are undesirably close to the given code word. Due to this closeness in weights, once in a while the channel noise causes the transmitted code word to be mistaken for a nearby code word. Indeed, it is this behavior that is responsible for the error floor (typically around a bit error rate of $10^{-5}$ to $10^{-6}$) that was mentioned earlier. In contrast, LDPC codes can be easily constructed so that they do not have such low-weight code words, and they can therefore achieve *vanishingly small* bit error rates. The error-floor problem in turbo codes can be alleviated by careful design of the interleaver.

Turning next to the issue of decoding complexity, we note that the computational complexity of a turbo decoder is dominated by the BCJR algorithm, which operates on the trellis for the convolutional code used in the encoder. The number of computations in each recursion of the BCJR algorithm scales linearly with the number of states in the trellis. Commonly used turbo codes employ trellises with 16 states or more. In contrast, LDPC codes use a simple parity-check trellis that has just two states. Consequently, the decoders for LDPC codes are significantly simpler than those for turbo decoders. Moreover, being parallelizable, LDPC decoding may be performed at greater speeds than turbo decoding.

However, a practical objection to the use of LDPC codes is that for large block lengths, their encoding complexity is high compared to turbo codes.

## ■ CONSTRUCTION OF LDPC CODES

LDPC codes are specified by a parity-check matrix denoted by **A**, which is *sparse*; that is, it consists mainly of 0s and a small number of 1s. In particular, we speak of $(n, t_c, t_r)$ LDPC codes, where $n$ denotes the block length, $t_c$ denotes the weight (i.e., number of 1s) in each column of the matrix **A**, and $t_r$ denotes the weight of each row with $t_r > t_c$. The rate of such a LDPC code is

$$r = 1 - \frac{t_c}{t_r} \tag{10.88}$$

whose validity may be justified as follows. Let $\rho$ denote the *density* of 1s in the parity-check matrix **A**. Then, following the terminology introduced in Section 10.3, we may set

$$t_c = \rho(n - k)$$

and

$$t_r = \rho n$$

where $(n - k)$ is the number of rows in **A** and $n$ is the number of columns (i.e., the block length). Therefore, dividing $t_c$ by $t_r$, we get

$$\frac{t_c}{t_r} = 1 - \frac{k}{n}$$

By definition, the code rate of a block code is $k/n$, hence the result of Equation (10.88) follows. For this result to hold, however, the rows of **A** must be *linearly independent.*

The structure of LDPC codes is well portrayed by *bipartite graphs*. Figure 10.31 shows such a graph for the example code of $n = 10$, $t_c = 3$, and $t_r = 5$. The left-hand nodes in the graph of Figure 10.31 are *variable nodes*, which correspond to elements of the code word. The right-hand nodes of the graph are *check nodes*, which correspond to the set of parity-check constraints satisfied by code words in the code. LDPC codes of the type exemplified by the graph of Figure 10.31 are said to be *regular* in that all the nodes of a similar kind have exactly the same degree. In the example graph of Figure 10.31, the degree of the variable nodes is $t_c = 3$, and the degree of the check nodes is $t_r = 5$. As the block length $n$ approaches infinity, each check node is connected to a vanishingly small fraction of variable nodes, hence the term *low-density*.

The matrix **A** is constructed by putting 1s in **A** at *random*, subject to the *regularity constraints*:

- Each column contains a small fixed number, $t_c$, of 1s.
- Each row contains a small fixed number, $t_r$, of 1s.

In practice, these regularity constraints are often violated slightly in order to avoid having linearly dependent rows in the parity-check matrix **A**.

Unlike the linear block codes discussed in Section 10.3, the parity-check matrix **A** of LDPC codes is *not* systematic (i.e., it does not have the parity-check bits appearing in diagonal form), hence the use of a symbol different from that used in Section 10.3. Nevertheless, for coding purposes, we may derive a generator matrix **G** for LDPC codes by means of *Gaussian elimination* performed in modulo-2 arithmetic; this procedure is illustrated later in Example 10.9. Following the terminology introduced in Section 10.3, the 1-by-$n$ code vector **c** is first partitioned as

$$\mathbf{c} = [\mathbf{b} \vdots \mathbf{m}]$$

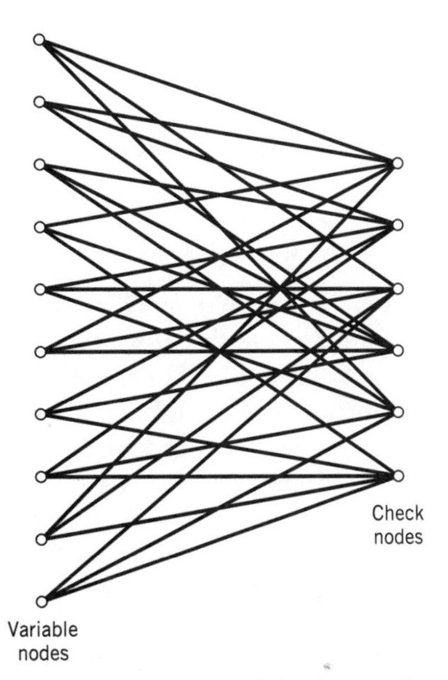


**FIGURE 10.31** Bipartite graph of the (10, 3, 5) LDPC code.

where $\mathbf{m}$ is the $k$-by-1 message vector, and $\mathbf{b}$ is the $(n\text{-}k)$-by-1 parity vector; see Equation (10.9). Correspondingly, the parity-check matrix $\mathbf{A}$ is partitioned as

$$\mathbf{A}^T = \begin{bmatrix} \mathbf{A}_1 \\ \cdots\cdots \\ \mathbf{A}_2 \end{bmatrix} \tag{10.89}$$

where $\mathbf{A}_1$ is a square matrix of dimensions $(n\text{-}k) \times (n\text{-}k)$, and $\mathbf{A}_2$ is a rectangular matrix of dimensions $k \times (n\text{-}k)$; transposition symbolized by the superscript $T$ is used in the partitioning of matrix $\mathbf{A}$ for convenience of presentation. Imposing the constraint of Equation (10.16) on the LDPC code, we may write

$$[\mathbf{b} \,\vdots\, \mathbf{m}] \begin{bmatrix} \mathbf{A}_1 \\ \cdots\cdots \\ \mathbf{A}_2 \end{bmatrix} = \mathbf{0}$$

or, equivalently,

$$\mathbf{b}\mathbf{A}_1 + \mathbf{m}\mathbf{A}_2 = \mathbf{0} \tag{10.90}$$

Recall from Equation (10.7) that the vectors $\mathbf{m}$ and $\mathbf{b}$ are related by

$$\mathbf{b} = \mathbf{m}\mathbf{P}$$

where $\mathbf{P}$ is the coefficient matrix. Hence, substituting this relation into Equation (10.90), we readily find that, for any nonzero message vector $\mathbf{m}$, the coefficient matrix of LDPC codes satisfies the condition

$$\mathbf{P}\mathbf{A}_1 + \mathbf{A}_2 = \mathbf{0}$$

which holds for *all* nonzero message vectors and, in particular, for $\mathbf{m}$ in the form $[0 \cdots 0\ 1\ 0 \cdots 0]$ that will isolate individual rows of the generator matrix.

Solving this equation for matrix $\mathbf{P}$, we get

$$\mathbf{P} = \mathbf{A}_2\mathbf{A}_1^{-1} \tag{10.91}$$

where $\mathbf{A}_1^{-1}$ is the *inverse* of matrix $\mathbf{A}_1$, which is naturally defined in modulo-2 arithmetic. Finally, the generator matrix of LDPC codes is defined by

$$\begin{aligned} \mathbf{G} &= [\mathbf{P} \,\vdots\, \mathbf{I}_k] \\ &= [\mathbf{A}_2\mathbf{A}_1^{-1} \,\vdots\, \mathbf{I}_k] \end{aligned} \tag{10.92}$$

where $\mathbf{I}_k$ is the $k$-by-$k$ identity matrix; see Equation (10.12).

It is important to note that if we take the parity-check matrix $\mathbf{A}$ for some arbitrary LDPC code and just pick $(n - k)$ columns of $\mathbf{A}$ at random to form a square matrix $\mathbf{A}_1$, there is *no* guarantee that $\mathbf{A}_1$ will be *nonsingular* (i.e., the inverse $\mathbf{A}_1^{-1}$ will exist), even if the rows of $\mathbf{A}$ are linearly independent. In fact, for a typical LDPC code with large block length $n$, such a randomly selected $\mathbf{A}_1$ is highly unlikely to be nonsingular, because it is very likely that at least one row of $\mathbf{A}_1$ will be all 0s. Of course, when the rows of $\mathbf{A}$ are linearly independent, there will be *some* set of $(n - k)$ columns of $\mathbf{A}$ that will make a nonsingular $\mathbf{A}_1$, as illustrated in Example 10.9. For some construction methods for LDPC codes the first $(n - k)$ columns of $\mathbf{A}$ may be guaranteed to produce a nonsingular $\mathbf{A}_1$, or at least do so with a high probability, but that is *not* true in general.

### ▶ EXAMPLE 10.9 (10, 3, 5) LDPC Code

Consider the bipartite graph of Figure 10.31 pertaining to a (10, 3, 5) LDPC code. The parity-check matrix of the code is defined by

$$\mathbf{A} = \begin{bmatrix} 1 & 1 & 0 & 1 & 0 & 1 & \vdots & 0 & 0 & 1 & 0 \\ 0 & 1 & 1 & 0 & 1 & 0 & \vdots & 1 & 1 & 0 & 0 \\ 1 & 0 & 0 & 0 & 1 & 1 & \vdots & 0 & 0 & 1 & 1 \\ 0 & 1 & 1 & 1 & 0 & 1 & \vdots & 1 & 0 & 0 & 0 \\ 1 & 0 & 1 & 0 & 1 & 0 & \vdots & 0 & 1 & 0 & 1 \\ \underbrace{0 \;\; 0 \;\; 0 \;\; 1 \;\; 0 \;\; 0}_{\mathbf{A}_1^T} & & & & & & \vdots & \underbrace{1 \;\; 1 \;\; 1 \;\; 1}_{\mathbf{A}_2^T} & & & \end{bmatrix}$$

which appears to be random, while maintaining the regularity constraints: $t_c = 3$ and $t_r = 5$. Partitioning the matrix $\mathbf{A}$ in the manner described in Equation (10.89):

$$\mathbf{A}_1 = \begin{bmatrix} 1 & 0 & 1 & 0 & 1 & 0 \\ 1 & 1 & 0 & 1 & 0 & 0 \\ 0 & 1 & 0 & 1 & 1 & 0 \\ 1 & 0 & 0 & 1 & 0 & 1 \\ 0 & 1 & 1 & 0 & 1 & 0 \\ 1 & 0 & 1 & 1 & 0 & 0 \end{bmatrix}$$

$$\mathbf{A}_2 = \begin{bmatrix} 0 & 1 & 0 & 1 & 0 & 1 \\ 0 & 1 & 0 & 0 & 1 & 1 \\ 1 & 0 & 1 & 0 & 0 & 1 \\ 0 & 0 & 1 & 0 & 1 & 1 \end{bmatrix}$$

To derive the inverse of matrix $\mathbf{A}_1$, we first use Equation (10.90) to write

$$\underbrace{[b_0, b_1, b_2, b_3, b_4, b_5]}_{\mathbf{b}} \underbrace{\begin{bmatrix} 1 & 0 & 1 & 0 & 1 & 0 \\ 1 & 1 & 0 & 1 & 0 & 0 \\ 0 & 1 & 0 & 1 & 1 & 0 \\ 1 & 0 & 0 & 1 & 0 & 1 \\ 0 & 1 & 1 & 0 & 1 & 0 \\ 1 & 0 & 1 & 1 & 0 & 0 \end{bmatrix}}_{\mathbf{A}_1} = \underbrace{[u_0, u_1, u_2, u_3, u_4, u_5]}_{\mathbf{u} = \mathbf{m}\mathbf{A}_2}$$

where we have introduced the vector $\mathbf{u}$ to denote the matrix product $\mathbf{m}\mathbf{A}_2$. By using Gaussian elimination, the matrix $\mathbf{A}_1$ is transformed into *lower diagonal form* (i.e., all the elements above the main diagonal are zero), as shown by

$$\mathbf{A}_1 \rightarrow \begin{bmatrix} 1 & 0 & 0 & 0 & 0 & 0 \\ 1 & 1 & 0 & 0 & 0 & 0 \\ 0 & 1 & 1 & 0 & 0 & 0 \\ 1 & 0 & 1 & 1 & 0 & 0 \\ 0 & 1 & 0 & 1 & 1 & 0 \\ 1 & 0 & 0 & 1 & 0 & 1 \end{bmatrix}$$

This transformation is achieved by the following modulo-2 additions performed on the columns of square matrix $\mathbf{A}_1$:

- Columns 1 and 2 are added to column 3.
- Column 2 is added to column 4.
- Columns 1 and 4 are added to column 5.
- Columns 1, 2 and 5 are added to column 6.

Correspondingly, the vector $\mathbf{u}$ is transformed as

$$\mathbf{u} \rightarrow [u_0, u_1, u_0 + u_1 + u_2, u_1 + u_3, u_0 + u_3 + u_4, u_0 + u_1 + u_4 + u_5]$$

Accordingly, premultiplying the transformed matrix $\mathbf{A}_1$ by the parity vector $\mathbf{b}$, using successive eliminations in modulo-2 arithmetic working backwards, and putting the solutions for the elements of the parity vector $\mathbf{b}$ in terms of the elements of the vector $\mathbf{u}$ in matrix form, we get

$$\underbrace{[u_0, u_1, u_2, u_3, u_4, u_5]}_{\mathbf{u}} \underbrace{\begin{bmatrix} 0 & 0 & 1 & 0 & 1 & 1 \\ 1 & 0 & 1 & 0 & 0 & 1 \\ 1 & 1 & 1 & 0 & 0 & 0 \\ 1 & 1 & 0 & 0 & 1 & 0 \\ 0 & 1 & 0 & 0 & 1 & 1 \\ 1 & 1 & 1 & 1 & 0 & 1 \end{bmatrix}}_{\mathbf{A}_1^{-1}} = \underbrace{[b_0, b_1, b_2, b_3, b_4, b_5]}_{\mathbf{b}}$$

The inverse of matrix $\mathbf{A}_1$ is therefore

$$\mathbf{A}_1^{-1} = \begin{bmatrix} 0 & 0 & 1 & 0 & 1 & 1 \\ 1 & 0 & 1 & 0 & 0 & 1 \\ 1 & 1 & 1 & 0 & 0 & 0 \\ 1 & 1 & 0 & 0 & 1 & 0 \\ 0 & 1 & 0 & 0 & 1 & 1 \\ 1 & 1 & 1 & 1 & 0 & 1 \end{bmatrix}$$

The matrix product $\mathbf{A}_2\mathbf{A}_1^{-1}$ is (using the given value of $\mathbf{A}_2$ and the value of $\mathbf{A}_1^{-1}$ just found)

$$\mathbf{A}_2\mathbf{A}_1^{-1} = \begin{bmatrix} 1 & 0 & 0 & 1 & 1 & 0 \\ 0 & 0 & 0 & 1 & 1 & 1 \\ 0 & 0 & 1 & 1 & 1 & 0 \\ 0 & 1 & 0 & 1 & 1 & 0 \end{bmatrix}$$

Finally, using Equation (10.92), the generator of the (10, 3, 5) LDPC code is

$$\mathbf{G} = \left[ \underbrace{\begin{matrix} 1 & 0 & 0 & 1 & 1 & 0 \\ 0 & 0 & 0 & 1 & 1 & 1 \\ 0 & 0 & 1 & 1 & 1 & 0 \\ 0 & 1 & 0 & 1 & 1 & 0 \end{matrix}}_{\mathbf{A}_2\mathbf{A}_1^{-1}} \begin{matrix} \vdots \\ \vdots \\ \vdots \\ \vdots \end{matrix} \underbrace{\begin{matrix} 1 & 0 & 0 & 0 \\ 0 & 1 & 0 & 0 \\ 0 & 0 & 1 & 0 \\ 0 & 0 & 0 & 1 \end{matrix}}_{\mathbf{I}_k} \right]$$

It is important to recognize that the LDPC code described in this example is intended only for the purpose of illustrating the procedure involved in the generation of such a code. In practice, the block length $n$ is orders of magnitude larger than that considered in this example. Moreover, in constructing the matrix $\mathbf{A}$, we may constrain all pairs of columns to have a *matrix overlap* (i.e., inner product of any two columns in matrix **A**) not to exceed 1; such a constraint, over and above the regularity constraints, is expected to improve the performance of LDPC codes. Unfortunately, with a small block length as that considered in this example, it is difficult to satisfy this additional requirement. ◀

## MINIMUM DISTANCE OF LDPC CODES

In practice, the block length of a LDPC code is large, ranging from $10^3$ to $10^6$, which means that the number of code words in a particular code is correspondingly large. Consequently, the algebraic analysis of LDPC codes is rather difficult. It is much more productive to perform a *statistical analysis* on an ensemble of LDPC codes. Such an analysis permits us to make statistical statements about certain properties of member codes in the ensemble. Moreover, an LDPC code with these properties can be found with high probability by a random selection from the ensemble.

Among these properties, the minimum distance of the member codes is of particular interest. From Section 10.3 we recall that the minimum distance of a linear block code is, by definition, the smallest Hamming distance between any pair of code vectors in the code. Over an ensemble of LDPC codes, the minimum distance of a member code is naturally a random variable. Elsewhere[20] it is shown that as the block length $n$ increases, for fixed $t_c \geq 3$ and $t_r > t_c$ the probability distribution of the minimum distance can be overbounded by a function that approaches a unit step function at a fixed fraction $\Delta_{t_c t_r}$ of the block length $n$. Thus, for large $n$, practically all the LDPC codes in the ensemble have a minimum distance of at least $n\,\Delta_{t_c t_r}$. Table 10.10 presents the rate $r$ and $\Delta_{t_c t_r}$ of LDPC codes for different values of the weight-pair $(t_c, t_r)$. From this table we see that for $t_c = 3$ and $t_r = 6$ the code rate $r$ attains its highest value of 1/2 and the fraction $\Delta_{t_c t_r}$ attains its smallest value, hence the preferred choice of $t_c = 3$ and $t_r = 6$ in the design of LDPC codes.

## PROBABILISTIC DECODING OF LDPC CODES

At the transmitter, a message vector **m** is encoded into a code vector **c** = **mG**, where **G** is the generator matrix for a specified weight-pair $(t_c, t_r)$ and therefore minimum distance $d_{\min}$. The vector **c** is transmitted over a noisy channel to produce the received vector

$$\mathbf{r} = \mathbf{c} + \mathbf{e}$$

where **e** is the error vector due to channel noise; see Equation (10.17). By construction, the matrix **A** is a parity matrix of the LDPC code; that is, $\mathbf{AG}^T = \mathbf{0}$. Given the received

**TABLE 10.10**[a] ***The rate r and fractional term $\Delta_{t_c t_r}$ of LDPC codes for varying weights $t_c$ and $t_r$***

| $t_c$ | $t_r$ | *Rate* r | $\Delta_{t_c t_r}$ |
|---|---|---|---|
| 5 | 6 | 0.167 | 0.255 |
| 4 | 5 | 0.2 | 0.210 |
| 3 | 4 | 0.25 | 0.122 |
| 4 | 6 | 0.333 | 0.129 |
| 3 | 5 | 0.4 | 0.044 |
| 3 | 6 | 0.5 | 0.023 |

[a]Adapted from Gallager (1962) with permission of the IEEE.

vector $\mathbf{r}$, the bit-by-bit decoding problem is to find the most probable vector $\hat{\mathbf{c}}$ that satisfies the condition $\hat{\mathbf{c}}\mathbf{A}^T = \mathbf{0}$.

In what follows, a bit refers to an element of the received vector $\mathbf{r}$, and a check refers to a row of matrix $\mathbf{A}$. Let $\mathcal{J}(i)$ denote the set of bits that participate in check $i$. Let $\mathcal{I}(j)$ denote the set of checks in which bit $j$ participates. A set $\mathcal{J}(i)$ that excludes bit $j$ is denoted by $\mathcal{J}(i)\backslash j$. Likewise, a set $\mathcal{I}(j)$ that excludes check $i$ is denoted by $\mathcal{I}(j)\backslash i$.

The decoding algorithm has two alternating steps: horizontal step and vertical step, which run along the rows and columns of matrix $\mathbf{A}$, respectively. In the course of these steps, two probabilistic quantities associated with nonzero elements of matrix $\mathbf{A}$ are alternately updated. One quantity, denoted by $P_{ij}^x$, defines the probability that bit $j$ is symbol $x$ (i.e., symbol 0 or 1), given the information derived via checks performed in the horizontal step, except for check $i$. The second quantity, denoted by $Q_{ij}^x$, defines the probability that check $i$ is satisfied, given that bit $j$ is fixed at the value $x$ and the other bits have the probabilities $P_{ij'} : j' \in \mathcal{J}(i)\backslash j$.

The LDPC decoding algorithm then proceeds as follows:[21]

### *Initialization*

The variables $P_{ij}^0$ and $P_{ij}^1$ are set equal to the *a priori* probabilities $p_j^0$ and $p_j^1$ of symbols 0 and 1, respectively, with $p_j^0 + p_j^1 = 1$.

### *Horizontal Step*

In the horizontal step of the algorithm, we run through the checks $i$. Define

$$\Delta P_{ij} = P_{ij}^0 - P_{ij}^1$$

For each weight-pair $(i, j)$, compute

$$\Delta Q_{ij} = \prod_{j' \in \mathcal{J}(i)\backslash j} \Delta P_{ij'}$$

Hence, set

$$Q_{ij}^0 = \frac{1}{2}(1 + \Delta Q_{ij})$$

$$Q_{ij}^1 = \frac{1}{2}(1 - \Delta Q_{ij})$$

### *Vertical Step*

In the vertical step of the algorithm, the values of the probabilities $P_{ij}^0$ and $P_{ij}^1$ are updated using the quantities computed in the horizontal step. In particular, for each bit $j$, compute

$$P_{ij}^0 = \alpha_{ij} p_j^0 \prod_{i' \in \mathcal{I}(j)\backslash i} Q_{i'j}^0$$

$$P_{ij}^1 = \alpha_{ij} p_j^1 \prod_{i' \in \mathcal{I}(j)\backslash i} Q_{i'j}^1$$

where the scaling factor $\alpha_{ij}$ is chosen to make

$$P_{ij}^0 + P_{ij}^1 = 1$$

In the vertical step, we may also update the *pseudo-posterior probabilities*:

$$P_j^0 = \alpha_j p_j^0 \prod_{i \in \mathcal{I}(j)} Q_{ij}^0$$

$$P_j^1 = \alpha_j p_j^1 \prod_{i \in \mathcal{I}(j)} Q_{ij}^1$$

where $\alpha_j$ is chosen to make

$$P_j^0 + P_j^1 = 1$$

The quantities obtained in the vertical step are used to compute a tentative estimate $\hat{\mathbf{c}}$. If the condition $\hat{\mathbf{c}}\mathbf{A}^T = \mathbf{0}$ is satisfied, the decoding algorithm is terminated. Otherwise, the algorithm goes back to the horizontal step. If after some maximum number of iterations (e.g., 100 or 200) there is no valid decoding, a decoding failure is declared. The decoding procedure described herein is a special case of the general low-complexity *sum-product algorithm*.

Simply stated, the sum-product algorithm passes probabilistic quantities between the check nodes and variable nodes of the bipartite graph. By virtue of the fact that each parity-check constraint can be represented by a simple convolutional coder with one bit of memory, we find that LDPC decoders are simpler to implement than turbo decoders, as stated earlier.

In terms of performance, however, we may say the following in light of experimental results reported in the literature: Regular LDPC codes do not appear to come as close to Shannon's limit as do their turbo code counterparts.

## 10.11 *Irregular Codes*

The turbo codes discussed in Section 10.8 and the LDPC codes discussed in Section 10.10 are both regular codes, each in its own individual way. The error-correcting performance of both of these codes over a noisy channel can be improved substantially by using their respective irregular forms.

In a standard turbo code with its encoder as shown in Figure 10.25, the interleaver maps each systematic bit to a unique input bit of convolutional encoder 2. In contrast, *irregular turbo codes*[22] use a special design of interleaver that maps some systematic bits to multiple input bits of the convolutional encoder. For example, each of 10 percent of the systematic bits may be mapped to eight inputs of the convolutional encoder instead of

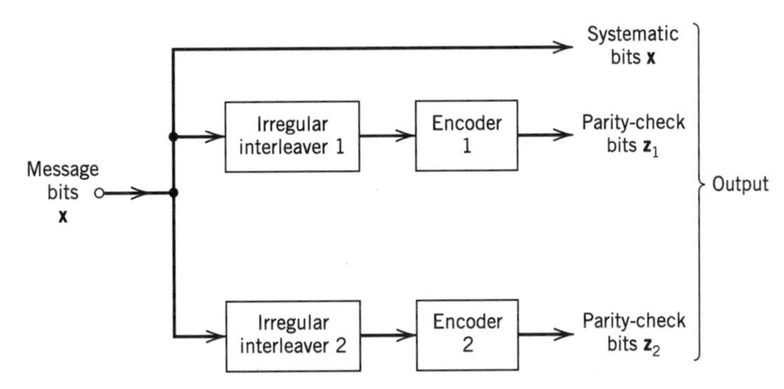


**FIGURE 10.32** Block diagram of irregular turbo encoder.

a single one. As shown in Figure 10.32, similar *irregular interleavers* are used in both convolutional encoding paths to generate the parity-check bits $\mathbf{z}_1$ and $\mathbf{z}_2$ in response to the message bits $\mathbf{x}$. Irregular turbo codes are decoded in a similar fashion to regular turbo codes.

To construct an *irregular LDPC code,*[23] the degrees of the variable and check nodes in the bipartite graph are chosen according to some distribution. For example, we may have an irregular LDPC code with the following graphical representation:

- One half of the variable nodes have degree 5 and the other half of the variable nodes have degree 3.
- One half of the check nodes have degree 6 and the other half of the check nodes have degree 8.

For a given block length and a given degree sequence, we define an ensemble of codes by choosing the edges (i.e., the connections between the variable and check nodes) in a random fashion. Specifically, the edges emanating from the variable nodes are enumerated in some arbitrary order, and likewise for the edges emanating from the check nodes.

Figure 10.33 plots the error performances of the following codes:[24]

- Irregular LDPC code: $k = 50{,}000$, $n = 100{,}000$, rate $= 1/2$
- Turbo code (regular): $k = 65{,}536$, $n = 131{,}072$, and rate $= 1/2$
- Irregular turbo code: $k = 65{,}536$, $n = 131{,}072$, and rate $= 1/2$

where $k$ is the number of message bits and $n$ is the block length. The generator polynomials for the two convolutional encoders in the regular/irregular turbo codes are as follows:

$$\text{Encoder 1: } g(D) = 1 + D^4$$
$$\text{Encoder 2: } g(D) = 1 + D + D^2 + D^3 + D^4$$

Figure 10.33 also includes the corresponding theoretical limit on channel capacity for code rate $r = 1/2$.

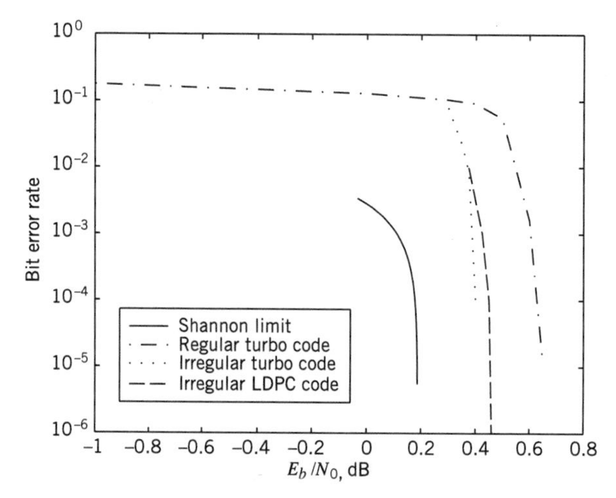


**FIGURE 10.33** Noise performances of regular turbo code, irregular turbo code and irregular low-density parity-check (LDPC) code, compared to the Shannon limit for code rate $r = 1/2$.

Based on the results presented in Figure 10.33, we may make the following observations:

- The irregular LDPC code outperforms the regular turbo code in that it comes closer to Shannon's theoretical limit by 0.175 dB.
- Among the three codes displayed therein, the irregular turbo code is the best in that it is just 0.213 dB away from Shannon's theoretical limit.

## 10.12 *Summary and Discussion*

In this chapter, we studied error-control coding techniques that have established themselves as indispensable tools for reliable digital communication over noisy channels. The effect of errors occurring during transmission is reduced by adding redundancy to the data prior to transmission in a controlled manner. The redundancy is used to enable a decoder in the receiver to detect and correct errors.

Error-control coding techniques may be divided into two broadly defined families:

1. *Algebraic codes*, which rely on abstract algebraic structure built into the design of the codes for decoding at the receiver. Algebraic codes include Hamming codes, maximal-length codes, BCH codes, and Reed-Solomon codes. These particular codes share two properties:

   *Linearity property*, the sum of any two code words in the code is also a code word.

   *Cyclic property*, any cyclic shift of a code word is also a code word in the code.

   Reed-Solomon codes are very powerful codes, capable of combatting both random and burst errors; they find applications in difficult environments such as deep-space communications and compact discs.
2. *Probabilistic codes*, which rely on probabilistic methods for their decoding at the receiver. Probabilistic codes include trellis codes, turbo codes, and low-density parity-check codes. In particular, the decoding is based on one or the other of two basic methods, as summarized here:

   *Soft input–hard output*, which is exemplified by the Viterbi algorithm that performs maximum likelihood sequence estimation in the decoding of trellis codes.

   *Soft input–soft output*, which is exemplified by the BCJR algorithm that performs maximum *a posteriori* estimation on a bit-by-bit basis in the decoding of turbo codes, or a special form of the sum-product algorithm in the decoding of low-density parity-check codes.

Trellis codes combine linear convolutional encoding and modulation to permit significant coding gains over conventional uncoded multilevel modulation without sacrificing bandwidth efficiency. Turbo codes and low-density parity-check codes share the following properties:

- Random encoding of a linear block kind.
- Error performance within a hair's breadth of Shannon's theoretical limit on channel capacity in a physically realizable fashion.

In practical terms, turbo codes and low-density parity-check codes have made it possible to achieve coding gains on the order of 10 dB, which is unmatched previously. These coding gains may be exploited to dramatically extend the range of digital communication receivers, substantially increase the bit rates of digital communication systems, or significantly decrease the transmitted signal energy per symbol. These benefits have significant implications for the design of wireless communications and deep-space communications, just to mention two important applications of digital communications. Indeed, turbo codes have already been standardized for use on deep-space communication links and wireless communication systems.

## NOTES AND REFERENCES

1. For an introductory discussion of error correction by coding, see Chapter 2 of Lucky (1989); see also the book by Adámek (1991), and the paper by Bhargava (1983). The classic book on error-control coding is Peterson and Weldon (1972). Error-control coding is also discussed in the classic book of Gallager (1968). The books of Lin and Costello (1983), Micheleson and Levesque (1985), MacWilliams and Sloane (1977), and Wilson (1998) are also devoted to error-control coding. For a collection of key papers on the development of coding theory, see the book edited by Berlekamp (1974).
2. For a survey of various ARQ schemes, see Lin, Costello, and Miller (1984).
3. In medicine, the term *syndrome* is used to describe a pattern of symptoms that aids in the diagnosis of a disease. In coding, the error pattern plays the role of the disease and parity-check failure that of a symptom. This use of *syndrome* was coined by Hagelbarger (1959).
4. The first error-correcting codes (known as Hamming codes) were invented by Hamming at about the same time as the conception of information theory by Shannon; for details, see the classic paper by Hamming (1950).
5. For a description of BCH codes and their decoding algorithms, see Lin and Costello (1983, pp. 141–183) and MacWilliams and Sloane (1977, pp. 257–293). Table 10.6 on binary BCH codes is adapted from Lin and Costello (1983).
6. The Reed-Solomon codes are named in honor of their inventors: see their classic 1960 paper. For details of Reed-Solomon codes, see MacWilliams and Sloane (1977, pp. 294–306). The book edited by Wicker and Bhargava (1994) contains an introduction to Reed-Solomon codes, a historical overview of these codes written by their inventors, Irving S. Reed and Gustave Solomon, and the applications of Reed-Solomon codes to the exploration of the solar system, the compact disc, automatic repeat-request protocols, and spread-spectrum multiple-access communications, and chapters on other related issues.
7. Convolutional codes were first introduced, as an alternative to block codes, by P. Elias (1955).
8. The term *trellis* was introduced by Forney (1973).
9. In a classic paper, Viterbi (1967) proposed a decoding algorithm for convolutional codes that has become known as the *Viterbi algorithm*. The algorithm was recognized by Forney (1972, 1973) to be a maximum likelihood decoder. Readable accounts of the Viterbi algorithm are presented in Lin and Costello (1983), Blahut (1990), and Adámek (1991).
10. Catastrophic convolutional codes are discussed in Benedetto, Biglieri, and Castellani (1987). Table 10.8 is adapted from their book.
11. For details of the evaluation of asymptotic coding gain for binary symmetric and binary-input AWGN channels, see Viterbi and Omura (1979, pp. 242–252) and Lin and Costello (1983, pp. 322–329).
12. Trellis-coded modulation was invented by G. Ungerboeck; its historical evolution is described in Ungerboeck (1982). Table 10.9 is adapted from this latter paper.

    Trellis-coded modulation may be viewed as a form of *signal-space coding*—a viewpoint discussed at an introductory level in Chapter 14 of the book by Lee and Messerschmitt (1994). For an extensive treatment of trellis-coded modulation, see the books by Biglieri, Divsalar, McLane, and Simon (1991), and Schlegel (1997).

13. Turbo codes were originated by C. Berrou and A. Glavieux. Work on these codes was motivated by two papers on error-correcting codes: Battail (1987), and Hagenauer and Hoecher (1989). The first description of turbo codes using heuristic arguments was presented at a conference paper by Berrou, Glavieux, and Thitimajshima (1993); see also Berrou and Glavieux (1996). For reflections on the early work on turbo codes and subsequent developments, see Berrou and Glavieux (1998).

    For a book on the basics of turbo codes, see Heegard and Wicker (1999). Using a procedure reminiscent of random coding (see Note 15), Benedetto and Montorosi (1996) have provided partial explanations for the impressive performance of turbo codes.

    In two independent studies reported in the papers by McEliece, MacKay, and Cheng (1998), and Kschischang and Frey (1998), it is shown that turbo decoding duplicates an algorithm in artificial intelligence due to Pearl (1982), which involves the propagation of belief. The term *belief* is another way of referring to *a posteriori* probability. These two papers have opened a new avenue of research, which links turbo decoding and learning machines. For an insightful discussion of turbo codes, see the book by Frey (1998).

    A pseudo-random interleaver is basic to the operation of turbo codes. Denenshgaran and Mondin (1999) present a systematic procedure for designing interleavers (i.e., permuters) for turbo codes.

14. The dual termination of turbo codes is discussed in Guinand and Lodge (1996).

15. Random coding is discussed in Cover and Thomas (1991), Section 8.7.

16. The plots presented in Fig. 10.27 follow those in Fig. 6.8 of the book by Frey (1998).

17. In the early 1960s, Baum and Welch derived an iterative procedure for solving the parameter estimation problem, hence the name *Baum-Welch algorithm* (Baum and Petrie (1966); Baum et al. (1970)). In the *BCJR algorithm*, named after Bahl, Cocke, Jelinek, and Raviv (1974), the Baum-Welch algorithm is applied to the problem of soft output, maximum likelihood decoding of convolutional codes.

18. The proof that the estimate $\hat{\mathbf{x}}$ in Eq. (10.76) approaches the MAP solution as the number of iterations approaches infinity is discussed in the paper by Moher and Gulliver (1998).

19. Low-density parity-check (LDPC) codes were originally discovered by Gallager (1962, 1963). They were rediscovered independently by MacKay and Neal (1995); see also MacKay (1999).

    In the 1960s and for a good while thereafter, the computers available at that time were not powerful enough to process the long block lengths that are needed to achieve excellent performance with LDPC codes, hence the lack of interest in their use for over twenty years.

20. For a detailed treatment of the statement that the probability distribution of the minimum distance of an LDPC code approaches a unit step function of the block length for certain values of weight-pair $(t_c, t_r)$, see Gallager (1962, 1963).

21. The decoding algorithm of LDPC codes described herein follows MacKay and Neal (1996, 1997).

22. Irregular turbo codes were invented by Frey and MacKay (1999).

23. Irregular LDPC codes were invented independently by MaKay et al. (1999) and Richardson et al. (1999).

24. The codes, whose performances are plotted in Fig. 10.34, are due to the following originators:
    - Regular turbo codes: Berrou and Glavieux (1996); Berrou et al. (1995).
    - Irregular turbo codes: Frey and MacKay (1999).
    - Irregular LDPC codes: Richardson et al. (1999).

# PROBLEMS

## Soft-Decision Coding

10.1 Consider a binary input $Q$-ary output discrete memoryless channel. The channel is said to be symmetric if the channel transition probability $p(j|i)$ satisfies the condition:

$$p(j|0) = p(Q - 1 - j|1), \qquad j = 0, 1, \ldots, Q - 1$$

Suppose that the channel input symbols 0 and 1 are equally likely. Show that the channel output symbols are also equally likely; that is,

$$p(j) = \frac{1}{Q}, \qquad j = 0, 1, \ldots, Q - 1$$

10.2 Consider the quantized demodulator for binary PSK signals shown in Fig. 10.3$a$. The quantizer is a four-level quantizer, normalized as in Fig. P10.2. Evaluate the transition probabilities of the binary input-quarternary output discrete memoryless channel so characterized. Hence, show that it is a symmetric channel. Assume that the transmitted signal energy per bit is $E_b$, and the additive white Gaussian noise has zero mean and power spectral density $N_0/2$.

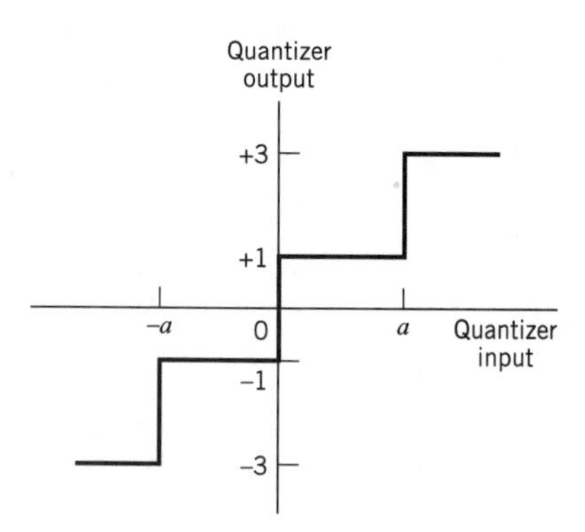


FIGURE P10.2

10.3 Consider a binary input AWGN channel, in which the binary symbols 1 and 0 are equally likely. The binary symbols are transmitted over the channel by means of phase-shift keying. The code symbol energy is $E$, and the AWGN has zero mean and power spectral density $N_0/2$. Show that the channel transition probability is given by

$$p(y|0) = \frac{1}{\sqrt{2\pi}} \exp\left[-\frac{1}{2}\left(y + \sqrt{\frac{2E}{N_0}}\right)^2\right], \qquad -\infty < y < \infty$$

## Linear Block and Cyclic Codes

10.4 In a *single-parity-check code*, a single parity bit is appended to a block of $k$ message bits $(m_1, m_2, \ldots, m_k)$. The single parity bit $b_1$ is chosen so that the code word satisfies the *even parity rule*:

$$m_1 + m_2 + \cdots + m_k + b_1 = 0, \qquad \text{mod } 2$$

For $k = 3$, set up the $2^k$ possible code words in the code defined by this rule.

10.5 Compare the parity-check matrix of the (7, 4) Hamming code considered in Example 10.2 with that of a (4, 1) repetition code.

**10.6** Consider the (7, 4) Hamming code of Example 10.2. The generator matrix G and the parity-check matrix H of the code are described in that example. Show that these two matrices satisfy the condition

$$\mathbf{H}\mathbf{G}^T = \mathbf{0}$$

**10.7** (a) For the (7, 4) Hamming code described in Example 10.2, construct the eight code words in the dual code.

(b) Find the minimum distance of the dual code determined in part (a).

**10.8** Consider the (5, 1) repetition code of Example 10.1. Evaluate the syndrome s for the following error patterns:

(a) All five possible single-error patterns

(b) All 10 possible double-error patterns

**10.9** For an application that requires error detection *only*, we may use a *nonsystematic* code. In this problem, we explore the generation of such a cyclic code. Let $g(X)$ denote the generator polynomial, and $m(X)$ denote the message polynomial. We define the code polynomial $c(X)$ simply as

$$c(X) = m(X)g(X)$$

Hence, for a given generator polynomial, we may readily determine the code words in the code. To illustrate this procedure, consider the generator polynomial for a (7, 4) Hamming code:

$$g(X) = 1 + X + X^3$$

Determine the 16 code words in the code, and confirm the nonsystematic nature of the code.

**10.10** The polynomial $1 + X^7$ has $1 + X + X^3$ and $1 + X^2 + X^3$ as primitive factors. In Example 10.3, we used $1 + X + X^3$ as the generator polynomial for a (7, 4) Hamming code. In this problem, we consider the adoption of $1 + X^2 + X^3$ as the generator polynomial. This should lead to a (7, 4) Hamming code that is different from the code analyzed in Example 10.3. Develop the encoder and syndrome calculator for the generator polynomial:

$$g(X) = 1 + X^2 + X^3$$

Compare your results with those in Example 10.3.

**10.11** Consider the (7, 4) Hamming code defined by the generator polynomial

$$g(X) = 1 + X + X^3$$

The code word 0111001 is sent over a noisy channel, producing the received word 0101001 that has a single error. Determine the syndrome polynomial $s(X)$ for this received word, and show that it is identical to the error polynomial $e(X)$.

**10.12** The generator polynomial of a (15, 11) Hamming code is defined by

$$g(X) = 1 + X + X^4$$

Develop the encoder and syndrome calculator for this code, using a systematic form for the code.

**10.13** Consider the (15, 4) maximal-length code that is the dual of the (15, 11) Hamming code of Problem 10.12. Do the following:

(a) Find the feedback connections of the encoder, and compare your results with those of Table 7.1 on maximal-length codes presented in Chapter 7.

(b) Find the generator polynomial $g(X)$; hence, determine the output sequence assuming the initial state 0001. Confirm the validity of your result by cycling the initial state through the encoder.

10.14 Consider the (31, 15) Reed-Solomon code.

(a) How many bits are there in a symbol of the code?

(b) What is the block length in bits?

(c) What is the minimum distance of the code?

(d) How many symbols in error can the code correct?

## Convolutional Codes

10.15 A convolutional encoder has a single-shift register with two stages, (i.e., constraint length $K = 3$), three modulo-2 adders, and an output multiplexer. The generator sequences of the encoder are as follows:

$$g^{(1)} = (1, 0, 1)$$
$$g^{(2)} = (1, 1, 0)$$
$$g^{(3)} = (1, 1, 1)$$

Draw the block diagram of the encoder.

*Note: For Problems* 10.16–10.23, *the same message sequence* 10111 . . . *is used so that we may compare the outputs of different encoders for the same input.*

10.16 Consider the rate $r = 1/2$, constraint length $K = 2$ convolutional encoder of Fig. P10.16. The code is systematic. Find the encoder output produced by the message sequence 10111. . . .

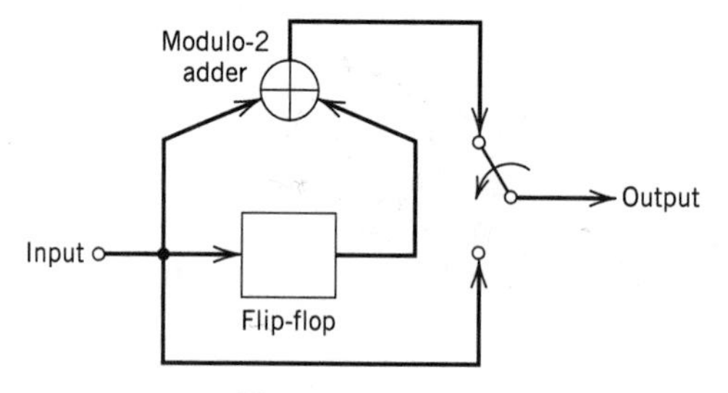


FIGURE P10.16

10.17 Figure P10.17 shows the encoder for a rate $r = 1/2$, constraint length $K = 4$ convolutional code. Determine the encoder output produced by the message sequence 10111. . . .

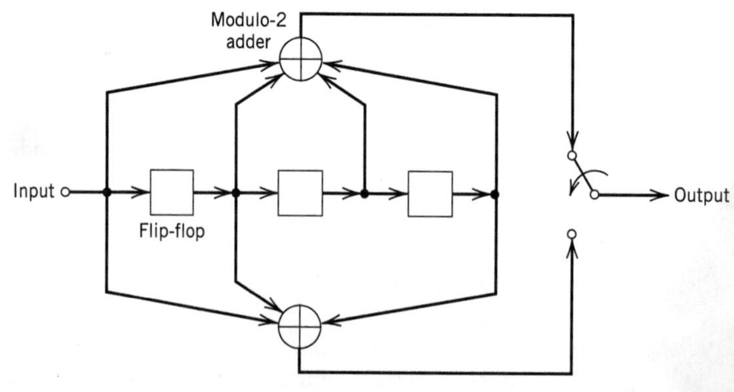


FIGURE P10.17

**10.18** Consider the encoder of Fig. 10.13*b* for a rate $r = 2/3$, constraint length $K = 2$ convolutional code. Determine the code sequence produced by the message sequence 10111. . . .

**10.19** Construct the code tree for the convolutional encoder of Fig. P10.16. Trace the path through the tree that corresponds to the message sequence 10111 . . . , and compare the encoder output with that determined in Problem 10.16.

**10.20** Construct the code tree for the encoder of Fig. P10.17. Trace the path through the tree that corresponds to the message sequence 10111. . . . Compare the resulting encoder output with that found in Problem 10.17.

**10.21** Construct the trellis diagram for the encoder of Fig. P10.17, assuming a message sequence of length 5. Trace the path through the trellis corresponding to the message sequence 10111. . . . Compare the resulting encoder output with that found in Problem 10.17.

**10.22** Construct the state diagram for the encoder of Fig. P10.17. Starting with the all-zero state, trace the path that corresponds to the message sequence 10111 . . . , and compare the resulting code sequence with that determined in Problem 10.17.

**10.23** Consider the encoder of Fig. 10.13*b*.

(a) Construct the state diagram for this encoder.

(b) Starting from the all-zero state, trace the path that corresponds to the message sequence 10111. . . . Compare the resulting sequence with that determined in Problem 10.18.

**10.24** By viewing the minimum shift keying (MSK) scheme as a finite-state machine, construct the trellis diagram for MSK. (A description of MSK is presented in Chapter 6.)

**10.25** The trellis diagram of a rate-1/2, constraint length-3 convolutional code is shown in Figure P10.25. The all-zero sequence is transmitted, and the received sequence is 100010000. . . . Using the Viterbi algorithm, compute the decoded sequence.

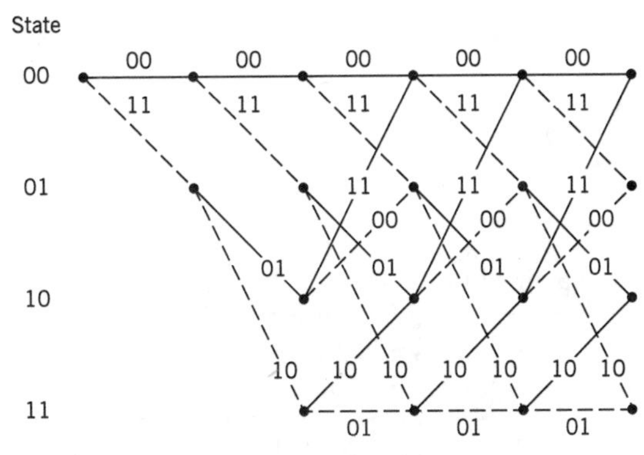


**FIGURE P10.25**

**10.26** Consider a rate-1/2, constraint length-7 convolutional code with free distance $d_{\text{free}} = 10$. Calculate the asymptotic coding gain for the following two channels:

(a) Binary symmetric channel

(b) Binary input AWGN channel

**10.27** In Section 10.6 we described the Viterbi algorithm for maximum likelihood decoding of a convolutional code. Another application of the Viterbi algorithm is for maximum likelihood demodulation of a received sequence corrupted by intersymbol interference due to a dispersive channel. Figure P10.27 shows the trellis diagram for intersymbol interference, assuming a binary data sequence. The channel is discrete, described by the finite impulse response (1, 0.1). The received sequence is (1.0, −0.3, −0.7, 0, . . .). Use the Viterbi algorithm to determine the maximum likelihood decoded version of this sequence.

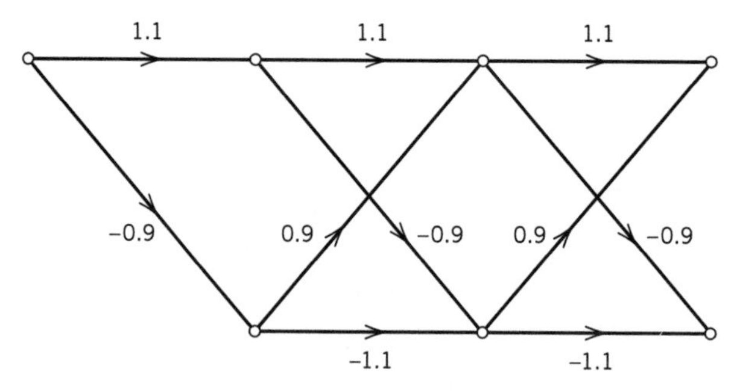


**FIGURE P10.27**

10.28 Figure P10.28 depicts 32-QAM cross constellation. Partition this constellation into eight subsets. At each stage of the partitioning, indicate the within-subset (shortest) Euclidean distance.

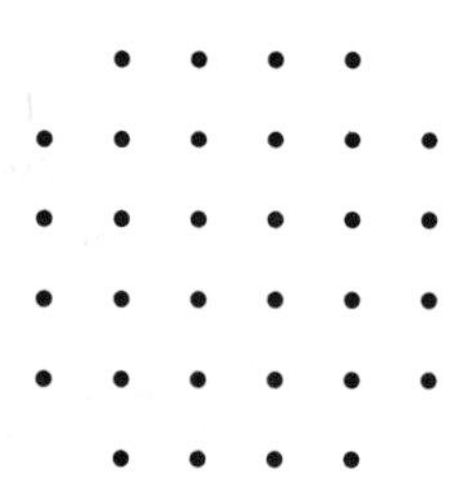

**FIGURE P10.28**

10.29 As explained in the Introduction to this chapter, channel coding can be used to reduce the $E_b/N_0$ required for a prescribed error performance or reduce the size of the receiving antenna for a prescribed $E_b/N_0$. In this problem we explore these two practical benefits of coding by revisiting Example 8.2 in Chapter 8 on the downlink power calculations for a domestic satellite communication system. In particular, we now assume that the design of the downlink includes the use of a coding scheme consisting of a rate-1/2 convolutional encoder with length $K = 7$ and Viterbi decoding. The coding gain of this scheme is 5.1 dB, assuming the use of soft quantization. Hence do the following:

(a) Recalculate the required $E_b/N_0$ ratio of the system.

(b) Assuming that the required $E_b/N_0$ ratio remains unchanged, calculate the reduction in the size of the receiving dish antenna that is made possible by the use of this coding scheme in the downlink.

10.30 Unlike the convolutional codes considered in this chapter, we recall from Chapter 6 that the convolutional code used in the voiceband modem V.32 modem is *nonlinear*. Figure P10.30 shows the circuit diagram of the convolutional encoder used in this modem; it uses modulo-2 multiplication and gates in addition to modulo-2 additions and delays. Explain the reason for nonlinearity of the encoder in Fig. P10.30, and use an example to illustrate your explanation.

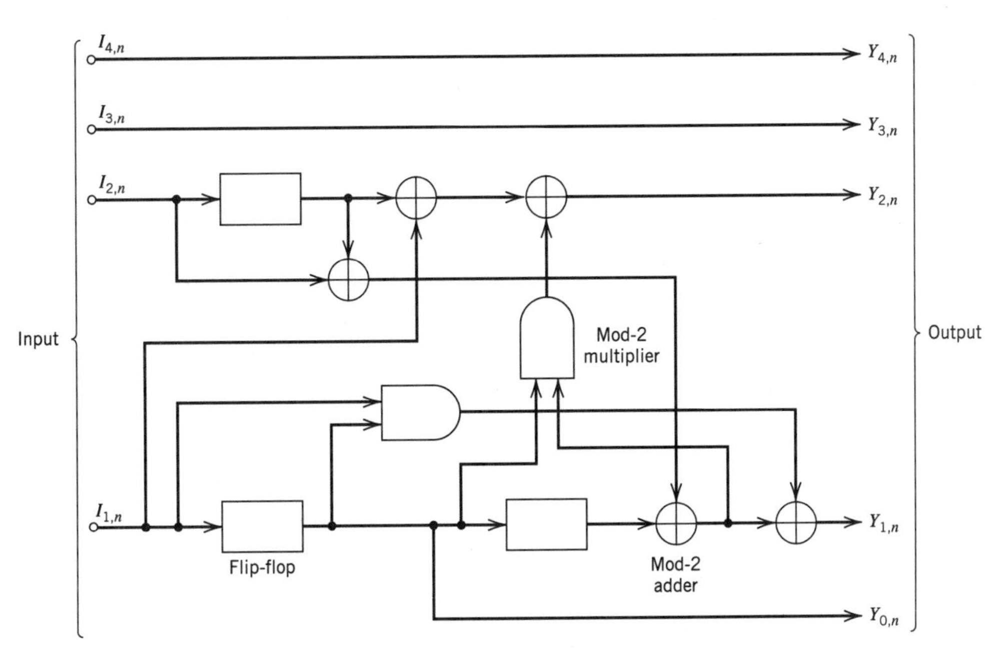


**FIGURE P10.30**

## Turbo Codes

**10.31** Let $r_c^{(1)} = p/q_1$ and $r_c^{(2)} = p/q_2$ be the code rates of RSC encoders 1 and 2 in the turbo encoder of Fig. 10.25. Find the code rate of the turbo code.

**10.32** The feedback nature of the constituent codes in the turbo encoder of Fig. 10.25 has the following implication: A single bit error corresponds to an infinite sequence of channel errors. Illustrate this phenomenon by using a message sequence consisting of symbol 1 followed by an infinite number of symbols 0.

**10.33** Consider the following generator matrices for rate 1/2 turbo codes:

$$\text{4-state encoder:} \qquad \mathbf{g}(D) = \left[1, \frac{1 + D + D^2}{1 + D^2}\right]$$

$$\text{8-state encoder:} \qquad \mathbf{g}(D) = \left[1, \frac{1 + D^2 + D^3}{1 + D + D^2 + D^3}\right]$$

$$\text{16-state encoder:} \qquad \mathbf{g}(D) = \left[1, \frac{1 + D^4}{1 + D + D^2 + D^3 + D^4}\right]$$

(a) Construct the block diagram for each one of these RSC encoders.

(b) Setup the parity-check equation associated with each encoder.

**10.34** The turbo encoder of Fig. 10.25 involves the use of two RSC encoders.

(a) Generalize this encoder to encompass a total of $M$ interleavers.

(b) Construct the block diagram of the turbo decoder that exploits the $M$ sets of parity-check bits generated by such a generalization.

**10.35** Turbo decoding relies on the feedback of extrinsic information. The fundamental principle adhered to in the turbo decoder is to avoid feeding a decoding stage information that stems from the stage itself. Explain the justification for this principle in conceptual terms.

**10.36** Suppose a communication receiver consists of two components, a demodulator and a decoder. The demodulator is based on a Markov model of the combined modulator and channel, and the decoder is based on a Markov model of a forward error correction code. Discuss how the turbo principle may be applied to construct a joint demodulator/decoder for this system.

### Computer Experiment

10.37 In this experiment we continue the investigation into turbo codes presented in Section 10.9 by evaluating the effect of block size on the noise performance of the decoder.

As before, the two convolutional encoders of the turbo encoder are as follows:

Encoder 1: [1, 1, 1]

Encoder 2: [1, 0, 1]

The transmitted $E_b/N_0$ is 1 dB. The block errors to termination are prescribed not to exceed 15.

With this background information, plot the bit error rate of the turbo decoder versus the number of iterations for two different block (i.e., interleaver) sizes: 200 and 400.

# APPENDIX 1

# PROBABILITY THEORY

## A1.1 *Probabilistic Concepts*

*Probability theory* is rooted in phenomena that, explicitly or implicitly, can be modeled by an experiment with an outcome that is subject to *chance*. Moreover, if the experiment is repeated, the outcome can differ because of the influence of an underlying random phenomenon or chance mechanism. Such an experiment is referred to as a *random experiment*. For example, the experiment may be the observation of the result of tossing a fair coin. In this experiment, the possible outcomes of a trial are "heads" or "tails."

To be more precise in the description of a random experiment, we ask for three features:

1. The experiment is repeatable under identical conditions.
2. On any trial of the experiment, the outcome is unpredictable.
3. For a large number of trials of the experiment, the outcomes exhibit *statistical regularity*; that is, a definite *average* pattern of outcomes is observed if the experiment is repeated a large number of times.

### ■ RELATIVE-FREQUENCY APPROACH

Let *event A* denote one of the possible outcomes of a random experiment. For example, in the coin-tossing experiment, event $A$ may represent "heads." Suppose that in $n$ trials of the experiment, event $A$ occurs $N_n(A)$ times. We may then assign the ratio $N_n(A)/n$ to the event $A$. This ratio is called the *relative frequency* of the event $A$. Clearly, the relative frequency is a *nonnegative real number less than or equal to one*. That is to say,

$$0 \leq \frac{N_n(A)}{n} \leq 1 \tag{A1.1}$$

If event $A$ occurs in none of the trials, $N_n(A)/n = 0$. If, on the other hand, event $A$ occurs in all the $n$ trials, $N_n(A)/n = 1$.

We say that the experiment exhibits *statistical regularity* if for *any* sequence of $n$ trials the relative frequency $N_n(A)/n$ converges to the same limit as $n$ becomes large. It thus seems natural for us to define the *probability of event A* as

$$P(A) = \lim_{n\to\infty}\left(\frac{N_n(A)}{n}\right) \tag{A1.2}$$

The limit shown in Equation (A1.2) should not be viewed in a mathematical sense. Rather, we think of Equation (A1.2) as a statement that the probability of an event is the long-term proportion of times that a particular event $A$ occurs in a long sequence of trials. For example, in the coin-tossing experiment, we may expect that out of a million tosses of a fair coin, about one half of them will show up heads.

The probability of an event is intended to represent the *likelihood* that a trial of the experiment will result in the occurrence of that event. For many engineering applications and games of chance, the use of Equation (A1.2) to define the probability of an event is acceptable. However, for many other applications, this definition is inadequate. Consider, for example, the statistical analysis of the stock market: How are we to achieve repeatability of such an experiment? A more satisfying approach is to state the properties that any measure of probability is expected to have, postulate them as *axioms*, and then use relative-frequency interpretations to justify them.

## ■ AXIOMS OF PROBABILITY

When we perform a random experiment, it is natural for us to be aware of the various outcomes that are likely to arise. In this context, it is convenient to think of an experiment and its possible outcomes as defining a space and its points. With the $k$th outcome of the experiment, say, we associate a point called the *sample point*, which we denote by $s_k$. The totality of sample points corresponding to the aggregate of all possible outcomes of the experiment is called the *sample space*, which we denote by $\mathsf{S}$. An event corresponds to either a single sample point or a set of sample points. In particular, the entire sample space $\mathsf{S}$ is called the *sure event*; the null set $\varnothing$ is called the *null* or *impossible event*; and a single sample point is called an *elementary event*.

Consider, for example, an experiment that involves the throw of a die. In this experiment there are six possible outcomes: the showing of one, two, three, four, five, and six dots on the upper face of the die. By assigning a sample point to each of these possible outcomes, we have a one-dimensional sample space that consists of six sample points, as shown in Figure A1.1. The elementary event describing the statement "a six shows" corresponds to the sample point {6}. On the other hand, the event describing the statement "an even number of dots shows" corresponds to the subset {2, 4, 6} of the sample space. Note that the term *event* is used interchangeably to describe the subset or the statement.

We are now ready to make a formal definition of probability. A *probability system* consists of the triple:

1. *A sample space* $\mathsf{S}$ *of elementary events (outcomes).*
2. *A class* $\mathscr{E}$ *of events* that are subsets of $\mathsf{S}$.
3. *A probability measure* $P(\cdot)$ assigned to each event $A$ in the class $\mathscr{E}$, which has the following properties:

   (i)
$$P(\mathsf{S}) = 1 \tag{A1.3}$$

   (ii)
$$0 \leq P(A) \leq 1 \tag{A1.4}$$

   (iii) If $A + B$ is the *union of two mutually exclusive events* in the class $\mathscr{E}$, then
$$P(A + B) = P(A) + P(B) \tag{A1.5}$$

Properties (i), (ii), and (iii) are known as the *axioms of probability*. Axiom (i) states that the probability of the sure event is unity. Axiom (ii) states that the probability of an event

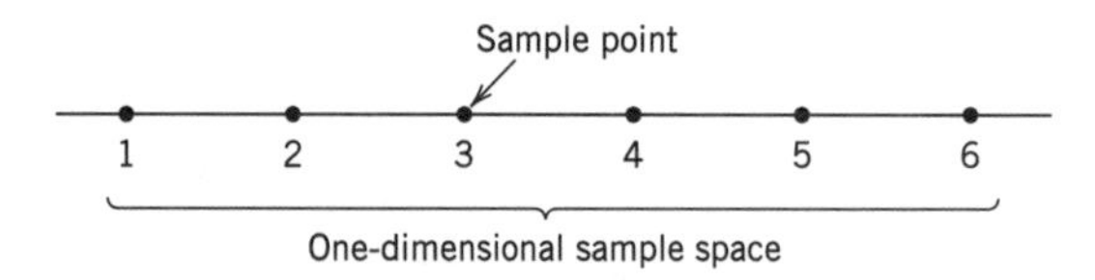


**FIGURE A1.1** Sample space for the experiment of throwing a die.

is a nonnegative real number that is less than or equal to unity. Axiom (iii) states that the probability of the union of two mutually exclusive events is the sum of the probabilities of the individual events. These three axioms are sufficient to deal with experiments with finite sample spaces.

Although the axiomatic approach to probability theory is abstract in nature, all three axioms have relative-frequency interpretations of their own. Axiom (ii) corresponds to Equation (A1.1). Axiom (i) corresponds to the limiting case of Equation (A1.1) when the event $A$ occurs in all the $n$ trials. To interpret axiom (iii), we note that if event $A$ occurs $N_n(A)$ times in $n$ trials and event $B$ occurs $N_n(B)$ times, then the union event "A or B" occurs in $N_n(A) + N_n(B)$ trials (since $A$ and $B$ can never occur on the same trial). Hence, $N_n(A + B) = N_n(A) + N_n(B)$, and so we have

$$\frac{N_n(A + B)}{n} = \frac{N_n(A)}{n} + \frac{N_n(B)}{n}$$

which has a mathematical form similar to that of axiom (iii).

Axioms (i), (ii), and (iii) constitute an implicit definition of probability. We may use these axioms to develop some other basic properties of probability, as described next.

**Property 1**

$$P(\overline{A}) = 1 - P(A) \tag{A1.6}$$

*where $\overline{A}$ (denoting "not A") is the complement of event A.*

The use of this property helps us investigate the *nonoccurrence of an event*. To prove it, we express the sample space $\mathsf{S}$ as the union of two mutually exclusive events $A$ and $\overline{A}$:

$$\mathsf{S} = A + \overline{A}$$

Then, the use of axioms (i) and (iii) yields

$$1 = P(A) + P(\overline{A})$$

from which Equation (A1.6) follows directly.

**Property 2**

*If M mutually exclusive events $A_1, A_2, \ldots, A_M$ have the exhaustive property*

$$A_1 + A_2 \cdots + A_M = \mathsf{S} \tag{A1.7}$$

*then*

$$P(A_1) + P(A_2) + \cdots + P(A_M) = 1 \tag{A1.8}$$

To prove this property, we first use axiom (i) in Equation (A1.7), and so write

$$P(A_1 + A_2 + \cdots + A_M) = 1$$

Next, we generalize axiom (iii) by writing

$$P(A_1 + A_2 + \cdots + A_M) = P(A_1) + P(A_2) + \cdots + P(A_M)$$

Hence, the result of Equation (A1.8) follows. When the $M$ events are *equally likely* (i.e., they have equal probabilities of occurrence), then Equation (A1.8) simplifies as

$$P(A_i) = \frac{1}{M}, \qquad i = 1, 2, \ldots, M$$

**Property 3**

*When events A and B are not mutually exclusive, then the probability of the union event "A or B" equals*

$$P(A + B) = P(A) + P(B) - P(AB) \tag{A1.9}$$

*where P(AB) is the probability of the joint event "A and B."*

The probability $P(AB)$ is called a *joint probability*. It has the following relative-frequency interpretation:

$$P(AB) = \lim_{n\to\infty}\left(\frac{N_n(AB)}{n}\right) \tag{A1.10}$$

where $N_n(AB)$ denotes the number of times the events $A$ and $B$ occur simultaneously in $n$ trials of the experiment. Axiom (iii) is a special case of Equation (A1.9); when $A$ and $B$ are mutually exclusive, $P(AB)$ is zero, and Equation (A1.9) reduces to the same form as Equation (A1.5).

## ▪ CONDITIONAL PROBABILITY

Suppose we perform an experiment that involves a pair of events $A$ and $B$. Let $P(B|A)$ denote the probability of event $B$, given that event $A$ has occurred. The probability $P(B|A)$ is called the *conditional probability of B given A*. Assuming that $A$ has nonzero probability, the conditional probability $P(B|A)$ is defined by

$$P(B|A) = \frac{P(AB)}{P(A)} \tag{A1.11}$$

where $P(AB)$ is the joint probability of $A$ and $B$.

We justify the definition of conditional probability given in Equation (A1.11) by presenting a relative-frequency interpretation of it. Suppose that we perform an experiment and examine the occurrence of a pair of events $A$ and $B$. Let $N_n(AB)$ denote the number of times the joint event $AB$ occurs in $n$ trials. Suppose that in the same $n$ trials, the event $A$ occurs $N_n(A)$ times. Since the joint event $AB$ corresponds to both $A$ and $B$ occurring, it follows that $N_n(A)$ must include $N_n(AB)$. In other words, we have

$$\frac{N_n(AB)}{N_n(A)} \leq 1$$

The ratio $N_n(AB)/N_n(A)$ represents the relative frequency of $B$ given that $A$ has occurred. For large $n$, this ratio equals the conditional probability $P(B|A)$; that is,

$$P(B|A) = \lim_{n\to\infty}\left(\frac{N_n(AB)}{N_n(A)}\right)$$

or, equivalently,

$$P(B|A) = \lim_{n\to\infty}\left(\frac{N_n(AB)/n}{N_n(A)/n}\right)$$

Recognizing that

$$P(AB) = \lim_{n\to\infty}\left(\frac{N_n(AB)}{n}\right)$$

and

$$P(A) = \lim_{n\to\infty}\left(\frac{N_n(A)}{n}\right)$$

the result of Equation (A1.11) follows.

We may rewrite Equation (A1.11) as

$$P(AB) = P(B|A)P(A) \tag{A1.12}$$

It is apparent that we may also write

$$P(AB) = P(A|B)P(B) \tag{A1.13}$$

Accordingly, we may state that *the joint probability of two events may be expressed as the product of the conditional probability of one event given the other, and the elementary probability of the other*. Note that the conditional probabilities $P(B|A)$ and $P(A|B)$ have essentially the same properties as the various probabilities previously defined.

Situations may exist where the conditional probability $P(A|B)$ and the probabilities $P(A)$ and $P(B)$ are easily determined directly, but the conditional probability $P(B|A)$ is desired. From Equations (A1.12) and (A1.13), it follows that, provided $P(A) \neq 0$, we may determine $P(B|A)$ by using the relation

$$P(B|A) = \frac{P(A|B)P(B)}{P(A)} \tag{A1.14}$$

This relation is a special form of *Bayes' rule*.

Suppose that the conditional probability $P(B|A)$ is simply equal to the elementary probability of occurrence of event $B$, that is,

$$P(B|A) = P(B)$$

Under this condition, the probability of occurrence of the joint event $AB$ is equal to the product of the elementary probabilities of the events $A$ and $B$:

$$P(AB) = P(A)P(B)$$

so that

$$P(A|B) = P(A)$$

That is, the conditional probability of event $A$, assuming the occurrence of event $B$, is simply equal to the elementary probability of event $A$. We thus see that in this case a knowledge of the occurrence of one event tells us no more about the probability of occurrence of the other event than we knew without that knowledge. Events $A$ and $B$ that satisfy this condition are said to be *statistically independent*.

## A1.2 *Random Variables*

It is customary, particularly when using the language of sample space, to think of the outcome of an experiment as a variable that can wander over the set of sample points and whose value is determined by the experiment. *A function whose domain is a sample space and whose range is some set of real numbers is called a random variable of the experiment.* However, the term *random variable* is somewhat confusing. First, the word *random* is not used in the sense of equal probability of occurrence, for which it should be reserved. Second, the word *variable* does not imply dependence (on the experimental outcome), which is an essential part of the meaning. Nevertheless, the term is so deeply imbedded in the literature of probability that its usage has persisted.

When the outcome of an experiment is $s$, the random variable is denoted as $X(s)$ or simply $X$. For example, the sample space representing the outcomes of the throw of a die is a set of six sample points that may be taken to be the integers $1, 2, \ldots, 6$. Then if we identify the sample point $k$ with the event that $k$ dots show when the die is thrown, the function $X(k) = k$ is a random variable such that $X(k)$ equals the number of dots that show when the die is thrown. In this example, the random variable takes only a discrete set of values. In such a case, we say that we are dealing with a *discrete random variable.* More precisely, the random variable $X$ can take only a finite number of values in any finite observation interval. If, however, the random variable $X$ can take any value in a whole observation interval, $X$ is called a *continuous random variable.* For example, the random variable that represents the amplitude of a noise voltage at a particular instant of time is a continuous random variable because it may take any value between plus and minus infinity.

To proceed further, we need a probabilistic description of random variables that works equally well for discrete as well as continuous random variables. Let us consider the random variable $X$ and the probability of the event $X \leq x$. We denote this probability by $P(X \leq x)$. It is apparent that this probability is a function of the *dummy variable* $x$. To simplify the notation, we write

$$F_X(x) = P(X \leq x) \tag{A1.15}$$

The function $F_X(x)$ is called the *cumulative distribution function* (cdf) or simply the *distribution function* of the random variable $X$. Note that $F_X(x)$ is a function of $x$, not of the random variable $X$. However, it depends on the assignment of the random variable $X$, which accounts for the use of $X$ as subscript. For any point $x$, the distribution function $F_X(x)$ expresses a probability.

The distribution function $F_X(x)$ has the following properties, which follow directly from Equation (A1.15):

1. The distribution function $F_X(x)$ is bounded between zero and one.
2. The distribution function $F_X(x)$ is a nondecreasing function of $x$; that is,

$$F_X(x_1) \leq F_X(x_2) \qquad \text{if } x_1 < x_2 \tag{A1.16}$$

An alternative description of the probability of the random variable $X$ is often useful. This is the derivative of the distribution function, as shown by

$$f_X(x) = \frac{d}{dx} F_X(x) \tag{A1.17}$$

which is called the *probability density function* (pdf) of the random variable X. Note that the differentiation in Equation (A1.17) is with respect to the dummy variable $x$. The name, density function, arises from the fact that the probability of the event $x_1 < X \leq x_2$ equals

$$\begin{aligned} P(x_1 < X \leq x_2) &= P(X \leq x_2) - P(X \leq x_1) \\ &= F_X(x_2) - F_X(x_1) \\ &= \int_{x_1}^{x_2} f_X(x)\, dx \end{aligned} \tag{A1.18}$$

The probability of an interval is therefore the area under the probability density function in that interval. Putting $x_1 = -\infty$ in Equation (A1.18), and changing the notation somewhat, we readily see that the distribution function is defined in terms of the probability density function as follows:

$$F_X(x) = \int_{-\infty}^{x} f_X(\xi)\, d\xi \tag{A1.19}$$

Since $F_X(\infty) = 1$, corresponding to the probability of a certain event, and $F_X(-\infty) = 0$, corresponding to the probability of an impossible event, we readily find from Equation (A1.18) that

$$\int_{-\infty}^{\infty} f_X(x)\, dx = 1 \tag{A1.20}$$

Earlier we mentioned that a distribution function must always be nondecreasing. This means that its derivative or the probability density function must always be nonnegative. Accordingly, we may state that a *probability density function must always be a nonnegative function, and with a total area of one.*

Thus far we have focused attention on situations involving a single random variable. However, we find frequently that the outcome of an experiment requires several random variables for its description. We now consider situations involving two random variables. The probabilistic description developed in this way may be readily extended to any number of random variables.

Consider two random variables X and Y. We define *the joint distribution function $F_{X,Y}(x, y)$ as the probability that the random variable X is less than or equal to a specified value x and that the random variable Y is less than or equal to a specified value y.* The variables X and Y may be two separate one-dimensional random variables or the components of a single two-dimensional random variable. In either case, the joint sample space is the $xy$-plane. The joint distribution function $F_{X,Y}(x, y)$ is the probability that the outcome of an experiment will result in a sample point lying inside the quadrant $(-\infty < X \leq x, -\infty < Y \leq y)$ of the joint sample space. That is,

$$F_{X,Y}(x, y) = P(X \leq x, Y \leq y) \tag{A1.21}$$

Suppose that the joint distribution function $F_{X,Y}(x, y)$ is continuous everywhere, and that the partial derivative

$$f_{X,Y}(x, y) = \frac{\partial^2 F_{X,Y}(x, y)}{\partial x \partial y} \tag{A1.22}$$

exists and is continuous everywhere. We call the function $f_{X,Y}(x, y)$ the *joint probability density function* of the random variables X and Y. The joint distribution function

$F_{X,Y}(x, y)$ is a nondecreasing function of both $x$ and $y$. Therefore, from Equation (A1.22) it follows that the joint probability density function $f_{X,Y}(x, y)$ is always nonnegative. Also the total volume under the graph of a joint probability density function must be unity, as shown by

$$\int_{-\infty}^{\infty} \int_{-\infty}^{\infty} f_{X,Y}(\xi, \eta)\, d\xi\, d\eta = 1 \tag{A1.23}$$

The probability density function for a single random variable ($X$, say) can be obtained from its joint probability density function with a second random variable ($Y$, say) in the following way. We first note that

$$F_X(x) = \int_{-\infty}^{\infty} \int_{-\infty}^{x} f_{X,Y}(\xi, \eta)\, d\xi\, d\eta \tag{A1.24}$$

Therefore, differentiating both sides of Equation (A1.24) with respect to $x$, we get the desired relation:

$$f_X(x) = \int_{-\infty}^{\infty} f_{X,Y}(x, \eta)\, d\eta \tag{A1.25}$$

Thus the probability density function $f_X(x)$ is obtained from the joint probability density function $f_{X,Y}(x, y)$ by simply integrating it over all possible values of the undesired random variable $Y$. The use of similar arguments in the other dimension yields $f_Y(y)$. The probability density functions $f_X(x)$ and $f_Y(y)$ are called *marginal densities*. Hence, the joint probability density function $f_{X,Y}(x, y)$ contains all the possible information about the joint random variables $X$ and $Y$.

Suppose that $X$ and $Y$ are two continuous random variables with joint probability density function $f_{X,Y}(x, y)$. The *conditional probability density function* of $Y$ given that $X = x$ is defined by

$$f_Y(y|x) = \frac{f_{X,Y}(x, y)}{f_X(x)} \tag{A1.26}$$

provided that $f_X(x) > 0$, where $f_X(x)$ is the marginal density of $X$. The function $f_Y(y|x)$ may be thought of as a function of the variable $y$, with the variable $x$ arbitrary, but *fixed*. Accordingly, it satisfies all the requirements of an ordinary probability density function, as shown by

$$f_Y(y|x) \geq 0$$

and

$$\int_{-\infty}^{\infty} f_Y(y|x)\, dy = 1$$

If the random variables $X$ and $Y$ are *statistically independent*, then knowledge of the outcome of $X$ can in no way affect the distribution of $Y$. The result is that the conditional probability density function $f_Y(y|x)$ reduces to the marginal density $f_Y(y)$, as shown by

$$f_Y(y|x) = f_Y(y)$$

In such a case, we may express the joint probability density function of the random variables $X$ and $Y$ as the product of their respective marginal densities, as shown by

$$f_{X,Y}(x, y) = f_X(x) f_Y(y)$$

In words, we may state that if the joint probability density function of the random variables $X$ and $Y$ equals the product of their marginal densities, then $X$ and $Y$ are statistically independent.

## A1.3 *Statistical Averages*

Having discussed probability and some of its ramifications, we now seek ways for determining the *average* behavior of the outcomes arising in random experiments.

The *expected value* or *mean* of a random variable $X$ is defined by

$$\mu_X = E[X] = \int_{-\infty}^{\infty} x f_X(x)\, dx \tag{A1.27}$$

where $E$ denotes the *statistical expectation operator*. That is, the mean $\mu_X$ locates the center of gravity of the area under the probability density curve of the random variable $X$. To interpret the expected value $\mu_X$, we write the integral in the defining Equation (A1.27) as the limit of an approximating sum formulated as follows. Let $\{x_k | k = 0, \pm 1, \pm 2, \ldots\}$ denote a set of uniformly spaced points on the real line:

$$x_k = \left(k + \frac{1}{2}\right)\Delta, \qquad k = 0, \pm 1, \pm 2, \ldots \tag{A1.28}$$

where $\Delta$ is the spacing between adjacent points. We may then rewrite Equation (A1.27) as the limiting form of a sum:

$$\begin{aligned} E[X] &= \lim_{\Delta \to 0} \sum_{k=-\infty}^{\infty} \int_{k\Delta}^{(k+1)\Delta} x_k f_X(x)\, dx \\ &= \lim_{\Delta \to 0} \sum_{k=-\infty}^{\infty} x_k P\left(x_k - \frac{\Delta}{2} < X \le x_k + \frac{\Delta}{2}\right) \end{aligned} \tag{A1.29}$$

For a physical interpretation of the sum on the right-hand side of Equation (A1.29), suppose that we make $n$ independent observations of the random variable $X$. Let $N_n(k)$ denote the number of times that the random variable $X$ falls inside the $k$th bin:

$$x_k - \frac{\Delta}{2} < X \le x_k + \frac{\Delta}{2}, \qquad k = 0, \pm 1, \pm 2, \ldots$$

Then, as the number of observations, $n$, is made large, the ratio $N_n(k)/n$ approaches the probability $P(x_k - \Delta/2 < X \le x_k + \Delta/2)$. Accordingly, we may approximate the expected value of the random variable $X$ as

$$\begin{aligned} E[X] &\simeq \sum_{k=-\infty}^{\infty} x_k \left(\frac{N_n(k)}{n}\right) \\ &= \frac{1}{n} \sum_{k=-\infty}^{\infty} x_k N_n(k), \qquad n \text{ large} \end{aligned} \tag{A1.30}$$

We now recognize the quantity on the right-hand side of Equation (A1.30) simply as the *sample average*. The sum is taken over all the values $x_k$, each of which is weighted by the number of times it occurs; the sum is then divided by the total number of observations to give the sample average. Indeed, Equation (A1.30) provides the basis for computing the expected value $E[X]$.

We next consider a more general situation. Let $X$ denote a random variable, and let $g(X)$ denote a function of $X$ defined on the real line. The quantity obtained by letting the argument of the function $g(X)$ be a random variable is also a random variable, which we denote as

$$Y = g(X) \tag{A1.31}$$

To find the expected value of the random variable $Y$, we could of course find the probability density function $f_Y(y)$ and then apply the standard formula

$$E[Y] = \int_{-\infty}^{\infty} y f_Y(y)\, dy$$

A simpler procedure, however, is to write

$$E[g(X)] = \int_{-\infty}^{\infty} g(x) f_X(x)\, dx \tag{A1.32}$$

Indeed, Equation (A1.32) may be viewed as generalizing the concept of expected value to an arbitrary function $g(X)$ of a random variable $X$.

## ■ MOMENTS

For the special case of $g(X) = X^n$, using Equation (A1.32) we obtain the $n$th *moment* of the probability distribution of the random variable $X$; that is,

$$E[X^n] = \int_{-\infty}^{\infty} x^n f_X(x)\, dx \tag{A1.33}$$

By far the most important moments of $X$ are the first two moments. Thus putting $n = 1$ in Equation (A1.33) gives the mean of the random variable as shown in Eq. (A1.27), whereas putting $n = 2$ gives the *mean-square value* of $X$:

$$E[X^2] = \int_{-\infty}^{\infty} x^2 f_X(x)\, dx \tag{A1.34}$$

We may also define *central moments*, which are simply the moments of the difference between a random variable $X$ and its mean $\mu_X$. Thus, the $n$th central moment is

$$E[(X - \mu_X)^n] = \int_{-\infty}^{\infty} (x - \mu_X)^n f_X(x)\, dx \tag{A1.35}$$

For $n = 1$, the central moment is, of course, zero, whereas for $n = 2$ the second central moment is referred to as the *variance* of the random variable $X$, which is written as

$$\text{var}[X] = E[(X - \mu_X)^2] = \int_{-\infty}^{\infty} (x - \mu_X)^2 f_X(x)\, dx \tag{A1.36}$$

The variance of a random variable $X$ is commonly denoted as $\sigma_X^2$. The square root of the variance, namely, $\sigma_X$, is called the *standard deviation* of the random variable $X$.

The variance $\sigma_X^2$ of a random variable $X$ in some sense is a measure of the variable's "randomness." By specifying the variance $\sigma_X^2$, we essentially constrain the effective width of the probability density function $f_X(x)$ of the random variable $X$ about the mean $\mu_X$.

A precise statement of this constraint is due to Chebyshev. The *Chebyshev inequality* states that for any positive number $\epsilon$, we have

$$P(|X - \mu_X| \geq \epsilon) \leq \frac{\sigma_X^2}{\epsilon^2} \tag{A1.37}$$

From this inequality we see that the mean and variance of a random variable give a *partial description* of its probability distribution, hence their common use in practice.

We note from Equations (A1.34) and (A1.36) that the variance $\sigma_X^2$ and mean-square value $E[X^2]$ are related by

$$\begin{aligned} \sigma_X^2 &= E[X^2 - 2\mu_X X + \mu_X^2] \\ &= E[X^2] - 2\mu_X E[X] + \mu_X^2 \\ &= E[X^2] - \mu_X^2 \end{aligned} \tag{A1.38}$$

where, in the second line, we have used the *linearity* property of the statistical expectation operator $E$. Equation (A1.38) shows that if the mean $\mu_X$ is zero, then the variance $\sigma_X^2$ and the mean-square value $E[X^2]$ of the random variable $X$ are equal.

## ■ Characteristic Function

Another important statistical average is the *characteristic function* $\phi_X(v)$ of the probability distribution of the random variable $X$, which is defined as the expectation of the complex exponential function $\exp(jvX)$, as shown by

$$\begin{aligned} \phi_X(v) &= E[\exp(jvX)] \\ &= \int_{-\infty}^{\infty} f_X(x) \exp(jvx)\, dx \end{aligned} \tag{A1.39}$$

where $v$ is real and $j = \sqrt{-1}$. In other words, the characteristic function $\phi_X(v)$ is (except for a sign change in the exponent) the Fourier transform of the probability density function $f_X(x)$; the Fourier transform is reviewed in Appendix 2. In this relation we have used $\exp(jvx)$ rather than $\exp(-jvx)$, so as to conform with the convention adopted in probability theory. Recognizing that $v$ and $x$ play analogous roles to the variables $2\pi f$ and $t$ of Fourier transforms, respectively, we deduce the following inverse relation from analogy with the inverse Fourier transform:

$$f_X(x) = \frac{1}{2\pi} \int_{-\infty}^{\infty} \phi_X(v) \exp(-jvx)\, dv \tag{A1.40}$$

This relation may be used to evaluate the probability density function $f_X(x)$ of the random variable $X$ from its characteristic function $\phi_X(v)$.

## ■ Joint Moments

Consider next a pair of random variables $X$ and $Y$. A set of statistical averages of importance in this case is the *joint moments*, namely, the expected value of $X^i Y^k$, where $i$ and $k$ may assume any positive integer values. We may thus write

$$E[X^i Y^k] = \int_{-\infty}^{\infty} \int_{-\infty}^{\infty} x^i y^k f_{X,Y}(x, y)\, dx\, dy \tag{A1.41}$$

A joint moment of particular importance is the *correlation* deifned by $E[XY]$, which corresponds to $i = k = 1$ in Equation (A1.41).

The correlation of the centered random variables $X - E[X]$ and $Y - E[Y]$, that is, the joint moment

$$\text{cov}[XY] = E[(X - E[X])(Y - E[Y])] \tag{A1.42}$$

is called the *covariance* of $X$ and $Y$. Letting $\mu_X = E[X]$ and $\mu_Y = E[Y]$, we may expand Equation (A1.42) to obtain the result

$$\text{cov}[XY] = E[XY] - \mu_X \mu_Y \tag{A1.43}$$

Let $\sigma_X^2$ and $\sigma_Y^2$ denote the variances of $X$ and $Y$, respectively. Then the covariance of $X$ and $Y$, normalized with respect to $\sigma_X \sigma_Y$, is called the *correlation coefficient* of $X$ and $Y$:

$$\rho = \frac{\text{cov}[XY]}{\sigma_X \sigma_Y} \tag{A1.44}$$

We say that the two random variables $X$ and $Y$ are *uncorrelated* if and only if their covariance is zero, that is, if and only if

$$\text{cov}[XY] = 0$$

We say that they are *orthogonal* if and only if their correlation is zero, that is, if and only if

$$E[XY] = 0$$

From Equation (A1.43) we observe that if one of the random variables $X$ and $Y$ or both have zero means, and if they are orthogonal, then they are uncorrelated, and vice versa. Note also that if $X$ and $Y$ are statistically independent, then they are uncorrelated; however, the converse of this statement is not necessarily true.

APPENDIX 2

# REPRESENTATION OF SIGNALS AND SYSTEMS

## A2.1 Fourier Analysis

Let $g(t)$ denote a *nonperiodic deterministic signal*, expressed as some function of time $t$. By definition, the *Fourier transform* of the signal $g(t)$ is given by the integral

$$G(f) = \int_{-\infty}^{\infty} g(t) \exp(-j2\pi ft)\, dt \tag{A2.1}$$

where $j = \sqrt{-1}$, and the variable $f$ denotes *frequency*. Given the Fourier transform $G(f)$, the original signal $g(t)$ is recovered exactly using the formula for the *inverse Fourier transform*:

$$g(t) = \int_{-\infty}^{\infty} G(f) \exp(j2\pi ft)\, df \tag{A2.2}$$

Note that in Equations (A2.1) and (A2.2) we have used a lowercase letter to denote the time function and an uppercase letter to denote the corresponding frequency function. The functions $g(f)$ and $G(f)$ are said to constitute a Fourier-transform pair.

For the Fourier transform of a signal $g(t)$ to exist, it is sufficient but not necessary that $g(t)$ satisfies three conditions known collectively as *Dirichlet's conditions*:

1. The function $g(t)$ is single-valued, with a finite number of maxima and minima in any finite time interval.
2. The function $g(t)$ has a finite number of discontinuities in any finite time interval.
3. The function $g(t)$ is absolutely integrable, that is,

$$\int_{-\infty}^{\infty} |g(t)|\, dt < \infty$$

We may safely ignore the question of the existence of the Fourier transform of a time function $g(t)$ when it is an accurately specified description of a physically realizable signal. In other words, physical realizability is a sufficient condition for the existence of a Fourier transform. Indeed, we may go one step further and state that all energy signals, that is, signals $g(t)$ for which

$$\int_{-\infty}^{\infty} |g(t)|^2\, dt < \infty$$

are Fourier transformable.

The Fourier transform provides the mathematical tool for measuring the frequency content, or spectrum, of a signal. For this reason, the terms *Fourier transform* and *spectrum* are often used interchangeably. Thus, given a signal $g(t)$ with Fourier transform $G(f)$, we may refer to $G(f)$ as the spectrum of the signal $g(t)$. By the same token, we refer to $|G(f)|$ as the *magnitude spectrum* of the signal $g(t)$, and refer to $\arg\{G(f)\}$ as its *phase spectrum*.

## ■ PROPERTIES OF THE FOURIER TRANSFORM

It is useful to have insight into the relationship between a time function $g(t)$ and its Fourier transform $G(f)$, and also into the effects that various operations on the function $g(t)$ have on the transform $G(f)$. This may be achieved by examining certain properties of the Fourier transform, which are summarized in Table A6.2.

## ■ DIRAC DELTA FUNCTION

Strictly speaking, the theory of the Fourier transform is applicable only to time functions that satisfy the Dirichlet conditions. Such functions include energy signals. However, it would be highly desirable to extend this theory in two ways:

1. To combine the Fourier series and Fourier transform into a unified theory, so that the Fourier series may be treated as a special case of the Fourier transform.
2. To include power signals (i.e., signals for which the average power is finite) in the list of signals to which we may apply the Fourier transform.

It turns out that both of these objectives can be met through the "proper use" of the *Dirac delta function*, or *unit impulse*.

The Dirac delta function or just delta function, denoted by $\delta(t)$, is defined as having zero amplitude everywhere except at $t = 0$, where it is infinitely large in such a way that it contains unit area under its curve; that is,

$$\delta(t) = 0, \qquad t \neq 0 \tag{A2.3}$$

and

$$\int_{-\infty}^{\infty} \delta(t)\, dt = 1 \tag{A2.4}$$

An implication of this pair of relations is that the delta function $\delta(t)$ must be an even function of time $t$, which is centered at $t = 0$.

For the delta function to have meaning, however, it has to appear as a factor in the integrand of an integral with respect to time and then, strictly speaking, only when the other factor in the integrand is a continuous function of time. Let $g(t)$ be such a function, and consider the product of $g(t)$ and the time-shifted delta function $\delta(t - t_0)$. In light of the two defining equations, Equations (A2.3) and (A2.4), we may express the integral of this product as follows:

$$\int_{-\infty}^{\infty} g(t)\, \delta(t - t_0)\, dt = g(t_0) \tag{A2.5}$$

The operation indicated on the left-hand side of this equation sifts out the value $g(t_0)$ of the function $g(t)$ at time $t = t_0$, where $-\infty < t < \infty$. Accordingly, Equation (A2.5) is referred to as the *sifting property* of the delta function. This property is sometimes used as the defining equation of a delta function; in effect, it incorporates Equations (A2.3) and (A2.4) into a single relation.

Noting that the delta function $\delta(t)$ is an even function of $t$, we may rewrite Equation (A2.5) so as to emphasize its resemblance to the convolution integral, as shown by

$$\int_{-\infty}^{\infty} g(\tau)\, \delta(t - \tau)\, d\tau = g(t) \tag{A2.6}$$

In words, the convolution of any function with the delta function leaves that function unchanged. We refer to this statement as the *replication property* of the delta function.

It is important to realize that no function in the ordinary sense has the two properties of Equations (A2.3) and (A2.4) or the equivalent sifting property of Equation (A2.5). However, we can imagine a sequence of functions that have progressively taller and thinner peaks at $t = 0$, with the area under the curve remaining equal to unity, whereas the value of the function tends to zero at every point except $t = 0$, where it tends to infinity. That is, we may view the delta function as *the limiting form of a pulse of unit area as the duration of the pulse approaches zero*. It is immaterial what sort of pulse shape is used.

## FOURIER TRANSFORMS OF PERIODIC SIGNALS

It is well known that by using the Fourier series, a periodic signal can be represented as a sum of complex exponentials. Also, in a limiting sense, Fourier transforms can be defined by complex exponentials. Therefore, it seems reasonable to represent a periodic signal in terms of a Fourier transform, provided that this transform is permitted to include delta functions.

Consider then a periodic signal $g_{T_0}(t)$ of *period* $T_0$. We can represent $g_{T_0}(t)$ in terms of the *complex exponential Fourier series*:

$$g_{T_0}(t) = \sum_{n=-\infty}^{\infty} c_n \exp(j2\pi n f_0 t) \tag{A2.7}$$

where $c_n$ is the *complex Fourier coefficient* defined by

$$c_n = \frac{1}{T_0} \int_{-T_0/2}^{T_0/2} g_{T_0}(t) \exp(-j2\pi n f_0)\, dt \tag{A2.8}$$

and $f_0$ is the *fundamental frequency* defined as the reciprocal of the period $T_0$; that is,

$$f_0 = \frac{1}{T_0} \tag{A2.9}$$

Let $g(t)$ be a pulselike function, which equals $g_{T_0}(t)$ over one period and is zero elsewhere; that is

$$g(t) = \begin{cases} g_{T_0}(t), & -\dfrac{T_0}{2} < t \le \dfrac{T_0}{2} \\ 0 & \text{elsewhere} \end{cases} \tag{A2.10}$$

The periodic signal $g_{T_0}(t)$ may now be expressed in terms of the function $g(t)$ as an infinite summation, as shown by

$$g_{T_0}(t) = \sum_{m=-\infty}^{\infty} g(t - mT_0) \tag{A2.11}$$

Based on this representation, we may view $g(t)$ as a *generating function*, which generates the periodic signal $g_{T_0}(t)$.

The function $g(t)$ is Fourier transformable. Accordingly, we may rewrite the formula for the complex Fourier coefficient as follows:

$$\begin{aligned} c_n &= f_0 \int_{-\infty}^{\infty} g(t) \exp(-j2\pi n f_0 t)\, dt \\ &= f_0 G(n f_0) \end{aligned} \tag{A2.12}$$

where $G(nf_0)$ is the Fourier transform of $g(t)$ evaluated at the frequency $nf_0$. We may thus rewrite the formula for the reconstruction of the periodic signal $g_{T_0}(t)$ as

$$g_{T_0}(t) = f_0 \sum_{n=-\infty}^{\infty} G(nf_0) \exp(j2\pi n f_0 t) \tag{A2.13}$$

or, equivalently, in light of Equation (A2.11)

$$\sum_{m=-\infty}^{\infty} g(t - mT_0) = f_0 \sum_{n=-\infty}^{\infty} G(nf_0) \exp(j2\pi n f_0 t) \tag{A2.14}$$

Equation (A2.14) is one form of *Poisson's sum formula*.

It is of interest to observe that the function $g(t)$, which constitutes one period of the periodic signal $g_{T_0}(t)$, has a continuous spectrum defined by $G(f)$. On the other hand, the period signal $g_{T_0}(t)$ itself has a discrete spectrum. We conclude, therefore, that *periodicity in the time domain has the effect of changing the frequency-domain description or spectrum of the signal into a discrete form defined at integer multiples of the fundamental frequency.*

## ■ FOURIER-TRANSFORM PAIRS

Table A6.3 presents a listing of some commonly used Fourier-transform pairs, the derivations of which follow from the material just presented.

## ■ TRANSMISSION OF SIGNALS THROUGH LINEAR SYSTEMS

A *system* refers to any physical device that produces an output signal in response to an input signal. It is customary to refer to the input signal as the *excitation* and to the output signal as the *response*. In a *linear* system, the *principle of superposition* holds; that is, the response of a linear system to a number of excitations applied simultaneously is equal to the sum of the responses of the system when each excitation is applied individually.

In the time domain, a linear system is described in terms of its *impulse response*, which is defined as *the response of the system (with zero initial conditions) to a unit impulse or delta function $\delta(t)$ applied to the input of the system.* If the system is *time invariant*, then the shape of the impulse response is the same no matter when the unit impulse is applied to the system. Thus, assuming that the unit impulse or delta function is applied at time $t = 0$, we may denote the impulse response of a linear time-invariant system by $h(t)$. Let this system be subjected to an arbitrary excitation $x(t)$. The response, $y(t)$, of the system is defined in terms of the impulse response $h(t)$ by

$$y(t) = \int_{-\infty}^{\infty} x(\tau)h(t - \tau)\, d\tau \tag{A2.15}$$

which is called the *convolution integral*. Equivalently, we may write

$$y(t) = \int_{-\infty}^{\infty} h(\tau)\, x(t - \tau)\, d\tau \tag{A2.16}$$

Hence, convolution is *commutative*.

In the convolution integral, three different time scales are involved: *excitation time* $\tau$, *response time* $t$, and *system-memory time* $t - \tau$. This relation is the basis of time-domain analysis of linear time-invariant systems. According to Equation (A2.15), the present value of the response of a linear time-invariant system is a weighted integral over the past history of the input signal, weighted according to the impulse response of the system. Thus the impulse response acts as a *memory function* for the system.

## ■ Frequency Response of Linear Time-Invariant Systems

Consider a linear time-invariant system of impulse response $h(t)$ driven by a complex exponential input of unit amplitude and frequency $f$, that is,

$$x(t) = \exp(j2\pi ft)$$

Using this excitation in Equation (A2.16), the response of the system is obtained as

$$\begin{aligned} y(t) &= \int_{-\infty}^{\infty} h(\tau) \exp[j2\pi f(t - \tau)] \, d\tau \\ &= \exp(j2\pi ft) \int_{-\infty}^{\infty} h(\tau) \exp(-j2\pi f\tau) \, d\tau \end{aligned} \tag{A2.17}$$

Define the *frequency response* of the system as the Fourier transform of its impulse response, as shown by

$$H(f) = \int_{-\infty}^{\infty} h(t) \exp(-j2\pi ft) \, dt \tag{A2.18}$$

The integral in the last line of Equation (A2.17) is the same as that of Equation (A2.18), except that $\tau$ is used in place of $t$. Hence, we may rewrite Equation (A2.17) in the form

$$y(t) = H(f) \exp(j2\pi ft) \tag{A2.19}$$

The response of a linear time-invariant system to a complex exponential function of frequency $f$ is, therefore, the same complex exponential function multiplied by a constant coefficient $H(f)$.

The frequency response $H(f)$ is, in general, a complex quantity, so we may express it in the form

$$H(f) = |H(f)| \exp[j\beta(f)] \tag{A2.20}$$

where $|H(f)|$ is called the *magnitude response*, and $\beta(f)$ is the *phase*, or *phase response*. In the special case of a linear system with a real-valued impulse response $h(t)$, the frequency response $H(f)$ exhibits conjugate symmetry, which means that

$$|H(f)| = |H(-f)|$$

and

$$\beta(f) = -\beta(-f)$$

That is, the magnitude response $|H(f)|$ of a linear system with real-valued impulse response is an even function of frequency, whereas the phase $\beta(f)$ is an odd function of frequency.

In some applications, it is preferable to work with the logarithm of $H(f)$ expressed in polar form rather than with $H(f)$ itself. Define the natural logarithm

$$\log H(f) = \alpha(f) + j\beta(f) \tag{A2.21}$$

where

$$\alpha(f) = \log|H(f)| \tag{A2.22}$$

The function $\alpha(f)$ is called the *gain* of the system. It is measured in *nepers*, whereas $\beta(f)$ is measured in *radians*. Equation (A2.21) indicates that the gain $\alpha(f)$ and phase $\beta(f)$ are the real and imaginary parts of the (natural) logarithm of the frequency response $H(f)$, respectively. The gain may also be expressed in *decibels* (dB) by using the definition

$$\alpha'(f) = 20 \log_{10} |H(f)| \tag{A2.23}$$

The two gain functions $\alpha(f)$ and $\alpha'(f)$ are related by

$$\alpha'(f) = 8.69\alpha(f) \tag{A2.24}$$

That is, 1 neper is equal to 8.69 dB.

## A2.2 Bandwidth

The time-domain and frequency-domain descriptions of a signal are *inversely* related. In particular, we may make the following important statements:

1. If the time-domain description of a signal is changed, the frequency-domain description of the signal is changed in an *inverse* manner, and vice versa. This inverse relationship prevents arbitrary specifications of a signal in both domains. In other words, *we may specify an arbitrary function of time or an arbitrary spectrum, but we cannot specify both of them together.*
2. If a signal is strictly limited in frequency, the time-domain description of the signal will trail on indefinitely, even though its amplitude may assume a progressively smaller value. We say a signal is *strictly limited in frequency* or *strictly band limited* if its Fourier transform is exactly zero outside a finite band of frequencies. The sinc pulse

$$\text{sinc}(t) = \frac{\sin(\pi t)}{\pi t}$$

is an example of a strictly band-limited signal. It is also *asymptotically limited in time*, which confirms the opening statement we made for a strictly band-limited signal. In an inverse manner, if a signal is *strictly limited in time* (i.e., the signal is exactly zero outside a finite time interval), then the spectrum of the signal is infinite in extent, even though the amplitude spectrum may assume a progressively smaller value. This behavior is exemplified by a rectangular pulse. Accordingly, we may state that a *signal cannot be strictly limited in both time and frequency.*

The *bandwidth* of a signal provides a measure of the *extent of significant spectral content of the signal for positive frequencies.* When the signal is strictly band limited, the bandwidth is well defined. For example, the sinc pulse $\text{sinc}(2Wt)$ has a bandwidth equal to $W$. However, when the signal is not strictly band limited, as is generally the case, we encounter difficulty in defining the bandwidth of the signal. The difficulty arises because the meaning of "significant" attached to the spectral content of the signal is mathematically imprecise. Consequently, there is no universally accepted definition of bandwidth. Nevertheless, there are some commonly used definitions for bandwidth, as discussed next.

When the spectrum of a signal is symmetric with a *main lobe* bounded by well-defined *nulls* (i.e., frequencies at which the spectrum is zero), we may use the main lobe as the basis for defining the bandwidth of the signal. Specifically, if the signal is *low-pass* (i.e., its spectral content is centered around the origin), the bandwidth is defined as one half the total width of the main spectral lobe since only one half of this lobe lies inside the positive frequency region. For example, a rectangular pulse of duration $T$ seconds has a main spectral lobe of total width $2/T$ hertz centered at the origin. Accordingly, we may define the bandwidth of this rectangular pulse as $1/T$ hertz. If, on the other hand, the signal is *band-pass* with main spectral lobes centered around $\pm f_c$, where $f_c$ is large enough, the bandwidth is defined as the width of the main lobe for positive frequencies. This definition of bandwidth is called the *null-to-null bandwidth*. For example, an RF pulse of duration $T$ seconds and frequency $f_c$ has main spectral lobes of width $2/T$ hertz centered around $\pm f_c$, where it is assumed that $f_c$ is large compared to $1/T$. Hence, we may define the null-to-null bandwidth of this RF pulse as $2/T$ hertz. On the basis of the definitions presented here, we may state that shifting the spectral content of a low-pass signal by a sufficiently large frequency has the effect of doubling the bandwidth of the signal; such a frequency translation is attained by using modulation.

Another popular definition of bandwidth is the *3-dB bandwidth*. Specifically, if the signal is low-pass, the 3-dB bandwidth is defined as the separation between zero frequency, where the magnitude spectrum attains its peak value, and the *positive frequency*, at which the amplitude spectrum drops to $1/\sqrt{2}$ of its peak value. For example, the decaying exponential $\exp(-at)$ has a 3-dB bandwidth of $a/2\pi$ hertz. If, on the other hand, the signal is band-pass, centered at $\pm f_c$, the 3-dB bandwidth is defined as the separation (along the positive frequency axis) between the two frequencies at which the magnitude spectrum of the signal drops to $1/\sqrt{2}$ of the peak value of $f_c$. The 3-dB bandwidth has the advantage in that it can be read directly from a plot of the magnitude spectrum. However, it has the disadvantage in that it may be misleading if the magnitude spectrum has slowly decreasing tails.

Yet another measure for the bandwidth of a signal is the *root mean square (rms) bandwidth*, which is defined as the square root of the second moment of a properly normalized form of the squared magnitude spectrum of the signal about a suitably chosen point. We assume that the signal is low-pass, so that the second moment may be taken about the origin. As for the normalized form of the squared magnitude spectrum, we use the nonnegative function

$$\frac{|G(f)|^2}{\int_{-\infty}^{\infty} |G(f)|^2\, df}$$

in which the denominator applies the correct normalization in the sense that the integrated value of this ratio over the entire frequency axis is unity. We may thus formally define the rms bandwidth of a low-pass signal $g(t)$ with Fourier transform $G(f)$ as follows:

$$W_{\text{rms}} = \left( \frac{\int_{-\infty}^{\infty} f^2 |G(f)|^2\, df}{\int_{-\infty}^{\infty} |G(f)|^2\, df} \right)^{1/2} \tag{A2.25}$$

An attractive feature of the rms bandwidth $W_{\text{rms}}$ is that it lends itself more readily to mathematical evaluation than the other two definitions of bandwidth, but it is not as easily measurable in the laboratory.

## ■ Time-Bandwidth Product

For any family of pulse signals that differ by a time-scaling factor, the product of the signal's duration and its bandwidth is always a constant, as shown by

$$(\text{duration} \times \text{bandwidth}) = \text{constant}$$

The product is called the *time-bandwidth product* or *bandwidth-duration product*. The constancy of the time-bandwidth product is another manifestation of the inverse relationship that exists between the time-domain and frequency-domain descriptions of a signal. In particular, if the duration of a pulse signal is decreased by reducing the time scale by a factor $a$, the frequency scale of the signal's spectrum, and therefore the bandwidth of the signal, is increased by the same factor $a$, by virtue of the *time-scaling property* of the Fourier transform, and the time-bandwidth product of the signal is thereby maintained constant; see item 2 of Table A6.2. For example, a rectangular pulse of duration $T$ seconds has a bandwidth (defined on the basis of the positive-frequency part of the main lobe) equal to $1/T$ hertz, making the time-bandwidth product of the pulse equal unity. Whatever definition we use for the bandwidth of a signal, the time-bandwidth product remains constant over certain classes of pulse signals. The choice of a particular definition for bandwidth merely changes the value of the constant.

To be more specific, consider the rms bandwidth defined in Equation (A2.25). The corresponding definition for the *rms duration* of the signal $g(t)$ is

$$T_{\text{rms}} = \left( \frac{\int_{-\infty}^{\infty} t^2 |g(t)|^2 \, dt}{\int_{-\infty}^{\infty} |g(t)|^2 \, dt} \right)^{1/2} \tag{A2.26}$$

where it is assumed that the signal $g(t)$ is centered around the origin. It may be shown that, using the rms definitions of Equations (A2.25) and (A2.26), the time-bandwidth product has the following form:

$$T_{\text{rms}} W_{\text{rms}} \geq \frac{1}{4\pi} \tag{A2.27}$$

where the constant is $1/4\pi$. The Gaussian pulse $\exp(-\pi t^2)$ satisfies this condition with the equality sign.

## ■ NOISE EQUIVALENT BANDWIDTH

The definitions of bandwidth just presented (i.e., 3-dB bandwidth, null-to-null bandwidth, and rms bandwidth) are all formulated in terms of deterministic signals. Another definition of bandwidth that presents itself in the study of random signals and systems is the noise equivalent bandwidth. Suppose that a white noise source of power spectral density $N_0/2$ is connected to the input of the simple RC low-pass filter of Figure A2.1; the corresponding value of the average output noise power is equal to $N_0/(4RC)$. For this filter, the half-power or 3-dB bandwidth is equal to $1/(2\pi RC)$. Here again we find that the average output noise power of the filter is proportional to the bandwidth.

We may generalize this statement to include all kinds of low-pass filters by defining a noise equivalent bandwidth as follows. Suppose that we have a source of white noise of

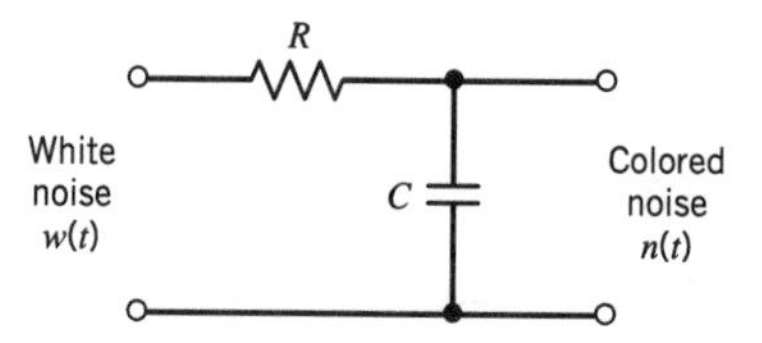


**FIGURE A2.1** RC low-pass filter.

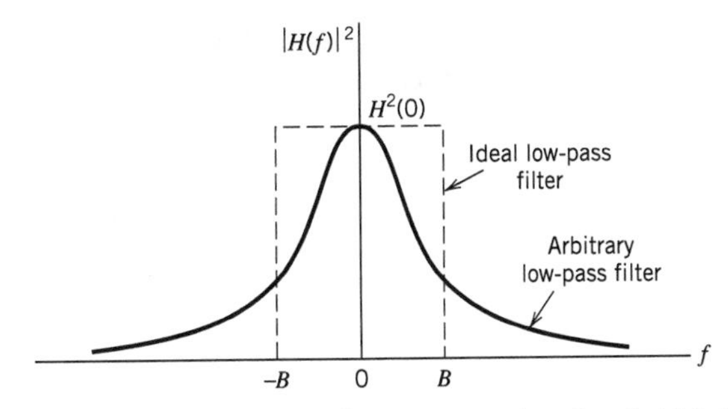


**FIGURE A2.2** Illustrating the definition of noise-equivalent bandwidth for a low-pass filter.

zero mean and power spectral density $N_0/2$ connected to the input of an arbitrary low-pass filter of transfer function $H(f)$. The resulting average output noise power is therefore

$$\begin{aligned} N_{\text{out}} &= \frac{N_0}{2}\int_{-\infty}^{\infty} |H(f)|^2\, df \\ &= N_0 \int_0^{\infty} |H(f)|^2\, df \end{aligned} \tag{A2.28}$$

where, in the last line, we have made use of the fact that the magnitude response $|H(f)|$ is an even function of frequency.

Consider next the same source of white noise connected to the input of an *ideal* low-pass filter of zero-frequency response $H(0)$ and bandwidth $B$. In this case, the average output noise power is

$$N_{\text{out}} = N_0 B H^2(0) \tag{A2.29}$$

Therefore, equating this average output noise power to that in Equation (A2.28), we may formally define the *noise equivalent bandwidth* as

$$B = \frac{\int_0^{\infty} |H(f)|^2\, df}{H^2(0)} \tag{A2.29}$$

Thus the procedure for calculating the noise equivalent bandwidth consists of replacing the arbitrary low-pass filter of transfer function $H(f)$ by an equivalent ideal low-pass filter of zero frequency response $H(0)$ and bandwidth $B$, as illustrated in Figure A2.2. In a similar way, we may define a noise equivalent bandwidth for bandpass filters.

## A2.3 Hilbert Transform

The Fourier transform is particularly useful for evaluating the frequency content of an energy signal or, in a limiting sense, that of a power signal. As such, it provides the mathematical basis for analyzing and designing frequency-selective filters for the separation of signals on the basis of their frequency content. Another method of separating signals is based on *phase selectivity*, which uses phase shifts between the pertinent signals to achieve the desired separation. The simplest phase shift is that of 180 degrees, which is merely a polarity reversal in the case of a sinusoidal signal. Shifting the phase angles of all components of a given signal by 180 degrees requires the use of an *ideal transformer*. Another phase shift of interest is that of $\pm 90$ degrees. In particular, when the phase angles of all components of a given signal are shifted by $\pm 90$ degrees, the resulting function of time is known as the Hilbert transform of the signal.

To be specific, consider a signal $g(t)$ with Fourier transform $G(f)$. The *Hilbert transform* of $g(t)$, which we shall denote by $\hat{g}(t)$, is defined by

$$\hat{g}(t) = \frac{1}{\pi}\int_{-\infty}^{\infty} \frac{g(\tau)}{t - \tau}\, d\tau \tag{A2.31}$$

Clearly, the Hilbert transformation of $g(t)$ is a linear operation. The *inverse Hilbert transform*, by means of which the original signal $g(t)$ is recovered from $\hat{g}(t)$, is defined by

$$g(t) = -\frac{1}{\pi}\int_{-\infty}^{\infty} \frac{\hat{g}(\tau)}{t - \tau}\, d\tau \tag{A2.32}$$

The functions $g(t)$ and $\hat{g}(t)$ are said to constitute a *Hilbert-transform pair*. A short table of Hilbert-transform pairs is given in Table A6.4.

We note from the definition of the Hilbert transform that $\hat{g}(t)$ may be interpreted as the convolution of $g(t)$ with the time function $1/\pi t$. We also know from the convolution theorem that the convolution of two functions in the time domain is transformed into the multiplication of their Fourier transforms in the frequency domain; see item 12 of Table A6.2. For the time function $1/\pi t$, we have (see Table A6.3)

$$\frac{1}{\pi t} \rightleftharpoons -j \operatorname{sgn}(f) \tag{A2.33}$$

where $\operatorname{sgn}(f)$ is the *signum function* defined in the frequency domain as

$$\operatorname{sgn}(f) = \begin{cases} 1, & f > 0 \\ 0, & f = 0 \\ -1, & f < 0 \end{cases} \tag{A2.34}$$

It follows therefore that the Fourier transform $\hat{G}(f)$ of $\hat{g}(t)$ is given by

$$\hat{G}(f) = -j \operatorname{sgn}(f) G(f) \tag{A2.35}$$

Equation (A2.35) states that given a signal $g(t)$, we may obtain its Hilbert transform $\hat{g}(t)$ by passing $g(t)$ through a linear two-port device whose frequency response is equal to $-j \operatorname{sgn}(f)$. This device may be considered as one that produces a phase shift of $-90$ degrees for all positive frequencies of the input signal and $+90$ degrees for all negative frequencies, as in Figure A2.3. The amplitudes of all frequency components in the signal, however, are

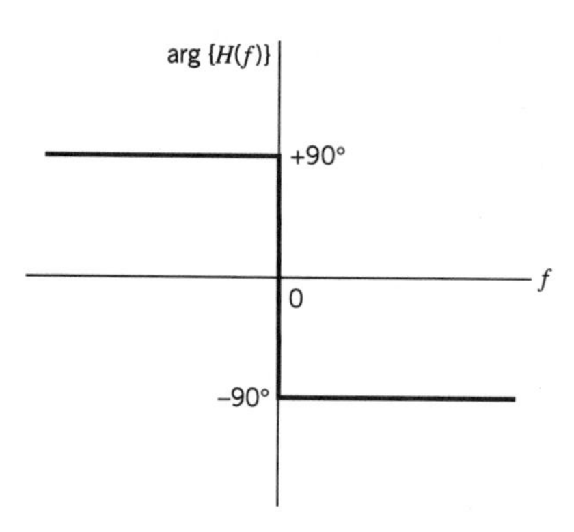


**FIGURE A2.3** Phase characteristic of linear two-port device for obtaining the Hilbert transform of a real-valued signal.

unaffected by transmission through the device. Such an ideal device is referred to as a *Hilbert transformer.*

### ■ PROPERTIES OF THE HILBERT TRANSFORM

The Hilbert transform differs from the Fourier transform in that it operates exclusively in the time domain. It has a number of useful properties, some of which are listed next. The signal $g(t)$ is assumed to be real valued, which is the usual domain of application of the Hilbert transform. For this class of signals, we may state the following:

1. A signal $g(t)$ and its Hilbert transform $\hat{g}(t)$ have the same magnitude spectrum.
2. If $\hat{g}(t)$ is the Hilbert transform of $g(t)$, then the Hilbert transform of $\hat{g}(t)$ is $-g(t)$.
3. A signal $g(t)$ and its Hilbert transform $\hat{g}(t)$ are orthogonal over the entire time interval $(-\infty, \infty)$, as shown by

$$\int_{-\infty}^{\infty} g(t)\hat{g}(t)dt = 0$$

Proofs of these properties are left as exercises for the reader; the proofs follow from Equations (A2.31), (A2.32) and (A2.35).

## A2.4 *Complex Representation of Signals and Systems*

### ■ PRE-ENVELOPE

Consider a real-valued signal $g(t)$. We define the *pre-envelope*, or *analytic signal*, of the signal $g(t)$ as the complex-valued function

$$g_+(t) = g(t) + j\hat{g}(t) \tag{A2.36}$$

where $\hat{g}(t)$ is the Hilbert transform of $g(t)$. We note that the given signal $g(t)$ is the real part of the pre-envelope $g_+(t)$, and the Hilbert transform of the signal is the imaginary part of the pre-envelope. Just as the use of phasors simplifies manipulations of alternating currents and voltages, so we find that the pre-envelope is particularly useful in handling band-pass signals and systems.

One of the important features of the pre-envelope $g_+(t)$ is the behavior of its Fourier transform. Let $G_+(f)$ denote the Fourier transform of $g_+(t)$. Then we may write

$$G_+(f) = G(f) + \text{sgn}(f)G(f)$$

from which we readily find that

$$G_+(f) = \begin{cases} 2G(f), & f > 0 \\ G(0), & f = 0 \\ 0, & f < 0 \end{cases} \tag{A2.37}$$

where $G(0)$ is the value of $G(f)$ at frequency $f = 0$. This means that the pre-envelope of a signal has no frequency content (i.e., its Fourier transform vanishes) for all negative frequencies.

From the foregoing analysis it is apparent that for a given signal $g(t)$ we may determine its pre-envelope $g_+(t)$ in one of two equivalent ways:

1. We determine the Hilbert transform $\hat{g}(t)$ of the signal $g(t)$, and then use Equation (A2.36) to compute the pre-envelope $g_+(t)$.
2. We determine the Fourier transform $G(f)$ of the signal $g(t)$, use Equation (A2.37) to determine $G_+(f)$, and then evaluate the inverse Fourier transform of $G_+(f)$ to obtain

$$g_+(t) = 2\int_0^\infty G(f)\exp(j2\pi ft)\,df \tag{A2.38}$$

For a particular signal $g(t)$ of Fourier transform $G(f)$, one of these two ways may be better than the other.

Equation (A2.36) defines the pre-envelope $g_+(t)$ for positive frequencies. Symmetrically, we may define the pre-envelope for *negative frequencies* as

$$g_-(t) = g(t) - j\hat{g}(t) \tag{A2.39}$$

The two pre-envelopes $g_+(t)$ and $g_-(t)$ are simply the complex conjugate of each other, as shown by

$$g_-(t) = g_+^*(t) \tag{A2.40}$$

where the asterisk denotes complex conjugation. The spectrum of the pre-envelope $g_+(t)$ is nonzero only for *positive* frequencies, as emphasized in Equation (A2.37); hence, the use of a plus sign as the subscript. In contrast, the spectrum of the other pre-envelope $g_-(t)$ is nonzero only for *negative* frequencies, as shown by the Fourier transform

$$G_-(f) = \begin{cases} 0, & f > 0 \\ G(0), & f = 0 \\ 2G(f), & f < 0 \end{cases} \tag{A2.41}$$

Thus the pre-envelopes $g_+(t)$ and $g_-(t)$ constitute a complementary pair of complex-valued signals. Note also that the sum of $g_+(t)$ and $g_-(t)$ is exactly twice the original signal $g(t)$.

## ■ CANONICAL REPRESENTATIONS OF BAND-PASS SIGNALS

Consider a band-pass signal $g(t)$ whose Fourier transform $G(f)$ is nonnegligible only in a band of frequencies of total extent $2W$, say, centered about some frequency $\pm f_c$. This is illustrated in Figure A2.4*a*. We refer to $f_c$ as the *carrier frequency*. In the majority of communication signals, we find that the bandwidth $2W$ is small compared with $f_c$, and so we refer to such a signal as a *narrowband signal*. However, a precise statement about how small the bandwidth must be for the signal to be considered narrowband is not necessary for our present discussion.

Let the pre-envelope of a narrowband signal $g(t)$, with its Fourier transform $G(f)$ centered about some frequency $\pm f_c$, be expressed in the form

$$g_+(t) = \tilde{g}(t)\exp(j2\pi f_c t) \tag{A2.42}$$

We refer to $\tilde{g}(t)$ as the *complex envelope* of the signal. Equation (A2.42) may be viewed as the basis of a definition for the complex envelope $\tilde{g}(t)$ in terms of the pre-envelope $g_+(t)$. We note that the spectrum of $g_+(t)$ is limited to the frequency band $f_c - W \le f \le f_c + W$, as illustrated in Figure A2.4*b*. Therefore, applying the frequency-shifting property of the

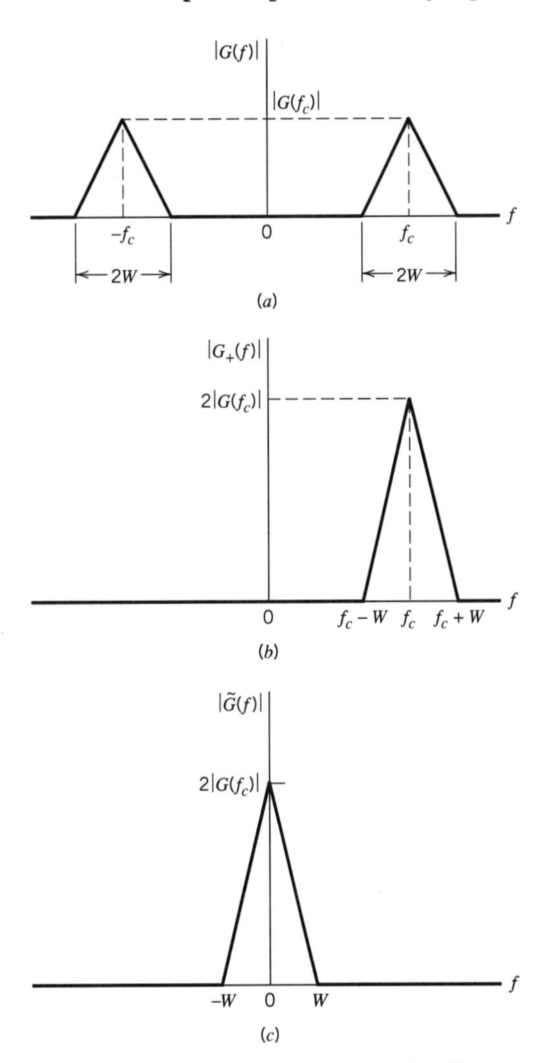


**FIGURE A2.4** (*a*) Magnitude spectrum of band-pass signal $g(t)$. (*b*) Magnitude spectrum of pre-envelope $g_+(t)$. (*c*) Magnitude spectrum of complex envelope $\tilde{g}(t)$.

Fourier transform to Equation (A2.42), which is described as item 5 in Table A6.2, we find that the spectrum of the complex envelope $\tilde{g}(t)$ is limited to the band $-W \leq f \leq W$ and centered at the origin as illustrated in Figure A2.4*c*. That is, the complex envelope $\tilde{g}(t)$ of a band-pass signal $g(t)$ is a *low-pass signal*, which is an important result.

By definition, the given signal $g(t)$ is the real part of the pre-envelope $g_+(t)$. We may thus express the original band-pass signal $g(t)$ in terms of the complex envelope $\tilde{g}(t)$ as follows:

$$g(t) = \text{Re}[\tilde{g}(t) \exp(j2\pi f_c t)] \tag{A2.43}$$

In general, $\tilde{g}(t)$ is a complex-valued quantity; to emphasize this property, we may express it in the form

$$\tilde{g}(t) = g_I(t) + jg_Q(t) \tag{A2.44}$$

where $g_I(t)$ and $g_Q(t)$ are both real-valued low-pass functions; their low-pass property is inherited from the complex envelope $\tilde{g}(t)$. We may therefore use Equations (A2.43) and (A2.44) to express the original band-pass signal $g(t)$ in the *canonical*, or *standard*, form:

$$g(t) = g_I(t)\cos(2\pi f_c t) - g_Q(t)\sin(2\pi f_c t) \tag{A2.45}$$

We refer to $g_I(t)$ as the *in-phase component* of the band-pass signal $g(t)$ and to $g_Q(t)$ as the *quadrature component* of the signal; this nomenclature recognizes that $\sin(2\pi f_c t)$ [i.e., the multiplying factor of $g_Q(t)$] is in phase-quadrature with respect to $\cos(2\pi f_c t)$ [i.e., the multiplying factor of $g_I(t)$] and $\cos(2\pi f_c t)$ is viewed as the reference.

According to Equation (A2.44), the complex envelope $\tilde{g}(t)$ may be pictured as a *time-varying phasor* positioned at the origin of the $(g_I, g_Q)$-plane, as indicated in Figure A2.5*a*. With time $t$ varying, the end of the phasor moves about in the plane. Figure A2.5*b* shows the phasor representation of the complex exponential $\exp(j2\pi f_c t)$. In the definition given in Equation (A2.43), the complex envelope $\tilde{g}(t)$ is multiplied by the complex exponential $\exp(j2\pi f_c t)$. The angles of these two phasors therefore add and their lengths multiply, as shown in Figure A2.5*c*. Moreover, in this latter figure, we show the $(g_I, g_Q)$-plane rotating with an angular velocity equal to $2\pi f_c$ radians per second. Thus, in the picture portrayed here, the phasor representing the complex envelope $\tilde{g}(t)$ moves in the $(g_I, g_Q)$-plane and at the same time the plane itself rotates about the origin. The original band-pass signal $g(t)$ is the projection of this time-varying phasor on a *fixed line* representing the real axis, as indicated in Figure A2.5*c*.

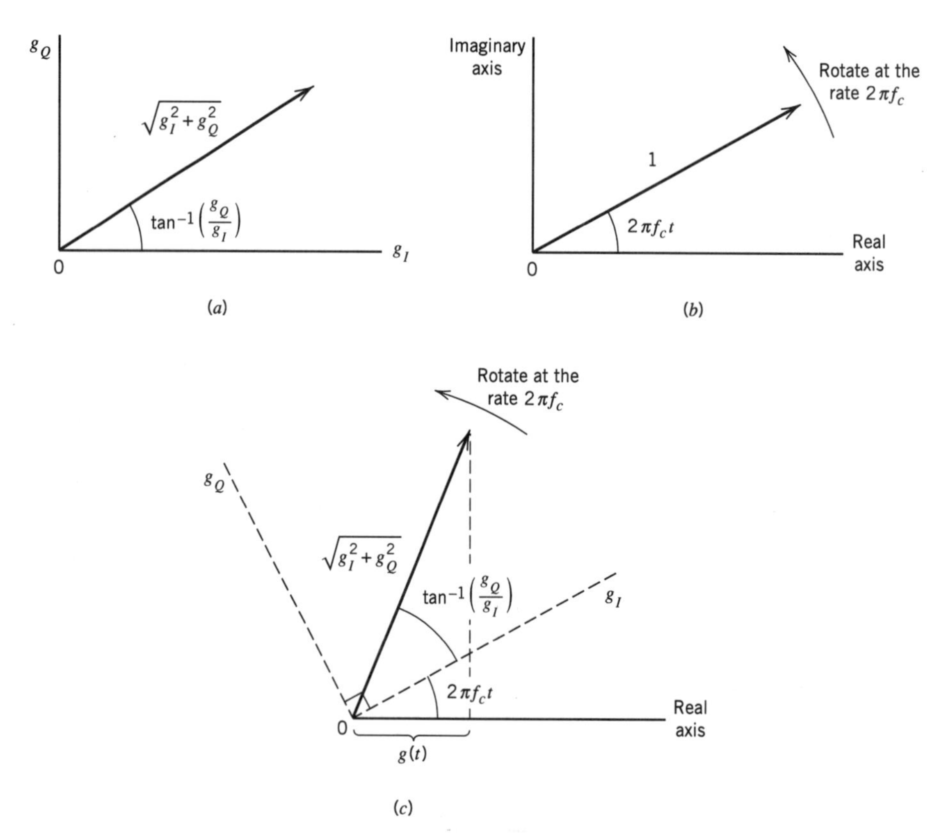


**FIGURE A2.5** Illustrating an interpretation of the complex envelope $\tilde{g}(t)$ and its multiplication by $\exp(j2\pi f_c t)$.

Since both $g_I(t)$ and $g_Q(t)$ are low-pass signals limited to the band $-W \leq f \leq W$, they may be derived from the band-pass signal $g(t)$ using the scheme shown in Figure A2.6*a*. Both low-pass filters in this figure are identical, each of which has a bandwidth equal to $W$. To reconstruct $g(t)$ from its in-phase and quadrature components, we may use the scheme shown in Figure A2.6*b*.

The two schemes shown in Figure A2.6 are basic to the study of *linear modulation systems*. The multiplication of the low-pass in-phase component $g_I(t)$ by $\cos(2\pi f_c t)$ and the multiplication of the low-pass quadrature component $g_Q(t)$ by $\sin(2\pi f_c t)$ represent linear forms of modulation. Given that the carrier frequency $f_c$ is sufficiently large, the resulting band-pass function $g(t)$ defined in Equation (A2.45) is referred to as a *passband signaling waveform*. Correspondingly, the mapping from $g_I(t)$ and $g_Q(t)$ into $g(t)$ is known as *passband modulation*.

Equation (A2.44) is the Cartesian form of expressing the complex envelope $\tilde{g}(t)$. Alternatively, we may express it in the polar form

$$\tilde{g}(t) = a(t) \exp[j\phi(t)] \tag{A2.46}$$

where $a(t)$ and $\phi(t)$ are both real-valued low-pass functions. Based on this polar representation, the original band-pass signal $g(t)$ is defined by

$$g(t) = a(t) \cos[2\pi f_c t + \phi(t)] \tag{A2.47}$$

We refer to $a(t)$ as the *natural envelope* or simply the *envelope* of the band-pass signal $g(t)$ and to $\phi(t)$ as the *phase* of the signal. Equation (A2.47) represents a *hybrid form of amplitude modulation and angle modulation*; indeed, it includes amplitude modulation, frequency modulation, and phase modulation as special cases.

From this discussion it is apparent that, whether we represent a band-pass (modulated) signal $g(t)$ in terms of its in-phase and quadrature components as in Equation (A2.45) or in terms of its envelope and phase as in Equation (A2.47), the information content of the signal $g(t)$ is completely preserved in the complex envelope $\tilde{g}(t)$.

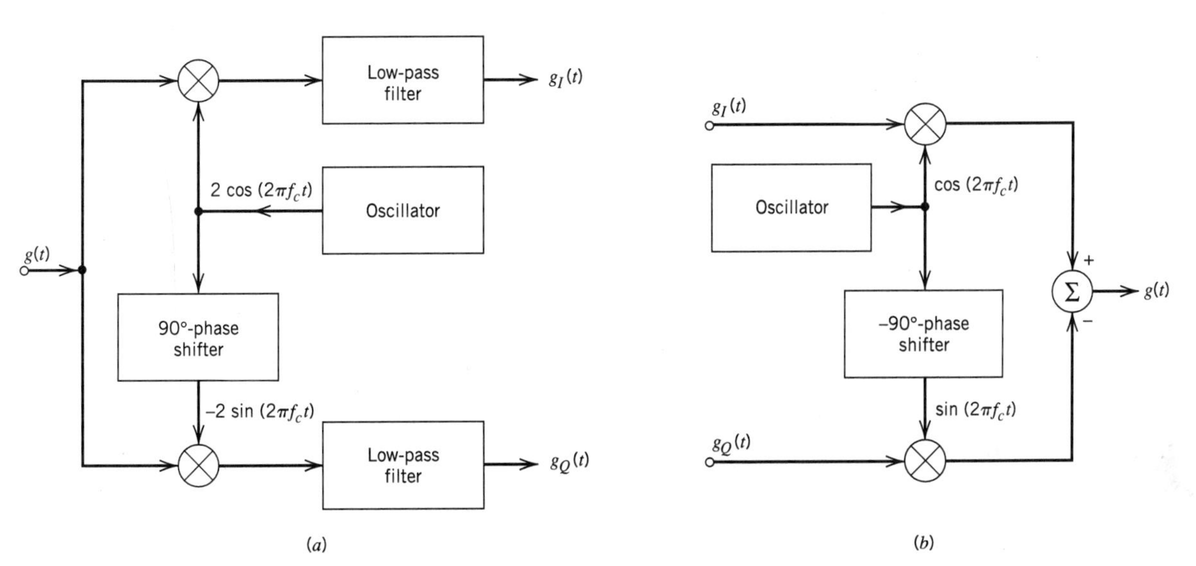


**FIGURE A2.6** (*a*) Scheme for deriving the in-phase and quadrature components of a band-pass signal. (*b*) Scheme for reconstructing the band-pass signal from its in-phase and quadrature components.

## ■ TERMINOLOGY

The distinctions among the three different envelopes that we have introduced to describe a band-pass signal $g(t)$ should be carefully noted. We summarize their definitions here:

1. The pre-envelope $g_+(t)$ for positive frequencies is defined by

$$g_+(t) = g(t) + j\hat{g}(t)$$

where $\hat{g}(t)$ is the Hilbert transform of the signal $g(t)$. According to this representation, $\hat{g}(t)$ may be viewed as the quadrature function of $g(t)$. Correspondingly, in the frequency domain we have

$$G_+(f) = \begin{cases} 2G(f), & f > 0 \\ G(0), & f = 0 \\ 0, & f < 0 \end{cases}$$

2. The complex envelope $\tilde{g}(t)$ equals a frequency-shifted version of the pre-envelope $g_+(t)$, as shown by

$$\tilde{g}(t) = g_+(t) \exp(-j2\pi f_c t)$$

where $f_c$ is the carrier frequency of the band-pass signal $g(t)$.

3. The envelope $a(t)$ equals the magnitude of the complex envelope $\tilde{g}(t)$ and also that of the pre-envelope $g_+(t)$, as shown by

$$a(t) = |\tilde{g}(t)| = |g_+(t)|$$

Note that for a band-pass signal $g(t)$, the pre-envelope $g_+(t)$ is a complex band-pass signal whose value depends on the carrier frequency $f_c$. On the other hand, the envelope $a(t)$ is always a real low-pass signal and, in general, the complex envelope $\tilde{g}(t)$ is a complex low-pass signal; the values of the latter two envelopes are independent of the choice of the carrier frequency $f_c$. This property gives the complex envelope $\tilde{g}(t)$ an analytic advantage over the original signal $g(t)$.

The envelope $a(t)$ and phase $\phi(t)$ of $g(t)$ are related to the quadrature components $g_I(t)$ and $g_Q(t)$ as follows (see the time-varying phasor representation of Figure A2.5*a*):

$$a(t) = \sqrt{g_I^2(t) + g_Q^2(t)}$$

$$\phi(t) = \tan^{-1}\left(\frac{g_Q(t)}{g_I(t)}\right)$$

Conversely, we may write

$$g_I(t) = a(t)\cos[\phi(t)]$$

$$g_Q(t) = a(t)\sin[\phi(t)]$$

Thus, each of the quadrature components of a band-pass signal contains both amplitude and phase information. Both components are required for a unique definition of the phase $\phi(t)$, modulo $2\pi$.

## ■ BAND-PASS SYSTEMS

Now that we know how to handle the complex low-pass representation of band-pass signals, it is logical that we develop a corresponding procedure for handling the analysis of band-pass systems. Specifically, we wish to show that the analysis of band-pass systems can be greatly simplified by establishing an analogy (or, more precisely, an isomorphism) between low-pass and band-pass systems. This analogy is based on the use of the Hilbert transform for the representation of band-pass signals.

Consider a narrowband signal $x(t)$, with its Fourier transform denoted by $X(f)$. We assume that the spectrum of the signal $x(t)$ is limited to frequencies within $\pm W$ Hz of the carrier frequency $f_c$. Also, we assume that $W < f_c$. Let this signal be represented in terms of its in-phase and quadrature components as follows:

$$x(t) = x_I(t)\cos(2\pi f_c t) - x_Q(t)\sin(2\pi f_c t) \tag{A2.48}$$

where $x_I(t)$ is the in-phase component and $x_Q(t)$ is the quadrature component. Then, using $x(t)$ to denote the complex envelope of $x(t)$, we may write

$$\tilde{x}(t) = x_I(t) + jx_Q(t) \tag{A2.49}$$

Let the signal $x(t)$ be applied to a linear time-invariant band-pass system with impulse response $h(t)$ and frequency response $H(f)$. We assume that the frequency response of the system is limited to frequencies within $\pm B$ of the carrier frequency $f_c$. The system bandwidth $2B$ is usually narrower than or equal to the input signal bandwidth $2W$. We wish to represent the band-pass impulse response $h(t)$ in terms of two quadrature components, denoted by $h_I(t)$ and $h_Q(t)$. Thus, by analogy to the representation of band-pass signals, we may express $h(t)$ in the form

$$h(t) = h_I(t)\cos(2\pi f_c t) - h_Q(t)\sin(2\pi f_c t) \tag{A2.50}$$

Define the *complex impulse response* of the band-pass system as

$$\tilde{h}(t) = h_I(t) + jh_Q(t) \tag{A2.51}$$

Hence, we have the complex representation

$$h(t) = \text{Re}[\tilde{h}(t)\exp(j2\pi f_c t)] \tag{A2.52}$$

Note that $h_I(t)$, $h_Q(t)$, and $\tilde{h}(t)$ are all low-pass functions limited to the frequency band $-B \le f \le B$.

We may determine the complex impulse response $\tilde{h}(t)$ in terms of the quadrature components $h_I(t)$ and $h_Q(t)$ of the band-pass impulse response $h(t)$ by using Equation (A2.51). Alternatively, we may determine it from the band-pass frequency response $H(f)$ in the following way. We first note from Equation (A2.52) that

$$2h(t) = \tilde{h}(t)\exp(j2\pi f_c t) + \tilde{h}^*(t)\exp(-j2\pi f_c t) \tag{A2.53}$$

where $\tilde{h}^*(t)$ is the complex conjugate of $\tilde{h}(t)$. Therefore, applying the Fourier transform to Equation (A2.53), and using the complex-conjugation property of the Fourier transform, which is described in item 10 in Table A6.2, we get

$$2H(f) = \tilde{H}(f - f_c) + \tilde{H}^*(-f - f_c) \tag{A2.54}$$

where $H(f)$ is the Fourier transform of $h(t)$, and $\tilde{H}(f)$ is the Fourier transform of $\tilde{h}(t)$. Equation (A2.54) satisfies the requirement that $H^*(f) = H(-f)$ for a real impulse response $h(t)$. Since $\tilde{H}(f)$ represents a low-pass frequency response limited to $|f| \le B$ with $B < f_c$, we deduce from Equation (A2.54) that

$$\tilde{H}(f - f_c) = 2H(f), \qquad f > 0 \tag{A2.55}$$

Equation (A2.55) indicates that for a specified band-pass frequency response $H(f)$, we may determine $\tilde{H}(f)$ by taking the part of $H(f)$ corresponding to positive frequencies, shifting it to the origin and then scaling it by the factor 2. To determine the complex impulse response $\tilde{h}(t)$, we take the inverse Fourier transform of $\tilde{H}(f)$, obtaining

$$\tilde{h}(t) = \int_{-\infty}^{\infty} \tilde{H}(f) \exp(j2\pi ft)\, df \tag{A2.56}$$

The representations just described for band-pass signals and systems provide the basis of an efficient method for determining the output of a band-pass system driven by a band-pass signal. We assume that the spectrum of the input signal $x(t)$ and the frequency response $H(f)$ of the system are both centered around the same frequency $f_c$. In practice, there is no need to consider a situation in which the carrier frequency of the input signal is not aligned with the midband frequency of the band-pass system, since we have considerable freedom in choosing the carrier or midband frequency. Thus, changing the carrier frequency of the input signal by an amount $\Delta f_c$, say, simply corresponds to absorbing (or removing) the factor $\exp(\pm j2\pi\, \Delta f_c t)$ in the complex envelope of the input signal or the complex impulse response of the band-pass system. We are therefore justified in proceeding on the assumption that $X(f)$ and $H(f)$ are both centered around $f_c$. Suppose then we use $y(t)$ to denote the output signal of the system. It is clear that $y(t)$ is also a band-pass signal, so that we may represent it in terms of its low-pass complex envelope $\tilde{y}(t)$, as follows:

$$y(t) = \text{Re}[\tilde{y}(t) \exp(j2\pi f_c t)] \tag{A2.57}$$

The output signal $y(t)$ is related to the input signal $x(t)$ and impulse response $h(t)$ of the system in the usual way by the convolution integral

$$y(t) = \int_{-\infty}^{\infty} h(\tau) x(t - \tau)\, d\tau \tag{A2.58}$$

In terms of pre-envelopes, we have $h(t) = \text{Re}[h_+(t)]$ and $x(t) = \text{Re}[x_+(t)]$. We may therefore rewrite Equation (A2.58) in terms of the pre-envelopes $x_+(t)$ and $h_+(t)$ as follows:

$$y(t) = \int_{-\infty}^{\infty} \text{Re}[h_+(\tau)]\, \text{Re}[x_+(t - \tau)]\, d\tau \tag{A2.59}$$

To proceed further, we make use of a basic property of pre-envelopes that is described by the following relation (presented here without proof):

$$\int_{-\infty}^{\infty} \text{Re}[h_+(\tau)]\, \text{Re}[x_+(\tau)]\, d\tau = \frac{1}{2} \text{Re}\left[\int_{-\infty}^{\infty} h_+(\tau) x_+^*(\tau)\, d\tau\right] \tag{A2.60}$$

where we have used $\tau$ as the integration variable to be consistent with that in Equation (A2.59). Next, we note that using $x(-\tau)$ in place of $x(\tau)$ has the effect of removing the complex conjugation on the right-hand side of Equation (A2.60). Hence, bearing in mind the algebraic difference between the argument of $x_+(\tau)$ in Equation (A2.60) and that of $x_+(t - \tau)$ in Equation (A2.59), and using the relationship between the pre-envelope and complex envelope of a band-pass function, we get

$$\begin{aligned} y(t) &= \frac{1}{2} \text{Re}\left[\int_{-\infty}^{\infty} h_+(\tau) x_+(t - \tau)\, d\tau\right] \\ &= \frac{1}{2} \text{Re}\left[\int_{-\infty}^{\infty} \tilde{h}(\tau) \exp(j2\pi f_c \tau) \tilde{x}(t - \tau) \exp(j2\pi f_c (t - \tau))\, d\tau\right] \\ &= \frac{1}{2} \text{Re}\left[\exp(j2\pi f_c t) \int_{-\infty}^{\infty} \tilde{h}(\tau) \tilde{x}(t - \tau)\, d\tau\right] \end{aligned} \tag{A2.61}$$

Thus comparing the right-hand sides of Equations (A2.57) and (A2.61), we readily deduce that for a large enough carrier frequency $f_c$, the complex envelope $\tilde{y}(t)$ of the output signal is related to the complex envelope $\tilde{x}(t)$ of the input signal and the complex impulse response $\tilde{h}(t)$ of the band-pass system as follows:

$$2\tilde{y}(t) = \int_{-\infty}^{\infty} \tilde{h}(\tau)\tilde{x}(t - \tau)\, d\tau \tag{A2.62}$$

or, using the shorthand notation for convolution,

$$2\tilde{y}(t) = \tilde{h}(t) \bigstar \tilde{x}(t) \tag{A2.63}$$

where ★ denotes convolution. In other words, *except for the scaling factor 2, the complex envelope $\tilde{y}(t)$ of the output signal of a band-pass system is obtained by convolving the complex impulse response $\tilde{h}(t)$ of the system with the complex envelope $\tilde{x}(t)$ of the input band-pass signal.* Equation (A2.63) is the result of the isomorphism, for convolution, between a band-pass function and the corresponding low-pass function.

The significance of this result is that in dealing with band-pass signals and systems, we need only concern ourselves with the low-pass functions $\tilde{x}(t)$, $\tilde{y}(t)$, and $\tilde{h}(t)$, representing the excitation, the response, and the system, respectively. That is, the analysis of a band-pass system, which is complicated by the presence of the multiplying factor $\exp(j2\pi f_c t)$, is replaced by an equivalent but much simpler low-pass analysis that completely retains the essence of the filtering process. This procedure is illustrated schematically in Figure A2.7.

The complex envelope $\tilde{x}(t)$ of the input band-pass signal and the complex impulse response $\tilde{h}(t)$ of the band-pass system are defined in terms of their respective in-phase and quadrature components by Equations (A2.49) and (A2.51), respectively. Substituting these relations in Equation (A2.63), we get

$$2\tilde{y}(t) = [h_I(t) + jh_Q(t)] \bigstar [x_I(t) + jx_Q(t)] \tag{A2.64}$$

Because convolution is *distributive*, we may rewrite Equation (A2.64) in the equivalent form

$$2\tilde{y}(t) = [h_I(t) \bigstar x_I(t) - h_Q(t) \bigstar x_Q(t)] + j[h_Q(t) \bigstar x_I(t) + h_I \bigstar x_Q(t)] \tag{A2.65}$$

Let the complex envelope $\tilde{y}(t)$ of the response be defined in terms of its in-phase and quadrature components as

$$\tilde{y}(t) = y_I(t) + jy_Q(t) \tag{A2.66}$$

Comparing the real and imaginary parts in Equations (A2.65) and (A2.66), we have for the in-phase component $y_I(t)$ the relation

$$2y_I(t) = h_I(t) \bigstar x_I(t) - h_Q(t) \bigstar x_Q(t) \tag{A2.67}$$

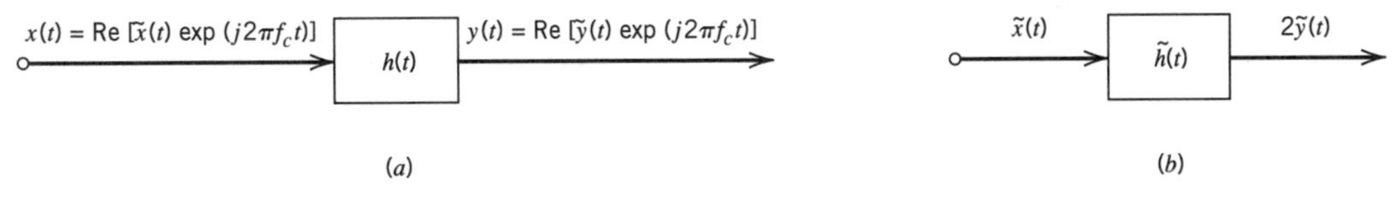


**FIGURE A2.7** (*a*) Narrowband filter of impulse response $h(t)$ with narrowband input signal $x(t)$. (*b*) Equivalent low-pass filter of complex impulse response $\tilde{h}(t)$ with complex low-pass input $\tilde{x}(t)$.

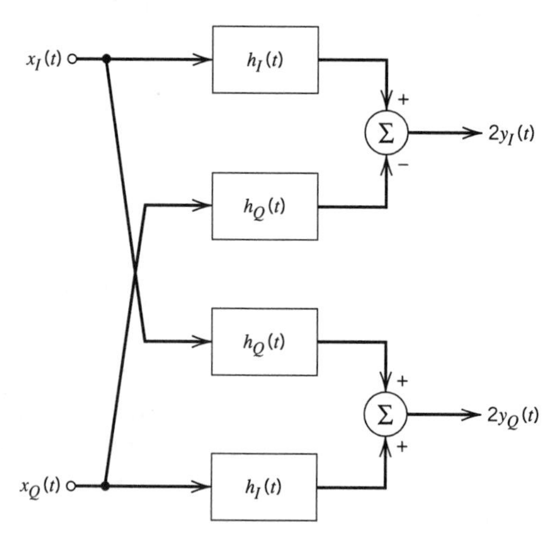


**FIGURE A2.8** Block diagram illustrating the relationships between the in-phase and quadrature components of the response of a band-pass filter and those of the input signal.

and for the quadrature component $y_Q(t)$ the relation

$$2y_Q(t) = h_Q(t) \bigstar x_I(t) + h_I(t) \bigstar x_Q(t) \tag{A2.68}$$

Thus, for the purpose of evaluating the in-phase and quadrature components of the complex envelope $\tilde{y}(t)$ of the system output, we may use the *low-pass equivalent model* shown in Figure A2.8. All the signals and impulse responses shown in this model are real-valued low-pass functions. Accordingly, this equivalent model provides a practical basis for the efficient simulation of band-pass filters or communication channels on a digital computer.

To sum up, the procedure for evaluating the response of a band-pass system (with mid-band frequency $f_c$) to an input band-pass signal (of carrier frequency $f_c$) is as follows:

1. The input band-pass signal $x(t)$ is replaced by its complex envelope $\hat{x}(t)$, which is related to $x(t)$ by

$$x(t) = \text{Re}[\hat{x}(t)\exp(j2\pi f_c t)]$$

2. The band-pass system, with impulse response $h(t)$, is replaced by a low-pass analog, which is characterized by a complex impulse response $\tilde{h}(t)$ related to $h(t)$ by

$$h(t) = \text{Re}[\tilde{h}(t)\exp(j2\pi f_c t)]$$

3. The complex envelope $\tilde{y}(t)$ of the output band-pass signal $y(t)$ is obtained by convolving $\tilde{h}(t)$ with $\tilde{x}(t)$, as shown by

$$2\tilde{y}(t) = \tilde{h}(t) \bigstar \tilde{x}(t)$$

4. The desired output $y(t)$ is finally derived from the complex envelope $\tilde{y}(t)$ by using the relation

$$y(t) = \text{Re}[\tilde{y}(t)\exp(j2\pi f_c t)]$$

# APPENDIX 3

# BESSEL FUNCTIONS

## A3.1 Series Solution of Bessel's Equation

In its most basic form, *Bessel's equation of order n* is written as

$$x^2 \frac{d^2y}{dx^2} + x\frac{dy}{dx} + (x^2 - n^2)y = 0 \tag{A3.1}$$

which is one of the most important of all variable-coefficient differential equations.[1] For each $n$, a solution of this equation is defined by the power series

$$J_n(x) = \sum_{m=0}^{\infty} \frac{(-1)^m\left(\frac{1}{2}x\right)^{n+2m}}{m!(n+m)!} \tag{A3.2}$$

The function $J_n(x)$ is called a *Bessel function of the first kind of order n*. Equation (A3.1) has two coefficient functions, namely, $1/x$ and $(1 - n^2/x^2)$. Hence, it has no finite singular points except the origin. It follows therefore that the series expansion of Equation (A3.2) converges for all $x > 0$. Equation (A3.2) may thus be used to numerically calculate $J_n(x)$ for $n = 0, 1, 2, \ldots$. Table A6.5 presents values of $J_n(x)$ for different orders $n$ and varying $x$. It is of interest to note that the graphs of $J_0(x)$ and $J_1(x)$ resemble the graphs of $\cos x$ and $\sin x$, respectively; see the graphs of Figure 2.23 in Chapter 2.

The function $J_n(x)$ may also be expressed in the form of an integral as

$$J_n(x) = \frac{1}{\pi}\int_0^{\pi} \cos(x \sin\theta - n\theta)\, d\theta \tag{A3.3}$$

or, equivalently,

$$J_n(x) = \frac{1}{2\pi}\int_{-\pi}^{\pi} \exp(jx \sin\theta - jn\theta)\, d\theta \tag{A3.4}$$

## A3.2 Properties of the Bessel Function

The Bessel function $J_n(x)$ has the following properties:

1. $$J_n(x) = (-1)^n J_{-n}(x) \tag{A3.5}$$

To prove this relation, we replace $\theta$ by $(\pi - \theta)$ in Equation (A3.3). Then, noting that $\sin(\pi - \theta) = \sin\theta$, we get

$$\begin{aligned} J_n(x) &= \frac{1}{\pi}\int_0^{\pi} \cos(x \sin\theta + n\theta - n\pi)\, d\theta \\ &= \frac{1}{\pi}\int_0^{\pi} [\cos(n\pi)\cos(x \sin\theta + n\theta) + \sin(n\pi)\sin(x\sin\theta + n\theta)]\, d\theta \end{aligned}$$

For integer values of $n$, we have

$$\cos(n\pi) = (-1)^n$$
$$\sin(n\pi) = 0$$

Therefore,

$$J_n(x) = \frac{(-1)^n}{\pi} \int_0^{\pi} \cos(x \sin\theta + n\theta)\, d\theta \tag{A3.6}$$

From Equation (A3.3), we also find that by replacing $n$ with $-n$:

$$J_{-n}(x) = \frac{1}{\pi} \int_0^{\pi} \cos(x \sin\theta + n\theta)\, d\theta \tag{A3.7}$$

The desired result follows immediately from Equations (A3.6) and (A3.7).

2.
$$J_n(x) = (-1)^n J_n(-x) \tag{A3.8}$$

This relation is obtained by replacing $x$ with $-x$ in Equation (A3.3), and then using Equation (A.3.6).

3.
$$J_{n-1}(x) + J_{n+1}(x) = \frac{2n}{x} J_n(x) \tag{A3.9}$$

This *recurrence formula* is useful in constructing tables of Bessel coefficients; its derivation follows from the power series of Equation (A3.2).

4. For small values of $x$, we have

$$J_n(x) \simeq \frac{x^n}{2^n n!} \tag{A3.10}$$

This relation is obtained simply by retaining the first term in the power series of Equation (A3.2) and ignoring the higher-order terms. Thus, when $x$ is small, we have

$$\begin{aligned} J_0(x) &\simeq 1 \\ J_1(x) &\simeq \frac{x}{2} \\ J_n(x) &\simeq 0 \qquad \text{for } n > 1 \end{aligned} \tag{A3.11}$$

5. For large values of $x$, we have

$$J_n(x) \simeq \sqrt{\frac{2}{\pi x}} \cos\left(x - \frac{\pi}{4} - \frac{n\pi}{2}\right) \tag{A3.12}$$

This shows that for large values of $x$, the Bessel function $J_n(x)$ behaves like a sine wave with progressively decreasing amplitude.

6. With $x$ real and fixed, $J_n(x)$ approaches zero as the order $n$ goes to infinity.

7.
$$\sum_{n=-\infty}^{\infty} J_n(x) \exp(jn\phi) = \exp(jx \sin\phi) \tag{A3.13}$$

To prove this property, consider the sum $\sum_{n=-\infty}^{\infty} J_n(x) \exp(jn\phi)$ and use the formula of Equation (A3.4) for $J_n(x)$ to obtain

$$\sum_{n=-\infty}^{\infty} J_n(x) \exp(jn\phi) = \frac{1}{2\pi} \sum_{n=-\infty}^{\infty} \exp(jn\phi) \int_{-\pi}^{\pi} \exp(jx \sin\theta - jn\theta)\, d\theta$$

Interchanging the order of integration and summation:

$$\sum_{n=-\infty}^{\infty} J_n(x) \exp(jn\phi) = \frac{1}{2\pi} \int_{-\pi}^{\pi} d\theta \exp(jx \sin \theta) \sum_{n=-\infty}^{\infty} \exp[jn(\phi - \theta)] \quad \text{(A3.14)}$$

We now invoke the following relation from Fourier transform theory:

$$\delta(\phi) = \frac{1}{2\pi} \sum_{n=-\infty}^{\infty} \exp[jn(\phi)], \qquad -\pi \leq \phi \leq \pi \quad \text{(A3.15)}$$

where $\delta(\phi)$ is a delta function. Therefore, using Equation (A3.15) in (A3.14) and then applying the sifting property of the delta function, we get

$$\begin{aligned} \sum_{n=-\infty}^{\infty} J_n(x) \exp(jn\phi) &= \int_{-\pi}^{\pi} \exp(jx \sin \theta)\, \delta(\phi - \theta)\, d\theta \\ &= \exp(jx \sin \phi) \end{aligned}$$

which is the desired result.

8. $$\sum_{n=-\infty}^{\infty} J_n^2(x) = 1 \qquad \text{for all } x \quad \text{(A3.16)}$$

To prove this property, we may proceed as follows. We observe that $J_n(x)$ is real. Hence, multiplying Equation (A3.4) by its own complex conjugate and summing over all possible values of $n$, we get

$$\sum_{n=-\infty}^{\infty} J_n^2(x) = \frac{1}{(2\pi)^2} \sum_{n=-\infty}^{\infty} \int_{-\pi}^{\pi} \int_{-\pi}^{\pi} \exp(jx \sin \theta - jn\theta - jx \sin \phi + jn\phi)\, d\theta\, d\phi$$

Interchanging the order of double integration and summation:

$$\begin{aligned} &\sum_{n=-\infty}^{\infty} J_n^2(x) = \\ &\frac{1}{(2\pi)^2} \int_{-\pi}^{\pi} \int_{-\pi}^{\pi} d\theta\, d\phi \exp[jx(\sin \theta - \sin \phi)] \sum_{n=-\infty}^{\infty} \exp[jn(\phi - \theta)] \end{aligned} \quad \text{(A3.17)}$$

Using Equation (A3.15) in (A3.17) and then applying the sifting property of the delta function, we finally get

$$\sum_{n=-\infty}^{\infty} J_n^2(x) = \frac{1}{2\pi} \int_{-\pi}^{\pi} d\theta = 1$$

which is the desired result.

Many of these properties of the Bessel function $J_n(x)$ may also be illustrated in numerical terms by referring to Table A6.5.

## A3.3 *Modified Bessel Function*

The *modified Bessel equation of order n* is written as

$$x^2 \frac{d^2y}{dx^2} + x \frac{dy}{dx} - (x^2 + n^2)y = 0 \quad \text{(A3.18)}$$

With $j^2 = -1$, where $j$ is the square root of $-1$, we may rewrite this equation as

$$x^2 \frac{d^2y}{dx^2} + x\frac{dy}{dx} + (j^2x^2 - n^2)y = 0$$

From this rewrite it is evident that Equation (A3.18) is nothing but Bessel's equation, namely, Equation (A3.1), with $x$ replaced by $jx$. Thus replacing $x$ by $jx$ in Equation (A3.2), we get

$$\begin{aligned} J_n(jx) &= \sum_{m=0}^{\infty} \frac{(-1)^m \left(\dfrac{jx}{2}\right)^{n+2m}}{m!(n+m)!} \\ &= j^n \sum_{m=0}^{\infty} \frac{\left(\dfrac{x}{2}\right)^{n+2m}}{m!(n+m)!} \end{aligned}$$

Next we note that $J_n(jx)$ multiplied by a constant will still be a solution of Bessel's equation. Accordingly, we multiply $J_n(jx)$ by the constant $j^{-n}$, obtaining

$$j^{-n}J_n(jx) = \sum_{m=0}^{\infty} \frac{\left(\dfrac{1}{2}x\right)^{n+2m}}{m!(n+m)!}$$

This new function is called the *modified Bessel function of the first kind of order n*, denoted by $I_n(x)$. We may thus formally express a solution of the modified Bessel equation, Equation (A3.18), as

$$\begin{aligned} I_n(x) &= j^{-n}J_n(jx) \\ &= \sum_{m=0}^{\infty} \frac{\left(\dfrac{1}{2}x\right)^{n+2m}}{m!(n+m)!} \end{aligned} \quad \text{(A3.19)}$$

The modified Bessel function $I_n(x)$ is a monotonically increasing real function of the argument $x$ for all $n$, as shown in Figure A3.1 for $n = 0, 1, 2$.

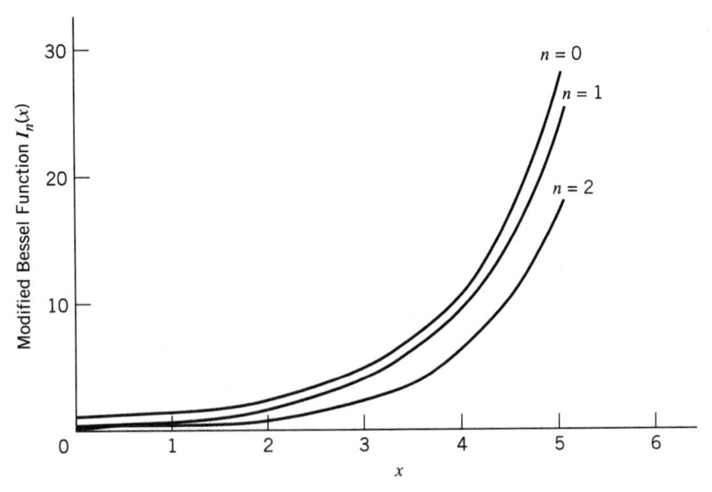


**FIGURE A3.1** Modified Bessel function $I_n(x)$ of varying order $n$.

The modified Bessel function $I_n(x)$ is identical to the original Bessel function $J_n(x)$ except for an important difference: The terms in the series expansion of Equation (A3.19) are all positive, whereas they alternate in sign in the series expansion of Equation (A3.2). The relationship between $J_n(x)$ and $I_n(x)$ is analogous to the way in which the trigonometric functions $\cos x$ and $\sin x$ are related to the hyperbolic functions $\cosh x$ and $\sinh x$.

An interesting property of the modified Bessel function $I_n(x)$ is derived from Equation (A3.13). Specifically, replacing $x$ by $jx$ and the angle $\phi$ by $\theta - \pi/2$ in this equation, and then invoking the definition of $I_n(x)$ in the first line of Equation (A3.19), we obtain

$$\sum_{n=-\infty}^{\infty} I_n(x) \exp(jn\theta) = \exp(x \cos \theta) \tag{A3.20}$$

From this relation it follows that

$$I_n(x) = \frac{1}{2\pi} \int_{-\pi}^{\pi} \exp(x \cos \theta) \cos(n\theta)\, d\theta \tag{A3.21}$$

This integral formula for $I_n(x)$ may, of course, also be derived from Equation (A3.4) by making the appropriate changes.

When the argument $x$ is small, we obtain the following asymptotic estimates directly from the series representation of Equation (A3.19):

$$I_0(x) \to 1 \qquad \text{for } x \to 0 \tag{A3.22}$$

and

$$I_n(x) \to 0 \qquad \text{for } n \geq 1 \text{ and } x \to 0 \tag{A3.23}$$

For large values of $x$ we have the following asymptotic estimate for $I_n(x)$, which is valid for all integers $n \geq 0$:

$$I_n(x) \simeq \frac{\exp(x)}{\sqrt{2\pi x}} \qquad \text{for } x \to \infty \tag{A3.24}$$

Note that this asymptotic behavior of $I_n(x)$ is independent of the order $n$ for large values of $x$.

## Notes and References

1. Equation (A3.1) is named for the German mathematician and astronomer Friedrich Wilhelm Bessel (1784–1846). For detailed treatments of the solution to this equation and related issues, see Wylie and Barrett (1982) and Watson (1966).

# APPENDIX 4

# CONFLUENT HYPERGEOMETRIC FUNCTIONS

## A4.1 *Kummer's Equation*

The *confluent hypergeometric function*[1] is a solution of *Kummer's differential equation*:

$$x \frac{d^2y}{dx^2} + (b - x)\frac{dy}{dx} - ay = 0 \tag{A4.1}$$

where, in general, the parameters $a$ and $b$ are complex numbers. For the case when $b \neq 0, -1, -2, \ldots$, the solution of Kummer's equation is defined by the series

$${}_1F_1(a; b; x) = 1 + \frac{a}{b}\frac{x}{1!} + \frac{a(a+1)}{b(b+1)}\frac{x^2}{2!} + \cdots \tag{A4.2}$$

where ${}_1F_1(a; b; x)$ denotes a confluent hypergeometric function parameterized by $a$ and $b$. In this notation, the first subscript denotes the number of factorials in the numerator of the general term in Equation (A4.2), the second subscript denotes the number of factorials, apart from $n!$, in the denominator. In Equation (A4.2), both subscripts are clearly 1.

## A4.2 *Properties of the Confluent Hypergeometric Function*

**Property 1**

*For small values of $x$, the confluent hypergeometric function approximates as*

$${}_1F_1(a; b; x) \simeq 1 + \frac{a}{b}x \qquad \text{for } x \to 0 \tag{A4.3}$$

This property follows directly from the series expansion of Equation (A4.2).

**Property 2**

*For $a = -1$ and $b = 1$ we have the exact identity:*

$${}_1F_1(-1; 1; x) = 1 - x \qquad \text{for all } x \tag{A4.4}$$

This property also follows directly from the series expansion of Equation (A4.2).

**Property 3**

*The confluent hypergeometric function for $a = -1/2$ and $b = 1$ is related exactly to the modified Bessel function for all $x$ as follows:*

$$ {}_1F_1\left(-\frac{1}{2}; 1; -x\right) = \exp\left(-\frac{x}{2}\right)\left((1 + x)I_0\left(\frac{x}{2}\right) + xI_2\left(\frac{x}{2}\right)\right) \qquad \text{(A4.5)} $$

*where $I_n(x)$ is the modified Bessel function of order $n$.*

A special case of Equation (A4.5) occurs when $x$ is large. From the definition of the modified Bessel function given in Appendix 3, we have the following asymptotic formula for large $x$:

$$ I_n(x) \simeq \frac{\exp(x)}{\sqrt{2\pi x}} \qquad \text{for } x \to \infty \qquad \text{(A4.6)} $$

Hence, combining Equations (A4.5) and (A4.6), we obtain the simple result

$$ {}_1F_1\left(-\frac{1}{2}; 1; -x\right) \simeq 2\sqrt{\frac{x}{\pi}} \qquad \text{for } x \to \infty \qquad \text{(A4.7)} $$

## Notes and References

1. For a discussion of confluent hypergeometric functions, see Jeffreys and Jeffreys (1956). Tabulated values of these functions are presented in Abramowitz and Stegun (1965).

# APPENDIX 5

# CRYPTOGRAPHY

Secrecy is certainly important to the security or integrity of information transmission. Indeed, the need for secure communications is more profound than ever, recognizing that the conduct of much of our commerce, business, and personal matters is being carried out today through the medium of computers, which has replaced the traditional medium of papers.

*Cryptology* is the umbrella term used to describe the science of secret communications; it is derived from the Greek *kryptos* and *logos* which mean "hidden" and "word," respectively.[1] The subject matter of cryptology may be partitioned neatly into *cryptography* and *cryptanalysis*. Cryptography deals with the transformations of a message into coded form by *encryption* and the recovery of the original message by *decryption*. The original message to be encrypted (enciphered) is called the *plaintext*, and the result produced by encryption is called a *cryptogram* or *ciphertext*; the latter two terms are used interchangeably. The set of data transformations used to do the encryption is called a *cipher*; normally, the transformations are parameterized by one or more *keys*. *Cryptanalysis*, on the other hand, deals with how to undo cryptographic communications by breaking a cipher or forging coded signals that may be accepted as genuine.

Cryptographic systems offer three important services:

1. *Secrecy*, which refers to the denial of access to information by unauthorized users.
2. *Authenticity*, which refers to the validation of the source of a message.
3. *Integrity*, which refers to the assurance that a message was not modified by accidental or deliberate means in transit.

A conventional cryptographic system relies on the use of a single piece of private and necessarily secret information known as the *key*; hence, conventional cryptography is referred to as *single-key cryptography* or *secret-key cryptography*.[2] This form of cryptography operates on the premise that the key is known to the encrypter (sender) and by the decrypter (receiver) but to no others; the assumption is that once the message is encrypted, it is (probably) impossible to do the decryption without knowledge of the key.

*Public-key cryptography*,[3] also called *two-key cryptography*, differs from conventional cryptography in that there is no longer a single secret key shared by two users. Rather, each user is provided with key material of one's own, and the key material is divided into two portions: a public component and a private component. The public component generates a public transformation, and the private component generates a private transformation. But, of course, the private transformation must be kept secret for secure communication between the two users.

## A5.1 *Secret-Key Cryptography*

Basically, the flow of information in a secret-key cryptographic system is as shown in Figure A5.1. The message source generates a plaintext message, which is encrypted into a cryptogram at the transmitting end of the system. The cryptogram is sent to an *authorized user* at the receiving end over an "insecure" channel; a channel is considered insecure if

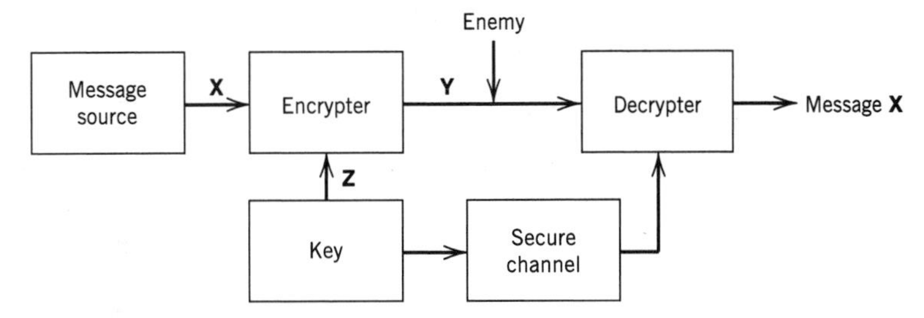


**FIGURE A5.1** Block diagram of secret-key cryptographic system.

its security is inadequate for the needs of its users. It is assumed that in the course of transmission the cryptogram may be intercepted by an *enemy cryptanalyst*[4] (i.e., would-be intruder into a cryptographic system). The requirement is to do the encryption in such a way that the enemy is prevented from learning the contents of the plaintext message.

In abstract terms, a *cryptographic system or cipher (for short) is defined as a set of invertible transformations of the plaintext space (i.e., the set of possible plaintext messages) into the cryptogram space (i.e., the set of all possible cryptograms)*. Each particular transformation corresponds to encryption (enciphering) of a plaintext with a particular key. The invertibility of the transformation means that unique decryption (deciphering) of the cryptogram is possible when the key is known. Let **X** denote the plaintext message, **Y** denote the cryptogram, and **Z** denote the key. Let **F** denote the invertible transformation producing the cryptogram **Y**, as follows:

$$\mathbf{Y} = \mathbf{F}(\mathbf{X}, \mathbf{Z}) = \mathbf{F}_z(\mathbf{X}) \tag{A5.1}$$

The transformation is intended to make the cryptogram **Y** useless to the enemy. At the receiving end of the system, the cryptogram **Y** is decrypted with the inverse transformation $\mathbf{F}^{-1}$ to recover the original plaintext message **X**, as shown by

$$\mathbf{F}^{-1}(\mathbf{Y}, \mathbf{Z}) = \mathbf{F}_z^{-1}(\mathbf{Y}) = \mathbf{F}_z^{-1}(\mathbf{F}_z(\mathbf{X})) = \mathbf{X} \tag{A5.2}$$

In physical terms, the cryptographic system consists of a set of instructions, a piece of physical hardware, or a computer program. In any event, the system is designed to have the capability of encrypting the plaintext (and, of course, decrypting the resulting cryptogram) in a variety of ways; the particular way chosen to do the actual encryption is determined by the specific key.

The security of the system resides in the secret nature of the key, which requires that the key must be delivered to the receiver over a *secure channel* (e.g., registered mail, courier service) as implied in Figure A5.1. The cryptographic system depicted in this figure provides a solution to the *secrecy problem*, preventing an enemy from extracting information from messages transmitted over an insecure communication channel. Cryptography also provides a solution to the *authentication problem*, preventing an enemy cryptanalyst from impersonating the message sender. In this second situation, the enemy cryptanalyst is the one who originates a "fraudulent" cryptogram **Y′** that is delivered to the receiver (decrypter), as shown in Figure A5.2. The authentic cryptogram **Y** is shown as a dashed input to the enemy cryptanalyst, indicating that the enemy produces the fraudulent cryptogram **Y′** without ever seeing the authentic one. The receiver may be able to recognize **Y′** as fraudulent by decrypting it with the correct key **Z**; hence, the line from the receiver output to the destination is shown dashed to suggest rejection of the fraudulent cryptogram **Y′** by the receiving user.

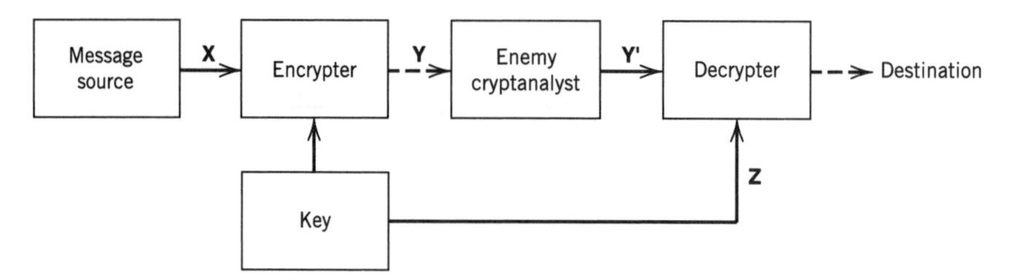


**FIGURE A5.2** Illustrating the intrusion of an enemy cryptanalyst.

## A5.2 *Block and Stream Ciphers*

Much as error-correcting codes are classified into block codes and convolutional codes, cryptographic systems (ciphers) may be classified into two broad classes: *block ciphers* and *stream ciphers*. Block ciphers operate in a purely combinatorial fashion on large blocks of plaintext, whereas stream ciphers process the plaintext in small pieces (i.e., characters or bits).

Figure A5.3 shows the generic form of a block cipher. The plaintext (consisting of serial data) is divided into large blocks, each of which is usually made up of a fixed number of bits. Successive blocks of the plaintext are enciphered (encrypted) using the same secret key, otherwise independently; the resulting enciphered blocks are finally converted into serial form. Thus, a particular plaintext block identical to a previous such block gives rise to an identical ciphertext block. Specifically, each bit of a particular ciphered block is chosen to be a function of all the bits of the associated plaintext block and the key; the goal of a block cipher is to have no specific bit of the plaintext ever appear in the ciphertext directly.

Block ciphers operate with a fixed transformation applied to large blocks of plaintext data, on a block-by-block basis. In contrast, a stream cipher operates on the basis of a time-varying transformation applied to individual bits of the plaintext. The most popular stream ciphers are the so-called *binary additive stream ciphers*, the generic form of which is shown in Figure A5.4. In such a cipher, the secret key is used to control a *keystream generator* that emits a binary sequence called the *keystream*, whose length is much larger than that of the key. Let $x_n$, $y_n$, and $z_n$ denote the plaintext bit, ciphertext bit, and keystream bit at time $n$, respectively. The ciphertext bits are then determined by simple modulo-2 addition of the plaintext bits and the keystream bits, as shown by

$$y_n = x_n \oplus z_n, \qquad n = 1, 2, \ldots, N \tag{A5.3}$$

where $N$ is the length of the keystream. Because addition and subtraction in modulo-2 arithmetic are exactly the same, Equation (A5.3) also implies the following relation

$$x_n = y_n \oplus z_n, \qquad n = 1, 2, \ldots, N \tag{A5.4}$$

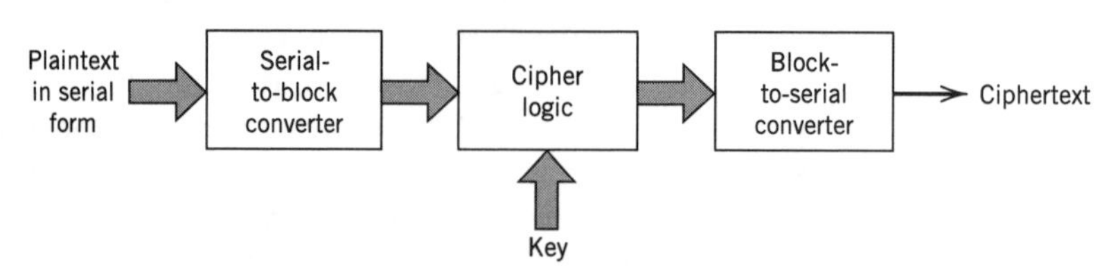


**FIGURE A5.3** Block diagram of a block cipher.

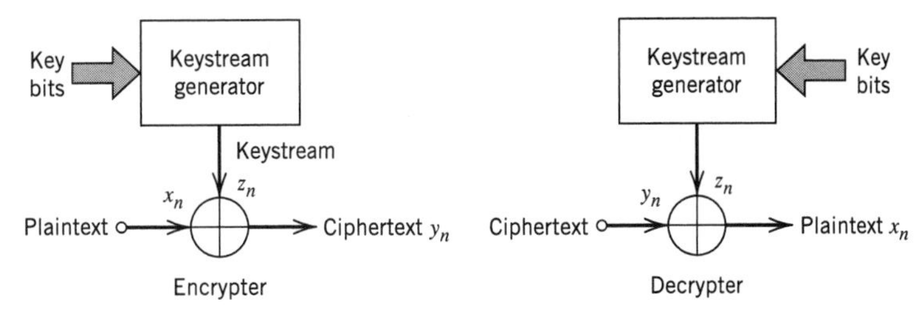


**FIGURE A5.4** Binary additive stream cipher.

We thus see that in binary additive stream ciphers, identical devices can be used to perform encryption and decryption, as shown in Figure A5.4. The secret key is chosen according to some probability distribution. To provide secure encryption, the keystream should resemble a coin-tossing (i.e., completely random) sequence as closely as possible.

Block ciphers are normally designed in such a way that a small change in an input block of plaintext produces a major change in the resulting output. This *error propagation* property of block ciphers is valuable in authentication in that it makes it improbable for an enemy cryptanalyst to modify encrypted data, unless knowledge of the key is available. On the other hand, a binary additive stream cipher has *no* error propagation; the decryption of a distorted bit in the ciphertext affects only the corresponding bit of the resulting output.

Stream ciphers are generally better suited for the secure transmission of data over error-prone communication channels; they are used in applications where high data rates are a requirement (as in secure video, for example) or when a minimal transmission delay is essential.[5]

## ■ REQUIREMENT FOR SECRECY

In cryptography, a fundamental assumption is that an enemy cryptanalyst has knowledge of the entire mechanism used to perform encryption, except for the secret key. We may identify the following forms of attack that may be attempted by the enemy cryptanalyst, depending on the availability of additional knowledge:

1. *Ciphertext-only attack* is a cryptanalytic attack in which the enemy cryptanalyst has access to part or all of the ciphertext.
2. *Known-plaintext attack* is a cryptanalytic attack in which the enemy cryptanalyst has knowledge of some ciphertext–plaintext pairs formed with the actual secret key.
3. *Chosen-plaintext attack* is a cryptanalytic attack in which the enemy cryptanalyst is able to submit any chosen plaintext message and receive in return the correct ciphertext for the actual secret key.
4. *Chosen-ciphertext attack* is a cryptanalytic attack in which the enemy cryptanalyst is able to choose an arbitrary ciphertext and find the correct result for its decryption.

A ciphertext-only attack occurs frequently in practice. In this form of attack, an enemy cryptanalyst uses only knowledge of the statistical structure of the language in use (e.g., in English the letter *e* occurs with a probability of 13 percent, and the letter *q* is always followed by *u*) and knowledge of some probable words (e.g., a letter probably begins with "Dear Sir/Madam:"). A known-plaintext attack may take place by virtue of the standard computer formats used in programming languages and data generation. In any case, the ciphertext-only attack is viewed as the weakest threat to which a cryptographic system can be subjected, and any system that succumbs to it is therefore considered totally insecure. Thus, for a cryptographic system to provide secrecy, at the minimum it should be immune to ciphertext-only attacks; ideally, it should also be immune to known-plaintext attacks.

## A5.3 *Information-Theoretic Approach*

In the *Shannon model of cryptography*, named in recognition of Shannon's 1949 landmark paper on the information-theoretic approach to secrecy systems, the enemy cryptanalyst is assumed to have unlimited time and computing power. But the enemy is presumably restricted to a ciphertext-only attack. Cryptanalysis in the Shannon model is defined as the process of finding the secret key, given the cryptogram (ciphertext) and the *a priori* probabilities of the various plaintexts and keys. The secrecy of the system is considered *broken* when the enemy cryptanalyst performs decryption successfully, obtaining a *unique* solution to the cryptogram.[6]

Let $\mathbf{X} = (X_1, X_2, \ldots, X_N)$ denote an $N$-bit plaintext message, and $\mathbf{Y} = (Y_1, Y_2, \ldots, Y_N)$ denote the corresponding $N$-bit cryptogram; that is, both the plaintext and the cryptogram have the same number of bits. It is assumed that the secret key $\mathbf{Z}$ used to construct the cryptogram is drawn according to some probability distribution. The uncertainty about $\mathbf{X}$ is expressed by the entropy $H(\mathbf{X})$, and the uncertainty about $\mathbf{X}$ given knowledge of $\mathbf{Y}$ is expressed by the conditional entropy $H(\mathbf{X}|\mathbf{Y})$. The *mutual information* between $\mathbf{X}$ and $\mathbf{Y}$ is defined by

$$I(\mathbf{X}; \mathbf{Y}) = H(\mathbf{X}) - H(\mathbf{X}|\mathbf{Y}) \tag{A5.5}$$

The mutual information $I(\mathbf{X}; \mathbf{Y})$ represents a basic measure of security (secrecy) in the Shannon model.

### ■ PERFECT SECURITY

Assuming that an enemy cryptanalyst can observe only the cryptogram $\mathbf{Y}$, it seems appropriate that we define the *perfect security* of a cryptographic system to mean that the plaintext $\mathbf{X}$ and the cryptogram $\mathbf{Y}$ are statistically independent. In other words, we have

$$I(\mathbf{X}; \mathbf{Y}) = 0 \tag{A5.6}$$

Then, using Equation (A5.5), we find that the condition for perfect security may be rewritten as

$$H(\mathbf{X}|\mathbf{Y}) = H(\mathbf{X}) \tag{A5.7}$$

Equation (A5.7) states that the best an enemy cryptanalyst can do, given the cryptogram $\mathbf{Y}$, is to guess the plaintext message $\mathbf{X}$ according to the probability distribution of all possible messages.

Given the secret key $\mathbf{Z}$, we recognize that

$$\begin{aligned} H(\mathbf{X}|\mathbf{Y}) &\le H(\mathbf{X}, \mathbf{Z}|\mathbf{Y}) \\ &= H(\mathbf{Z}|\mathbf{Y}) + H(\mathbf{X}|\mathbf{Y}, \mathbf{Z}) \end{aligned} \tag{A5.8}$$

The conditional entropy $H(\mathbf{X}|\mathbf{Y}, \mathbf{Z})$ is zero if, and only if, $\mathbf{Y}$ and $\mathbf{Z}$ together uniquely determine $\mathbf{X}$; this is indeed a valid assumption when the decryption process is performed with knowledge of the secret key $\mathbf{Z}$. Hence, we may simplify Equation (A5.8) as follows:

$$\begin{aligned} H(\mathbf{X}|\mathbf{Y}) &\le H(\mathbf{Z}|\mathbf{Y}) \\ &\le H(\mathbf{Z}) \end{aligned} \tag{A5.9}$$

Thus, substituting Equation (A5.9) into (A5.7), we find that for a cryptographic system to provide perfect security, the following condition must be satisfied:

$$H(\mathbf{Z}) \geq H(\mathbf{X}) \qquad \text{(A5.10)}$$

The inequality of Equation (A5.10) is *Shannon's fundamental bound for perfect security*; it states that for perfect security, the uncertainty of a secret key **Z** must be at least as large as the uncertainty of the plaintext **X** that is concealed by the key.

For the case when the plaintext and key alphabets are of the same size, the use of Shannon's bound for perfect security yields the following result: *The key must be at least as long as the plaintext.* The conclusion to be drawn from this result is that the length of the *secret key* needed to build a perfectly secure cryptographic system may be impractically large for most applications. Nevertheless, perfect security has a place in the practical picture: It may be used when the number of possible messages is small or in cases where the greatest importance is attached to perfect security.

A well-known, perfectly secure cipher is the *one-time pad*[7] (sometimes called the *Vernam cipher*), which is used for unconventional applications such as two users communicating on a hotline with high confidentiality requirements. The one-time pad is a stream cipher for which the key is the same as the keystream, as shown in Figure A5.5. For encryption the input consists of two components: a message represented by a sequence of message bits $\{x_n | n = 1, 2, \ldots\}$, and a key represented by a sequence of statistically independent and uniformly distributed bits $\{z_n | n = 1, 2, \ldots\}$. The resultant cipher $\{y_n | n = 1, 2, \ldots\}$ is obtained by the modulo-2 addition of the two input sequences, as shown by

$$y_n = x_n \oplus z_n, \qquad n = 1, 2, \ldots$$

Consider, for example, the binary message sequence 00011010 and the binary key sequence 01101001. The modulo-2 addition of these two sequences is written as follows:

| | |
|---|---|
| Message: | 00011010 |
| Key: | 01101001 |
| Cipher: | 01110011 |

In the encryption rule described here, key bit 1 interchanges 0s and 1s in the message sequence, and key bit 0 leaves the message bits unchanged. The message sequence is recovered simply by modulo-2 addition of the binary cipher and key sequences, as shown by

| | |
|---|---|
| Cipher: | 01110011 |
| Key: | 01101001 |
| Message: | 00011010 |

The one-time pad is perfectly secure, because the mutual information between the message and the cipher is zero; it is therefore completely undecipherable.

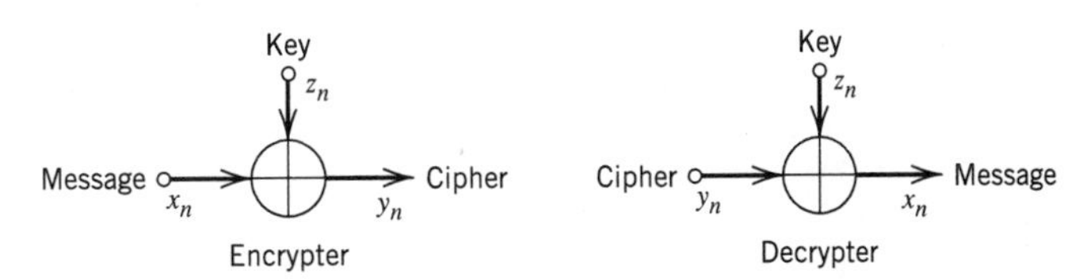


**FIGURE A5.5** One-time pad (Vernam cipher).

## ■ UNICITY DISTANCE

Consider now the practical case of an imperfect cipher and ask the question: When can an enemy cryptanalyst break the cipher? As the amount of intercepted text increases, intuitively we expect that a point may be reached at which it becomes possible for an enemy cryptanalyst with unlimited time and computing power to find the key and thus break the cipher. This critical point in the Shannon model is called the *unicity distance*, which is formally defined as the smallest $N$ such that the conditional entropy $H(\mathbf{Z} | Y_1, Y_2, \ldots, Y_N)$ is approximately zero. For a particular kind of "random cipher," the unicity distance is approximately given by[8]

$$N_0 \simeq \frac{H(\mathbf{Z})}{r \log L_y} \tag{A5.11}$$

where $H(\mathbf{Z})$ is the entropy of the key $\mathbf{Z}$, and $L_y$ is the size of the ciphertext alphabet. The parameter $r$ is the *percentage redundancy* of the message information contained in the $N$-bit ciphertext; it is itself defined by

$$r = 1 - \frac{H(\mathbf{X})}{N \log L_y} \tag{A5.12}$$

where $H(\mathbf{X})$ is the entropy of the plaintext $\mathbf{X}$. In most cryptographic systems, the size $L_y$ of the ciphertext alphabet is the same as the size $L_x$ of the plaintext alphabet; in such a case, $r$ is just the percentage redundancy of the plaintext itself. Although the derivation of Equation (A5.11) assumes a certain well-defined "random cipher," it can be used to estimate the unicity distance for ordinary types of ciphers, which is the routine practice today.

Let $K$ be the number of digits in the key $\mathbf{Z}$ that are chosen from an alphabet of size $L_z$; then we may express the entropy of the key $\mathbf{Z}$ as follows:

$$H(\mathbf{Z}) \leq \log(L_z^K) = K \log L_z \tag{A5.13}$$

with equality if and only if the key is completely random. Let the size $L_z$ of the key alphabet be the same as the size $L_y$ of the ciphertext alphabet, and let the key be chosen completely at random to maximize the unicity distance. Then, substituting Equation (A5.13) with equality into Equation (A5.11), we get the simple result

$$N_0 \simeq \frac{K}{r} \tag{A5.14}$$

To illustrate the application of Equation (A5.14), consider a cryptographic system with $L_x = L_y = L_z$, which is used for the encryption of English text. The percentage redundancy $r$ for typical English text is about 75 percent. Hence, according to Equation (A5.14), an enemy cryptanalyst can break the cipher after intercepting only about $1.333K$ bits of ciphertext data, where K is the key size.

However, it is important to note that an imperfect cipher that is potentially breakable can still be of practical value. When the intercepted ciphetext contains sufficient information to satisfy Equation (A5.11), there is no guarantee that an enemy cryptanalyst with limited computational resources can actually break the cipher. Specifically, it is possible for the cipher to be designed in such a way that the task of the cryptanalysis, though known to be attainable with a finite amount of computation, is so overwhelming that it will literally exhaust the physical computing resources of the universe. In such a case, the imperfect cipher is said to be *computationally secure*.

## ROLE OF DATA COMPRESSION IN CRYPTOGRAPHY

Lossless data compression or data compaction is a useful tool in cryptography. We say this because data compaction removes redundancy, thereby increasing the unicity distance $N_0$ in accordance with Equation (A5.11). To exploit this idea, data compaction is used prior to encryption in the transmitter, and the redundant information is reinserted after decryption in the receiver; the net result is that the authorized user at the receiver output sees no difference, and yet the information transmission has been made more secure. It would be tempting to consider the use of perfect data compaction to remove all redundancy, thereby transforming a message source into a completely random source and resulting in $N_0 = \infty$ with any key size. Unfortunately, we do not have a device capable of performing perfect data compaction on realistic message sources, nor is it likely that there will ever be such a device. It is therefore futile to rely on data compaction alone for data security. Nevertheless, limited data compaction tends to increase security, which is the reason why cryptographers view data compression as a useful trick.

## DIFFUSION AND CONFUSION

In the Shannon model of cryptography, two methods suggest themselves as general principles to guide the design of practical ciphers. The methods are called *diffusion* and *confusion*, the aims of which (by themselves or together) are to frustrate a statistical analysis of ciphertext by the enemy and therefore make it extremely difficult to break the cipher.

In the method of diffusion, the statistical structure of the plaintext is hidden by spreading out the influence of a single bit in the plaintext over a large number of bits in the ciphertext. This spreading has the effect of forcing the enemy to intercept a tremendous amount of material for the determination of the statistical structure of the plaintext, since the structure is evident only in many blocks, each one of which has a very small probability of occurrence. In the method of confusion, the data transformations are designed to complicate the determination of the way in which the statistics of the ciphertext depend on the statistics of the plaintext. Thus, a good cipher uses a combination of diffusion and confusion.

For a cipher to be of practical value, however, it must not only be difficult to break the cipher by an enemy cryptanalyst, but also it should be easy to encrypt and decrypt data given knowledge of the secret key. We may satisfy these two design objectives using a *product cipher*, based on the notion of "divide and conquer." Specifically, the implementation of a strong cipher is accomplished as a succession of simple component ciphers, each of which contributes a modest amount of diffusion and confusion to the overall makeup of the cipher. Product ciphers are often built using substitution ciphers and transposition ciphers as basic components; these simple ciphers are described next.

1. *Substitution cipher.*
In a substitution cipher each letter of the plaintext is replaced by a fixed substitute, usually also a letter from the same alphabet, with the particular substitution rule being determined by the secret key. Thus the plaintext

$$\mathbf{X} = (x_1, x_2, x_3, x_4, \ldots)$$

where $x_1, x_2, x_3, \ldots$ are the successive letters, is transformed into the ciphertext

$$\begin{aligned} \mathbf{Y} &= (y_1, y_2, y_3, y_4, \ldots) \\ &= (f(x_1), f(x_2), f(x_3), f(x_4), \ldots) \end{aligned} \tag{A5.15}$$

| | |
|---|---|
| Plaintext letters | ABCDEFGHIJKLMNOPQRSTUVWXYZ |
| Ciphertext letters | YDUBHNACSVXELPFMKQJRWGOZIT |

**FIGURE A5.6** Substitution cipher.

where $f(\cdot)$ is a function with an inverse. When the substitutes are letters, the key is a permutation of the alphabet. Consider, for example, the ciphertext alphabet of Figure A5.6, where we see that the first letter $Y$ is the substitute for $A$, the second letter $D$ is the substitute for $B$, and so on. The use of a substitution cipher results in confusion.

2. *Transposition cipher.* In a transposition cipher, the plaintext is divided into groups of fixed period $d$ and the same permutation is applied to each group, with the particular permutation rule being determined by the secret key. For example, consider the permutation rule described in Figure A5.7, for which the period is $d = 4$. According to this cipher, letter $x_1$ is moved from position 1 in the plaintext to position 4 in the ciphertext. Thus, the plaintext

$$\mathbf{X} = (x_1, x_2, x_3, x_4, x_5, x_6, x_7, x_8, \ldots)$$

is transformed into the ciphertext

$$\mathbf{Y} = (x_3, x_4, x_2, x_1, x_7, x_8, x_6, x_5, \ldots)$$

Although the single-letter statistics of the ciphertext **Y** are the same as those of the plaintext **X**, the higher-order statistics are changed. The use of a transposition cipher results in diffusion.

By interleaving the simple substitutions and transpositions and repeating the interleaving process many times, it is possible to build a strong cipher equipped with good diffusion and confusion.

### ▶ EXAMPLE A5.1

Consider the plaintext message

THE KING IS DEAD LONG LIVE THE KING

Using the permuted alphabet described in Figure A5.6 for the substitution cipher, this plaintext is transformed into the ciphertext

RCHXSPASJBHYBEFPAESGHRCHXSPA

Suppose next we apply the permutation rule described in Figure A5.7 for the transposition cipher; accordingly, the ciphertext resulting from the substitution cipher is further transformed into

HXCRASPSHYBJFBEBSGEACHRHPASX

which has no resemblance to the original plaintext. ◀

| | | | | |
|---|---|---|---|---|
| Plaintext letters | $x_1$ | $x_2$ | $x_3$ | $x_4$ |
| Ciphertext letters | $x_3$ | $x_4$ | $x_2$ | $x_1$ |

**FIGURE A5.7** Transposition cipher.

## A5.4 Data Encryption Standard

The *data encryption standard* (DES)[9] is certainly the best known, and arguably the most widely used, secret-key cryptoalgorithm; the term *algorithm* is used to describe a sequence of computations. The basic DES algorithm can be used for both data encryption and data authentication. It is the standard cryptoalgorithm for data storage and mail systems, electronic funds transfers (retail and wholesale), and electronic business data interchange.

The DES algorithm is a strong block cipher that operates on 64-bit blocks of plaintext data and uses a 56-bit key; it is designed in accordance with Shannon's methods of diffusion and confusion. Essentially the same algorithm is used for encryption and decryption. The overall transformations employed in the DES algorithm may be written as $P^{-1}\{F[P(\mathbf{X})]\}$, where $\mathbf{X}$ is the plaintext, $P$ is a certain permutation, and the function $F$ combines substitutions and transpositions. The function $F$ is itself obtained by cascading a certain function $f$, with each stage of the cascade referred to as a *round*.

The flow-chart of Figure A5.8 shows the details of the DES algorithm for encryption. After a certain initial permutation, a plaintext of 64 bits is divided into a left-half $L_0$ and a right-half $R_0$, each of which is 32 bits long. The algorithm then performs 16 rounds of a key-dependent computation, with the $i$th round of the computation described as follows:

$$L_i = R_{i-1} \qquad i = 1, 2, \ldots, 16 \tag{A5.16}$$

$$R_i = L_{i-1} \oplus f(R_{i-1}, Z_i) \qquad i = 1, 2, \ldots, 16 \tag{A5.17}$$

On the right-hand side of Equation (A5.17), the addition is modulo-2 and each $Z_i$ is a different 48-bit block of the key used in round $i$. The function $f(\cdot, \cdot)$ is a function with a 32-bit output. The result of the 16th round is reversed, obtaining the sequence $R_{16}L_{16}$. This 32-bit sequence is input into a final permutation $P^{-1}$ to produce the 64-bit ciphertext. The aim is that after 16 rounds of key-dependent computations, the patterns in the original plaintext are undetectable in the ciphertext. From Equations (A5.16) and (A5.17), we note that for decryption the function $f(\cdot, \cdot)$ need not be invertible, because $(L_{i-1}, R_{i-1})$ can be recovered from $(L_i, R_i)$ simply as follows:

$$R_{i-1} = L_i \qquad i = 1, 2, \ldots, 16 \tag{A5.18}$$

$$L_{i-1} = R_i \oplus f(L_i, Z_i) \qquad i = 1, 2, \ldots, 16 \tag{A5.19}$$

Equation (A5.19) holds even if the function $f(\cdot, \cdot)$ is a many-to-one function (i.e., it does not have a unique inverse).

Figure A5.9 shows the flowchart for computing the function $f(\cdot, \cdot)$. The 32-bit block $R$ is first expanded into a new 48-bit block $R'$ by repeating the edge bits of each successive 4-bit word (i.e., the bits numbered 1, 4, 5, 8, 9, 12, 13, 16, . . . , 28, 29, 32). Thus, given the 32-bit block $R$ written as

$$R = \underbrace{r_1r_2r_3r_4}_{\substack{\text{first}\\ \text{4-bit word}}} \quad \underbrace{r_5r_6r_7r_8}_{\substack{\text{second}\\ \text{4-bit word}}} \quad \cdots \quad \underbrace{r_{29}r_{30}r_{31}r_{32}}_{\substack{\text{eighth}\\ \text{4-bit word}}}$$

we construct the expanded 48-bit block $R'$ as follows:

$$R' = \underbrace{r_{32}r_1r_2r_3r_4r_5}_{\substack{\text{first}\\ \text{6-bit word}}} \quad \underbrace{r_4r_5r_6r_7r_8r_9}_{\substack{\text{second}\\ \text{6-bit word}}} \quad \cdots \quad \underbrace{r_{28}r_{29}r_{30}r_{31}r_{32}r_1}_{\substack{\text{eighth}\\ \text{6-bit word}}}$$

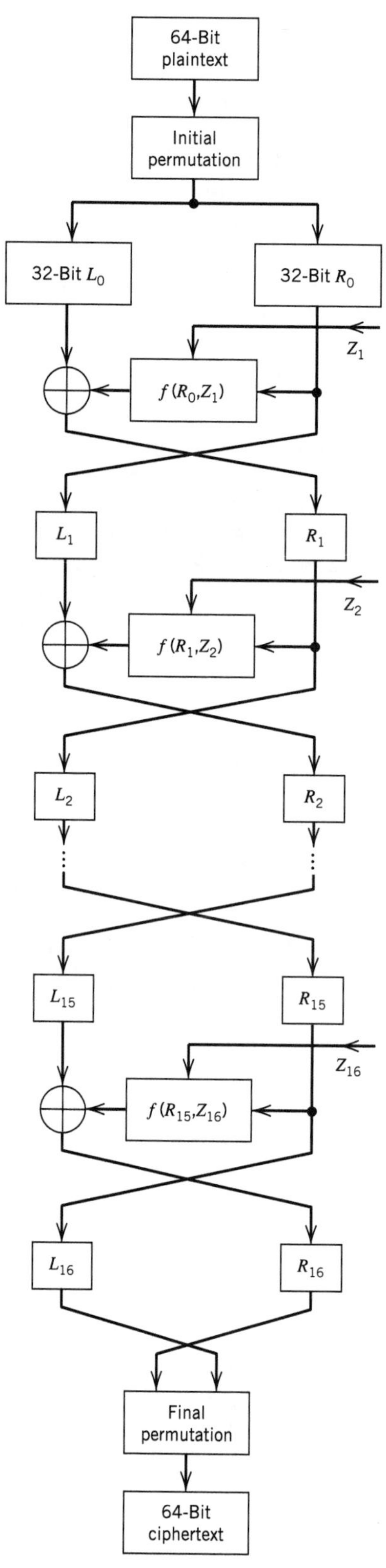


**FIGURE A5.8** Data encryption standard. (From Diffie and Hellman, 1979, with permission of the IEEE.)

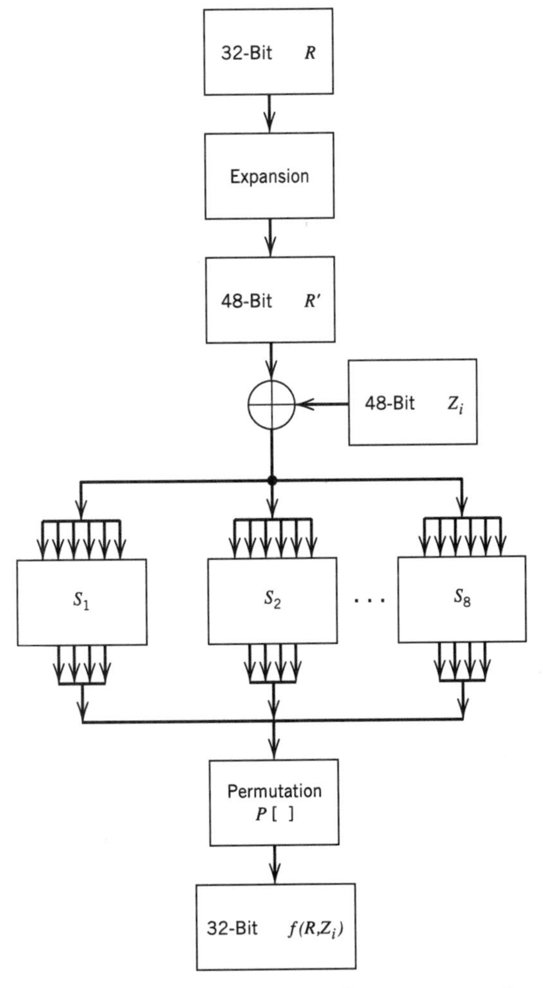


**FIGURE A5.9** $f(R, K)$ flowchart. (From Diffie and Hellman, 1979, with permission of the IEEE.)

The 48-bit blocks $R'$ and $Z_i$ are added modulo-2, and the resultant is divided into eight 6-bit words. Let these words be denoted by $B_1, B_2, \ldots, B_8$. We thus write

$$B_1 B_2 \cdots B_8 = R' \oplus Z_i \tag{A5.20}$$

Each 6-bit word $B_i$ is input to a substitution box $S_i$ in the form of a look-up table, producing a 4-bit output $S_i(B_i)$. Each output bit of the substitution box $S_i$ $(B_i)$ is a Boolean function of the 6-bit word $B_i$. The eight outputs $S_1(B_1), S_2(B_2), \ldots, S_8(B_8)$ are arranged into a single 32-bit block that is input to the permutation box denoted by $P[\cdot]$. The permuted output so produced is the desired 32-bit function $f(R, Z_i)$, as shown by

$$f(R, Z_i) = P[S_1(B_1)S_2(B_2) \cdots S_8(B_8)] \tag{A5.21}$$

The 48-bit block $Z_i$ for the $i$th iteration uses a different subset of the 64-bit key $Z_0$. The procedure used to determine each $Z_i$ is called the *key-schedule calculation*, the flowchart of which is shown in Figure A5.10. The key $Z_0$ has eight parity bits in positions 8, 16, . . . , 64, which are used for error detection in their respective 8-bit bytes; the errors

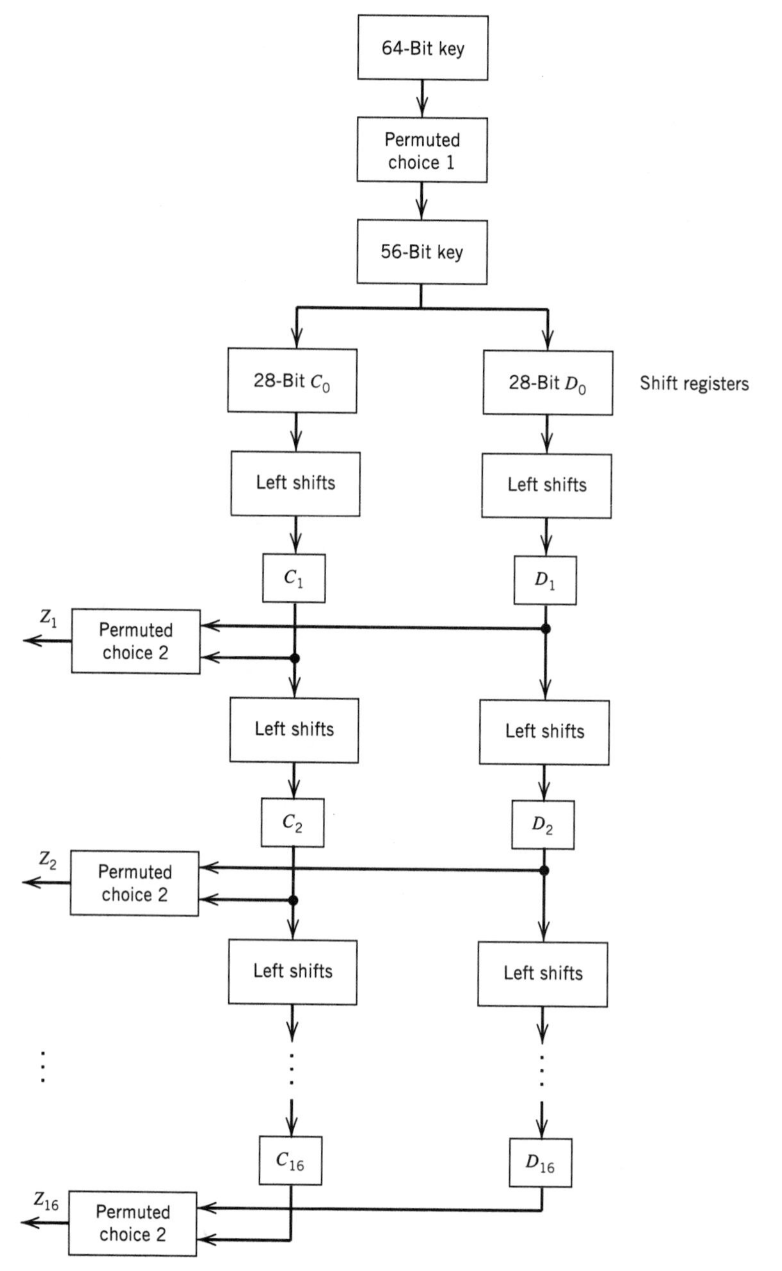


**FIGURE A5.10** Flowchart for the key-schedule calculation. (From Diffie and Hellman, 1979, with permission of the IEEE.)

of concern may arise in the generation, distribution, and storage of the key $Z_0$. The permuted choice 1 disregards the parity bits of $Z_0$ and then permutes the remaining 56 bits that are loaded into two 28-bit shift registers, each with 24 taps. The 48 taps of the two shift registers are subjected to 16 iterations of computation, with each iteration involving one or two cyclic left shifts followed by a permutation, referred to as permuted choice 2.

The outputs resulting from these 16 iterations provide the different 48-bit blocks $Z_1$, $Z_2, \ldots, Z_{16}$ of the key used in iteration 1, 2, . . . , 16, respectively.

Despite all the claims to the contrary, it appears that no one has yet demonstrated a fundamental weakness of the DES algorithm. Notwithstanding all the controversy surrounding its use, perhaps the most significant contribution of the DES algorithm is the fact that it has been instrumental in raising the level of interest in using cryptography as a mechanism for secure computer networks.

## A5.5 *Public-Key Cryptography*[10]

For a pair of users to engage in cryptographic communication over an insecure channel, it is necessary for the users to exchange key information prior to communication. The requirement for a secure distribution of keys among authorized users applies to all cryptographic systems, regardless of their type. In conventional cryptography, the users employ a physically secure channel (e.g., courier service or registered mail) for key distribution. However, the use of such a supplementary channel points to a major limitation of conventional cryptography. Needless to say, the use of courier service or registered mail for key distribution is costly, inconvenient, low-bandwidth, and slow; also, it is not always secure.

The problem of key distribution is particularly accentuated in large communication networks, where the number of possible connections grows as $(n^2 - n)/2$ for $n$ users. For large $n$, the cost of key distribution becomes prohibitive. Thus, in the development of large, secure communication networks, we are compelled to rely on the use of insecure channels for both exchange of key information and subsequent secure communication. This constraint raises a fundamental question: How can key information be exchanged securely over an insecure channel? In *public-key cryptography*, this seemingly difficult issue is resolved by making some key material "public" and thereby considerably simplifying the task of key management. This is in direct contrast to conventional cryptography, where the key is kept completely secret from an enemy cryptanalyst.

A public key cryptographic system is described by two sets of algorithms that compute invertible functions (transformations). Let these two sets of algorithms be denoted by $\{E_z\}$ and $\{D_z\}$ that are indexed by $z$. The invertible transformations computed by these algorithms may be written as follows

$$E_z\colon f_z(x) = y \tag{A5.22}$$

$$D_z\colon f_z^{-1}(y) = x \tag{A5.23}$$

where $x$ is a certain input message in the domain of some function $f_z$ indexed by $z$, and $y$ is the corresponding cryptogram in the range of $f_z$. A fundamental requirement of the system is that the function $f_z$ must be a *trapdoor one-way function*. The term "one-way" refers to the fact that for $x$ in the domain of $f_z$, it must be easy to compute $f_z(x)$ from knowledge of the algorithm $E_z$, but for a certain cryptogram $y$ in the range of $f_z$, an enemy cryptanalyst must find it extremely difficult to compute the inverse $f_z^{-1}(y)$. On the other hand, an authorized user in possession of the associated algorithm $D_z$ would find it easy to compute the inverse $f_z^{-1}(y)$. Thus the *private key* (algorithm) $D_z$ provides a "trapdoor" that makes the problem of inverting the function $f_z$ appear extremely difficult from the viewpoint of the cryptanalyst, but easy for the (sole authorized) possessor of $D_z$. Since knowledge of the key (algorithm) $E_z$ does not by itself make it possible to compute the inverse of $f_z$, it may be made *public*; hence, the name "public-key cryptography."

The notion emerging from the description of a public-key cryptographic system presented herein is that the keys come in inverse pairs (i.e., public key and private key), and that each pair of keys has two basic properties:

1. *Whatever message is encrypted with one of the keys can be decrypted with the other key.*
2. *Given knowledge of the public key, it is computationally infeasible to find the secret key.*

The use of public-key cryptography as described herein makes it possible to solve the secrecy problem as follows. Subscribers to a secure communication system list their public keys in a "telephone directory" along with their names and addresses. A subscriber can then send a private message to another subscriber simply by looking up the public key of the addressee and using the key to encrypt the message. The encrypted message (i.e., ciphertext) can only be read by the holder of that particular public key. In fact, should the original message (i.e., plaintext) be lost, even its sender would find it extremely difficult to recover the message from the ciphertext.

The key management of public-key cryptography makes it well suited for the development of large, secure communication networks. Indeed, it has evolved from a simple concept to a mainstay of cryptographic technology.

## ■ DIFFIE–HELLMAN PUBLIC KEY DISTRIBUTION

In a simple and yet elegant system known as the *Diffie–Hellman public key-distribution system*, use is made of the fact that it is easy to calculate a discrete exponential but difficult to calculate a discrete logarithm. To be more specific, consider the *discrete exponential function*

$$Y = \alpha^X \bmod p \qquad \text{for } 1 \leq X \leq p - 1 \tag{A5.24}$$

where the arithmetic is performed modulo-$p$. The $\alpha$ is an integer that should be *primitive* (i.e., all powers of $\alpha$ generate all the elements mod $p$ relatively prime to $p - 1$). Correspondingly, $X$ is referred to as the *discrete logarithm* of $Y$ to the base $\alpha$, mod $p$, as shown by

$$X = \log_\alpha Y \bmod p \qquad \text{for } 1 \leq Y \leq p - 1 \tag{A5.25}$$

The calculation of $Y$ from $X$ is easy, using the trick of square-and-multiply. For example, for $X = 16$ we have

$$Y = \alpha^{16} = \{[(\alpha^2)^2]^2\}^2$$

On the other hand, the problem of calculating $X$ from $Y$ is much more difficult.

In the Diffie–Hellman public key-distribution system, all users are presumed to know both $\alpha$ and $p$. A user $i$, say, selects an independent random number $X_i$ uniformly from the set of integers $\{1, 2, \ldots, p\}$ that is kept as a *private secret*. But the discrete exponential

$$Y_i = \alpha^{X_i} \bmod p \tag{A5.26}$$

is deposited in a *public directory* with the user's name and address. Every other user of the system does the same thing. Now, suppose that users $i$ and $j$ wish to communicate privately. To proceed, user $i$ fetches $Y_j$ from the public directory and uses the private secret $X_i$ to compute

$$\begin{aligned} K_{ji} &= (Y_j)^{X_i} \bmod p \\ &= (\alpha^{X_j})^{X_i} \bmod p \\ &= \alpha^{X_j X_i} \bmod p \end{aligned} \tag{A5.27}$$

In a similar way, user $j$ computes $K_{ij}$. But we have

$$K_{ji} = K_{ij} \tag{A5.28}$$

Accordingly, users $i$ and $j$ arrive at $K_{ji}$ as the *secret key* in a conventional cryptosystem. Another user must compute $K_{ji}$ using the information $Y_i$ and $Y_j$ obtained from the public directory, applying the alternative formula

$$K_{ji} = (Y_j)^{\log Y_i} \bmod p \tag{A5.29}$$

Apparently, there is no other method for an enemy to find the secret key $K_{ji}$; however, there is no proof for it. In light of what we said earlier, Equation (A5.29) is difficult to calculate as it involves a discrete logarithm, whereas Equation (A5.27) is easy to calculate as it involves a discrete exponential. Thus, security of the system depends on the difficulty encountered in computing a discrete logarithm.

The Diffie–Hellman public key-distribution system is the oldest system in its class; nevertheless, it is still generally considered to be one of the most secure and practical public key-distribution systems.

## A5.6 Rivest–Shamir–Adleman System

To develop a public-key cryptographic system is no easy task. Indeed, numerous such systems have been proposed in the literature, but unfortunately most of them have proven to be insecure. To date, the most successful implementation of public-key cryptography is the *Rivest–Shamir–Adleman (RSA) system,*[11] which uses ideas from classical number theory. It is considered to be one of the most secure cryptographic systems in that it has withstood many attempts by experts in the field to break it.

The RSA algorithm is a block cipher based on the fact that finding a random prime number of large size (e.g., 100 digit) is computationally easy, but factoring the product of two such numbers is currently considered computationally infeasible. Specifically, the computation of parameters specific to the RSA algorithm proceeds as follows:

1. Choose two *very large prime numbers*, $p$ and $q$, at random; the prime numbers have to be fairly carefully chosen as some prime numbers lead to a very weak system.
2. Multiply the numbers $p$ and $q$, obtaining the product

$$pq = n \tag{A5.30}$$

Find the *Euler totient function* of $n$, using the formula

$$\phi(n) = (p - 1)(q - 1) \tag{A5.31}$$

Equation (A5.31) follows from the definition of the Euler totient function $\phi(n)$ as the number of positive integers $i$ less than $n$, such that the *greatest common divisor* of $i$ and $n$ is equal to one.

3. Let $e$ be a positive integer less than $\phi(n)$, such that the greatest common divisor of $e$ and $\phi(n)$ is equal to one. Hence, find a positive integer $d$ less than $\phi(n)$, such that

$$de = 1 \bmod \phi(n) \tag{A5.32}$$

The RSA trapdoor one-way function is then defined simply by computing the *discrete exponentiation*

$$f_z(x) = x^e = y \bmod n \tag{A5.32}$$

The values of $n$ and $e$ constitute the public key; hence, publishing the easy-to-find algorithm $E_z$ to compute the function $f_z$ amounts just to publishing the numbers $n$ and $e$.

The prime numbers $p$ and $q$ constitute the private key. Since $d$ is related to $p$ and $q$, possession of the easy-to-find (when one knows the trapdoor $z$) algorithm $D_z$ to compute the inverse function $f_z^{-1}$ amounts just to knowing $p$ and $q$. In particular, the inverse function is defined by

$$f_z^{-1}(y) = y^d \bmod n \tag{A5.34}$$

The decrypting exponent $d$ is found using Equation (A5.32), which is equivalent to the statement (in ordinary integer arithmetic) that

$$de = \phi(n)Q + 1 \tag{A5.35}$$

for some integer $Q$. Note that $\phi(n)$ is itself related to $p$ and $q$ by Equation (A5.31). Since $y = x^e$, we may use Equations (A5.32) and (A5.33) to write

$$\begin{aligned} y^d &= x^{de} \\ &= x^{\phi(n)Q+1} \\ &= ((x^{\phi(n)})^Q)x \end{aligned} \tag{A5.36}$$

We now make use of a celebrated theorem of Euler, which states that for any positive integers $x$ and $n$ with $x < n$, we have

$$x^{\phi(n)} = 1 \bmod n \tag{A5.37}$$

Hence, the use of Equation (A5.37) in (A5.36) yields the desired decryption:

$$y^d = x \tag{A5.38}$$

We thus see that finding the inverse function $f_z^{-1}$ is easy, given knowledge of the prime numbers $p$ and $q$.

The *security* of the RSA cryptoalgorithm rests on the premise that any method of inverting the function $f_z$ is *equivalent to factoring* $n = pq$. This equivalence raises the question: Is an attack by factoring $n$ computationally feasible? It appears that the answer is no, provided that the prime numbers $p$ and $q$ are on the order of 100 decimal digits each and that there is no revolutionary breakthrough in factoring algorithms.

## ■ DIGITAL SIGNATURES[12]

For an electronic mail system to replace the use of ordinary paper mail for business transactions, it must be possible for a user of the system to "sign" an electronic message. The use of a *digital signature* provides proof that the message did originate from the sender. To satisfy this requirement, the digital signature must have the following properties:

- The receiver of an electronic message is able to verify the sender's signature.
- The signature is not forgeable.
- The sender of a signed electronic message is unable to disclaim it.

To implement digital signatures using the RSA algorithm, we may proceed as follows. A user in possession of the private key $d$ may sign a given message block $m$ by forming the signature

$$s = m^d \bmod n \tag{A5.39}$$

It is difficult to compute $s$ unless the private key $d$ is known. Hence, a digital signature defined in accordance with Equation (A5.39) is difficult to forge. Moreover, the sender of message $m$ cannot deny having sent it, since no one else could have created the signature $s$. The receiver proceeds by using the public key $e$ to compute

$$\begin{aligned} s^e &= (m^d)^e \bmod n \\ &= m^{de} \bmod n \\ &= m \bmod n \end{aligned} \tag{A5.40}$$

where, in the last line, use is made of Equation (A5.32). Hence, the receiver is able to validate the sender's signature by establishing that the computation of $s^e \bmod n$ produces the same result as the deciphered message $m$. Thus, the RSA algorithm satisfies all the three necessary properties of a digital signature.

## A5.7 Summary and Discussion

Cryptography is a "hot" research area. This statement should not come as a surprise. Considering the fact that we are in an *information society*, the importance of cryptography as a security mechanism will continue to grow. In this appendix, we have presented an introductory treatment of this highly important subject.

We may classify cryptography into secret-key cryptography and public-key cryptography, depending on whether the key used for the encryption of a message and its decryption is completely secret or partly public. Alternatively, we may classify a cryptographic system into a block cipher or stream cipher, depending on the method of implementation. A block cipher exhibits error propagation, which can prove highly valuable in authentication.

Among the many cryptographic systems developed to date, the data encryption standard (DES) and the Rivest–Shamir–Adleman (RSA) algorithms stand out as the most successful ones. Both of these cryptoalgorithms are block ciphers. They differ from each other in that the DES algorithm involves the use of a secret key whereas the RSA algorithm involves the use of a public key. In a secret-key system, the same key is shared both by the sender and the receiver. On the other hand, in a public-key system, the key is split into two parts: a public key located in the transmitter and a private (secret) key located in the receiver; in the latter system, it is computationally infeasible to recover the plaintext message from its encrypted version without knowledge of the private key.

Although public-key cryptosystems such as RSA provide an effective method for key management, they are inefficient for the bulk encryption of data due to low bandwidths. In contrast, conventional cryptosystems such as DES provide better throughput, but they require key management. This suggests the possible use of a hybrid approach exploiting the best elements of both cryptosystems as the basis for the practical design of a secure communication system. For example, the RSA algorithm may be used for authentication, and the DES algorithm for encryption.

## NOTES AND REFERENCES

1. For an introductory treatment of cryptography, see Chapter 15 of the book by Adámek (1991). For a comprehensive treatment of the many facets of cryptology, see the book edited by Simmons (1992); this book is an expanded edition of a Special Issue of the *Proceedings of the IEEE* (1988) on cryptology. The chapter contributions of the book by Simmons are written by leading authorities on the subject of cryptology. A nice treatment of cryptology is also presented in the book by van Tilborg (1988).
2. The era of scientific secret-key cryptography was ushered in with the publication of a landmark paper by Shannon (1949), which established the connection between cryptography and information theory.
3. The era of public-key cryptography was established with the publication of another landmark paper by Diffie and Hellman (1976), which showed for the first time that it is possible to have secret communications without any transfer of a key between sender and receiver. It was the paper by Diffie and Hellman that sparked the explosion of research interest in cryptology, which has continued ever since.
4. The term *enemy cryptanalyst* is commonly used in cryptology to refer to a cryptogram interceptor (eavesdropper); its usage originates from military applications.
5. For a comprehensive treatment of stream ciphers, see Chapter 2 written by R. A. Rueppel in the book *Contemporary Cryptology*, edited by Simmons (1992).
6. For a highly readable account of the Shannon model of cryptography, see the opening chapter by J. L. Massey in the book edited by Simmons (1992).
7. The one-time pad derives its name from its use (shortly before, during, and after World War II) by spies of several governments, who were given a pad of paper with a randomly chosen key and told to use it only for a single encryption. The one-time pad is also known as Vernam's cipher, so named in recognition of its originator, G. S. Vernam.
8. For a derivation of Equation (A5.11), see the original paper by Shannon (1949).
9. The history of the DES algorithm is recounted by M. E. Smid and D. K. Branstad in Chapter 1 of the book edited by Simmons (1992). For a description of the DES algorithm, see Diffie and Hellman (1979). See also the books by Meyer and Matyas (1982) and Torrieri (1992, Chapter 6).
10. For a comprehensive treatment of public-key cryptography, see Chapter 4 by J. Nechvatal in the book edited by Simmons (1992). This book also includes a chapter contribution by W. Diffie that describes the several attempts to devise secure public-key cryptoalgorithms and the gradual evolution of a variety of protocols based on them.
11. The RSA system is patented; it is named in recognition of its originators R. L. Rivest, A. Shamir, and L. Adleman. The original reference for this cryptosystem is Rivest, Shamir, and Adleman (1978).
12. The idea of a digital signature was first discussed by Diffie and Hellman (1976). Its implementation using the RSA algorithm is described by Rivest, Shamir, and Adleman (1978). For a detailed treatment of digital signatures, see Chapter 6 by C. J. Michell, F. Piper, and R. Wild in the book edited by Simmons (1992).

# APPENDIX 6

# TABLES

The twelve tables compiled in this final appendix cover the following:

- *ASCII code*
- *Fourier and Hilbert transforms*
- *Bessel functions*
- *Error function*
- *Selected modem standards*
- *Trigonometric identities, series expansions, and integrals*
- *Useful constants and recommended unit prefixes*

## TABLE A6.1 *ASCII code*

*Bit Position*

| | | | | | | | | | | | | |
|---|---|---|---|---|---|---|---|---|---|---|---|---|
| | | | | *7* | *0* | *0* | *0* | *0* | *1* | *1* | *1* | *1* |
| | | | | *6* | *0* | *0* | *1* | *1* | *0* | *0* | *1* | *1* |
| *4* | *3* | *2* | *1* | *5* | *0* | *1* | *0* | *1* | *0* | *1* | *0* | *1* |
| 0 | 0 | 0 | 0 | | NUL | DLE | SP | 0 | @ | P | \ | p |
| 0 | 0 | 0 | 1 | | SOH | DC1 | ! | 1 | A | Q | a | q |
| 0 | 0 | 1 | 0 | | STX | DC2 | ″ | 2 | B | R | b | r |
| 0 | 0 | 1 | 1 | | ETX | DC3 | # | 3 | C | S | c | s |
| 0 | 1 | 0 | 0 | | EOT | DC4 | $ | 4 | D | T | d | t |
| 0 | 1 | 0 | 1 | | ENQ | NAK | % | 5 | E | U | e | u |
| 0 | 1 | 1 | 0 | | ACK | SYN | & | 6 | F | V | f | v |
| 0 | 1 | 1 | 1 | | BEL | ETB | ' | 7 | G | W | g | w |
| 1 | 0 | 0 | 0 | | BS | CAN | ( | 8 | H | X | h | x |
| 1 | 0 | 0 | 1 | | HT | EM | ) | 9 | I | Y | i | y |
| 1 | 0 | 1 | 0 | | LF | SUB | * | : | J | Z | j | z |
| 1 | 0 | 1 | 1 | | VT | ESC | + | ; | K | [ | k | { |
| 1 | 1 | 0 | 0 | | FF | FS | ' | < | L | \ | l | : |
| 1 | 1 | 0 | 1 | | CR | GS | – | = | M | ] | m | } |
| 1 | 1 | 1 | 0 | | SO | RS | . | > | N | ^ | n | ~ |
| 1 | 1 | 1 | 1 | | SI | US | / | ? | O | — | o | DEL |

| | |
|---|---|
| ACK | Acknowledge |
| BEL | Bell or alarm |
| BS | Backspace |
| CAN | Cancel |
| CR | Carriage return |
| DC1 | Device control 1 |
| DC2 | Device control 2 |
| DC3 | Device control 3 |
| DC4 | Device control 4 |
| DEL | Delete |
| DLE | Data link escape |
| EM | End of medium |
| ENQ | Enquiry |
| EOT | End of transmission |
| ESC | Escape |
| ETB | End of transmission block |
| ETX | End of text |
| FF | Form feed |
| FS | File separator |
| GS | Group separator |
| HT | Horizontal tab |
| LF | Line feed |
| NAK | Negative acknowledge |
| NUL | Null or all zeros |
| RS | Record separator |
| SI | Shift in |
| SO | Shift out |
| SOH | Start of heading |
| SP | Space |
| STX | Start of text |
| SUB | Substitute |
| SYN | Synchronous idle |
| US | Unit separator |
| VT | Vertical tab |

(From Couch, 1990, with permission of Macmillan.)

**TABLE A6.2** ***Summary of properties of the Fourier transform***

| *Property* | *Mathematical Description* |
|---|---|
| 1. Linearity | $ag_1(t) + bg_2(t) \rightleftharpoons aG_1(f) + bG_2(f)$ where $a$ and $b$ are constants |
| 2. Time scaling | $g(at) \rightleftharpoons \frac{1}{\lvert a \rvert} G\left(\frac{f}{a}\right)$ where $a$ is a constant |
| 3. Duality | If $g(t) \rightleftharpoons G(f)$, then $G(t) \rightleftharpoons g(-f)$ |
| 4. Time shifting | $g(t - t_0) \rightleftharpoons G(f) \exp(-j2\pi f t_0)$ |
| 5. Frequency shifting | $\exp(j2\pi f_c t) g(t) \rightleftharpoons G(f - f_c)$ |
| 6. Area under $g(t)$ | $\int_{-\infty}^{\infty} g(t)\, dt = G(0)$ |
| 7. Area under $G(f)$ | $g(0) = \int_{-\infty}^{\infty} G(f)\, df$ |
| 8. Differentiation in the time domain | $\frac{d}{dt} g(t) \rightleftharpoons j2\pi f G(f)$ |
| 9. Integration in the time domain | $\int_{-\infty}^{t} g(\tau)\, d\tau \rightleftharpoons \frac{1}{j2\pi f} G(f) + \frac{G(0)}{2} \delta(f)$ |
| 10. Conjugate functions | If $g(t) \rightleftharpoons G(f)$, then $g^*(t) \rightleftharpoons G^*(-f)$ |
| 11. Multiplication in the time domain | $g_1(t) g_2(t) \rightleftharpoons \int_{-\infty}^{\infty} G_1(\lambda) G_2(f - \lambda)\, d\lambda$ |
| 12. Convolution in the time domain | $\int_{-\infty}^{\infty} g_1(\tau) g_2(t - \tau)\, d\tau \rightleftharpoons G_1(f) G_2(f)$ |

**TABLE A6.3** ***Fourier-transform pairs***

| *Time Function* | *Fourier Transform* |
|---|---|
| $\text{rect}\left(\frac{t}{T}\right)$ | $T\,\text{sinc}(fT)$ |
| $\text{sinc}(2Wt)$ | $\frac{1}{2W}\,\text{rect}\left(\frac{f}{2W}\right)$ |
| $\exp(-at)u(t), \quad a > 0$ | $\frac{1}{a + j2\pi f}$ |
| $\exp(-a\lvert t\rvert), \quad a > 0$ | $\frac{2a}{a^2 + (2\pi f)^2}$ |
| $\exp(-\pi t^2)$ | $\exp(-\pi f^2)$ |
| $\begin{cases} 1 - \frac{\lvert t\rvert}{T}, & \lvert t\rvert < T \\ 0, & \lvert t\rvert \geq T \end{cases}$ | $T\,\text{sinc}^2(fT)$ |
| $\delta(t)$ | $1$ |
| $1$ | $\delta(f)$ |
| $\delta(t - t_0)$ | $\exp(-j2\pi f t_0)$ |
| $\exp(j2\pi f_c t)$ | $\delta(f - f_c)$ |
| $\cos(2\pi f_c t)$ | $\frac{1}{2}[\delta(f - f_c) + \delta(f + f_c)]$ |
| $\sin(2\pi f_c t)$ | $\frac{1}{2j}[\delta(f - f_c) - \delta(f + f_c)]$ |
| $\text{sgn}(t)$ | $\frac{1}{j\pi f}$ |
| $\frac{1}{\pi t}$ | $-j\,\text{sgn}(f)$ |
| $u(t)$ | $\frac{1}{2}\delta(f) + \frac{1}{j2\pi f}$ |
| $\sum_{i=-\infty}^{\infty} \delta(t - iT_0)$ | $\frac{1}{T_0}\sum_{n=-\infty}^{\infty} \delta\left(f - \frac{n}{T_0}\right)$ |

*Notes:* $u(t)$ = unit step function
$\delta(t)$ = delta function, or unit impulse
$\text{rect}(t)$ = rectangular function of unit amplitude and unit duration centered on the origin
$\text{sgn}(t)$ = signum function
$\text{sinc}(t)$ = sinc function

**TABLE A6.4** *Hilbert transform pairs*[a]

| *Time Function* | *Hilbert Transform* |
|---|---|
| $m(t)\cos(2\pi f_c t)$ | $m(t)\sin(2\pi f_c t)$ |
| $m(t)\sin(2\pi f_c t)$ | $-m(t)\cos(2\pi f_c t)$ |
| $\cos(2\pi f_c t)$ | $\sin(2\pi f_c t)$ |
| $\sin(2\pi f_c t)$ | $-\cos(2\pi f_c t)$ |
| $\frac{\sin t}{t}$ | $\frac{1-\cos t}{t}$ |
| $\text{rect}(t)$ | $-\frac{1}{\pi}\log\left\lvert\frac{t-\frac{1}{2}}{t+\frac{1}{2}}\right\rvert$ |
| $\delta(t)$ | $\frac{1}{\pi t}$ |
| $\frac{1}{1+t^2}$ | $\frac{t}{1+t^2}$ |
| $\frac{1}{t}$ | $-\pi\delta(t)$ |

[a]In the first two pairs, it is assumed that $m(t)$ is band-limited to the interval $-W \le f \le W$, where $W < f_c$.

Notes: $\delta(t)$: delta function
rect($t$): rectangular function of unit amplitude and unit duration centered on the origin
log: natural logarithm

**TABLE A6.5** *Table of Bessel functions*[a]

| | $J_n(x)$ | | | | | | | | |
|---|---|---|---|---|---|---|---|---|---|
| n\x | *0.5* | *1* | *2* | *3* | *4* | *6* | *8* | *10* | *12* |
| 0 | 0.9385 | 0.7652 | 0.2239 | −0.2601 | −0.3971 | 0.1506 | 0.1717 | −0.2459 | 0.0477 |
| 1 | 0.2423 | 0.4401 | 0.5767 | 0.3391 | −0.0660 | −0.2767 | 0.2346 | 0.0435 | −0.2234 |
| 2 | 0.0306 | 0.1149 | 0.3528 | 0.4861 | 0.3641 | −0.2429 | −0.1130 | 0.2546 | −0.0849 |
| 3 | 0.0026 | 0.0196 | 0.1289 | 0.3091 | 0.4302 | 0.1148 | −0.2911 | 0.0584 | 0.1951 |
| 4 | 0.0002 | 0.0025 | 0.0340 | 0.1320 | 0.2811 | 0.3576 | −0.1054 | −0.2196 | 0.1825 |
| 5 | — | 0.0002 | 0.0070 | 0.0430 | 0.1321 | 0.3621 | 0.1858 | −0.2341 | −0.0735 |
| 6 | | — | 0.0012 | 0.0114 | 0.0491 | 0.2458 | 0.3376 | −0.0145 | −0.2437 |
| 7 | | | 0.0002 | 0.0025 | 0.0152 | 0.1296 | 0.3206 | 0.2167 | −0.1703 |
| 8 | | | — | 0.0005 | 0.0040 | 0.0565 | 0.2235 | 0.3179 | 0.0451 |
| 9 | | | | 0.0001 | 0.0009 | 0.0212 | 0.1263 | 0.2919 | 0.2304 |
| 10 | | | | — | 0.0002 | 0.0070 | 0.0608 | 0.2075 | 0.3005 |
| 11 | | | | | — | 0.0020 | 0.0256 | 0.1231 | 0.2704 |
| 12 | | | | | | 0.0005 | 0.0096 | 0.0634 | 0.1953 |
| 13 | | | | | | 0.0001 | 0.0033 | 0.0290 | 0.1201 |
| 14 | | | | | | — | 0.0010 | 0.0120 | 0.0650 |

[a]For more extensive tables of Bessel functions, see Watson (1966, pp. 666–697), and Abramowitz and Stegun (1965, pp. 358–406).

**TABLE A6.6** *The error function*[a]

| $u$ | $erf(u)$ | $u$ | $erf(u)$ |
|---|---|---|---|
| 0.00 | 0.00000 | 1.10 | 0.88021 |
| 0.05 | 0.05637 | 1.15 | 0.89612 |
| 0.10 | 0.11246 | 1.20 | 0.91031 |
| 0.15 | 0.16800 | 1.25 | 0.92290 |
| 0.20 | 0.22270 | 1.30 | 0.93401 |
| 0.25 | 0.27633 | 1.35 | 0.94376 |
| 0.30 | 0.32863 | 1.40 | 0.95229 |
| 0.35 | 0.37938 | 1.45 | 0.95970 |
| 0.40 | 0.42839 | 1.50 | 0.96611 |
| 0.45 | 0.47548 | 1.55 | 0.97162 |
| 0.50 | 0.52050 | 1.60 | 0.97635 |
| 0.55 | 0.56332 | 1.65 | 0.98038 |
| 0.60 | 0.60386 | 1.70 | 0.98379 |
| 0.65 | 0.64203 | 1.75 | 0.98667 |
| 0.70 | 0.67780 | 1.80 | 0.98909 |
| 0.75 | 0.71116 | 1.85 | 0.99111 |
| 0.80 | 0.74210 | 1.90 | 0.99279 |
| 0.85 | 0.77067 | 1.95 | 0.99418 |
| 0.90 | 0.79691 | 2.00 | 0.99532 |
| 0.95 | 0.82089 | 2.50 | 0.99959 |
| 1.00 | 0.84270 | 3.00 | 0.99998 |
| 1.05 | 0.86244 | 3.30 | 0.999998 |

[a]The error function is tabulated extensively in several references; see for example, Abramowitz and Stegun (1965, pp. 297–316).

## TABLE A6.7 *Selection of ITU voiceband (telephone line) modem standards*

| | *ITU Standard*[a] | *Type of modulation* | *Bit rate, b/s* | *Symbol rate, bauds* |
|---|---|---|---|---|
| (a) Symmetric modems: | V.21 | Binary FSK | 300 | 300 |
| | V.22 bis | QPSK | 1,200 | 600 |
| | V.26 | QPSK | 2,400 | 1,200 |
| | V.27 | 8-PSK | 4,800 | 2,400 |
| | V.32 | 16-QAM | 9,600 | 2,400 |
| | V.34 | 1024-QAM | 28,800 | 3,429 |
| | V.34 High Speed | Nested-constellation of four 960-QAM constellations | 33,600 | |
| (b) Asymmetric modems: | V.90: Downstream | Digital | 56,000 | |
| | Upstream | V.34 High Speed | 33,600 | |

[a]The suffix "bis" designates the second version of a particular standard.

## TABLE A6.8 *Trigonometric identities*

$$\exp(\pm j\theta) = \cos\theta \pm j\sin\theta$$
$$\cos\theta = \tfrac{1}{2}[\exp(j\theta) + \exp(-j\theta)]$$
$$\sin\theta = \frac{1}{2j}[\exp(j\theta) - \exp(-j\theta)]$$
$$\sin^2\theta + \cos^2\theta = 1$$
$$\cos^2\theta - \sin^2\theta = \cos(2\theta)$$
$$\cos^2\theta = \tfrac{1}{2}[1 + \cos(2\theta)]$$
$$\sin^2\theta = \tfrac{1}{2}[1 - \cos(2\theta)]$$
$$2\sin\theta\cos\theta = \sin(2\theta)$$
$$\sin(\alpha \pm \beta) = \sin\alpha\cos\beta \pm \cos\alpha\sin\beta$$
$$\cos(\alpha \pm \beta) = \cos\alpha\cos\beta \mp \sin\alpha\sin\beta$$
$$\tan(\alpha \pm \beta) = \frac{\tan\alpha \pm \tan\beta}{1 \mp \tan\alpha\tan\beta}$$
$$\sin\alpha\sin\beta = \tfrac{1}{2}[\cos(\alpha - \beta) - \cos(\alpha + \beta)]$$
$$\cos\alpha\cos\beta = \tfrac{1}{2}[\cos(\alpha - \beta) + \cos(\alpha + \beta)]$$
$$\sin\alpha\cos\beta = \tfrac{1}{2}[\sin(\alpha - \beta) + \sin(\alpha + \beta)]$$

## TABLE A6.9 *Series expansions*

**Taylor series**

$$f(x) = f(a) + \frac{f'(a)}{1!}(x-a) + \frac{f''(a)}{2!}(x-a)^2 + \cdots + \frac{f^{(n)}(a)}{n!}(x-a)^n + \cdots$$

where

$$f^{(n)}(a) = \left.\frac{d^n f(x)}{dx^n}\right|_{x=a}$$

**MacLaurin series**

$$f(x) = f(0) + \frac{f'(0)}{1!}x + \frac{f''(0)}{2!}x^2 + \cdots + \frac{f^{(n)}(0)}{n!}x^n + \cdots$$

where

$$f^{(n)}(0) = \left.\frac{d^n f(x)}{dx^n}\right|_{x=0}$$

**Binomial series**

$$(1+x)^n = 1 + nx + \frac{n(n-1)}{2!}x^2 + \cdots, \qquad |nx| < 1$$

**Exponential series**

$$\exp x = 1 + x + \frac{1}{2!}x^2 + \cdots$$

**Logarithmic series**

$$\log(1+x) = x - \tfrac{1}{2}x^2 + \tfrac{1}{3}x^3 - \cdots$$

**Trigonometric series**

$$\sin x = x - \frac{1}{3!}x^3 + \frac{1}{5!}x^5 - \cdots$$

$$\cos x = 1 - \frac{1}{2!}x^2 + \frac{1}{4!}x^4 - \cdots$$

$$\tan x = x + \frac{1}{3}x^3 + \frac{2}{15}x^5 + \cdots$$

$$\sin^{-1} x = x + \frac{1}{6}x^3 + \frac{3}{40}x^5 + \cdots$$

$$\tan^{-1} x = x - \frac{1}{3}x^3 + \frac{1}{5}x^5 - \cdots, \qquad |x| < 1$$

$$\operatorname{sinc} x = 1 - \frac{1}{3!}(\pi x)^2 + \frac{1}{5!}(\pi x)^4 - \cdots$$

## TABLE A6.10 *Integrals*

Indefinite integrals

$$\int x \sin(ax)\, dx = \frac{1}{a^2}[\sin(ax) - ax\cos(ax)]$$

$$\int x \cos(ax)\, dx = \frac{1}{a^2}[\cos(ax) + ax\sin(ax)]$$

$$\int x \exp(ax)\, dx = \frac{1}{a^2}\exp(ax)(ax - 1)$$

$$\int x \exp(ax^2)\, dx = \frac{1}{2a}\exp(ax^2)$$

$$\int \exp(ax)\sin(bx)\, dx = \frac{1}{a^2 + b^2}\exp(ax)[a\sin(bx) - b\cos(bx)]$$

$$\int \exp(ax)\cos(bx)\, dx = \frac{1}{a^2 + b^2}\exp(ax)[a\cos(bx) + b\sin(bx)]$$

$$\int \frac{dx}{a^2 + b^2x^2} = \frac{1}{ab}\tan^{-1}\left(\frac{bx}{a}\right)$$

$$\int \frac{x^2\, dx}{a^2 + b^2x^2} = \frac{x}{b^2} - \frac{a}{b^3}\tan^{-1}\left(\frac{bx}{a}\right)$$

Definite integrals

$$\int_0^\infty \frac{x\sin(ax)}{b^2 + x^2}\, dx = \frac{\pi}{2}\exp(-ab), \qquad a > 0,\ b > 0$$

$$\int_0^\infty \frac{\cos(ax)}{b^2 + x^2}\, dx = \frac{\pi}{2b}\exp(-ab), \qquad a > 0,\ b > 0$$

$$\int_0^\infty \frac{\cos(ax)}{(b^2 - x^2)^2}\, dx = \frac{\pi}{4b^3}[\sin(ab) - ab\cos(ab)], \qquad a > 0,\ b > 0$$

$$\int_0^\infty \operatorname{sinc} x\, dx = \int_0^\infty \operatorname{sinc}^2 x\, dx = \frac{1}{2}$$

$$\int_0^\infty \exp(-ax^2)\, dx = \frac{1}{2}\sqrt{\frac{\pi}{a}}, \qquad a > 0$$

$$\int_0^\infty x^2\exp(-ax^2)\, dx = \frac{1}{4a}\sqrt{\frac{\pi}{a}}, \qquad a > 0$$

**TABLE A6.11 *Useful constants***

## Physical Constants

| | |
|---|---|
| Boltzmann's constant | $k = 1.38 \times 10^{-23}$ joule/degree Kelvin |
| Planck's constant | $h = 6.626 \times 10^{-34}$ joule-second |
| Electron (fundamental) charge | $q = 1.602 \times 10^{-19}$ coulomb |
| Speed of light in vacuum | $c = 2.998 \times 10^{8}$ meters/second |
| Standard (absolute) temperature | $T_0 = 273$ degrees Kelvin |
| Thermal voltage | $V_T = 0.026$ volt at room temperature |
| Thermal energy $kT$ at standard temperature | $kT_0 = 3.77 \times 10^{-21}$ joule |
| One hertz (hz) = 1 cycle/second; 1 cycle = $2\pi$ radians | |
| One watt (W) = 1 joule/second | |

## Mathematical Constants

| | |
|---|---|
| Base of natural logarithm | $e = 2.7182818$ |
| Logarithm of $e$ to base 2 | $\log_2 e = 1.442695$ |
| Logarithm of 2 to base $e$ | $\log 2 = 0.693147$ |
| Logarithm of 2 to base 10 | $\log_{10} 2 = 0.30103$ |
| Pi | $\pi = 3.1415927$ |

**TABLE A6.12 *Recommended unit prefixes***

| *Multiples and Submultiples* | *Prefixes* | *Symbols* |
|---|---|---|
| $10^{12}$ | tera | T |
| $10^{9}$ | giga | G |
| $10^{6}$ | mega | M |
| $10^{3}$ | kilo | K ($k$) |
| $10^{-3}$ | milli | m |
| $10^{-6}$ | micro | $\mu$ |
| $10^{-9}$ | nano | n |
| $10^{-12}$ | pico | p |

# GLOSSARY

## Conventions and Notations

1. The symbol | | means the absolute value, or magnitude, of the complex quantity contained within.
2. The symbol arg( ) means the phase angle of the complex quantity contained within.
3. The symbol Re[ ] means the "real part of," and Im[ ] means the "imaginary part of."
4. Unless stated otherwise, the natural logarithm is denoted by log. Logarithms to bases 2 and 10 are denoted by $\log_2$ and $\log_{10}$, respectively.
5. The use of an asterisk as superscript denotes complex conjugate, e.g., $x^*$ is the complex conjugate of $x$.
6. The symbol $\rightleftharpoons$ indicates a Fourier-transform pair, e.g., $g(t) \rightleftharpoons G(f)$, where a lowercase letter denotes the time function and a corresponding uppercase letter denotes the frequency function.
7. The symbol $F[\ ]$ indicates the Fourier-transform operation, e.g., $F[g(t)] = G(f)$, and the symbol $F^{-1}[\ ]$ indicates the inverse Fourier-transform operation, e.g., $F^{-1}[G(f)] = g(t)$.
8. The symbol ★ denotes convolution, e.g.,

$$x(t) \star h(t) = \int_{-\infty}^{\infty} x(\tau)h(t - \tau)\, d\tau$$

9. The symbol $\oplus$ denotes modulo-2 addition, except in Chapter 10 where binary arithmetic is used and modulo-2 addition is denoted by an ordinary plus sign throughout that chapter.
10. The use of subscript $T_0$ indicates that the pertinent function $g_{T_0}(t)$, say, is a periodic function of time $t$ with period $T_0$.
11. The use of a hat over a function indicates one of two things:
    (a) the Hilbert transform of a function, e.g., the function $\hat{g}(t)$ is the Hilbert transform of $g(t)$, or
    (b) the estimate of an unknown parameter, e.g., the quantity $\hat{\alpha}(\mathbf{x})$ is an estimate of the unknown parameter $\alpha$, based on the observation vector $\mathbf{x}$.
12. The use of a tilde over a function indicates the complex envelope of a narrowband signal, e.g., the function $\tilde{g}(t)$ is the complex envelope of the narrowband signal $g(t)$. The exception to this convention is in Section 10.8, where, in the description of turbo decoding, the tilde is used to signify extrinsic information and thereby distinguish it from log-likelihood ratio.
13. The use of subscript + indicates the pre-envelope of a signal, e.g., the function $g_+(t)$ is the pre-envelope of the signal $g(t)$. We may thus write $g_+(t) = g(t) + j\hat{g}(t)$, where $\hat{g}(t)$ is the Hilbert transform of $g(t)$. The use of subscript $-$ indicates that $g_-(t) = g(t) - j\hat{g}(t) = g_+{}^*(t)$.
14. The use of subscripts $I$ and $Q$ indicates the in-phase and quadrature components of a narrowband signal, a narrowband random process, or the impulse response of a narrow-band filter, with respect to the carrier $\cos(2\pi f_c t)$.

15. For a low-pass message signal, the highest frequency component or message bandwidth is denoted by $W$. The spectrum of this signal occupies the frequency interval $-W \le f \le W$ and is zero elsewhere. For a band-pass signal with carrier frequency $f_c$, the spectrum occupies the frequency intervals, $f_c - W \le f \le f_c + W$ and $-f_c - W \le f \le -f_c + W$, and so $2W$ denotes the bandwidth of the signal. The (low-pass) complex envelope of this band-pass signal has a spectrum that occupies the frequency interval $-W \le f \le W$.

    For a lowpass filter, the bandwidth is denoted by $B$. A common definition of filter bandwidth is the frequency at which the magnitude response of the filter drops by 3 dB below the zero-frequency value. For a band-pass filter of mid-band frequency $f_c$ the bandwidth is denoted by $2B$, centered on $f_c$. The complex low-pass equivalent of this band-pass filter has a bandwidth equal to $B$.

    The transmission bandwidth of a communication channel, required to transmit a modulated wave, is denoted by $B_T$.
16. Random variables or random vectors are uppercase (e.g., $X$ or $\mathbf{X}$), and their sample values are lowercase (e.g., $x$ or $\mathbf{x}$).
17. A vertical bar in an expression means "given that," e.g., $f_X(x \mid H_0)$ is the probability density function of the random variable $X$, given that hypothesis $H_0$ is true.
18. The symbol $E[\ ]$ means the expected value of the random variable enclosed within; the $E$ acts as an operator.
19. The symbol var[ ] means the variance of the random variable enclosed within.
20. The symbol cov[ ] means the covariance of the two random variables enclosed within.
21. The average probability of symbol error is denoted by $P_e$.

    In the case of binary signaling techniques, $p_{10}$ denotes the conditional probability of error given that symbol 0 was transmitted, and $p_{01}$ denotes the conditional probability of error given that symbol 1 was transmitted. The *a priori* probabilities of symbols 0 and 1 are denoted by $p_0$ and $p_1$, respectively.
22. The symbol $\langle\ \rangle$ denotes the time average of the sample function enclosed within.
23. Boldface letter denotes a vector or matrix. The inverse of a square matrix $\mathbf{R}$ is denoted by $\mathbf{R}^{-1}$. The transpose of a vector $\mathbf{w}$ is denoted by $\mathbf{w}^T$. The Hermitian transpose of a complex-valued vector $\mathbf{x}$ is denoted by $\mathbf{x}^H$; Hermitian transposition involves both transposition and complex conjugation.
24. The length of a vector $\mathbf{x}$ is denoted by $\| \mathbf{x} \|$. The Euclidean distance between the vectors $\mathbf{x}_i$ and $\mathbf{x}_j$ is denoted by $d_{ij} = \| \mathbf{x}_i - \mathbf{x}_j \|$.
25. The inner product of two real-valued vectors $\mathbf{x}$ and $\mathbf{y}$ is denoted by $\mathbf{x}^T\mathbf{y}$; their outer product is denoted by $\mathbf{x}\mathbf{y}^T$. If the vectors $\mathbf{x}$ and $\mathbf{y}$ are complex valued, their inner product is $\mathbf{x}^H\mathbf{y}$, and their outer product is $\mathbf{x}\mathbf{y}^H$.
26. The vector product of two $M$-by-1 vectors $\boldsymbol{\alpha}$ and $\boldsymbol{\beta}$ is an $M$-by-1 vector defined by

$$\boldsymbol{\alpha} \cdot \boldsymbol{\beta} = \begin{bmatrix} \alpha_1\beta_1 \\ \alpha_2\beta_2 \\ \vdots \\ \alpha_M\beta_M \end{bmatrix}$$

where $\alpha_k$ and $\beta_k$ are the $k$th elements of $\boldsymbol{\alpha}$ and $\boldsymbol{\beta}$, respectively. The $L_1$ norm of the vector product $\boldsymbol{\alpha} \cdot \boldsymbol{\beta}$ is defined by

$$\| \boldsymbol{\alpha} \cdot \boldsymbol{\beta} \|_1 = \sum_{m=1}^{M} \boldsymbol{\alpha}_m \boldsymbol{\beta}_m$$

## Functions

1. Rectangular function: $\text{rect}(t) = \begin{cases} 1, & -\frac{1}{2} < t < \frac{1}{2} \\ 0, & |t| > \frac{1}{2} \end{cases}$

2. Unit step function: $u(t) = \begin{cases} 1, & t > 0 \\ 0, & t < 0 \end{cases}$

3. Signum function: $\text{sgn}(t) = \begin{cases} 1, & t > 0 \\ 0, & t = 0 \\ -1, & t < 0 \end{cases}$

4. (Dirac) delta function: $\delta(t) = 0, \quad t \neq 0$

   $$\int_{-\infty}^{\infty} \delta(t)\, dt = 1$$

   or, equivalently,
   $$\int_{-\infty}^{\infty} g(t)\, \delta(t - t_0)\, dt = g(t_0)$$

5. Sinc function: $\text{sinc}(x) = \dfrac{\sin(\pi x)}{\pi x}$

6. Sine integral: $\text{Si}(u) = \displaystyle\int_0^u \frac{\sin x}{x}\, dx$

7. Error function: $\text{erf}(u) = \dfrac{2}{\sqrt{\pi}} \displaystyle\int_0^u \exp(-z^2)\, dz$

   Complementary error function: $\text{erfc}(u) = 1 - \text{erf}(u)$.

8. Binomial coefficient $\dbinom{n}{k} = \dfrac{n!}{(n-k)!k!}$

9. Bessel function of the first kind of order $n$: $J_n(x) = \dfrac{1}{2\pi} \displaystyle\int_{-\pi}^{\pi} \exp(jx \sin\theta - jn\theta)\, d\theta$

10. Modified Bessel function of the first kind of zero order: $I_0(x) = \dfrac{1}{2\pi} \displaystyle\int_{-\pi}^{\pi} \exp(x \cos\theta)\, d\theta$

11. Confluent hypergeometric function $_1F_1(a; b; x) = 1 + \dfrac{a}{b}\dfrac{x}{1!} + \dfrac{a(a+1)}{b(b+1)}\dfrac{x^2}{2!} + \cdots$

## Abbreviations

A: ampere
AC: alternating current
ADC: analog-to-digital converter
ADM: adaptive delta modulation
ADPCM: adaptive differential pulse-code modulation
ADSL: asymmetric digital subscriber line
AM: amplitude modulation
ANSI: American National Standards Institute
APB: adaptive prediction with backward estimation
APF: adaptive prediction with forward estimation
AQB: adaptive quantization with backward estimation

| | |
|---|---|
| **AQF:** | adaptive quantization with forward estimation |
| **ARQ:** | automatic-repeat request |
| **ASCII:** | American National Standard Code for Information Interchange |
| **ASK:** | amplitude-shift keying |
| **ATM:** | asynchronous transfer mode |
| **AWGN:** | additive white Gaussian noise |
| **b/s:** | bits /second |
| **BER:** | bit error rate |
| **BISDN:** | broadband ISDN |
| **BPF:** | band-pass filter |
| **BSC:** | binary symmetric channel |
| **CAP:** | carrierless amplitude/phase modulation |
| **CCITT:** | Consultative Committee for International Telephone and Telegraph (Now renamed the ITU) |
| **CDM:** | code-division multiplexing |
| **CDMA:** | code-division multiple access |
| **CELP:** | code excited linear predictive (model) |
| **CO:** | central office |
| **codec:** | coder/decoder |
| **CPFSK:** | continuous-phase frequency-shift keying |
| **CRC:** | cyclic redundancy check |
| **CW:** | continuous wave |
| **DAC:** | digital-to-analog converter |
| **dB:** | decibel |
| **dBW:** | decibel referenced to 1 watt |
| **dBmW:** | decibel reference to 1 milliwatt |
| **DC:** | direct current |
| **DEM:** | demodulator |
| **DES:** | data encryption standard |
| **DFT:** | discrete Fourier transform |
| **DM:** | delta modulation |
| **DMT:** | discrete multitone |
| **DPCM:** | differential pulse-code modulation |
| **DPSK:** | differential phase-shift keying |
| **DSB-SC:** | double sideband–suppressed carrier |
| **DS/BPSK:** | direct sequence/binary phase-shift keying |
| **DSL:** | digital subscriber line |
| **exp:** | exponential |
| **FDM:** | frequency-division multiplexing |
| **FDMA:** | frequency-division multiple access |
| **FEXT:** | far-end crosstalk |
| **FFT:** | fast Fourier transform |
| **FH:** | frequency hop |

**FH/MFSK:** frequency hop/*M*-ary frequency-shift keying
**FMFB:** frequency modulator with feedback
**FSK:** frequency-shift keying
**GMSK:** Gaussian filtered MSK
**GSM:** global system for mobile communication
**HDTV:** high definition television
**Hz:** Hertz
**IDFT:** inverse discrete Fourier transform
**IF:** intermediate frequency
**I/O:** input/output
**IP:** internet protocol
**IS-95:** intermediate standard-95
**ISDN:** integrated services digital network
**ISI:** intersymbol interference
**ISO:** International Organization for Standardization
**ITU:** International Telecommunications Union
**JPEG:** joint photographic experts group
**LAN:** local-area network
**LDM:** linear delta modulation
**LMS:** least-mean-square
**log:** natural logarithm
**$\log_2$:** logarithm to base 2
**$\log_{10}$:** logarithm to base 10
**LPC:** linear predictive coding (model)
**LPF:** low-pass filter
**MAP:** maximum *a posteriori* probability
**ML:** maximum likelihood
**mmse:** minimum mean-square error
**modem:** modulator–demodulator
**MPEG:** motion photographic experts group
**ms:** millisecond
**$\mu$s:** microsecond
**MSK:** minimum shift keying
**NCO:** number-controlled oscillator
**NEXT:** near-end crosstalk
**nm:** nanometer
**NRZ:** nonreturn-to-zero
**NTSC:** National Television Systems Committee
**OC:** optical carrier
**OFDM:** orthogonal frequency-division multiplexing
**OOK:** on–off keying
**OSI:** open systems interconnection
**PAM:** pulse-amplitude modulation

| | |
|---|---|
| PCM: | pulse-code modulation |
| PDM: | pulse-duration modulation |
| PG: | processing gain |
| PLL: | phase-locked loop |
| PN: | pseudo-noise |
| POTS: | plain old telephone service |
| PPM: | pulse-position modulation |
| PSK: | phase-shift keying |
| PSTN: | public switched telephone network |
| PWM: | pulse-width modulation |
| QAM: | quadrature amplitude modulation |
| QoS: | quality of service |
| QPSK: | quadriphase-shift keying |
| RF: | radio frequency |
| rms: | root-mean-square |
| RS: | Reed-Solomon |
| RS-232 | Recommended standard-232 (port) |
| RSA: | Rivest-Shamir-Adelman |
| RSC: | recursive systematic convolutional (code) |
| RZ: | return-to-zero |
| s: | second |
| SDH: | synchronous digital hierarchy |
| SDMA: | space-division multiple access |
| SDR: | signal-to-distortion ratio |
| SNR: | signal-to-noise ratio |
| SONET: | synchronous optical network |
| STFT: | short-time Fourier transform |
| STM: | synchronous transfer mode |
| TC: | time compression |
| TCM: | trellis-coded modulation |
| TDM: | time-division multiplexing |
| TDMA: | time-division multiple access |
| TV: | television |
| UHF: | ultra high frequency |
| V: | volt |
| VCO: | voltage-controlled oscillator |
| VHF: | very high frequency |
| VLSI: | very-large-scale integration |
| W: | watt |
| WDM: | wavelength division multiplexing |

# Bibliography

## Books


M. Abramowitz and I.A. Stegun, *Handbook of Mathematical Functions with Formulas, Graphs, and Mathematical Tables* (New York: Dover Publications, 1965).

N. Abramson, *Information Theory and Coding* (New York: McGraw-Hill, 1963).

J. Adamék, *Foundations of Coding* (New York: Wiley, 1991).

Y. Akaiwa, *Introduction to Digital Mobile Communication* (New York: Wiley, 1997).

J.B. Anderson, T. Aulin, and C.E. Sundberg, *Digital Phase Modulation* (New York: Plenum Publishers, 1986).

J.B. Anderson and S. Mohan, *Source and Channel Coding: An Algorithmic Approach* (Boston, Mass: Kluwer Academic, 1991).

J.B. Anderson, *Digital Transmission Engineering* (Piscataway, N.J.: IEEE Press, 1999).

R.B. Ash, *Information Theory* (New York: Wiley, 1965).

Bell Laboratories Technical Staff, *A History of Engineering Science in the Bell System: The Early Years (1875–1925)*, (Books on Demand, Ann Arbor, Michigan: 1975).

J.C. Bellamy, *Digital Telephony*, Second Edition (New York: Wiley, 1991).

S. Benedetto and E. Biglieri, *Principles of Digital Transmission with Wireless Applications* (New York: Kluwer Academic/Plenum Publishers, 1999).

S. Benedetto, E. Biglieri, and V. Castellani, *Digital Transmission Theory* (Englewood Cliffs, N.J.: Prentice-Hall, 1987).

W.R. Bennett, *Introduction to Signal Transmission* (New York: McGraw-Hill, 1970).

K.B. Benson and J.C. Whitaker, *Television Engineering Handbook*, rev. ed. (New York: McGraw-Hill, 1992).

T. Berger, *Rate Distortion Theory: A Mathematical Basis for Data Compression* (Englewood Cliffs, N.J.: Prentice-Hall, 1971).

E.R. Berlekamp, *Algebraic Coding Theory* (New York: McGraw-Hill, 1968).

E.R. Berlekamp (editor), *Key Papers in the Development of Coding Theory* (Piscataway, N.J.: IEEE Press, 1974).

V.K. Bhargava, D. Haccoun, R. Matyas, and P. Nuspl, *Digital Communications by Satellite: Modulation, Multiple Access, and Coding* (New York: Wiley, 1981).

E. Biglieri, D. Divsalar, P.J. McLane, and M.K. Simon, *Introduction to Trellis-Coded Modulation with Applications* (New York: Macmillan, 1991).

J.A.C. Bingham, *The Theory and Practice of Modem Design* (New York: Wiley, 1988).

R.B. Blachman and J.W. Tukey, *The Measurement of Power Spectra, from the Point of View of Communication Engineering* (New York: Dover, 1958).

H.S. Black, *Modulation Theory* (Princeton, N.J.: Van Nostrand, 1953).

R.E. Blahut, *Principles and Practice of Information Theory* (Reading, Mass.: Addison-Wesley, 1987).

R.E. Blahut, *Digital Transmission of Information* (Reading, Mass.: Addison-Wesley, 1990).

G.E.P. Box and G.M. Jenkins, *Time Series Analysis: Forecasting and Control* (San Francisco: Holden Day, 1976).

R.N. Bracewell, *The Fourier Transform and Its Applications*, 2nd ed., rev. (New York: McGraw-Hill, 1986).

L. Brillouin, *Science and Information Theory*, 2nd ed. (New York: Academic Press, 1962).

K.W. Cattermole, *Principles of Pulse-code Modulation* (New York: American Elsevier, 1969).

W.Y. Chen, *DSL: Simulation Techniques and Standards Development for Digital Subscriber Line Systems* (Indianapolis, Ind.: Macmillan Technical Publishing, 1998).

J.M. Cioffi, *Digital Data Transmission*, EE379C Course Textbook, Stanford University, 1998.

G.C. Clark, Jr., and J.B. Cain, *Error-correction Coding for Digital Communications* (New York: Plenum Publishers, 1981).

L.W. Couch, *Digital and Analog Communication Systems*, 5th ed. (Englewood Cliffs, N.J.: Prentice-Hall, 1997).

T.M. Cover and J.A. Thomas, *Elements of Information Theory* (New York: Wiley, 1991).

H. Cramér and M.R. Leadbetter, *Stationary and Related Stochastic Processes: Sample Function Properties and Their Applications* (New York: Wiley, 1967).

W.B. Davenport, Jr., and W.I. Root, *An Introduction to the Theory of Random Signals and Noise* (New York: McGraw-Hill, 1958).

R.C. Dixon, *Spread Spectrum Systems*, 2nd ed. (New York: Wiley, 1984).

R.C. Dixon (editor), *Spread Spectrum Techniques* (New York, IEEE Press, 1976).

L.J. Doob, *Stochastic Processes* (New York: Wiley, 1953).

J.J. Downing, *Modulation Systems and Noise* (Englewood Cliffs, N.J.: Prentice-Hall, 1964).

W. Feller, *An Introduction to Probability Theory and its Application*, vol. 1, 3rd ed. (New York: Wiley, 1968).

T.L. Fine, *Theories of Probability: An Examination of Foundations* (New York: Academic Press, 1973).

L.E. Franks (editor), *Data Communication: Fundamentals of Baseband Transmission* (Dowden, Hutchison, and Ross, 1974).

L.E. Franks, *Signal Theory* (Englewood Cliffs, N.J.: Prentice-Hall, 1969).

R.L. Freeman, *Telecommunications Transmission Handbook*, 4th ed. (New York: Wiley, 1998).

R.M. Gagliardi, *Introduction to Communications Engineering*, 2nd ed. (New York: Wiley, 1988).

R.G. Gallager, *Information Theory and Reliable Communication* (New York: Wiley, 1968).

R.G. Gallager, *Low-Density Parity-Check Codes* (Cambridge, Mass.: MIT Press, 1963).

F.M. Gardner, *Phaselock Techniques*, 2nd ed. (New York: Wiley, 1979).

V.K. Gary and J.E. Wilkes, *Principles & Applications of GSM* (Englewood Cliffs, N.J.: Prentice-Hall, 1999).

A. Gersho and R.M. Gray, *Vector Quantization and Signal Compression* (Boston, Mass.: Kluwer Academic, 1992).

J.D. Gibson (editor), *The Mobile Communications Handbook* (Piscataway, N.J.: IEEE Press, 1996).

R.D. Gitlin, J.F. Hayes, and S.B. Weinstein, *Data Communications Principles* (New York: Plenum, 1992).

B. Goldberg and H.S. Bennett (editors), *Communication Channels: Characterization and Behavior* (New York: IEEE Press, 1976).

S.W. Golomb (editor), *Digital Communications with Space Applications* (Englewood Cliffs, N.J.: Prentice-Hall, 1964).

S.W. Golomb, *Shift Register Sequences* (San Francisco: Holden-Day, 1967).

R.M. Gray and L.D. Davisson, *Random Processes: A Mathematical Approach for Engineers* (Englewood Cliffs, N.J.: Prentice-Hall, 1986).

P.E. Green, Jr., *Computer Network Architectures and Protocols* (New York: Plenum, 1982).

P.E. Green, Jr., *Fiber Optic Networks* (Englewood Cliffs, N.J.: Prentice-Hall, 1993).

M.S. Gupta (editor), *Electrical Noise: Fundamentals and Sources* (New York: IEEE Press, 1977).

R.W. Hamming, *Coding and Information Theory* (Englewood Cliffs, N.J.: Prentice-Hall, 1980).

S. Haykin, *Communication Systems*, 3rd ed. (New York: Wiley, 1994).

S. Haykin, *Adaptive Filter Theory*, 3rd ed. (Englewood Cliffs, N.J.: Prentice-Hall, 1996).

S. Haykin and B. Van Veen, *Signals and Systems* (New York: Wiley, 1999).

C. Heegard and S.B. Wicker, *Turbo Coding* (Boston, Mass.: Kluwer Academic Publishers, 1999).

G. Held, *The Complete Modem Reference*, 3rd ed. (New York: Wiley, 1997).

C.W. Helstrom, *Statistical Theory of Signal Detection* (Elmsford, N.Y.: Pergamon Press, 1968).

C.W. Helstrom, *Probability and Stochastic Processes for Engineers*, 2nd ed. (New York: Macmillan, 1990).

K. Henney (editor), *Radio Engineering Handbook* (New York: McGraw-Hill, 1959).

J.K. Holmes, *Coherent Spread Spectrum Systems* (New York: Wiley, 1982).

W.C. Jakes, Jr. (editor), *Microwave Mobile Communications* (New York: Wiley, 1974).

N.S. Jayant and P. Noll, *Digital Coding of Waveforms: Principles and Applications to Speech and Video* (Englewood Cliffs, N.J.: Prentice-Hall, 1984).

N.S. Jayant (editor), *Waveform Quantization and Coding* (New York: IEEE Press, 1976).

H. Jeffreys, Sir, *Theory of Probability*, 3rd ed. (Oxford: Clarendon Press, 1967).

H. Jeffreys, Sir, and B.S. Jeffreys, *Methods of Mathematical Physics*, 3rd ed. (Cambridge University Press, 1956).

M.C. Jeruchim, B. Balaban, and J.S. Shanmugan, *Simulation of Communication Systems* (New York: Plenum, 1992).

E.C. Jordan and K.G. Balmain, *Electromagnetic Waves and Radiating Systems*, 2nd ed. (Englewood Cliffs, N.J.: Prentice Hall, 1968).

A. Khintchin, *Mathematical Foundations of Information Theory* (New York: Dover, 1957).

A.N. Kolmogorov, *Foundations of the Theory of Probability* (New York: Chelsea Publishing, 1956).

V.A. Kotel'nikov, *The Theory of Optimum Noise Immunity* (New York: McGraw-Hill, 1960).

J.D. Kraus, *Antennas* (New York: McGraw-Hill, 1950).

S. Kullback, *Information Theory and Statistics* (New York: Dover, 1968).

P. Lafrance, *Fundamental Concepts in Communication* (Englewood Cliffs, N.J.: Prentice-Hall, 1990).

B.P. Lathi, *Modern Digital and Analog Communication Systems*, 2nd ed. (Oxford University Press, 1995).

I. Lebow, *Information Highways and Byways* (Piscataway, N.J.: IEEE Press, 1995).

E.A. Lee and D.G. Messerschmitt, *Digital Communication*, 2nd ed. (Boston, Mass.: Kluwer Academic, 1994).

J.S. Lee and L.E. Miller, *CDMA Systems Engineering Handbook* (Boston, Mass.: Artech House Publishers, 1998).

Y.W. Lee, *Statistical Theory of Communication* (New York: Wiley, 1960).

W.C.(Y.) Lee, *Mobile Communications Engineering* (New York: McGraw-Hill, 1982).

A. Leon-Garcia, *Probability and Random Processes for Electrical Engineering*, 2nd ed. (Reading, Mass.: Addison-Wesley, 1994).

C.R. Lewart, *The Ultimate Modem Handbook* (Englewood Cliffs, N.J.: Prentice-Hall, 1998).

S. Lin and D.J. Costello, Jr., *Error Control Coding: Fundamentals and Applications* (Englewood Cliffs, N.J.: Prentice-Hall, 1983).

W.C. Lindsey, *Synchronization Systems in Communication and Control* (Englewood Cliffs, N.J.: Prentice-Hall, 1972).

W.C. Lindsey and M.K. Simon (editors), *Phase-locked Loops and Their Applications* (New York: IEEE Press, 1978).

W.C. Lindsey and M.K. Simon, *Telecommunication Systems Engineering* (Englewood Cliffs, N.J.: Prentice-Hall, 1973).

M. Loève, *Probability Theory* (Princeton, N.J.: Van Nostrand, 1963).

R.W. Lucky, *Silicon Dreams: Information, Man, and Machine* (New York: St. Martin's Press, 1989).

R.W. Lucky, J. Salz, and E.J. Weldon, Jr., *Principles of Data Communication* (New York: McGraw-Hill, 1968).

F.J. MacWilliams and N.J.A. Sloane, *The Theory of Error-correcting Codes* (Amsterdam: North-Holland, 1977).

V.K. Madisetti and D.B. Williams (editors), *The Digital Signal Processing Handbook* (Piscataway, N.J.: IEEE Press, 1998).

R.J. Marks II, *Introduction to Shannon Sampling and Interpolation Theory* (New York/Berlin: Springer-Verlag, 1991).

J.C. McDonald (editor), *Fundamentals of Digital Switching*, 2nd ed. (New York: Plenum, 1990).

R. McDonough and A.D. Whalen, *Detection of Signals in Noise*, 2nd ed. (New York: Academic Press, 1995).

R.J. McEliece, *The Theory of Information and Coding: A Mathematical Framework for Communication* (Reading, Mass.: Addison-Wesley, 1977).

A. Mengali and N. D'Andrea, *Synchronization Techniques for Digital Receivers* (New York: Plenum, 1997).

D.J.G. Mestdagh, *Fundamentals of Multiaccess Optical Fiber Networks* (Boston, Mass.: Artech House Publishers, 1995).

C.H. Meyer and S.M. Matyas, *Cryptography: A New Dimension in Computer Data Security* (New York: Wiley, 1982).

H. Meyr and G. Ascheid, *Synchronization in Digital Communications*, vol. 1 (New York: Wiley, 1990).

H. Meyr, M. Moeneclaey, and S.A. Fechtel, *Digital Communication Receivers: Synchronization, Channel Estimation and Signal Processing* (New York: Wiley, 1998).

A.M. Michelson and A.H. Levesque, *Error-control Techniques for Digital Communication* (New York: Wiley, 1985).

D. Middleton, *An Introduction to Statistical Communication Theory* (New York: McGraw-Hill, 1960).

J.G. Nellist, *Understanding Telecommunications and Lightwave Systems: An Entry Level Guide* (Piscataway, N.J.: IEEE Press, 1992).

C.F.J. Overhage (editor), *The Age of Electronics* (New York: McGraw-Hill, 1962).

P.F. Panter, *Modulation, Noise and Spectral Analysis, Applied to Information Transmission* (New York: McGraw-Hill, 1965).

A. Papoulis, *Probability, Random Variables, and Stochastic Processes*, 2nd ed. (New York: McGraw-Hill, 1984).

J.D. Parsons, *The Mobile Radio Propagation Channel* (New York: Wiley, 1992).

K. Pahlavan and A.H. Levesque, *Wireless Information Networks* (New York: Wiley, 1996).

J. Pearl, *Probabilistic Reasoning in Intelligent Systems: Networks of Plausible Inference* (San Mateo, Calif: Morgan Kaufman Publishers, 1988).

W.W. Peterson and E.J. Weldon, Jr., *Error Correcting Codes*, 2nd ed. (Cambridge, Mass.: MIT Press, 1972).

J.R. Pierce and A.M. Noll, *Signals: The Science of Telecommunications* (New York: Scientific American Library, 1990).

J.R. Pierce, *Symbols, Signals and Noise: The Nature and Process of Communication* (New York: Harper, 1961).

H.V. Poor, *An Introduction to Signal Detection and Estimation*, 2nd ed. (New York/Berlin: Springer-Verlag, 1994).

T. Pratt and C.W. Bostian, *Satellite Communications* (New York: Wiley, 1986).

J.G. Proakis, *Digital Communications*, 3rd ed. (New York: McGraw-Hill, 1995).

L.R. Rabiner and R.W. Schafer, *Digital Processing of Speech Signals* (Englewood Cliffs, N.J.: Prentice-Hall, 1978).

K.R. Rao and P. Yip, *Discrete Cosine Transform: Algorithms, Advantages, Applications* (New York: Academic Press, 1990).

T.S. Rappaport, *Smart Antennas* (Piscataway, N.J.: IEEE Press, 1998).

T.S. Rappaport, *Wireless Communications: Principles and Practice* (Piscataway, N.J.: IEEE Press, 1996).

S.O. Rice, "Noise in FM receivers," in M. Rosenblatt (editor), *Proceedings of the Symposium on Time Series Analysis* (New York: Wiley, 1963), pp. 395–411.

J.H. Roberts, *Angle Modulation: The Theory of Systems Assessment*, IEE Communication Series 5 (London: Institution of Electrical Engineers, 1977).

H.E. Rowe, *Signals and Noise in Communication Systems* (Princeton, N.J.: Van Nostrand, 1965).

D.J. Sakrison, *Communication Theory: Transmission of Waveforms and Digital Information* (New York: Wiley, 1968).

C. Schlegel, *Trellis Coding* (Piscataway, N.J.: IEEE Press, 1997).

M. Schwartz, W.R. Bennett, and S. Stein, *Communication Systems and Techniques* (New York: McGraw-Hill, 1966).

M. Schwartz, *Information Transmission, Modulation and Noise: A Unified Approach*, 3rd ed. (New York: McGraw-Hill, 1980).

M. Schwartz, *Telecommunication Networks: Protocols, Modeling, and Analysis* (Reading, Mass.: Addison Wesley, 1987).

K.S. Shanmugan, *Digital and Analog Communication Systems* (New York: Wiley, 1979).

C.E. Shannon and W. Weaver, *The Mathematical Theory of Communication* (Urbana: University of Illinois Press, 1949).

G.J. Simmons (editor), *Contemporary Cryptology: The Science of Information Integrity* (Piscataway, N.J.: IEEE Press, 1992).

M.K. Simon, J.K. Omura, R.A. Scholtz, and B.K. Levitt, *Spread Spectrum Communications*, vols. I, II, and III (New York: Computer Science Press, 1985).

B. Sklar, *Digital Communications: Fundamentals and Applications* (Englewood Cliffs, N.J.: Prentice-Hall, 1988).

D. Slepian (editor), *Key Papers in the Development of Information Theory* (New York: IEEE Press, 1974).

N.J.A. Sloane and A.D. Wyner, *Claude Shannon: Collected Papers* (Piscataway, N.J.: IEEE Press, 1993).

D.R. Smith, *Digital Transmission Systems* (Princeton, N.J.: Van Nostrand Reinhold, 1985).

I.S. Sokolnikoff and R.M. Redheffer, *Mathematics of Physics and Modern Engineering* (New York: McGraw-Hill, 1966).

J.J. Spilker, Jr., *Digital Communications by Satellite* (Englewood Cliffs, N.J.: Prentice-Hall, 1977).

W. Stallings, *ISDN and Broadband ISDN*, 2nd ed. (New York: Macmillan, 1992).

T. Starr, J.M. Cioffi, and P.J. Silverman, *Understanding Digital Subscriber Line Technology* (Englewood Cliffs, N.J.: Prentice-Hall, 1999).

R. Steele, *Delta Modulation Systems* (New York: Wiley, 1975).

R. Steele and L. Hanzo (editors), *Mobile Radio Communications*, 2nd ed. (New York: Wiley, 1999).

J.J. Stiffler, *Theory of Synchronous Communications* (Englewood Cliffs, N.J.: Prentice-Hall, 1971).

E.D. Sunde, *Communications Systems Engineering Theory* (New York: Wiley, 1969).

A.S. Tanenbaum, *Computer Networks*, 2nd ed. (Englewood Cliffs, N.J.: Prentice-Hall, 1995).

S. Tantaratana and K.M. Ahmed, *Wireless Applications of Spread Spectrum Systems: Selected Readings* (Piscataway, N.J.: IEEE Press, 1998).

T.M. Thompson, *From Error-correcting Codes Through Sphere Packings to Simple Groups* (The Mathematical Association of America, Washington D.C.: 1983).

D.J. Torrieri, *Principles of Military Communication Systems*, 2nd ed. (Boston, Mass.: Artech House Publishers, 1992).

G.L. Turin, *Notes on Digital Communications* (Princeton, N.J.: Van Nostrand-Reinhold, 1969).

A. Van der Ziel. *Noise: Source, Characterization, Measurement* (Englewood Cliffs, N.J.: Prentice-Hall, 1970).

H.F. Vanlandingham, *Introduction to Digital Control Systems* (New York: Macmillan, 1985).

H.C.A. van Tilborg, *An Introduction to Cryptology* (Boston, Mass.: Kluwer, 1988).

H.L. Van Trees, *Detection, Estimation, and Modulation Theory*, Part I (New York: Wiley, 1968).

A.J. Viterbi, *Principles of Coherent Communication* (New York: McGraw-Hill, 1966).

A.J. Viterbi and J.K. Omura, *Principles of Digital Communication and Coding* (New York: McGraw-Hill, 1979).

G.N. Watson, *A Treatise in the Theory of Bessel Functions*, 2nd ed. (New York: Cambridge University Press, 1966).

N. Wax (editor), Selected Papers on Noise and Stochastic Processes (New York: Dover Publications, 1954).

E.T. Whittaker and G.N. Watson, *A Course in Modern Analysis*, 4th ed. (New York: Cambridge University Press, 1952).

S.B. Wicker and V.K. Bhargava (editors), *Reed-Solomon Codes* (Piscataway, N.J.: IEEE Press, 1994).

B. Widrow and S.D. Stearns, *Adaptive Signal Processing* (Englewood Cliffs, N.J.: Prentice-Hall, 1985).

N. Wiener, *The Extrapolation, Interpolation, and Smoothing of Stationary Time Series, with Engineering Applications* (New York: Wiley, 1949).

S.G. Wilson, *Digital Modulation and Coding* (Englewood Cliffs, N.J.: Prentice-Hall, 1996).

E. Wong, *Stochastic Processes in Information and Dynamical Systems* (New York: McGraw-Hill, 1971).

P.M. Woodward, *Probability and Information Theory, with Applications to Radar*, 2nd ed. (Elmsford, N.Y.: Pergamon Press, 1964).

J.M. Wozencraft and I.M. Jacobs, *Principles of Communication Engineering* (New York: Wiley, 1965).

W.W. Wu, *Elements of Digital Satellite Communication*, vol. I (New York: Computer Science Press, 1984).

C.R. Wylie and L.C. Barrett, *Advanced Engineering Mathematics*, 5th ed. (New York: McGraw-Hill, 1982).

R.D. Yates and D.J. Goodman, *Probability and Stochastic Processes: A Friendly Introduction for Electrical and Computer Engineers* (New York: Wiley, 1999).

J.H. Yuen (editor), *Deep Space Telecommunications Systems Engineering* (New York: Plenum, 1983).

R.E. Ziemer and R.L. Peterson, *Digital Communications and Spread Spectrum Systems* (New York: Macmillan, 1985).

R.E. Ziemer and W.H. Tranter, *Principles of Communications*, 3rd ed. (Boston, Mass.: Houghton Miflin, 1990).

## PAPERS, REPORTS, PATENTS[1]

M.R. Aaron and D.W. Tufts, "Intersymbol interference and error probability," *IEEE Trans. on Information Theory*, vol. IT-12, pp. 26–34, 1966.

J.E. Abate, "Linear and adaptive delta modulation," *Proceedings of the IEEE*, vol. 55, pp. 298–308, 1967.

A.N. Akansu, P. Duhamel, X. Lin, and M. de Courville, "Orthogonal transmultiplexers in communications: A review," *IEEE Transactions on Signal Processing*, vol. 46, pp. 979–995, 1998.

Y. Akaiwa and Y. Nagata, "Highly efficient digital mobile communications with a linear modulation method," *IEEE Journal on Selected Areas in Communications*, vol. SAC-5, pp. 890–895, 1987.

O. Al-Shaykh, R. Neff, D. Taubman, and A. Zakhour, "Video sequence compression." In V.K. Madisetti and D.B. Williams (editors), *The Digital Signal Processing Handbook*, CRC Press, pp. 55-1–55-19, 1998.

F. Amoroso, "The bandwidth of digital data signals," *IEEE Communications Magazine*, vol. 18, no. 6, pp. 13–24, 1980.

J.B. Anderson and D.P. Taylor, "A bandwidth-efficient class of signal space codes," *IEEE Transactions on Information Theory*, vol. IT-24, pp. 703–712, 1978.

R.R. Anderson and J. Salz, "Spectra of digital FM," *Bell System Tech. J.*, vol. 44, pp. 1165–1189, 1965.

R. Arens, "Complex processes for envelopes of normal noise," *IRE Trans. on Information Theory*, vol. IT-3, pp. 204–207, 1957.

E.H. Armstrong, " A method of reducing disturbances in radio signaling by a system of frequency modulation," *Proceedings of the IRE*, vol. 24, pp. 689–740, 1936.

E. Arthurs and H. Dym, "On the optimum detection of digital signals in the presence of white Gaussian noise—A geometric interpretation and a study of three basic data transmission systems," *IRE Trans. on Communication Systems*, vol. CS-10, pp. 336–372, 1962.

B.S. Atal and J.R. Remde, "A new model of LPC excitation for producing natural-sounding speech at low bit rates," *Proc. ICASSP '82*, pp. 614–17, 1982.

B.S. Atal and M.R. Schroeder, "Stochastic coding of speech signals at very low bit rates," IEEE International Conference on Communications, May 1984.

M. Austin, "Decision-feedback equalization for digital communication over dispersive channels," *MIT Research Laboratory of Electronics Technical Report* 461, 1967.

E. Ayanoglu, N.R. Dagdeviren, J.E. Mazo, and R. Saltzberg, "High-speed modem synchronized to a remote codec," United States Patent 5,394,437, February 28, 1995.

E. Ayanoglu, N.R. Dagdeviren, G.D. Golden, and J.E. Mazo, "An equalizer design technique for the PCM modem: a new modem for the digital public switched network," *IEEE Transactions on Communications*, vol. 46, pp. 763–774, 1998.

L.R. Bahl, J. Cocke, F. Jelinek, and J. Raviv, "Optimal decoding of linear codes for minimizing symbol error rate," *IEEE Transactions on Information Theory*, vol. IT-20, pp. 284–287, 1974.

G. Battail, "Coding for the Gaussian channel: the promise of weighted output decoding," *International J. Satellite Communications*, vol. 7, pp. 183–192, 1989.

G. Battail, "Pondération des symbols décodés par l'algorithme de Viterbi," *Ann. Télécommunication*, vol. 42, pp. 31–38, 1987.

E. Bedrosian, "The analytic signal representation of modulated waveforms," *Proceedings of the IRE*, vol. 50, pp. 2071–2076, 1962.

P.A. Bello, "Characterization of randomly time-variant linear channels," *IEEE Transactions on Communication Systems*, vol. CS-11, pp. 360–393, 1963.

S. Benedetto and G. Montorsi, "Unveiling turbo codes: Some results on parallel concatenated coding schemes," *IEEE Transactions on Information Theory*, vol. 42, pp. 409–428, 1996.

W.R. Bennett, "Spectra of quantized signals," *Bell System Tech. J.*, vol. 27, pp. 446–472, 1948.

N. Benvenuto, et al., "The 32 kb/s ADPCM coding standard," *AT&T Technical Journal*, vol. 65, pp. 12–22, Sept./Oct. 1986.

C. Berrou and A. Glavieux, "Near optimum error correcting coding and decoding: turbo codes," *IEEE Transactions on Communications*, vol. 44, pp. 1261–1271, 1996.

C. Berrou and A. Glavieux, "Reflections on the Prize Paper: Near optimum error-correcting coding and decoding turbo codes," *IEEE Information Theory Society Newsletter*, vol. 48, no. 2, p. 1 and pp. 24–31, June 1998.

C. Berrou, A. Glavieux, and P. Thitmajshima, "Near Shannon limit error-correction coding and decoding: turbo codes," *International Conference on Communications*, pp. 1064–1090, Geneva, Switzerland, May 1993.

V.K. Bhargava, "Forward error correction schemes for digital communications," *IEEE Communications Magazine*, vol. 21, no. 1, pp. 11–19, 1983.

R.C. Bose and D.K. Ray-Chaudhuri, "On a class of error correcting binary group codes," *Information and Control*, vol. 3, pp. 68–79, 1960.

K. Brandenburg and G. Stoll, "ISO-MPEG-1 Audio: A generic standard for coding of high-quality digital audio," *Journal of the Audio Engineering Society*, vol. 42, pp. 780–792, 1994.

D.G. Brennan, "Linear diversity combining techniques," *Proceedings of the IRE*, vol. 47, pp. 1075–1102, 1959.

A. Buzo, A.H. Gray, Jr., R.M. Gray, and J.D. Markel, "Speech coding based upon vector quantization," *IEEE Transactions on Acoustics, Speech, and Signal Processing*, vol. ASSP-28, pp. 562–574, 1980.

C.R. Cahn, "Combined digital phase and amplitude modulation communication systems," *IRE Transactions on Communication Systems*, vol. CS-8, pp. 150–155, 1960.

J.R. Carson, "Notes on the theory of modulation," *Proceedings of the IRE*, vol. 10, pp. 57–64, 1922.

J.R. Carson and T.C. Fry, "Variable frequency electric circuit theory with application to the theory of frequency modulation," *Bell System Tech. J.*, vol. 16, pp. 513–540, 1937.

E.F. Casas and C. Leung, "OFDM for data communication over mobile radio FM channels," *IEEE Transactions on Communications*, vol. 39, pp. 783–793, 1991.

J.G. Chaffee, "The application of negative feedback to frequency-modulation systems," *Bell System Tech. J.*, vol. 18, pp. 404–437, 1939.

R.W. Chang, "Synthesis of band-limited orthogonal signals for multichannel data transmission," *Bell System Tech. J.*, vol. 45, pp. 1775–1796, 1996.

W.Y. Chen, G.H. Im, and J.J. Werner, "Design of digital carrierless AM/PM transceivers," *Standard Project, T1E1.4/92-149, AT&T and Bellcore*, August 19, 1992.

S. Chennakeshu and G.J. Sauliner, "Differential detection of $\pi$/4-shifted-DQPSK for digital cellular radio," *IEEE Transactions on Vehicular Technology*, vol. 42, pp. 46–57, 1993.

J.M. Cioffi, V. Oksman, J.-J. Werner, T. Pollet, P.M.P. Spruyt, J.S. Chow, and K.S. Jacobsen, "Very-high-speed digital subscriber lines," *IEEE Communications Magazine*, vol. 37, pp. 72–79, April, 1999.

L.J. Cimini, Jr., and Y. Li, "Orthogonal frequency division multiplexing for wireless communications," tutorial notes, TU18, International Conference on Communications '99, Vancouver, British Columbia, Canada, June, 1999.

A.C. Clarke, "Extraterrestrial relays," *Wireless World*, vol. 51, pp. 305–308, October 1945.

C.E. Cook and H.S. Marsh, "An introduction to spread spectrum," *IEEE Communications Magazine*, vol. 21, no. 2, pp. 8–16, 1983.

J.P. Costas, "Synchronous communications," *Proceedings of the IRE*, vol. 44, pp. 1713–1718, 1956.

J.P. Costas, "Poisson, Shannon, and the radio amateur," *Proceedings of the IRE*, vol. 47, pp. 2058–2068, 1959.

M.G. Crosby, "Frequency modulation noise characteristics," *Proceedings of the IRE*, vol. 25, pp. 472–514, April 1937.

C.C. Cutler, "Differential quantization of communication signals," United States Patent 2-505-361, 1952.

C.L. Dammann, L.D. McDaniel and C.L. Maddox, "D2 channel bank—Multiplexing and coding," *Bell System Tech. J.* vol. 51, pp. 1675–1699, 1972.

F. Daneshgaran and M. Mondin, "Design of interleavers for turbo codes: Iterative interleaver growth algorithms of polynomial complexity," *IEEE Transactions on Information Theory*, vol. 45, pp. 1845–1859, 1999.

R. deBuda, "Coherent demodulation of frequency-shift keying with low deviation ratio," *IEEE Trans. on Communications*, vol. COM-20, pp. 429–435, 1972.

F.E. DeJager, "Deltamodulation, a method of PCM transmission using the 1-unit code," *Phillips Research Reports*, vol. 7, pp. 442–46, 1952.

F.E. DeJager and C.B. Dekker, "Tamed frequency modulation: A novel method to achieve spectrum economy in digital transmission," *IEEE Transactions on Communications*, vol. COM-26, pp. 534–542, 1978.

J.A. Develet, "A threshold criterion for phase-lock demodulation," *Proceedings of the IEEE*, vol. 51, pp. 349–356, 1963.

W. Diffie and M.E. Hellman, "New directions in cryptography," *IEEE Transactions on Information Theory*, vol. IT-22, pp. 644–654, 1976.

W. Diffie and M.E. Hellman, "Privacy and authentication: An introduction to cryptography," *Proceedings of the IEEE*, vol. 67, pp. 397–427, 1979.

D. Divsalar, "Turbo codes," MILCOM 96 tutorial, San Diego, November 1996.

M.I. Doelz and E.H. Heald, "Minimum shift data communication system," U.S. Patent 2977417, March 1961.

R.M. Dolby, "An audio reduction system," *Journal of the Audio Engineering Society*, vol. 15, p. 383, 1967.

J. Dungundji, "Envelopes and pre-envelopes of real wave-forms," *IRE Transactions on Information Theory*, vol. IT-4, pp. 53–57, 1958.

P. Elias, "Coding for noisy channels," *IRE Convention Record*, Part 4, pp. 37–46, March 1955.

L.H. Enloe, "Decreasing the threshold in FM by frequency feedback," *Proceedings of the IRE*, vol. 50, pp. 18–30, 1962.

V.M. Eyuboglu, "Detection of coded modulation signals on linear, severely distorted channels using decision-feedback noise prediction with interleaving," *IEEE Transactions on Communications*, vol. COM-36, pp. 401–409, 1988.

D.D. Falconer, "Carrierless AM/PM," *Bell Laboratories, Internal Memorandum*, July 3, 1975.

K. Feher, "MODEMS for emerging digital cellular-mobile radio system," *IEEE Transactions on Vehicular Technology*, vol. 40, pp. 355–365, 1991.

J.L. Flanagan, M.R. Schroeder, B.S. Atal, R.E. Crochiere, N.S. Jayant, and J.M. Tribolet, "Speech coding," *IEEE Transactions on Communications*, vol. COM-27, pp. 710–737, 1979.

B. LeFloch, R. Halbert-Lassalle, and D. Castelain, "Digital sound broadcasting to mobile receivers," *IEEE Transactions of Broadcasting*, vol. 35, pp. 493–503, 1989.

G.D. Forney, Jr., "Maximum likelihood sequence estimation of digital sequences in the presence of intersymbol interference," *IEEE Transactions on Information Theory*, vol. IT-18, pp. 363–378, 1972.

G.D. Forney, Jr. "The Viterbi algorithm," *Proceedings of the IEEE*, vol. 61, pp. 268–278, 1973.

G.D. Forney, Jr., and M.V. Eyuboglu, "Combined equalization and coding using precoding," *IEEE Communications Magazine*, vol. 29, no. 12, pp. 25–34, 1991.

G.D. Forney, Jr., L. Brown, M.V. Eyuboglu, and J.L. Moran III, "The V.34 high-speed modem standard," *IEEE Communications Magazine*, pp. 28–93, December 1996.

L.E. Franks, "Carrier and bit synchronization in data communications—A tutorial review," *IEEE Transactions on Communications*, vol. COM-28, pp. 1107–1121, 1980.

B.J. Frey and D.J.C. MacKay, "Irregular turbocodes," *Proceedings of the 37th Annual Allerton Conference on Communication, Control, and Computing*, Allerton House, Illinois, September 1999.

H.T. Friis, "Noise figures in radio receivers," *Proceedings of the IRE*, vol. 32, pp. 419–422, 1944.

K.E. Fulz and D.B. Penick, "T1 carrier system," *Bell System Tech. J.*, vol. 44, pp. 1405–1451, 1965.

D. Gabor, "Theory of communications," *Journal of IEE* (London), vol. 93, Part III, pp. 429–457, 1946.

D.L. Gall, "MPEG: a video compression standard for multimedia applications," *Communications of the ACM*, vol. 34, pp. 47–58, 1991.

R.G. Gallager, "Low-density parity-check codes," *IRE Transactions on Information Theory*, vol. 8, pp. 21–28, 1962.

W.A. Gardner, "Introduction to Einstein's contribution to time-series analysis," *IEEE ASSP Magazine*, vol. 4, pp. 4–5, October 1987.

W.A. Gardner and L.E. Franks, "Characterization of cyclostationary random signal processes," *IEEE Transactions on Information Theory*, vol. IT-21, pp. 4–14, 1975.

D.A. George, "Matched filters for interfering signals," *IEEE Transactions on Information Theory*, vol. IT-11, pp. 153–154, 1965.

A. Gersho, "Adaptive equalization of highly dispersive channels for data transmission," *Bell System Tech. J.*, vol. 48, pp. 55–70, 1969.

R.A. Gibby and J.W. Smith, "Some extensions of Nyquist's telegraph transmission theory," *Bell Systems Tech. J.*, vol. 44, pp. 1487–1510, 1965.

R.D. Gitlin and E.Y. Ho, "The performance of staggered quadrature amplitude modulation in the presence of phase jitter," *IEEE Transactions on Communications*, vol. COM-23, pp. 348–352, 1975.

M.J.E. Golay, "Note on digital coding," *Proceedings of the IRE,* vol. 37, p. 657, 1949.

M.J.E. Golay, "Binary coding," *IRE Transactions on Information Theory,* vol. PGIT-4, pp. 23–28, 1954.

R. Gold, "Optimal binary sequences for spread spectrum multiplexing," *IEEE Transactions on Information Theory*, vol. IT-13, pp. 619–621, 1967.

R. Gold, "Maximal recursive sequences with 3-valued recursive cross correlation functions," *IEEE Transactions on Information Theory*, vol. IT-14, pp. 154–156, 1968.

B. Goode, "Scanning the issue: Special issue on global information infrastructure," *Proceedings of the IEEE*, vol. 85, pp. 1883–1886, 1997.

R.M. Gray, "Vector quantization," *IEEE ASSP Magazine*, vol. 1, no. 2, pp. 4–29, 1984.

W.J. Gruen, "Theory of AFC synchronization," *Proceedings of the IRE*, vol. 41, pp. 1043–1048, 1953.

P. Guinand and J. Lodge, "Trellis termination for turbo encoders," *Proceedings of 18th Biennial Symposium on Communications, Queen's University*, Kingston, Canada, June 1996.

D.W. Hagelbarger, "Recurrent codes: Easily mechanized, burst-correcting binary codes," *Bell System Tech. J.*, vol. 38, pp. 969–984, 1959.

J. Hagenauer, E. Offer, and L. Papke, "Iterative decoding of binary block and convolutional codes," *IEEE Transactions on Information Theory*, vol. 42, pp. 429–445, 1996.

J. Hagenauer and P. Hoeher, "A Viterbi algorithm with soft-decision outputs and its applications," *IEEE Globecom 89*, pp. 47.11–47.17, November 1989, Dallas, Texas.

R.W. Hamming, "Error detecting and error correcting codes," *Bell System Tech. J.*, vol. 29, pp. 147–160, 1950.

J.C. Hancock and R.W. Lucky, "Performance of combined amplitude and phase-modulated communication systems," *IRE Transactions on Communication Systems*, vol. CS-8, pp. 232–237, 1960.

H.H. Hanning and J.W. Pan, "D2 channel bank system aspects," *Bell System Tech. J.*, vol. 51, pp. 1641–1657, 1972.

H. Harashima and H. Miyakawa, "Matched-transmission technique for channels with intersymbol interference," *IEEE Transactions on Communications*, vol. COM-20, pp. 774–779, 1972.

T.V.L. Hartley, "Transmission of information," *Bell System Tech. J.*, vol. 7, pp. 535–563, 1928.

F.S. Hill, Jr., "On time-domain representations for vestigial sideband signals," *Proceedings of the IEEE*, vol. 62, pp. 1032–1033, 1974.

D.A. Huffman, "A method for the construction of minimum redundancy codes," *Proceedings of the IRE*, vol. 40, pp. 1098–1101, 1952.

P.A. Humblet and M.G. Troulis, "The information driveway," *IEEE Communications Magazine*, pp. 64–68, December, 1996.

G.-H. Im and J.-J. Werner, "Bandwidth-efficient digital transmission over unshielded twisted-pair wiring," *IEEE Journal on Selected Areas in Communications*, vol. 13, pp. 1643–1655, 1995.

H. Insoe, Y. Yasuda, and J. Murakami, "A telemetering system by code modulation: Δ-Σ modulation," *IRE Transactions on Space Electronics and Telemetry*, vol. SET-8, pp. 204–209, 1962.

M. Ishizuka and K. Hirade, "Optimum Gaussian filter and deviated-frequency locking scheme for coherent detection of MSK," *IEEE Transactions on Communications*, vol. COM-28, pp. 850–857, 1980.

I.M. Jacobs, "Practical applications of coding," *IEEE Transactions on Information Theory*, vol. IT-20, pp. 305–310, 1974.

N.S. Jayant, "Adaptive delta modulation with a one-bit memory," *Bell System Tech. J.*, vol. 49, pp. 321–342, 1970.

N.S. Jayant, "Digital coding of speech waveforms, PCM, DPCM and DM quantizers," *Proceedings of the IEEE*, vol. 62, pp. 611–632, 1974.

N.S. Jayant, "Coding speech at low bit rates," *IEEE Spectrum*, vol. 23, no. 8, pp. 58–63, 1986.

A.J. Jerri, "The Shannon sampling theorem—its various extensions and applications: A tutorial review," *Proceedings of the IEEE*, vol. 65, no. 11, pp. 1565–1596, 1977.

J.B. Johnson, "Thermal agitation of electricity in conductors," *Physical Review*, second series, vol. 32, pp. 97–109, 1928.

P. Kabal and S. Pasupathy, "Partial-response signaling," *IEEE Transactions on Communications*, vol. COM-23, pp. 921–934, 1975.

I. Kalet, "The multitone channel," *IEEE Transactions on Communications*, vol. 37, pp. 119–124, 1989.

I. Kalet, J.E. Mazo, and B.R. Saltzberg, "The capacity of PCM voiceband channels," *IEEE International Conference on Communications*, pp. 507–511, Geneva, Switzerland, 1993.

H. Kaneko, "A unified formulation of segment companding laws and synthesis of codes and digital companders," *Bell System Tech. J.*, vol. 49, pp. 1555–1588, 1970.

A.I. Khintchine, "Korrelationstheorie der stationören stochastischen Prozese," *Mathematiche Annalen*, vol. 1, 109, pp. 415–458, 1934.

H. Kobayashi, "Correlative level coding and maximum-likelihood decoding," *IEEE Transactions on Information Theory*, vol. IT-17, pp. 586–594, 1971.

R. Kohno, "Spatial and temporal communication theory using adaptive antenna array," *IEEE Personal Communications*, pp. 28–35, February, 1998.

E.T. Kretzmer, "Generalization of a technique for binary data communications," *IEEE Transactions on Communication Technology*, vol. COM-14, pp. 67–68, Feb. 1966.

F.R. Kschischang and B.J. Frey, "Interactive decoding of compound codes by probability propagation in graphical models," *IEEE Journal on Selected Areas in Communication*, vol. 16, pp. 219–230, 1998.

J.W. Lechleider, "Line codes for digital subscriber lines," *IEEE Communications Magazine*, vol. 27, pp. 25–32, September 1989.

B.M. Leiner, V.G. Cerf, D.D. Clark, R.E. Kohn, L. Kleinrock, D.C. Lynch, J. Postel, L.G. Roberts, and S. Wolff, "A brief history of the Internet," *Commun. ACM*, vol. 40, pp. 102–108, February 1997.

A. Lender, "The duobinary technique for high-speed data transmission," *IEEE Transactions on Communications and Electronics*, vol. 82, pp. 214–218, May 1963.

A. Lender, "Correlative digital communication techniques," *IEEE Transactions on Communication Technology*, vol. COM-12, pp. 128–135, 1964.

A. Lender, "Correlative level coding for binary-data transmission," *IEEE Spectrum*, vol. 3, no. 2, pp. 104–115, 1966.

N.-S. Lin and C.-P.J. Tzeng, "Full-duplex data over local loops," *IEEE Communications Magazine*, vol. 26, pp. 31–42, February 1988.

S. Lin, D.J. Costello, and M.J. Miller, "Automatic-repeat-request error control schemes," *IEEE Communications Magazine*, vol. 22, no. 12, pp. 5–16, 1984.

Y. Linde, A. Buzo and R.M. Gray, "An algorithm for vector quantizer design," *IEEE Trans. on Communications*, vol. COM-28, pp. 84–95, 1980.

D. Linden, "A discussion of sampling theorems," *Proceedings of the IRE*, vol. 47, pp. 1219–1226, 1959.

C.L. Liu and K. Feher, "Noncoherent detection of $\pi/4$-shifted systems in a CCI-AWGN combined interference environment," *Proceedings of the IEEE 40th Vehicular Technology Conference*, San Francisco, 1989.

S.P. Lloyd, "Least squares quantization in PCM," unpublished Bell Laboratories Technical Note, 1957. This report was reprinted in *IEEE Transactions on Information Theory*, vol. IT-28, pp. 129–137, 1982.

J. Lodge, R.Young, P. Hoeher, and J. Hagenauer, "Separable MAP 'filters' for the decoding of product and concatenated codes," *Proceedings of the IEEE International Conference on Communications*, pp. 1740–1745, Geneva, Switzerland, May 1993.

R. W. Lucky, "Automatic equalization for digital communication," *Bell System Tech. J.*, vol. 44, pp. 547–588, 1965.

R. W. Lucky, "Techniques for adaptive equalization of digital communication systems," *Bell System Tech. J.*, vol. 45, pp. 255–286, 1966.

R. Lugannani, "Intersymbol interference and probability of error in digital systems," *IEEE Transactions on Information Theory*, vol. IT-15, pp. 682–688, 1969.

V.H. MacDonald, "Advanced mobile phone service: the cellular concept," *Bell System Tech. J.*, vol. 58, pp. 15–41, 1979.

D.J.C. MacKay, "Good error-correcting codes based on very sparse matrices," *IEEE Transactions on Information Theory*, vol. 45, pp. 399–431, 1999.

D.J.C. MacKay and R.M. Neal, "Near Shannon limit performance of low density parity check codes," *Electronics Letters*, vol. 33, No. 6, pp. 457–458, 1997; and vol. 32, no. 18, pp. 1645–1646, 1996.

D.J.C. MacKay, S.T. Wilson, and M.C. Davey, "Comparison of constructions of irregular Gallager codes," *IEEE Transactions on Communications*, vol. 47, pp. 1449–1454, 1999.

J. Max, "Quantizing for minimum distortion," *IRE Transactions on Information Theory*, vol. IT-6, pp. 7–12, 1960.

K. Maxwell, "Asymmetric digital subscriber line: Interim technology for the next forty years," *IEEE Communications Magazine*, vol. 34, pp. 100–106, October 1996.

R.J. McEliece, D.J.C. MacKay, and J.-F. Cheng, "Turbo coding as an instance of Pearl's belief propagation algorithm," *IEEE Journal on Selected Areas of Communication*, vol. 16, pp. 140–152, 1998.

D. Mennie, "AM stereo: Five competing options," *IEEE Spectrum*, vol. 15, no. 6, pp. 24–31, 1978.

M.L. Moher, "Cross-entropy and iterative detection," Ph.D. thesis, Department of Systems and Computer Engineering, Carleton University, Ottawa, Canada, May 1997.

M.L. Moher and T.A. Gulliver, "Cross-entropy and iterative decoding," *IEEE Transactions on Information Theory*, vol. 44, pp. 3097–3104, 1998.

P. Monsen, "Feedback equalization for fading dispersive channels," *IEEE Transactions on Information Theory*, vol. IT-17, pp. 56–64, 1971.

K.H. Mueller and J.J. Werner, "A hardware efficient passband equalizer structure for data transmissions," *IEEE Transactions on Communications*, vol. COM-30, pp. 538–541, 1982.

K. Murota and K. Hirade, "GMSK modulation for digital mobile radio telephone," *IEEE Transactions on Communications*, vol. COM-29, pp. 1044–1050, 1981.

E. Murphy, "Whatever happened to AM stereo?" *IEEE Spectrum*, vol. 25, p. 17, 1988.

P. Noll, "MPEG digital audio coding standards." In V.K. Madisetti and D.B. Williams (editors), *The Digital Signal Processing Handbook*, Piscataway, N.J.: IEEE Press, pp. 40-1–40-28, 1998.

D.O. North, "An analysis of the factors which determine signal/noise discrimination in pulsed carrier systems," *Proceedings of the IEEE*, vol. 51, pp. 1016–1027, 1963; this paper is a reprint of a classified RCA Report published in 1943.

H. Nyquist, "Certain factors affecting telegraph speed," *Bell System Tech. J.*, vol. 3, pp. 324–346, 1924.

H. Nyquist, "Thermal agitation of electric charge in conductors," *Physical Review*, second series, vol. 32, pp. 110–113, 1928.

H. Nyquist, "Certain topics in telegraph transmission theory," *Transactions of the AIEE*, vol. 47, pp. 617–644, Feb. 1928.

M.W. Oliphant, "The mobile phone meets the Internet," *IEEE Spectrum*, vol. 36, pp. 20–28, August, 1999.

B.M. Oliver, J.R. Pierce, and C.E. Shannon, "The philosophy of PCM," *Proceedings of the IRE*, vol. 36, pp. 1324–1331, 1948.

D.Y. Pan, "Digital audio compression," *Digital Technical Journal*, vol. 5, pp. 1–14, 1993.

S. Pasupathy, "Nyquist's third criterion," *Proceedings of the IEEE,* vol. 62, pp. 860–861, 1974.

S. Pasupathy, "Correlative coding—A bandwidth-efficient signaling scheme," *IEEE Communications Magazine*, vol. 15, no. 4, pp. 4–11, 1977.

S. Pasupathy, "Minimum shift keying—A spectrally efficient modulation," *IEEE Communications Magazine*, vol. 17, no. 4, pp. 14–22, 1979.

A.J. Paulraj and B.C. Ng, "Space-time modems for wireless personal communications," *IEEE Personal Communications*, pp. 36–48, February, 1998.

A.J. Paulraj and C.B. Papadias, "Space-time processing for wireless communications," *IEEE Signal Processing Magazine*, pp. 49–83, November, 1997.

R.L. Pickholtz, D.L. Schilling, and L.B. Milstein, "Theory of spread-spectrum communications—A tutorial," *IEEE Transactions on Communications*, vol. COM-30, pp. 855–884, 1982.

R. Price, "Nonlinearly feedback-equalized PAM vs. capacity for noisy filter channels," International Conference on Communications, ICC '72, pp. 22.12–22.17, June 1972, Philadelphia.

R. Price and P.E. Green, Jr., "A communication technique for multipath channels," *Proceedings of the IRE*, vol. 46, pp. 555–570, 1958.

J.G. Proakis, "Advances in equalization for intersymbol interference," *Advances in Communications Systems*, edited by A.J. Viterbi, vol. 4, pp. 123–198, Academic Press, 1975.

S. Qureshi, "Adaptive equalization," *IEEE Communications Magazine*, vol. 20, no. 2, pp. 9–16, March 1982.

S. Qureshi, "Adaptive equalization," *Proceedings of the IEEE*, vol. 73, pp. 1349–1387, 1985.

T.A. Ramstad, "Still image compression." In V.K. Madisetti and D.B. Williams (editors), *The Digital Signal Processing Handbook*, Piscataway, N.J.: IEEE Press, pp. 52-1–52-27, 1998.

I.S. Reed and G. Solomon, "Polynomial codes over certain finite fields," *Journal of SIAM*, vol. 8, pp. 300–304, 1960.

A.H. Reeves, "The past, present and future of PCM," *IEEE Spectrum*, vol. 12, no. 5, pp. 58–63, 1975.

S.A. Rhodes, "Effect of noisy phase reference on coherent detection of offset-QPSK signals," *IEEE Transactions on Communications*, vol. COM-22, pp. 1046–1055, 1974.

S.O. Rice, "Mathematical analysis of random noise," *Bell System Tech. J.*, vol. 23, pp. 282–332, 1944; vol. 24, pp. 46–156, 1945.

S.O. Rice, "Statistical properties of a sine-wave plus random noise," *Bell System Tech. J.*, vol. 27, pp. 109–157, 1948.

S.O. Rice, "Envelopes of narrow-band signals," *Proceedings of the IEEE*, vol. 70, pp. 692–699, 1982.

T. Richardson, A. Shokrollahi, and R. Urbanke, "Design of provably good low-density parity check codes," submitted in 1999 to *IEEE Transactions on Information Theory*.

R.L. Rivest, A. Shamir, and L. Adleman, "A method for obtaining digital signatures and public key cryptosystems," *Communications of the ACM*, vol. 21, pp. 120–126, 1978.

W.L. Root, "Remarks, mostly historical, on signal detection and signal parameter estimation," *Proceedings of the IEEE*, vol. 75, pp. 1446–1457, 1987.

A. Ruiz, J.M. Cioffi, and S. Kasturia, "Discrete multiple tone modulation with coset coding for the spectrally shaped channel," *IEEE Transactions on Communications*, vol. 40, pp. 1012–1029, 1992.

W.D. Rummler, "A new selective fading model—Application to propagation data," *Bell System Tech. J.*, vol. 58, pp. 1037–1071, 1979.

B.R. Saltzberg, "Comparison of single-carrier and multitone digital modulation for ADSL applications," *IEEE Communications Magazine*, vol. 36, pp. 114–121, November, 1998.

B.R. Saltzberg, "Performance of an efficient parallel data transmission system," *IEEE Transactions on Communication Technology*, vol. COM-15, pp. 805–811, 1967.

S.D. Sandberg and M.A. Tzannes, "Overlapped discrete multitone modulation for high speed copper wire communications," *IEEE Journal on Selected Areas in Communications*, vol. 13, pp. 1571–1585, 1995.

D.V. Sarwate and M.B. Pursley, "Crosscorrelation properties of pseudorandom and related sequences," *Proceedings of the IEEE*, vol. 68, pp. 593–619, 1980.

B. Sayar and S. Pasupathy, "Nyquist 3 pulse shaping in continuous phase modulation," *IEEE Transactions on Communications*, vol. COM-35, pp. 57–67, 1987.

H.R. Schindler, "Delta modulation," *IEEE Spectrum*, vol. 7, no. 10, pp. 69–78, 1970.

R.A. Scholz, "The origins of spread-spectrum communications," *IEEE Transactions on Communications*, vol. COM-30, pp. 822–854, May 1982.

R.A. Scholz, "Notes on spread-spectrum history," *IEEE Transactions on Communications*, vol. COM-31, pp. 82–84, 1983.

J.S. Schouten, F. DeJager, and J.A. Greefkes, "Delta modulation, a new modulation system for telecommunication," *Phillips Technical Review*, vol. 13, pp. 237–245, 1952.

C.E. Shannon, "A mathematical theory of communication," *Bell System Tech. J.*, vol. 27, pp. 379–423, 623–656, 1948.

C.E. Shannon, "Communication theory of secrecy systems," *Bell System Tech. J.*, vol. 28, pp. 656–715, 1949.

C.E. Shannon, "Communication in the presence of noise," *Proceedings of the IRE*, vol. 37, pp. 10–21, 1949.

M.K. Simon and D. Divsalar, "On the implementation and performance of single and double differential detection schemes," *IEEE Trans. on Communications*, vol. 40, pp. 278–291, 1992.

B. Sklar, "A primer on turbo code concepts," *IEEE Communications Magazine*, vol. 35, pp. 94–102, December 1997.

B. Sklar, "A structural overview of digital communications—A tutorial review," Part I, *IEEE Communications Magazine*, vol. 21, no. 5, pp. 4–17, 1983; Part II, vol. 21, no. 7, pp. 6–21, 1983.

D. Slepian, "On bandwidth," *Proceedings of the IEEE*, vol. 64, pp. 292–300, 1976.

B. Smith, "Instantaneous companding of quantized signals," *Bell System Tech. J.*, vol. 36, pp. 653–709, 1957.

E.S. Sousa and S. Pasupathy, "Pulse shape design for teletext data transmission," *IEEE Trans. on Communications*, vol. COM-31, pp. 871–878, 1983.

S. Stein, "Unified analysis of certain coherent and noncoherent binary communication systems," *IEEE Transactions on Information Theory*, vol. IT-10, pp. 43–51, 1964.

C.E. Sundberg, "Continuous phase modulation," *IEEE Communications Magazine*, vol. 24, no. 4, pp. 25–38, 1986.

M. Tomlinson, "New automatic equaliser employing modulo arithmetic," *Electronics Letters*, vol. 7, pp. 138–139, March 1971.

D.W. Tufts, "Nyquist's problem—The joint optimization of transmitter and receiver in pulse amplitude modulation," *Proceedings of the IEEE*, vol. 53, pp. 248–259, 1965.

G.L. Turin, "An introduction to matched filters," *IRE Transactions on Information Theory*, vol. IT-6, pp. 311–329, 1960.

G.L. Turin, "An introduction to digital matched filters," *Proceedings of the IEEE*, vol. 64, pp. 1092–1112, 1976.

G. Ungerboeck, "Channel coding with multilevel/phase signals," *IEEE Transactions on Information Theory*, IT-28, pp. 55–67, 1982.

G. Ungerboeck, "Trellis-coded modulation with redundant signal sets," Parts 1 and 2, *IEEE Communications Magazine*, vol. 25, no. 2, pp. 5–21, 1987.

M.C. Valenti, "An introduction to turbo codes," EE Department, Virginia Polytechnic Institute & State University, Blacksburg, Virginia, unpublished, 1998.

B. van der Pol, "The fundamental principles of frequency modulation," *Journal of IEE* (London), vol. 93, part III, pp. 253–258, 1946.

J.H. Van Vleck and D. Middleton, "A theoretical comparison of visual, aural, and meter reception of pulsed signals in the presence of noise," *Journal of Applied Physics*, vol. 17, pp. 940–971, 1946.

A.J. Viterbi, "Error bounds for convolutional codes and an asymptotically optimum decoding algorithm," *IEEE Trans. on Information Theory*, vol. IT-13, pp. 260–269, 1967.

A.J. Viterbi, "Spread-spectrum communications—Myths and realities," *IEEE Communications Magazine*, vol. 17, no. 3, pp. 11–18, May 1979.

A.J. Viterbi, "When not to spread spectrum—A sequel," *IEEE Communications Magazine*, vol. 23, no. 4, pp. 12–17, 1985.

A.J. Viterbi, "Wireless digital communication: A view based on three lessons learned," *IEEE Communications Magazine*, vol. 29, no. 9, pp. 33–36, 1991.

G.K. Wallace, "The JPEG still picture compression standard," *Communications of the ACM*, vol. 34, pp. 31–44, 1991.

D.K. Weaver, Jr., "A third method of generation and detection of single-sideband signals," *Proceedings of the IRE,* vol. 44, pp. 1703–1705, 1956.

J. Weiss and D. Schremp, "Putting data on a diet," *IEEE Spectrum*, vol. 30, pp. 36–39, August 1993.

T.A. Welch, "A technique for high performance data compression," *Computer*, vol. 17, no. 6, pp. 8–19, 1984.

L.-F. Wei, "Rotationally invariant convolutional channel coding with expanded signal space—part I: 180°," *IEEE Journal on Selected Areas in Communications*, vol. SAC-2, pp. 659–671, 1984.

L.-F. Wei, "Rotationally invariant convolutional channel coding with expanded signal space—part II: nonlinear codes," *IEEE Journal on Selected Areas in Communications*, vol. SAC-2, pp. 672–686, 1984.

L.-F. Wei, "Trellis-coded modulation with multidimensional constellations," *IEEE Transactions on Information Theory*, vol. IT-33, pp. 483–501, 1987.

S.B. Weinstein, "Echo cancellation in the telephone network," *IEEE Communications Magazine*, vol. 15, no. 1, pp. 8–15, 1977.

S.B. Weinstein and P.M. Ebert, "Data transmission by frequency-division multiplexing using the discrete Fourier transform," *IEEE Transactions on Communications*, vol. COM-19, pp. 628–634, 1971.

J.J. Werner, "Tutorial on carrierless AM/PM—Part I: Fundamentals and digital CAP transmitter," *AT&T Bell Laboratories Report*, Minneapolis, June 23, 1992.

J.J. Werner, "Tutorial on carrierless AM/PM—Part II: Performance of bandwidth-efficient line codes," *AT&T Bell Laboratories Report*, Middletown, February 6, 1993.

B. Widrow and M.E. Hoff, Jr., "Adaptive switching circuits," *WESCON Convention Record*, Pt. 4, pp. 96–104, 1960.

J.H. Winters, "Smart antennas for wireless systems," *IEEE Personal Communications*, pp. 23–27, February, 1998.

J.H. Winters, "Adaptive antennas for wireless communications," International Conference on Communications '99, Tutorial Notes TU5, Vancouver, June 6, 1999.

A.D. Wyner, "Fundamental limits in information theory," *Proceedings of the IEEE*, vol. 69, pp. 239–251, 1981.

J.L. Yen, "On the non-uniform sampling of bandwidth-limited signals," *IRE Transactions on Circuit Theory*, vol. CT-3, pp. 251–257, 1956.

O.C. Yue, R. Luganani, and S.O. Rice, "Series approximations for the amplitude distribution and density of shot processes," *IEEE Transactions on Communications*, vol. COM-26, pp. 45–54, 1978.

N. Zervos and I. Kalet, "Optimized decision feedback equalization versus optimized orthogonal frequency division multiplexing for high-speed data transmission over the local cable network," International Conference on Communications, ICC '89, pp. 35.2.1–35.2.6, June 1989.

J. Ziv and A. Lempel, "A universal algorithm for sequential data compression," *IEEE Transactions on Information Theory*, vol. IT-23, pp. 337–343, 1977.

J. Ziv and A. Lempel, "Compression of individual sequences via variable-rate coding," *IEEE Transactions on Information Theory*, vol. IT-24, pp. 530–536, 1978.

W.Y. Zou and Y. Wu, "COFDM: An overview," *IEEE Transactions on Broadcasting*, vol. 41, pp. 1–5, 1995.

## NOTES

1. The following abbreviations are used for some of the journal papers:

   ACM: Association for Computing Machinery
   AIEE: American Institute of Electrical Engineers
   IEEE: Institute of Electrical and Electronics Engineers
   IEE: Institution of Electrical Engineers (London)
   IRE: Institute of Radio Engineers
   SIAM: Society for Industrial and Applied Mathematics

# INDEX

**B**

**E**

**G**

**Q**